Perspectives in Human Biology

Loren Knapp

University of South Carolina

Wadsworth Publishing Company
I(T)P® An International Thomson Publishing Company

Belmont, CA • Albany, NY • Bonn • Boston • Cincinnati • Detroit • Johannesburg • London • Madrid
Melbourne • Mexico City • New York • Paris • Singapore • Tokyo • Toronto • Washington

Biology Editor: *Jack C. Carey*
Assistant Editor: *Kristin Milotich*
Editorial Assistants: *Kerri Abdinoor, Michael Burgeen*
Marketing Manager: *Dave Garrison*
Print Buyer: *Karen Hunt*
Permissions Editor: *Peggy Meehan*
Production: *Brenda Owens, Tom Modl*
Text Design and Dummy: *Hespenheide Design*
Copy Editors: *Betty O'Bryant, Bridget Neumayr*
Illustrators: *John Clark, Deborah Cowder, Stan Maddock, Sandra McMahon, Randy Miyake, Tech-Graphics, Cyndie C. H. Wooley*
Cover Design: *Cuttriss & Hambleton*
Cover Image: *Stephen Simpson/FPG*
Compositor: *Carlisle Communications, Ltd.*

Printed in the United States of America
1 2 3 4 5 6 7 8 9 10

For more information, contact Wadsworth Publishing Company, 10 Davis Drive, Belmont, CA 94002, or electronically at http://www.thomson.com/wadsworth.html

International Thomson Publishing Europe
Berkshire House 168-173
High Holborn
London, WC1V 7AA, England

Thomas Nelson Australia
102 Dodds Street
South Melbourne 3205
Victoria, Australia

Nelson Canada
1120 Birchmount Road
Scarborough, Ontario
Canada M1K 5G4

International Thomson Publishing GmbH
Königswinterer Strasse 418
53227 Bonn, Germany

International Thomson Editores
Campos Eliseos 385, Piso 7
Col. Polanco
11560 México D.F. México

International Thomson Publishing Asia
221 Henderson Road
#05-10 Henderson Building
Singapore 0315

International Thomson Publishing Japan
Hirakawacho Kyowa Building, 3F
2-2-1 Hirakawacho
Chiyoda-ku, Tokyo 102, Japan

International Thomson Publishing
Southern Africa
Building 18, Constantia Park
240 Old Pretoria Road
Halfway House, 1685, South Africa

Library of Congress Cataloging-in-Publication Data

Knapp, Loren W., 1948–
Perspectives in human biology/Loren Knapp.
p. cm.
Includes bibliographical references and index.
ISBN 0-314-20110-6
1. Human biology. I. Title
QP34.5.K62 1997
612—dc21 97-22459
CIP

Books in the Wadsworth Biology Series

Biology: Concepts and Applications, 3rd, Starr

Biology: The Unity and Diversity of Life, 8th, Starr and Taggart

Human Biology, 2nd, Starr and McMillan

Molecular and Cellular Biology, Wolfe

Introduction to Cell and Molecular Biology, Wolfe

Marine Life and the Sea, Milne

Environment: Problems and Solutions, Miller

Environmental Science, 6th, Miller

Living in the Environment, 10th, Miller

Sustaining the Earth, 3rd, Miller

Oceanography: An Invitation to Marine Science, 2nd, Garrison

Essentials of Oceanography, Garrison

Oceanography: An Introduction, 5th, Ingmanson and Wallace

Plant Physiology, 4th, Salisbury and Ross

Plant Physiology Laboratory Manual, Ross

Plant Physiology, 4th, Devlin and Witham

Exercises in Plant Physiology, Witham et al.

Plants: An Evolutionary Survey, 2nd, Scagel et al.

Psychobiology: The Neuron and Behavior, Hoyenga and Hoyenga

Sex, Evolution, and Behavior, 2nd, Daly and Wilson

Introduction to Microbiology, Ingraham and Ingraham

Dimensions of Cancer, Kupchella

Evolution: Process and Product, 3rd, Dodson and Dodson

Fundamentals of Physiology: A Human Perspective, 2nd, Sherwood

Human Physiology: From Cells to Systems, 3rd, Sherwood

Biology: Science and Life, Cummings

Human Heredity, 4th, Cummings

Perspectives in Human Biology, Knapp

Environmental Science: A Systems Approach to Sustainable Development, 5th, Chiras

General Botany, Rost/Barbour/Murphy/Stocking/Thornton

Understanding Human Anatomy and Physiology, Stalheim-Smith and Fitch

An Introduction to Ocean Sciences, Segar

General Ecology, Krohne

Biotechnology, Barnum

CONTENTS IN BRIEF

DETAILED CONTENTS

CHAPTER 1

Science and a View of the World

CHAPTER 2

Chemistry and Life

CHAPTER 3

Cells and Life

CHAPTER 4

Cells into Tissues

CHAPTER 5

The Architecture of the Skin

CHAPTER 6

The Skeletal System

CHAPTER 7

The Muscles and Muscular Systems

CHAPTER 8

The Respiratory System

CHAPTER 9

The Digestive System and Human Nutrition

CHAPTER 10

The Kidneys and the Urinary System

CHAPTER 11

The Cardiovascular System and Blood

CHAPTER 12

The Immune System and Defending the Body

CHAPTER 13

The Endocrine System

CHAPTER 14

Sex and Life

CHAPTER 15

The Changing Form of New Life

CHAPTER 16

Sexually Transmitted Diseases

CHAPTER 17

The Nervous System

CHAPTER 18

The Senses

CHAPTER 19

Human Behavior and Learning

CHAPTER 20

Genetics and Human Heredity

CHAPTER 21

Cancer

CHAPTER 22

Evolution and Human History

CHAPTER 23

The Biosphere

PREFACE

The Author's Viewpoint

Perspectives in Human Biology begins with the assumption that there are many important ways to view human activities in the world in which we live, but that the most powerful perspective is that of science. It is interesting to note that the original meaning of the word science is knowledge. The focus of *Perspectives* is on the underlying principles of biological science and on how the discovery of these principles has allowed humans unparalleled capacity to describe themselves and their relationship to the physical world. This capacity has manifested itself in many ways and has resulted in the establishment of medical specialties, such as anatomy and physiology, and basic research fields, such as molecular biology and ecology. Scientific approaches to understanding the world and our place in it have led to the invention of the means to describe and manipulate the smallest particles of atomic matter, as well as the development of the tools needed to influence many large and diverse types of ecosystems. Although other nonscientific views provide vital perspectives on how members of our species relate to each other (psychology and sociology), organize into groups (politics and religion), and establish working relationships (economics) to the benefit of individuals, families, and nations, the most powerful modern vision of us and our place in the world is provided through the eyes of science.

Perspectives in Human Biology is organized by scale, from small to large. The goal is to establish a framework of thinking in which the smallest units of matter (atoms and molecules) are perceived as connected in logical stages to the largest (ecosystems), as students progress through the book. *Perspectives* starts out with a concise overview of what science is and how scientific thinking proceeds. It also undertakes to show that what scientists do in their work provides an intellectual basis for judging the ceaseless changes in the way we view the world. The book provides insights into the underlying foundation of thought by which we define what life is, and how and when life might have arisen during the history of the primordial Earth. *Perspectives* emphasizes the fact that science is a uniquely human endeavor and is required by its nature to be organized as an open-ended and argumentative process in which new discoveries build on, or replace, established ideas only when they can be rigorously supported with data. This leads to the constant updating of scientific knowledge and drives the development and refinement of all discoveries from the newest innovations in engineering and electronics to the clinical applications of biotechnology. Within this constantly changing knowledge base, *Perspectives* creates a framework for understanding the science of human biology.

The chemistry of life and the organization and function of cells and tissues are covered early in the book. These subjects establish the core of knowledge needed to bridge the conceptual nature of atoms and molecules to the actual behavior of cells. Cells are central to life, and they are the basic units that compose all living things. *Perspectives* goes into the details of how the collections of cells in multicellular organisms such as ourselves are organized into tissues and organs, which in turn have specialized roles to perform. The expansive middle section of *Perspectives in Human Biology* presents and integrates the anatomy and physiology of the major human organ systems, from skin to nerves. The particular emphasis of this section is on understanding the function of specialized cells and on explaining how this knowledge may be applied to maintaining our individual and collective health and well-being. But this book is not only about describing issues of human health and well-being, or depicting humans as organized masses of specialized molecules, cells, and tissues in organ systems, but rather, it shows humans as extremely complex and unique biological forms intimately embedded in the environments (habitats and niches) in which they live and function.

The concluding sections of *Perspectives* focus on a number of broad issues of importance in widening a student's view of him- or herself. These include issues related to developing a better understanding of human behavior, human evolution, and human genetics. It is my contention that, in the context of understanding human genetics and evolution, we come to see ourselves as closely related to one another and, by descent, ultimately to all life on Earth. I have sought to develop enduring connections between human genetics and human history and to extend the knowledge in these fields to understanding human behavior. The question of how we are influenced by our genes as opposed to how we are shaped by the environments in which we presently live is of tremendous scientific and social importance, and it has not yet been completely resolved. It is a question that all students should consider from a biological point of view as they come to terms with their own development and growth and the immense complexity and diversity of life around them.

Perspectives in Human Biology closes with a description of the biosphere. Students are here confronted with a brief but broad exposure to some of the main ideas associated with our understanding of the organization and function of the life-sustaining surface of Earth. It is within this context that a description of aquatic and terrestrial ecosystems is presented, and that a short journey across the ever-changing face of the planet is undertaken. Such a journey is a metaphor for life: one journey over, another begun. The delicate balance of forces that maintains many of the ecosystems on Earth is subject to disturbance from the destructive activities of humans, as in the case of no other species living or extinct. The success of human beings in science with their seemingly mindless application of technology is viewed by some as a potentially fatal flaw in human nature. In discussing the biosphere, the author asks the reader to pause and consider issues important to the future. How may we learn to protect the Earth for generations to come while still reaping the benefits of a human science aimed at enhancing our own quality of life? *Perspectives* ends as it began with the assumption that there are many ways to view human activities in the world. I hope it has led the student to consider that we should seek to develop the clearest view of science. Why? The answer lies on the horns of a dilemma: science is solely the province of humankind and the one endeavor that is capable of providing both the tools to destroy us and the tools to sustain us. It is to build knowledge of and about humans, from the fundamental particles of atoms composing them to the globe-spanning organization of ecosystems in which they live, that *Perspectives in Human Biology* was written.

In Support of Learning

Starting with the first chapter and continuing in all chapters throughout *Perspectives in Human Biology,* there is a pattern of organization to aid the student in understanding what is presented. This includes an abundance of illustrations, photos, and tables. Included also are definitions of *key terms* presented in the margins of the pages of the chapters in which the material is covered as well as in a general glossary. This will help reinforce general scientific and biological concepts by allowing terms associated with them to be emphasized as the student reads. Furthermore, at the end of each chapter is a series of *Questions for Critical Consideration.* These are formulated to invite the student to speculate on what they have read, as well as accurately to relate the facts presented in each chapter. To enrich the student beyond the scope of *Perspectives,* a number of books and articles of particular relevance are listed at the end of the chapter. This will allow the interested student to explore in more detail some of the information presented in *Perspectives.* In addition, there are many special sections throughout the book (called *Biosites*) that provide readers with interesting and intriguing material of historical or conceptual importance aimed at widening their view of the significance of the subject matter of the chapter.

I would like to acknowledge those who served as reviewers for this book. These individuals are: Daryl Adams, Mankato State University; Venita Allison, Southern Methodist University; David R. Anderson, Pennsylvania State University, Fayette Campus; Clyde Bottrell, Tarrant County Community College; J.D. Brammer, North Dakota State University; Alfred G. Buchanan, Santa Monica College; Sebastian Cherian, Mt. St. Clare College; Victor Chow, City College of San Francisco; Florence Dusek, Des Moines Area Community College; Stephen Freedman, Loyola University of Chicago, Lakeshore Campus; Albert Gordon, Southwest Missouri State University; Sheldon R. Gordon, Oakland University; Kenneth Gregg, Winthrop University; Laszlo Hanzely, Northern Illinois University; John P. Harley, Eastern Kentucky University; Debra Howell, Chabot College; Madelyn Hunt, Lamar University; Cran Lucas, Louisiana State University, Shreveport; David L. Mason, Wittenberg University; Charles Mays, Depauw University; John McCue, St. Cloud State University; Don Naber, University College, University of Maine; Joel Piperberg, University of Pennsylvania, Millersville; David Quadaguo, Florida State University; Robert S. Sullivan, Marist College; David Weisbrot, William Paterson University; Roberta Williams, University of Nevada-Las Vegas.

DEDICATION

Perspectives in Human Biology is dedicated to my wife Suzie and my children Laurie and Zann, whose enthusiasm, support, and timely review of the evolving text and art were invaluable from beginning to end. The book is also dedicated to all the students of my courses in human biology over the years, whose role as willing guinea pigs helped shape its content for future classes who, hopefully, will benefit from it. Special thanks to Peter Marshall, without whom I might never have been given the opportunity to present these perspectives at all.

CHAPTER 1

Science and a View of the World

INTRODUCTION

There are many ways to view the world in which we live. We see around us a continuous swirl of human activities based on changes introduced by political events, economics, social and cultural interactions, and advances in science and technology. Each is important in its own way in helping each of us organize a self-consistent interpretation of the multifaceted world in which we live. Perhaps the most powerful and unifying vision of the world in modern times is that of science, which has been of inestimable value in satisfying the human desire to understand how Nature operates. We routinely use scientific merit as a measure of the worth of actions and ideas. The special area of the sciences dealing with living organisms and their relationships to the environment is called biology, and the vast frontiers of science are nowhere more complicated than in the study of life.

WHAT IS SCIENCE?

This chapter provides a useful background of material for understanding human biology. We begin with brief descriptions and examples of what science is about, in general, and how scientists in different disciplines have attempted to make sense of the world around us. These descriptions and accompanying examples will provide a prelude to the discussions and descriptions focusing on human biology in later chapters of this book and will help provide an intellectual and cultural framework in which to fit information on human anatomy and physiology, human reproduction, human behavior and social activity, and human history.

First, we might ask: What is science and why is our species so intensely and uniquely curious about itself and its relationship to the physical and biological world? Second, who are scientists, what do they do, and how do they do it? These questions require not only explanation but also a context in which such activities may be viewed.

An unrestricted definition of the word *science* is knowledge. However, this only reveals part of the meaning of the word. Science in the twentieth century has a more restricted meaning, referring specifically to knowledge and study of the physical world (that is, physics, chemistry, biology). The sciences have as goals not only generating new knowledge through basic research but also developing practical applications of that knowledge. It is the combination of basic research and the applications developed from those discoveries that give us a sense of what it is that scientists do. For example, the discovery of X rays by physicists led to practical ways of examining the internal skeletal structure of the living body. The discovery of aspirin by chemists led to aspirin's general availability and use as a pain killer. The discovery of penicillin by biologists led to the first practical antibiotics, which have saved millions of lives over the last half century.

Some discoveries result from good fortune and are unintended. However, as the famous chemist and microbiologist Louis Pasteur is credited with saying, "chance favors the prepared mind." Scientists are prepared to discover and to connect ideas and facts that may have been previously considered as unrelated. Scientists ask questions and seek answers about many different aspects of nature, including human nature, and attempt to determine the underlying framework and mechanisms by which nature operates. For scientists, asking questions is a profession, an ongoing and often relentless process coupled to a search for solutions. A scientific search is essentially systematic, that is, methodical and well planned (though a little luck never hurts). Such organization is developed with the intent of evaluating **quantitative** and **qualitative** relationships between living organisms and the objects and forces with which they interact in the natural world. The discoveries of scientists are reported in books and journals and, when generally agreed upon, the findings enter the realm of facts. Once the facts are known and presented, humans in all walks of life easily may acquire the new knowledge and take advantage of the new techniques and products that result from the application of that knowledge.

To get the most out of the discoveries of science, and particularly the study of human biology, you must develop a working knowledge of many well-established areas of science, that is, be scientifically literate. Such literacy includes basic understanding of the facts and information derived from the study of the disciplines of physics, chemistry, and geology. Each of these disciplines, and many related subspecialties, directly and/or indirectly establish and/or support a sound basis for the study of life. Physics, chemistry, and geology find a place in the descriptions found in this chapter and surface regularly in discussions throughout the book. However, for all its power to describe the world in an objective and methodical fashion, science is not immune to criticism on a number of fronts. There are many people who vehemently disagree with the answers scientists provide to questions concerning the origins of life and, particularly, to the origin of human life. The controversy over this particular issue will be discussed in some detail later in this chapter in an effort to define what science is and what it is not.

THE WORLD OF PHYSICS AND CHEMISTRY

A knowledge of physics and chemistry is essential to an appreciation of biological systems. These two areas of science provide information on matter and energy. In the twentieth century, thanks to Albert Einstein and many other physicists, we know that matter and energy are different aspects of the same thing; that is, they are interchangeable. Matter is the tangible form of materials and objects and is composed of atoms. It has mass and/or weight and takes up space. We can feel it, taste it, see it, and describe it in terms of shape, hardness, and texture. Matter on Earth exists in one of three phases: solid, liquid, and gas (Figure 1–1). For example, water is normally a liquid but, if the temperature decreases enough (for water, 0° C), it becomes a hard, crystalline solid known as ice. If the temperature increases enough (again for water, 100° C), it becomes a gas or vapor. In these cases, temperature reflects a change in the heat energy of the molecules.

Quantitative relating to quantity or amount
Qualitative relating to quality or kind

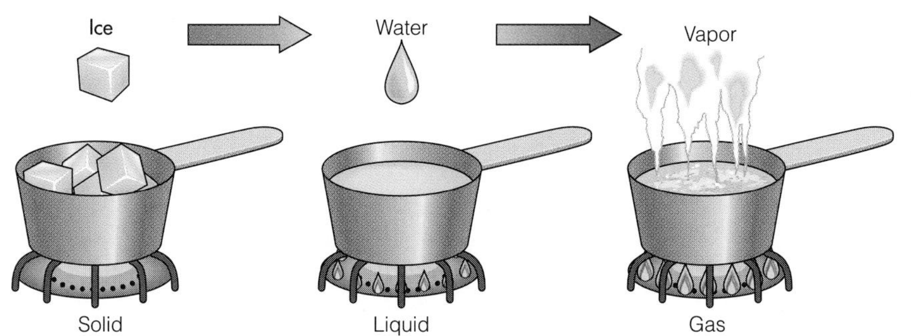

Figure 1–1 The Three Phases of Matter

Changes in states of matter occur in conjunction with an increase or decrease in the heat content, which is a reflection of the kinetic energy of the atoms or molecules.

Energy, on the other hand, is intangible. It cannot always be felt or seen directly, but its actions on material objects can be observed. Energy is defined as the capacity to do work. There are two basic forms: potential and kinetic (Table 1–1). **Potential energy** is stored energy; that is, ready to do work, but not activated or released to do so. **Kinetic energy** is the energy of motion, for example, the runner sprinting from the starting blocks or the crash of water tumbling over Niagara Falls. These two forms of energy are closely linked together. For example, a car battery has a great deal of potential electrical energy. When you start your car, the stored energy is released as current and flows to the electrical and mechanical parts of the engine where it is converted to kinetic energy. Likewise, gasoline has a great amount of chemical potential energy. When it is burned as fuel, it is converted explosively to the kinetic energy of molecular movement needed to allow your car engine to run. In terms of biological systems, human fuels, such as starches and sugar, contain the biochemical potential energy that allows an Olympic sprinter to burst from the starting line at the sound of the gun in a flurry of muscle activity. However, at the most fundamental physical level, matter and energy must be viewed as merely different aspects of the same thing. One of the keys to understanding how biological systems operate is to keep in mind that living systems are based on the same principles of structure and function as nonliving systems. Physics, chemistry , and biology are all subject to the same basic laws that govern the transformations of matter and energy.

Table 1–1 Forms of Energy

FORM	EXAMPLE
Electrical	Battery
Chemical	Gasoline
Mechanical	Wind-up spring
Gravitational	Water moving down hill
Nuclear	Splitting the atom (fission)
Heat	Absorption of sunlight

THE GEOLOGICAL WORLD

Geology is the study of the physical structure and dynamics of change of Earth. It deals with how Earth was formed and how it has been altered over time. This includes consideration of the building of mountains, of erosion, of oceanic and atmospheric dynamics, of volcanoes, and much more that pertains to the processes and forces that bring about a sense of global stability even in the face of continuous change. Geological Earth is the necessary physical framework on which life arose, evolved, and is maintained. Earth continues to provide the land, water, and air in and on which the diversity of life abounds.

Origins of Earth

With the present structural framework of Earth in mind, it is interesting to consider how this planet formed. Earth was formed approximately 5 billion years ago along with a star (our Sun), and a host of eight other planets, which presently orbit the Sun as part of the solar system (Figure 1–2). Earth is the third planet from the Sun, is an average of 93 million miles away from it, and takes a year to complete one orbit. Earth initially was very hot and inhospitable for life. It was built from the condensing gases and accretion of particulate matter. In addition, new material was added by bombardment from meteors, asteroids, comets, and other solar system objects with which it collided. It took Earth's surface millions of years to cool from a molten, primordial state to form a solid crust similar to the one we stand on today. Even today, however, a few miles down, Earth is extremely hot and fluid filled, a little like a warm jelly doughnut (Figure 1–3)—the deeper toward the core, the hotter the material is. A clear indication of this molten state is observed during volcanic activity.

Potential energy the energy of state or position, as a rock sitting at the top of a hill or an unconnected charged battery

Kinetic energy the energy of motion

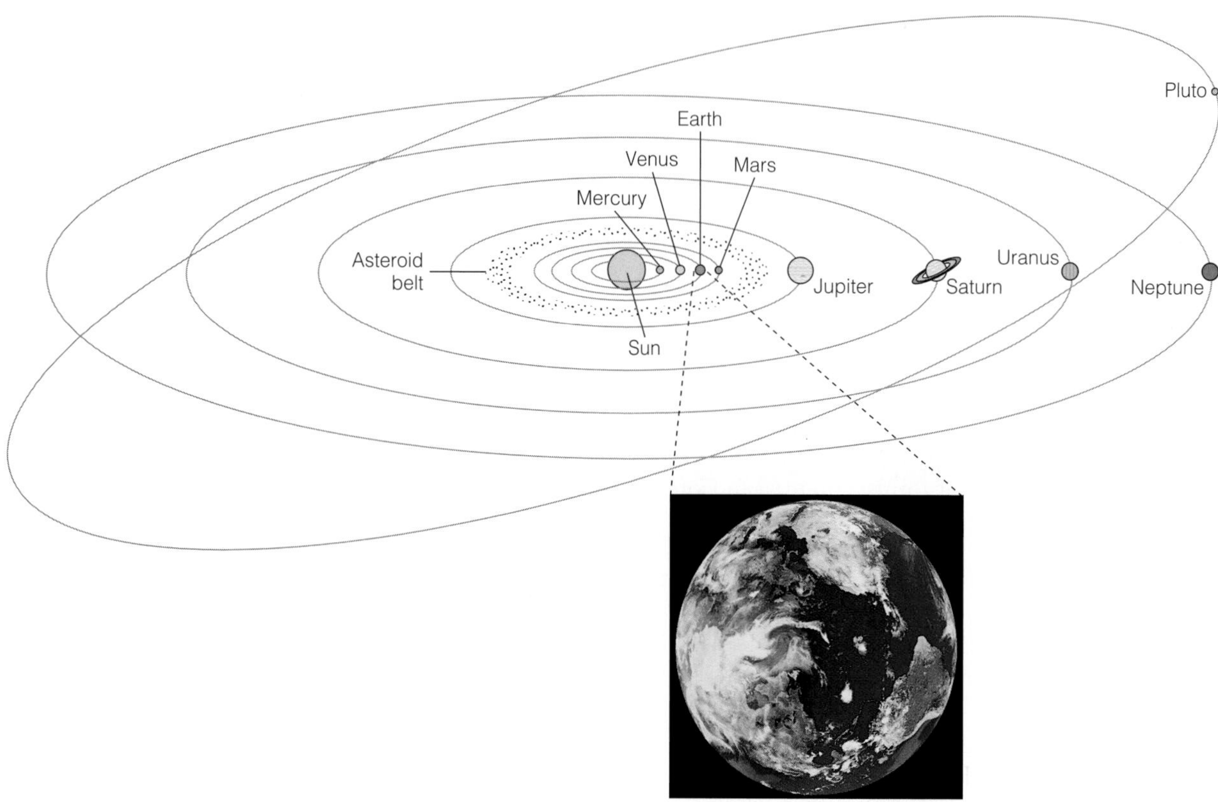

Figure 1–2 Our Solar System—The Third Planet from the Sun
Earth is one of nine planets that revolve around the Sun. It is the only planet that exhibits life and oceans of water.

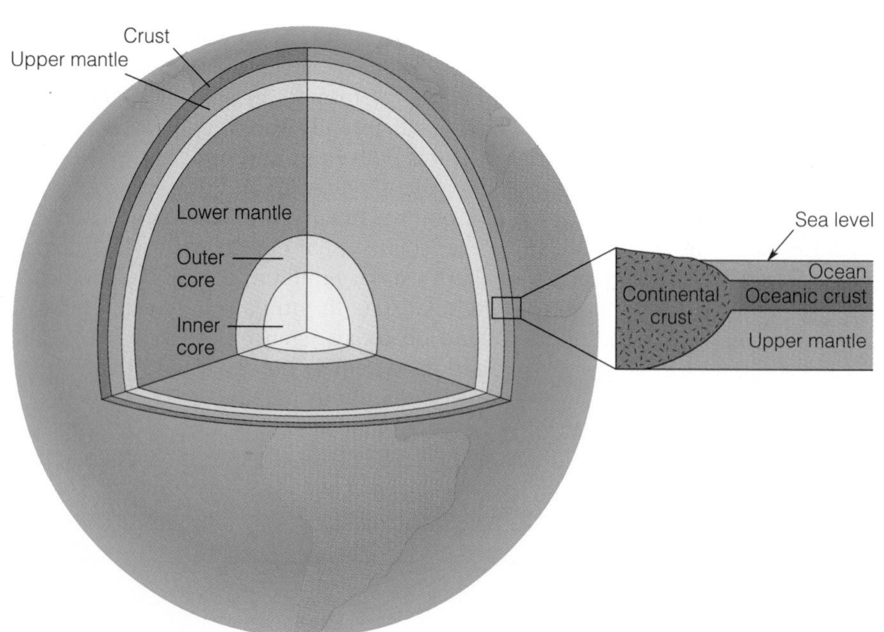

Figure 1–3 Earth's Crust and Mantle
A cutaway of Earth's crust shows the layers and their relative depths. The Earth is approximately 12,000 km in diameter. The continental crust is 20 km to 90 km thick. The oceanic crust is 5 km to 10 km thick.

For example, lava from the Mauna Loa volcano in the Hawaiian Islands is derived from mantle material that erupts onto the surface through vents in Earth's crust.

The Crust of the Planet

The crust of the planet is solid, but does not form a uniformly smooth and seamless surface. It is thicker under the continents (continental crust) and thinner under the oceans (oceanic crust). Earth's surface is not like a Ping-Pong ball but more like a leather soccer ball with creases, folds, and a patchwork of connected materials. Earth's crust is formed from a series of gigantic plates (called **tectonic plates**) that are in constant, slow motion relative to one another over the surface, always rubbing and pushing together (Figure 1–4). The separation of the modern continents has occurred by **continental drift** over the last 200 million years. These global movements did much to distribute and isolate the modern life forms we recognize on different continental and island land masses. For example, the kangaroos of Australia evolved on that gigantic continent in geological isolation and are found as natural populations nowhere else on Earth. The movement of the plates, and particularly the collisions they have with one another, is the cause of earthquakes and volcanic activity around the world. Cities in the United States, such as Los Angeles and San Francisco, and in Japan, such as Osaka, have had serious earthquake problems over the years because they are located very near the conjunction of Earth's dozen or so major plates.

The Early Atmosphere

When Earth was young (3 to 4 billion years ago), the atmosphere was very different from that of today (Table 1–2). Although it contained many of the same gases that are found in today's atmosphere, the relative proportion of these gases was different. Carbon dioxide, hydrogen, water vapor, and nitrogen were all present, but there was no gaseous oxygen and very little ammonia and methane. The oxygen that was present was found in combination with other elements, such as in minerals, or in water, or in carbon dioxide. These combinations are observed today in the atmosphere, as well as in the form of the silicon dioxides that compose sand at the beach or along a river bed. As a result of the lack of free oxygen in Earth's early history, the radiation from the Sun was much more intense on the surface of the planet. Present levels of oxygen in the atmosphere (approximately 20%) and its chemical cousin, **ozone,** prevent much of the harmful ultraviolet radiation produced by the Sun from reaching Earth's surface. Where did the free oxygen of the modern atmosphere come from? Oxygen in the atmosphere was generated by the activity of living organisms that thrived on Earth billions of years ago. The cycle of oxygen formation and consumption continues

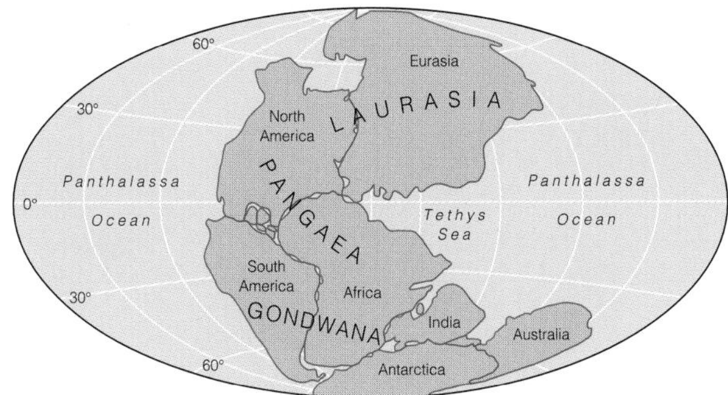

(a) Earth about 200 million years ago

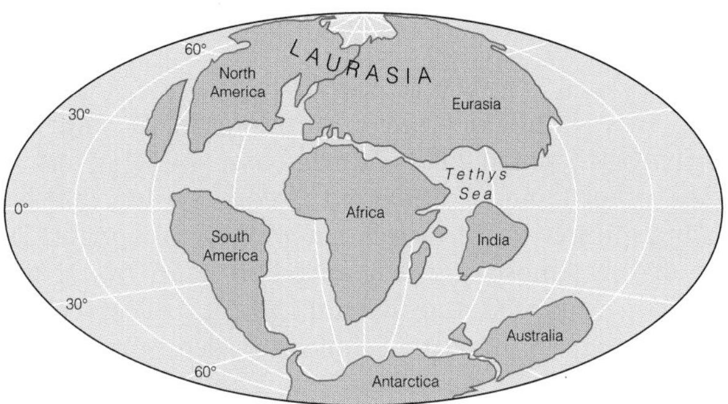

(b) Earth about 70 million years ago

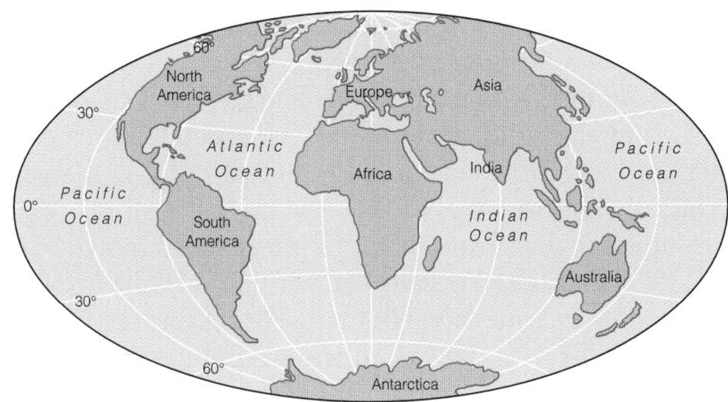

(c) Earth today

Figure 1–4 Continental Drift

Continental drift is the movement of land masses over long periods of time. Laurasia, Pangaea, and Gondwana are names of ancient continental masses. The Panthalassa Ocean and Tethys Sea are names of ancient bodies of water. (a) Earth about 200 million years ago (b) Earth about 70 million years ago.

Tectonic plates large areas of Earth's crust undergoing constant deformation

Continental drift slow movement of crustal land masses over the surface of Earth

Ozone 0_3 a highly reactive form of oxygen

Table 1-2 Earth's Atmosphere

MODERN EARTH (REDUCING ATMOSPHERE)		ANCIENT EARTH (NONREDUCING ATMOSPHERE)	
Gas	**Percent**	**Gas**	**Percent**
Nitrogen	79	Nitrogen	Unknown
Oxygen	20	Oxygen	0
Carbon dioxide	0.03–0.04	Carbon dioxide	Unknown
Methane	Trace	Methane	Transient
Hydrogen	0	Hydrogen	Unknown
Ozone in . . .	Trace	Ozone	0

Table 1-3 Elements with Biological Relevance

Carbon	Oxygen
Nitrogen	Hydrogen
Sulfur	Phosphorus
Sodium	Potassium
Iron	Magnesium
Calcium	Iodine
Chlorine	

today in conjunction with **photosynthesis** and **cell respiration,** which will be discussed in later chapters.

The other familiar ingredient needed for life that was missing initially on Earth was water in a liquid phase. The heat of the atmosphere of young Earth kept water in a gaseous or vaporized state. Eventually, as Earth cooled, and bombardment of the surface from space debris became less frequent, the water vapor condensed and rains drenched the land. Water filled up many of the low areas and turned them into oceans, seas, and lakes. The present oceans of Earth cover over two-thirds of the planetary surface and in most places are several miles deep.

The importance of geology to our understanding of biology is that the solid, fluid, and gaseous materials (earth, water, and air) provided the necessary physical framework not only for the origins of life but also for the continued existence of living organisms. The relentless slow changes that go on in the crust, the oceans and seas, and the atmosphere have provided a wide range of living conditions on Earth. It is within these continually changing environments that new types of organisms have arisen and older types, which have become ill-adapted or less competitive, have disappeared.

THE BIOLOGICAL WORLD AND THE ORIGINS OF LIFE

The First Organisms

Most of what has been described up to this point has been related to the origin, presence, location, and changes in nonliving materials of Earth—land, water, and air. There was no life on early Earth; however, the main chemical ingredients necessary for life as we know it today (carbon, nitrogen, oxygen, sulfur, hydrogen, and phosphorous) were present (Table 1–3). The best contemporary estimates suggest that life did not arise until about 3 to 3.5 billion years ago. Very little is known about what organisms were like in those primordial times. However, several things seem certain: the organisms were simple and lacked a nucleus and were microscopic in size. They came to be present in huge numbers and great diversity. Some of them developed the unique capacity to split molecules of water and release gaseous oxygen. This ancient biochemical reaction, in conjunction with others that allowed light energy to be captured directly from the Sun and utilized in cell metabolism (early photosynthesis), began the slow process of releasing oxygen into the atmosphere. It took billions of years to build up and sustain the 20% oxygen level in the air of modern-day Earth. Oxygen was of great importance to the evolution of higher forms of life, which, because of their size and complexity, needed oxygen for efficient energy **metabolism.** The biological record indicates that it was approximately 600 million years ago, during the onset of the Paleozoic era (or nearly 3 billion years after life first arose, see Biosite 1–1) that organisms we might recognize, in modern terms, appeared on Earth (Figure 1–5). This explosion of types of organisms included the appearance of arthropods, molluscs, and most types of marine invertebrates. In the final analysis, it was probably the availability of oxygen at a sufficiently high level in the atmosphere and dissolved in the oceans and seas of the world that made possible the evolution of advanced forms of life of many kinds, including our own ancient chordate and vertebrate ancestors over 500 million years ago. A description of the Kingdoms of Life will be presented in a subsequent section of this chapter and in Appendix A.

The Step from Nonliving to Living

Understanding the step from inanimate materials to living, animate organisms is an immense challenge to biologists and filled with controversy. Like the origin of the universe,

Photosynthesis capture and conversion of light energy to chemical energy

Cell respiration energy-forming activities in a cell that require oxygen

Metabolism the totality of chemical reactions taking place within an organism

1–1

CAMBRIAN EXPLOSION

Life began approximately 3.5 billion years ago on Earth. There are many traces of fossil bacteria in the rocks remaining from that period (in Southern Africa particularly). Prior to that time Earth itself was probably in an inhospitably hot state not conducive to life. Simple organisms flourished, but diversification of types was probably limited. For the next 2 billion years there was a slow, steady increase in diversity, particularly in the late Precambrian period when simple multicellular organisms evolved. The Edicara fauna of Australia contains soft-bodied organisms such as jellyfish, soft corals, and worms. The first consistent and continuous fossil record suddenly began in the Cambrian period approximately 600 million years ago. There was an explosion of diversity of types, with all the major invertebrate phyla evolving within a few million years of one another.

What is the explanation for the sudden rise in diversity and the development of hard tissues (so they are better preserved than soft-bodied forms)? The major arguments, or rather speculations, are that while multicellular organisms may have been abundant in the Precambrian oceans, the chemistry of Earth's waters prevented mineralization so we have a poor record of their existence. This is not the case, however, because evidence from unaltered sediments suggests that species were not highly diversified, mineralized or not. It has also been suggested that the barrier to animal and plant evolution and diversification was the low level of oxygen in the atmosphere. It was only when the free-oxygen level reached 20% or so that it was available for metabolic use and to block the dangerous UV radiation from the Sun. This would be fine, if there had not already been significant oxidation of materials in the preceding billion years. A final speculation is that the organisms evolving during the Cambrian transition suddenly developed sexual reproduction and this provided the mechanism needed for development of greater diversity. Unfortunately, there is evidence for sexual reproduction in our soft-bodied ancestors of the Precambrian. What are we left with to explain this sudden burst of evolutionary activity? We are left with precious little to provide a clear understanding of what happened during the Cambrian period that gave rise to such diversity, but we should be somewhat mollified that we are here as a result of diversity and can continue to consider it in the future.

evidence of the origin of life has been totally obscured by the passage of time. It is generally agreed among scientists, however, that although not sufficient for life, one of the first steps towards life was the production of organic molecules. How might this have occurred? One way to do this experimentally is to set up conditions similar to those envisioned as having existed on early Earth. In a fascinating experiment by Stanley Miller and Harold Urey in 1953, water, carbon dioxide, ammonia, hydrogen, and methane were placed in a flask and exposed to heat and an electrical discharge. The electricity was used to simulate lightning (Figure 1–6), thought to be prevalent in the early atmosphere. What came out of the flask after a few days were amino acids, short peptides, and other organic molecules characteristic of living organisms. Notice that no oxygen was put in the flask. As stated previously, geological evidence suggests that there was no free oxygen available. This type of oxygen-free environment is considered a **reducing atmosphere.** The presence of oxygen, on the other hand, tends to bring about the breakdown or oxidation of organic molecules, thus affecting their stability and longevity. It is no coincidence that the molecules necessary for the formation of primordial life forms emerged early in Earth's history, when there was no oxygen in the primitive atmosphere. They might not have formed otherwise.

The Miller–Urey experiments and dozens of others like them arose from the testing of hypotheses put forth by J.B.S. Haldane and A.I. Oparin in the 1920s. Their hypotheses suggested that a form of spontaneous generation occurred in the waters of early Earth. Oparin and Haldane were the sources of the ideas that were used to speculate on the existence of a reducing atmosphere and the presence of a "primordial" soup in the oceans. There is much controversy among scientists over the validity of the Haldane–Oparin

Reducing atmosphere conditions in which oxygen is absent from an environment

Time (millions of years ago)	Era	Period	Life		Physical Events
	Cenozoic	Tertiary	Modern life	Development of humans Large carnivores Monkeys Horses Development of mammals	Glacial advances and retreats Rise of Alps and Himalayas Rocky Mountain uplift
100 200	Mesozoic	Cretaceous Jurassic Triassic	Middle life	Extinction of dinosaurs Flowering plants Climax of dinosaurs Dinosaurs	Mountains form in western North America Present-day Atlantic Ocean forms Appalachian Mountains uplift
300 400 500	Paleozoic	Permian Carboniferous Devonian Silurian Ordorician Cambrian	Ancient life	Conifers Reptiles Coal-forming forests widespread Amphibians Earliest forests Land plants and animals Primitive fish Vertebrates Marine invertebrates	Appalachian and Ural mountains begin Seas drain from North America Extensive seas and coal swamps cover North America Mountains and volcanoes in eastern North America Mountain building in Europe Extensive seas cover continents
600		Precambrian		Primitive marine life	Earth's crust forms Free oxygen accumulates Total age about 5 billion years

Figure 1–5 The Appearance of Life on Earth
Most major, recognizable life forms appeared on Earth during the Cambrian period at the beginning of the Paleozoic era between 500 million and 600 million years ago.

view of early Earth and the relationship of the Miller–Urey experiments to the origin of life. Many arguments focus on what the composition of the early atmosphere really may have been. Geologists suggest that there was no methane nor ammonia in the atmosphere at that time, or what was there was short-lived because of the intense radiation from the Sun. The Miller–Urey experiments performed in the absence of methane and ammonia give rise to far fewer organic molecules, most with no relevance to biological systems. Many other scientists have developed ideas to account for the origins of life, some asserting the preeminence of RNA, and some claiming that proteins or proteinlike molecules were responsible. Work by Sidney Fox and his group has proposed that proteinlike molecules, formed under conditions of early Earth, are able to form cell-like structures that are claimed to have a variety of properties similar to microbial cells. The history of the ideas of the modern view of prebiotic Earth and the individuals who developed them is fascinating. It is recounted in Robert Shapiro's book, *Origins,* listed at the end of this chapter.

There is a great deal of uncertainty as to exactly what the compositions of Earth's atmosphere and oceans were, where conditions favoring generation of life might have existed, and how specific biologically relevant molecules were formed. However, it is clear that components present very early in Earth's history were capable of forming many types of relevant organic molecules, and organic molecules were the key to the development of life.

Criteria for Life

As might be guessed from the previous section, the earliest life on Earth was undoubtedly very simple in form and substantially different from modern organisms. Speculation about this often leads to the question of what constitutes a living organism and what are the attributes of life? What criteria might be used to define whether an object or material is living or nonliving? There are many criteria (Table 1–4). The most fundamental criterion for life is an organism's chemical composition. As previously discussed, a unique aspect of all living organisms is that they are composed of highly organized organic molecules (even the remnants of dead cells and tissues reflect a once living state). The cellular structure of living organisms is also a strict and valuable criterion for life, although determining the presence of cells or cellular structures may be difficult because of their small size. This difficulty may be overcome

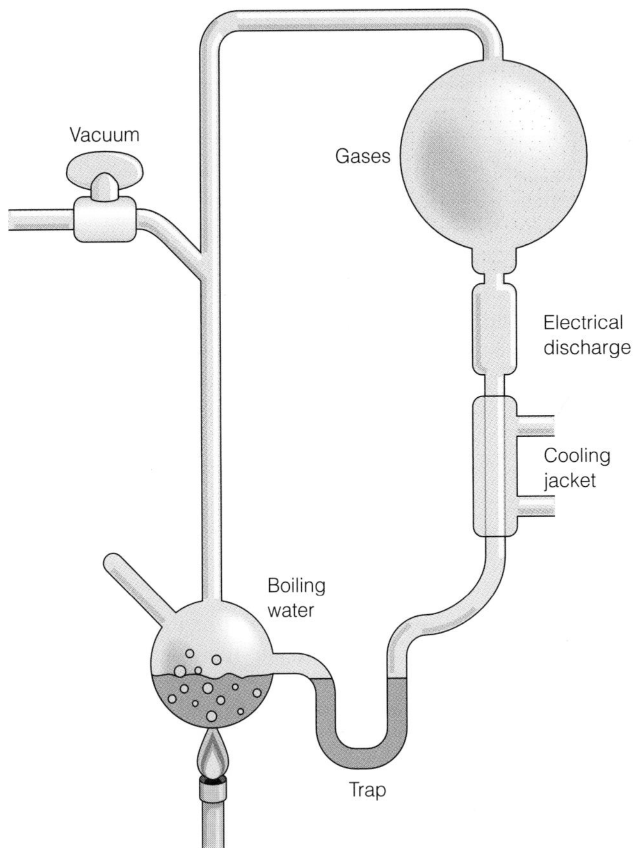

Figure 1–6 The Miller–Urey Experiment
Miller and Urey used an apparatus similar to this to produce and trap amino acids by passing an electric spark through a mixture of gases thought to represent Earth's early atmosphere.

to some extent if the cells can be shown to move or grow in a directed fashion. The capacity for active movement is an excellent defining characteristic of a living organism. However, care must be taken to ensure that the movement of objects, particularly very small objects such as cells, are not simply subject to **Brownian motion.** If an organism does not move outright, it may grow towards, or away from, a variety of chemical or physical stimuli (that is, nutrients, light, gravity). This selective responsiveness is called a **taxis,** if the entire organism moves, and a **trophism,** if part of the organism moves or grows directionally. The occurrence of such responses suggests that an organism may be excitable or irritable; that is, it can be stimulated or disturbed into a stereotypic action. Irritability or excitability can, therefore, be considered an important characteristic of life.

Yet another criterion applied to living things is the capacity for reproduction or replication at the cellular and multicellular level. For simple microbes that existed billions of years ago (and their present day counterparts), this means dividing equally from one cell into two by an **asexual reproduction** process called **binary fission,** or unequally by a process of **budding.** For complex multicellular organisms, such as humans, reproduction requires not only binary fission, but the formation of specialized cells, known as **gametes,** which are capable of interacting during a process of **fertilization.** The combining of two gametes, one from each parent, brings about the formation of a new individual and is called **sexual reproduction.**

Response to stimuli or irritants, active movement, and cell division all require energy. Energy production by organisms requires energy sources. Organisms have to be able to acquire and use materials from their environments to extract energy from them and to coordinate the use of that energy to copy their internal, highly organized structure. The acquisition of energy rich molecules to carry out the activities discussed above can be measured. The

Table 1–4 Criteria for Life

1. Chemical composition based on organic molecules
2. Structure based on cells
3. Capacity for active movement
4. Irritability
5. Growth
6. Acquisition and use of materials to produce energy
7. Reproduction (asexual or sexual)

Brownian motion random movement of microscopic particles influenced by kinetic energy of the molecules composing the liquid or gas in which they are suspended or dissolved

Taxis movement of (usually) simple organisms in response to stimulus

Tropism orientation by an organism or part of an organism to external stimulation

Asexual reproduction a form of reproduction not requiring specialized germ cells

Binary fission reproduction of a cell by division into two equal parts

Budding reproduction of a new individual by outgrowth from the parent organism

Gametes cells specifically associated with sexual reproduction

Fertilization the interaction of germ cells to form a new individual by sexual reproduction

Sexual reproduction a form of reproduction involving interaction of germ cells

mechanisms by which energy is obtained and used by organisms, from microbes to man, are clear and distinctive criteria of life.

Individually, these attributes are necessary, but may not be sufficient for life. For example, a candle flame may move, acquire and utilize energy, and reproduce itself continuously until it runs out of air or wax. Is a candle flame alive? Only in poetry. The occurrence of these characteristics in an object may be obvious (that's not an object; it's an organism) or not, but their collective appearance often represents a level of organization of matter and energy that is associated only with living things.

CONTROVERSY—ORIGINS AND EVOLUTION

The origins of life have so far been considered with respect to a scientific point of view. There are, however, other ways of approaching the problem of origins, and it is with these other ways that controversy and animosity often arise. The controversy results from the intellectual collision between two very different ways of thinking about the world in which we live. The antagonists in this philosophical conflict are, on the one hand, scientists, whose interpretations have been generally recounted in previous sections of this chapter, and, theologians or religious adherents, who hold to beliefs about the creation of the world by a supreme being or beings.

Just so there is no confusion on one point of great importance, there is really no way to be certain how the world began and how life arose. There are very persuasive scientific arguments about many aspects of this process that can be made based on what is known of physics, chemistry, and biology in today's world. However, there is no direct way to establish under what circumstances life began. No one was there to see the unfolding of these primordial events, and nothing remains that can be considered as irrefutable physical evidence as to the mode and method of biogenesis. Neither science nor religion(s) is able to provide a definitive, universally acceptable answer to this question.

The Claims

Be that as it may, the major religions claim, in one way or another, that the world was created by a supreme power beyond the comprehension of humankind using forces that may no longer be in operation. Many of the mythologies of peoples from around the world converge on the idea of creation by design and sustained by divine intervention. The basis for this type of belief is faith, a cornerstone of all world religions. In fact, faith in a set of beliefs is all that is needed for acceptance of this view of the world. A worldview such as this may be based on a codified system of oral or written histories or on personal or collective revelations. Once established, however, the system of beliefs is essentially unchangeable and, moreover, often unquestioned as to its veracity. Revealed truth, and the deep faith in it, is a powerful force in our society, and often at odds with the findings and even the methods of the sciences.

Science also seeks truth, in the form of understanding the operation of natural laws, but not necessarily *the* truth as defined by adherents of the various religions of the world. Science uses a very different process and travels along a very different road to discover the rules governing the Universe and establishing the potential for and occurrence of early life. Science is a way of accumulating, organizing, and applying knowledge about the physical world. These processes depend on objects that can be observed and forces that can be measured in a reproducible manner. These observations and measurements lend themselves to asking questions about how processes in nature occur. The questions that can be answered *objectively* may be used to make predictions that can be tested further to prove the assumptions. In short, this question-answer-question approach constitutes a scientific method (Figure 1–7). Because of the requirement for objectivity of this method there are, in principle, no outside imperatives imposed on the method, no higher authority to be accommodated, no faith required to accept the results. If the answers derived from the thought processes and methods of science employed are flawed or incomplete, or do not make physical, chemical, geological, or biological sense, scientists are forced back to the drawing board for new and better ideas and more critical experimental approaches.

There is a great deal of uncertainty in science—truth is a goal, not a given. This does not invalidate other approaches to questions of great moral and ethical significance to us as individuals and as a society. However, science and scientists focus on those aspects of nature that either can be tested directly or inferred from facts and information that are themselves subject to objective scrutiny. The methods of science are in many ways similar to those used by most of us every day to solve all kinds of ordinary problems. How? Consider the way you observe events or phenomena in everyday life, and how you may deal with the facts and circumstances surrounding them. This probably includes describing the phenomenon, posing questions about it for which answers are possible, and then applying the appropriate methods to resolve the problem. For example, everyone with a car has at one time or another had a problem with getting it started. If your car will not start, you would suspect problems from a number of sources—electrical, mechanical, fuel. The stepwise process of elimination of some of the possibilities increases the likelihood of others. Narrowing the possibilities by tests for the presence of gasoline, or the fact that the headlights burn brightly when you turn them on, may lead inevitably to the conclusion that the starter is the cause of the problem.

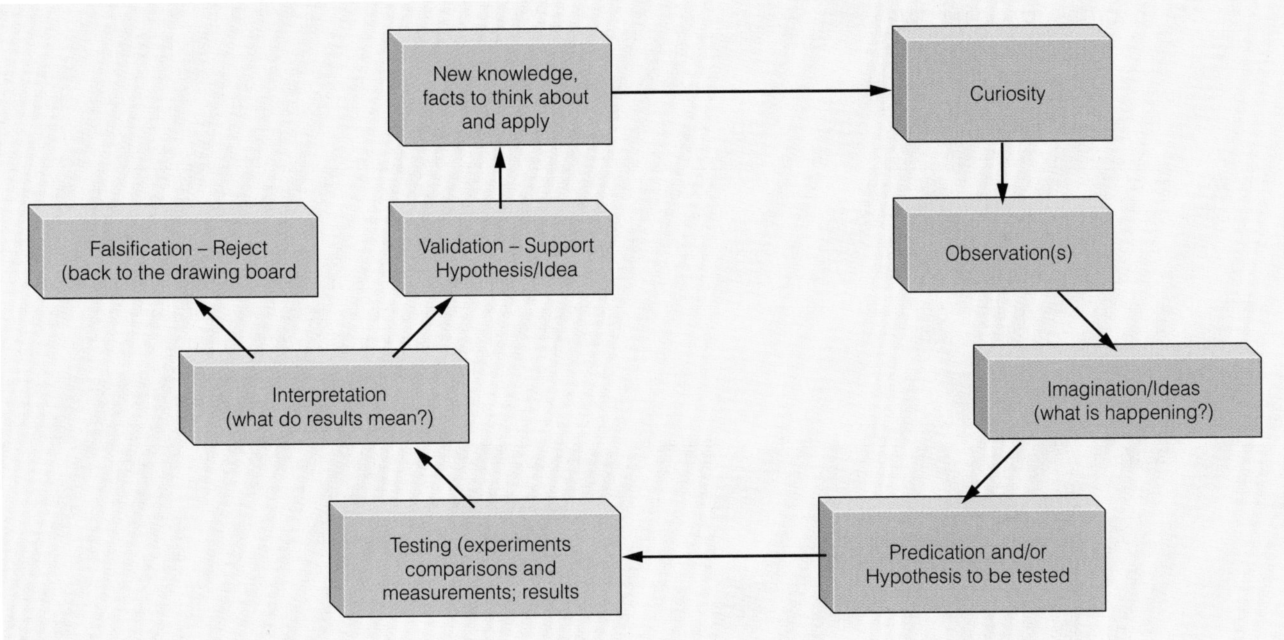

Figure 1–7 The Nature of Scientific Inquiry
The nature of a scientific inquiry is open-ended. The cycle leading from curiosity to new knowledge and facts, as indicated here, may be only one of many ways to apply the method of science.

At this point, you may call a tow truck and have the car transported to the service center, where, indeed, trained technicians discover that the starter is the cause of your car's problem. Their line of reasoning may well be identical to your own objective assessment.

For scientists in particular, the methods used must be able to produce the same results when used by others. Throughout this book, the results of objective thinking and scientific methods will be presented, with due consideration to all the unknowns and uncertainties that surround our present and imperfect knowledge of the world. However, keep in mind that the methods employed and the logic used is not so different from the way we all go about solving the ordinary, everyday problems we face.

KINGDOMS OF LIFE

To make sense out of the complexity of the physical world around us and the diversity of life forms, humans have always attempted to organize objects and materials, including organisms, into different classes, which can be compared and contrasted. The usefulness of this is often self-evident, as we order types of things into categories that have functional or structural utility. For example, the organization of items in a grocery store usually provides a reasonable association of related products. Fruits and vegetables are together in one part of the store, meat in another, and dairy products in yet another; or the arrangement of a library, in which books are organized by subject matter, author(s) or title, and cross-referenced to ensure easy access to the collection. Things are organized so we can find them easily, logically, and repeatedly. This provides us the opportunity to compare and contrast similarities and dissimilarities among the objects or concepts in question.

Unicellularity

Comparing and contrasting similarities and differences are also used for the categorization of life forms (Figure 1–8). Biologists have attempted (mostly successfully) to arrange organisms in coherent orders or groupings based on the degree of similarity to one another (a specialty known as **taxonomy**). Some types of organisms occur in nature as single cells and reproduce by binary fission or budding. These unicellular organisms may grow together in clusters or colonies, but they behave predominately as individuals

Taxonomy the study of the classification of living and extinct organisms

SOME CHARACTERISTICS OF THE FIVE KINGDOMS

Kingdom	Cell Type	Cell Number	Major Mode of Nutrition
Monera	Procaryotic	Unicellular	Absorb or photosynthesize
Protista	Eucaryotic	Unicellular	Absorb, ingest, or photosynthesize
Fungi	Eucaryotic	Most multicellar	Absorb
Plantae	Eucaryotic	Multicellular	Photosynthesize
Animalia	Eucaryotic	Multicellular	Ingest

Figure 1-8 The Kingdoms of Life: (a) Monera (b) Fungi (c) Protista (d) Plantae (e) Animalia

and have unique characteristics that separate them into natural groups. This is exemplified in the large and diverse group of **procaryotic organisms,** known as bacteria, which exist as single cells, and in protists, which are **eucaryotic,** single-celled, **organisms.** Bacteria often grow in colonies but can grow and flourish in isolation. The shape, size, nutrient requirements, and biochemical properties of these organisms help define the separate species.

Viruses—Living or Nonliving?

By definition, single cells are the smallest units of life. They can reproduce themselves and have attributes, as described earlier, characteristic of all living things. However, there is a group of very important particles or entities that do not qualify completely as living organisms, but do have profound effects on the growth and survival of cells. This group is known as viruses and they seem to qualify best as

Procaryotic organisms unicellular organisms with no definable nucleus.

Eucaryotic organisms a single or multicellular organism whose cells have a nucleus

quasi-life forms. Viruses are **intracellular parasites.** They penetrate into cells and take over part of the cellular machinery, diverting the work of the cell to the production of new viruses. In this way they often destroy or incapacitate cells.

Viruses cannot reproduce or function in the absence of a living host cell. However, many types of viruses can survive for long periods outside of cells. Even though viruses are very simple in their construction relative to cells, they are by no means simple in their functions and adaptability. Viruses occur in a wide variety of forms and may target specific cell types within an organism for infection. Unicellular organisms are not free from this invasion. Bacteria harbor viruses, called **bacteriophages,** which affect all classes of bacteria. There are plant viruses, whose effects are devastating on trees and crops and result in billions of dollars in agricultural losses each year. There is also a wide range of animal viruses, some of which have severe effects on human health. Among the best known are the viruses that are associated with the flu, cold sores and genital herpes, and human immunodeficiency (HIV). Some of the viruses that infect other animals infect us as well and can be the source of worldwide human epidemics. The origins of viruses that are agents of the flu are often discovered to be harbored in pigs and ducks in Asia. The natural hosts for viruses, that is, the organisms in which the virus may remain latent without killing, are often not known. Nor are the **vectors** of viruses, which spread the virus from its natural host to affected species, always known. Mosquitos are a common vector for a number of viruses, including those that cause **yellow fever** and **dengue fever.** The deadly **Ebola virus** of Africa, which has occurred in dangerous outbreaks in recent years, is lethal in humans (and probably other primates), but the natural resevoir and vectors for it have not yet been discovered.

Multicellularity

Organisms composed of many cells are called multicellular, and their cells, tissues, and organs act together in a coordinated fashion. Individual cell types, tissues, and organs of a species often have specialized structures and functions that can be used to differentiate one species from others. For example, we have little trouble differentiating birds, which have feathers, from mammals, which have hair, or plants, with leaves, from clams, with mineralized shells. In higher forms of animals and plants, in which two different, but related, forms exist (for example, male and female), reproduction involves sexual interactions between individuals to bring about contact between gametes from both types to give rise to new individuals of that species. In these cases, individuals that are identical to one another morphologically (other than sexual differences) and are capable of producing viable and fertile offspring are, by definition, a **species.**

One of the great steps forward in the history of biology was the rational organization of all known organisms with regard to the relationships in structure and function that they have with one another. The more characters, or **traits,** that two organisms share in common, the more closely they may be related to one another. This rationale also extends into the evolutionary past, such that a common ancestor of one or more present-day species may be recognized from the **fossil record.** The theory of evolution, particularly with reference to the appearance of new species, is based on the concept of descent with modification from a common ancestor.

With all the need for rational organization of organisms, past and present, into appropriate groupings, a problem sometimes arises in the naming of particular species. This was certainly a problem faced by seventeenth and eighteenth century biologists and taxonomists. For example, what if members of the same species found in different locations or regions around the world are called by different names? Should there be a separate entry for each of the organisms? This problem was exacerbated during the years of worldwide exploration and discovery during the fifteenth and sixteenth centuries. Species with similar, if not identical, characteristics, but different local names, were collected worldwide and brought back to Europe for study and cataloging. Confusion in naming made it difficult to know who was who, and what was what. If there was going to be a resolution to the confusion, it meant that a universally agreed-upon, systematic method of classification and naming had to be developed.

Quasi-life forms viruses and intracellular parasites that cannot reproduce or function outside a host

Intracellular parasites viruses or some classes of monerans that live inside cells and generally destroy cells

Bacteriophages a class of viruses that infect bacterial cells

Vectors carriers of infectious disease agents between natural host and organisms pathologically infected

Yellow fever lethal viral disease carried by mosquitoes to humans

Dengue fever a viral infection carried by mosquitos to humans and characterized by headache, severe joint pain, and rash

Ebola virus biological agent of a lethal disease in which hemorrhaging occurs in all organs of the body; vector unknown

Species a population of genetically similar organisms that is reproductively isolated, produces viable and fertile offspring, and shares an evolutionary history

Traits genetically inherited characteristics

Fossil record the imprint or mineralized remains of previously living things preserved in geological sediments

1–2

LINNAEUS AND TAXONOMY

Carolus Linnaeus began life as Carl Linne in 1707. His humble beginnings were overcome by his natural talent and intellect, which provided him a chance to study botany in Holland. When he returned to his native Sweden, he began organizing and naming plant specimens. The organization of living things was a passion to Linnaeus and to many scientists of the period. With an avalanche of new discoveries occurring all the time around the world, the organization of species was not only a major task but also absolutely imperative if there was to be order in the natural sciences. He was so caught up in naming things that he also renamed himself with a Latin equivalent.

Common sense tells us that it is easy to tell the difference between a cat and fish, but what about the differences between different kinds of fish, particularly those that are similar but not identical? The biggest problems are in the details. Linnaeus accomplished two objectives in his life; he developed the binomial naming system, and he organized life into hierarchical categories based on shared traits. He had problems and made mistakes, just as the rest of us, but his system of organization of species and of the naming of those species is still used today. Linnaeus suffered a stroke late in his life so severe that the "namer of all things" could not even remember his own.

The Linnean System

The solution to most of the problems in classification was provided by Carolus Linnaeus (1707–1778) and his students (Biosite 1–2). They amassed huge collections of organisms from around the world, unified the descriptions of the characteristics of species, and established a universal binomial naming system based on Latin nomenclature (a so-called dead, or unchanging, language). This has led, in modern times, to the development of well-accepted sets of criteria (morphological, historical, molecular) for organismal classification and a naming system that attempts to relate organisms to one another based on characteristics from the most general features shared among them to the most specific. The major types of organisms, with fundamental similarities to one another, are organized into categories known as **Kingdoms.** There are presently five Kingdoms recognized by most biologists, into which all living species may be placed. These Kingdoms are Monera, Protista, Fungi, Plantae, and Animalia. A brief description of the categories and types of life forms in each of the Kingdoms is presented in Appendix A, with special emphasis placed on the criteria used in determining the various types of animals in the Kingdom Animalia, to which our species belongs.

EXPECTATIONS AND PERSPECTIVES

What else can you expect to encounter as you begin your study of human biology? There are several chapters in this book that establish the underlying structural framework of all life. This first chapter has introduced some of the physical, chemical, and geological concepts and considerations involved in defining and understanding this framework and how living organisms fit into it. The levels of organization of matter (and the manner in which they are measured) on Earth range from the structure of atoms and molecules to the composition and operation of ecosystems in the **biosphere** (Figure 1–9). The organization of matter at the atomic and molecular levels constitutes the study of physics, chemistry, and biochemistry (Chapter 2). The organization of matter into organic and inorganic categories provides a basis for defining life. The cell is the basic, and smallest, unit of life, uniquely composed, as it is, of a combination of organic and inorganic molecules. All organisms, regardless of the Kingdom to which they belong, are either unicellular or multicellular. There is no noncellular form of life. Viruses may be considered quasi-life forms, because their capacity to reproduce is dependent on infection of cells as intracellular parasites. The criteria used to establish groupings of forms of life are based on the unique attributes of living things as individuals and as pop-

Kingdoms the most inclusive category of taxonomy; five Kingdoms generally recognized

Biosphere the zone formed by a combination of atmosphere, lithosphere, and hydrosphere in which life exists on Earth.

Matter	Example(s)	Scale of measurement
• Atom	Hydrogen Atom	$10^{-12} - 10^{-10}$ m
• Molecule	water, amino acid	$10^{-10} - 10^{-8}$ m
• Cell	bacteria, sperm, red blood cell	$10^{-7} - 10^{-4}$ m
• Tissue	skin (epidermis and dermis)	$10^{-3} - 10^{-2}$ m thick
• Organ	variable — some glands heart	10^{-3} m 10^{0} m
• Organism	Human	$>10^{0}$ but $<10^{1}$ m
• Population (variable scale)	Human	10^{6} m^2 or more (i.e., a town or city)
• Community (variable scale)	Human	10^{7} m^2 or more (nation)
• Ecosystem (variable scale)	Desert	10^{7} m^2 or more (region)
• Biosphere	Earth	10^{7} m in diameter $<10^{14}$ m^2
• Exosphere	Solar system galaxies known universe	10^{13} m in diameter 10^{21} m in diameter 10^{26} m in extent

Figure 1–9 Size Range of Matter

ulations. It is the structural and functional similarity among individuals, with allowance for some variation, that is reflected in the characterization of species presented in Appendix A.

Chapters 3 and 4 describe the internal structure, composition, and interactions among cells of multicellular organisms, with particular regard to human cells and tissues. The study of the function of cells and tissues constitutes the discipline of physiology. The study of the structure of cells, tissues, and organs constitutes the disciplines of microanatomy and anatomy. These functional and structural approaches to studying and describing biological systems open the way to the specific study of human cells, tissues, and organ systems. We will examine a range of human organ systems from the skin and its architecture to the nervous system and the basis for cellular communication and learning. Included in this presentation of organismal structure and function are detailed descriptions of how organs and organ systems develop during embryogenesis, how they change with age, and how they are affected by disease.

Having established a working knowledge of the nature of the complex structures and actions of the diverse organs and organ systems, the goals then will be to focus on higher levels of activities, or interactions, among humans and between humans and their environments. Human behavior is presented in the context of not only how we act and respond to the world around us but also how we learn, accumulate, and use knowledge. This is followed by a critical inquiry into genetics and human heredity, which underlie all aspects of life and its perpetuation. The genetic information of a species, including **Homo sapiens,** determines the nature of its structure and morphology, reproductive behavior, and its many and diverse adaptations to the environments in which it lives. The underlying capacity for change in biological systems, in the form of mutations in the genetic information of a species, allows for one of the most far-reaching capabilities of life on Earth—evolution. A chapter on evolution and human history will provide a backdrop in which to consider how mutation, sexual reproduction, and natural selection ensure the biological means to bring about changes in populations of organisms over many generations. The ideas associated with the theory of evolution will be presented and how humans fit into the historical changes of life on Earth will be described in detail.

Homo sapiens the name of a species within the genus Hominidiae; usually characterized as a grouping of bipedal primates; employed as a term for the human species

The most complex level of organization of matter and energy on Earth is observed in the workings of the biosphere, which combines living and nonliving systems of the planet together in complicated cycles, chains, and webs. Beyond Earth and the biosphere, which envelopes its surface, the focus of science has been outward into the essentially empty, lifeless reaches of space (except for the search for life beyond Earth in a discipline called **exobiology**). Our journey ends below the frontiers of space. The final chapter presents a description of the biosphere, which combines the unique physical, chemical, geological, and biological factors necessary to support the great diversity of life. Changes in the physical, chemical, and geological aspects of the biosphere bring about continuous change in life on Earth. Such a combination of factors on a planet is rare, and so far as is known, unique. Indeed, we know of no other life-bearing planet like Earth in any other part of the galaxy. It is in the context of understanding the continuity between, and among, the many levels of inorganic and organic organization and interactions within the biosphere, from atoms to ecosystems, that this book presents a perspective on life from a human scientific point of view.

Exobiology the study of life outside or beyond Earth

Questions for Critical Inquiry

1. What is science? How does it affect the way we, as humans, view the world? What other ways are there to view the world around us?
2. What is biology? Why is it important to know essential facts about physics, chemistry, and geology in order to get the most out of studying biology?
3. What advantage is there to organizing species into groups on the basis of their similarities to one another?
4. What are some of the social conflicts that have arisen (and might arise in the future) in the progress of science?

Questions of Facts and Figures

5. How old is the Earth? How was it formed?
6. What are some of the attributes of life?
7. Is the surface crust of Earth uniform and stable? What is the planet like underneath the crust?
8. What was the likely composition of Earth's atmosphere prior to the arising of life? What is the present composition? What gaseous molecules were missing? How were they formed?
9. What is the present best estimate of when life arose on Earth? When did most of the familiar forms of life on Earth arise (clams, arthropods, worms)?
10. How many Kingdoms of life are presently used by biologists to organize species? What are they?

References and Further Readings

Bronowski, J. (1965). *Science and Human Values.* New York: Harper & Row.

Ferris, T. (1988). *Coming of Age in the Milky Way.* New York: Anchor Books.

Gould, S. J. (1982). *The Panda's Thumb.* New York: Norton.

Hanson, R. W. (1986). Ed. *Science and Creation.* New York: Macmillan.

Hazen, R. , and Trifil, J. (1990). *Science Matters: Achieving Scientific Literacy.* New York: Anchor Books.

Shapiro, R. (1987). *Origins: A Skeptic's Guide to the Creation of Life on Earth.* New York: Bantam Books.

Ziman, J. (1976). *The Force of Knowledge.* New York: Cambridge University Press.

CHAPTER 2

Chemistry and Life

INTRODUCTION

Why should anyone studying human biology be interested in the organization of matter or the various forms of energy? What is to be gained from describing the structure of objects we may not be able to even see and forces we may not feel? One way to approach these questions is by restating them to make them personal. What am I made of? How am I organized? How do I interact with the environment? Is "what I am" connected with the other life on Earth? How? These questions have no simple scientific answers even though they have been asked, in one form or another, ever since humans began to think. In reconsidering the questions of why there should be interest in knowing the structure of matter and the various forms of energy, the answer lies in the fact that matter and energy are at the core and foundation of what we are, and, if we are ever to understand ourselves, we must know something of how we are put together.

FRAMEWORK FOR LIFE

In addition to the personal and practical reasons for understanding physics and chemistry and the properties of matter and energy, humans have a driving curiosity to try to understand all about ourselves and to see what makes us tick. Even our scientific name, *Homo sapiens,* translates from Latin to mean intelligent man. The most appropriate means of answering questions about ourselves and accurately perceiving and describing the world around us has always been to use mind over matter.

An appropriate starting point in this quest for knowledge is to consider the following three simple statements:

1. All matter is made of atoms.
2. Atoms interact to form molecules.
3. Molecules interact to form living cells.

The implications of these statements, individually and collectively, have had an enormous impact on science and particularly on biology. They also have had a tremendous influence on the attitudes and expectations of our society, as the ideas and concepts evolving from them are transformed into new and practical technologies. The statements suggest, as was discussed in Chapter 1, that the properties of life are contingent on an underlying framework of matter and energy.

Here, then, is another reason to study the properties of matter. The activities we may take for granted, such as breathing, vision, and the continuous beat of our hearts, all depend on special kinds of molecules working together properly in the cells and tissues of our bodies based on principles of physics and chemistry. However, the fact is that, even though atoms and molecules are necessary for life to exist, they are not sufficient for it.

Necessary and Sufficient

What does it mean that atoms and molecules are necessary, but not sufficient, for life? If it is known what a cell is made of, why is it not possible to simply construct new cells from the proper ingredients, similar to making a cake? Scientific experiments in which isolated molecules of living cells are simply mixed together do not result in anything alive. It is relatively easy to collect all the molecules of a cell or cells in a test tube. The disruption of their integrity is brought about in a number of ways, including the act of simply grinding them up in a mortar and pestle. This is a problem not unlike that faced by Humpty Dumpty. The egg man in this Mother Goose nursery rhyme example is broken into pieces that cannot be put back together again. The same holds for cells and organisms, and so far, neither all the king's horses, nor all the king's scientists have been able to recombine the parts.

The problem of reconstruction, after disruption of structure and loss of function, is also analogous to a watchmaker's dilemma. For example, a clock or wristwatch is a very intricate machine with many complicated parts and mechanisms. Using the right tools, however, it can be easily and completely dismantled into the screws, springs, gears, hands, and face from which it is made. The problem clearly is not in taking machines or instruments apart, but rather in putting the parts back together again to make them whole and functional. Have you ever tried to reassemble your own wristwatch from a collection of dismantled pieces? The reassembly is far more difficult and time consuming than disassembly. This is because it requires not only all the right pieces and the right tools but also a knowledge of the order of their assembly, orientation, and proper mechanical interactions. Enter the watchmaker and his/her experience and training. If that skill and knowledge is lacking, the chance to reassemble a functional watch is very low.

The Ultimate Intricacy

By the same token, the organization and interactions of the molecules that make up living cells are far more complicated and intricate than the pieces of a watch (as we will learn in the biochemistry section later in this chapter). The problem for the reassembly of a cell is that there is not a blueprint nor design to work from to restore the order and arrangement of all the separated molecular pieces that collectively make a cell a living entity. Similar to the parts of a watch, atoms and molecules must fit together perfectly and be arranged in special ways to manifest life. To learn more about how atomic and molecular pieces fit together in a living cell, we explore in this chapter some of the physical, chemical, and biochemical properties of atoms and molecules. It is intellectually quite humbling to realize how little we actually know about how living matter is organized and what life is.

CHEMISTRY AND ATOMS

What is an Atom?

The question of what is an atom has been the subject of much conjecture and experimentation for millenia. A fair definition is that an atom is the basic, universal building block of matter. Atoms are organized such that there is a dense, positively charged core and a negatively charged particle cloud surrounding the core. The core, or nucleus, is composed of two kinds of particles—protons, which carry a positive charge, and neutrons, which have no charge (Table 2–1). The cloud of particles surrounding the nucleus is composed of electrons (Figure 2–1). The number of protons in the core of an atom is a characteristic feature. The specific number of protons determines the atomic number,

Table 2-1 The Atom and Its Parts

ATOM OR SUBATOMIC PARTICLE	CHARGE	MASS (WEIGHT)*	SIZE*
Atom	No net charge	Variable	Variable
Proton	Positive	1	2×10^{-15}m
Neutron	No charge	1	2×10^{-15}m
Electron	Negative	0.0005	less than 1×10^{-16}m

*Variability of atomic size and mass depends on the number of subatomic particles.

which in turn identifies that atom, and all atoms like it, as a specific element. Atoms are the smallest possible material objects that retain identities associated with the elements they represent. In ancient Greece, where the concept of the atom arose over 2,500 years ago, the word meant indivisible.

To get a sense of the relative sizes of the particles involved, imagine single atoms of silicon and oxygen to be the size of poppy seeds. A poppy seed is small but clearly visible to the naked eye. If all the atoms found in a rock the size of your fist were poppy seeds, the rock would expand to occupy a volume the size of the earth itself. Although the "indivisible" atom has proven to be divisible in modern physics (for example, fission of uranium in nuclear reactors), under normal conditions of life, the structure and function of atoms are the basis for chemical and biological systems.

Elements and Atomic Number

An element is defined by the number of protons contained in the atomic nucleus. The number of protons in an atom is called the atomic number. Each of the 92 natural elements (and the elements beyond uranium as well) has a particular and distinct atomic number. For example, the element hydrogen has one proton. Helium, the gas we use to fill up balloons to float for parties, has two protons. Carbon has six protons, and oxygen has eight. The elements are numbered and organized in a Periodic Table of Elements according to the number of protons each contains (inside front cover). Hydrogen is number 1, helium number 2, carbon number 6, and oxygen number 8. The largest and final of the natural elements is uranium, which has 92 protons. Chemical relationships in the properties of the elements exist with respect to the column in which the elements are found. For example, in column I, lithium, sodium, and potassium have similar chemical properties. Likewise, in column VIII, helium, neon, argon, and xenon are related chemically. These repeated similarities in each column are the basis for calling the table "periodic."

Atomic Weight and Isotopes

The weight of an atom depends on how many protons and neutrons are contained in the nucleus. The more there are, the heavier the atom. A single type of element may have different numbers of neutrons (Figure 2–1), in which case it is known as an isotope. There are naturally occurring isotopes of all elements. The most common isotope of an element found in nature is used as a standard. For example, the most common isotope of carbon has six protons and six neutrons. Six neutrons plus six protons gives an atomic weight of 12 (known as carbon-12). But carbon atoms commonly have seven or even eight neutrons (carbon-13 and carbon-14, respectively). These are less abundant isotopes of carbon, but they are perfectly natural. Some isotopes are inherently unstable and have a tendency to disintegrate. This instability is the basis for the phenomenon of radioactivity. Isotopes of many elements are radioactive. Hydrogen-3, which is also known as tritium,

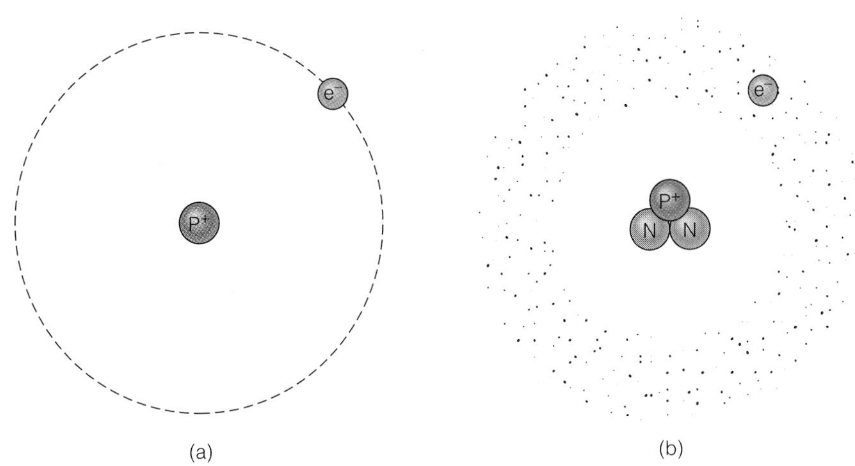

Figure 2–1 The Structure of an Atom
(a) In the old model of an atom the electron (e^-) orbits the nucleus (proton, P^+), similar to a satellite orbiting Earth in a fixed path. (b) In the new model the electron may position itself at any number of distances from the nucleus, but it is most probably located in a shell at a set distance from the nucleus (area of high-density stipling). The presence of neutrons (N) does not alter the relationship between P and e but does determine the type of isotope. For example, P and e (as in part a) is hydrogen; P + N + e is deuterium; P + N + N + e is tritium.

2–1

WHY SHOULD WE BE AFRAID OF RADIOACTIVE MATERIALS?

Radioactivity is a byproduct of the breakdown of an atomic nucleus. This kind of radiation is frightening because in most cases it cannot be seen or felt without special detectors. It is extremely hazardous to life because it is damaging to the molecules that are important for life. A short exposure to high levels of radioactivity can cause so much damage to living tissue that it results in immediate death. Long-term exposure to lower levels of radioactivity can lead to alterations in DNA and proteins and may result in cancer and birth defects. This is one of the reasons why people get so upset about potentially unsafe nuclear reactor facilities. Recent large-scale disasters in the nuclear energy industries of the United States (Three-Mile Island, 1982) and particularly in the Ukraine (Chernobyl, 1986) released huge amounts of radioactive isotopes into Earth's atmosphere. The magnitude of the short-term effects of the Chernobyl reactor meltdown were enormous; the entire region had to be evacuated and quarantined. Problems in the Ukraine will continue in the 1990s and into the twenty-first century as people attempt to move back into the contaminated region. The long-term effects of these radioactive releases on human life, and on the environment, may take generations to determine. It is difficult if not impossible to escape exposure to these materials because they diffuse extensively in the atmosphere, and all who have to breathe do so from a reservoir of shared air.

and carbon-14 are both radioactive isotopes. Uranium-235, with 92 protons and 143 neutrons, is also highly radioactive (Biosite 2–1).

 CHEMICAL BONDS

The Role of Electrons

An individual atom is a composite structure that consists of a nucleus, in which protons and neutrons form a very stable relationship, and electrons, which orbit around the nucleus and are less constrained. Electrons are not so strongly held in orbit around a nucleus that they cannot move or be exchanged between atoms. Electrons tend to be indiscriminate in their interactions between and among atoms; that is, they can be jarred loose, or pulled away, from their orbit around one nucleus and captured or shared with another. The field of chemistry is based in large part on understanding how the sharing, transfer, and interactions of electrons between atoms in molecules occurs. The transfer, or sharing, of electrons is the basis for chemical bonds between atoms. Molecules may have as few as two atoms, such as molecules of oxygen (O_2), or billions of atoms linked together, as found in a molecule of DNA, which carries the hereditary blueprint for organismal growth and development.

Covalent Bonds

Electrons are constantly in motion and their positions at any instant are determined by **probability functions.** The complex path of their movement around a nucleus is called an orbital and is somewhat similar to the moon orbiting the earth, or the earth orbiting the sun, except that an electron moves much more rapidly and radically, forming a cloudlike shell. An electron and a nucleus are like partners in a dance. This is a convenient way to visualize electrons in relation to one another and in relation to the atoms sharing them. When electrons are shared between atoms, they form what are called **covalent bonds.** Changing covalent bonds results in changes in the chemical properties of molecules, and patterned, orderly changes in molecules are characteristic of living organisms.

Electrons in the outermost shells of an atom are most accessible for exchanging partners with other atoms. For example, in a molecule of water (Figure 2–2) the oxygen atom has eight electrons, six of which are in the outermost shell, and each hydrogen atom has a single electron. The general rule for chemical bonds formed between atoms is that the atoms lose, gain, or share electrons in order to fill their outermost shells. An electron from each of the two hydrogen atoms is shared with the oxygen atom and fills the oxygen atom's outermost shell. Filling this shell makes

Probability functions mathematical terms used to predict statistical likelihood of events or spatial distributions

Covalent bonds chemical bonds between atoms in which electrons are shared

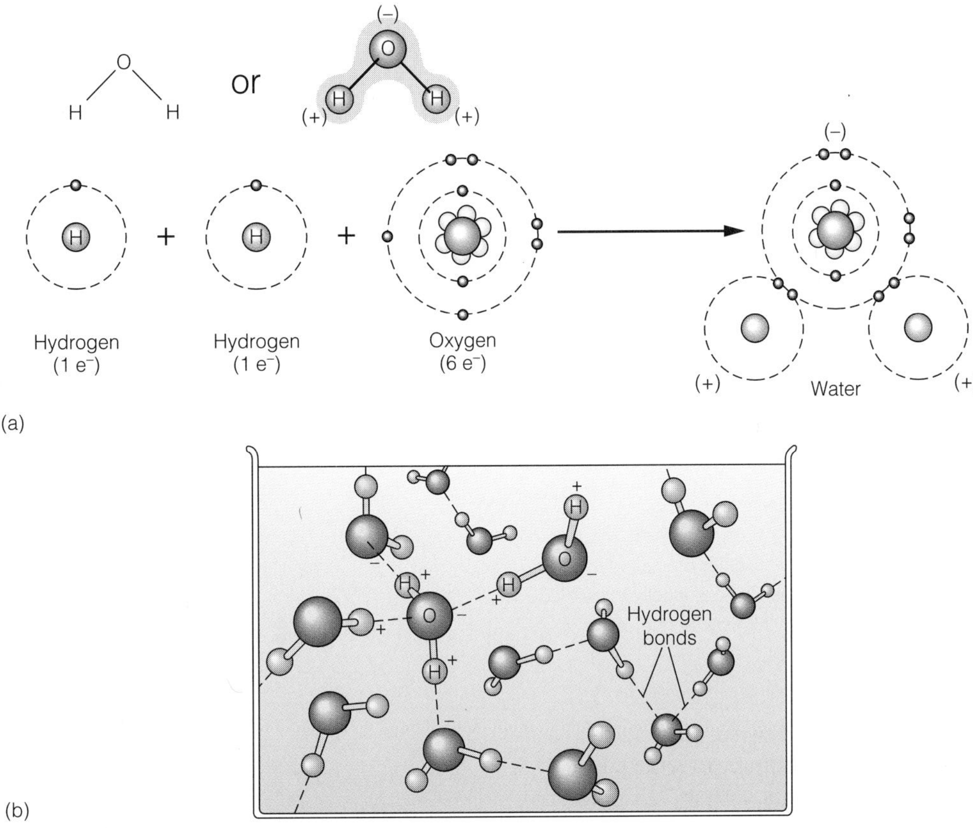

Figure 2–2 Water Molecules and Their Interactions

(a) Covalent bonds
Hydrogen and oxygen atoms interact to form water. The stick model (H-O-H) and the space-filling model next to it represent the polar structure of water molecules. Oxygen shares an e^- from each hydrogen atom to fill its outermost shell with $8e^-$. Each hydrogen atom shares an e^- from the oxygen atom to fill its outermost shell with $2e^-$.

(b) Hydrogen bonds in water
A hydrogen bond, represented by a dashed line, results from the attraction between the positive region of one water molecule and the negative region of another. Note that each water molecule is hydrogen-bonded to several other water molecules.

the association of hydrogen and oxygen very stable. It also influences the shape of the water molecule (Figure 2–2). Oxygen is a larger atom than the two hydrogen atoms combined, and the angle of the bonds between them is fixed by the electron orbitals. The angle of the bonds is crucial to the chemical properties of water. Water molecules are polarized with respect to shape and positive and negative charge distribution. This shape and charge distribution allows water to be cohesive and highly structured (Figure 2–2[b]), as well as to interact with other molecules that are polarized or carry a charge. This cohesiveness involves **hydrogen bonds,** which result from the attraction of the positive region of one water molecule with the negative part of another.

The number of covalent bonds a single atom can establish and the ease in which bonds are made and broken are key factors in the chemistry of life. A case in point is the ubiquitous element of life, carbon. Carbon atoms have four electrons in an outermost shell that can hold eight (Figure 2–3). Atoms tend to add or lose electrons to fill their outermost shells, which usually contain eight electrons (Figure 2–4). Each carbon atom can easily share electrons with up to four other elements at angles that optimize the separation of the linked atoms. Carbon atoms can potentially form multiple relationships with many other types of elements, making and breaking such associations with relative ease. The interactions of carbon with other elements

Hydrogen bond attractive bond between molecules by positive and negative charged regions, as in the case of interacting hydrogen and oxygen in water molecules

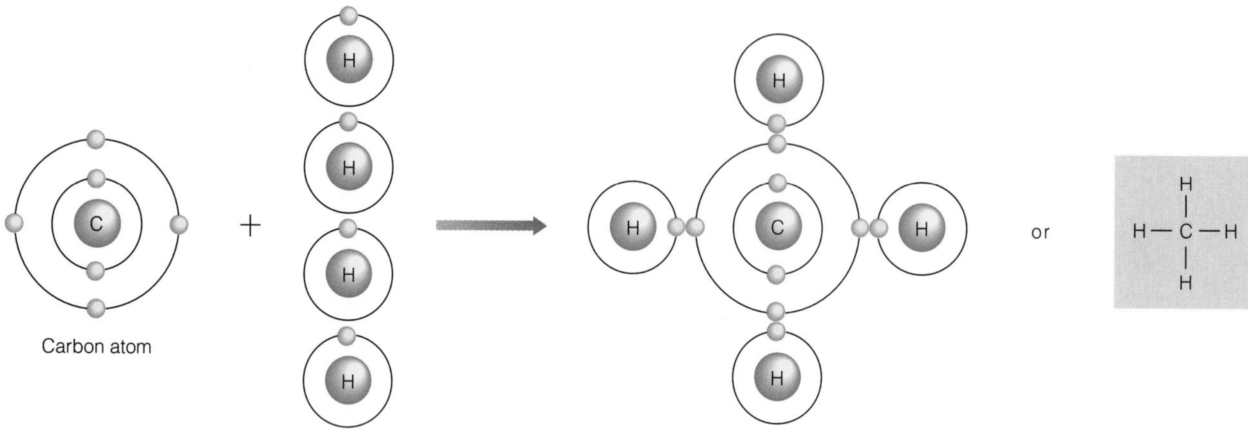

Figure 2–3 Carbon and Its Valency
The carbon atom has four electrons in its outermost shell. The tendency is for carbon to share electrons with other atoms to form four covalent bonds. As shown, four hydrogens share electrons with carbon to fill collectively their outer shells. CH_4 is methane, a common gas found in nature.

has proven to be extensive, diverse, and fairly easily altered. Hence, the chemistry of life is, in large part, based on understanding the chemistry of carbon.

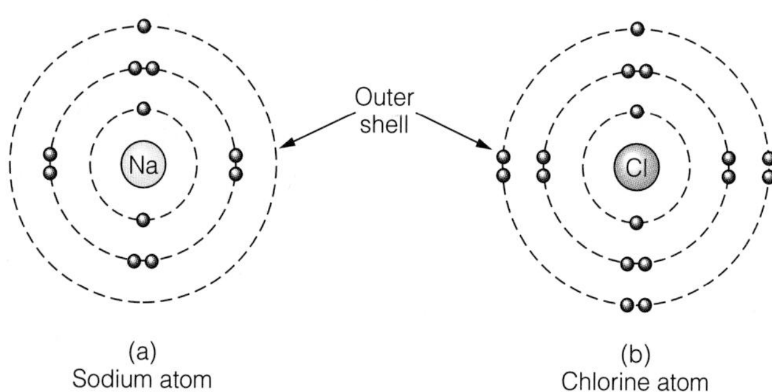

Figure 2–4 The Rule of Eight
The general rule of eight is that atoms tend to add or lose electrons to fill their outermost shells. In (a), sodium (Na) would tend to lose an electron from its outermost shell. This would result in a configuration of $8e^-$ in the newly formed outer shell and a positive charge of +1. In (b), chlorine (Cl) would tend to gain an electron to fill its existing outer shell. This would result in a configuration of $8e^-$ and a negative charge of −1.

Ionic Bonds

Another type of bond important in biological systems is the ionic bond (Figure 2–5). Unlike a covalent bond, an ionic bond requires interactions between atoms that carry either a positive or a negative charge. Such atoms are called ions. Formation of ions occurs by the loss of an electron from one element and the gain of that electron by another element. In ionic bonds there is no sharing of electrons—it is all or none. The loss of an electron leaves one atom with a net positive charge (because the number of protons is greater than the number of electrons). The gain of an electron by an atom produces in the other atom a net negative charge. Electrically charged ions occur in two types—**cations,** if positively charged, and **anions,** if negatively charged.

An example of ionic bonding is found in ordinary table salt, sodium chloride (NaCl) (Figure 2–5). The outer shell of sodium contains a single electron. The outermost shell of chlorine contains seven electrons. Sodium and chlorine atoms interact with each other by transference of the electron in sodium to chlorine, so that each ion attains a stable outermost shell with eight electrons. This tendency to fill an outermost shell (by losing or gaining an electron) is called the "rule of eight." A property of compounds with ionic bonds is that they readily dissolve in water. This is because of the ability of the polar water molecule to separate and surround sodium and chloride ions. Salt crystals added to water will disappear into solution in a matter of seconds. The ions thus dissolved may not be visible, but, if the water is allowed to evaporate, salt crystals will form once again.

Sodium and chloride ions are necessary for human survival. Elemental sodium, which is a silvery white solid, and chlorine, which is a yellow gas, are both highly toxic. It seems somewhat ironic that the loss, or gain, of a single electron in the outermost shell of an atom can transform that atom from a poison to a requirement for life. It should

Cations positively charged ions
Anions negatively charged ions

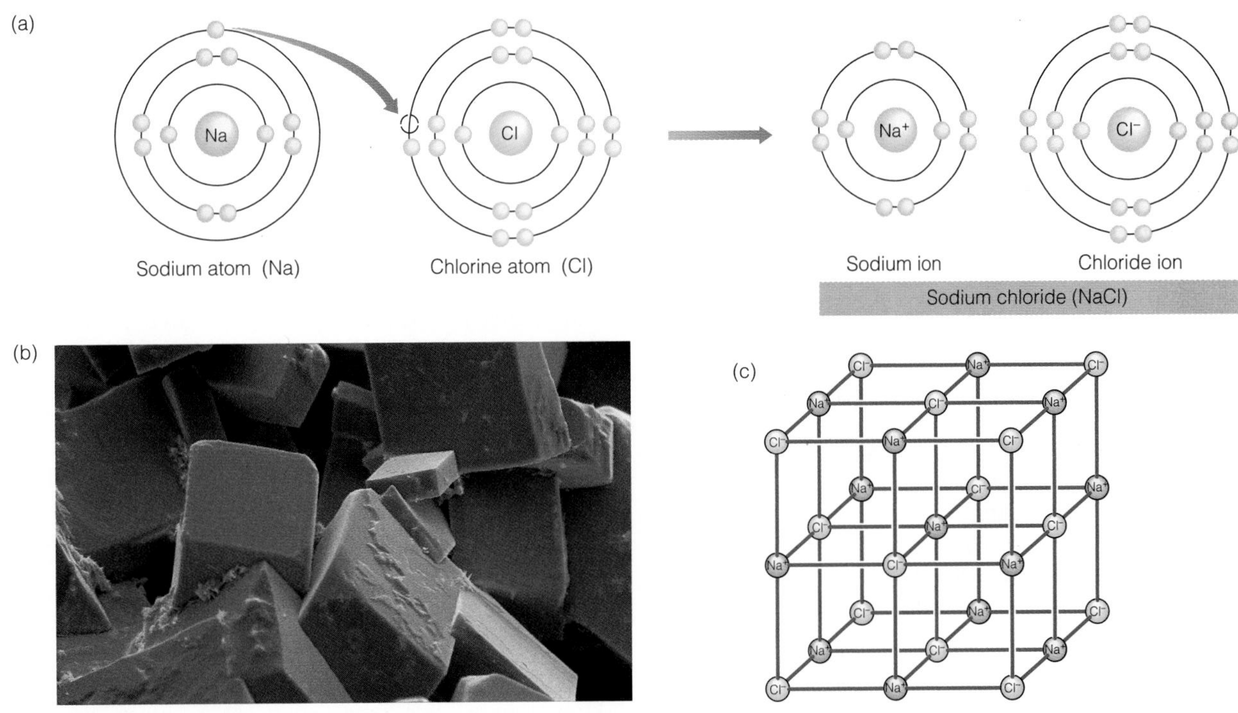

Figure 2–5 Ionic Bonds
(a) The electron of sodium is lost to chlorine, thus providing each atom with a filled outer shell. The Na is *positively charged* and the Cl is *negatively charged*. (b) The ions associate to form crystals. (c) NaCl ions are highly organized and form a crystal lattice, in this case common table salt.

be pointed out that some types of molecules held together by covalent bonds are easily dissolved in water as well. In this case the entire molecule dissolves in water even though the individual covalent bonds that hold the molecule together do not dissociate. For example, simple sugar molecules, such as the sucrose used to sweeten tea or coffee, do not dissociate as they dissolve. The individual sugar molecules composing the crystal are surrounded by water molecules and disappear into a clear solution.

Polarity

How does water surround molecules such as sucrose and dissolve them? To answer this question in more detail, a closer look at the structure of a water molecule is required (Figure 2–2). There are two end regions associated with the structure of a water molecule. These regions are called the poles of the molecule; thus, water is called a polar compound. Because electrons are naturally more attracted to oxygen (based on the size of the oxygen nucleus and configuration of electrons), the oxygen end of water is slightly more negatively charged. This results in the hydrogen end of the molecule having a slightly more positive charge (Figure 2–2). These features of electron distribution give water unique properties, one of which is its ability to interact with molecules that are polar or that have positive and/or negative charges. In this way, water has the capacity to surround and separate molecules, or ions from one another and, thus, to act as a solvent. Because so many different types of compounds have regions of positive and/or negative charge and are, therefore, polar, they can be dissolved in water.

There are many types of nonpolar molecules that cannot be dissolved easily in water. Such molecules generally lack any charge, positive or negative, and, therefore, have difficulty interacting with water molecules. Examples of nonpolar compounds include fats, waxes, oils, and petroleum products, which, when mixed with water or aqueous solutions, immediately separate into distinct **immiscible phases.** Failure of these types of molecules to dissolve is easy to demonstrate. The oil and vinegar of salad dressings do not stay mixed, and the fat and broth from chicken soup or beef stock separate quickly. The chemical basis for separation into different phases is the incompatability of polar and nonpolar molecules.

> **Immiscible phase** that which cannot be mixed; often refers to lack of solubility of a compound in water

ACIDS AND BASES

What is an acid? The image that comes to mind is a dangerous liquid that burns skin or eyes on contact. This is true in many cases but not in all. For instance, hydrochloric acid (HCl) is well known as a strong acid (Table 2–2). It is commonly used to etch metals and to acidify swimming pools to retard the growth of microorganisms. But perhaps surprisingly, it is also the acid produced by cells of the stomach that helps in chemical digestion of food (as described in Chapter 9). A strong acid, therefore, is not necessarily dangerous to health but, rather, may be essential for it. Sulfuric acid (H_2SO_4) is another very strong acid. It is not produced by living cells, but it has many valuable industrial uses. However, sulfuric acid can be a problem when it occurs in sufficient quantities in the natural environment. One of the main urban environmental concerns is conversion of sulfur-containing materials (such as those derived from the burning of sulfur-rich coal) to sulfuric acid in the atmosphere, which may in turn precipitate in the form of an acidic rain.

Not all acids are strong and/or dangerous. By comparison, **acetic acid** is a relatively weak acid. Its presence determines the sharp taste of vinegar (which is usually 4%–5% acetic acid). The active ingredient of aspirin tablets is **N-acetylsalicylic acid,** which is also a weak acid. Chemically, an acid is any molecule that releases a hydrogen ion (H^+) into solution as it dissolves. The molecule that remains after the hydrogen ion has separated from the parent molecule is called a **base.** Using dissociation of molecules of water, it has been possible to define a meaning and a measure of acidity and its base counterpart alkalinity. Water molecules dissociate to form H^+ ions (an acid) and OH^- ions (a base) and then reassociate again into a water molecule. Relatively few water molecules dissociate at any one instant, but all water molecules are capable of it (Biosite 2–2).

Table 2–2 Common Acids and Bases

ACID	BASE	APPLICATION OR SOURCE
Hydrochloric acid		Swimming pool, stomach acid
Acetic acid		Vinegar
Sulfuric acid		Industrial uses, acid rain from conversion of sulfur in coal
N-acetylsalicylic acid		Aspirin, analgesic
Nitric acid		Industrial uses, conversion of nitrogen oxides in air
	Sodium hydroxide	Drain cleaners, industrial uses
	Calcium hydroxide	Drain cleaners, industrial uses
	Aluminum hydroxide	Antacids
	Sodium hypochlorite	Bleach
	Ammonia	Fertilizers, cleansers, refrigerants

pH—A Measure of H^+ Concentration

The dissociation-association cycle of water to ions establishes what is called an equilibrium, or chemical balance; that is, for every water molecule that dissociates into ions, there is a pair of H^+ and OH^- ions associating into a water molecule. The measure of acidity is based on the concentration of H^+ ions in solution and is called the pH. The H, in pH, refers to the hydrogen ions that dissociate. pH is a familiar symbol and occurs on the labels of many products used every day (Biosite 2–2). Cosmetics, shampoos, soaps, antacids, and many types of food and drink have a pH value listed. Advertisers emphasize their products as "pH balanced" to assure us that they are concerned for our well-being, as well as our pocketbooks (Table 2–2).

In addition to acids, a variety of chemical bases are well known in a wide range of products found on market shelves (Table 2–2). The drain cleaners used to unplug household plumbing are composed of the hydroxides of sodium and potassium (NaOH and KOH). They are very strong bases (pH 13–14). The antacids used to relieve stomach upsets are composed of calcium and aluminum hydroxides, and they work to absorb acid in stomach juices. Knowing that excess stomach acid can be eliminated, or neutralized, by the addition of an appropriate counteractive chemical base is valuable. Compounds that resist, or reverse, changes in the pH of a solution are called **buffers.** Buffers act by chemically sponging up excess H^+ or OH^- ions (Biosite 2–2). An understanding of neutralization of acid comes directly from the principles of chemistry. Such principles can be applied beneficially to the human body, which is itself a chemical factory undergoing continuous and complex chemical reactions.

The pH of the Human Body

From a physiological point of view, controlling the pH of various fluids of the human body is no trivial matter. Blood, lymph, and digestive fluids each need to be main-

Acetic acid mild organic acid found commonly in vinegar

N-acetylsalicylic acid aspirin, a weak acid

Base opposite of an acid; hydrogen ion acceptor

Buffers compounds capable of neutralizing solutions of both acids and bases

BIO site 2–2

pH

The measurement of H^+ ion concentration on a pH scale has a range from 0 to 14. Seven (7) is the value that represents a neutral pH. This is based on the dissociation of water to H^+ and OH^-. There is an equilibrium established with no imbalance in the number of H^+ and OH^- ions. If the H^+ concentration increases, as with hydrochloric acid added to water, the pH goes down (as far as 0). If the H^+ concentration decreases, the pH value goes up (as far as 14). The scale for measuring pH is negatively logarithmic, which means that each number value on the scale, as it goes down from 14 to 0, represents 10 times more acid than the previous one. For example, an acidic solution with a pH of 5 has 10 times the number of H^+ of a solution at pH 6.

Buffers act to resist pH change in the face of an increase of H^+ or OH^-. They absorb the excess H^+ from acids or release H^+ in the presence of a base. Buffers usually consist of a mixture of a weak acid and its conjugate base. For example, in the blood interstitial fluids and cytoplasm a carbonic acid-bicarbonate buffer system is common. H_2CO_3 and HCO_3^- are the acid and base pair. The H_2CO_3 releases H^+ in the presence of base to form water ($H^+ + OH^- = H_2O$) and the HCO_3^- absorbs H^+ from acid to form the weak carbonic acid ($HCO_3^- + H^+ = H_2CO_3$).

tained at the appropriate pH for survival (Figure 2–6). The function of natural buffers in human blood is particularly relevant. The pH of blood normally is close to 7.4, which is considered slightly alkaline. If too much acid is present in the blood and the pH drops, a condition of **acidosis** arises. If too little acid is present, or there is too much base, the pH rises, establishing a condition called **alkalosis**. In either case, the body's pH balance is disrupted, and an afflicted individual may suffer serious harm, including coma and even death. The normal chemical components of blood, including large and small protein molecules as well as many types of ions, provide natural buffers in blood and keep fluctuations in pH to a minimum, and within safe limits. Diet and disease can affect significantly the presence and types of molecules produced by a body's tissues and, thus, alter the pH and the chemistry of the body.

To review the material covered up to this point, acids are hydrogen ion releasers, and bases are hydrogen ion acceptors. The strength of an acid depends on the ease with which it releases hydrogen ions, the number released, and their concentration in a solution. The increase in the presence of hydrogen ions is reflected in the measure of increased hydrogen ion concentration/activity, or pH. Awareness of simple chemical and biological principles, including those involving the functions and effects of acids and bases, provides a practical knowledge that may help protect us from ourselves.

SPECIAL REACTIONS

Getting Useful Energy Out of Molecules

The constant exchange of electrons among and between atoms brings about chemical changes that are a necessity of life. Many of these chemical reactions result in cellular energy production. However, not all molecules are equal in this regard; some kinds of molecules have more potential energy than others. Molecules with a relative abundance of carbon-hydrogen bonds, such as carbohydrates and fats, store much more useful and accessible energy for the human body than molecules that have fewer (or none), such as proteins, carbon dioxide, or alcohols.

Metabolic Reactions—A Slow Burn

The combustion or burning of natural gas, oil, coal, and the cellulose of wood, whose molecules are rich in C-H bonds, provide an example of energy-liberating reactions. Their combustion keeps our homes warm in winter and our cars running and helps cook our meals (Table 2–3). During the burning of gas, oil, coal, and wood, oxygen combines with hydrocarbons to form carbon dioxide, water, and a tremendous amount of heat energy. On an individual human scale, the cells of the body also produce energy, but they do so using chemical reactions to bring about the breakdown of carbohydrates (such as table sugar sucrose, which is composed of glucose and fructose) and fat, which are energy-rich molecules. Fortunately, the oxidation of the sugar that supplies the body with a major portion of its energy, as the candy

Acidosis a condition of blood or body fluids in which the pH becomes abnormally acidic

Alkalosis a condition of blood or body fluids in which the pH becomes abnormally basic

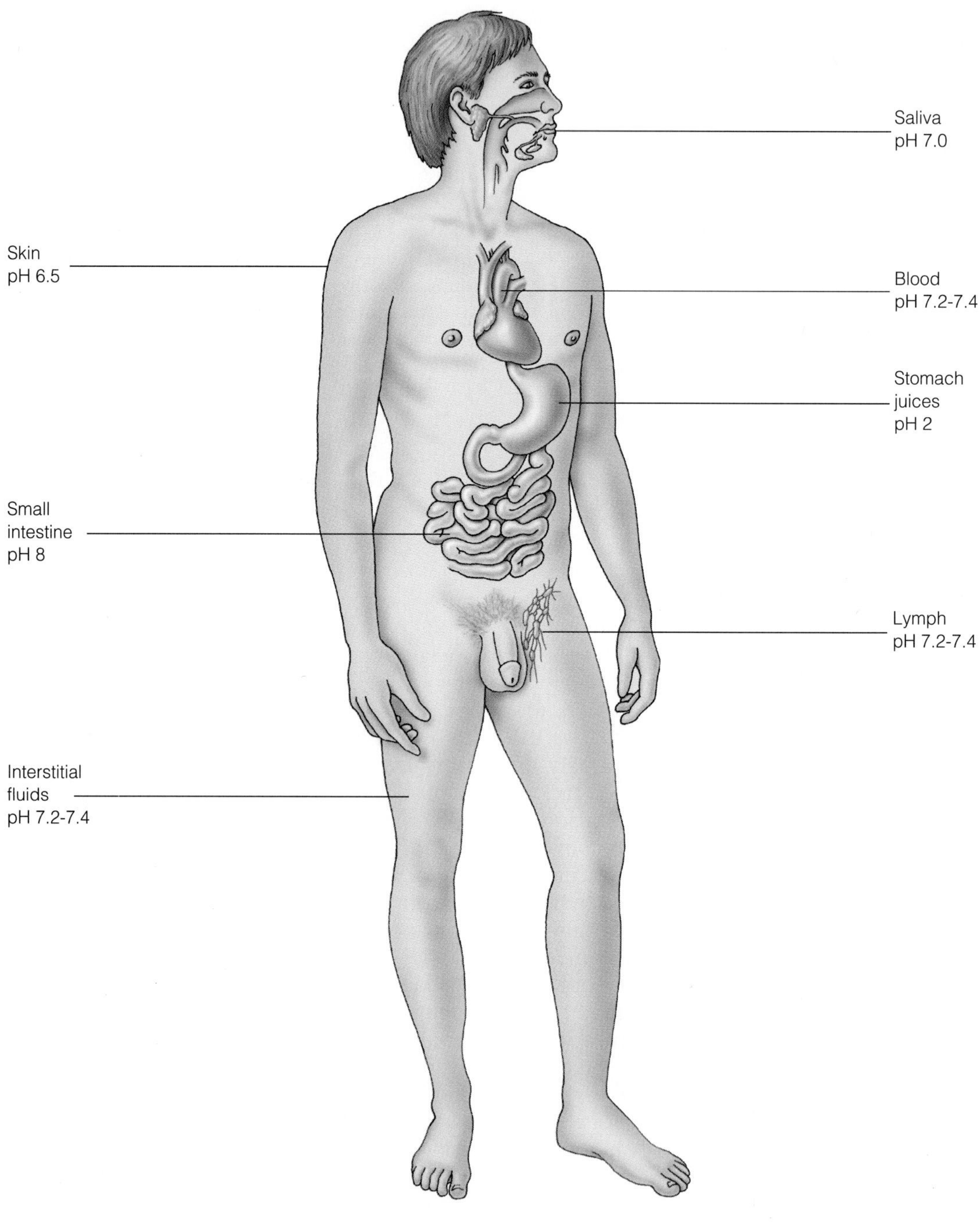

Figure 2–6 The pH of Different Regions of the Human Body

Table 2-3 Energy-Rich Compounds

COMPOUND	USE AND RATE OF ENERGY RELEASE
Coal	Heating, relatively rapid release (combustion)
Gasoline	Engines, rapid release
Oil	Heating, rapid release
Natural gas(es)	Heating, rapid release
methane	
propane	
butane	
Carbohydrates	Metabolic energy, slow release (biological)
glucose	
sucrose	
glycogen	
fats	

bar we may need before class, is a relatively slow process and does not release the total potential energy contained in the molecules of sugar all at once, as in the combustion of natural gas. Otherwise, our lives would be explosively short and sweet. For a glucose molecule, whether derived from cellulose or sucrose, the reaction is as follows:

$$C_6H_{12}O_6 + 6O_2 \rightarrow 6CO_2 + 6H_2O + \text{Heat} + \text{Useful energy}$$

The reactions that bring about the conversion of sugars to carbon dioxide, water, and energy take place within every cell in the human body. The study of this class of chemical reactions that take place within living cells has created a discipline that combines biology and chemistry in a specialty called **biochemistry.**

BIOCHEMISTRY—THE CHEMISTRY OF LIFE

From a biological point of view, the molecules that are most important for human life are found in four major classes—carbohydrates, lipids, proteins, and nucleic acids. Each of these classes of compounds represents a diverse family of related organic molecules that are involved in the biochemistry of cells.

ORGANIC MOLECULES

What is an organic molecule? By definition an organic molecule is a molecule that contains carbon. The definition of organic molecules is very broad. The exceptions to the rule are carbon dioxide and carbon monoxide (CO_2 and CO), which are generally considered to be inorganic molecules. Regardless of these exceptions, the special category of organic molecules serves to underscore the importance of carbon in biological systems. Carbon, with its capacity to form a multiplicity of covalent bonds, is central to the structure and function of each of the four major types of organic molecules listed above.

Common Elements

Carbon alone is insufficient to provide the diversity of molecules needed for life. The most common elements composing living organisms in addition to carbon are oxygen, hydrogen, nitrogen, sulfur, and phosphorous. Be aware, however, that not all biologically important molecules are organic. Some molecules, such as water, oxygen, and nitrogen, and many types of ions, such as sodium, calcium, potassium, and iron, are inorganic but clearly necessary for human life. It should come as no surprise that both organic and inorganic compounds in appropriate combinations are needed for all forms of life.

CARBOHYDRATES

Sugars and starches are in a class of molecules known as carbohydrates. They are composed primarily of carbon, hydrogen, and oxygen in a ratio of approximately 1:2:1. For example, a molecule of glucose, which is one of the most important sugars in the body, has a chemical formula of $C_6H_{12}O_6$, a ratio of C:H:O of roughly 1:2:1.

Carbohydrates are essential energy resources as well as building blocks for larger molecules produced throughout the body. Carbohydrates are categorized into two groups—simple and complex. Simple carbohydrates have one or two sugar units combined together. Complex carbohydrates may have thousands of sugar units strung together in long branched and unbranched chains (Figure 2–7).

Simple Sugars

The simple sugars are called **monosaccharides** and **disaccharides.** There is a large family of these compounds, whose chemical compositions are nearly identical. For instance, the formula for glucose, $C_6H_{12}O_6$, is the same as for many other biologically important six-carbon sugars. The six-carbon sugars are known as **hexoses** (Figure 2–7). There are two forms of monosaccharides "open" and "ring-structure." The ring structure is the most common form.

Biochemistry the chemistry of the products of living organisms

Monosaccharides single simple sugar

Disaccharides covalent combinations of two monosaccharides

Hexoses six-carbon sugars, i.e., glucose

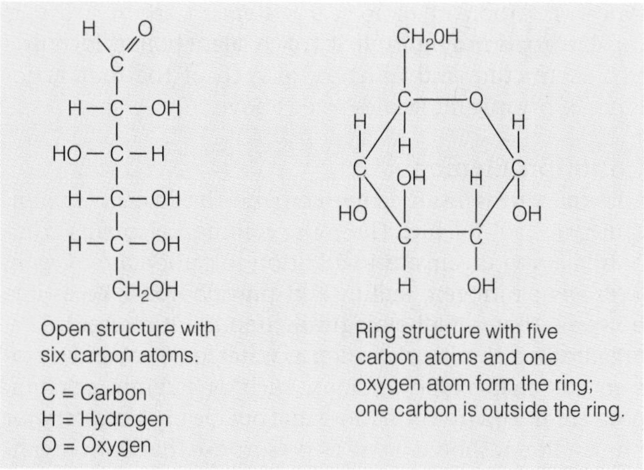

Open structure with six carbon atoms.

C = Carbon
H = Hydrogen
O = Oxygen

Ring structure with five carbon atoms and one oxygen atom form the ring; one carbon is outside the ring.

Figure 2–7 The Shape of Monosaccharides

(a) Monosaccharide (glucose, shown, or galactose)

(b) Disaccharide (lactose, maltose, or sucrose)

. . . glucose – glucose – glucose – glucose – glucose – . . .

(c) Polysaccharide

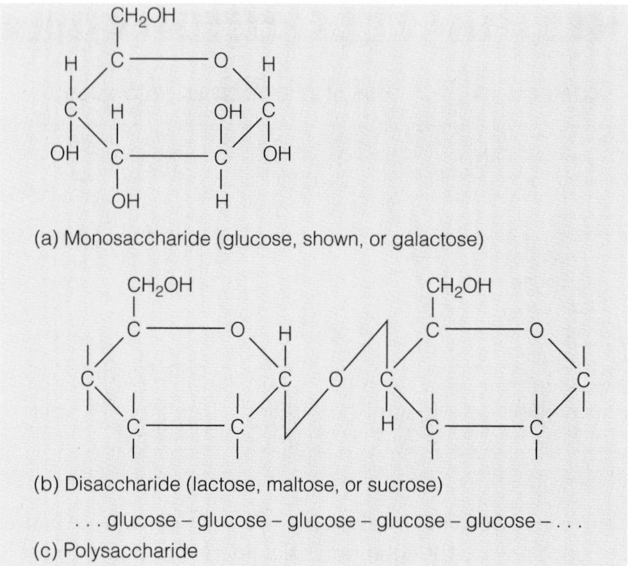

Figure 2–8 Monosaccharides, Disaccharides, and Polysaccharides

Many monosaccharides linked together form a polysaccharide (for example, glycogen). Polysaccharides are used as storage molecules in animals and plants. Some polysaccharides in plants are used structurally, such as for cellulose.

This also includes fructose and galactose. The differences between different hexoses are in how the atoms in each molecule are arranged. For humans, glucose is the primary molecular fuel driving cellular activities. Cells may be likened to tiny engines where glucose is equivalent to gasoline for cars. Monosaccharides are also important as building blocks in the synthesis of diverse biological molecules.

In addition to monosaccharides, there are many biologically important and naturally occurring disaccharides. These compounds are composed of two sugars covalently linked together (Figure 2–8). Those most commonly consumed by humans are sucrose, which is derived from green plants, lactose, which is part of mammalian mother's milk, and maltose, which comes from germinating seeds and the digestion of starch, a polysaccharide. These specific disaccharides are composed of different combinations of monosaccharides. Sucrose is a combination of glucose and fructose; lactose is a combination of glucose and galactose; and maltose is made of two glucose molecules (Figure 2–8[b]). However, a cell does not generally have the capacity to take up all types of disaccharides. Usually such sugars must be broken down into their monosaccharide constituents and then brought into the cell. In addition, nonglucose monosaccharides must be converted to glucose before being utilized as an energy source. This is true for all sugars that enter the body, including the complex sugars that we consume in the form of starches.

Complex Sugars—Cellulose, Starches, and Glycogen

When large numbers of single sugars, particularly glucose, are covalently linked together in long chains, they form **polysaccharides** (Figure 2–8[c]). Molecules such as glycogen, various types of starch, and the cellulose of plants are polysaccharides. How is it that glucose can be assembled into long polymers? From a structural viewpoint, glucose may be envisioned as having many chemical arms with molecules capable of forming covalent bonds with other sugars in a variety of different arrangements. The variety of covalent bonds is like a long, straight chain of human beings formed by holding hands. (Figure 2–9). If the right hand of one person were used to hold the left hand of another (the way we hold hands to walk together), all the individuals would end up facing the same direction no matter how many people formed the human chain. If hands were held left to left and right to right, the direction faced by each individual would alternate back to front—same people, different arrangement. Glucose molecules are basically analogous to this, except that each molecule has more "hands" than do humans. What makes polysaccharides different from one another, therefore, is not necessarily the presence of different monosaccharides but the pattern by which the identical sugar molecules are linked together.

Polysaccharides many monosaccharides covalently linked together to form long chains, i.e., glycogen or cellulose

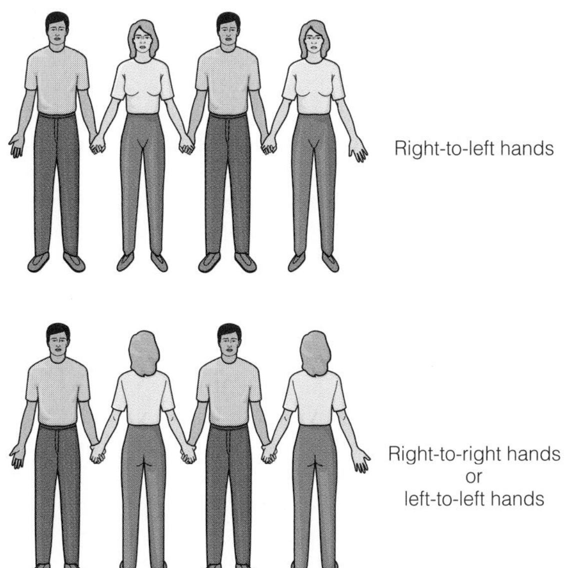

Figure 2–9 Polysaccharides and Holding Hands Analogy
The organization of polysaccharides allows for variation in the covalent linkages made between carbon atoms in monosaccharides. If right and left hands are used to form bonds between people, all individuals face the same direction. If right-to-right or left-to-left hands are used, it is clear that the orientation between individuals is very different.

Cellulose Glucose molecules can be linked in long straight chains, chains with alternating faces, short branched structures, as well as highly branched and extremely long molecules. The way in which the molecules are arranged is extremely important because cells of the human body do not have the ability to synthesize, or digest, all types of polysaccharides. A good example of this is the polysaccharide cellulose that is one of the main structural molecules synthesized by plants from glucose and, probably, one of the most abundant organic molecules on Earth. We ingest this material in many of the foods we eat, but our bodies lack the essential enzymes necessary for its digestion. Therefore, the glucose molecules trapped in cellulose are unavailable as sources of energy or as molecular building blocks for the cells and tissues of the human body.

This is not the case for other groups of mammals, such as cows and horses, who, because they are herbivores, depend exclusively on plants for food and have the capacity to digest cellulose. One might assume that a cow has digestive enzymes that we humans lack. Oddly enough, neither of these plant-eating mammals synthesizes the proper enzymes to break down or hydrolyze cellulose. How, then, do herbivores breakdown cellulose? The secret of a cow's success is the microbes that live in its complex, multicompartment stomach. These specialized, microscopic protists produce enzymes capable of breaking down cellulose to simple sugars, which in turn are used by the host animal. In return for the metabolic favor, the host cow provides a home for these guest microbes. This special host-guest relationship is called a **symbiosis.** This particular kind of host-microbe relationship does not occur in humans.

However, not all is lost with respect to the functional use of cellulose. In humans, many fibrous, and indigestible polysaccharides, such as cellulose, do have a valuable physiological significance. They provide fiber, or bulk, material needed to support the general movement, or motility, of ingested materials within and along the intestinal tract. It is fiber that helps give structure to the mass of material that passes through the intestinal tract. This process in turn promotes movement of the entire content of the gut, possibly prevents constipation, stimulates muscles of the digestive tract to retain tone, and may actually improve the body's ability to utilize glucose.

Starches and Glycogen Starches are also plant polysaccharides. They are derived from potatoes, rice, corn, and grains and are a main source of complex sugars in the human diet. The key storage polysaccharide synthesized by humans and other mammals is glycogen. Like cellulose, starch and glycogen are polymers of glucose. Unlike cellulose, however, both starch and glycogen are easily broken down by the enzymes of the human digestive tract and elsewhere in cells of the body. When more starch is ingested and converted to glucose than is immediately required for the energy needs of the body, (for example, the extra helping of spaghetti at dinner), the resulting excess of glucose is transported into cells, converted to glycogen, and stored (Figure 2–10). Cells of the liver are the main storehouse of this material, but it is also stored in other tissues, such as muscle. Between meals, when energy needs are not met with incoming supplies, glycogen reserves are broken down again to glucose. The glucose molecules are transported out of cells in the liver and into the blood for distribution to cells throughout the entire body. By controlling the recruitment of glucose into and out of storage, the human body is able to satisfy its energy needs on a continuous basis.

It is important to keep in mind that the human body is a massive chemical factory and the chemistry of carbohydrates is only one of many different processes that go on simultaneously within us. What makes carbohydrates so

Symbiosis living together of two dissimilar organisms in a mutually beneficial relationship

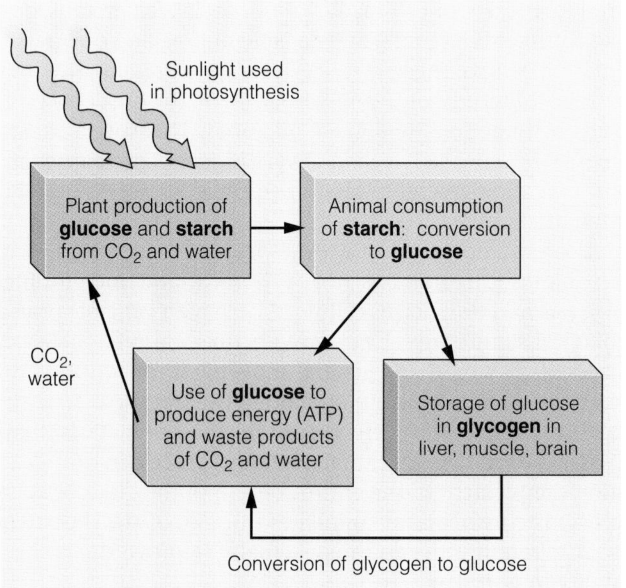

Figure 2–10 Utilization of Glucose, Glycogen, and Starch Formation of glucose and glucose-based polysaccharides in plants depends on a source of energy (sunlight), CO_2, and water. The energy stored in glucose is released in animal cell metabolism and produces CO_2 and water. Storage of glycogen is temporary.

vitally important is the fact that chemical energy is stored in these molecules and that this energy is readily accessible for cell metabolism. This situation is similar to having a savings account at a bank. You put extra money (if there is such a thing) into your account, so that when you need cash between paychecks you can withdraw money from savings. The human body has its own savings plan and it stores sugars and many other types of molecules for later use. The polysaccharides that can be digested by humans, such as glycogen and starches, are readily broken down into simple sugars.

While complex carbohydrates are used principally as sources of energy and fiber, they also may play an important role as chemical building blocks in other types of molecules. The presence of carbohydrates is also observed in covalent linkages to molecules of protein, lipid, and nucleic acids.

Glycoproteins, Glycolipids, and Nucleic Acids

Glycoproteins and **glycolipids** are composite molecules and are very important in the structure and function of cells. Glycoproteins are proteins to which sugars are covalently linked. Included in this group of molecules are hormones, enzymes, and structural proteins. One such structural protein is collagen, which makes up a large portion of human bones, connective tissues, such as tendons and ligaments, and the skin. Glycoproteins may be important in protecting the cells of the stomach and intestines from the acids and digestive enzymes used to degrade the food we eat. The class of molecules known as glycolipids provide another important example of molecular partnership. Glycolipids are composed of fats and carbohydrates. Glycolipids have a crucially important role in the structure of the plasma membrane. The basis for human blood types is determined in part by differences in glycolipids on the surfaces of red blood cells.

Another important group of molecules in which sugars play a fundamentally important structural and functional role are the **nucleic acids.** Among the most familiar molecules constituting this family of compounds are the **nucleosides** and **nucleotides** of the purine and pyrimidine bases found in DNA and RNA. The sugars ribose and deoxyribose are five-carbon sugars, or pentoses. They form an integral part of nucleosides and nucleotides and are necessary for the formation of polymers of DNA and RNA, as will be discussed in detail later in this chapter.

LIPIDS AND FATS

Lipids and fats are used as structural molecules for the assembly of the plasma membrane, intracellular membranes, and in the synthesis of storage materials. The classification of lipids is based largely on their insolubility, as opposed to relatedness in chemical structure (Figure 2–11). A variety of different types of molecules are considered to be lipids. In general, lipids and fats are nonpolar and insoluble in water. Many lipids have fluid properties at body temperature. The lipids of plants, such as corn oil and olive oil, which are different in composition from lipids found in animals, are clear, thick liquids at normal room temperatures.

Triglycerides

Over 90% of the human body's fats are found in the form of **triglycerides,** the remaining 10% is made up of phospho-

Glycoprotein a combination of lipid and carbohydrates

Glycolipid a combination of lipid and carbohydrates

Nucleic acid DNA and RNA; composed of covalently linked nucleotides

Nucleoside a purine or pyrimidine base with covalently linked ribose or deoxyribose sugar

Nucleotide a nucleoside with one or more phosphate groups covalently linked to the ribose or deoxyribose sugar

Lipids class of hydrophobic molecules composing fats, waxes, and sterols

Triglycerides a class of lipids characterized by linkage of three fatty acids to a glycerol molecule

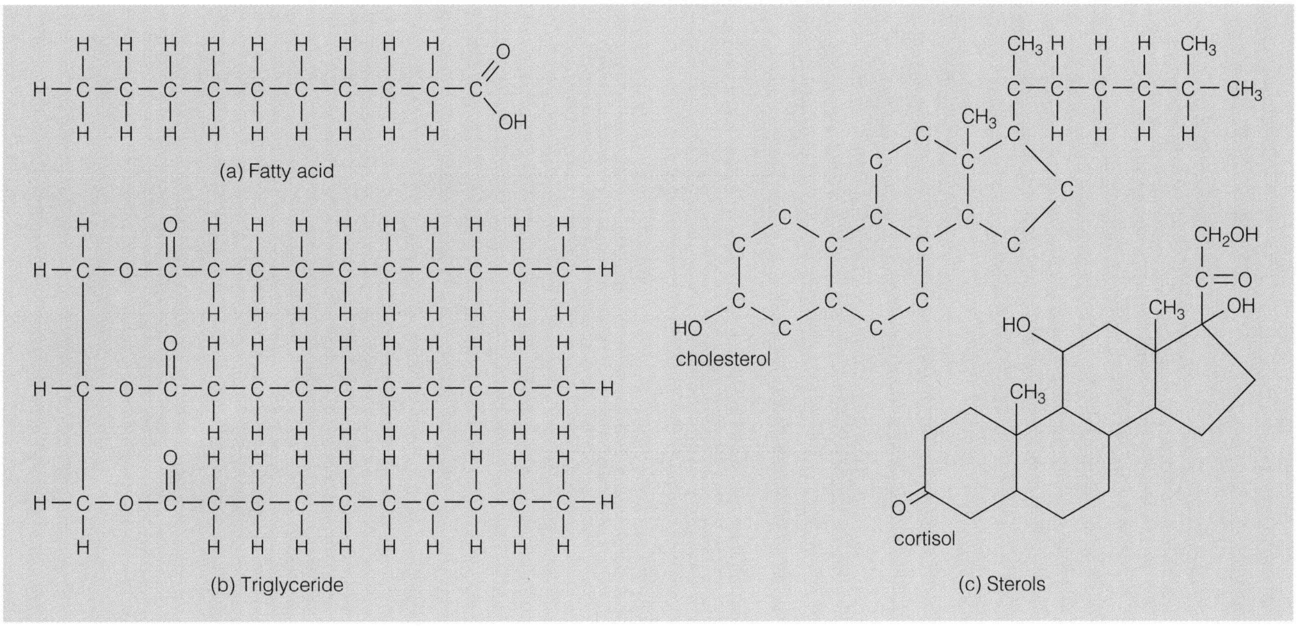

Figure 2–11 Types of Lipids
(a) Fatty acids are long-chain hydrocarbons with an acidic group (C=O)OH at one end of the molecule. (b) Three fatty acids are linked together to form a triglyceride. (c) Sterols are multi-ring compounds with different chemical groups covalently linked to the rings.

lipids, glycolipids and **sterols** (Figure 2–11). Triglycerides are the body's major storage form of fat. Triglycerides are composed of three fatty acids (Figure 2–12), which are generally long, nonpolar hydrocarbon chains, linked to a molecule known as **glycerol.** Glycerol is a three-carbon compound with a fatty acid chain linked to each carbon. Triglycerides are stored in the **adipose,** or fatty, **tissues** of the body.

Triglycerides are also found circulating in the blood unassociated with membranes or adipose tissue. Because of their long hydrocarbon chains, these molecules have a higher ratio of hydrogen to carbon than do carbohydrates of similar size. This is important in the energy economy of an organism because oxidation of fats releases nearly twice as much chemical energy as does the oxidation of equivalent amounts of starch.

Storing Energy-Rich Molecules—Carbohydrates Versus Fats

How is it that cells store two different types of energy-rich molecules? Why not just carbohydrates? Why not just fats? The problem is one of space and metabolic recruitment. Polysaccharides, such as glycogen, take up a large volume of space within a cell, but they are readily available for conversion to glucose. The reason for a large space requirement is that polysaccharides have a high affinity for water, with which they surround themselves. The association of water molecules with glycogen results in a composite that takes up a great deal more volume than if it were waterless (Figure 2–13[a]). Imagine a box full of dry, compressed sponges. They take up a limited amount of space. However, if water is added, the sponges expand dramatically to many times their original volume. This is similar to the situation for the storage of polysaccharides within cells, in which the complex sugar molecules absorb water in the cytoplasm and expand to fill the space needed to contain them.

What about the storage of fats? Fats are spatially more economical as reserve materials because they are nonpolar and do not associate readily with water. In addition, the molecular composition of fats is such that they store up to twice the chemical potential energy as an equivalent number of polysaccharide molecules. Also, fats are commonly compartmentalized into specialized fat cells, which in turn

Sterols a class of lipids with multi-ring structure to which steroids and cholesterol belong

Glycerol a three-carbon compound used as a backbone for the formation of triglycerides and phospholipids

Adipose tissue fatty storage tissues of the body

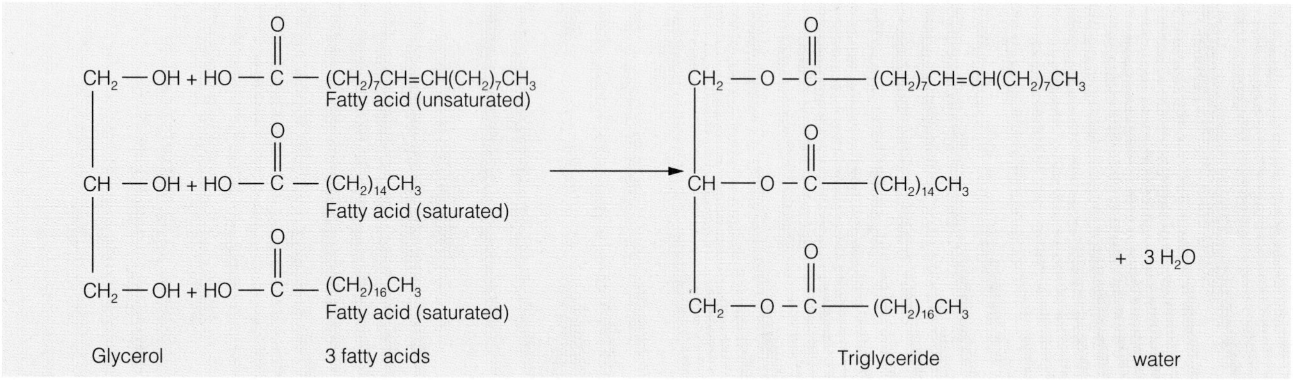

$$
\begin{array}{llll}
CH_2-OH + HO-\overset{O}{\overset{\|}{C}}-(CH_2)_7CH{=}CH(CH_2)_7CH_3 & & CH_2-O-\overset{O}{\overset{\|}{C}}-(CH_2)_7CH{=}CH(CH_2)_7CH_3 & \\
\quad\quad\quad\quad\quad\quad\quad\quad\quad\quad\text{Fatty acid (unsaturated)} & & & \\
CH-OH + HO-\overset{O}{\overset{\|}{C}}-(CH_2)_{14}CH_3 & \longrightarrow & CH-O-\overset{O}{\overset{\|}{C}}-(CH_2)_{14}CH_3 & \\
\quad\quad\quad\quad\quad\quad\quad\quad\quad\quad\text{Fatty acid (saturated)} & & & + \; 3\,H_2O \\
CH_2-OH + HO-\overset{O}{\overset{\|}{C}}-(CH_2)_{16}CH_3 & & CH_2-O-\overset{O}{\overset{\|}{C}}-(CH_2)_{16}CH_3 & \\
\quad\quad\quad\quad\quad\quad\quad\quad\quad\quad\text{Fatty acid (saturated)} & & & \\
\text{Glycerol} \quad\quad\quad \text{3 fatty acids} & & \text{Triglyceride} & \text{water}
\end{array}
$$

Figure 2–12 Formation of Triglycerides
Triglycerides are composed of three fatty acid molecules (saturated or unsaturated) and a glycerol molecule. The linkage of the three fatty acids generates water. This reaction can proceed in both directions—one to synthesize triglycerides for storage in adipose tissue and the other to degrade fats to supply energy to the cells.

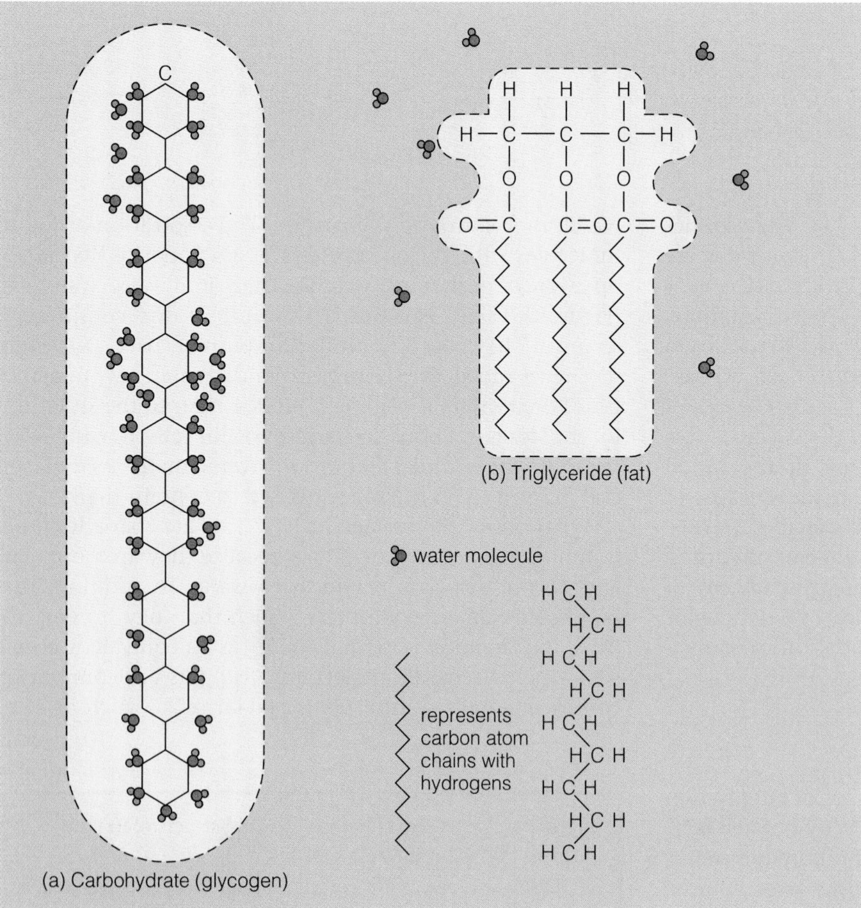

Figure 2–13 Storage Molecules—Space and Energy
The volume taken up by equivalent size molecules of polysaccharides and fat is determined by water surrounding the two types of molecules. Because carbohydrates bind water and lipids do not, the volume taken up by carbohydrates is much larger. Energy available in triglycerides is nearly two times as much as in an equivalent amount of carbohydrate. Thus, lipids are more space saving and energy rich than carbohydrates. However, they are not as readily utilized as are carbohydrates.

are organized into fatty tissues. The drawback to the exclusive storage of fats as an energy resource is that fats are not as easily or as rapidly recruited to supply energy as are the polysaccharides. This establishes the well-balanced, dual storage phenomenon observed in the human body. A limited supply of complex carbohydrates (glycogen in cells) is broken down first to supply immediate chemical energy as a greater reserve of fats is more slowly recruited. Thus, to serve both short-term and long-term needs, the body stores two fundamentally different types of energy-rich molecules.

A Nonenergy Role for Fats

Fatty tissues also serve in nonenergy related roles in the body. For instance, they act as natural "shock absorbers," or cushioning devices, for many organs. The fat pad around the kidneys helps prevent damage from bumps, falls, and bodily traumas that might otherwise jar these organs loose from their attachments to the body wall (see Chapter 10). The fatty layers of the skin help insulate organisms from the cold and give buoyancy in water. Mammals of all kinds use fat for insulation. Whales, porpoises, and seals, for example, have thick layers of body fat to protect them from the cold of the open seas.

Saturation of Fats

The hydrocarbon structure of a fatty acid molecule is chemically characterized by the degree of its saturation, a term that denotes the number of carbon-hydrogen bonds in the molecule. This characteristic is often listed on the label of food products. For instance, let us compare butter, which is animal fat, and margarine, which is produced from plant fats (Figure 2–14). There are fewer hydrogen atoms per carbon atom in plant fats. As a result of this chemical difference, fatty acids of plants are considered to be less

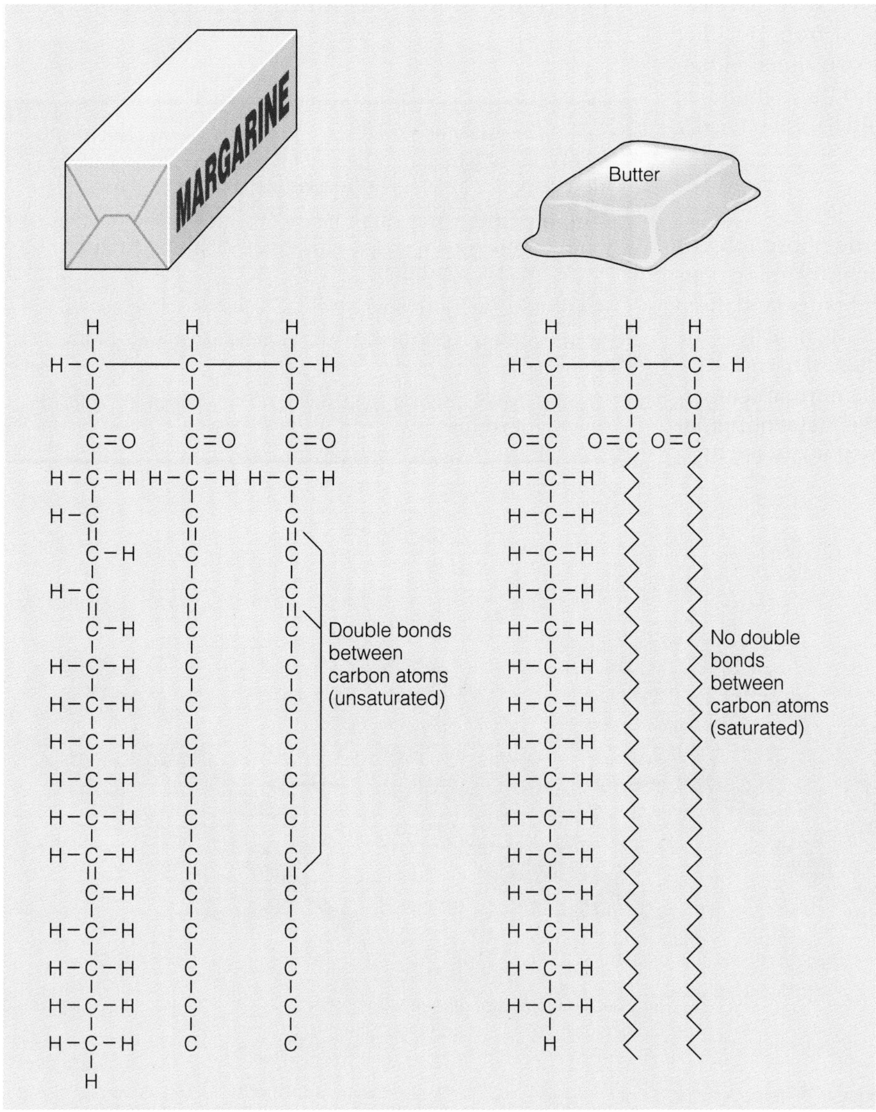

Figure 2–14 Saturated Versus Unsaturated Fats (Butter Versus Margarine)

The presence of double bonds in margarine is characteristic of unsaturated fats. These fats have a lower melting temperature and fewer hydrogen bonds.

saturated. The fatty acids in butter, on the other hand, are usually completely saturated. If one hydrogen-carbon bond is eliminated in a fatty acid hydrocarbon chain (a chemical reaction called **dehydrogenation**), the fatty acid is considered to be monounsaturated. If two or more hydrogens are eliminated, the molecule is considered to be polyunsaturated.

In terms of human health, there is a correlation between the level of saturation of dietary fats and cardiovascular disease. One of the major contributors in arteriosclerosis (hardening of the arteries) is the association of saturated fats and cholesterol into plaque, which develops along and within the walls of blood vessels. This type of arteriosclerosis is called atherosclerosis. During the process of atherosclerosis plaque may increase in amount along the artery wall to such an extent that it interferes with or even completely blocks the flow of blood. The blockage of blood vessels in the heart or brain can lead to serious heart disease, strokes, and death (Figure 2–15). However, it is not clear that all saturated fat should be eliminated from our diet and that only unsaturated fats should be consumed. A balanced diet should include both types, and they should be derived from natural as opposed to artificial, chemically modified sources.

Phospholipids and Glycolipids

Two other important groups of fats are the phospholipids and glycolipids. They are found predominantly in the plasma membrane and other cellular membranes and help establish a semipermeable barrier that demarcates the inside of a cell from the outside environment (Figure 2–16). Phospholipids tend to be semisolid at the normal temperature of the human body (on average 38° C). The nonpolar properties and viscosity of phospholipids at relatively high temperatures account in part for the functional and structural integrity of cellular membranes. Phospholipids and glycolipids have basically the same organization as triglycerides, except that phosphate groups associated with organic molecules such as choline, or carbohydrate molecules are covalently linked to the glycerol backbone. These types of alterations of lipids generally make the molecule **amphipathic,** which means it is **hydrophilic** (soluble in water) at one end, and **hydrophobic** (insoluble in water) at the other. These are precisely the properties needed to establish and maintain the lipid bilayers of cell membranes, as will be presented in Chapter 3.

Sterols and Steroids

Sterols, of which **cholesterol** is among the best known and most notorious, are found in combination with phospholipids and glycolipids in membranes and in combination with triglycerides in stored fats. The structure of sterols is quite different from the structure of triglycerides, phos-

Dehydrogenation removal of hydrogen atoms, specifically removal of hydrogen in fatty acids to form carbon-carbon double bonds and less saturated fats

Amphipathic chemical property of a molecule in which both positive and negative charges exist simultaneously under certain conditions

Hydrophilic "water loving"

Hydrophobic property of molecules that are not soluble in water

Cholesterol common lipid substance in mammals; part of the sterol family

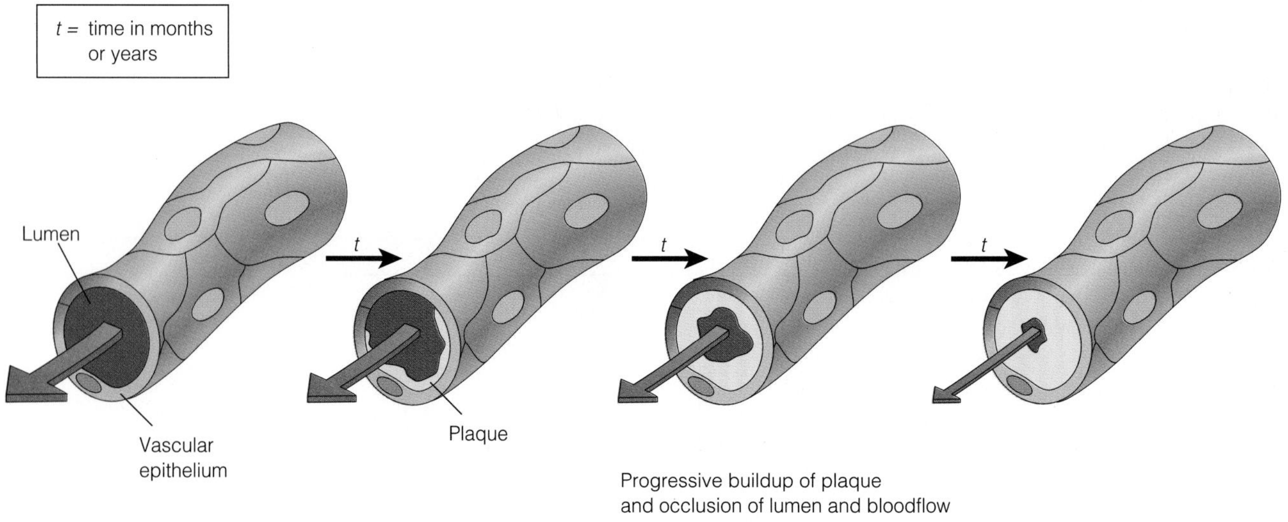

Figure 2–15 Progressive Buildup of Plaque in Atherosclerosis

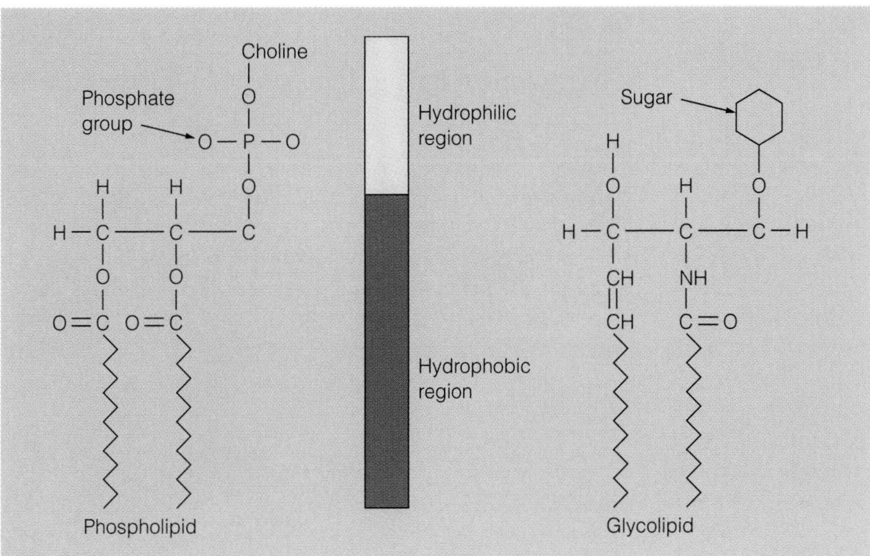

Figure 2–16 Phospholipids and Glycolipids
The phospholipids and glycolipids are found in the cellular membrane. There are some differences in structure between these two lipids, but both have a hydrophilic region and a hydrophobic region that allow them to form lipid bilayers.

pholipids, and glycolipids. Sterols and steroids form multiring structures (Figure 2–11). Cholesterol is by far the most commonly recognized member of the sterol chemical family and is the precursor for many other sterols, including all the steroid hormones. Cholesterol molecules are particularly important in plasma membranes where they are capable of inserting themselves between phospholipids and glycolipids.

A wide range of different types of steroid hormones is essential to regulate growth and development and to repair of tissue damage in the human body. In recent years, steroids have also gained notoriety through their misuse by athletes and bodybuilders (Table 2–4). Certain types of steroids, particularly those based on the structure of the male hormone testosterone, provide a chemical signal for rapid and profound increases in muscle mass and, thus, physical strength and endurance. The cost in human terms of such Herculean changes in body structure are increased probabilities of heart and kidney failure, impotence, and potentially drastic and antisocial changes in behavior.

PROTEINS AND AMINO ACIDS

Proteins are arguably the most influential and important molecules in the body. The name protein comes from the Greek word *proteios,* which means first. Proteins and protein-containing molecules form both the bulk of the structural materials that make up the body (bones, connective tissue, skin, and muscle) and the preponderance of enzymes, which act as catalysts in most chemical reactions in living organisms and control the rates of those reactions. The only class of molecules to challenge the preeminence of proteins are the nucleic acids, which in the form of DNA and RNA ultimately control the production of proteins. However, it should be pointed out that the synthesis of both DNA and RNA depends on enzymes. The arguments over which types of molecules are the most important is largely moot, however, because the two groups are absolutely required for all living systems.

Amino Acids—Building Blocks of Proteins

Proteins are constructed from a family of molecules known as amino acids. There are 20 different amino acids commonly used to make human proteins. Each of the amino acids shares some characteristics with all the others, and at the same time each is unique. The unique portion of an amino acid always occurs at the same carbon atom in the amino acid backbone and is generally referred to as a reactive group or R group (Figure 2–17). The R group may be as simple as a single hydrogen atom or a sulfur atom linked to a hydrogen atom (SH group), or as complex as the long-chained structures of Lysine (Figure 2–17). These molecules are classified as acids because, when they are dissolved in water, they release hydrogen ions into solution.

Table 2–4 Steroids and Their Biological Effects

STEROID(S)	EFFECT(S)
Sex steroids	Metabolic, developmental, sex determination
testosterone	Male secondary sex characteristics, fertility
estrogen and progesterone	Female secondary sex characteristics, fertility, ovarian cycle, menstruation
Mineralocorticoids	Influence salt and water regulation and bodily fluid balances
Glucocorticoids	Influence general cell metabolism, involved in inflammatory reactions

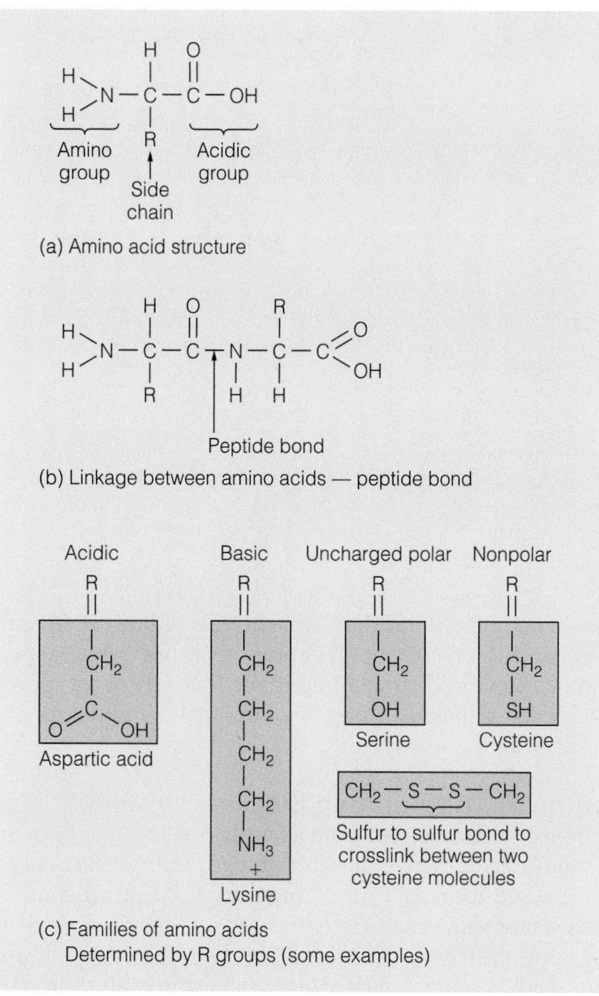

Figure 2–17 Amino Acids

Table 2-5 Amino Acids and Their Symbols

SYMBOL	ABBREVIATION	AMINO ACID
A	Ala	Alanine
C	Cys	Cysteine
D	Asp	Aspartic acid
E	Glu	Glutamic acid
F	Phe	Phenylalanine
G	Gly	Glycine
H	His	Histidine
I	Ile	Isoleucine
K	Lys	Lysine
L	Leu	Leucine
M	Met	Methionine
N	Asn	Asparagine
P	Pro	Proline
Q	Gln	Glutamine
R	Arg	Arginine
S	Ser	Serine
T	Thr	Threonine
V	Val	Valine
W	Trp	Tryptophan
Y	Tyr	Tyrosine

Amino acids are building blocks for the synthesis of proteins (Table 2-5). Chemical reactions that allow individual amino acids to be linked together, as if beads in a necklace, involve the formation of covalent peptide bonds. Dipeptides, tripeptides, and polypeptides are so named because they involve the linkage together of two, three, or many amino acids, respectively. Polypeptides may form extremely long chains of hundreds or thousands of amino acids that twist and fold in characteristic shapes.

Primary, Secondary, Tertiary, and Quaternary Structure

Primary Structure Large polypeptides (that is, peptides containing more than 100 amino acids) and/or associations of two or more separate polypeptides are called proteins. The primary structure of a protein is the sequence of covalently linked amino acids. Such a sequence may be hundreds of units long and is unique to a specific type of protein (Figure 2–18[a]).

Secondary Structure The secondary structure of a protein is the arrangement of the chain of amino acids in three dimensions. All objects in the real world have 3-D geometry, including proteins. There are three basic types of geometric arrangements for proteins observed in nature. The two major arrangements are the sheetlike beta configuration and the helically twisted arrangement known as the alpha helix (Figure 2–18[b]). The third type of arrangement lacks alpha and beta structure and is called a random coil. Alpha and beta protein geometries depend on hydrogen bonds, which, although weaker than covalent or ionic bonds, are strong enough to stabilize the shape of a polypeptide. Hydrogen bonds are electrostatic in nature; that is, they depend on the attraction of positive and negative charges carried on atoms within a polypeptide. Such bonds occur predominantly between hydrogen atoms and oxygen atoms in different amino acids within the polypeptide chain.

Tertiary Structure The third level, or tertiary structure, of a protein involves even more folding and twisting, often into a spherical or globular shape. The interaction of the R groups on different amino acids of a polypeptide chain, or between different parts of a polypeptide chain, accounts for tertiary structure (Figure 2–18[c]).

$^{+}H_3N$– His Ser Glu Gly Thr Phe Thr Ser Asp Tyr Ser Lys Tyr Leu Asp Ser Arg Arg Ala Gln Asp Phe Val Gln Trp Leu Met Asn Thr – COO^{-}

(a) Primary Structure

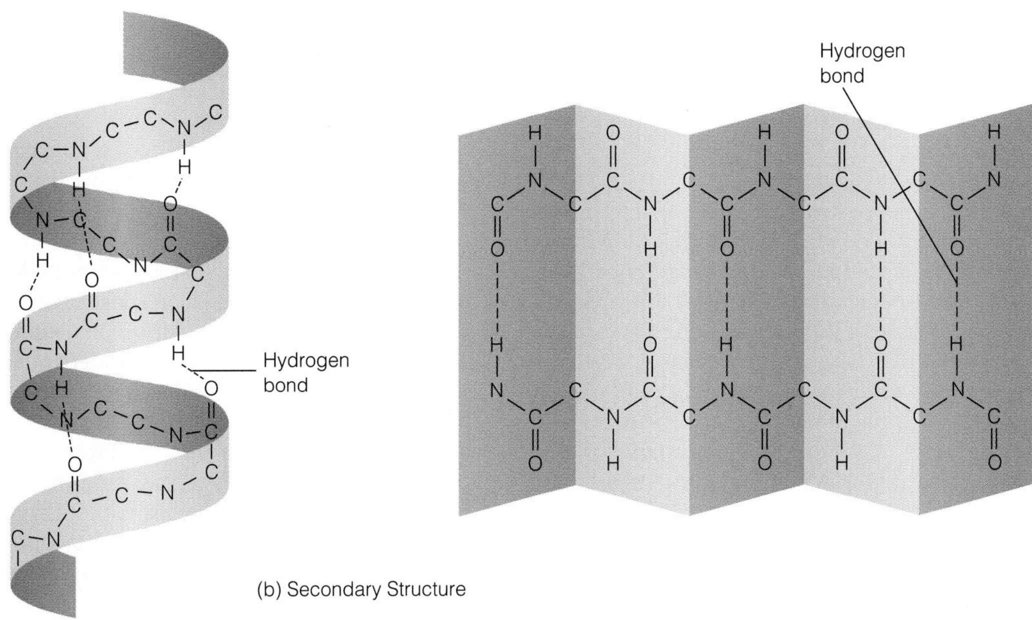

(b) Secondary Structure

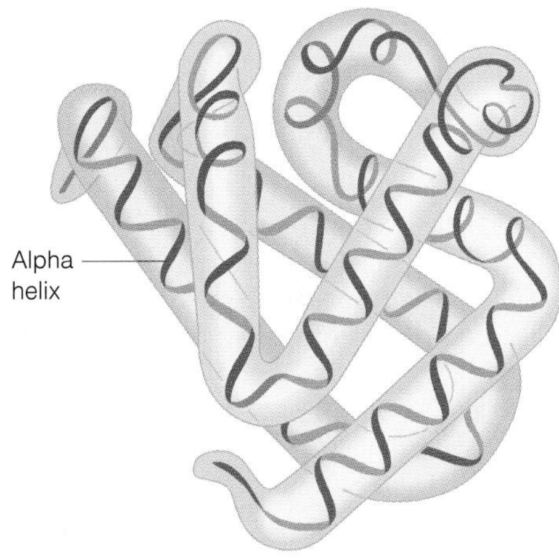

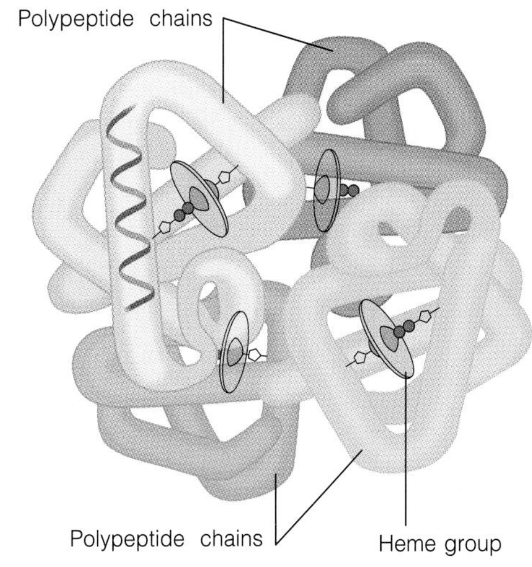

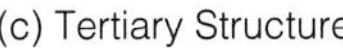
(c) Tertiary Structure

(d) Quarternary Structure

Figure 2–18 The Structure of Proteins

(a) The primary structure of a protein or polypeptide is its sequence. (b) The secondary structure involves folding of the protein into alpha or beta structures. (c) The tertiary structure involves further folding in space and establishment of domains in which alpha, beta, or random structures are formed. (d) Quaternary structure is the association of two or more proteins together, as shown for the protein hemoglobin in this diagram.

Sulfur Bonds Sulfur-sulfur covalent bonds (S-S) (Figure 2–17[c]) restrict the shape of proteins by cross-linking different regions of the polypeptide. To envision this type of linkage, imagine your whole body is a single protein molecule. What happens if you tie your hands and feet together with a short rope? Try running. Your new shape, as in the shape of a cross-linked protein, affects the way in which your whole body is able to function (Figure 2–19). This is roughly equivalent to the S-S linkages that occur between two cysteines (S-H) within a polypeptide chain. The shape and flexibility of a protein are very important to the function of a protein.

A Twisted Rope Tertiary structure does not alter primary or secondary structure. The organization of a protein molecule at each higher structural level is constrained by the limitations of the previous level. Envision a protein as a piece of rope with a series of knots tied along the rope one-inch apart (Figure 2–20). The number of knots represents the number of amino acids in a polypeptide and each knot can be any one of 20 different amino acids. The order of the representative amino acid knots is the primary structure. The secondary structure of such a model is represented by wrapping the rope around a flexible rubber tube to form a helix. It can be wrapped either left or right, but in nature nearly all helical proteins are right-handed or alpha helices. Coiling of the flexible rubber tube with the polypeptide rope wrapped around it throws the protein into an even more complex three-dimensional array. This category of coiled coils (the rope is coiled and the tube is coiled) is analogous to the tertiary structure of a protein. R group interactions (such as sulfur-sulfur bonds in cysteine) act to stabilize or initiate the folds and twists at special sites along the primary sequence. This model is somewhat oversimplified because it leaves out consideration of the actual sequence of amino acids, which determines the position of hydrogen bonds, sulfur-sulfur bonds, and the points of turning in the polypeptide chain based on a specific sequence. However, the model does offer a tangible way to visualize the complex three-dimensional geometry of a polypeptide.

Quaternary Structure There is a final level of organization in proteins, called quaternary structure. This fourth level of organization takes into consideration the interaction between or among several independent polypeptides (Figure 2–18[d]). Many proteins are made up of two or more polypeptide chains. In these proteins each of the individual polypeptides is called a subunit. Proteins made up of two, three, four, or more polypeptide subunits are common. For example, hemoglobin, the principal oxygen-binding protein of red blood cells has four polypeptide chains (called globin) associated together. The globin polypeptides fit neatly into a quaternary arrangement through interactions

Figure 2–19 Protein as Body, Body as Protein

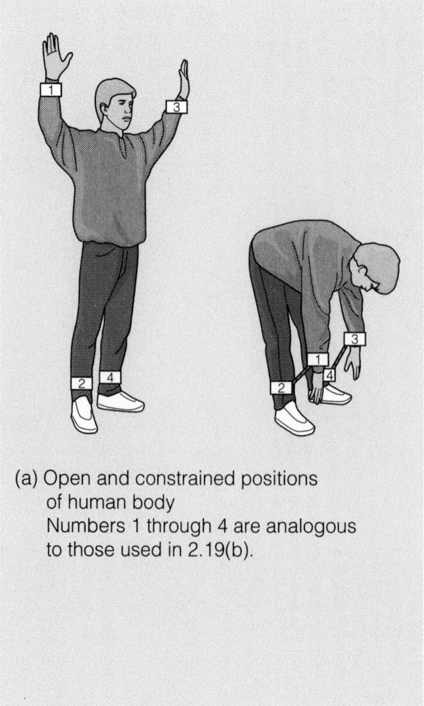

(a) Open and constrained positions of human body
Numbers 1 through 4 are analogous to those used in 2.19(b).

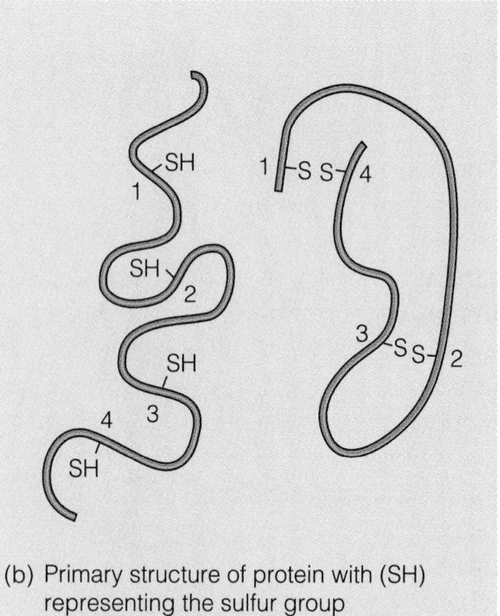

(b) Primary structure of protein with (SH) representing the sulfur group of cysteine. Numbers 1 through 4 indicate relative position of cysteines in the protein. Cross linkage of S-S constrains the shape of the protein (see also Figure 2.17[c]).

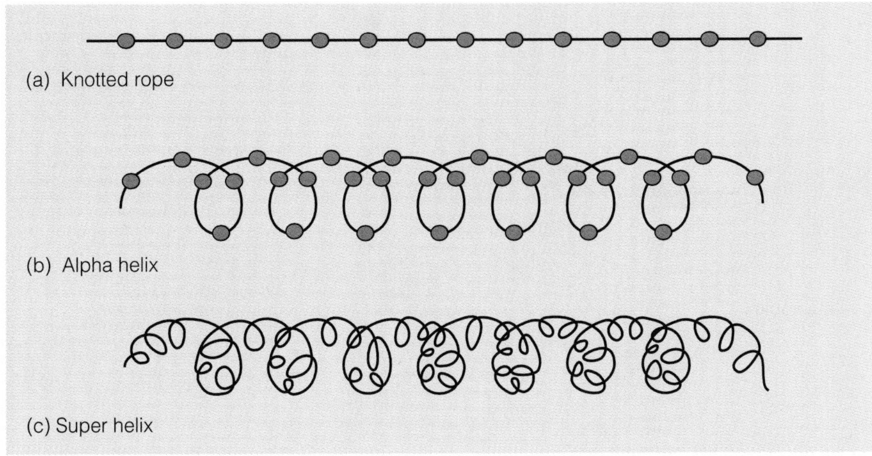

Figure 2–20 Model of the Twisted Rope
The twisting of a knotted rope introduces helical structure. (a) Each dot represents a knot and an amino acid with no twists. (b) Uniform twisting brings about an alpha helix formation. (c) Twists of twists

with an iron-containing complex called heme. The combination of the iron–heme complex and the globin polypeptides is hemoglobin. Slight changes in the sequence of amino acids in globin may change the shape of the proteins, which in turn may lead to an altered oxygen-binding function. This is potentially disasterous. A good example of this is sickle-cell anemia. This disease state results from a genetic alteration in DNA coding for a single amino acid in globin. The binding of oxygen to sickle-cell hemoglobin is altered, the cell shape is altered, and the individual suffers serious physical and physiological disability.

STRUCTURAL PROTEINS AND ENZYMES

There are two major categories of proteins—structural proteins, such as globin described above, and enzymes, to be considered in the next section. Structural proteins account for the majority of our body mass. Structural proteins are major components of skin, bone, and muscle, which together make up 75%–80% of human body weight. Structural proteins may function in purely structural protection, such as the keratins in the skin, in dynamic physical support, as is the case for collagen in the bones, ligaments, and tendons, and in cycles of contraction and relaxation, as observed in the action of actin and myosin in muscles. Table 2–6 lists well-known structural proteins, what they do, and where they are located in human tissues.

Enzymes and Catalysts

The other category of proteins, the enzymes, is involved in controlling the rate of chemical reactions. In this context, enzymes are biological catalysts. A **catalyst** is a compound that participates in a chemical reaction in such a way as to increase the chance (and, therefore, the rate) for that reaction to take place (Figure 2–21). In addition, the catalyst itself remains unchanged by the overall chemical reaction. Therefore, a single protein enzyme molecule may be used over and over in a cycle that results in the continuous formation of products.

The word *enzyme* comes from the Greek word meaning "in leaven," which was essentially an agent used to make bread rise. The action of enzymes has been the focus of a great deal of scientific research in the nineteenth and twentieth centuries (Biosite 2–3). Enzymes were first identified in fermentation reactions involving yeast. This process results in the production of alcohol, acetic acid (as in vinegars), and carbon dioxide gas. Historically, humankind has always been interested in fermented products either for ritualistic purposes or to enliven social interactions. Beers and

Catalyst an agent that increases the rate of a chemical reaction without itself undergoing permanent change

Table 2–6 Structural Proteins and Their Tissue and Cellular Distribution

PROTEIN	LOCATION IN TISSUE OR CELL TYPE
Actin	Muscle sarcomeres, cytoskeleton
Myosin	Muscle sarcomeres
Keratin	Epithelial cells, (for example, in the cytoskeleton of epidermal cells)
Collagen	Extracellular matrix of connective tissue, (for example, in bone)
Elastin	Extracellular matrix of connective tissue, (for example, of cartilage, large arteries)
Tubulin	Axonemes of cilia, flagella, cytoskeleton

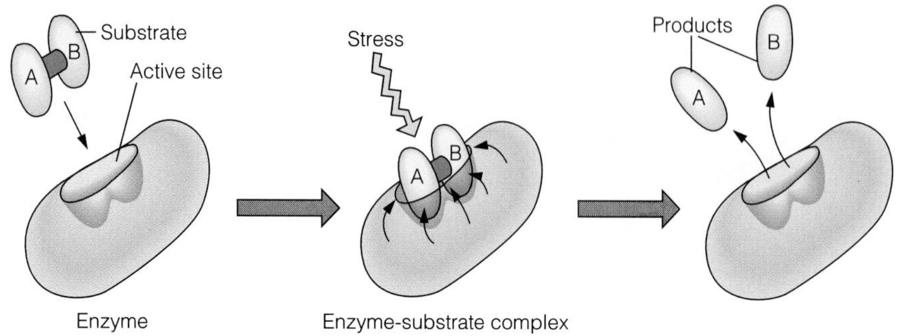

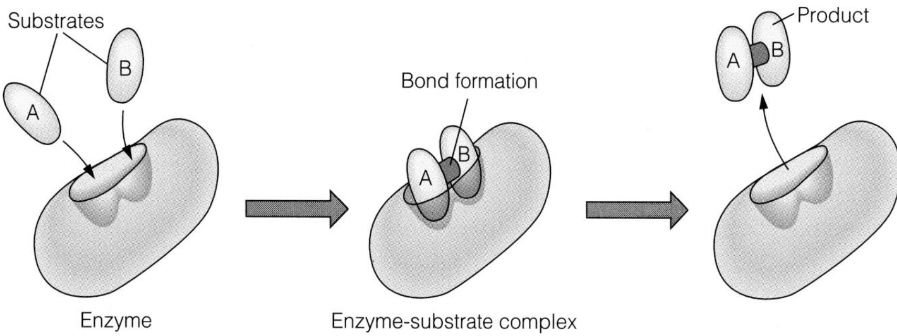

Figure 2–21 Enzyme Binding to Substrate and Processing
(a) The enzyme binds to the substrate in the active site. Induced fit puts stress on the substrate, converting it to products which are released. (b) The process is reversible, such that A and B may become substrates, which are linked together to form the AB product.

wines of many types have been made and served since ancient times. Some things do not change much, do they?

Catalysts are particularly important with respect to the **activation energy** of chemical reactions. The energy hill associated with the initiation of biochemical reactions is lowered considerably by the presence of enzymes (Figure 2–22). There are many different kinds of enzymes inside and outside of cells, each catalyzing a specific type of chemical reaction. It is the coordination of multienzyme reaction sequences, for example, that produces a polysaccharide from monosaccharides, or triglycerides from fatty acids and glycerol, or polypeptides from amino acids. Neither synthesis of a large molecule from smaller precursor molecules nor the breakdown of large molecules into smaller ones takes place in a single step. Many different types of enzymes are required to do these jobs. The processes are sequential and incremental, each enzyme involved in a sequence may alter the carbohydrate, lipid, or protein by adding, subtracting, or rearranging one molecule or atom at a time in a carefully orchestrated series of steps.

Enzyme-Catalyzed Reactions

An enzyme-catalyzed reaction can be represented most simply in the following way

$$E + S \rightleftharpoons ES \rightleftharpoons E + P$$

where E is the enzyme (with hands), S is the substrate molecule or molecules (with handholds) at the start of the reaction, and P is the product (Figure 2–23). ES represents a temporary combination of the enzyme and the substrate, which rapidly regenerates the enzyme with release of the product. The rate at which these reactions take place is called the turnover rate. Different enzymes have different turnover rates (Table 2–7). As the arrows in Figure 2–23 show and the hands demonstrate, the reaction can go in either direction. A multitude of enzymes is involved in thousands of chemical reactions that are going on all the time inside and outside the cells of all living organisms.

Getting Over the Energy Hill

Under what conditions do chemical reactions in this giant chemical factory we call a body take place? Most biochemical reactions do not occur readily at low temperature. In general, if the reactants of a human biochemical reaction were simply mixed in a test tube without the appropriate enzyme(s), the temperature needed to start the reaction would be considerably higher than body temperature. This

Activation energy the energy needed to initiate a chemical reaction

2–3 ENZYMES

Enzymes catalyze chemical reactions and are responsible for the high rate, specificity, and efficiency of human cellular biochemical reactions. Until the 1980s it was considered that only proteins had the capacity to act as enzymes. This was because of the complex 3-D structures that proteins could assume (see primary, secondary, tertiary, and quaternary structure in text). Pits and grooves on the surface of globular proteins served as sites for the binding of substrates and conversion of them into products. However, recent discoveries have shown that RNA molecules, called ribozymes, can also act as enzymes and perform these catalytic activities. In some of these reactions the ribozymes act as both the enzyme and the substrate!

One of the most important aspects of cellular processes to remember is that cells can only do what their enzymes enable them to do. There are six major categories of enzyme function.

ENZYME	ACTIVITY
1. Oxidoreductases	Electron transfer
2. Transferases	Transfer groups between substrates
3. Hydrolyases	Hydrolysis of substrates
4. Lyases	Add atoms to double bonds
5. Isomerases	Change organization of molecule
6. Ligases	Condensation reactions using ATP

constitutes the energy barrier, or hill, that must be overcome for the chemical reaction to start (Figure 2–22). A serious drawback for spontaneous chemical reactions in living systems is that excessive heat tends to destroy or denature protein molecules. You cannot simply put a flame to a part of the body and expect the reactions in the heated area to run faster or more efficiently.

Enzymes are nature's solution to the problem of high energy hills for biochemical reactions in a living organism. Enzymes lower the energy needed for a reaction to take place, so that chemical changes can occur at temperatures within biological limits. Most chemical reactions in humans occur at or near 38° C, which is the average temperature of the human body. In nature there is a wide range of temperature conditions in which organisms of many different types normally grow and thrive. Enzyme-catalyzed chemical reactions are able to take place normally in fish at below freezing temperatures in arctic oceans, or near boiling in the thermophilic bacteria found in geothermal hot springs and geysers.

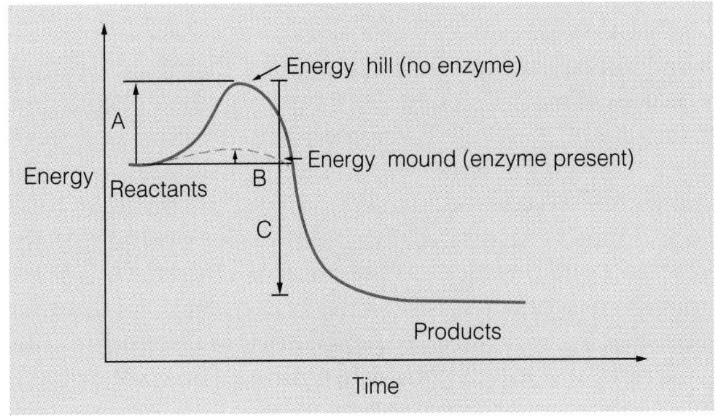

Figure 2–22 Energy Hills and Valleys
A Energy of activation, with no enzyme to catalyze the reaction, takes greater kinetic energy for the reaction to take place.
B Energy of activation is reduced with an enzyme to catalyze the reaction, and the probability that the reaction to form products will occur goes up.
C Energy of product formation is when products have less energy than reactants.

The Structure of Enzymes—Inducing a Fit

With all the different requirements for survival for different organisms in different environments, how do these catalytic molecules actually work? First, let us consider geometry. Most enzymes are globular molecules; that is, they are roughly spherical in shape with flexible folds and extensions on their surfaces. Second, only a limited portion of the enzyme is directly involved in the chemical reaction, not the whole protein. In fact, the regions of activity are often associated with the folds on the surface of the enzyme (Figure 2–21). These pits and grooves act as binding sites to hold the reacting molecules in place and as catalytic sites to alter a substrate into a product. This means that an enzyme somehow "recognizes" the proper reactant(s), as in a lock and key, with the added feature that in fitting together a change in the shape of the enzyme is induced. This induced fit puts a strain on the reactants (the ES complex and the hands separating in Figure 2–23),

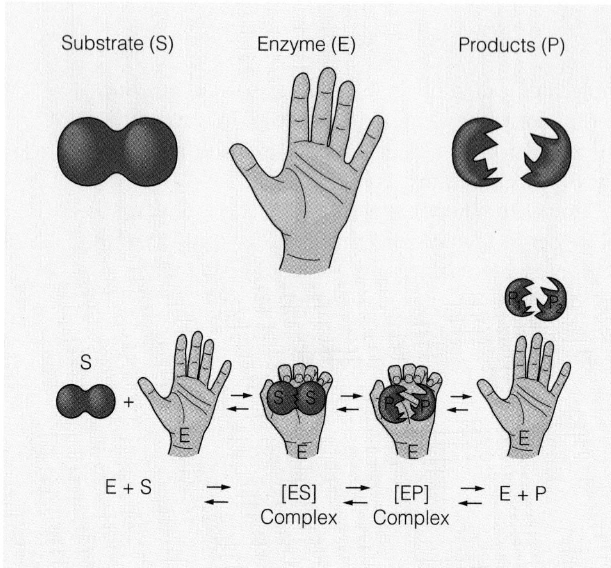

Figure 2–23 The Gripping Hand—Another Way of Looking at Enzyme Action

which brings about chemical changes that make or break covalent bonds. Reactant S is bound by the enzyme and, thus, P will be formed. Alternatively, because these reactions can go in either direction, P associates with the enzyme and is converted into S. Making and breaking covalent bonds is facilitated by proximity of position of the reactants and products on the surface of the enzyme. What may be a product in one case, is a reactant in another. Enzymes are very versatile molecules and essential for life.

One of the stipulations concerning an enzyme is that it not be permanently changed by the reaction of substrate to product. This means that one enzyme can sequentially and separately catalyze many identical chemical reactions. In fact, some enzymes operate at thousands or hundreds of thousands of cycles per second. Others are much slower. The rate at which an enzyme churns out products is called the turnover rate (Table 2–7). One of the most important roles for enzymes in cells is in the replication of our genetic material, DNA, a process in which both speed and accuracy are essential. The enzymes involved have fairly low turnover rates, but in conjunction with other factors (such as mistake correction mechanisms) generate a remarkably low number of errors.

NUCLEIC ACIDS

DNA

There are two types of nucleic acids, **DNA** and **RNA.** It was not until the mid-1940s that nucleic acids were seriously considered as the molecules of heredity. The work of Oswald Avery and his colleagues established DNA as the principle molecule of heredity. The structure of the DNA molecule was not fully described until 1953, by James Watson and Francis Crick (Biosite 2–4). Yet, in the span of just over four decades, biologists and biochemists have deciphered the genetic code, sequenced genes, reconstructed genes, transferred genes from humans to bacteria (and other species) and are beginning the sequencing and analysis of the entire human genome. A genome is the total DNA found in the chromosomes within the nucleus of a cell. Nucleic acid molecules are composed of purines and pyrimidines and are clearly of fundamental biological importance (Figure 2–24). But, what exactly are nucleic acids? How are they organized into the huge molecules of DNA and the smaller RNA molecules that establish the genetic code and the means to properly express it?

The Structure of DNA DNA is a long double-stranded molecule (Figure 2–25) that looks as if it were a ladder that has been held at each end and twisted to the right to form a helix. It may well be thought of as the "ladder of life," because contained in the organization and sequence of molecules in DNA in the chromosomes of each cell of the human body is the hereditary information necessary to produce a human being. All types of organisms on Earth use the same code to specify their growth and development. The formation of long, information-containing molecules of DNA is an excellent example of the importance of biochemical polymerization. Proteins are also polymers, but in proteins the subunits used to construct them are the group of 20 amino acids discussed previously (Figure 2–17). In the case of DNA, the polymers are formed by covalent linkage of nucleotides containing specific purine and pyrimidine bases.

DNA deoxyribonucleic acid; the genetic material
RNA ribonucleic acid; formed during transcription; a carrier of specific genetic information

Table 2–7 Turnover Rates of Selected Enzymes

ENZYME	RATE OF TURNOVER
Carbonic anhydrase	$0.6–1 \times 10^6$/sec
Catalase	1×10^5/sec
Penicillinase	2×10^3/sec
DNA polymerase	$1–2 \times 10^3$/sec
Galactosidase	2×10^2/sec
Tryptophan synthetase	2/sec

2–4

DNA AND ITS DISCOVERERS

There have been a number of important discoveries occurring over nearly one hundred years that led to the discovery of the structure and function of DNA and the mechanisms by which it is utilized as genetic material. Johann Miescher named the nucleic acids in the 1860s following its characterization from pus in wounds. However, Oswald Avery and his coworkers first showed that DNA was the active, transforming principle in bacteria. It was believed up to (and even after Avery's work) that nucleic acids were too simple to contain the genetic information. Avery isolated pure DNA from virulent strains of bacteria and exposed nonvirulent strains to it. These benign strains were transformed permanently to virulence. Alfred Hershey and Martha Chase later showed this to be true for viruses that infect bacteria. DNA was the key to heredity. Erwin Chargaff analyzed DNA from many different organisms and determined that the ratio of A:T and G:C was always one to one. This is referred to as Chargaff's Rule. It was with this background that the X-ray studies used by James Watson and Francis Crick (via Maurice Wilkins and Rosalind Franklin) led to the basic structure of DNA. The DNA molecule is a double-stranded alpha helix. Subsequent discoveries by Crick, Marshall Nirenberg, and Severo Ochoa led to the cracking of the genetic code. The history of these discoveries and the development of molecular biology are presented in Freeman Judson's book, *The Eighth Day of Creation,* and recounted in *The Alpha Helix* by James Watson.

The Composition of DNA The purines found in DNA are adenine (symbolized as A), and guanine (G). The pyrimidines are cytosine (C) and thymine (T) (Figure 2–24). These molecules are referred to as bases and are characteristically found in DNA in associations known as base pairs. Before bases link together covalently to form a DNA molecule, they must undergo two major modifications. The first is the addition of a sugar molecule, known as deoxyribose (the D in DNA), which is a pentose or five-carbon sugar. This converts a base, either a purine or a pyrimidine, to what is called a nucleoside. The nucleoside in turn has to be altered by the addition of phosphate groups. The base-sugar-phosphate combination is called a nucleotide, or a nucleoside phosphate. The importance of these chemical modifications is that polymerization of DNA depends on the covalent linkage between the phosphate group of one nucleotide and the sugar group of an adjacent nucleotide (Figure 2–25). The sugars and phosphate groups establish a "backbone" from which the individual purines and pyrimidines (ATGC) extend.

The long linear structure that emerges from the construction of this backbone is only part of a complete double-stranded DNA molecule. How is a helical, ladderlike arrangement of the complete two-stranded molecule formed? It is formed by hydrogen bonding between the bases in two different strands of DNA. The strength and stability of hydrogen bonds is dependent on their number and complementarity between the bases (Figure 2–25). Covalent bonding between sequential nucleotides accounts for one-half of the DNA molecule (a left or right half of a ladder). The complementarity of nucleic acids is critical in this regard because the bases of each strand are directed toward one another and toward the central axis of the helix, as if two halves of a ladder were coming together to form the whole structure. Effective hydrogen bonding, therefore, matches an adenine with a thymine, or a guanine with cytosine at each step of the ladder. The stability of these particular match-ups of base pairs, A with T and G with C, is determined by the chemical structure of purines and pyrimidines and the number of hydrogen bonds they form. Normally, there are two hydrogen bonds between A and T and three between G and C. This was the structure of DNA first described by Watson and Crick, a discovery so important to understanding heredity and genetics that it eventually led to their sharing the Nobel prize.

A Template for Replication Once it was understood that the DNA molecule was a double helix, whose complementary strands were held together by hydrogen bonds between bases, several previously unknown mechanisms were explained. First, a single strand of the DNA molecule, freed from its partner, could act as a template to form a new second strand and synthesize a new double-stranded DNA molecule composed of one original strand and one new

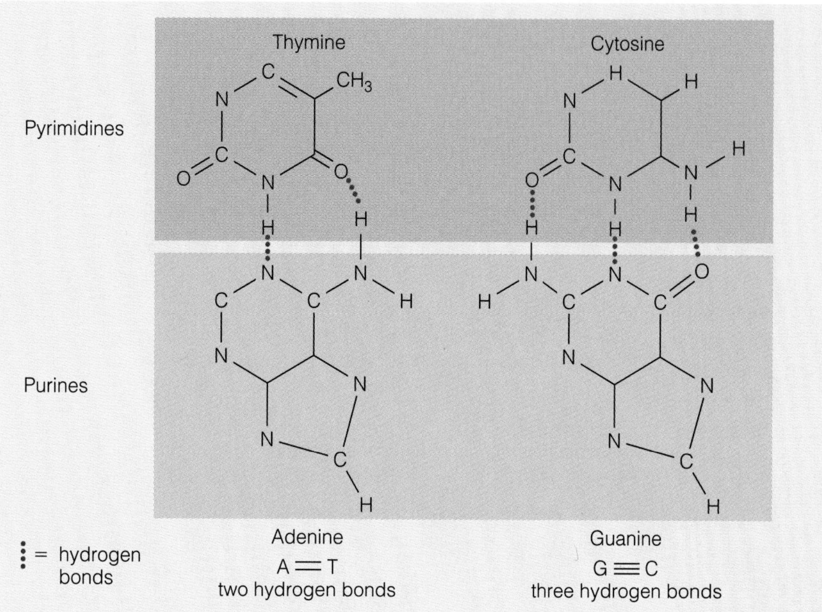

Figure 2–24 Purines and Pyrimidines—The Bases
The purines and pyrimidines found in DNA are adenine, guanine, thymine, and cytosine. These compounds are nitrogen-containing, ring-organized molecules whose relationships depend on size, shape, and hydrogen bonds. Purines have two rings; pyrimidines have one.

strand. This use of one strand as a template to make the other is called **semiconservative replication** (Figure 2–26). Second, such structural and functional features suggested a mechanism to account for the systematic doubling of the entire complement of chromosomal DNA observed in dividing cells. Third, it provided a way to explain, at the molecular level, the genetic basis for inheritance. After all, each new cell must have a copy of the information from which it was made in order to reproduce itself in turn. Also, each egg and sperm must have genetic information that can be combined to form a new, genetically unique individual. The question is, how is the information necessary for maintaining a complex, dynamic living cell encoded into a double-stranded molecule that itself is composed of only four different nucleotide bases?

Lifecode

A Four-Letter Alphabet The secret to the complexity of a living cell is a code, an order, and an arrangement of the small number of nucleotide bases in DNA in a way that carries information and ultimately has biological meaning. The English alphabet is a code. It is constructed so as to use 26 units (letters) to build words that, as a result of order and arrangement, carry information and have meaning. Substitutions of a single letter in a word can make subtle or profound differences in its meaning and the meaning of the sentences in which it is used. For instance, in the sentence "The book was green," there is little doubt about its meaning. By changing one letter in the word *book,* a *c* for a *b,* an entirely new sentence and meaning is generated.—"The cook was green." Although we might not mind reading a green book, we might choose not to eat food prepared

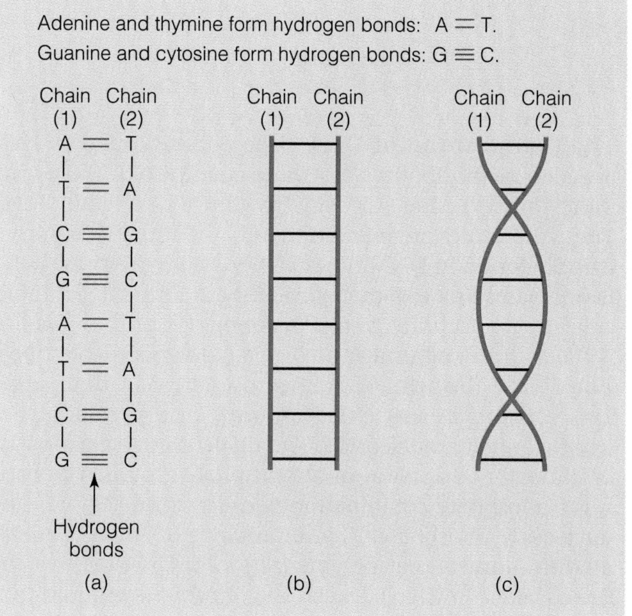

Figure 2–25 DNA Chains and Double Strands
Individual chains of covalently linked purine and pyrimidine bases (chains 1, 2) form hydrogen bonds that hold them together (a). The ladder representation (b) provides a model that may be twisted (c) to show the nature of the double helix.

Semiconservative replication synthesis of DNA in which one strand of a double-stranded molecule is used to copy and make a new strand

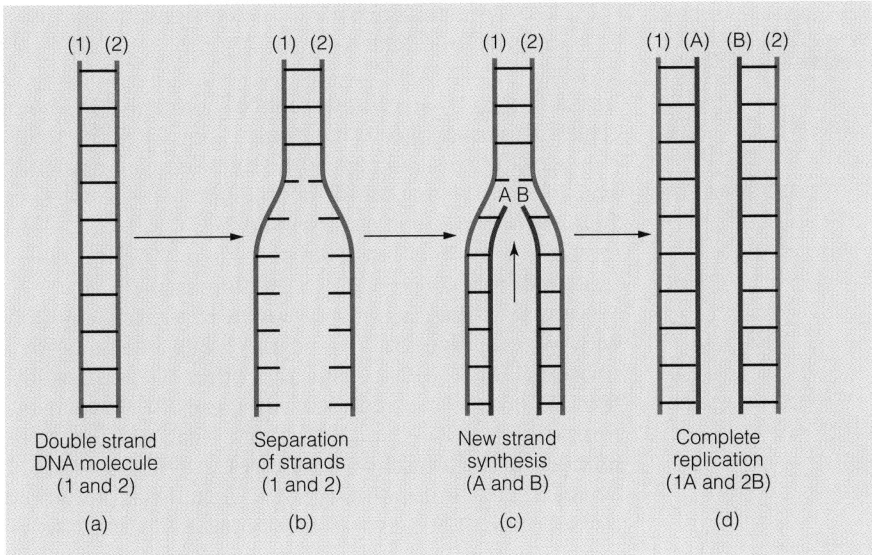

Figure 2–26 DNA Replication—A Semiconservative Process
DNA synthesis takes place on both strands of a double-stranded molecule (a) after separation of the strands (b). Synthesis (c) produces two new strands (A, B) that are complementary to the parent strands (1, 2). When synthesis is complete (replication), there are two identical strands, one original and one new strand are linked together with hydrogen bonds.

by a green cook! Total nonsense arises, if uninterpretable changes occur, such as "The zook was green" or "The book was greed." Either subtle differences of meaning or total nonsense in a code may bring about disaster. Unlike the English alphabet with its 26 letters, the biological lifecode uses only four letters in the form of the purine and pyrimidine bases: A, T, C, and G; and substitutions or alterations in the order in which they appear in a DNA molecule can be lethal.

Deciphering the Sequence of Nucleotides—A Magic Number of Three The sense of the genetic code depends on the order in which the nucleotides are arranged in a DNA molecule. How might such a code be deciphered? As already discussed in the previous section on proteins, the information in the DNA must eventually be translated into the sequence of amino acids in polypeptides. A major part of the definition of a **gene,** is the sequence of DNA that codes for a complete and functional protein. To better understand this, let's crack the code. Certain information about nucleic acids and amino acids is needed. First, there are four bases and 20 amino acids. We need a mechanism to convert from an alphabet of four letters to an alphabet of 20. Obviously, there is not a one-to-one correspondence between nucleotides and amino acids. Is there a two-to-one correspondence? Are combinations of two nucleic acids sufficient to code for 20 amino acids? No. Mathematically, four things taken two at a time (represented as 4^2) give rise to only 16 different possibilities, which is not enough to work in the translation of the lifecode. How about combinations of three of the four nucleotides? Four things taken three at a time (4^3) give rise to 64 different possibilities. This is more than enough to code for 20 amino acids, with some redundancy built into the coding system. In fact this triplet combination is the basis for information coded in DNA: units of three nucleotides in sequence form a codon. During transcription, as will be described in the next section, DNA is used to form RNA. RNA also contains a triplet code. Each codon is represented by sequences of three nucleotides and, as Table 2–8 shows, there are specific sequences of nucleotides that code for specific amino acids.

Transcription and Translation This is all well and good, but how do we convert the useful information coded in chromosomal DNA in the nucleus of a cell, a gene, to proteins, which are produced in the cytoplasm of the cell? For this we need a messenger service that shuttles specific genetic information to the part of the cell where synthesis of proteins actually takes place. The identity and function of messenger molecules was discovered several years after the structure of DNA was established. These molecules are known as **messenger RNAs** (mRNA) and individual messages are transcribed directly from genes contained in a strand of DNA (Figure 2–27). The process of forming RNA from DNA is called **transcription** and is analogous to copying, by hand, a small part of a manuscript for study or use by someone else. The act of copying or transcribing may take place in a library, but the copy can be transported and used by others at a distant location. In molecular transcription, DNA is the manuscript and RNA is the copy.

Gene genetic unit of inheritance; composed of DNA

Messenger RNA one of three major types of RNA; contains genetic information used in translation

Transcription the synthesis of RNA from a DNA template

Table 2–8 The Genetic Code—Converting Nucleic Acid Sequence to Protein

FIRST POSITION	SECOND POSITION				THIRD POSITION
↓	U	C	A	G	↓
U	Phe	Ser	Tyr	Cys	U
	Phe	Ser	Tyr	Cys	C
	Leu	Ser	STOP	STOP	A
	Leu	Ser	STOP	Trp	G
C	Leu	Pro	His	Arg	U
	Leu	Pro	His	Arg	C
	Leu	Pro	Gln	Arg	A
	Leu	Pro	Gln	Arg	G
A	Ile	Thr	Asn	Ser	U
	Ile	Thr	Asn	Ser	C
	Ile	Thr	Lys	Arg	A
	Met	Thr	Lys	Arg	G
G	Val	Ala	Asp	Gly	U
	Val	Ala	Asp	Gly	C
	Val	Ala	Glu	Gly	A
	Val	Ala	Glu	Gly	G

The genetic code translates RNA sequence (using A, U, G, and C, in which U in RNA equals T in the original DNA) into amino acid sequence. For example, the RNA sequence UUU translates to phe (phenylalanine), while the sequence CCC translates to pro (proline). The triplets UAA, UAG, and UGA do not code for any amino acid and act to stop protein synthesis.

RNA is similar, but not identical, to DNA in its chemical properties. The following are some of the significant differences:

1. RNA is a single-stranded molecule, not double-stranded, as is DNA.
2. Nucleotides in RNA have a different sugar molecule associated with them, that is, ribose instead of deoxyribose.
3. RNA contains the pyrimidine uracil as a substitute for thymine (Table 2–8).

In addition, RNA molecules are much shorter than DNA molecules. This is because, as mentioned above, RNA molecules represent only a small part of the total genetic information available in the long, linear double-stranded chromosomal DNA housed in the nucleus (Figure 2–27). In fact, a mRNA molecule represents but a single gene among the tens of thousands that are present in the chromosomal DNA.

Messenger RNA is transported out of the nucleus along with two other types of RNA (Figure 2–28). One is known as ribosomal RNA (rRNA) and the other as transfer RNA (tRNA). Both types represent different classes of genes from mRNA. rRNAs and tRNAs are final and functional products of the transcription of rRNA and tRNA genes. This expands the definition of a gene to include all products of transcription that serve as functional RNAs. mRNAs are unique in that they serve as templates for synthesis of proteins. All three types of RNA are related functionally and are used to assemble the protein-synthesizing machinery of the cell, which are **ribosomes.** The synthesis of protein on ribosomes in the cytoplasm is called **translation** (Figures 2–28 and 2–29). Ribosomes have the capacity to use the triplet coding sequence of nucleotides in the mRNA (AUGC) and translate them into a polypeptide chain containing the specific sequence of amino acids coded originally by the DNA.

Ribosomes intracellular structures involved in the synthesis of protein during the process of translation

Translation the synthesis of protein from an RNA template held in a ribosome

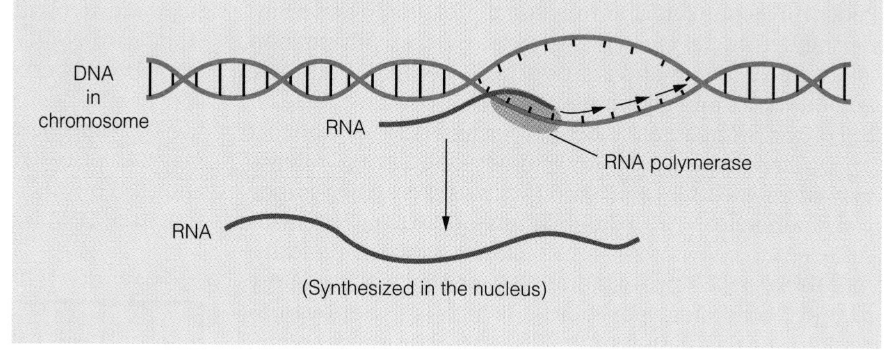

Figure 2–27 Transcription
Transcription is the process by which RNA is made from a DNA template. The enzyme involved is known as RNA polymerase. DNA strands separate to allow RNA polymerase to bind to one of them and use the DNA sequence to act as a template to make a complement of any strand of RNA. RNAs are single-stranded molecules and are much shorter than the DNA molecule from which they are synthesized. There are three major types of RNAs coded for in the genome—messenger RNA (mRNA), ribosomal RNA (rRNA), and transfer RNA (tRNA).

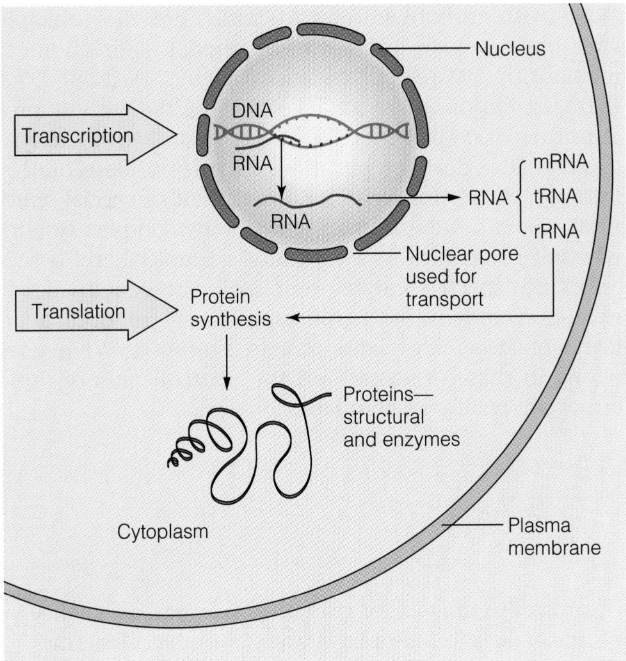

Figure 2–28 Transport of RNA From the Nucleus
RNA is synthesized in the nucleus. Once formed the RNA types are actively transported through nuclear pores into the cytoplasm. They combine to form a protein synthesis complex that releases proteins into the cytoplasm or into the endoplasmic reticulum.

What is it that provides the bridge between an RNA sequence and a protein sequence? The bridges are provided by tRNAs, which are specially adapted to simultaneously recognize a codon on the mRNA and to carry an amino acid to that site. Individual amino acids are attached to specific tRNAs. Glycine has its own tRNA, as do lysine, cysteine, and the rest of the animo acids. There is space for two tRNAs at a time in an assembled ribosome, and the binding of the tRNAs is by means of an anticodon. For example, the codon for phenylalanine is UUU. The anticodon which allows the tRNA with the correct amino acid to link with the mRNA is AAA. A and U form hydrogen bonds that hold them together within the ribosome. When they are both present and bound to their specific codon, an enzyme reaction in the ribosome transfers an amino acid from one of the tRNAs to the other. A tRNA that has given up its amino acid is released from the ribosome, the ribosome moves along the length of the mRNA, and a new "loaded tRNA" binds the next codon. The synthesis of peptide bonds continues until the message ends, at which time the new protein is released and the ribosomes disassemble. Unlike transcription (DNA to RNA), which essentially stays in the same basic chemical language (purines and pyrimidines), translation, as the name implies, requires a change from one language to another. The translation must be true to the original nucleic acid sequence. Every protein made from a specific type of mRNA has to be the same. A key to understanding genetics and inheritance is knowing that

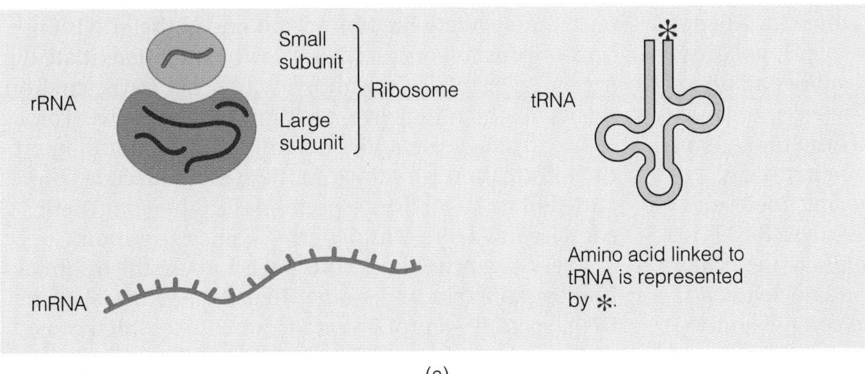

(a)

tRNA *
in
tRNA
out
Stop
mRNA
Start
P site
A site
New protein being translated

(b)

Figure 2–29 Translation
In (a) the RNA components formed in the nucleus assemble into a protein synthesizing complex known as a ribosomal complex in the cytoplasm. In (b) ribosomes synthesize protein from sequences of nucleic acid in the mRNA. tRNAs bring in amino acids, bind to the correct sequence, and then discharge the amino acid and leave the ribosome. rRNA is associated with nearly 100 proteins to form the two subunits of the ribosome.

the DNA sequence of a gene, transcribed into a functionally equivalent mRNA and utilizing ribosomes and adaptor tRNA molecules, produces not only a functional polypeptide but also the identical polypeptide time after time. Without this stability of information transfer from genome to cytoplasm, cell structure and function could not be maintained.

Processing the Genetic Code

The basic code of life is written in the specific sequence of nucleotides in DNA. However, for this information to be useful to a cell and ultimately to an organism, it must be transcribed to RNA and then translated into proteins (Figures 2–27 and 2–28). Proteins are the cellular workhorses, the engines, and frameworks of living cells. They, along with carbohydrates and lipids, are the principle structural materials of life. Clearly, there is a direct, linear relationship between DNA and proteins. Without DNA there is no information to make proteins, but without proteins there is no way to make new DNA (or RNA). The synthesis of DNA during replication, and the synthesis of RNA during transcription, requires a multitude of special structural proteins and enzymes. As with any complex system, whether mechanically, electrically, or molecularly based, errors are expected and do occur. Biological systems are no exception, and low but significant rates of error occur at all levels of DNA, RNA, and protein synthesis. What goes wrong in these processes and the consequences of these errors will be discussed in later chapters.

Summary

In this chapter the following statements were considered:

1. Matter is made of atoms.
2. Atoms interact to form molecules.
3. Molecules interact to form living cells.

In the context of these questions the atomic structure of the elements was outlined and their properties discussed. *All matter is made of atoms.*

Atoms are the building blocks of molecules. The chemical properties of molecules were discussed, which included such topics as covalent bonds, ionic bonds, pH, polarity, oxidation-reduction reactions, and availability and utilization of chemical energy. Differences between organic and inorganic chemistry provided a backdrop for a discussion of animate and inanimate objects and materials.

Molecules interact to form living cells. Organic molecules found in and produced by living cells were described. This included descriptions of many familiar types of molecules including carbohydrates, fats, proteins, and nucleic acids. The structure and function of simple and complex carbohydrates, as well as the structure and properties of fats and lipids, were discussed and compared. The organization and structure of proteins were described in detail, and the function of enzymes and their roles in chemical reactions in cells and living organisms were considered. Purines and pyrimidines and nucleosides and nucleotides were shown to provide the building blocks for an information-rich lifecode or genetic code. Hereditary information, that is, that information that is passed from one generation of living organisms to the next, is contained in specific sequences of nucleotides found in DNA and RNA molecules. This hereditary information in humans is contained in each of 46 chromosomes located in the nucleus of every cell in the body. The genetic code is represented as a triplet code that can be utilized to translate sequence information in RNA into proteins.

With all that is known from the study of modern biology, however, life is still a mystery. This is observed in our incomplete knowledge of how cells organize their contents and regulate the timing and rate of their division, how organisms control their growth and differentiate during embryogenesis, how individuals adapt to the environments in which they live, and how species evolve. For a cell to divide, grow, and differentiate, it has to call upon the information contained in the specific lifecode contained in its DNA. Every species has a different genetic content and a different pattern of gene expression.

Every organism must both have and use the information contained in its genome. To use the information in DNA, specialized proteins are required. So, which came first, proteins (particularly enzymes), which make it possible to synthesize and assemble new DNA, or DNA, which provides the code that makes it possible to synthesize new proteins? These questions of the origin and perpetuation of life cannot at present be answered fully. A description of some of these problems and issues was presented in Chapter 1, and more detailed information is presented in Chapter 3, which focuses on the structure and function of the fundamental unit of all living organisms, the cell.

Questions for Critical Inquiry

1. What is the watchmaker's dilemma? How does it relate to living organisms? Can it be solved? How?
2. Why might it be important to know about the nature of the materials from which living organisms are constructed?
3. Should human DNA, and the lifecode it contains, be open to manipulation by biologists and other scientists?
4. What is the difference between necessary and sufficient conditions with respect to life?

Questions of Facts and Figures

5. What is the structure of an atom?
6. How are atomic number and atomic weight arrived at?
7. What are the two major types of chemical bonds and how do they differ with respect to electrons?
8. Describe and differentiate between the four basic types of organic molecules.
9. pH represents a measure of what ions in a solution?
10. What is the basis for polarity and nonpolarity in molecules?
11. How do enzymes act to increase the rate of a chemical reaction?
12. What is transcription? What is translation? How are the two processes linked?
13. What are primary, secondary, and tertiary structure in proteins?
14. Why is it more efficient to store fats than it is to store carbohydrates?
15. Who discovered the structure of DNA? What were some of the implications of the structure as described?

References and Further Readings

Alberts, B., et al. (1994). *Cell and Molecular Biology.* New York: Garland Publishing.

Judson, H. F. (1979). *The Eighth Day of Creation: The Makers of the Revolution in Biology.* New York: Simon and Schuster.

McCarty, M. (1985). *The Transforming Principle: Discovering that Genes Are Made of DNA*. New York: W.W. Norton & Co.

Ohanian, H. (1995). *Modern Physics,* 2nd ed. New York: Prentice Hall.

Roger, D., Goode, S. and Mercer E. (1997). *Chemistry: Principles and Practice*. Fort Worth, TX: Saunders College Publishing.

Stryer, L. (1988). Biochemistry, 3rd ed. New York: W. H. Freeman and Co.

Watson, J. D. (1968). *The Double Helix: A Personal Account of the Discovery of the Structure of DNA*. New York: Atheneum.

CHAPTER 3

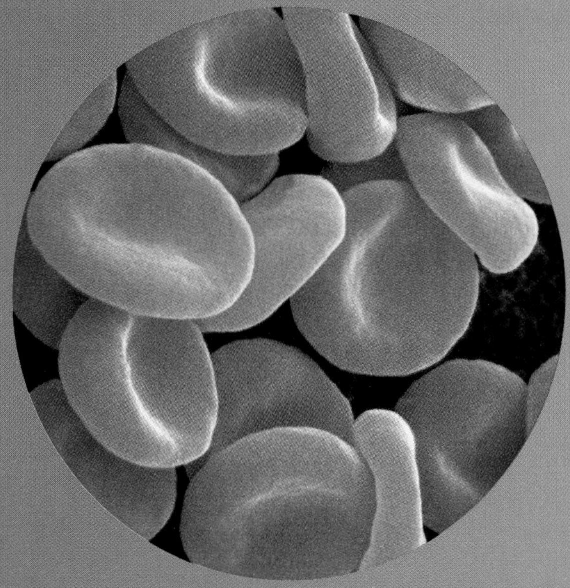

Cells and Life

INTRODUCTION

A cell is the most basic biological entity (or unit) of form and function capable of carrying out all the processes necessary for life. This may seem to be a somewhat oversimplified definition, but it serves to focus us on the entity itself, a cell, which, regardless of appearances, is an intricate and complicated system of dynamic structural and functional elements. Cells have the capacity to synthesize, degrade, modify, and organize organic and inorganic molecules, to regulate transport of organic and inorganic materials into and out of themselves, and to acquire, produce, and use energy-rich molecules. There are enormous gaps in our knowledge and understanding of cells, even though they are the simplest of life forms. This chapter presents an analysis of the parts of cells and the processes they carry out as a first step in an effort to make sense of the vastly more complicated tissues and organs of multicellular organisms such as humans.

THE HUMAN BODY AND ITS CELLS—THE COMPLEXITY OF MULTICELLULARITY

There is great diversity in the kinds of cells composing a multicellular organism and great variation in the total number of cells composing any particular individual. Tiny nematode worms may have fewer than 1000 cells; a massive organism such as an elephant has perhaps one hundred trillion. Between these extremes are we humans, who are composed of perhaps tens of trillions of cells. The numbers alone, however, are misleading. It is not how many cells an organism has but how they are organized and function together. Furthermore, it is particularly important to know what cells can do independently and what they can do only through interactions with one another. In this chapter we confront questions of how cells perform all the activities necessary for their survival. For example, which parts of a cell are involved in transport of molecules and ions? How are organic molecules synthesized and organized? How are molecules stored, degraded, and exported? One way to find out how the many parts of a cell work is to separate them from one another and study them independently. Biologists have used this approach successfully for decades and have become very good at separating and determining the nature and activities of the parts of cells. Maybe Humpty Dumpty was broken into pieces because he was pushed off the wall by the king's scientists who were interested to see what he was made of.

MORPHOLOGICAL CONSIDERATIONS—A DIVERSITY OF CELL TYPES

The Neighborhood Plan

The human body has so many different kinds of cells that in an intact individual it is often difficult to tell one type from another or to determine precisely what any one cell is doing at any particular time in a tissue or organ. For purposes of comparison, this is somewhat analogous to describing the changes in the active and diverse neighborhoods in which each of us lives. Neighborhoods have different types of houses and businesses being built or torn down, different families with different numbers of children and pets moving in and out, and dynamic patterns of traffic crisscrossing in the streets under telephone and power lines. To describe a neighborhood precisely from one minute to the next is a formidable task. Likewise, human cells exist among neighbors, and many of the cells are highly specialized and interactive and organized into functional units known as *tissues,* which will be discussed in Chapter 4. Different cell types have different size populations, different locations within the body, and different timing of activities such as cell migration, cell division, and cell death. For example, nerve cells interconnect, as do telephone lines, carrying information rapidly between different parts of the body. Muscle fibers contract and relax to allow the movement of bones so we can walk, run, and jump. Cells of the skin die as they reach the surface, yet seal tightly together to envelope the body and protect it from the outside environment.

Human cells are specialized in so many different ways that there is no one cell type that can be singled out as a representative. However, specialization does not mean that cells are altogether different from one another. In fact, they share many features that might be referred to as housekeeping functions. These shared characteristics reflect common needs for metabolic pathways, supply of common structural materials, and transport functions that allow the cell to survive and adapt to change from minute to minute. Emphasis on shared properties and processes of cells lends itself to model building. A model allows one to combine all the features of cells into one hybrid, generalized cell. We will do this in an effort to establish the relationships of all the parts to each other and to the whole.

A MODEL CELL

To build a model, one needs the appropriate building materials. The following are four basic structural components to be used in the architecture of a model cell (Figure 3–1):

1. Membranes
2. Organelles
3. Complexes of macromolecules
4. Polymer systems.

Those aspects of the model that represent key cellular structures include:

1. Cellular membranes
2. Cytoplasm
3. Nucleus, nuclear envelope, and chromosomes
4. Golgi apparatus, lysosomes, and vesicles
5. Peroxisomes
6. Cytoskeleton
7. Centrioles, cilia, and flagella
8. Mitochondria (and chloroplasts in plant cells).

This list encompasses a great many structures and components that perform a range of required functions in a cell. Cells are complicated entities and have to have specialized and generalized parts to carry out all the processes required of them. Knowledge of the many parts will allow us to bring their relationships to one another and to the whole into sharper focus.

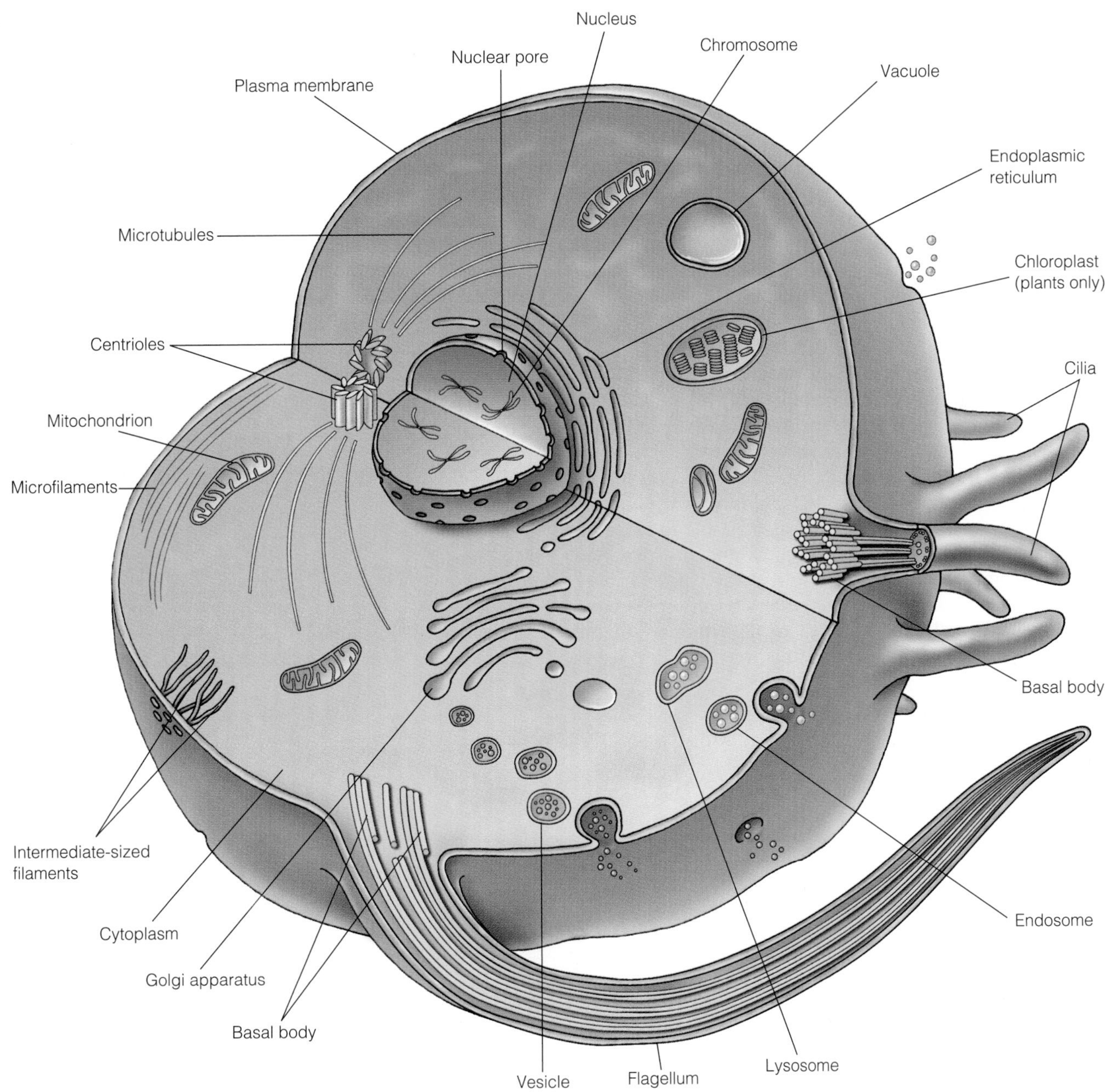

Figure 3–1 A Model Cell

The major compartments of the cell are the cytoplasm, which is enclosed by the plasma membrane, and the nucleus, which contains the chromosomes. Several filament or tubule systems occur in the cytoplasm, including microfilaments, microtubules, and intermediate-sized filaments. Specialized extensions of the plasma membrane include cilia and flagella, which utilize microtubules and basal bodies to allow cell movement. The endoplasmic reticulum is a series of membrane-enclosed channels that are involved in protein synthesis and transport. The endoplasmic reticulum communicates with the Golgi apparatus, which in turn produces vesicles for secretion. The Golgi apparatus also forms the lysosome, which interacts with endosomes as part of the process of cell eating and digestion. Mitochondria produce ATP for use as cellular energy, and, in plants, the chloroplast converts sunlight to chemical energy during photosynthesis and produces organic molecules for storage and use. Vacuoles store waste products of cells internally. Centrioles organize the microtubules in an area close to the nucleus. The nucleus has nuclear pores, which allow RNA to be transported to the cytoplasm and proteins to be imported into the nucleus.

Cell Membranes and the Cytoplasm

Every cell has a distinct boundary, a structural barrier capable of separating the internal materials from the extracellular environment. This boundary is called the **plasma membrane.** It is composed of a phospholipid bilayer and is slightly viscous and extremely thin (Figure 3–2). In addition to providing a thin physical framework of envelopment, the plasma membrane also plays an important role in controlling the movement or transport of materials into and out of a cell. The cytoplasm is the internal, gel-like fluid of a cell. **Cytoplasm** is mostly water (70%–80%), with a great number and variety of different types of molecules, ions, and other solutes dissolved within it (Figure 3–3). Most of the components of the cell, excluding the plasma membrane and nucleus, are either dissolved or suspended in the cytoplasm. The nucleus has a membrane envelope of its own (the nuclear envelope), which encloses and separates the primary genetic material (DNA in the chromosomes) from the cytoplasm. The total content of a cell within the plasma membrane including the nucleus is called **protoplasm.**

In addition to phospholipids, plasma membranes are composed of cholesterol, glycolipids, and proteins. The plasma membrane is between 8 nm and 10 nm (nanometers) thick (Figure 3–2). A membrane that thin, spread over a large area, might be expected to be extremely fragile. However, plasma membranes of this thickness in cells are relatively strong, durable, and quite stable. This is because of the small size of most cells (10μm–20μm, micrometers). Stability and strength in this case are a matter of proportion, the greater the volume of a cell enclosed within a thin membrane, the more susceptible that membrane is to *lysis.* The phospholipids and proteins of a plasma membrane form a self-contained, continuous, and flexible layer around a cell. For the purposes of our ongoing comparisons, imagine that a balloon partially filled with water represents a cell. The elastic layer is the membrane, the water is the enclosed cytoplasm. As does a water-filled balloon, living cells conform to their physical surroundings without being broken, even if this means stretching, contracting, floating, or flattening.

A plasma membrane is a two-dimensional fluid. This

Plasma membrane complex phospholipid bilayer membrane composing the barrier surrounding and defining a cell

Cytoplasm the viscous fluid found within a cell, excluding the nucleus

Protoplasm viscous fluid and all components of a cell contained within the plasma membrane, including the nucleus (*see* cytoplasm)

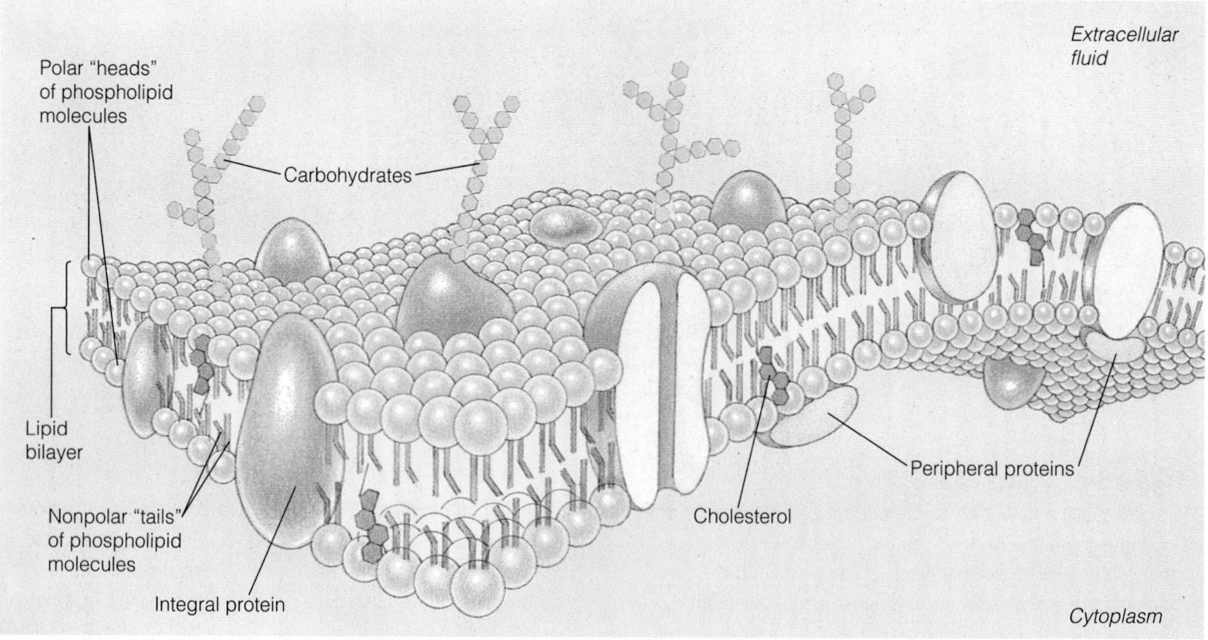

Figure 3–2 Membrane Structure
The plasma membrane is composed primarily of phospholipids (the lipid bilayer) and proteins embedded in the bilayer lipids. Proteins may extend through the entire bilayer (integral proteins) or be associated with only one side of the membrane (peripheral proteins).

results from the preferential **hydrophobic interactions** of phospholipids with one another to avoid direct contact with water. A plasma membrane has a viscosity about the consistency of light oil. This fluidity allows cells the capacity to change shape in direct response to internal and external conditions. Thus, deformability of membranes serves to protect the cell and its contents. However, this does not mean that a cell is indestructible. On the contrary, damage to the plasma membrane may easily result in cell lysis and lead to cell death. Perforation of a plasma membrane allows the cytoplasm to leak out, leaving only the ghostly remnant of the lipid sack that once contained it. Normally, however, the plasma membrane of a cell is durable and adaptive to changes in the environments in which cells normally operate.

A Fluid Mosaic

The description of the properties of the phospholipid bilayer as a two-dimensional fluid provides an interesting picture of how lipids are organized in plasma membranes. What about proteins? Proteins and phospholipids within the membrane interact directly with one another and have a relationship somewhat similar to the relationship icebergs have with the waters in which they float. The phospholipid bilayer forms a fluid sea in which proteins are embedded. Because the plasma membrane is essentially a thin sheet, some of the proteins (and there are many of them) actually extend completely through it and poke into the cytoplasm. Proteins that traverse completely through the plasma membrane are called **integral proteins.** Those that do not span the bilayer, but are associated with one side or the other, are called **peripheral proteins** (Figure 3–4). Proteins are the functional partners in the plasma membrane and affect the way a cell transports materials into itself, migrates, connects to its neighbors, and communicates. Both proteins and phospholipids are capable of moving in the plane of the membrane by diffusion or under the influence of outside forces.

Endoplasmic Reticulum

There is also a network of phospholipid membranes within the cytoplasm (Figure 3–5). These internal membranes form an elaborate network called the endoplasmic reticu-

> **Hydrophobic interactions** "water hating"; interactions between molecules to the exclusion of water
>
> **Integral proteins** proteins that are embedded in a cell membrane and exposed on both sides of the phospholipid bilayer
>
> **Peripheral proteins** proteins associated with one or the other surface of a cell membrane

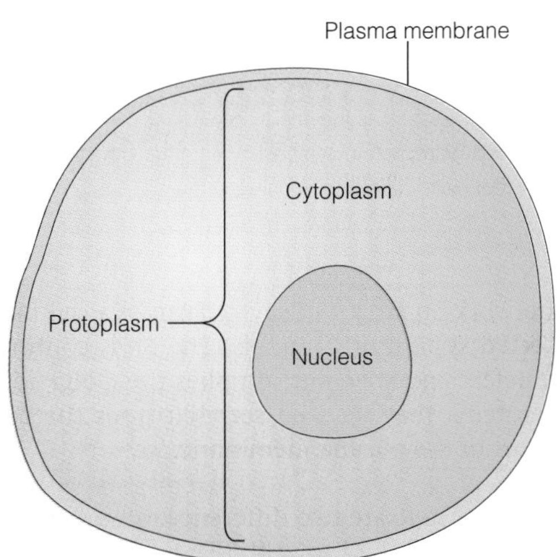

Figure 3–3 Cytoplasm and Protoplasm
Protoplasm is the entire content of a cell including the nucleus and the cytoplasm in this diagram. The cytoplasm is the viscous material within the plasma membrane but excluding the nucleus. Cytoplasm is composed of 70%–80% water plus proteins, salts, and carbohydrates. The activities that take place in the cytoplasm include protein synthesis, energy production, intracellular transport of materials and biochemical reactions that change the cell shape and allow for movement or motility.

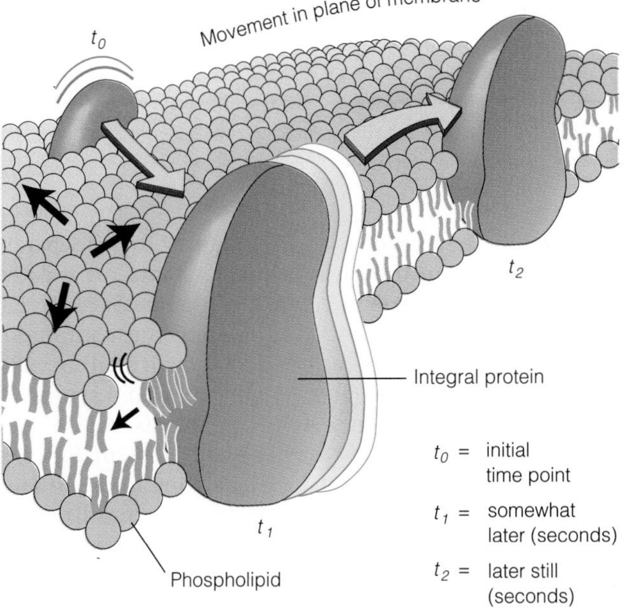

Figure 3–4 Fluid Mosaic Model of the Plasma Membrane
The fluid mosaic model presents a dynamic picture of the plasma membrane, in which integral proteins and phospholipid molecules are in constant "fluid" motion in the plane of the membrane.

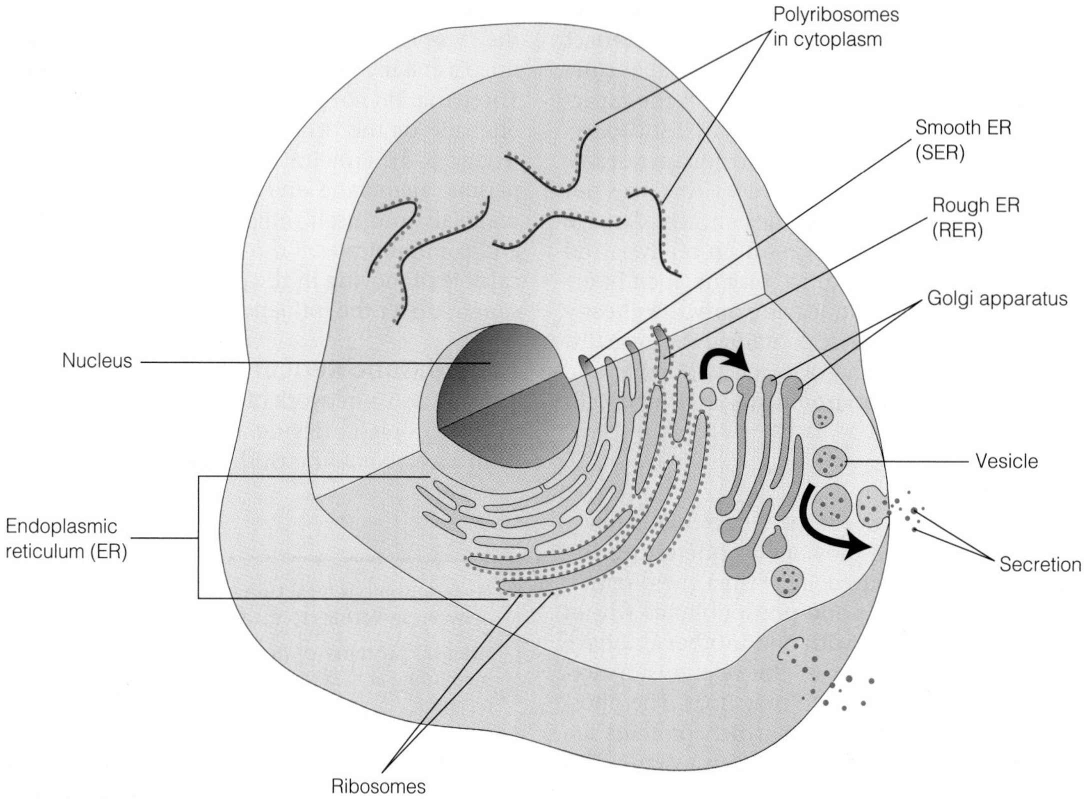

Figure 3–5 Rough and Smooth Endoplasmic Reticulum and the Golgi Apparatus
The nucleus, endoplasmic reticulum (ER) and Golgi apparatus are related to one another. The outer surface of the nucleus is continuous with the ER and the ER produces vesicles that communicate with the Golgi apparatus. The Golgi apparatus produces vesicles that are involved in cellular secretion. The SER is the site of lipid synthesis. The RER is the site of protein synthesis, because of its association with ribosomes. The Golgi apparatus is the site of glycosylation. Ribosomes are also associated with protein synthesis within the cytoplasm and form polyribosomes with mRNA molecules.

lum, or ER. They surround a compartment that is separate and distinct from the cytoplasm. This forms an isolated space called the **cisternal space** or the *lumen of the ER.* The ER is a dynamic structure, undergoing continuous changes in amount, organization, and activity within cells. The membranes of the ER are involved in lipid synthesis, protein synthesis, glycosylation of proteins, and in the internal cellular transport of proteins. Different types of cells have different amounts of ER, ranging from virtually none in red blood cells to massive amounts in cells of the liver. The distribution and organization of ER is directly related to the synthetic needs of the cell. Some regions within a single cell have more ER than others. For example, most of the ER in a nerve cell is in the body of the cell where protein synthesis takes place and not in the **dendrites** and **axon** that grow out from it. The endoplasmic reticular network organizes the cytoplasm of a cell into an interconnected system of channels. The ER is continuous with the nuclear envelope and supplies the Golgi apparatus with proteins that may be secreted from the cell or become a part of the plasma membrane.

RER and SER There are two different kinds of endoplasmic reticulum—rough and smooth (Figure 3–5). Rough en-

> **Cisternal space** volume within the membranes of an organelle such as the endoplasmic reticulum or Golgi apparatus
>
> **Dendrites** extensions arising from a nerve cell body
>
> **Axon** extension from the surface of a neuron

doplasmic reticulum (RER) has ribosomes attached to it. The ribosomes are actively involved in protein synthesis. As new proteins are synthesized on RER, they are inserted through the ER membrane into the lumen. At that point the proteins are no longer considered to be in the cytoplasm proper, because the ER compartment is separate and *topologically distinct* from it. Proteins synthesized in the RER are **glycosylated** to a limited extent and undergo further addition of carbohydrates in the Golgi apparatus.

The other type of ER is called smooth endoplasmic reticulum (SER), because it lacks ribosomes. The principal role of the SER is phospholipid synthesis. All the major phospholipid bilayers of the cell, including the plasma membrane, the ER, and the Golgi apparatus and its vesicles, are synthesized by the smooth endoplasmic reticulum. In addition, the SER is important in detoxification of the cell. Enzymes located in the SER of liver cells serve to chemically modify a number of types of toxins and, thus, make them easier for the body to **excrete** in the **urine** through the action of the kidney.

Ribosomes, ER, and Cytoplasmic Protein Synthesis

For proteins to be synthesized in cells, ribosomes must be present and active (Figure 3–6). As discussed briefly in Chapter 2, ribosomes are macromolecular complexes composed of dozens of polypeptides and several types of rRNA that provide the machinery to translate information from mRNA into protein. Ribosomes are generally found in close association with the endoplasmic reticulum and free in the cytoplasm. Cytoplasmic ribosomes may be assembled into polyribosomes, which are responsible for the synthesis of proteins that will remain within the cell (Figure 3–6 [b]). Cellular proteins contain short sequences of amino acids that act as targeting signals and determine when, and to what location, a newly synthesized protein will go within the cell. These signals target proteins to other intracellular compartments, such as mitochondria and peroxisomes, as well as to the nucleus. Proteins destined for the Golgi apparatus, lysosomes, and vesicles, or secreted from the cell are synthesized on ribosomes attached to the ER (Figure 3–6[c]). In this case, signal sequences on the proteins direct them through the ER into the cisternae. The rough endoplasmic reticulum helps to coordinate the activities of ribosomes in vital synthetic processes and to control the protein traffic patterns within cells.

Nucleus and Nuclear Envelope

The nucleus is the single most important organelle found in a cell (Figure 3–7). The genetic information for cell growth and development is encoded in extremely long, linear molecules of DNA, which in turn is organized within the nucleus into structures known as chromosomes (see Figure 3–7). Each of the 46 chromosomes of a human cell contains a single DNA molecule and is usually too thin and extended within the nucleus to be observed directly with a microscope. However, if the single DNA molecule of every chromosome were to be lined up together end to end, the DNA would extend nearly two meters! Chromosomes and the information they contain constitute the **genome.** In addition to DNA, chromosomes are composed of several types of proteins. One group, known as **histone proteins,** provide the means for DNA to be condensed to take up less space within the nucleus. This combination of proteins and DNA is called **chromatin.**

The nucleus is surrounded by a special membrane called the nuclear envelope. The nuclear envelope is composed of a double layer of phospholipid membranes (Figure 3–7). The outer nuclear membrane is often continuous with the ER. The inner nuclear membrane of the envelope is in contact with the inside of the nucleus and the chromatin. The double envelope is penetrated by channels called nuclear pores, which regulate the transport of molecules into and out of the nucleus.

Nuclear pores are actively opened to let large molecules into and out of the nucleus. The main sources of genetic information in a cell are DNA and RNA, which are synthesized within the nucleus. The main structural and functional molecules, the proteins, are synthesized exclusively in the cytoplasm. Without the nuclear pores there would be no communication between the nuclear compartment and the cytoplasm. However, nuclear pores have their limits as to the size of molecules that can get through them. Chromosomes and, therefore, DNA are much too big to pass through a nuclear pore. But smaller molecules, such as the various types of RNA, move easily from nucleus to cytoplasm, and proteins move readily from the cytoplasm to the nucleus. These smaller molecules, such as the messenger

Glycosylated with sugars covalently linked to a molecule, as in the case of glycoproteins or glycolipids

Excrete to eliminate waste from a cell, tissue, or organ

Urine excreted waste product in mammals containing nitrogen and produced in kidneys and eliminated through the urinary system

Genome the full set of genes found in the chromosomes of an individual, which specify the development, growth, and maintenance of that individual

Histone proteins special proteins associated with DNA in chromosomes

Chromatin DNA and associated proteins in the nucleus

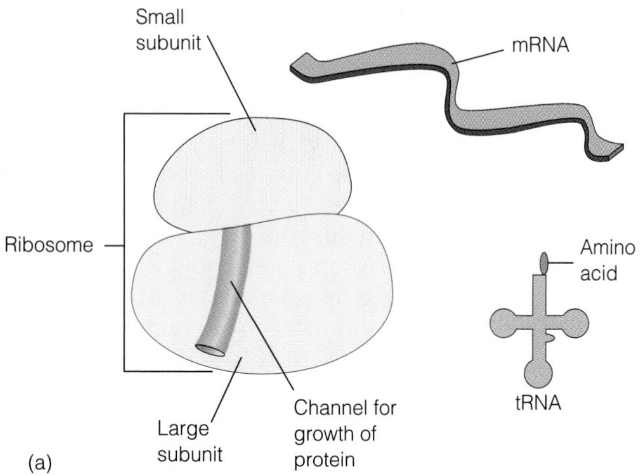

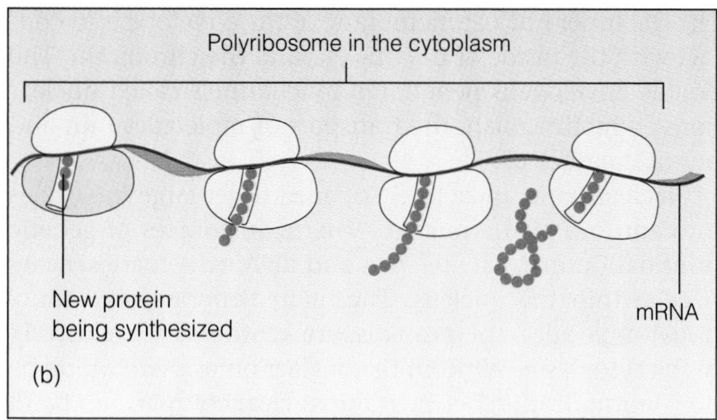

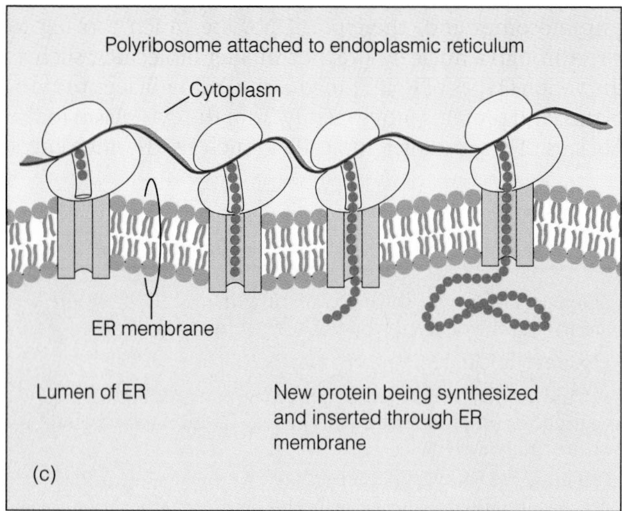

Figure 3–6 The Ribosome and Its Function
The components of a functional ribosome are shown in part (a). The attachment of many ribosomes to a single mRNA is called a polyribosome. Protein synthesis takes place continuously as ribosomes move along the mRNA and proteins get longer and longer (b) and (c). In part (b) the protein will be released into the cytoplasm. In part (c) the ribosome is attached to the endoplasmic reticulum (ER) and proteins are inserted into the ER lumen as they are formed.

RNAs and certain cytoplasmic proteins, are involved in controlling how and when the information coded in the chromosomal DNA is used. Messenger RNA molecules are transported into the cytoplasm, associate with ribosomes, and direct the process of protein synthesis. Thousands of nuclear pores can be produced to bridge the gap between the inner and outer membrane of the nuclear envelope, and they can be opened and closed in response to the cell's need to exchange materials in the cytoplasm and nucleus. In this way pores can be used to regulate not only what gets exchanged between the nucleus and the cytoplasm but also how much, when, and where the exchange takes place (Figure 3–8).

Golgi Apparatus, Vesicles, Lysosomes, and Vacuoles

Golgi Apparatus and Vesicles The Golgi apparatus is derived from the ER but has an independent status once formed. It is made up of a series of flattened, platelike structures that are in reality membranes folded around internal, cisternal compartments. The membrane compartments are flattened in the middle and expanded at the edges to give the impression of being a stack of dishes (Figure 3–9). There are three main regions in the Golgi apparatus:

The *cis region,* which is nearest to the ER

A **medial** or middle **region**

The *trans region,* which is furthest away from the ER.

The membranes of the Golgi are similar in composition to the endoplasmic reticulum. In fact, membrane vesicles bud off from the ER and are transported to the Golgi where they fuse and become part of the Golgi membrane (Figure 3–9). The Golgi apparatus communicates with the ER and with the lysosomes and the plasma membrane through exchange of vesicles.

There are several important functions for the Golgi apparatus, each of which is related to structural needs of the cell. The Golgi apparatus is involved in carbohydrate synthesis, assembling new plasma membrane and recycling old, making lysosomes, and producing and packaging

Medial region of or relating to the middle plates of a Golgi apparatus

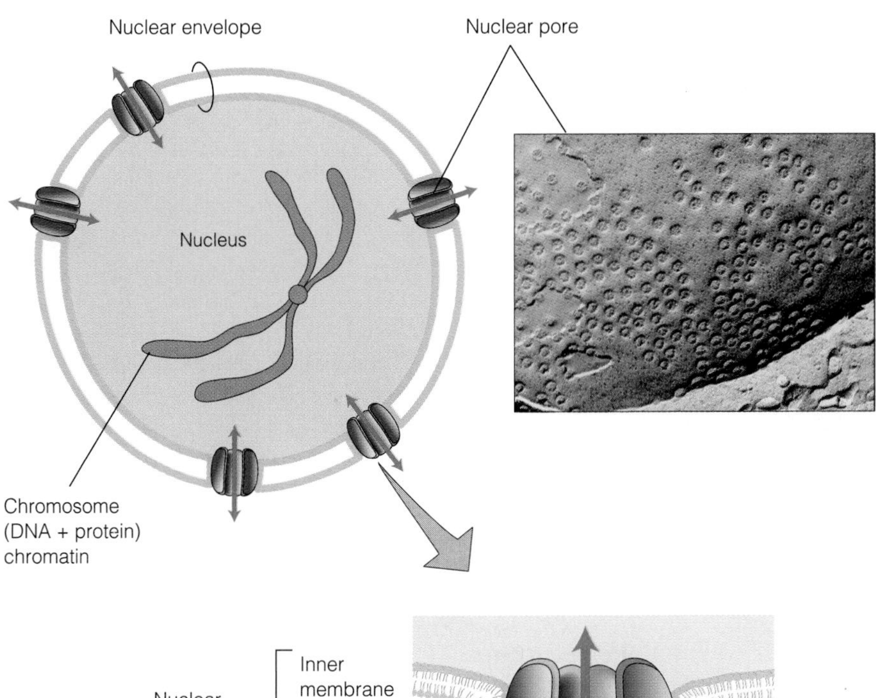

Figure 3–7 The Nucleus
The nucleus is surrounded by a double membrane known as the nuclear envelope and penetrated by channels known as nuclear pores. The nucleus contains the chromosomes, which are too large to move through the pores into the cytoplasm. Inset is a scanning electron microscope image of nuclear pores.

Figure 3–8 Genetic Information Transfer
The genetic information of the cell is in the nucleus in the form of nuclear DNA. In Step 1 transcription of DNA to RNA occurs. In Step 2 transport of RNAs of all types (mRNA, rRNA, and tRNA) occurs through the nuclear pores. In Step 3 ribosomes are assembled and in Step 4 translation of mRNA to protein occurs. Many types of proteins return to the nucleus by transport through nuclear pores and influence DNA and RNA transcription (Step 5).

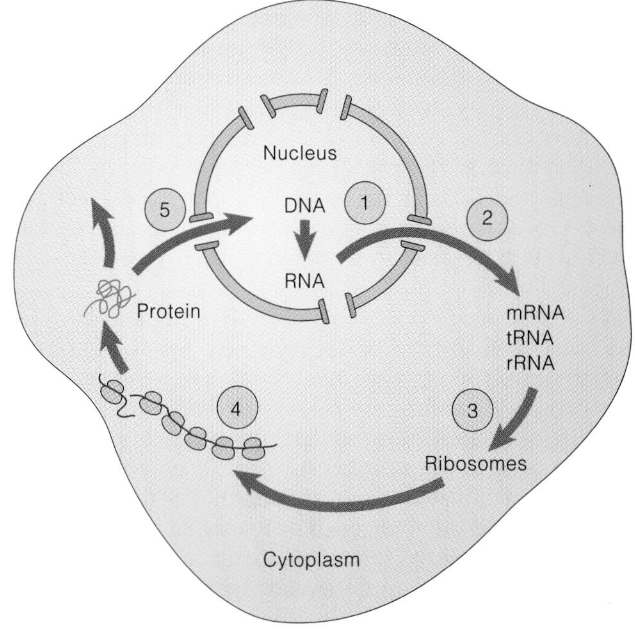

proteins and glycoproteins for storage or secretion from the cell. Most cells are continuously making new membrane to replace the existing one and, in the process, removing existing plasma membrane for recycling. The origin of new plasma membrane is principally the membranes of Golgi vesicles. Vesicles bud from the surface of the trans Golgi, migrate to the inner surface of the plasma membrane and fuse (Figure 3–9). The reverse of this process involves vesicles budding from the plasma membrane and entering the cytoplasm where they are incorporated back into the internal membranes of the cell. In this way a cell stays roughly the same size throughout its lifetime. This dynamic balance is mediated by the Golgi apparatus through the regulation of vesicular traffic to and from the cell surface.

Lysosomes Lysosomes are specialized vesicles produced by the Golgi apparatus. They have a unique role in phagocytosis and cellular digestion. The lysosome is a veritable storehouse of enzymes known as acid hydrolases that are capable of digesting a wide range of types of molecules (Figure 3–10). This provides an interesting paradox. If lysosomes have enzymes capable of digesting all kinds of

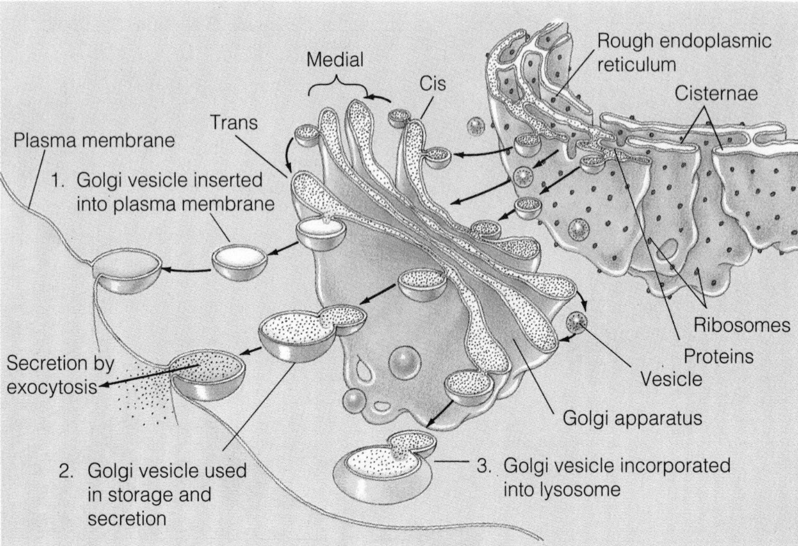

Figure 3–9 The Golgi Apparatus and Secretion
The Golgi apparatus is a stacked membrane structure with cis, medial and trans plates. Materials (proteins, glycoproteins) are transported by vesicles to the Golgi from the endoplasmic reticulum. Vesicles are also used to transport material between the Golgi plates. Secretion occurs by packaging of materials in the Golgi into secretory vesicles which insert into the plasma membrane and release their contents. The lysosome is also a vesicular product of the Golgi apparatus. The Golgi is responsible for most of the glycosylation in a cell.

molecules, including those present in the cell itself, why is it that the cell is not destroyed by these enzymes? The answer is that the cell is protected because the digestive enzymes are sequestered safely inside the lysosome. However, when a cell eats or drinks, through the processes of **phagocytosis** (cell eating) and **pinocytosis** (cell drinking), it is able to actively bring in material from outside the cell (a general process known as **endocytosis**). It does this by wrapping extracellular particles up in a plasma membrane sack and forming a cytoplasmic vesicle called an **endosome.** This important process is the reverse of **exocytosis,** which occurs in conjunction with secretion and the addition of new membrane to the plasma membrane (see Figure 3–9). Following endocytosis, the lysosome fuses with endosomes to form a *phagosome.* This exposes the inside of the endosome to the battery of powerful digestive enzymes found within the lysosome. The content of the endosome is broken down enzymatically into small molecules that can be transported through the phagosome membrane and into the cytoplasm, thus, providing building-block molecules and nutrients needed for the cell to grow.

Vacuoles Vacuoles are a class of vesicles that are basically the storage containers of cells. They are thought to be related to lysosomes and are particularly prevalent in plant cells. Vacuoles are constructed to package materials produced by the cells for later use or to sequester molecules for which the cell has no use or may be otherwise toxic. Many of these materials represent materials brought into the cell by endocytosis from outside and undigested by lysosomal enzymes. The materials are packaged inside membrane sacs, like kitchen trash bags, and stored until they are exported as trash.

Peroxisomes

Peroxisomes are a class of vesicles that are involved in oxidative reactions utilizing hydrogen peroxide (H_2O_2). Peroxides are familiar to most of us as powerful oxidizers used to disinfect wounds or to bleach hair or fabric. They are so powerful that special enzymes within the protected environment of the peroxisome must be used to convert water and oxygen to hydrogen peroxide, which can then be used to degrade toxic substances such as formaldehydes and phenolic compounds into less toxic forms. Peroxisomes are not derived from the ER but rather are self-replicating organelles within the cytoplasm.

Cytoskeleton—Microfilaments, Intermediate-Sized Filaments, and Microtubules

The interwoven network of filaments found in cells are made up of basically three types of proteins. These pro-

Phagocytosis active, energy-requiring cellular transport process (cell eating); used to bring large molecules or particles into a cell

Pinocytosis active, energy-requiring cellular transport process (cell drinking); used to bring fluids and dissolved substances into a cell

Endocytosis active, energy-requiring transport process involving invagination and vesiculation of the plasma membrane

Endosome vesicle arising from endocytosis

Exocytosis active, energy-requiring transport process involving evagination or budding of plasma membrane

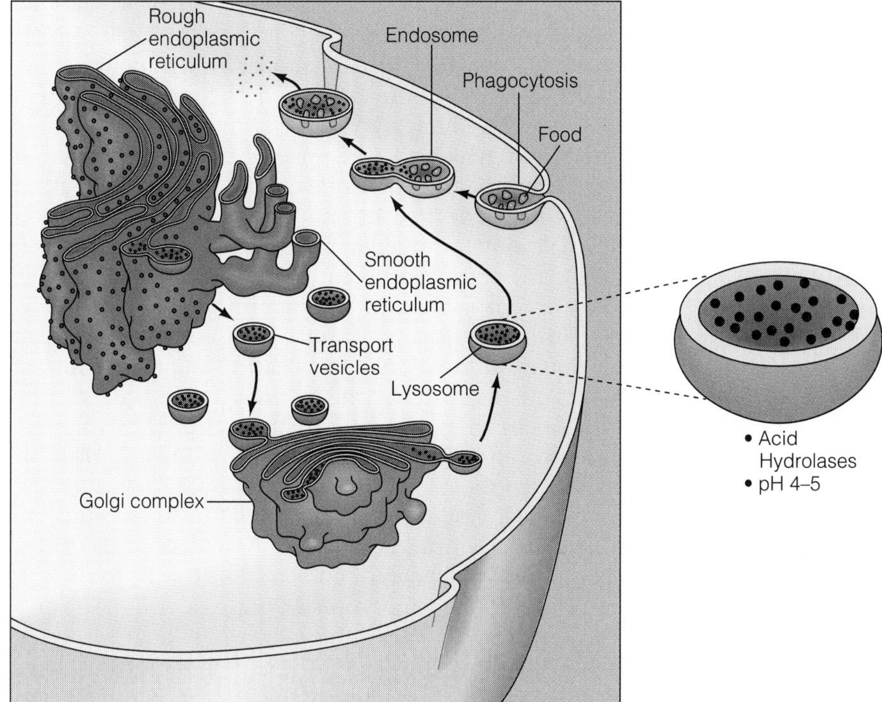

Figure 3–10 The Lysosome and Phagocytosis
The lysosome is formed by the Golgi apparatus. The content of the lysosome is highly acidic and includes enzymes called hydrolases, which are active at pH 4–5. Lysosomes fuse with endosome(s) and digest their content, which is then transported into the cytoplasm of the cell for use.

teins are capable of being assembled (a process known as polymerization) into long filamentous and tubular structures called **microfilaments, intermediate-sized filaments** (IF) and **microtubules** (Figure 3–11). The names provide the best description of their physical character, relative size, and shape. Microfilaments are small filaments (5 nm–6 nm in diameter) found within the cytoplasm of all cell types and composed of the protein *actin.* Likewise, microtubules are small tubes, but they are thicker than microfilaments (22 nm–25 nm in diameter) and composed of the protein **tubulin.** Intermediate-sized filaments are intermediate between microfilaments and microtubules (8 nm–11 nm in diameter) and are composed of proteins known as intermediate-filament proteins (Figure 3–11). These three types of polymers occur in a variety of combinations and arrangements in different types of cells. They may interact directly or indirectly with the plasma membrane.

Together microfilaments, microtubules, and intermediate-sized filaments make up a complex cytoplasmic skeleton, or **cytoskeleton.** Similar to the poles of a tent, cytoskeletal filaments and tubules are very important in determining the shape of cells. Cell motility and the positioning and movement of organelles also depend on the function of the cytoskeleton. The organization of the cytoskeleton can, and does, change dramatically within a cell, and, when this occurs, the cell shape changes. Rearrangements of the cytoskeleton are a natural, and necessary, part of cell growth, cell division, and cell movement. Alterations in cell morphology often reflect changes in the relationship between the cytoskeleton, the plasma membrane, and extracellular environment.

Centrioles, Cilia, and Flagella

Centrioles, cilia, and flagella share in common an underlying framework of microtubules, which suggests a relationship to the cytoskeleton. In fact, centrioles play a significant role in the organization of the cytoskeleton and in establishing the *spindle fibers* used to separate chromosomes during mitosis (discussed later in this chapter). Centrioles are a collection of nine short, triplet bundles of

Microfilaments ubiquitous components of the cytoskeleton of eucaryotic cells; composed of actin; 5 nm–6 nm in diameter

Intermediate-sized filaments family of filaments composing the cytoskeleton; 8 nm–11 nm in diameter

Microtubules ubiquitous components of the cytoskeleton of eucaryotic cells; composed of tubulin; 22 nm–25 nm in diameter

Tubulin protein composing microtubules

Cytoskeleton filaments or tubules within the cytoplasm of a cell; affect cell shape

microtubules (Figure 3–12[a]). There are normally two to four of these organelles in a cell, depending on the stage of the **cell cycle.**

Centrioles are related to cilia and flagella through their capacity to act as base plates or basal bodies for the attachment of microtubules. Some types of cells have specialized fingerlike extensions projecting out from their cell surfaces called **cilia** and **flagella.** A specialized microtubular structure called an **axoneme** extends throughout the length of a cilium or flagellum.

In cells on which cilia are present, they usually occur in large numbers, and their movement is wavelike, with a strong force generated in only one direction. The unidirectional motion of cilia provides propulsion for cell movement, or for movement of materials over a cell surface. Cilia play a prominent role in the directed movement of many species of unicellular eucaryotes, which may have hundreds of cilia beating out a rhythm of movement that propels them through the watery environments in which they live. Cilia in human tissues are found on cells that are themselves not able to move. For example, epithelial cells lining the passageways into the lungs have a vast array of cilia, which beat forcefully in one direction, outward. Any foreign particles, or debris, encountering this lawn of cilia are moved out of the respiratory tract. Imagine yourself lying on the lawn at a local park and having millions of individual blades of grass beating forcefully in one direction under your body. You might well be lifted and moved over the grassy surface without having to take a step. What a way to keep people off the grass!

A flagellum is basically a form of specialized cilium. Usually, there is only one flagellum per cell. The only flagellated cells found in humans are sperm cells. The whipping motion of their long tails allows sperm to move rapidly over relatively great distances. Sperm take advantage of this mechanism of motility in moving through the female reproductive tract. This greatly increases the chances of a sperm encountering an egg, which in turn may be slowly moving over the epithelial lining of a Fallopian tube under the influence of cilia.

(a)

(b)

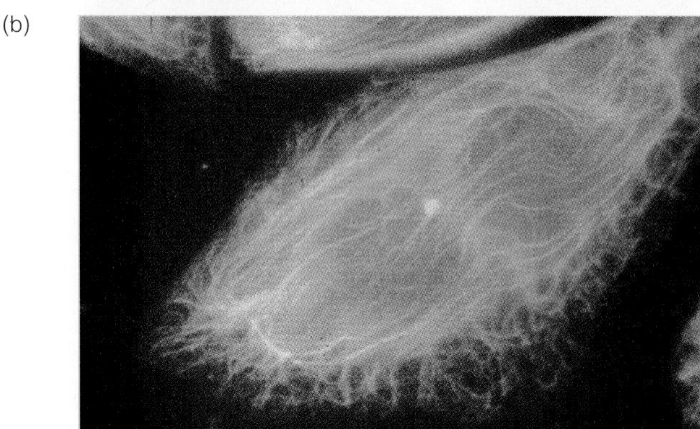

(c)

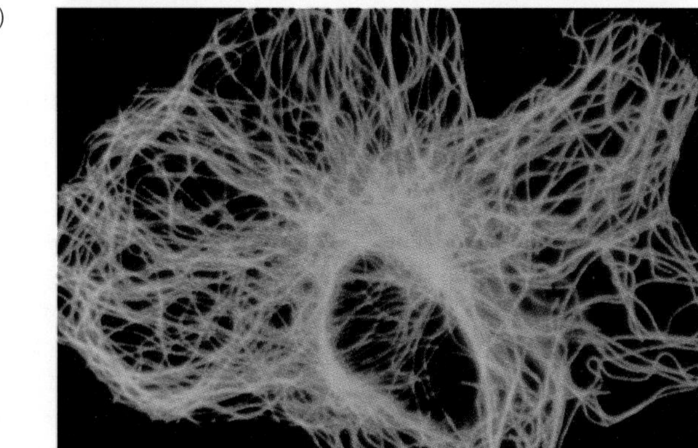

Figure 3–11 Filaments, Tubules, and the Cytoskeleton
The cytoskeleton is composed principally of microfilaments (a), intermediate-sized filaments (b), and microtubules (c). Microfilaments are composed of actin, intermediate-sized filaments (b) are represented by keratin, and microtubules (c) are composed of tubulin. In many cells all three types of cytoskeletal elements are present simultaneously.

Mitochondria

Mitochondria are the energy producers of eucaryotic cells. They are plentiful in the cytoplasm of most cell types and range in shape from spherical to elongated cylinders. However, regardless of shape, they share a characteristic double-layered envelope made up of a highly convoluted inner membrane and a smooth-appearing outer membrane.

Cell cycle the repetitious process by which a cell prepares for and completes division

Cilia cell surface extensions surrounding an axoneme; involved in cell movement

Flagella extensions of specialized cells (such as sperm) containing axonemes

Axoneme microtubular structure of cilia and flagella

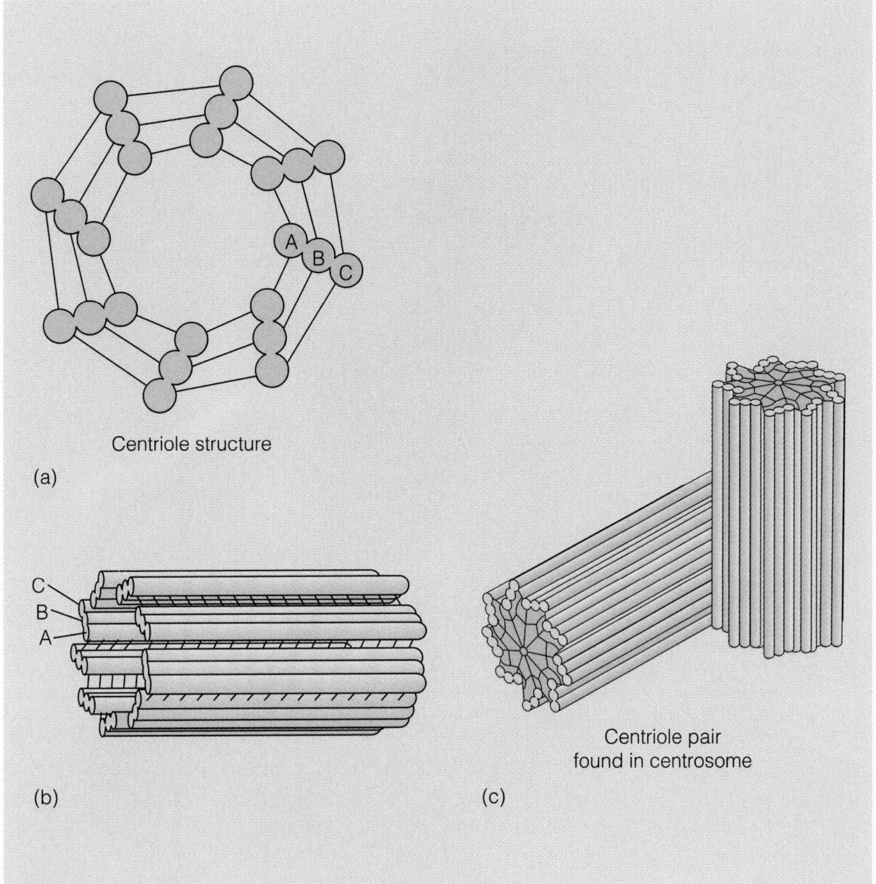

Figure 3–12 The Centriole
Centrioles are composed of nine sets of three microtubules (designated A,B,C) and organized into a cylindrical shape, parts (a) and (b). Centrioles usually occur in pairs (c) within the centrosome.

These two membranes are separated by an intermembrane space. Located within the envelope is a cytoplasmlike material known as the *matrix* (Figure 3–13). Dissolved or suspended in the matrix are a variety of proteins, enzymes, and RNAs, as well as mitochondrial DNA. The chemical reactions that produce ATP take place predominantly in the matrix and the inner membrane (Figure 3–13, inset).

The energy produced in the mitochondria of a cell supplies the majority of energy needed for that cell to function. DNA and protein synthesis, cell movement, cell division, and exo- and endocytosis all require a great deal of readily available energy. Energy used by cells to perform work is produced by a process called **cellular respiration.** A mitochondrion is the cellular equivalent to a portable generator. However, instead of producing electricity, a mitochondrion produces millions of ATP molecules each second in active cells. Thus, mitochondria are unique in many ways (Figure 3–13):

1. They have a double membrane, similar to that observed in the nucleus.
2. They produce and export energy-rich ATP molecules critical to the needs of the cell.
3. They have their own DNA, separate and distinct from that found in the nucleus.

The discovery that DNA is present in mitochondria has generated much speculation. If mitochondria have their own DNA, might they really be cells within cells? What is the origin of this organelle? Mitochondrial DNA is a closed circular molecule and, thus, is distinctly different from the long, stringlike linear DNA molecules found in chromosomes of the host cell nucleus. In fact, the mitochondrial DNA is similar to bacterial DNA. Studies have shown that mitochondrial DNA contains genes, but mitochondria cannot live very long outside a cell. This means that even though they have their own DNA, they need assistance from the host cell to function and survive. Evidence suggests that hundreds of millions of years ago a fortuitous encounter occurred between a bacteriumlike organism (ancestor to present day mitochondria) and an early eucaryotic cell (one having a

Cellular respiration energy-producing reactions in a cell

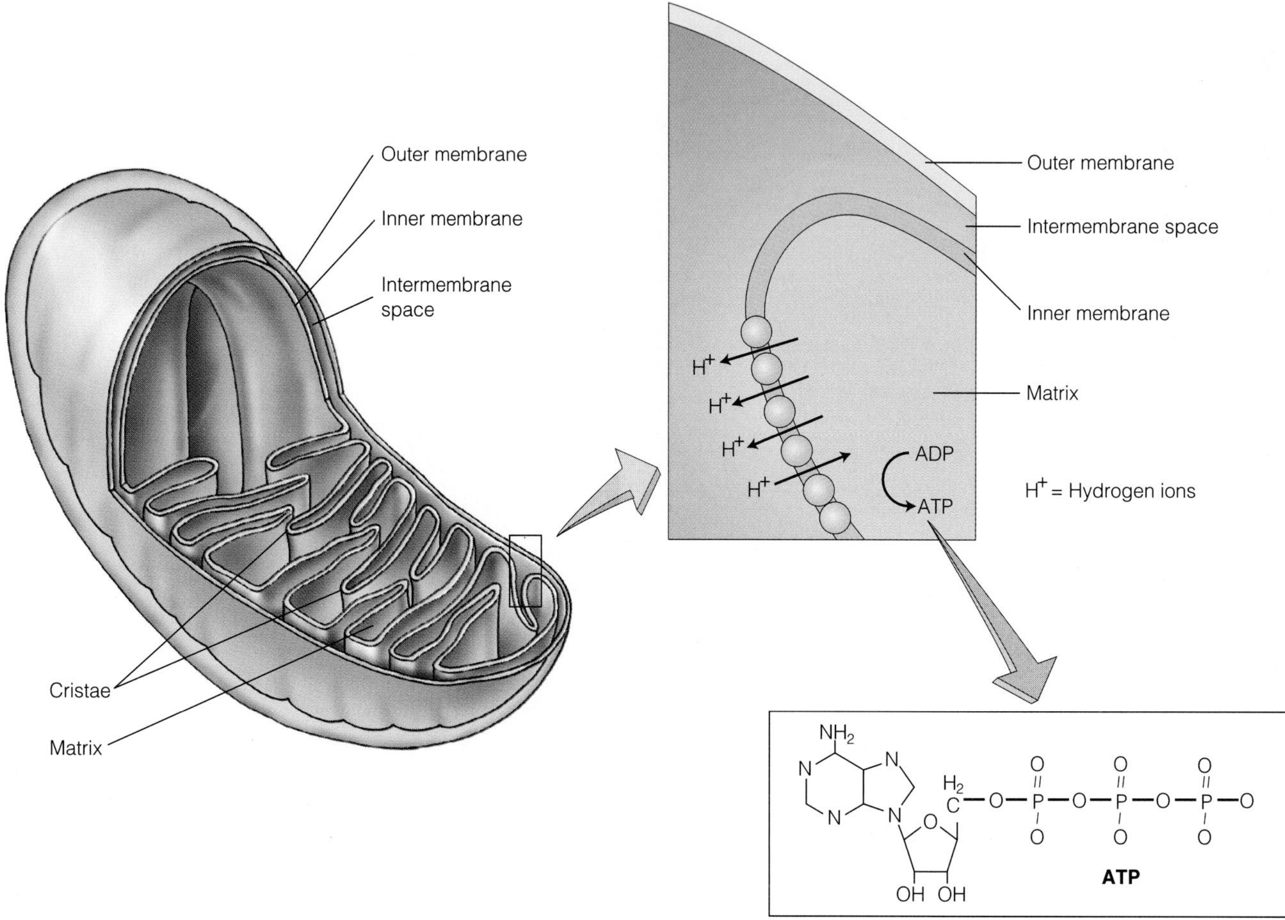

Figure 3–13 The Mitochondrion
The mitochondrion is a double-membrane organelle. The outer membrane is relatively flat and porous. The inner membrane is embedded with proteins involved in electron transport and ATP formation. Hydrogen ions (H^+) are transported out of the matrix into the intermembrane space during electron transport. ATP is formed when H^+ move back into the matrix. ATP is adenosine triphosphate (inset) and is the most commonly used high energy molecule used by cells.

nucleus and linear chromosomes) (Figure 3–14). This evolved into a **symbiotic relationship,** which is maintained today in all groups of eucaryotic organisms, including protists, fungi, and plants, as well as animals. These tiny, energy-producing organelles are permanent guests of all types of eucaryotic cells, paying their way with the formation of energy-rich ATP molecules needed to ensure cellular and, ultimately, organismal survival (Biosite 3–1).

FUNCTIONAL CONSIDERATIONS

Cellular Respiration and the Production of ATP

In terms of utilizing its energy resources, cells spend ATP as we spend money. The cell earns this spendable income through cellular respiration. This means that the food we eat must provide a source of molecules capable of being used to generate ATP. The energy of carbohydrates and fats, for instance, is converted and stored in the chemical bonds of ATP to pay the energy price for the physical and chemical activities of cells that underlie all human activity and behavior.

Cellular respiration involves a complex series of enzymatic reactions that sequentially release chemical energy from the food we eat. In the presence of oxygen, cellular

Symbiotic relationship a relationship in which two independent entities coexist together to their mutual benefit

3–1

THE SUSPECTED ORIGIN OF MITOCHONDRIA

The origins of mitochondria are in the distant past, perhaps a billion years ago. They were crucial to the evolution of eucaryotes. The ancestors of modern-day mitochondria were procaryotic cells that were somehow taken up by a pre-eucaryotic cell (perhaps by endocytosis that did not lead to degradation of the ingested cell). This is depicted in Figure 3–14. The outer membrane of the present-day mitochondrion is similar to the plasma membrane of the host cell. The inner membrane of the mitochondrion is distinctly different from it. The DNA of the mitochondria is circular and found suspended free in the matrix. mRNA, rRNA, and tRNA are transcribed from the mitochondrial genome, and proteins are translated on mitochondrial ribosomes. However, the genetic information contained in the mitochondrial DNA is not sufficient to allow the organelle to survive separately from the host cell.

Based on the energy requirements of eucaryotic cells, it is doubtful that eucaryotes as we know them today, and particularly multicellular organisms, could have evolved without the symbiotic relationship that was established in the ancient seas of Earth a billion or more years ago.

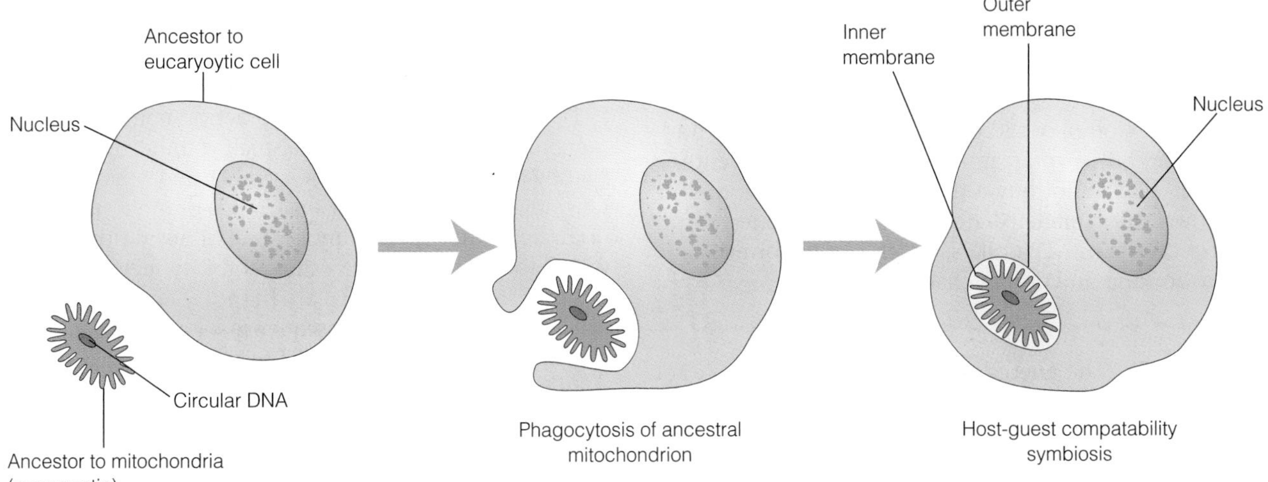

Figure 3–14 The Origin of Mitochondria in Eucaryotic Cells
The ancestors to the current mitochondria were probably procaryotic organisms, which were captured by phagocytosis but not lysed. The inner membrane of the mitochondrion is of procaryotic origin as is the presence of circular DNA. The outer membrane of the mitochondrion is of host cell origin. The host-guest relationship is mutually beneficial, or symbiotic.

respiration is highly efficient and is called **aerobic cellular respiration.** In the absence of oxygen, respiration is called **anaerobic cellular respiration** and is very inefficient. The biochemical reactions involved in aerobic cellular respiration take place initially in the cell cytoplasm. These reactions are part of the process of **glycolysis** and result in the breakdown of glucose, which is a six-carbon sugar, into two three-carbon molecules called *pyruvic acid or*

Aerobic cellular respiration energy-producing reactions requiring oxygen

Anaerobic cellular respiration energy-producing reaction occurring in the absence of oxygen

Glycolysis metabolic pathways within a cell converting glucose to pyruvate

pyruvate. These reactions produce several ATP molecules; however, they require the cell to use ATP to generate ATP. The net result is that glycolysis does not produce much extra ATP. However, the process does not end there. Pyruvate is transported into a mitochondrion, where in the presence of oxygen 15 or more ATP molecules are formed for each pyruvate oxidized (30 or more ATPs for each glucose molecule). The oxidation of pyruvate occurs in a series of reactions known as the **Krebs cycle** and ends up forming water and carbon dioxide as waste products (Figure 3–15). The sequences of enzymatic reactions in glycolysis and the Krebs cycle are highly organized and integrated. In general, the breakdown of glucose in reactions involving oxygen and water may be represented as follows:

$$C_6H_{12}O_6+H_2O+6CO_2 \rightarrow 6CO_2+6H_2O+\text{Energy}$$
(net of 30 or more ATPs)

The fats and complex carbohydrates that we ingest as part of our diet are energy-rich molecules. It is the oxidation of the carbon-hydrogen bonds in these compounds (as discussed in Chapter 2) that ultimately produces energy for cellular work. But the collective potential energy of complex carbohydrates and fats is not directly available for use by a cell. It must be released in small amounts in a sequential manner. Complex carbohydrates are a little like hundred dollar bills—you cannot use them in a vending machine. You cannot buy a thing until you break the big bills down into smaller denominations. The energy currency of cells is in ATP-dollars, which can be formed only if mitochondria function as change makers.

Uses of ATP The energy stored in ATP is accessible for cell use by the breaking of chemical bonds. The structure of ATP includes three phosphate groups linked to a deoxyribose sugar, which is attached to adenine (Figure 3–13, inset). The final two of these phosphates are called *pyrophosphate groups,* and it is the hydrolysis of these bonds during the biochemical reactions carried out by cells that releases the bond energy for use. ATP is considered the fundamental source of energy in cells because nearly all activities of cells use it. In addition, after transfer of energy from ATP to other molecules, the ADP that results is recycled to form new ATP. There are a few other energy-rich molecules used by cells to carry out specific activities, including nicotinamide dinucleotide (NADH), Flavin adenine dinucleotide ($FADH_2$), and Guanosine triphosphate (GTP) (Figure 3–15), but ATP is by far the most common.

Aerobic cellular respiration connects the activities of the mitochondria to the energy needs of cells and, ultimately, to the energy needs of the entire body. The evolution of efficient aerobic respiration in cells occurred when a symbiotic relationship between mitochondria and host cells was established over a billion years ago. In fact, this ancient partnership was an essential step in the evolution of all eucaryotic organisms.

The Movement of Molecules

Passive Transport and Diffusion To better understand how cells are able to perform what appear to be some amazing feats of growth, changes in shape, transport, and movement, an examination of a few of the mechanisms that make such feats possible is in order. The physical and chemical basis for cellular transport are absolutely critical to understanding the functions of cells. The simplest of processes for movement of materials into and out of cells is **diffusion.** Diffusion is a passive phenomenon based on the inherent kinetic energy of molecules. It is the familiar process by which the smell of perfume drifts over the still night air, or cloud shapes change against the clear blue sky, or colored dyes disperse in undisturbed water. It is a fundamental physical process by which molecules of all kinds of materials move from areas of higher concentration to those of lower concentration (Figure 3–16).

From a theoretical as well as practical point of view, the more molecules contained in a small volume, the more likely they are to encounter one another and bounce off in a new direction. Bounces and rebounds of molecules tend to occur such that the net movement is away from areas in which there are more molecules and into regions in which there are fewer (Figure 3–16[a]). In the case of perfume in air, diffusion occurs as a consequence of random molecular interactions. The probability of interactions between molecules increases with increased concentration. Open a bottle of perfume in the middle of a closed room and the fragrance molecules begin to evaporate and escape the bottle in which they were contained. If the air is kept absolutely still, it may take hours for the molecules to diffuse throughout the room. Eventually, however, the room will fill uniformly with molecules that were once concentrated in the bottle. This uniformity of distribution of molecules does not mean that the movement of individual molecules ceases (Figure 3–16[d]). It means simply that the probability and outcome of colli-

Krebs cycle pathway of sequential enzymatic reactions occurring in mitochondria utilizing acetyl-CoA to generate carbon dioxide and ATP

Diffusion movement of a substance from an area of higher concentration to one of lower concentration

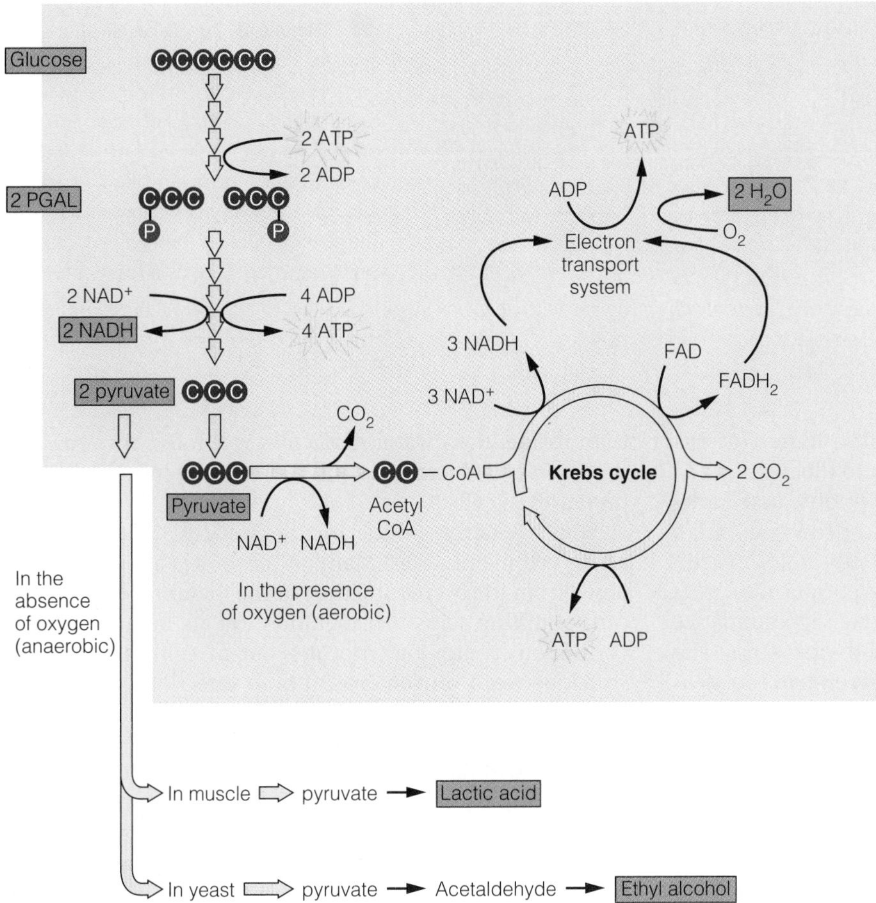

Figure 3–15 Glycolysis and the Krebs Cycle—Energy Production in Cells
Energy production begins with the breakdown of glucose during glycolysis. The six-carbon glucose is broken down into two three-carbon molecules known as PGAL (phosphoglyceraldehyde), which utilize 2 ATP. PGAL is converted to pyruvate forming two molecules of NADH and four molecules of ATP. This produces a net gain of 2 ATP molecules, formation of 2 pyruvates and ends the glycolysis stage of energy production. In the absence of oxygen (anaerobic conditions) pyruvate is converted to lactic acid in muscle cells. Pyruvate is converted to acetaldehyde and then to ethyl alcohol in yeast, commonly referred to as fermentation. In the presence of oxygen pyruvate enters the mitochondrion and is converted to CO_2 and a two-carbon compound (acetyl) that is linked to a molecule known as coenzyme A (CoA) or acetyl CoA. The acetyl CoA enters the Krebs cycle in the matrix of the mitochondria and the acetyl group is transferred to molecules of the cycle. Ultimately, the two carbons are degraded to CO_2 and released as waste. In the process NADH and $FADH_2$ are produced and enter the electron transport system (ETS) of the inner membrane of the mitochondria. $FADH_2$ is flavin adenine dinucleotide and NADH is nicotin amide dinucleotide. Both $FADH_2$ and NADH have the capacity to undergo chemical changes needed to carry energy from chemical reactions in the Krebs cycle into the ETS. NADH and $FADH_2$ transfer their hydrogens and electrons to molecules of the ETS, which pump them into the intermembrane space. The hydrogen ions re-enter the matrix and in the process synthesize ATP (see Figure 3–13 for details). Oxygen is the final acceptor of electrons in the ETS and forms H_2O as a final product. The net gain of ATP for each glucose molecule that undergoes glycolysis and participates in the Krebs Cycle and ETS is 30 or more (usually 36).

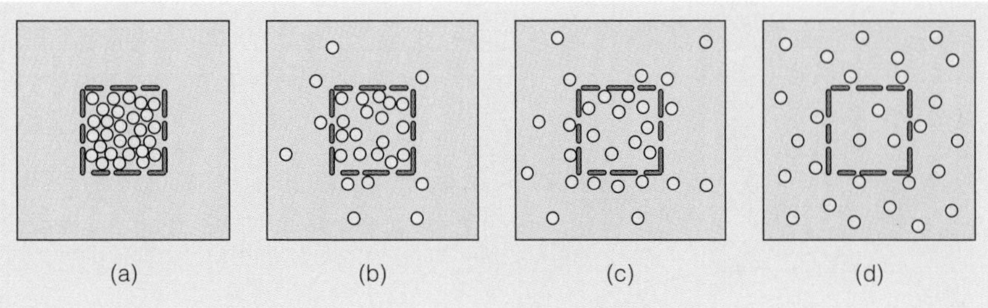

Figure 3–16 Diffusion
In (a) the concentration of particles is contained exclusively within the inner, perforated box. With time (b) particles move out into the larger volume of the closed box. Eventually, particles diffuse out (c) and (d) to uniformly fill the large volume. The kinetic energy of the particles forces their separation in (a) as a result of a higher rate of collision and a greater probability that particles will move away from one another.

sions between molecules results in a distribution of fragrance indistiguishable from one instant to the next.

Diffusion also occurs in aqueous solutions, and molecules dissolved in the tissues and cells of the body are a good example. Molecules dissolved in body fluids distribute themselves from areas of higher concentration to areas of lower concentration. In this way materials in abundance outside of cells can move into cells and vice versa. The process and rate of diffusion along a **concentration gradient** requires no expenditure of cellular energy. It depends solely on the differences in concentration between two areas or locations and an unhindered path along which to migrate.

Not all movement of molecules into and out of cells is governed by passive diffusion because not all pathways of molecular or ionic migration are unhindered. For example, ions and molecules often require an active process utilizing cellular energy to be transported across the plasma membrane. Movement of molecules under these circumstances involves mechanisms of **active transport.** However, before discussing active transport, we need to establish the nature of the movement of water molecules across membranes.

Osmosis and Tonicity Human beings are composed of 60% water. Cells are filled with it and surrounded by it. Human survival depends on water. Water is a universal component of life and of the aqueous fluids that compose the cytoplasm of cells and the interstitial fluids that surround them. However, these facts are a bit misleading because, if human cells are exposed to pure water, they will burst and be destroyed. Clearly, something in the water of our cells and body fluids prevents this from happening. That something is dissolved molecules and ions.

Water molecules, as any other molecules that are free to move, are subject to diffusion along a concentration gradient. For example, if there are more water molecules within a cell than in the same volume outside that cell, water will move outward (Figure 3–17). If, on the other hand, there is a higher concentration of water molecules outside the cell than inside, then water molecules will move inward. This process is called **osmosis** and is one of the most important processes in cells.

Water is fairly unique in the freedom it has to move across cell membranes. Many other solutes are greatly hindered from freely crossing the plasma membrane, and some cannot diffuse across the membrane at all. Because water is continuously moving into and out of cells, a balance is struck between movements in both directions. This establishes a no-net-gain situation called *equilibrium*. Alterations in this equilibrium may have dire consequences, including dramatic shrinkage and/or bursting of cells. Changes in equilibrium are brought about by changes in the amounts or types of substances dissolved in the cytoplasm, or in the external environment, which cannot pass through the plasma membrane. These dissolved, impermeable substances determine the property of a solution known as its *tonicity*. Solutions containing different amounts of solutes per unit volume of water, when separated by a semipermeable membrane, undergo a net movement of water molecules and create an **osmotic pressure.**

The most spectacular example of the consequences of alterations in tonicity and the resultant changes in osmotic pressure involves experiments with red blood cells. Human red blood cells have a characteristic biconcave shape (Figure 3–17[b]). When cells are placed into a saline solution containing approximately 1% sodium chloride, these cells do not change shape at all, indicating no net

Concentration gradient difference in the concentration of a substance from one region or compartment to another

Active transport cellular transport processes requiring ATP

Osmosis the movement or diffusion of water from an area of higher concentration or activity to a lower one across a semipermeable membrane barrier

Osmotic pressure the force or pressure produced by osmosis

movement of water into or out of the cells. This solution is called **isotonic.** On the other hand red blood cells suspended in distilled water rapidly round up, similar to inflating balloons, and burst. This is because of the high osmotic pressure of water relative to the cell cytoplasm. It is the net movement of water molecules along a concentration gradient into the red blood cells that causes them to explode. Water containing a relatively low concentration of salt (or pure water), which causes bursting, is called a **hypotonic solution.** At the other extreme, cells shrivel up into a spiked, or crenulated, shape when placed into a solution containing 2%–3% or more of salt (Figure 3–17[b]). The net movement of water under the influence of this osmotic pressure is out of the cell. A high-concentration salt solution is called **hypertonic.**

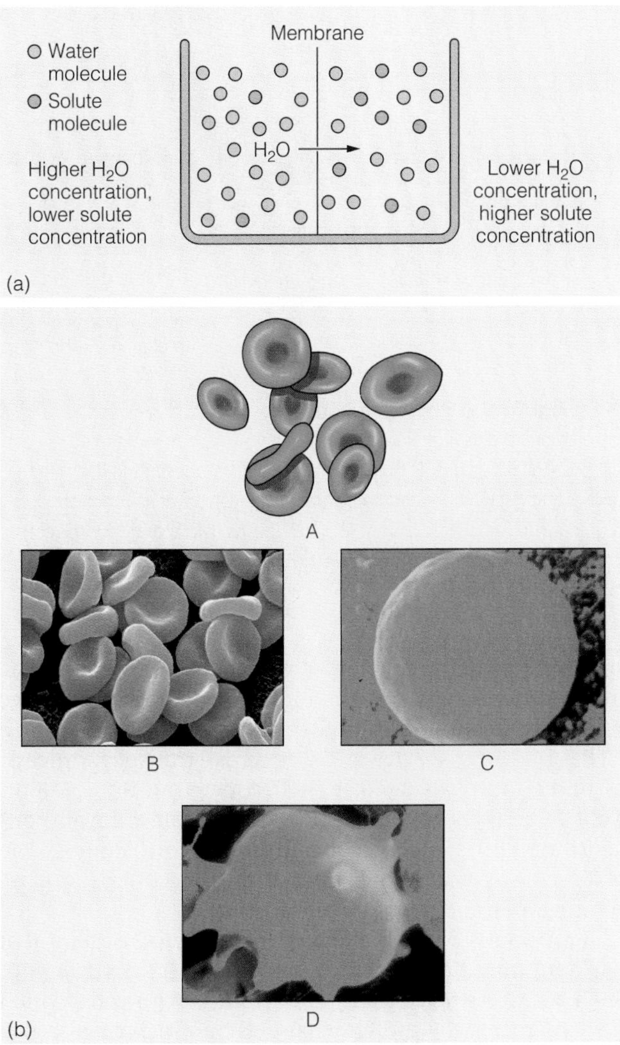

Figure 3–17 Osmosis and Red Blood Cells
(a) Osmosis is the movement of water across a membrane in response to differences in solute concentrations. (b) Mammalian red blood cells are normally biconcave in shape (A). When placed in an isotonic solution, the cells do not change shape (B). When placed in a hypotonic solution (such as distilled water), the solute concentration inside the cells results in water movement into the cells, until they lyse or burst, as if balloons (C). When placed in a hypertomic solution, the cells are exposed to a higher concentration of solutes outside the cell and water moves out of the cell, shrinking the cells dramatically (D).

Facilitated Diffusion, Cotransport, and Active Transport Passive transport of molecules into cells does not require cellular energy, but cells often have to bring in molecules and ions that (1) do not move freely across cellular membranes and (2) must be moved against their concentration gradient. Cells have special mechanisms to carry out these processes. The transport of molecules that cannot simply diffuse through the membrane uses a process called **facilitated diffusion.** In this case, special channel-forming proteins in the plasma membrane (called *permeases*) can be opened, or closed, to allow for movement of molecules and/or ions into a cell (Figure 3–18). The channel proteins have the capacity to bind selectively to particular types of molecules or ions located at the cell surface. For example, the binding of a molecule of glucose to its specific permease opens a channel through the membrane. The glucose molecule then passes through the channel and into the cytoplasm. The channel closes and is ready to bind another glucose molecule for transport. This process is considered passive, simply because it does not directly require ATP. However, it does require the binding of molecules or ions to specific channel proteins, which, in turn, open to form a channel extending through the plasma membrane.

Another type of transport carried out by cells is called *cotransport.* In cotransport, two different types of ions or molecules can be brought into the cell simultaneously, or exchanged—one from inside to outside, the other from outside to inside. Cotransport takes advantage of ionic gradients established across the plasma membrane of a cell.

Isotonic solution a solution containing solutes matching physiological concentrations; of or relating to conditions of equal tension or tonicity

Hypotonic solution a solution containing a lower concentration of solutes relative to physiological norm

Hypertonic solution a solution containing a higher concentration of solutes relative to physiological norm

Facilitated diffusion use of specialized membrane channels to make the transport of materials into and out of cells easier; no cellular energy is required

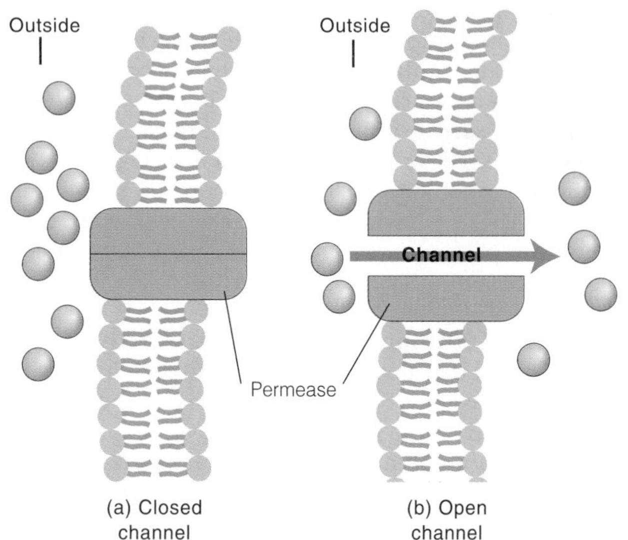

Figure 3–18 Facilitated Diffusion
Facilitated diffusion depends on permease molecules embedded in the plasma membrane. The permeases form channels that open in response to binding specific molecules. When open, the channels allow molecules to diffuse into the cell from a region of higher concentration (outside) to a region of lower concentration (inside).

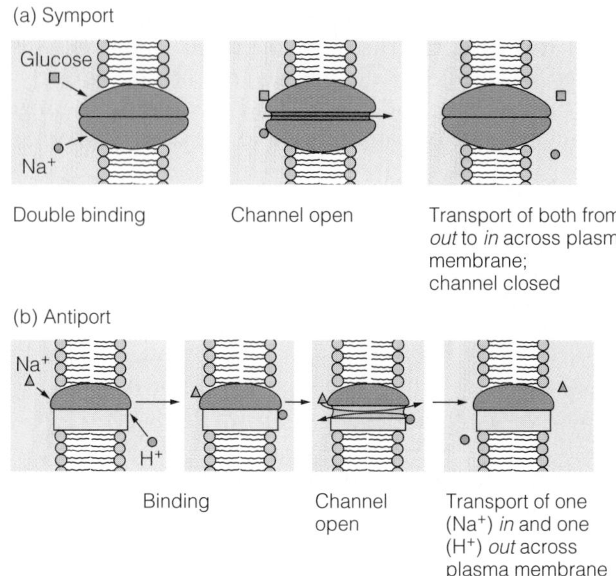

Figure 3–19 Cotransport Processes
(a) A symport channel is one in which two different molecules or ions (glucose and sodium [Na^+] as shown) bind to the channel protein and are transported in the same direction. (b) An antiport channel is one in which two different molecules or ions (Na^+ and Hydrogen [H^+] as shown) bind on opposite sides of the membrane and move in opposite directions, in or out.

One example of how this works is in the cotransport of sodium ions and glucose molecules. There are generally about 40 times as many sodium ions outside a cell as inside, providing a steep chemical gradient potential across the plasma membrane. When a sodium ion *and* a glucose molecule bind simultaneously to the sodium-glucose cotransport protein, a channel opens (somewhat similar to the two keys needed to open a safe deposit box at a bank) to allow both substances to cross the membrane. This type of channel is called a **symport** and is *unidirectional* (Figure 3–19[a]). Piggybacking the transport of two substances is very important to the cell and does not directly use precious cellular energy currency (ATP) to operate. An **antiport** functions in a similar fashion but, instead of molecules or ions moving in the same direction across the plasma membrane, they move in opposite directions, or *bidirectionally* (Figure 3–19[b]). Again, as in the case of symports, a gradient of ions is involved in driving the movement of one type of ion inward while allowing the outward flow of a different type of ion against its own concentration gradient.

Processes of transport that require ATP are called active transport. In some ways active transport is similar to facilitated transport. For example, molecules or ions must be bound to, or associated with, a specific membrane-spanning protein receptor in the plasma membrane. The major difference is that the molecule bound to an active transport receptor does not translocate unless, and until, ATP is hydrolyzed (Figure 3–20). Active transport processes move materials against their concentration gradient. That is to say, they move substances from a lower concentration to a higher concentration. For example, the sodium ions that move inward during cotransport are pumped out again by active transport.

Active transport is analogous to pumping the roomful of perfume, mentioned earlier, back into the bottle again. Such a process takes energy, and so it is with cells, which have many special protein channels by which they actively bring in molecules. Most often, the price of transport is paid in hydrolysis of ATP molecules. Of course, if the substances that are transported are themselves energy rich and capable of being used to produce ATP, such as glucose, the energy of transport is well spent.

Symport transport channel in plasma membrane

Antiport mechanism of cellular transport involving exchange in both directions across the cell membrane

THE CELL CYCLE AND CELL DIVISION

It should be remembered that the DNA of eucaryotic cells is organized into a number of long, linear strands called chromosomes. All the genes of a cell are encoded in chromosomal DNA sequences and are collectively called the genome. The DNA of chromosomes is specifically associated with histone proteins, and together they compose the functional form of genetic material found in the nucleus called chromatin (refer to Figure 3–7). As you might imagine, it is no trivial matter for one cell to become two. A cell not only has to replicate the DNA of all chromosomes exactly but also to supply all the proteins, lipids, and carbohydrates needed to nearly double the mass and volume of the cell. Cells are also required to closely follow a programmed series of events that control progression through the cycle (Figure 3–21). Timing is everything in cell division.

There are two types of cell division. The first involves the equal division of all the genetic material in somatic cells and is called **mitosis.** The second is **meiosis,** or reduction division, which results in equal division of genetic material and a decrease in the total amount of DNA in each of the progeny cells (Figure 3–22). Meiosis is the process by which eggs and sperm are formed. Abnormalities in the timing, rate, or position of chromosomes during mitosis or in the location in the body of dividing cells can be lethal. Uncontrolled cell division is a primary attribute of cancer cells.

Interphase, G_1, S, and G_2

Starting with **somatic cells** and mitosis, one might ask, how does a cell go about dividing? There is a great deal of preparation for this event, which in reality is only one of a series of steps involved in a complete cell cycle (Figure 3–21). The journey through a cell cycle described here begins at a point just after cell division has been completed. This stage of the cycle is called G_1 and it represents the beginning of **interphase,** which is the time between cell divisions. It is during interphase that a cell begins to grow. The *G* in G_1 stands for gap and represents the period of time between the completion of cell division and the beginning of the synthesis of DNA and the assembly of chromatin. Once DNA synthesis is initiated, the cycle enters the S phase, or synthesis phase. To successfully divide later in the cycle, the cell must have precisely the right amount and composition of DNA so that it can endow what will be two new daughter cells with identical genetic information at the end of the cycle.

In the S phase, the cell exactly doubles the amount of DNA in each of its chromosomes. In humans this means complete duplication of 23 pairs of chromosomes. Once the synthesis of DNA is completed, the cell changes gears into the third phase of the cell cycle, known as G_2. This is the second gap in the cycle and represents the period between the end of DNA synthesis and the initiation of cell division. During G_2, the cell continues to grow and prepare itself for division. When interphase is finished (G_1, S, G_2), the cell is triggered to enter mitosis.

A great deal is known about the regulation of events in the cell cycle, but many facets of the process are still a mystery. For example, how is it that different cell types have different rates at which they divide (Table 3–1)? Precursors to human blood cells in the bone marrow divide extremely rapidly, finishing a complete cell cycle within a few hours. On the other hand, it may take days for a skin cell to go through a complete cell cycle. Some types of cells cease dividing altogether, as is the case for many neurons in the human brain and for muscle fibers. Cells that withdraw naturally from the cell cycle (called G_0) often do not return to it, even when tissue is damaged by accident or dis-

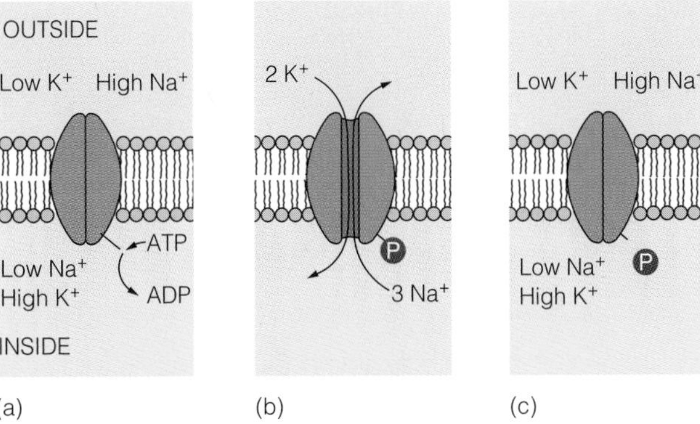

Figure 3–20 Active Transport and ATP
In (a) the active transport channel is closed. The concentration of Na^+ is high outside the cell while the concentration of K^+ is low. The reverse is true inside the cell. When ATP gives up one of its phosphates (a) to alter the channel proteins, the channel opens (b). This allows Na^+ to move out against its concentration gradient and K^+ to move in. In (c) the phosphate is removed from the channel protein, the channel closes and transport of Na^+ and K^+ ceases.

Mitosis phase of cell cycle in which chromosomes are divided equally into two daughter cells; consists of four stages: prophase, metaphase, anaphase, and telophase

Meiosis reduction division of cells (particularly in formation of gametes) in which chromosome number is reduced by half (*see* haploid)

Somatic cells from the Greek soma or body; cells of the body excluding the gametes

Interphase portion of the cell cycle in which the cell is not dividing

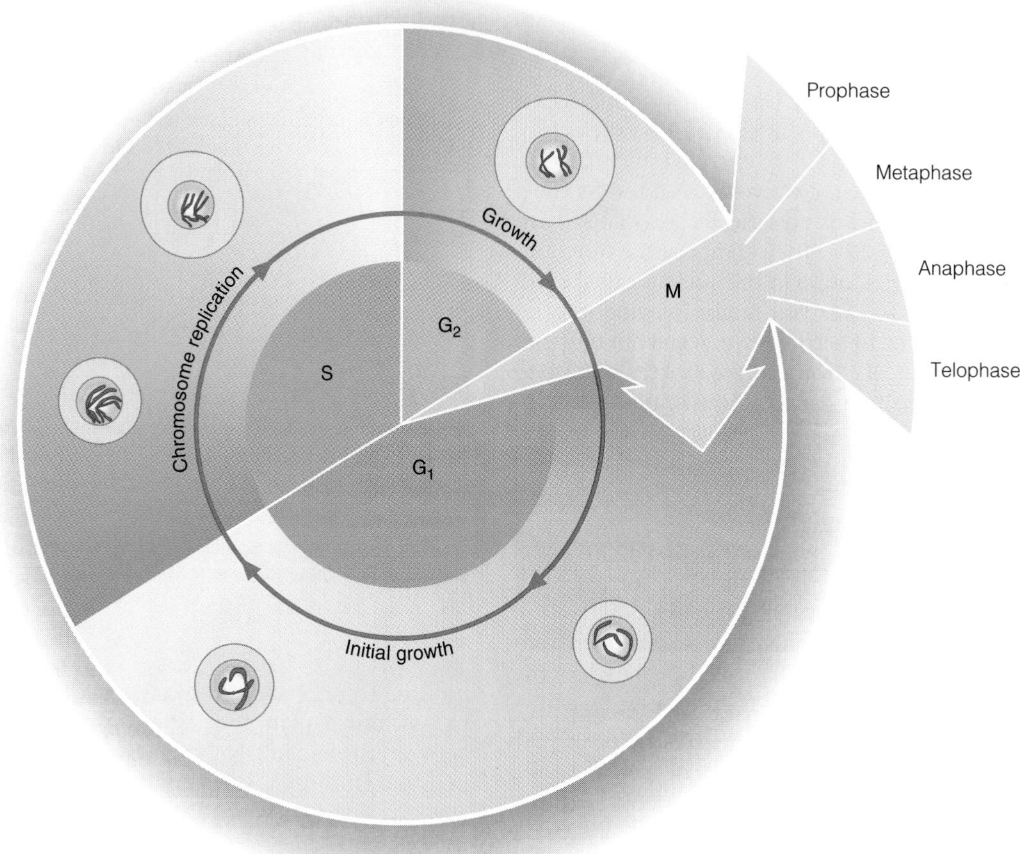

Figure 3–21 The Cell Cycle
The cell cycle of a dividing cell involves four phases: G_1, S, G_2, and M. G_1 is the initial growth phase, S is the phase in which DNA is replicated, G_2 is a second growth phase, and M is mitosis. Mitosis has four stages: prophase, metaphase, anaphase, and telophase. It is during the M phase that chromosomes separate and the cell divides. Prophase is characterized by condensation of the chromosomes, their attachment to the spindle fibers, and the disappearance of the nuclear envelope. Metaphase events align the chromosomes at the metaphase plate. Anaphase events separate the chromosomes at the centromere and move the separated chromosomes to opposite poles of the cell. Telophase events reverse the losses occurring in prophase by reestablishing the nuclear envelope, reorganizing the spindle fibers, and decondensing the chromosomes. Cell cleavage or cytokinesis produces two new daughter cells. As long as cells are dividing, they follow the set pattern described here as the cell cycle.

ease and new cells may be needed to replace the losses.

In addition to differences in total cycle time, the amount of time spent in any one phase of the cell cycle is also highly variable. For example, an average cell from human connective tissue (known as a fibroblast), may take 16 to 24 hours to divide. The time spent in each phase of the fibroblast cell cycle has been determined. A fibroblast may spend 5 to 6 hours in G_1, 8 to 10 hours in S, 6 to 7 hours in G_2 and about an hour to complete mitosis. The duration of mitosis is generally fairly short in most types of dividing cells. The preparations for mitosis that have been made in G_1, S, and G_2 allow mitosis to follow in a fairly rapid and stereotyped series of changes. The cell cycle is similar to being in a boat traveling on a long river headed towards a waterfall. It may take hours (or days) to get to the falls (G_1 + S + G_2), depending on the cell type, but once the falls are reached the time of descent over them (mitosis, M) is pretty much the same.

Mitosis

Mitosis itself is subdivided into four stages. They are *prophase, metaphase, anaphase,* and **telophase** (Figure

Telophase the final stage of mitosis

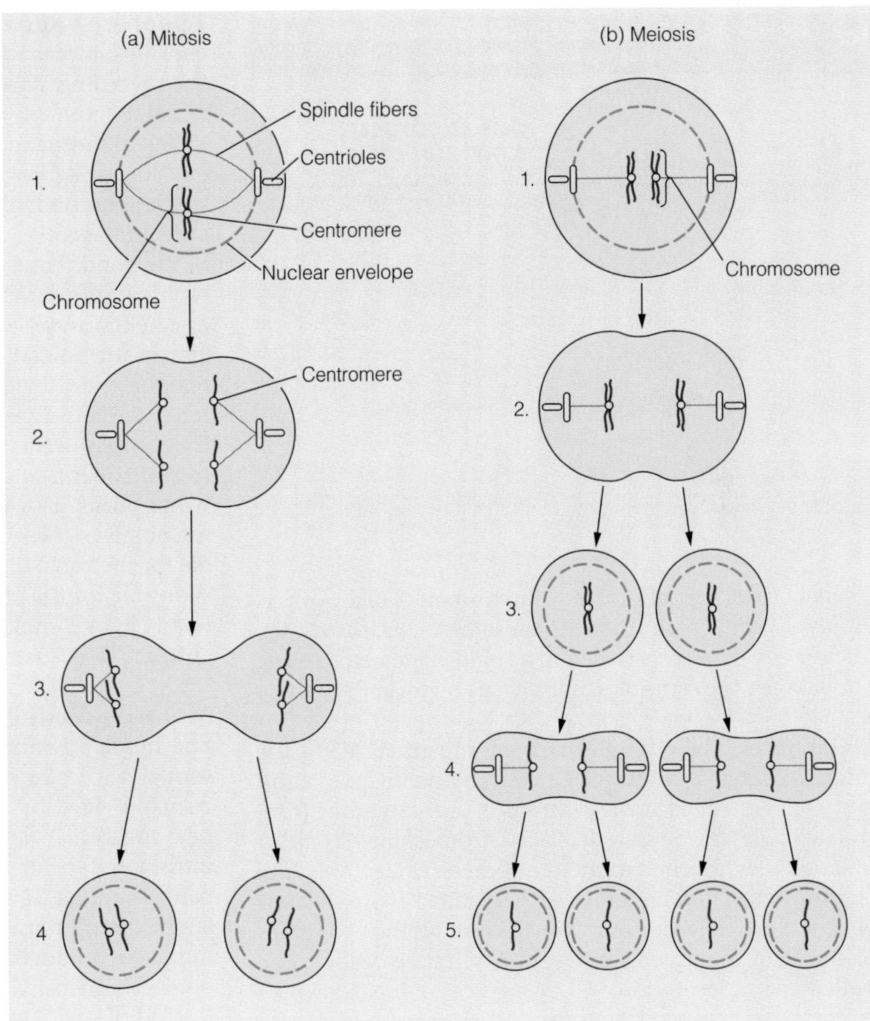

Figure 3–22 Mitosis and Meiosis
The events of mitosis and meiosis are different but use the same mechanisms. In mitosis (a1), the chromosomes condense and associate with the spindle fibers as the nuclear envelope disappears. Chromosomes are pulled apart (a2) at the centromeres and travel to opposite poles of the cell. In (a3) the chromosomes have reached the poles and the cell begins to cleave. The division of the cell (a4) results in two identical daughter cells. In meiosis (b1), the first cell division is called a reduction division because it separates homologous chromosomes from one another intact and not by splitting at the centromere (b2). This results in two daughter cells with one each of the original pair (b3). The second meiotic division (b4) takes place in a fashion identical to mitosis (a2), with the chromosome splitting at the centromere. Subsequent cleavage results in a total of four (b5) new cells, each with half the number of chromosomes and half the amount of DNA.

3–22). Each stage blends into the next succeeding one in a continuous fashion. Division of mitosis into four parts is a convenient way to conceptualize the process and lends itself to biological analysis. The first stage of mitosis is prophase.

Prophase Prophase is the period during which the long, linear strands of chromatin in each chromosome are condensed (Figures 3–21 and 3–22). One way to visualize the condensation process is to imagine a fishing line that has been cast out over the surface of a pond. The line is so thin that it cannot be seen. The process of condensation is accomplished by reeling in the fishing line, adding layer upon layer of line to the reel. One of the unique attributes of the long, linear chromatin is that it is self-reeling. As condensation progresses, more and more of the chromatin is coiled in on itself and takes up less and less space. Finally, it is compacted into a recognizable condensed chromosome. There are two arms to each of the chromosomes (representing the duplicated DNA strands produced in S) and they are connected at a site known as the **centromere.** The condensation occurs nearly simultaneously during prophase for all 46 chromosomes in the nucleus of human cells.

At the same time that chromosomes are condensing during prophase, the nuclear envelope that surrounds them is disintegrating. This removes an obstacle to the eventual separation of chromosomes into daughter cells. The centrioles change position and function during this time. As discussed earlier, centrioles are crucial for organizing the microtubules that will eventually form a tightropelike framework known as the **spindle apparatus.** It is along the spindle micro-

Centromere region of a chromosome at which it separates during anaphase; attachment site for spindle microtubules

Spindle apparatus assembly of microtubules used to translocate chromosomes to the poles during anaphase

Table 3-1 Rate of Cell Division in the Cell Cycle of Selected Cell Types

CELL TYPE	TIME TO COMPLETE A DIVISION CYCLE
Epithelial cells lining the gut	11–24 hours
Fibroblasts *in vitro*	18–24 hours
Red blood cell precursors in human bone marrow	15–20 hours
Cells in growing hair follicle(mouse)	12 hours
Malignant cancer cell	10–12 hours
Mature neurons, muscle fibers	No cell division
Especially rapid rates observed in procaryotes and animal embryonic cells	
Embryonic blastomeres of frog	30 minutes
Escherichia coli (bacteria)	20 minutes
Embryonic cells of fruit fly	As short as 7 minutes!

tubules that migration of chromosomes takes place. Prophase ends with the chromosomes condensed, the nuclear envelope dispersed, and a spindle apparatus strung out between the centrioles as if a circus highwire.

Metaphase The chromosomes are organized along the midline or equator of the cell by the competing pull of microtubules attached at the centromere and extending from the two poles of the spindle. Each chromosome has many microtubules attached to the centromere. Collectively, the chromosomes line up along what is called the *metaphase plate* and each chromosome, which contains the two copies of the DNA replicated during the S phase, is ready to be pulled apart. Separation will take place by breakdown of connections between DNA strands at the centromere.

Anaphase As the chromosomes begin to separate and move to opposite poles, the period of metaphase ends and the next stage of mitosis, anaphase, begins. This phase presents evidence for an interesting balancing act, as equal parts of each chromosome slowly migrate away from the other. Experimental intervention aimed at destroying or disrupting microtubules within the spindle apparatus results in halting the movement of chromosomes. Naturally occurring abnormalities in chromosome migration may result in a phenomenon called **nondisjunction.** In this process the segregation of one of the two chromosomes of a pair fails to take place. This leaves one of the daughter cells with an extra chromosome and the other daughter cell with a deficit. Cells missing a complete chromosome cannot survive because important and different genetic information is contained in each of the 23 pairs of human chromosomes. The presence of a complete extra chromosome can also be disastrous. This is because many of the products of genes are highly regulated and, if they are present in double the amount, they may alter the behavior of the cell.

Telophase—Karyokinesis and Cytokinesis Once the chromosomes reach the poles of the spindle, the actions of the preceding stages starts to go in reverse. At the end of telophase, the spindle fibers along which the chromosomes traveled during anaphase disappear, and two new nuclear envelopes begin to appear. The chromosomes themselves begin to decondense again into thin strands within the nucleus, unraveling, as if cast out once again over the waters of the pond. The events of the first three phases of mitosis were concerned with the reorganization of the cytoplasm, organelles, and the nucleus in preparation for **karyokinesis** or the migration of the chromosomes. Also during telophase, the dividing cell begins to cleave into two daughter cells. This aspect of mitosis is known as **cytokinesis.** Cytokinesis occurs in conjunction with the contraction of microfilaments and, similar to pulling the cord of a drawstring purse, it slowly divides and pinches off the plasma membrane between the two cells. Karyokinesis is responsible for the equal distribution of DNA, and cytokinesis is responsible for the cleavage of the two cells. Cleavage brings to an end a complete cell cycle, with each of the new cells entering G_1. What goes around, comes around.

Meiosis—Reduction Division

The human genome is composed of 23 pairs of chromosomes, with one member of each pair having originally been provided by either the mother or the father of the new individual (a son or daughter). This means that the information contained in each matching member of the 23 pairs is similar, but not identical, having arisen from different family lineages. This is a very important factor for human reproduction and the formation of gametes. This process is meiosis and involves two successive cell divisions instead of one (Figure 3–23). In the first meiotic cell division each member of the pair of homologous chromosomes is separated from the other into daughter cells. Unlike mitosis, individual chromosomes are not separated at the centromere, so each cell gets either the chromosome originally provided by the mother or the father, but not both! During the second (and final) division, the chromosomes divide at the centromere and equivalent genetic information is segregated into the gametes. The outcome of these divisions is a reduction in the number of chromosomes (amount of DNA), and the distribution of the unique maternal or paternal information contained in the original genome of the cells. Each gamete has one each of

Nondisjunction failure of a chromosome to separate during mitosis or meiosis and enter one of the two daughter cells

Karyokinesis separation of chromosomes during mitosis; specifically during anaphase

Cytokinesis physical cleavage of one cell into two

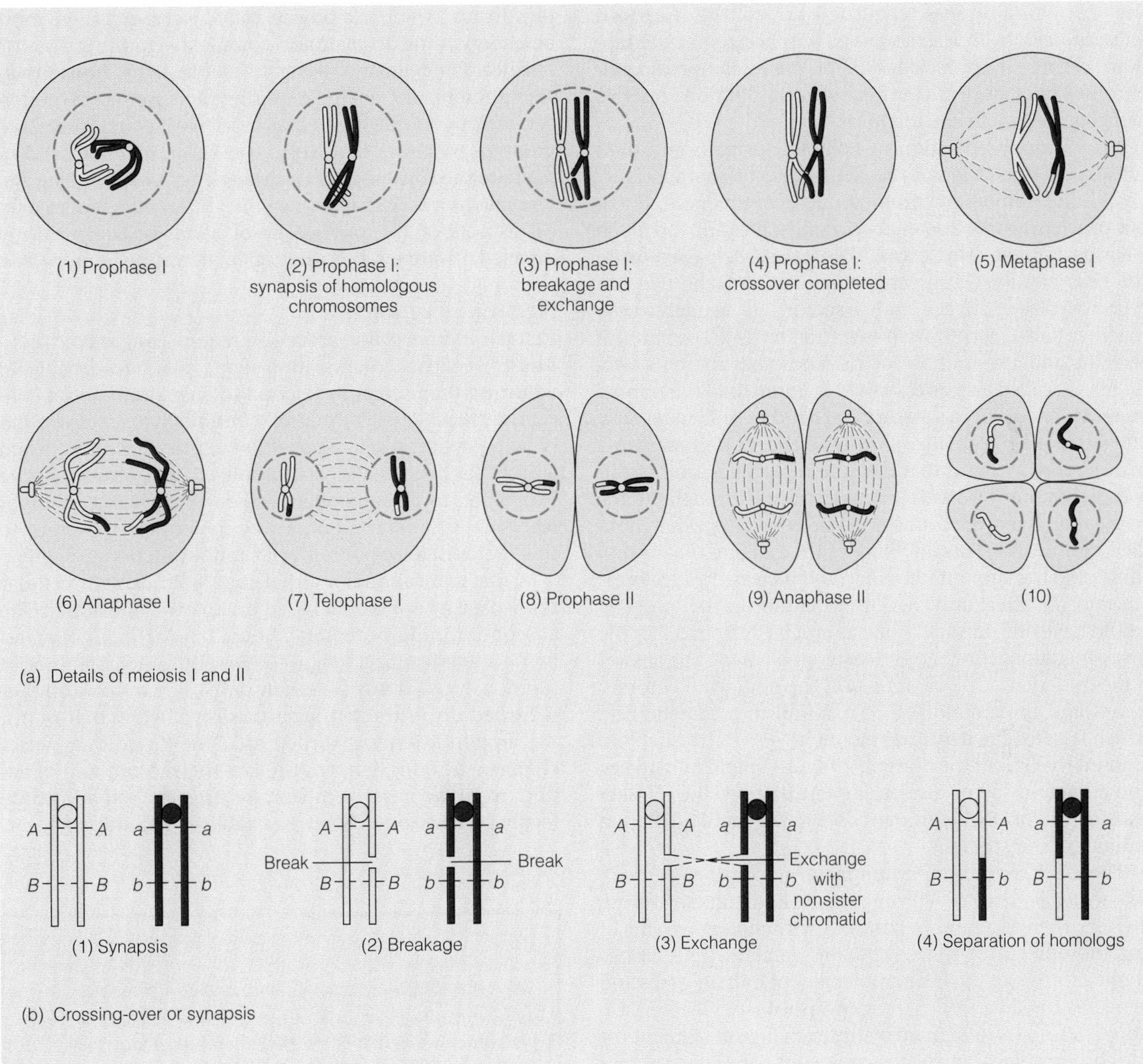

Figure 3–23 Details of Meiosis—Reduction Division, Synapsis, and Exchange
In (a) homologous chromosomes (black and white) undergo synapsis and exchange of genetic material (a1–4). Steps (a5–a10) complete the first and second meiotic division as shown in Figure 3–22(a)1 and (a)2. In (b) specific regions of the homologous chromosomes are aligned during synapsis (Aa, Bb), undergo breakage, and exchange (b2,3) and then complete the crossover (b4). Equivalent but not necessarily identical genetic information is exchanged during this process.

the 23 chromosomes characteristic of the **haploid** amount of DNA in *Homo sapiens,* but, because of the random nature of the separation of parental homologous chromosomes during the first meiotic division, no two gametes are genetically identical.

Why should separation of homologous pairs and their subsequent separation at the centromere be so important? First, if the number of chromosomes is not reduced, a serious problem arises—too many chromosomes and too much DNA per gamete. This excess DNA would be passed on to the cells of succeeding generations, because the two cell types involved, an egg and a sperm, are actually fused together during fertilization (see Chapter 14). The amount of DNA and the number of chromosomes in the newly fused cell, called a **zygote,** would be more than the normal amount for our species. Assuming that that individual grew and developed normally, it would have 92 chromosomes instead of 46. Extend that into the next generation of offspring (assuming this doubling occurred throughout the population), and the number of chromosomes would double again to 184. This doubling of the number of chromosomes and the amount of DNA would increase in a *geometric progression.* After only seven generations, the progeny would have one hundred times as much DNA per cell as the first generation! How is this disaster prevented? The answer is meiosis. Meiotic division reduces the number of chromosomes and the amount of DNA found in a somatic cell, which is called the **diploid** amount of DNA, to the haploid amount that is half the normal. When a haploid human egg and a haploid sperm fuse during fertilization, the number and amount of DNA is brought back to the standard diploid amount of our species.

There is a second important factor to consider in meiosis—*recombination* of chromosomes and genes. During meiosis, there is a randomizing of the homologous pairs of chromosomes into the developing gametes. This brings about a sorting of the full complement of human genes and increases the diversity of the **gene pool** (see Chapter 20). This diversity is augmented by the exchange of segments of the chromosomes between homologous pairs prior to their separation in the first meitoic division (called **crossing-over** and **synapsis**) (Figure 3–23). Each of the billions of humans, past and present, has been and is a totally genetically unique individual (other than *identical twins*), and it is the events of meiosis that provide the mechanisms to generate such diversity.

HOMEOSTASIS—MAINTAINING BALANCES

It is a commonly held view that for complex systems (living as well as nonliving) to appear to stay the same, they must undergo continuous change. To counter the natural tendency of complex structures to break down to less complex forms, which is a process called **entropy,** there must be energy input to maintain complexity to bring about a balance. For cellular and organismal systems, **homeostasis** represents the energy-requiring processes that counteract entropy. Homeostasis can be viewed as the integrated group of processes that work together to maintain and/or regulate the constancy of the integrity of a system. The system may not stay exactly the same, but it often appears to. All aspects of the physiology of a cell or organism are directed toward establishing and maintaining homeostasis.

Negative Feedback

Cellular homeostatic processes require **negative feedback.** Feedback occurs in systems that are capable of adjusting themselves to respond to changes in preset conditions. Mechanistically, a cell continually reassesses what is to be synthesized, repaired, or replaced within it and adjusts its behavior biochemically and in terms of gene expression to respond effectively to those needs. Without feedback information and a way to connect the input to the appropriate response a cell cannot function properly.

A mechanical analogy that may help explain cellular homeostasis, is a model based on an air conditioner. The air-conditioning system in a house or apartment operates by negative feedback (Figure 3–24). If the temperature in the house gets above a certain point, the air conditioner is turned on. If the temperature drops below that point, the air conditioner is turned off. The device that senses changes in room temperature is a thermostat, and variation from a preset temperature triggers self-adjusting, feedback reactions. What is established under such cir-

Haploid one complete set of chromosomes represented by one member of each pair of a diploid set, 23 in humans. This set is also called the lN number of chromosomes

Zygote product of fusion of two gametes; a new genetically distinct individual of a sexually reproducing species

Diploid the full complement of 46 chromosomes (23 pairs in humans) Also known as the 2N number of chromosomes

Gene pool full complement of genes (alleles) found in a population of interbreeding organisms (a species)

Crossing-over exchange of DNA between homologous chromosomes usually during meiosis

Synapsis exchange of parts of chromosomes during recombination events of meiosis

Entropy in thermodynamics, the tendency towards disorder

Homeostasis the tendency for a system or organism to reach a state of physiological equilibrium

Negative feedback condition in which variance from a set point induces changes in a system that returns that system to the set point

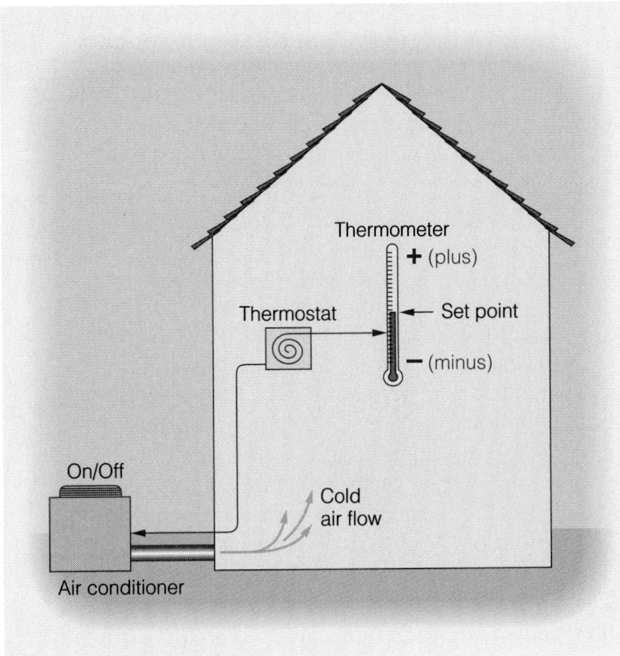

Figure 3–24 Negative Feedback
Negative feedback is the process by which a system regulates itself around a specific set point. Using an air conditioning system as a model, as the temperature increases above the set point, the thermostat is triggered to turn on the air conditioner. The air conditioning sends cold air to the house or office until the temperature drops to or slightly below the set point. The thermostat then turns off the air conditioner. The cycle repeats to negatively counteract upward or downward deviation from the set point.

cumstances is called a **feedback loop,** a continuous cycle of changes that keep the temperature of a room or house very near the set point. There is always some variation from one minute to the next, but in a well-adjusted and efficient air-conditioning system this variation is always only slightly above or below the set point.

Thermoregulation in Humans This description of negative feedback is representative of homeostatic mechanisms in general and is analogous to the way in which biological functions are regulated. We might be asked, for example, how does the human body regulate its heat content? Human beings, as all mammals, are warm blooded or **endothermic.** This means that a human body actively (that is, using energy) maintains a fairly constant temperature in the face of changes in internal heat and in the external environment. If it is colder outside than inside the body, heat is generated (that is, muscle contractions, shivering) to warm up. If it is warmer outside than the preset temperature of the body, there are means to cool down. The stipulation in this case is that the temperature of the human body must be maintained within certain limits at all times, or it will suffer from either **hypothermia** (too low a body temperature) or **hyperthermia** (too high a body temperature). Neither of these conditions is healthy, and extremes in either condition may lead to unconsciousness and/or death.

How does the body determine when its temperature or heat content is too high or too low? It does so through a cellular thermostat that monitors the heat content of the blood. The thermostat in humans is located in the brain, specifically in the **hypothalamus.** As blood changes temperature during its journey through the body, variations in heat content are picked up by special nerve sensors in the hypothalamus that continuously monitor them. These nerves send feedback signals to other parts of the brain, which, in turn, activate or deactivate the function of organs and tissues (muscles contracting, sweat glands secreting), which themselves are capable of generating or dissipating heat. For example, in response to a rising body temperature, blood circulates very near the surface of the skin from deep within the body core, giving up heat to the environment. Sweat glands are activated, and evaporation of their watery secretions carries heat away from the body. Blood, as water, is a suitable medium through which heat can be carried and exchanged, as well as being used as a medium to monitor very subtle changes in the body's heat content. The general processes involved in warming and cooling a body are **conduction, convection, evaporation,** and **radiation** (Biosite 3–2).

With these mechanisms of gaining or losing heat in mind, consider the physiological homeostatic changes that occur when the temperature of the environment in which we find ourselves changes from a comfortable, air-conditioned 24° C inside an office building to an uncom-

Feedback loop the ability of an end-product of a process to alter the process

Endothermic warm-blooded; maintenance of constant body temperature

Hypothermia loss of heat from a body beyond normal limits

Hyperthermia overheating of a body above normal limits

Hypothalamus region of the brain involved in controlling involuntary body functions, e.g., thermal regulation

Conduction relating to exchange of heat between materials in direct contact

Convection transfer of heat by circulation of gas or fluid

Evaporation change of state of a substance from liquid to vapor

Radiation energy emitted in the form of waves and particles (i.e., lightwaves and photons)

3–2

GAINING AND LOSING HEAT: CONDUCTION, CONVECTION, EVAPORATION, AND RADIATION

Conduction of heat from one object to another depends on direct contact. Heat will be transferred by diffusion until both objects have reached the same temperature, which, depending on the individual masses of the objects in question, will be less than the initially higher temperature of the one but greater than the initially lower temperature of the other. Convection refers to heat that is transferred by the circulation or movement of a fluid or gas, such as water or air. Convective exchange depends on a temperature differential. For example, cool air blowing over the surface of the human body carries heat away. In conjunction with evaporation, this is a very effective means to cool the body.

Evaporation means literally to convert to vapor. Vaporization of water and movement of the vapor away from the surface of the body brings about cooling. Water molecules heated by conduction and radiation absorb energy and, as they evaporate, carry away heat that would otherwise be maintained. Thus, a body is cooled by the evaporation of water molecules and their dissipation by convection. Radiation is the fourth type of heating process. The use of the term refers to electromagnetic radiation, such as visible light, X rays, ultraviolet light, and microwaves. Such radiation can travel through a vacuum, unlike the processes of conduction and convection. Radiation transfers energy to atoms and molecules within objects when it impinges on them. This causes a local increase in kinetic energy and increases the overall heat content of the object. A good example of this radiative heating process is putting your food in a microwave oven. Microwave radiation increases the kinetic energy of water molecules, which, in turn, transfer their heat by conduction to other molecules with which they come in contact. The same principle is involved when the human body is heated by the sun. Radiation is absorbed by the skin, which heats up and, thereby, the heat is transferred to the rest of the body.

fortable 37° C outside. As heat content of the body increases from a combination of conductive, convective, and/or radiative effects, the temperature of blood rises. The hypothalamus monitors these changes and sends controlling signals to the parts of the brain that activate the release of fluids from sweat glands. Sweat is 99% water, so, as it is secreted and reaches the surface of the skin, it begins to evaporate. Evaporation, in conjunction with convection, leads to cooling. As long as the environment remains hot enough, and we do not run out of internal water, the body will keep sweating—the more evaporation that occurs, the more effective the cooling. In addition to sweating, the body also responds to elevation of heat content by changing the routes of blood within the circulatory system. This rerouting of blood to the surface results in conduction of heat away from the body core. In this circumstance, homeostasis is working on two levels: one with respect to the activity of sweat glands, and the other in regulating the routing of blood to the surface of the body. Now that the body has adapted to heat stress, what happens if the temperature suddenly drops?

As we reenter the air-conditioned office building, the thermal regulating system of the body reacts appropriately to the **stress.** Stress is defined as any factor that changes or alters an existing equilibrium. The environment in which we find ourselves is now cooler than body temperature. Heat content of the body decreases due to conduction and convection. As body heat content is dissipated, the temperature of the blood decreases. The hypothalamus responds to cooling by deactivating the majority of sweat glands. This reduces loss of water and, therefore, aids in retaining heat within the body. Circulating blood is rerouted so that less flows to the surface and more is retained in the core. Again, this reduces heat loss. If the temperature is too cool, yet another heat balancing activity may begin, shivering. As the temperature of the body drops further below the set point, neural signals to muscles may trigger

Stress physiological state resulting from factors that tend to alter an existing equilibrium

their contraction. In shivering, the rapid, repetitious pattern of contraction of muscles generates heat. It is helpful to remember that, if it is cold enough to start shivering, it is wise not to try to stop it, you probably need the heat that is being generated.

There are many other types of feedback mechanisms involved in controlling the many and diverse chemical and mechanical functions of the body. Many homeostatic mechanisms will be described in later chapters of this book, when the functions and activities of specific tissues and organs of the human body are discussed. Without negative feedback loops of these kinds to regulate the body's physiological activities in response to stress, a human being could not adapt in the ever-changing environment in which he or she lives.

Positive Feedback

In **positive feedback,** the signal that results from a change in the set-point conditions of the body does not return the system to its former balance. Rather, it increases, or enhances, the deviation from equilibrium even further, either up or down (Figure 3–25). The disadvantage of positive feedback in the normal operation of the body is fairly obvious. A signal to relentlessly increase body heat, as when we have a fever, or to increase the rate of the heart beat may be disastrous in a very short time. The same is true for the regulation of the rate of breathing, or any of the hundreds of subtle, biochemical, enzymatic, and physiological balancing acts that the cells, tissues, and organs of the body perform continuously as part of homeostasis. However, there are a few instances in which positive feedback plays an important role in natural body functions. The most profound instance of this is observed in childbirth. During labor and delivery, the rate and intensity of contractions of the uterus are linked to the release of hormones. Each contraction sends a signal to release more hormone. The strength and number of these contractions continues to increase until the baby is pushed from the birth canal. The resolution to this state occurs when the positive feedback loop is interrupted as signals from the uterus cease following delivery.

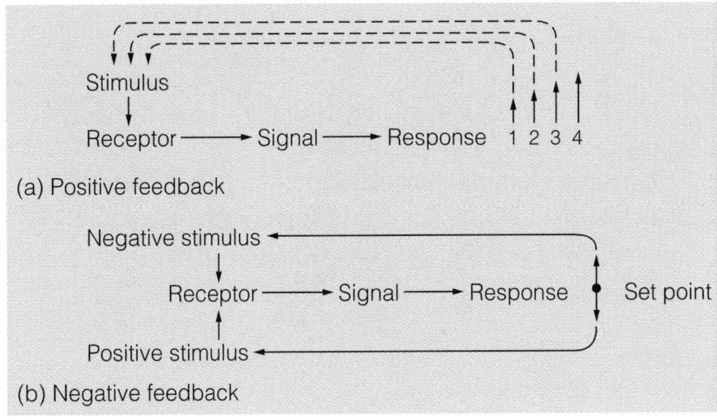

Figure 3–25 Positive Feedback
In positive feedback a stimulus results in a response that enhances subsequent responses (a). If arrow 1 indicates the initial response, it then acts as a stimulus that results in arrow 2. The height of the arrows indicates increased amplitude of effect in each cycle. This is compared to a negative feedback (b) in which the arrow up results in a negative stimulus, which reduces the response, or a positive stimulus which increases the response to the set point.

> **Positive feedback** a condition in which variance from a set point induces further variance of that system from a set point

Summary

In this chapter, structures and functions of cells have been described in detail from inside and out. The structure and activities of cellular membranes, organelles, macromolecular complexes, and cytoskeletal elements were shown to be integrated in their performances as cells grow, divide, transport materials across their membranes, and otherwise adapt to the stresses induced by their environments. Cells in complex, multicellular organisms have immediate neighbors and extracellular materials with which they interact directly, but they are also metabolically linked to their more distant parts of a body. Diffusion, osmosis, passive and active transport, and cell division, as well as the processes of cellular respiration and metabolism, all play roles in maintaining cell and tissue structure and function. Homeostasis is the result of all the balancing acts carried on within and among cells to adapt to stress.

In some ways cells are independent of their neighbors and, because of independence, are deserving of their designation as the smallest units of life. In other ways cells, particularly those of multicellular organisms, are clearly dependent on one another. Without the integrated activity of trillions of cells, a multicellular organism such as *Homo sapiens* would not survive long. A single cell in the human body acts to some degree in accordance with its individual nature but must also conform to its role as a part of a complex society of cells. In the next chapter, we will discuss the organization of cells into tissues and describe the structures and functions of the complex society of cells from which the human body is made.

Questions for Critical Inquiry

1. What advantage or disadvantage is there to being multicellular?
2. How are each of the components of the model cell described in this chapter influenced by negative feedback on their activities?
3. In what ways might a cell be considered more than the sum of its parts?
4. What might be the significance of the observation that the genome is located in its own special compartment, the nucleus?

Questions of Facts and Figures

5. What is a cell?
6. What Kingdoms are composed of single-cell or unicellular organisms?
7. What are the main categories of cellular substructures within cells?
8. What is the plasma membrane of a cell composed of? What model describes its function?
9. What are the three types of cytoskeletal elements?
10. In what compartment is the endoplasmic reticulum found? What are its functions?
11. In what organelle is genomic DNA found? What is unique about the membranes of this organelle?
12. Intracellular digestion involves what organelle? From what other organelle is this digestive structure derived?
13. Cilia and flagella are specialized cellular structures. With what activities are they associated? What cytoskeletal element is found within cilia and flagella?
14. ATP is the energy currency of cells. In what organelle is this molecule synthesized? Which of the chemical bonds in ATP are considered high-energy bonds?
15. What is cellular respiration?
16. What are the two major types of transport across the plasma membrane of a cell?
17. What is osmosis? How does the tonicity of a solution affect the morphology of cells?
18. How is facilitated diffusion different from passive diffusion? How is facilitated diffusion different from active transport?
19. What is the energy source for active transport? Passive diffusion?
20. What are the four phases of the cell cycle?
21. What are the four stages of mitosis?
22. What is homeostasis?
23. How do negative and positive feedback differ from one another? How does each relate to homeostasis?
24. Describe the processes of conduction, convection, evaporation, and radiation.

References and Further Readings

Alberts, B., Bray, D., Lewis, J., Raff, M., Roberts, K., Watson, J. (1994). *Molecular Biology of the Cell*. 3rd ed. New York: Garland Publishing.

Lehninger, A. (1971). *Bioenergetics: The Molecular Basis of Biological Energy Transformations*. 2nd ed. Menlo Park, CA: Benjamin-Cummings.

CHAPTER 4

Cells into Tissues

INTRODUCTION

Individual human cells are fully capable of carrying out the basic biochemical activities or housekeeping processes necessary to function effectively and to grow, differentiate, and divide successfully. This includes the cellular respiration and metabolic processes needed to produce and use energy; the structural means to replace and maintain membranes; organelles, and cytoskeleton; the activities involved in controlling the transport of molecules and ions into and out of the cell; and the regulation of the fundamental processes of cell division. However, the housekeeping processes of individual cells are not sufficient to satisfy the exceptional demands for energy and activity to operate a massive multicellular organism. Such demands are satisfied through cellular specialization and the coordination possible only if cells work together.

TISSUES

The human body is composed of a complex, totally integrated society of cells whose organization into different and highly specialized forms and functions is required. Any reference to a generalized single cell is of limited value in explaining higher-order structure and function. Cells in all higher organisms function in multicellular arrays, with different cell types working together in basic organizational units known as "tissues," and it is with tissues and their composition that we concern ourselves in this chapter.

How is it that the organization of cells into tissues came to be and to what end? From an evolutionary point of view multicellularity provided the chance for cells to specialize and compete with unicellular organisms for new and different niches in the environment. It is clear that a variety of types of unicellular organisms live solitary and successful existences. An enormous number of species of organisms in two of the five Kingdoms, the Monera and the Protista, exist throughout their lifetimes as single cells (Figure 4–1). Each unicellular organism is autonomous in its activities. Although procaryotic and eucaryotic organisms are capable of solitary existence, they also are limited in the degree of their specialization. The two Kingdoms are characterized by cells of very small size, a requirement for aqueous environments, and a limited amount of genetic information in their genomes.

Specialization

Unlike more primitive forms of animals, such as sponges or corals, the cells of vertebrates are an organized and integrated society of trillions of cells. For example, all the activities of humans and other higher organisms are carried out by specialized cells in dozens of different types of tissues functioning together. Thinking, reading, touching, breathing, swallowing, and general body movements all require an astounding level of coordinated activity. Each of the activities listed here may occur simultaneously, even as you read this paragraph. Underlying these observable behaviors are the specific and integrated activities of different and often distantly separated types of cells.

Homo sapiens *Homo sapiens* is a prime example of an organism in which specialization without integration is useless. The human body cannot coordinate movement without muscle contractions, and muscles cannot be controlled and coordinated without nerves, and neither muscles nor nerves can survive without oxygen delivered by blood cells. A recurring theme of this chapter is the value of integration—integration of specialized cells into tissues and of specialized tissues with one another. What is presented are descriptions of special groups of tissues and the cells composing them, whose basic structural and functional attributes set them clearly apart from one another. The formal studies of tissue organization is *histology,* or microanatomy, and *pathology.* Furthermore, it should be noted that different tissue types occur combined with one another to form organs, which, in turn, are integrated together into organ systems. The structures and functions of organs and organ systems represent major characteristic features of human anatomy and physiology. Discussion of each of the major organ systems of the human body will be taken up successively in separate chapters of this book.

BASIC TISSUE TYPES

There are four basic tissue types found in animals and they are derived from embryonic germ cell layers known as the ectoderm, mesoderm and endoderm. They are the epithelial tissue, connective tissue, muscle, and nerve (Figure 4–2). Each tissue derived from its embryonic progenitor type is identified by its own special cellular types and biochemical characteristics. The basic types are described ini-

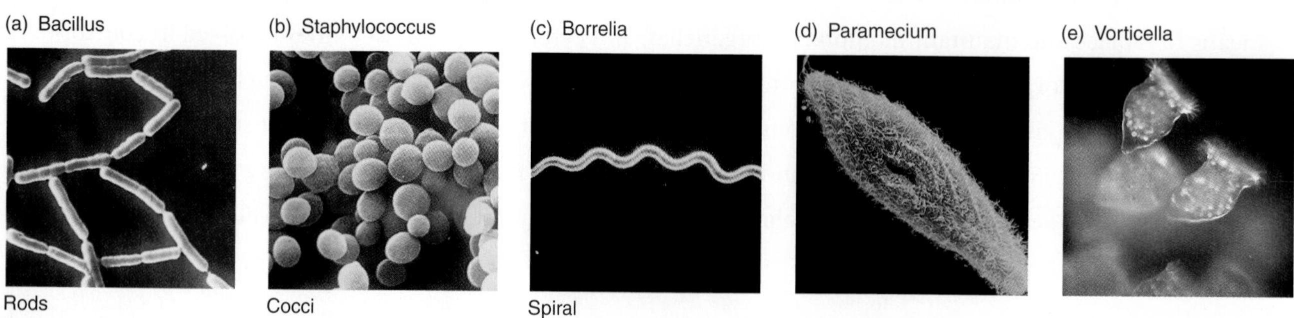

Figure 4–1 Unicellular Organisms
Unicellular organisms are found in the Kingdoms Monera and Protista. The shapes of bacteria (a)—(c), which are monerans, are diverse. Paramecium (d) and vorticella (e) are protists but also show dramatic differences in shape.

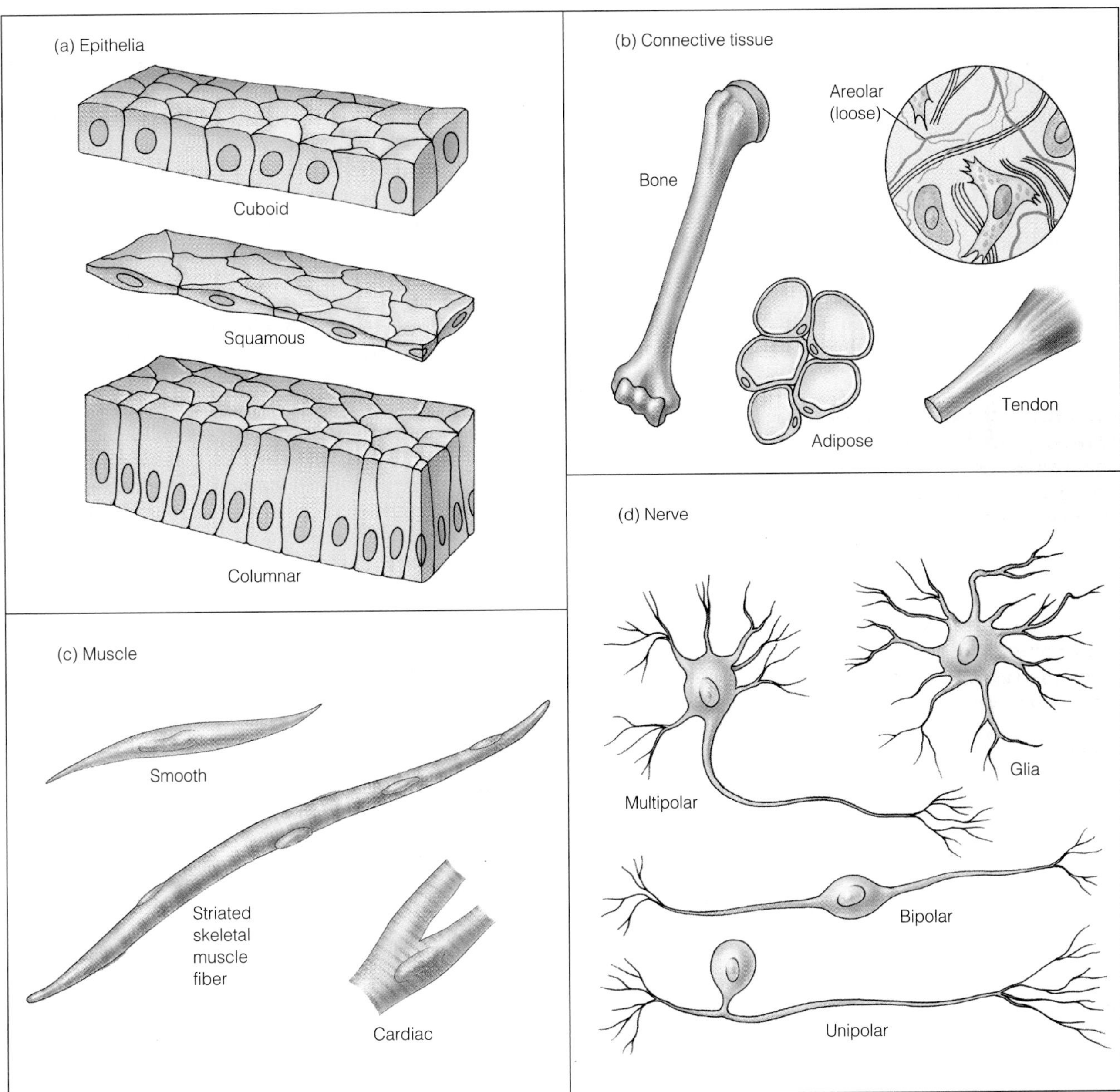

Figure 4–2 Basic Tissue Types
There are four basic tissue types found in animals. Epithelia (a) are tight-fitting cells that form barriers. Connective tissues (b) hold other tissues together and are structural in form. Muscle (c) is the contractile tissue and nerve (d) is the excitable tissue. Each type is composed of many different cell types.

tially in an overview and then described in detail. There are many subtypes of each of these basic tissues and the subtypes may be composed of cells that play specific roles only in particular regions, or within specific organs of the body (Table 4–1).

EPITHELIA—AN OVERVIEW

Epithelial tissues or **epithelia** are tight-fitting cells that form sheets that cover the surfaces of the body and its many organs, inside and outside. Wherever there is a surface on the body or an organ, be it skin, lungs, or heart, there is a specialized epithelium to cover and protect it.

Table 4–1 Germ Layer Origins of Tissue Types and Selected Cellular Subgroupings

ECTODERM	MESODERM	ENDODERM
Nerve—neurons and glia	Skeletal muscle	Liver
Skin—epidermis and hair, glands and nails	Skin—dermis	Pancreas
Facial bones—neural crest derived	Most other bones of skeleton	Lungs, trachea
Pigment cells (melanocytes)	Cartilage	Gall bladder
	Kidney and urinary system	
Lens of eye	Heart	Gut—stomach, intestines

Thus, there are many different types of epithelia. For example, on the outside of the body is skin (Figure 4–3), which is composed of a tough, relatively dry, thick epithelium called the **epidermis** and an underlying connective tissue called the **dermis.** The mouth, throat, digestive tract, lungs, and heart, on the other hand, are lined with softer, more delicate mucus epithelia, which are moist and lubricated. The thick epidermis of the body's external surface and the thin moist internal epithelia provide excellent examples of how form fits function, all the while maintaining the structural integrity of the organism.

Epithelia and the Cells That Compose Them

There are three basic types of epithelial cells found in animal epithelial tissues, and their designations are based on shape—cuboidal, columnar, and squamous (Figure 4–4). These cells occur separately in simple epithelia and some forms of stratified epithelia and in combination in more complex types of stratified epithelia. A simple epithelium is a layer of epithelial cells only one cell-layer thick (Figure 4–4 [a]). The cells are closely attached to one another laterally (side to side) and form a sheet or membrane that is

Epithelia a group of tissues composed of tightly adherent cells that form barriers to the movement of materials into and out of the body

Epidermis specialized epithelium of skin; overlies dermis

Dermis the component of skin that underlies the epidermis and is connective in character

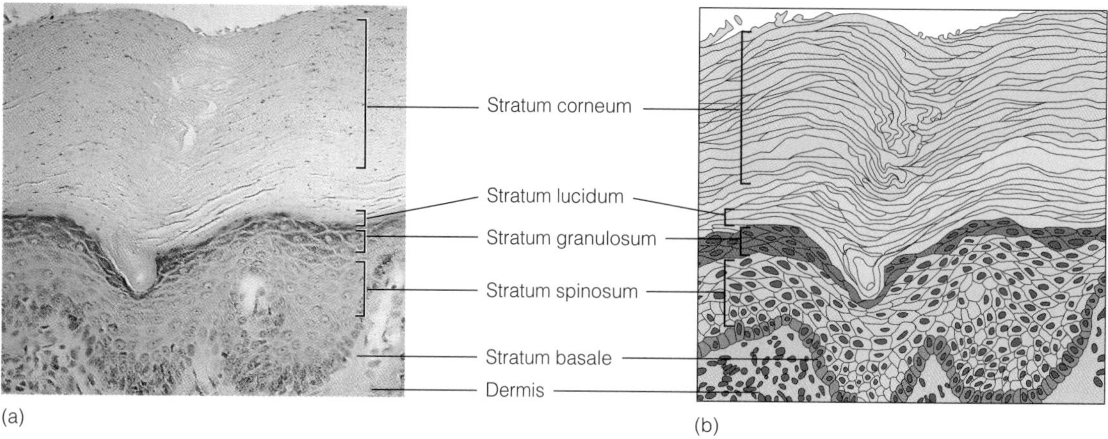

Figure 4–3 The Structure of Skin
The skin is a combination of tissues including an epithelium (epidermis) and connective tissue (dermis). The layers of the epidermis are built up from cells dividing in the stratum basale or germinativum. Subsequent cell layers (stratum spinosum and stratum granulosum) undergo programmed cell death. The outer layer of the palm skin (a) is the stratum corneum made up of corneocytes and is a nonliving layer. In thick skin a stratum lucidum is present between the stratum granulosum and stratum corneum.

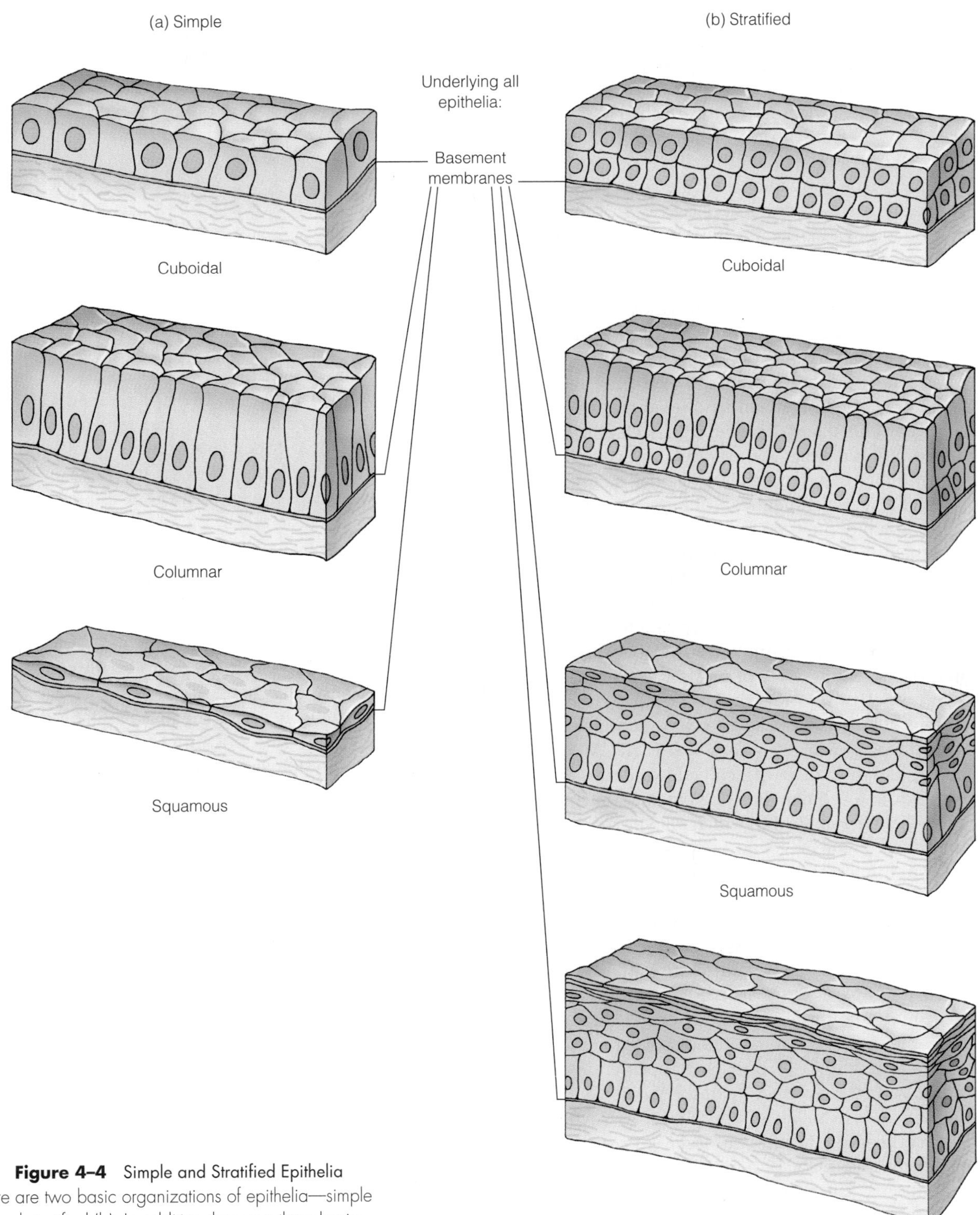

Figure 4–4 Simple and Stratified Epithelia
There are two basic organizations of epithelia—simple (a) and stratified (b). In addition there are three basic morphologies of epithelial cells—cuboidal, columnar, and squamous. All types of epithelia are attached to an underlying connective tissue basement membrane composed of the extracellular matrix (ECM).

continuous over a large area of the body surface both externally and internally.

A stratified epithelium (Figure 4–4 [b]) has two (or more) distinctive layers of cells arranged one atop the other. Cells of a stratified epithelium are strongly attached to each other laterally, but they are also connected vertically as well. The tight packing of cells to one another is a characteristic of all epithelia. One consequence of the close positioning and strong adherence of epithelial cells to each other is that epithelia function as barriers in the control of the movement of molecules and ions into and out of the body. Epithelia act as virtually impermeable seals over the internal and external surfaces of a body. The epidermis is a good general example. It provides the primary physical barrier to protect against potential loss of body water from evaporation, water gain from absorption, and the encroachment into the body of microorganisms and toxic substances.

The cellular characteristics of epithelia change as one examines different regions of the body. However, different types of epithelia blend naturally together, maintaining their intercellular attachments and barrier functions. For example, the epidermal epithelium transitions into a mucus epithelium as it progresses from the cheek, over the lip, and into the mouth. In addition to the continuity of attachments holding epithelial cells together, all epithelia must be strongly attached to an underlying foundation or substratum. This foundation is called the **basement membrane** and is composed of substances that form part of the **extracellular matrix** (or ECM). Extracellular matrix molecules are produced by cells and secreted into the environment in which the cells live. The ECM forms a complex gel-like substance with embedded fibrous materials, which acts as a cellular glue. The basement membrane is a very special type of ECM to which all types of epithelia are attached (see Figure 4–4, and Table 4–2).

There is an interesting dilemma associated with describing surfaces from an *anatomical* and *physiological* point of view. What is considered outside and what is considered inside the body with respect to surfaces lined with epithelia? If epithelia are ultimately continuous with one another and form an impermeable barrier to molecules and ions, how are inside and outside of the body demarcated? If you were given the task of painting the outside surface and all directly connected surfaces of a giant open-ended pipe, how would you accomplish your task? Would you cover only the "apparent" outside of the tube? If not, where, as you painted the ends, would you decide to stop? As you painted the ends, you would notice that the surfaces curving to the interior of the pipe are connected, one folded back into itself in space. In order to paint the entire surface, and all connected surfaces, you would have to paint the "apparent" inside of the pipe as well.

This situation is similar to what occurs over the surface of the human body. As the epidermis transforms into the moist epithelium of the mouth, where does the outside end and inside begin? Changes in the structural appearance of the cells of the epithelia that line the oral region and then the different regions of the digestive tract are all connected to one another. This is also true for the other end of the digestive tract, where the epithelium of the rectum transforms back to an epidermis in the region around the anus. The conclusion to be drawn from this anatomical continuity in epithelial connection over the entire human body is that the apparent inside of the body is really not. We are led to the conclusion that for molecules or ions to be "inside" the body, these materials must pass into and/or through the various types of epithelia. To accomplish this, individual epithelia cells have physiological mechanisms to transport these materials actively across the plasma membrane. This cellular transport activity constitutes a highly effective and well-regulated physiological barrier to the movement of molecules and ions into the body proper.

Table 4–2 Molecules Found in the Extracellular Matrix

MOLECULES	TYPES	GENERAL FUNCTION
Fibronectin	Glycoprotein	Structural
Laminin	Glycoprotein	Structural
Collagen	Glycoprotein	Fibrous structural
Proteoglycans	Proteins/ carbohydrates	Structural, space filling
Glycosaminoglycans	Carbohydrates	Structural, space filling
Proteinases	Glycoproteins	Enzymatic (degradation of proteins)

Cell Types Found in Epithelia

Simple Epithelium As previously stated, epithelial cells come in three basic shapes or morphologies. Cubes, cylinders, and flat, platelike structures known as squames (Figure 4–4). Simple epithelia may be composed of a single layer of any of these types of cells. Thus, a simple epithelium can be defined as any one of several types, including a simple cuboidal epithelium, a simple columnar epithelium, or a simple squamous epithelium. This description depends solely on the morphology of cells from which it is composed.

Basement membrane extracellular matrix organized to support an epithelium

Extracellular matrix (ECM) a complex, multicomponent material produced by surrounding cells; composed predominantly of proteins and carbohydrates

Stratified Epithelium Stratified epithelia are characterized in a similar fashion (Figure 4–4). A stratified cuboidal epithelium is composed solely of cuboidal cells. This type of epithelium is rare, as is the occurrence of a stratified columnar epithelium. However, stratified layers combining cuboidal or columnar cells with overlying squamous cells are common, for example, the epidermis. The epidermis is considered to be a stratified squamous epithelium.

Pseudostratified Epithelium There are two other types of epithelia worthy of special note. One type is known as a **pseudostratified epithelium,** the other as a **transitional epithelium** (Figure 4–5). As the name implies, a pseudostratified epithelium looks as if it is a stratified epithelium, but, in actuality, it is not. To be legitimately considered stratified, an epithelium must have two or more distinct horizontally arranged cell layers. Close examination of a pseudostratified epithelium reveals that all the cells composing the tissue are actually attached at their lower or basal surfaces to a basement membrane. In a true stratified epithelium the cells must be in separate and distinct layers in which only the underlying layer of cells is attached to the basement membrane. What is misleading is the fact that the position of nuclei of some epithelial cells of a pseudostratified epithelium are clearly above others, giving the false impression of two separate and distinct layers, when in fact there is only one.

Transitional Epithelium A transitional epithelium undergoes changes in appearance as a result of the stretching of the epithelial cell layers due to some applied force or pressure. The epithelium found in the urinary bladder is a good example of a transitional epithelium. The designation transitional is based on the observation that as the urinary bladder stretches to accommodate the pressure of the temporary storage of urine, the epithelial cells are altered from a polygonal shape to a flattened and extended squamous shape. This distended shape is reversed when the bladder contracts as it empties (Figure 4–5). The stretched appearance of urinary bladder epithelial cells is a direct response to pressure induced by increase in volume.

Intercellular Junctions

The adhesion between epithelial cells has been mentioned as being important in maintaining the barrier functions and continuity of the layers. To modify a phrase from a speech by President Abraham Lincoln at the onset of the Civil War, an epithelium divided cannot stand. The strength of an epithelium is in union, and the union of the body's epithelial tissues depends on cells being held together very tightly (Figure 4–6). The key factor in preventing leaks is an extensive array of strong and resilient adhesions between the surface membranes of neighboring epithelial cells. The adhesive sites have special proteins associated with them and occur at specific locations between adjacent plasma membranes. The specialized regions are called *adhesive junctions* (Figure 4–6). The formation and maintenance of these intercellular junctions bind the epithelial cells together into a continuous sheet. This strikes at the very heart of what any tissue in a multicellular organism is required to do—maintain stable cell associations in order to function. If cells fail to cohere, tissues dissociate and function is lost or diminished. Epithelial cells are among the most cohesive of all cell types and the tissues they form are the most spatially extended.

Pseudostratified epithelium specialized epithelium that appears to be stratified but in which all cells are directly attached to the basement membrane

Transitional epithelium a specialized epithelium in which cells undergo a change from polygonal to elongated as the epithelium is subjected to stretching

(a)

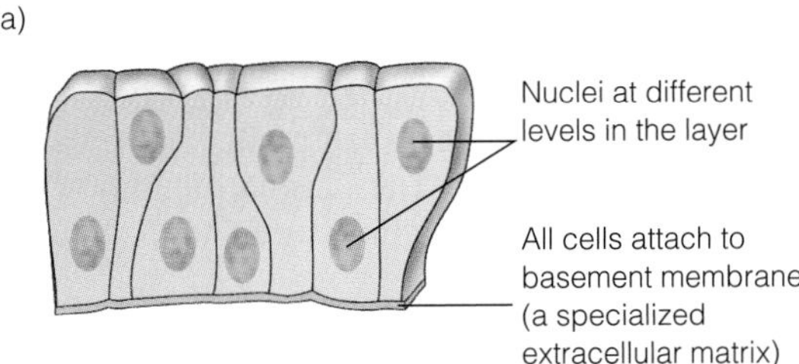

(b)

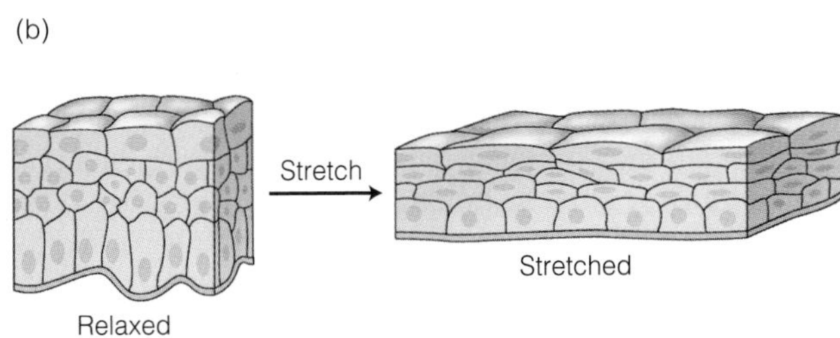

Figure 4–5 Pseudostratified and Transitional Epithelia
Pseudostratified epithelia (a) are characterized by the appearance of being stratified when in fact they are not. Transitional epithelia undergo shape changes in response to stretching. These changes allow for expansion of the epithelia to accommodate pressure.

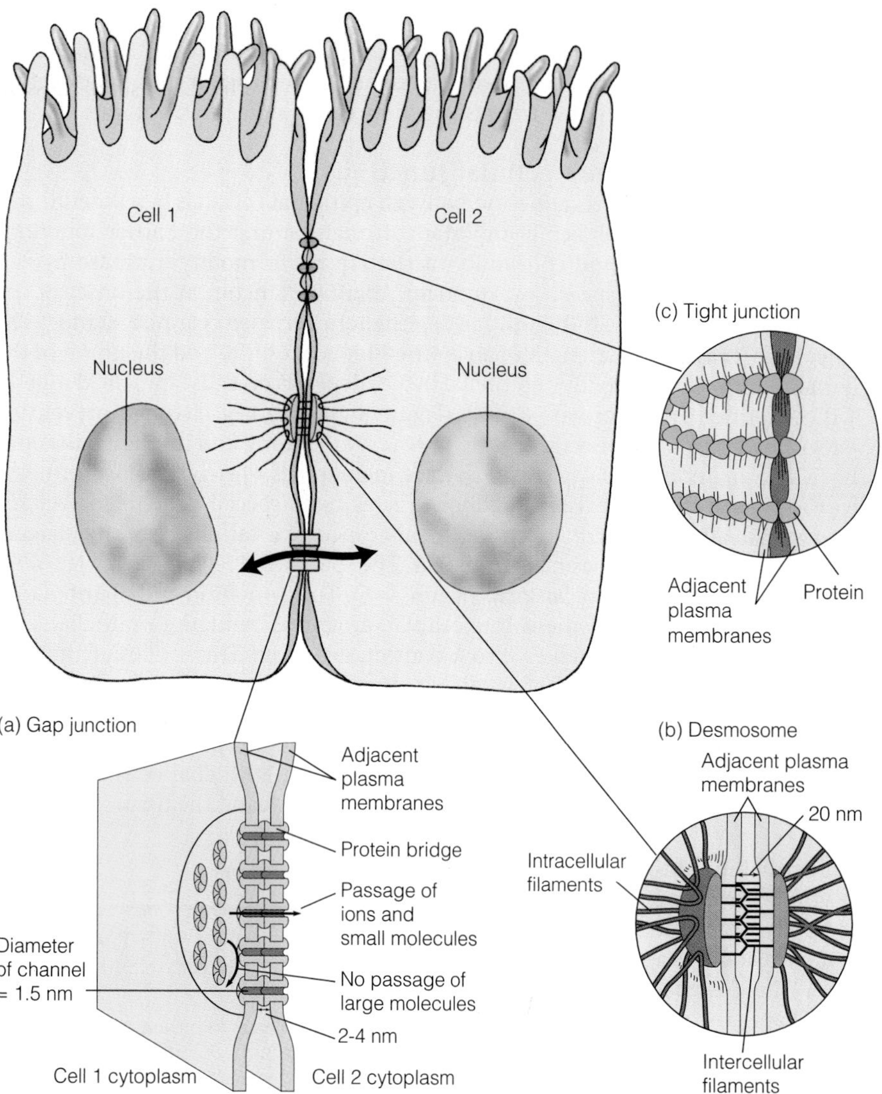

Figure 4–6 Junctions Between Cells

Epithelial cells are held together and communicate through a number of specialized junctions. Gap junctions (a) allow passage of small molecules between cells for communication. Desmosomes (b) hold cells together and are called adhesive junctions. Tight junctions (c) are barriers to the passage of substances between adjacent cells.

Epithelia have basically three types of intercellular junctions, two of which are **occlusive junctions** and/or adhesive junctions, that is, tight junctions and adhesive junctions (Figure 4–6), and one of which forms channels between cells (**gap junction**). To simplify envisioning adhesion among epithelial cells, think of an epithelial cell as a cylindrical object, similar to a can of soup, and the epithelium, composed of epithelial cells, as cans packed as closely together as possible. Such an epithelium is very uniform in composition and thickness. Even when the cans are pushed together as closely as physically possible, there are still significant spaces between each can. These spaces represent "leaks" in the soup can epithelium. Anything smaller than the size of the spaces between cans will pass through unobstructed. But suppose the top edges of all the cans are stretched so they may be welded together? This would change the shape of the top of the cans from circular to hexagonal and allow them to be held tightly with no spaces for leakage. It is tight junctions (Figure 4–6) that form bands of adhesive connections between neighbor cells utilizing integral proteins that serve to block the movement of materials between cells. This is why such junctions are called occlusive. The plasma membranes of cells readily accommodate stretching and shape changes,

Occlusive junctions form barriers to movement of substances between cells; tight junctions

Gap junction an intercellular junction that establishes a direct connection or channel between two adjoining cells

so tight junctions between epithelial cells, similar to the welds between cans, are able to establish and maintain a hexagonal shape.

Continue the soup can analogy one step further. With only the top edge of the cans held together, the bottoms are relatively free to move around. If each can is welded to each of its neighbors at a few places on their shared lateral surfaces, the whole collection of cans now forms a strongly bound unit, indeed, the equivalent of an epithelial membrane. The second major type of adhesive junction, responsible for lateral adhesions between cells, is called a spot junction, or **desmosome** (Figure 4–6). Tight junctions and desmosomes help keep an epithelium leakproof and epithelial cells tightly bound together, respectively.

The third type of junction commonly found between epithelial cells is called a gap junction (Figure 4–7). Gap junctions are composed of proteins that form channels through the plasma membrane and are specialized to provide a means for cells to communicate directly among themselves. Communication through intercellular channels allows cells to coordinate their metabolic activities. Small but important molecules, such as ATP, *cyclic AMP,* and glucose, as well as ions, such as Ca^{+2}, pass easily between cells connected by gap junctions. The exchange of molecules between cells is involved in coordination of cell and tissue activities. For example, the rate of growth and differentiation of cells may be regulated by exchange of small messenger molecules able to pass through the 2 nm channel formed by the gap junction. In addition, if one of a group of epithelial cells is damaged or destroyed, damage to closely interacting cells is prevented by closing gap junctions.

CONNECTIVE TISSUE—GLUES, FRAMEWORKS, AND HARD TISSUES

The group of tissues known as connective tissues have two different but related properties. One type of connective tissue produces biological glues, which are used to hold other cells and tissues together. The other type produces rigid structures or frameworks, such as those composed of bone

Desmosome symmetrical adhesive junction formed between cells

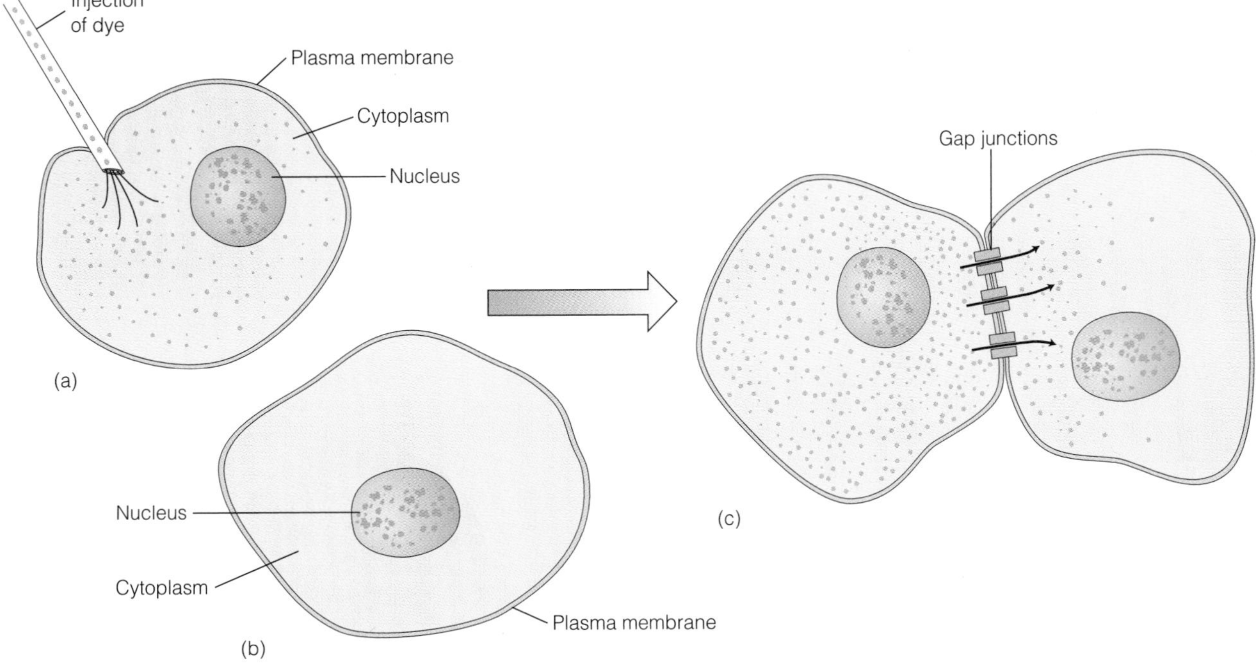

Figure 4–7 Cell-Cell Communication—Gap Junctions
To demonstrate that cells communicate directly, a cell can be injected with small dye molecules (a). When the labeled cell is brought into contact with an unlabeled cell (b) and (c), dye diffuses from the labeled cell into the unlabeled cell through gap junctions formed between them. This may be observed using a microscope.

and cartilage, which provide physical support for the attachment of other human tissues and organs. Imagine the problems you might have in moving your body if your muscles were not attached to your bones.

Connective Tissue—Details

Connective tissues, are a diverse and heterogenous group of related tissues made up of a number of types of cells. In fact, connective tissues are heterogeneous enough to be put into two general categories: **connective tissue proper** and *special connective tissues*. Although connective tissues are structurally diverse in appearance, they have many shared functional features (Figure 4–8).

Connective Tissue Proper Some types of connective tissue proper (Figure 4–8), appear at first glance to be the

> **Connective tissue proper** fibrous type connective tissues

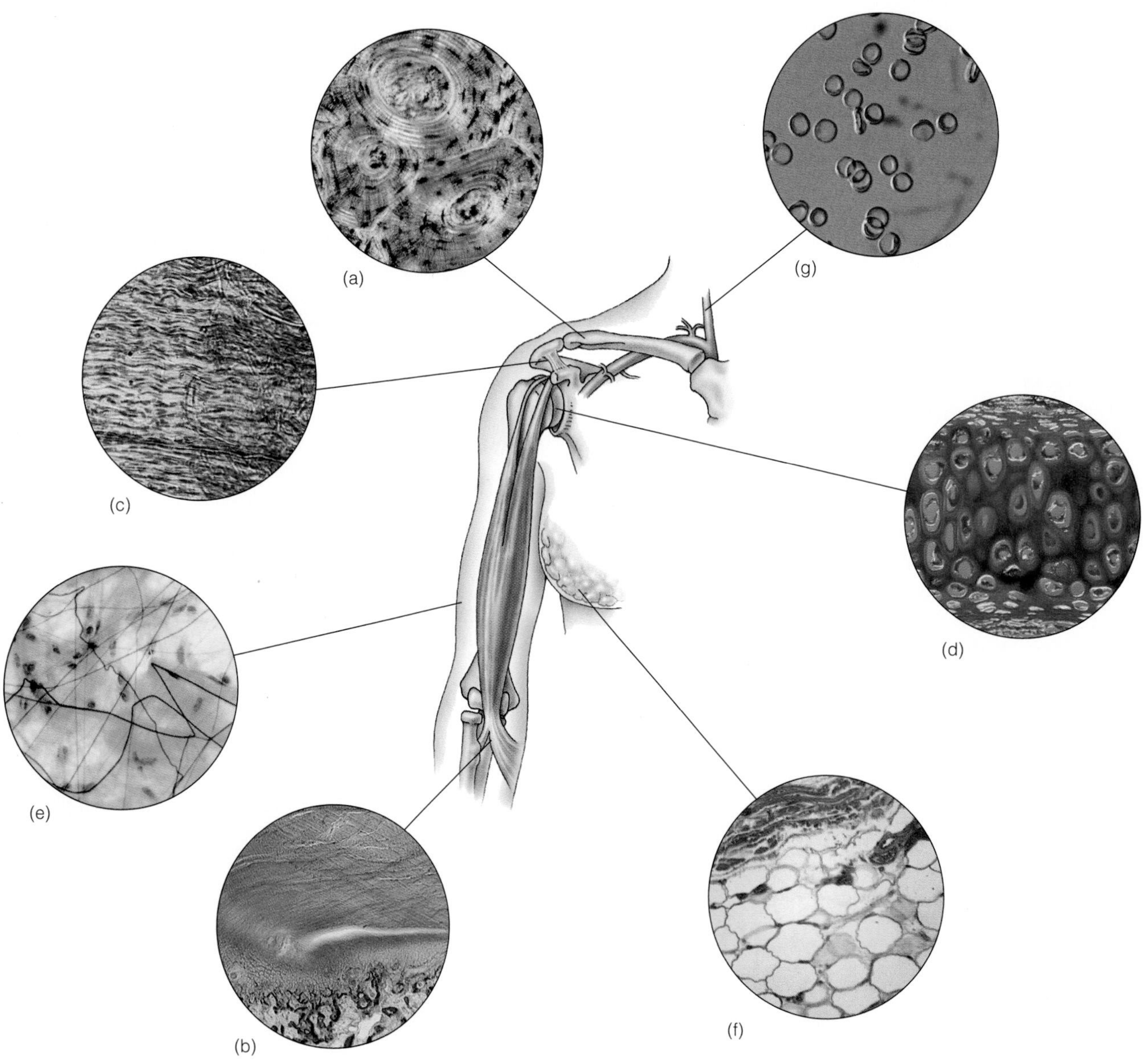

Figure 4–8 Diversity of Connective Tissues
Connective tissue is found in all regions of the body: (a) Bone (b) Tendon (c) Ligament (d) Cartilage (e) Areolar (f) Adipose (g) Blood.

structural opposite of an epithelial tissue. This includes **areolar tissues** and fatty, **adipose tissues,** which are often loosely arranged and have space between cells. Others are very highly organized in appearance and are tough, resilient, and inelastic. Examples of some of the toughest connective tissues include **tendons** and **ligaments** [Figure 4–8 (b) and (c)]. Tendons have to be tough to effectively connect muscles to bone, and ligaments, likewise, must have great strength to maintain the connection of bone to bone.

The cells of connective tissue proper are generally irregular in morphology and have relatively large intercellular spaces between them. They are often only loosely held together by associations with the extracellular materials (ECM) surrounding them. The ECM is composed of many different kinds of molecules, many of which are sticky, gluelike, space-filling substances (Table 4–2). In fact, many types of glue sold commercially are produced from the most abundant of animal extracellular matrix molecules, *collagen*. In the days of horse-drawn wagons, it was no joke when a horse got too old to pull the cart. The animal might easily have been sent to the glue factory.

The principal cell type composing connective tissue proper is called a **fibrocyte.** The progenitors of fibrocytes are called **fibroblasts.** Fibrocytes get their name from the fibrous, extracellular matrix molecules, including collagen, fibronectin, and laminin that they produce in abundance and in which they are embedded (Figure 4–9). These versatile cells produce and secrete a variety of other proteins, glycoproteins, and proteoglycans that alter the composition and structural properties of the environment around them. Connective tissue proper is meant to be leaky to some extent. Instead of the fibroblasts, fibrocytes, and other cell types being attached directly to one another, they may be attached only to ECM (Figure 4–9). Production of ECM molecules often forces open spaces between cells and promotes the movement of cells and materials. This is particularly important with respect to the diffusion of water and molecules dissolved in water.

Unlike epithelial cells, fibrocytes do not maintain a consistent shape. One reason for this is because fibrocytes actively migrate, and, to do so, they need to change shape. They do so by pushing and pulling themselves through, over, and around the extracellular obstacles they find in their way. Fibrocytes and fibroblasts produce most of the extracellular molecules of connective tissue proper. With respect to the growth and maintenance of connective tissue in the human body, connective tissue is produced, repaired, and rebuilt constantly throughout a lifetime. It is fibrocytes, fibroblasts, and related connective tissue cells that constantly produce, remodel, and repair the materials in the microenvironments in which they reside.

Some types of connective tissue proper have more cells in proportion to extracellular materials and some have less. The connective tissue underlying the epidermis, called dermis, is a good example of this (Figure 4–4). The dermis is rich in fibroblasts and ECM, but there are regions of the dermis in which the ratio of cells to ECM is greater than others. Variations in the cellular and extracellular composition and density of the dermis are important factors in the structure of the skin. Epidermal cell attachment to the dermis must be strong but flexible. The dermis is deformable and modestly compressible. These properties of connective tissue are observable in the form of ridges under the epidermis of the fingers and hands. Fingerprints

Areolar tissues loose connective tissue

Adipose tissues fatty connective tissue

Tendons tough, resilient connective tissues connecting muscle to bone

Ligaments tough, resilient connective tissue connecting bones

Fibrocyte a mature, less active form of fibroblast

Fibroblast common connective tissue cell type, secretes ECM molecules

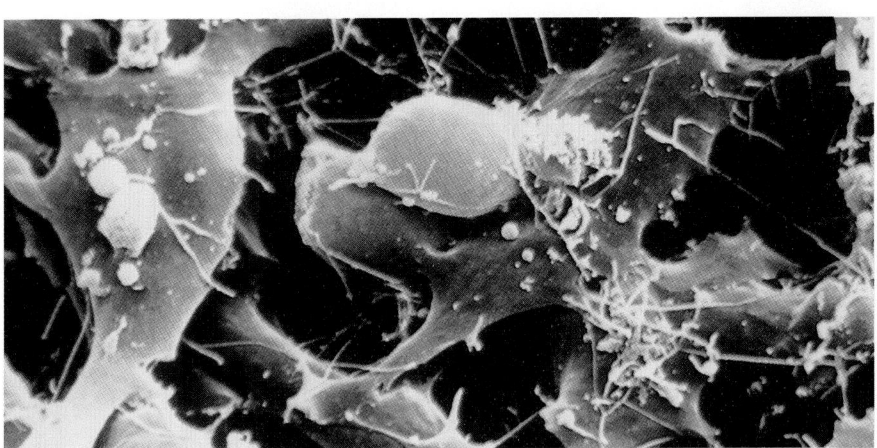

Figure 4–9 Fibroblasts and the Extracellular Matrix

Fibroblasts produce and associate with extracellular matrix molecules.

represent patterns of dermal connective tissue in the skin. The inherent strength, flexibility, and deformability of the ridges of fingerprints allow humans to pick up and manipulate small objects more easily than if the surfaces of fingers were smooth (Figure 4–10).

Blood Certain types of specialized body fluids are also considered to be connective tissues. For example, blood is a specialized connective tissue. Red and white blood cells in circulation are suspended in a fluid matrix and shuttled rapidly throughout a closed circuit of vascular tubing (Figure 4–11). In this case, it is necessary to consider not only the relative proportion of cells to ECM but also the nature of the fixed or fluid relationships of specific types of cells to their extracellular environments. Blood cells are produced in the red marrow of bones, an environment that is more highly structured than the fluids in which the mature red blood cells flow. Their origins in the center of bones is testimony to their connective tissue affiliations.

Bones, Cartilage, and the Skeleton The specialized connective tissues that compose the skeleton are unique in the qualities of their hardness. The skeleton is principally an articulated set of bones that are organized in specific patterns during embryonic development. The skeleton is organized to protect and support many internal organs, particularly the central nervous system, which is encased in the skull and spine, and to provide a framework for the muscle attachments required for movement (see Chapters 6 and 7 for more details). There are basically two types of special connective tissues associated with the skeleton—cartilage and bone (Figure 4–12). They are different from one another in chemical composition, as well as in the degree of their hardness and flexibility. Bones are mineralized structures and are harder than cartilage. Cartilage is tough and resilient but much more flexible than bone. Both connective tissue types are extremely tough and resilient. The cells that produce cartilage are called **chondrocytes,** and the cells that produce bone are called **osteocytes.**

Chondrocytes and Osteocytes Chondrocytes produce cartilage (Figures 4–8 and 4–12). They secrete collagen as well as extracellular materials that have unique elastic properties. In addition to the protein collagen, cartilage is composed of an abundance of a protein called **elastin,**

Chondrocytes cartilage forming cells
Osteocytes mature, bone-producing cells
Elastin elastic extracellular matrix protein found in many types of connective tissue

Figure 4–10 Fingerprints
Fingerprints are patterned elevations of the epidermis resulting from the buildup of connective tissue. The patterns found in humans are unique to each individual and are used in identification.

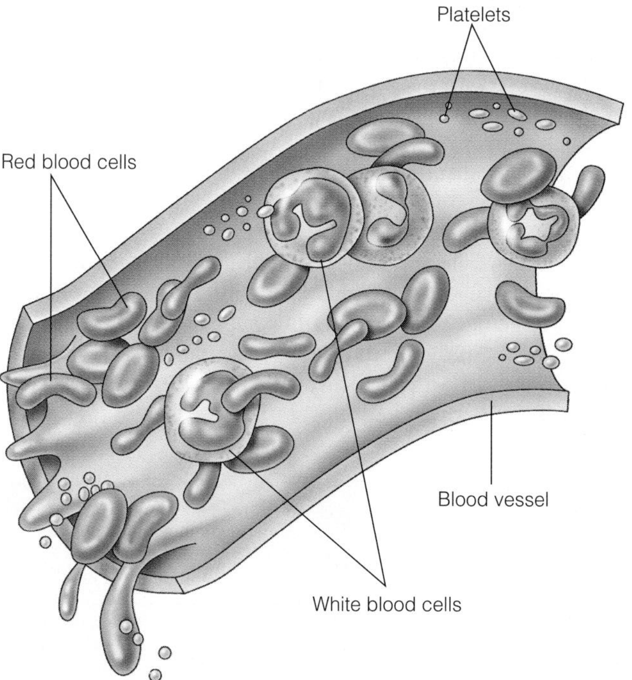

Figure 4–11 Blood as Connective Tissue
Human blood is composed of approximately 40% cells or formed elements (red blood cells, white blood cells, and platelets) and 60% fluid. The fluid is known as plasma and contains water, protein, lipid, salts, and dissolved gases, such as carbon dioxide and nitrogen.

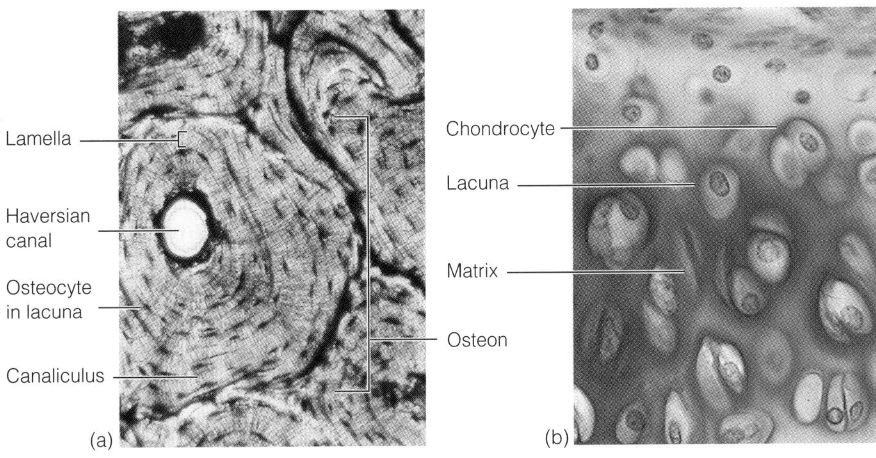

Figure 4–12 Skeletal Elements—Morphology, Structure, and Chemistry
The unit structure of compact bone (a) is the osteon, which is a concentrically organized structure surrounding a Haversian canal. The lamellar structure of the osteon has within it spaces for osteocytes (lacunae), which are connected to one another via a canaliculus. Cartilage (b) is less mineralized than bone, however, chondrocytes are located in spaces (lacunae) and surrounded by a tough matrix.

which is rubbery and flexible. Cartilagenous tissues help in the operation of the skeleton, specifically in the case of the moveable joints. The articulating ends of bones are covered with smooth, nearly frictionless cartilage capsules to ensure easy, sliding motion. In another region of the body, cartilage is found to be the main structural material in the **trachea.** The trachea is the connecting tube through which air passes from the **pharynx** to the **lungs** (your **Adam's apple** is made out of this same material). Cartilage in the trachea, and in the bronchial passageways of the lungs, is tough, slightly deformable for flexibility, and essential to stabilizing the opening of the airways. Cartilage also acts as a shape modifying material. Did you ever wonder why a dried human skeleton has no protruding nose or attached ears? Feel the end of your own nose. It is solid, but it is not bone. It feels rubbery and flexible because it is constructed of cartilage. So is the structural material of your ears. In fact, during embryogenesis, the human skeleton is initially constructed of cartilage and only later in fetal development and continuing after birth does bone replace the cartilage model of the skeleton. The many types of cartilage and their special structural characteristics and distribution in the body are listed in Table 4–3.

Osteocytes produce mineralized materials from within bone (Figure 4–12). Osteocytes secrete collagen, but in addition they are capable of secreting mineralized materials. In this capacity, osteocytes are distinctly different from chondrocytes. Bone is harder and stronger than cartilage but less flexible. The basic strength in bone is provided by minerals. The particular kind of mineralization occurring in bones in the human body is called *calcification.* Calcification is the process by which the calcium containing compounds, **calcium phosphate,** is incorporated with collagen into bone, as part of the bone matrix. This process of bone building is called **ossification** and is the direct result of the mineralizing activities of osteoblasts, which are the precursors to the mature bone-forming cells, the osteocytes.

Special secretory vesicles containing calcium phosphate are released by osteocytes within forming bone. The minerals are deposited, along with collagenous materials, into the extracellular matrix. Thus, bone is composed of a unique combination of two-thirds inorganic (calcium phosphate) and one-third organic substances (collagen and other proteins). The combination of minerals and molecules is amazingly strong, but still flexible enough so that slight bending can occur. The same principle of combining fibers and solids applies to building materials for bridges

Table 4-3 Types of Cartilage

CARTILAGE TYPE	LOCATION(S)
Elastic	External ear, auditory tube, epiglottis, surrounding largest of arteries
Hyaline	Joint surfaces at end of bones, trachea, bronchi
Fibrocartilage	Intervertebral discs, pubic symphsis, attachment areas of tendons and ligaments

Trachea cartilagenous tube connecting pharynx to lungs and allowing uninterrupted passage of air

Pharynx region of the mouth and throat that extends from the nasal cavity to the larynx and esophagus

Lungs paired organs of the thorax in which oxygen is taken in and carbon dioxide is expelled from the body

Adam's apple the cartilage of the larynx, which protrudes in human males

Calcium phosphate the major mineral component of bone

Ossification the process of bone formation

and highways, which use steel wire embedded in concrete to withstand the stress and strain of motor vehicles.

Osteocytes are located in regions of bone known as **osteons,** or **Haversian systems** (Figure 4–12), which are also described in more detail in Chapter 6. In an intact Haversian system, individual osteocytes are found in each of millions of tiny compartments within bone known as **lacunae.** Osteocytes in adjacent lacunae interact with one another through interconnected channels known as **canaliculi.**

MUSCLE CELLS AND FIBERS—A CAPACITY TO CONTRACT

Muscle cells and muscle fibers have the unique capacity to contract (Figure 4–13). Changes in the length of groups of muscles attached to bones generates the force required to move the skeleton, inflate the lungs during breathing, and squeeze the chambers of the heart to change their volume during the pumping of blood. The timing of the contractions of muscles is controlled by neurochemical stimulation directed by nerves.

Muscle Types

There are three different types of muscle cells found in the human body (Figure 4–13)—smooth muscle cells, cardiac muscle cells, and striated skeletal muscle fibers, which compose the skeletal muscles of the body. The most familiar are the muscle fibers composing striated skeletal muscle that allow the human body to move. The smooth muscles are associated with more subtle control of internal organs, such as the constriction and dilation of blood vessels, and the process of **peristalsis** in the digestive tract. Cardiac muscle cells compose the heart and contract in an integrated and coordinated fashion in response to internal and external signals that regulate heartbeat.

Striated Skeletal Muscle and Muscle Fibers Striated skeletal muscle is formed from specialized cells known as muscle fibers that contain many nuclei. Muscle fibers are formed early in human embryonic development from the fusion of hundreds of individual embryonic muscle cells (called **myoblasts**) into one large multinucleated muscle fiber. Muscle fibers have a characteristic internal organization of striations or crosshatchings (Figure 4–14). These striations represent cytoskeletal elements called **sarcomeres** and are composed principally of microfilaments and myosin filaments organized into special contractile units within muscle fibers.

Cardiac Muscle The heart is composed of a specialized type of muscle cell known as a cardiac **myocyte** (Figure 4–15). Cardiac myocytes are single cells with a single nucleus, not multicellular fibers. Each cardiac myocyte is tightly adhered to its neighbors and densely striated. The heart is basically one solid muscle, surrounding a series of blood-filled chambers. Cardiac muscle contraction is an involuntary action; that is, it occurs automatically and does not require nor respond to conscious control (try to stop your heartbeat by force of will). Both the involuntary nervous system and a special group of cells inherent to the

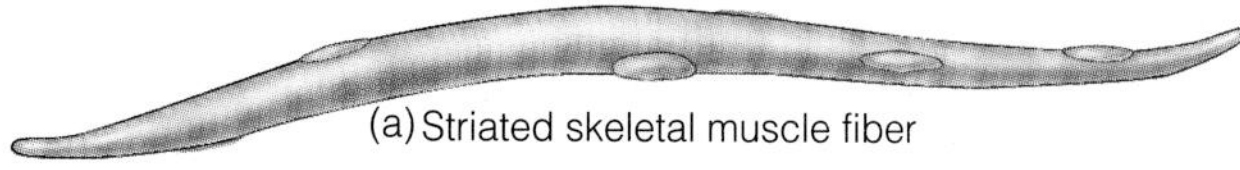

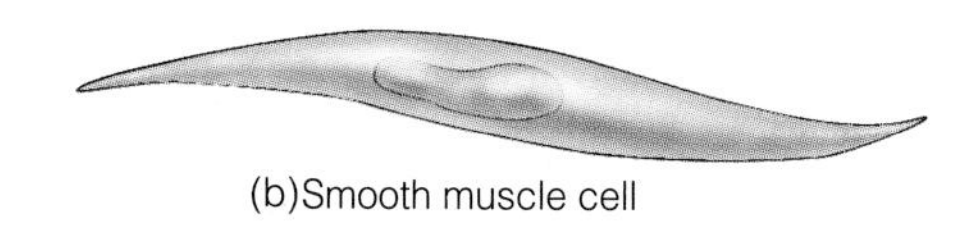

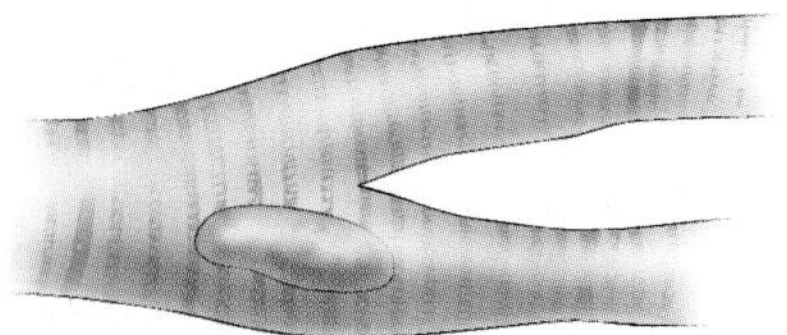

Figure 4–13 Muscle Types
There are three types of muscle in the human body with specialized cells associated with each type. Striated skeletal muscle is composed of muscle fibers (a); smooth muscle is composed of smooth muscle cells (b); the heart is composed of cardiac myocytes (c).

Osteons small oval structural units repeated within compact bone, composed of a central canal surrounded by concentric rings of bone

Haversian systems *see* osteons

Lacunae small spaces within bone and cartilage in which cells (osteocytes or chondrocytes) reside

Canaliculi small canals connecting cellular domains within bone

Peristalsis waves of muscular contractions that move materials within a hollow tubular structure or organ, e.g., material movement within the digestive tract

Myoblast embryonic form of muscle cells that will fuse together to form a syncytium that will become muscle fiber

Sarcomeres units of contraction in some types of muscle cells; composed principally of actin and myosin

Myocyte a mature muscle cell

heart, known as the **pacemaker,** control the beat of the heart. A single heartbeat represents the coordinated contraction of billions of cardiac myocytes. Individual cardiac muscle cells are connected by special adhesive junctions known as **intercalated discs** which contain desmosomes and gap junctions. The persistent, uninterrupted beat of a human heart continues for a lifetime. In fact, the beat of a heart defines a lifetime, for when cardiac myocytes cease contracting, human life ends.

Smooth Muscle The third type of muscle cell is smooth muscle cells (Figure 4–16). Each smooth muscle cell

Pacemaker (1) the sinoatrial node of the heart. (2) an implanted electrical device to control the beat of the heart

Intercalated discs a junctional complex containing desmosomes and gap junctions; prevalent in cardiac muscle

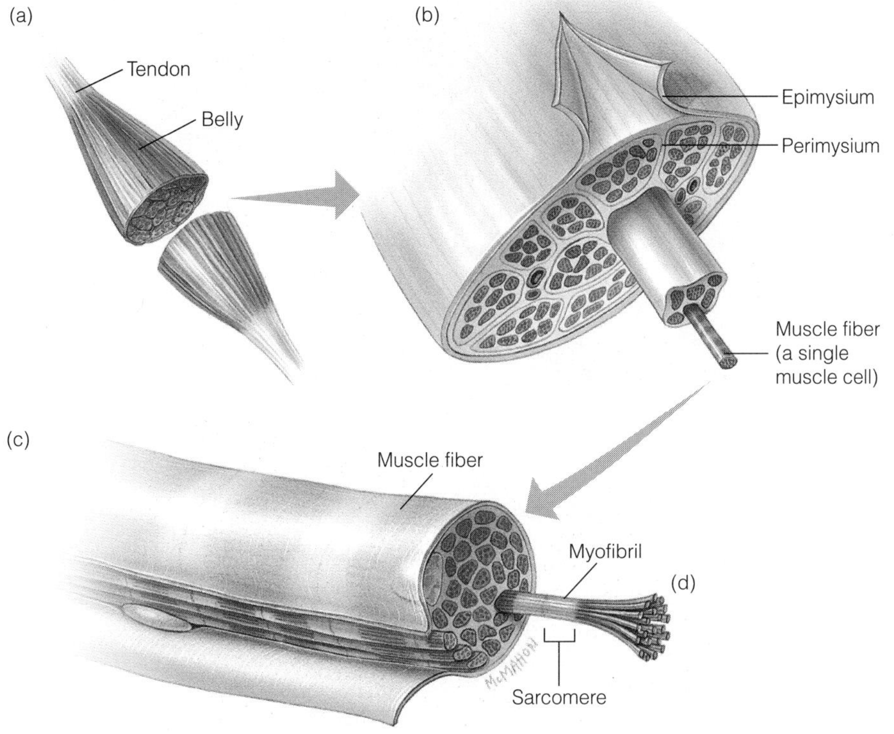

Figure 4–14 Striated Muscle
There are several levels of skeletal muscle organization. An entire muscle with tendon is shown in part (a). A cross section of the muscle indicates the bundles of muscle groups and shows a single muscle fiber (b). In part (c) a single muscle fiber is dissected to examine a single myofibril (d), which contains the contractive units of this muscle type, the sarcomere.

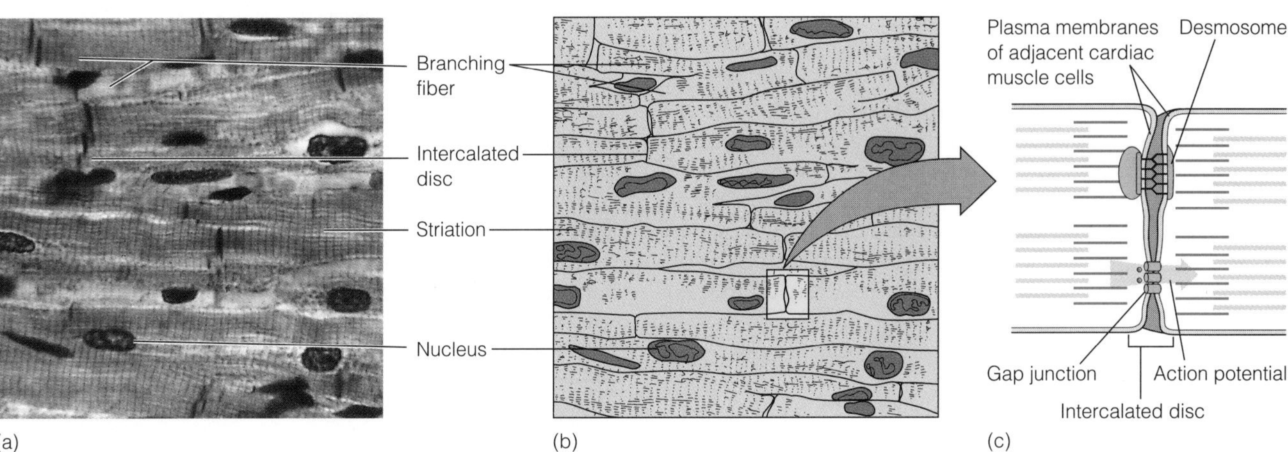

Figure 4–15 Cardiac Muscle
Cardiac muscle cells are held together tightly by intercalated discs, which contain desmosomes and gap junctions. Cardiac muscle cells have a single nucleus and are densely striated.

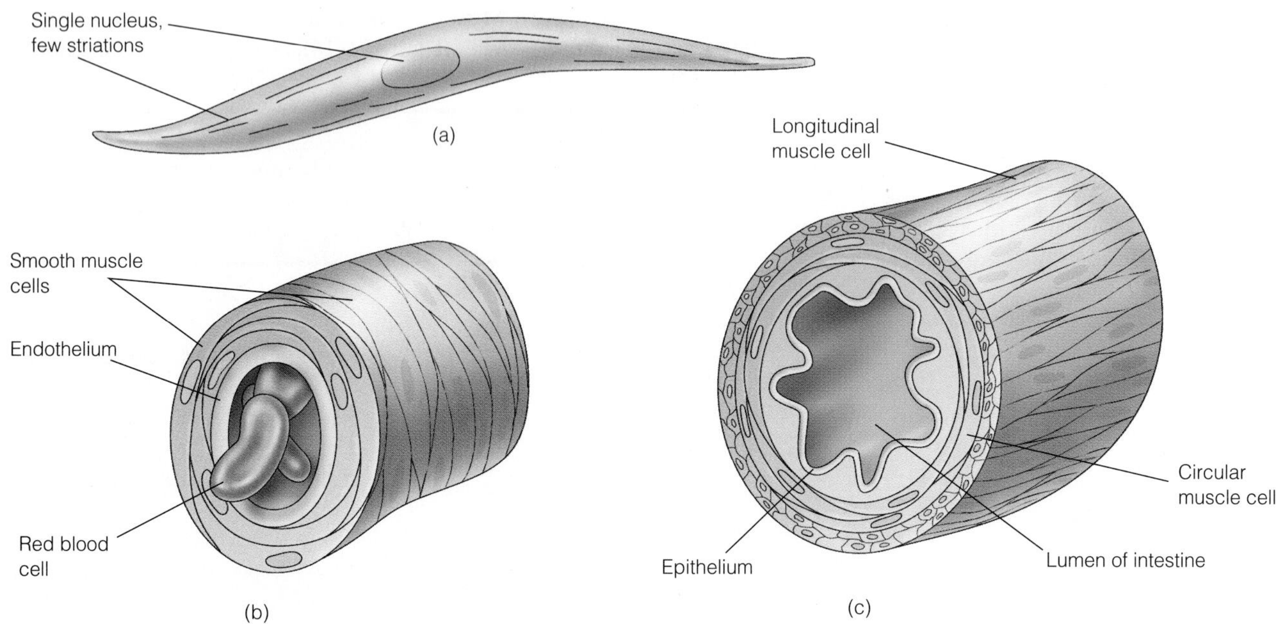

Figure 4–16 Smooth Muscle
The smooth muscle cell (a) may be organized in several ways surrounding internal hollow structures or organs. In part (b) smooth muscle cells are organized in a circular fashion around a blood vessel. In part (c) smooth muscle cells are organized in both a circular fashion and in longitudinal bands. The contraction of the smooth muscle constricts flow and aids in peristalsis.

contains a single nucleus but has very few, if any, obvious striations, although it is fully capable of strong contractions. Smooth muscle cells coordinate their activities and often collectively surround many internal structures involved in the movement of materials through or within the body. Peristalsis, which is the muscular process by which materials are moved within the intestinal tract, is dependent on the coordinated contractions of millions of smooth muscle cells all along the tubular digestive tract. The control of blood flow in **capillaries** (for oxygen and carbon dioxide exchange) and throughout the rest of the cardiovascular system is controlled by smooth muscles constricting **arterioles** and **arteries** that lead into them. Likewise, blood pressure in the vascular system is controlled by constriction and dilation of blood vessels in conjunction with the contraction or relaxation of the surrounding smooth muscle cells.

NERVE CELLS AND TISSUES—THE EXCITABLE CELLS

Nerve cells are the generators and carriers of electrical signals in the principal communication network of the body, the nervous system. Nerve cells connect with and coordinate the activities of many different cell types, including other nerve cells, as well as muscles and glands. Nerve cells provide the body with nearly instantaneous communication through the control of signals to and from nerve cells within the network (Figure 4–17). The physical and chemical mechanisms involved in nerve signaling and intercellular communication underlie the function of the brain and manifest themselves in all aspects of human behavior, including intellectual activities. Even in these times of high technology electronics and supercomputers, the functional capacity and sophisticated processing in the human brain is unmatched.

Nerves and the Cells that Compose Them

The cell types composing nerve tissue are functionally the most complicated group of cells in the human body. There are two major types of cells found in the nervous system.

Capillaries the smallest vascular elements of the circulatory system

Arterioles small arteries

Arteries vascular elements of the circulatory system carrying blood away from the heart; usually oxygenated

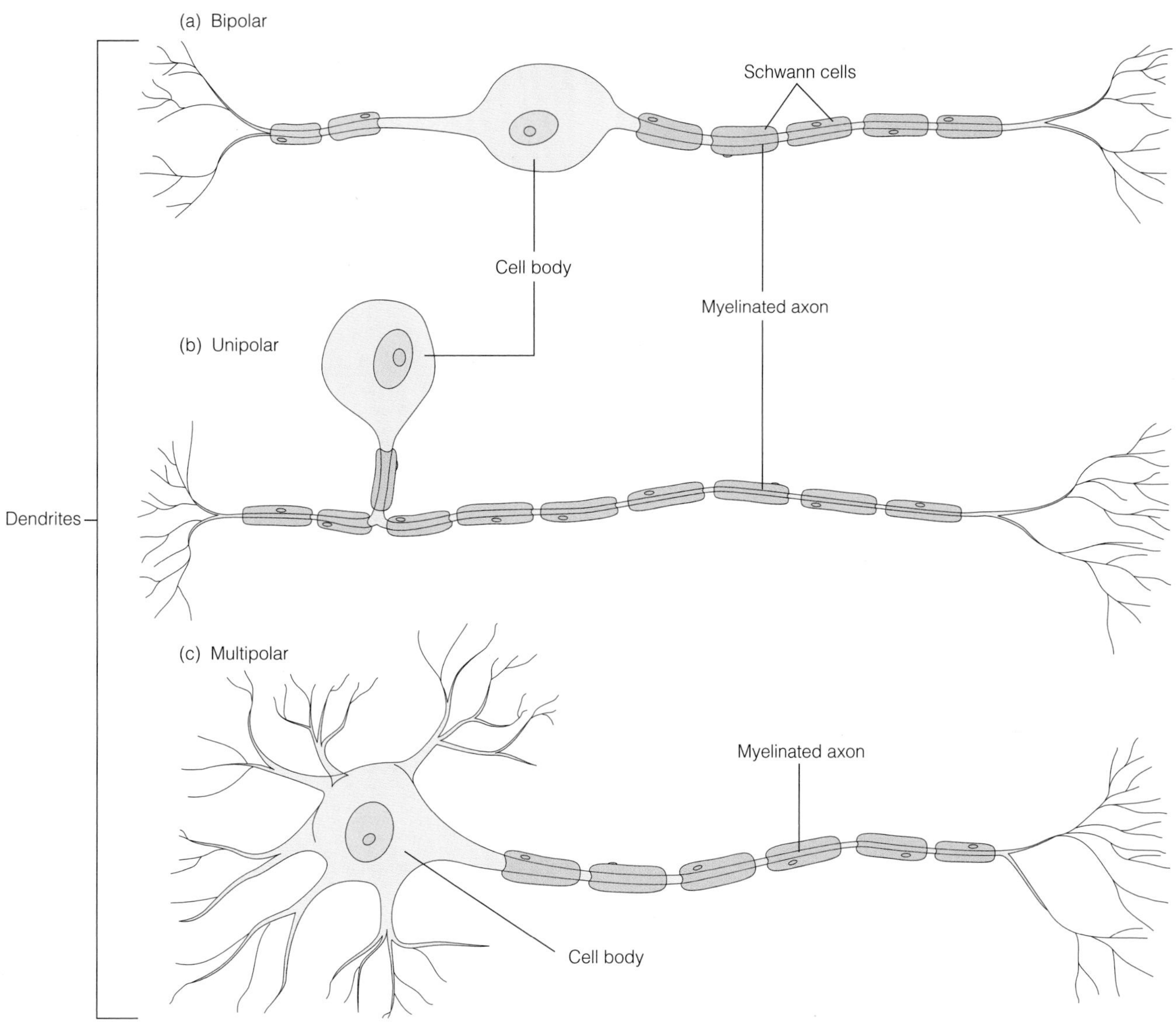

Figure 4–17 Nerve Cell Morphology
There are three basic morphologies associated with neurons: unipolarity, bipolarity, and multipolarity. The determination of this character is based on the number of dendrites a neuron has.

One group is **neurons,** the other is **glia** (Figures 4–17 and 4–18). Neurons have the inherent capacity to generate and transmit electrical signals among and between themselves and other target cell types. Neurons are found in characteristic sizes and shapes, which are correlated to the region of the nervous system in which they are found and the functions they serve (Figure 4–18). The signaling among neurons is considered both chemical and electrical (that is, **neurotransmitters,** ions, and *voltage potential*) in nature. The electrical signals are conducted by ion fluxes through the plasma membrane of a nerve cell, similar to a wave, from one end of the cell to the other. Contacts between neurons occur in regions known as a **synapse** by means of signaling molecules known as neurotransmitters. A single neuron may contact one or many other neurons and may

Neurons electrically excitable cell type of nerve tissue

Glia a class of cells found in nerve tissue; plays a supportive, structural role in nervous system

Neurotransmitters substances released from neurons that influence the activities of other cells including other neurons

Synapse junction between excitable cells; a gap across which neurotransmitters diffuse between nerve cells or between nerve cells and muscle cells or fibers

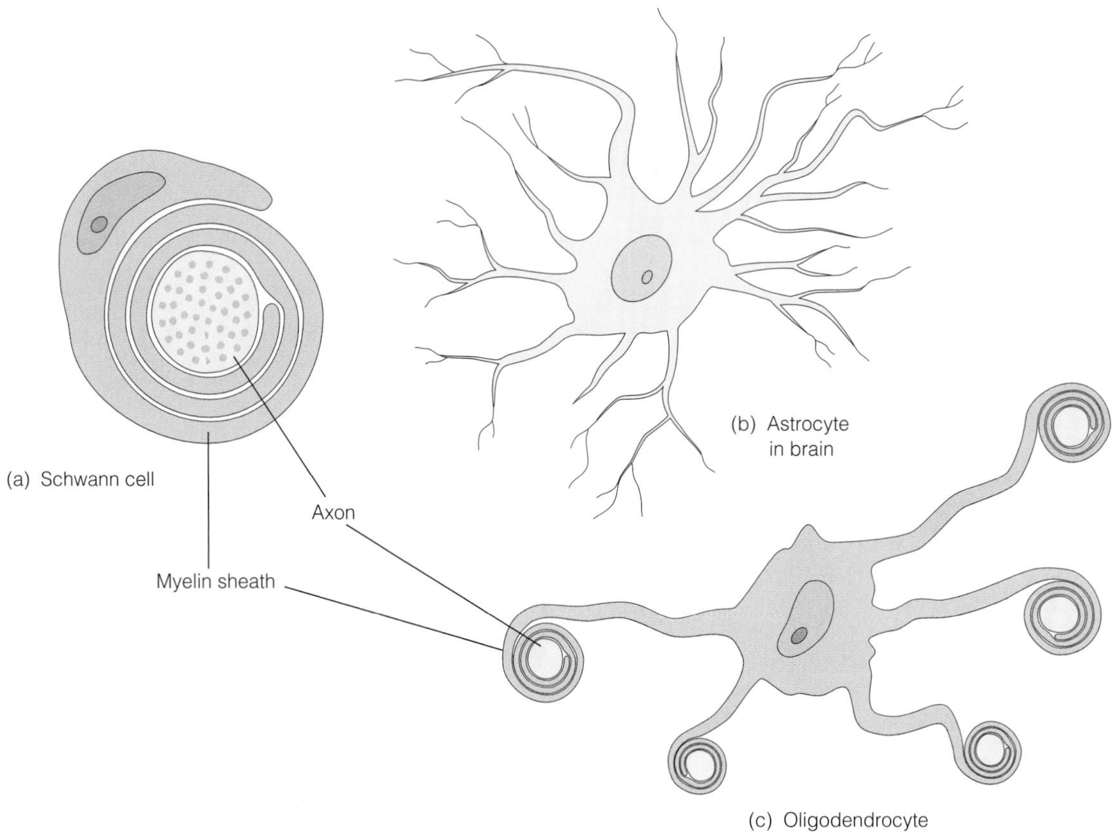

Figure 4–18 Glial Cell Morphology
There are three main types of glial cells in the nervous system: the Schwann cells (a), which produce myelin in the peripheral nervous system; the astrocyte (b), which is involved in establishing the blood-brain barrier; and the oligodendrocyte (c), which myelinates neurons in the central nervous system.

itself receive signals through synaptic contacts with many others. Nerve cells in the brain may have thousands of synaptic contacts (Figure 4–19).

Types of Neurons The generation of signals and their long-distance transmission through the nervous system is carried out solely by neurons. Neuron cell morphology suggests much about the way in which they operate to integrate the functions of the nervous system. Neurons occur in three basic shapes—unipolar, bipolar, and multipolar (Figure 4–17). All three types have features characteristic of all neurons, including an axon, dendrite(s), and a *cell body.*

The first of these cell types, the unipolar cells, are relatively rare in the human nervous system, as will be discussed more fully in Chapter 17. Unipolar cells are characterized by a cell body from which a single cellular extension arises. The extension operates as both an axon and a dendrite. Bipolar cells have both an axon and a dendrite. Multipolar cells, which are the most common of the three types found in the human nervous system, have an axon and many dendrites. Regardless of the type of neuron, however, no neuron has more than one axon. It is principally through the interactions of neuronal cells by means of axons and dendrites, that the neurological circuitry of the nervous system is established and maintained.

Glial Cells Glial cells are quite different structurally and functionally from neurons (Figure 4–18), but they too, occur in several different sizes and shapes. Glial cells are not capable of receiving nor transmitting electrical signals. The presence of large numbers of glial cells and the range of their activities are fundamentally important to the structure of the nervous system tissue, as well as to the function of individual neurons.

Functionally, glial cells serve two main purposes. First, they act as structural elements to support and to insulate neurons from one another, and second, they provide nutrients for the continued function and survival of neurons. Structurally, their gluelike properties help organize and hold the nervous system together. The word *glia* is derived from the Greek for glue. Without glial cells to perform these functional and structural roles in the nervous system, the neuron network does not work properly, and the organism cannot behave in an integrated, normal fashion. In the absence or reduction in number of glia, the brain would probably be an elaborate, anatomical short circuit, with neurons acting as bare wires in contact with one another. Evidence that this occurs is apparent in diseases such as **multiple sclerosis,** in which the body destroys its own glial cells and exposes individual neurons to one another in the brain.

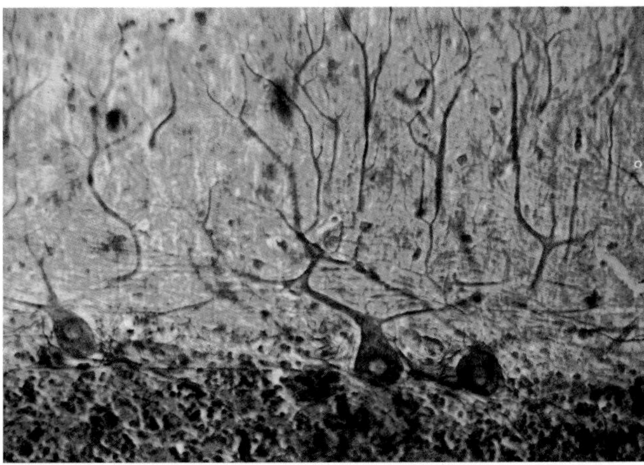

Figure 4–19 Purkinje Cell
Purkinje cells are located in the cerebellum where they are involved in coordination of complex muscle activity. A single purkinje cell may establish thousands of synapses.

Glial cells occur in three morphological types—**astrocytes,** which are shaped like stars; **oligodendrocytes,** which have extensively branched surfaces, and **Schwann cells,** which are organized, similar to a jelly roll, around a neuron (Figure 18). Each cell type is capable of interacting closely with neurons by wrapping around axons and dendrites. This intimacy provides the structural glue, the insulation, and the nutrient exchanging capability necessary to keep the nervous system working.

Multiple sclerosis degenerative disease of the nervous system; involves destruction of myelin surrounding neurons

Astrocytes class of glial cells; part of the blood-brain barrier

Oligodendrocytes glial cells responsible for myelination in the central nervous system

Schwann cells glial cells responsible for myelination in the peripheral nervous system

Summary

There are four distinctive types of tissues in animals. The cells of each tissue are distinctive in structure and function but also depend on associations among themselves and with extracellular materials surrounding them. The four basic tissue types are epithelia, connective tissue, muscle, and nerve. Epithelia are tightly adhered cells that form the protective layers of the surface of the human body. Connective tissues are diverse, ranging from the loose areolar types to compact types such as dermal tissues, tendons, and ligaments and to the flexible and hard tissues of cartilage and bone. Muscles are the contractile elements of the human body and their cells come in three types: striated fibers, cardiac myocytes, and smooth muscle cells. Each has distinctive characteristics fitted to its functions in using contraction as a force to generate motion. Nerve tissues are composed of neurons, which conduct electrical signals, and glia, which support and sustain the neurons.

The mechanisms by which cells organize and specialize have yet to be completely resolved. However, as recounted in this chapter, the diversity of cellular form and function of each of the basic tissue types lends itself to unique combinations in the construction of organs and organ systems. We now turn our attention to the description of individual organ systems, starting with the integument and the architecture of the skin.

Questions for Critical Inquiry

1. Why is multicellularity considered an important step in evolution?
2. Why do we need to know about the form and function of cells in tissues?
3. Why is it important to clearly differentiate between inside and outside the body? What are the criteria you used?
4. What are some of the consequences for an organism of
   (a) An epithelial cell that loses its capacity to adhere to its neighbors?
   (b) A connective tissue cell that cannot produce an ECM?
   (c) A muscle cell that contracts and will not relax?
   (d) A reduction in the number of glial cells in the brain?

Questions of Facts and Figures

5. What are the four basic tissue types?
6. What are the three types of muscle and in what capacities do they function?
7. Name four major types of epithelia.
8. In which type of tissue would you expect to find the most extracellular matrix?
9. What are three types of junctions found between epithelial cells? How do they differ from one another?
10. When examining a dried human skeleton, why are there no nose or ears?
11. What is the composition of bone? To what do we attribute its strength?
12. Name the three types of neurons based on their shapes.
13. How many axons does the average neuron have?
14. The word *glia* is from the Greek for glue. Why would an entire class of cells be given this name?
15. To what underlying structure(s) does an epithelium attach?
16. Blood is considered to be in what basic tissue group? Why?
17. What properties of cartilage make it important to the structure and function of the trachea?

References and Further Readings

Leeson, T. S., and Leeson, C. R. (1981). *Histology,* 4th ed. Philadelphia: W B Saunders.

Sherwood, L. (1993). *Human Physiology,* 2nd ed. St. Paul, MN: West Publishing.

Stalheim-Smith, A., and Fitch, G. K. (1993). *Understanding Human Anatomy and Physiology.* St. Paul, MN: West Publishing.

Tortora, G. (1986). *Principles of Human Anatomy,* 4th ed. New York: Harper Collins.

CHAPTER 5

The Architecture of the Skin

INTRODUCTION

Examining the thick, densely scaled skin of an adult alligator reminds one of the value of an armored body. This tough exterior is certainly useful in protecting the reptile from damage incurred in its predatory activities, but it also protects against a variety of other less violent environmental insults. Skin need not be overly thick and tough to be protective. Water dwelling species, such as fish and amphibians have skin that is relatively thin and simple in comparison to other vertebrates, such as reptiles, birds, and mammals. Under the influence of the relative dryness of Earth's atmosphere and wide daily fluctuations in temperature and radiation from the sun, these latter groups evolved protective layers to seal off their internal structures. This chapter focuses on the organization and growth of human skin and the specialized structures that serve as the human body's armor against the ravages of the environment.

PROTECTIVE COVERINGS

Skin is a principal type of protective covering, or integument, for the bodies of all vertebrates. An integument is a fairly generalized biological structure, encompassing a wide range of different and specialized natural coverings found in animals and plants (for example, seed coats). With regard to animal skin, integument also includes such specialized features as feathers, scales, quills, and hair. Human skin is composed primarily of two layers of cells that are tightly attached to one another, the epidermis and the dermis (see Figure 5–1 and Chapter 4). The epidermis is an epithelial tissue and the dermis is a connective tissue type. As simple as human skin may appear from the surface, it is actually very complex structurally and functionally. The epidermis composes the outermost protective layer of the body, forming a **keratinized stratified epithelium** that efficiently excludes what is outside the body and ensures that what is inside is protected. The dermis incorporates a number of other anatomical structures including blood vessels and nerves, as well as a diverse array of extracellular matrix molecules. Thus, the dermis provides strong physical and nutritional support for the attachment and maintenance of the epidermis.

Shared Characteristics of Vertebrate Skin

There is a high degree of similarity in the skin of all groups of vertebrates, across all the classes (Figure 5–2). Among mammals, for example, there are specializations of many different but related types. Hair, fur, quills, and whiskers are all anatomically equivalent in terms of their tissue origins, how they grow, and much of their biochemical composition. Each of the structures is a derivative of the epidermis, grows from a follicle, and shares in common epidermal proteins known as **keratins.** Keratins are abun-

Keratinized stratified epithelium the epidermis of skin

Keratins a class of proteins forming intermediate-sized filaments and bundles; the structural proteins of keratinocytes and many other types of epithelial cells

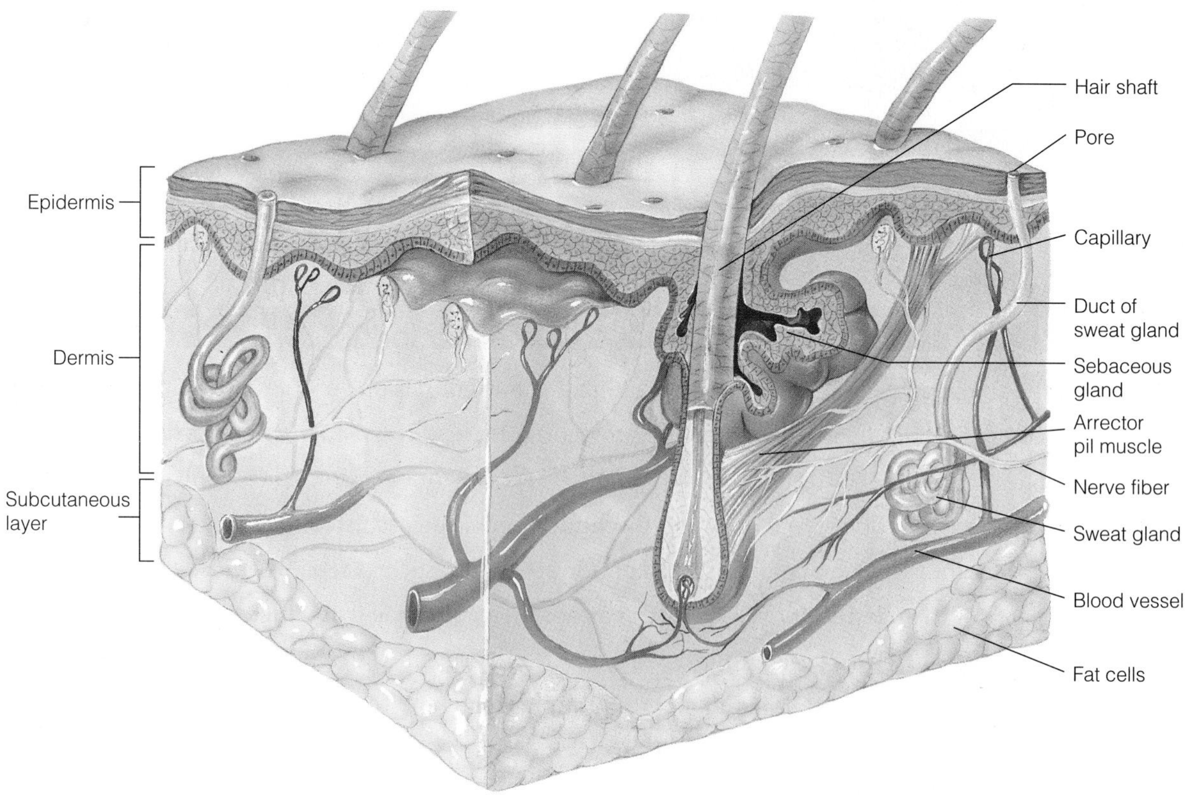

Figure 5–1 Human Skin

Human skin is composed of epidermis and dermis. This diagram shows a number of structures associated with the skin including hair, glands, nerve, muscle, and vascular elements.

dant in the skin of vertebrates from fish to humans, and they are related biochemically and genetically in all cases. Structural, biochemical, and genetic similarities in skin among such diverse species attests to the evolutionary success of this tissue architecture as a means of keeping separate the inside of the body from the outside environment. However, skin has many other important attributes that go far beyond its barrier function. This chapter presents a description of the anatomy and physiology of the skin in relation to the diverse functions it has in maintaining the integrity of the human body.

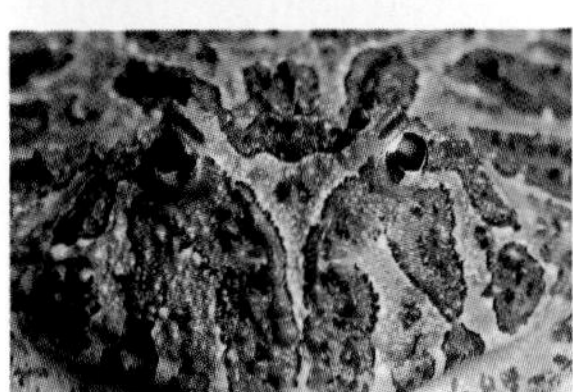

Figure 5–2 Comparative Anatomy of Skin
The organization of skin is similar in all vertebrates. Specializations of the integument characterize class differences. Amphibians and fish have the simplest skin; reptiles and birds have scales; birds have feathers. Mammals have hair and nails/or claws.

THE ANATOMY OF SKIN

When we encounter another living human being, the first part of his or her body we see is the skin. The smiling face, the waving hands, the moving arms and legs are each a familiar indicator of life. Yet, the fact is, that what we see on the surface of an individual is not alive at all, it is composed almost exclusively of dead cells (Figure 5–3). How is it that dead cells compose the entire surface of the living human body? First, it should be noted that the cells of the skin are continuously growing and differentiating and that there are living cells in many layers of the epidermis beneath the superficial layers of dead cells. Epidermal cells differentiate and migrate outward to the surface and, as they do so, they die. Dead cells accumulate in thick layers on the surface. This facility for differentiation and cell death is also true for hair and nails, which are essentially nonliving structures. The living epidermal cells are plentiful beneath the tough, nonliving exterior. The deepest of these layers is attached to a basement membrane and is the source of all new epidermal cells, replenishing the constant loss of dead cells at the surface. How does such a dynamic, living layer of cells give rise to layers of cells that end up dead at the surface? What tissues are involved, and where are the different cell types and specialized proteins that characterize human skin found?

Living Skin

Human skin is composed of cells representative of all the basic tissue types. However, as stated previously, the main tissue components, and the two that team up to form the basic structures of the skin, are the epidermis and dermis (Figure 5–1). These layers, and the molecules they produce, are responsible for an amazing range of integrated functions associated with skin. Indeed, the epidermis and the dermis have a special relationship with one another. The epidermis effectively covers the exterior surface of the body, but it has to have a substratum or foundation to hold on to. The dermis provides the foundation for attachment, as well as the biochemical and nutritional components necessary for growth of the epidermis.

The Epidermis The epidermis is a keratinizing, stratified epithelium and is renewing itself continuously (Figure 5–3). It is composed predominantly of cells that synthesize tough and durable alpha keratin proteins. Differentiating epidermal cells are generally referred to as **keratinocytes.** Epidermal cells are arranged in successive layers from deep

Keratinocytes cells producing keratin in the epidermis

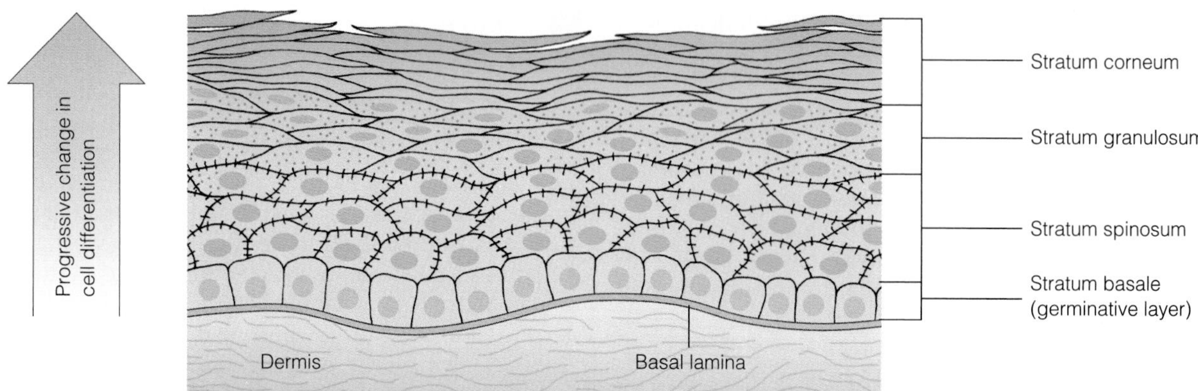

Figure 5–3 The Epidermal Structure
The stratum basale is the source of all cells in the epidermis. As the cells in this layer divide they move up and differentiate. The second layer forms a multitude of desmosomes (spines in spinosum). The third layer produces a multitude of granules (granulosum). The fourth layer is a nonliving layer made of corneocytes (corneum), which is hard and durable and is slowly lost from the skin surface as new cells grow up to renew the epidermis.

(that is, adjacent to the dermis) to superficial, which is the layer of dead cells that composes the outer surface, where they are referred to as **corneocytes.** The thickness of the human epidermis varies from 0.8 mm–1.4 mm. This is about the thickness of a dozen pieces of notebook paper. Considering the extent to which we must rely on the epidermis to protect us from the outside world, this thickness does not seem like very much.

The deepest layer of epidermal cells is called the **germinative,** or **basal cell layer.** These cells are attached to the underlying dermis in a well-defined region referred to as the basement membrane or **basal lamina.** The epidermis is a dynamic layer of cells, and cell growth is continuous throughout our lifetimes. The entire epidermis of human skin is replaced on average every four to five weeks. The loss of dead cells from the surface of the skin keeps the thickness of the epidermis about the same at all times. However, the thickness of the epidermis does vary somewhat in different regions of the body. For instance, the epidermis on the palms of the hands and the soles of the feet is thicker than the epidermis on top of the hands and feet. The replacement rates for regions of the epidermis that are subject to relatively constant mechanical abrasion, such as the palms of the hands and the soles of the feet, are generally higher, whereas the replacement rates for areas such as the back and arms are lower. In general, the skin surface is lost by physical contact with material objects in the environment. The cells may be rubbed off by the fabric of our clothes or washed off in the shower. One estimate suggests that the human body may lose 0.3 g–0.4 g of dead skin each day.

There are also external conditions that can bring about local changes in the thickness of the skin. Persistent use, or abuse, of skin in a particular region of the body often leads to a local thickening of the epidermis into what is called a callous. This indicates that the skin is not only dynamic in its normal pattern of growth but also can change that pattern in response to changes in the environment. Callouses are common, and particularly prominent, on hands and feet, which are most often used in human encounters with the physical world.

Another characteristic of keratinoctyes is that they produce lipids that are water resistant (similar to the wax or oil used to prevent wetting of wood or metal). Lipids prevent the incursion of water into the body and reduce loss of water out of the body. Imagine that each time you took a shower or bath your body soaked up the water as if a sponge or that the water in your tissues simply leaked out through the skin. The delicate balance of the water content of bodily fluids would be disrupted totally in either case, and neither we nor any other mammal could survive such fluctuations. The dual capacity of keratinocytes to provide both intercellular adhesion and a lipid-based water repellency makes the epidermis uniquely fitted to protect us against the extremes of wet and dry in the external environment.

Corneocytes dead cells of the outer layers of the epidermis

Germinative cell layer in skin, the deepest living cell layer from which cells of other layers are derived

Basal lamina an assemblage of extracellular matrix materials found in conjunction with epithelia (*see* basement membrane)

When all the characteristics of the epidermis are considered together, it is apparent that this tissue provides the structural elements (cells, proteins, and lipids) and functional attributes (resistance to mechanical disruption, chemical damage, and leakage) that make our skin flexible, durable, and, for the most part, impenetrable to water. However, no tissue (living or dead) is indestructible, not even a tough, membranous envelope such as the skin.

The Dermis Directly beneath the epidermis is the dermis. Two tissue types could not be more different from one another. The cells of the epidermis are held together tightly with little or no space between them. A prominent character of the dermis is that it is fibrous in nature and the cells are associated relatively loosely with one another (see Chapter 4). Dermal cells produce, secrete, and are surrounded by the extracellular matrix. In addition, the dermis has a rich endowment of blood vessels. The main functions of the dermis are to act as an attachment site for the epidermis and as a source for the nutrients and oxygen that go to the epidermis, (Figure 5–4).

The dermis also connects to other types of tissues, ensuring that the skin is attached strongly to the underlying anatomical structures, including muscles and bones. It certainly would be disconcerting to have our skin slipping around over the surface of the body. If the skin, as a separate entity and composed of the epidermis and dermis, were removed from the rest of the body, it could be folded into a neat pile of sheetlike material (the ultimate body suit) and would weigh approximately 2.5 kg–3.0 kg. At its thickest points, probably on the bottoms of the feet, the skin is 5mm–6mm thick, or about the thickness of a telephone cord. The familiar adage about beauty being skin deep is more meaningful with these facts in mind.

Wounds and Wound Healing In the face of physical or chemical damage, the skin has mechanisms to repair itself. Puncture wounds that penetrate the epidermis expose the inside of the body to components of the outside environment. This is always a potentially dangerous situation, involving as it does possible infection and loss of blood and other bodily fluids. For minor cuts, abrasions, and burns, the skin does a remarkable job of sealing itself up and undergoing reconstruction. If a small area of the epidermis is totally destroyed, keratinocytes of adjacent regions will grow into the wound area and establish a new, contiguous epidermis, which in turn will reestablish the stratified cell layers representative of the original tissue (Figure 5–5).

Hair and Nails

So far we have considered the skin as having a fairly uniform sheetlike morphology, but there are a number of

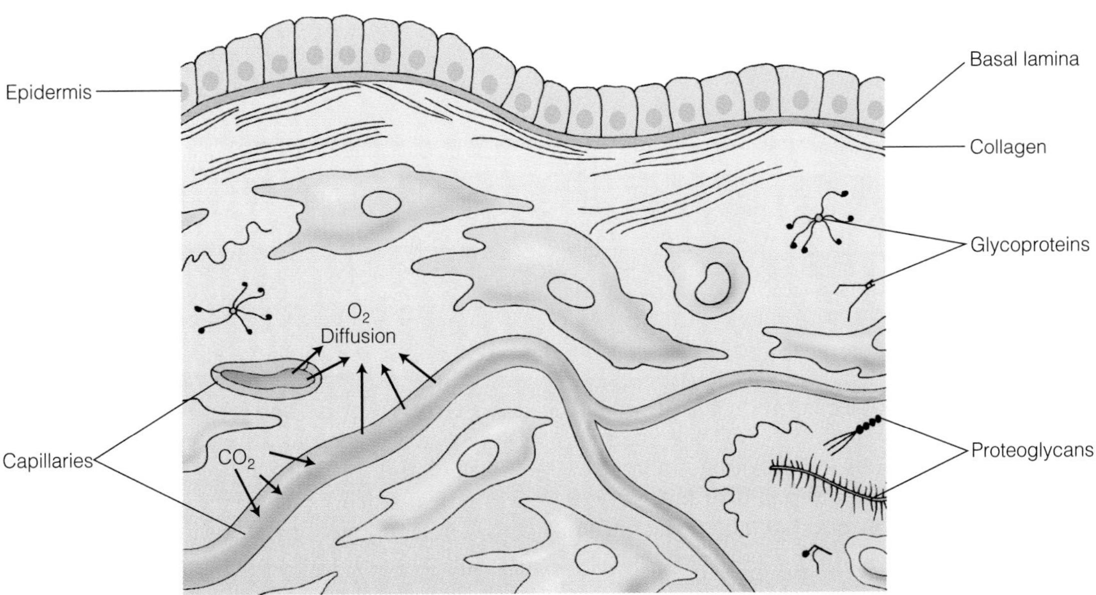

Figure 5–4 The Structure of the Dermis
The dermis is composed of connective tissue cells and the extracellular matrix molecules they produce. This includes collagen, proteoglycans, and glycoproteins. Capillaries traverse the dermis and deliver and exchange oxygen (O_2) and carbon dioxide (CO_2).

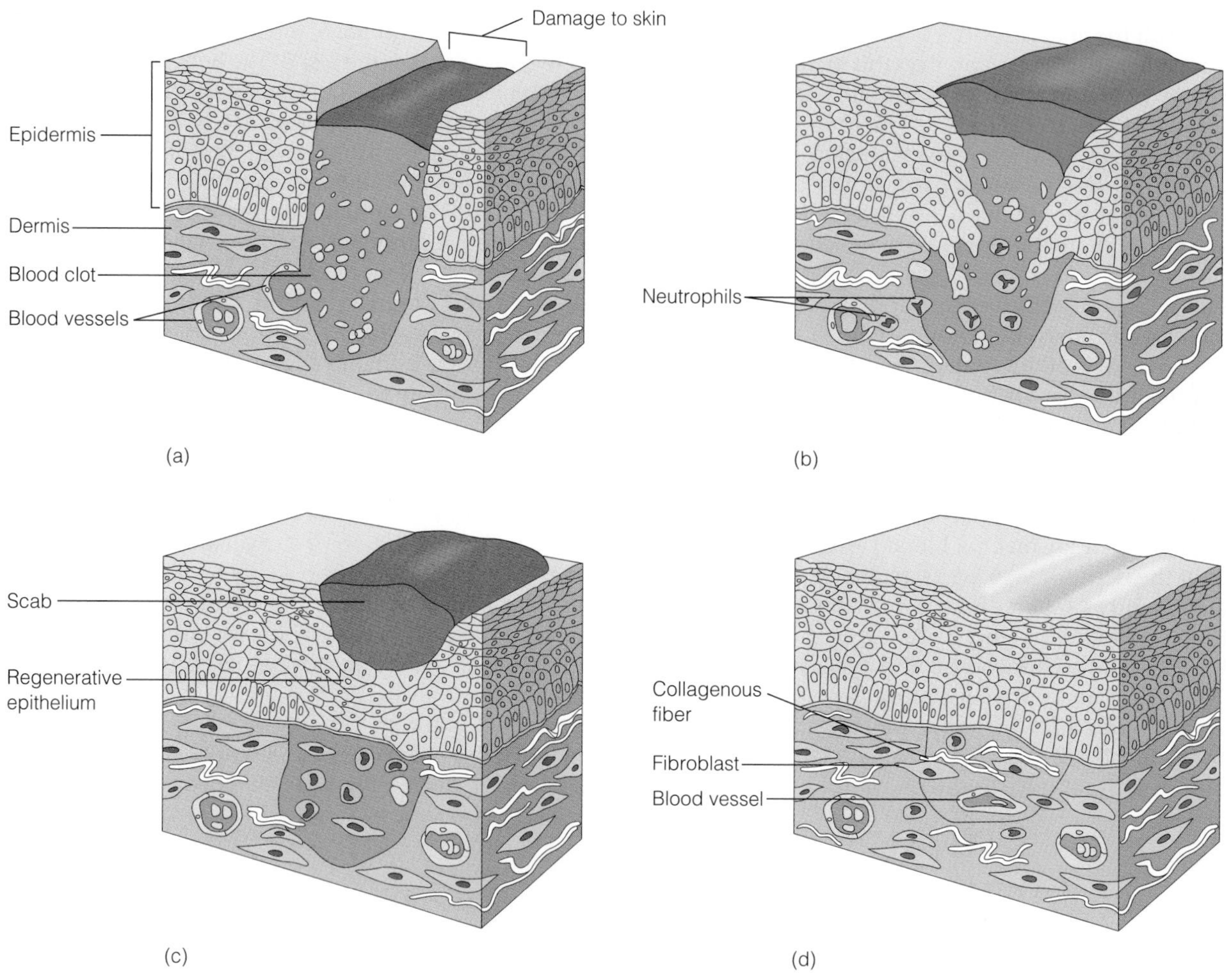

Figure 5–5 Wound Healing
In (a) damage to the skin results in formation of a blood clot (scab). In (b) within a few hours white blood cells (neutrophils) enter the area to destroy bacteria and clear cellular debris. Regeneration of the epidermis occurs (c) from the edges of the wound. In (d) re-establishment of an intact epidermis and dermis once again protects the body from encroaching bacteria or environmental exposure.

specialized structures that do not conform to this image. For example, hair and nails are derivatives of the epidermis, yet these structures are not obviously skinlike in appearance. There is a pattern to hair and nail distribution over the skin surface. An interesting question about the formation of these structures is how the pattern is established. The initiation of hair formation occurs during the first few weeks of human embryonic development. If epidermal-dermal interactions in the skin fail to occur properly during this time, an individual may end up with patchy regions of hair, or no hair at all. This also occurs with fingernails and affects an individual for his or her lifetime.

Hair Growth and Development One of the most easily recognized specializations of the epidermis is hair. As the skin itself, a hair cannot develop or be maintained without interactions between the epidermis and the dermis. These interactions result in the downward growth of the epidermis into the underlying dermis to form a hair follicle. Over the lifetime of a healthy human being, hair will form cyclically from a follicle (Figure 5–6). The physical structure we call a hair, particularly that part of the hair we see on the surface of the skin, is composed of dead cells. However, during its initial growth, a hair is derived from living epidermis. The cells within the growing region undergo differentiation as they move away from the follicle and along the path of the hair shaft. The living layers of a hair associate with pigment-producing cells known as **melanocytes** (to be discussed later in this chapter), which pass pigment into the epidermal cells and give hair its color. Hair is composed principally of keratin, which is one of the hardest and most durable proteins found in humans. If you have ever attempted to shave your face, legs, or underarms with a dull razor, you can appreciate just how tough hair is.

In humans, high-density hair growth is restricted to certain regions of the body, notably the head, underarms, and genital area. This is not true for other mammals, who are usually covered with dense hair or fur. In addition to differences in the density of hair over the human body, the types of hair found in different regions is often of quite different quality (that is, thickness, potential length, and color) as well. The hair on the head tends to be thick and long (or at least each hair has the potential to grow quite long if it is not cut), and hair on the arms tends to be thin and relatively short. The hair in the genital and underarm areas tends to be thick, heavily pigmented, and relatively short. The hair of peoples from different geographical regions and racial backgrounds may also be quite different with respect to cross-sectional shape. The extremely curly hair of individuals of African ancestry tends to be flat in shape, and the very straight hair of some Asiatic groups tends to be round. Wavy hair is often intermediate between the extremes, being ellipsoidal in shape (Figure 5–7).

Melanocytes pigment-producing cells of the body that produce melanin

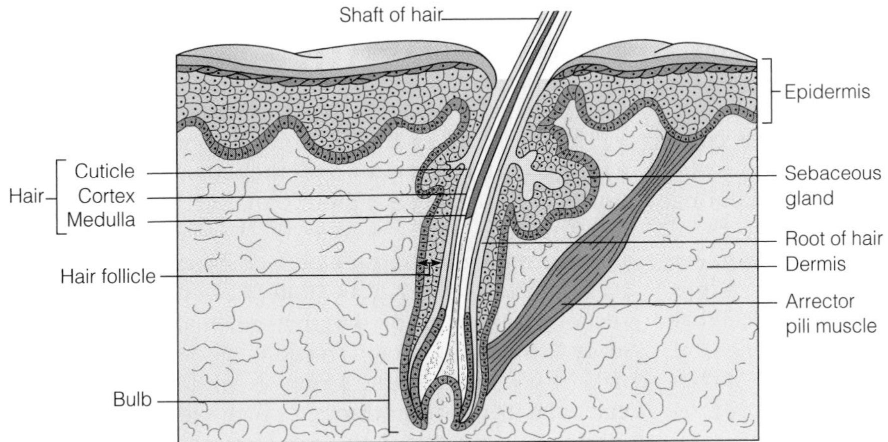

Figure 5–6 Hair and Hair Follicle

Hair is a derivative of the epidermis. This diagram shows a hair, its follicle, and associated structures, including the arrector pili muscle that allows a hair to "stand on end."

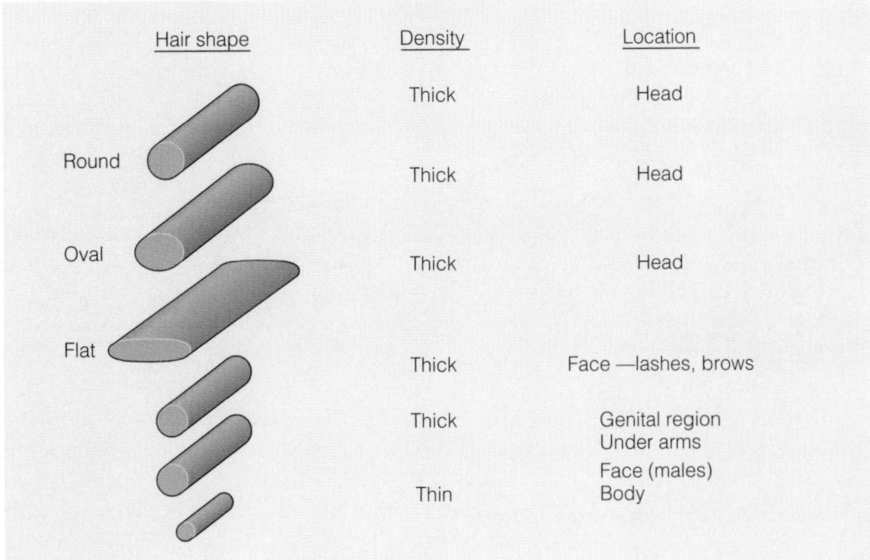

Figure 5–7 Hair Shape, Density, and Distribution

Hair shape plays a role in determining straight, wavy, or curly hair (round, oval, flat, respectively).

Unlike animal fur, human body hair does not serve as body insulation or protection. Our rather sparse hair is mostly vestigial in this regard, and we can certainly live quite normally without it. This is not to say that there is no significance to body hair on human beings. After all, hair mirrors many of the sexual, behavioral, and cultural influences of our place and time. Some types of hair do serve specific functions. The hairs of the nostrils are important because they help filter out particulates as we breathe. The hairs of the eyelashes and brows help reduce the amount of particulate material that could potentially fall into the eyes.

Nail Growth and Development The growth of nails on the ends of fingers and toes of humans is essentially an evolutionary modification of claws found on other types of mammals. Nails are also remarkably similar to virtually all such structures found on the ends of the digits of terrestrial and semiterrestrial vertebrates including birds, reptiles, and amphibians. Nails, as is hair, are derived from the epidermis. In the case of nails, the epidermis and dermis on the top, or dorsal, surface of the fingers and toes interact during embryonic development to specify nail growth in these regions. Unlike hairs, which grow out in the shape of a filamentous cylinder, nails grow out in the shape of a gently curved, relatively large plate (Figure 5–8). A nail is composed almost entirely of keratin, which, as in hair, is the principal protein expressed by nail epidermal cells during terminal differentiation.

The epidermis of the top surface of a finger curves under at the cuticle and is in contact with the supportive dermis of that region. It is underneath the area of the cuticle that the nail is made. Nails grow outward along the length of the dorsal finger tips. The region of growth from which a nail extends is called the nail matrix, and the surface of the finger beneath an existing nail is referred to as the nail bed. The nail is translucent because no pigment is included in cells during nail formation. For anyone who has lost a nail (remember the hammer stroke that missed or the chest of drawers that caught your finger between it and the wall?), the rate and process of growth can be observed. It takes weeks for a nail to grow back completely. The slow growth of a nail during regeneration demonstrates the continuous nature and orderliness of normal nail growth. Nails are growing all the time, damaged or not. This is why we have to trim our nails occasionally lest they interfere with the use of our fingers.

One of the most important functions of nails is to protect the tips of our fingers and toes. Because our hands are always involved in the manipulation of objects around us, the tips of our fingers tend to get in harm's way more often than other parts of the body. The nails also provide a rigid, thin edge that helps in picking up small objects that would otherwise resist the softer, more deformable skin of the finger tips. Human beings have nails on opposable fingers and thumbs, an important structural characteristic of the human hand. Opposable digits have been of immense significance in human evolution, and nails at the ends of fingers enhance the range of size and composition of materials that we are able to manipulate.

The skin, hair, and nails comprising our body's surface are structural manifestations of the action of natural selection favoring this type of protection. Each of these anatomical structures helps prevent mechanical and chemical insults from being transmitted to the interior, where real harm to the body could be done. However, these are not the only features of the skin that are important physiologically to the human body. The skin is also a mediator of signals from outside the body to inside and provides glandular mechanisms through which substances from the inside are secreted onto the outer surface.

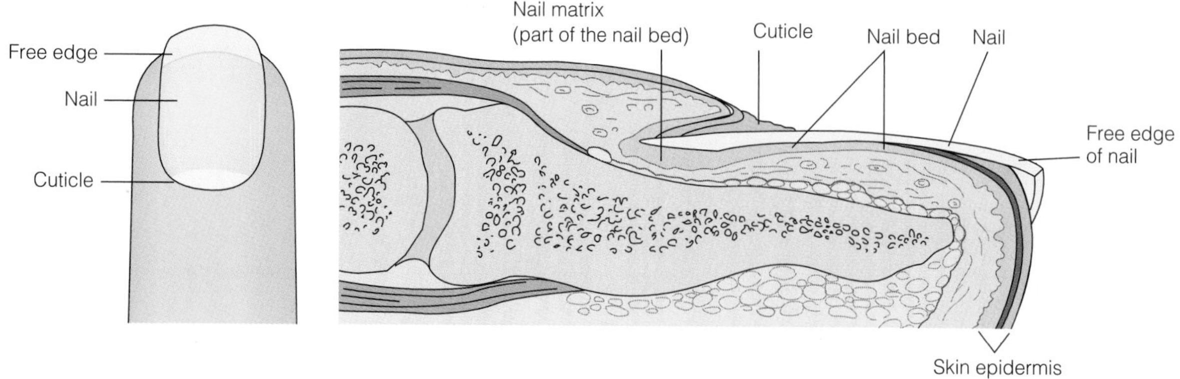

Figure 5–8 Fingernail
The fingernail is a derivative of the epidermis. Each nail grows from the nail matrix, out over the nail bed. The free edge protects the finger tips.

Glands of the Skin

There are basically two types of glands associated with the skin, **sweat glands** and **sebaceous glands.** (Figure 5–9). Sweat glands and sebaceous glands are both **exocrine glands.** Exocrine glands produce secretions that are exuded onto an epithelial surface, in this case the skin. The products of these glands reach the surface of the body by a duct or tubular system (Figure 5–9). The epithelial cells of these glands are derived from and continuous with the epidermis so the sealing of the body surface is not compromised. Sweat glands are involved predominantly in the secretion of water and solutes and sebaceous glands secrete thick, oily substances that provide lubrication for and water repellency to hair and skin. Both types of glands are distributed over nearly the entire body surface. Sebaceous glands are found predominantly in association with the epidermis of the hair canal.

Sweat Glands The two types of sweat glands are **eccrine sweat glands** (the most common type) and **apocrine sweat glands.** There are tens of thousands of these glands, each embedded deep in the dermis and reaching the surface through a tiny twisted tube (Figure 5–9). The distribution of sweat glands is not uniform over the body surface. Some regions of the skin have many more sweat glands than others. The palms of the hands and the soles of the feet have gland densities reaching hundreds of glands per square centimeter. Other regions, such as the back and buttocks, have far fewer. Some regions, such as the lips, nail beds, clitoris, and penis, have no sweat glands at all. Apocrine sweat glands are associated with the hair follicles principally in the skin of the arm pit. Apocrine sweat glands begin to function around puberty and their secretions are thicker and less watery than those of the eccrine type. Eccrine sweat glands develop independently of other specialized structures of the skin.

One of the most important functions of eccrine sweat glands is in the homeostatic temperature regulation of the body (discussed in Chapter 3). The surface area of the human body is quite large (approximately 2 m^2–3 m^2). This surface is subjected continuously to temperature changes in the environment. The body is either giving up heat to the environment or accumulating heat from the environment by conduction, convection, evaporation, and/or radiation. Sweat glands control the amount and rate of secretion of water from the body and provide an evaporative mechanism by which the internal heat can be altered. By retaining water the body is essentially holding on to an internal source of heat that helps in maintaining a nearly constant body temperature.

The human body responds to an increase in heat by stimulating the sweat glands to secrete water onto the surface of the skin. As the water in sweat begins to evaporate, heat is carried away from the body. As an example, anyone who has ever come out of a swimming pool wet and dripping on a hot day can appreciate the initial coolness associated with the water clinging to the skin. It is the evaporation of this water, carrying away heat from the body surface, that makes us feel cool. When the water is gone, the body begins to heat up once again.

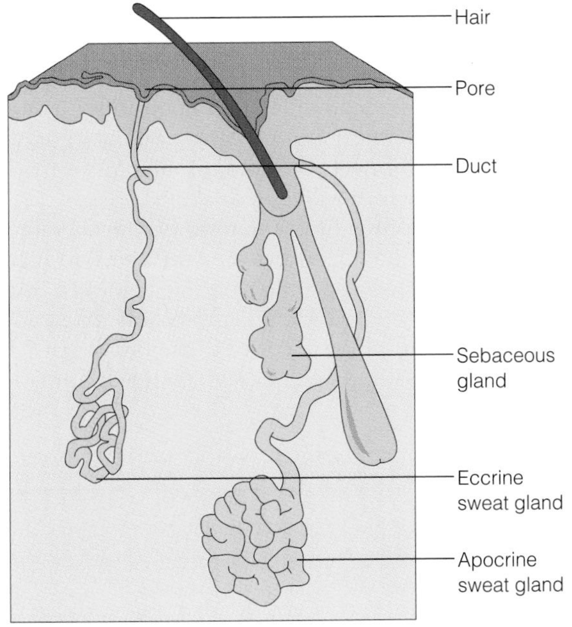

Figure 5–9 Glands of Skin
There are two types of skin glands—sweat glands and sebaceous glands. Eccrine sweat glands produce watery sweat, apocrine sweat glands produce a thicker fluid and are often associated with hair follicles.

Sebaceous Glands Sebaceous glands are of less critical physiological importance than the sweat glands, but they

Sweat glands exocrine glands secreting sweat or other substances (*see* apocrine and eccrine sweat gland)

Sebaceous glands exocrine glands of skin that produce oily secretion released into hair follicle

Exocrine glands glands with ducts that deliver secretions to the surface of an epithelium

Eccrine sweat glands exocrine glands producing watery secretion of sweat commonly distributed over the surface of the entire body

Apocrine sweat glands exocrine gland producing sweat and a thick secretion

do help lubricate and protect skin and hair (Figure 5–9). These glands produce oils that soften the skin and lubricate hair as it emerges from the hair canal. This oily product is called **sebum** and is composed of a variety of substances including lipids, fatty acids, cholesterol, and dead cells. Pimples, blackheads, whiteheads, and zits are all names given to mild forms of skin problems associated with sebaceous glands. In many cases, these conditions can be corrected with diligent application of over-the-counter medications and good skin hygiene. However, there are more serious problems associated with abnormal function of sebaceous glands. One of them is **acne.** In this condition, there is a tendency for the sebaceous glands and pores to become plugged with sebum. Because the gland continues to produce its secretions, regardless of whether the channel to the surface is blocked or not, there is a net buildup of these substances. Bacteria can grow very well on the waxy materials trapped in the gland, so infection and inflammation can ensue. This causes an increase in the size of the gland and distorts and displaces the skin surrounding it. This situation is worst during puberty, particularly in males because sebaceous gland activity and the production of sebum is enhanced by the male hormone **testosterone.** Generally, when puberty ends, so does the acute spread of acne. However, the scars remain, mute testimony to the uneasy relationship between skin and hormones during human growth and development.

Pigmentation of the Skin

Four major types of pigments are found to influence the color of human skin. Each type gives a different coloration, so the combination of these types in different amounts and locations in skin provides a wide range of color possibilities. This is easily observed in a population of human beings, in which a natural range of colors from black to white occurs, with brown, yellow, and red hues in between. The classes of pigments involved are (1) red pigments of oxygenated hemoglobin, (2) blue pigments of deoxygenated hemoglobin, (3) the yellow carotenoids, and (4) the brown-black melanins (Table 5–1).

Oxygenated and Deoxygenated Hemoglobin The colors imparted by the hemoglobins are associated with the arterial, venous, and capillary blood flow in the vessels of the dermis, which, unless masked by the epidermal distribution of the other two pigment types, give a pink (arterial) or blue (venous) hue to a translucent skin. The rosy color of a blush on our cheeks, elicited by embarassment or excitement, represents the increased flow of oxygenated hemoglobin in blood cells through capillary beds of the skin. Because veins tend to be close to the body surface, that is, just beneath the skin, there can be an eerie blue hue to the skin of extremely pale individuals due to deoxygenated hemoglobin in the blood cells. This used to be a sign of aristocracy, clearly emblematic of a "blue-blood."

Carotenoids The carotenoids are water-insoluble, unsaturated hydrocarbon molecules and must be obtained by the body from the diet. They are related in chemical structure to vitamin A, for which they are the precursor molecules. Vitamin A is a very important factor in the health of the skin. A prime example of a carotenoid is beta carotene. This pigment gives rise to the yellow-orange color of carrots. The carotenoid molecule is a precursor for the yellow pigments found in the human skin. A color range of yellow to light red is associated with different types and concentrations of these pigments in the skin.

Melanin and Melanocytes Without doubt, the most important pigment of the human skin with respect to both color and protection is **melanin** (Table 5–1). An explanation of why this is so provides the opportunity to describe a complex biochemical and physiological chain of events that ultimately connects the color of humans to the environments in which they live.

Probably one of the most interesting and generally misunderstood phenomena associated with the skin pigmentation is tanning. Most people have a tendency to think of it as simply a way to enhance their physical appearance or attractiveness and not as a critically important part of the function of the skin. As previously described, human skin

Table 5–1 Pigments of the Skin and Body

PIGMENT	LOCATION
Oxyhemoglobin	Oxygenated blood cells (arteries)
Deoxyhemoglobin	Deoxygenated blood cells (veins)
Carotenoids	In skin, vitamin A-related
Melanin	In skin, formed by melanocytes and pigment epithelial cells of the eye

Sebum oily secretion of sebaceous gland

Acne condition of skin in which sebaceous glands are infected and inflamed

Testosterone steroid hormone, associated with development of male secondary sexual characteristics

Melanin brown/black pigment of skin produced by melanocytes

is composed of two basic cell layers, the epidermis and the dermis. However, there are special cell types within these layers that provide unique services to the skin. One of these is the melanocyte. Melanocytes are melanin pigment producers, and they are found in both the dermis and epidermis. Melanin is packaged inside melanocytes in an organelle called the **melanosome.** The melanocyte has long, fingerlike cellular extensions that can pass through the basement membrane that separates dermis and epidermis and between epidermal cells (Figure 5–10). These extensions directly contact basal cells and keratinocytes. Pigment-bearing melanosomes are transferred from melanocytes to epidermal cells and imbue them and the entire skin with characteristic color. One melanocyte may serve several keratinocytes, forming an epidermal melanin unit. (Figure 5–10).

What role does this pigment serve in the function of the skin? A major role is protection of the body from radiation. There are a number of specific kinds of radiation in sunlight that are damaging. This includes X rays, ultraviolet (UV) rays, and microwaves. These particular types of radiation can be very dangerous to living organisms. Any radiation that can penetrate all the way through a human body (X rays) is certainly a potential hazard, especially if it also damages cells and molecules as it passes through. Likewise, any radiation that selectively heats up water molecules to the boiling point (microwaves) can have disasterous effects on a body composed of 60%–70% water. Fortunately, direct exposure to these particular wavelengths of radiation is very infrequent and, in the case of X rays, usually only under the watchful eye of trained technicians.

However, this is not the case for ultraviolet radiation. UV rays are a natural part of the sun's light and impinge on us through Earth's atmosphere each day, even through a cloud cover. It is this type of radiation that exposes skin to a serious problem. UV radiation is energetic enough to cause cell and tissue damage (remember the last time you were sunburned?), and there is essentially no way to avoid it if we are to participate in outdoor activities during the day. Enter the melanocyte. A fundamental molecular property of the pigment melanin is that it can absorb UV light. This means that the more melanin you have in your skin, the more protection you have against the harmful effects of sunlight. The response to melanocytes to sunlight, particularly to the intense sunlight of a midsummer's day, is to increase synthesis of melanin, increase the number of melanosomes, and increase the rate of transfer of melanosomes into epidermal cells.

Tanning is a targeted process. If you hang your arm out of the window of your car for more than a few minutes as you drive on a sunny day, you will end up with a tan arm. The unexposed portions of your body will remain unaffected. The more you expose, the more you tan. The skin is simply responding to overexposure to UV light. This process of tanning is also time dependent. As long as an individual is consistently exposed to sunlight (hours a day for some) the body will remain tan. When a tan individual reduces or eliminates exposure time in the sun, the melanocytes revert to a less active state. Less pigment is made, fewer melanosomes are transferred to the epidermis, and gradually the skin pigmentation returns to a paler, baseline density. The pigment is lost as cell death of maturing epidermal cells occurs and the top layers of the skin are sloughed off. There is danger associated with tanning. Extended exposure to UV radiation causes long-term damage to the skin and may lead to cancer. As with many activities in which we partake, tanning should be done in moderation and with the use of appropriate sunscreens to protect the skin.

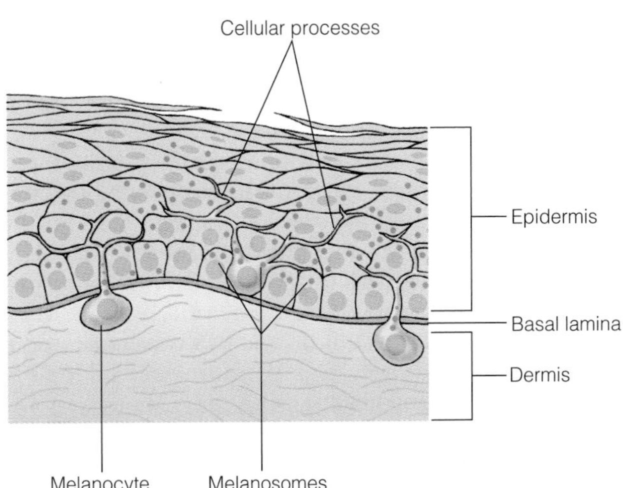

Figure 5–10 Melanocytes and Melanin in the Skin
The dark color of skin is determined by the amount of melanin deposited in the epidermis by melanocytes.

What about people whose skin is naturally heavily pigmented? This is obviously the case for people indigenous to Africa and their descendants in other parts of the world. The difference in color between a "black" person and a "white" person does not correspond to the number of

Melanosome organelle found in melanocytes in which melanin is stored; passed from a melanocyte to epidermal cells

melanocytes found in the skin. Rather, the melanocytes of a black individual produce more melanin and transfer more melanosomes to the epidermis than do melanocytes of white individuals. In addition, the transfer occurs regardless of light conditions.

The natural regime of sunlight in equatorial Africa (now, as in the past) is one in which the intensity of radiation is very high. A naturally high level of pigmentation in the skin of people living in that environment was then, and certainly is now, beneficial because of the UV radiation barrier it provides. Because *Homo sapiens* and its predecessor species originally evolved in Africa, this constant or **constitutive condition** of pigment formation must have been altered in subpopulations of early humans as the northern migrations of people out of Africa occurred during the last several hundred thousand years. There is an interesting conjecture as to how the constant expression of high levels of melanin in skin changed to an expression that is regulated by exposure to sunlight in human populations separated over time and geographic distance. That conjecture involves the production of vitamin D.

Skin and Vitamin D Vitamin D is a **hormone** crucial for calcium utilization and bone growth in humans. Biochemically, it is derived from reactions between sunlight and molecules related to cholesterol found in the skin. If sunlight is consistently intense, as it is around the equator of this planet, enough light can penetrate even a heavily pigmented skin to stimulate the formation of vitamin D from its cholesterol-like precursor molecule. If sunlight is much less intense or seasonal, as is found in the northern latitudes of Earth, heavy pigmentation of the skin could effectively block the passage of light to the living cell layers and reduce or eliminate the production of vitamin D. Natural selection would favor the survival of individuals with less pigment, because light would be able to penetrate and convert precursor molecules in the skin to vitamin D, a necessary ingredient of good health.

Assuming that this reduction in pigment synthesis and utilization reflected genetic differences among those individuals, this advantageous trait could be passed on to offspring and so, over time, result in the evolution of less pigmented subpopulations of humans. While this scheme is only speculative, it does offer us a chance to think about how human populations are related to each other. In the long run, many of the differences we see among people today, such as skin color or the morphology of hair, can be attributed to the natural selection of what, by chance, turned out to be the advantageous adaptions of some of our ancestors to the complex, changing environments in which they lived.

Sensory Systems of the Skin

There are probably few who would now doubt the importance of the skin in the general protection of the body from environmental insult or its role in preventing loss of our internal fluid contents. However, there are properties of skin that are more subtle in nature and require long-range interactions with other types of tissues. The sensory systems of the skin fall into this category. Close your eyes. How is it that we know when and where we contact an object in the space around us, or when and where we sense heat, cold, or air movement in our proximity? We do so using special sense organs and cells that are distributed in the skin over the entire surface of the body.

These sense organs and cells are receptors for and transmitters of information from the environment. Different types of receptors respond specifically to touch, light, and heavy pressure, and can discriminate between hot and cold over a wide range of temperatures. The receptors are able to do this by transforming the physical signals they receive from the environment into nerve impulses sent to the brain. Connections and integration of the information in the brain determine whether or not we cry out in pain or murmur in pleasure. The types of receptors found in human skin are shown in Figure 5–11 and are described in detail in Chapter 18.

THE SKIN AND IMMUNITY

The primary defense of our body begins with the skin. The skin contains a wide range of mechanical and chemical features for this purpose, from waterproofing to radiation protection. The skin also has a secondary set of characteristics related to defense, which involves specialized cells of the immune system. The cells that are responsible for this immunological defense are called **Langerhans cells** (Figure 5–12). They are found in the epidermis among the keratinocytes, but they are not of epidermal origin. Their origin is the bone marrow and, during their differentia-

Constitutive condition a condition in which the expression of a trait is continuous and not controlled by external stimulation, e.g., pigmentation of skin

Hormone a substance produced by endocrine glands and secreted directly into bloodstream, which elicits specific responses from target cells

Langerhans cells immune reactive cells of the skin; derived from bone marrow cells

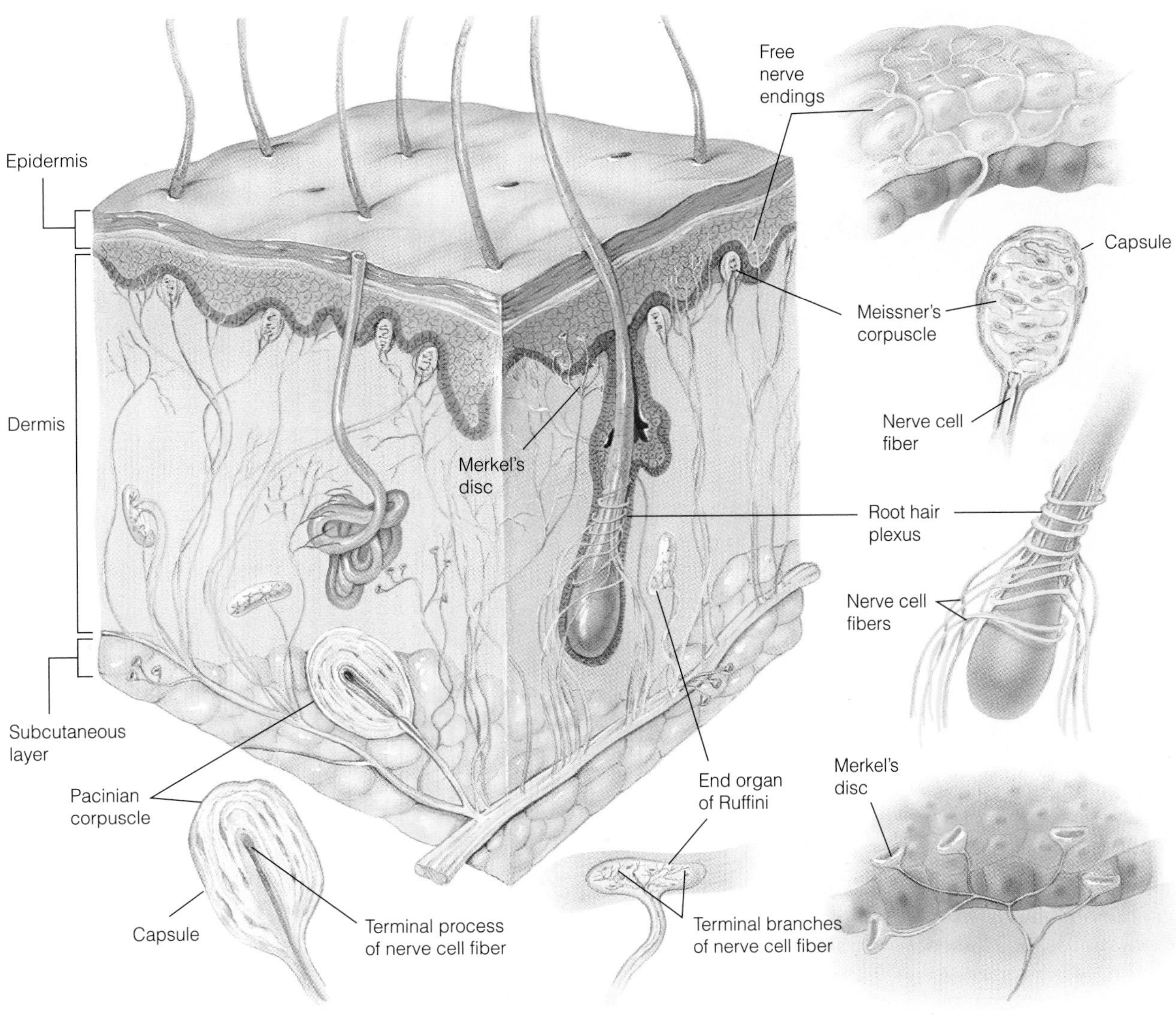

Figure 5-11 Receptors in Skin
The skin has within it a multitude of nerve receptors. They receive stimulation from pressure and touch (free nerve endings, Meissner's corpuscle, hair root plexus, pacinian corpuscle) and hot/cold (end organ of Ruffini, Merkel's disc).

tion, they migrate to and from the skin where they associate with the epidermis and specialize in detecting foreign materials. In this capacity, they recognize invading microorganisms and toxic substances and act to inform and activate other defensive cells of the immune system. The immune system is composed of a multitude of cell types and is continuously poised to respond to nonself substances by mediating reactions that will lead to the destruction of these invaders and eliminate them from the body.

Diseases

Abnormalities of Growth When things go wrong in the skin, it usually does not take long to find out. A scratch or minor burn that becomes infected is obvious and can be

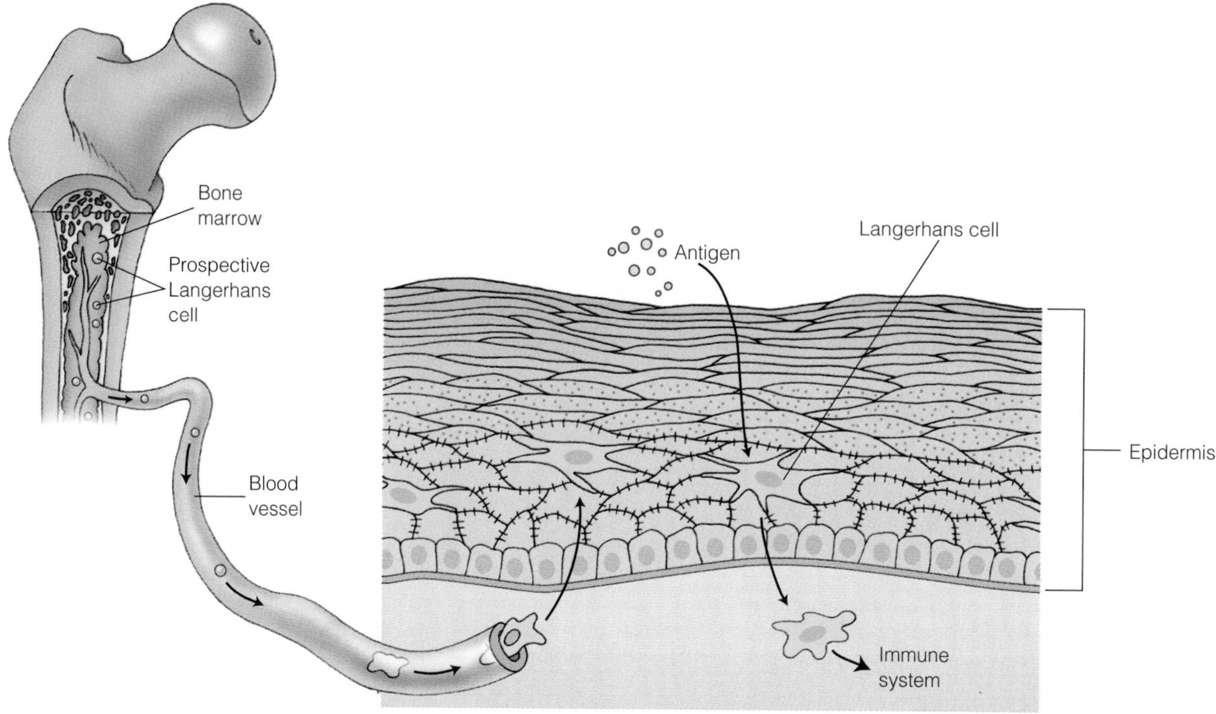

Figure 5–12 Langerhans Cells
Langerhans cells have their origins in bone marrow. They are part of the human body's immune system. They migrate to the skin where they locate in the epidermis. Upon exposure to foreign antigens, they return to the immune system as part of the body's defenses.

treated promptly and effectively. An allergic response to chemicals or objects in contact with our skin, such as pollen, fabrics, or cosmetics, is quickly recognized and remedied. Some conditions of the skin, such as acne, are generally bothersome but not life threatening. When something goes seriously wrong, however, we usually need immediate medical attention. Let us consider some of the well-known and serious problems associated with the skin (Table 5–2).

Cancer Skin cancer is one of the most prevalent types of serious skin disease. Different types of cancers develop from different kinds of skin cells. Cancerous transformation of the cells from the epidermis are, in general, referred to as **carcinomas.** If the basal cells of the germinative layer of the epidermis transform to become cancerous, the cancer is called a **basal cell carcinoma.** If other, more superficial, cells of living layers of epidermis are transformed, they may form **squamous cell carcinomas.** When melanocytes are transformed, they become **melanomas,** which, although they are not carcinomas, are among the most dangerous types of cancers. The hallmark of a cancer is the rate at which the cells composing it grow and the extent to which that growth results in the invasion or spread of cells into adjoining normal tissues. Cell proliferation is rapid in many types of cancer, and the cells are usually abnormal in morphology. Fortunately, skin cancers are among the easiest for us to detect, and their quick removal is often a simple process (Figure 5–13).

Carcinomas malignant cancers originating from an epithelial cell type

Basal cell carcinoma malignant cancer cell arising from basal cells of the epidermis

Squamous cell carcinomas malignant cancer cells arising from the transformation of squamous or elongated cells of the epidermis

Melanomas malignant cancer cells derived from the transformation of a melanocyte

Table 5-2 Damage to and Diseases of the Skin

DISEASE OR DAMAGE	SYMPTOMS
Cuts and scrapes	Bleeding, infection, inflammation
Allergic responses	Itching and inflammation
Acne	Blockage of sebaceous gland, infection, and inflammation
Cancer (carcinomas and melanomas)	Rapid, abnormal growth, tumors, invasion, and metastasis
Basal cell carcinoma	
Squamous cell carcinoma	
Melanomas	
Burns (first, second and third degree)	Damage to epidermis, dermis, and underlying connective tissue and muscle
Congenital malformations	Development of abnormal skin, hair, abnormal nails, and abnormal skin glands

Burns

The skin is truly multifaceted in its capacity to adapt and repair itself. Without the protective embrace of the skin a human cannot long survive. One of the most dangerous situations for the body is extensive destruction of the skin, as occurs with bad burns. The extent to which skin is damaged by burning is classified as first degree, second degree, or third degree.

First degree burns affect only the epidermis. **Second degree burns** affect both the epidermis and dermis but not the underlying connective tissue or muscle. **Third degree burns** involve the deep penetration and destruction of the integument including underlying layers of non-skin tissues. The percentage of the body's surface that is burned is also a crucial factor. Third degree burns over a large percentage of the body expose unprotected regions to just the kinds of problems that the skin was meant to prevent. Aside from the pain, the most serious problems are **dehydration** and the potential introduction into the body of dangerous chemical, radiation and/or microbiological agents. Assuming that an individual was to initially survive a severe, near full-body burn, there is another problem. Can the skin grow back? The answer to this question is yes, but very slowly. This is because the new skin must grow from the edges of the skin that was not damaged in the first place (Figure 5–5).

Burns and Test Tube Skin Enormous strides have been made in burn therapy in the past few years. One of the most interesting is the use of keratinocytes grown in culture in the laboratory to replace the ones lost in the burn. Pieces of skin from the victim can be used as a source of cells grown in the laboratory. The growth of these cells in culture (called *in vitro* growth) results in large sheets of cells held together by the same kinds of adhesive junctions discussed in Chapter 4 in reference to the strength and tightness of an intact epidermis. These sheets of keratinocytes can be transplanted back onto the burn victim and can help reestablish the protective barrier separating the hostile

First degree burns damage to skin from heat directly affecting only the epidermis

Second degree burns damage to skin from heat affecting both the epidermis and underlying dermis

Third degree burns severe damage to skin and underlying tissues (connective tissue, muscle) from heat

Dehydration loss of water; drying out

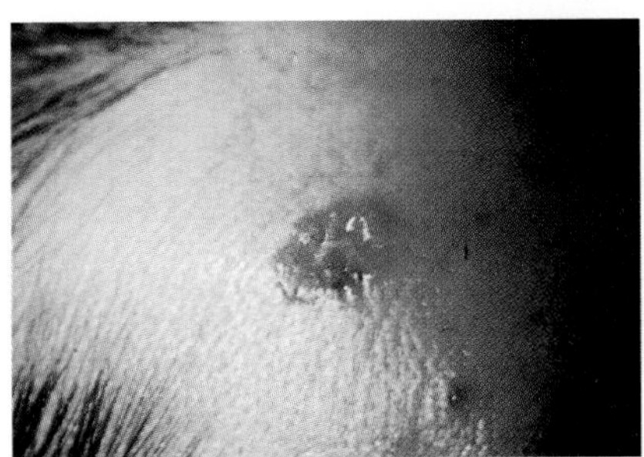
(a) Basal cell carcinoma

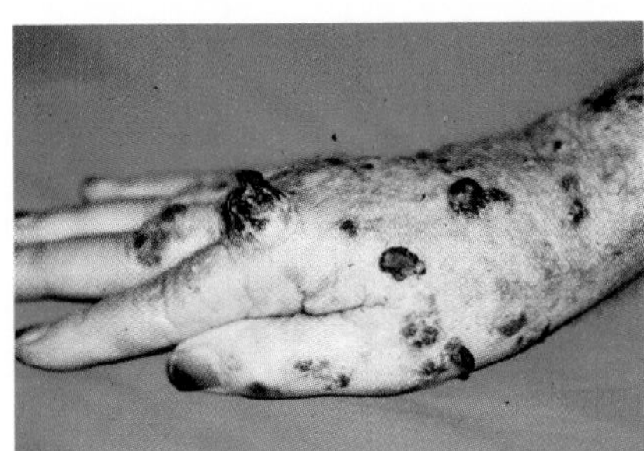
(b) Squamous cell carcinoma

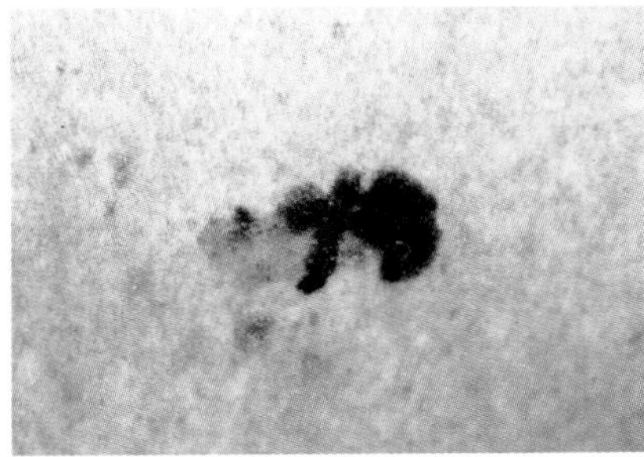
(c) Melanoma

Figure 5-13 Skin Cancers
The cancers of skin are dangerous if left untreated. Severe damage to skin occurs with rapid growth of these types of cancer. Melanomas, if left untreated, are lethal.

outside environment and the life-sustaining inside of the body (Figure 5–14). The advantage of this procedure is that it uses healthy cells of the individual burn victim, and thus, avoids the complications that arise in the potential immunological rejection of foreign skin cells from donors.

CULTURAL CONSIDERATIONS

For all the importance that the skin and its derivatives have for the biological integrity of the body, we are generally more concerned with how they look than how they work. This is certainly not unexpected, nor unusual. If the skin or any other part of the body is in generally good condition, our concerns are focused less on health and more on other aspects of our lives. The various tissue and organ systems of the body normally take care of themselves quite well. However, skin is unique among organ systems because it is exposed on the outside of the body, is exposed to our own inspection, and is in easy view of others. This fact has provided us with the opportunity to develop ways to change its appearance without interfering with its function. The cultural significance of personal appearance cannot be overstated. This is reflected historically in a diversity of traditions, arising first in ancient cultures in which body ornamentation, social behavior (for example, mating and combat), and religious ritual were linked (Figure 5–15).

Modern societies, East and West, are the children of these traditions as well. This is demonstrated in the economic success and influence of industries that cater to the

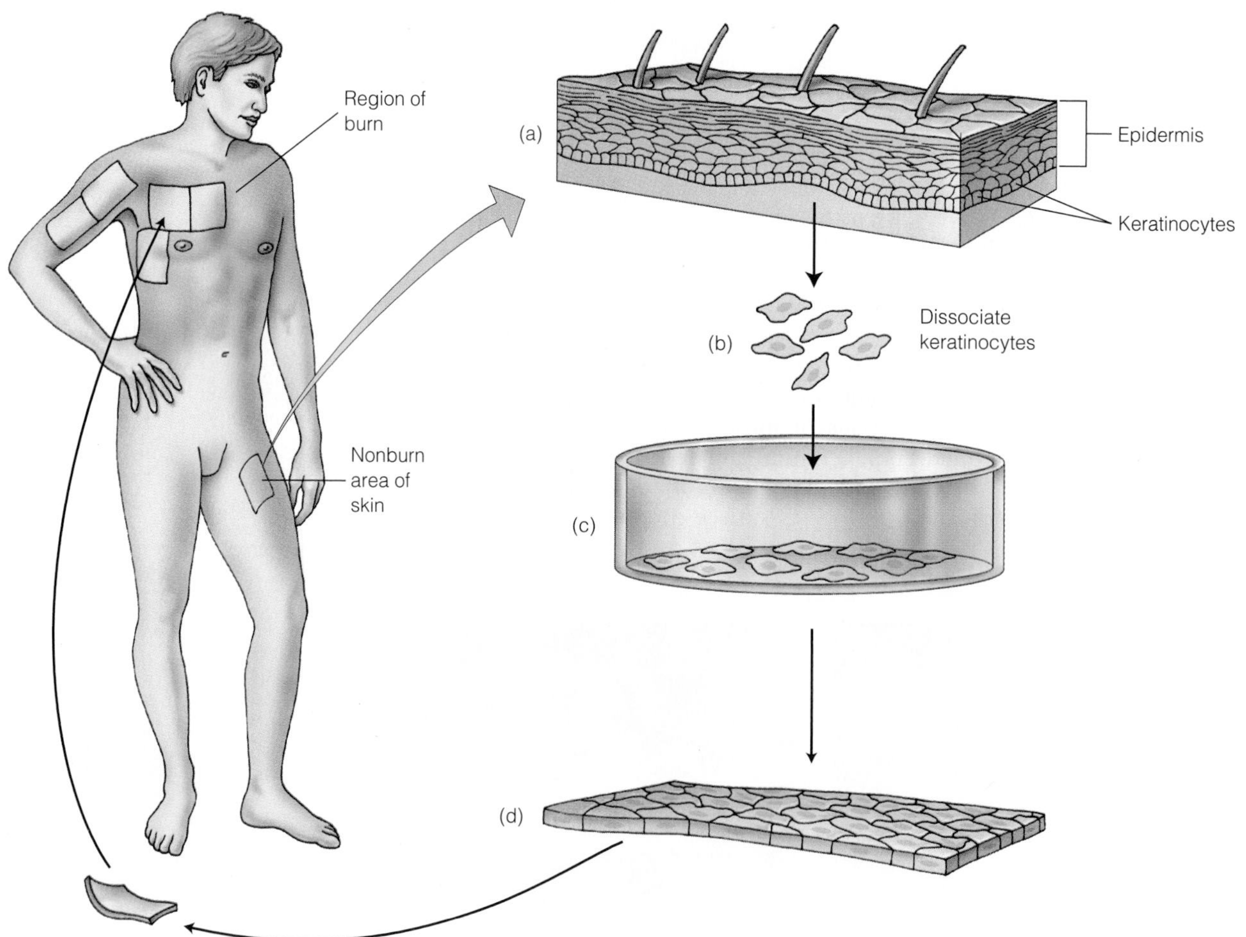

Figure 5–14 Replacing Damaged Skin with Cells Grown *in Vitro:* Replacement of the epidermis of skin begins with a sample of the victim's own normal, undamaged skin (a). The epidermis is dissociated (b) and keratinocytes are grown in culture (*in vitro*) for 1–3 weeks (c). The sheet of cells that forms in culture (d) is placed back over the burned areas and grows again on the body.

health and beauty of skin, hair, and nails. Cosmetics are used to color the skin, lotions to soften and heal it, soaps to clean it, and clothes and jewelry to accentuate and style it. Hair and nails are likewise colored, curled, clipped, painted, and styled to fit our desires and personalities (or, at least what we hope fits our personalities). All these manipulations and changes, which are so important to individual self-esteem and social standing, take place on the surface of an organic canvas of dead cells only a few millimeters thick. It is through our skin that we meet the world both as biological and social entities, and it is fortunate that this thin, translucent barrier is up to the task.

Figure 5-15 Skin Modifications—Cultural Considerations

The skin and its derivatives are an organic canvas for coloring, curling, camouflaging, decorating, and protecting in many and unusual ways.

Summary

The architecture of the skin is multifaceted. Skin is composed of two layers, the epidermis and the dermis. The epidermis is a keratinized stratified epithelium that serves a major role in the protection of the body. It does so by establishing a flexible, durable, and self-renewing layer that prevents water loss, radiation damage, and penetration by outside agents. The dermis is a connective tissue type. It serves as a substratum for the attachment of the epidermis and is invested with blood vessels and nerves, which allow it to play a supportive role in gas exchange, nutrient delivery, and sensory reception.

Human skin has a number of specializations, including

hair follicles and hair, nails, sweat glands, and sebaceous glands. Each of these specializations is a derivative of the epidermis, but each also requires direct interactions with the underlying dermis. Damage to the skin, particularly the epidermis, is repaired by local cell proliferation and migration into the site of damage. Extensive damage is more difficult to repair and takes a longer time to accomplish. Burn damage to the skin may be augmented by new techniques of *in vitro* cell culture, which replaces damaged epidermis with sheets of cells grown in test tubes and culturing flasks. Cancers of the epidermis are generally referred to as carcinomas. Other types of cancer can occur in the skin. One of the most common is melanoma. Skin cancers are usually treated successfully because they are easy to detect.

The phenomenon of tanning is the result of the production and transfer of the pigment melanin into the epidermis. Other pigments, including carotenoids and hemoglobin contribute to the coloration of human skin. A suggested evolutionary significance for pigmentation is protection from UV radiation. As a result of the protective functions of skin, skin plays a major role in general human health. The skin has both biological and sociocultural significance.

Questions for Critical Inquiry

1. How has skin evolved differently among different animals? What is the basis for these differences?
2. What mechanisms does skin utilize to protect against the following
   (a) Water loss (or incursion)?
   (b) Radiation?
   (c) Contact with rough, abrasive surfaces?
3. How is that skin is both alive and dead? Of what advantage might this be?
4. How might skin color among different geographically isolated groups of humans be changed and for what reasons?

Questions of Facts and Figures

5. What are the two main layers of the skin?
6. What cells are responsible for formation of pigment within the epidermis?
7. What are the principal components of the extracellular matrix of the dermis?
8. What are the functions of human hair?
9. What are the two major types of glands in the skin? How do they function? What do they produce?
10. What senses are associated with the skin?
11. Does the skin have an immune function? If so, what cell or cells are involved?

References and Further Readings

Bereiter-Hahn, J., Matoltsy, A. G., and Richards, K. S. Eds. (1986). *Biology of the Integument, Vol. II, Vertebrates*. Berlin: Springer-Verlag.

Brown, C. H. (1975). *Structural Materials in Animals*. Belfast, Ireland: Pitman Publishing.

Sherwood, L. (1993). *Human Physiology,* 2nd ed. St. Paul, MN: West Publishing.

CHAPTER 6

The Skeletal System

INTRODUCTION

Imagine the excitement of being at the opening of an ancient Egyptian tomb. For weeks you have watched all the work at the excavation site focused on clearing a path to the doors. The blocks of the stone doors are enormous and intricately laid, suggesting workmanship of great skill and artistry. What were the ancient people entombed in this burial site like? How did they live? How did they die? Much of what will be discovered about these ancient human beings will be derived from the anatomy of the remains, and the most permanent elements of the human body are the hard and durable bones of the skeleton. This chapter introduces details of the development, anatomy, and physiology of bones and their integration into the remarkable framework of the vertebrate skeleton.

THE PAST AS PROLOGUE

If this were an adventure story, the scene might be taking place in a north African desert at the doors of a long-forgotten chamber in a pyramid. The fantasy of such tales has certainly been told before, and the discovery of tombs, such as that of the Egyptian pharaoh Tutankhamen early in the twentieth century, have an aura of romance and danger surrounding them (Figure 6–1). But the questions of what ancient people were like, and how they lived and died, are important to us for many reasons. More than simple curiosity about the past and our origins has motivated much of the intellectual inquiry of biology, anthropology, and history. Keys to the origins of life, the beginnings of civilization and culture, and the advance of human science and technology are revealed in the rare and unique sites where pieces of the past are brought to light. What do we know of the anatomy and physiology of the human beings of the past? How are they different? Not much remains of the soft tissues of a human body even a few weeks after death (unpreserved by modern techniques), let alone after hundreds or thousands of years. What does remain is very important, because it allows scientists and scholars to cast an inquiring, analytical light into the dark and otherwise inaccessible corners of the human past. The sole remains of the bodies from ancient tombs and grave sites may well be only dry and brittle bones.

Figure 6–1 Ancient Bones

SKELETON

The organization of the human body (and the body of all vertebrates) is dependent on a rigid internal framework of bones organized into a skeleton. Bones are composed chemically of a combination of mineral (calcium phosphate) and organic materials (principally collagen) (Figure 6–2). There are normally 206 bones in the human body (Table 6–1), though occasionally extra fingers or toes are formed in an individual (Figure 6–3). Many of the bones occur in matching pairs, one on each side of the body axis. Thus, the overall **symmetry** of the skeleton is two-sided or bilateral. This symmetry is easily demonstrated. If you stand at the edge of a full-length mirror and divide yourself in half lengthwise into a composite that is half-real body and half-mirror image, your right and left sides will be equivalent. If you orient yourself in any other way other than along the longitudinal midline, the reflection will not be symmetrical.

The bones of the skeleton of *Homo sapiens* are divided into two large groups—those that establish the long axis of the body (the axial skeleton) and those that compose the appendages (appendicular skeleton) (Figure 6–2). The axial skeleton includes the *skull, vertebrae,* and *ribs*. The remaining bones of the skeleton compose the appendages (arms and legs) or bones that make up the pelvic girdle and pectoral or shoulder girdle.

Axial Skeleton

The axial skeleton constitutes the body's long axis as well as the anatomical center line for bilateral organization. The bones of the axial skeleton provide the rigid, internal armor that protects the central nervous system and principal

Symmetry division of structure into exactly similar parts with respect to shape, size, and position

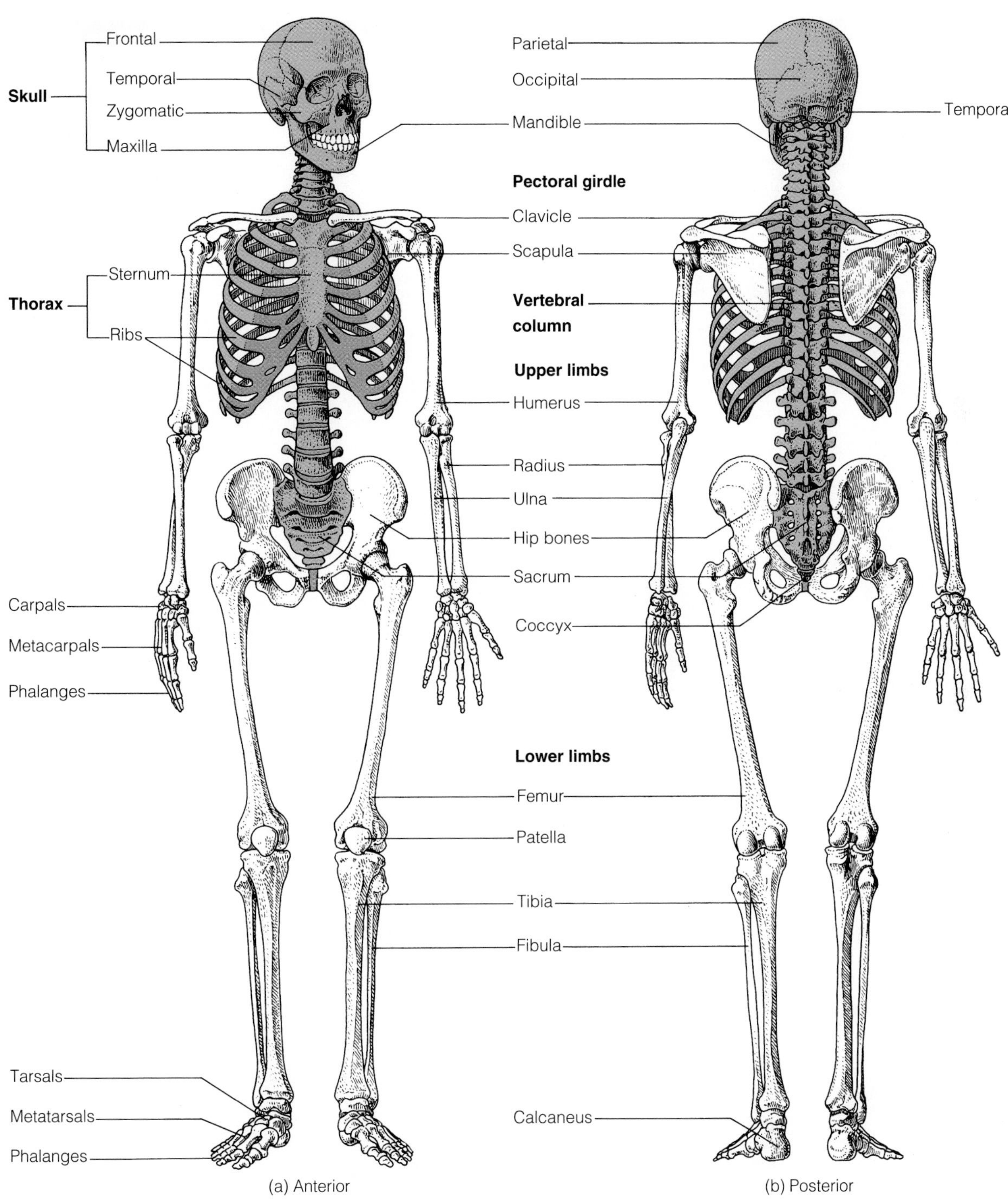

Figure 6–2 The Human Skeletal System
The axial and appendicular skeletons are differentiated by shading (axial) or unshaded structure.

Table 6–1 Bones of the Human Skeleton

AXIAL SKELETON		APPENDICULAR SKELETON	
Region of Skeleton	**Number of Bones**	**Region of Skeleton**	**Number of Bones**
Skull	22	Pectoral girdle	4
Cranial bones		Clavicle (2)	
Frontal (1)		Scapula (2)	
Occipital (1)		Upper limbs	60
Sphenoid (1)		Humerus (2)	
Ethmoid (1)		Ulna (2)	
Parietal (2)		Radius (2)	
Temporal (2)		Carpal (16)	
Sutural (variable)		Metacarpal (10)	
Facial bones		Phalanx (28)	
Vomer (1)		Sesamoid (variable)	
Mandible (1)		Pelvic girdle	2
Maxilla (2)		Coxal (2)	
Zygomatic (malar) (2)		Lower limbs	60
Nasal (2)		Femur (2)	
Lacrimal (2)		Patella (2)	
Inferior nasal concha (2)		Tibia (2)	
Palatine (2)		Fibula (2)	
Auditory ossicles	6	Tarsal (14)	
Malleus (2)		Metatarsal (10)	
Incus (2)		Phalanx (28)	
Stapes (2)		Sesamoid other than patella (variable)	
Hyoid	1		
Vertebral column	26		
Cervical vertebra (7)		Total	126
Thoracic vertebra (12)			
Lumbar vertebra (5)			
Sacrum (1)		Axial + Appendicular	206
Coccyx (1)			
Rib	24		
Sternum	1		
Total	80		

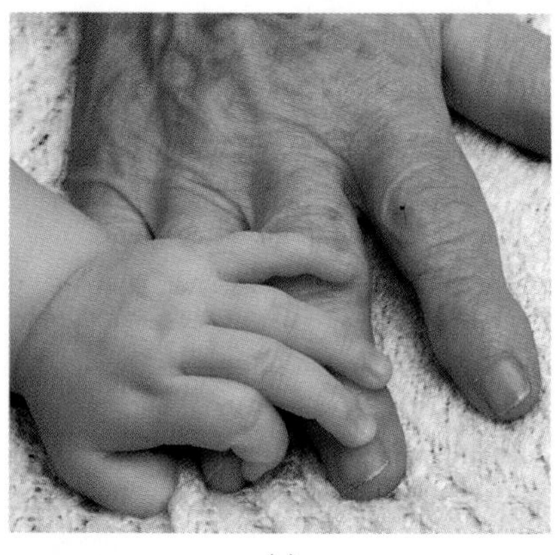
(a)

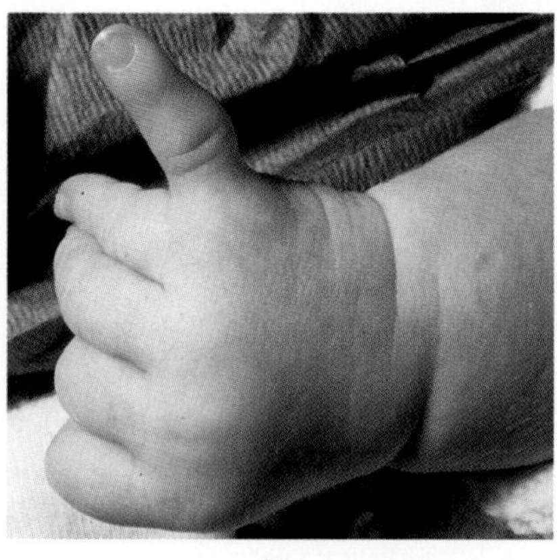
(b)

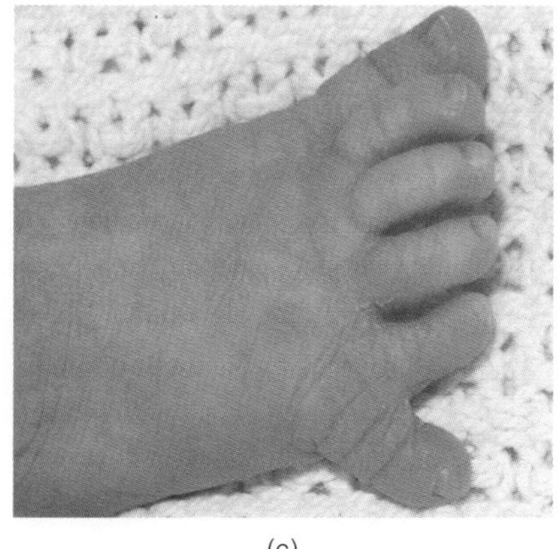
(c)

Figure 6–3 Abnormal Skeletal Development
The development of extra fingers and toes is uncommon but not unusual. The normal hands in (a) are compared to a six-fingered hand in (b) or a six-toed foot (c).

organs of the thoracic cavity. The bones of the skull and vertebrae surround the brain and spinal cord, and the ribs form a cage around the heart and lungs (Figure 6–2).

Skull The relatively spherical human skull is actually a complex, hollow structure constructed of bones held together rigidly by special joints called sutures (Figure 6–4) which give the appearance of the bones having been sewn together. Sutures will be discussed in more detail in the section of this chapter dealing with the structure of joints. The upper portion of the skull is called the *cranium,* and it creates a cavity within which the brain is housed (Figure 6–4). Like the metal plates of a battleship welded together to form the hull, the eight bones of the cranium are fused around the cranial cavity. These bones are often penetrated by holes through which nerves and blood vessels enter or exit the cavity. Of particular importance for the passage of the spinal cord into the skull is the *foramen magnum,* or large hole, of the occipital bone (Figure 6–4).

There are 14 bones that compose the face (Table 6–1 and Figure 6–5). These bones are fused or hinged, as in the case of the jaw, and, when mature, they form a rigid framework for the orbits of the eyes, the cheeks, the nasal cavity, and the oral cavity. Within many bones of the face are air spaces lined with mucous membranes, which are referred to as the *paranasal sinuses* (Figure 6–5). The hardest structures produced by the human body are the teeth (Figure 6–5). They grow out from the *maxilla* and *mandible* bones, which form the upper and lower jaws, respectively.

Vertebrae The vertebral column or spinal column is made up of a series of bones called vertebrae (Figure 6–6). They create a hollow tubular space for the spinal cord. Each individual vertebra has a central canal. Unlike the cranium, which is constructed of rigidly fused bones, the spinal column is a flexible unit with the vertebrae stacked one atop the other. Individual vertebrae are separated from one another by softer, cushioning *intervertebral discs* made from

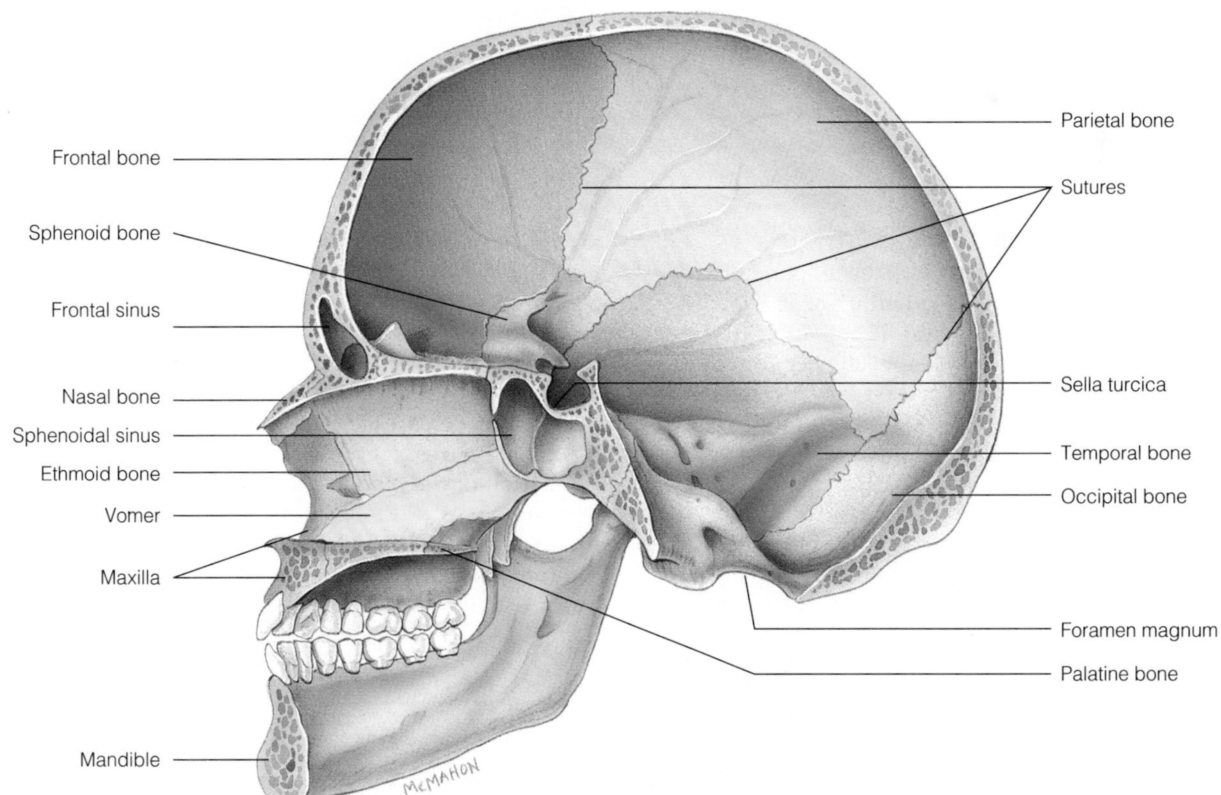

Figure 6–4 The Human Skull
This side, cutaway, view of the human skull allows us to identify the principal bones and structures that make up the adult skull.

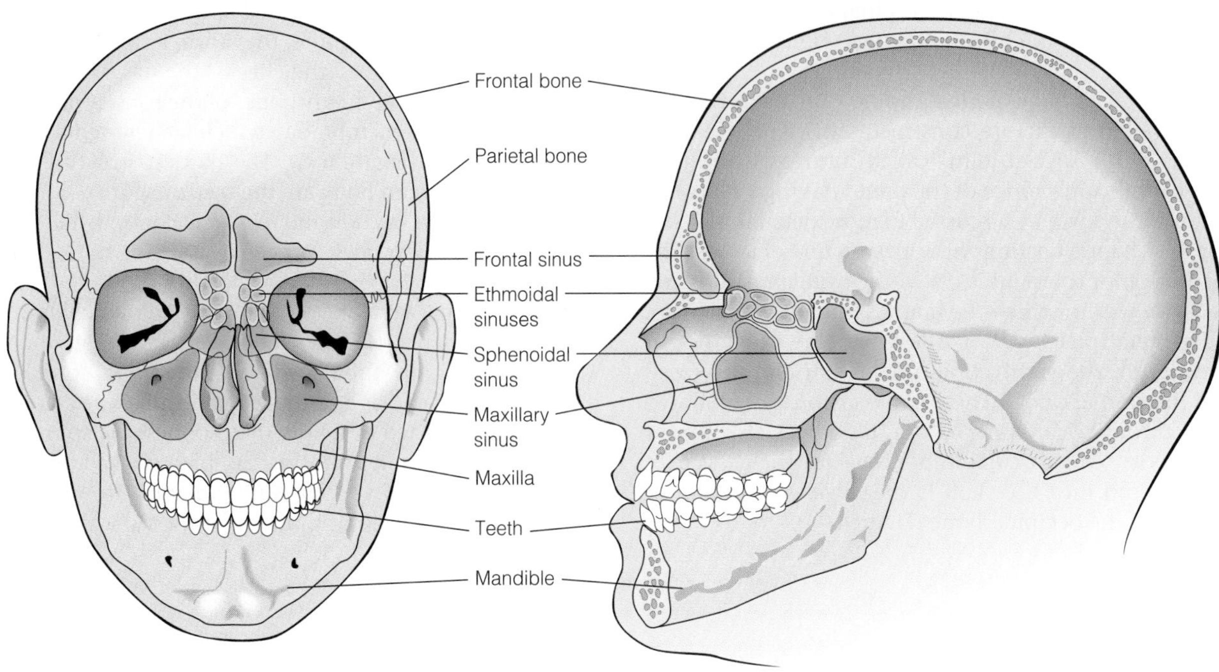

Figure 6–5 Bones of the Face and Paranasal Sinuses
The major bones of the face have hollow, mucous-membrane lined spaces called paranasal sinuses. They occur in the frontal, ethmoid, sphenoid, and maxilla bones.

cartilage (Figure 6–6). There are five types of vertebrae named for the regions in which they are found. The first three types are referred to as **cervical, thoracic** and **lumbar** vertebrae. There are 7 cervical, 12 thoracic, and 5 lumbar vertebrae. The other two types of vertebrae compose the *sacrum* and the *coccyx*. Sacral and coccygeal vertebrae are fused together (Figure 6–6) and form parts of the pelvic girdle. Of particular importance are the first cervical vertebra of the spine, known as the *atlas,* and the second, called the *axis*. The skull rests upon and is supported by the atlas, which, in turn, allows the head to rotate on the axis.

The position of the foramen magnum is very important with regard to the human capacity to stand comfortably upright. In most mammals, and all other vertebrates, this hole in the occipital bone is positioned at the back of the head instead of underneath it, as in humans (Figure 6–4). For example, the heads of cats and dogs extend roughly along a horizontal line from the spine, which itself is horizontal in a body supported by four legs. Most terrestrial mammals other than *Homo sapiens* (and higher primates to some degree) are functionally *quadrupeds* (walk on all four limbs) with the spinal column oriented horizontally. The vertical organization of the human skeleton has required evolutionary reworking not only of the shape, position, and holes in bones but of the position and attachment of internal organs and muscles. The pelvis in humans is altered to stabilize the vertical stacking of vertebrae, the suspension of reproductive organs and the attachment of back and leg muscles. Human organs are attached to bones and to the body wall. These attachments help the organs resist the tendency of gravity to pull them down and out of position. The vertical stance of the human body has made circulation of the blood somewhat more complicated and demanding. As we discuss in Chapter 11, the human heart has to work relatively harder than those of most mammals to lift blood from the lower extremities and push it to the upper extremities (for example, the brain). Circulation in

Cervical of, or referring to, the region of the neck; the first 7 vertebrae

Thoracic of, or referring to, the cavity enclosed by the ribs in the upper torso; the set of 12 vertebrae immediately below the cervical vertebrae

Lumbar of, or referring to, the 5 vertebrae in the region of the lower back, immediately below the thoracic vertebrae

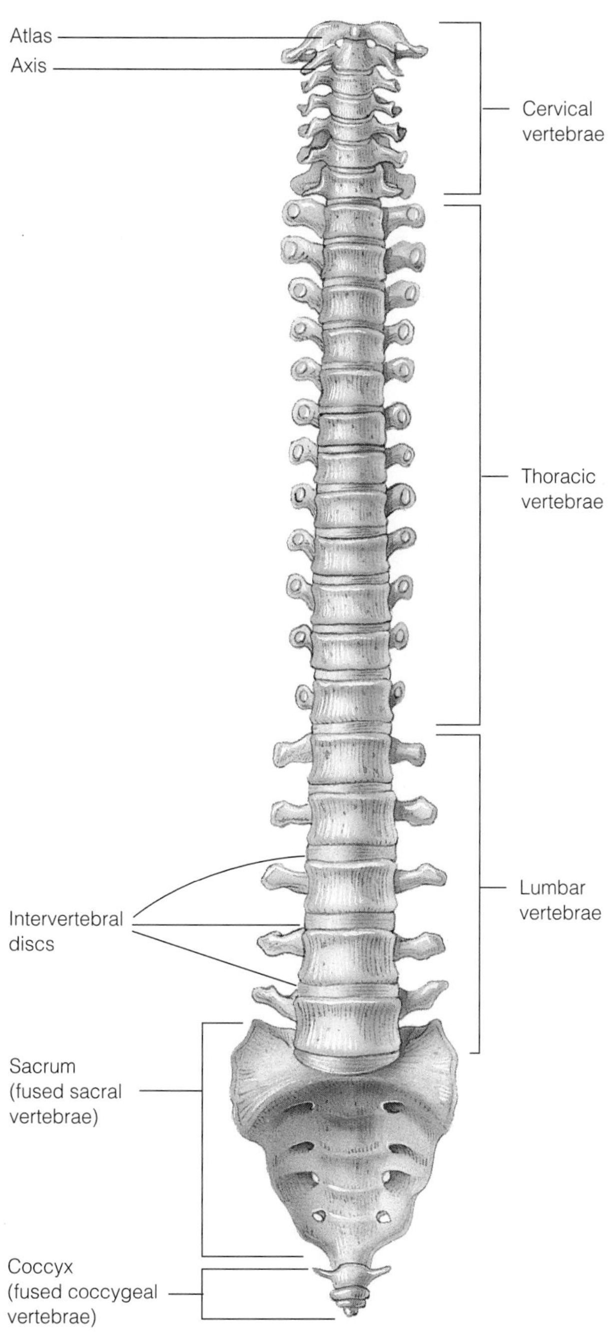

Figure 6–6 Vertebrae
This anterior view of the vertebrae shows five types—cervical, thoracic, lumbar, sacral, and coccygeal. Intervertebral discs are found between each unfused vertebra.

all animals is aided by valves in the veins, which make it easier to transport blood back to the heart, but they are particularly important in humans.

Ribs The final group of bones of the axial skeleton are the ribs (Figure 6–2). These 12 pairs of ribs form a cage around the thoracic cavity. The major organs of the upper torso are located in the thoracic cavity. The heart and lungs, along with the major large blood vessels and passageways for air (*trachea*) and food (*esophagus*), are surrounded, and protected, by the rib enclosure (Figure 6–7). The ribs are attached in the back (posterior) to the thoracic vertebrae of the spine and arc around to the front (anterior or the ventral surface) of the body to nearly completely enclose the chest and parts of the upper abdomen (Figures 6–2 and 6–7). The upper 10 pairs of ribs attach in front to the *sternum* either directly or indirectly. The first seven pairs are directly attached to the sternum and are called *true ribs*. The next three pairs are attached by shared common cartilage to the sternum and are called *false ribs*. The final two pairs of ribs are not attached to the sternum and they are called *floating ribs* (you can feel them move by pushing on the lowest part of your rib cage). The ribcage is further stabilized through the attachment of the sternum to the clavicles or collar bones (Figures 6–2 and 6–7), which in turn are attached to the scapulae or *shoulder blades*. This combination of articulations gives strength to the framework of bones surrounding the thoracic cavity. This stability is also important for muscle attachments to these bones, which are involved in breathing and in the movement of the head and arms.

Appendicular Skeleton

The appendicular skeleton is composed of bones that make up the appendages (arms and legs) and bones that integrate the appendage attachment to the axial skeleton (the pelvic and pectoral girdles). The bones of the arms and legs follow the same pattern in number and arrangement. This pattern of bone organization is observed in all mammals and most vertebrates, particularly terrestrial vertebrates. The similarity in the structure of the vertebrate skeleton is not accidental; it is lineal and lends strong support to the theory of evolution, that is, the concept of descent with modification. Based on skeletons, humans are related to all other vertebrate species by descent from a common vertebrate ancestor. Equivalently positioned bones in all limbed vertebrates are considered to be *homologous;* that is, they arise from the same origins during embryonic development. The shape of homologous bones is often different because they have evolved to be adapted for different functions (Figure 6–8). When we study the bones of the human skeleton, we are in effect studying one of many different outcomes of the skeletal evolution of vertebrates.

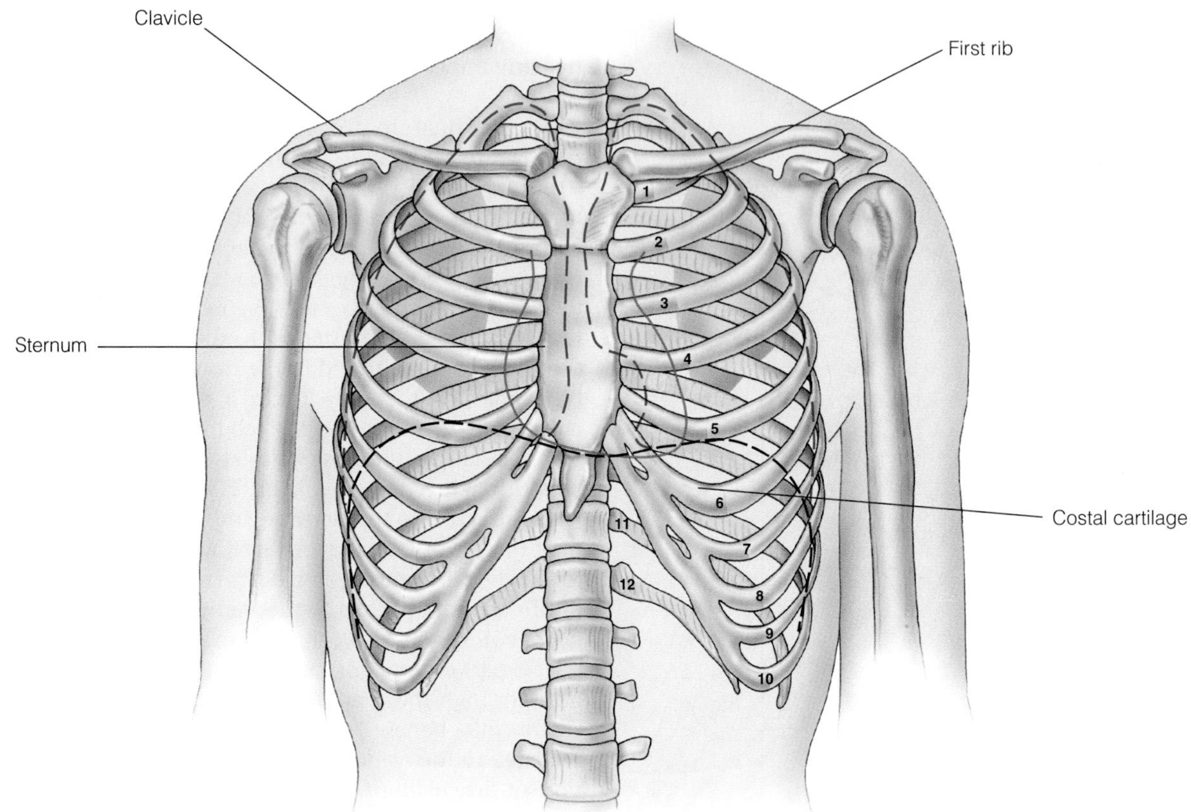

Figure 6–7 Ribcage and Thoracic Organs
There are 12 pairs of ribs attached posteriorly to each of the thoracic vertebrae. The first 7 pairs are called true ribs and the next three pairs are called false ribs. The final 2 pairs are called floating ribs because they are not attached anteriorly to the sternum.

Bones of the Arms and Shoulder Girdle The appendages of the upper body are composed of 60 bones, matched side to side, and four bones that establish the shoulder girdle (Figure 6–2 and Table 6–1). The largest of these bones, the humerus, forms the upper arm. The humerus is attached to the shoulder girdle at its **proximal end** by articulations with the shoulder blade and collar bone (Figure 6–2). The **distal end** of the humerus is attached to the two bones of the forearm, the ulna and the radius. These three bones (humerus, ulna, and radius) form the elbow, which is capable of a wide range of movements (Figure 6–9).

The ulna is the inner bone of the forearm (nearest the little finger at the wrist) and the radius (nearest the thumb) is the outer bone when viewed in a skeleton in the standard anatomical position (Figures 6–2 and 6–9). During the rotation of the arm, these bones cross over one another to provide the twist that is familiar in the movements of the arms. The distal end of the ulna and radius interact with several of the bones of the wrist. The bones of the wrist are called *carpals* (Figure 6–9). The bones of this region are organized in two rows and are capable of gliding movements.

The hand is made up of 19 bones (Figure 6–9), called *metacarpals* and *phalanges* (fingers). The metacarpals connect to phalanges, forming the knuckles. The fingers (digits 2–5) have three bones each and the thumb (digit 1) has two. The hand is remarkably adept at gripping and in bringing the ends of the fingers into contact with the

Proximal end the end nearest the long axis of the body

Distal end the end farthest away from the long axis of the body

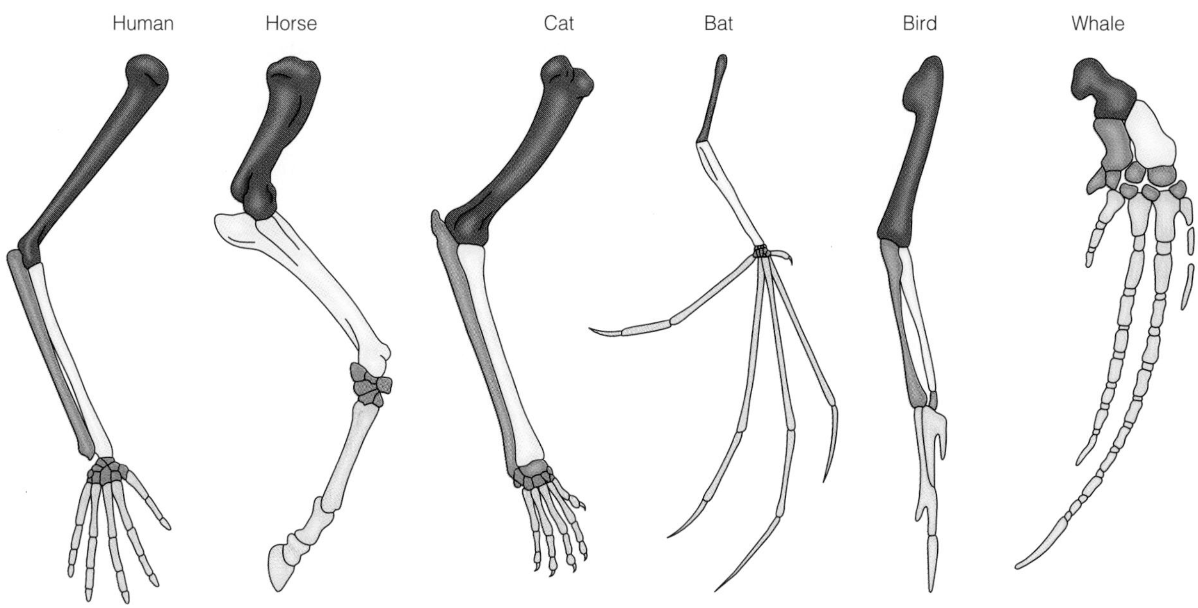

Figure 6–8 Homologous Bones
The bones of the forelimbs of vertebrates are considered homologous. They may serve different functions, but they are composed of the same bones (color coding) and have similar embryological origins.

thumb. This capacity is called *opposition* and is unique to primates and particularly well developed in humans (Figure 6–10).

Bones of the Legs and Pelvic Girdle In an organization nearly identical to that of the bones in the arm, forearm, wrist, and hand, each leg is organized into thigh, calf, ankle, and foot. The largest bone of the human body is the upper leg bone, the *femur.* The proximal end of the femur forms a joint with the pelvic girdle. During early childhood, the pelvic girdle forms a basket-shaped structure resulting from the fusion of two pairs of three bones (Figure 6–2). The femur articulates with two bones of the lower leg, the *tibia* and *fibula,* and with the kneecap (*patella*) to form the knee. The combination of these four bones of the leg and the ligaments that hold them together makes the knee one of the most complicated joints in the human body (Figure 6–11).

The tibia is the larger of the two bones of the lower leg. The front surface of the tibia is commonly referred to as the shin and can easily be felt as a ridge extending down the front of the lower leg. It is certainly felt when one bumps into a chair in the dark. Although the organization of the leg bones is similar to the arm bones, different types of joints between them affect the specific range of movement. The distal end of the tibia and fibula interact with the ankle. The ankle is composed of *tarsal* bones that connect to the long *metatarsal* bones of the foot that articulate with the small bones of the toes. This is similar to the organization of the hand, but the foot is built to support the weight of the body and the toes are not capable of opposition in gripping or manipulating objects.

There are no differences in the number and type of bones found in males and females of our species. However, in general, males have larger, heavier bones. A significant difference in the shape, width, and density of the pelvis is observed between males and females. The angle of the pubic arch (Figure 6–2 and Biosite 6–1) in females is greater than 90° and in males it is less than 60°. This is related anatomically to the capacity of females to give birth.

PROPERTIES OF BONES

Sizes and Shapes

Human bones range in size from a few millimeters in length for the tiny bones of the middle ear to half a meter or more in length for the long bones of the humerus and femur (Figure 6–2). The large, long bones of the body support attachment of the massive muscles of the arms, shoulders, pelvic girdle, and legs that are necessary for movement and lifting (Biosite 6–2).

There are four main shape categories of bones. Bones that are longer than they are wide are called *long bones.* Bones that tend toward a cuboidal shape are called *short bones. Flat bones* are thin and platelike structurally, with compact bone surfaces separated by a spongy layer of bone. Bones that do not fit into any of the previous three categories are called *irregular bones* (Figure 6–12).

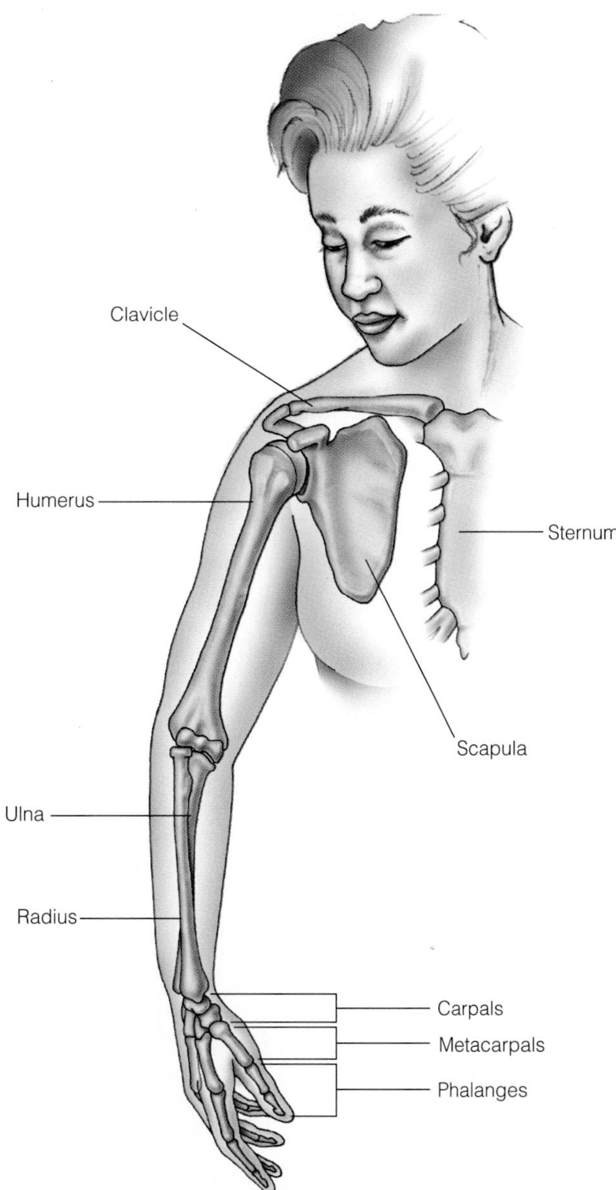

Figure 6–9 Arm Bones
The bones of the arm are organized from proximal to distal as the humerus, ulna, radius, carpals, metacarpals, and phalanges.

Long Bones The long bones of the human skeleton include the bones of the arms and hands, as well as those of the legs (femur, tibia, fibula) and feet (Figures 6–2 and 6–12). Many of the weight-bearing long bones are somewhat curved to accommodate the pressures and stresses associated with resisting the pull of gravity, body movement, and lifting. The curvature provides limited flexibility, similar to the action of a leaf spring found in cars, and, thus, reduces the chances for fractures and breaks under heavy loads or sudden jumps and landings.

Short Bones and Flat Bones The length and width of short bones is roughly equal, so they appear cuboidal in shape. The bones of the wrist and ankles fit into this category. Flat bones are characterized by their thinness. The platelike structure of these bones serves in areas of the body where great bulk is not needed, but protection of underlying organs is imperative. This is the case for the several bones that form the cranial cavity (Figure 6–4). In this category also are the ribs and sternum, which provide the protective framework that surrounds the thoracic cavity and protects the heart, lungs, and other important anatomical structures.

Irregular Bones This is a catch-all category for all the bones that do not fit into long, short, or flat groups. The individual shapes of these bones are varied. For example, the shape of many of the bones of the skull are incredibly complex. The sphenoid bone that forms the side of the face and the back of the orbits of the eyes is a butterfly shape with spread wings (Figure 6–12). Taken out of the context in which the sphenoid is integrated with the other bones of

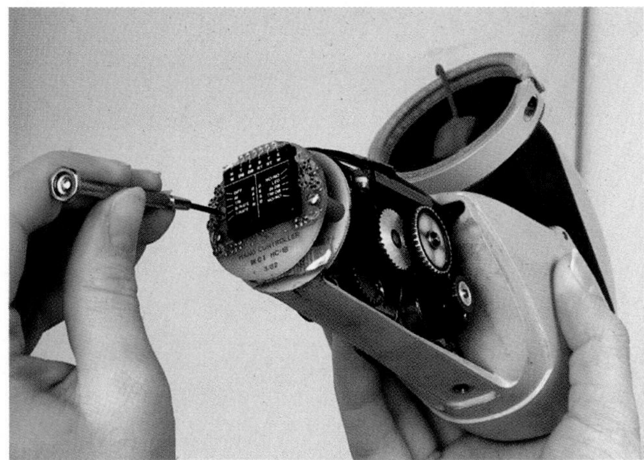

Figure 6–10 Gripping Hand
The bones of the hand are organized such that the thumb can be directly opposed to each of the other digits.

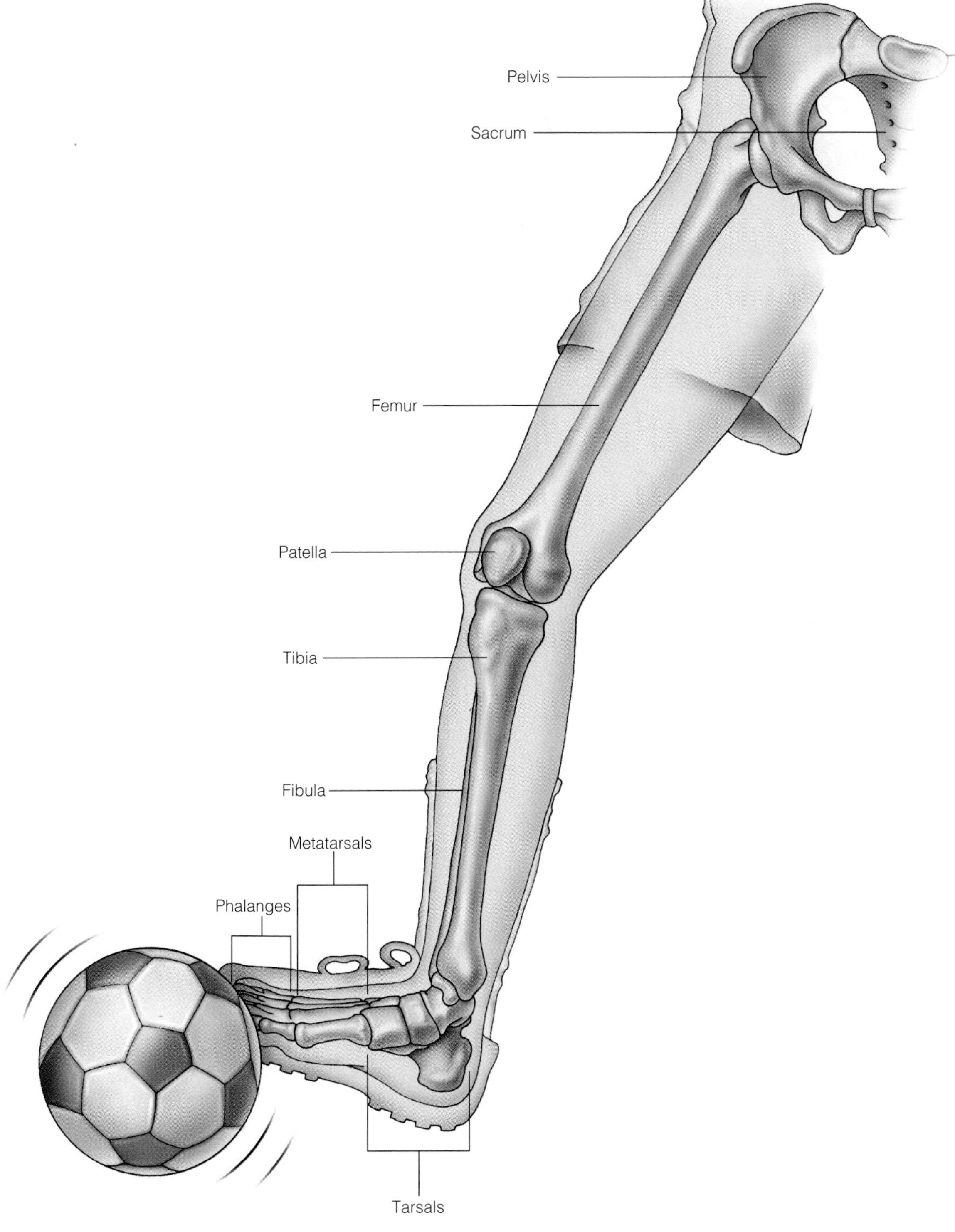

Figure 6–11 Leg Bones
The bones of the leg are organized, from proximal to distal, similarly to the organization of the arm. A single femur is attached to the patella and to the tibia and fibula, which in turn attach to the tarsal bones. The metatarsal bones and phalanges complete the foot.

6–1

THE HUMAN PELVIS: DIFFERENCES IN MALES AND FEMALES

In general, males have larger bones than females, but the bones are otherwise comparable in form and function. However, this is not the case for the human pelvis, in which several differences occur. First, the angle of the pubic arch in females (90°) is greater than the pubic arch in males (55°). Second, the pelvic inlet of females is also larger. Third, the pubic symphysis in females undergoes temporary changes during late pregnancy that make the joint more pliable to allow for the necessary expansion of the pelvis during childbirth. These fundamental differences between males and females have been selected by nature through evolution to accommodate reproduction.

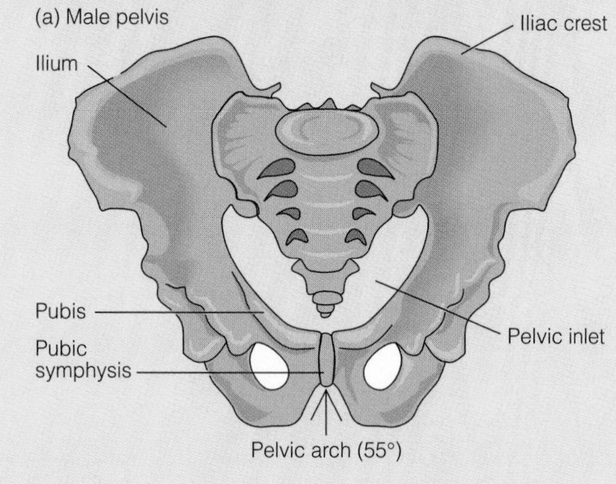

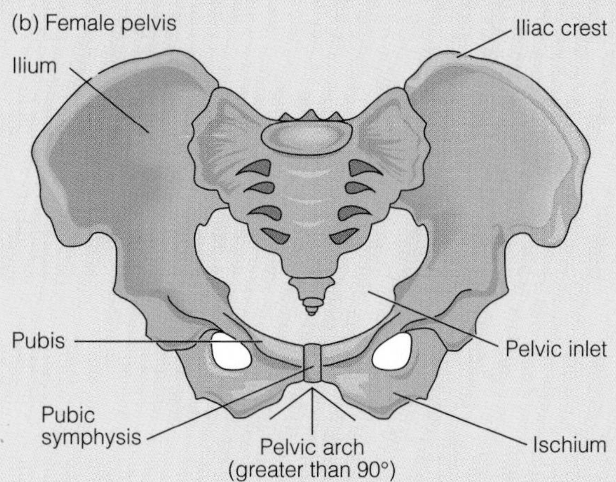

the skull, it appears more as a work of abstract art. Vertebrae are also irregular-shaped bones.

Structure of Bone

As previously mentioned, human bones are composed of a combination of organic molecules and hard minerals (Figure 6–2). The organic material is mostly proteinaceous in character and includes long, fibrous proteins such as collagen. The mineral or inorganic materials of bone are predominantly calcium phosphate. The minerals alone form crystals of great hardness, but they shatter easily into powder under pressure or stress. The organic materials are strong and tough, but they are flexible. If bones were made of organic materials only, they would be too flexible to allow us to stand, work, lift, and/or move effectively for very long. If bones were composed of calcium phosphate alone, they would be hard and rigid but extremely brittle.

This brittle state, not unlike that observed in the dried bones of skeletons, arises from dessication and the lack of most organic materials. This is the case of the bones of skeletons from an ancient tomb or gravesite. In fact, the origin of the word *skeleton* comes from the Greeks and refers to the dry and brittle state of the bones of long dead bodies. Worse for the living is the abnormal process known as **osteoporosis** in which the organic content of bone is reduced significantly with respect to the mineral content (Figure 6–13). This condition results from complications associated with reduced steroid hormone levels in the human body and is often observed in women who have undergone menopause. Osteoporosis results in bones that fracture easily under what would otherwise be normal stress. Vertebrae and bones of the pelvic girdle are particularly

Osteoporosis a disease state in which bones become brittle from loss of organic materials; such bones are particularly susceptible to breakage

BIO site 6–2

A COMPARISON OF BONE SIZE AND BODY MASS

As the human body gets larger and more massive, so the bones must grow to support it. But there are limits. The strength of bones is based on two factors: (1) the materials from which they are made—organic compounds (principally collagen) and inorganic materials (principally calcium phosphate), and (2) the size and shape of the bone. The largest bones of the body are those that support the most weight. This is true for all vertebrates, as may be observed in the size of leg bones in humans, horses, elephants, and even in the extinct dinosaurs. Why can we (or any other animal) not attain the size of the giant organisms of science fiction novels?

The problem for vertebrates, and the human body in particular, is that the mass of an individual increases with the volume of an individual, and volume increases as a function of a factor of 3 (m^3). The strength of a bone (or any structural material) on the other hand, increases as a function of cross-sectional area, or a factor of 2 (m^2). What would happen if a 10 kg (220 pounds), 1.8 m (6 feet) individual were suddenly to become twice as large? This individual would be 3.6 m tall (12 feet) and weigh 800 kg. Under these circumstances the bone strength (area) would increase by a factor of 2^2 (4), but body mass would increase by 2^3 (8). The major skeletal bones supporting such a mass would be insufficient and the bones would crack and break with the first step. This relationship between area and volume is why it is virtually impossible for any extant animal (spider, ant, moth, or human) to be as large as they are often portrayed in the movies!

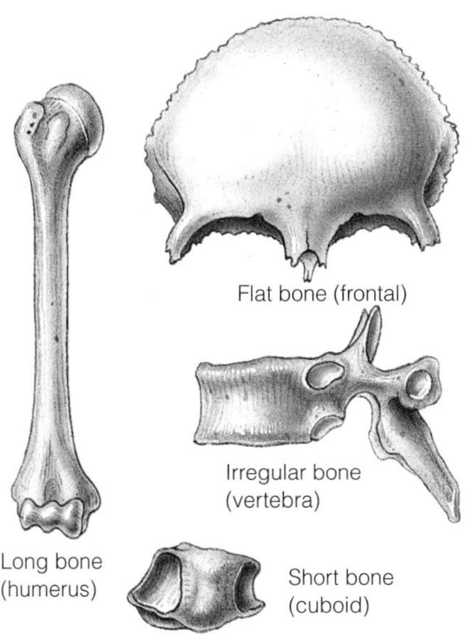

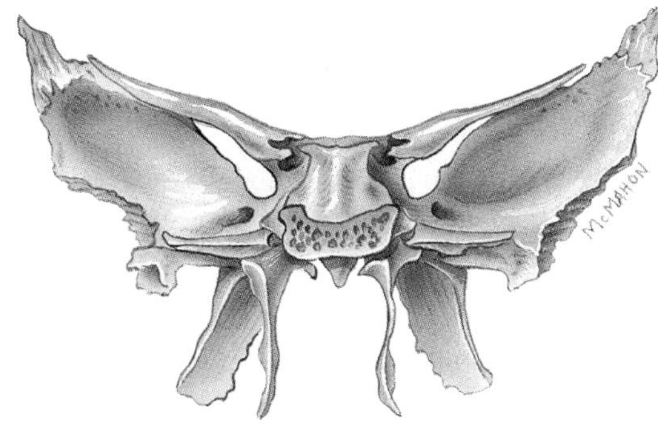

Figure 6–12 Shapes of Bones
Bones are classified as long, short, flat, and irregular in shape (a). The complexity of the shape of the sphenoid bone (b) is highly irregular, but it integrates with the several bones of the face and cranium.

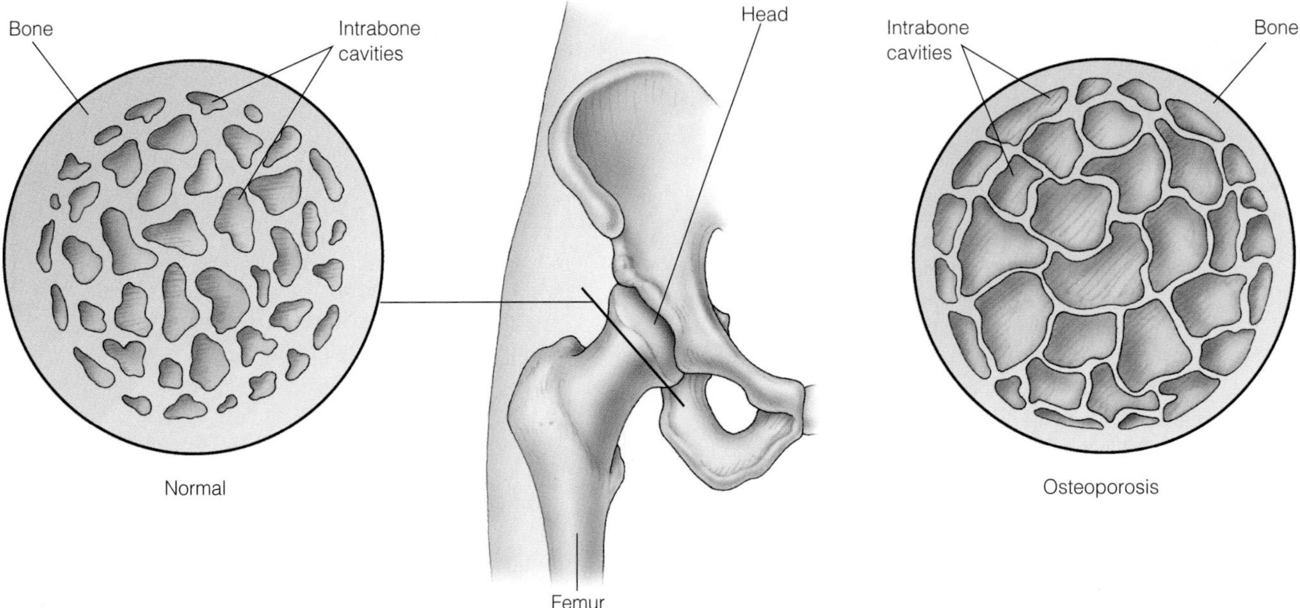

Figure 6–13 Normal and Osteoporotic Bone
The head of the femur is susceptible to breaking in individuals with osteoporosis. The bones in these individuals become less dense and brittle.

susceptible. Fortunately, normal bones are neither too dry and hard nor overly flexible, but rather, a combination of the two—hard and flexible.

MAKING BONES—OSTEOGENESIS

Osteoblasts and Osteocytes

New bone is made both on the surface of existing bone and within the bone (Figure 6–14). The surface of bones is covered by an epithelium called the *periosteum.* This membrane has two layers, the outermost of which is a tough fibrous sheet and the inner layer of which contains bone-forming cells known as **osteoblasts.** The term *blast* means bud or form and combined with *osteo* means bone forming. Within the dense matrix of compact bone are specialized cells known as *osteocytes,* which produce bone from inside. The suffix *cyte* means cell, in this case a mature bone cell. Osteocytes are located within small cavities known as *lacunae* (Figure 6–15). There are millions of lacunae within each bone of the body. These tiny pockets are arranged in a radial, or circular, fashion around a blood vessel. Together the lacunae and the blood vessel they surround are known as a *Haversian system* or *osteon.* The lacunae of a single osteon are perforated so that cellular extensions of osteocytes from different lacunae within the system can directly communicate with each other. The channels between lacunae are called *canaliculi,* or little canals.

Types of Bone—Compact and Spongy

Structurally, there are two types of bone—compact and spongy. *Compact bone* is highly organized and composed of repeating units of Haversian systems. Each of the osteons is a long, thin rodlike structure made up of concentric rings of mineralized material called *lamellae* in which the lacunae and canaliculi are organized (Figures 6–14 and 6–15). Compact bone is characterized by bundles of the rodlike osteons tied together in a dense, rigid, and strong material composite. *Spongy bone* also contains lacunae that enclose osteocytes. However, spongy bone is not organized into osteons but rather into a network of interwoven rods or bars called *trabeculae* that have spaces between them (see Figure 6–14).

Remodeling of Bone

All types of bones in the human body are broken down into their constituent molecular components and rebuilt on a continuous basis throughout our lifetime. This process of turnover of bone brings about new growth,

Osteoblasts immature bone-forming cells

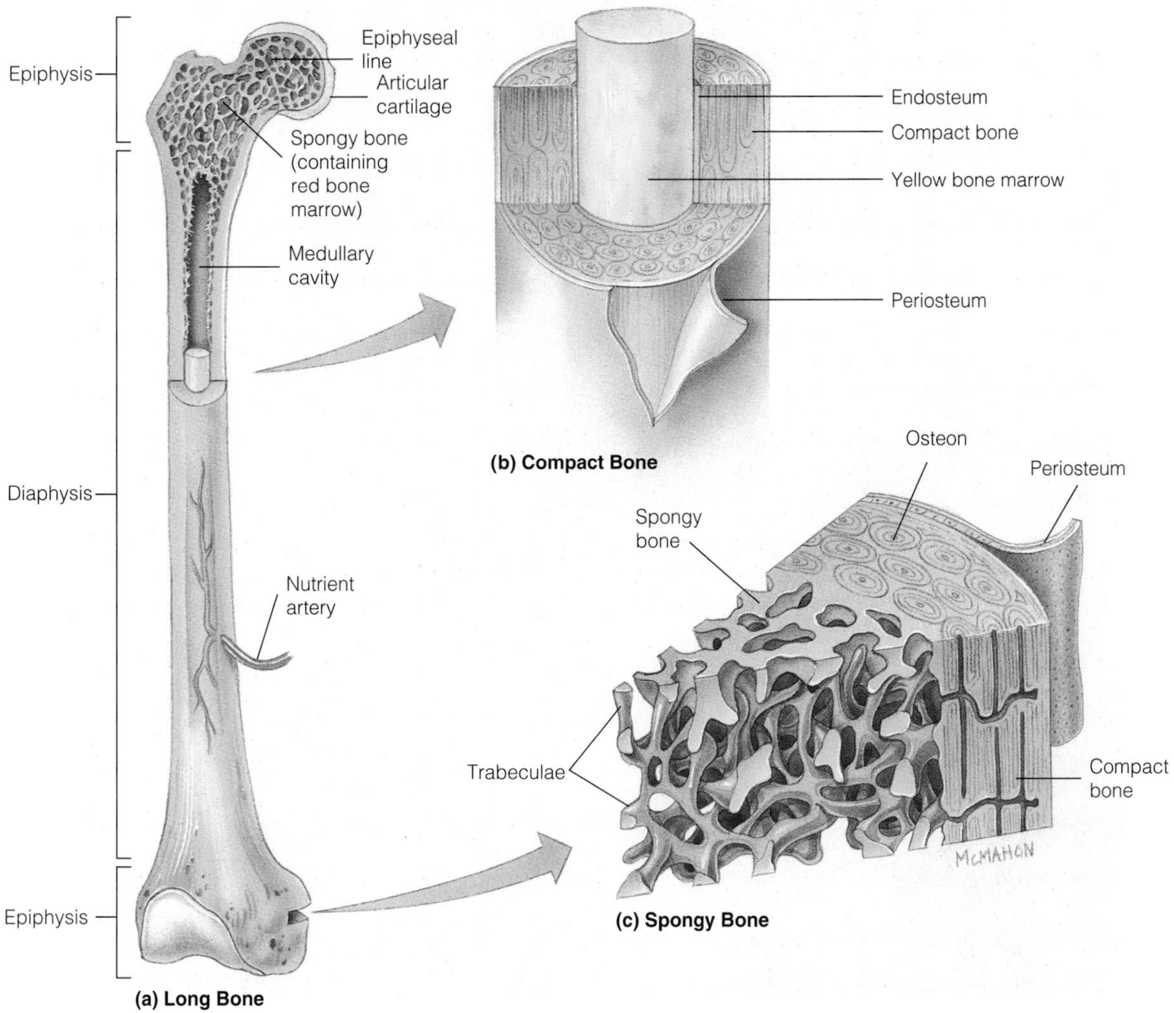

Figure 6–14 Structure of Bone
A long bone is shown in (a), with a central diaphysis and two epiphyses at the ends. Compact bone (b) is dense mineralized material surrounded on the inside by an endosteum and on the outside by a periosteum. Spongy bone (c) is less dense containing hollow spaces bridged by trabeculae.

which, in turn, maintains the overall strength and integrity of the skeleton and provides the opportunity for change in the shape or density of bones in response to stress or damage. **Remodeling** of bone is homeostatic in nature and is highly regulated by feedback loops. The homeostasis reflects the balance between the amount of bone that is removed and the amount of bone that is produced. Bone remodeling and the maintenance of the skeleton by a balance of replacement and renewal is called turnover, a process that fulfills the familiar adage that "the only way to stay the same is to keep changing."

Bones have a unique way of being rebuilt, which involves osteoblasts and **osteoclasts** (Figure 6–15 and next section). Osteoblasts are prevalent in the osteogenic periosteum and *endosteum* surrounding and lining bones. Osteoblasts differentiate and grow to form new bone, as described earlier, and eventually differentiate into osteocytes encased within a lacuna.

Osteoclasts An osteoclast is an entirely different type of cell from an osteoblast or an osteocyte. The origin of

> **Remodeling** with respect to bone, the constant turnover of old bone and its replacement with new bone
>
> **Osteoclasts** cells involved in the breakdown of bone

osteoclasts is in the blood-forming tissues of the marrow. Immature osteoclasts migrate from within the marrow directly to the surface of bones where they carry out their activities. They are giant multinucleated cells. The suffix *clast* means to break or destroy. For example, a more common usage of clast is in the word *iconoclast,* defined as an individual who attacks established beliefs or institutions. The osteoclast attacks and destroys established compact and spongy bone (Figure 6–15). It does so using enzymes to dissolve the organic components of bone and solubilize the calcium phosphate. Calcium and phosphate ions released in this way are recirculated to the rest of the body through the vascular system, through which the ions are made available to the osteocytes and osteoblasts to make new bone. The balance of actions between bone dissolution and bone formation is the basis for skeletal homeostasis, which includes the processes of turnover and remodeling. The bones of the human skeleton are in a constant state of flux and a living human skeleton is replaced completely many times during a lifetime.

DEVELOPMENT OF BONES

Embryo and Fetus

The formation of bones is initiated early in human fetal development and continues for many years after birth. The rate of bone formation in the fetus is more rapid than in an adult, but it is incomplete at the time of birth. Ossification of the embryonic skeleton starts at approximately two months of gestation and defines the transition from embryo to the fetus. The skeleton of the embryo initially consists of collagenous connective tissue produced by fibroblasts. This connective tissue is replaced by cartilage, which establishes a **cartilage model** (Figure 6–16). The cartilage is replaced by mineralized materials as the bones

Cartilage model embryonic skeleton prior to ossification

Figure 6–15 The Osteon
The unit of structure of compact bone is called the osteon. An osteon is composed of osteocytes within lacunae connected in lamellae by canaliculi. Each osteon surrounds a Haversian canal with blood vessels and nerves inside.

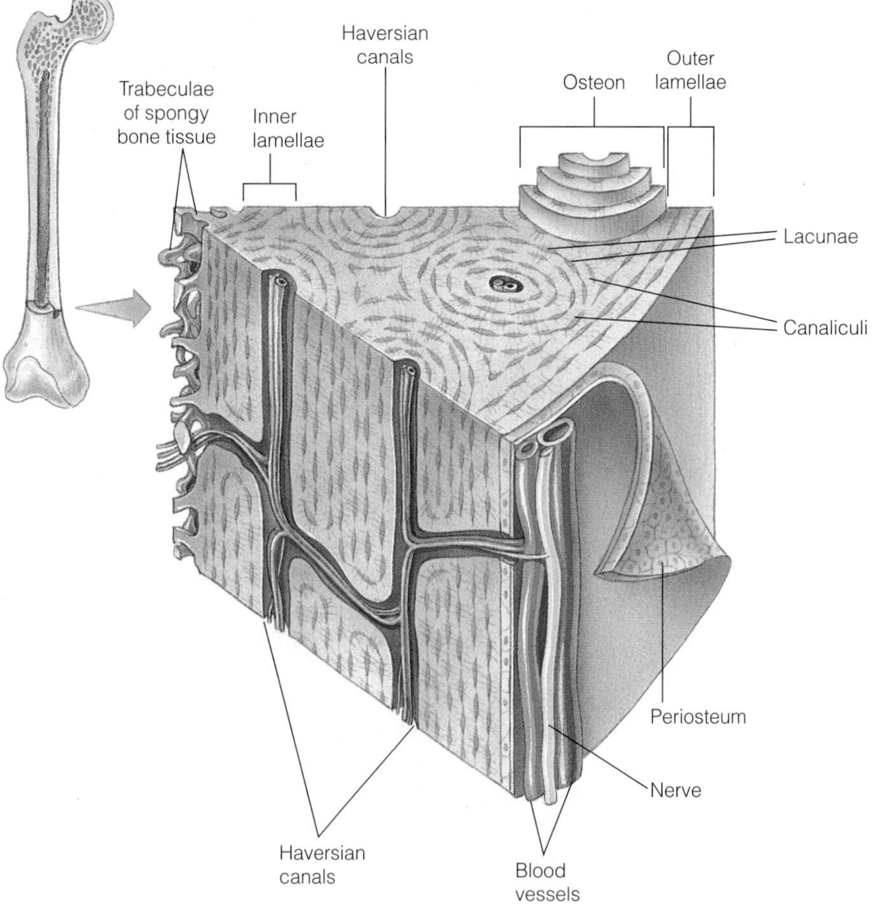

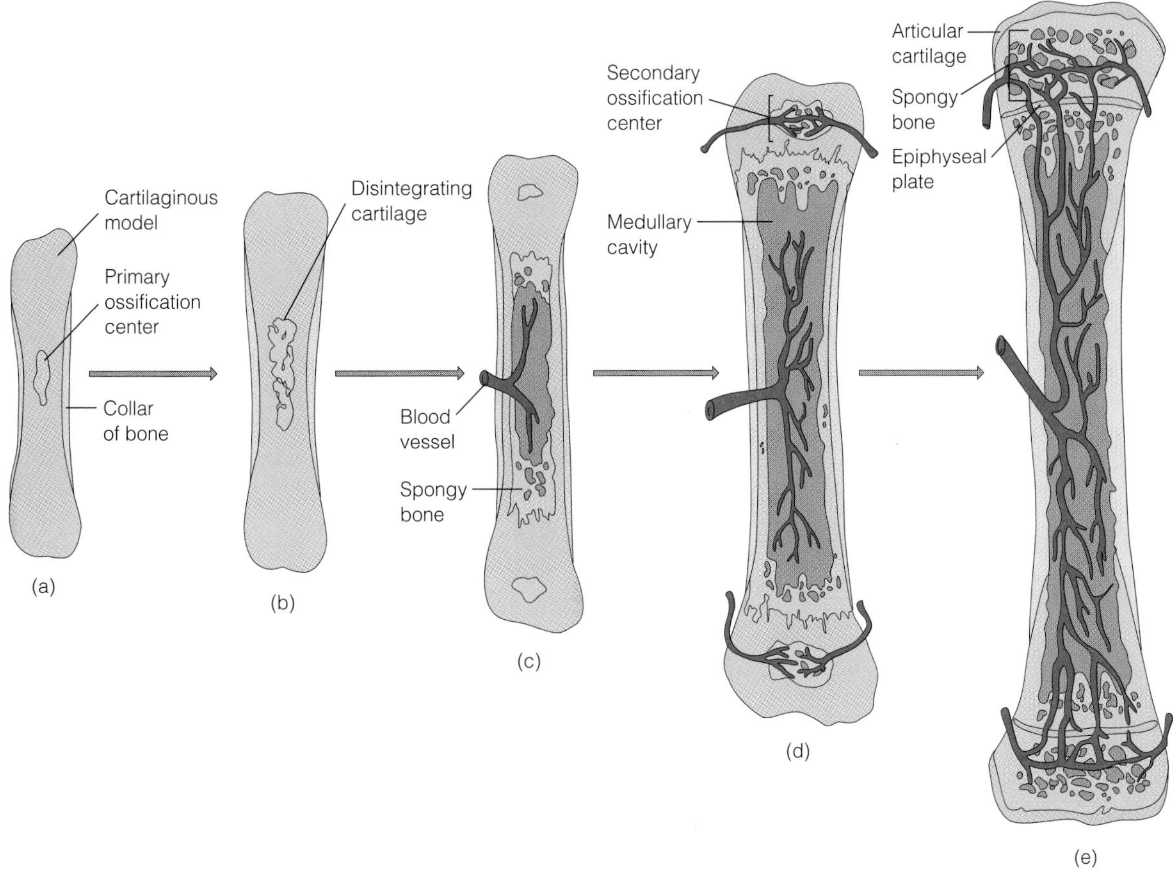

Figure 6–16 Bone Formation During Development and Growth

A cartilagenous model of a developing long bone (a) is replaced from within by spongy bone (b) and (c). This growth is supported by invasion of blood vessels into the cartilage. Osteoblasts invade secondary ossification sites and establish epiphyseal growth plates (d). Osteoclasts break down some bone to form a medullary cavity. Bone replaces cartilage in all areas except for the joint surface, which is hyaline cartilage, and in the nongrowing epiphyseal plate (e).

grow during fetal development and by birth a large portion (but not all) of a newborn baby's bones are ossified.

Pattern of Growth—Formation of the Diaphysis and Epiphysis

The growth of a long bone provides a good example of the pattern of bone growth. The ends of a long bone are called the epiphyses and the middle region between them is called the diaphysis (Figure 6–14). The area first undergoing ossification during fetal development is the central region of the diaphysis. This is followed by the development of ossification centers in the epiphyseal ends. Ossification of the diaphysis and epiphyses of bones occurs in late fetal development and continues after birth. The zones of bone growth in the epiphyses progress in two directions, thus accounting for the length of the bone. Growth in the region of the diaphysis occurs radially, thus increasing the thickness of the bone. In humans, growth in the epiphyses is essential for attaining full height and continues for many years. Bone growth is regulated by a variety of factors, including dietary protein and calcium and hormones, particularly the pituitary *growth hormone* (Table 6–2 and Chapter 13).

The growth zone of long bones shrinks as the human body moves from adolescence to adulthood, and, eventually, the region of growth is restricted to the *epiphyseal growth plate.* When the growth plate ossifies, the elongation of bones ceases altogether. This cessation of growth usually occurs by late teens in humans, earlier in females than in males. After the growth plate ossifies, humans grow no taller.

Table 6–2 Factors Affecting Bone Growth

NUTRITION	HORMONES AND GROWTH FACTORS	INHERITANCE	VITAMINS
Organic protein and amino acid intake Inorganic calcium, phosphate, and carbonate intake and production	Growth hormone, thyroxin, calcitonin, parathyroid hormone, and fibroblast growth factor	A number of genes affecting synthesis of collagen, production of calcium phosphate, and hormone synthesis and reception	C in collagen assembly, D (calciferol) in calcium absorption

Abnormalities—Nutrients and Teratogens

Proper nourishment during human fetal and postnatal growth is very important to the formation of fully functional and structurally sound bones. An undernourished child may not form a strong healthy skeleton and as a result may suffer from deficiencies in bone size and strength for a lifetime. Congenital and/or genetic problems in bone development may even prevent bones from forming in their normal anatomical positions. Drug-induced alterations in the development of the embryonic and fetal skeleton are often predictable, as exemplified in babies affected by the drug thalidomide.

Thalidomide is a mild tranquilizer and was initially targeted to be given to pregnant women to relieve anxiety associated with the stress and responsibilities of child bearing. Unfortunately, when ingested during the first trimester of a pregnancy, this drug interferes with and prevents the growth of bones of the limbs. Thalidomide is not unique in this property. There are a multitude of chemical and biological agents (drugs, viruses, toxins) that cause serious abnormalities during embryonic and fetal development (Table 6–3). They are technically called teratogens, for the monstrous abnormalities they induce.

REPAIR OF BONES

Accidents happen and bones may be broken. The most common forms of breaks are the simple fracture, the complete fracture, and the compound fracture (Figure 6–17). Simple fractures are cracks in the bone that do not result in physical separation of the pieces. A more complicated and damaging break, the complete fracture, is one in which the bone is completely separated or displaced and the pieces damage the surrounding tissue to various extents. The worst possible breaks are compound fractures in which the separated pieces of the bone or bones protrude through the skin. It is very important to have proper treatment after breaking a bone, particularly if displacement has occurred. Splints, pins, and casts are used to bring the bone pieces back into proper register and alignment and hold them in place during the healing process. Maladjustments in the way in which bones are allowed to repair themselves may seriously limit their future performance as part of an integrated skeletal system. In recent years electrical stimulation of bones has been shown to improve the rate of healing.

The structural repair of bones is accomplished using some of the mechanisms that are involved in the normal remodeling of bones. The first step in the repair is to form a connective tissue scar, or *callous*. Callous formation results from the action of connective tissue cells, such as fibroblasts, which produce gluelike extracellular matrix molecules to hold the bones and tissues together. The initial material is then replaced by cartilage, not unlike the early cartilage model established before bone mineralization takes place in the embryo. The cartilage, in turn, is invaded by periosteal osteoblasts and compact and spongy bone formation takes place. Because of the constant remodeling of bone under normal circumstances, breaks in bones early in life may be entirely healed and show little or no evidence of ever having occurred. This capacity to heal bone breaks declines significantly with age.

BONES AND BLOOD FORMATION

In human adults blood is formed by **hematopoietic cells** located in the marrow of many (but not all) bones of the skeleton (Figure 6–18). The blood forming tissues are in the *medullary cavities* of bones and constitute what is called the *red marrow*. In bones that are not involved in the formation of blood, the medullary cavity contains fatty tissue called *yellow marrow*. However, the formation of blood cells in adults is quite different from that during embryonic and

Hemopoietic cells progenitor cells of blood cells

fetal development. During embryonic development, the progenitor blood-forming cells arise outside the embryo and migrate first to the spleen, then to the liver before finally colonizing the marrow of the developing bones. Early in fetal development many more of the bones have red marrow, which becomes more restricted in distribution as we mature.

JOINTS—ARTICULATIONS OF BONES

All but a few of the 206 bones of the human body are connected to one another to form what is called an articulated skeleton. The articulated bones are physically held together by tough cords of connective tissue known as ligaments (Figure 6–19). Inherent in this arrangement is the movement of bones relative to one another under the force of muscle contractions, as will be discussed in Chapter 7. The relative movement of bones under these conditions is made possible by specialized articulating surfaces called **joints,** which occur between bones. Ligaments, which are composed of tough and inelastic connective tissue, attach to protrusions, or processes, on one bone, extend over the joint between bones, and attach to processes on the adjacent bone.

Table 6–3 Teratogens and Their Effects

TERATOGEN	EFFECTS
Viruses	
Rubella, roseola, herpes	bone and limb deformities
Cytomegalovirus, parvoviruses	brain and nervous system abnormalities
Alcohol	brain defects
Drugs	
Tranquilizers (such as thalidomide)	limb defects
Cocaine	brain and heart defects
Vitamins	
Vitamin A and derivatives	limb, brain and nervous system defects
Antibiotics	limb and nervous system defects
Bacteria and their toxins	
Mycoplasma	brain defects; lethal to embryo
Trepomema pallidum (syphilis)	bone, brain, and heart defects
Hormones and their derivatives	
Steroids	reproductive system, skeletal, and nervous system defects

Joints regions of attachment between bones

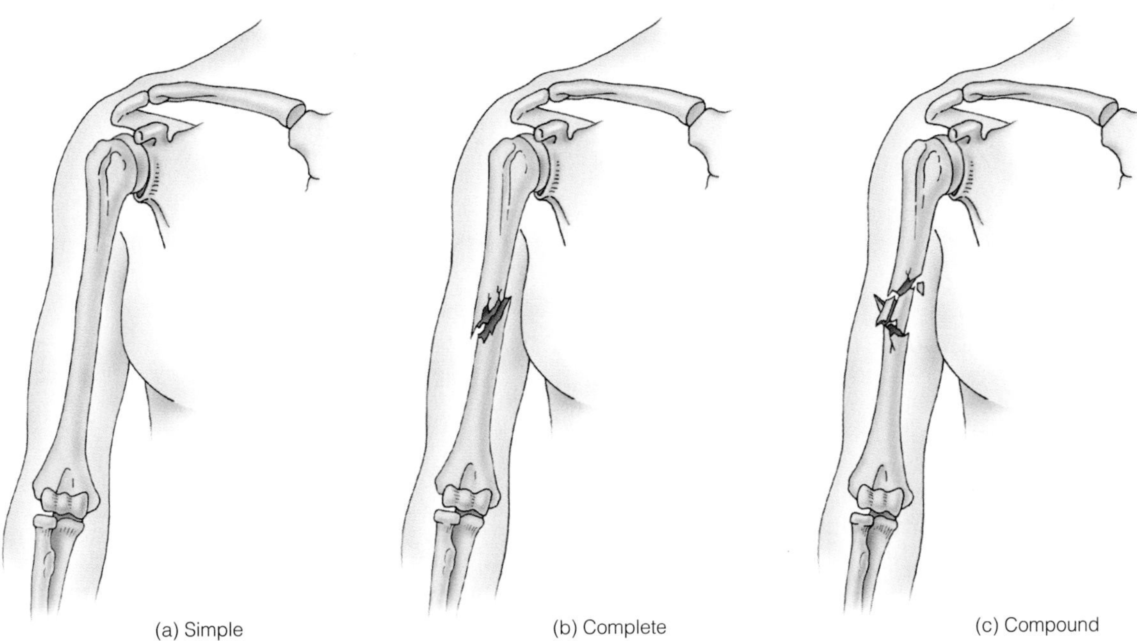

Figure 6–17 Broken Bones
Of the three types of breaks shown, the complete fracture (b) and compound fracture (c) are the most serious. Improper treatment of any break can result in impaired future performance of the bone in the skeletal system.

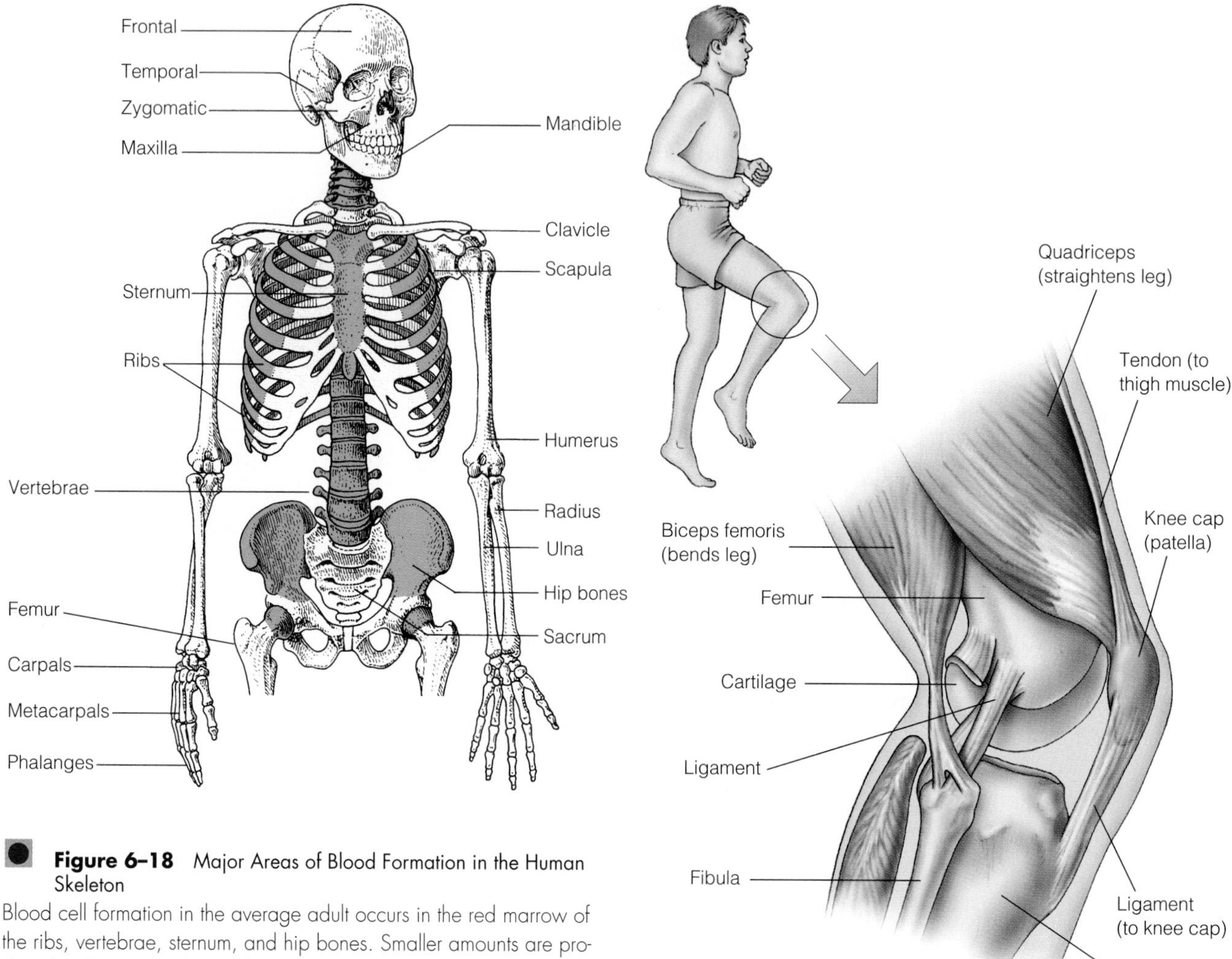

Figure 6–18 Major Areas of Blood Formation in the Human Skeleton
Blood cell formation in the average adult occurs in the red marrow of the ribs, vertebrae, sternum, and hip bones. Smaller amounts are produced in the proximal end of some long bones, such as the femur.

Figure 6–19 Ligaments
Ligaments attach bone to bone. One of the most complex joints in which bones are attached is the knee. Both the tibia and fibula are attached to the femur, and the tibia and femur are attached to the patella.

Types of Joints Allowing Movement

What types of joints are involved in accommodating movements? There are several types, including *gliding joints, hinge joints,* and *pivot joints,* as well as *ellipsoidal, saddle,* and *ball-and-socket joints* (Figure 6–20). Gliding joints accommodate movements in two dimensions. Wrist and ankle joints are included in this group. The hinge joint allows movement in a single plane and involves principally extension and flexion movements as observed in the movements around the elbow and knee. A pivot joint requires that the rounded end on one bone fit tightly and precisely into a complementary depression in the other. Rotation occurs around such a joint, and examples of this type of movement include head-twisting interactions of the atlas and the axis or the rotation of the forearm. Side-to-side and back and forth movements are often mediated by the ellipsoidal joint. This is reflected in the complex movements of the wrist. A saddle joint is one in which the surfaces of both the interacting bones are saddle shaped. The interaction of the thumb metacarpal and its associated carpal is an example of this type of joint. The ball-and-socket joint involves one bone with a ball-shaped end and the other with a cuplike receiving surface. The hip joint is a fine example of this type of joint, which allows movement in three planes.

The movements that occur between bones, as was the case for joints, fall into several categories. There are *gliding movements, angular movements, rotational movements,*

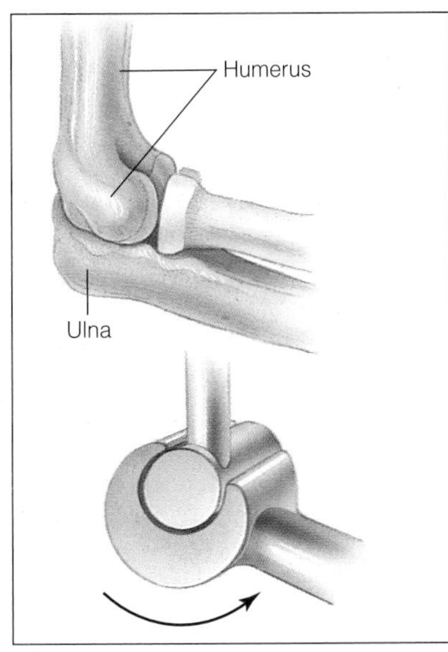

(a) Hinge Joint

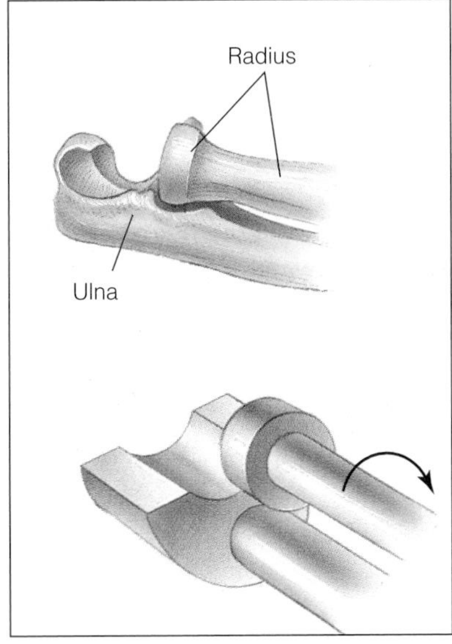

(b) Pivot Joint

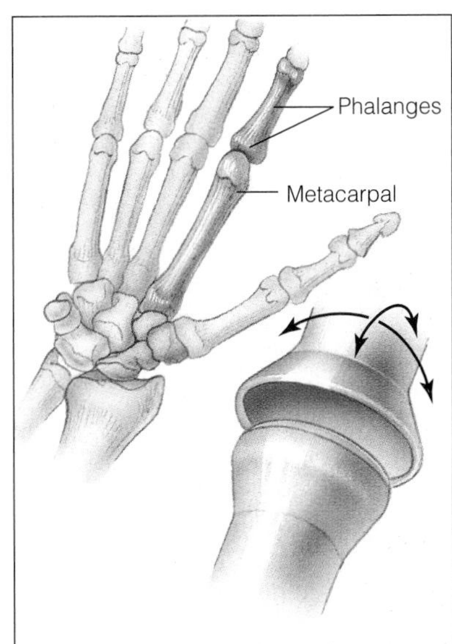

(c) Ellipsoid Joint

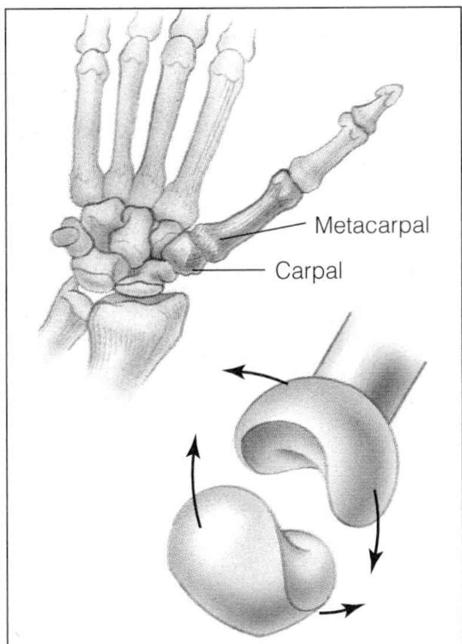

(d) Saddle Joint

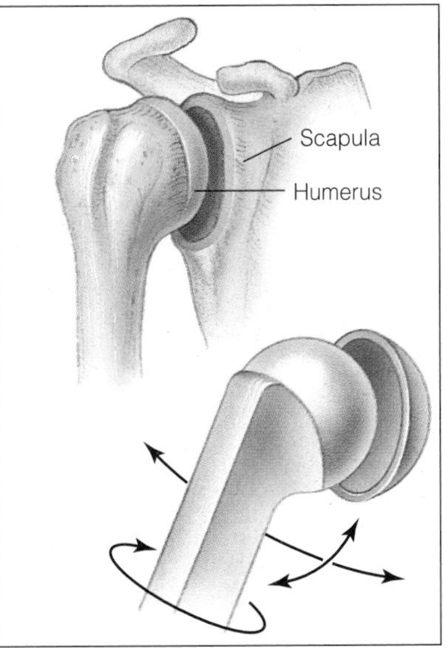

(e) Ball-and-Socket Joint

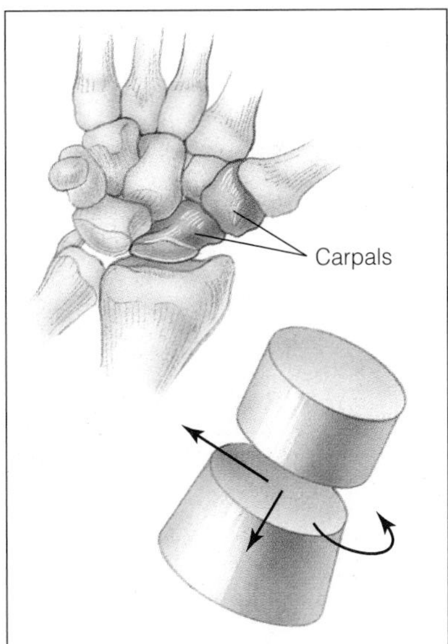

(f) Gliding Joint

Figure 6–20 Movement of Bones at Articulations
The hinge joint in (a) allows flexion and extension movements. The pivot joint in (b) , at the elbow, allows rotational movement. The ellipsoidal joint in (c) allows adduction and abduction of the hand. The saddle joint in (d) allows movement of the thumb. The ball-and-socket joint of the shoulder (e) provides rotational and abductive/adductive movement. The gliding joints of the wrist (f) allow for complex movement of bones in the wrist.

circumductive movements, and a set of special movements that combine features of the other categories. In gliding movements, two flat or slightly curved surfaces simply glide over one another, as exemplified in the interactions of the wrist and ankle bones of the hands and feet (Figure 6–20 [f]). Angular movements increase or decrease the angle between adjacent bones. Such movements as lifting your arm away from your body or bringing it closer to the body are considered angular movements, as are the flexion and extension of the forearm. Some of these movements are depicted in Figure 6–20. Rotation is the movement of a bone around its axis. If you shake your head to signify "no," it is a rotational movement of the head on the vertebral axis. If you change the position of the palms of the hands from up to down and back again, it is also a rotational movement. Circumduction is a special type of rotational movement, as occurs in the motion of the arm around the shoulder joint.

To Move or Not to Move

There are several different types of joints, some allowing movement, some not. The three major classes of joints are called *fibrous, cartilagenous,* and *synovial joints* (Figure 6–21). The **synovial joints,** the form involved in movement of the skeleton, are the most familiar type. These joints establish a fluid-filled space between bones that allows nearly frictionless freedom of movement. Synovial joints allow a wide range of different movements including pivots, rotations, and glides.

To Move—Synovial Joints The defining feature of a freely moveable joint is a *synovial cavity* between the articulating bones. This joint is surrounded by a special membrane and filled with fluid (Figure 6–21). Lining the walls of the synovial cavity is a specialized synovial membrane. This membrane produces a slippery and highly lubricating material referred to as *synovial fluid* (Figure 6–21). The articulating bones essentially float over one another in a nearly frictionless manner. The joint is encased in a strong, fibrous capsule that integrates with the periosteum of each of the adjacent bones and helps hold the two bones together. The shape of bones at each joint is critical to their function. Articulating bone surfaces must be complemen-

Synovial joint a fully moveable joint

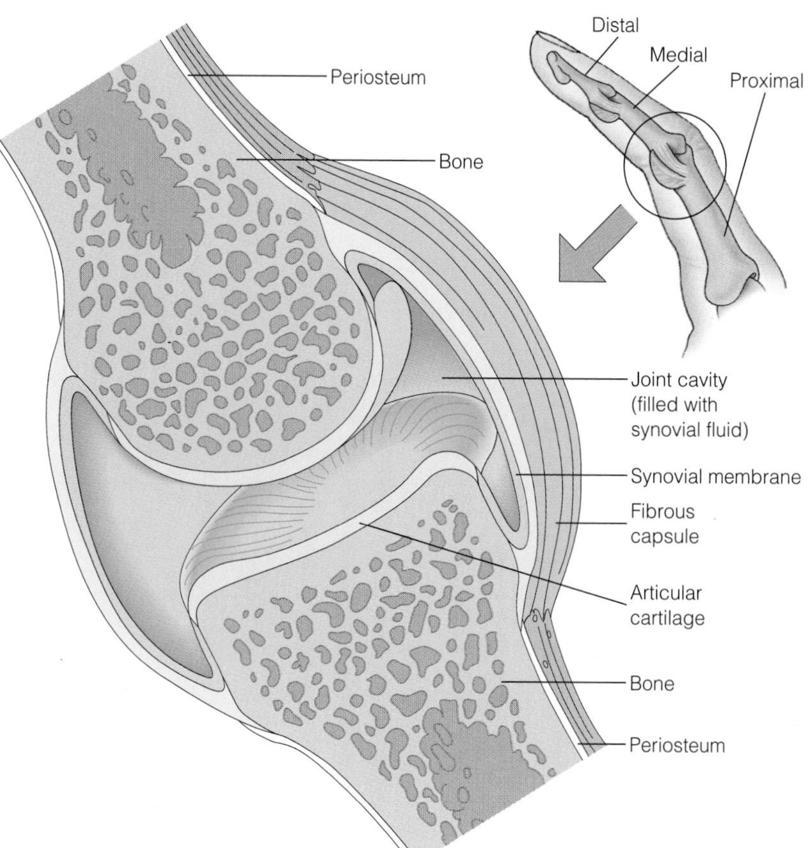

Figure 6–21 A Synovial Joint
A synovial joint has a number of characteristic features. It possesses a joint cavity filled with synovial fluid. The cavity is lined with synovial membranes that secrete the fluid. The ends of each articulating bone are covered by hyaline cartilage and surrounded by a fibrous capsule.

tary to perform the types of movements that occur between them.

To Move a Little—Cartilagenous Joints These joints are constructed of cartilage and allow slight movement between the bones they connect. This is characteristic of a class of cartilaginous joints known as *symphyses*. For example, the symphysis between the surfaces of the bones connecting the front of the pelvic girdle is a flat, broad connective tissue plate that allows slight movement. In fact, during the late stages of pregnancy, the *pubic symphysis* is altered and becomes more pliable and stretchable (Biosite 6–1). This allows for maximal expansion of the pelvis during delivery of a baby. Another example of this type of joint is the intervertebral disc, which is composed of cartilage and cushions the contact between vertebrae (Figure 6–6).

Not to Move—Fibrous Joints Fibrous joints are connections between bones that do not allow movement. The classic example of this type of joint is the **suture.** Sutures are found only in the skull between the bones forming the cranium (Figure 6–4). The shape of the suture is often interdigitating in appearance. The bones of the skull of a newborn infant are incomplete and the zones between them are flexible (Figure 6–22). The soft spots of cartilage between unfused bones are called **fontanels,** or little fountains, because the infant's blood flow can be observed pulsing underneath them. As in the pubic symphysis of the mother, the incomplete sutures of the neonate skull allow the baby's head to be compressed so as to pass more easily through the birth canal.

DISORDERS OF BONES AND JOINTS

Dietary

There are a number of deficiencies and diseases that are specifically associated with bones and joints (Table 6–2). Dietary deficiencies are particularly telling on the structure of bone. Calcium and phosphorus (as phosphate ions) are required to provide the mineral building blocks for bones. As was discussed in the section on osteoblasts and osteoclasts, bones are in a constant state of remodeling, so imbalances in supplies of these inorganic materials can be serious. This is particularly important for women during pregnancy, when the extra burden of supplying organic and inorganic materials to the skeleton of the fetus can deplete the mother's bones of calcium and other ions and molecules found in bone. Sufficient protein in the diet (and the *essential amino acids* of which they are composed) are also necessary. Amino acids furnish the organic building blocks from which collagen and proteoglycans needed for bone formation are synthesized. Sufficient supplies of

Suture fibrous joint particularly evident between the bones of the cranium

Fontanels meaning tiny fountains; regions between bones of the cranium that have yet to form sutures

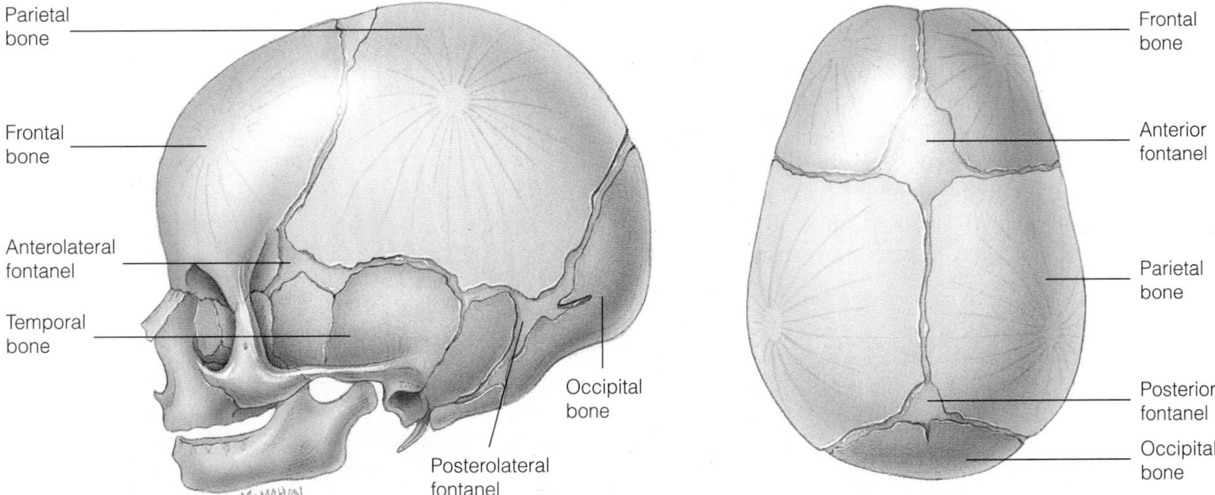

Figure 6–22 The Fontanels of the Fetal Skull
The different fontanels in the fetal skull allow the bones of the fetal skull to be displaced during delivery. In adults, these joints (sutures) between skull bones are rigid.

these organic components ensure that the right ingredients for bone growth, remodeling, and repair are available to the osteogenic cells.

Vitamins and hormones are also very important for bone formation. *Vitamin D* (*calciferol*) is essential for the utilization of calcium and phosphate. If minerals are not recruited for mixing the concrete of the bone matrix, the resulting bones will be subject to bowing and bending. *Rickets* is a condition in which a deficiency of vitamin D results in soft, overly pliable bones in children. *Osteomalacia* is a similar condition in adults and also results in softening of the bones. A consequence of this softening (demineralization) is revealed by the effect of Earth's gravity, which is sufficient to cause bending of the weight-bearing bones. Osteomalacia is the opposite of osteoporosis, which is a condition in which the organic components of bone are depleted and the affected bone becomes brittle and far more susceptible to fractures than normal bone (Figure 6–13).

Diseases

Conditions such as *rheumatism, bursitis,* and *tendinitis* refer to conditions of inflammation that affect supporting structures of bones and joints (Table 6–4). Rheumatism is a painful condition affecting bones, ligaments, joints, and other tissues. **Arthritis** is a form of rheumatism, and one major form of this disease that affects joints is the disabling *rheumatoid arthritis.* This disease is the most common type of arthritic inflammation of joints. The disease affects the synovial membranes and, if left unchecked, will progressively destroy the capsular articular cartilage and lead to excessive mineralization in the joint and the fusion of adjacent decapsulated bones. This, in turn, immobilizes the fingers and has crippling effects on the hands. Bursitis is an inflammation of the membranes surrounding joints and bones. Tendinitis is the inflammation of tendons (which connect muscles to bones) and synovial membranes surrounding joints such as the wrist, shoulders, fingers, and elbow (as in tennis elbow).

Cancers of bone cells are called **osteomas** (Table 6–4). They result from the transformation of osteogenic cells and the subsequent overgrowth of the normal bone by cancer cells. Other types of cancer cells can grow in bones.

Table 6–4 Diseases and Abnormalities Affecting Bones and Joints

Cancer
Osteomas- arise from transformed osteoblasts
Chondromas- arise from transformed chondroblasts
Lymphomas and Myelomas- arise from transformed bone marrow cells
Metastasized cancer cells from carcinomas and other cancer types
Bacterial infections
Syphilis
Osteoporosis
Brittle bones; associated with reduced estrogen
Rickets
Deficiency of vitamin D induces soft bones in children
Osteomalacia
Soft bones in adults
Osteopetrosis
Excessively thick dense bones
Scurvy
Deficiency of vitamin C in diet affecting collagen synthesis
Rheumatism
Inflammation or pain in joints
Bursitis
Inflammation of a bursa or membrane between a tendon or a bone
Arthritis
Inflammation of joints due to infection or metabolic causes
Tendinitis
Inflammation or pain in tendons attaching muscles to bones

Transformed blood cells give rise to *lymphomas* and *myelomas,* whose origin is the bone marrow. In fact, most types of malignant cancers undergo *metastasis,* which is a form of translocation of transformed cancer cells from the site of origin of the cancer to other sites. Such metastatic cells can and do spread to bones, where they form cancerous foci that are difficult, if not impossible, to eradicate by contemporary medical procedures (see Chapter 21 for more details on cancer).

Arthritis an inflammation of a joint or joints
Osteoma cancer of bone

Summary

The bones of the human body are organized into an articulated skeleton, which supports and protects, in one way or another, all other organ systems. Bones provide a rigid framework connected by ligaments and to which tendons and muscles attach to bring about movement of the body. Moveable joints provide hinges for muscle-driven movements, but other types of joints connect bones into immobile configurations that strengthen the skeleton, similar to the welded plates of the hull of a ship.

Bones are remodeled continuously by the combined efforts of two types of cells—osteoblasts and osteoclasts. Osteocytes are found in lacunae within osteons and are capable of synthesizing and assembling organic molecules and inorganic compounds, such as collagen and calcium phosphate, respectively, to form the strong and durable substance of bone. Of all the organ systems of the human body the skeleton remains intact the longest after life has ceased.

Some of the most revealing remnants of the human past are found in the brittle bones of bodies buried in long forgotten graves. Trapped in the protective arms of dry, silent tombs, or held in the preserving waters of peat bogs or permafrost, the bones of *Homo sapiens* tell a story of us. As the skeletons of these long dead men, women, and children are examined and reassembled, so we reconstruct the history of our kind.

Questions for Critical Inquiry

1. What purposes does an internal skeleton serve in animals? What about animals such as insects and clams that do not have an internal skeleton? How might they be supported in structure and function?
2. Are the human skeleton and skeletons of other mammals related to one another? How?
3. In what ways has the human skeleton (*Homo sapiens*) evolved over time to accommodate erect posture?
4. Is there a skeleton in an embryo? Are there bones in an embryo? If so, when and where do they form; if not, when and where will they form?

Questions of Facts and Figures

5. What does it mean to be bilaterally symmetrical? Are we perfectly bilaterally symmetrical?
6. How many bones are there in a normal human body?
7. What are the two major categories of skeletal organization? How are they related to the symmetry of the body?
8. What cells make bone? What cells "unmake" bone?
9. What types of joints are there in the human body? Which of them allow for movement of the skeleton?
10. Where is blood formed in a human adult? Is it formed there during embryonic and fetal development? If not, where else is it formed?
11. What is the organic and inorganic composition of bone?
12. What are the cancers of bone called?

References and Further Readings

Parker, S. (1993). *The Body Atlas.* London: Dorling Kindersley.

Stalheim-Smith, A., and Fitch, G. K. (1993). *Understanding Human Anatomy and Physiology.* St. Paul, MN: West Publishing.

Tortora, G. J. (1986). *Principles of Human Anatomy.* New York: Harper & Row.

CHAPTER 7

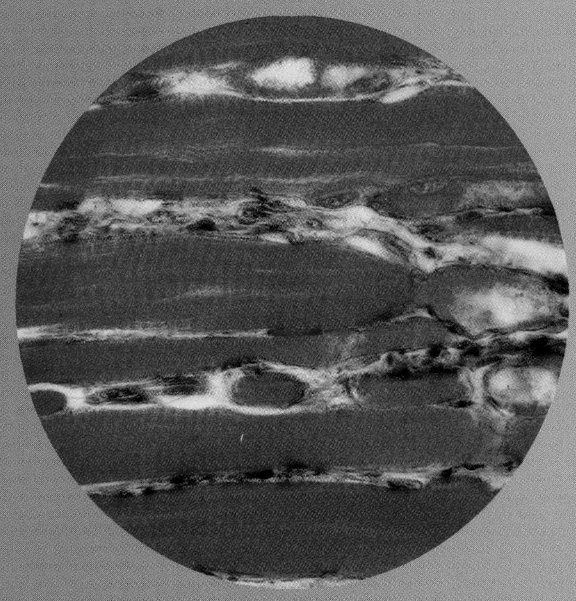

The Muscles and Muscular Systems

INTRODUCTION

Have you ever watched people move? Have you simply observed them in ordinary activities such as walking, skipping, jumping, lifting, and riding? It is quite amazing how many different movements a human body performs in carrying out what seem to be the most ordinary of tasks. As observers, we tend to watch and appreciate the graceful movements of a dancer, or the endurance of a long distance runner, or the awesome strength of a weightlifter. They are impressive and often inspiring. However, in an objective appraisal, the specialized talents of the athlete and the dancer reflect abilities inherent to all human beings. We share in common a part of our anatomy, the muscular system, that makes bodily movements not only possible but often elevates these movements to something quite exceptional.

THREE TYPES OF MUSCLES—STRIATED SKELETAL, SMOOTH, AND CARDIAC

The muscles of the human body with which we are most familiar are actually large groups of muscle cells bundled together and connected in specific patterns to the bones of the skeleton. As described in Chapter 4, muscles are composed of several types of muscle cells or fibers whose key attribute is the ability to contract when stimulated to do so. The controlled contraction of muscles of all types is essential for active movement. In this chapter we focus on the relationships between the structure and function of muscles and the cells of which they are composed. The formal study of muscles is called myology. The prefix *myo* means muscle and will in this chapter appear often in conjunction with the naming of cells (myocytes and myofibers), and the specialized structural components within them (myofilaments).

Muscle cells occur in three basic types—striated skeletal, smooth, and cardiac (Figure 7–1). The cellular aspects of the three muscle cell types provide a way to differentiate them from one another. Skeletal muscle cells are long and fibrous and are referred to as muscle fibers or myofibers. This designation applies because myofibers are formed from the fusion of many individual myoblast cells. Myoblasts are immature muscle cells found in the embryo and are precursors to myofibers. Myofibers retain the many nuclei resulting from this fusion and have a highly organized, internal cytoplasmic structure with the appearance of striations or bands. Smooth muscle cells are just that, cells not fibers. They lack obvious striations within their cytoplasm and have but a single nucleus. Cardiac muscle cells are also single cells with a single nucleus, but they are densely striated. Cardiac myocytes are held together by unique and powerful intercellular adhesive junctions.

The ability of muscles to contract is the key to several important anatomical functions, and the different types of muscles and muscle cells are involved in different functions. In general, skeletal muscle fibers are used to move the skeleton. This is brought about when the nervous system stimulates skeletal muscle fibers to contract while connected to bones of the skeleton by tendons (Figure 7–2). The skeletal muscles are often massive and overlap one another. The thickness and overlap of muscles gives the body its shape and contours.

Smooth muscles are composed of highly organized groups of individual muscle cells. For example, such groupings are found surrounding the intestinal tract, blood vessels, and the lungs and provide the force required to move solids, fluids, and air through the internal channels of these anatomical structures (Figure 7–3). The action of smooth muscles in contracting around and constricting the tubelike structures of the body provides a means to (1) aid in the motility of food by means of peristalsis, (2) move blood, and (3) control and restrict the movement of air entering and leaving the alveoli. The location of smooth muscles within the body's visceral organs gives them a second, and common, designation as visceral muscle. Smooth muscle action is controlled involuntarily by the autonomic subdivision of the central nervous system.

Cardiac muscle cells (referred to as *cardiac myocytes*) differentiate and organize exclusively in the heart. The functional communication, tight adhesion, and coordinated group action of cardiac myocytes during contraction control the volume of the chambers of the heart. This contraction brings about the movement of blood through the chambers and into the **pulmonary** and **systemic** human **circulatory system** (Figure 7–4). Contraction of cardiac muscles, as that of visceral muscles, is controlled by the autonomic nervous system and is, thus, largely involuntary in its actions.

With these properties of muscle cells in mind, this chapter will focus on the structure and function of the three types of muscles, the means by which they contract, and their anatomy, histology, and intracellular organization. We start with striated skeletal muscle.

STRIATED SKELETAL MUSCLE

When one views the musculature of the human body, it is apparent that there is a massive amount of skeletal muscle tissue and that there are many different and independently acting muscles or muscle groups involved in movement and posture. In fact, the human body is by mass 50% muscle and is composed of over 600 different muscles connected with the skeleton (Figure 7–5). Some muscles are quite small, such as those that control the movement of the eyes, and others are large by comparison, as the *Rectus femoris* of the thigh. What all the skeletal muscles have in common is that they are composed of bundles of muscle fibers. Individual muscle fibers form large groups or bundles, many of which, in turn, represent the commonly named muscles of human anatomy.

Bundles of Muscles

How are groups of muscle fibers held together? This is accomplished by the formation of connective tissue sheaths, which interact with muscle tissue to form larger and larger bundles (Figure 7–6). The three main types of connective tis-

> **Pulmonary circulatory system** blood vessels leading to and away from the lungs; associated with the right side of the heart
>
> **Systemic circulatory system** blood vessels leading to the entire body other than the lungs and back to the heart; associated with the left side of the heart

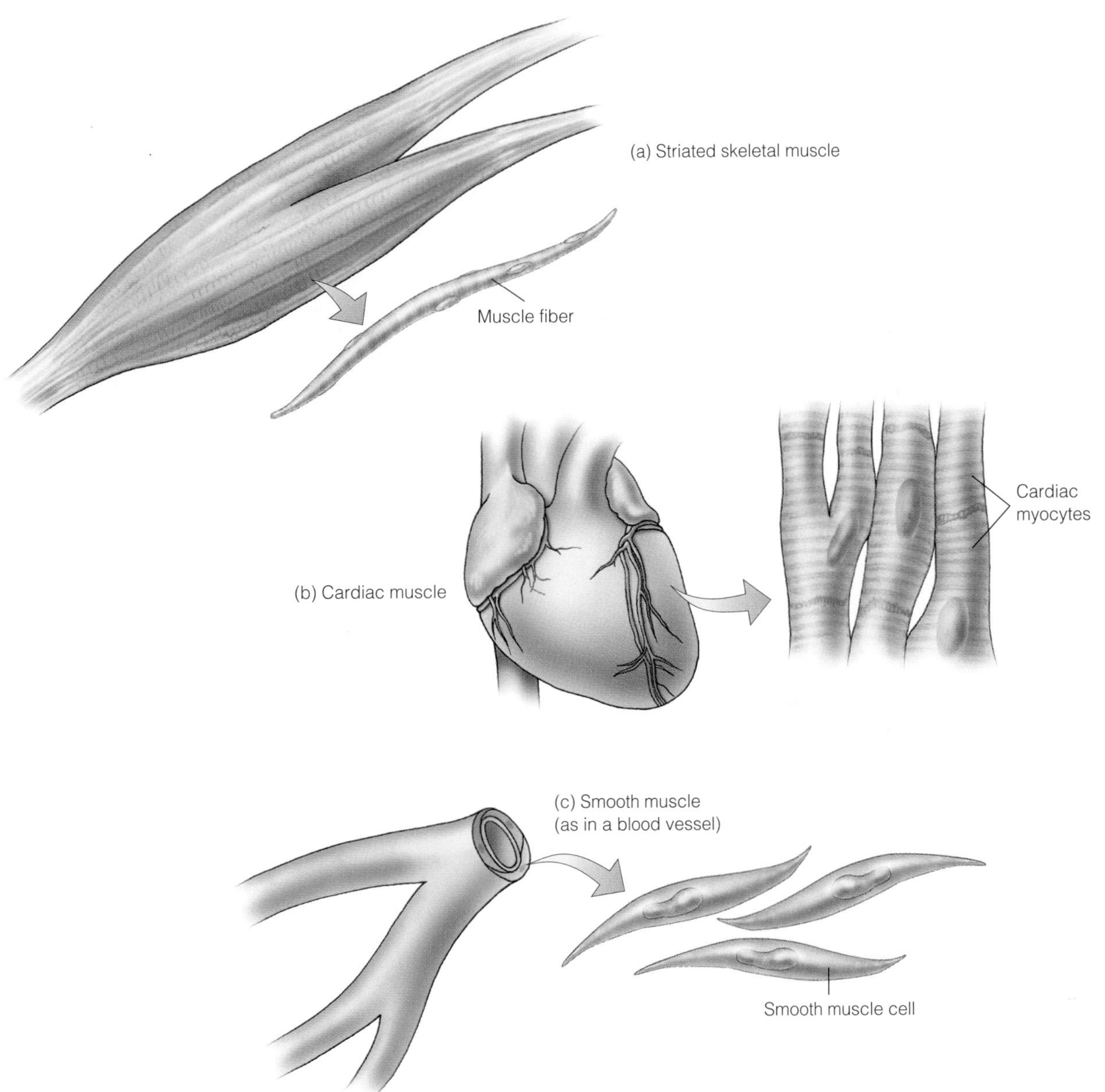

Figure 7–1 Three Types of Muscle
The three types of muscle are (a) striated skeletal muscle, (b) cardiac muscle, and (c) smooth muscle. Each provides special contraction to move the body (a), pump blood (b), and control the circulation of blood to all regions of the body (c).

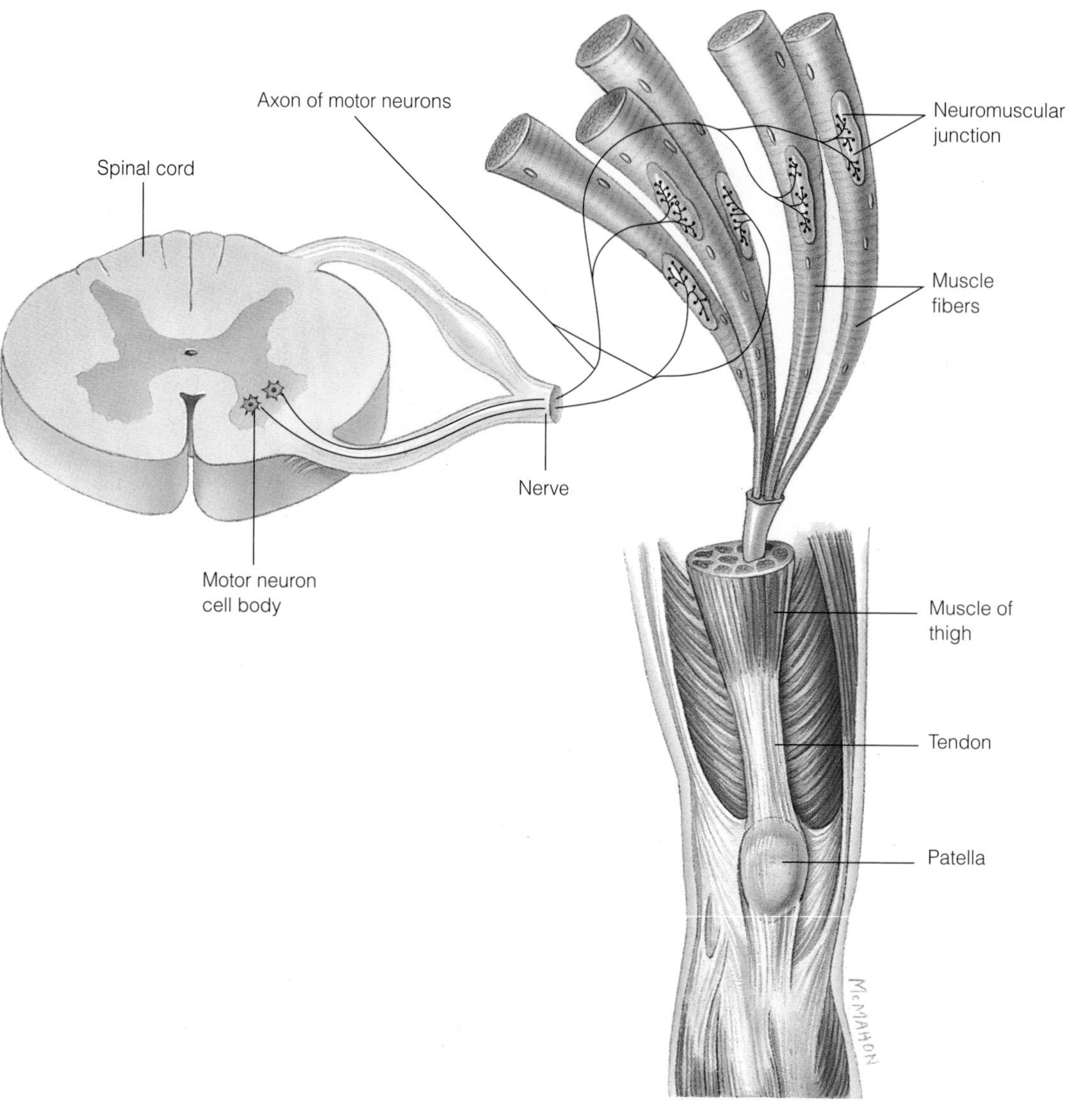

Figure 7–2 Integration—Nerve, Muscle, Bone, and Tendon
The action of muscles is controlled by the nervous system. The integration of muscle and bone is mediated by tendons. Movement of the skeleton depends on proper connections among these elements.

sue involved in surrounding and interconnecting these bundles are the *epimysium, perimysium,* and *endomysium.* These specialized connective tissues are differentiated from one another principally based on the size and number of myofibers they envelope. The epimysium surrounds an entire muscle and incorporates within it other structural elements, including blood vessels and nerves. The perimysium surrounds smaller bundles of muscle fibers within the muscle and includes tiny blood capillaries that deliver oxygen and nutrients and remove waste from the individual muscle fibers. The endomysium is the most delicate of the three types and surrounds individual muscle fibers. Collectively the connective tissues tie the myofibers, bundles, and entire muscle group together in a strongly bound unit. Tendons, which are tough, flexible, fibrous connective tissue, integrate with all three layers and provide a means to extend beyond the end of a muscle to attach to bone.

Leverage—Attachment of Muscles with Bones

The attachment of muscles to bones occurs via tendons. As described in Chapter 6, the synovial joints between bones provide for movement of the bones of the skeleton relative to one another. The sites at which a muscle is actually attached via tendons to a bone is an important determi-

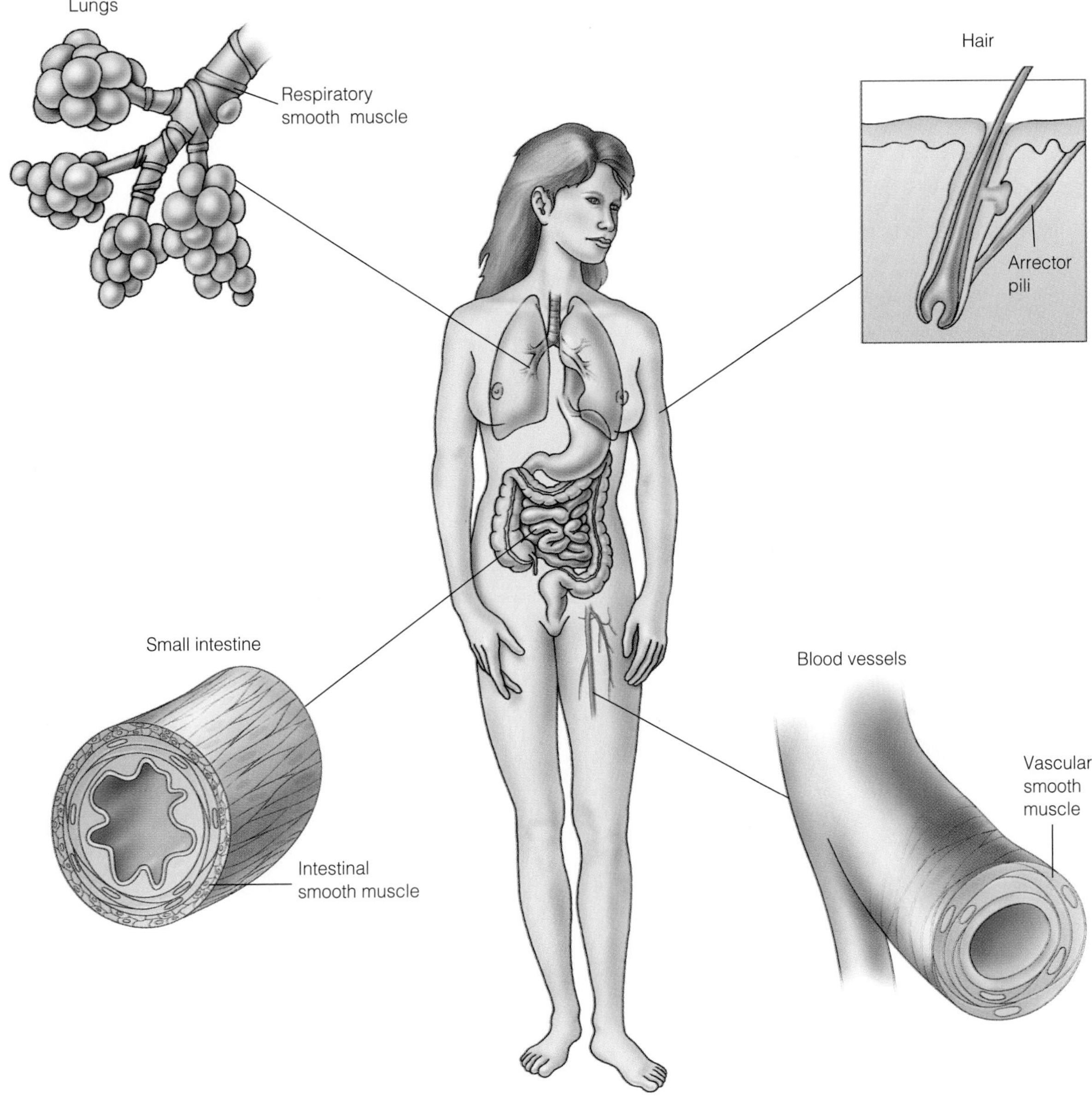

Figure 7–3 Smooth Muscles and Tubular Elements of the Body
Smooth muscles are associated with many different organs. Contraction of smooth muscles restricts flow of air to the lungs, flow of blood to the body, and movement of material in the digestive tract. Smooth muscle also controls movement of body hair.

nant of the direction and strength of skeletal movements around the joints. Muscles have two types of attachment points, referred to as **origins** and **insertions** (Figure 7–7). For movement to occur, striated skeletal muscles must be attached to at least two different bones. One end of a muscle forms an attachment that anchors the muscle to an

Origins sites on bone at which tendons attach and stabilize muscles

Insertions sites on bone at which tendons attach to allow muscle contraction to move bones around a joint

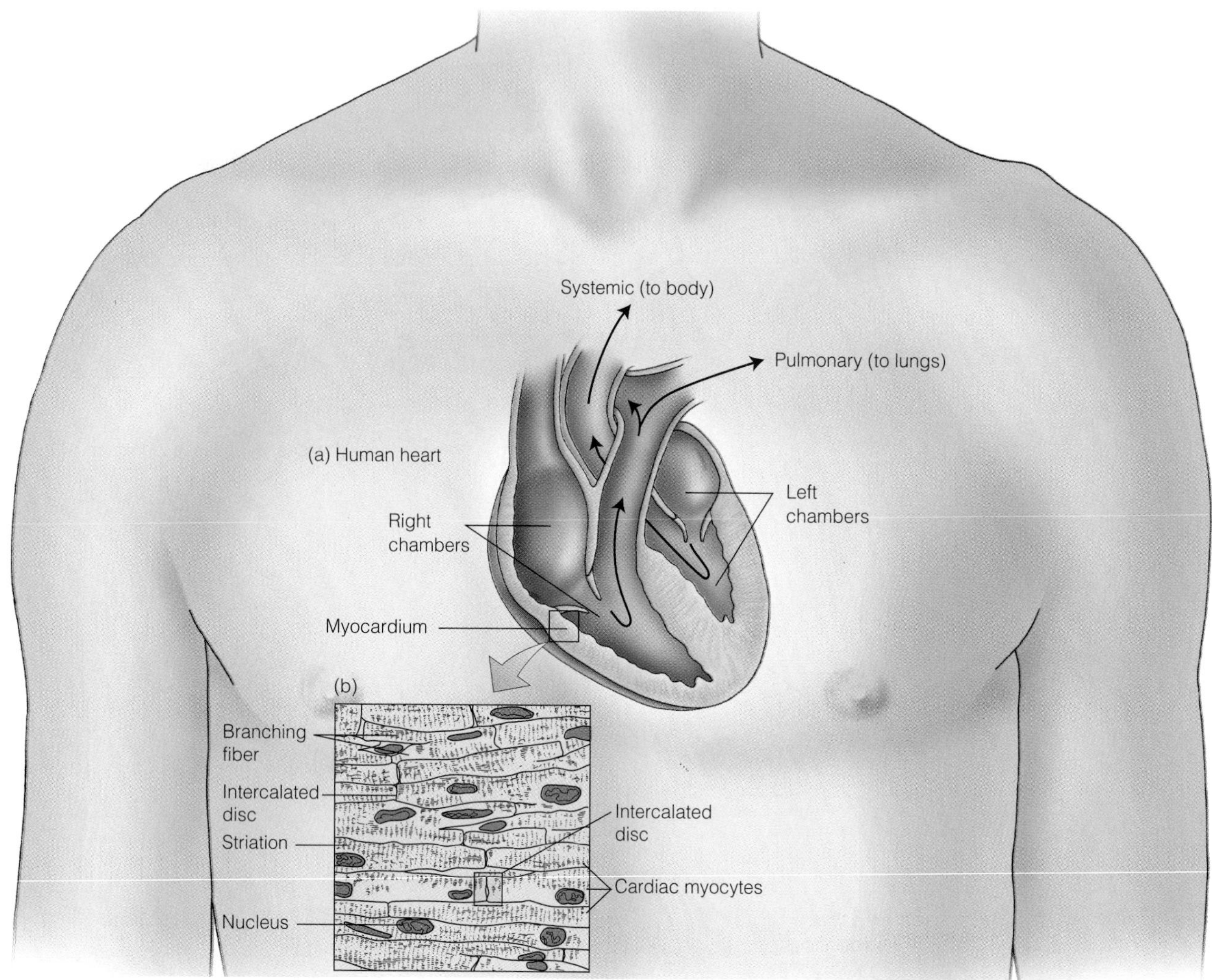

Figure 7–4 The Heart and the Organization of Cardiac Myocytes
The action of the heart (a) is to pump blood using the contraction of the cardiac myocytes around the chambers. The muscle layer of the heart is the myocardium, which is composed of densely packed cardiac myocytes (b).

immobile bone and establishes the origin. The other end attaches to the bone that is to be moved and establishes the insertion. Implicit in this scheme is the notion of muscles working together in a coordinated fashion. In fact, all active movement of the skeleton requires the coordinated contraction and/or relaxation of many different muscle groups.

The attachment of muscles on different bones and across the joints connecting those bones provides *leverage*. The physical description of levers and how they act is important in understanding the musculoskeletal system from an engineering point of view (Figure 7–8). There are three classes of levers, whose nature depends on the relationship between the position of a *fulcrum* (F) and the application of the force (E) with respect to the resistance (R) to be overcome. Using a simple teeter-totter model, the first class of levers can be explained. The fulcrum is positioned midway between the two riders. When one rider is much heavier than the other, the lighter rider is propelled upward. The only way to balance the teeter-totter is to move the fulcrum nearer to the heavier rider. Thus, in principle, if enough pressure is put on one end of the lever, the other end (with a mass or resistance associated with it) will rise. The placement of the fulcrum is important, and, as the ancient Greek mathematician and inventor Archimedes once claimed, given a sufficiently long lever

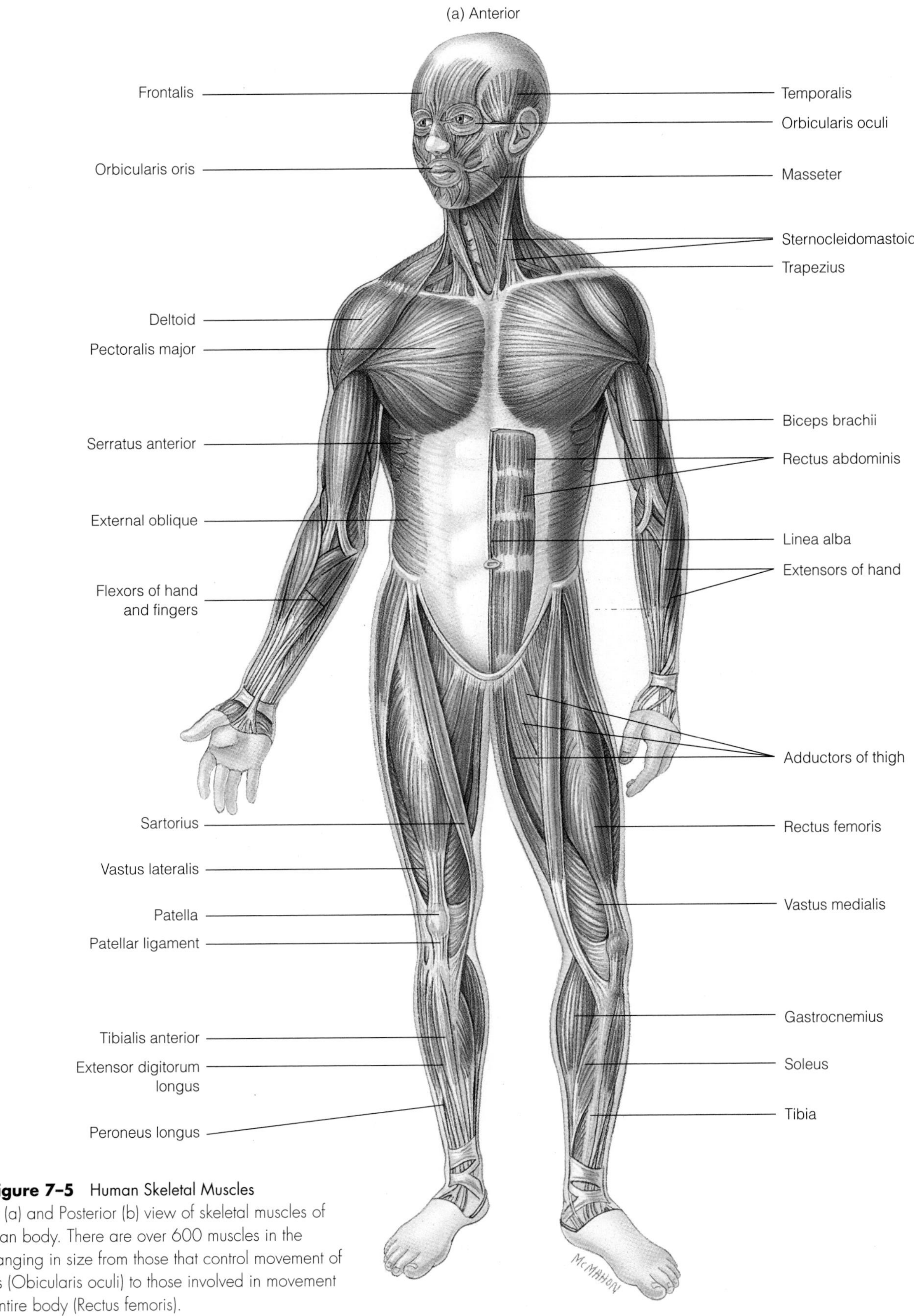

Figure 7–5 Human Skeletal Muscles
Anterior (a) and Posterior (b) view of skeletal muscles of the human body. There are over 600 muscles in the body, ranging in size from those that control movement of the eyes (Obicularis oculi) to those involved in movement of the entire body (Rectus femoris).

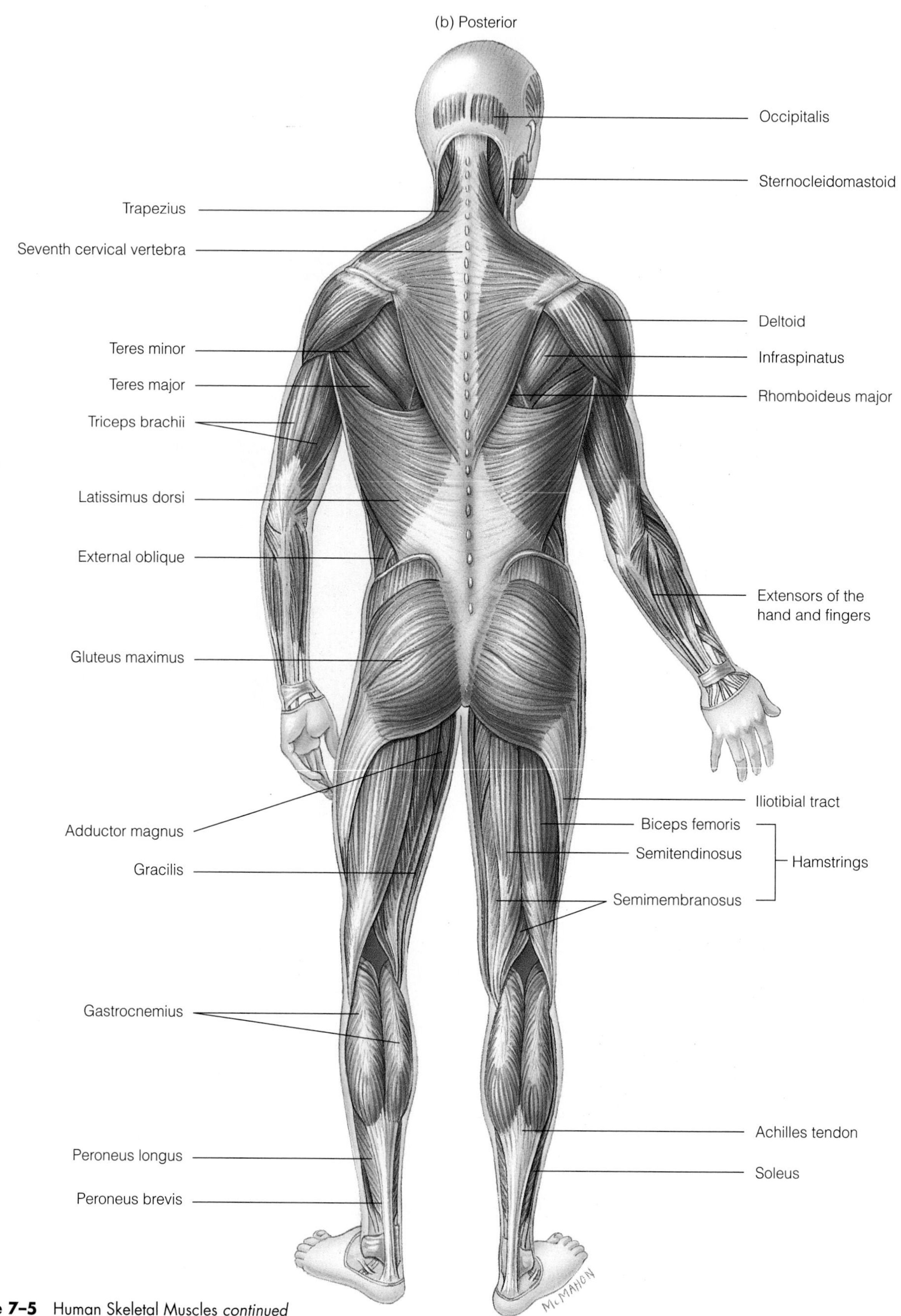

Figure 7–5 Human Skeletal Muscles *continued*

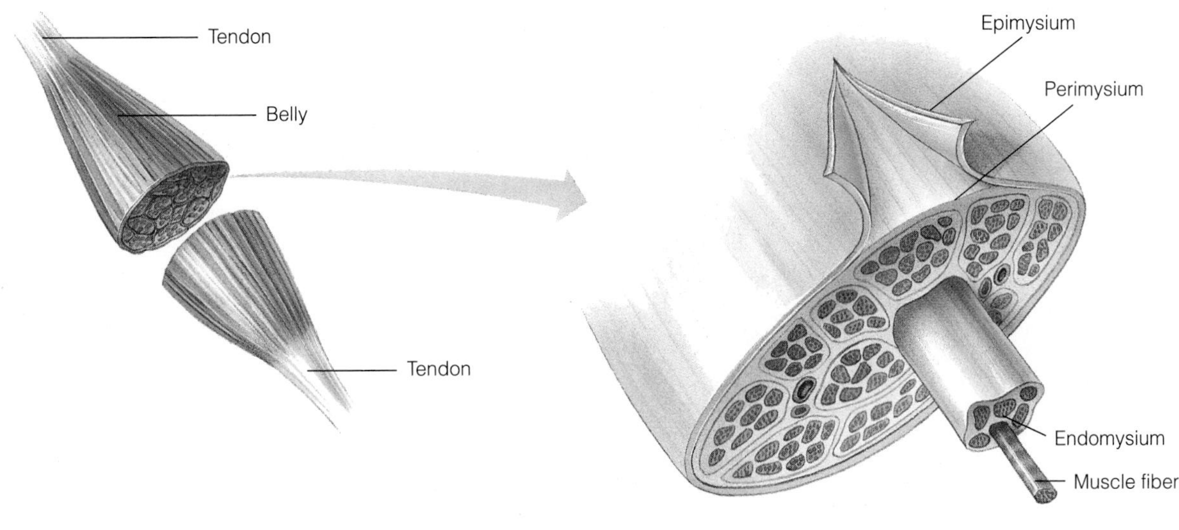

Figure 7–6 Muscle Sheaths
To function properly, muscle fibers must be tightly bound together. Individual muscle fibers are held together by an endomysium (b). Small bundles of fibers are connected by a perimysium, and the entire muscle is wrapped in an epimysium [(a) and (b)].

(and an appropriately placed fulcrum), he could move the world. An example of this type of movement in the human body is the nodding motion of the head. The fulcrum is the atlas of the spine, the force is applied by the contraction of the muscles of the neck, and the head is the resistance at the other end of the lever (Figure 7–8[a]).

Change the position of the fulcrum and the action of the muscles are changed. In a second class of levers the fulcrum is at one end of the lever and the force is applied at the other end. The resistance is in between (Figure 7–8[b]). A human body lever of this type is exemplified in the lifting of the body on the balls of the feet. The final type of lever is one in which the fulcrum is at one end of the lever, the resistance is at the other end, and the force is applied somewhere between [Figure 7–8(c)]. The movement of the lower arm by the action of the *Biceps brachii* muscle is a case in point. The origin(s) of the muscle are on the scapula, and the insertion is on the radius bone of the forearm. The action of this movement is called *flexion*. The opposite action is called *extension* and, as will be discussed, it involves a different muscle, the *Triceps brachii*.

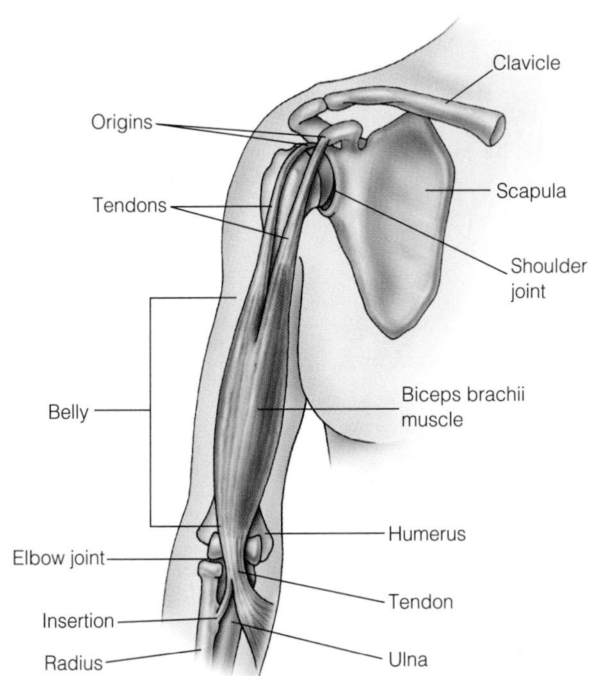

Figure 7–7 Origins and Insertions of a Muscle
The origin and insertion of the *Biceps brachii* muscle are shown here. Attachment of muscles across a joint (the elbow) provides the means to move the lower arm in a flexing motion.

Muscles of the Head and Neck

The muscles of the head (Figure 7–9) control facial expressions, **mastication** (chewing), opening and closing of

Mastication the action of chewing

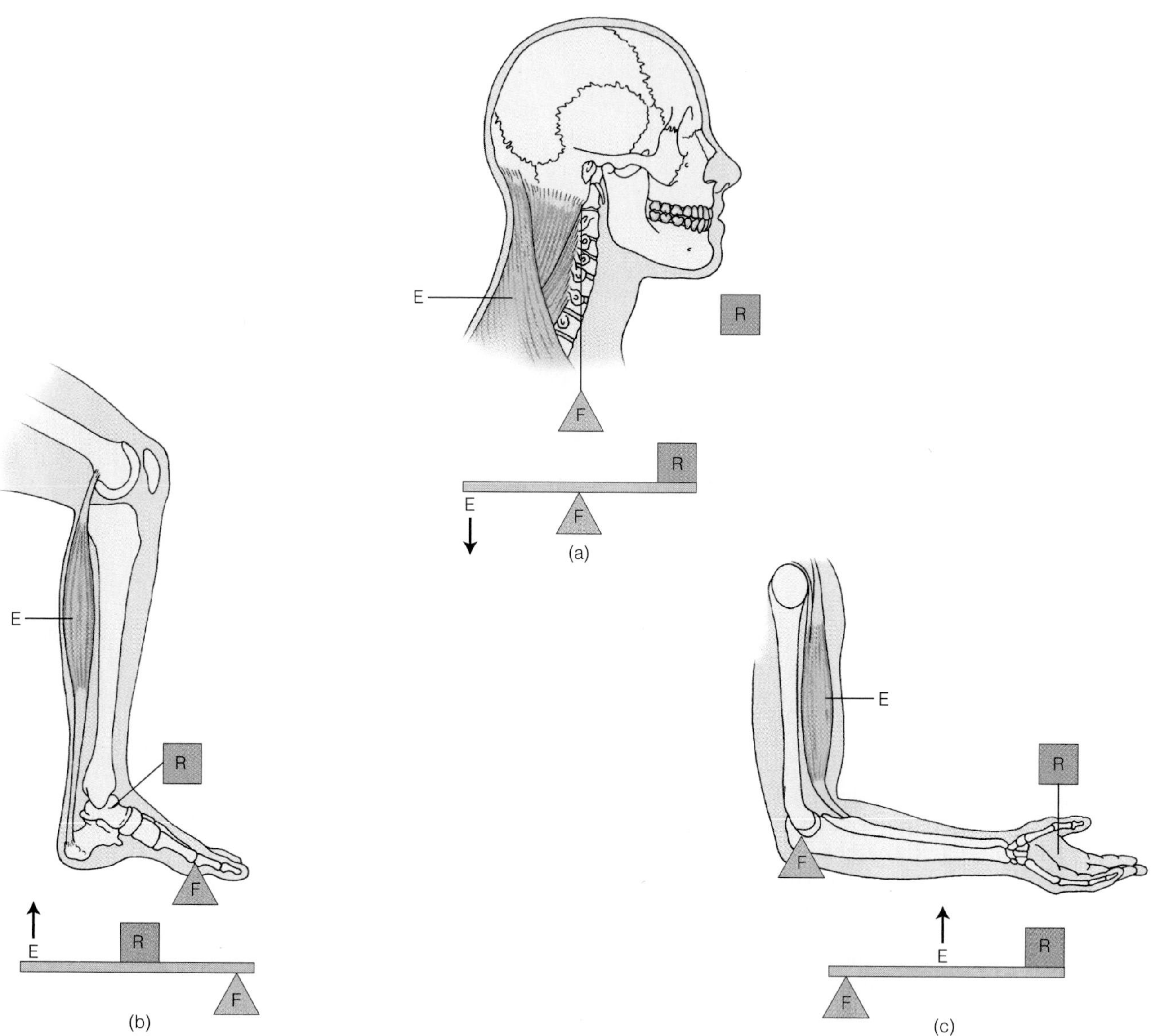

Figure 7–8 Levers and Skeletal Movement
Three types of levers are represented. In (a) the fulcrum (F) is between the resistance (R) and the application of force (E). In (b) the resistance is between the force and the fulcrum, and in (c) the application of force occurs between the fulcrum and the resistance. Examples of the action of these levers are nodding of the head (a), lifting of the body on the balls of the feet (b), and flexing the arm (c).

the eyes, and movements of the head up and down and side to side. Smiling and frowning involves the coordination of many different muscles of the face including the orbicular muscles of the eyes and mouth and those of the chin, cheeks, and forehead. The tiniest muscles of the body are found within the middle ear. They are attached to the three small bones that transduce sound waves hitting the *tympanic membrane*. These mechanical forces generate signals that are picked up by the nerve receptors of the inner ear.

Muscles of the Arm

The principal muscles of the arm are the *Deltoid, Biceps brachii, Triceps brachii, Brachialis, Pronator teres, Supinator,* and *Brachioradialis,* as well as flexor and extensor muscles. These groups of muscles are involved in the movement of the upper and lower arm as well as the movement of the fingers (Figure 7–10). The actions of the muscles of the arm undergo a finely regulated combination of contractions and relaxations to precisely control the movement of the arms, hands, and fingers. The enhancement or complementary action of muscles working together is called **synergism.** Synergistic muscles help control the direction of a movement by adjusting the angle of the contraction of the main muscle group and by stabilizing the other muscles involved (Figure 7–11). This is similar in outcome to controlling the airflow over the wings and tail of an airplane. The wings and tail themselves are set at a permanent angle, but portions of them (ailerons, flaps, and rudder) allow adjustments for takeoffs, maneuverings, and landings. Synergistic muscles are our rudders.

Synergism working together

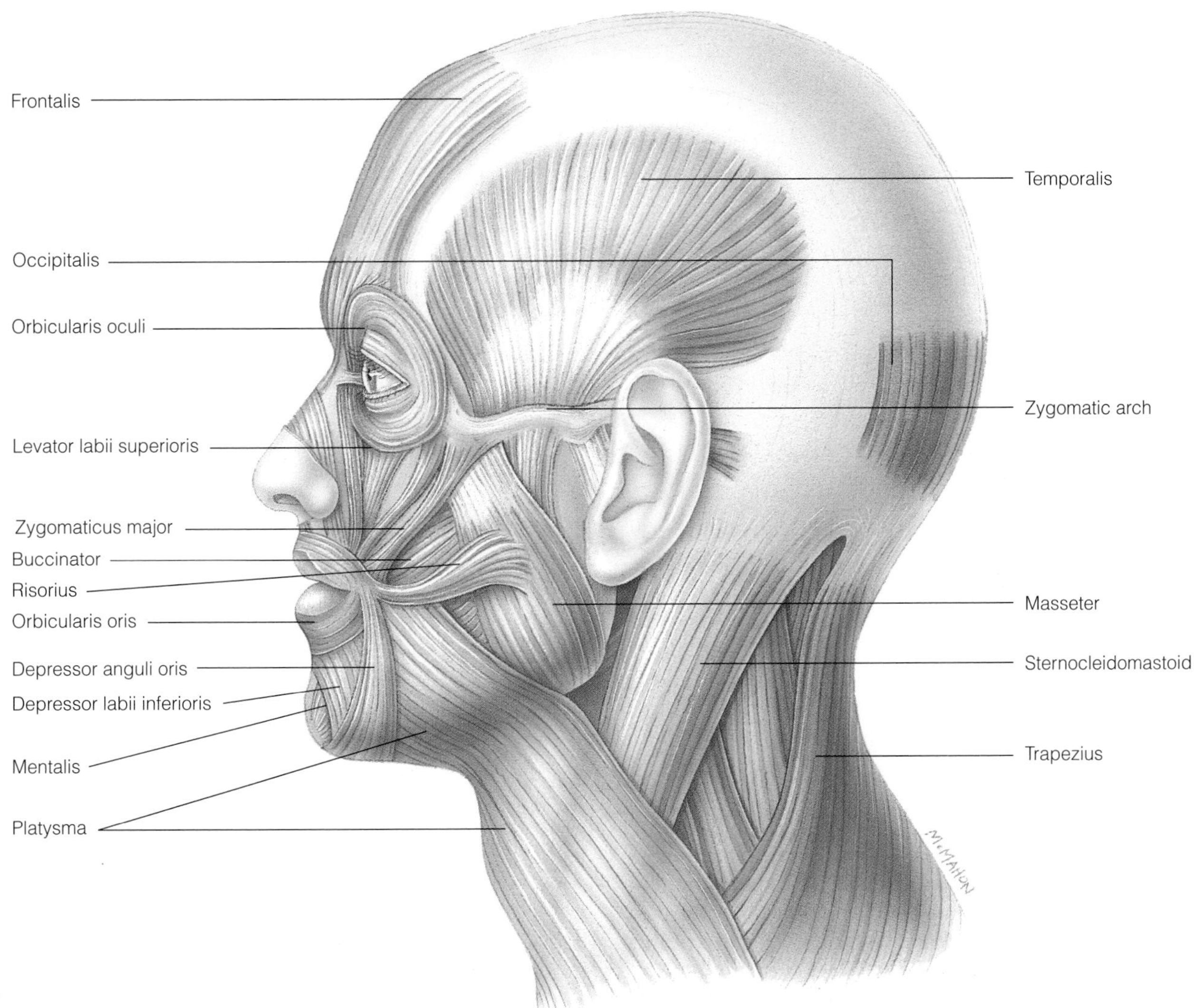

Figure 7–9 Muscles of the Head and Neck
Muscles of the head and neck provide a number of types of movement including nodding (Sternocleidomastoid), blinking (Orbicularis), chewing (Masseter), and smiling (Risorius).

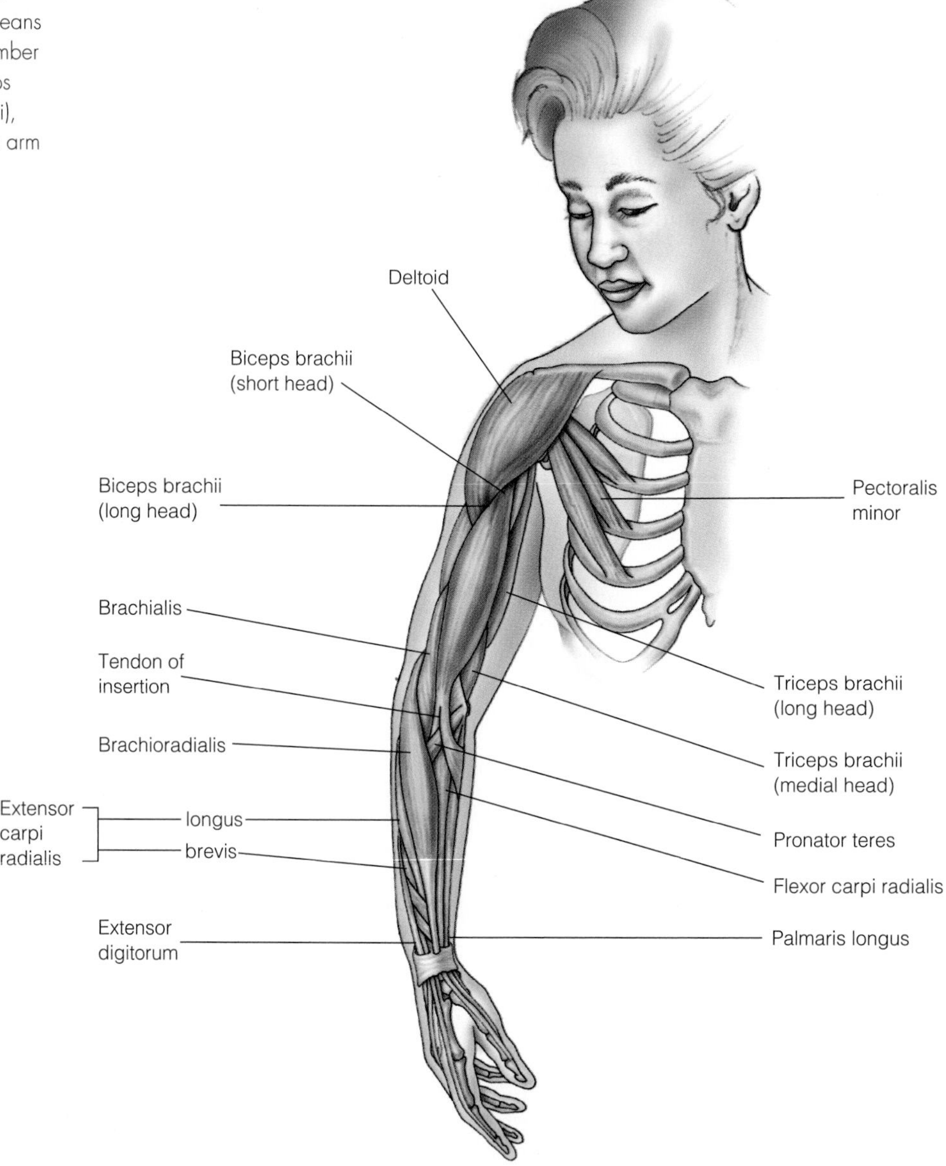

Figure 7–10 Muscles of Arm
The muscles of the arm provide the means to move the skeletal elements in a number of movements including flexion (Biceps brachii) and extension (Triceps brachii), finger movements, and rotation of the arm at the shoulder and elbow.

The opposite of synergism in muscle interactions is called **antagonism** and is important in developing tension between muscle groups. For example, the contraction of the *Biceps brachii* muscle on the front of the arm results in flexion. The contraction of the *Triceps brachii* muscle on the back of the arm results in the opposite motion, extension. During these reciprocal movements of muscles, one is contracting, while the other is relaxing (Figure 7–11). This dynamic opposition between muscle groups allows one to control the spatial position of the arm and the rate at which it moves.

The reciprocal nature of these muscles is an important consideration, because muscle contraction is an active process. Muscle extension, however, is not an active process

Antagonism working against one another

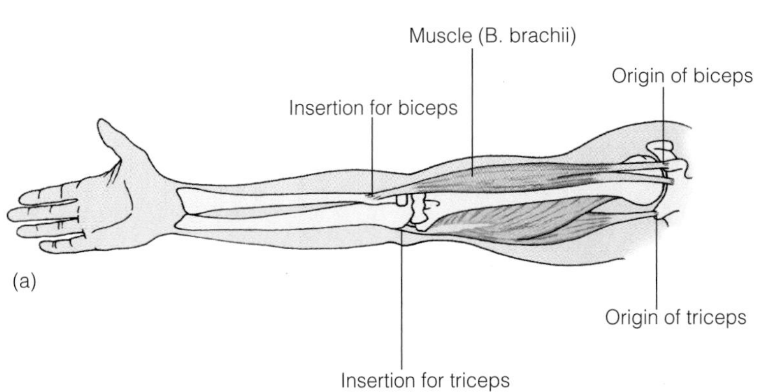

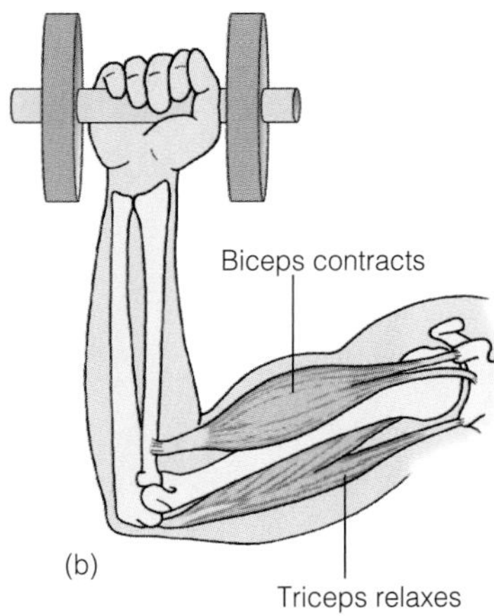

Figure 7–11 Synergism and Antagonism
Muscle contraction and relaxation are coordinated in specific muscle groups to allow movement. The muscle pair Biceps brachii and Triceps brachii are antagonistic. Muscles that contract with either Biceps or Triceps are synergistic.

(Figure 7–11). Muscles do not extend actively and, as a result, they must be pulled back into a stretched, extended position by the action of other muscles that are contracting. This, then, is the situation between the antagonistic muscles of the front and back of the arm. When the *Biceps brachii* contracts, it flexes the radius towards the humerus using the elbow as a hinge. The bones of the upper arm and shoulder provide the framework against which the pull of the *Biceps* and synergistic muscles are stabilized. It ordinarily would seem ridiculous if the upper arm and body were thought of as moving towards the lower arm instead of the other way around. However, suppose you are climbing up the face of a cliff or doing pull-ups on a horizontal bar. Under those circumstances one uses the grip of the hands to stabilize the lower arm and the entire body is pulled up by moving the upper arm (and attached body) towards the lower arm. Either way, the end result is the same.

To bring the action of antagonistic muscles into sharper focus, let us consider what might happen next. The *Biceps brachii* are completely contracted during the pull up, establishing a full flexing of the arm. The *Triceps brachii* are relaxed. One certainly would not want the *Biceps brachii* to remain contracted for too long, at least no longer than necessary to perform whatever action is intended. Keep in mind that contraction is active, extension is passive. If the *Biceps brachii* were to suddenly relax, the arms would be subject to potential injury. There would be no control of the rate of angular separation of the arm bones and an individual would snap down rapidly. To offset this catastrophic relaxation, antagonistic muscles are employed, in this case the *Triceps brachii*. The extension of the arms from a flexed position is controlled by the contraction of the *Triceps*. Working in combination, this flexion-extension cycle allows the human body to move dynamically and safely.

Muscles of the Chest, Shoulders, and Back

The muscles of the chest integrate with those of the shoulders and neck to support the movement of the arms. Principal muscles of this group are the *Deltoid, Pectoralis major, Trapezius,* and *Serratus anterior* (Figure 7–12). The *diaphragm* and the *intercostal muscles* between the ribs, which are essential for breathing, underlie the superficial muscles of the abdominal wall and chest. Muscles of the back include the *Latissimus dorsi* (which circle around to the front of the body as well) and the *Teres major* muscle. There are also a number of muscles that move the spinal column, referred to as *Spinalis muscles,* based on their position and region of origin and insertion. This entire set of muscles provides for movements associated with breathing, position changes of the upper body, and motion of head and arms and helps establish a dynamic tension between the front of the body and the back, which maintains human posture.

Muscles of the Abdomen

The muscles of the abdominal region stretch over areas that have no skeletal elements beneath them (Figure 7–13).

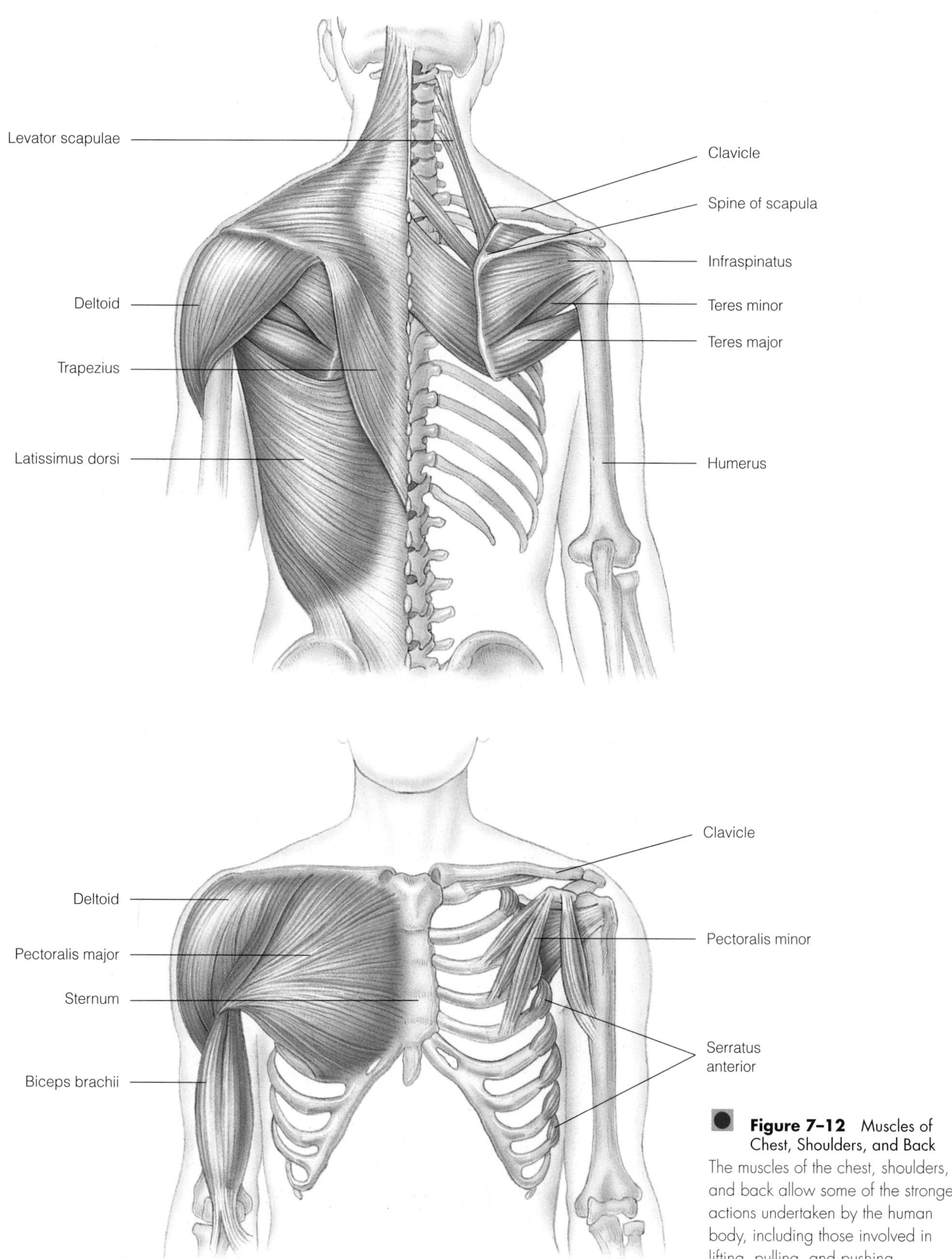

Figure 7–12 Muscles of Chest, Shoulders, and Back
The muscles of the chest, shoulders, and back allow some of the strongest actions undertaken by the human body, including those involved in lifting, pulling, and pushing.

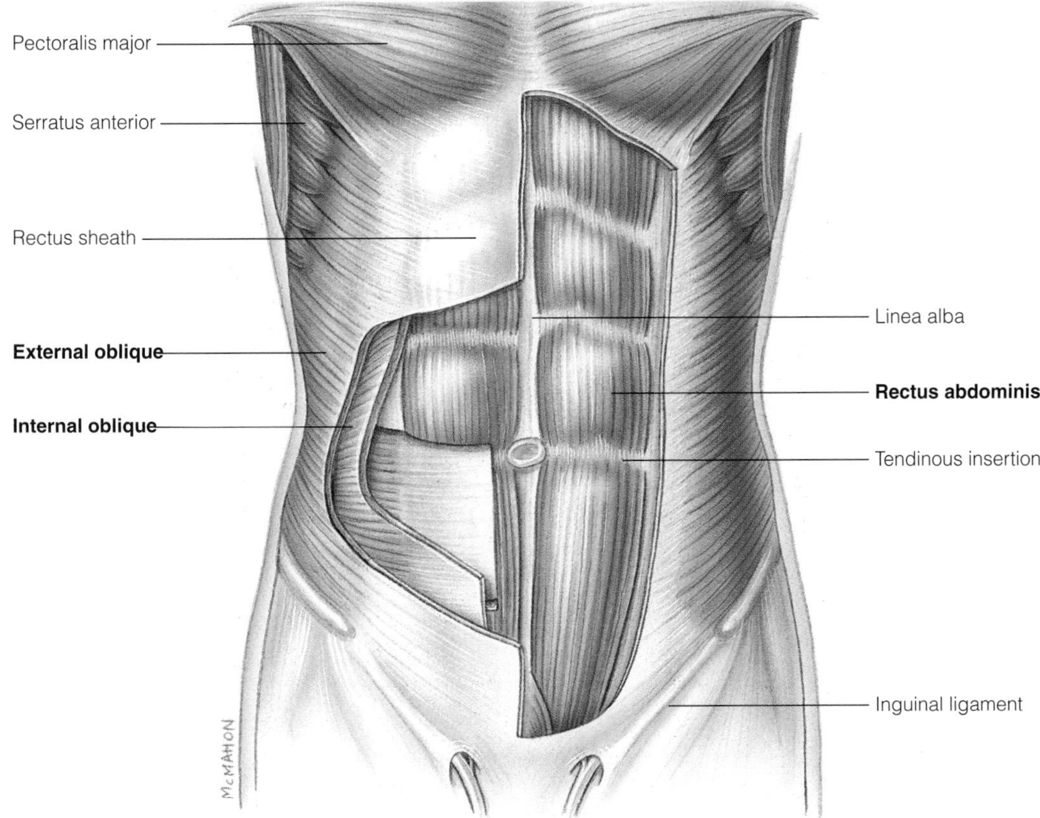

Figure 7–13 Muscles of the Abdomen
Muscles of the abdomen provide protection for underlying organs, as well as the capacity to control the movement of the upper body, such as bending.

The muscles are thick and strong and protect the organs of the abdominal cavity, while maintaining the inherent flexibility of the central body. Bending over (and its opposite, straightening up) requires elaborate synergistic and antagonistic muscle group interactions. The principal muscles of the abdomen are the *Rectus abdominus* and the *External* and *Internal oblique* muscles. The Rectus muscle has its origin along the pubic symphysis and its insertion onto ribs and the sternum. The appearance of ripples in the stomach muscles represents the *tendinous intersections* between the Rectus muscle. The midline indentation is also tendinous and is called the *Linea alba* (Figure 7–13).

Muscles of the Pelvis and Legs

In a manner similar to the attachment of major muscles to the bones of the arms and shoulders, major muscles of the legs are attached to the pelvic girdle and lower spine (Figure 7–14). The principal muscles of the legs and hips are the *Gluteus maximus,* which gives shape to the human buttocks, and the *Psoas major, Rectus femoris, Adductor,* and *Vastus* muscles of the upper leg. Origins of these muscles are on the pelvis or the spine and insertions are on the femur, patella, and tibia.

The major muscles of the lower leg are the *Gastrocnemius, Soleus, Peroneus,* and *Tibialis* muscles. There are also a number of flexor and extensor muscles, which are involved in the movement of the toes. The Gastrocnemius muscle is well known from its attachment to the heel by the *Achilles tendon* (Figure 7–14). Mythologically, the Greek warrior-hero Achilles had only one physical weakness that could bring about his demise. It was the tendon that connected his *Gastrocnemius* to his foot. According to the legend, his mother dipped him in the river Styx as a child to impart invulnerability to him. Unfortunately, she held him by the tendons that bear his name and the waters did not touch them. They were his only weak spot. Separation of the *Gastrocnemius* muscle from the heel resulting from damage to the Achilles tendon during strenuous activities, such as participation in sports, is seriously debilitating and requires prolonged recovery time.

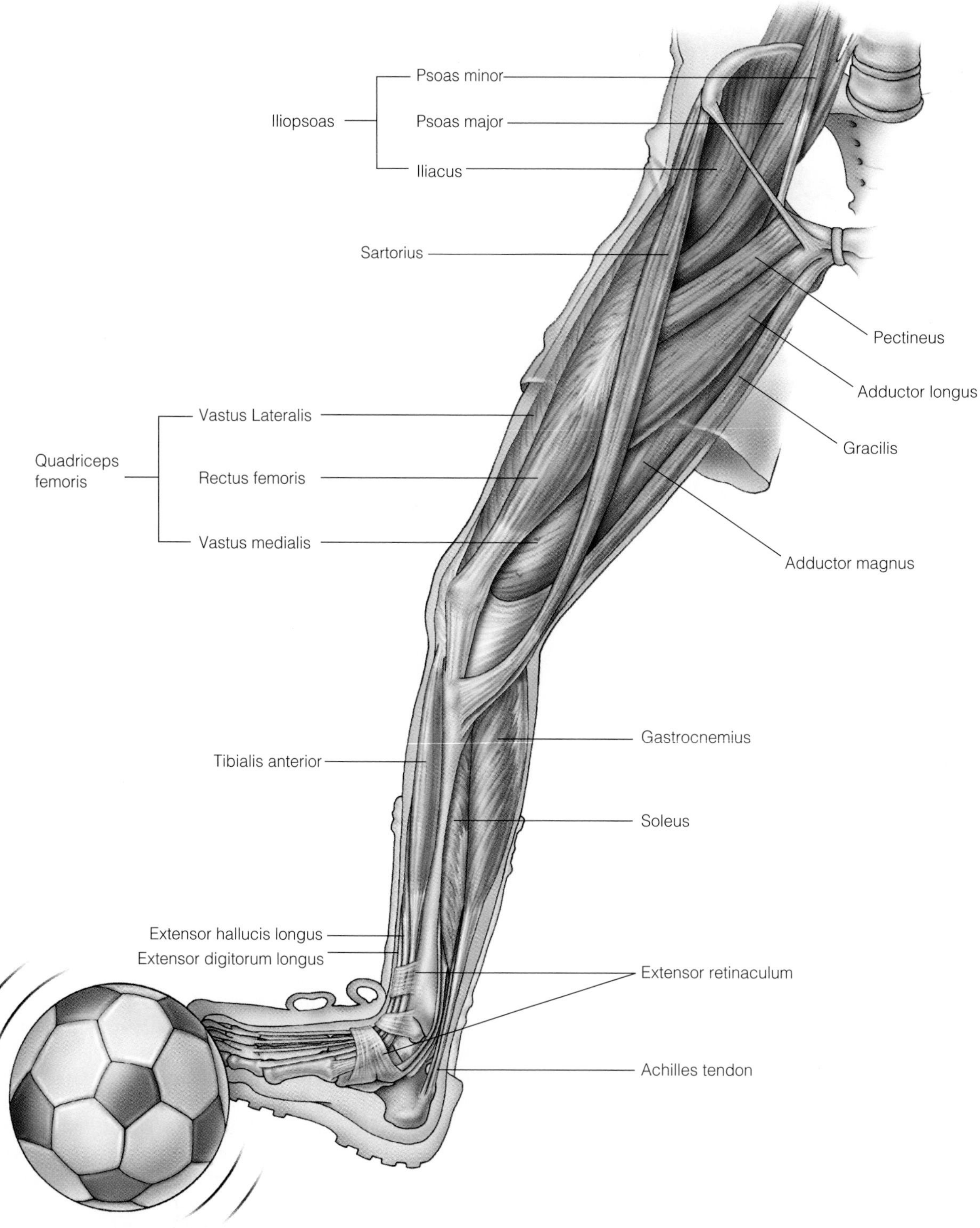

Figure 7–14 Muscles of the Pelvis and Legs
The muscles of the pelvis and legs are among the most powerful in the human body. Aside from allowing coordinated movements of the legs and feet, they support the erect stature and lifting capacity of the body.

Isotonic and Isometric Actions of Muscles

The actions of muscles during their contractions fall into two general categories—**isotonic** and **isometric.** These categories differentiate actions in which a muscle does or does not fully contract (Figure 7–15). Isotonic interactions are those that involve the uniform contraction of a muscle while moving an object. For example, lifting a glass of water or a book is an isotonic movement. The tone or tension of the muscle stays the same throughout the movement. Try it for yourself. Close your left hand over your right arm (so you can feel the action of the *Biceps* and *Triceps* muscles) as you lift this book in your right hand. The tension of the contracting muscle remains the same as the action is performed.

Isometric actions are those in which the tension on muscles increases during the action because the muscle cannot fully contract against the resistance encountered. Clasping your hands together across your chest and pulling one arm against the other is an example of an isometric action. In this case there is a constant increase in the tension of opposing muscles as they attempt to complete the movement. Quite often, the actions performed in common activities blend the two forms of movement. A combination of the two types of actions is observed in opening a jar whose lid is stuck. The initial force of gripping and twisting the lid is isometric because the muscles involved meet with significant resistance. When the lid is turned hard enough, however, the resistance is overcome and the lid twists off. The action is finished with an isotonic flourish.

SMOOTH MUSCLE

Smooth muscle is composed of individual muscle cells that work together in groups, rather than as a result of fusion as exhibited in myofibers. The individual cells are usually much smaller than their larger, striated skeletal muscle counterparts. (Figure 7–16). A single smooth muscle cell may range in size from 5μ–10μ in diameter, and 100μ–300μ in length. They have a single nucleus and no obvious striations, although they use the same principles to contract as observed in all muscles. The control of

> **Isotonic** uniform contraction of a muscle during movement of an object
>
> **Isometric** nonuniform contraction of increasing tension against a resistance

Figure 7–15 Isotonic and Isometric Actions

Isotonic movements (a) involve uniform contraction of muscles while undergoing an action. Lifting weights, sit-ups, and walking up stairs are isotonic actions. Isometric activities (b) result in increased tension on muscles and disallow full contraction. Pulling one arm against another is an isometric action.

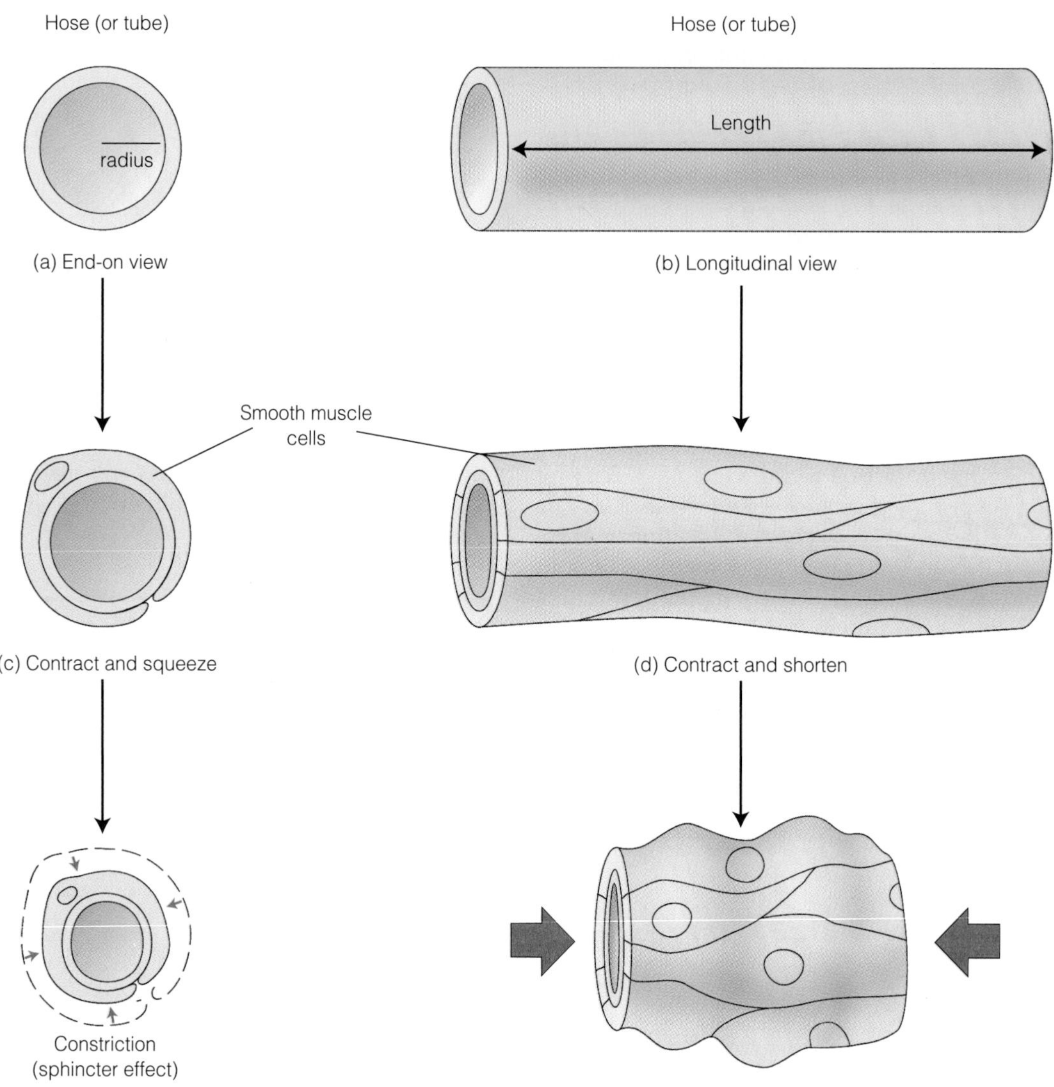

Figure 7–16 Arrangement of Smooth Muscle Cells
The arrangement or orientation of smooth muscle cells is associated with their function. A circular orientation (a) provides the capacity to constrict (c). A longitudinal arrangement (b) provides the capacity to shorten (d).

smooth muscle cell activity is mediated predominantly by input from the autonomic nervous system and, thus, is involuntary in nature.

The activities controlled by smooth muscles are the key to understanding their roles in the human body. Contraction of smooth muscles is carried out automatically, and, therefore, the processes in which they are involved are intimately associated with homeostasis. For example, when smooth muscle cells that surround blood vessels contract, they reduce or cut off blood flow to the regions of the body downstream from the constriction. This is analogous to a crimp in a garden hose, the flow from the nozzle ceases but the hose is still filled with water under pressure. An important manifestation of this type of action occurs when part of the human body is seriously damaged and the vascular system is compromised. The body responds rapidly by restricting blood flow to the area of injury by contracting smooth muscle cells associated with blood vessels in critical upstream regions. This reduces blood pressure in damaged arteries, diverting blood flow to other areas.

Effective clotting of blood depends, in part, on the capacity to redirect blood flow and reduce pressure on the developing clot. One way to think about this is to envision what traffic cops do at the site of a multicar pileup that stalls traffic for blocks down a busy street. They redirect incoming traffic around and away from the accident to prepare to clear the obstacles. Only then can ambulances, tow trucks, and the fire department get to the accident scene and take care of the people and vehicles involved. Smooth muscles play a role similar to the traffic cop by redirecting the flow of circulation so that homeostatic mechanisms of blood clotting and wound sealing can begin.

Based on a consideration of their cellular dimensions, smooth muscle cells are unusually long and tapered (see Figure 7–1 and 7–16). They wrap themselves around the various types of tubular and sacklike structures inside the body (blood vessels, intestine, stomach, exocrine glands) and contract (squeeze) or relax appropriately with respect to timing and duration. However, smooth muscle cells are not solely oriented around a tubular structure; they are also aligned along the length of such structures. How does this affect the function of the tissues with which smooth muscles are associated? Once again a garden hose analogy will provide a view of how flow and movement within tubes is controlled.

There are principally two ways to view the structure of a hose—in cross section (an end-on view) and in long section (a side view) (Figure 7–16). In the end-on view the hose has a circular profile. From the side it appears as a long tube. Using the hose as a model for a blood vessel, smooth muscle cells may be envisioned as oriented either around the circumference of the hose or stretched along the longitudinal axis on the surface of the hose. When contraction of muscle cells in the circumferential orientation occurs, it constricts the diameter of the hose, like the draw string on a purse. When contraction of the muscle cells occurs along the length of the hose, it pulls and shortens it. Imagine squeezing the hose to constrict flow in one area and contracting the longitudinal cells in front of the constriction. This is possible in cellular and tissue circumstances because cells in the two orientations can be coordinately controlled by the nervous system. When the longitudinal cells are activated, whatever material is within the tube will be moved forward. This activity is not isolated to one event but to many similar actions in sequence—constriction, forward movement, constriction, forward movement, and on and on. This is essentially how the combined action of smooth muscle cells in different orientations is able to move fluids along vascular and glandular elements and solids and liquids through the intestinal tract of the body.

Voluntary and Involuntary Actions

A voluntary act in human behavior is one that proceeds from a conscious decision to do something and uses the voluntary pathways of the central nervous system. An involuntary act is one that is controlled unconsciously and uses the involuntary pathways of the central nervous system. One act occurs by our own volition, the other automatically. With respect to striated skeletal muscle it is normal to consciously activate muscles and move. If you desire to move a hand to scratch your head, you simply think about it deliberately and do it. The actual steps in the process by which you carry out such an action is incredibly complex and not well understood (involving the brain, spinal cord, motor neurons, motor end plates, synapses, and neurotransmitters), but the voluntary nature of the action is easy to demonstrate.

As important as it is to have conscious control of many of the activities of muscles, the body cannot take chances. There are some basic physiological activities associated with homeostasis that are too important to be left to deliberate or intentional control. Heart beat, blood flow, digestion, body temperature regulation, the subtle control of appetites and drives, and sleep and wakefulness are all examples of activities or states in which humans have some but not final control. Involuntary actions may be initiated and completed contrary to, or in the absence of choice, and so are not subject to the control of will. Try stopping your heart beat. No luck? Try to stop the peristaltic movement of food actively translocating through your intestinal tract. Any success? These actions and the regulation of the muscles involved in them are for the most part beyond conscious, voluntary control, and it is well that they are. There is some evidence that **biofeedback** can have a direct influence on involuntary actions, but such control is limited.

Voluntary and involuntary actions are directed by different and specialized parts of the central nervous system, which sets up natural divisions of labor in the body. In a worst-case imaginary scenario, one in which voluntary control is required for every single activity that might go on in the body, what would happen when we slept? Further, if we decided not to sleep, would we simply burn out from lack of rest, instead of passing away as consciousness fled? Be glad

Biofeedback control of unconscious or involuntary bodily processes through conscious thought.

that the body is given no conscious choice in such important matters of physiology concerning the operation of the many systems that maintain the homeostatic balances.

CARDIAC MUSCLE

The adult human heart is about the size of a closed fist. It resides in its own special cavity within the chest, behind the protective rib cage and sternum and is tilted slightly to the left of the longitudinal midline of the body (Figure 7–4). The heart muscle tissue is organized to contract against itself and, thus, has the capacity to reduce the volume of its internal chambers. This is done transiently and at a regular interval, or beat. With such a rhythmic beat, the heart acts as an in-line pump for continuous flow of the blood. The heart beat occurs repeatedly for a lifetime in a highly regulated effort to displace blood through the pulmonary and systemic circulatory system (see Chapter 11) to ensure the delivery of oxygen to all the tissues. The complex network of blood vessels is thousands of miles long and requires powerful contractions to move blood through it. It is the collection of billions of cardiac muscle cells, organized within the walls of the heart that carry out this essential activity.

Cardiac Myocytes

The heart muscle is composed of a unique type of myocyte called a cardiac myocyte. Cardiac myocytes are found only in the heart and have attributes of both striated skeletal muscle fibers and smooth muscle cells. As in skeletal muscle fibers, cardiac myocytes are striated. Striations within muscle cells are an indicator of the potential for powerful contractions. Like smooth muscle cells, cardiac myocytes are single cells with only one nucleus. The heart is controlled involuntarily by both internal pacemakers and externally by the central nervous system. However, unlike either skeletal or smooth muscle, cardiac myocytes have specialized intercellular adhesive structures known as *intercalated discs* which hold the cells together strongly and gap junctions that allow the cells to function together as a unit (Figure 7–4 inset).

Intercalated Discs

There are hundreds to thousands of intercalated discs on the surface of each cardiac myocyte, and each cell has multiple interactions with neighboring cells in the heart. These attachments between cells are what gives strength and unity of action to the entire heart muscle. The heart must beat in a highly coordinated fashion to function at all. The intercalated discs are the spot welds between cardiac myocytes and are similar structurally and functionally to the desmosomes found in and between epithelial cells (see Chapter 4).

The organization, adhesion, and orientation of cardiac myocytes is similar in importance to the circular and longitudinal arrangements of smooth muscle cells in the intestinal tract and the longitudinal alignment of muscle fibers in striated skeletal muscle. The direction of the force of heart muscle contraction must be controlled and maintained to accomplish its purpose. The work of the heart muscle is to squeeze the chambers within, so that the blood that fills them is propelled through the circulatory system. The timing of the squeezing beat is tightly regulated both internally and externally.

An Inherent Beat

The heart myocytes have an inherent beat of their own. If a few heart cells are removed from a human being or any other animal with a heart, those cells individually will continue to beat (Figure 7–17). They can be observed under a microscope and the rate and degree of their contractions can be measured. If two or more of these isolated cardiocytes are brought into contact with one another and adhere, the group of cells will communicate directly and beat as one. This suggests that heart cells share signals with one another in creating the rhythm of their beat. As previously described, communication between heart cells is mediated by gap junctions, which form tiny channels between cells. The adherence of the cells is based on desmosomal junctions. Because the mass of the heart and the number of cardiac myocytes is so large, there is an additional mechanism to coordinate the beat. There is a specialized region of the heart that controls the basic, resting heart beat rate. This region is called the **sinoatrial** (or **SA) node** and it has the inherent property of initiating a contraction cycle in the heart muscle approximately every 0.8 seconds. A second region of the heart is also involved in controlling the beat. This region is called the **atrioventricular** (or **AV) node,** and it helps distribute the signal from the SA node so that the contraction is timed appropriately in different regions of the heart (see Chapter 11 for details). This capacity to regulate the beating of heart muscle cells is observed even in an intact, isolated heart, which, if removed from the human body under the appropriate conditions, will go on beating in the surgeon's hands. This capacity is important for the success of organ donor programs, in which a living, beating heart may be transferred from a dead accident victim into a living recipient in dire need of a new heart.

Sinoatrial node also known as SA node; controls pace of the heart beat

Atrioventricular node also known as AV node, propagates signal from SA node in pace of heart beat

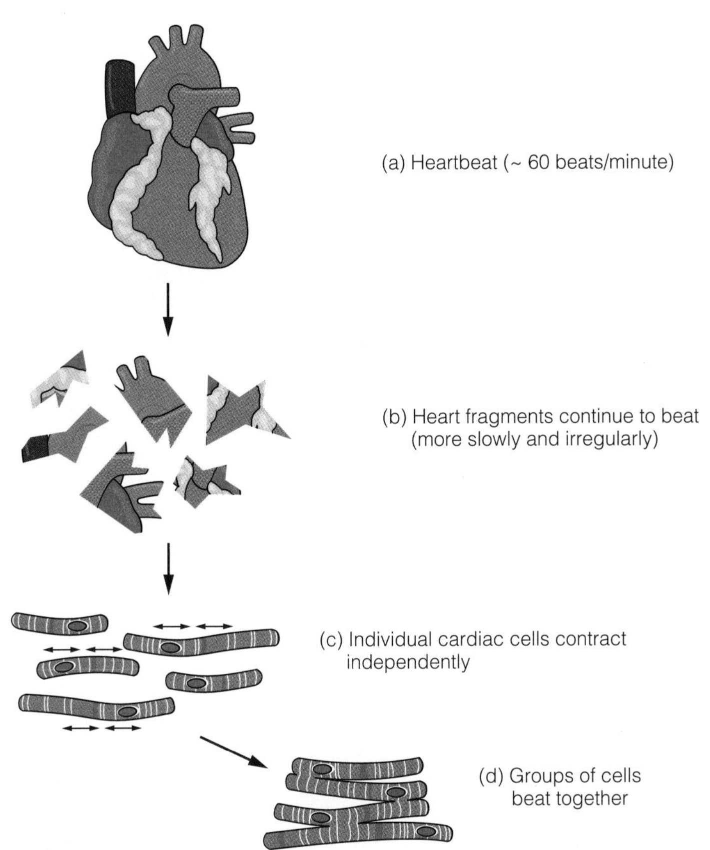

Figure 7–17 The Beat Goes On There is an inherent capacity of the heart (a) and fragments (b) to beat. Even when heart tissue has been reduced to single cells they continue to beat (c). As cells re-associate in small groups the beat of the cells is coordinated (d).

External Control of the Heartbeat

The heart is also controlled externally. The inherent beat of the heart is not sufficient for all conditions in which it functions. When the human body is physically active, the rate of the heartbeat increases. When we get excited or afraid, the heartbeat also goes up. When we relax or feel safe, it decreases. The ability to regulate changes in the heartbeat is not only inherent to the muscles of the heart but also requires outside control and coordination. Coordinating changes in the rate of the heartbeat is controlled by the autonomic nervous system. This interaction ultimately connects the beat of the heart to the world of the senses and perceptions mediated by the central nervous system. The involuntary control of the action of the human heart is mediated specifically by the *Vagus nerve* and, thus, links heartbeat to states of mind (anxiety, fear, pleasure) as well as to the body's state of physical activity (running, swimming, dancing).

NEUROMUSCULAR INTERACTIONS

The nervous system plays a fundamental role in activating the muscles. There will be more coverage of this topic in Chapter 17, but a brief description here is warranted because muscles, as the targets of nerve control, are coordinated by them in their actions. Nerves are composed of bundles of axons arising from the spinal cord, which connect to muscles at *synapses* on the individual myofibers known as **neuromuscular junctions** (Figure 7–18). In electrical terms this is similar to plugging in an extension cord to a kitchen appliance. Flipping the on-off switch activates or deactivates the appliance. Nerves use cellular extension cords to activate molecular switches as input to muscle fibers of the body. The agents of stimulation are chemical signals, referred to as **neurotransmitters.** The principal neurotransmitter of neuromuscular interactions is *acetylcholine.* Acetylcholine is released from vesicles in the nerve and binds to receptors in the surface of myofibers. As will be discussed in detail, the binding of

Neuromuscular junction synapse between nerve cell and muscle cell

Neurotransmitters chemical signal molecules involved in regulating the electrical activity of nerve and muscle

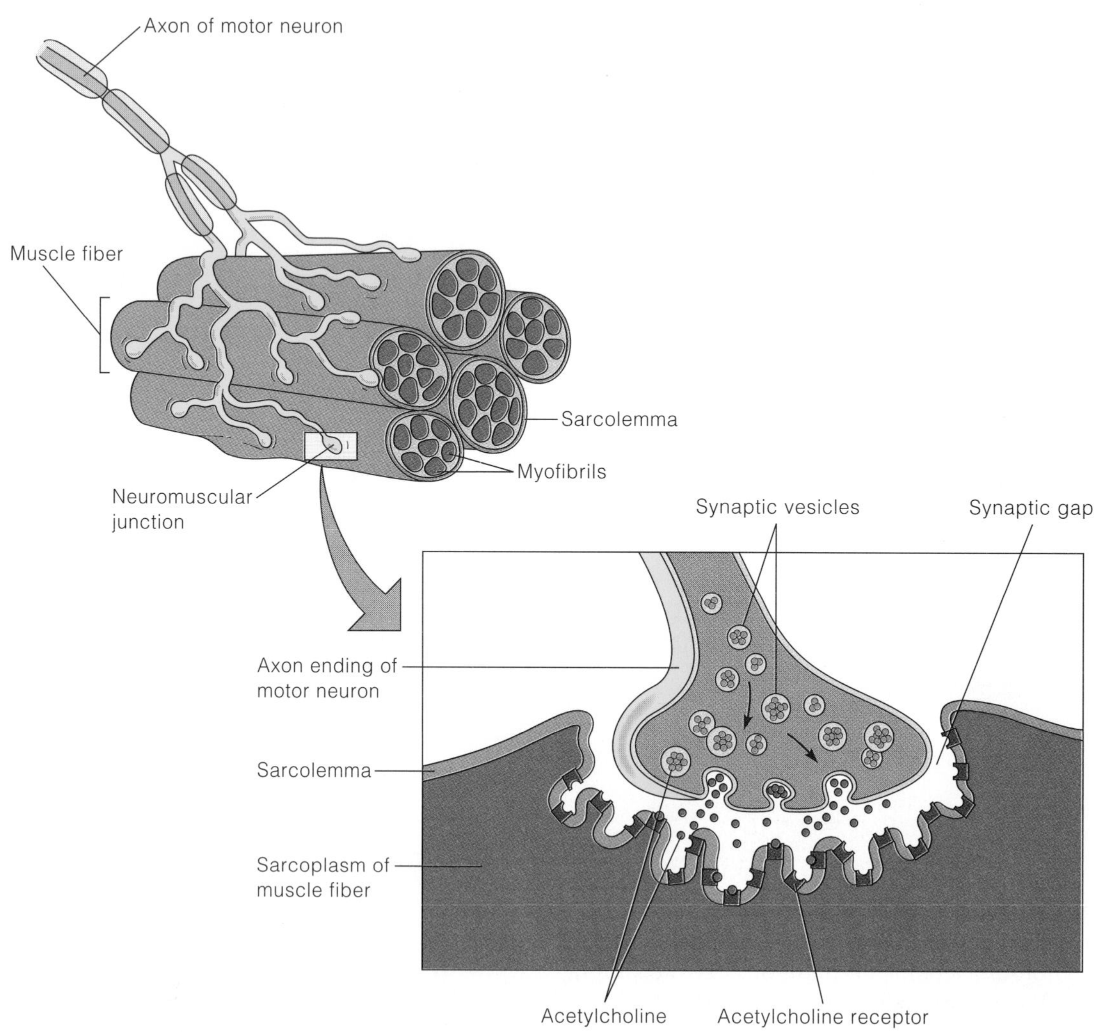

Figure 7–18 Neuromuscular Junction
The interaction between a nerve and a muscle cell or fiber is based on a synapse. The stimulation of a muscle to contract depends on the release and reception of the neurotransmitter acetylcholine.

acetylcholine activates the rapid release of Ca^{+2} from the endoplasmic reticulum of a muscle fiber and triggers the cell to contract. When acetylcholine is withdrawn or inactivated, the affected muscle relaxes.

THE MECHANISMS OF MUSCLE CONTRACTION

The mechanisms of contraction of any muscle cell or fiber are nearly the same regardless of whether skeletal, smooth, or cardiac myocytes are considered. For contraction, it is what is inside the muscle cell that counts, particularly with respect to the *cytoskeleton* (Figure 7–19). The action of the cytoskeletal striations and the changes they undergo allows myocytes to contract. What do we know about the cytoplasm of muscle cells? As was presented in Chapter 2, the major parts of a cell are the membranes, the organelles, the specialized macromolecular complexes, and the cytoskeleton. In the case of a muscle cell, the cytoskeleton plays a prominent role in the capacity of myocytes to contract. Specific organelles and membranes are also important, but their presence and activity are far less obvious than the internal *myofilament-based cytoskeleton* that constitutes the striations observed in cardiac muscle cells and skeletal muscle fibers.

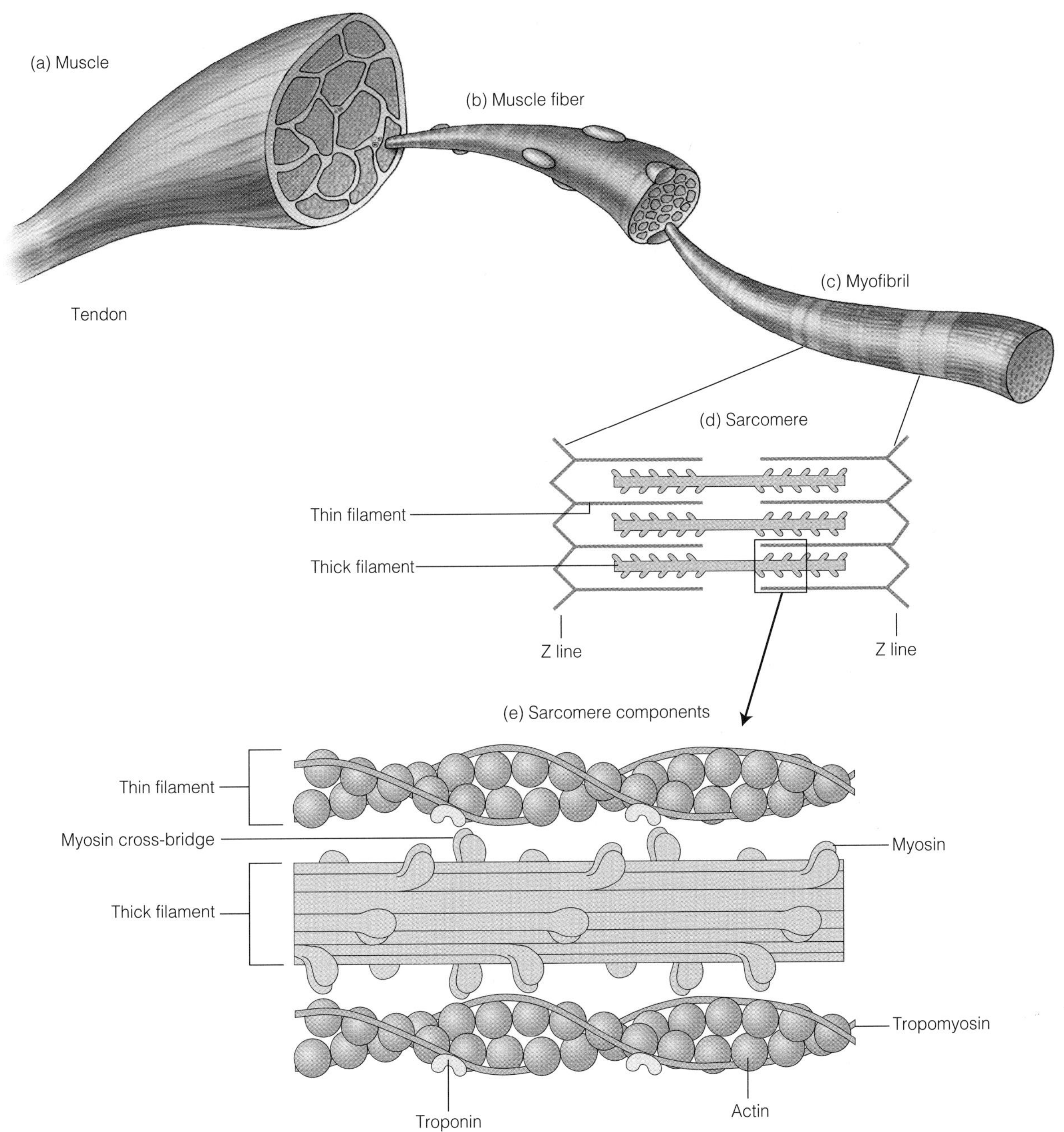

Figure 7–19 Levels of Organization in Muscle Contraction
Skeletal striated muscle (a) is composed of muscle fibers (b). The fibers are constructed with myofibrils (c) arranged into repeated units of structure known as sarcomeres (d). Sarcomeres are composed of actin, myosin, and other regulatory proteins (e), such as tropomyosin and troponin.

Myofibrils and Sarcomeres

There are basically two states in which muscle cells are found—relaxed and contracted. Figure 7–20 shows a comparison of the relative lengths of a relaxed and a contracted muscle. How does a muscle transition between these two states? To answer this question, we need to describe the structures involved. Striated skeletal muscles are composed of myofibers. The basic structural elements found within each muscle fiber are called *myofibrils,* which are, in turn, composed of a linear series of specialized elements known as sarcomeres. The sarcomeres are themselves composed of organized arrays of thick and thin filaments,

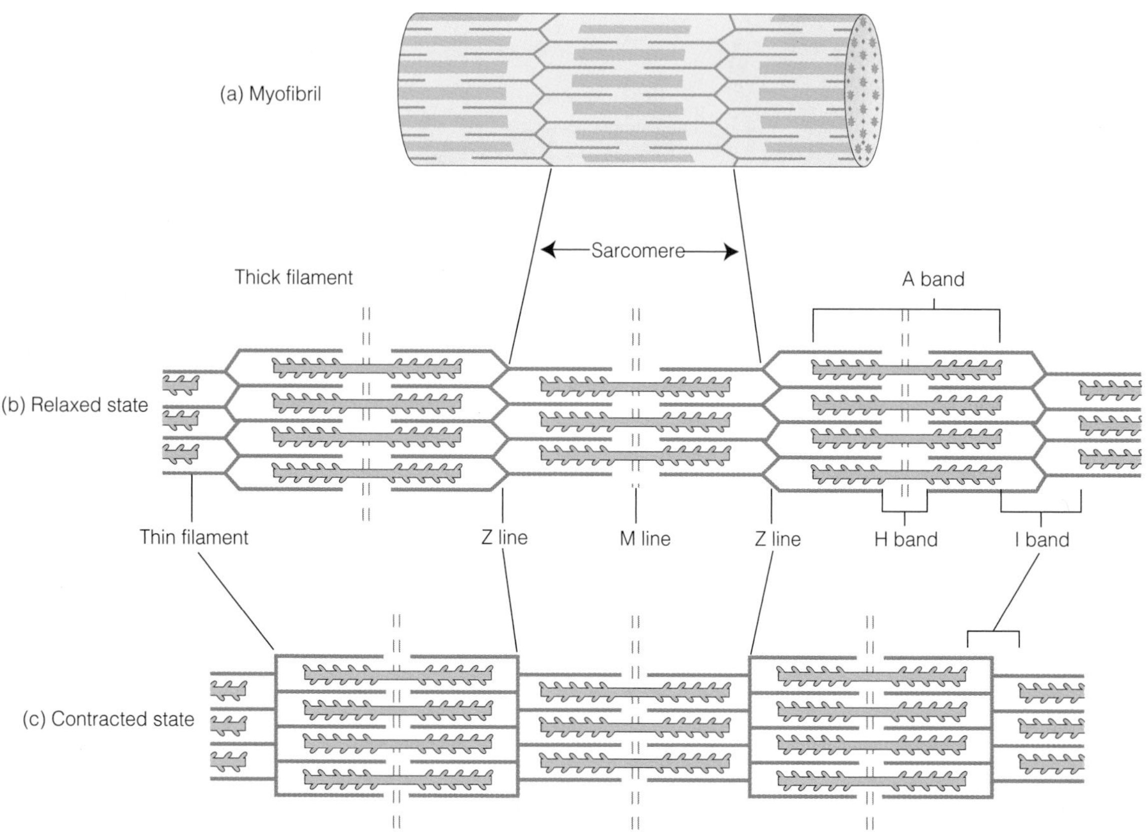

Figure 7–20 Sliding-Filament Model
A sarcomere (a) is composed of many actin and myosin filaments that overlap one another. In a relaxed state (b) myosin and actin filaments in a sarcomere do not completely overlap. In a contracted state (c) the filaments slide over one another and shorten.

which overlap one another like interdigitated fingers (Figure 7–19).

The thick filaments are formed from the protein *myosin.* The thin filaments are composed of polymers of the protein actin and of other associated proteins known as *tropomyosin* and *troponin* (Figure 7–19). The myosin and actin filaments interact along a portion of their length and it is the overlap of these cytoskeletal elements that represents the essential structure needed for muscle cells to contract. The thin filaments run in two directions from attachments at their ends to plates known as *Z lines.* The attachment of these filaments running in opposite directions over the thick filaments in conjunction with the Z-line attachments provide the framework against which muscle contraction occurs. This is, then, the sarcomere, and it is the repeated pattern of sarcomeres connected end to end (and side to side) that constitutes a myofibril (Figure 7–19). The sarcomeres of a muscle fiber are shown in a relaxed state but are poised to contract. The diagram shows a sarcomere that has contracted.

Interaction of Thick and Thin Filaments

There is considerable change in the arrangement of thick and thin filaments and in the length of each of the repeated sarcomeres during muscle contraction. Ultimately, the sarcomeres of myofibrils are connected to the plasma membrane of the muscle fiber (called the *sarcolemma*) so that, when they contract, the entire muscle fiber is shortened dramatically (Figure 7–20). The collection of muscle fibers in turn is bundled together to form a muscle so that the collective contraction of many myofibers is coordinated and powerful. What needs to be explained is how muscle contraction works at the molecular level. This calls for designing a molecular machine and a molecular motor from the parts described above and depicted in diagrams and from information available to us from studies of the biochemistry of muscle cells.

The Sliding-Filament Model

To explain the difference between the sarcomeres of the contracted and relaxed muscle fiber depicted in Figure

7–20 some assumptions are made. The first assumption for the model is that the thin and thick filaments within a sarcomere are able to slide over one another to shorten the sarcomere during contraction. This sliding is an all-or-nothing reaction and is nearly instantaneous. The fact that actin filaments run in opposite directions over a centrally organized bundle of myosin filaments makes this fairly easy to visualize. This organization and potential for sliding action constitutes the basis for what is known as the **sliding-filament model.**

The second assumption is that this is a reversible process. It should be remembered that muscles actively contract and passively extend. Thus, the reversibility of the contracted state requires that a muscle fiber once contracted must be pulled into its original extended state by the action of some outside force. This could be the pull from another muscle contracting in the opposite direction or the pull of gravity. A third assumption is that the process requires, and has plenty of energy (ATP) and other factors, particularly calcium ions, in order to occur. Figure 7–19 and 7–20 presents the model at several levels of organization from muscle tissue to myofiber to molecules, with a continuity of structure based on current knowledge.

The thick myosin filaments are arranged together and attached or stabilized at the center line of each sarcomere (called the *M line*). Myosin filaments are bipolar and extend toward the ends of each sarcomere but not far enough to actually attach to these poles. Regions at the end of each sarcomere are the Z lines and, as previously described, the thin actin filaments are attached to them and extend towards the M line. The actin filaments do not directly attach to the M line, but they do overlap extensively with the bipolar myosin filaments. This, then, is the situation in the relaxed state of the muscle fiber. The areas where the two filament types overlap are called the *A bands*. The regions in which the two types of filaments do not overlap are called the *I bands*. The A and I stand for **anisotropic** and **isotropic,** respectively. As we will see, the A band and I band refer to areas of the sarcomere that change (A) or do not change in appearance (I) as a result of contraction.

The dynamics of the sliding filament model can be presented in the following sequence of events. The interaction between myosin filaments and actin filaments is mediated by the actin-associated protein tropomyosin. Tropomyosin lies along the length of the actin filament and in the relaxed state prevents thin and thick filaments from contacting each other. Tropomyosin interacts with a calcium-binding protein known as troponin, which also interacts with actin. Calcium ions are generally in very low abundance in the cytoplasm of cells. In muscle cells of all types, calcium ions are sequestered in the endoplasmic reticulum, and in muscle fibers this forms a special system of membranes known as the *sarcoplasmic reticulum.* An elaborate system of invaginations of the sarcolemma gives rise to the *T tubules*. The T tubules are closely associated with the ER and are, thus, able to relay the signal from the binding of acetylcholine rapidly and effectively to the cell interior. As long as the calcium ions are prevented from entering the cytoplasm (or the *sarcoplasm* as it is called in muscle fibers) the muscle will remain relaxed. However, when calcium ions are released and become available in the sarcoplasm, the troponin binds them and changes its relationship to tropomyosin. This in turn induces tropomyosin to slide over the surface of the actin filament and uncovers a region that can bind to myosin. When actin and myosin are connected, the thick and thin filaments are ready to slide over one another. The only thing needed for this to occur is ATP.

Myosin is a motor molecule and acts also as an enzyme known as *ATP'ase*. Myosin has a special ability to bind ATP and use it as a source of energy. ATP is produced in mitochondria during conditions of aerobic respiration (see Chapter 3). Mitochondria are particularly abundant in muscle cells and provide the tremendous quantities of ATP needed for muscle activities. The ATP bound to myosin undergoes a conversion to ADP when myosin binds to an actin filament. This conversion of ATP to ADP changes the shape of the myosin molecule and pulls the actin filament over it towards the M line, thus, shortening the sarcomere. This is why myosin is considered a motor molecule, because it actively moves the actin filaments to which it is attached. At this point we have all the elements needed for a muscle to contract and we know how the system operates, but how are the events initiated and controlled? After all, it is important that a muscle cell not remain contracted, nor relaxed, but be able to cycle appropriately between the two states.

For the proper regulation of muscle contraction we need input from the nervous system, as mentioned earlier. Muscle contraction is activated by a neurotransmitter-mediated signal released from a neuron at a neuromuscular junction. The neurotransmitter involved in signaling muscle contraction is acetylcholine. Acetylcholine binds to a receptor on the surface of the muscle fiber and alters the T tubules so that calcium ions are released from the sarcoplasmic reticulum into the sarcoplasm. This is a massive release of ions that are immediately bound by troponin. Troponin alters tropomyosin and allows myosin and actin to connect together. In the presence of adequate amounts of ATP the thick and thin filaments rapidly and completely slide over one another in each of thousands of

Sliding-filament model model used to explain the interaction of myosin and actin during the shortening of sarcomeres in muscle contraction

Anisotropic able to polarize light

Isotropic nonpolarizing

sarcomeres and bring about dramatic shortening of the fiber.

Relaxation of the network of filaments comes about when the stimulus to contract is removed or when the cells run out of the ATP or calcium to maintain a state of contraction. When the neurons stop releasing acetylcholine, the T system of tubules reverses its action and the sarcoplasmic reticulum rapidly pumps the free calcium ions back into the ER. In the absence of calcium, troponin is altered and induces tropomyosin to cover the myosin-actin binding site. This prevents their interaction and halts contraction. Relaxation is a passive process and, as such, filaments slide back over one another effectively only under the influence of a pull from other muscles or gravity. The A and I bands appear differently in a relaxed and a contracted sarcomere. The A band does not change much in appearance during contraction, but the I band nearly disappears. The distance that a single sarcomere contracts is equal to the difference between the length of the A band and the length of the I band in a relaxed muscle fiber.

The Force of Numbers

There is force in numbers. The more sarcomeres there are that are tied together in a series, the greater the total length of muscle contraction. The more muscle fibers there are that are aligned together laterally, the greater the total strength of contraction (Figure 7–21). This helps to explain why striations are so easily observed in striated skeletal muscle and cardiac myocytes. There are so many sarcomeres along the multitude of myofibrils found in these cells that they are readily apparent even at low magnification under a microscope. The striations are huge collections of highly organized sarcomeres. This also helps explain why contraction is an active process and relaxation is a passive process. The sliding of the filaments over each other, based on the motor molecule attributes of myosin, is unidirectional and bipolar. The racheting of the thick and thin filaments takes energy and calcium ions. The relaxation begins when these filaments are uncoupled and represents essentially the release of tension on the molecular machinery as contraction ends. Smooth muscle cells also have these molecular contractile elements within them but in low enough numbers and somewhat altered organization that they are not readily apparent. However, the molecular mechanisms of contraction are basically the same for all three types of muscle cells.

Hypertrophy and Atrophy

With some idea of how a muscle functions, we can ask how do muscles grow? Most of us are aware that Mr. Olympia and Ms. Olympia are competitive titles bestowed each year on the individual male and female judged to have the biggest and most symmetrically muscular bodies in the world. How do these males and females become so muscular? They were certainly not born that way. Even children of body building parents are not themselves any more muscular than children of parents who are not athletes. The answer to how muscles grow is simple; they grow by being used. The more work that you do and the more often you do it (weight lifting, pull-ups, push-ups, running), the bigger and stronger the muscles involved in those activities become. Muscles grow in response to the work demanded of them.

However, this does not directly answer the question of how a muscle actually grows. What is the cellular or tissue basis for it? Do our bodies increase the number of muscle cells to increase muscle mass? This would require an increase in the rate of cell proliferation, a condition known as **hyperproliferation,** and in many kinds of tissues this type of growth does account for increased cellular and tissue mass. However, this is not the case for skeletal striated muscle growth. There is very little increase in the number of skeletal muscle fibers after birth. The only way for the mass of a muscle to increase, with no increase in the number of cells, is for each individual muscle fiber to become larger. This type of growth is called **hypertrophy** and is the mechanism by which muscles of the human body grow and strengthen.

Hypertrophic growth is induced by work, which adaptively increases the number of myofibrils and sarcomeres within muscle fibers. If an individual works hard enough and long enough, the size and strength of muscles can become enormous. As we humans change jobs and activities over the course of our lives, we often see changes in the size and tone of muscles in different regions of the body. The runner develops powerful legs, the swimmer strong arms and legs, the carpenter strong arms and shoulders. Much of this activity is also good for the heart, which grows in size and strength in response to many types of exercise. Physical activity is good for the overall health and performance of the human body.

Reverse the situation. What happens when we cannot, or will not, use our muscles? Under these conditions muscle fibers undergo **atrophy,** or shrinkage, from disuse or damage. This process results in a decrease in the number of myofibrils and sarcomeres, decrease in mass, and decrease in strength. Why should this occur? The body is attuned to the physical demands placed upon it, if muscles are not used, or are affected by some disease that interferes with their normal function, cellular economy is such that less energy and nutritional resources are committed to them. The turnover and unreplaced loss of a portion of the contractile elements of a muscle fiber results in a decrease in muscle size.

Hyperproliferation excessive cell division; usually associated with increase in tissue or organ size

Hypertrophy increase of cell size to accommodate growth in tissue or organ size

Atrophy a wasting away

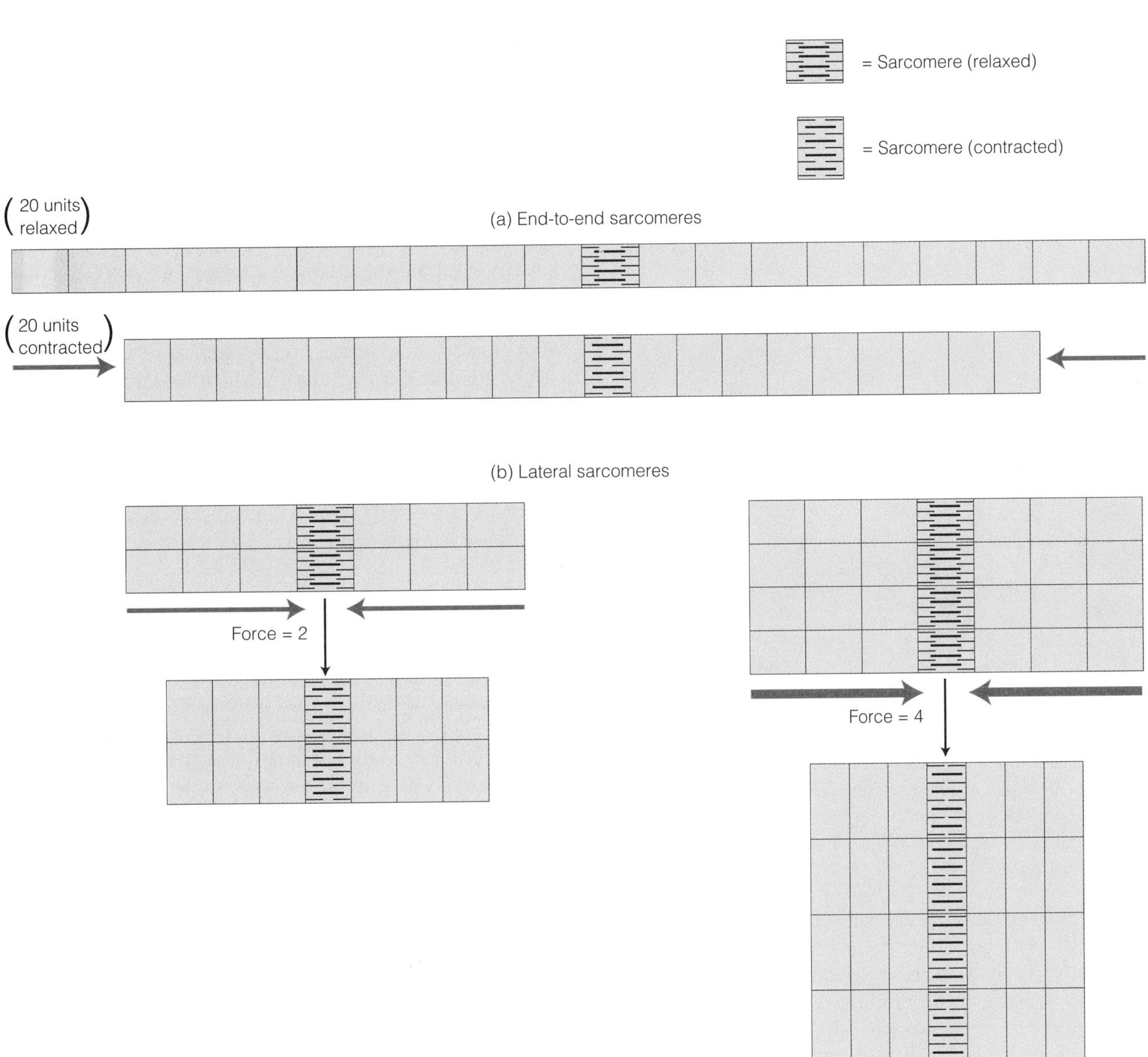

Figure 7–21 Force in Numbers
If the muscle fiber is long, there are more end-to-end associations of sarcomeres. Thus, total length of contraction is greater with longer fibers (a). If more sarcomeres are organized laterally, the force of contraction is stronger (b).

EMBRYOGENESIS AND THE FORMATION OF MUSCLES

All three types of muscles are derived from the *mesoderm* during embryonic development (Figure 7–22). This so-called middle layer formed during embryogenesis also gives rise to cells that form all types of connective tissue including bones, tendons, ligaments, and fibrous tissues that surround, support, and protect muscles. Smooth and cardiac muscle cells are single cells throughout their development. Muscle cells migrate to and increase in number at the sites in which they will differentiate into their final forms. However, the muscle fibers found in striated skeletal muscle are somewhat different. Early in embryogenesis the cells that will give rise to skeletal muscles are single cells with a single nucleus. At that stage they are called

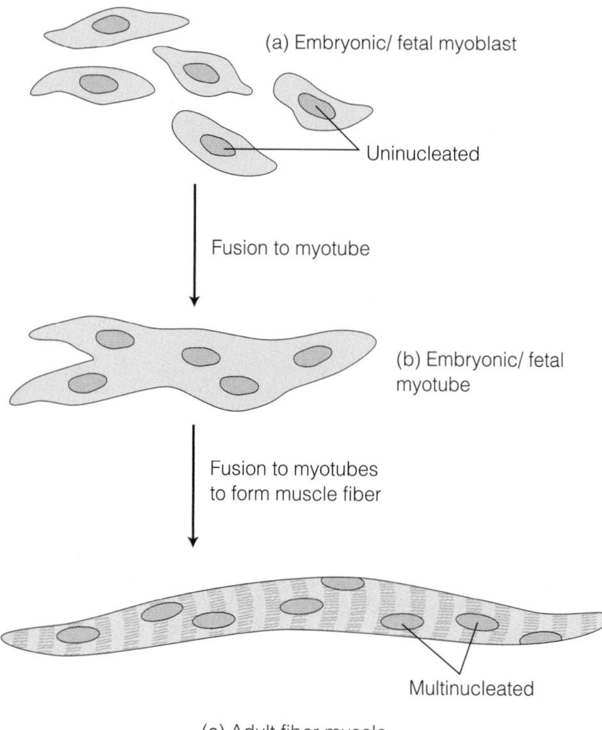

Figure 7–22 Myoblast Fusion
In the embryo, myoblast cells have a single nucleus (a) as they begin to fuse. Fusion results initially in the formation of myotubes (b) which align and fuse together to form muscle fibers in adults (c).

myoblasts, which are precursor cells to what will become myofibers. Myoblasts migrate to locations in which they will eventually form muscles. Myoblasts align and associate together along the developing skeleton during embryogenesis and at these sites they undergo a process called *cell fusion* (Figure 7–22). The initial fusion of uninuclear myoblasts gives rise to relatively small multicellular and multinuclear structures called *myotubes*. Myoblasts continue to fuse and incorporate into developing myotubes, and myotubes fuse with other myotubes to eventually give rise to muscle fibers that include hundreds or thousands of cells and their nuclei.

MUSCLE TERMINOLOGY, MUSCLE DAMAGE, AND MUSCLE DISEASES

Terminology

Myology generally refers to the medical study of muscles. The prefix myo has been used often in this chapter in association with descriptions of muscles. **Myogenesis** is the term used to describe the formation of muscles. *Myopathy* combines the prefix with pathology and, thus, refers to the study of muscle damage or the outcome of diseases that affect the muscular system (Table 7–1). *Myotonia* refers to a condition in which a muscle has a tendency to remain contracted and is unable to relax, as in the case of muscle spasms. When skeletal or cardiac muscles are physically damaged, they cannot repair themselves by proliferating new muscle cells. Instead, they undergo a process called *fibrosis,* which brings about scarring as it replaces the damaged or dead muscle tissue with nonmuscle connective tissue. This fibrous scar material is derived from fibroblasts that invade the damaged tissue and secrete collagen and other gluelike extracellular matrix molecules. Connective tissue of this type is neither contractile, nor very elastic, so overall muscle performance is diminished. In general, when damage occurs to the cells and tissues of the body, including muscles, a condition of inflammation results. This condition is characterized by local redness, heat, and pain at the site of damage and is part of the process of repair.

Diseases

Muscular Dystrophy There are a number of heritable diseases in which muscles undergo degeneration and may be ultimately destroyed. They are collectively called *muscular dystrophy.* Individual muscle fibers are affected in the process of degeneration brought about by this class of disease, which is usually progressive and lethal. However, not all types of muscles in the body are equally affected by muscular dystrophy. The wasting away of striated skeletal muscles in the periphery of the body, such as the arms and legs, is more pronounced than that of internal muscles, such as the diaphragm. There is no known cure for muscular dystrophic diseases, but physical activity and exercise seem to help retard their progress.

Myasthenia Gravis *Myasthenia gravis* is a disease of muscles that is for unknown reasons more common in females than in males. This condition results in muscle weakness. The major problem encountered in myasthenia gravis is the failure of motor neurons to communicate properly, or effectively, with the muscles that they enervate. For muscles to stay strong and retain tone, they must be stimulated by nerves on a continuous basis. In cases of myasthenia gravis adequate stimulation does not occur. This disease is progressive and severely debilitating. The underlying cause of the failure of nerve-muscle communication is related to an autoimmune destruction of neuro-

> **Myoblasts** immature muscle cells, particularly those fated to fuse and form muscle fibers
>
> **Myogenesis** the development or genesis of muscles

Table 7-1 Muscle Diseases and Damage

Muscular dystrophy
lethal childhood genetic disorder in which skeletal muscle degenerates and there is loss of muscle function (see pictures [a] and [b])

(a)

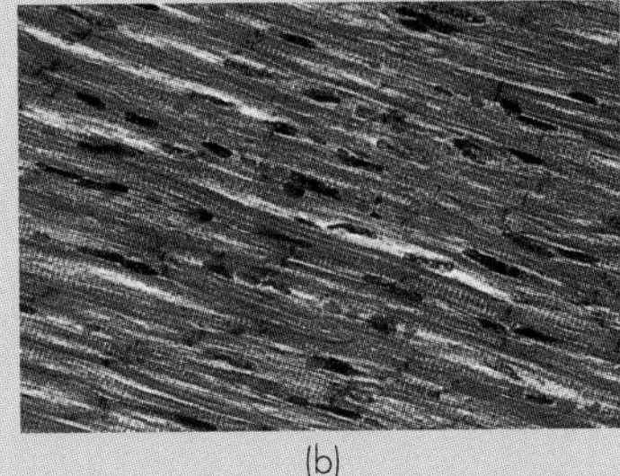
(b)

Myasthenia gravis
autoimmune disease in which there is a progressive decrease in muscle strength due to failure of interaction between nerve and muscle

Cancer
sarcomas and myomas resulting in abnormal growths and tissue damage (see picture [c])

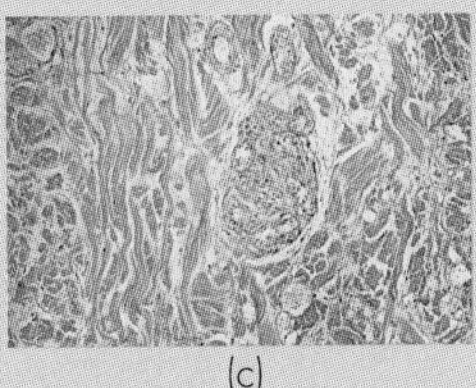
(c)

Tetanus
also known as lockjaw, a condition resulting from a bacterial toxin affecting muscles such that they stay abnormally in a state of contraction (see pictures [d] and [e])

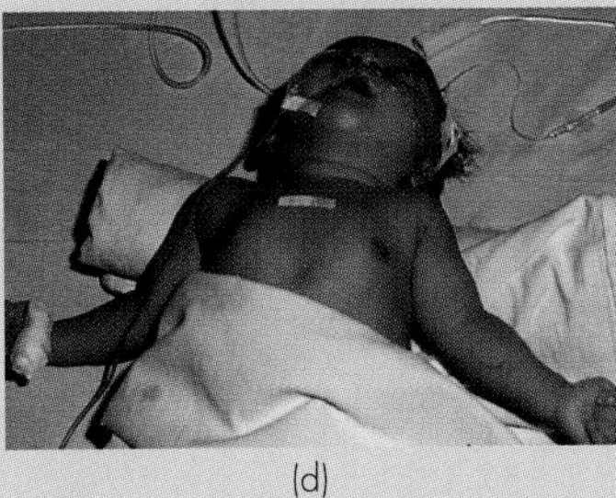
(d)

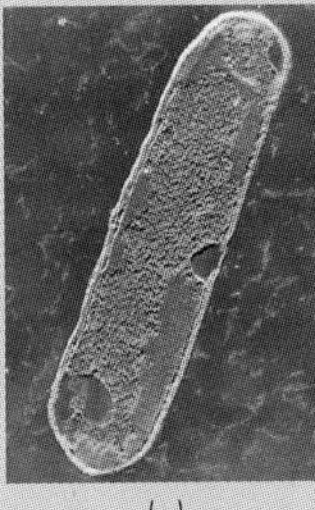
(e)

Poisons
compounds that affect acetylcholine neurotransmitter function at a neuromuscular junction; many of the poisons in pesticides inhibit an essential enzyme known as acetylcholine esterase

Physical damage
accident related; damage to skeletal muscle is naturally repaired by the body, but the replacement connective tissue results in scarring that decreases muscle performance

transmitter receptors on muscle fibers. As will be discussed in Chapter 12, **autoimmunity** is an abnormal condition in which the human body's immune system recognizes itself as foreign and begins to attack its own cells and products. This antiself attack eventually interferes so extensively with the neuromuscular junctions that muscles become completely functionless.

Tetanus *Tetanus* is a condition in which a muscle, or muscles, is irreversibly trapped in a contracted state. One form is induced by a toxin produced by a common soil bacterium known as *Clostridium tetani,* which enters the body through a cut or puncture caused by a contaminated nail, wire, or other sharp object. The common name for this condition is *lockjaw,* a reference to the effect it has on the jaw muscles of livestock that have been infected with *C. tetani.* Humans are affected similarly and, as all mammals, ultimately die from the toxin's effect on the diaphragm, the muscle that controls expansion of the thoracic cavity during breathing. Antitetanus treatments are available and provide immunity for humans and animals exposed to the toxin.

In comparing myasthenia gravis and tetanus, we see two extremes of the effects of disease on muscles. In myasthenia there is a failure to contract, in tetanus there is a failure to relax. In both cases, which are clearly opposite and extreme conditions affecting muscles, the dynamic balance between contraction and relaxation is lost, and the muscle cells cannot do their job.

Cancer—Sarcomas and Connective Tissue Cancers

Transformation of cells is the process by which cells lose control of their normal capacity to grow and divide. Unregulated proliferation of cells can lead to cancer (see Chapter 21 for details). Cancers of muscle cells and within muscles are called *sarcomas.* The origin of these types of cancers is usually within the connective tissues associated with muscles. Because skeletal muscle fibers do not normally divide, transformation is rare. However, this is not the case for smooth muscle cells, which are capable of cell division throughout their lives. The abnormal, rapid growth of muscles or the cells that compose the surrounding connective tissue should always be taken seriously.

Autoimmunity abnormal condition in which the body's immune system attacks and destroys the body or its parts

Summary

Muscles compose approximately 50% of the adult human body weight. There are three types of muscle cells—smooth, cardiac, and striated skeletal—and all are specialized for contraction. Smooth and cardiac muscle cells occur as single cells and are mononucleated. Skeletal muscle cells are actually multinucleated fibers arising during embryonic development from the fusion of hundreds of individual myoblast cells. Cardiac and skeletal muscles contain elaborate cytoskeletal organizations composed of sarcomeres and are, thus, striated; smooth muscles do not have so elaborate intracellular organization and are not striated. The nervous system mediates the activity of all types of muscles, though the heart has an internal regulation as well. Striated skeletal muscle is, in general, under voluntary control; smooth muscle and cardiac muscle are under involuntary control.

There are over 600 different skeletal muscles in the human body, ranging in size from the tiny muscles that control the movement of an eye or transduction of sound to the inner ear to the large muscles that control the movements of the arms and legs. Tendons attach muscles to bones. Different muscle groups attached to different parts of the skeleton, either working together as synergists or in opposition as antagonists, bring about the controlled and often elegant movements of which the human body is capable.

Cardiac muscle cells are found only in the heart and are held together by specialized adhesive junctions within structures known as intercalated discs. Within the intercalated discs are found gap junctions. Cardiac muscle cells communicate directly by gap junctions. Cardiac myocytes have an inherent capacity to beat. This is true from very early in embryonic development, during which these cells establish not only the structure of the heart but also both internal and external means of controlling the rate and depth of contractions. The collective contraction of cardiac myocytes brings about a squeezing of the chambers within, and forces blood through the thousands of miles of vascular elements of the human body's circulatory system.

Smooth muscle cells are found within and surrounding blood vessels, lung tissue, the intestinal tract, and many other sites. Their orientation around and/or along internal networks of tubing in the body provides a means to move liquids and solids within the tubes. Peristalsis within the gut is a prime example of the action of smooth muscles. Each type of muscle, regardless of category, has complexes of thin and thick filaments that provide the molecular machinery to contract. Muscles of all types actively contract, but the extension to their original shape occurs passively in response to other muscles or gravity.

Questions for Critical Inquiry

1. What is the primary function of muscles? Why is voluntary control of muscle action so important to human activities?
2. What are involuntary muscles? In what ways are they essential to human body functions?
3. How do groups of skeletal striated muscles work together to control movement in a precise way?
4. What role does leverage play in body movement? How do muscles and bones interact to provide leverage?

Questions of Facts and Figures

5. What is a sarcomere? What are the molecular and biochemical requirements for it to operate effectively?
6. What are the three types of muscle cells or fibers?
7. In what type of muscle are intercalated discs found?
8. Where are the smallest muscles of the body located?
9. What is the *Achilles tendon* and why is it important?
10. Where do muscles and nerves interact? The release of what neurotransmitter stimulates a muscle cell or fiber to contract? What intracellular ion is involved in the capacity of a muscle cell to contract?
11. Explain the sliding filament model of the sarcomere.
    (a) What are the I and A bands?
    (b) What is the Z line?
    (c) Where are actin and myosin found in the sarcomere?
    (d) Where are tropomyosin and troponin found?
    (e) What is the T system of tubules?

References and Further Readings

Alberts, B., Bray, D., Lewis, J., Raff, M., Roberts, K., and Watson, J. D. (1994). *Molecular Biology of the Cell,* 3rd ed. New York: Garland Publishing.

Stalheim-Smith, A., and Fitch, G. K. (1993). *Understanding Human Anatomy and Physiology.* St. Paul, MN: West Publishing.

Sherwood, L. (1993). *Human Physiology,* 2nd ed. St. Paul, MN: West Publishing.

CHAPTER 8

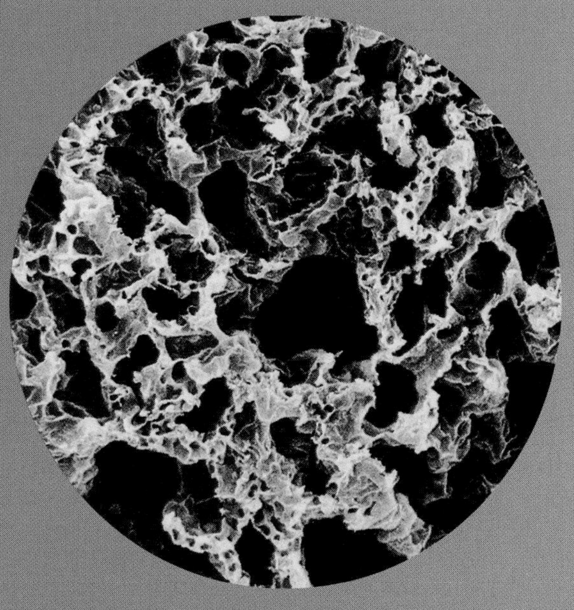

The Respiratory System

INTRODUCTION

Survival of the body is dependent on the minute-to-minute effectiveness of the lungs to carry out efficient gas exchange. The respiratory system of humans is capable of exchanging massive amounts of gases between the body and the atmosphere, particularly oxygen and carbon dioxide. This chapter presents the anatomy and physiology of the respiratory system in the context of survival. Interference with the function of the lungs has dire consequences. This chapter presents information on the composition of air and the effects on the lungs and body of ever increasing amounts of pollutants. Because we take in so large a volume of air, so rapidly and continuously, upwards of 20,000 liters a day, the quality of the atmosphere we breathe is very important. Because human industry and automobile traffic produce an enormous amount of atmospheric pollution, it easily might be said that we are our own worst enemy.

EVERY BREATH WE TAKE

We often take for granted many of the homeostatic processes going on in our bodies that act to maintain our physical and physiological well-being. Although it is true that the only way that we appear to stay the same is to undergo constant change, most physiological changes in the anatomical systems of the human body occur under circumstances that are beyond our abilities to control and/or capacity to perceive. Bones grow and replace themselves at a very slow rate, smooth and cardiac muscles continuously contract and relax, skin is constantly replaced by a nearly endless turnover of epidermal cells. However, for the most part, we do not feel these changes or activities taking place; we simply accept that they must be happening.

In this chapter, the focus is on the anatomy and physiology of the human lungs and respiratory system, the system of tissues that brings in air, absorbs oxygen, and exchanges gases (particularly oxygen and carbon dioxide) with the atmosphere around us. It will become clear after reading this chapter why breathing is not only easy to perceive but also urgent from one minute to the next. If we do not breathe in a continuous effort to bring in new oxygen and eliminate toxic levels of carbon dioxide, we do not survive. It is as simple as that.

Air is approximately 20% *oxygen* (O_2) and 79%–80% *nitrogen* (N_2) (Table 8–1). A very small but important percentage of air is also composed of *carbon dioxide* (CO_2) (0.03%–.04%) and other gases and particles. Oxygen is the essential ingredient of air and must reach the blood to then be delivered to the tissues of the body to form ATP during aerobic respiration (see Chapter 3). It is the job of the respiratory system to carry out gas exchange for the body. Blood cells carry oxygen to all the tissues. In as few as five to ten seconds without oxygen the brain shuts down and we lose consciousness. Deprivation of oxygen for longer than that can damage brain cells, often irreparably and result in **coma** or death. What is this respiratory system that provides our bodies with the ability to exchange atmospheric gases needed to survive from minute to minute? How does it work? What organs and tissues are involved in its performance? What is its anatomical structure? Let's take a deep breath and find out.

Table 8–1 Composition of the Atmosphere

79%–80% nitrogen
20% oxygen
0%–5% water (as humidity)
0.03%–0.04% carbon dioxide
Trace amounts of carbon monoxide, argon, ozone, methane, and other hydrocarbons, nitrogen oxides, sulfur oxides and air-borne particles (dust and pollen)

ANATOMY OF THE RESPIRATORY SYSTEM

Much of the structure of the human respiratory system might easily be compared to the input of air through the air conditioning ducting of a large building (Figure 8–1). The ducts and tubing that lead from the air conditioners, which might be considered the "nose" of the building, are large and free from obstructions. There are certainly filters to clear particles from the incoming air, but they are changed often so as not to impede the smooth flow of fresh air being pumped into the building. The ducting becomes smaller and smaller as the air progresses into different parts of the building, and smaller yet as they divide and provide circulation of conditioned air to each individual room. Ultimately, it is the air reaching rooms occupied by workers in the various offices of the building that is important. After all, this is where the main human activities take place. Because this is a recycling system, the incoming air must eventually be expelled from the building, in a fashion similar to our breathing in and breathing out.

Nose and Nasal Cavity

How is this architectural model related to the structure of the human respiratory system? By comparison, the air we breathe is conditioned during its initial passage into the body. The normal route into the body is through the nose, which has well-known external features, but less well-known internal structures. The shape of the nose is determined largely by cartilage, which composes the structural material of the soft parts of the nose. The cartilage is attached to the bones of the face with the ridge of the nose roughly between the eyes. There are two openings on the underside of the nose, each called a *nostril* or *external naris* (Figure 8–2). They open into an internal nasal cavity. The nasal cavity is anatomically important because it corresponds to spaces created by the organization of the bones of the skull and face. The inner end of the nasal cavity forms an *internal naris*, which opens to the *pharynx* at back of the mouth.

The nasal cavity communicates with hollow spaces within the frontal, sphenoidal, maxillary, and ethmoidal bones (see Figure 6–5). These intrabone spaces are known as the paranasal sinuses (Figure 8–2). There are two routes

Coma a persistent abnormal state of unconsciousness

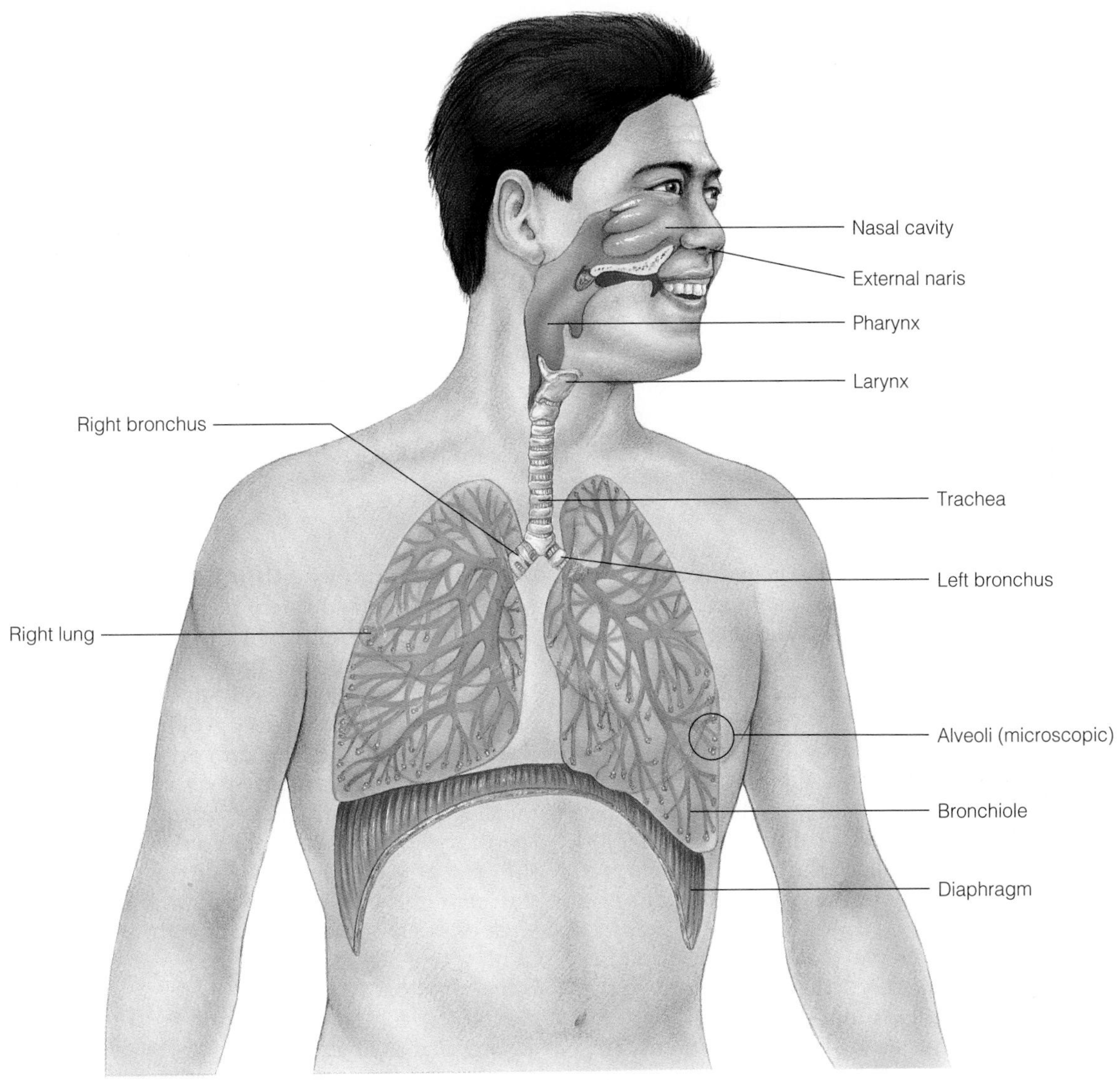

Figure 8–1 The Respiratory System
The respiratory system is involved in delivering oxygen to the body and eliminating carbon dioxide.

air can take through the head. The first is the oral cavity, which is separated from the nasal cavity by the roof of the mouth (palatine and maxillary bones). The oral cavity leads to the pharynx, but, as will be discussed, there are many reasons that breathing through the mouth is not preferred. The second route is through the nose and nasal cavity. This route also leads to the pharynx but is better than the oral route for several reasons.

The nose forms early during fetal development as a combination of bone and cartilage along the midline of the face. Most of the external shape of the nose is constructed of cartilage, so it feels softer and more flexible than the base of the nose, which is bone. The nostrils are physically separated from one another by a ***nasal septum.*** Hair grows within the initial chamber, or *vestibule,* of the nose near the external nares. These hairs are the initial filtering system for incoming air and block the entrance of the nose against passage of large particles of dust and dirt, so that such materials are impeded from entering the nasal cavity, paranasal sinuses, throat, and, ultimately, the lungs.

The nasal cavity provides a staging area for controlling the quality of air entering the body. First, it alters the

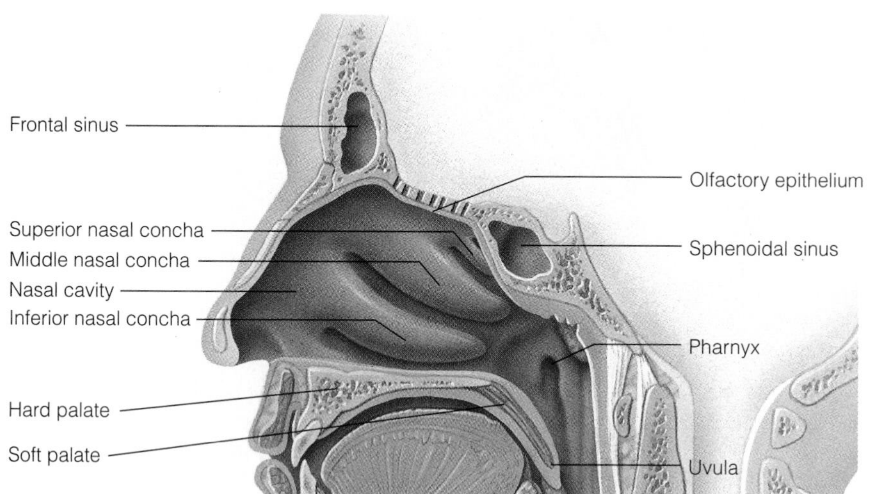

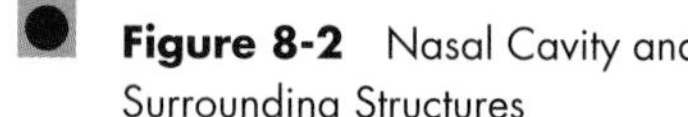

Figure 8-2 Nasal Cavity and Surrounding Structures

The nasal cavity is surrounded by bones. Air enters through the external naris and passes into the pharynx through the internal naris. Paranasal sinuses are indicated as the frontal and sphenoidal sinuses.

temperature of the entering air. If the incoming air is cold, it mixes with air already in the nasal cavity and is warmed substantially before it moves towards the lungs. If it is initially hot, it is cooled. If you breathe deeply on very cold days in winter, it provides an interesting lesson in the importance of the nasal cavity (Table 8–2). If you breathe through your mouth, the air has very little chance to warm up before being sucked down into the lungs. After a few inhalations, the cold air infusion into your lungs can be felt quite distinctly and may even be painful. If you breathe slowly through your nose, it allows the incoming air to mix with the warmer air already in the nasal cavity and prewarms it as it passes through.

Second, the nasal cavity serves to moisten the incoming air. The lining of the respiratory system, particularly in the lungs, is very moist. The exchange of gases necessary for respiratory function depends on maintaining moist surfaces in the lungs, particularly in the alveoli. On very dry days (under condition of low relative humidity), breathing can dessicate otherwise moist surfaces throughout the respiratory system. Breathing through your mouth on such days leaves a dry, unpleasant feeling throughout the oral cavity, over the surface of the tongue, and in the throat.

Third, the nasal cavity is effective in slowing down the flow of air. This provides a chance for molecules in the air that we can smell (called **odorants**) to interact with receptors in the **olfactory epithelium,** which lines the upper portion of the nasal cavity (Figure 8–3). *Olfactory receptor cells* have nerve endings in this region that send signals directly to the brain to inform us not only that we are smelling but also what specifically we are smelling. For better or for worse we can hardly avoid odorants be they freshly popped, buttered popcorn or week-old hot summer garbage. An accurate and sensitive sense of smell may under special circumstances be crucial. Responsiveness to specific types of odorant molecules warns of dangers, as in the case of smoke from fires or noxious chemicals. Without the mixing and slowing of airflow in the nasal cavity, our sense of smell would not work nearly as well.

Paranasal Sinuses

During the cold and flu season we are bombarded with advertisements about remedies for aches, pains, and fever. Many of these remedies are touted for their ability to reduce inflammation and swelling of the sinuses. Sinus swelling can cause headaches, toothaches, and facial discomfort. We hear about sinuses all the time, but what are sinuses? Sinuses are air-filled spaces within the bones of the face that are lined with mucous epithelia. The sinuses of the face are called the paranasal sinuses and the epithelia of these sinuses are continuous with the epithelium of the nasal cavity (Figures 8–2 and 8–4). The sinus spaces are

Table 8-2 Functions of the Nose
Filtration of particles from air
Temperature regulation of incoming air
Moistening of incoming air
Slowing of the flow of air to allow the air mixing required for olfaction

Odorants molecules or substances that interact with nerve receptors and that we perceive as smells

Olfactory epithelium specialized epithelium located along the upper surface of the nasal cavity in which nerve receptors for odorants are found

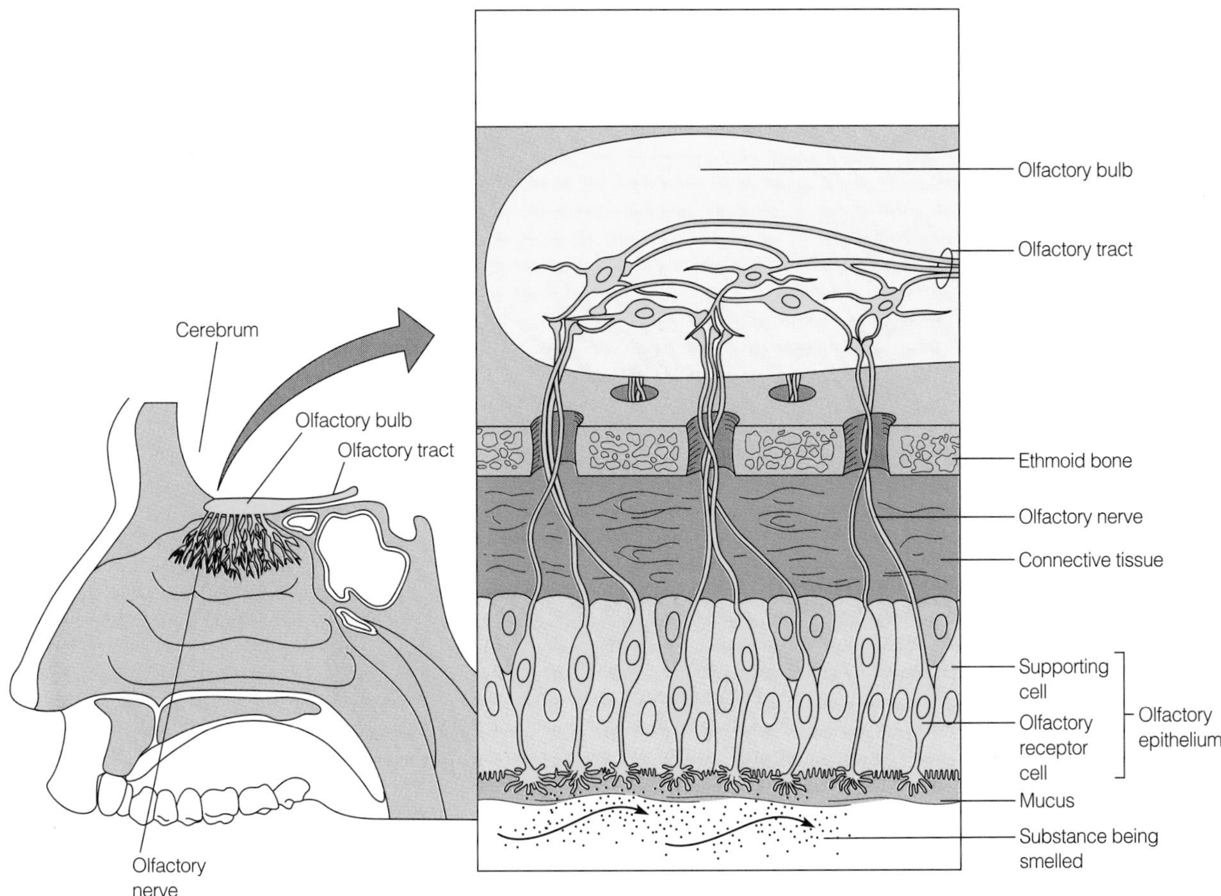

Figure 8–3 Olfaction
The sense of smell depends on nerve receptors in the olfactory epithelium. The nerves pass through the ethmoid bone and directly into the brain.

quite narrow but extensive and increase the surface area of epithelia and the capacity for air inside the head. In humans, an allergic response or an infection often cause sinus tissues to swell, which puts pressure on surrounding tissue and the entire face may hurt.

Pharynx

Once the preconditioned air leaves the nasal cavity, it is pulled down through an internal naris and into a region of the mouth known as the pharynx (Figure 8–4). This region communicates directly with the oral cavity. Therefore, the mouth serves as a crossroads, one pathway used in breathing and the other in swallowing. Thus, the pharynx is a relatively small but important transitional region that is eventually divided into two separate channels, one leading to the *lungs,* the other to the *stomach.* The one leading to the lungs is called the *trachea.* The one leading to the stomach is called the *esophagus.* This pathway to the stomach will be the topic of Chapter 9 on the digestive system.

Larynx

The first part of the pathway leading from the pharynx to the trachea is the **larynx** (Figure 8–5). The larynx is made of extremely tough cartilagenous connective tissue, such that it is not easily distorted or collapsed. A collapse of the larynx, or any other part of the ducting that leads to the lungs, would be disastrous. Of course, an interruption of a few seconds without air and the oxygen it contains results

Larynx the region above the trachea in which sounds are made; also known as the voice box

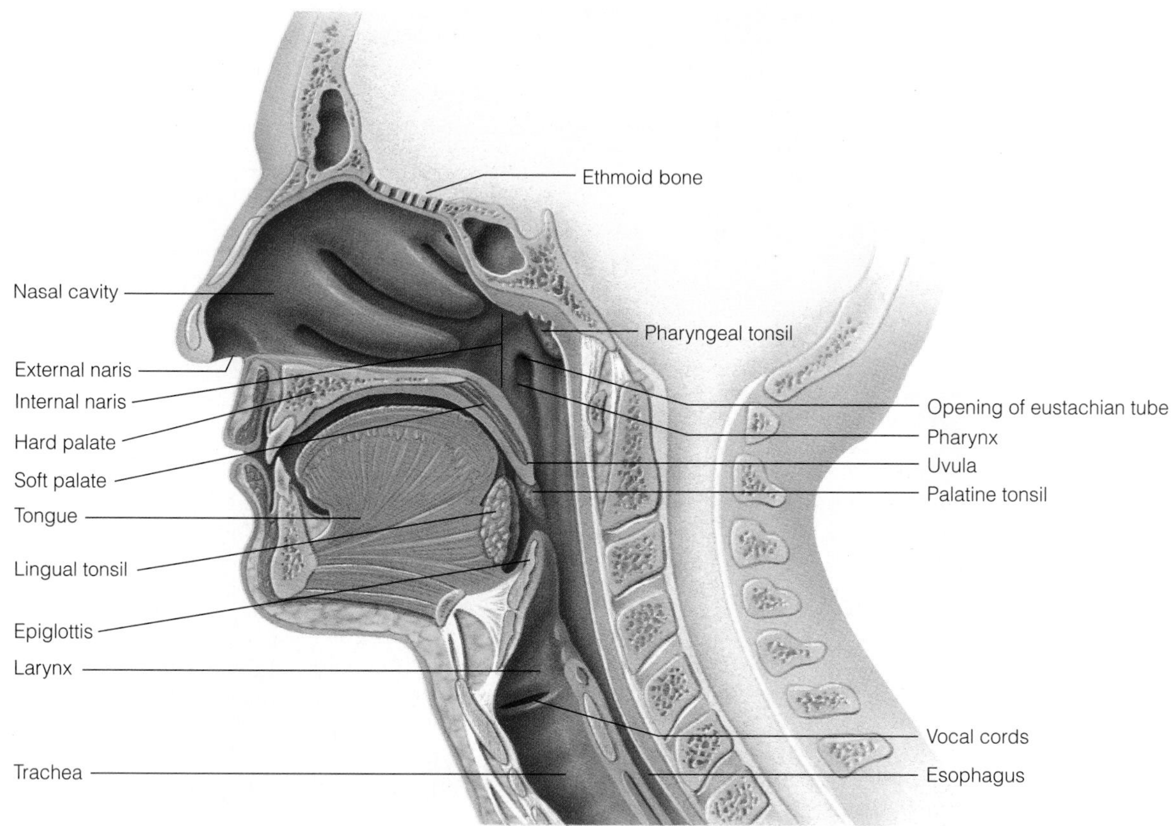

Figure 8–4 The Upper Regions of the Respiratory System
The upper regions of the respiratory system include the nasal cavity, the pharynx, and the larynx.

in unconsciousness. The larynx is also called the voice box and contains the *vocal cords* (Biosite 8–1). The vocal cords vibrate in response to the passage of air over them. A by-product of this vibration of the cords is sound. The mouth, pharynx, and, particularly, the larynx are useful in producing the special sounds needed for human communication.

The range of the frequency of vibrations of the vocal cords is wide (from a few **hertz** to several thousand hertz). Precise control of changes in length and tension on the cords determines the **pitch** of the sound emitted. In humans, as in no other animals, the ability to control the vocal cords has resulted in the extensive manipulation of sound. Changes in the shape of the mouth and positions of tongue, teeth, and lips are important as well. Many of the sounds we make are learned and patterned in their utterance (that is, "the rain in Spain falls mainly on the plain," as Liza Dolittle was forced to learn in the musical *My Fair Lady*), forming a system of communication that we call a *language*. However, nearly all sound in humans that is derived from the larynx, whether it is screaming in fear, groaning in pain, or grunting in disgust, contains interpretable information. Inflammation of the larynx, which may arise as a consequence of screaming for hours at the most important football game of the season or at a rock concert, can result in a loss of voice. This voiceless condition is called *laryngitis*. In these particular situations the vocal cords may have been strained and need time to recover. Oh, blissful silence.

Trachea

The trachea is constructed of the same tough cartilagenous material as the larynx. However, there does need to be some flexibility in the tubing of the passageway leading to the lungs; otherwise, we could not move our head effectively nor bend over. The trachea is not bone because bone would be too rigid. The cartilage of the trachea is a compromise. It is tough and flexible. It will not collapse easily, but it is not rigidly constrained. The anatomy of the tra-

Hertz units of frequency, in cycles per second
Pitch the frequency of vibrations of a sound

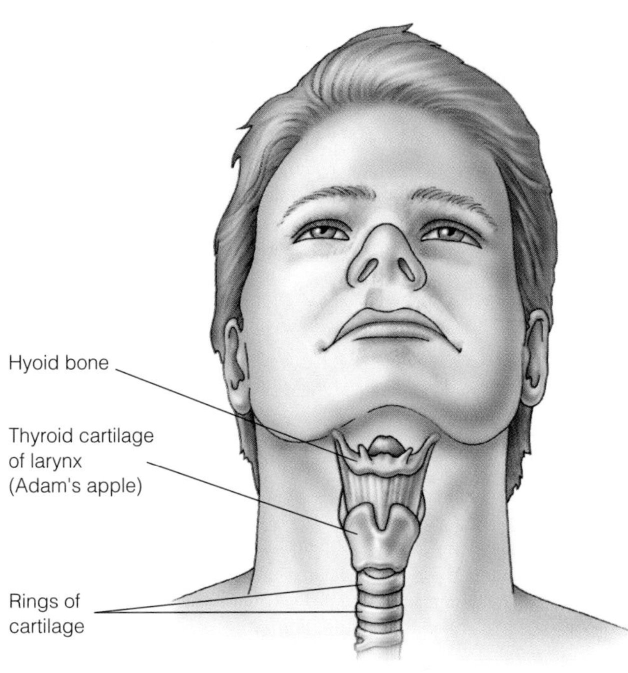

Figure 8–5 Larynx
The larynx is the region in which air is directed to the trachea and in which the vocal cords are located.

chea is described as a series of hard cartilagenous rings (Figure 8–5), linked together by smooth muscle and elastic connective tissue along its entire length. This stacked-ring structure allows for toughness and flexibility—the two essential mechanical features of the initial passageway for air on its way to the lungs.

It is important to note that, because the pharynx is a space common to both the trachea and the esophagus, there needs to be an effective way of preventing the flow of materials destined for the stomach from entering the lungs, and vice versa. The airway leading to the trachea passes between the vocal cords through a space called the *glottis*. There is a flap just above this space, called the *epiglottis* (Figure 8–6). When we swallow, the epiglottis folds down over the glottis and prevents food from entering the larynx and the trachea beyond. Occasionally, food does enter into the larynx or trachea, but a cough or gag reflex usually occurs immediately to expel the material. Sometimes this gag reflex is insufficient and human intercession is needed to dislodge the object from the air passageways.

Bronchial Tree

The trachea is several inches long and runs down from the larynx into the thoracic cavity as a single tube. The trachea branches to form two somewhat smaller channels that enter each lung. Each of the smaller tubes is called a *bronchus* (Figure 8–7). In contrast to the rigid bronchi, the

8–1

THE VOICE BOX OF HUMAN MALES AND FEMALES

Why do males have a larger voice box than females? The differential growth of the larynx between sexes is linked to relatively high levels of testosterone in males and to the influence of this hormone on male growth at puberty. During the development and growth of the larynx, or voice box, a portion of this structure, known as the thyroid cartilage, grows differentially larger in males and is called the Adam's apple. It stands out in the front of the throat and moves up and down as we swallow. Females also have an Adam's apple, but it is just not as big nor as obvious as in males. Both the size of the voice box and the thickness and length of the vocal cords are significant in determining the pitch of the voice. On average, males have deeper voices than females because they have a larger larynx and thicker and longer vocal cords.

This structural change can be seen as well as heard in the throats and voices of young men growing through puberty; their voices transition from a higher pitch to a lower pitch within a year or two. The larynx is only one of many organs or structures of the human body that are influenced by levels of this hormone. Adult males have roughly 10 times as much testosterone as females and the larger Adam's apple of males is one of the manifestations of its anatomical growth effects. If females take testosterone, or testosterone derivatives (for instance, female bodybuilders or athletes who choose to use anabolic steroids), many aspects of their anatomy and physiology change dramatically. These changes include an increase in the size of the larynx and the accompanying deepness of the voice.

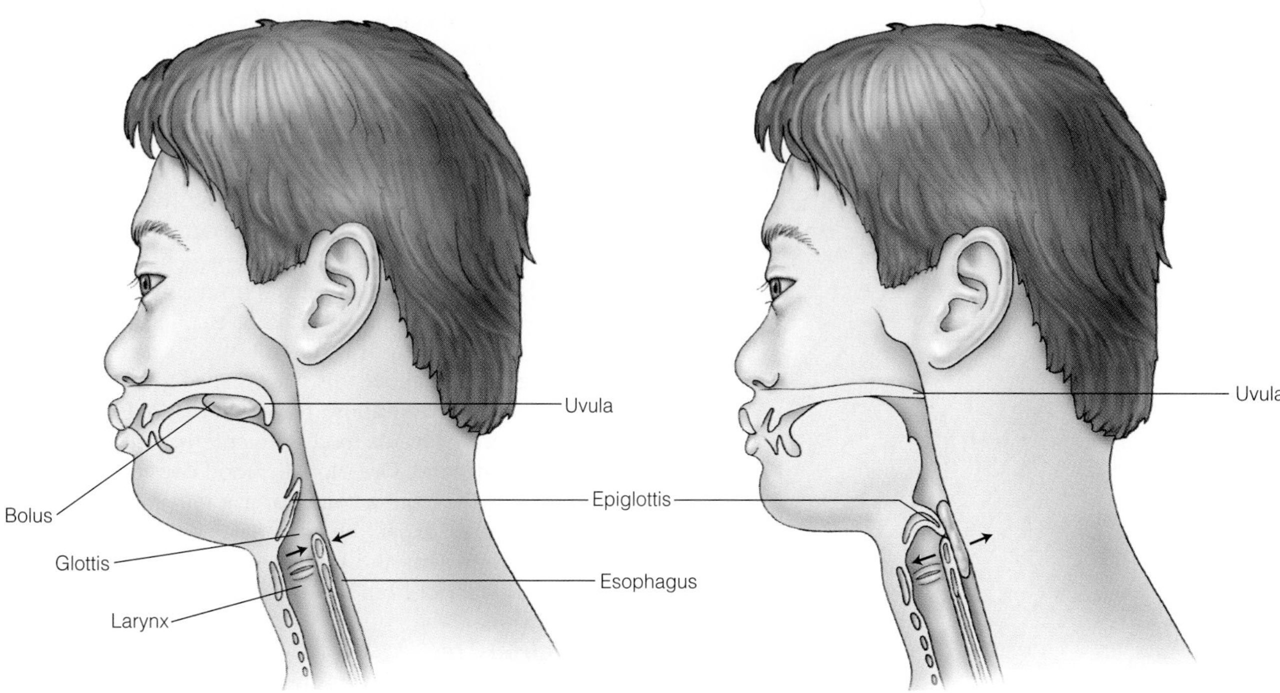

Figure 8–6 Epiglottis and Glottis
The coordinated action of the epiglottis to cover the glottis prevents unintended passage of food into the larynx during swallowing.

tissues of the lungs are very delicate and spongy in substance and appearance to accommodate effective gas exchange. Much of the volume of the lungs is composed of air-filled cavities called *alveoli,* which are well suited for the exchange of gases. However, to get the air into the terminal cavities of the lungs, the tubes bringing air in must remain fairly rigid, even as they become smaller and smaller, just like the branches of a tree.

Alveoli

A bronchus divides into a system of *bronchioles* as it passes within each lung and lung lobe. The bronchioles themselves subdivide into even smaller tubes, known as *respiratory bronchioles.* Eventually, all the branching of the respiratory tree ceases, and the pathways of air terminate in grapelike clusters of alveoli (Figure 8–7[b]). There are approximately a million alveoli in each lung, and it is the alveolar space that gives the spongelike structure to the lungs. The delivery of new air into the alveoli is essential, but the volume of air needed varies with the physiological state of the body. When a body is relaxed, it may take in and expel an average of 15 or so breaths a minute and take in about half a liter of air with each breath. The total volume inhaled measures out to be approximately 12 thousand liters of air a day. This volume estimate does not take into consideration differences in the rate and depth of breathing when the body is hard at work, participating in athletic events, or exercise.

LUNGS

Humans have two lungs, but they are not equivalent in size or in the number of lobes they contain (Figure 8–7[a]). The right lung is larger than the left and contains three lobes, while the left contains two. There is a reasonable explanation for this disparity in the size and location of the lungs. The difference in size and volume of the left lung allows space for the heart to fit in a region of the thoracic cavity called the **mediastinum.** The heart and lungs are the major occupants of the thoracic cavity, but they share it with a number of large arteries and veins (for example, the *aorta* and *superior vena cava*) that deliver blood from the heart to the lungs and systemically throughout the body.

Mediastinum space in which the heart resides surrounded by the left lung

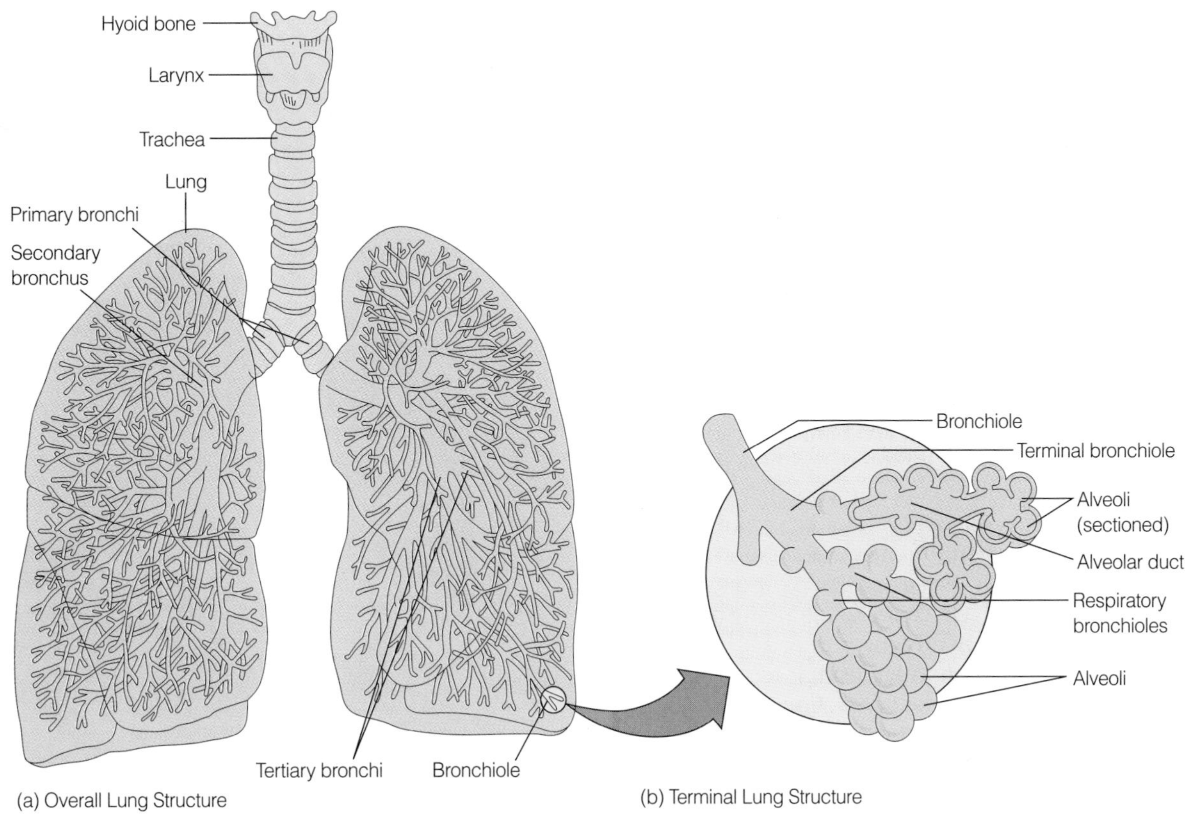

Figure 8–7 Bronchial Tree
From the trachea, incoming air enters the primary bronchi (a). The bronchus of each lung divides into smaller and smaller diameter tubes called bronchioles (b). The respiratory bronchioles (c) divide and reduce in size and air enters the terminal structures of the branching tree, the alveolus (d).

The thoracic cavity is lined with a membranous layer, known as the *pleura,* which separates the heart and lungs from the underlying abdominal cavity and gastrointestinal tract. With the heart and lungs occupying the same cavity, it is extremely important that they are isolated effectively from one another.

Surrounding the Lungs—The Pleura

The lungs are surrounded by two layers of membranes, known as pleural membranes or the **pleura.** The outer layer is called the *parietal pleura.* This membrane is attached to the wall of the pleural cavity. The inner, *visceral pleura* encloses the lungs themselves. The space between the visceral and parietal pleura is coated with a lubricating fluid, which allows them to move almost frictionlessly over one another as the lungs expand and contract. If there were friction or resistance between the organs, or if individual independent movements were restricted, the membranes might become irritated and inflamed. This happens in certain types of diseases and results in what is referred to as *pleurisy.* Pleurisy makes every breath taken a painful experience. The heart is also surrounded by membranes, called *epicardial membranes,* which help isolate it from the left lung and other structures in the thoracic cavity.

Based on the anatomical and physiological descriptions presented so far, the respiratory tubing that delivers air to the lungs is tough, flexible, and decreases in diameter the further into the lungs it penetrates. The endpoint of the branching is the alveolus, which contains little, if any, of the rigid structural materials present in the duct system. The structure is appropriate because its function is gas exchange with the blood in the capillaries. The surface area of the lungs, as estimated from the number and size of

Pleura the membranes surrounding the lungs

alveoli, is very large. Spread out, as if a sheet, the surface area of the lungs is in the range of $60m^2$–$70m^2$, roughly equal to the playing area of a tennis court (Figure 8–8). Across this surface area thousands of liters of gases are exchanged each day between the lungs and the atmosphere. This surface must stay moist and warm to function properly. We now turn our attention to the functional aspects of gas exchange in the lungs.

PHYSIOLOGY OF THE RESPIRATORY SYSTEM

If the cells and tissues throughout the body are to remain alive, they must receive oxygen and eliminate carbon dioxide. The presence of oxygen is essential to drive the engines of efficient energy production in mitochondria within cells. Without sufficient oxygen, ATP cannot be produced in adequate quantities to support the energy demands of the complex organisms such as *Homo sapiens*. Without ATP, the metabolic machinery of cells comes to a grinding halt. At the point at which oxygen is no longer available, respiration changes from an aerobic (oxygen-using), to an anaerobic (oxygen-lacking) pathway. The production of ATP requires fuel as well as oxygen. As with any energy-producing activity in the body, waste is produced along with energy. Glucose provides most of the fuel for human cells. This metabolic breakdown of glucose gives rise to ATP and results in the production of carbon dioxide. Excessive concentration of carbon dioxide in the body results in the poisoning of cells. The principal function of the lungs, then, is to exchange gases—in with oxygen, out with carbon dioxide.

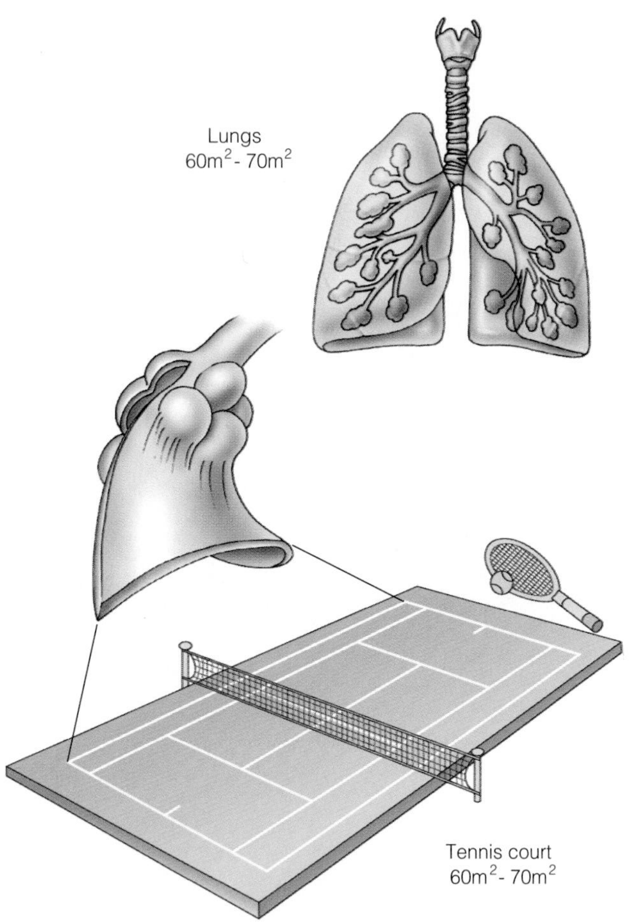

Figure 8–8 Surface Area of the Lungs
The surface area for exchange of gases in the alveoli of the lungs is approximately the area of a tennis court.

Oxygen and Carbon Dioxide

Oxygen The movement of oxygen into the body takes place across the plasma membranes of epithelial cells lining the alveoli of the lungs and through the endothelial cells composing the walls of the capillary blood vessels with which they are in contact (Figure 8–9). The epithelium of the alveolar cavities is one cell thick, as is the endothelium of the blood capillaries. Both types of epithelia are permeable to oxygen and CO_2 or have cellular mechanisms that facilitate gas exchange. The double membrane interface is thin and simple. This simplicity in cellular organization is required for efficient function, because gas exchange carried out in the lungs operates predominantly by *diffusion*. The thinner the boundary is between alveoli and capillaries, the faster and more efficient the diffusion of gases is. Diffusion is easily interrupted by barriers (for example, extra layers of cells). Incoming oxygen must ultimately diffuse across the plasma membrane of red blood cells and be bound to the oxygen-binding protein, *hemoglobin*.

Carbon Dioxide Carbon dioxide is handled somewhat differently than oxygen, though it too must pass through red blood cell membranes, the capillary endothelium, and the alveolar epithelium to be expelled from the lungs. As previously stated, the origin of carbon dioxide is the breakdown of glucose to CO_2 in the metabolically active tissues of the body. CO_2 diffuses through the interstitial fluids between the cells in which it is produced and into the blood (Figure 8–10). Some of the carbon dioxide is dissolved in the blood plasma, and some enters red blood cells by active means. Most of the carbon dioxide in blood cells is converted to bicarbonate ions and is carried to the lungs in that form. Conversion of bicarbonate ions back into gaseous carbon dioxide in the lungs allows it to diffuse from red blood cells, through the capillary wall, through the alveolar epithelium, and into an alveolus.

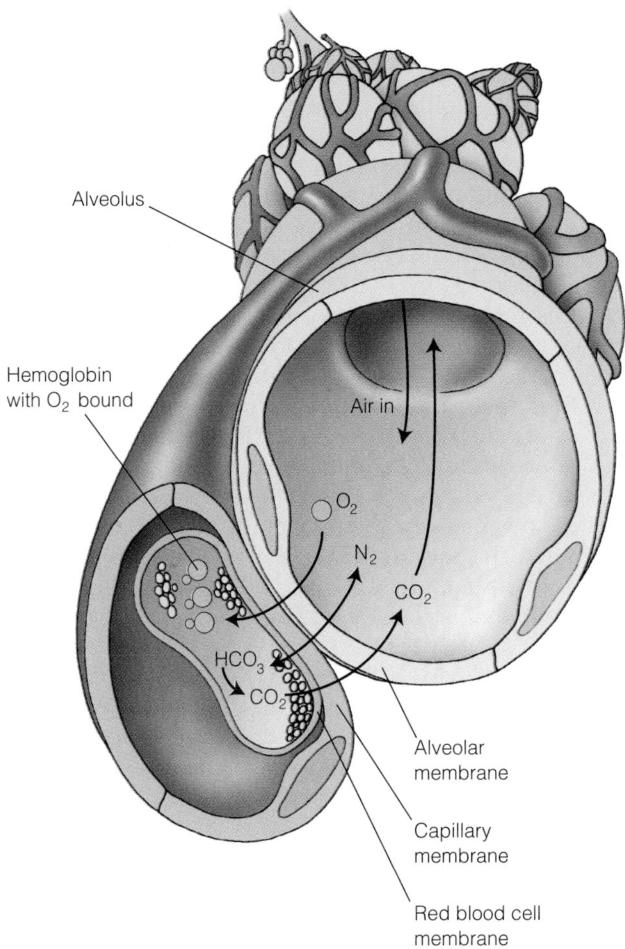

Figure 8–9 The Alveolus, the Capillary, and the Red Blood Cell

Close conjunction of a red blood cell within a pulmonary capillary and an alveolus in the lungs is needed for effective gas exchange.

Exchange of Gases

Several important factors affect the exchange of gases within the lungs. First, the concentration of oxygen in the alveoli is much higher than the concentration of oxygen in the blood in the surrounding blood capillaries. Thus, there is a natural concentration gradient and a tendency for oxygen to pass through the lung and capillary membrane barriers and to enter the blood. Second, the concentration of carbon dioxide is much higher in the blood than in alveoli, so carbon dioxide moves out of the blood, crosses the membrane barriers into the alveoli and exits the body with each breath. An additional requirement for effective gas exchange is that the epithelium lining alveoli must remain moist. The molecules of oxygen and carbon dioxide are essentially dissolved in aqueous fluids as they undergo exchange across the membranes in both directions. The effectiveness of this process assumes that the alveoli remain open, and expanded, on a continuous basis. Because there is little or no structural support for maintaining alveolar expansion, a phenomenon known as **surface tension** may bring about their collapse. Surface tension in the lungs is reduced by special molecules known as **surfactants.** These surface tension-relieving molecules are produced by cells in the lungs and ensure that alveoli remain open.

Binding and Unbinding of Oxygen to Hemoglobin

There is little question of the importance of getting oxygen through the lungs and into red blood cells. However, how does it bind within a red blood cell and how is it released when it reaches its destination in peripheral tissues? If we follow a molecule of oxygen from an alveolus into a blood cell, the process reveals that it crosses the cellular membranes of the alveoli, capillaries, and red blood cell plasma membrane efficiently. Once oxygen passes into a red blood cell, it is bound tightly by molecules of hemoglobin. Hemoglobin is formed of four polypeptides and an iron complex known as heme. Each hemoglobin, in turn, binds four oxygen molecules. There are approximately 300 million hemoglobin molecules in each red blood cell and an average of 10 trillion to 15 trillion circulating red blood cells in a human body. Multiplying the number of red blood cells by the number of hemoglobin molecules in each cell shows that our blood is capable of carrying more than 10×10^{21} (a billion trillion!) oxygen molecules. In addition, because red blood cells circulate within a closed vascular system, hemoglobin is continuously alternating between binding oxygen, as the cells move through the lungs, and releasing oxygen to tissues in the peripheral parts of the body.

It is interesting to note that, although oxygen specifically binds to hemoglobin, oxygen also is released in the appropriate place in the body and under the right conditions to reach the tissues and cells that need it. It would do us very little good if the red blood cells bound oxygen and then would not unbind it. In the oxygen-rich environment of the lungs hemoglobin holds tight to oxygen, but in the oxygen-poor, carbon dioxide-rich environments distant from the lungs hemoglobin readily gives up oxygen. Thus, metabolically active cells receive a constant supply of

Surface tension the property of liquids that makes them resistant to penetration at their surface due to the cohesive interactions of the molecules that compose them

Surfactant a wetting agent or detergent that reduces surface tension

oxygen to make ATP. However, this is only the first half of the gas exchange journey. The potentially toxic CO_2 must be removed (see Figure 8–10).

How is carbon dioxide eliminated? Fortunately, carbon dioxide is considerably more soluble in aqueous solutions than oxygen is, and human blood is composed of nearly 90% water. Therefore, carbon dioxide, which is at high concentration in the metabolically active tissues, is able to diffuse into capillary blood. In addition, as has been described, carbon dioxide has several means of transport from peripheral tissues back to the lungs. It may be dissolved in the plasma (9%–10%), bound to hemoglobin (27%–30%), or carried by blood in the form of dissolved bicarbonate ions (60%–64%).

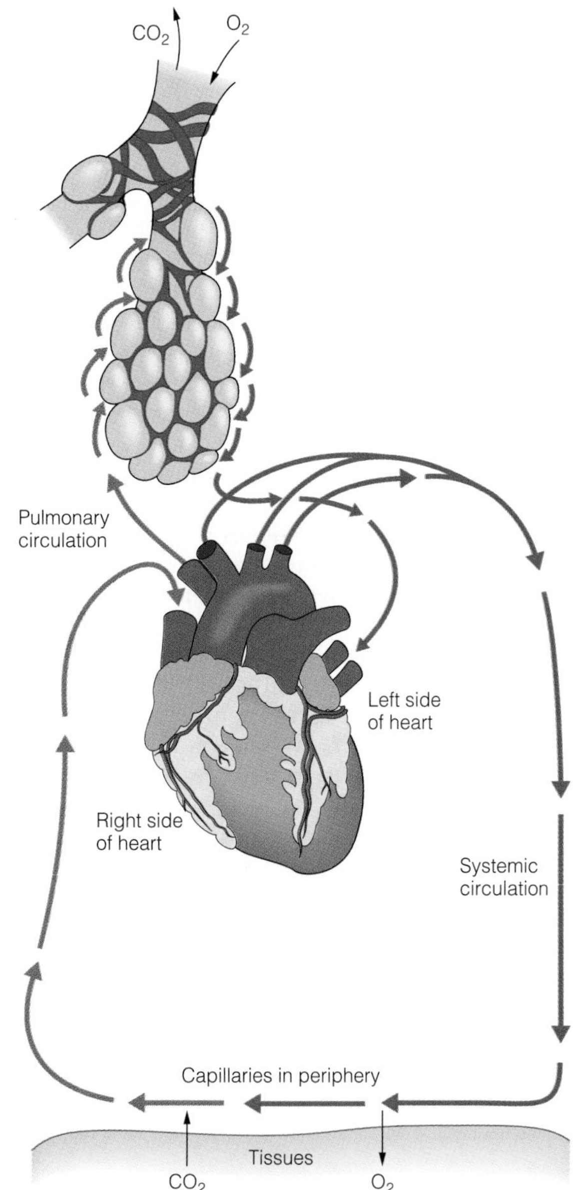

Figure 8–10 Oxygen and Carbon Dioxide—The Pathways In and Out of the Body
Oxygen enters the body through the alveoli of the lungs. Carbon dioxide exits the body from this site also. Oxygen is delivered to the body's tissues through arteries. Carbon dioxide is delivered to the lungs through veins.

The exchange cycle of gases in the blood follows an exacting pathway—alveolar capillary blood cells enriched with oxygen from the lungs circulate into regions of the body that require oxygen. The blood delivers a load of oxygen from the red blood cells and takes on a new cargo of dissolved or bound carbon dioxide for the return trip to the lungs. In the lungs, carbon dioxide is unloaded by red blood cells, crosses the membranes of the capillaries and alveoli and enters the air space of the lungs. Oxygen, once again in high concentration in the lungs, crosses the membranes in the opposite direction and becomes bound to the hemoglobin molecules of red blood cells.

The Other Gas—Nitrogen

The role of atmospheric oxygen in the function of the respiratory system as well as the exchange of carbon dioxide that arises from animal cell metabolic activity have been discussed. However, there is another gas in Earth's atmosphere, which is the major component of air, nitrogen (N_2). Nitrogen composes nearly 80% of Earth's atmosphere. At normal atmospheric pressures, nitrogen enters the body through the lungs and is present in blood and tissues, but it is basically chemically unreactive. However, nitrogen is important physiologically for the part it plays in maintaining air pressure sufficient for the lungs to operate efficiently. If the percentage of nitrogen dropped, it would mean that the percentage of the other gases would increase. The most obvious candidates for increase are oxygen and carbon dioxide. An increase in the percentage of oxygen would be a real disaster for the long-term health of humans and other air-breathing organisms. Exposure to too much oxygen, for too long, damages cells and tissues by oxidative degradation. A significant increase in the percentage of carbon dioxide would also be disastrous. Not only is it a metabolic poison capable of killing rapidly but also it has long-term effects on global warming through the enhancement of the greenhouse effect. Nitrogen serves a valuable purpose in maintaining the breathability of air precisely because it is chemically inert, yet abundant, under normal atmospheric conditions.

Barometric Pressure and Nitrogen in the Blood

Although it is not used metabolically, nitrogen enters the blood and is carried throughout the circulatory system, entering all tissues of the body. However, if we breathe air containing nitrogen under conditions of high pressure (*hyperbaric conditions*), such as during **scuba diving** at great depths, nitrogen can present a serious problem (Figure 8–11). The principle is fairly simple. If the pressure to which a liquid (including blood) is exposed is increased, the number of gas molecules, such as oxygen, carbon dioxide, and nitrogen, that can be dissolved in the liquid increases. Everyone has observed the fizz of gas released upon opening a carbonated drink. The bubbles that rise are formed from carbon dioxide molecules. CO_2 is initially dissolved in the liquid and held there under pressure, instantly escaping when the top is popped.

Deep Water The problem for the human body under the pressure of deep water is analogous to a carbonated canned drink. In a deep sustained dive, nitrogen molecules become dissolved in the blood in greater than normal concentration, proportional to the pressure (Figure 8–11). Oxygen is not as abundant, nor as soluble as nitrogen, so nitrogen accounts for the predominance of gas dissolved. A diver's body is confronted with greater and greater pressure as he or she goes deeper and deeper, and more and more gas is dissolved in the body's fluids. As a general rule, for every 10 meters of depth in water there is a unit increase in barometric pressure (see Figure 8–11). If a diver rises to the surface from a great depth too rapidly, he or she is exposed too dramatically to lower pressure and, as a carbonated drink when the cap is popped off, gas bubbles begin to form. Under these circumstances, nitrogen bubbles form within the blood and tissues of the body leading to painful and often fatal consequences. A doubling of the body from abdominal pain induced by nitrogen gas expansion results in what is called **the bends.** Bubbles in the blood may block blood flow and, thus, adversely affect the heart and brain. How is such a fate to be avoided? The bends, and other problems with bubbles of gas in the blood, can be avoided by rising slowly to the surface. A slow rise from a great depth allows the nitrogen in the body to equilibrate slowly with changing pressure and allows time for excess nitrogen to be ventilated. Gas mixtures used in oxygen tanks for deep diving often substitute helium for nitrogen. Helium is not as soluble as nitrogen in the blood. It was only when sustained deep diving became feasible that this problem arose. Under normal conditions, nitrogen is "breather friendly" in its chemical inertness.

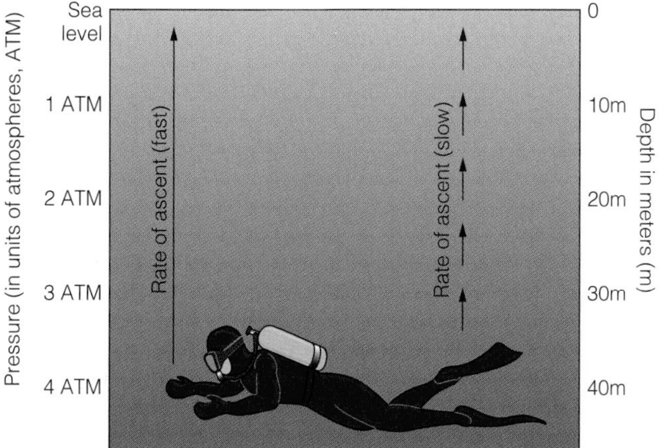

Figure 8–11 Nitrogen and High Pressure
Under the pressure of great depth in water (hyperbaric) nitrogen gas is forced to dissolve in blood at a higher concentration. If a diver rises too rapidly from a great depth, the nitrogen gas escapes the blood and forms dangerous bubbles in the circulatory system. This may lead to the bends.

MECHANICS OF BREATHING

How do we breathe? What are the mechanics of inhaling and exhaling. The groundwork for breathing has been laid out in the discussion of the exchange of gases across the epithelium of the lungs, and into and out of the vascular circulation, but how does air get moved into the lungs in the first place? And then how do we get it out again?

The lungs themselves do not have the capacity for self-expansion; that is, there are no muscles organized internally to allow the lungs to expand or contract in a controlled fashion. However, the lungs are composed of elastic connective tissues and, once they have been forcefully expanded, they tend to return to a low volume resting state. The lungs are like balloons. It takes an outside force to inflate them, but because they are elastic and stretched under pressure, when the pressure is released they empty by elastic rebound. This rebound in conjunction with muscle-driven changes in the shape of the thoracic cavity accounts for the force of exhalation, or *expiration,* of air. How are the forces for expanding the volume of the lungs applied?

The answer to this question lies in the interactions of muscles, bones, and membranes surrounding the lungs. The lungs are located inside a protective cage of bones, the ribs, and surrounded by thin epithelia, the pleural membranes. Along the lower surface of the thoracic cavity and adjacent to the lungs is a large muscle known as the diaphragm. The diaphragm is extensible, such that its contraction affects the shape of the thoracic cavity. Other

Scuba diving swimming under water with scuba gear; scuba is short for *s*elf-*c*ontained *u*nderwater *b*reathing *a*pparatus

The bends a condition associated with formation of gas bubbles within blood vessels as a result of decreased pressure from rising too rapidly from great depths in water

groups of muscles affecting the shape and volume of the thoracic cavity are located between the true ribs. This group of muscles is known as *intercostal muscles*. If you touch your rib cage while taking a deep breath, you can feel your chest expand. This expansion of the thoracic cavity is associated with the breathing in of air. Breathing out occurs when the rib cage and diaphragm return to their original configuration and reduce the size of the thoracic cavity.

To initiate the actions required for inhalation, the muscles involved require controlling signals from the brain. Nerve impulses to the diaphragm and intercostal muscles surrounding the thoracic cavity provide a means of coordinating and controlling the timing and forcefulness of the muscle contractions. Signals may be generated voluntarily, which means conscious control (go ahead, breathe faster or slower), or involuntarily, which autonomically controls breathing (go ahead, try to stop yourself from breathing for a few minutes). The regulation and control of breathing ultimately resides in the *brain stem* (Figure 8–12). Specialized nerve receptor cells in the brain stem have the capacity to sense small changes in the concentration of gases dissolved in the blood, particularly carbon dioxide. The neural signals generated by these cells are sent to other parts of the brain and then routed through motor neurons in the spinal cord to the muscles associated with breathing. Nerve endings in the connective tissue of the lungs sense when the lungs are expanding and, to avoid overexpansion, send signals back to the brain to cease muscle contraction. With no signals for muscle contraction to maintain the inflated state of the lungs, the balloonlike structures begin to deflate. The balance of on-off signaling to and from the brain and lungs provides a well-balanced homeostatic negative feedback that regulates breathing and protects the lungs from damage.

Diaphragm and Expansion

The major force driving the expansion of the lungs is the contraction of the diaphragm to alter the shape of the thoracic cavity (Figure 8–13). The consequence of this change in shape is the enlargement of the thoracic cavity in which

Figure 8–12 Nerve Connections and the Lungs

Nerves control the rate and depth of breathing. Muscles of the diaphragm and between the ribs contract to change the volume of the thoracic cavity and to expand the lungs.

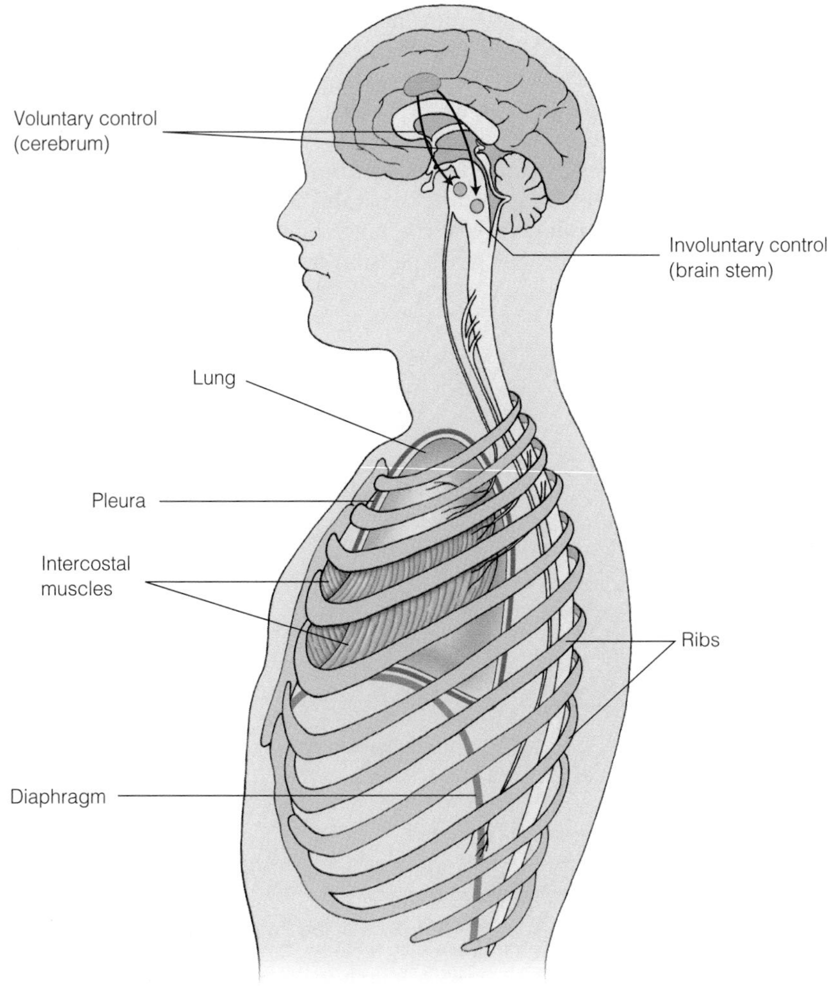

the lungs are located. A model for this action of expansion of the lungs is observed in the "sealed jar" experiment. As the volume of the jar is increased by stretching the elastic membrane at its base, the balloons within are forced to expand and fill with air. Likewise, because the pleural membranes surrounding the lungs form a tight seal over each lung, the lungs themselves are expanded as the diaphragm contracts and the volume of the thoracic cavity is increased. As the lungs begin to expand into the enlarging thoracic cavity, the pressure within them changes negatively with respect to the outside air. Thus, the pressure inside the lungs is less than the pressure of the outside air. This difference causes movement of outside air into the lungs to equalize the pressure filling the newly available expanding lung volume.

The intercostal muscles of the ribs also serve an important function in the expansion and contraction of the thoracic cavity. The contraction of one set of intercostal muscles (***inspirational muscles***) helps change the position of the ribs to accommodate expansion. During inspiration, the muscles help lift and separate the ribs to increase the size of the thoracic cavity. During expiration, the opposite-acting ***expirational muscles*** contract to pull the ribs back together. This serves to accommodate the reduction in size of the thoracic cavity as we exhale. The combination of the contraction and relaxation of the diaphragm and the coordinated contraction of the intercostal muscles account for the forces that change the shape of the thoracic cavity, produce a negative pressure, and bring about the influx of air.

The reverse of inhalation is exhalation. As described earlier, exhalation occurs largely by rebound of the expanded elastic tissues of the lungs in conjunction with the changes in the contraction of the groups of muscles described above. Cessation of expansion is initiated by the negative feedback signals arising from nerve cells, referred to as *stretch receptors* located in the lungs. The air pressure within the expanded lungs is greater than the barometric pressure of the air outside. As a result, the lungs are poised to empty by elastic rebound. This occurs as the diaphragm relaxes and expirational intercostal muscles contract to bring about a return to the resting shape of the thoracic cavity.

LUNG VOLUME AND RESERVES

The discussion now turns to the parameters associated with the variations in the useful volume of the lungs and how they relate to the coordination of lung activity and physiology during strenuous physical activity (Figure 8–14). There are several things to consider in answering this question. Keep in mind that the lungs are only partially emptied during exhalation. Likewise, they are usually only partially filled by inhalation. Why should this be the case? First, some statistics: the *total capacity* of the lungs in an average adult is approximately 5.5 liters, or about a gallon and a half (Table 8–3). The *working volume* of the lungs, that is, the amount of volume that can be maximally inhaled or exhaled, is somewhat less than the total, around 4.5 liters. The difference between the two (approximately a liter) is called the *residual volume*. The residual volume keeps the airways open for gas exchange on a continuous basis and helps prevent the collapse of alveoli from surface tension. An inactive or relaxing human generally takes in about 0.5 liters of air with each breath. This is called the *tidal volume*. If we inhale deeply, we can pull in a further 2.5 liters of air. This is called the *inspiratory reserve*. If we forcefully exhale at the end of a normal breath, we can eliminate approximately 1.5 liters. This is called the *expiratory reserve*. The lungs are sufficiently elastic to adjust the volume of air intake to the level of activity of the body. Less oxygen (and, thus, less air intake) is needed when we are resting than when we are working or exercising vigorously.

Even though an individual may consciously control the working volume of air that he or she inhales or exhales, there are limits to self-regulation. We cannot stop breathing for very long on our own volition. Likewise, heavy exercise, excitement, and/or danger often induce involuntary increases in the rate and depth of breathing. What are the mechanisms underlying involuntary changes in breathing behavior? Is there a specific chemical basis for the control of breathing?

CONTROL OF BREATHING

In this regard, an initial consideration of how breathing may be controlled and regulated is quite logically focused on oxygen. It might be assumed that the need for oxygen would be the primary determinant in controlling when and how deeply we breathe. Certainly, more oxygen is needed when we exercise, because muscles require oxygen to generate the level of ATP necessary for cellular and tissue activities. If individuals hold their breath long enough, they will simply run out of oxygen. Therefore, one might assume that the body is telling us to breathe because the oxygen supply is low. This is quite logical, and quite incorrect. The mechanism by which breathing is controlled depends primarily on brain receptors monitoring the level of carbon dioxide in the blood.

As has been described, carbon dioxide is the waste product of cell metabolism. CO_2 is produced from the catalysis of glucose. Glucose is oxidized to CO_2 during the process

Inspirational muscles muscles used to inhale
Expirational muscles muscles used to exhale

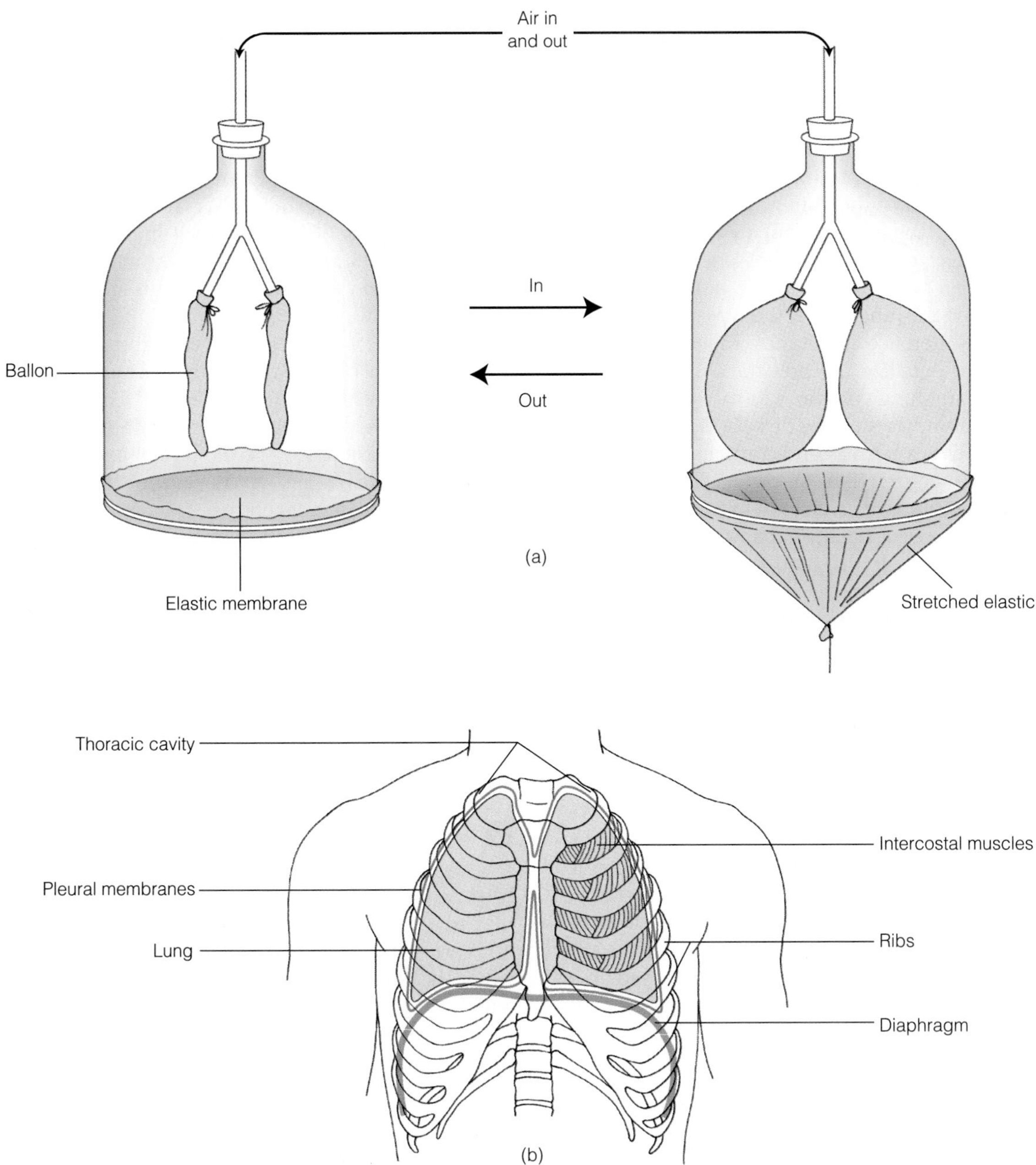

Figure 8–13 Lungs and Expansion—The Balloon Model
In (a) expansion of the elastic membrane at the base of a sealed jar pulls air into the balloons and expands them. Likewise, in (b) expansion of the pleural membranes of the thoracic cavity by contraction of the diaphragm results in expansion of the lungs.

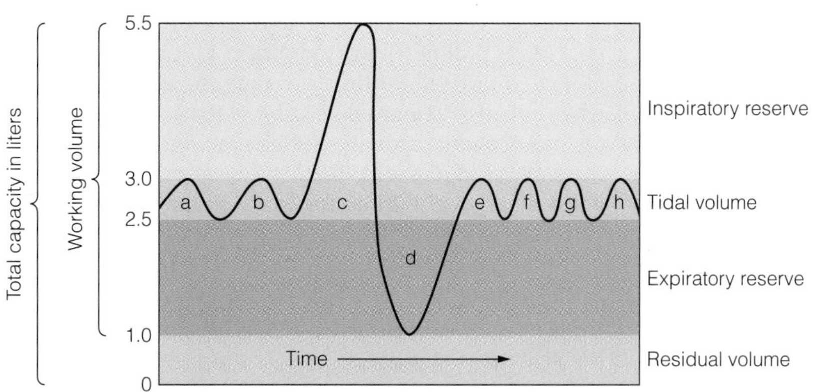

Figure 8–14 Lung Volume and Breathing
This series of breaths (a–g) indicates normal tidal volume, inhaling (upward curve), and exhaling (downward curve). In (c) a maximal deep breath uses the entire inspiratory reserve. In (d) a maximal exhalation expels the expiratory reserve. Then the breath during (e) returns to the normal tidal volume and is maintained in (f), (g), and (h).

Table 8-3 Lung Capacity and Volumes in Breathing

MEASUREMENT	VOLUMES*
Total capacity	5.5 liters
Working volume	4.5 liters
Residual volume	1.0 liters
Tidal volume	0.5 liters
Inspiratory reserve	2.5 liters
Expiratory reserve	1.5 liters

*based on an average adult individual

that produces ATP. If the cells are more active, more glucose is oxidized, and more carbon dioxide is produced. A test of the preeminence of carbon dioxide, and not oxygen, in the regulation of breathing is reasonably simple. If a person breathes into a closed container, such as a tightly sealed bag, and rebreathes the air in the bag continuously, his or her rate of breathing will increase with time. Under these circumstances, oxygen is being continuously depleted and carbon dioxide is building up.

How do we test for the specific effects of oxygen depletion and carbon dioxide increase? If we construct a more complicated apparatus through which to rebreath used air, we can add chemicals to the equipment that will absorb the carbon dioxide as we exhale but will not interact with oxygen. In such a case, it has been shown that there is no increase in the rate of breathing even though O_2 is depleted rapidly. Therefore, it must not be the level of oxygen that is critical. In another experimental approach, it was found that simply increasing the amount of carbon dioxide in the air increased the rate of breathing, even when oxygen was present at a normal concentration. From these results the inference is that the level of carbon dioxide in the blood is critical in the control of breathing. To reiterate, the changes in the level of carbon dioxide are continually monitored by special receptor cells in the central nervous system, which, in turn, signal other areas of the brain (and then the appropriate muscles) to undergo a cycle of breathing.

However, the monitoring of oxygen is not completely excluded from the regulatory process. There are receptors for oxygen in the *carotid arteries* and the *aortic arch,* which are the major blood vessels of the thoracic cavity. The carotid artery delivers blood to the brain. These receptors act as a backup system for chemically monitoring blood. A decrease in oxygen level does send stimulatory signals to the brain, but this is not the principal pathway for control of breathing. In addition, there are receptors that inhibit and interrupt the breathing cycle. A variety of noxious chemicals are able to induce signals from special nerve receptors located in the larynx to shut down breathing to avoid the inhalation of potentially dangerous substances into the lungs. This instantaneous stop-breathing response gives us a chance to escape airborne hazardous materials without breathing more than necessary.

AIR AND THE ATMOSPHERE

Structure and function go hand in hand in the human body, and the relationship between the anatomical structure and physiological functions of the lungs has been discussed at great length. For a human body, the map of the respiratory system starts in the nose and mouth and charts an inward path for the molecules of air that lead first through the nasal cavity and sinuses, through the pharynx and larynx, and then down the trachea into the bronchi and their ever-branching bronchioles into the tiny, moist recesses of the alveoli. The trip back simply reverses the pathway. It all takes about five seconds or less for the average, relaxed deep breath of air. Gas exchange is the dividend of the process, with oxygen taken up by red blood cells and CO_2 expelled. But outside the body is a huge reservoir of air, the **atmosphere,** which disperses what is

Atmosphere the specific group of gases surrounding Earth that compose the air we breathe

exhaled and replenishes what is inhaled during the next breath. We now turn our attention to the the atmosphere, which may be likened to an invisible gaseous envelope that surrounds Earth.

A Slice of the Sky

To describe the gases and their distribution around the planet, we must leave the body and examine the physical environment in which we live. Note that the term *in* is used with respect to the atmosphere. We live *on* Earth's surface, that is, supported by the ground, and we occasionally immerse ourselves *in* water, but air is the most important medium for us, and we live *in* it. Atmosphere is an expansive term, used to describe much more than the mixture of gases in the air that surrounds us. The atmosphere stretches miles up into the sky and encompasses the entire planet. The atmosphere is similar to a great invisible envelope which protects us from and warms us against the onslaught of the vast cold emptiness of deep space.

A view from space reveals the beauty and dynamics of the air. Clouds float in the atmosphere, partially covering the lands and seas in their shadows. Dust storms disturb the desert sands sending particles high above the solid land. Volcanos occasionally erupt sending vapors and particles high into the atmosphere. Lightning flashes in a thousand different places on Earth's surface each day. The entire surface of the planet is in a state of constant flux and activity. The study of these changes in atmospheric conditions is called **meteorology,** and the variations associated with the changes result in local conditions we call **weather.** As much as the atmosphere changes, there are some things in the atmosphere that change very little. One of those is its overall molecular composition (see Table 8–1).

Ninety-nine plus percent of the air we breathe is composed of two gases, nitrogen (79%–80%) and oxygen (20%) or nearly four molecules of nitrogen for each molecule of oxygen. Carbon dioxide and water vapor make up important constituents of the air, but they are present in lower concentrations (0.03%–0.04% carbon dioxide; 0%–5% water vapor). There are normally trace amounts of other gases as well (Table 8–1), but those constituents are of little consequence, unless they are particularly toxic. There is a significant difference in the absolute amounts of gases present in the atmosphere at different altitudes—greater amounts are near the surface of the planet than away from it. However, the ratio of nitrogen to oxygen at any altitude stays the same throughout. Analysis of the gases of the atmosphere is evaluated normally at sea level.

Pressure and Altitude—As Light as Air?

The air we breathe is under pressure, referred to as *barometric pressure.* This is because gases, as all matter on Earth, are influenced by gravity. This is what gives us and all other material objects **weight.** The atmosphere also has weight, because it is made of molecules that are attracted to the enormous pull of Earth's gravity. Envision a column of air several miles tall resting on each square centimeter of Earth's surface (Figure 8–15). Air is light, but several miles of atmosphere actually weighs a significant amount, roughly 1.1 kg/cm^2 (15 lbs/in^2) at sea level. This pressure is equivalent to 11,000 kg for every square meter of Earth's surface at sea level. One consequence of this is that barometric pressure varies with altitude. The higher one goes up in the atmosphere, the less pressure there is and, because there are fewer gas molecules per unit volume, the thinner the atmosphere becomes.

Living at Altitude Living at different altitudes affects breathing and gas exchange. Assume you live at sea level. There are plenty of oxygen molecules to breathe and our bodies are well adapted to the concentration. If you were to fly to Mexico City, you would end up nearly 3000 meters above sea level. The change in pressure is from 1.1 kg/cm^2 (or 11,000 kg/m^2) to around 0.8 kg/cm^2 (or 8,000 kg/m^2) (Figure 8–15). This means not only that the pressure is different but also that with each breath you take there are far fewer molecules of oxygen entering the lungs to bind to hemoglobin in red blood cells. As a result, red blood cells deliver less oxygen to peripheral tissues, particularly the brain, and in a very short time you may feel nauseated, exhausted by any physical activity, and have a terrible headache. The higher in altitude one goes, the more severe are the effects. Exposure of an unadapted individual to the atmosphere at the peak of Mt. McKinley in Alaska (6,000 meters) would lead to severe mental disorientation and drowsiness. Higher altitudes can induce collapse, coma, and death.

Adapting Not all is lost. The body is capable of compensating for changes in atmospheric pressure over a fairly short time and to accommodating to the reduced availability of oxygen molecules. How do we adapt to changes in altitude and the thinning of the atmospheric envelope? The human body responds to deprivation of oxygen by producing more red blood cells and more hemoglobin per red blood cell and by more rapidly venting air into and out of the lungs. The increase in the oxygen-carrying capacity of the blood offsets the reduction in the number of oxygen

Meteorology study of weather and atmospheric conditions that give rise to it

Weather local conditions of the atmosphere; such as, rain, wind, hot, and cold

Weight the attraction of mass to the Earth resulting from gravity

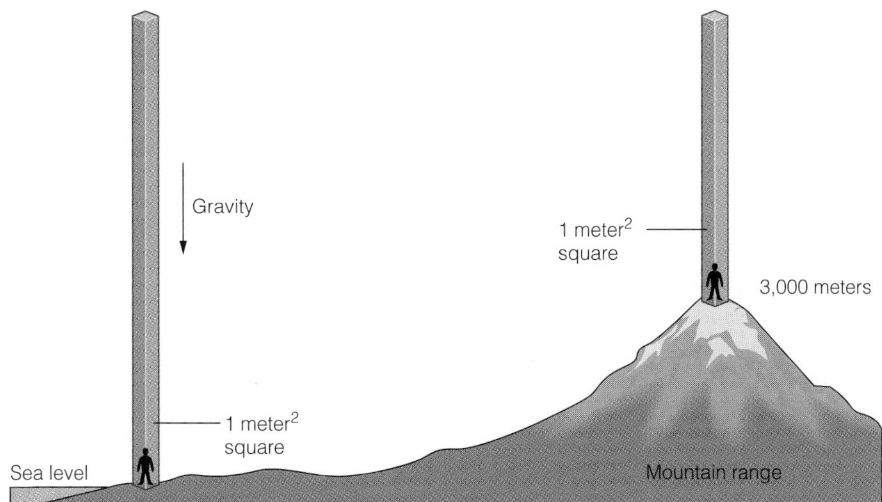

Figure 8–15 An Atmospheric Column of Air—The Pressure on Us All
At sea level the pressure of a column of air 15 miles high is 11,000 kg/m^2. Because the column is shorter at higher altitudes, the air pressure is reduced and the density of the air is decreased. At 3,000 m above sea level the pressure is approximately 8,000 kg/m^2. This pressure difference and the change in air density has profound effects on the performance of the lungs.

molecules taken in with each breath. **Acclimation** of the human body takes some time at high altitudes, usually several days to several weeks, depending on altitude (Biosite 8–2). However, humans are not the only organisms that change altitudes as part of their normal behavior and activity. The most obvious of the nonhuman species to undergo rapid changes in altitude are birds. In amazing feats of adapting to altitude some high-flying birds, such as condors, routinely fly as high as 10,000 to 11,000 meters (higher than Mt. Everest, the highest peak in the world) with apparent ease. The efficiency of oxygen exchange, differences in lungs, addition of other air-retaining structures, and efficiency of energy use make birds the consummate creatures of the atmosphere.

DISEASES OF THE RESPIRATORY SYSTEM

There are many diseases that seriously affect the respiratory system. Some of them are quite prevalent and constitute serious problems for human health. Because the respiratory system is so complex, and its functions so

Acclimation adapting to a new environment or set of conditions

8–2

ACCLIMATING TO ALTITUDE

An interesting application of knowledge about acclimating to high altitudes is observed in the training of Olympic athletes. Many countries have national training facilities located at relatively high altitudes. For example, the U.S. Olympic Training Facility is located at an altitude of greater than 1500 meters in the state of Colorado. The advantage of this location is that when an athlete's body adapts to altitude (that is, more red blood cells, more hemoglobin/cell), he or she can return to lower altitudes and perform demonstrably better. The atmosphere at lower altitudes provides more oxygen, and the increased oxygen-carrying capacity of the athlete's blood is a real boon. The disadvantage is that along with the increase in the number of red blood cells that occurs with adaptation to high altitude comes an increase in the viscosity or thickness of the blood. This makes it slightly more difficult for the heart to pump blood through the vascular system and increases the chance for blood clots to form.

important and varied, you find distinct and characteristic problems associated with many of its different anatomical regions (Table 8–4).

Nose, Nasal Cavity, and Paranasal Sinuses

The epithelial linings of the nose, nasal cavity, and paranasal sinuses are affected by viruses (colds and flu), which usually result in swelling, inflammation, and irritation of the mucous membranes. In response to infection, or inflammation, the mucosa may accumulate fluids and swell. There is an extensive array of environmental agents that affect the mucosa and underlying connective tissues of the sinuses. Many of them induce allergic responses (**allergens**). Tissue swelling and runny noses abound during peak pollen seasons, and pressure from swelling can lead to painful headaches. Bacterial and fungal infections can also develop in the nasal cavity and sinuses, leading to serious damage to the mucous membranes if left untreated.

Pharynx and Larynx

The pharynx contains the *tonsils* and *adenoids*. These lymphoid tissues can become infected by bacteria or viruses and become inflamed. These types of infections (for example, *tonsillitis*) are predominantly associated with childhood. If the problem is persistant, the lymphoid tissues involved may be surgically removed (*tonsillectomy*), though, in general, physicians try to see that they are retained through the judicious use of antibiotics. Inflammation and infection in the adjacent region of the larynx can lead to hoarseness or loss of speech.

As mentioned earlier in this chapter, laryngitis is quite common and can be attributed not only to infections but also to stress and over use of the vocal cords. Professional singers and vociferous sports fans often have to rest their voices between performances. Even politicians suffer. President Clinton required special care to keep his voice during the 1996 presidential campaign.

Blockage of the larynx with food or other objects is a relatively common occurrence. If the blockage persists, a person can suffocate. One way in which a blockage can be removed is by using the Heimlich maneuver (Biosite 8–3).

Bronchial Tree

Asthma and bronchitis occur deeper within the respiratory system and affect the tiny branches of the respiratory tree, the bronchioles. Asthma is a complex allergic-type reaction, which results in the contraction of smooth muscles and constriction of bronchioles. When these muscles contract, the bronchioles are narrowed and the flow of air into and out of the lungs is impeded. Asthmatics have a terrible time breathing. Bronchitis is an inflammation of the mucosa of bronchioles and often results in the excessive production of mucus, which, in turn, blocks the air passageways. Persistent bronchitis, and bronchiole inflammation, can lead to serious infections and further degradation of the ducting of the lungs.

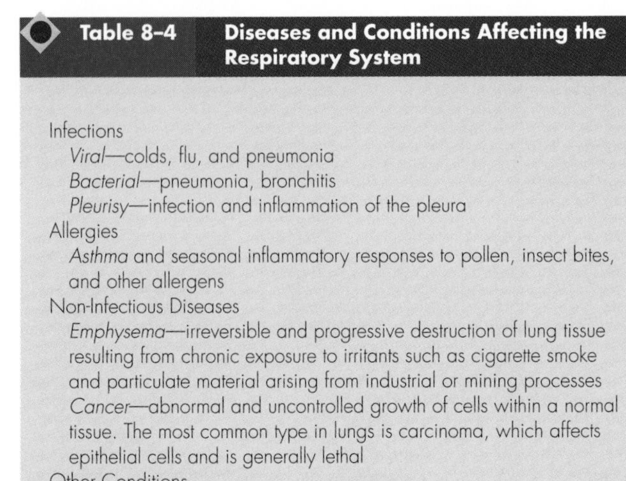

Table 8–4 Diseases and Conditions Affecting the Respiratory System

Infections
- *Viral*—colds, flu, and pneumonia
- *Bacterial*—pneumonia, bronchitis
- *Pleurisy*—infection and inflammation of the pleura

Allergies
- *Asthma* and seasonal inflammatory responses to pollen, insect bites, and other allergens

Non-Infectious Diseases
- *Emphysema*—irreversible and progressive destruction of lung tissue resulting from chronic exposure to irritants such as cigarette smoke and particulate material arising from industrial or mining processes
- *Cancer*—abnormal and uncontrolled growth of cells within a normal tissue. The most common type in lungs is carcinoma, which affects epithelial cells and is generally lethal

Other Conditions
- *Pulmonary edema*—build up of fluids in the lungs from poor cardiac function

Alveoli

The final destination of air in the lungs is the small, grapelike clusters of alveoli. In the alveoli, oxygen, carbon dioxide, and other atmospheric gases are exchanged. The alveoli are affected by such diseases as *pneumonia* and **emphysema.** Pneumonia may be caused by viruses, bacteria, or fungi, but all of these agents lead to the destruction of lung tissue and with it the capacity to respire effectively. Bacterial pneumonia can be treated with antibiotics, yet, it is often more dangerous than the *walking pneumonia* caused by viruses. Some forms of bacterial and fungal infections are difficult to treat. Individuals with AIDS may contract serious fungal pneumonia infections, which ultimately claim their lives.

Another devastating disease affecting the alveoli is emphysema. Emphysema is the result of chronic exposure of the lungs to irritants, such as cigarette smoke, particulate matter from industrial and manufacturing processes, and coal dust that inhibits the *cilia* of lung epithelial cells and reduces their ability to move particulate materials out of the lungs. This disease results in the loss of elasticity and

Allergens materials or agents that elicit an allergic immune response

Emphysema irreversible and progressive deterioration of the lungs

8–3

THE HEIMLICH MANEUVER

The Heimlich maneuver was developed to dislodge materials and objects from blocking the trachea and preventing the passage of air. As shown in (a) a rescuer positions his or her fist directly over the upper abdomen and below the rib cage. With arms wrapped around the victim's waist and one hand holding the closed fist, the rescuer rapidly squeezes inward and upward several times in rapid succession (b). This forces out the air remaining in the lungs and dislodges (hopefully) the object (usually food) from the trachea. The process may be repeated as necessary.

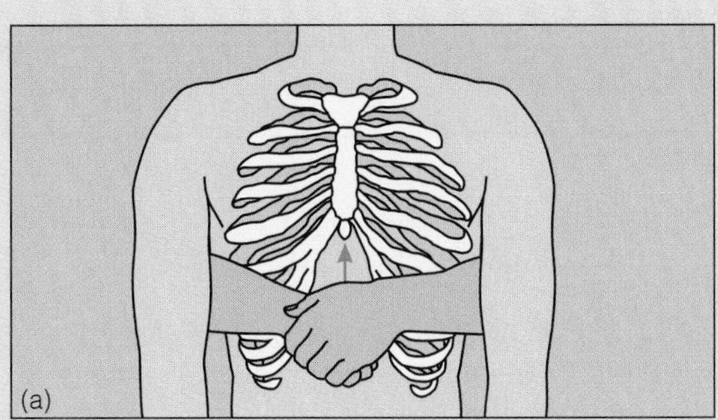

destruction of the connective tissue of alveolar walls. Instead of contracting and pushing out the air within them, the alveoli remain filled with stale air and are inert with respect to elastic rebound. This permanent inflation of the lungs leads to a buildup of fibrous connective tissue, which, in turn, inhibits the diffusion of oxygen and carbon dioxide across the alveolar epithelial cell membranes. These changes lead to damage to the capillaries that surround the alveoli and a blockage of blood flow. This makes the heart work harder and leads to a general decline in heart-lung efficiency. Emphysema is a progressive disease, which means that, once it develops, it continues to get worse. Ultimately the disease destroys the alveoli. Severe cases restrict almost all physical activity for an individual because of lack of ventilation due to respiratory system destruction. Without sufficient gas exchange death is inevitable.

Pleura

The pleural membranes covering the lungs are very important to respiration. They seal the surface of the lungs so that lung expansion is possible in conjunction with expansion of the thoracic cavity. Pleurisy is characterized by an inflammation of these membranes. If the pleura are severely infected or inflamed, every breath may be painful. This is because the otherwise smooth, moist, and frictionless interaction of the pleura with the lungs is altered.

Pulmonary Edema

The heart plays a role in keeping fluids from building up in the lungs. Strong cardiac pumping helps keep these fluids circulating and prevents their buildup in lung tissue. **Pulmonary edema** is the buildup of tissue fluids in the lungs and is usually a result of a weakened heart. Less efficiency in cardiopulmonary functions, that is, after a heart attack or in cases of emphysema, is a dangerous situation. The buildup of fluids in the lungs often results from infections and inflammations of the alveoli, bronchioles, and respiratory mucosa. The fluid buildup reduces the available surface area for gas exchange and leaves an individual short of breath.

Pulmonary edema buildup of watery fluids in the lungs; associated with poor heart function

Cancer

Of all diseases of the lungs, the most frightening and debilitating is cancer. Lung cancer is one of the deadliest of lung diseases. The most common types of cancers of the lungs are **carcinomas.** Carcinomas arise from the transformation of normal epithelial cells lining the lungs into cancerous cells that grow out of control. The causes of this transformation of cells to a cancerous state are many and diverse and not completely understood (see Chapter 21 for details). Smoking cigarettes is a leading cause of lung cancer. Industrial air pollutants, such as asbestos and organic solvents, are also agents that have been shown by laboratory tests to be capable of inducing cancer. Exposure to airborne radioactive materials is also a known agent of cancer of the lungs. Although a great deal is known about cancer and its causes, we do not know nearly enough.

Cancers are not a single disease with a single cure. More knowledge of how the compounds mentioned above (and thousands of others) may transform cells to a cancerous state and how the body responds to the changes they bring about over the course of a lifetime are of great importance for human health and longevity. It is too often the case that the presence of cancerous cells in the lungs cannot be detected until too late to remedy the situation. Surgery, radiation treatments, and chemotherapy are still the best weapons in use to combat the spread of cancers. However, the effectiveness of preventive measures such as changes in lifestyle (stopping smoking), diet, or geographic location of residence may be important. An ounce of prevention is worth a pound of cure.

Carcinomas malignant cancers of epithelial origin

Summary

This chapter has described the great sea of air that surrounds us all and the journey these gases make into the lungs and out again. In this round trip the internal territory of the respiratory system has been mapped and the mechanics of breathing described. The structure of the respiratory system is uniquely fitted to its performance—structure underlies function. As in the air-conditioning system of a skyscraper, the human respiratory tree divides again and again to deliver a cargo of precious molecular oxygen from the air to the red blood cells in the capillaries surrounding the millions of alveoli of the lungs. Exchange of gases in the lungs carries away one of the major waste products of our cellular metabolism, carbon dioxide. The change in the level of CO_2 in the blood is monitored by the brain stem to regulate and control breathing. It is quite sobering to realize how perilously close we are to death should our next breath be interfered with. Life is maintained from minute to minute with every breath you take.

Questions for Critical Inquiry

1. How is the quality of the air we breathe important to respiratory function? Should we be concerned with the quality of air we breathe? Why?
2. What roles do O_2 and CO_2 have in human metabolic processes?
3. How are the rate and depth of breathing affected by strenuous activity? What is the brain-muscle-lung connection?

Questions of Facts and Figures

4. What are some of the functions of the respiratory system other than gas exchange?
5. What is the composition of air? How does it change with altitude?
6. Where does gas exchange take place in the lungs?
7. What purposes are served by the nose and nasal cavity?
8. In what way is the structure of the trachea and the bronchial tree important to respiratory function?
9. In what anatomical region is the voice box located?
10. What are the membranes that surround the lungs?
11. Is nitrogen metabolically important under normal atmospheric conditions at sea level? Under what conditions does it become harmful?

12. What groups of muscles are involved in breathing?
13. (a) What is the total air volume capacity (on average) of adult lungs?
    (b) What is the residual volume?
    (c) What is the tidal volume?
    (d) What are the inspirational and expirational reserves?
14. What is emphysema?
15. What type of cancer commonly occurs in the lungs?

References and Further Readings

Hazen, R. H., and Trefil, J. (1990). *Science Matters.* New York: Anchor Books.

For information on the Heimlich maneuver,

Heimlich, H. J., and Uhley, M. H. (1979). "The Heimlich Maneuver," *Clinical Symposia* 31:1–32.

Monroe, J. S., and Wicander, R. (1994). *The Changing Earth.* St. Paul, MN: West Publishing.

Stalheim-Smith, A., and Fitch, G. K. (1993). *Understanding Human Anatomy and Physiology.* St. Paul, MN: West Publishing.

CHAPTER 9

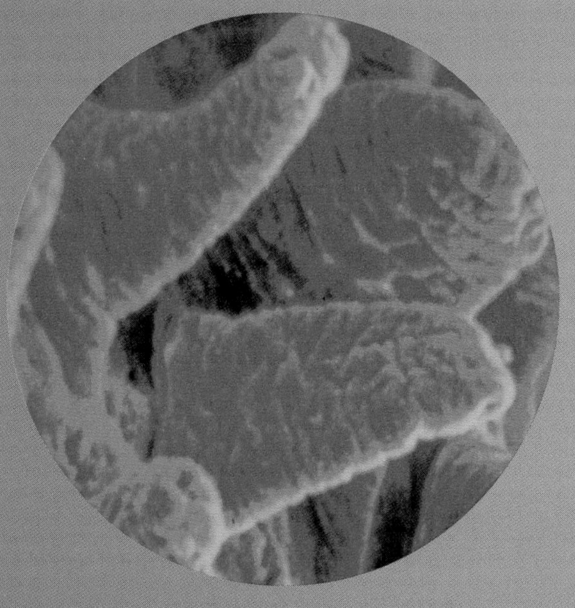

The Digestive System and Human Nutrition

INTRODUCTION

The collective balance or homeostasis of the body's internal physiology is kept within specific norms through the consumption of energy-rich molecules and their conversion to ATP. Different organ systems in the body use energy to integrate different aspects of this biological balancing process. However, only two organ systems are fundamentally involved in providing the molecules needed to produce energy and nutrients as building blocks for growth. They are the respiratory system and the digestive system. Each functions to bring in molecular resources from outside the body and utilize them for production of ATP and synthesis of organic molecules. The respiratory system helps ensure that the body is supplied with oxygen, through the exchange of gases with the atmosphere. However, oxygen is only one of the ingredients of energy production by animal cells. Now, we turn to the other ingredients and investigate the digestive system, which brings in and converts nutrients to cellular fuel.

AN OPEN-ENDED TUBE

The digestive system is basically the long, tubular structure (about 9 m–10 m) through which raw materials and energy-rich nutrients enter and leave the body (Figure 9–1). The digestive system provides an anatomical and physiological arena in which the raw materials we consume from a diversity of sources in the world around us are moved, broken down, and made available for use by the body's cells and tissues. The digestive system is constructed in the form of a long, convoluted tube, which is subdivided, both structurally and functionally, into specialized compartments and/or regions. The tube is continuous, starting with the mouth at the anterior end and terminating with the anus at the posterior end.

THE DISASSEMBLY LINE

One might relate the digestive process to a factory assembly line, or better, a "disassembly" line, in which the large particles of food ingested during a meal, or snack, are systematically dismantled and converted to a form that can be

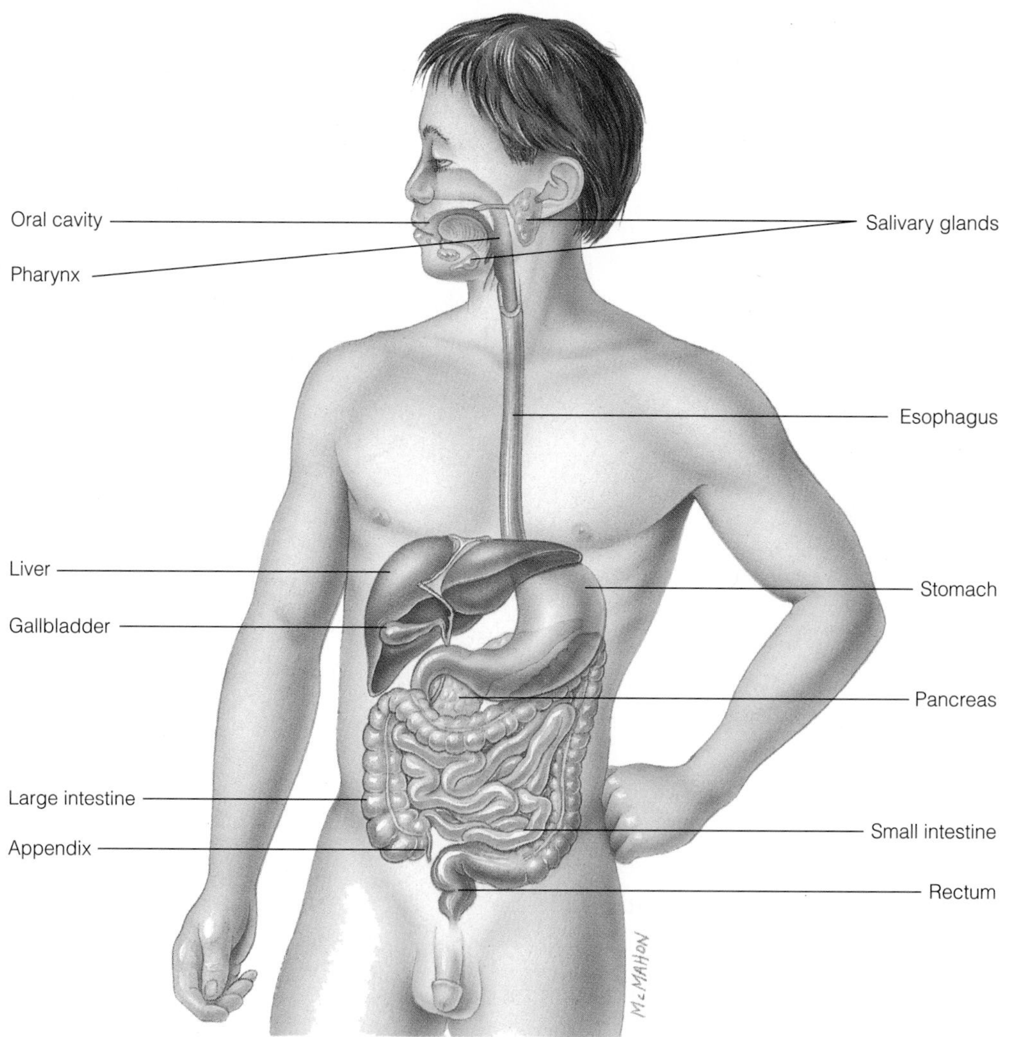

Figure 9–1 Digestive System
Input to the digestive system is through the mouth and output is through the rectum and anus. Between the entrance and exit are 25 to 30 feet of tubular elements.

utilized by the body (Figure 9–2). The process of digestion works more slowly than respiration, taking hours and not seconds or minutes, but it provides the second key to cellular energy production, that is, a source of fuel. Complex carbohydrates and fats, for example, are energy-rich components of our diet and are easily and rapidly broken down in the digestive process. It is only after they have been broken down and absorbed into a cell that they can be used in combination with oxygen to produce ATP. In addition, the digestive process supplies organic and inorganic construction materials for use by the cells and tissues of the body. Carbohydrates, lipids, and amino acids are combined with water, vitamins, and minerals to satisfy metabolic needs (see also the human nutrition section of this chapter).

It is worth remembering that cells are the basic units of function within the body. They are microscopic factories that are in constant need of energy and materials to synthesize proteins, lipids, complex carbohydrates, and genetic materials, such as DNA. Cellular metabolism is a never-ending process. The human body's needs are often measured in terms of minimum daily requirements for specific types of nutrients. This will be discussed in more detail later in the chapter. The body does have some energy reserve storage capacity for use in emergencies, but it does not take long for the body to begin digesting itself when intake is reduced or cut off.

CONVERTING RAW MATERIALS

The problems faced by the digestive system are formidable ones. This is because most of the material we ingest as food is in a form that cannot be used directly by cells in the body. The food we eat is, in fact, foreign material clearly produced originally to have a structural role, or function, in another species of animal or plant. These materials were produced to function in that animal or plant and not in us. It is not only foreign in its composition but also much too large and complex to be utilized by individual cells. Raw materials that cannot be transported through the plasma membrane and into the cytoplasm are basically useless metabolically. As with the proverbial sailor adrift alone at sea in a lifeboat, there is water all around, but not a drop to drink. So it would be with most classes of nutrients, except,

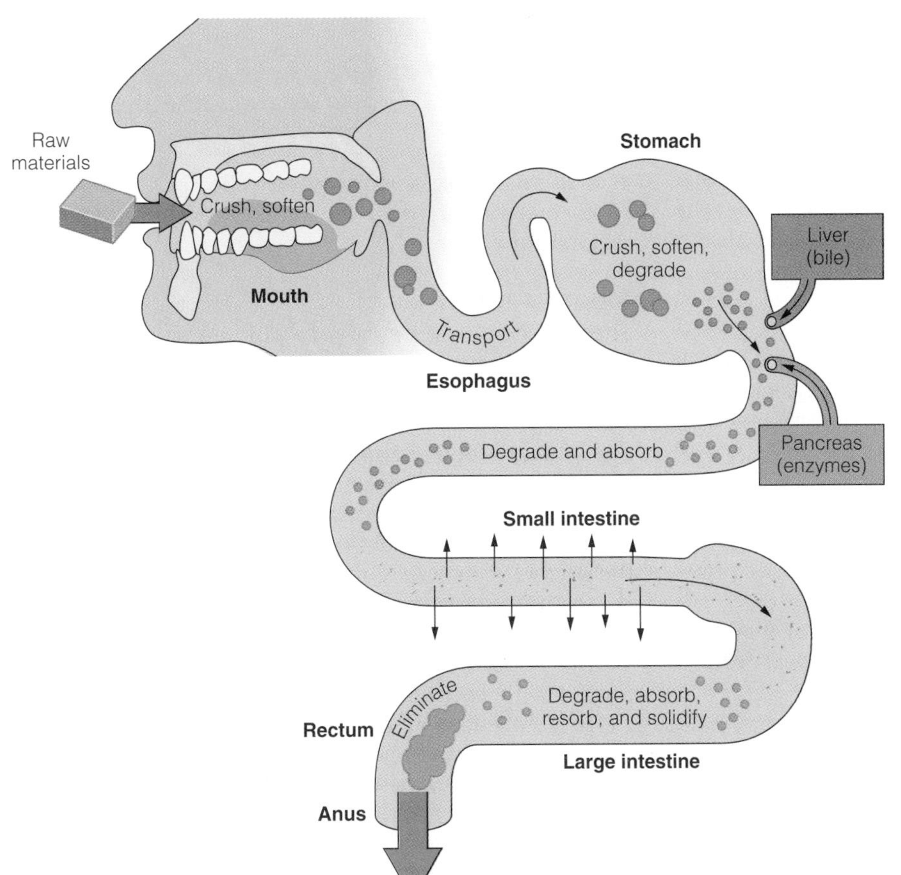

Figure 9–2 Disassembly Line
The disassembly of raw materials begins in the mouth. Large, soft material is produced, which is transported through the esophagus to the stomach. Large material is crushed and degraded to smaller, more fluid material, which enters the small intestine where it is enzymatically degraded to molecules and absorbed into the body. Indigestible material passes into the large intestine where it is re-solidified as water is resorbed into the body. The solid waste, feces, is eliminated through the anus following temporary storage in the rectum.

of course, that these materials can be efficiently broken down into their constituent molecular parts, transported into cells, and used as fuel and building supplies. It is the process of digestion and the anatomical system in which these processes occur that we will now explore.

THE MAP

A map of the anatomical terrain of the digestive tract may help us make sense of the processes going on along its length (Figures 9–1 and 9–2). Science fiction writers and movie producers have created worlds in which the characters in their stories are shrunk down to a size small enough to enter the blood stream, or airways, of the body, or to interact with a part of the outside environment that we take for granted as being too small to directly concern us. Isaac Asimov's *The Fantastic Journey,* and the movies, *Inner Space* and *Honey, I Shrunk the Kids,* are all part of this imaginative world of book and film. The idea of the microcosm or "small world view" with respect to the parts of the body is a valuable one, because in real life there is complete spatial continuity between small and large. It is only a matter of scale. Imagine what it might be like to encounter microscopic regions of the body as if they were the regular-sized world around us. A tooth might appear the size of a mountain. The flow of saliva might seem as some mighty river. To recognize unfamiliar objects or terrain, we would need a map with a starting place, clear signposts, and a goal. For the digestive system, the map is a modified anatomical diagram (Figure 9–2). The starting place is the mouth, and the goal is to understand how we ingest, digest, and absorb nutrients and eliminate the unused waste.

Setting the Course

There are four basic types of activities carried out by the digestive system. They are the mechanical breakdown of the large, potentially tough or hard materials we ingest in the presence of saliva; the biochemical modification of these materials from larger to smaller forms (ultimately to basic, small, building-block molecules); the absorption of molecules into cells (and their dispersion by way of the blood to all parts of the body); and the elimination of unusable, undigested, and waste materials.

So, the overall requirements for processes of digestion include the following in spatial and chronological order:

1. Mechanical breakdown
2. Biochemical modification
3. Absorption
4. Elimination

To propel the materials through the system on schedule, there is an inherent muscular activity controlling movement operating within the digestive tract from the esophagus to the *rectum.* The process is called peristalsis, and it keeps materials moving in a timely fashion from one region to the next.

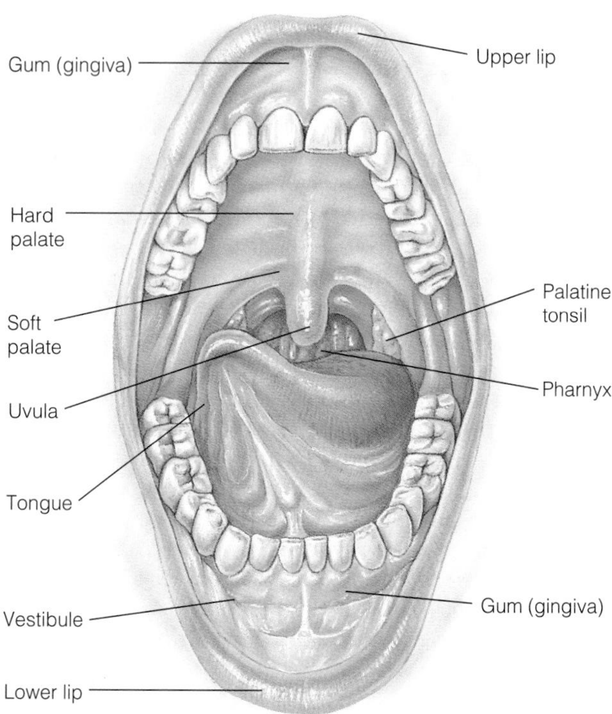

Figure 9–3 The Cavernous Mouth
An anterior view of the oral cavity and structures within it.

Mouth

The mouth is a cavernous space with numerous specialized structures organized within it. It is a warm, moist environment (Figure 9–3) and is the principal compartment in which mechanical disruption of bulk foodstuffs takes place. The intake of food or **ingestion** through the mouth is the first step in eating. After actually placing food in the mouth and sensing its warmth, texture, and flavor, the initial mechanical action is chewing or mastication. This reduces the size of the materials by means of the grinding action of the teeth. So starts the mechanical breakdown of food.

Teeth A look inside an open mouth reveals two, U-shaped rows of teeth, each solidly embedded in bone. One group of 14–16 is embedded in the maxilla, and a matching set is embedded in the mandible (Figure 9–4). Teeth are the hardest structures in the human body. They are made of proteins,

Ingestion intake of food, eating

specifically enamel and dentin, in which minerals of calcium phosphate are incorporated. Like bone, these dental proteins and minerals are produced by cells, but unlike bone, teeth that are broken cannot repair themselves (Figure 9–4). A maximum of 32 teeth are found in the adult human mouth though in many individuals fewer are found. Teeth grow from and remain strongly embedded in the bones of the jaw, and the pattern of their distribution allows the upper and lower rows of teeth to be brought into register when the mouth is closed. This conjunction of top and bottom teeth is called the **bite.** There are several different types of teeth, but basically they fall into two categories—teeth that tear and teeth that grind. Teeth that tear, the *incisors* and *canines,* are sharp edged or pointed. Teeth that grind, the *cuspids* and *bicuspids,* are flat or ridged. The function of the first group is to divide up the food into portions capable of being ground up by the second.

Saliva Chewing mixes pieces of food with the fluids in the mouth. This fluid is called *saliva.* Saliva is 99% water and is produced by several different glands, particularly in the regions of the mandible and maxilla. The major source of these fluids is the salivary glands, which are a group of glands located in both upper and lower areas of the mouth.

> **Bite** the alignment of the upper and lower jaws, particularly with respect to teeth

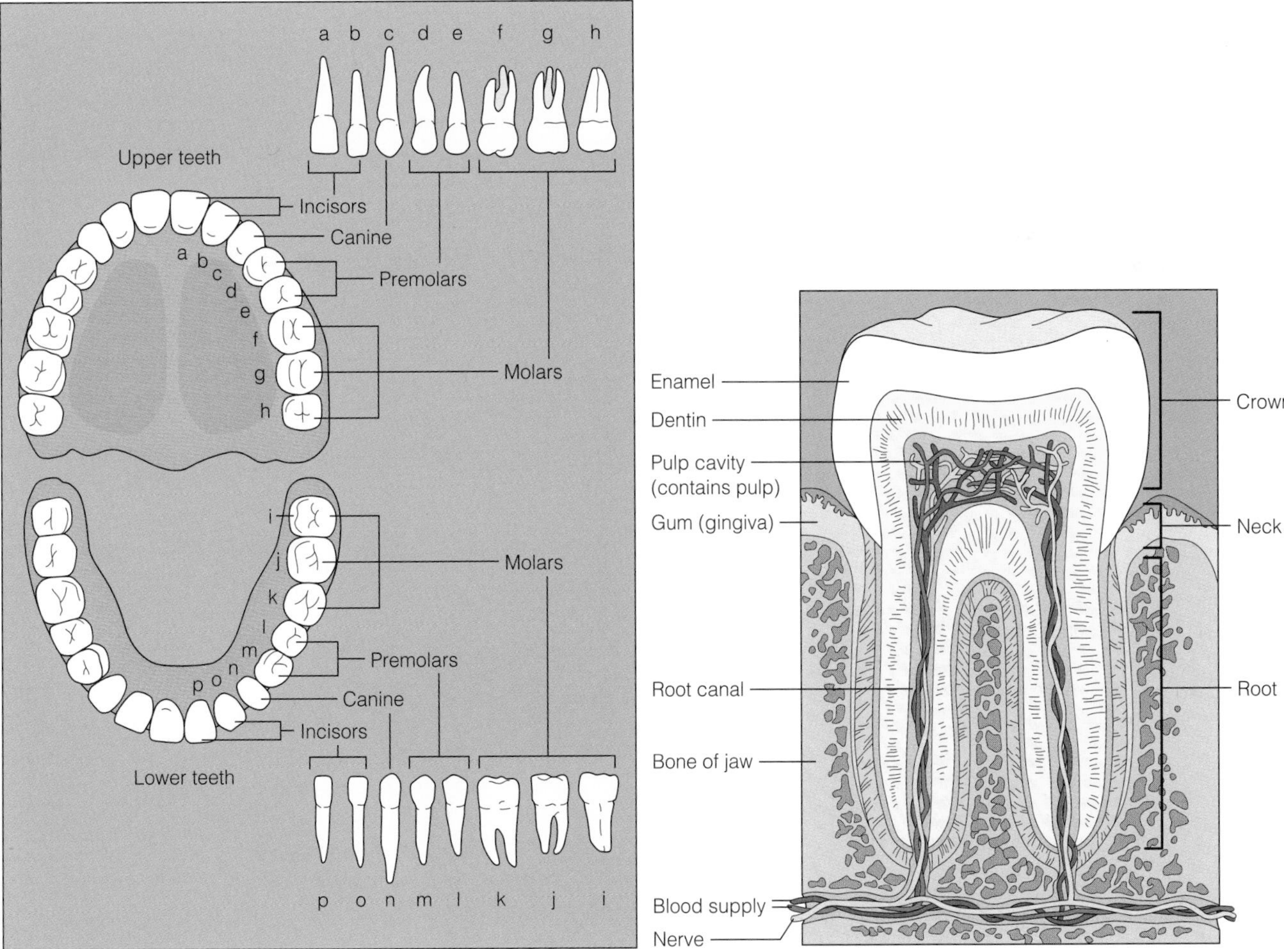

(a) Permanent Teeth **(b) Tooth Structure**

Figure 9–4 Pattern and Organization of Teeth
In (a) the pattern and organization of permanent teeth are shown for the upper and lower jaws. The shape of teeth is important to their function. In (b) the structure of a tooth is shown.

They are the *parotid, sublingual,* and *submandibular glands.* Together, they produce a liter or more of saliva each day (Figure 9–5). They also produce an enzyme, known as **salivary amylase,** that begins the breakdown of polysaccharides. By the time we finish chewing each bite of food, it is a rounded mass of moist, crushed material called a *bolus.* Food in this condition is usually soft and pliable and does not damage the tissues lining the oral cavity. The oral cavity is lined with a thin, moist epithelium covering various ridges and folds. This type of flexible mucous membrane originates on the inside of the lips and extends throughout the mouth and back to the pharynx. With regional variation in structure and specialization mucous membranes are found commonly throughout the digestive system.

Palate The upper surface of the mouth is called the **palate** (Figure 9–3). The anterior part of the palate is hard and ridged. It is essentially bone covered with mucous membrane. The posterior part of the palate, which lacks bone, is referred to as soft. The hanging flap of tissue at the end of the soft palate in the back of the mouth is a muscular extension called the *uvula* (open your mouth and say "aah" while looking down your throat in the mirror). Behind, and to the sides of, the uvula are the *palatine tonsils,* and at the base of the tongue are the *lingual tonsils.* Tonsils are lymphoid tissue and, thus, are part of the immune system, which is important in defense against infection and disease. The palatine tonsils are probably best known by the pain and problems they inflict on us when they swell up in response to infection. Their swelling makes it hurt to swallow and, if the condition persists, they are often surgically removed.

Tongue Positioned within the U-ring of teeth in the space surrounded by the lower jaw is the **tongue** (Figure 9–6). This organ is muscular and flexible. It is intimately involved in the process of swallowing, which requires a com-

Salivary amylase an enzyme produced by salivary glands that digests starches

Palate the upper surface of the mouth

Tongue a flexible, flattened muscular structure of the mouth used in eating, swallowing, and speech

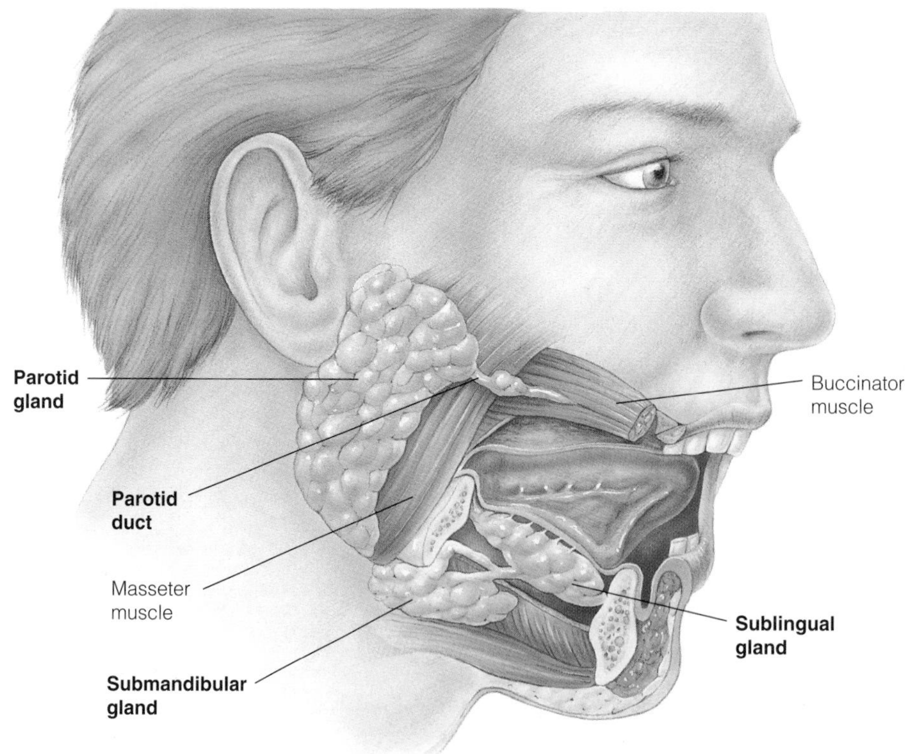

Figure 9–5 Salivary Glands
Salivary glands produce liters of saliva each day to soften and begin the breakdown of raw materials we ingest.

plex set of tongue movements. The tongue also plays a key role in speech and the formation of intelligible sounds. In conjunction with the vocal cords, and the teeth and lips, the variety and quality of sounds that come out of our mouths is amazing. For fun, read outloud any sentence in this book while visualizing the relative position of your tongue, teeth, and lips, as you do so.

Papillae and Taste Buds The surface of the tongue is covered with a variety of specialized, elevated structures located on its upper surface and edges. These elevations are called **papillae.** There are three kinds of papillae found in humans: *filiform, fungiform,* and *circumvallate.* (Figure 9–6). **Taste buds** are found in fungiform and circumvallate papillae and are formed in association with sensory cells that are part of the nervous system. Taste buds are capable of discriminating between different types of tastable molecules, and this is the basis for how we sense sweetness, sourness, saltiness, and bitterness in the food we eat (Figure 9–6). The sensory cells have receptors on their surfaces that act in a lock-and-key mechanism. A cell surface "lock" is opened by a flavor "key," and a signal is sent to the brain for integration. However, taste buds can be fooled. For example, the sweet taste of sugar is based on the molecular structure of a carbohydrate, as in the case of the table sugar sucrose; however, sugar-substitutes, such as Nutrasweet™, elicit a similar sensation even though it is composed of amino acids. The shape of taste is a fascinating phenomenon.

The mouth is usually a moist environment and made more so by the fluids released by the salivary glands in anticipation of, and during eating. Consequently, there is little problem in dispersing the flavor-eliciting compounds throughout the mouth. However, if you dry your tongue (with a paper napkin, for instance) and then place on it a crystal of salt, or sugar, for a short time, there will be no sensation of taste at all. Only when the crystal dissolves are the taste bud receptors stimulated. One way in which the

> **Papillae** specialized raised areas of the tongue and mouth
>
> **Taste buds** specialized chemoreceptors in supporting structures located on fungiform and circumvallate papillae

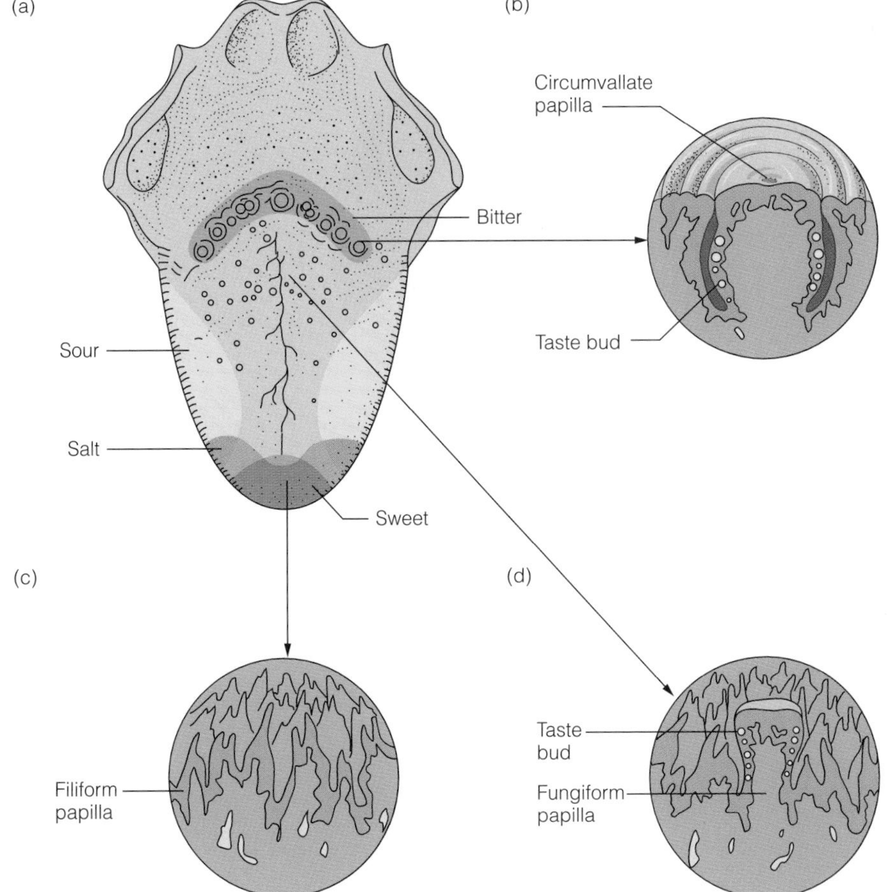

Figure 9–6 The Tongue, Papillae, and Taste Buds

The human tongue is a muscular structure with a complex epithelial surface architecture (a). Covering the upper surface of the tongue are a number of different types of elevations known as papillae. The circumvallate papilla (b), for example, is a large dome-shaped structure that has taste buds associated with it (c). Taste buds allow us to discriminate between the different taste sensations of substances we eat.

tongue can be mapped for distribution of different types of taste buds is to introduce tiny quantities of crystallized materials onto specific regions and measure the responses of individuals with respect to the taste sensations they receive.

Another role for papillae on the surface of the tongue is related to the roughness they impart to that surface. This, in turn, serves in the mechanical control of fluids as well as in the positioning of the bolus. An irregular or rough surface works better than a smooth one in this case, much as the ridges found on finger tips help grip smooth objects. In many infectious diseases of the mouth, scales are shed from papillae at a slow rate, and a mixture of these scales and bacteria accumulate to cause the feeling of a "coated tongue."

Pharynx

The back of the mouth and the nasal cavity converge in a region known as the pharynx (Figures 9–1 and 9–3). This region is essential as a staging area for activities that begin in the nose and/or mouth—breathing and swallowing. The anatomy of breathing was described in Chapter 8 and involves the passage of air through the pharynx into the larynx. Swallowing occurs at the finish of chewing, as the bolus passes through the pharynx to the esophagus. After softening up the food, we are ready to swallow it. This requires the bolus to be positioned on the tongue at the entrance to the pharynx. The mechanical events associated with the act of swallowing automatically close off the entrance into the larynx, by folding the epiglottis over the glottis. This leaves the channel into the esophagus open (Figure 9–7). Commitment of materials into the digestive tract begins in earnest as we swallow, because at that point, there is generally no turning back.

Esophagus

The esophagus is a muscular, flexible tube approximately 30 cm long, depending on the height of the individual, and lined with a stratified epithelium. It connects the pharynx with the stomach and is capable of peristaltic control of the movement of food within it. The musculature of the esophagus is quite interesting and fundamentally important (Figure 9–8). The upper third of the esophagus is surrounded by voluntary muscle, which allows us to partially control a swallow. The middle third has a mix of voluntary and involuntary muscle, and the muscles surrounding the final third are entirely involuntary. In other words, we progressively lose the capacity to control the food we swallow as it moves down the esophagus. As mentioned earlier,

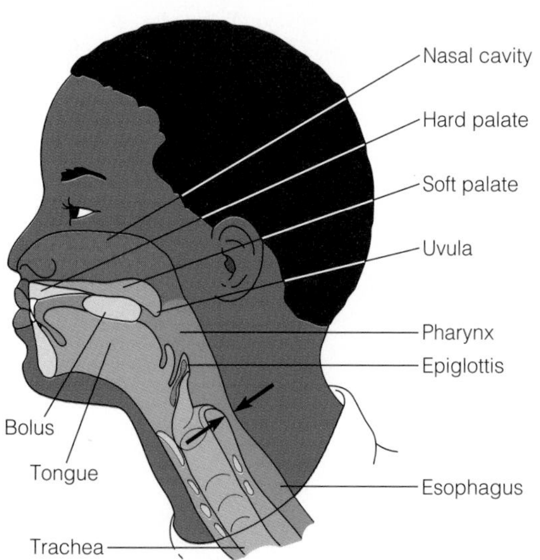

(a) Bolus of food entering pharynx initiates swallowing reflex; upper esophageal sphincter (arrows) is closed

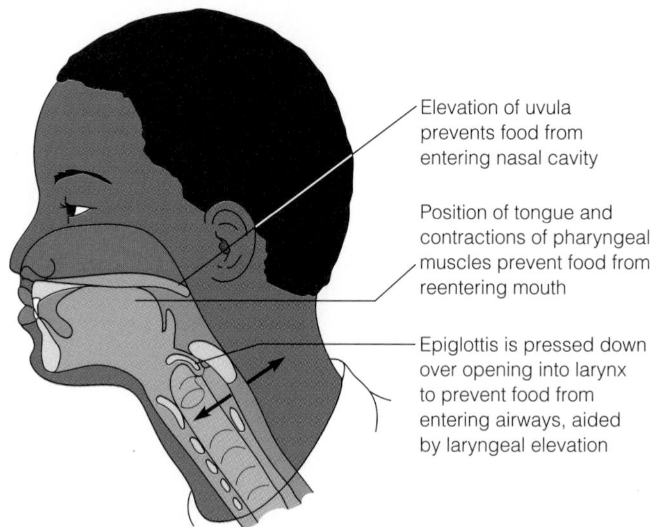

(b) Involuntary stage of swallowing begins; upper esophageal sphincter (arrows) opens

Figure 9–7 The Act of Swallowing

The act of swallowing involves a complicated series of events. The changing position of the tongue is crucial to effective swallowing and ensuring that food travels down the esophagus and not the trachea.

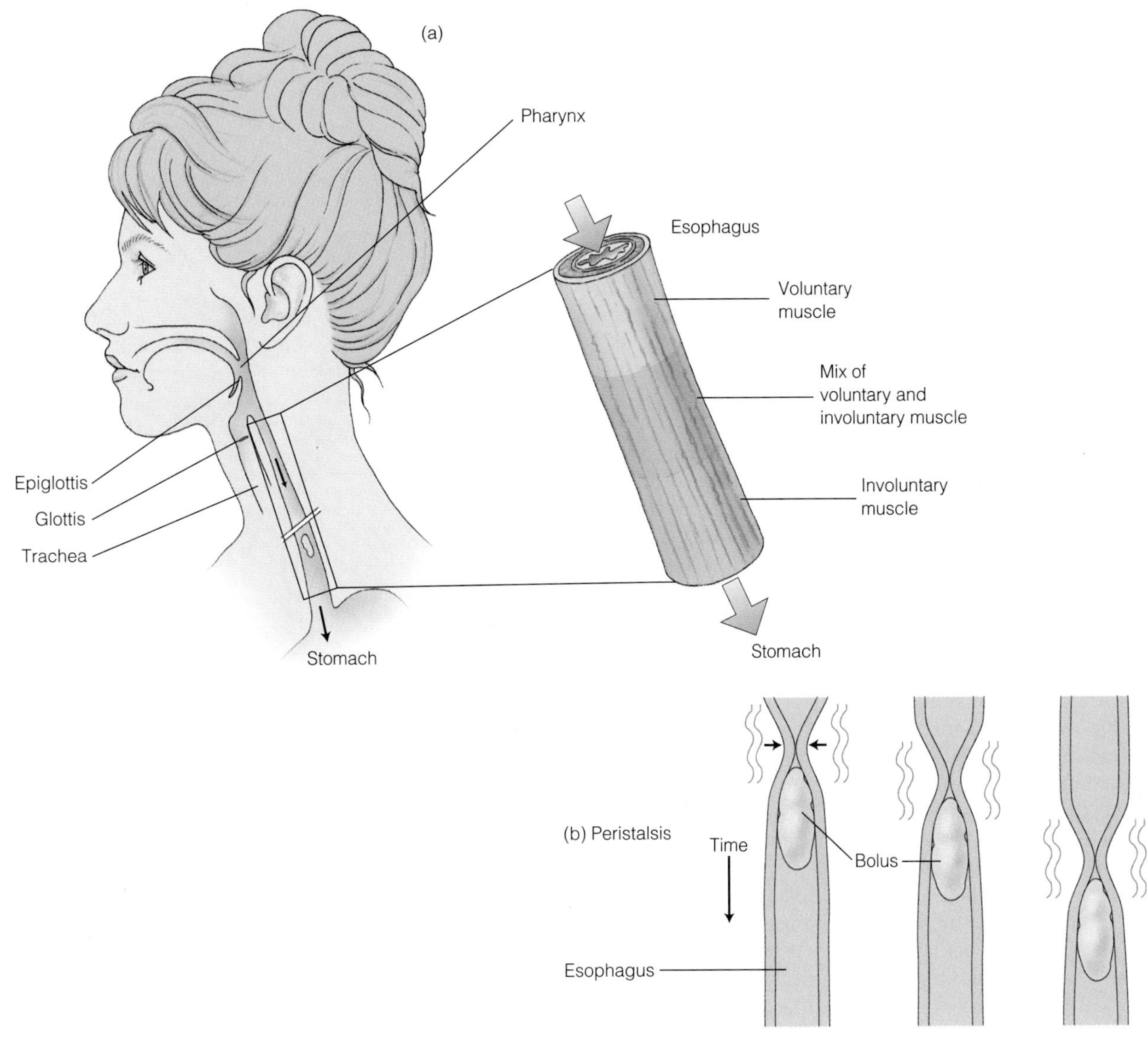

Figure 9–8 The Esophagus
Once material has entered the esophagus it begins to move downward towards the stomach. Both voluntary and involuntary muscles are involved (a) and movement is by peristalsis (b).

peristalsis provides a way to move materials within the esophagus. This is akin to squeezing a long tube filled with paste. Where we squeeze, the material inside is displaced within the tube in both directions. However, if we move our fingers down just slightly and squeeze again, we can sequentially and progressively move all the paste out of the end of the tube (Figure 9–8[b]). The other action that the squeezing of the muscles surrounding the digestive tract allows is *segmentation,* which will be discussed in a later section. This tends to divide up, or segment, the total mass of material within the tract for easier handling. Peristalsis begins in the esophagus and is so efficient that we can even swallow upside down. This ability is not observed in some species of animals, such as birds and reptiles. Take a moment when the opportunity arises to watch a bird drink and notice that it has to raise its head and let the water run down its throat under the influence of gravity.

The digestive tract usually functions unidirectionally. The bolus moves down from the mouth, through the esophagus, and into the stomach. The exception to the rule is *vomiting,* which results in the violent expulsion of the contents of the stomach out of the body. Although physically distressing, vomiting is a protective mechanism by which the body rids itself of potentially harmful materials.

The esophagus traverses the thoracic cavity and passes through the diaphragm to enter the abdominal cavity. Within the abdomen the esophagus opens into the stomach. There is a valvelike flap at this conjunction, called the *gastroesophageal* or *cardiac sphincter,* which allows material to pass into the stomach, but fairly effectively prevents backflow of material from the stomach into the esophagus. This is an important control point because the chemical environment of the stomach is very different from, and damaging to, that of the esophagus (Figure 9–9).

Stomach

The stomach is an expandable, muscular organ capable of stretching to accommodate variable amounts of liquids and semisolids. Because of its unique musculature (Figure 9–9), the stomach is capable of powerful and persistent contractions, which result in the thorough mixing of fairly large volumes of food. The inner surface of the stomach has deep folds within it called **rugae** which diminish as the stomach expands to accommodate the incoming food. The expanded volume of the stomach can attain several liters, though expansion to this degree is usually uncomfortable. However, some people have considerably larger holding capacity than others.

What makes the conditions in the stomach harsher than that of the mouth and esophagus is that cells lining the stomach produce and secrete hydrochloric acid (HCl) (Figure 9–10). This production of acid is regulated by hormones and behavior and is linked to the actual consumption, or anticipation of consumption, of food. HCl plays a role in activating enzymes that are present in inactive forms in the stomach and, thus, initiates the biochemical breakdown or digestion of large molecules into smaller ones. The churning of stomach muscle contraction helps bring about the rapid, thorough mixing of food and aids in digestion by means of more complete and rapid exposure of nutrients to enzymes and acid.

Under these highly acidic, and strongly digestive conditions, what is it that protects the stomach from digesting itself? The answer is mucus. The stomach is protected from its own acid by mucus produced and secreted by numerous glandlike cells known as *goblet cells* (Figure 9–9), which are an integral part of the gastric epithelium. So, in addition to cells that produce acid, there are also cells that produce a counteractive mucus. The lining of the stomach is thick and highly infolded, but the gastric epithelium itself is only a single cell layer thick. Mucus protection, however, is not universal throughout the digestive tract. The esophagus is a stratified epithelium and does not produce mucus. This is one of the reasons why the cardiac sphincter is so

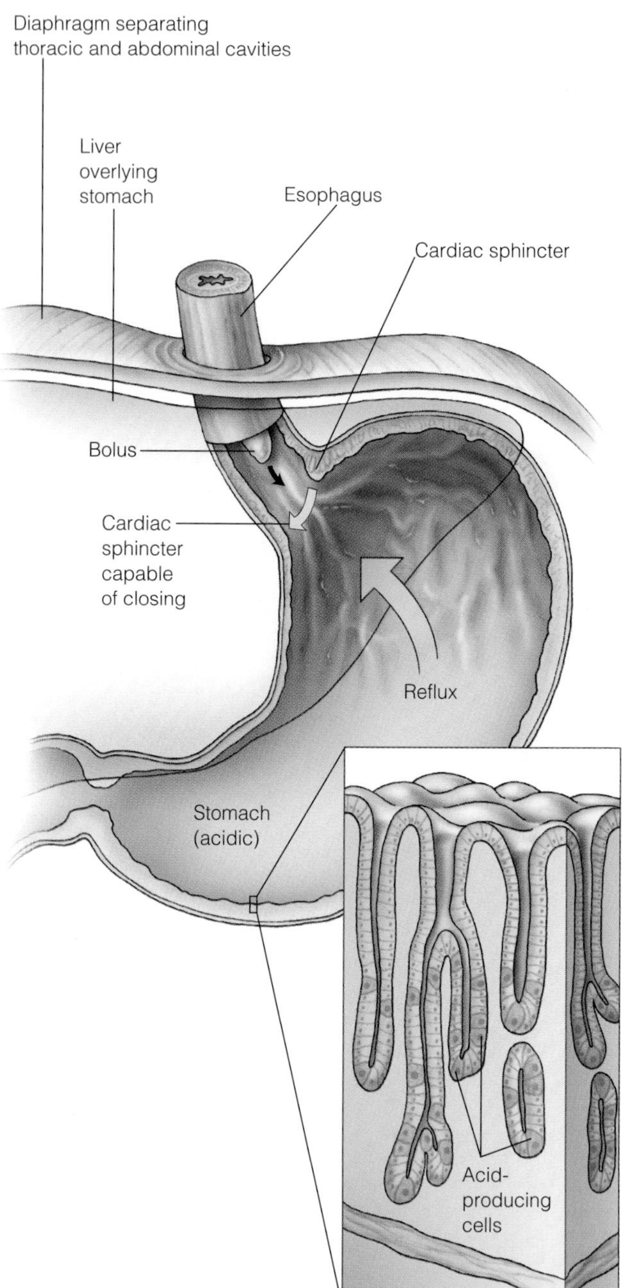

Figure 9–9 The Stomach
Entrance to the stomach is through the cardiac sphincter, which is located near the region where the esophagus passes through the diaphragm. The stomach is an acidic environment, which is produced by cellular synthesis of hydrochloric acid.

Rugae the deep folds in the stomach lining that allow for volume expansion

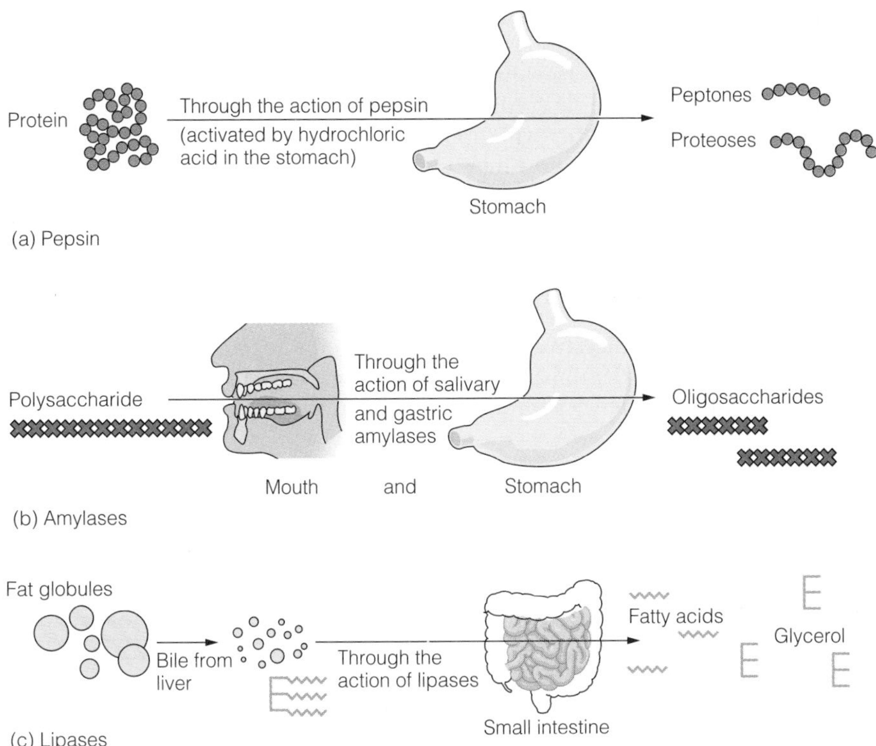

Figure 9–10 Enzymes of the Mouth, Stomach, and Intestines
Enzymes throughout the digestive system act to degrade large molecules or complexes of molecules into smaller ones. Pepsin and other proteases (a) break down proteins. Amylases and other glycosidases break down polysaccharides (b), and lipases break down triglycerides (c).

important. The escape of stomach acid back into the esophagus with partially digested food, a condition known as **reflux,** can cause considerable distress and/or damage ranging from heartburn to esophageal ulcers.

Although the digestive process initially begins in the mouth with the breakdown of complex carbohydrates by the salivary enzyme amylase, it is in the stomach that digestion really begins in earnest. As stated earlier, the stomach establishes an extremely acidic environment (pH 2) for the bolus. In fact it is through the production of hydrochloric acid and the acid activation of enzymes that the stomach functions in the degradation of proteins. It is the acidic pH of the stomach that activates gastric enzymes, such as *pepsin,* to become functional (Figure 9–10). When maintained at the low pH of the stomach, this enzyme acts as a very efficient catalyst in the breakdown of proteins into smaller fragments, known as *proteoses* and *peptones.*

Saliva in the mouth and hydrochloric acid in the stomach are released as we anticipate and then actually begin eating. HCl activates enzymes for the degradation of proteins, but there are other types of enzymes activated as well. Fat digesting enzymes, known as *lipases,* are also present in the stomach and begin to break down fats into fatty acids (Figure 9–10). The mechanical churning of the stomach makes more efficient the acidic and enzymatic breakdown of the bolus we have ingested and leads to the formation of a viscous fluid known as **chyme.**

Although food is subjected to acid and many types of enzymes for several hours in the stomach, the breakdown of food is not complete. Further degradation is necessary for nutrients to be completely digested, and complete digestion is a prerequisite for absorption of nutrients into cells. This aspect of the overall digestive process is not carried out in the stomach. Because the nutrients are not yet completely broken down, there is little point to trying to get them transported into and through the epithelial cells that line the stomach. However, the stomach is capable of absorbing molecules. For example, aspirin, which is a mild acid, and ethyl alcohol, which is a component of fermented and distilled drinks, are both small molecules and are directly absorbed through the stomach lining into the vascular circulation. This is why each of these compounds has relatively rapid effects—pain relief or intoxication—after we ingest them. With these, and a few other exceptions in mind, however, the fact is that most of the absorption of nutrients (as well as most of the absorption of alcohol and aspirin) takes place in the small intestine and not in the stomach.

Reflux movement of material from the stomach into the esophagus

Chyme the semisolid material resulting from partial digestion of food in the stomach

Small Intestine and Accessory Organs

Duodenum, Jejunum, and Ileum At the appropriate time, usually several hours after a main meal, the churned up chyme of the stomach is released into the adjacent region of the digestive tract, the small intestine (Figure 9–11). The chyme is further liquified and emulsified into what is referred to as **chyle.** When in a relaxed state, the small intestine is roughly 6 m–7 m long and is divided into three main regions. The initial 0.5 m, or so, is called the *duodenum.* The second 2 m–2.5 m is called the *jejunum,* and the final 3 m–4 m is called the *ileum.* Aside from the functional differences among these regions, they share many anatomical and physiological characteristics and are largely indistinguishable from one another. Each region is actively involved in the absorption of nutrients and the resorption of water from chyle. A large volume of water is needed to dissolve and transport the nutrients within the digestive system. However, most of the water must then be resorbed into the body to prevent dehydration.

For these absorptive and resorptive processes to be efficient, a large surface area in the intestines is required. Evidence of this is observed along the inner surface of the small intestine, which is covered with a number of large folds and fingerlike extensions, known as **villi,** protruding into the lumen of the tract. The intestines are lined with a simple epithelium covered with **microvilli.** The presence of microvilli dramatically increases the surface area of this long intestinal tube. Each cell of the epithelium has thousands of these tiny projections. Thus, the folds, villi, and microvilli along the entire 6 m–7 m length of the small intestine provide a surface area of absorption in excess of 200 m^2 (Figure 9–12).

To further understand the progressive nature of the digestive process, we can examine the contents of the duodenum and see how the once large molecules that we ate a few hours ago are fairing as they enter the small intestine.

Chyle milky emulsified fats absorbed from small intestine into lacteals

Villi folds in the epithelium lining the intestines

Microvilli fine, filamentous extensions of the cells lining the intestines and other bodily sites

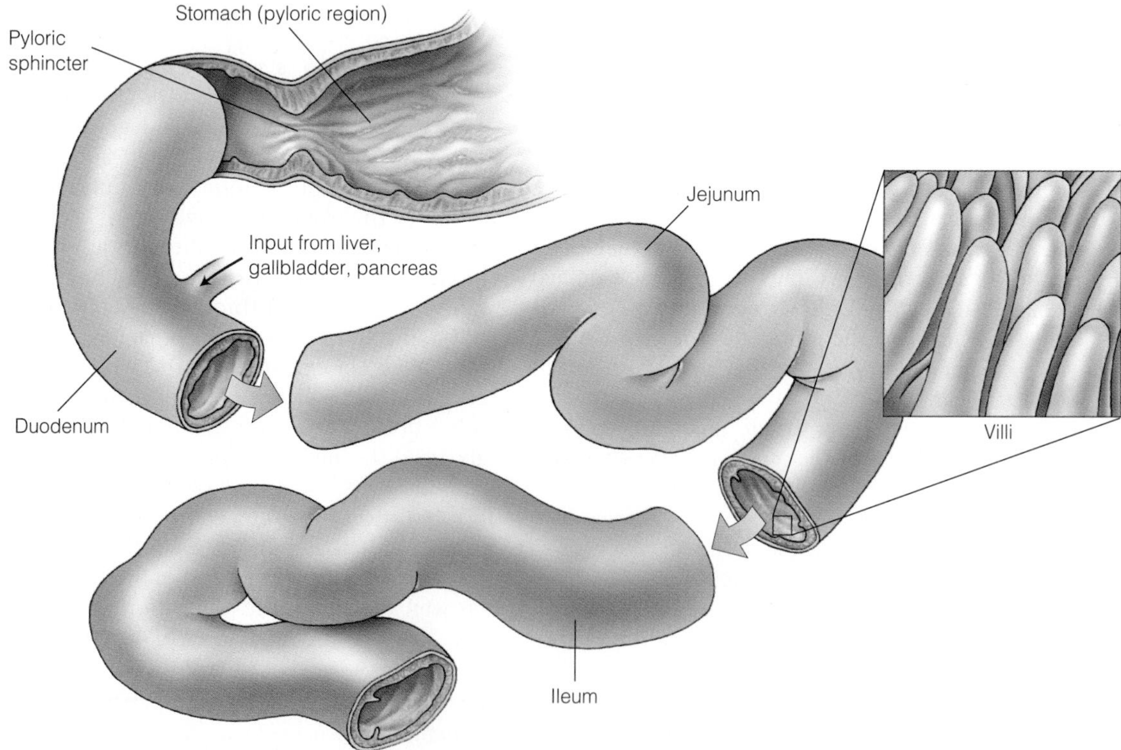

Figure 9–11 Small Intestine
From the stomach, digesting material moves into the small intestine for further degradation and absorption of nutrients. The small intestine has three regions: the duodenum, the jejunum, and the ileum.

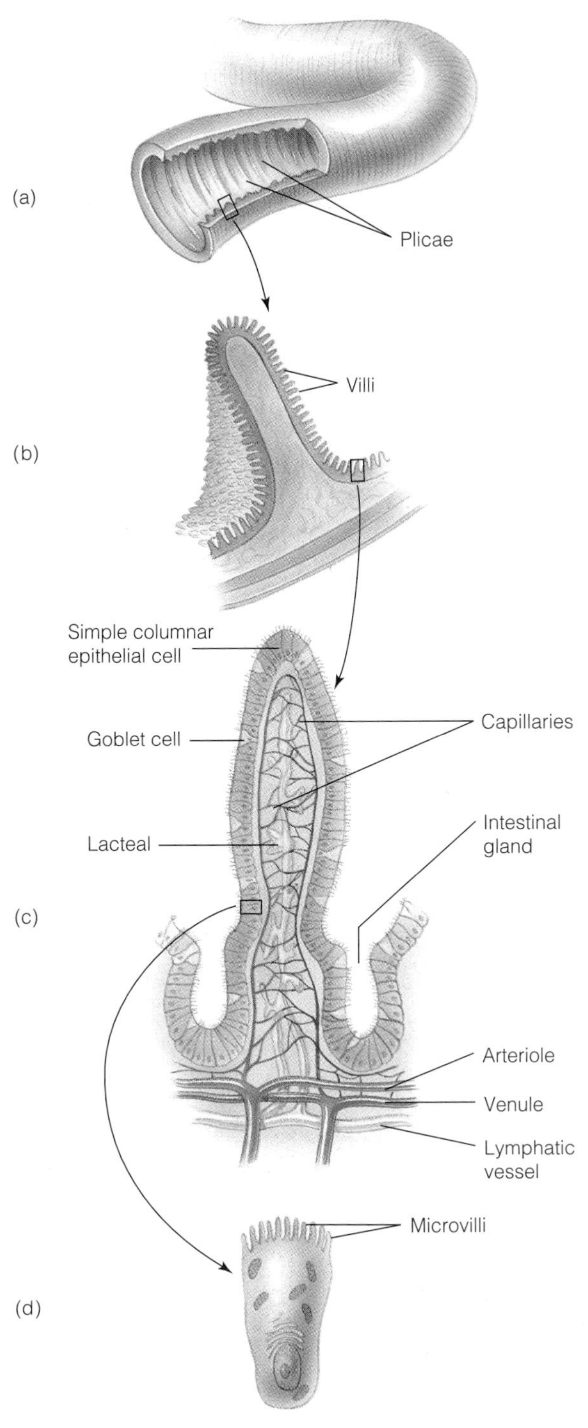

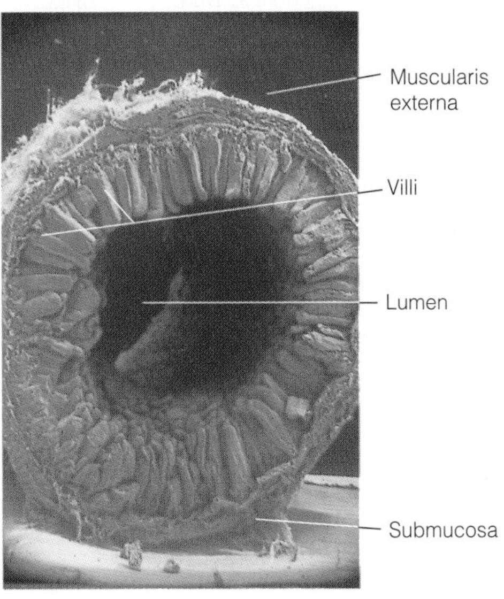

Figure 9–12 Surface Area of the Intestine
The surface area of the internal surface of the intestine is folded into plicae (a), which are further folded into villi (b) and (c). The cells of a villus have microvilli extending from them (d). Absorption of nutrients takes place across the cells of the villi where nutrients are then transported into capillaries of the vascular system and lacteals of the lymphatic system.

All the proteins, carbohydrates, and fats that were consumed in the form of meat, potatoes, and butter have been broken down into simpler forms. For example, complex carbohydrates have been digested into oligosaccharides and proteins have been digested into small peptides. The stomach has certainly taken care of the initial phases of food breakdown. However, digestion is far from complete.

Just as the conditions in the stomach are very different from those in the esophagus, the conditions in the small intestine are very different from those of the stomach. The small intestine is alkaline in pH. The stomach acids are neutralized by the buffers of the small intestine to ensure high levels of activity of the intestinal enzymes present. Unlike pepsin in the stomach, most of the intestinal enzymes cannot tolerate an acidic environment. There is a great variety of enzymes in the small intestine (Table 9–1). The collective activities of these enzymes catalyze the complete digestion of larger molecules into their component parts. At this point it is fair to ask, where do all these enzymes come from? This is a question of great significance and to answer it we need to backtrack a bit to discuss a group of organs that play an accessory role in the digestive process.

Accessory Organs of Digestion The liver, gallbladder, and pancreas are considered here as accessory organs in the digestive process, only because they do not directly participate in the digestion and absorption of the food we eat (Figure 9–13). However, their indirect involvement is, to varying degrees, imperative for proper system function. The liver and pancreas have other important roles separate and distinct from digestion. The gallbladder is predominantly a storage organ and functions in conjunction with the liver to regulate the availability of bile.

Liver The liver is a large, multilobed organ located within the abdominal cavity and is positioned asymmetrically on the right side of the body axis. It is located just under the ribs and diaphragm and has ducts that connect to the gallbladder. The liver has many important functions including principal roles in **detoxification** of the blood, turnover of red blood cells, distribution of nutrients, and production of serum proteins such as *albumin* (Figure 9–13, Biosite 9–1). With regard to the digestive system, the liver produces special molecules, known as **bile salts,** which act as **emulsifiers** of fats. Emulsification is the process by which fat globules ingested during a meal are broken down into smaller fat vesicles within the duodenum (giving rise to chyle), so that the enzymes associated with hydrolysis of lipids (lipases) can do a more efficient job of converting fats into their constituent fatty acids.

The bile molecules are a by-product of hemoglobin degradation. The role of the liver in the breakdown of old red blood cells, which are essentially tiny bags of hemoglobin, is, thus, linked to the digestive process. In addition, the color of human feces derives in large part from a pigment in bile called *bilirubin.* Bilirubin arises from the breakdown of hemoglobin from red blood cells resorbed in the liver. Waste not, want not.

Gallbladder The products of the liver that are targeted for the digestive tract are shuttled to the duodenal region of the small intestine by two channels known as the right and left *hepatic ducts* (Figure 9–13). These ducts combine to form the *bile duct,* which, in turn, connects to the intestine. The flow of these biliary secretions is timed to occur in conjunction with the migration of chyme from the stomach into the duodenum. In anticipation of this process, bile is also stored for later use. Much of the bile produced by the liver is stored in the gallbladder. This ensures a reservoir of these molecules for mass release during digestion. The gallbladder does not make bile, it only stores it. In the event of malfunction of the gallbladder, such as infection, blockage, or physical damage, digestion can be seriously affected. This is a particularly acute problem for those who consume a large amount of fat in their diets. The undigested portion of fat may pass through the intestines unabsorbed, denying the body of needed lipid nutrients. In addition this may also alter the motility of all the materials moving through the digestive system and result in diarrhea.

Pancreas The other important accessory digestive organ is the pancreas. The pancreas is a long, lobated organ that is oriented transversely across the abdomen (Figures 9–14), and drains from left to right into the duodenum. This organ is also a dual purpose gland. On the one hand, the pancreas acts as an exocrine gland and produces and secretes an enormous number of different kinds of digestive enzymes. These catalytic proteins are secreted directly into the duodenum through the *pancreatic ducts* (Figure 9–14). In its second role, the pancreas serves as an endocrine gland and produces the hormones *insulin* and *glucagon,* as will be discussed in Chapter 13. With respect to digestion, however, only the exocrine functions of this multipurpose gland concern us here.

Table 9–1 Classes of Digestive Enzymes

CLASS	FUNCTION
Proteinases or Proteases	Breakdown proteins into peptides
Peptidases	Breakdown peptides into amino acids
Lipases	Breakdown triglycerides into fatty acids
Nucleases	Breakdown DNA and RNA into nucleotides
Glycosidases	Breakdown complex carbohydrates into simple sugars

The enzymes that are produced by the pancreas fall into several different categories (Table 9–1). It is clear that, for all the different types of food that we consume to be broken down in preparation for cellular absorption, there must be sets of very specific enzymes capable of doing so. Proteins are reduced to amino acids by *proteases* and *peptidases.* Complex sugars, such as starch, are reduced to simple sugars by glycosidases. DNA and RNA are broken down into individual nucleosides by *nucleases.* Fats and other lipids are cleaved into their component parts by lipases. Pancreatic enzymes are very powerful in their digestive and degradative actions, so powerful in fact that one may wonder why they do not digest the cells and tissues of the organ that produces them. The reason for this is that they are initially produced by the pancreas in inactive forms. In this inactive form, in which they are called *proenzymes,* it takes special conditions to activate them. Guess where such conditions are found in the body? The small intestine. Because the pancreatic digestive enzymes are activated upon arrival in the

Detoxification processes used by cells to make toxic compounds less toxic

Bile salts emulsifying molecules formed by the liver by degradation of hemoglobin

Emulsifiers molecules that disperse large fat globules into smaller lipid vesicles

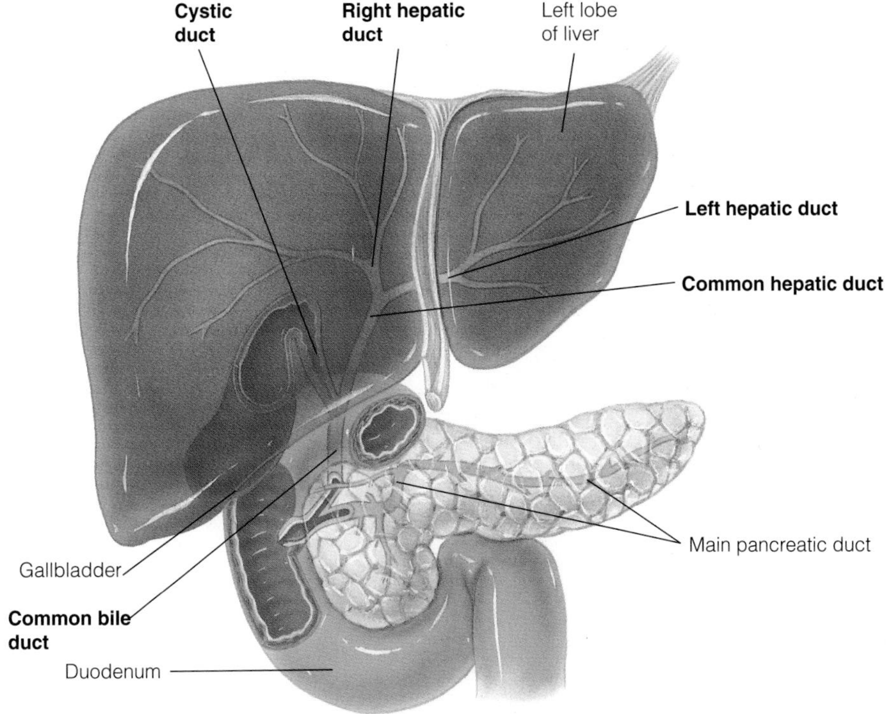

Figure 9–13 Location and Function of Accessory Organs of the Digestive System
The liver and the pancreas are the principal accessory organs of the digestive system. The liver produces bile and the pancreas produces proenzymes that are activated only when they reach the lumen of the duodenum. The gallbladder stores bile produced in the liver.

9–1

THE PATTERNS OF SYNTHESIS AND DELIVERY OF NUTRIENTS

Following the absorption of nutrients through the epithelium of the small intestine and passage into vascular and lymphatic systems, blood-borne nutrients are processed in the liver. Amino acids, glucose, and fats enter into the liver and into the liver cells, where they are used to produce proteins and glycogen or are converted to energy. The liver is the center of a distribution network that delivers glucose to the muscles, brain, and fatty tissues. The liver supplies amino acids for the muscles and other tissues and directs fats to and from the fatty tissues. As shown in the diagram, the liver is the nexus for the pathways of use and storage necessary to maintain the homeostatic balances or energy-rich molecules and molecular building blocks for the entire body.

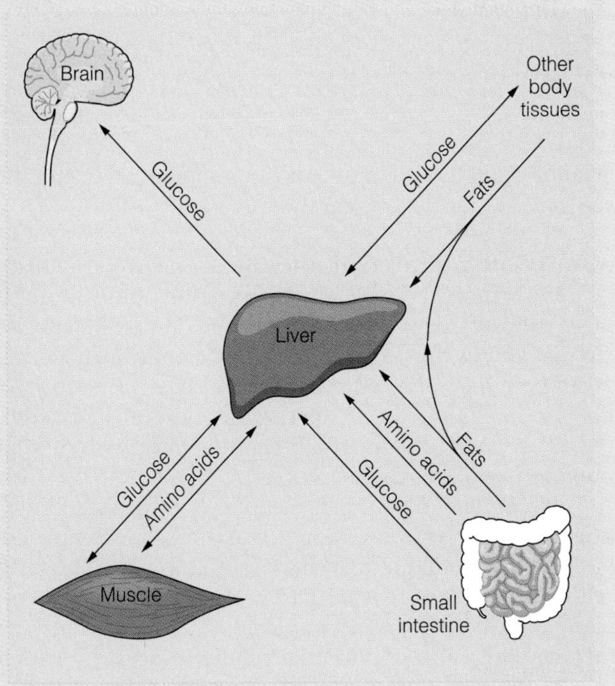

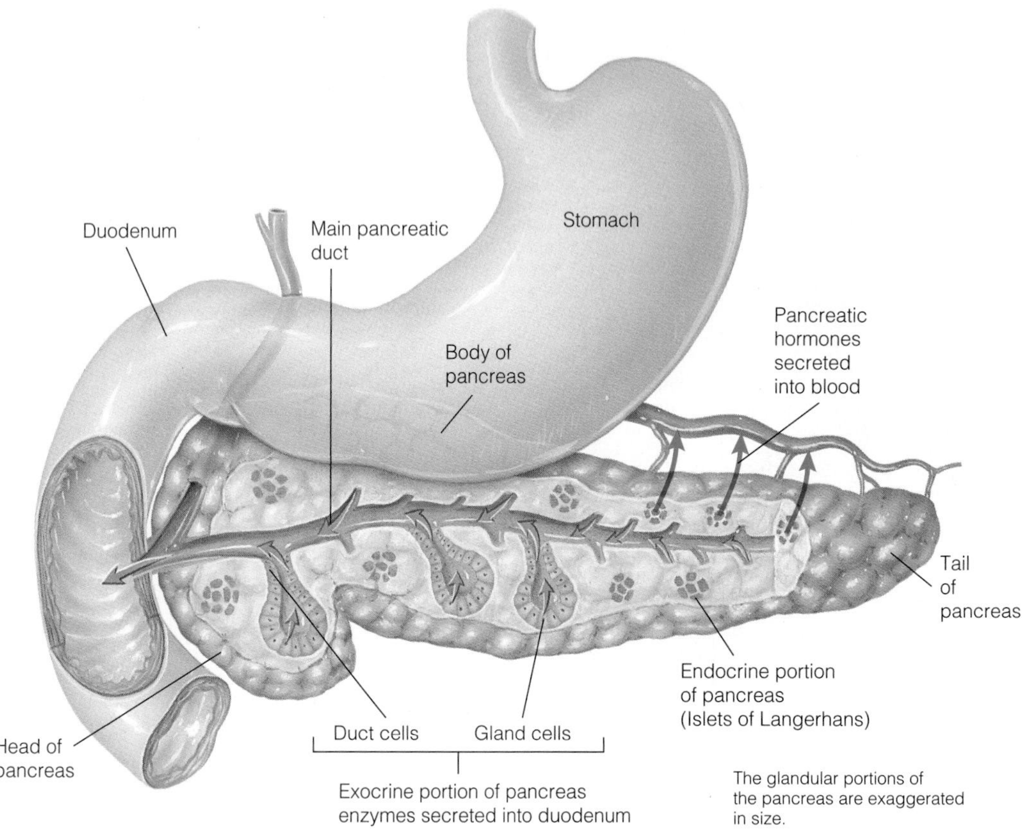

Figure 9–14 The Pancreas—Exo- and Endocrine Functions
The pancreas has two types of glandular cells—exocrine and endocrine. Exocrine glands (gland cells and duct cells) produce and secrete enzymes used for digestion. Endocrine glands (islets of Langerhans), produce hormones.

alkaline small intestine, their actions are limited to specific substrates available to them there.

Absorption in the Small Intestine The liver, gall bladder, and pancreas disgorge their secretions into the right place (the duodenum) at the right time (in the presence of chyme) and in the properly protected form (proenzymes). Even though the liver and the pancreas have other roles to play in bodily functions, the coordination of their activities as accessory organs in the digestive process is of vital importance to homeostasis.

The process of digestion is completed in the small intestine. The carbohydrates, proteins, and fats are completely broken down into monosaccharides, amino acids, and fatty acids. Each of these can now pass through the plasma membrane of cells by a variety of passive and active transport mechanisms. Absorption is the cellular process by which cells of the small intestine take in nutrients (Figure 9–12). However, the epithelial lining of the small intestine should also be thought of as a conduit through which nutrients are brought into the body and not as a final destination.

Where do nutrients go once they pass through the intestinal epithelium? As seen in Figure 9–15, the intestinal tract is held in place by the *mesentery* and served by an extensive vascular network. The blood vessels form an intricate and dense network of capillary beds associated with the intestines. At the microscopic level, there is a knot of vascular elements closely associated with each villus along the entire length of the small intestine (Figure 9–15). The nutrients that are transported through the intestinal cell lining are taken up by the vascular system and carried by the blood and lymph to the rest of the body. Amino acids, sugars, and nucleic acids are readily taken up into the vascular capillaries through facilitated and active transport. One of the first organs through which the blood flows is the liver, which helps purify the blood (detoxification) and absorbs glucose to store as glycogen. Fatty acids are handled in specialized lymphatic vessels known as *lacteals.*

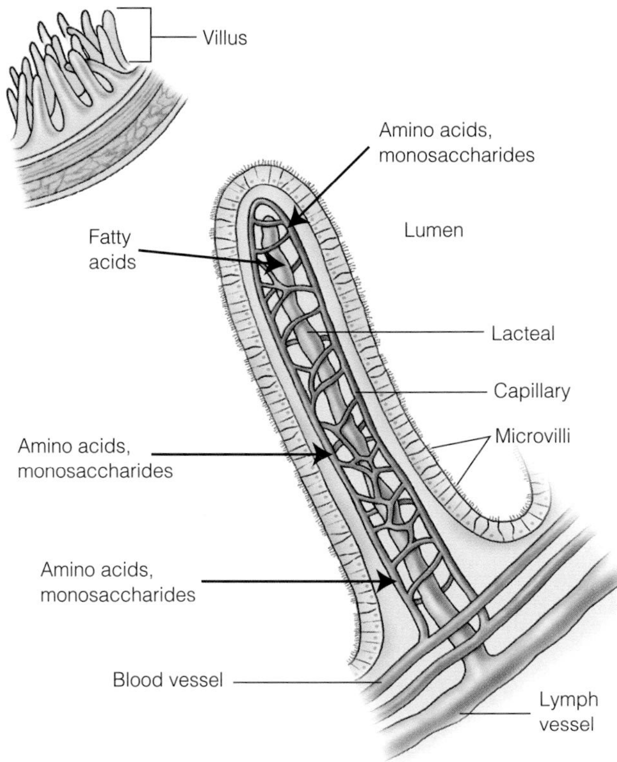

Figure 9–15 Passage of Nutrients into the Body
Nutrients are transported across the epithelial cells of the intestine and are absorbed by capillaries (for amino acids, monosaccharides) or lacteals (for fatty acids). It is only after uptake and transport to the blood or lymph that nutrients are made available for other cell types in the body.

Lacteals are unique blind-ended capillaries of the lymphatic system that have their origins in the villi. The fats taken up by the lacteals are transported through the lymph system, as opposed to the blood vasculature, but they are eventually combined with the blood through a duct emptying into the *subclavian vein.*

Indigestible Materials—Fiber In terms of uptake of molecules from the lumen of the small intestine, the epithelium does most of the job of absorption. However, not everything that we eat is broken down by the combination of the mechanical, chemical, and enzymatic arsenal to which it is exposed. If it is not digested, the potential nutrients cannot be absorbed. One of the most important indigestible substances for the human body is the polysaccharide **cellulose.** Cellulose is a principal structural material of plants and one of the most abundant molecules on Earth. It is found in the cell walls of plant tissues, from leaves to roots. When we eat a salad made of lettuce, tomatoes, and onions, we are consuming a significant amount of cellulose. What happens to cellulose if the human body has no means to digest it? Does it simply pass through the intestine unutilized? No. Cellulose (as well as other indigestible materials) provide **fiber,** which may be invisible with respect to nutrients, but is important in other ways (Figure 9–16).

What is the indigestible fiber material good for? As it turns out these materials are very important for removing cells of intestinal lining and maintaining the process of motility. Fibrous materials help ensure the solidity of feces, particularly in the large intestine where formation of feces takes place. Segmentation of this material (Figure 9–16[b]) distributes it relatively uniformly within the intestine. Passage of the fibrous bulk through the intestine also helps in the absorption of nutrients, probably by removing old epithelial cells. In fact, ingested fibrous materials combine with the shed cells. Because the equivalent of the entire epithelial lining of the intestines is replaced every few days, this shedding process accounts for several pounds of cells a week. In addition, bacteria, which grow in the intestinal tract, are also part of the endogenous indigestible material. They represent what is called the *natural flora* of the gut (Table 9–2). Bacteria, as do the cells of the epithelial lining of the intestines, account for much of the solid mass of material that is continually produced within and must pass through the intestines. Bacteria are also very important in other ways to the proper function of the digestive process. They are the principal participants in several phases of nutrient production, digestion, and absorption.

Large Intestine

Chyle is the combination of digestible and indigestible materials. Feces, however, even though moist, is normally quite solid. The solidification process, which requires resorption of water, takes place in the large intestine (Figure 9–17). At the junction between the small and large intestine there is a blind outpocketing of the ascending colon, known as the *cecum.* The vestigial, but often troubling, **appendix** is located at the end of the cecum. The appendix can give humans a serious problem on occasion as a result of infection and inflammation (*appendicitis*). The

Cellulose fibrous polysaccharide produced by plants and indigestible in vertebrates

Fiber indigestible materials in a diet that help in movement and absorption of nutrients in the intestinal tract, for example, cellulose

Appendix the blind end of the cecum, an outpocketing of the intestine

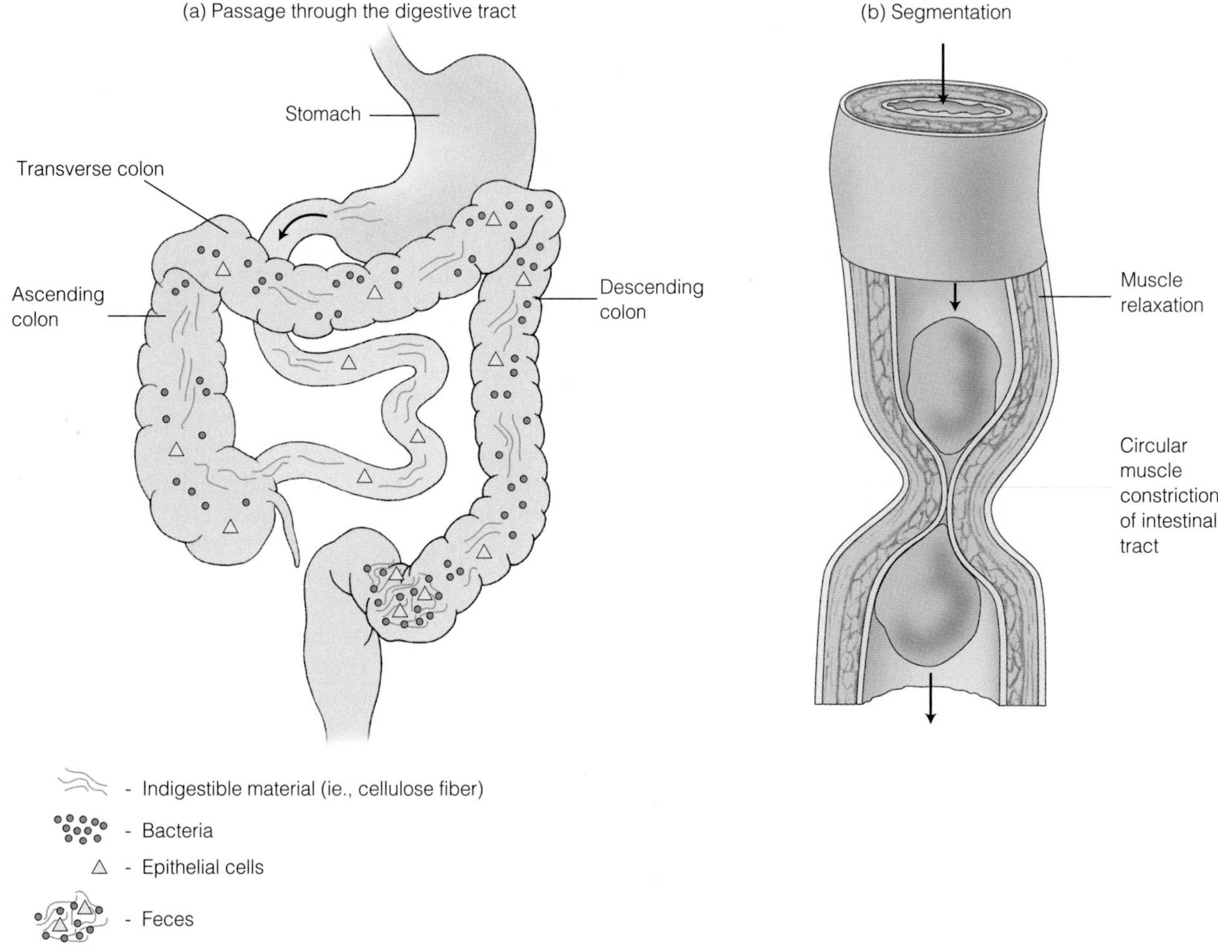

Figure 9–16 Fiber and the Intestines
Fiber is material within the digestive tract (a) that is not broken down. This material moves through the intestines by peristalsis and segmentation (b). Fiber is solidified with bacteria and sloughed epithelial cells to form feces.

region of transition from the small to the large intestine is called the ileocecal junction. The two regions are separated by a valve (the *ileocecal valve*) that controls movement of materials into the large intestine.

Beyond the ileocecal valve is the large intestine or **colon.** The large intestine is approximately two-meters long, and is divided into five major colonic regions (Figure 9–17). The first of these is called the *ascending colon.* Material within this region travels upward to the *transverse colon,* which crosses the abdomen just under the stomach from right to left. Just to keep things properly balanced (after all, everything that goes up must come down!), the last leg of the journey of material within the large intestine is through the *descending colon* and into an S-shaped region known as the *sigmoid colon.* The final compartment of the large intestine is the rectum, in which the compacted combination of fibrous and cellular waste products known as *feces* are stored prior to expulsion through the anus during the act of *defecation.*

Functionally, the large intestine has several important jobs to perform before the materials within it are eliminated. First, it resorbs water from the chyle introduced from the small intestine. Second, it resorbs minerals, particular-

Colon the large intestine

ly calcium ions. Third, the large intestine must remain hospitable to bacteria that, in turn, produce essential vitamins (Table 9–2). Vitamins, such as *vitamin K*, are produced by bacteria and absorbed through the wall of large intestines. And finally, the large intestine, by means of the action of surrounding muscles, helps control the progressive peristaltic movement of materials within it at the proper pace and then controls their expulsion from the body at the appropriate time.

Table 9–2 Bacteria and Other Microorganisms Commonly Found in the Human Intestinal Tract

CLASS	ORGANISM	COMMENT
Bacteria	*Escherichia coli, Streptococcus faecalis, Bacteriodes species*	These organisms occur in large numbers (billions per gram of feces), produce enzymes that degrade molecules, produce vitamins and help solidify feces.
Fungi	*Candida* and *Torulopsis*	The yeasts.
Protista	*Balantidium, Entamoeba*	The proportion of the fungi and protista are dependent largely on the diet of the individual

See book by Stanier, R. et al. listed at end of chapter.

Figure 9–17 The Large Intestine
Material from the small intestine enters the large intestine through the ileocecal valve. There are several regions of the large intestine including the ascending colon, the transverse colon, the descending colon, the sigmoid colon, and the rectum. Feces exit the body through the anus.

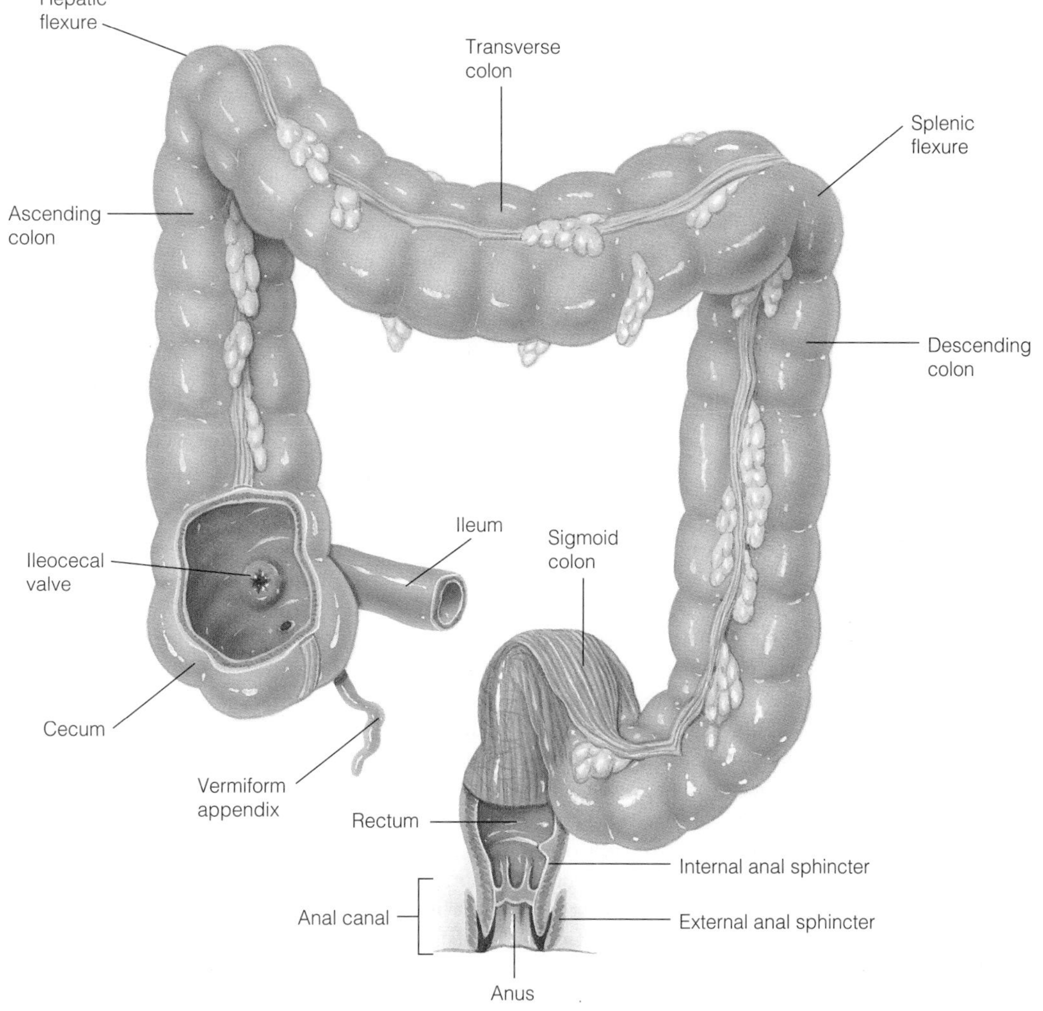

Resorption of Water The resorption of water that occurs in the large intestine provides a means to dry up the mass of material to be eliminated (Figure 9–18). The body is really quite stingy with its water. Homeostatic mechanisms associated with several major organ systems, including the urinary, respiratory, integumentary, and endocrine systems, in addition to the digestive system, are intimately involved in maintaining the water balance of the body. The proper composition and viscosity of body fluids must be closely maintained. Excessive loss of water leads to dehydration, which can be disastrous. The large intestine usually finishes the job of water resorption for the body. Occasionally, however, bacterial and viral infections or toxins cause the motility of materials within the intestinal tract to increase. If the materials move more rapidly through the intestines, there is less time for resorption of water. The failure to resorb water results in materials that are loose and unformed. A common manifestation of high motility is *diarrhea*. If diarrhea persists, water is not able to be resorbed efficiently through the large intestine and dehydration becomes a problem. This is particularly problematic in babies and small children, whose body masses are small and are more rapidly affected by loss of water.

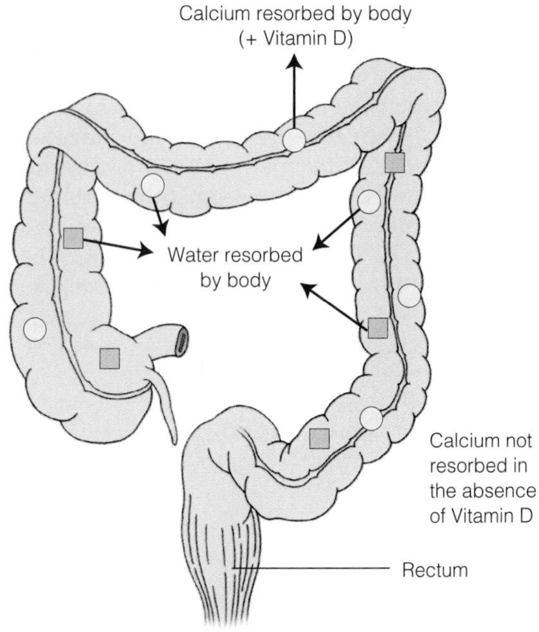

Figure 9–18 Water and Mineral Resorption
Water is preserved for the body by resorption in the large intestine. Minerals, such as calcium, are also resorbed by the large intestine. The process of resorption is dramatically affected by the presence (+) or absence (−) of vitamin D.

Minerals Along with the resorption of water, the absorption of minerals also takes place in the large intestine (Figure 9–18). While sodium and potassium are regulated in the kidney, calcium ions are absorbed into the body by transport across the wall of the large intestine. The role of calcium in the body is multifaceted. Calcium is needed for formation of bones and teeth, muscle contraction, nerve cell function, and enzyme activity, to name but a few of its uses. The absorption of calcium depends upon the presence of vitamin D. Lack of this vitamin (normally formed in the skin by the action of sunlight, but also a supplement of commercially available cow's milk) leads to the disease rickets. This condition results in a softening of bones by decalcification and their "bowing," particularly in the legs in response to the weight of the body.

Intestinal Flora—Friendly Bacteria Bacteria also help finish the job of digestion in the large intestine (see Table 9–2). The epithelial cells of the large intestine itself do not produce enzymes for this part of the digestive process, and there are no accessory organs that connect to it. It is bacteria themselves that produce and secrete degradative enzymes. The bacterial enzymes are similar in activity to host enzymes found in the small intestine. The intestinal microbes (for example, *E. coli*) are the so-called good bacteria of the gut. They divide rapidly, so most of the time there is no shortage of them or their products in the intestine. Because of this rapid rate of proliferation and accumulation, up to 50% of our feces may be bacterial mass. The distinctive smell of feces arises from some of the chemical byproducts of the breakdown of the materials in the large intestine by bacterial action. Bacteria are friendly to the human body for another reason as well. As mentioned earlier, they produce essential vitamins (Table 9–2). Many of the B vitamins are also produced by bacteria and absorbed through the walls of the large intestine. Vitamin K, which is a factor in the clotting of blood, is produced by human bacterial flora. The cells of the human body cannot synthesize vitamins, but neither can they survive without them.

Motility The final attribute of the large intestine that we will examine, is the control it has over the pace of movement of fibrous material or bulk within the tract. As in the rest of the intestinal tract, the large intestine undergoes peristalsis and segmentation (Figures 9–8 and 9–16). The muscle contractions controlling the position and movement of the solidifying mass are very important. After the nutrients are extracted from the food we eat, what is left over has to be eliminated. The passage of water, minerals, and vitamins back into the body leaves basically solid, or semisolid, waste products behind. Muscle activity of the large intestine moves these materials to the end of the line and collects and stores them for a relatively short time before they are expelled. The rectum is the storage area for feces. The timing of contractions of the muscles within the large intestine are periodic, with cycle times occurring over several hours. Usually, the contractions of muscles in the large intestine are stimulated by ingestion of food. So after eating, there is a peristaltic episode and we have the urge to defecate.

The mechanical aspects of defecation are by no means simple (Figure 9–19). The contractions of the large intestine are powerful; however, for materials to be finally forced out of the body, the sphincter muscles surrounding the anus must also be relaxed. The muscles of this sphincter are marginally under voluntary control, that is to say, we can prevent defecation for a while by voluntary action but not indefinitely. We learn to control this early in life, because it is not socially acceptable to defecate whenever the urge comes over us. Social conventions and hygiene establish for the most part the appropriate time and place to defecate.

PROBLEMS IN THE DIGESTIVE TRACT

There are a number of infections and diseases associated with the digestive tract that cause varying degrees of problems for us. Gastric distress after eating or drinking too much, or consuming bad food or drink, have plagued us all at one time or another. A few of the more serious and persistent problems of the digestive system are discussed here with respect to the regions in which they occur (Table 9–3).

Mouth

The bacterial destruction of the teeth give rise to what is called **dental caries.** Bacteria dissolve the hard surfaces of the teeth and form cavities in them. If left untreated, loss of teeth can alter seriously the ability to chew and, therefore,

Dental caries bacteria-induced deterioration of teeth; cavities

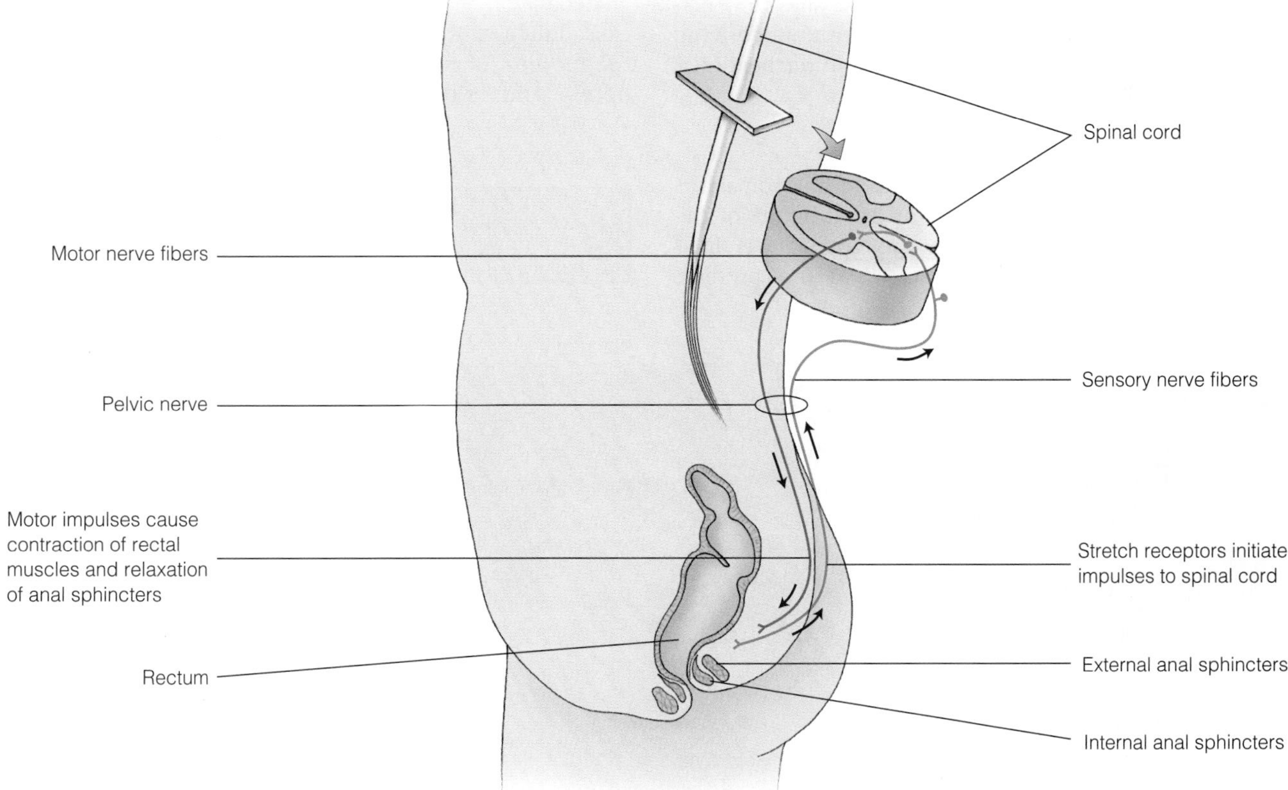

Figure 9–19 Defecation

The impulse to defecate results from stretch receptors in the rectum that stimulate neurons in the spinal cord.

Table 9-3 Problems in the Digestive System

PROBLEM	SOURCE OR ORIGIN OF PROBLEM
Indigestion and reflux	Diet, ulcers, excessive acid production, esophageal malfunction
Ulcers	Diet, bacterial infection
Tooth decay	Diet and bacterial infection
Gingivitis	Diet and bacterial infection
Diarrhea	Toxins, bacterial infections, viral infections
Peritonitis	Bacterial infection
Colitis	Diet, genetic predisposition
Polyps	Diet, genetic predisposition
Cancer	Carcinomas occurring in all regions of the digestive system (for example, colon cancer)

interfere with the first step in the process of digestion, that of mechanical grinding and softening of food. Just as important are diseases of the gums, particularly *gingivitis.* This type of infection erodes the gums, or *gingiva,* which support and surround the teeth. The gingiva is continuous with the oral mucosa and its degradation can lead to the same result as dental caries and worse. After all, the teeth are only as healthy and functional as the tissues in which they are embedded, including the gingiva and underlying bone (Figure 9–20). There are many problems further down the tract as well.

Esophagus and Stomach

The transition between the esophagus and the stomach is characterized by a distinct change in the structure of the epithelium lining the two regions. The esophagus is lined with a stratified epithelium, which does not produce mucus. The stomach is lined by an epithelium that produces copious amounts of mucus. Reflux from the stomach, which is highly acidic, can erode the unprotected epithelium of the esophagus and form an **ulcer** or open wound. Ulcers occur most frequently in the esophageal, stomach, and duodenal regions because of the acidity of the stomach content. The degree of seriousness of an ulcer is dependent on how much damage it does to the lining of the gut. If the acidity of the fluids erodes the epithelium alone, it can be extremely painful and debilitating. If it penetrates further into the connective tissue and muscle layers underlying the epithelium, it can cause bleeding and greater pain and distress. Really bad ulcers may even penetrate the entire gut wall and perforate the peritoneal cavity. This type of ulcer is most dangerous because of the chance of serious infection. Leakage of gut content into the abdominal cavity may lead to *peritonitis.* Most types of ulcers can be controlled with stress reduction, changes in diet, and special drugs that control the level of acid production by the cells of the stomach lining. In some cases surgery may be performed to remove a portion of the gastric epithelium containing hydrochloric acid producing cells.

Intestines

Bacterial and viral infections of the intestinal tract can cause much pain and distress. These infections often alter the motility of materials within the tract by altering the uptake and output of water into the intestines. This

Ulcer an open wound, particularly in the digestive tract, in which the epithelial lining is eroded

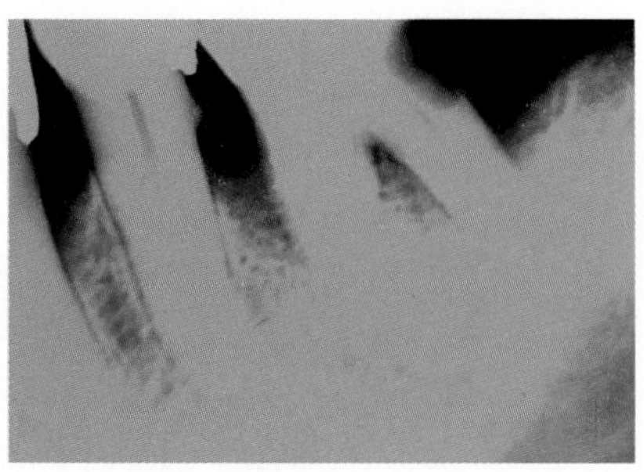

(a) Dental caries

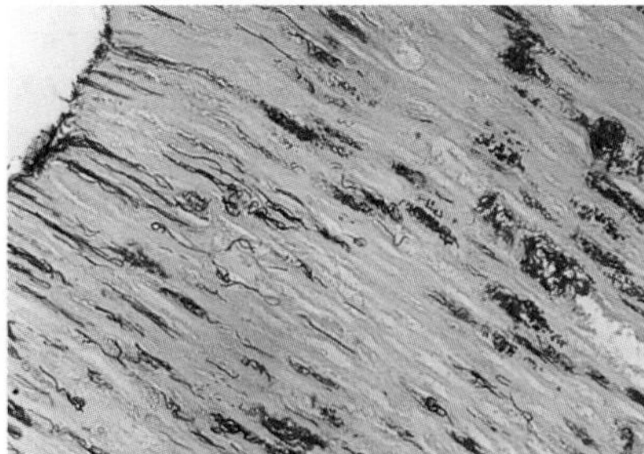
(b) Gingivitis

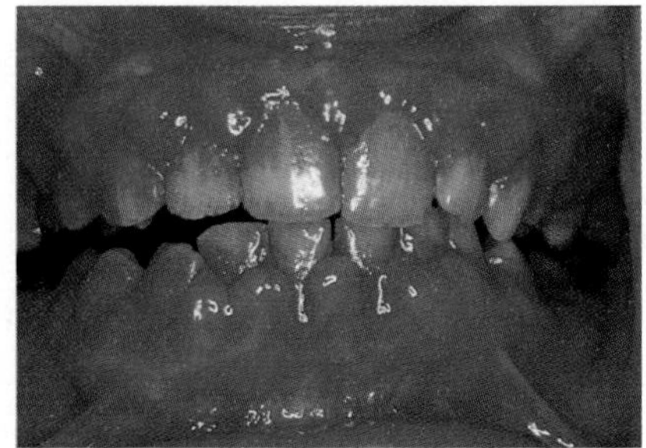
(c) Cancer

Figure 9–20 Diseases of the Mouth

In (a) the destruction of the teeth, or dental caries, is shown. Commonly called cavities, this destruction arises through the action of bacteria. Likewise, gingivitis (b) is gum disease and also arises from bacterial infection. Cancer (c) can arise in any of the tissues of the mouth and in any location. Most forms are either carcinomas or sarcomas.

changes the entire digestive process, which can be thrown into chaos for a long period of time. Certain bacterial toxins, particularly **cholera toxin,** causes massive water loss from the body. This leads to morbid dehydration and, in many cases, to death. With regard to bacterial infections, they can often be effectively treated with antibiotics. However, there are drawbacks to using antibiotics, one being that they kill the good bacteria as well as the bad. This leaves the body deprived of the digestive functions of the good bacteria, which aid in bulk formation and vitamin synthesis. This is potentially true for any situation in which antibiotics are taken for days or weeks during an infection. We should always be aware of the consequences of taking these medications and seek medical attention and advice in their use. Appendicitis also falls into this class of bacterial infections, and there can be serious consequences if medical treatment is delayed.

Cancer

The most dangerous of diseases of the digestive tract are cancers. Unfortunately, cancers of the mouth, esophagus, stomach, and intestines are fairly common in occurrence. Carcinomas arise as a result of transformation of epithelial cells that line the gut. Smoking and chewing tobacco have been implicated in human mouth and lip cancers. Sarcomas arise as a result of transformation of connective tissue and muscle cells that underlie the epithelial layers. There are obvious signs that indicate grave problems in the digestive tract. One of them is the appearance of blood in the feces, which should always be treated seriously because it may be an indication of cancer, precancer, or other potentially severe damage (for example, ulcers or *polyps*) to a portion of the digestive tract.

HUMAN NUTRITION—HOW YOU ARE WHAT YOU EAT

As we have discussed, the human digestive system is a very capable and efficient disassembly line; it is involved in the breakdown of food (complex organic molecules) into simple compounds, such as monosaccharides, amino acids, and fatty acids, that can be absorbed into the body. The digestive process also involves many nonorganic and noncaloric materials as well, including water, vitamins, and minerals. All these materials considered together are **nutrients,** which supply energy, building blocks, or cofactors for chemical reactions in the human body. Some of these molecules are synthesized from precursors by the cells as part of their normal metabolism and, thus, are considered nonessential in the diet. Others are not produced by the human body and are considered *essential nutrients.* Essential nutrients must be obtained from the foods we consume. If the body does not obtain (or has an imbalance in the amounts of) essential nutrients for energy and building blocks, the result is malnutrition. The average body (if there is such a thing) is composed largely of water (about 60%), fat (about 20%), protein, carbohydrate, the bone mineral calcium phosphate (about 20%), and a very small combined amount of other minerals and vitamins (less than 1%). The dynamic balance among the intake of nutrients associated with digestion, their conversion to energy (ATP) and building blocks for growth, and their elimination as solid, liquid, or gaseous waste is essential for health.

Nutrients and Energy

The nutrients in the food we eat fall into six classes. As the human body composition analysis above shows, they are water, fat, protein, carbohydrates, minerals, and vitamins. Some of these nutrients, particularly carbohydrates and lipids but also proteins, yield energy in the form of ATP, which is measured in the commonly used unit of heat, the *calorie.* The caloric unit usually applied to the energy content of a molecule is the kilocalorie (1,000 calories) and is designated as a Calorie (with a capital C). The measure of these units is in grams, that is, Calories per gram. In this way, we are able to estimate the caloric content of food based on the number of grams in particular classes of organic molecules. For carbohydrates, there are 4 Calories/gram, for fats there are 9 Calories/gram, and for proteins there are 4 Calories/gram. The energy from food is used to produce heat, transport molecules and ions across cell membranes, move muscles, and carry out many of the enzyme catalyzed chemical reactions of cells.

A common recommendation for intake of nutrients is published every five years or so by the U.S. government to help advise people on their dietary intake. The Recommended Dietary Allowances (RDA) are set for energy (Calories), protein, vitamins (including A, D, E, and C), and minerals (including calcium, phosphorous, iron, and zinc). The full set of recommendations is presented in Table 9–4(a). These recommendations are for healthy people only and not requirements. Individual needs may vary, and any preexisting medical condition should be

Cholera toxin a deadly toxin produced by the bacteria *Vibrio cholera,* which causes massive dehydration through water loss from the intestines

Nutrients substances from which we derive sustenance; consist of six classes: water, fat, protein, carbohydrates, minerals, and vitamins

Table 9–4 U.S. Recommended Daily Allowance

NUTRIENT	RDA FOR AN ADULT MALE (1968)	RDA FOR AN ADULT FEMALE (1968)	U.S. RDA FOR ADULTS[A]
Nutrients that must appear on the label.[b]			
Protein (g), PER ≥ casein[c]	45	—	45
Protein (g) PER < casein	65	55	65
Vitamin A (RE)	1000[d]	800[d]	1000[d]
Vitamin C (ascorbic acid) (mg)	60	55	60
Thiamin (vitamin B_1 (mg)	1.4	1.0	1.5
Riboflavin (vitamin B_2) (mg)	1.7	1.5	1.7
Niacin (mg)	18	13	20
Calcium (g)	0.8	0.8	1.0
Iron (mg)	10	18	18
Nutrients that may appear on the label:			
Vitamin D (IU)	—	—	400[e]
Vitamin E (IU)	30[f]	25[f]	30[f]
Vitamin B_6 (mg)	2.0	2.0	2.0
Folic acid (folacin) (mg)	0.4	0.4	0.4
Vitamin B_{12} (μg)	6	6	6
Phosphorus (g)	0.8	0.8	1.0
Iodine (μg)	120	100	150
Magnesium (mg)	350	300	400
Zinc (mg)	—	—	15
Copper (mg)	—	—	2
Biotin (mg)	—	—	0.3
Pantothenic acid (mg)	—	—	10

[a]Separate tables of U.S. RDA are published for infants, children, and pregnant and lactating women.
[b]Must appear whenever nutrition labeling is required.
[c]PER is an index of protein quality explained in Chapter 5. Casein is milk protein.
[d]1000 RE was originally expressed as 5000 IU. 800 RE was originally expressed as 4000 IU.
[e]400 IU vitamin D is the same as 10 μg; see inside front cover.
[f]30 IU vitamin E is the same as 30 mg. 25 IU vitamin E is the same as 25 mg. The RDA for vitamin E has since been lowered.

Note: As of 1980, the U.S. RDA numbers used on labels were still those taken from the 1968 RDA. There was no great need to update them, because they still would be judged generous by any standard, and because the expense of converting labels to a different set of numbers would be too great to warrant the change. The circled numbers are those chosen for the U.S. RDA from the adult male and female recommendations. In each case, the higher number is chosen. In the cases of thiamin, niacin, iodine, and magnesium, the RDA for an adolescent boy are used, because these are even higher than the adult RDA. In the cases of calcium and phosphorus, 1 g/day is used, more than the adult RDA. Pregnant and lactating women and rapidly growing teenagers have RDA even higher than this, but 1 g was considered generous enough for use as a standard for labels.

In the cases of the last four nutrients—zinc, copper, biotin, and pantothenic acid—RDA had not been set as of 1968, but these nutrients were known to be essential. The agency set "guestimates" for these so that labels showing percentages of U.S. RDA could include them. As of 1980, all four of these nutrients were included in the RDA tables, but the U.S. RDA values were not changed to correspond; they were considered close enough already.

Source: Adapted from *Food Technology* 28, no. 7 (1974):5.

taken into consideration before altering nutrient intake (Biosite 9–2).

The production of most of the energy for the body takes place in the mitochondria. In this organelle, glucose is utilized most efficiently to produce ATP during aerobic cellular respiration. Without oxygen (anaerobic respiration) a cell is nearly 20 times less efficient in producing ATP. Without glucose the body is forced to use fat and proteins as sources of energy. The persistent lack of glucose as an energy source can spell trouble for the body because the breakdown of lipids results in a buildup of reactive and damaging compounds known as ketones in the blood, and the breakdown of proteins eliminates muscle tissue for structural and function activities.

Classes of Nutrients

Carbohydrates Sugars and starches are the common forms of carbohydrates ingested in the food we eat. In general, *complex carbohydrates,* or polysaccharides such as starch and glycogen, are broken down into glucose, which in turn, is used as the principal energy source of cells (Figure 9–21). But some forms of complex carbohydrates are not degraded by the body and fall into a class of materials referred to as fiber. Cellulose, hemicellulose, and pectin are the best known of these polysaccharides. As described earlier in this chapter, fiber is very important in the motility of solid and semisolid materials in the intestines and may actually help in the absorption of simple sugars (Figure 9–16).

BIO site 9–2

WHAT IS AN IDEAL DIET?

It has been said that if you succeed in diet planning you will have established not only a healthy regimen for nutritional intake but also an enjoyable variety of tasty and interesting foods so you do not get bored. There is an old adage that claims, "we eat with our eyes," so in addition to food being healthful and tasty, it should also look attractive and appetizing. An excellent diet will have the following characteristics—adequacy, balance, calorie control, moderation, variety, and visual or sensory appeal.

Adequacy of diet will provide enough of each of the essential food groups, and balance will ensure that no one type is overrepresented. The energy derived from the food we eat is what runs our metabolism, so it is imperative that we have enough energy-rich nutrients to fulfill our metabolic needs. Because these needs depend on the activities we are involved in, the caloric input should be sufficient to meet demands. If you work out in a gym daily, or run 25 miles a week, you will need more energy than a sedentary individual.

Over 2000 years ago Aristotle first proclaimed that we should live in moderation. It was a wise insight then and continues to be today. A moderate diet is one that does not provide excess fat, sugar, or salt (or any other nutrients that in excess may adversely affect health). Variety in the choice of foods used to fulfill your dietary needs reduces the chance of monotonous and repetitive meals. Variety, combined with the appealing preparation and presentation of food, makes each meal special and circumvents the boredom associated with eating the same old thing, no matter how healthful it may be.

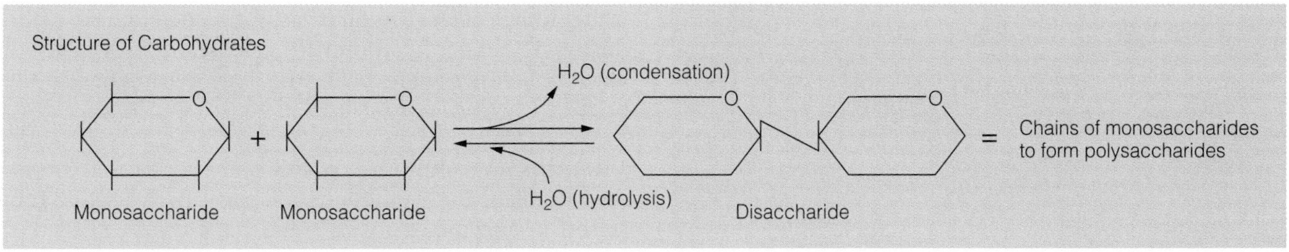

Figure 9–21 Structure of Carbohydrates

Carbohydrates have an undeserved reputation for being the fatteners of the human body. In fact, the contrary is the case. Consumption of complex carbohydrates is recommended by a number of studies including those of the U.S. Senate and the U.S. National Academy of Sciences. This includes potatoes, pasta, and rice. The sugars to avoid in large amounts are the simple sugars, such as glucose, sucrose, and dextrose, which offer plenty of calories but no fiber.

Proteins The structure and function of proteins was presented in detail in Chapter 2, but it should be reiterated that proteins, and the amino acids of which they are composed, are among the most important structural and functional molecules of the body (Figure 9–22). Most of the organic mass of the muscles and bones is protein, as is the preponderance of enzymes, which carry out the chemical reactions that make life possible at biologically relevant temperatures. The breakdown of protein in the digestive tract results in the release of the 20 individual types of amino acids. These molecules are absorbed by cells and used to make new proteins in a pattern specific for the individual. Most of the amino acids used to make proteins are synthesized *de novo* inside cells. However, there are several types that are not made by cells and have to be imported preformed. These are referred to as the *essential amino acids* (Table 9–5). In terms of dietary intake of protein, consideration should be taken of the fact that some proteins are incomplete with regard to one or more essential amino acids, and others are complete. Complete proteins are found in meat, fish, and poultry as well as in milk, cheese, and eggs. The diet should supply all the essential amino acids simultaneously, so that protein synthesis is not interrupted or incomplete because of the lack of any one type.

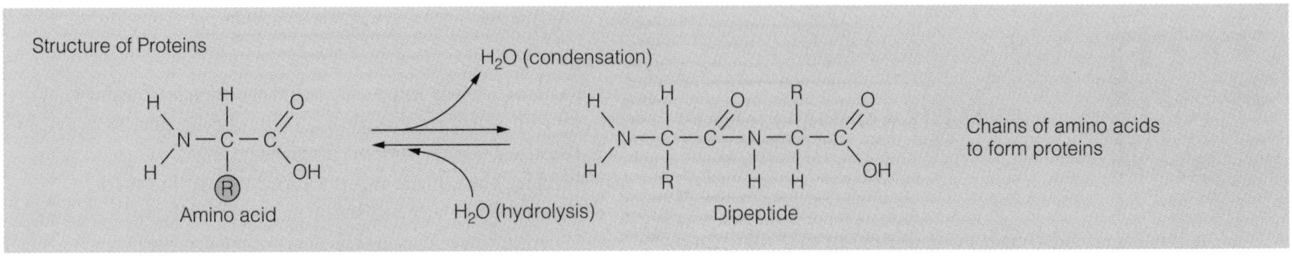

Structure of Proteins

H_2O (condensation)

Amino acid

H_2O (hydrolysis)

Dipeptide

Chains of amino acids to form proteins

Figure 9–22 Structure of Proteins

Table 9–5 Amino Acids

Amino Acids
Alanine
Arginine**
Aspartic acid
Cysteine**
Cystine
Glutamic acid
Glutamine
Glycine
Histidine*
Isoleucine*
Leucine*
Lysine*
Methionine*
Phenylalanine*
Proline
Serine
Threonine*
Tryptophan*
Tyrosine**
Valine*

*Essential amino acids for adults
**Conditionally essential amino acids

Fats Fats and oils are an essential part of any diet. All cells in the body require lipids to form cellular membranes, to build up energy reserves, and to establish insulation for thermal regulation and cushioning for protection of organs. The plasma membrane of all cells is composed of phospholipids and glycolipids, which form a bilayer that separates the outside and inside of a cell. Another important class of lipids is cholesterol. Cholesterol is made by cells as well as utilized from dietary intake. This sterol is inserted into cell membranes and alters their fluidity and resistance to disruption. Without lipids there can be no cells. Problems with cholesterol arise in many forms of vascular disease.

Fatty acids are a particularly important part of the diet. These long-chained molecules come in forms that are either saturated or unsaturated (Figure 9–23). Some fatty acids, such as *linoleic acid* and *linolenic acid* are essential fatty acids. They are considered polyunsaturated as a result of the reduced number of hydrogen-carbon bonds. A distinction must be made with regard to total fat intake and risk of cardiovascular disease. It is the intake of saturated fat, not cholesterol, that is the major dietary factor in raising blood cholesterol to dangerous levels. Trigylcerides are made from the combination of fatty acids and the three-carbon compound glycerol (Figure 9–23). *Triglycerides* are the chief storage products arising from overconsumption of fats and carbohydrates. Their structure makes them one of the most efficient ways to store such metabolic excess. Adipose tissue is composed of cells whose primary activity is the storage of fat. This fat is easily recruited for use in producing ATP in conjunction with glucose. In the absence of glucose, ketones are formed, which are detrimental to the body.

Vitamins Vitamins are organic compounds that are indispensable to body functions and particularly useful as cofactors in enzyme catalyzed reactions. They are needed in extremely small amounts and have no caloric value. There are 13 known vitamins (Table 9–6), of which nine are soluble and the remainder insoluble. Vitamins are not synthesized by human cells; they must be obtained preformed in the foods we eat. Serious problems can arise in the body in the absence of sufficient vitamins (Table 9–4).

Insoluble Vitamins Vitamins A, D, E, and K are insoluble. Each of these vitamins is stored in the fatty tissues of the body and, therefore, may last longer than the soluble vitamins, which tend to be excreted fairly rapidly. The sources of these vitamins are many (Table 9–6), including a variety of foods, bacteria, and light-induced conversion of precursor molecules. Vitamin A is involved in a number of important cellular phenomena, including photoreceptor cell function in vision and epithelial cell growth and maintenance. The epithelia of all areas of the body, from the skin to the lungs and from the gut to the vagina and urinary tract, require vitamin A. In recent years it has also

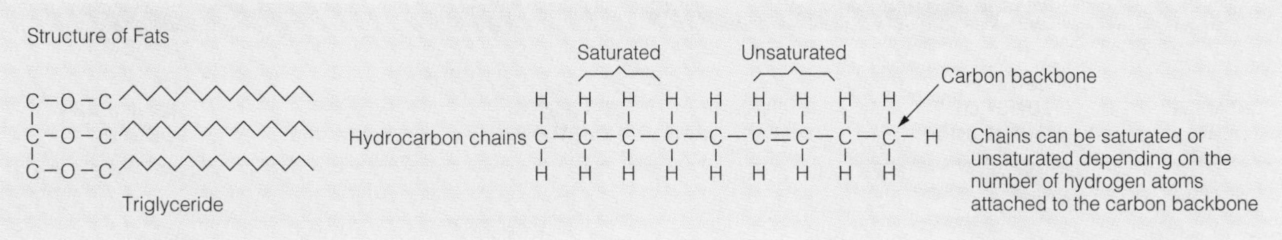

Structure of Fats

Triglyceride

Hydrocarbon chains

Saturated

Unsaturated

Carbon backbone

Chains can be saturated or unsaturated depending on the number of hydrogen atoms attached to the carbon backbone

Figure 9–23 Structure of Fats

been implicated in embryonic development as a **morphogen.** Vitamin A can also act as a **teratogen** if introduced in too high an amount or at the wrong time in development. Vitamin D regulates calcium to maintain bone. The conversion of cholesterol-like molecules to vitamin D takes place in skin exposed to sunlight (for as little as 30 minutes a day). Vitamin E serves as an **antioxidant,** preventing other molecules from being chemically degraded by oxygen. This is true for the protection of polyunsaturated fats, which are particularly susceptible to oxidation. Vitamin E protects the membranes of the lungs, where exposure to oxygen is a necessary function in gas exchange. Vitamin K is produced by bacteria in the human intestinal tract. It plays a valuable role in enzyme reactions involved in the formation of blood clotting factors and participates with Vitamin D in the regulation of calcium.

Soluble Vitamins The soluble vitamins fall into two families, the B vitamins and vitamin C. Vitamins of both groups are excreted rapidly from the body and need to be replaced continuously. The sources of these vitamins are listed in Table 9–6 and include a variety of meats, fish, and poultry as well as seeds, fruits, vegetables, and milk. The B vitamins are cofactors or coenzymes. They interact with protein enzymes to activate them in their performance as catalysts. The B vitamins act in a large number of enzyme pathways affecting the synthesis of DNA and RNA, amino acids, glycolysis and the Krebs cycle and electron transfer reactions of the mitochondria. Some diseases or conditions associated with deficiencies in the B vitamins are listed in Table 9–6.

Minerals Minerals are inorganic compounds, ions, or elements. They are required in conjunction with water and organic compounds to maintain the health and structure of the human body. The major minerals used by the body are calcium, chloride, magnesium, sodium, sulfur, phosphorus, and potassium. In addition to these elements, trace levels of many other minerals, including such commonly recognized atoms as iron, zinc, and iodine, are required (Table 9–7). Many of these minerals are dissolved in water and form the **electrolytes** of the body fluids. Sodium ions are the single most abundant species of ion found in the human body, and they are fundamentally important in transport phenomena associated with the plasma membranes of cells. Sodium, calcium, and potassium are essential for the function of nerves and muscles, as they move passively and/or actively across the membranes of these cell types. As discussed in Chapter 6, calcium phosphate is the principal component of bones, imparting the strength and rigidity required to support movement of the human body.

Trace Minerals Some of the trace minerals are as important as their more abundant relatives. For example, iron is essential for the function of hemoglobin and the binding of oxygen. Zinc is associated with dozens of enzymes, which cannot function in its absence. Iodine is needed in only millionths of a gram to serve its role in the formation of *thyroxine* in the *thyroid gland,* but in its absence **goiters** form and behavior may be severely altered. A list of the 25 essential minerals is presented in Table 9–7.

Water It goes without saying that the single most important molecule for life is water. The body is 60% water, the blood is 90% water, and an average cell is also 90% water. Water is the principal cellular and bodily solvent and

Morphogen a molecule that influences growth and morphogenesis of an organism

Teratogen a molecule or agent that causes deformities in development and growth of organisms

Antioxidant a molecule that retards or prevents oxidation

Electrolytes the water soluble charge-bearing molecules and ions found in bodily fluids

Goiters growths of the thyroid gland resulting from insufficient levels of dietary iodine

Table 9–6 Vitamins

VITAMIN NAMES	CHIEF FUNCTIONS IN THE BODY	DEFICIENCY DISEASE NAME	DEFICIENCY SYMPTOMS	TOXICITY SYMPTOMS	SIGNIFICANT SOURCES
			Blood/Circulatory System		
Biotin	Part of a coenzyme used in energy metabolism, fat synthesis, amino acid metabolism, and glycogen synthesis	(No name)	Abnormal heart action	(No toxicity symptoms reported)	Widespread in foods
			Digestive System		
			Loss of appetite, nausea		
			Nervous/Muscular Systems		
			Depression, muscle pain, weakness, fatigue		
			Skin		
			Drying, rash, loss of hair		
			Blood/Circulatory System		
Vitamin C (ascorbic acid)	Collagen synthesis (strengthens blood vessel walls, forms scar tissue, matrix for bone growth), antioxidant, thyroxine synthesis, amino acid metabolism, strengthens resistance to infection, helps in absorption of iron	Scurvy	Anemia (small-cell type)[a] atherosclerotic plaques, pinpoint hemorrhages	Blood cell breakage in certain racial groups[b]	Citrus fruits, cabbage-type vegetables, dark green vegetables, cantaloupe, strawberries, peppers, lettuce, tomatoes, potatoes, papayas, mangos
			Digestive System		
				Nausea, abdominal cramps, diarrhea	
			Immune System		
			Depression, frequent infections		
			Mouth, Gums, Tongue		
			Bleeding gums, loosened teeth.		
			Muscular/Nervous Systems		
			Muscle degeneration and pain, hysteria, depression		
			Skeletal System		
			Bone fragility, joint pain		
			Skin		
			Rough skin, blotchy bruises		
			Other		
			Failure of wounds to heal	Interference with medical tests; aggravation of gout symptoms; deficiency symptoms may appear at first on withdrawal of high doses	
			Blood/Circulator System		
Thiamin (vitamin B_1)	Part of a coenzyme used in energy metabolism, supports normal appetite and nervous system function	Beriberi	Edema, enlarged heart, abnormal heart rhythms, heart failure	Rapid pulse	Occurs in all nutritious foods in moderate amounts; pork, ham, bacon, liver, whole grains, legumes, nuts
			Nervous/Muscular Systems		
			Degeneration, wasting, weakness, pain, low morale, difficulty walking, loss of reflexes, mental confusion, paralysis	Weakness, headaches, insomnia, irritability	

[a]Small-cell type anemia is *microcytic anemia;* large-cell type is *macrocytic* or *megaloblastic anemia.*
[b]Groups susceptible to vitamin C toxicity are Sephardic Jews, Africans, and Asians.

(Continued)

Table 9–6 Vitamins *(continued)*

VITAMIN NAMES	CHIEF FUNCTIONS IN THE BODY	DEFICIENCY DISEASE NAME	DEFICIENCY SYMPTOMS	TOXICITY SYMPTOMS	SIGNIFICANT SOURCES
			Mouth, Gums, Tongue		
Riboflavin (vitamin B_2)	Part of a coenzyme used in energy metabolism supports normal vision and skin health	Ariboflavinosis	Cracks at corners of mouth,[c] magenta tongue	(No symptoms ordinarily reported)	Milk, yogurt, cottage cheese, meat, leafy green vegetables, whole-grain or enriched breads and cereals
			Nervous System and Eyes		
			Hypersensitivity to light,[d] reddening of cornea		
			Other		
			Skin rash	Interference with anticancer medication	
			Digestive System		
Niacin (nicotinic acid, nicotinamide, niacinamide, vitamin B_3, vitamin G); precursor is dietary tryptophan	Part of a coenzyme used in energy metabolism; supports health of skin, nervous system, and digestive system	Pellagra	Diarrhea	Diarrhea, heartburn, nausea, ulcer irritation, vomiting	Milk, eggs, meat, poultry, fish, whole grain and enriched breads and cereals, nuts, and all protein-containing foods
			Mouth, Gums, Tongue		
			Black, smooth tongue[e]		
			Nervous System		
			Irritability, loss of appetite, weakness, dizziness, mental confusion progressing to psychosis or delirium	Fainting	
			Skin		
			Skin rash on areas exposed to sun	Painful flush and rash	
			Other		
				Abnormal liver function, low blood pressure	
			Blood/Circulatory System		
Vitamin B_6 (pyridoxine, pyridoxal, pyridoxamine)	Part of a coenzyme used in amino acid and fatty acid metabolism, helps convert tryptophan to niacin, helps make red blood cells	(No name)	Anemia (small-cell type)[a]	Bloating	Green and leafy vegetables, meats, fish, poultry, shellfish, legumes, fruits, whole grains
			Digestive System		
			Mouth, Gums, Tongue		
			Smooth tongue[e]		
			Nervous/Muscular Systems		
			Abnormal brain wave pattern, irritability, muscle twitching, convulsions	Depression, fatigue, irritability, headaches, numbness, damage to nerves, difficulty walking	
			Skin		
			Irritation of sweat glands, rashes		
			Other		
			Kidney stones		
			Blood Circulatory System		
Folacin (folic acid, folate, pteroylglutamic acid)	Part of a coenzyme used in new cell synthesis	(No name)	Anemia (large-cell type)[a]		Leafy green vegetables, legumes, seeds, liver
			Digestive system		
			Heartburn, diarrhea, constipation	Diarrhea	
			Immune System		
			Depression, frequent infections		
			Mouth, Gums, Tongue		
			Smooth red tongue[e]		

[c]Cracks at the corners of the mouth are termed *cheilosis* (kee-LOH-SIS).
[d]Hypersensitivity to light is *photophobia*.
[e]Smoothness of the tongue is caused by loss of its surface structures and is termed *glossitis* (gloss-EYE-tis).

(Continued)

Table 9–6 Vitamins *(continued)*

VITAMIN NAMES	CHIEF FUNCTIONS IN THE BODY	DEFICIENCY DISEASE NAME	DEFICIENCY SYMPTOMS	TOXICITY SYMPTOMS	SIGNIFICANT SOURCES
			Nervous System		
			Depression, mental confusion, fainting	Insomnia, irritability	
			Other		
				Masking of vitamin B_{12} deficiency symptoms	
			Blood/Circulatory System		
Vitamin B_{12} (Cyanocobalamin)	Part of a coenzyme used in new cell synthesis, helps maintain nerve cells	(No name[f])	Anemia (large-cell type)[a]	(No toxicity symptoms known)	Animal products (meat, fish, poultry, shellfish, milk, cheese, eggs)
			Mouth, Gums, Tongue		
			Smooth tongue[e]		
			Nervous System		
			Fatigue, degeneration progressing to paralysis		
			Skin		
			Hypersensitivity		
			Digestive System		
Pantothenic acid	Part of a coenzyme used in energy metabolism	(No name)	Vomiting, intestinal distress	Occasional diarrhea	Widespread in foods
			Nervous System		
			Insomnia, fatigue		
			Other		
				Water retention (infrequent)	
			Blood/Circulatory System		
Vitamin A (retinol, retinal, retinoic acid); precursor is provitamin A carotenoids such as beta carotene	Vision; maintenance of cornea, epithelial cells, mucous membranes, skin; bone and tooth growth; reproduction; hormone synthesis and regulation; immunity; cancer protection	Hypovitaminosis A	Anemia (small-cell type)[a]	Red blood cell breakage, nosebleeds	Retinal: fortified milk, cheese, cream, butter, fortified margarine, eggs, liver
			Digestive System		
			Diarrhea, general discomfort	Abdominal cramps and pain, nausea, vomiting, diarrhea, weight loss	
			Immune System		
			Depression; frequent respiratory, digestive bladder, vaginal, and other infections	Overreactivity	Beta carotene: Spinach and other dark leafy greens, broccoli, deep orange fruits (apricots, peaches, cantaloupe) and vegetables (squash, carrots, sweet potatoes, pumpkin)
			Mouth, Gums, Teeth		
			Abnormal tooth and jaw alignment		
			Nervous/Muscular Systems		
			Night blindness (retinal)	Blurred vision, pain in calves, fatigue, irritability, loss of appetite, bone pain	
			Skin and Cornea		
			Keratinization, corneal degeneration leading to blindness,[g] rashes	Dry skin, rashes, loss of hair	
			Other		
			Kidney stones, impaired growth	Cessation of menstruation, growth retardation, liver and spleen enlargement	

[f]The name *pernicious anemia* refers to the vitamin$_{12}$ deficiency caused by lack of intrinsic factor, but not to that caused by inadequate dietary intake.
[g]Corneal degeneration progresses from *keratinization* (hardening) to *xerosis* (drying) to *xerophthalomia* (thickening, opacity, and irreversible blindness).

(Continued)

Table 9–6 Vitamins *(continued)*

VITAMIN NAMES	CHIEF FUNCTIONS IN THE BODY	DEFICIENCY DISEASE NAME	DEFICIENCY SYMPTOMS	TOXICITY SYMPTOMS	SIGNIFICANT SOURCES
			Blood/Circulatory System		
Vitamin D (calciferol, cholecalciferol, dihydroxy-vitamin D); precursor is the body's own cholesterol	Mineralization of bones (raises calcium and phosphorus blood levels by increasing absorption from digestive tract, withdrawing calcium from bones, stimulating retention by kidneys)	Rickets osteomalacia		Raised blood calcium	Self-synthesis with sunlight; fortified milk, fortified margarine, eggs, liver, fish
			Digestive System		
				Constipation, weight loss	
			Nervous System		
				Excessive thirst, headaches, irritability, loss of appetite, weakness, nausea	
			Other		
			Abnormal growth, joint pain, soft bones	Kidney stones, stones in arteries, mental and physical retardation	
			Blood/Circulatory System		
E (alphatocopherol, tocopherol)	Antioxidant (detoxification of strong oxidants), stabilization of cell membranes, regulation of oxidation reactions, protection of PUFA and vitamin A	(No name)	Red blood cell breakage, anemia	Interference with anticlotting medication	Plant oils (margarine, salad dressings, shortenings), green and leafy vegetables, wheat germ, whole grain products, butter, liver, egg yolk, milk fat, nuts, seeds
			Digestive System		
				General discomfort	
			Nervous/Muscular Systems		
			Degeneration, weakness, difficulty walking, intermittent claudication		
			Other		
			Fibrocystic breast disease		
			Blood/Circulatory System		
Vitamin K (phylloquinone, naphthoquinone)	Synthesis of blood-clotting proteins and a blood protein that regulates blood calcium	(No name)	Hemorrhaging	Interference with anticlotting medication; vitamin K analogues may cause jaundice	Bacterial synthesis in the digestive tract; liver, green leafy vegetables, cabbage-type vegetables, milk

dissolved in it are proteins, sugars, minerals, and vitamins. Most chemical reactions in the body take place using water (*hydrolysis reactions*) and/or forming water (*condensation reactions*) (for example, see Figures 9–21 and 9–22). It also provides a medium for suspension of hydrophobic materials such as lipids and cells (as in blood). The origins of life over 3.5 billion years ago took place in water and all living things since have required a continuous supply of this essential compound. We drink it to replenish ourselves, we bathe in it to clean ourselves, its evaporation cools us and forms the clouds in the sky that shade us. Water is the ultimate nutrient.

Table 9–7 The 25 Essential Minerals

- Carbon
- Hydrogen
- Oxygen
- Nitrogen
- Major minerals
  - Calcium
  - Chlorine
  - Magnesium
  - Phosphorus
  - Potassium
  - Sodium
  - Sulfur
- Trace minerals
  - Chromium
  - Cobalt
  - Copper
  - Fluorine
  - Iodine
  - Iron
  - Manganese
  - Molybdenum
  - Nickel
  - Selenium
  - Silicon
  - Tin
  - Vanadium
  - Zinc

Summary

The anatomical map used in this chapter has plotted a course through the digestive system from the mouth to the anus. It has pinpointed the areas in which specific activities of this system take place and where specific problems may develop. If we stretched the entire digestive tract into a straight tube (seen partially in Figures 9–1 and 9–2), we would see that it is very like a well-organized, compartmentalized, and efficient factory. The main difference in this factory comparison is that the digestive machinery forms a disassembly line instead of an assembly line. The cutting and grinding of teeth divides and pulverizes the food and softens it with water, so it can be easily swallowed. The stomach churns, mixes, and hydrolyzes the bolus into fluidlike consistency (chyme) and passes it on in spurts to the small intestine, where the final disassembly of molecules from the milky, emulsified chyle takes place. Once the food is totally dismantled to its component molecular pieces by emulsifiers and enzymes, it is actively transported through the epithelial cells of the wall of the small intestine and into the blood and lymph vessels of the body, to supply the ceaseless demands for energy and chemical building blocks. The remains move into the large intestine where the job of reclaiming water and absorbing minerals and vitamins is accomplished. The waste is stored, briefly, and then shipped out in compact, solidified packets known as feces. From a structural and functional point of view, it is clear that the disassembly line of the digestive tract plays a fundamental role in supplying nutrients and energy sources needed for homeostasis and that we are, indeed, what we eat.

Questions for Critical Inquiry

1. How is the digestive process related to energy supplies for the body? What are the sources of this energy?
2. In what way is it that, "We are what we eat?"
3. Compare the complementary roles of the digestive system and the respiratory system in supplying the molecules needed to produce energy for the body's needs.
4. What role should government take in assuring the quality, safety, and healthfulness of food products sold to the public?

Questions of Facts and Figures

5. What are the four basic types of activities carried out by the digestive system? Where do each of these activities take place?
6. What is the function of fiber in the digestive process? What are some sources of fiber?
7. How much saliva is produced each day by the human body? What is it used for? How is the water from which it is made recovered by the body?
8. What are the three types of papillae found on the human tongue?
9. What is peristalsis? Segmentation?
10. How is a bolus formed? Where?
11. What acid is produced by the stomach? What is the pH of the stomach during digestion?
12. Where are the digestive enzymes of the small intestine produced? How do they get to the duodenum? Where is bile produced? Stored?
13. Where does most of the absorption of nutrients occur in the digestive tract?
14. What role do bacteria play in digestion?
15. What types of materials are considered to be food?
16. What do the initials RDA stand for? What are some of those recommendations?

References and Further Readings

Alberts, B., Bray, D., Lewis, J., Roff, M., Roberts, K., and Watson, J. D. (1994). *Molecular Biology of the Cell,* 3rd ed. New York: Garland Publishing.

Hamilton, E. M. N., Whitney, E. N., and Sizer, F. S. (1988). *Nutrition: Concepts and Controversies,* 4th ed. St. Paul, MN: West Publishing.

Sherwood, L. (1993). *Human Physiology,* 2nd ed. St. Paul, MN: West Publishing.

Stanier, R., Doudoroff, M., and Adelberg, E. (1970). *The Microbial World,* 3rd ed. Englewood Cliffs, NJ: Prentice Hall.

CHAPTER 10

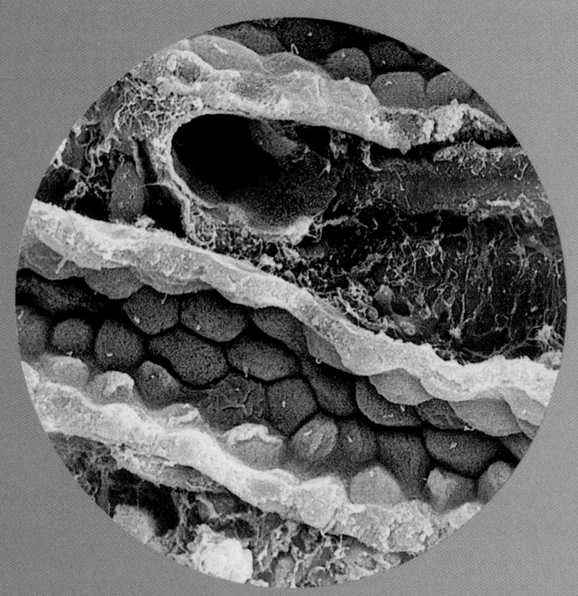

The Kidneys and the Urinary System

INTRODUCTION

Quenching one's thirst with a tall glass of water is both refreshing and essential. The human body, which is mostly water, tightly regulates its intake and output, as well as how it is used. The kidneys and urinary system control the amount of water in the body and the means by which water and metabolic wastes are eliminated. What might happen if you ran out of water and no more was available? Within hours the situation could become serious. Imagine the following scenario, in which a passenger on an ill-fated trip recounts an adventure in the desert.

Act. 1. Scene 1. *On a bus on a hot desert highway. Protagonist contemplating.*

It's only a matter of minutes until the bus runs out of gas. Murphy's law predicts it should happen right in the middle of the desert at high noon. And this just about fulfills expectations. I still don't know why the driver didn't fill up on gasoline before we headed out. Well, at least we have enough water to survive in this inferno until someone realizes we are missing and sends help. I think we have enough water. I wonder if somebody checked on the water supply before we left? The last thing I want is to dehydrate and die of thirst. I guess my kidneys will have to work overtime on this trip.

ANATOMY OF THE URINARY SYSTEM

Everyone has some awareness of what the kidneys do. The job of the kidneys and urinary system is a complex one, filtering the blood continuously to eliminate toxic nitrogenous waste products, such as urine. But it also has a significant role in reclaiming beneficial nutrients, and regulating the balance of water and various ions in the fluids of the body. The human body is nearly two-thirds water to begin with and maintains a uniform concentration of many types of solutes. The human body has two kidneys, one on each side of the spine at about the level of the lower ribs (Figure 10–1), as well as a system of tubing and reservoirs to transport and contain the urine. Each kidney is shaped like a kidney bean, convex on one side and concave on the other (Figure 10–2). A kidney is about the size of a human fist and is attached strongly to the dorsal abdominal body wall, behind the peritoneal cavity. The kidneys are supplied with blood by major arteries under high blood pressure. These arteries, known as **renal arteries,** recirculate 6,000–7,000 liters of blood each day to the kidneys to be filtered and refiltered, in the endless progression of continuous circulatory flow. The body can function well with a single kidney, so two kidneys provide us with more efficiency and a backup for any unilateral problems that may arise.

A ONE-WAY STREET

To understand how the kidneys function in filtration and water-solute-nitrogenous waste balance, we first need to

Renal arteries the large arteries that split off the aorta to supply the left and right kidneys with blood under high pressure

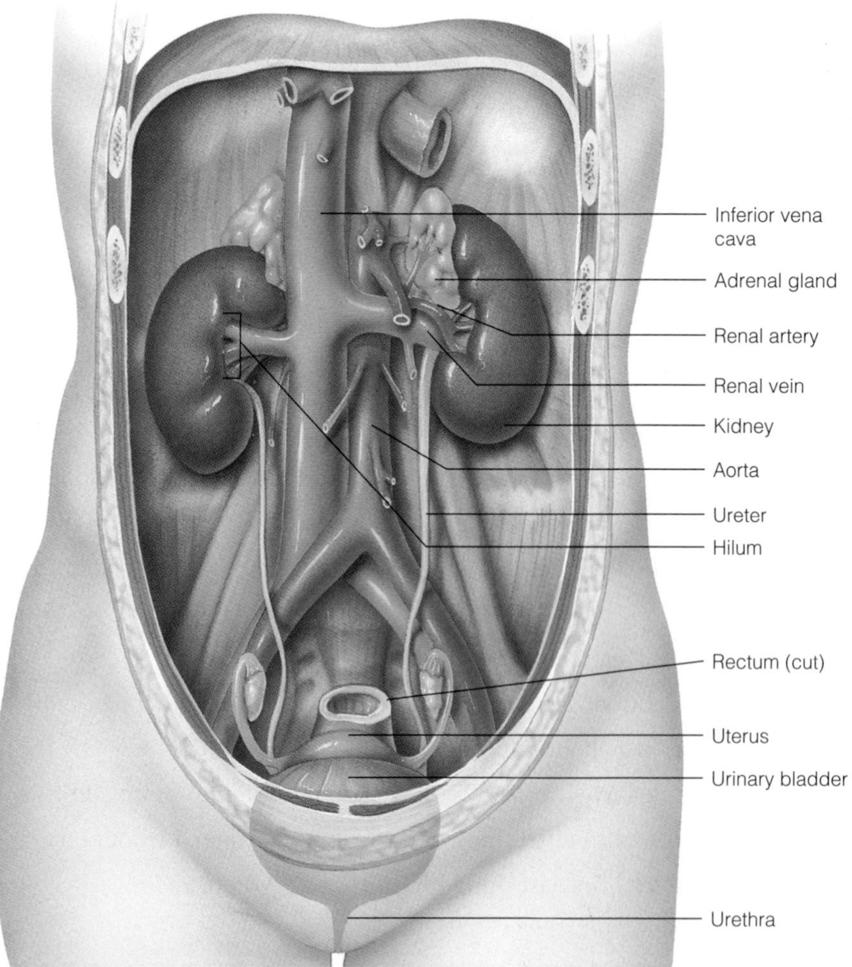

Figure 10–1 The Urinary System
The urinary system includes the kidneys, which produce urine, and a means to deliver urine to the outside of the body. The ureter, urinary bladder, and urethra allow for the removal of urine.

know about the structure and anatomy of the kidney (Figure 10–2) and the route taken by materials from the blood in the kidney and urinary system, which is basically a one-way street. There are cellular channels and carriers directing the movement of molecules of different kinds into and out of the cells composing the tubular system (Figure 10–3). The kidney employs a number of specialized mechanisms to retain useful molecules and ions that pass through it following the initial filtration step. An overview of the urinary system (Figure 10–1) shows that each kidney is connected to the *urinary bladder* by a large tube known as a *ureter.* The ureter from each kidney funnels the urine from the **renal pelvis** to the bladder. The urinary bladder is a very stretchable, hollow organ. It is capable of expanding, temporarily, to hold an ever-increasing volume of urine arriving from the kidneys. At some point in the "filling up" process, stretch receptors in the wall of the bladder are triggered to send signals to the brain that are interpreted as "it's time to eliminate urine." In adults the timing of elimination is under predominantly voluntary control. The urine exits the bladder through the *urethra* and is voided. The urethra in males is somewhat longer than in females (20 cm versus 4 cm–5 cm), because it passes through the entire length of the penis.

This quick sketch of the structures and processes of fluid production and elimination shows that there are specific regions of the urinary system in which urine is made and other regions in which storage and transport out of the body are involved.

Kidney

The kidney ultimately gives us the freedom we have as mobile, independent organisms. Getting rid of nitrogenous metabolic wastes on a continuous basis keeps us free from the effects of these accumulating toxins (a condition known as **uremia**). Not all animals produce urine (Table 10–1). Most nonmammalian organisms that live in the sea, for example, depend upon the huge volume of water in which they live to dissipate nitrogenous wastes such as ammonia.

Four Zones of the Kidney

The kidney itself has four major zones (Figure 10–4). The outside of the kidney is called the *capsule* and it forms a

Renal pelvis the final cavity of a kidney to which urine flows from the pyramids and papillae and from which the ureters emerge

Uremia a toxic buildup of urea in blood

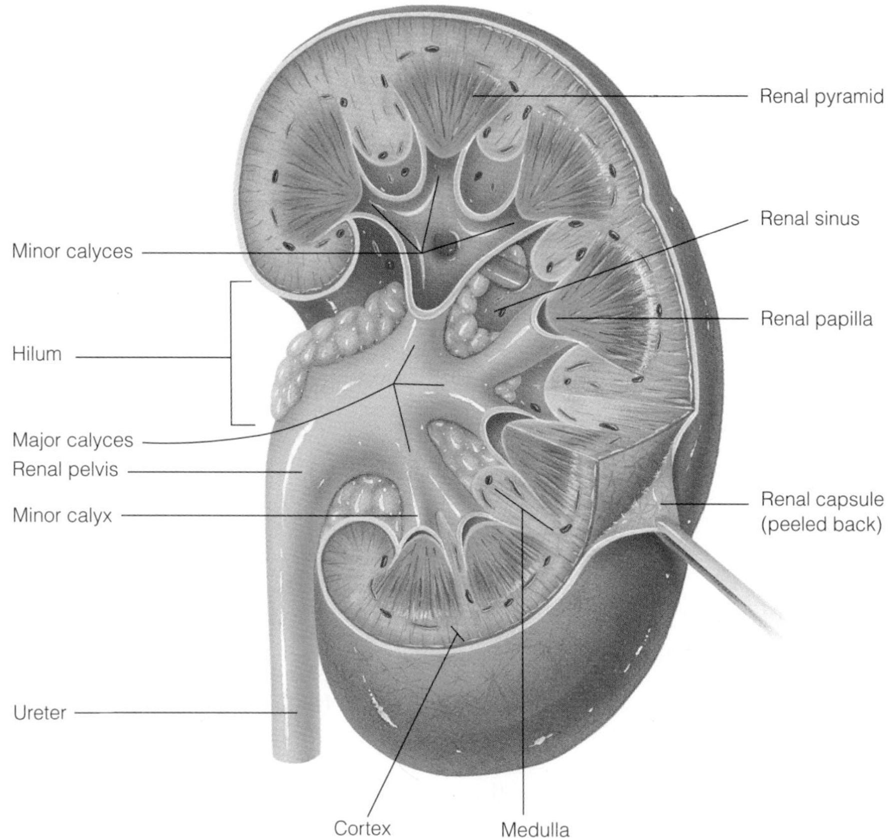

Figure 10–2 The Kidney
The kidney is organized in layers. The functional units of the kidney (the nephrons) are located in the cortex and medulla. The flow of urine is from the cortex and medulla, through the renal pyramids and into the calyces, and, finally, to the renal pelvis.

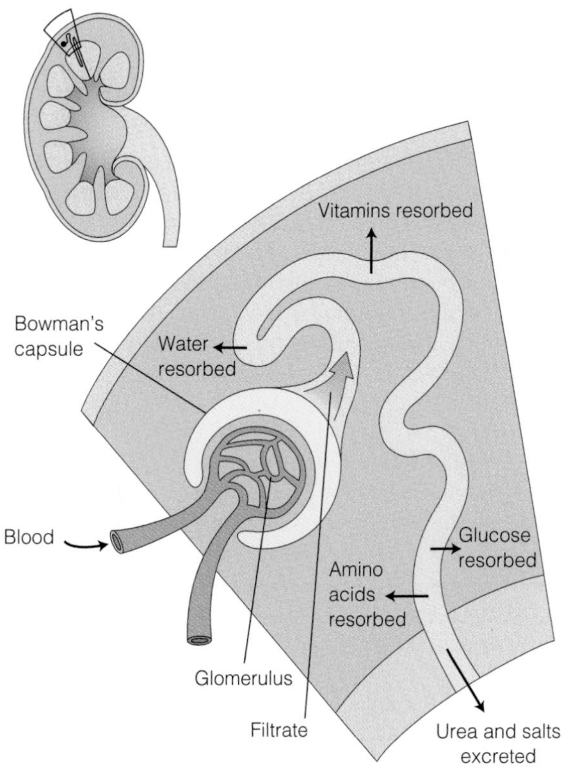

Figure 10–3 Resorption of Nutrients and Water
Most of the nutrients lost from the blood during filtration are resorbed in the first portion of the nephron. Vitamins, minerals, glucose, amino acids, and water are transported back into the body.

Table 10–1 Nitrogenous Wastes and Vertebrates

Fish and amphibians release ammonia directly into water.
Birds and reptiles convert ammonia to uric acid.
Mammals (humans) convert ammonia to urea.

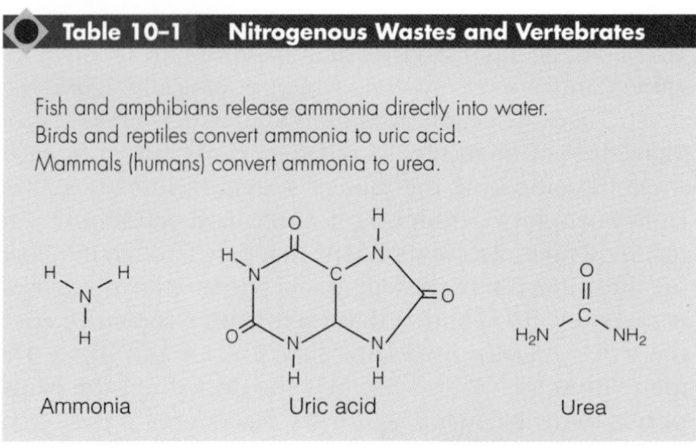

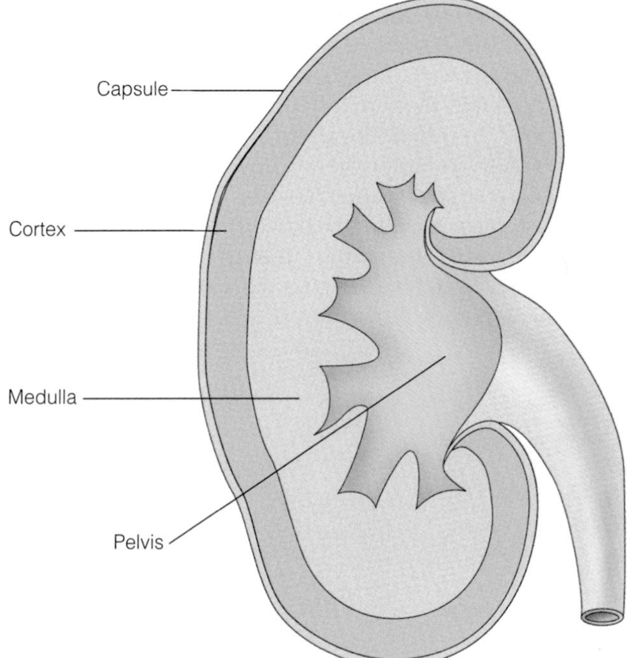

Figure 10–4 Layers of the Kidney
The kidney has four main layers. The capsule is tough connective tissue; the cortex and medulla contain the nephrons and collecting ducts; and the pelvis is the repository of urine as it begins to flow out of the kidney.

tough connective tissue layer that supports and protects the inner, more delicate structures. The capsule also functions to attach the kidney to the body wall, so it will not slip out of place or collapse, a serious condition known as *ptosis.* A dislodged kidney can twist and turn and cut off the flow of urine through a ureter, similar to putting a kink in a garden hose. This can be very dangerous.

Beneath the capsule in the outer, or more superficial, region of the kidney proper is a zone called the *cortex,* and underlying the cortex is a region called the *medulla* (Figure 10–5). The medulla surrounds a cavernous area known as the renal pelvis, through which the urine passes as it is shuttled via the ureter to the urinary bladder. More specifically, the renal pelvis is organized with a series of internal structures known as *pyramids.* The pyramids have a striated appearance and terminate in *papillae,* or fingerlike structures, that are directed towards the nominal center of the kidney pelvis. Each of the papillae is surrounded by a cuplike process from the renal pelvis called a *calyx.* The calyces help in collecting, and directing, the urine produced by the kidneys into the large cavity of the renal pelvis (Figure 10–2).

Passage of Blood

The start of the journey of blood through the urinary system begins with blood flow from the renal arteries that lead to the left and right kidneys (Figure 10–1). This is a high flow rate, high-pressure system that directs the entire blood volume of the body through the kidneys more than 150 times each day. The renal arteries enter a kidney through a region known as the **hilum** on the concave surface. Blood exits the kidney through the same region, via the **renal vein.** Upon entering through the hilum, the renal artery divides into branches that form a series of vascular arcs over the cortex of the kidney proper and returns along adjacent pathways to exit through the renal veins. The ureter also departs the kidney through the hilum, as well (Figure 10–2). The arterial blood flows into the cortex of the kidney to deliver blood to the **nephrons.**

Nephron

The Glomerulus and Bowman's Capsule The nephron is the structural and functional unit of the kidney (Figure 10–5). In humans, there are about a million nephrons in each kidney. Each tiny nephron has a filtration segment, which surrounds a mass of special blood capillaries known as the **glomerulus.** The glomerulus is located within a double-thick envelope of epithelial cell layers known as

> **Hilum** the region of exit of the renal pelvis from the kidney
>
> **Renal vein** the large vein that carries deoxygenated blood away from the kidneys
>
> **Nephron** the functional unit of the kidney wherein blood is filtered and nutrients are resorbed
>
> **Glomerulus** an array of arterioles through which blood is filtered into Bowman's capsule

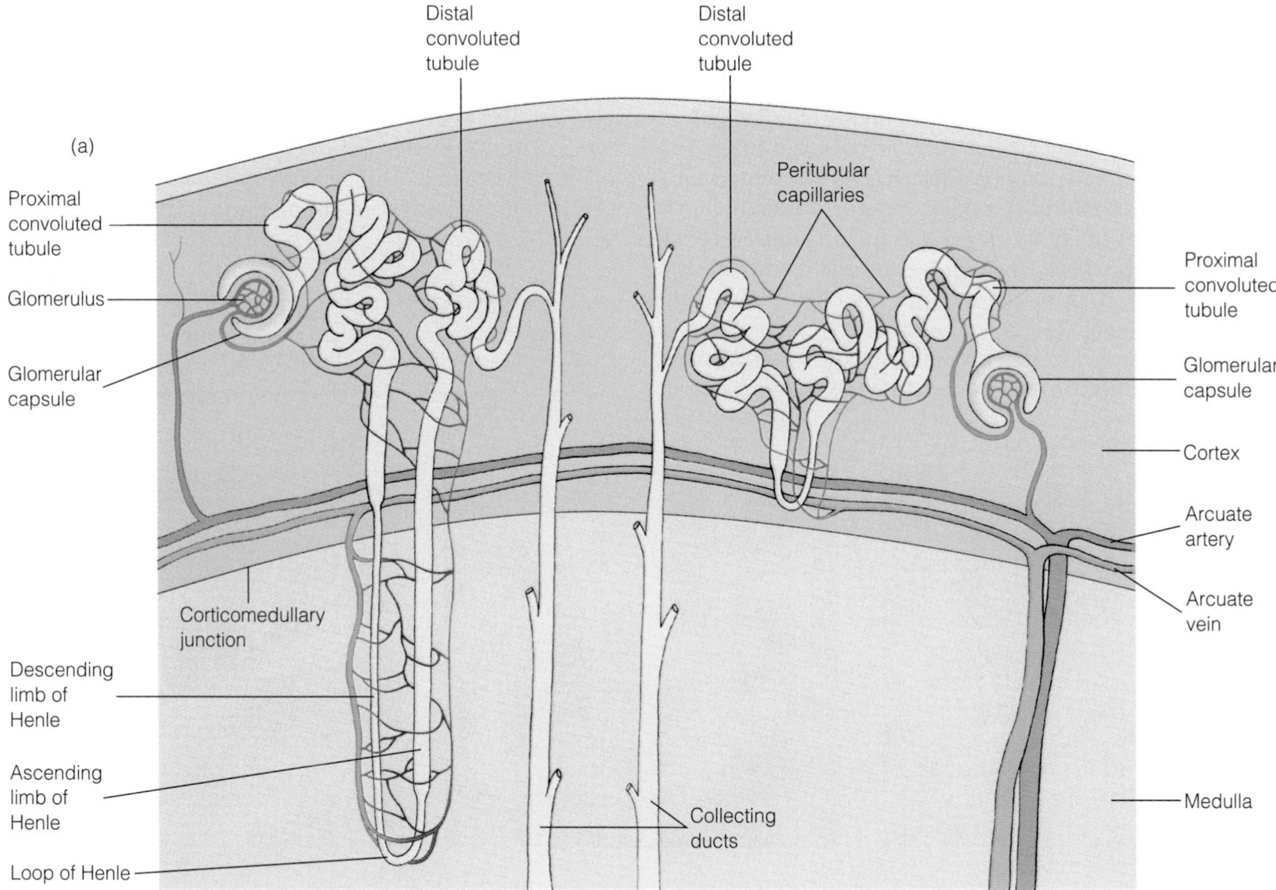

Figure 10–5 Cortex and Medulla
Nephrons are organized within the cortex and medulla. Some nephrons have a loop of Henle that passes into the medulla (a). Others are located entirely within the cortex (b). Both types connect to collecting ducts that pass toward the renal pyramids.

Bowman's capsule (Figure 10–5). This initial segment of the nephron is usually located within the cortex, or right at the boundary of the cortex and the underlying medulla (Figure 10–5). Filtration of blood initially occurs in the capsular part of the nephron, but as freshly filtered fluids pass through the tubes of the nephron, selective reclamation of useful materials (that is, water, glucose, amino acids, vitamins) takes place. By the end of the journey through a nephron, urine has been produced and is collected for elimination. The amount of urea composing the urine is closely regulated.

Bowman's capsule is an epithelial structure that encloses the glomerulus and results in two separated layers with a cavity, or space, between them. Imagine the glomerulus as a vascular fist punching its way into an inflated bag. The bag collapses around the fist in a double layer. This is analogous to the anatomy of a Bowman's capsule. This double envelopment is very important. The driving force for the filtration in the nephron is the systemic blood pressure, and the first major organs through which the abdominal systemic blood passes are the kidneys. In addition the glomerulus itself restricts the flow of blood, increasing filtration pressure by having smaller diameter blood vessels at the end of its tortuous track than at the beginning. The remainder of the nephron is a relatively long twisted tube arising from the end of Bowman's capsule, through which the filtrate eventually flows through the medulla and back up to the cortex before being diverted again through the medulla on its way to the renal pelvis (Figure 10–5 and 10–6). The tubular system of the kidneys is surrounded by a closely associated system of capillaries known as *peritubular capillaries*. It is through these capillaries that most or all of the useful molecules and ions of the filtrate are eventually resorbed back into the body.

Leaky Endothelium The endothelial cells lining the glomerular capillaries form tiny slits between them called *pedicels*. Water and small dissolved materials in the blood can be forced or *sieved* through the pedicels. These openings provide a kind of controlled leakage of the blood to the outer epithelial layer of Bowman's capsule, which also allows the diffusion of these molecules. Urea, glucose, amino acids, water, and various other solutes all move easily from the glomerular capillaries, through the inner epithelial layer of Bowman's capsule, and into the space between the capsule's inner and outer layers. Blood cells and large proteins are too big to pass through the capillary pores. This means that, as blood is forced into a porous, cul-de-sac of capillaries, it has a substantial amount of water and dissolved substances removed from it. The components of blood that do not pass into the Bowman's capsule exit the glomerulus and continue their circulation through the rest of the kidney. Keep in mind that it is arterial oxygenated blood that is being filtered because high pressure provides the force needed to push molecules through the pores. The blood that passes through the glomerulus remains oxygenated. Unlike conventional capillaries, oxygen is not exchanged in the nephron. Upon exiting the

Bowman's capsule a double-walled structure surrounding the glomerulus and into which the filtrate of blood flows

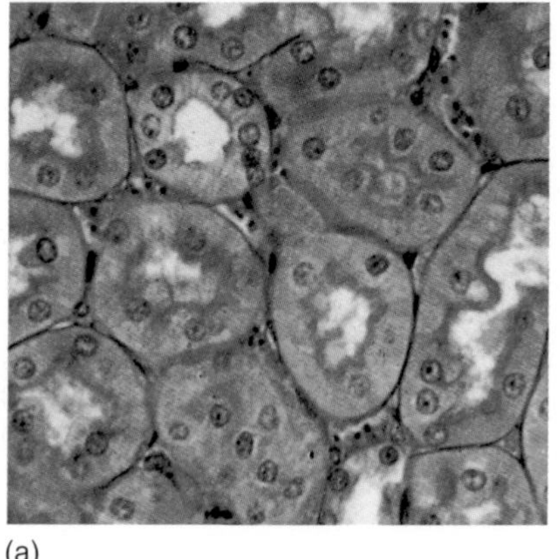
(a)

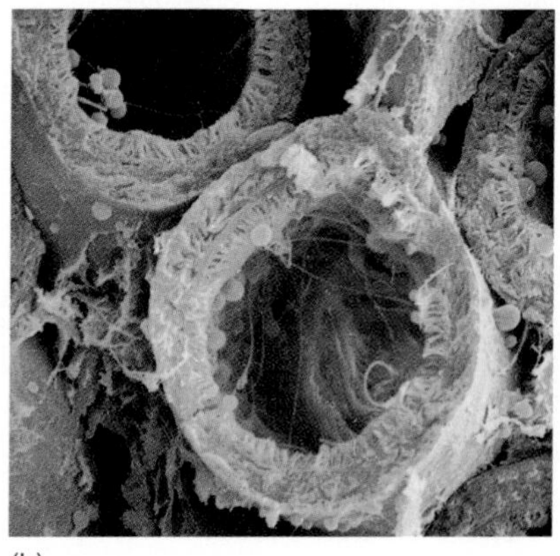
(b)

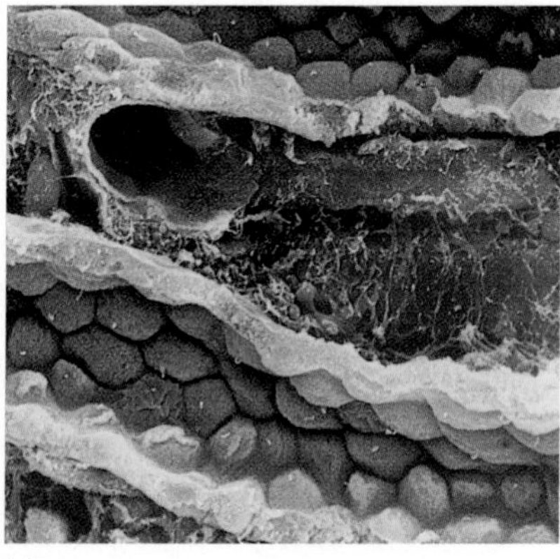
(c)

Figure 10–6 Tissue Organization of the Kidney
(a) Cross Section of Tubules (b) Magnified View of Tubules
The system of tubules is complex within the kidney, as shown in this cross section (a) of proximal and distal tubules and collecting ducts. The tubes twist and turn as indicated in the scanning electron micrographs shown in (b) and (c).

glomerulus, the blood goes on to deliver oxygen in the conventional manner to the cells in all regions of the kidney.

Proximal Convoluted Tubule Bowman's capsule is a double-walled epithelium. The filtrate from the glomerulus is trapped between the walls, but there is an exit at one end that forms a tube. The filtrate begins its flow through this tube, which constitutes the beginning of the rest of the nephron. The first region of this tubing is called the **proximal convoluted tubule.** (Figures 10–5 and 10–6). Proximal describes a relative anatomical position; in this case the proximal tubule is in closest proximity to Bowman's capsule. Many of the activities involved in the reclamation of useful molecules and solutes takes place in the proximal convoluted tubule. The initial filtration step is nonspecific in nature. The pores in the glomerulus and Bowman's capsule let everything of the right size pass through regardless of their physiological importance. The human body certainly has to get rid of the toxic materials to maintain homeostasis, but for the same reason it needs to retain glucose, amino acids, vitamins, water, and minerals for bodily well-being. The reclamation steps are very specific and very efficient. The epithelial cells of the proximal tubules selectively resorb beneficial materials from the filtrate, as do the epithelial cells of the other distal regions of the nephron. This material is transferred into the peritubular capillaries that surround the nephric tubules and, thus, is retained for the body. Out of the hundreds of liters of filtrate that are produced each day in the kidneys only about 1% exits the body as urine. Reclamation in the nephron and neighboring collecting tubules accounts for the 99% that is retained.

Reclamation What molecules do the cells of the proximal convoluted tubule actually reclaim? All of the glucose in the filtrate is resorbed in this region of the nephron. There should be little, or no glucose, in the urine that is excreted from the body. The presence of sugar in the urine is a sign of kidney malfunction and is observed in individuals with diabetes mellitus, for example. Vitamins and amino acids are also reabsorbed into the peritubular capillaries, following passage through the epithelia of proximal tubules, and returned directly to the blood. Eighty to ninety percent of the sodium and chloride ions, which are important in nerve and muscle activity, and most of the bicarbonate ions, which help maintain the pH of the blood, are also reclaimed for the body by the cells of this region of the nephron.

Balancing the pH of the blood is another important function of the kidneys. Fluctuations of blood pH beyond the normal, slightly alkaline range of 7.2–7.4 can be dangerous for the body. Acidosis (excessively low pH) and alkalosis (excessively high pH) are indications of serious problems in homeostatic mechanisms of the body. pH is determined in part by the presence of bicarbonate ions in the blood. Cells in the kidneys have specific transport machinery to regulate the movement of these ions.

By what mechanisms do all these ions and molecules get back across the membranes of the epithelial cells and into the body? It takes energy and active transport across plasma membranes to reclaim them. The energy is supplied in the form of ATP. The hydrolysis of ATP is coupled with the function of protein transmembrane carriers embedded in the plasma membranes of cells. The carriers specifically recognize, bind, and selectively transport molecules including glucose, amino acids, vitamins, and bicarbonate ions, as well as sodium and chloride ions. Sodium ions in particular are regulated closely. The transport processes are energy expensive but are essential as well as efficient. (Figure 10–7).

Water, on the other hand, traverses the plasma membrane passively by means of osmosis (Figure 10–8). Net movement of water molecules occurs in large part because sodium is transported through the cells and into the interstitial tissues and peritubular capillary system. The proximal and distal tubule cell membranes are readily permeable to water, as are the epithelial cells in other parts of the nephron. The epithelial cells of the lumen of the proximal tubules have a dense lawn of microvilli. The microvilli increase the surface area of absorption, which in turn influences the rapid and efficient reclamation of many beneficial constituents back into the body.

Loop of Henle The process of selective reclamation continues in the **loop of Henle** (Figure 10–7), which is the next region of the nephron. One of the main functions of the cells of the loop is to regulate the transport of sodium ions and water molecules out of the loop and into the interstitial fluids of the medulla. Keep in mind that during the kidney's reclamation of useful components there is no transfer of nitrogenous wastes out of the nephron. The retention and concentration of urine is the principal role of the kidney, even though it has many other functions as well. As indicated in Figures 10–7 and 10–8, the concentration of sodium ions drives osmosis. This part of the tubular system of the nephron descends down through the cortex, and into the medulla, and then ascends back up through the medulla into the cortex (Figure 10–5). It is particularly important that both the descending and ascending arms of the loop of Henle have different permeabilities to water molecules.

Proximal convoluted tubule tubule immediately following Bowman's capsule and responsible for resorption of nutrients

Loop of Henle an extended region following the proximal convoluted tubule in which urine is concentrated

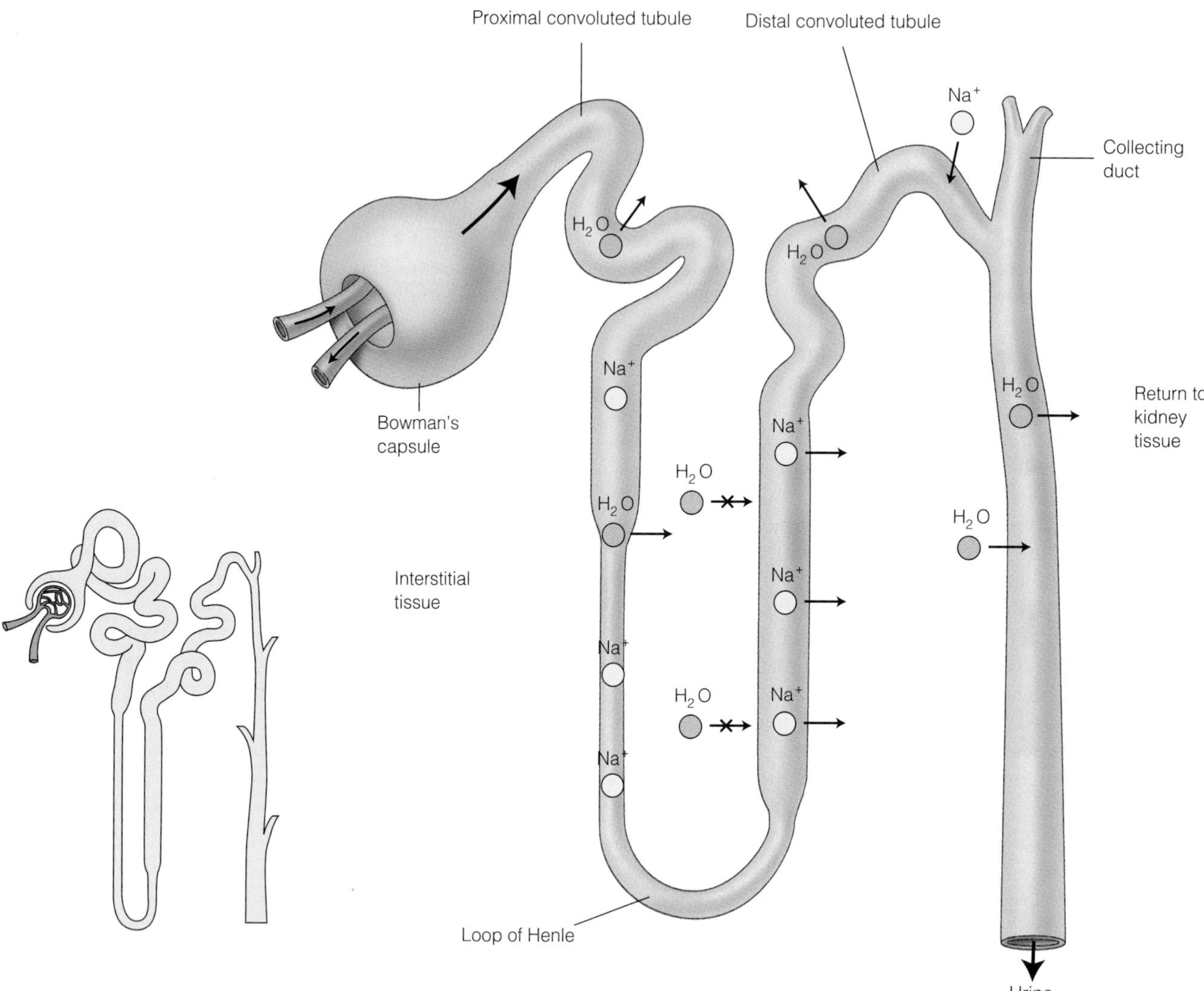

Figure 10–7 Sodium Ion and Water Movement in the Nephron

Sodium ions and water molecules are transported into and out of the tubular system of the nephron. Kidney cells in different regions are selectively permeable to water and sodium ions. Of particular importance is the movement of water out of the descending loop of Henle and the absence of active pumping of Na^+ in this region (→ indicates direction of movement; ↮ indicates a block to movement). The situation is reversed in the ascending loop. Water molecules are blocked and Na^+ is actively pumped into the interstitial tissue.

Water readily moves passively out of the descending loop, but is blocked from entering the ascending loop. However, sodium ions are actively transported out of the ascending loop. The sodium ions that are pumped out of the ascending loop move into the tissues surrounding the entire loop, and water molecules in the descending loop diffuse out in response to them. This osmosis increases the sodium concentration in the descending loop. The filtrate within the descending loop flows continuously into the region of the ascending loop, which continues to pump out more sodium ions. This sets up a physiological situation in which an ever-increasing concentration of sodium ions builds up in the entire loop. Ultimately, this buildup of salt concentration controls the amount of water available for resorption into the body.

Distal Convoluted Tubule The ascending part of the loop connects with the **distal convoluted tubule** in the

> **Distal convoluted tubule** follows the loop of Henle and continues the job of resorption of nutrients, especially water

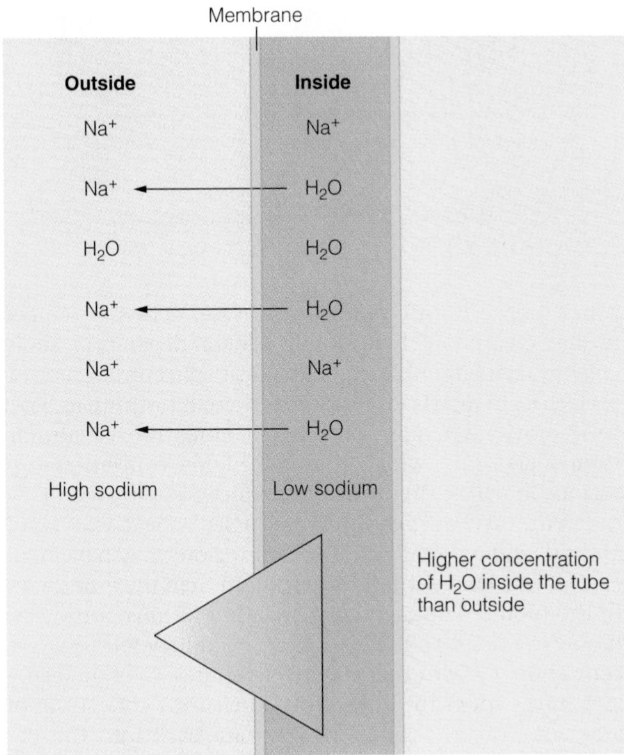

Figure 10–8 Osmosis
Osmosis is the movement of water molecules from a region of higher concentration (inside the tube as shown) to one of lower concentration. The movement of molecules creates an osmotic pressure. In this case, the membrane is permeable to water but impermeable to sodium ions.

kidney cortex (Figures 10–5 and 10–6). The twisting and turning of convoluted regions increases the length of tubing and the overall volume and surface area of the nephron. The increase in the inner surface area of exchange makes the resorption and concentration process more effective. The distal tubule is involved in sodium regulation and water resorption. The cells of the distal tubule pump out sodium ions and are permeable to water.

Collecting Tubules The end of the distal convoluted tubule connects to *collecting tubules* (Figure 10–5). Each collecting duct services several nephrons, and in the collecting tubules the process of concentrating the urine is carried out in earnest. Cells of the collecting tubule are usually impermeable to sodium and other ions. However, high ion concentrations in the medulla force water molecules to diffuse through the epithelial cells lining the duct and back into the body tissues by osmosis. This serves finally to reclaim needed water and to concentrate the urine with respect to urea and other molecules and ions. Throughout this process urea has remained in the filtrate within the lumen of the nephrons and collecting tubules.

FILTRATION AND OSMOSIS

The anatomical and physiological aspects of kidney function of concern to us in this chapter are **filtration,** *osmotic balance,* and *excretion* from the body. Filtration involves the removal of toxic substances from the blood. Osmotic balance ensures the body's capacity to remain hydrated. Excretion provides the means to remove urine from the body. The filtration occurs under the high blood pressure of the renal arteries, which branch from the aorta, as the aorta passes down through the abdominal cavity (Figure 10–1). The renal arteries carry up to 20% of the heart's systemic blood output during any one cycle. The filtration of the blood by the kidneys is brought about as the plasma of blood is forced through sievelike cells that allow passage of water and small molecules but retain blood cells and large molecules in normal circulation. Without a high blood pressure, the kidneys do a less efficient job of filtration. In fact, if blood pressure gets too low, the kidneys may fail, and toxic levels of waste products will build up in the blood.

As if it were a factory operating at high capacity, the human body produces an enormous amount of waste. This is to be expected from an active organism, whose metabolic needs demand the constant breakdown of energy-rich molecules to supply energy and nutrients to maintain itself. Even in its most relaxed state, the human body burns fuel to keep the heart beating, the lungs breathing, the digestive process going, and the brain functioning. The burning of any kind of fuel creates waste. Car engines burn gasoline, or other hydrocarbon fuels, and in the process produce carbon dioxide and carbon monoxide along with many other types of waste molecules. Nuclear reactors use the energy of atoms, but produce dangerous radioactive and thermal waste in the process. The fuels of the human engine are carbohydrates, lipids, and proteins. The catalytic breakdown of these molecules results in the formation of carbon dioxide and nitrogenous waste products, such as urea. Carbon dioxide and urea have adverse effects on cellular metabolism, if they become too concentrated. It is imperative that they be eliminated from the body, lest they do us harm. As discussed in Chapter 8, carbon dioxide is eliminated by release into the atmosphere with every breath we take. The nitrogenous wastes are predominantly

Filtration the action of separating components of the blood on the basis of size

the product of the breakdown of nucleic acids and proteins. These wastes must be disposed of in a different way. The kidney has a major role in this process.

A second essential function of the kidneys is to maintain the body's fluid and osmotic balance. This means that water is either retained or eliminated depending on body demands. As we have seen, the processes associated with water regulation take place in a part of the kidney relatively distant from the location at which initial filtration of blood occurs and is subject to relatively low-pressure conditions. The control of water is linked to the level of solutes in the blood and body fluids. The regulation of solute levels in the body, particularly sodium ions, by the kidney is reflected in the volume of urine produced by us each day. Urine is a combination of water, solutes, and nitrogenous waste products, the major component of which is urea. Urea is concentrated by the action of the kidney, and flushed from the body through the urinary system, which provides the leak-free plumbing necessary to carry waste out of the body.

Table 10–2 Sources of Nitrogenous Wastes

Protein
Amino groups and nitrogen containing side groups of amino acids are broken down into ammonia and converted into urea and/or uric acid.
Nucleic acids
Nitrogen atoms in the rings and side groups of nucleotides are broken down to uric acid.

URINE PRODUCTION

The amount of urine produced each day depends upon many factors, including how much water we consume, relative to how much water the body needs. It is also influenced by what we eat (proteins, salt), by the temperature of the environment in which we live, and by the level of physical activity (Table 10–2). These last two factors influence body mechanisms used for evaporative cooling of the body. Heavy sweating on a hot day puts demands on us to consume more water. This increased consumption does not necessarily increase the volume of urine. To provide sufficient water to accommodate continuous evaporative cooling of the body, the kidney must claim more of the water passing through the nephrons and collecting duct system. Normally, however, the greater the water intake, the larger the volume of urine produced. It also follows that the larger the volume of urine produced, the less concentrated it is. Less concentrated urine is lighter in color. Urine ranges in color from nearly colorless to dark yellow. The yellow color of urine results from the small amount of bile pigments contained in it, which are more concentrated in a smaller total volume. On average, humans excrete one to two liters of fluid a day, but the kidney may produce as much as 100 to 200 liters of filtrate during the same period. That accounts for an absorption efficiency of 99% or more.

WATER AND SOLUTES

Most of the movement of water molecules in the tubules and collecting ducts of the kidney is by osmosis, which is characterized by the diffusion of water across a semipermeable membrane. Diffusion in general depends on molecular movement and on a gradient of concentration of the molecules in question. As long as there is nothing to block their movement, molecules of all kinds move naturally from a region in which they are highly concentrated to regions in which they are not as concentrated. Water is no different. As has just been discussed, the tissue fluids around kidney tubules in different regions may have higher, or lower, concentrations of sodium than the fluids within the tubule. The high concentration of ions outside the tubule in the interstitial tissues of the kidney sets up a concentration gradient that effectively dilutes water molecules and forces them to move down their own concentration gradient from the glomerular filtrate back into the peritubular capillaries and the interstitium. As observed in the proximal tubule, sodium ions are actively transported out of the filtrate and into the body fluids. As a result of the movement of the salt into the interstitial tissues, water diffuses out of the tubules also. The relationship between sodium ion concentration and water movement is very important for kidney function.

In the case of the ascending arm of the loop of Henle, the epithelial cells are impermeable to water, even though sodium ions are actively pumped out of the tubule. This all changes in the distal convoluted tubule, where the direction of both sodium and water movement is into the interstitial tissues. The final arbiter of sodium and water movement is the collecting tubule. Sodium ion transport is blocked in this region, but water can move out (Figures 10–7 and 10–8). Because the fluids in the tissues surrounding the collecting tubule are of higher osmotic strength than the filtrate, water generally flows out and leaves behind a concentrated urine.

But this is not always the case. For instance, water will be retained in the filtrate if we drink a lot of water over a short period of time. Alternatively, if we do not have any water or conditions are such that we need to retain body water to survive, water will be efficiently resorbed. Consider the desert scenario presented at the beginning of the chapter, in which the prospects of getting water to drink are low. The way in which the body regulates the amount of water that is retained or eliminated by the kidneys depends on hormones.

HORMONE REGULATION OF WATER, SODIUM, AND POTASSIUM

ADH

There are special nerve cells in the brain that have receptors capable of monitoring the osmotic strength of blood as it circulates. These nerve cells are located in the **hypothalamus,** which is a small collection of nerve cells located in the middle of the brain, in the neighborhood of the **pituitary gland** (Figure 10–9). In fact, the nerve cells of the hypothalamus have cellular extensions or processes that pass directly to the posterior lobe of the pituitary gland. Stored in the hypothalamic cell processes is a hormone known as *antidiuretic hormone,* or *ADH.* This hormone is an oligopeptide and has profound effects on the *resorption* of water in the distal convoluted and collecting tubules.

Consider the following sequence of events. As the body becomes dehydrated (loses water) and the salt concentration of the blood increases (sodium ions), the nerves in the hypothalamus are stimulated to release ADH from the posterior lobe of the pituitary gland. The hormone enters the blood circulation and travels in the blood stream throughout the body. In the kidney, ADH interacts with cell surface receptors on the tubule cells and changes their behavior. This change alters the permeability of the tubule cells to water. Permeability goes up dramatically under the influence of ADH and water is resorbed by the body much more efficiently than in the absence of the hormone. The urine becomes very concentrated, deep yellow in color, and reduced in volume.

Hypothalamus region of the brain controlling vital involuntary functions, particularly the regulation of hormone secretion of the pituitary gland

Pituitary gland the primary endocrine gland; consists of two lobes, anterior and posterior, that secrete a variety of protein or peptide hormones.

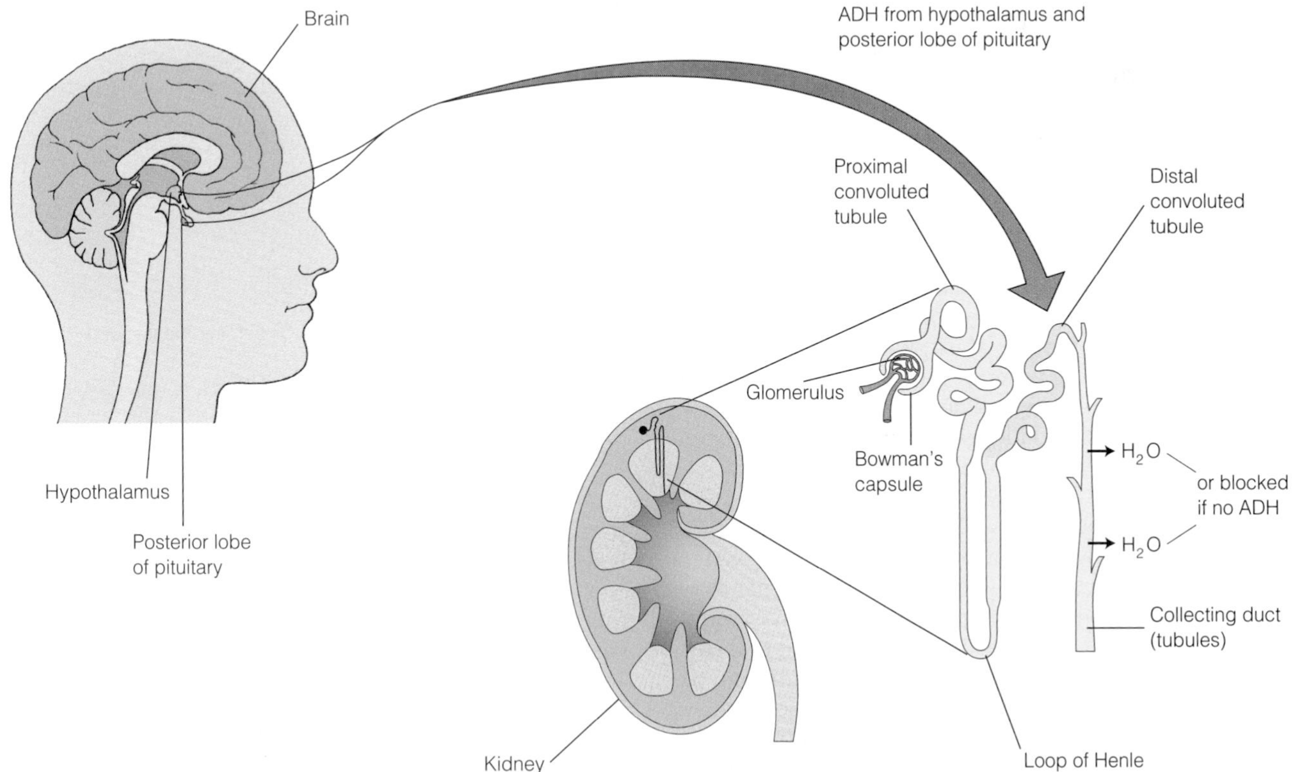

Figure 10–9 The Brain, the Kidney, and Antidiuretic Hormone (ADH)
ADH is produced and released from the brain. ADH increases the resorption of water into the body, and in its absence water is excreted in dilute urine.

In the reverse situation, imagine that you have just had two big glasses of water and your body is well hydrated. The nerve cells in the hypothalamus remain silent, because the concentration of ions in the blood is below the threshold level to stimulate the release of ADH. Without elevated ADH, the kidney tubules are less efficient at resorbing water (or do not resorb water at all). The water in the urine in the distal and collecting tubules remains there and the urine is relatively dilute, pale yellow in color, and greater in volume. This hormone-regulated responsiveness to changes in the **osmolarity** of the blood is key to maintaining homeostasis in the body. The negative feedback pathway between kidney and hypothalamus, mediated by osmolarity and hormone release, controls the composition of the body's interstitial and circulatory fluids, which vary little from moment to moment.

To summarize this feedback loop, we observe that as the osmolarity of the blood increases, the brain releases a hormone that causes the kidney to transport water in the tubules and collecting ducts back into the body. When the osmotic strength of the blood decreases, as a result of more water coming into the body tissues, the hypothalamus turns off the flow of ADH, and more water is retained in the tubules and eliminated from the body. When things go wrong with the regulation of this process, as when the production of ADH is abnormally inhibited, or eliminated, the body is in serious trouble.

Diabetes Insipidus One such situation is a rare disease known as diabetes insipidus. ADH in these individuals is found to be in insufficient supply to activate the resorption machinery in the kidney tubules (Figure 10–9). In the absence of ADH, the body does not resorb water efficiently. There is a tremendous increase in the production of dilute urine and a general dehydration of the body. The dehydration leads to an increase in thirst, which in turn leads to consumption of more water to satisfy the thirst and, further, the overproduction of urine. It is not unusual for a diabetic of this type to produce gallons of urine a day instead of the normal volume of approximately one liter!

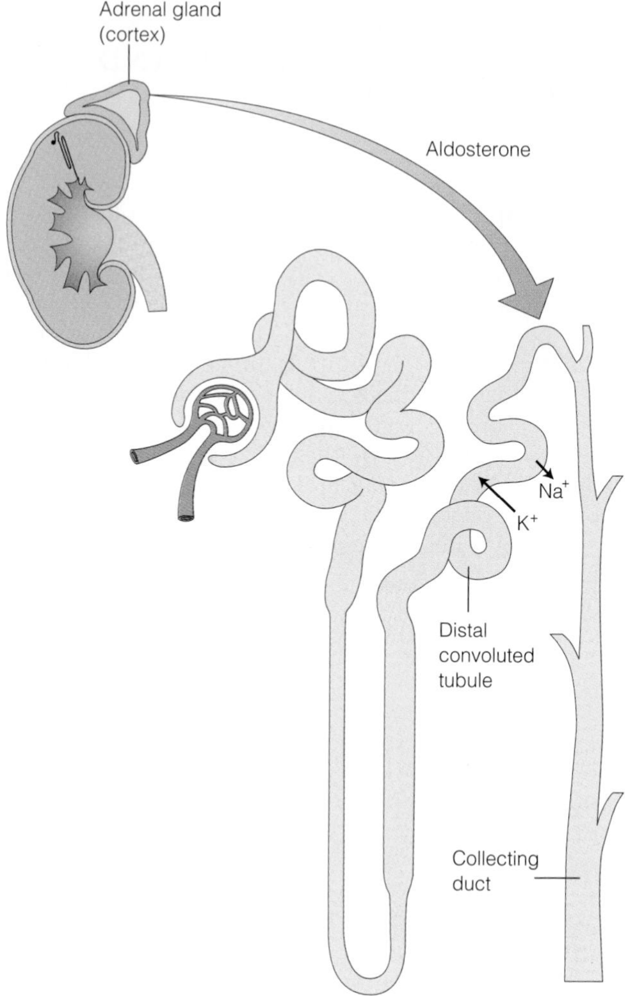

Figure 10–10 The Adrenal Gland, the Kidney, and Aldosterone

Sodium and potassium are important ions in the human body. Aldosterone regulates the resorption of sodium and the excretion of potassium to maintain ionic balance in the body.

Aldosterone, Sodium, and Potassium Ions

The movement of water into and out of the tubule system of the kidneys is tied to a number of different demands on the body. One of the most important, as we have seen, is the amount of sodium in the filtrate of the blood that passes through the kidneys and in the tissues that surround the kidney tubules. Water moves along an osmotic gradient, and sodium ions provide the basis for establishing that gradient. Sodium is the most abundant cation in human tissue fluids, making up 90%, or so, of all ions in these fluids. As you might have already suspected, there is also a hormonal mechanism to control the level of sodium ions transported out of or retained by the kidneys for use by the body (Figure 10–10).

The body needs sodium ions for a variety of cellular functions, but of particular importance are tissue water balance and the electrical activity of the nervous system. Too little sodium in the body can lead to headache and

Osmolarity the property of a solution determined by the concentration of ions dissolved in it

muscular weakness and, if sufficiently low, to shock, mental aberrations, and coma. Too much sodium can lead to hypertension and cardiovascular problems. The body maintains a concentration of sodium ions in the blood and tissues to within 1%–2% of a set, normal value at all times. This means that the excess salt that we take in with the food we eat, the content of which can be very high in some foods, must be eliminated quickly and efficiently. The hormone *aldosterone* plays a primary role in this process.

Aldosterone is a steroid hormone produced by cells of the cortex of the **adrenal glands** (Figure 10–10). Sodium loss from the body occurs from the activities of the kidneys and sweat glands as already described (see also Chapter 5). As the level of sodium in the interstitial tissues and blood drops, aldosterone is released into circulation and targets the cells of the distal convoluted tubule. Aldosterone elicits an increase in the efficiency of the export of sodium ions from the glomerular filtrate and, thus, resorption of these ions back into the body. Aldosterone is also released in response to low blood volume, which increases sodium ion concentration. This results in enhanced water reclamation by the kidneys for the body.

Aldosterone release is also stimulated by an increase in the level of potassium ions. Potassium ions are also of great physiological importance for the body, and their concentration in blood is closely controlled. Excessive fluctuations of this ion can lead to paralysis, mental confusion, and changes in the electrical activity of the heart. Potassium, like sodium, is also regulated by aldosterone, but in just the opposite way that sodium is. If potassium levels in the body go up, aldosterone will increase the excretion of potassium from the cells of the distal convoluted tubules into the filtrate, while at the same time increasing the resorption of sodium ions into the body. The actions of aldosterone are not completely understood in humans but are of fundamental importance, as is apparent in the case of its reciprocal effects on potassium and sodium ions in the body.

POLLUTING THE WATERS

Uremia

What happens if the kidneys do not work properly? We have seen how hormones can influence the function of these organs, but what if the urea that the kidneys are supposed to concentrate in the urine is not efficiently eliminated? Such conditions are devastating to the health of the body. Urea, ammonia, and other nitrogen-rich molecules are the waste products of the body's metabolism and are particularly toxic to us. (Biosite 10–1). If these compounds are allowed to build up in the blood, they cause a condition known as uremia. Uremia is a serious problem for human beings (and all mammals), because we carry around a fixed volume private ocean wherever we go. It cannot be allowed to become polluted with toxic metabolic wastes. Marine and fresh water organisms (from sponges to fish) pump wastes out of their bodies and into the surrounding sea to be diluted to inconsequentially low levels. However, when aquatic organisms are restricted to relatively small volumes of water, their environments may also become toxic to them. This is why we periodically change the water in our

Adrenal glands endocrine glands located atop the kidneys; produce steroids and catecholamines such as adrenaline

10–1

NITROGENOUS WASTE

Ammonia, urea, and uric acid are the three most common forms of nitrogenous waste produced by animals. These compounds are toxic in even low amounts in the bodies of the species that produce them. Therefore, they must be removed on a continuous basis. Ammonia is the most toxic of the three. The human body expends considerable energy preventing buildup of ammonia through the formation of urea, which is less toxic, and, therefore, easy to store. The kidney is the main organ of excretion of urea and uric acid in vertebrates. The same is true for uric acid, which is eliminated by many organisms in a semi-solid form. The origin of all three of these molecules is the *catabolism* (or breakdown) of proteins and nucleic acids, which are rich in nitrogen atoms. The body has a specific metabolic pathway, referred to as the urea cycle, that converts nitrogen-rich molecules into urea for elimination by means of the kidneys.

fish tanks. Survival of our finned friends depends on massive dilution of the toxic wastes they produce. On a larger scale, a polluted lake or bay is hazardous to the health of all its inhabitants, as well as to humans who consume or use its waters. This is certainly analogous to what happens to a body with the buildup of internal toxic wastes. It, too, becomes uninhabitable. Uremic poisoning results in lethargy, mental confusion, coma, and ultimately death.

How does uremic poisoning occur? Generally, uremia arises when the kidney fails to function properly. If insufficient blood volume is filtered or the filtering capacity of the nephrons is reduced, urea simply passes out of the glomerulus and continues along the circulatory path with everything else in the blood. Diseases that directly affect the function of nephrons within the cortex and medulla of the kidney are referred to as **nephritis.** There are many kinds of nephritis, the worst being those that completely shut down the filtration system. These cases, if left untreated, result in the relatively rapid demise (within days) of a human being. However, with the help of modern medical technology not all is lost. There is a procedure that offers a chance to filter the blood without having normal kidney function. This procedure involves vascular hook-up of an individual to an artificial kidney machine, and a process called **dialysis** (Figure 10–11).

Artificial Kidneys and Dialysis

Dialysis takes advantage of several important features of the blood and circulatory system. First of all, urea is a very small molecule. Its size is the basis for passing through the glomerular capillary pores in the first place. *Pores* is the operative word here. Imagine the construction of a material that had extremely small pores in it that would allow molecules of a very small size to go through but would not allow bigger molecules or cells to pass. There is such a material, and it is known as *dialysis membrane* (often in the form of tubing). Imagine further that we can provide a detour for blood to the outside of the body; and then deliver it back again. That's easy enough: simply insert a needle into a fairly large artery (the radial artery of the arm, for instance) to withdraw blood, and insert a needle further downstream to return blood to the circulatory system.

In this scenario the blood flows from the body through the dialysis tubing and back into the body after the urea and other small molecules have escaped through the pores. Whoops. This is where a problem arises. We need most of the small molecules and ions back in the body. That's what the kidney does so well; it actively transports, or uses osmosis, to pull all the good stuff back and leaves the harmful materials to be eliminated. For this to occur in the artificial kidney machine, the dialysis tubing must be bathed in a solution that contains all the good ingredients of the blood, and none of the bad. The dialysis tubing is semipermeable, which means that small molecules can move in, or out, but not large ones. In addition, molecules and ions will diffuse from a high concentration (in the tubing) to a low concentration (outside the tubing). That is how the urea is eliminated. However, if glucose, sodium, potassium, vitamins, amino acids, bicarbonate ions, and many other types of molecules and ions are put into the dialysis solution at the same concentration that they are found in blood, and at a physiologically appropriate pH, then the fluid that is returned to a body after dialysis will be nearly exactly the same as the fluid that left. What is lost? Urea and nitrogenous wastes. Kidney dialysis has saved many lives over the last few decades and made the quality of life much better for those who suffer from debilitating kidney diseases, such as nephritis, which destroy the kidney's capacity to filter the blood.

PLUMBING AND THE FLOW OF URINE OUT OF THE BODY

The Ureter

Now that the function of the kidneys has been described in some detail, let's turn our attention to the system by which urine is collected, stored, and removed from the body. Specifically, what happens to the urine that flows out of the collecting ducts? After beneficial substances such as glucose, vitamins, water, sodium, and potassium have been resorbed into the body, the remaining highly concentrated urine begins its journey out. The distal convoluted tubules drain into the collecting tubules that traverse the kidney cortex and medulla. The collecting ducts pass through the renal pyramids and open, by means of the papillary ducts, into the renal pelvis. The urine that has collected within the pelvic cavity of each kidney drains into the ureters, which connect them to the urinary bladder (Figure 10–12). Each ureter is approximately 25 cm–30 cm long. The cellular lining of the pelvis and ureter form a tight and occluding epithelium that allows none of the urine to diffuse into the surrounding tissues. This would clearly defeat the purpose of the previous filtration process. The epithelium of the urinary system is one of the least permeable in the human body. Peristaltic movements and gravity help carry the urine down the ureter, where it empties into and is stored within the expandable urinary bladder.

Urinary Bladder

The urinary bladder also is lined with a specialized epithelium, which forms a tight epithelial seal. The epithelial lin-

Nephritis a disease associated with the kidney function, particularly with respect to the nephron

Dialysis procedure in which blood is cleansed in an artificial kidney machine

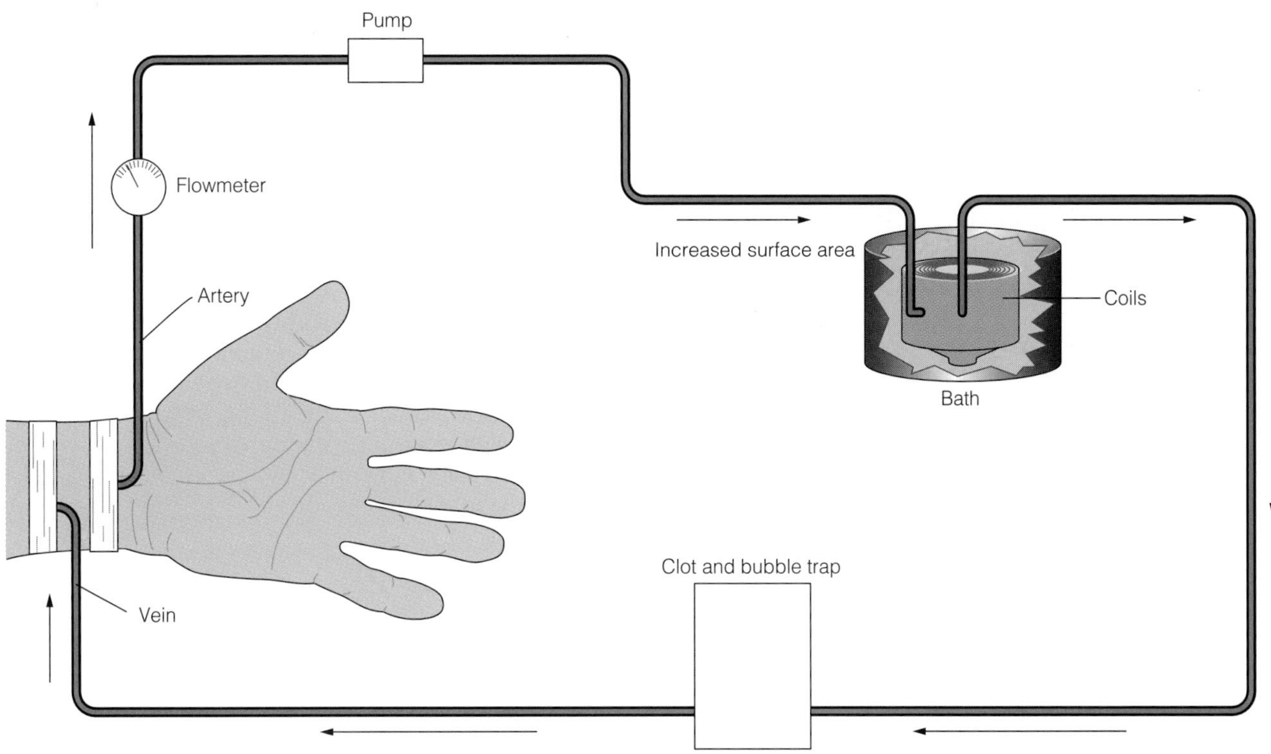

(b)

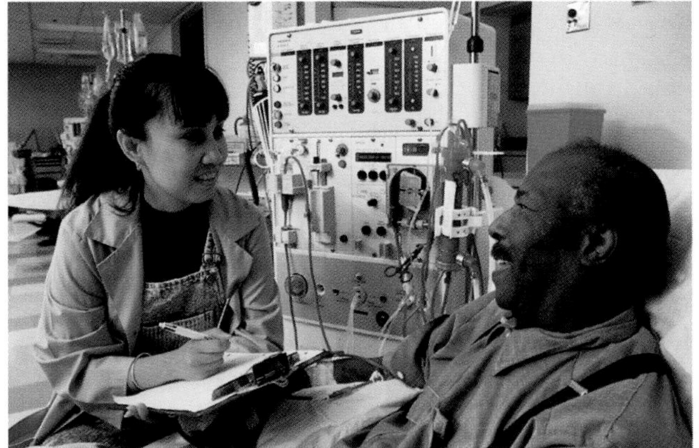

Figure 10–11 Kidney Dialysis
Kidney dialysis redirects blood from an artery into a machine that separates urea and other metabolic wastes from the blood. Nutrients and needed ions are replaced and blood flows back into body circulation (a). In (b) an individual hooked up to a kidney dialysis machine is shown.

ing of the bladder is called a **transitional epithelium** (see Chapter 4 for details). The unique aspect of the epithelium of this organ is that it is capable of considerable expansion, without damage. The cells of this stratified epithelium are roughly columnar in shape when the bladder is empty (Figure 10–13). However, as the bladder expands under the pressure of urine volume building up, the epithelium changes appearance. The cells that appeared columnar become elongated and squamous in shape. The empty urinary bladder looks something like a deflated balloon. It becomes spherical as it fills up and stretches the transitional epithelium. The overall capacity of the bladder is approximately 800 ml, but pressure is felt

Transitional epithelium specialized epithelium whose cells undergo shape change (flattening) during stretching, as in the urinary bladder

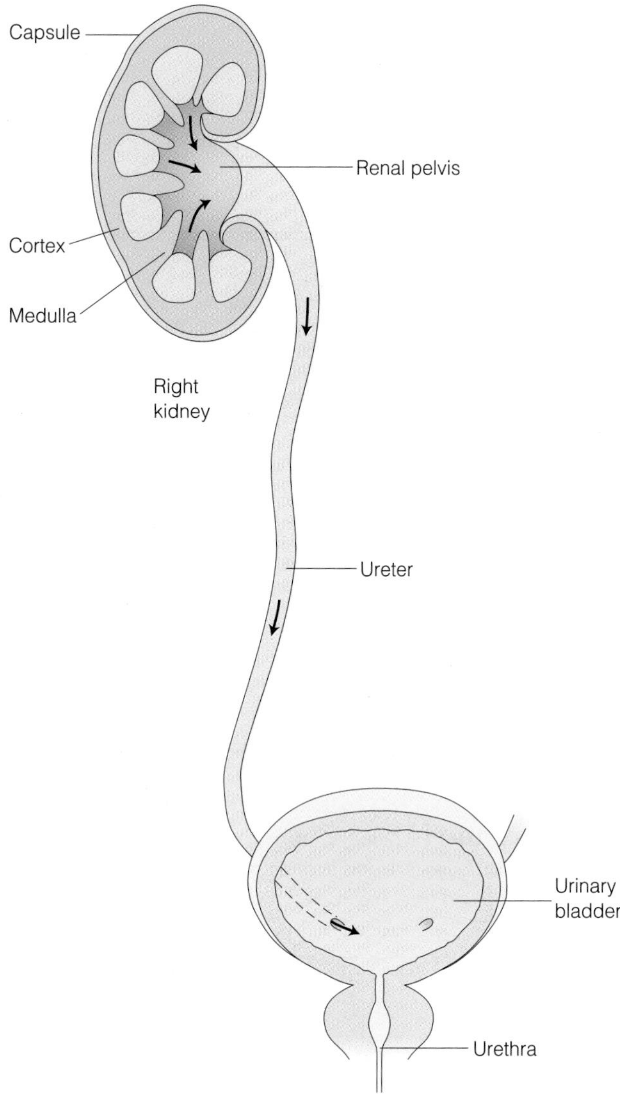

Figure 10–12 The Ureter and Urinary Bladder
Urine produced in the kidney flows from the renal pelvis into the ureter and is stored in the urinary bladder. Controlled release from the urinary bladder occurs through the urethra.

with a fluid build up of about 200 ml. A signal to the brain from stretch receptors in the bladder wall initiates the desire to urinate.

Internal and External Sphincters

There are two muscular sphincters associated with the output side of the bladder, which are connected to the urethra. There is an *internal urethral sphincter* and an *external urethral sphincter* (Figure 10–13). The internal sphincter is under involuntary control. The muscles contract under the influence of nerve signals generated in response to the stretch receptors in the urinary bladder wall. As a result, the internal sphincter opens. An individual has no control over this activity and, if it were the only means by which urine retention was controlled, our lifestyles would certainly be more complicated. As it is, however, the external sphincter is under voluntary control. Control of these muscles is what we learned in infancy, when we were potty trained. The voluntary control may be lost, or reduced in efficiency from nerve damage, such as back injury or stroke, or as a function of aging. For whatever reason, the inability to voluntarily control the release of urine from the bladder results in a condition known as **incontinence.** While this is not life threatening, it is certainly embarrassing and undesirable. Normal control of the external sphincter gives us considerable latitude in the timing of when we urinate. We can usually wait until appropriate social and hygienic circumstances are present. However, as is said so often, "when you gotta go, you gotta go."

Urethra

The final part of the journey out of the human body takes place through the urethra, which is a small tube exiting the urinary bladder and connecting to the outside (Figure 10–13). The passage of urine along this single exit tube from the urinary bladder is controlled by the voluntary and involuntary muscles of the urethral sphincters. Once both the internal and external sphincters are open, the pressure from the filled urinary bladder initiates and sustains the flow of urine out of the body. In females, as depicted in Figure 10–13(a), the urethra is relatively short and exits the body in the upper region of the *vestibule,* above and separate from the *vagina.* In males, the urethra is considerably longer (Figure 10–13[b]). It passes through the entire length of the *penis,* and shares that pathway, at different times and under different conditions, with *seminal fluids* produced by the male reproductive system.

DISEASES AND PROBLEMS IN THE PLUMBING

Obstruction

Problems of blockage of urination are serious and painful. If the urinary bladder cannot be emptied, for instance, as a result of physical blockage, it screams out a painful message to the brain. In fact, medical intervention may be necessary

Incontinence inability to control the urethral sphincters and, thus, to control urine flow from the urinary bladder

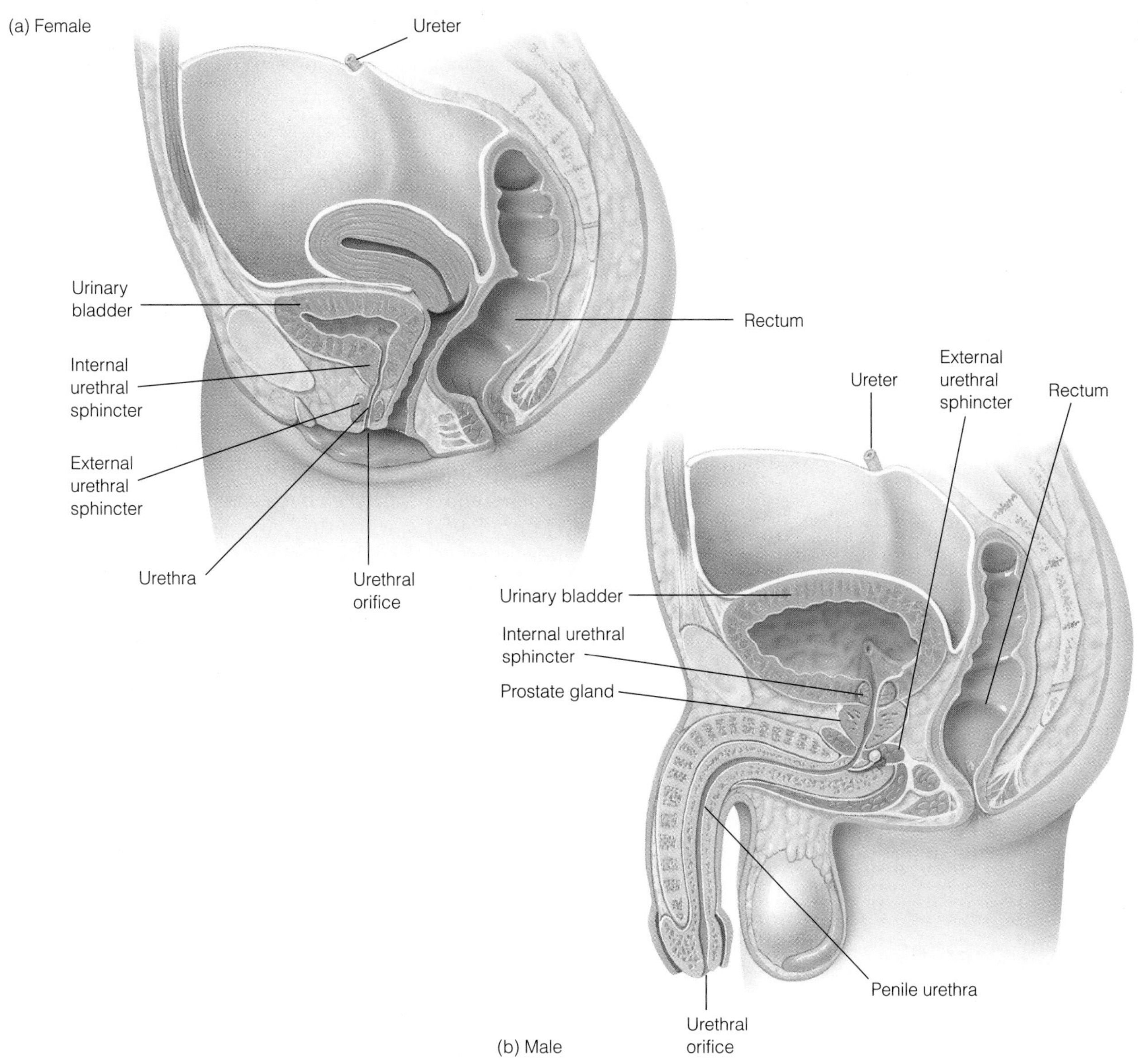

Figure 10–13 The Urinary Bladder and Urethra
The urinary bladder is the repository of stored urine. Periodically, the bladder is emptied through the urethra. The urethra in males is approximately 20 cm long. The urethra in females is approximately 4 cm long.

to drain the bladder, lest it burst, or tear open. **Catheterization** (Figure 10–14) is the procedure by which the urinary bladder is drained. To carry out this procedure, a thin tube is inserted into the urethra, gently forced through the area of the obstruction and into the bladder. An external obstruction in the flow of urine often occurs in older males as a result of abnormal growth of the *prostate gland,* which compresses, or pinches, the urethra as it extends from the urinary bladder. The prostate gland is part of the male reproductive tract, and it surrounds the urethra at this point. Swelling of the prostate due to infection, inflammation, or

> **Catheterization** the insertion of a tube into the urethra with the intent of draining the urinary bladder

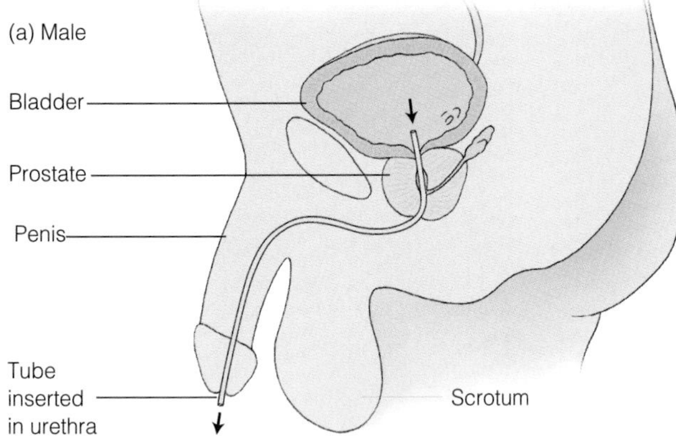

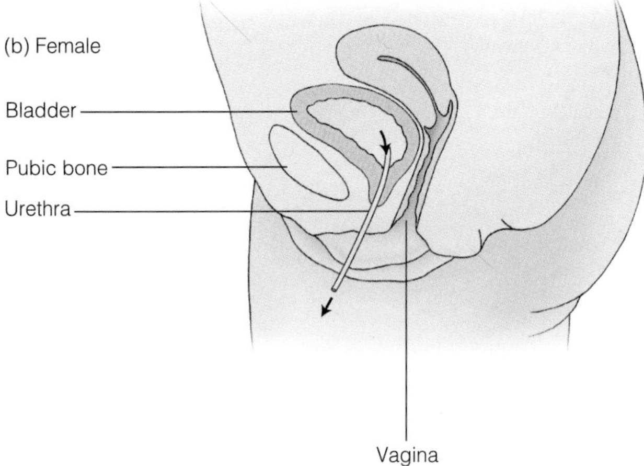

Figure 10–14 Catheterization in Males and Females
Catheterization is the process by which a tube is inserted through the urethra into the urinary bladder to collect or release urine. The urethra of males is longer than in females, so catheterization is somewhat more complicated.

cancer (see Chapter 21) may squeeze the urethra shut, like a most unnatural sphincter. Surgical removal of part or all of the prostate gland is often required to permanently eliminate the obstruction.

Stones

There are also internal disorders of both males and females that are capable of painfully blocking the flow of urine. These usually occur farther up the urinary tract in the region of the renal pelvis and ureters. This blockage is caused by *renal calculi*, or **kidney stones** (Figure 10–15). Kidney stones form as a result of crystallization or precipitation of various types of minerals in the urine. These crystals usually occur as a result of high levels of mineral intake, a decrease in water intake, metabolic changes in the pH of urine, and/or altered hormonal regulation of calcium and other minerals. The stones may have different mineral compositions, but they usually involve uric acid, calcium oxalate, and/or calcium phosphate. The stones are basically insoluble and, when they grow large enough, can block flow of urine. In addition, kidney stones can move within the tubules and cavities of the kidney and, when they do, they can cause tremendous pain.

Lithotripter

Stones usually form in the pelvis, where they can fill up much of the surface of the cavity. The *staghorn calculus* is an example of this type of stone formation (Figure 10–15b). Some forms of stones are jagged in shape, and, as they move into lower parts of the urinary system, that is, pelvis to ureter, they can jab the epithelial lining and cause pain, damage, and bleeding. Surgery may be called for in some cases, but in other cases nonsurgical techniques have

> **Kidney stones** mineralized structures formed in the kidney that may block urine flow, damage delicate epithelia, and cause pain; also known as renal calculi

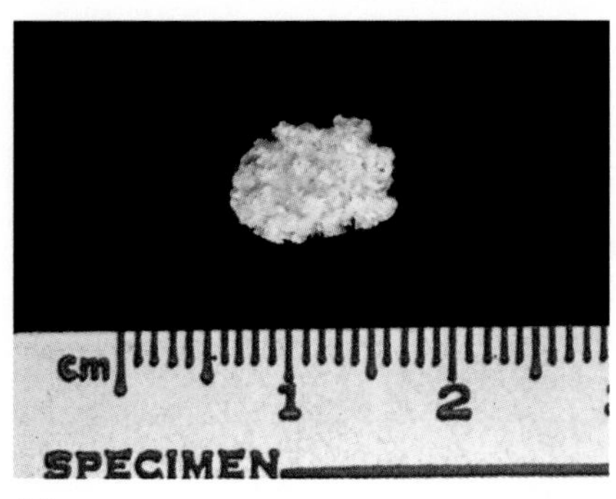

(a)

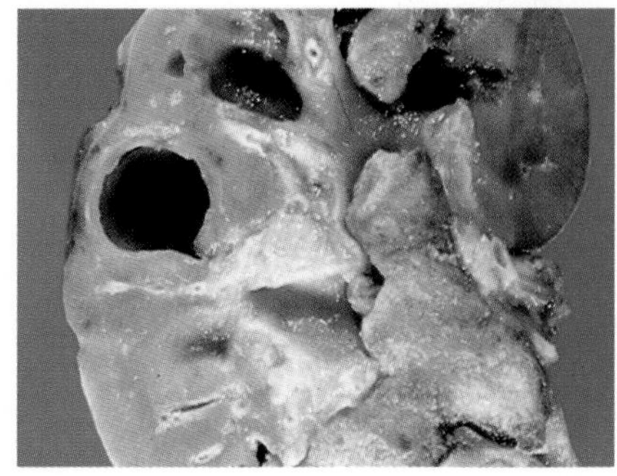

(b)

Figure 10–15 Kidney Stones
Kidney stones form when minerals do not remain dissolved in urine. Stones can be small and smooth (a) and pass naturally out of the kidney. Larger stones or stones that are not smooth may not be passed easily and may be extremely painful when they are passed. Layers of calcified material (calculus) may form in the renal pelvis and block urine flow.

proven successful. One nonsurgical procedure involves the shattering of stones within the intact kidney by the application of ultrasound waves generated by a machine known as a **lithotripter** or stone-smasher (Figure 10–16). A patient to be treated in this way is immersed in water, and the location of the kidney stones are targeted by computer imaging. The ultrasound waves, directed at the stones, shatter them into small pieces, which may then pass through the pelvis, ureter, bladder, and urethra, and exit the body. Oh what a relief it is! However, people who have had stones tend to be predisposed to forming them. It behooves those who have suffered from kidney stones to change their drinking habits or diets to reduce the chances for their reformation.

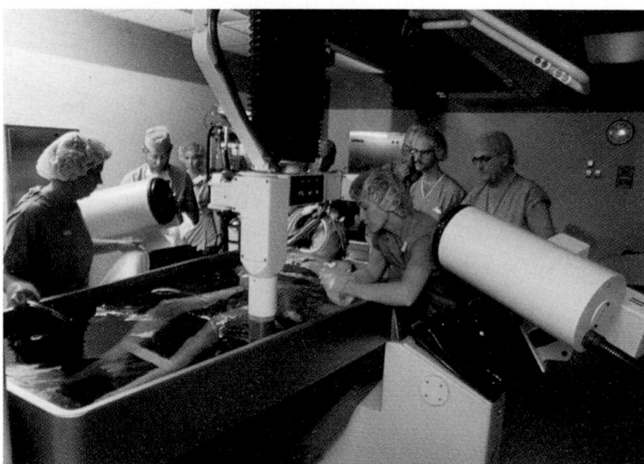

Figure 10–16 Lithotripter
The lithotripter, or stone smasher, works by using sound waves to break stones within the kidney without surgical intervention. This procedure only works on certain types of kidney stones, however.

KIDNEY TRANSPLANTS

Transplantation of major organs from one individual to another nearly always suffer the problems of *tissue rejection*. The body of an individual receiving an organ from a donor reacts against that tissue, or organ, as if it were a foreign invader. This usually leads to the complete destruction of the organ (that is, rejection) and the demise of the host individual. Advances in the understanding of tissue and organ **histocompatibility,** however, have greatly increased the success of organ transplants. One of the real success stories of organ transplantation has been in kidney transplants. For the purposes of this discussion, assume that we have found a nearly perfect histocompatibility match for an individual with severe kidney disease. This means that the molecular characteristics of the cells of the organ to be transplanted are nearly identical to those of the host (see Chapter 12 for more details). Because the human body can function normally on just one kidney, this sets up the possibility of receiving a kidney from a live donor (something that is not possible for heart or lung transplants). The donor has only to be willing to give up one of his or her kidneys (Figure 10–17). As you might imagine, such a donation is the subject of serious consideration for the donor. It is a major operation to remove a kidney, and as with any such procedure, it can be dangerous. For this reason the donor is usually a family member or a close friend.

Lithotripter literally stone smasher; a machine that uses vibrations to destroy certain types of kidney stones

Histocompatibility the mutual tolerance among tissues of two individuals that allows transplanting; tissue differences establish the molecular basis for tissue rejection

Figure 10–17 Kidney Transplant Surgery
Kidney failure is lethal. A short-term measure that can be taken is kidney dialysis, but ultimately, and when possible, a kidney must be replaced. Transplantation surgery is often successful and the individual receiving the transplant can function normally within a short period of time.

You may ask, however, "If dialysis works so well, why does a person with bad kidneys need a transplant?" This is a good question because dialysis can work quite well—for a time. The problem that arises is that the continuous penetration of arteries (ultimately using up sites in both the arms and legs) eventually leads to their degradation and destruction. Replacement vessels can be and often are used in transplantation, but this too becomes more and more difficult over time. Imagine dialysis every three to four days for the rest of your life. This means poking large needles into blood vessels a 100 times or more per year. Not only that, but the procedure ties you to a machine for several hours each time you hook up, and you may only feel good (toxins removed) for a day or two. Uremia is not a pleasant condition. The replacement of a kidney provides the individual with independence once again, freedom from the metabolic problems of toxic buildup. We often take such freedom for granted. The cleansing of the body's inner ocean on a continuous basis is a boon to the quality of life. The success of kidney, and other organ, transplants is a tribute to both basic and medical research. It is nice to know that scientists and physicians can accomplish so much through understanding the many diverse organ systems of the human body.

Summary

The kidneys are especially vital organs in the system of organs that make up the human body. They provide the proper fluid balance, waste product removal, and ion regulation necessary for all the other organ systems to function properly. The kidneys are organized in four zones (capsule, cortex, medulla, and renal pelvis) and are composed of functional units referred to as nephrons.

Each kidney has a million or more nephrons and they filter the blood constantly. Each nephron is divided into several parts based on anatomical and physiological characteristics. The initial segment includes the glomerulus and Bowman's capsule, through which high-pressure filtration of blood begins. There are many desirable materials in this filtrate and the remaining regions of the nephron provide a means to reclaim them. The proximal convoluted tubule reclaims nutrients such as glucose, vitamins, amino acids, and water. The loop of Henle, the distal convoluted tubule, and the collecting tubules regulate the resorption of water and ions (sodium and potassium) into the body. Urea is concentrated in the filtrate and called urine, which is passed into collecting tubules and removed from the kidneys via the ureters. The ureters connect to the urinary bladder, in which urine is stored until released from the body during urination. The system is highly efficient, reclaiming all nutrients and important ions as well as 99% of the water initially filtered through the glomerulus and Bowman's capsule. The processes of water resorption and ion flux are regulated by the hormones ADH and aldosterone, respectively.

With these elaborate anatomical and physiological mechanisms in place, we can be sure that, even if we run out of water for an extended period of time in an arid desert at high noon, our kidneys will work overtime to maintain the purity and healthy composition and volume of the inner ocean that makes our mobile, independent terrestrial existence possible.

CODA

Act 1. Scene 2. *Sometime later, the protagonist is suffering the summertime-hot desert in the shade of the bus, going nowhere fast.*

Hopefully, someone will come along real soon with water and gasoline, so I can get out of this desert predicament.

Questions for Critical Inquiry

1. How is it that the kidneys offer us an independent existence in a terrestrial environment?
2. How do the major functions of the kidney relate to long-term survival of the individual?
3. What is the source of nitrogenous waste in the human body?
4. How has understanding the structure and function of the kidney provided a better quality of human life?

Questions of Facts and Figures

5. Where are the fluid wastes of the human body stored? How are they eliminated? Does elimination of nitrogenous waste occur in all animals in this manner?

6. Why is high blood pressure (arterial blood) necessary for the proper functioning of the kidneys?
7. What is the composition of urine?
8. A nephron is composed of several regions with different anatomy and function. Name the regions, describe their anatomy, and explain their function.
9. What is the pH of blood? How is it affected by kidney function?
10. What hormones are involved in the regulation of water and sodium and potassium ions?
11. Name and describe two human diseases in which the major symptom is excessive water loss.
12. What is the most prevalent type of ion found in the fluids of the human body?
13. What is kidney dialysis and how does it work?
14. How do humans control the timing of urination?
15. What are kidney stones?
16. Why are kidney transplants so important to the health of an individual undergoing regular dialysis? What problems arise in any transplant procedure?

References and Further Readings

Alberts, B., Bray, D., Lewis, J., Raff, M., Roberts, K., and Watson, J.D. (1994). *Molecular Biology of the Cell,* 3rd ed. New York: Garland Publishing.

Sherwood, L. (1993). *Human Physiology,* 2nd ed. St. Paul, MN: West Publishing.

Stalheim-Smith, A., and Fitch, G.K. (1993). *Understanding Human Anatomy and Physiology.* St. Paul, MN: West Publishing.

CHAPTER 11

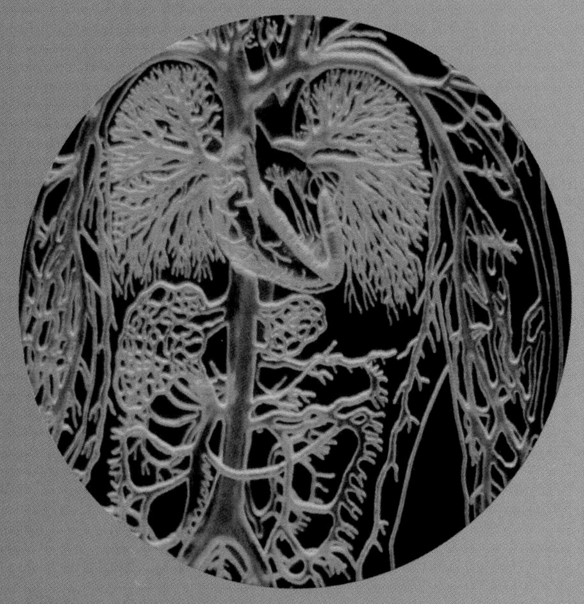

The Cardiovascular System and Blood

INTRODUCTION

They told me I collapsed into the arms of a very surprised mailman. I had just opened the front door to personally collect my letters from him. I was excited and expecting the pictures that my son took of all the family on his last visit. Last visit, indeed. I don't remember my fall into the oblivion of darkness, but I do remember the instantaneous and agonizing mountain of pain in my chest. Everything else was blocked from my mind. It felt as if someone or something was squeezing my heart with the unyielding grip of an iron fist.

DEFINING THE HEART

From an anatomical and physiological point of view, the heart is the centerpiece of the cardiovascular system. It is the muscular organ that pumps the body's blood. It can be physically damaged, often beyond repair. It may cease to function completely for some short length of time, as in the case of the individual described in the initial paragraph of this chapter, or its inactivity may disconnect us painfully, and permanently, from life. On the other hand, from a psychological point of view the human concept of the heart goes well beyond the anatomical. It has never been a simple matter to describe the metaphorical heart. We often say we want to get "to the heart of a matter," or that certain relationships are "matters of the heart" or, even more poetically, that "the heart is a lonely hunter." Emotionally, we think of the heart as breakable, perhaps by loss of, or separation from, a loved one. Spiritually, we may think of the heart as the center of our being, a source of courage, bravery, and honor. Psychologically, we may be softhearted, reflecting a type of personality that is inherently kind and gentle, or perhaps, gullible. The range of feelings we have about the heart is intimately connected to the way we use and apply the word to so many aspects of our lives. In the totality of the human experience there are clearly many ways in which to define heart.

In this chapter the focus is on the anatomical and physiological aspects of the heart and the system of vascular elements that carries the blood it pumps. The heart and vasculature are integrated into a cardiovascular system, which is a most efficient means of delivering blood to all tissues and organs of the body (Figure 11–1).

THE ANATOMY OF THE HEART

The heart is a relatively small organ, approximately the size of your fist. The heart is positioned at an angle, slightly off center, within the left side of the thoracic cavity, just under the region where the ribs come together in the front of the body at the sternum (Figure 11–2). It fits snugly within a space known as the mediastinum, which is surrounded by the left lung. The heart itself is composed predominantly of muscle and connective tissue.

The major function of the heart is contraction. However, unlike the striated skeletal muscles that move bones, the contraction of the heart muscle serves instead to reduce the internal volume of the chambers within it and to force blood to flow unidirectionally throughout the vascular circulation. The human heart is constructed around four chambers. Each chamber has a specific role in propelling blood within the cardiovascular system. The muscular elements of the heart are cardiac myocytes (Figure 11–3). As discussed in Chapter 7, this particular type of muscle cell (with a single nucleus and dense cytoplasmic striations) is found in no other organ of the body. The heart muscle, or **myocardium,** is lined and surrounded by layers of epithelial cells known as the *endocardium* (on the inside) and the *epicardium* (on the outside). The epicardium faces the *pericardial cavity.* The heart is enveloped by a *pericardial sac,* which is composed of two membranous layers (Figure 11–4). The outer layer is tough and fibrous and helps hold the heart in position within the chest through attachments to the sternum, to large blood vessels connecting to the heart, to the diaphragm, and to the pleura. This *fibrous pericardium,* as it is called, also helps prevent overexpansion of the heart. The inner, or *serous pericardium,* is delicate and moist. The fluid-filled space between the serous pericardium and the epicardial layer of the heart comprises the pericardial cavity. The fluid in this cavity acts as a lubricant to eliminate excessive friction between the membranes as the heart beats.

The maintenance of friction-free movement of the heart is extremely important. The heart beats an average of 60 to 80 times a minute, depending on the individual and his or her activities, and does so for an average of 75 years or more. Simple calculations reveal that the heart beats 3,600 to 4,800 times an hour and in excess of 86,000 times a day. Because the heart beats so often and so consistently (over 2,000,000,000 beats in an average life time), you can imagine what would happen if there was any unnecessary friction. Every heart beat would be a painful reminder of how important the proper function of the membranes of the heart really is. For many people this is a reality, because, in a condition known as *pericarditis,* the membranes surrounding the heart become painfully inflamed.

THE CHAMBERS

The heart is the pump of the cardiovascular system. It is integrated into the closed system of tubes in a linear fashion, essentially an in-line pump, to give the blood a periodic push in the proper direction. If we were to map the territory of the heart, we would find that the input and output sides of the heart are like open doors into and out of a four-room house (Figure 11–5). It is functionally important that there are two chambers each on the left and right sides of the heart. Each side has an **atrium** above and a **ventricle**

> **Myocardium** the muscular layer of the heart
>
> **Atrium** the upper chamber of vertebrate heart that passes blood to ventricle
>
> **Ventricle** the lower chamber of vertebrate heart that pumps blood out of the heart

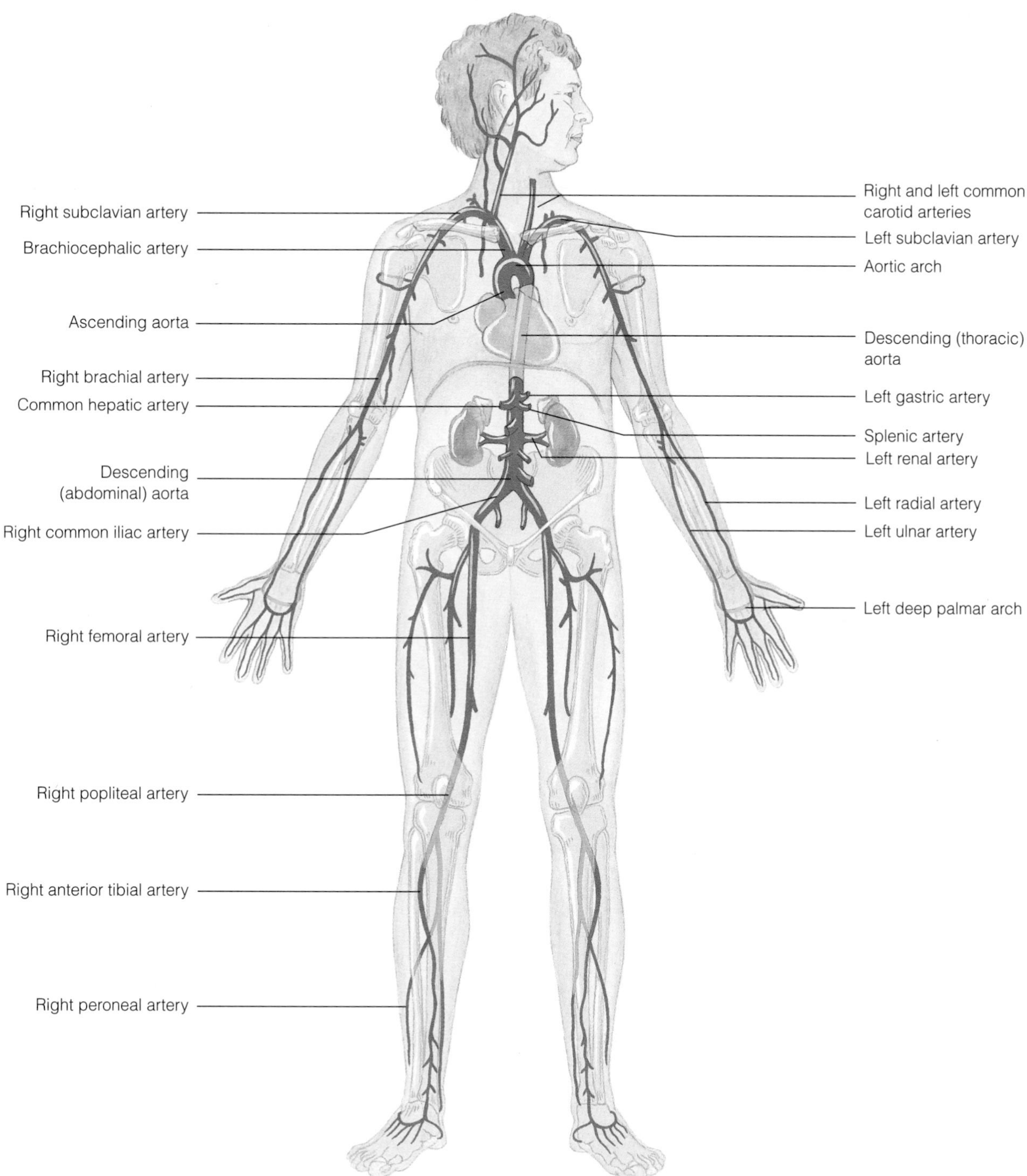

Figure 11-1 Cardiovascular System
The cardiovascular system is composed of arteries and veins. Shown here are the distributing arteries that carry oxygenated blood to the peripheral tissues of the body.

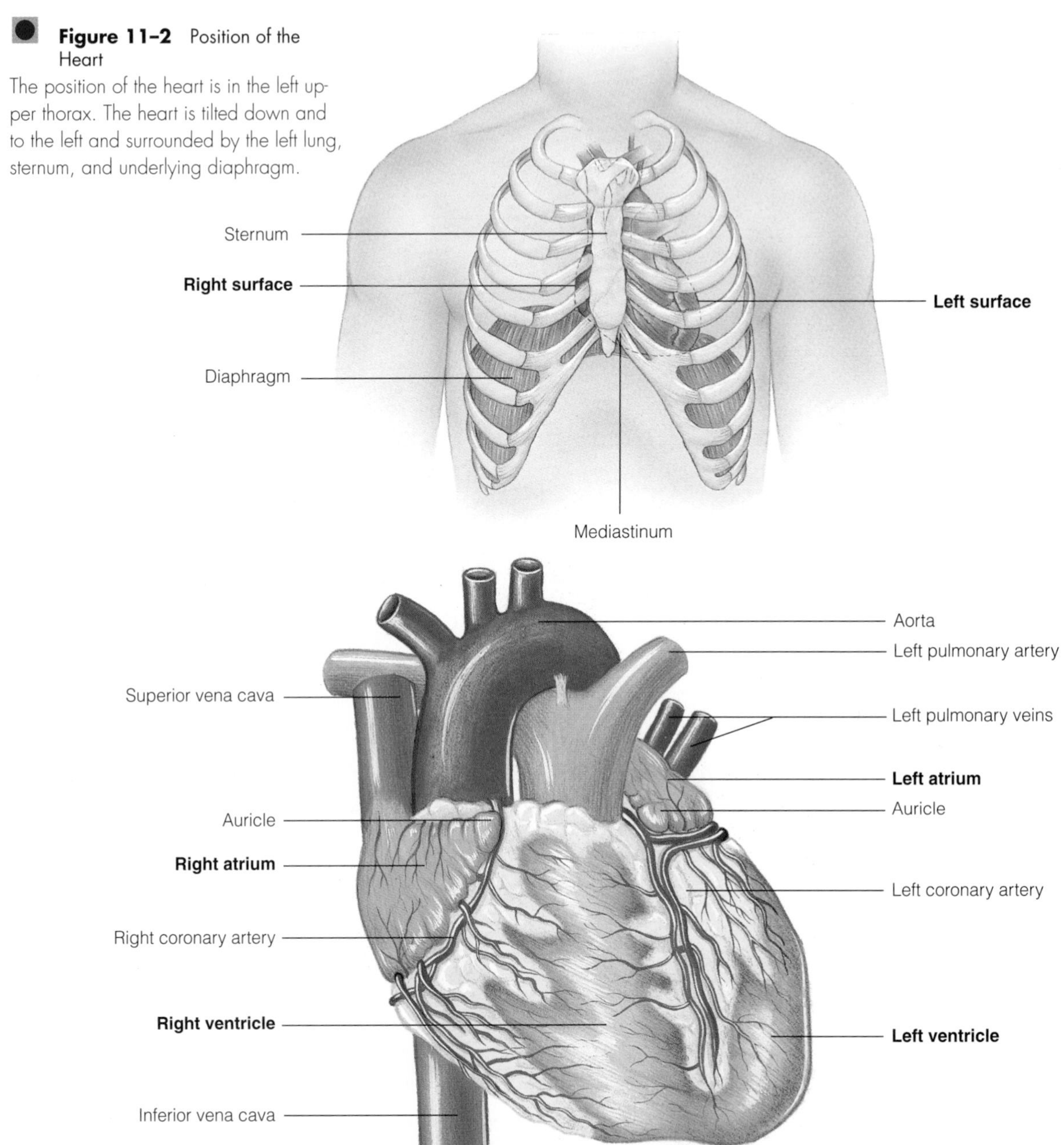

Figure 11–2 Position of the Heart
The position of the heart is in the left upper thorax. The heart is tilted down and to the left and surrounded by the left lung, sternum, and underlying diaphragm.

below. The main entrance to the heart is through the right atrium. The blood entering this chamber comes from the largest vein in the body, the vena cava. The vena cava has two branches, the upper or superior vena cava and the lower or inferior vena cava. Therefore, the blood arriving at the heart is drained from the upper and lower parts of the body, respectively. The blood at this point in the system is deoxygenated, the red blood cells having already delivered their cargo of oxygen to the tissues of the body. The right side of the heart will pump the returning red blood cells to the lungs for more oxygen. The color of deoxygenated venous blood is a deep, dark shade of red. The oxygenated blood from an artery, on the other hand, is bright red. The difference in color depends on the amount of oxygen that is bound to the hemoglobin in red blood cells.

The chambers of the heart are separated from one another side-to-side by a thick wall, or *septum* (Figure 11–3). The chambers on each side of the heart communi-

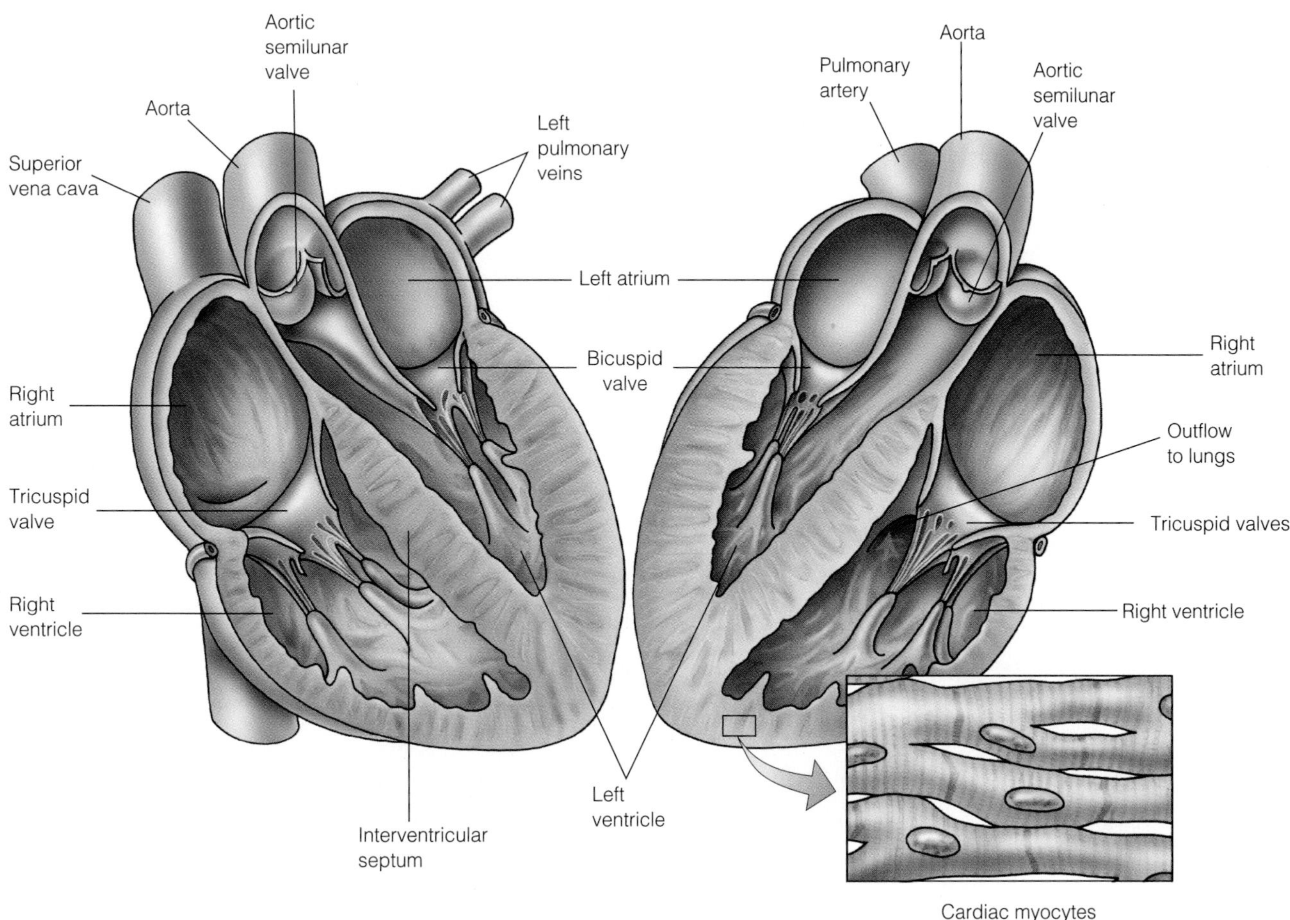

Figure 11–3 Muscular Chambers of the Heart
The heart of all mammals is composed of four chambers. Each chamber is surrounded by muscular layers whose contraction helps pump the blood.

cate by special one-way valves. This increases cardiac efficiency by preventing the mixing of blood from one chamber back into the other. This arrangement serves to direct the traffic flow of blood in one direction. As mentioned, the pattern of blood flow in the heart occurs along preset pathways. The staging of the flow of blood through the heart means that blood enters and leaves the heart twice: once to enter the lungs, and again to enter the circulation to the rest of the body.

VASCULAR FLOW

The two pathways leading to and from the heart are called the **pulmonary circulation** and the systemic circulation (Figures 11–1 and 11–5). Pulmonary circulation is the province of the right side of the heart and it directs deoxygenated blood into the lungs and oxygenated blood back to the heart. Systemic, or system wide, circulation is the province of the left side of the heart. The contraction of the heart muscle begins at the top of the heart and progresses to the bottom, or apex. This means that the contraction of the heart is *asymmetrical* (Figure 11–6). The overall contraction of the heart from start to finish takes less than a second to occur. During this time the atria are squeezed by the cardiac muscle that surrounds them just before the ventricles are squeezed by their surrounding muscle. The blood

Pulmonary circulation the flow of blood from the right ventricle through lung circulation and returning to the left atrium

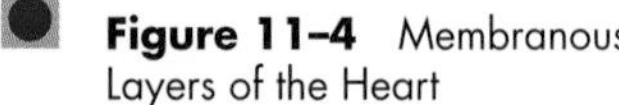

Figure 11–4 Membranous Layers of the Heart

In addition to a thick muscular layer, the heart is lined (endocardium) and covered (epicardium) with epithelial membranous layers.

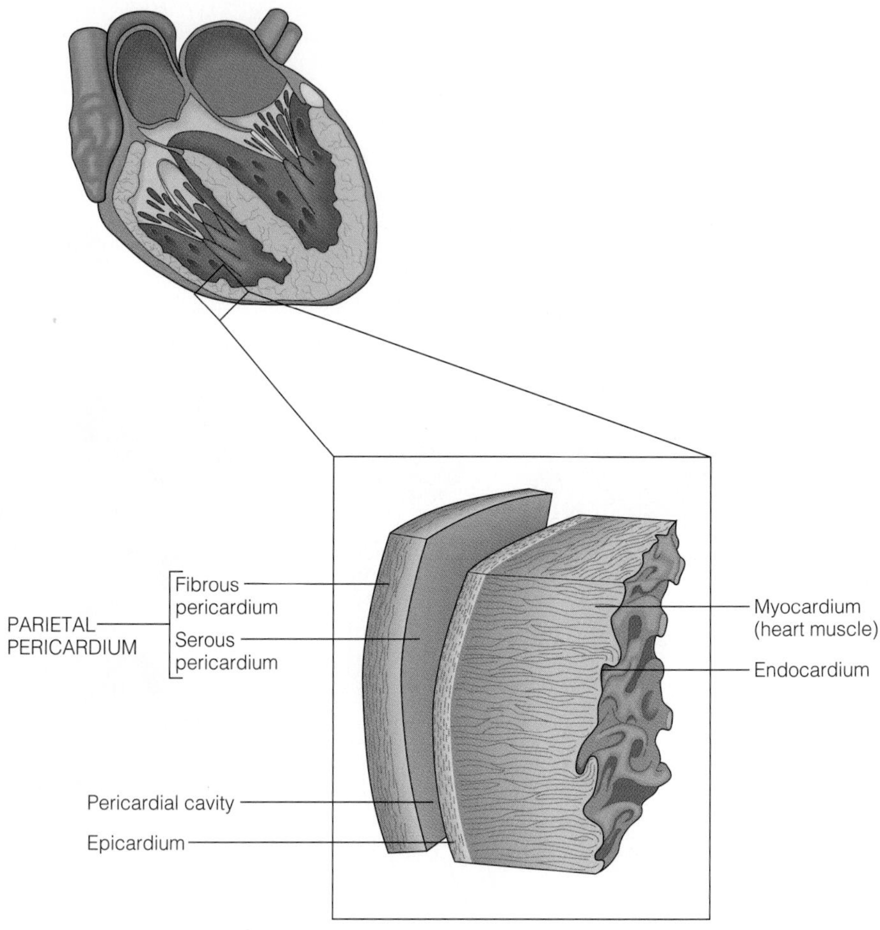

from the right atrium is pushed first through a one-way *tricuspid valve* into the lower right ventricular chamber (Figure 11–7). Only then does the muscle around the ventricle contract and squeeze. Because of the construction of the tricuspid valve, the pressure from the squeezing of the ventricular muscles forces the valve closed, so there is no backflow. The flow of blood from the right ventricle is directed through the *pulmonary semilunar valve* and into the lungs through the **pulmonary artery.** The vascular system is normally filled to capacity with blood, which puts continuous pressure on the entire system. The heart's chambers and valves compartmentalize, and squeeze, only a part of the total volume of blood during any one beat. The openness of the system allows blood to move uninterrupted throughout the thousands of miles of tubing that make up the vascular plumbing of the body.

Once in the vascular network of the lungs, the blood vessels divide again and again, becoming smaller and smaller, and eventually form an incredibly fine meshwork of capillaries. It is this meshwork of capillaries that surrounds the millions of alveoli in the lungs. The capillaries surrounding the alveoli have a tremendous surface area of absorption, which in turn is necessary to accommodate the rapid, high-volume exchange of gases. In the chapter on the respiratory system (Chapter 8), the flow of oxygen into the red blood cells and the outflow of carbon dioxide from the blood fluids was described. We breathe in tens of thousands of liters of air each day, and it is the constant flow of blood through the pulmonary circulation, driven by the pumping action of the right side of the heart, that makes possible the efficient exchange of the gases necessary to supply the body's metabolic need for oxygen. This is akin to the old time bucket brigade for putting out a fire, when a line of men and women would use a small number of buckets refilled with water over and over to put out the fire. The red blood cells are tiny oxygen-carrying buckets,

Pulmonary artery the artery directing deoxygenated blood from heart to lungs

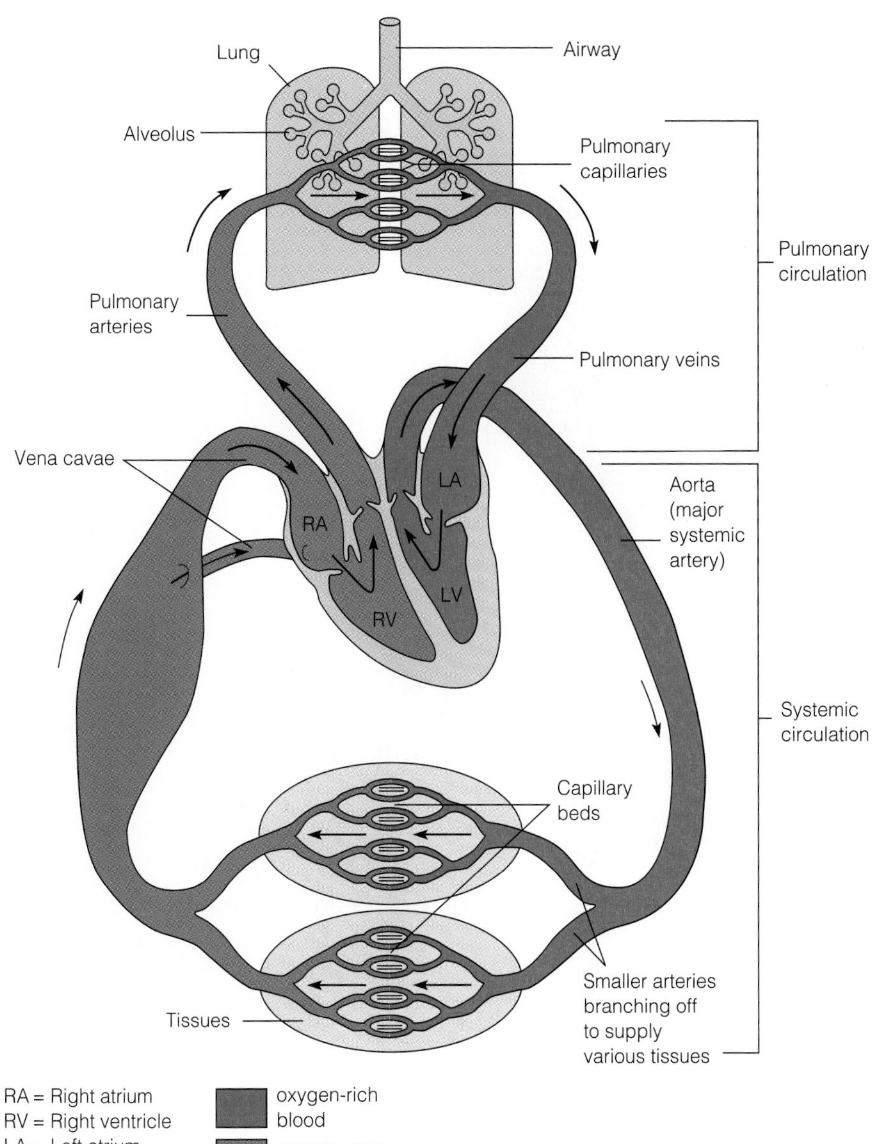

Figure 11–5 Circulation Within the Heart

There are two pathways of circulation to and from the heart. Pulmonary circulation takes blood to the lungs and utilizes the right atrium and ventricle. Systemic circulation takes blood to the body and utilizes the left atrium and ventricle.

cycling over and over again through the lungs and carrying their precious molecular cargo not to put the fires out but to keep the body's metabolic fires burning!

Once hemoglobin molecules in blood cells have bound oxygen in the lungs, the blood flows back to the heart. The tiny capillaries combine to form larger and larger vessels, similar to streams flowing into creeks and creeks into rivers, until the blood they contain passes through the **pulmonary vein** and is delivered to the left atrium (Figure 11–5). The blood is now oxygenated and ready for distribution to tissues of the body. The heart has not missed a beat in this process. The contraction of the left atrium, which by the way contracts at the same time as the right atrium, forces blood through a *bicuspid valve,* known as the *mitral valve,* and into the left ventricle (Figure 11–7). The muscle surrounding the left ventricle is thicker and stronger than any other part of the heart. This is because the force of the contraction of the left ventricle is used to push blood

Pulmonary vein the vein directing oxygenated blood from the lungs to the heart

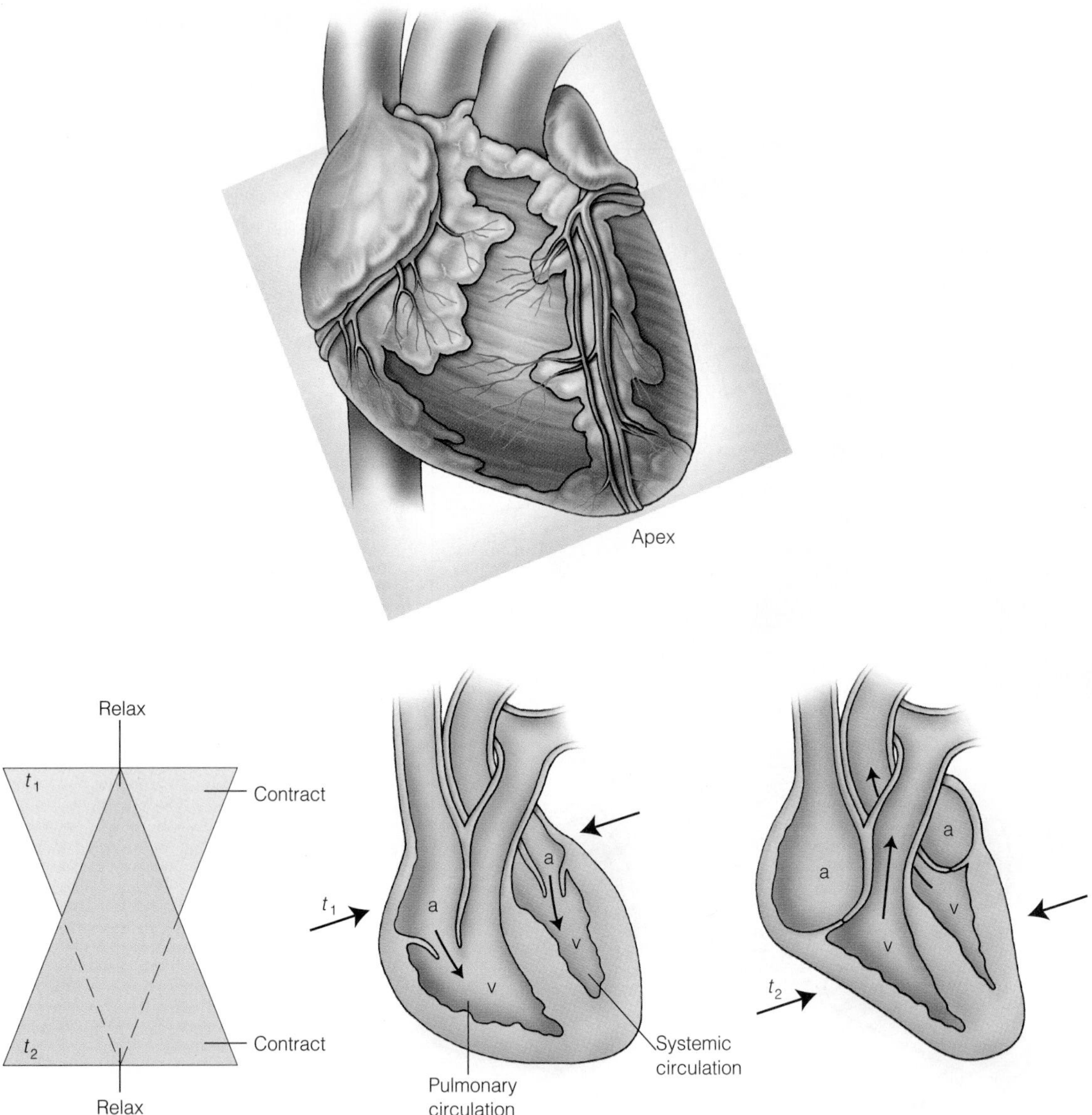

Figure 11-6 Asymmetry of Heart Muscle Contraction
The timing of muscle contraction in different regions of the heart is asymmetrical and highly coordinated. Initial contraction begins in the atria (a) and spreads to the ventricles (v).

through all the arteries and veins of the entire body (other than the lungs). This is a lot of work for the left side of the heart, and so it must be relatively stronger than the right side, which pumps blood solely through the vascular elements connecting the heart and lungs. When the muscle around the left ventricle contracts, the mitral valve is slammed shut and the blood flows out through the only other pathway open to it, the *aortic semilunar valve*. Blood is pushed at high velocity and high pressure through the back door of the heart and into the systemic circulation beyond.

Arteries

Elastic Arteries The blood vessels that allow flow away from the heart and carry oxygenated blood to the tissues of the body are called arteries. As the blood exits the left ventricle, it enters into the largest arterial structures of the hu-

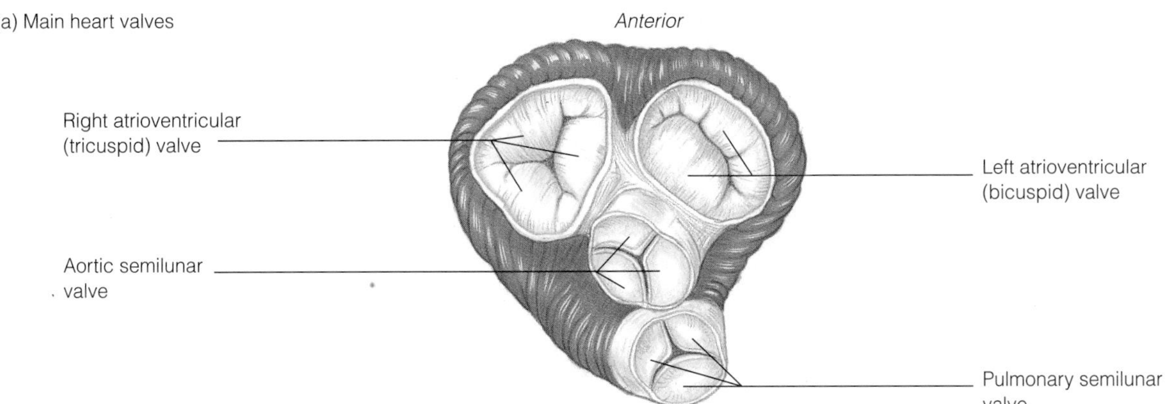

(b) Aortic semilunar valve

(c) Artificial replacement valve

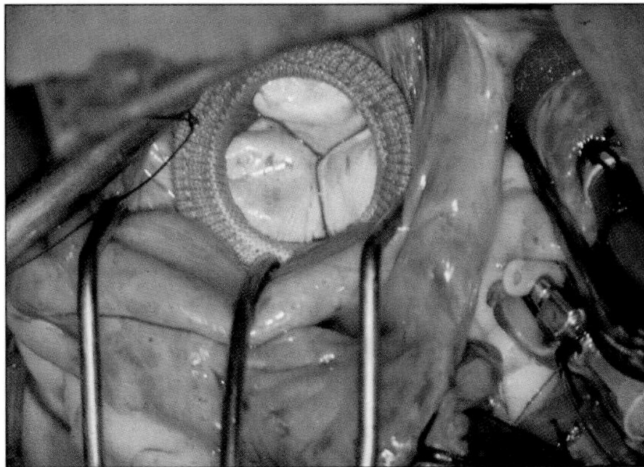

Figure 11-7 Valves of the Heart
The valves of the heart control the direction of blood flow and prevent backflow. Some of the main valves are shown diagrammatically in (a). An artificial valve is shown in (b) before implantation in a heart (c).

man body, the *ascending aorta* and the vessels of the aortic arch. These arteries, and several other large arteries that branch from them, are called **elastic arteries** because they have as part of their structure a great deal of elastic connective tissue (Figure 11–8). These largest of arteries are forced to expand under the pressure of the blood flow from the left ventricle. Blood pressure is highest in this part of the vascular system. Between heartbeats, which is the fraction of a second between the finish of ventricular contraction and the beginning of the next atrial contraction, the elastic rebound of the aortic vessels provides a continuity in the pressure of blood throughout the thousands of miles of arterial tubing. This process is somewhat similar to the air pressure used to inflate a bagpipe that is so important in sustaining its evocative sound between the player's breaths. The energy stored in the expansion of the elastic arteries, as that of the air pressure in the bagpipe, can be converted to maintaining high pressure for blood flow, or making music, as the case may be.In addition to being elastic, arteries are muscular and have connective tissue sheaths or *tunics* surrounding them. If we examine a cross section of an artery, we observe that it is constructed in many layers (Figure 11–9). The inner layer of epithelium is called the **endothelium.** This epithelial layer is continuous

Elastic arteries arteries invested with elastic connective tissue to allow expansion and contraction under the pressure of blood flow

Endothelium the epithelial lining of blood vessels

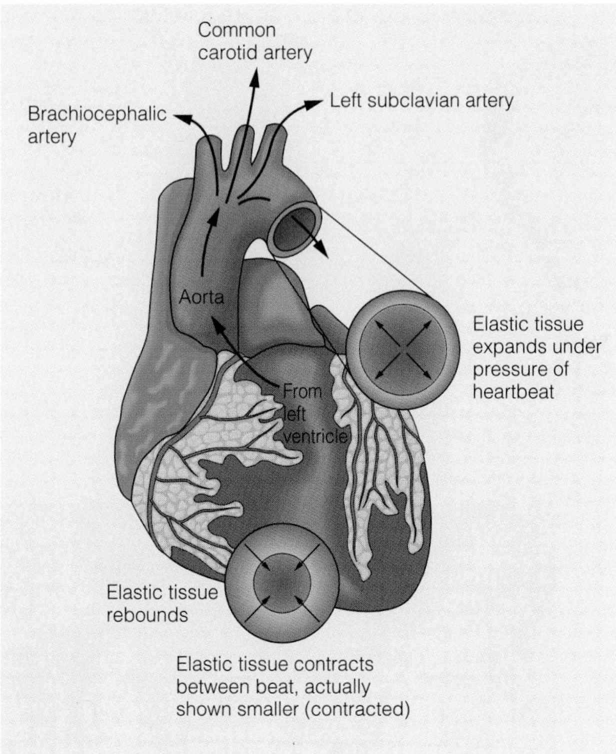

Figure 11–8 Aorta and Elastic Arteries
Elastic arteries expand under the stress of pressure from a heartbeat and contract between beats to maintain arterial blood pressure.

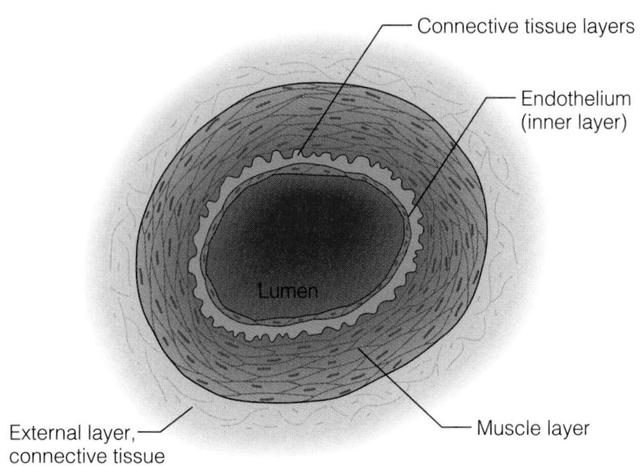

Figure 11–9 Structure of a Blood Vessel
Blood vessels are composed of several layers, which give strength and resilience under conditions of constant blood flow.

throughout the entire circulatory system. It may be likened to wall-to-wall carpeting on the floor of a house, except the carpeting would cover the walls, windows, and ceilings as well! The endothelium ensures that the blood and oxygen do not leak out into the tissues of the body in areas where it should not. The volume surrounded within the space enclosed by the endothelium of the circulatory system is called the **lumen.** This space is equivalent to the space within plastic and metal tubing that is part of the plumbing of our houses or apartments.

Underlying the endothelium is a resilient layer of connective tissue. The endothelium adheres to this connective tissue, which in turn is attached to a thick layer of smooth muscle cells that surround the artery in a continuous band. To hold the muscle cells, connective tissue, and endothelium in place, there is yet another layer of connective tissue that envelopes the entire structure. This external tuniclike layer is thick and strong and functions to prevent distortions of the artery under pressure. There are abnormal conditions under which the weakened wall of an artery will bulge out like a balloon. This is called an **aneurysm** and can be very dangerous (Figure 11–10). The artery may burst, which is disastrous, or it may put pressure on neighboring structures such as other arteries, veins, or nerves and cause local problems of blood delivery and loss of nerve sensation.

Arteries carrying oxygenated blood are under relatively high pressure. You can feel the pulsing flow of blood in arteries in many regions of your body. Place your fingertips on the area next to your Adam's apple and you will feel the blood coursing through the carotid artery on its way to the brain. If the flow of this blood is stopped, even for a few seconds, you may lapse into unconsciousness.

As stated previously, the largest arteries of the human body are elastic in nature. These arteries are too big and impermeable for the exchange gases. The aortic arch is similar to a freeway interchange in a major city at the end of the business day. The densely packed automobile traffic, like cells in an artery, initially moves in a continuous flow in one direction, branching off as the expressways diverge. There is no direct access to the neighborhoods through which the drivers of the cars pass, and there is really no way to stop. Offramps provide the needed exits from the flow of traffic along the expressways and connect these major highways with the thoroughfares of the city's gridwork of streets and avenues. Finally, after many twists and turns, you enter your neighborhood, pull up in the driveway, and sigh in relief. You are home at last.

Lumen a hollow space within blood vessels (or other tubular structures) through which blood flows

Aneurysm a bulge or outpocketing of a blood vessel resulting from weakness or damage

Distributing Arteries and Arterioles The circulation of blood within the body is analogous to the expressway. To distribute the blood to the tissues, the arterial vessels must decrease in size. This decrease in size of arteries helps keep the blood pressure consistently high in the face of an ever-increasing number of smaller and smaller blood vessels. The intermediate-sized arteries are called **distributing arteries** and are the ones with which most of us are probably familiar (Figure 11–11). They lead to the head, arms, legs, and internal organs. For example, the renal arteries that supply the high-pressure blood to the kidneys fall into this category. They are among the first arteries to branch off from the *descending aorta* (see Chapter 10). The gridwork of vessels distributes blood to all organs and regions. The division continues, and the smaller arteries fan out to deliver blood to nearly every nook and cranny of the body. The distributing arteries divide and give rise to arterioles, or little arteries. An arteriole is generally of microscopic proportions. Arterioles are the last step in the division of the arteries prior to forming capillaries.

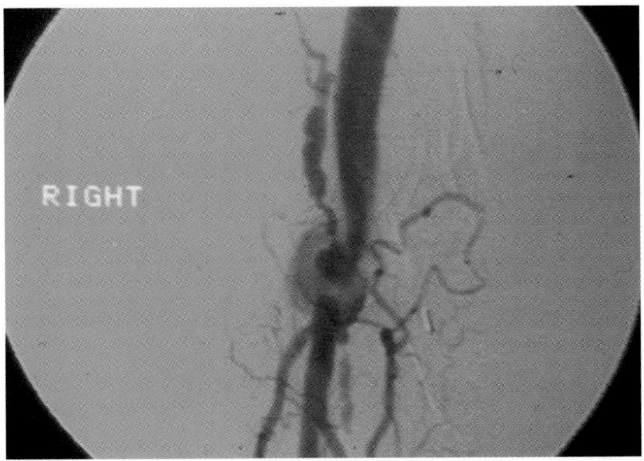

Figure 11–10 An Aneurysm
An aneurysm occurs when the walls of a blood vessel (usually an artery) are weakened and bulge outward under the pressure of blood flow.

Capillaries Capillaries are the smallest and most delicate of the vessels of the circulatory system. It is within the fine networks of capillaries, which penetrate to nearly all regions of the body, that oxygen and carbon dioxide are

> **Distributing arteries** arteries of midsize that distribute blood to all regions of the body; the commonly named arteries, such as the carotid or brachial arteries

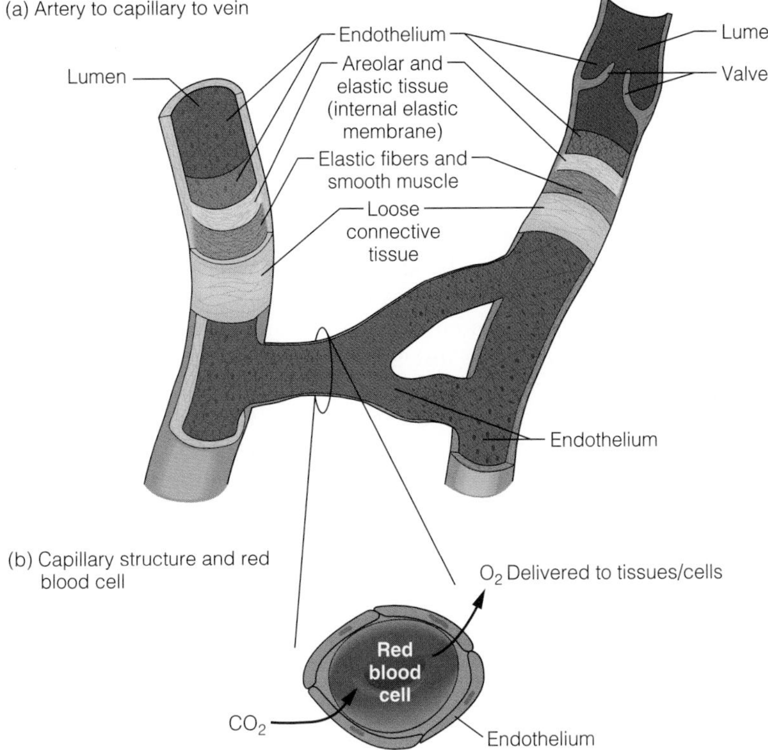

Figure 11–11 Capillary Structure and Connections
Capillaries connect the arterial blood vessels with the venous blood vessels (a). Within the capillaries oxygen and carbon dioxide are exchanged (b).

exchanged. The simple, minimalist structure of a capillary allows gases to move by diffusion or other transport mechanisms from the red blood cells and fluids of the circulatory system into the surrounding tissues and from those tissues back into the capillaries (Figure 11–11). For example, oxygen in the lungs diffuses into the capillaries from the oxygen-rich alveoli and carbon dioxide migrates out. The direction of movement of these two gases in other tissues is in the reverse direction; oxygen is released from hemoglobin and diffuses out of the red blood cells, while carbon dioxide diffuses in. Diffusion works well and efficiently over short distances (nanometers and micrometers) along concentration gradients. The higher concentration of a substance in one area results in the spread of that substance into areas in which it is less concentrated. High oxygen concentration in the lungs sets up a gradient with the blood. The net result is movement of oxygen into red blood cells. The lower oxygen tension in the rest of the tissues of the body reverses that movement and oxygen is supplied to the active cells and tissues of the body.

Blood flow throughout the peripheral regions of the body is controlled by regulating the flow through the capillaries. This is where the tiny muscle cells surrounding arterioles come into play (Figure 11–12). Smooth muscle cells are wrapped around the precapillary arterioles, and their contraction provides the mechanism to constrict and close them. When this happens, the blood is prevented from entering the capillaries. A good example of this type of control is the restricted flow of blood to the surface of the body on a cold day. The capillary beds that serve the skin carry a substantial amount of body heat with them. On a cold day, an exposed human body can ill afford to lose **core body heat.** So what happens? The brain senses surface temperature changes through cooling of the circulating blood and autonomically restricts the volume of blood flow to the surface by constricting the *precapillary sphincters*.

As blood cells enter the capillaries, they are individually squeezed tightly against the endothelial cells lining the tiny vessels. This causes a deformation in red blood cell shape and establishes the conditions necessary for discharge of oxygen. The dissociation of oxygen from hemoglobin continues on a cell-by-cell basis, until all of the hemoglobin molecules become deoxygenated. The release of oxygen molecules from the grip of hemoglobin and its diffusion across cell membranes supply the local region with the oxygen necessary to run the metabolic machinery of cells. Red blood cells exit the capillaries, their oxygen spent, and start the journey back to the heart.

A Rest Stop Let's review our trip so far, as blood moves from the doorways of the heart to the outlands of the peripheral capillaries. The right side of the heart pumps blood into the lungs through the pulmonary circulation. Red blood cells are charged up with oxygen and pushed back into the heart through the left atrium. Unidirectional flow is enforced by judiciously placed valves, which prevent backflow from one chamber to another. The left ventricle is the strongest part of the heart-pump. It pushes the blood out of the heart and into the systemic circulation. The largest of the elastic arteries carry the high-pressure fluid upward to the head and outward to the torso and limbs, dividing and reducing the diameter of blood vessels at ever-increasing distances from the heart. Distributing arteries bring the blood to major organs and regions, and smaller arteries and arterioles serve local tissue neighborhoods. Precapillary sphincters of arterioles, controlled by contracting smooth muscle cells, help regulate flow into the capillaries, where gas exchange takes place and the red blood cells are relieved of their oxygen load.

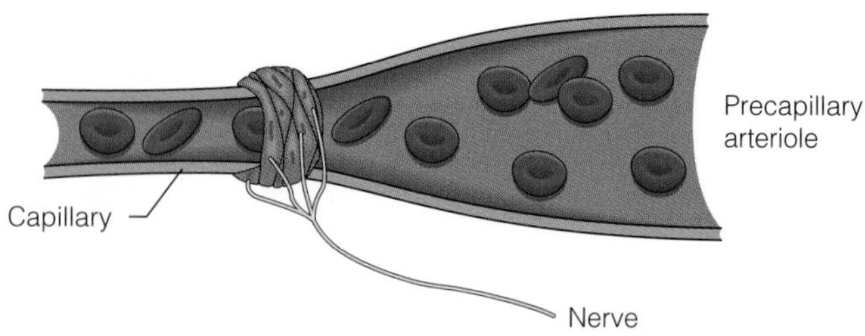

(a) Muscle-activated sphincter open or dilated

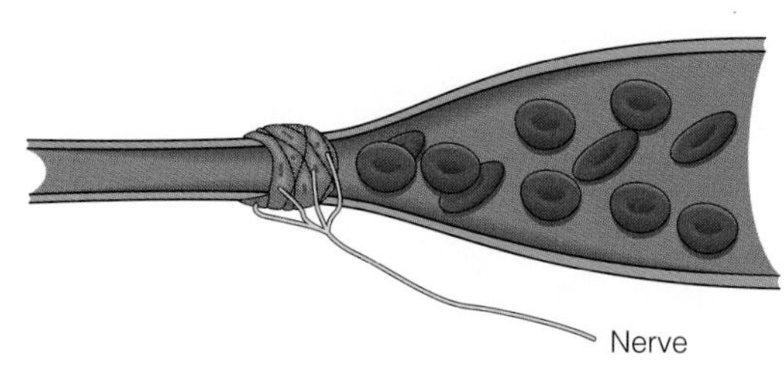

(b) Sphincter closed or constricted

Figure 11–12 Smooth Muscle Control of Blood Flow
Smooth muscle occurring around the precapillary arterioles can be dilated to allow flow of blood into a capillary or can be constricted to prevent flow of blood (b).

Core body heat the heat of the central portion of the human body; in humans maintained at approximately 38° C.

Veins

Venules As the flow of blood exits the capillaries, it enters vascular elements known as *venules* and the blood pressure decreases. The venules are the small tributaries of the venous system through which blood begins its flow back to the heart. Postcapillary blood vessels are different from their precapillary counterparts. This is reflected in the structural differences between arteries and veins. The thickness of the walls of the venous blood vessels is reduced with respect to the diameter of equivalently sized arteries. The structural change is similar to a series of high-pressure, small-diameter waterpipes emptying into a much larger diameter tube. Once the rapid outflow of water in the small pipes is purged (from arteries), it mingles with the slower waters of the wider channels (that is, into venules and then intermediate-size veins) and meanders through the wide aqueducts under low pressure. The blood pressure drop across the capillary beds is dramatic, but not all pressure is lost. However, this pressure drop presents a problem for the circulatory system. Without sufficiently high pressure, how does the blood flow back to the heart? This situation is further complicated by the fact that humans stand erect and must be able to move the blood against the force of gravity, which is always tugging us with a downward force. The pressurized system of arteries is up to the task, but the lower pressure of veins is not—at least, not without some innovative compensatory mechanisms.

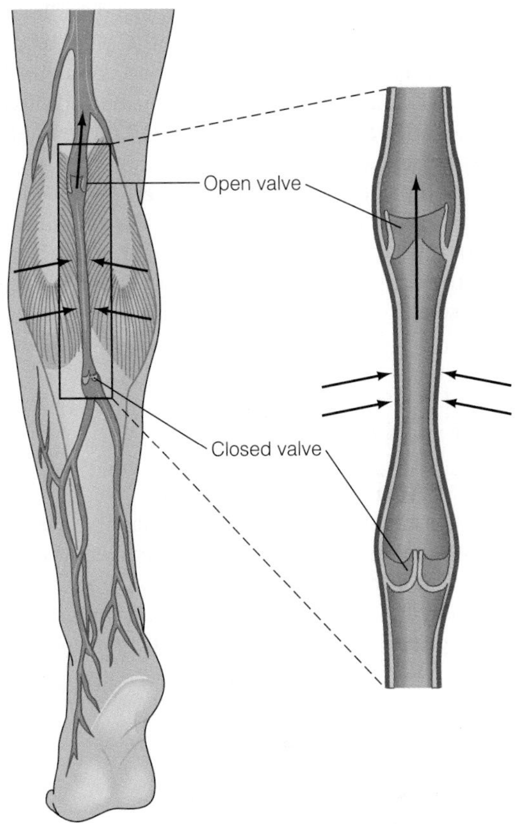

Figure 11–13 Movement of Blood in Veins—Valves and Muscles
Muscle contraction helps move low-pressure blood back toward the heart. One-way valves prevent backflow of blood.

Valves and Squeezing How does the blood climb back up from the lower extremities and the distant regions of the arms to reach the heart? There are two complementary mechanisms by which this occurs. The first is one-way valves, and the second is the squeezing of veins by contracting skeletal muscles through which the veins pass (Figure 11–13). Contraction of skeletal muscles during various human activities such as walking, running, and other exercise promotes the movement of fluids upward through the veins to the heart. This is similar to the peristaltic movements of the materials within the digestive tract but not quite as concerted and automatic. However, the mode of muscle-driven movement combined with the one-way valve system is quite effective.

Valves are crucial to the return of blood to the heart, because they trap small volumes of blood and prevent backflow, avoiding potential pooling in the lower extremities. It is not desirable for feet and legs to fill up and swell as a result of poorly circulating blood. This can happen in people who are on their feet a lot but do not move around very much. Painful swelling of veins, particularly in the lower legs, is common and can lead to **varicose veins** (Figure 11–14). *Varicosities* usually occur in veins close to the surface of the skin or other epithelium, where, under pressure from gravity, they essentially pop out of the connective tissues that normally hold them in place. This is not to be confused with an aneurysm, which is the breakdown of the tunic of an artery, but rather a varicosity is the detachment of an intact vessel from the connective tissue that generally holds it in place.

As the vessels become larger in diameter, the walls become thicker and stronger and are invested with more and more valves. The largest veins are those nearest the heart, such as the superior and inferior vena cava. Circulating blood has, thus, returned from a circuit of the cardiovascular system to the entrance of the heart once again.

Varicose veins abnormally swollen veins that have broken away from the connective tissues that surround them and protrude outward on a surface

The pattern of circulation of blood just described for the human body was first discovered in the seventeenth century by an English anatomist named William Harvey (Biosite 11–1). The intricacy of the distribution of vascular elements is spectacular. It is estimated that the total length of blood vessels in an average-sized human, if the vessels were cut and pasted into one long tube, would be in excess of 50,000 miles! This puts an enormous pressure on the heart, not only in terms of its muscular strength but also in terms of the consistency in timing of the contraction of the heart muscle. How is the heartbeat coordinated?

THE PHYSIOLOGY OF THE HEART

Electrical Activity of the Heart

The heart is bioelectrically wired. It has its own internal electrical signal generator called a pacemaker, its own internal electrical conducting system (*Purkinje fibers*), and an outside source of electrical stimulation (**sympathetic** and **parasympathetic nerves**), which all coordinate the beat of the heart with events going on in other parts of the body, or respond to events occurring outside the body.

SA Node The existence of an internal stimulating capacity in the heart is observed in the ability of a heart to beat even after removal from the body (Figure 11–15). This capacity is inherent to the cardiac myocytes themselves, which have an intrinsic pace at which they beat even when there is no functionally intact heart in which to beat. However, for the heart to pump blood, it has to be an intact unit, and there must be an internal control mechanism. This control is provided by a pacemaker. Pacemaker is the common name given to the cellular structure in the heart called the sinoatrial (SA) node. This node is located in the region of the right atrium and embedded in the thick

Sympathetic nerves nerves arising from spinal ganglia that control organ function; generally excitatory

Parasympathetic nerves nerves arising from the brain and brain stem (cranial nerves) and lower spine that control organ function; generally inhibitory

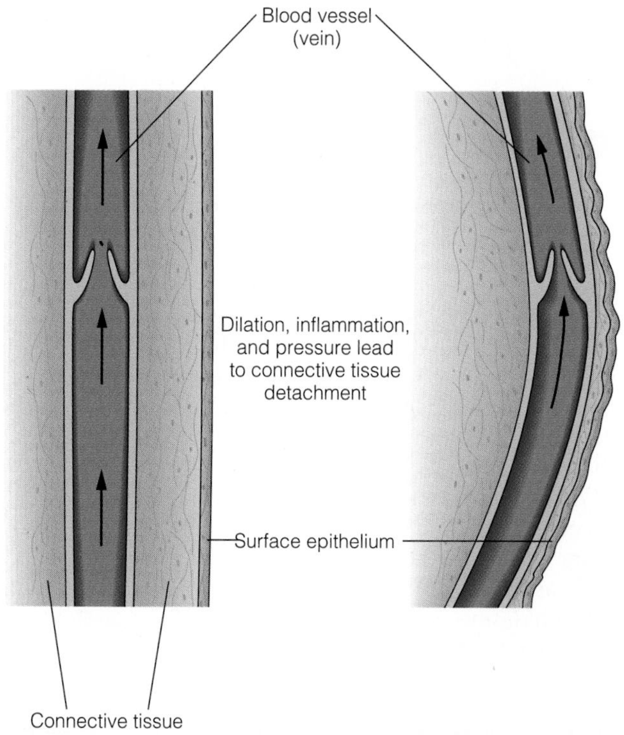

Figure 11–14 Varicose Veins
Varicose veins arise when veins near the surface of the body become inflamed and dilated and pull away from the connective tissue surrounding them.

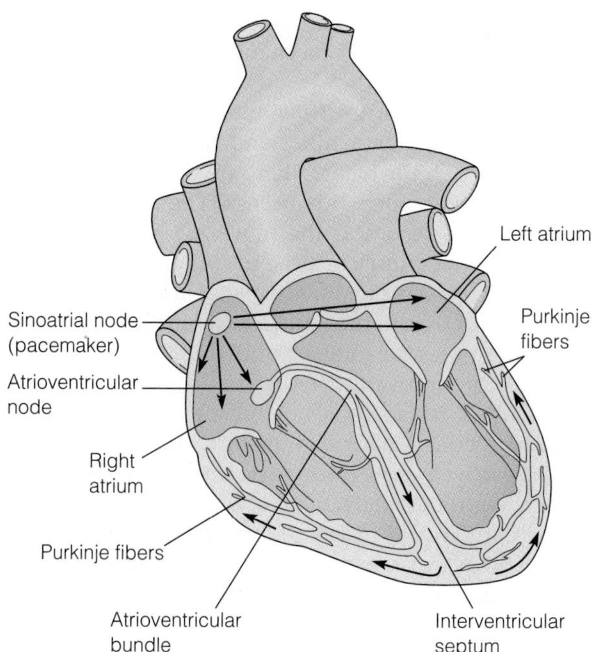

Figure 11–15 The Heart and Its Nodes
The inherent electrical activity of the heart is controlled by the sinoatrial (SA) node. The signals from this node are conducted to the atrioventricular (AV) node and then to the Purkinje fibers to the ventricles of the heart.

11–1

WILLIAM HARVEY AND THE CIRCULATION OF THE BLOOD

Just prior to 1620, William Harvey discovered the circulation of the blood. He did this with all the odds of ignorance stacked against him and with a certain amount of luck. The rational basis of medicine had been established by the Greeks, particularly by Hippocrates, whose oath is still repeated by physicians as they enter their profession. The problem was that the nature of medicine had not changed up to Harvey's time. Nearly 2000 years had passed without serious revision in the way physicians thought about and treated human body functions.

During Harvey's time in Italy, the study of anatomy in medical school was based on a few corpses (usually men condemned to death) dissected in an open auditorium for the medical students to watch. This was not conducive to new discoveries about human anatomy and physiology. In addition, the microscope had yet to be invented and the existence of microorganisms was not even guessed. How better to describe the cause of diseases but as imbalances in the humors representing the elements of earth, water, air, and fire. The arteries were thought to contain air (for which they are named), and it was the heart that produced a vacuum to pull air into the body through the lungs.

Harvey was not satisfied with this description. He began a series of experiments with organisms of all kinds because he was convinced that the heart was "the foundation of life" and wanted to know how it worked. The circulation of blood as we know it today is not an obvious fact. We can see some of the blood pulsing through the body, but we cannot see it all. Harvey was able use animal models to establish that circulation occurs through the continuous movement of blood through the circulatory system. The key to his discoveries was the presence of membranous flaps in the veins. What could they be? He determined they were valves. He made the leap of imagination that the valves were there to prevent the flow of blood backward into the heart! Once he showed this, he extended his findings to humans. It all fit together perfectly. For his efforts William Harvey permanently changed the way that scientists were to look at the heart and vascular circulation. He moved us from a 2000-year-old tradition of accepting the authority of the ancients into the beginnings of the modern era of experimental biology that characterizes contemporary research in anatomy and physiology.

muscular wall of the heart below the opening of the superior vena cava (Figure 11–15). The electrical activity of the SA node initiates each cardiac contraction cycle and sets the basic pace of the beat.

AV Node The signal generated by the SA node is sent along a cellular conduction pathway that leads from the top part of the heart (the atrial region) to the apex of the heart, which is in the ventricular region. To boost the signal to the lower part of the heart, a second node, known as the atrioventricular (AV) node, participates in the conduction process. Conduction fibers, referred to as Purkinje fibers, spread throughout the muscular wall of the ventricles to deliver a uniform stimulation over the surface of the heart (Figure 11–15). This helps explain, in part, the asymmetrical pumping action of the heart. The muscles of the atria, both right and left, start contracting first under the influence of the SA node. The AV node boosts the signal to the ventricles, after a slight delay, so the ventricular muscles contract secondarily. Then, the entire heart relaxes for a short period, as it awaits the next signal from the SA node. The whole cycle takes about 0.8 seconds.

ECG The electrical activity of the heart may be measured from outside the body. Electrodes placed on the chest, side, and back can be used to collect electrical information from the contracting heart in what is called an electrocardiogram (ECG; Biosite 11–2). This measurement may be routinely conducted during a physical examination at the doctor's office and provides the physician with a great deal of valuable information on the current health of your heart. The ECG of a single heartbeat is presented in Figure 11–16, as part of an explanation of the process. All one needs to know to start with is part of the English alphabet. At time zero the graph is flat. When the SA node initiates a signal, there is a slight blip in the electrical map called the *P-wave*. Then, there is a tiny downward blip, called Q, and an immediate, strong upward wave called R. This is followed

11–2

AN ELECTROCARDIOGRAM—MAPPING THE HEART'S ELECTRICAL ACTIVITY

One of the best ways to test the function of the heart is to monitor its electrical activity. The stimulating influence at the beginning of each heartbeat is an electrical impulse from the sinoatrial node. This signal is carried along a special conduction system and reiterated by a second impulse from the atrioventricular node. When this all works perfectly, the heartbeat occurs asymmetrically from the auricles to the ventricles, and then returns for an instant to a neutral state. This occurs approximately every 0.8 seconds in a resting heart.

What if the heart is damaged? This occurs, for example, through lack of oxygen resulting from an embolism or atherosclerosis. Cardiac myocytes may die under these conditions and the heart may never recover full function. In these cases the electrical activity of the heart is also altered. If a series of electrodes are placed on the front, side, and back of the body to monitor and map local heart muscle electrical activity, then you may find that regions of the heart are electrically blank. If there is no electrical activity, there is no contraction of muscle cells. If the area affected is large enough, the efficiency of the heart contraction is reduced and human health is altered.

An electrocardiogram measures the electrical activity of the heart and helps determine if, and to what extent, a heart may have been damaged by myocardial infarction, **ischemia,** and other types of damage.

by a plummeting, downward deflection called S. This group of changes is called, as common sense suggests, the *QRS complex.* The graph is flat again for a short interval, and then finishes the cycle with a small upward deflection known as a *T-wave.* PQRST takes about 0.6 seconds. With a 0.2 second rest for the heart, the cycle repeats itself. What does this electrical rhythm mean in terms of the beat of a heart? P waves indicate the immanent contraction of both the right and left atria. The electrical signal stimulates the atrial muscle cells to contract in concert. Then, the action spreads to the lower part of the heart. QRS represents the signal spread to the ventricles, the muscles of which contract immediately thereafter. The T-wave resets the ventricular musculature for the next round of contraction. Simple as ABC . . . , sorry, not quite, PQRST.

Blood Pressure and Hypertension

Systole and Diastole There are two terms that are closely associated with the beat of the heart and the pressure of the flow of blood. They are listed on the cardiovascular monitoring machines in department stores and supermarkets across the country. These machines provide easy access to measurement of your personal systolic and diastolic blood pressure. The states of **diastole** and **systole** represent phases of contraction and relaxation of the heart muscle (Figure 11–17). The contraction (systolic phase) of the muscles surrounding the atria sends blood into the relaxed ventricles. The atria themselves then relax (a diastolic phase), and the blood-filled ventricles prepare for contraction. This cardiac preparation is completed between P and R in the ECG (Figure 11–16). It is by the contraction and relaxation of the muscles surrounding the ventricles (particularly the left ventricle) that blood pressure is monitored. Contraction of the ventricles, as in the case of the atria, is also systole, and relaxation of the ventricles results in diastole. The left ventricle, however, pumps blood to the body, so its output can be measured peripherally. For example, the pulse we can feel in an artery of the wrist reflects the rhythm of ventricular systole and diastole.

Monitoring Systemic Circulation The machines in the drugstores and supermarkets measure the pressure in the arteries of the arm (particularly the *brachial artery*) during left ventricular systolic and diastolic phases. These machines do so by compressing the arteries of the upper arm

Diastole the period in a heartbeat during which both atria and ventricles are relaxed

Ischemia a condition in which blood flow is restricted and low or no oxygen is delivered to an organ

Systole the period in a heartbeat during which both atria and ventricles are contracting

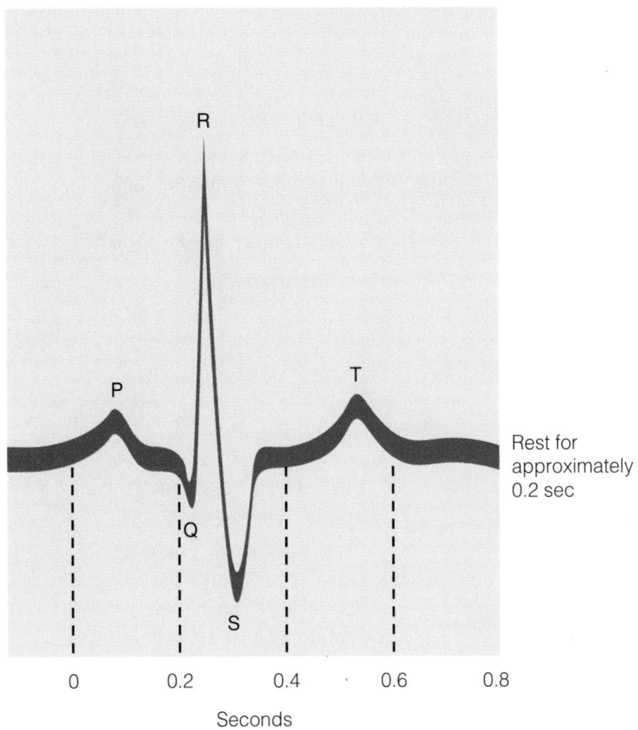

Figure 11–16 An Electrocardiogram (ECG)
The average heartbeat is completed in approximately 0.6 sec. The heart rests for 0.2 sec before it beats again. The electrical activity is measured during an ECG and results in a characteristic pattern of blips (P, Q, R, S, and T).

so that no flow of blood can occur within them (Figure 11–17 [a]). The cuff, which is the band that shrinks around the arm to restrict blood flow, slowly releases the upper arm from the external pressure and monitors when blood flow returns. The same principle applies when a garden hose is squeezed to stop the water. More pressure is on the hose from the outside than the water pressure applies from the inside. If we had a way to measure the pressure needed to stop the water, we would know the water pressure within. In the process of squeezing the arteries of our arm, we (or the physician/nurse) may determine that we have a blood pressure of 125 over 82 (125/82). This means that contractive pressure of the ventricles (in this case, the left ventricle, which supplies the force for systemic circulation), creates a systolic pressure equal to 125 millimeters of mercury in a manometer. The relaxation of the ventricular heart muscles drops the pressure within the ventricles to zero. So why does our blood pressure not drop to zero? A diastolic pressure of 82 represents the latent pressure within the system and reflects, in part, the rebound of elastic arteries, such as the aorta. Values in Table 11–1 indicate that a reading of 125/82 falls into a normal range of values for healthy male and female humans. Many people have blood pressures well beyond the normal range. Higher than normal blood pressure is called **hypertension** and has many causes. Hypotension in which blood pressure is too low also is a problem and can affect organ functions, for example, in filtration of blood through the kidneys.

Blood Pressure and Health Blood pressure readings provide an indication of the health of the cardiovascular system. Blood pressure may vary considerably, depending on a great many variables. The age and sex of an individual play a part in determining the average blood pressure of that individual, which by the way should represent the outcome of many readings, not just one, and should be taken at different times and under different conditions (see Figure 11–17 and the *sphygmomanometer*). Anxiety, excitement, activity, general physical condition, and existing cardiovascular disease all play a role in determining the blood pressure, even from minute to minute. If blood pressure is too high, there may be excessive physical stress on the blood vessels and the heart, which can lead to irregularities in beat, valve damage, or ruptures. If blood pressure is too low, the supply of blood to organs may be inadequate. In the case of the kidney, which requires a considerable amount of pressure to filter the blood, it may shut down completely if the overall pressure of the cardiovascular system gets too low.

Essential Hypertension Every human body needs the physiological capability to temporarily increase the pressure of the blood in the vascular system. Emergency or dangerous situations alter blood pressure and redirect blood-flow pathways within the body to enhance physical performance (for example, run, lift, push, pull) and protect us and/or those around us. Anxiety or stress of various types, of which we may not even be aware, can cause transient elevations of blood pressure. Simply having your blood pressure measured can induce its elevation. The term used to define the generally unexplainable upward fluctuations in blood pressure is *essential hypertension*. Most people have this type of hypertension at one time or another.

Blockage of Arteries

Sustained hypertension, however, is a different problem. One of the main causes of high blood pressure in humans is the reduction in the diameter of arteries. This occurs in individuals who, for reasons of diet or genetic predisposition

Hypertension high blood pressure

(a) Measurement of Blood Pressure

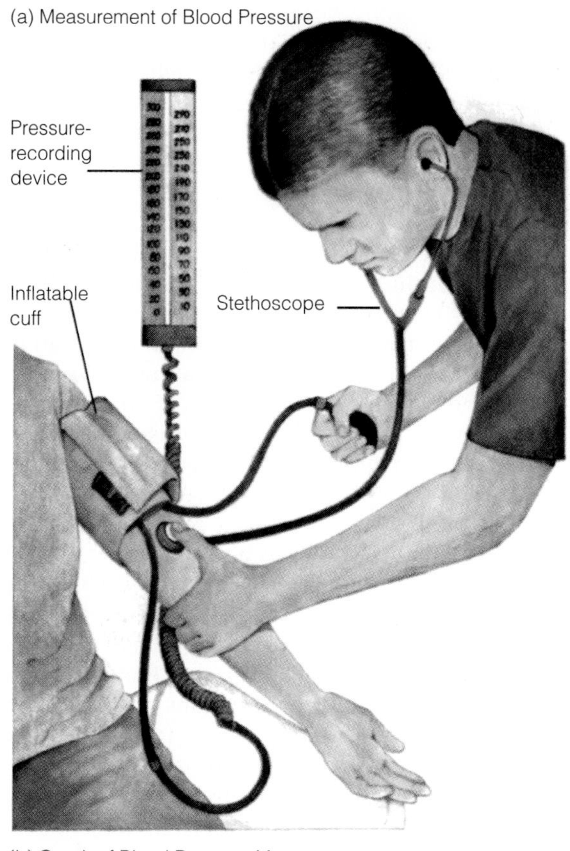

(b) Graph of Blood Pressure Measurement

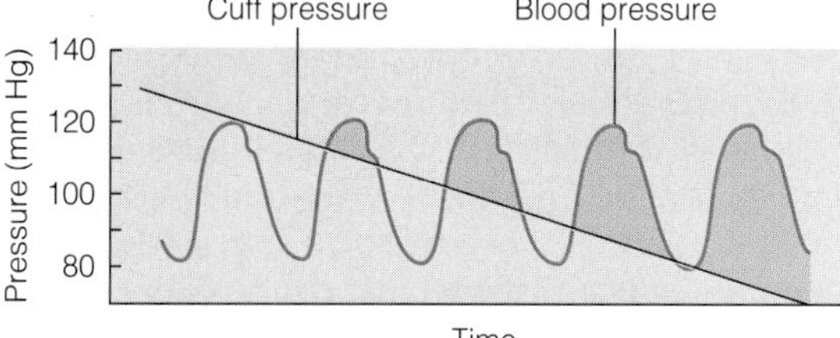

(c) When Blood Pressure is 120/80

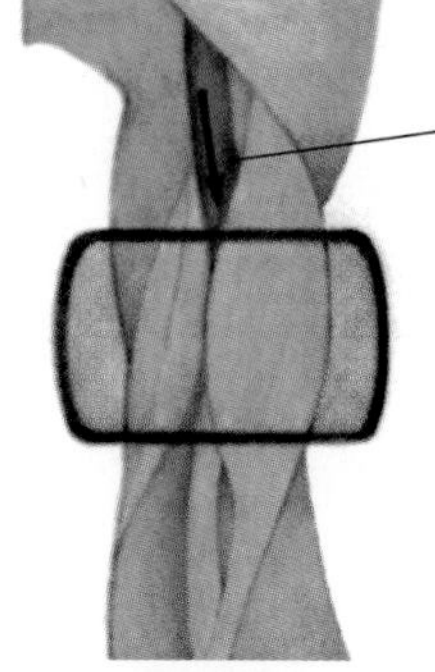

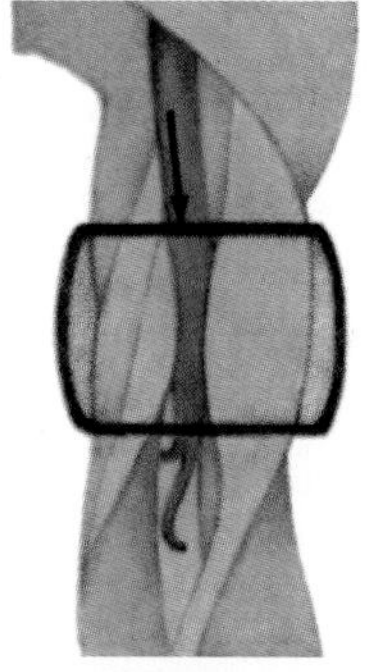

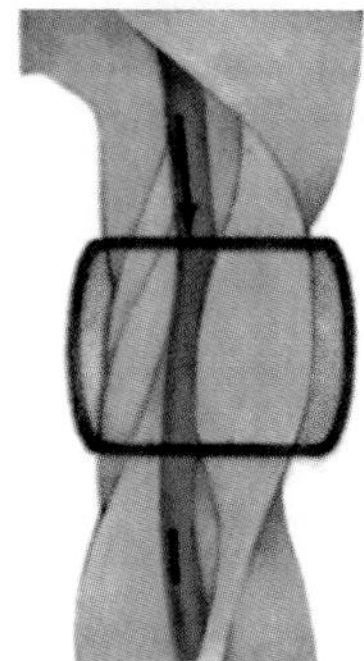

(d)

Figure 11-17 Diastole and Systole—Blood Pressure
A sphygmomanometer is generally used to measure blood pressure (a). Diastole is a brief period of rest when both atria and ventricles are relaxed. Systole is when both atria and ventricles are contracted. The highest reading in a measure of blood pressure (b) and (c) is systolic pressure; the lowest is diastolic pressure. Diastolic pressure does not drop to zero because of elastic rebound. A sphygmomanometer, commonly seen in places such as drugstores, is shown in (d).

Table 11-1 Normal Values for Human Blood Pressure (mm of Hg)

	SYSTOLIC	DIASTOLIC
Normal	100–140	60–90
Hypotension	Less than 100	Less than 60
Hypertension	Greater than 140	Greater than 90

or both, build up abnormal layers of materials (called **plaque**) along the wall of an artery (Figure 11–18). This process takes a relatively long time (years) and is somewhat like plastering the walls, floor, and ceiling of a hallway in your home with layer upon layer of cement until the hall is completely blocked. As the diameter of the arteries shrinks, the pressure goes up. Remember that the volume of blood in the body stays the same, so if the diameter of some of the vessels decreases, the pressure in the system must go up. Eventually, in the worst cases, the arteries simply close off and no blood can flow through them. Without blood there can be no oxygen delivered, and oxygen deprivation is a cellular death warrant.

This type of blockage, or **occlusion,** does not necessarily take place everywhere in the vascular system at once. However, you can imagine that such blockage in critical regions can mean the difference between life and death. Blockage of the *coronary arteries* (Figure 11–2), which bring oxygen to the heart muscle itself, may cause what is called a **myocardial infarction.** Without oxygen, cardiac myocytes die. If enough of them are destroyed, the heart beat becomes weak and/or irregular, or may fail to continue to beat at all. An ECG is capable of mapping silent regions of the heart, where muscle cells have died (Figure 11–19). Usually, there is some early indication that there is a problem developing. Chest pains that run down the left arm should suggest to anyone who has them to seek medical attention. This condition is known as *angina pectoris* and is symptomatically treated with drugs, such as *nitroglycerin,* which dilate the blood vessels and return blood flow to afflicted regions. However, this treatment does not solve the long-term problem.

Blockage of arteries in the brain results in deprivation of oxygen to brain cells (Figure 11–20). This condition may result in **stroke.** As in the muscle cells of the heart, the nerve cells of the brain cannot replace themselves after being damaged. Depending on the extent and region of oxygen deprivation, a stroke may have permanently debilitating effects on a wide range of human body functions and activities. Often these effects occur only on one side of the body. As will be discussed in Chapter 17 on the nervous system, the left side of the brain influences what goes on in

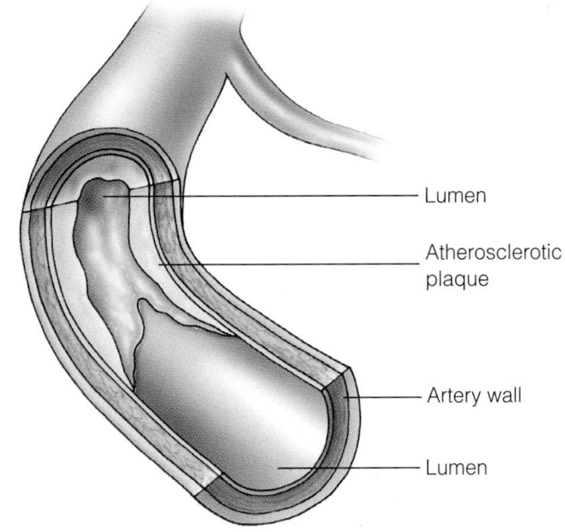

Figure 11-18 Plaque and Arteries
Buildup of atherosclerotic plaque along the walls of arteries reduces blood flow and increases blood pressure. Extensive buildup of plaque may occlude the artery completely.

the right side of the body and vice versa. Among the most obvious *unilateral effects* observed in victims of stroke are imbalances in facial muscle tension and loss of ability to control the movement of an arm or a leg.

Prevention and Intervention

There are a number of ways to prevent vascular system problems. The first is diet. Much of the plaque material that builds up along the walls of arteries is lipid in nature. This includes cholesterol and saturated fats as well as high- and low-density lipids (Table 11–2). When the body takes in more of some of these lipid materials than it can use, it tends to store them. The buildup of plaque in the artery (resulting from a condition known as atherosclerosis; see Figure 11–18) can be reversed in many cases simply by reducing the intake of foods rich in these substances.

Plaque fatty materials deposited along the surface of blood vessels, often partially or completely occluding the flow of blood

Occlusion a blockage of flow in the circulatory system

Myocardial infarction a blockage of blood flow to the myocardium; commonly associated with heart attacks

Stroke the loss of brain function (particularly motor control) due to oxygen deprivation caused by blockage of blood flow

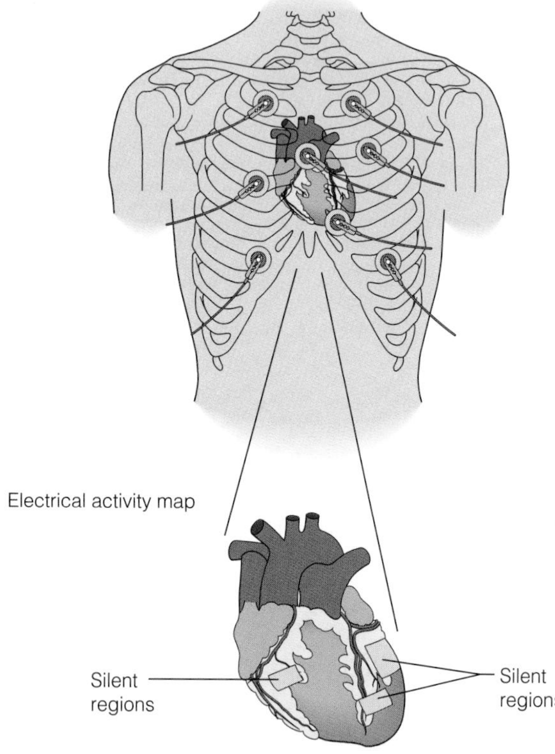

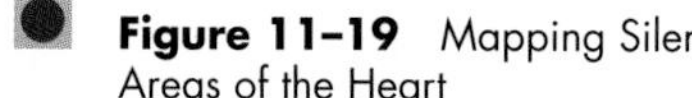

Figure 11–19 Mapping Silent Areas of the Heart

Damage to the heart from lack of oxygen to the heart muscle can be measured using electrodes to map electrical activity. Areas having no electrical signals arising from them are "silent regions," in which muscles are not contracting during a heart beat.

Low-cholesterol/low-saturated fat diets can reverse the buildup of plaque. However, it's not always as easy as that. If the atherosclerotic plaque is extensive, it may need more than a new diet to reduce or correct it. There are a number of drugs that have been introduced recently into the market that are effective in reducing plaque.

The most drastic measure to correct atherosclerosis is surgery. When diet and drugs do not work, or the situation is critical for the coronary arteries serving the heart itself, bypass surgery is needed to replace the blood vessels themselves. This is a dangerous and expensive operation, but it has saved lives and enhanced the quality of life for many of those who have undergone it. Somehow it seems easier, and wiser, to change a fat-rich diet and add some exercise-rich activities to our daily lives early on to prevent these problems from arising in the first place.

BLOOD

The anatomy and physiology of the system of blood vessels through which blood is pumped has been described, as has been the structure and function of the heart, arteries, veins, and capillaries. But we have not yet discussed blood itself. What is blood? How much of it do we have, and how and where is it made?

The Tissue that Flows

Blood is the tissue that flows (Figure 11–21). Blood contains both formed elements, such as cells, and nonformed elements, such as the fluid components of **plasma.** The measure of packed cells in any volume of blood is called the hematocrit. Blood is mostly water, but has within it a variety of very important ingredients, dissolved and suspended. The fluid part of blood is called plasma and contains molecules needed for maintaining a balanced pH, nutrient delivery, immune defenses (see Chapter 12), and all the factors needed for clotting of the blood (Table 11–3). Plasma is closely related in composition to *serum,* which is the straw-colored liquid that is separated from clotting blood minus the various factors associated with the clot itself. There are a variety of ions and other solutes dissolved in blood plasma. This includes sodium and chloride ions, as

Plasma the nonformed or fluid material of blood

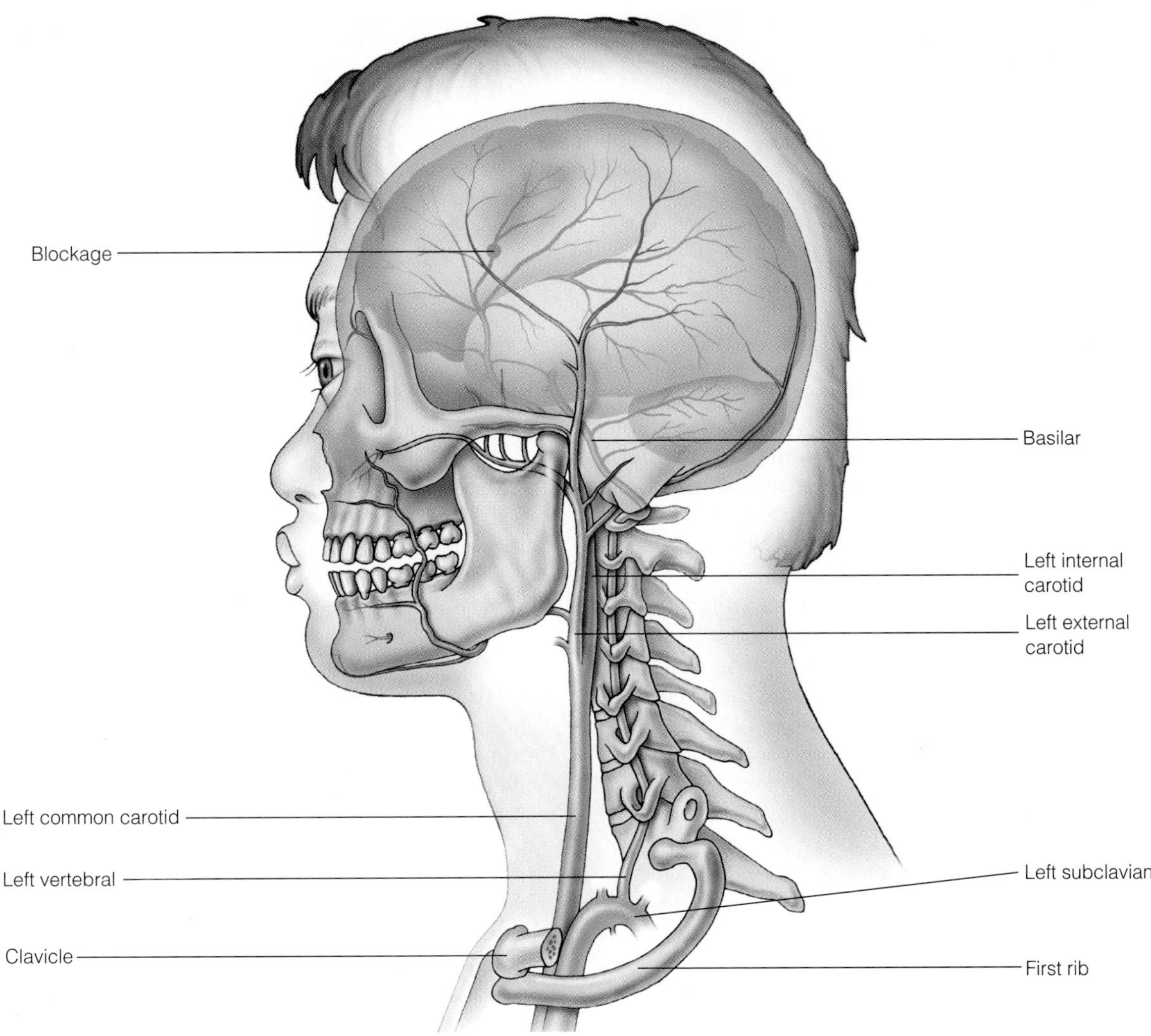

Figure 11–20 Brain Arteries
A lateral view of the brain is shown with blood entering the brain primarily from the carotid and vertebral arteries. Blockage of blood vessels denies oxygen to brain cells, which may die. This is the basis for stroke.

well as salts of calcium, potassium, magnesium, and many others. The blood also has significant levels of dissolved protein molecules, as well as carbohydrates and lipids. All of these combine to give the blood fluid considerable *viscosity* or thickness.

The blood is also replete with cells and cell-like particles, which are referred to as **formed elements.** There are many different kinds of formed elements found suspended and flowing in the internal river of blood including red blood cells (also referred to interchangeably as *erythrocytes*), white blood cells, and platelets. The predominant type of formed element by far is the erythrocyte. Erythrocytes are the fundamental cellular elements of the blood because they carry oxygen to body tissues and aid in the exchange of carbon dioxide. White blood cells include lymphocytes, monocytes, macrophages, and granulocytes, all of which are involved in defense of the body. Platelets are cell-like particles derived from cells in the bone marrow.

Formed elements the cellular components of blood, including red and white blood cells and platelets

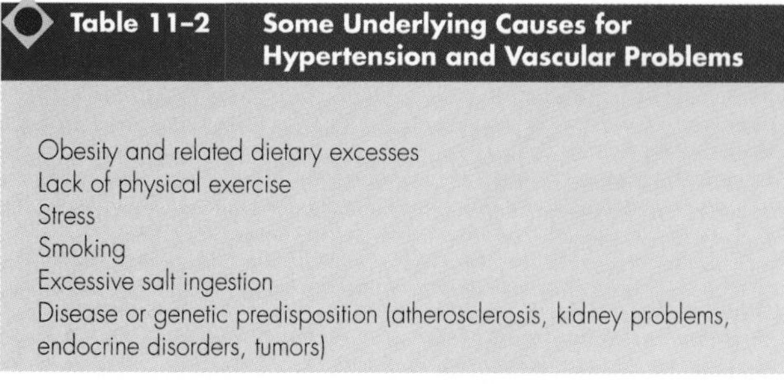

Table 11-2 Some Underlying Causes for Hypertension and Vascular Problems

Obesity and related dietary excesses
Lack of physical exercise
Stress
Smoking
Excessive salt ingestion
Disease or genetic predisposition (atherosclerosis, kidney problems, endocrine disorders, tumors)

Blood Volume

Blood is a fluid connective tissue, so it can be measured relatively easily in terms of its volume. What is the volume of blood for humans? The absolute volume varies somewhat between individuals of different size and sex, but blood represents approximately 7% of body weight. This works out at about 5–6 liters of blood (about 4 kg) for a 70 kg person. The number of red blood cells in this volume is enormous, easily 12 trillion to 14 trillion. Erythrocytes are among the smallest cells of the human body. They have a unique biconcave shape, which is reminiscent of a donut that did not quite finish getting a hole. Each cell is only .000007 meter in diameter. The number of cells in a tiny drop of blood is roughly equal to the population of New York city on a busy day. Males have an average of 5.4 million to 5.5 million cells per cubic millimeter. Females average about 4.8 million cells per cubic millimeter. Babies of both sexes average 4.5 million to 4.6 million. Owing to the importance of erythrocytes in gas exchange, it is imperative for medical science to know where these cells come from, how they are formed, and how their numbers in an individual are maintained.

Table 11-3 The Components of Plasma

Water
Proteins
  Albumins, alpha and beta globulins (e.g., clotting factors), fibrinogen, gamma globulin (antibodies)
Lipids
  Triglycerides, cholesterol
Carbohydrates
  Glucose
Ions
  Sodium, potassium, bicarbonate, chloride, calcium
Hormones
Gases
  Dissolved carbon dioxide and nitrogen

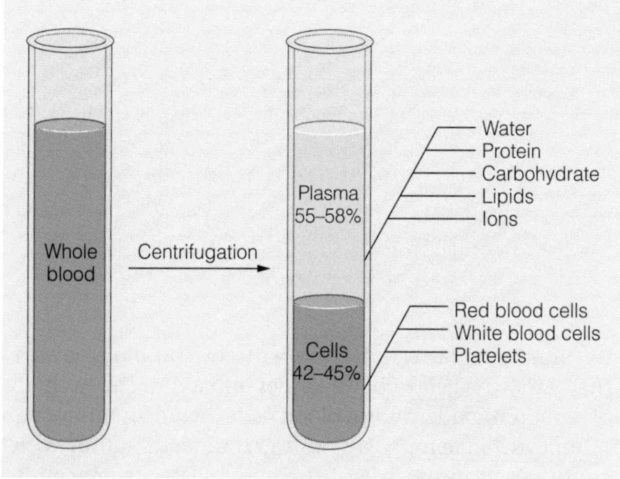

Figure 11-21 Blood and Its Components
Blood is composed of both formed elements (cells and platelets) and nonformed elements (plasma components). The percentage of formed to nonformed elements in blood is called the hematocrit. The value is easily determined by spinning a volume of blood in a centrifuge and measuring the volumes of the two elements. The average hematocrit value for an adult female is 42%; the average value for an adult male is 45%.

Sources of Blood

Blood cells in adults are formed in bones. In fact, in adult humans, blood cells are formed in special regions of a relatively few bones in the body. The bones of the ribs, sternum, vertebrae, and long bones are the principal sources of new blood cells. Within these bones is found the blood-producing tissue known as red marrow (also called hematopoietic tissue). The production of red blood cells is stimulated by a hormone known as **erythropoietin,** which is produced in the kidney. Erythropoietin is extremely specific in its effects on target cells, and the cells that produce it are very sensitive to factors controlling blood volume and erythrocyte concentration. For example, when we donate blood, this hormone is synthesized and released into vascular circulation and directly stimulates the increase in production of erythrocytes necessary to replace those lost. Although it is popularly thought that marrow in all bones forms blood, most of the bones in the adult human body, other than those mentioned above, have yellow marrow, which is predominantly fatty in nature and not blood forming (see Chapter 6 for more information on bones).

Turnover of Red Blood Cells

The average lifetime of a human red blood cell is 120 days. Erythrocytes born and mobilized in the marrow of a rib, for

Erythropoietin a hormone stimulating the production of erythrocytes

example, have only four months to circulate before they are demobilized. To maintain 12 trillion to 14 trillion red blood cells in the body on a continuing basis, over a million new red blood cells must be formed in the marrow every second of every day in our lives! The liver has the job of eliminating aging red blood cells. This selective removal of old erythrocytes occurs in a manner that is not completely understood. The cells of the liver somehow read a message in the molecules on the red blood cell's surface that advertizes their age. Wear and tear on these cells after a quarter of a million or more trips around the circulatory system makes them ripe for retirement. The old red blood cells are pulled out of circulation and their content resorbed by the liver. An important part of the job of the liver is to recycle the useful components of the red blood cells for the body. Hemoglobin is processed and the iron is saved for reutilization in new red blood cells. The globin proteins are transformed into *bile* for use in emulsification of fats for digestion—waste not, want not.

Embryo and Fetus

The sites at which red blood cells are made is not the same in an embryo or fetus as it is in the adult. The origin of the first red blood cells is actually from a region outside the embryo itself, in the *extraembryonic tissues*. As will be covered in detail in Chapter 15, there is a separation that occurs very early in development between the cells that will become the embryo proper and the cells that will form the membranes that surround or connect to the embryo (for example, *yolk sac, amnionic sac* and *chorion*). The precursors of the first human red blood cells arise in the extraembryonic tissues and migrate into the embryo itself. These migratory cells first populate the developing *spleen* and later in development migrate to the developing liver. Only much later do they arrive at their permanent home in the marrow of bones. In addition, fetal and early childhood blood formation in bone marrow is much more extensive than in adults. Many more bones are involved in making red blood cells early in our lives.

STOPPING LEAKS—THROMBOSIS

Among the many and diverse components of the blood is a special group involved in formation of blood clots, a process called **thrombosis.** When we cut ourselves sufficiently severely that we bleed, these components provide the molecular sealing mechanism. The players in this scenario are principally erythrocytes, platelets, and *fibrin* (Figure 11–22). Red blood cells play a passive role in this situation, as we will see later. As mentioned previously, platelets (which are also called *thrombocytes*) are not exactly cells, but they are specialized fragments of a cell. They are produced by budding from a massive progenitor cell, known as a *megakaryocyte*. This giant cell is located in bone marrow and releases tiny fragments of itself as membrane packets (the platelets), which in turn enter the circulation. There are about 200 million platelets in every milliliter of our blood.

Fibrin, on the other hand, is not cellular at all, it is a protein. A form of it circulates in the blood as a normal part of the plasma. You might be asking yourself, "If these components are already in the blood, why doesn't clotting occur all the time?" Fortunately, fibrin enters the circulation initially in an unreactive form of the active protein known as *fibrinogen*. In addition, intact platelets are unreactive until they break open and release their contents. This breakdown usually does not occur until a blood vessel is compromised in some way, such as following a cut or other physical damage.

Embolism

An intact blood vessel is lined with a vascular endothelium, which forms an epithelial sheet of cells that is contiguous throughout the entire vascular system. The potential for blood clotting has to be very carefully controlled. Clotting has to occur at the right place and at the right time. Forming a clot, which is commonly referred to as a **thrombus,** within a normal, intact system of vessels can be dangerous because it may break loose into the circulation. The release of a clot into the circulatory system is called an **embolus.** An embolus may stop the flow of blood to critical regions of the body, such as the heart or brain (Figures 11–20 and 11–22). Heart attacks and strokes are two of the most serious consequences of embolic blockage. Being unable to form clots at all is equally serious. Failure to stop bleeding means certain death, because there is no way to arrest the flow of blood from even the tiniest of cuts. In its most deadly form this condition is called *hemophilia*. Normally, the factors involved in clotting are inactive. What is it then that stimulates the platelets to react and the fibrinogen to convert to fibrin and join the process to stop the bleeding?

A Cascade of Events

Damage to a blood vessel exposes blood and blood vessels to a new environment. No longer isolated from the connective tissue that underlies the endothelium, the platelets become attached to the proteins of the extracellular matrix, particularly collagen, and burst open. The disintegration of

Thrombosis the process of blood clotting; involves a cascade of complex interactions

Thrombus a blood clot

Embolus a bulging out of a blood vessel from weakened wall structure

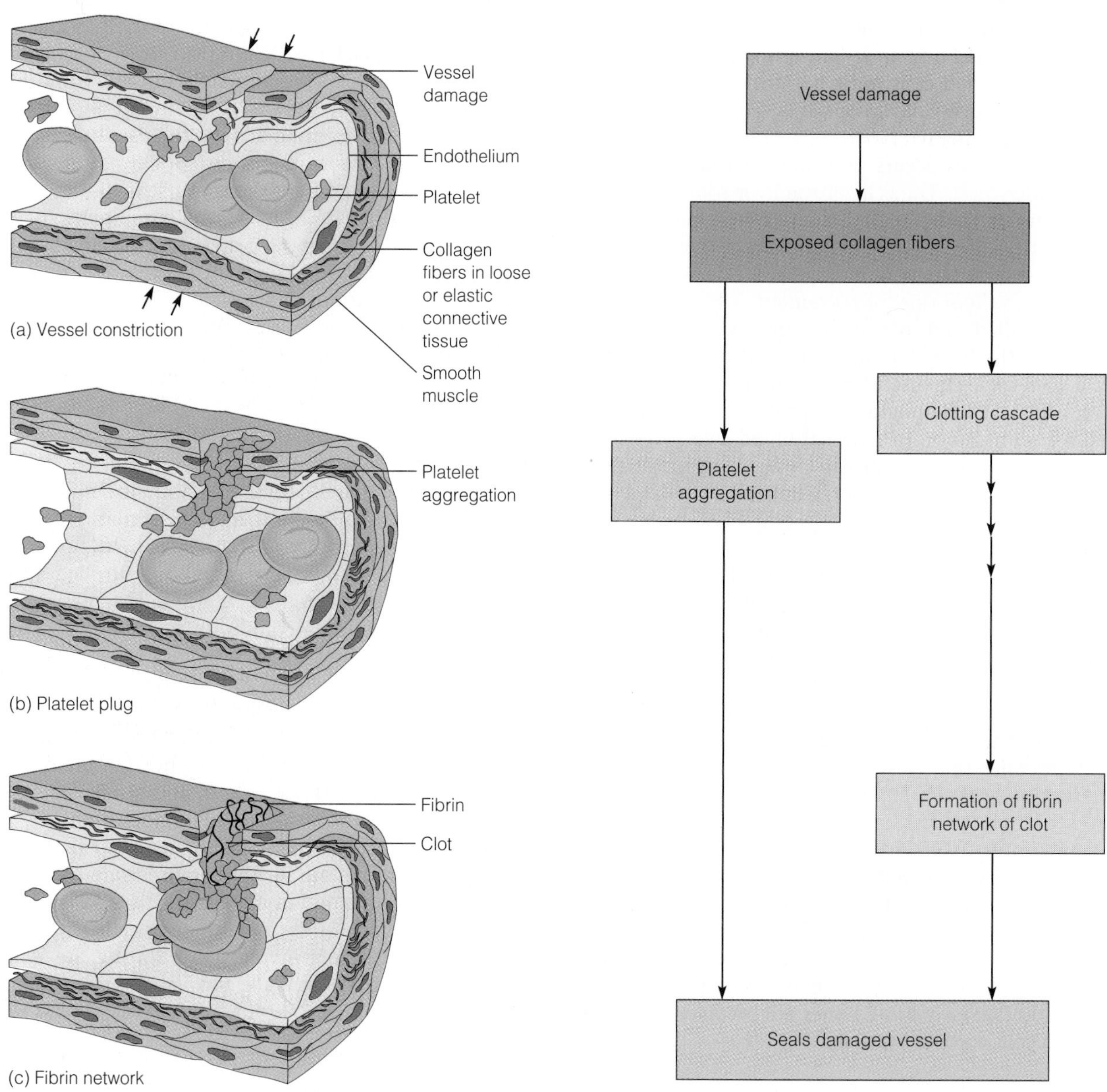

Figure 11-22 Blood Clotting—A Cascade of Events
The cascade of events in blood clotting is rapid and self-limiting. This ensures that vessel damage is repaired but that the clot does not spread to uninvolved areas.

platelets releases a number of stored molecules, including a very important one known as *adenosine diphosphate* or ADP. ADP is a signal molecule that attracts more platelets to the damaged site. A cascade of events is underway. The building blocks for sealing the hole begin to accumulate. In addition to platelet responses, a factor is released by the damaged tissue as well. In conjunction with plasma, calcium, and other circulating molecules, an enzyme called *thrombin* is formed from an inactive precursor known as *prothrombin*. Many of the protein factors involved in clotting are produced and maintained in an inactive form. This is essential because, as in the straw that broke the camel's back, they are ready to initiate clotting in an instant. Thrombin catalyzes the conversion of fibrinogen to fibrin. Fibrin forms the gluey fibrous meshwork that holds the clot together. Fibrin interacts with platelets and red blood cells by tying them into a big clot at the site of injury. If this occurs on the surface of the skin from a cut or scrape, what forms is commonly known as a *scab*. The cascade of blood-clotting events occurs rapidly and is very localized in extent. Once the cut or puncture is sealed, there are anti-clotting factors, such as *heparin,* that prevent the spread of the clot by limiting the clotting reaction. The clot ultimately protects the injured vessel by sealing it up so blood is not lost and entrance of foreign organisms and molecules is prevented (Figure 11–22).

Summary

The cardiovascular system delivers blood to all the organs of the human body and is essential in distributing adequate amounts of oxygen and nutrients to the cells that compose them. The key to the system is the in-line pump, the heart, that provides the force needed to push blood through the various types of blood vessels in the body. The heart is a muscular, four-chambered organ about the size of a human fist. It is surrounded by specialized membranes to protect and isolate it from other organs and structures within the thoracic cavity. Within the heart itself, there are two atria (upper chambers) connected by one-way valves to each of two ventricles (lower chambers), which in turn pump the blood through either the pulmonary circulation of the lungs or the systemic circulation of the rest of the body.

The heart has a highly regulated capacity to contract upon itself and reduce the volume of its chambers, thus, providing the motive force to push the blood through thousands of miles of circulatory tubing, commonly referred to as arteries, veins, and capillaries. Contraction of the heart is dependent on the electrical activity of the SA and AV nodes, which produce an asymmetrical contraction of the heart muscle. The contraction and relaxation of the heart, known as the states of systole and diastole, produce blood pressure within the system that can be measured quantitatively to assess cardiovascular heath. Although blood pressure in humans can vary from individual to individual, general ranges for males and females have been established. Diet and physical activity play an important role in preventing and treating abnormal conditions, such as hypertension, and problems associated with atherosclerosis.

The vascular system is organized as a one-way street that connects back to itself. Arteries carry blood away from the heart (usually oxygenated blood), and veins carry blood back to the heart after oxygen has been depleted from it. Large arteries are usually elastic in nature and help maintain blood pressure between heartbeats. Arterioles are the smallest of the arterial vessels and, as a consequence of their locations at the entrance to capillaries, control much of the flow of blood to the periphery and surface of the body. It is within the capillaries that red blood cells release the oxygen bound to hemoglobin to replenish the body's cells and tissues. Venules (the smallest of the veins) are connected to capillaries on the opposite side from the arterioles. All types of veins have one-way valves that aid in the return of relatively low-pressure blood to the heart.

The source of blood cells of all types in adult humans is bone marrow, specifically red bone marrow. However, the embryonic origins of blood cells occur outside the embryo proper. Blood cells migrate first to the spleen and from there to the liver before finally coming to reside in the marrow. Blood is composed of formed (cellular) and nonformed (water, proteins, salts) elements. The predominant type of formed element is the red blood cell or erythrocyte, which plays an essential role not only in oxygen delivery but in the process of thrombosis, or blood clotting. The formed and nonformed elements of the blood are poised to respond to damage (cuts, punctures) by sealing up the leaks using fibrin, platelets, and erythrocytes to form blood clots. Abnormal clotting can result in blood clots that may break loose within the cardiovascular system, which may give rise to blockages (embolisms) leading to heart attack and stroke.

Questions for Critical Inquiry

1. How does the anatomical organization of the heart ensure efficient flow of blood? What might go wrong with these arrangements that would alter their effectiveness?
2. How is heart muscle different from all other types of muscle? How is its contraction controlled and regulated?
3. What is the functional relationship between the SA and AV nodes? Why is the relationship important?
4. What are some of the causes of essential hypertension? What other factors might come into play to increase blood pressure?
5. What is a thrombosis? Under what conditions does one occur normally? What is the relationship between a thrombus and an embolism?

Questions of Facts and Figures

6. Is the flow of blood unidirectional? If so, why?
7. What is the molecule within red blood cells that binds oxygen? Where does oxygen binding take place? Where does unbinding take place?
8. What is the estimated total length of blood vessels in the human body?
9. What are the differences between arteries and veins?
10. How do valves function throughout the vascular system?
11. What are the two circulatory pathways of blood?
12. How do smooth muscles help control the pattern of blood flow on the human body?
13. What are the systolic and diastolic phases of the heart?
14. Where are red blood cells made in the adult human body? How many per second? What is the total number of red blood cells in the body? How much volume of blood do the average human male and female contain?

References and Further Readings

Cummings, M. R. (1996). *Biology: Science and Life* St. Paul, MN: West Publishing.

Cunningham, A. (1988). "William Harvey: The Discovery of the Circulation of the Blood," *Man Masters Nature: Twenty-Five Centuries of Science*. Porter, R., ed. New York: George Braziller.

Sherwood, L. (1993). *Human Physiology,* 2nd ed. St. Paul, MN: West Publishing.

Stalheim-Smith, A., and Fitch, G. K. (1993). *Understanding Human Anatomy and Physiology* St. Paul, MN: West Publishing.

CHAPTER 12

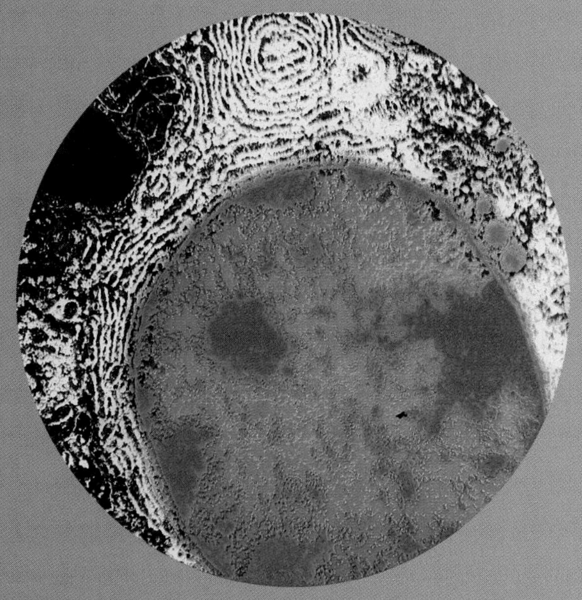

The Immune System and Defending the Body

INTRODUCTION

It has been said many times in the context of athletic competition and military campaigns that often the best defense is a good offense. This certainly describes systems of natural protection and immunity in the human body. The cells and molecules involved in providing immune protection arise as part of the same system that continuously gives rise to red blood cells. However, in the case of cells involved in defense and immunity, the immune system depends on the growth and development of a large family of cell types known as white blood cells. This class of cells is involved in establishing and maintaining the inherent defenses of the human body. These cellular- and molecular-based defenses against foreign invasion fall into three basic categories—fixed, mobile, and acquired.

FIXED DEFENSES

The human body is constantly confronted with a vast array of living and nonliving foreign agents and organic materials that potentially may do it damage. Viruses, bacteria, fungi, and toxins are all potential agents of tissue destruction. However, the body has an amazing array of defenses to protect itself. The first lines of defense against most of these agents are the epidermis of the skin and the mucous membranes lining the internal surfaces of the body, such as those found in the lungs and digestive tract (Figure 12–1). These are the truly fixed defenses, analogous to walls, bunkers, and moats to keep the enemy out. Epithelial tissues tightly seal and cover the external and internal surfaces of the body. For example, the skin provides a tough, water-tight surface composed of a continually replenished layer of nonliving material known as keratin. Skin is quite good at its job, but it is not perfect. Cuts and abrasions open the skin and provide a chance for an armada of organisms from outside to enter the body.

The mucous membranes of the respiratory and digestive tracts are thinner and more moist than the skin, but they serve well to trap and remove, or destroy, foreign materials that have entered the body. The stomach is protective because it produces an extremely acidic environment in which the hydrolysis of many kinds of foreign organic materials takes place all the time. Fluids produced by *lacrimal glands* located above the orbit of each eye drain as tears onto the surface of the eye (Figure 12–2). Tears act to wash debris and foreign materials of all types from the surface of the eye into the nasal cavity and into the pharynx through the act of swallowing, down the esophagus, and, ultimately, to destruction in the stomach. In addition to simply washing debris away, tears contain an enzyme known as **lysozyme,** which is capable of digesting bacterial cell walls and destroying many kinds of microorganisms. However, these epithelia, and their acidic or enzymatic secretions, are not perfect barriers. Once a bacterium or a virus penetrates the first line of defenses, the body requires backup mechanisms to seek them out and destroy them. The second line of defense, complementary to the fixed defenses, is composed of the cells and cellular products that compose a complex and multifaceted immune system.

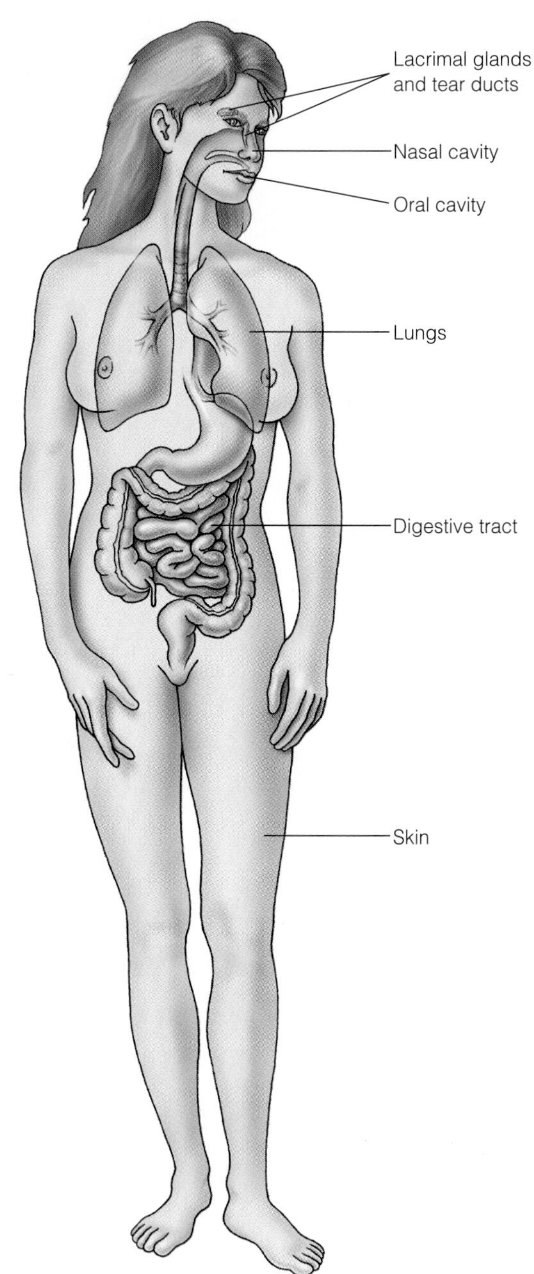

Figure 12–1 Fixed Defenses
The fixed defenses provide barriers to the entry of foreign materials or organisms into the body.

Lysozyme an enzyme of tears with antibacterial action

MOBILE DEFENSES

To make sense of what is meant by the designation mobile defenses, we will take a careful look at several different cell types involved. These cells fall into the category of blood cells known as white blood cells or WBCs. White blood cells along with red blood cells (RBCs) are produced in the red marrow of bones. WBCs are probably best known from the

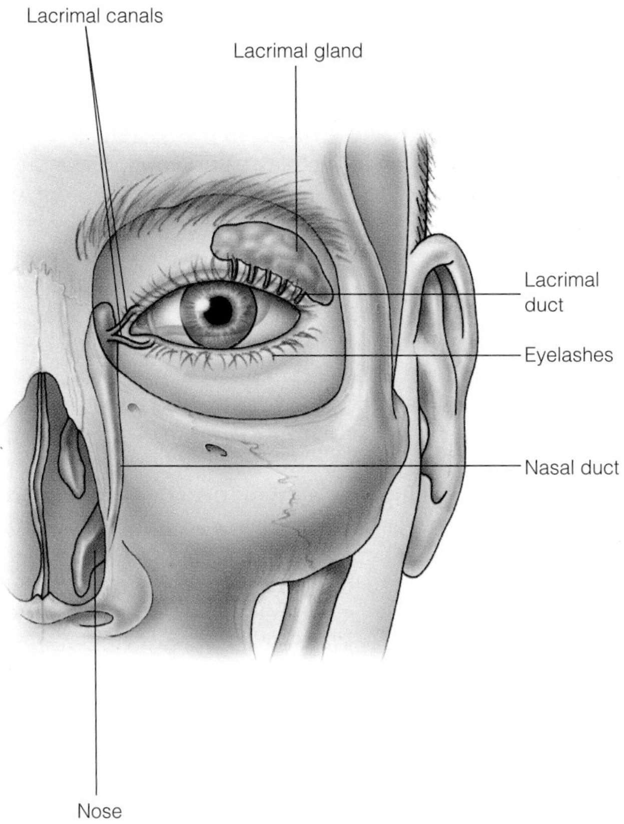

Figure 12-2 Eyes and Tears
The lacrimal glands produce a fluid that washes over the eye and flushes debris through the lacrimal canals and into the nasal cavity.

morphological appearance they have when circulating within the vascular system of the body. However, it should be kept in mind that their characteristic functions are manifested predominantly in the interstitial tissues of the body where they behave as phagocytes. WBCs have the unique ability to exit the circulatory system by migrating between endothelial cells in the capillaries and into surrounding tissues. There are two major groups of white blood cells, **granulocytes** and **agranulocytes** (Figure 12–3). In the most general sense, this means that the cells of the first group have obvious and characteristic granules within their cytoplasm and that the cells of the second group do not. However, this is oversimplified in terms of the actual structural and functional differences between the various cell types. As will be described, each of the types of WBCs follows a very different pathway of development and specialization.

Granulocytes

The categories of granulocytes are the *neutrophils, eosinophils,* and the *basophils* (Figure 12–3[a]). Neutrophils are the most prevalent of all white blood cell types, composing about 60–65% of the total. Eosinophils make up 2%–4% of this population and basophils are 1% or less. WBCs are represented

Granulocytes a class of white blood cells containing obvious cytoplasmic granules

Agranulocytes a class of white blood cells containing no obvious granules

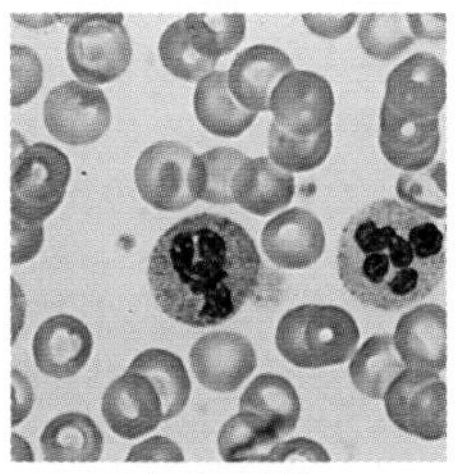
Neutrophils

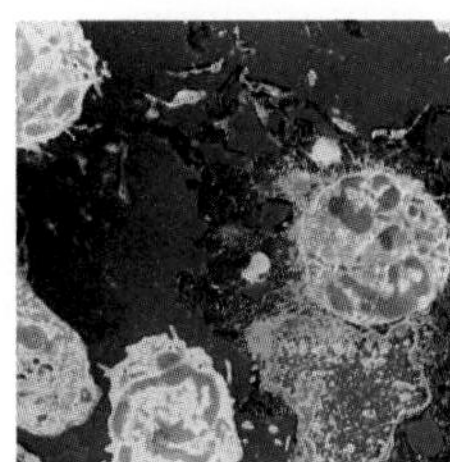
Eosinophils

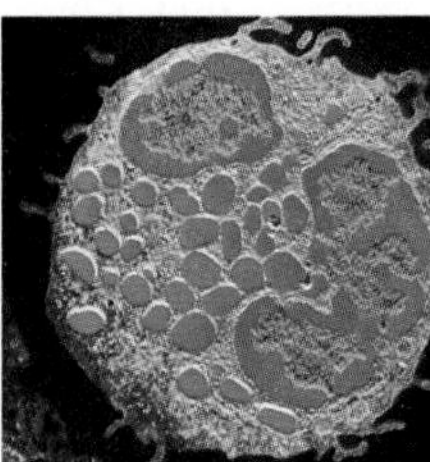
Basophils

(a) Granulocytes

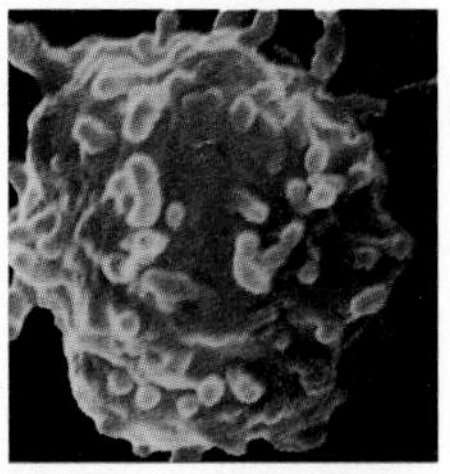
Lymphocyte T-Cell

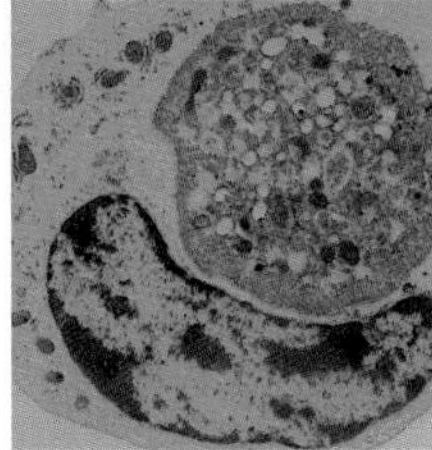
Monocytes

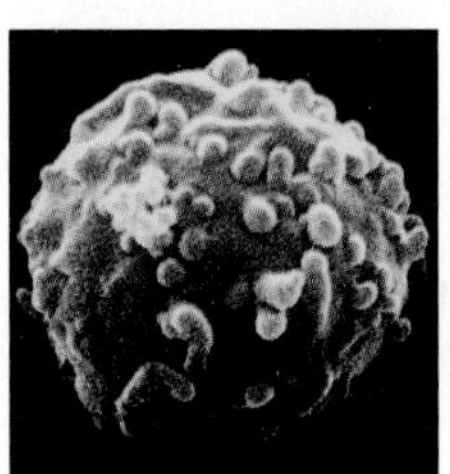
Lymphocyte B-Cell

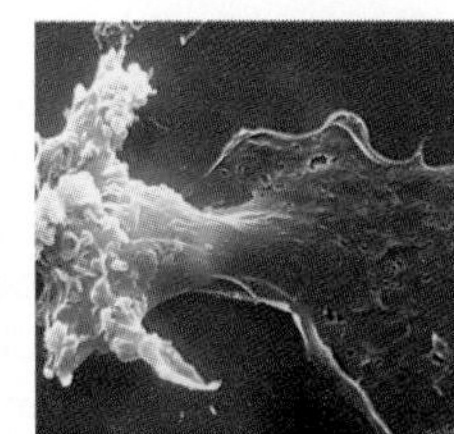
Macrophages

(b) Agranulocytes

Figure 12-3 Mobile Defenders—White Blood Cells
The two classes of white blood cells (WBCs) are (a) granulocytes and (b) agranulocytes, which include lymphocytes (B- and T-cells) and monocytes/macrophages.

at levels a thousand times or so less than RBCs (Figure 12–4). The difference in total number of RBCs versus WBCs is dramatic. RBCs occur in blood at 4 million to 5 million per cubic millimeter, whereas WBC types combined normally occur at approximately 10 thousand per cubic millimeter. Examination of a sample of blood using a microscope indicates the rarity of the occurrence of white blood cells. However, the numbers alone are misleading. The low percentage of white blood cells in blood belies their biological importance.

Elements of the mobile defense systems are also associated with certain specified areas of the body. They are not fixed in place in the sense of the epithelial tissues, but rather they reside in more or less restricted outposts. This elaborate, regionally organized defense is part of what is called the **reticuloendothelial system.** For the most part the cells of the reticuloendothelial system are phagocytic. This means that they are capable of recognizing and devouring microbes and other materials that enter the body. These phagocytic cells are stationed in special sites within the bone marrow, liver, spleen, thymus gland, and lymph nodes. As we will discuss shortly, the thymus gland, lymph nodes, and spleen are part of a system of circulatory elements known as the *lymphatics.*

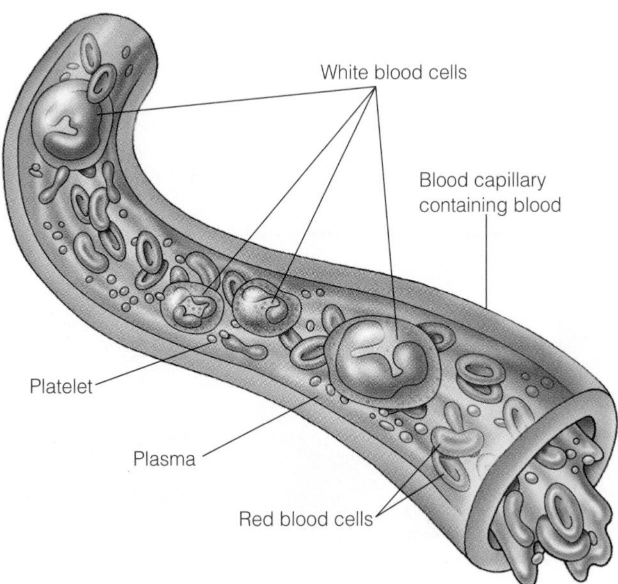

Figure 12–4 Blood Cells in Circulation
The formed elements of blood are red blood cells, white blood cells, and platelets. White blood cells are the defenders of the body. Several types of white blood cells are able to exit the circulation and enter surrounding tissues. White blood cells are represented in the following numbers in each cubic millimeter of blood: neutrophils, 5000; lymphocytes, 3000; monocytes, 1000; eosinophils, 400; and basophils, 100. For comparison, red blood cells are represented at 4 million to 5 million cells per cubic millimeter.

The lymphatic system is a second major circulatory system in the body. The fluid flowing in this system is known as *lymph* and is related to the plasma of blood. The system interconnects specialized lymph nodes and lymph organs, which together house enormous numbers of lymphocytes (Figure 12–5). The system of interconnected lymph nodes swells markedly during viral infections, a clear indication that the body is mounting an immune defense reaction. The lymphatics have several roles in addition to being essential for immune defenses. This system returns fluids that have leaked from the blood capillaries to the blood. It also transports fat derived from digestion back into the blood, and it acts as part of the fixed defenses that filter body fluids. The lymphatic system interconnects all parts of the body and eventually retrieves and collects lymph that re-enters blood circulation through portals in the subclavian veins (Figure 12–6).

It is worth noting that the reticuloendothelial cells of the liver, cells known as *Kupffer cells,* are probably responsible for capturing aged red blood cells and beginning the process of recycling their contents (Figure 12–7). The recognition of self and non-self (foreignness) changes with time, even with respect to the identification of cells of our own body. As will be described later, a number of important human diseases occur as a consequence of the failure of the mechanisms involved in self-recognition.

Neutrophils What if the microbes escape the clutches of the stationary, reticuloendothelial phagocytes? For this we send out the cavalry. Granulocytes are quick to counterattack the invasion of microbes. A bacterial infection of the skin, for example, may become inflamed and swollen. Neutrophils are attracted to these sites of damage or encroachment by special chemical factors released from the tissues affected. These chemical attractants act as a homing signal, which is picked up by receptors on the surface of circulating neutrophils. The neutrophils follow the vascular street map to the afflicted region, using the gradient of chemical attractants. Employing special adhesive molecules on their cell surfaces, the neutrophils attach to the endothelial cells of the capillary and push their way into the surrounding tissue. From there, they migrate to the target area and begin to ingest the foreign invaders (usually bacteria) and engulf inflammatory substances (Figure 12–8).

> **Reticuloendothelial system** macrophages and phagocytic cells circulating in blood, connective tissue, and capillaries capable of removing and destroying foreign cells

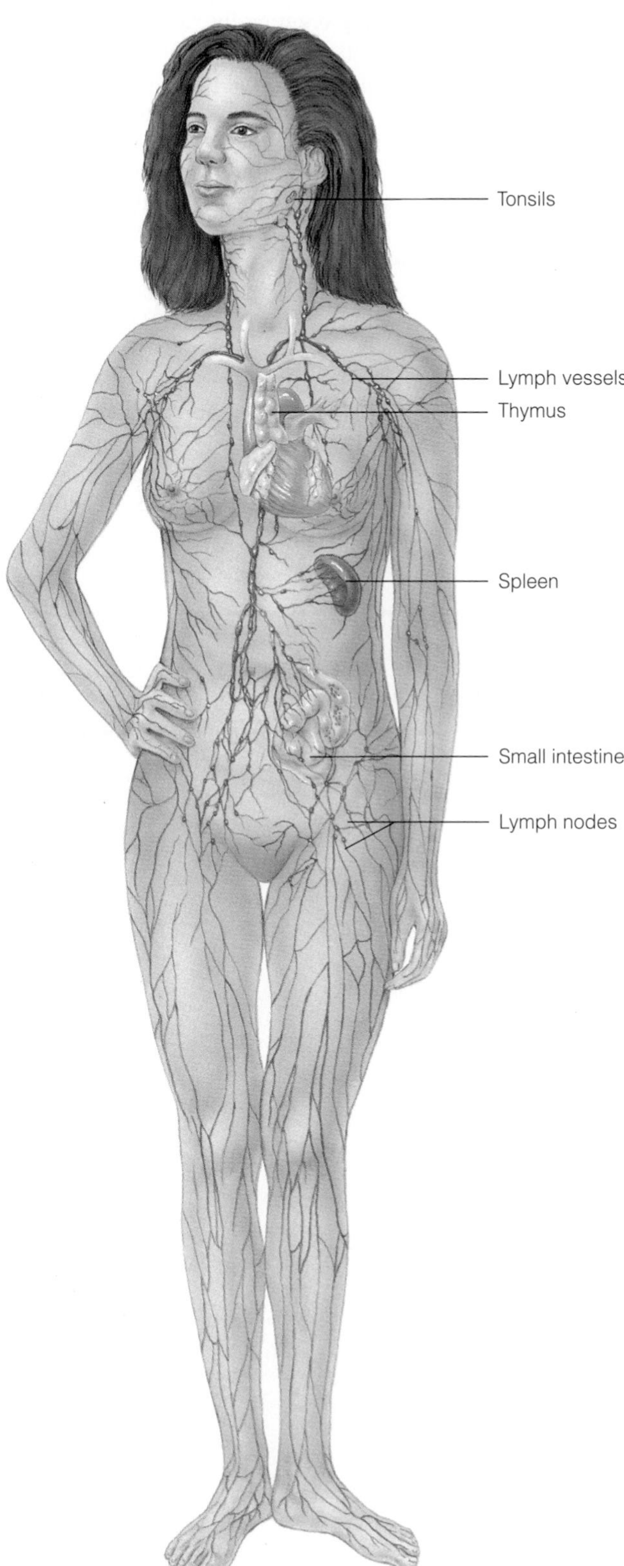

Figure 12-5 Lymphatic System
The lymphatic system includes several specialized organs (thymus, spleen) and a large network of lymph vessels and lymph nodes.

Figure 12-6 Lymph Vessels
Fluid enters the lymphatic system through lymph capillaries. The lymph flows within lymphatic vessels, through nodes and ducts, and empties into the vascular circulatory system through the subclavian vein.

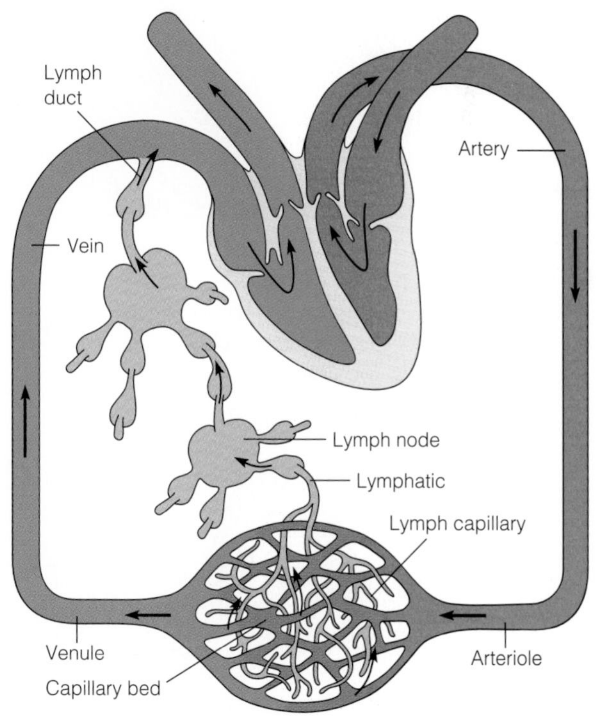

Once inside the neutrophil, the ingested material becomes the target of lysosomes. These organelles are essentially bags of enzymes of many different types that break down the foreign materials. As more and more neutrophils respond to the chemical attractant and migrate to the target area, more and more debris from the battle builds up. The neutrophils have a dual role in the process. The first is to attack and eliminate the infectious (or toxic) agents involved. The second is to clean up the mess afterwards, which they do with the help of *macrophages*. Pus in a wound arises through the action of neutrophils. Pus is a mixture of dead tissue, dead bacteria, and neutrophils stuffed by phagocytic consumption.

Eosinophils Eosinophils participate in this process of eliminating foreign intruders in a fashion similar to

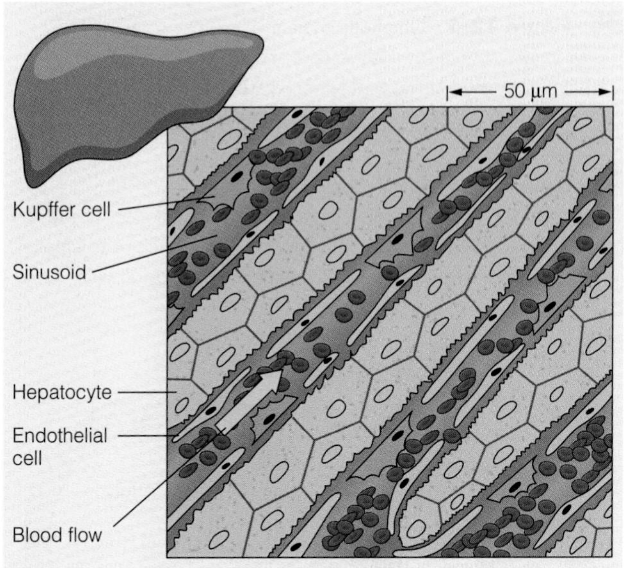

Figure 12–7 The Liver and Kupffer Cells
The liver contains macrophage-like cells known as Kupffer cells. These cells are part of the reticuloendothelial system and help remove debris, foreign materials, and old red blood cells from circulation.

neutrophils, except they are generally slower to arrive at the inflamed site and a bit more specific in what they respond to. The eosinophils are the principal aggressors against **parasites** (for example, worms and flukes), as compared to neutrophils, which are particularly sensitive to bacterial invasion. This differential response to different types of invaders is an important clue for a physician in determining treatment for infections that are not readily eliminated by host defenses. Increase (or decrease) in the number of specific types of white blood cells in the blood may indicate the type of infection you have (Table 12–1).

Basophils Basophils are rare in the blood, making up less than 1% of all WBCs. However, they are prevalent in the interstitial tissues of the body (Figure 12–9). Basophils are approximately the size of neutrophils and display granules that contain molecules with special properties. In addition

Parasites organisms living within other organisms obtaining nutrients at the host's expense; eosinophils function to destroy many types of parasites

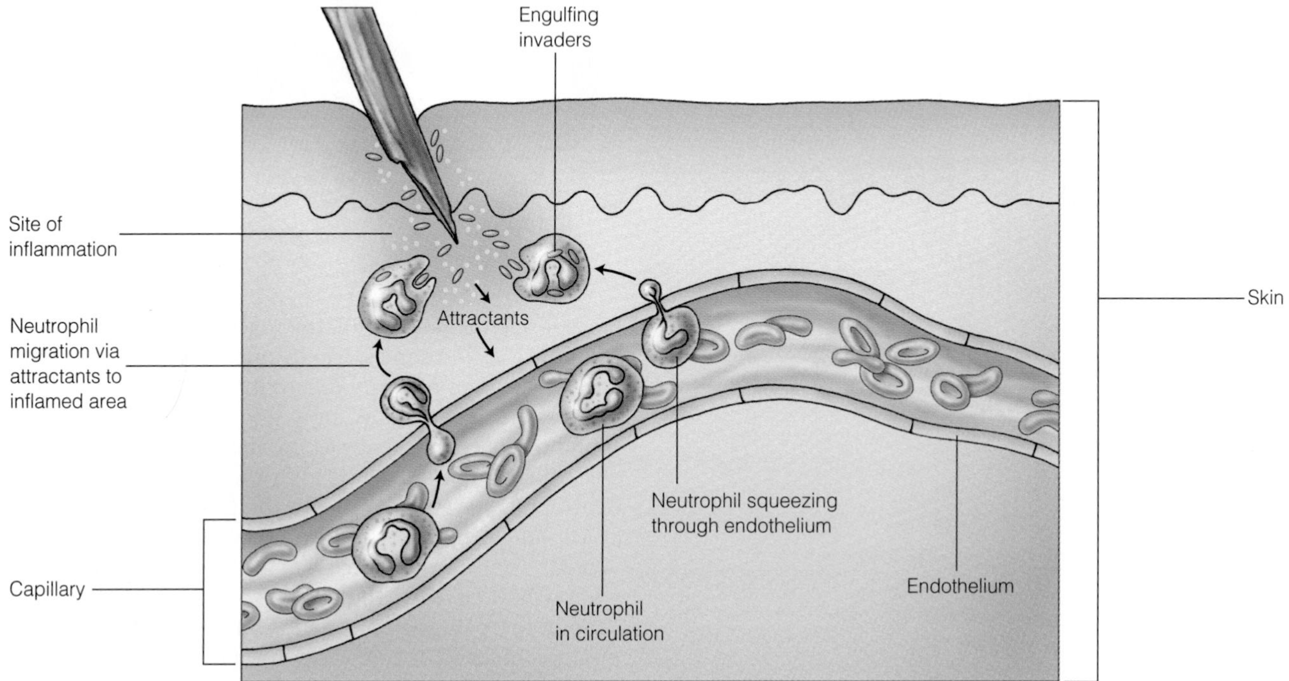

Figure 12–8 Neutrophils and Tissue Inflammation
Neutrophils are rapidly deployed to the site of damage (e.g., splinter) and inflammation. They exit the circulation and migrate to the site, where they attack invading organisms and eliminate debris.

Table 12-1 White Blood Cells and Disease

CELL TYPES	NORMAL PERCENTAGE OF TOTAL	HIGHER THAN NORMAL CONDITIONS	LOWER THAN NORMAL CONDITIONS
Overall WBC cell counts (4,500–11,000 mm^3 blood)	100%	Leukemia, bacterial infection, ulcers, uremia, pregnancy	Hepatitis, measles, cirrhosis of liver, typhoid fever
Granulocytes			
Neutrophils	60–65	Appendicitis, bacterial infection	Not known
Eosinophils	2–4	Parasitic infections, scarlet fever, allergies	Not known
Basophils	0–1	Not known	Not known
Agranulocytes			
Lymphocytes (B/T)	20–30	Viral infections, mononucleosis	AIDS (T-lymphocytes selectively destroyed)
Monocytes	5–10	Tuberculosis, Hodgkin's disease	

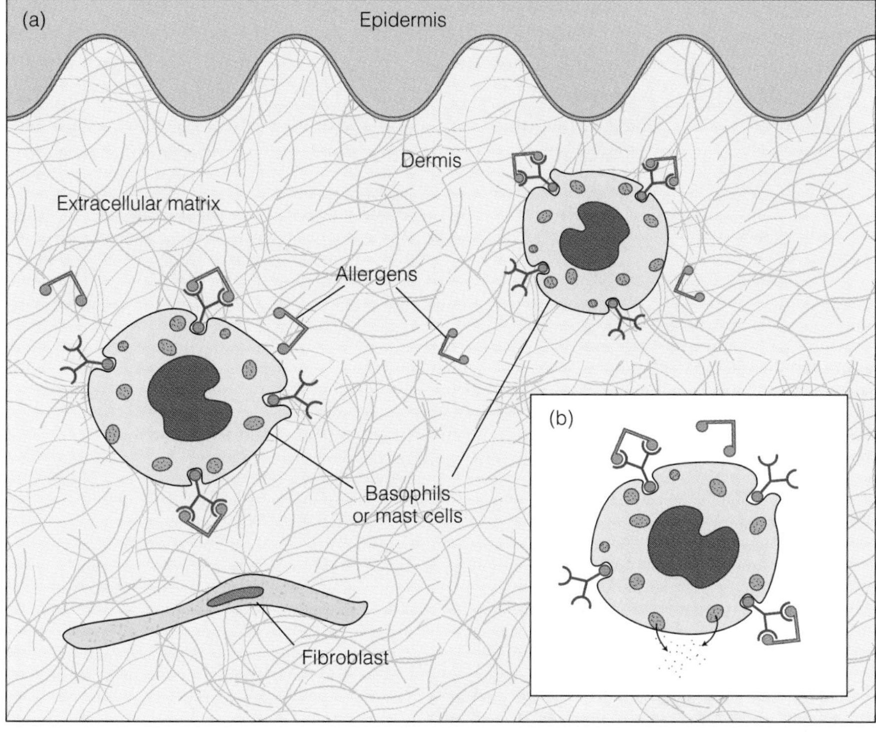

Figure 12-9 Basophils Within Tissues

Basophils (or mast cells) are located in tissues throughout the body as indicated for skin in (a). Basophils bind allergens (b) using surface-located IgE antibodies and respond to such binding by releasing histamines and other chemicals. This is the basis for an allergic response.

to enzymes, the granules store **histamine** and heparin. Histamine is released in tissues that have been injured or chemically stimulated and initiate inflammatory responses. The effects of histamine are to loosen the adhesive contacts between endothelial cells in capillaries and shift water from blood to tissues to bring about swelling. The loosening of the endothelium allows more rapid entry of neutrophils, eosinophils, and macrophages into the affected area.

Heparin is also released by basophils, but its function is quite different from that of histamine. Heparin interferes with blood clotting and, thus, is considered an *anticoagulant.* As discussed in Chapter 11, heparin may help limit the extent of thrombosis to damaged blood vessels. Basophils in the interstitial tissues are similar and probably identical to mast cells, which also release histamines. Basophils will be discussed again later in the chapter in conjunction with allergic responses and hypersensitivity.

Histamine a compound produced and released by basophils during an allergic response

Agranulocytes

Agranulocytes include B- and T-lymphocytes (also known as B-cells and T-cells) and the monocyte/macrophage lineage of cells. Lymphocytes are the second most abundant type of white blood cells composing 20%–30% of the total. Monocytes and macrophages are actually different forms of the same cell type and represent an average of about 5%–10% of WBCs (Figure 12–4 and Table 12–1). Monocytes are found in circulation, while macrophages are found migrating within tissues. Monocytes attach to and translocate themselves out of the blood vessels by pushing their way between the endothelial cells lining the capillaries, as do neutrophils and eosinophils. The percentages of different WBCs in the blood vary depending on a number of conditions in the body. Infections of different types cause fluctuations in the relative numbers of the different kinds of white blood cells, as do certain kinds of noninfectious diseases. Lymphocytes are very important defenders of the body's integrity and their increase in number is an excellent indicator of an existing infection. High white blood cell counts in the blood, particularly lymphocytes and monocytes, are often an indication of a viral infection. It is with regard to lymphocytes that we see the greatest range and the most complexity of human defensive responses to foreign invasion.

ACQUIRED DEFENSES

The acquisition of human immunity against specific agents, such as viruses and toxins, is ultimately based on recognition of the molecules of which the invader is composed. It is in this role as recognizers of foreignness that T- and B-cells function. B- and T-cells are both derived from precursor cells within the bone marrow, but each type undergoes differentiation in a different part of the lymphatic and blood systems. B-cells differentiate in the bone marrow and T-cells can differentiate only in the thymus gland, hence, B and T letter designations. As described earlier, there are many specialized and interconnected lymphoid organs involved in immunological defense and each has resident phagocytic cells and lymphocytes present. In this section we will first focus on B-lymphocytes and in the section following we will describe T-lymphocytes.

B-Lymphocytes and Antibodies

B-Lymphocytes are indirectly involved in recognizing foreign substances. There are two main types of B-cells. The first is the plasma cell, which produces and secretes antibodies that circulate in the body fluids. Secreted antibodies are also referred to as *humoral antibodies* and the response of B-lymphocytes to stimulation is called a **humoral response.** The second type of B-cell is the *memory cell* (Figure 12–10). The memory cell remains after the initial humoral response has declined and serves as a reservoir of cells that can rapidly produce more plasma cells should the body ever be exposed to the same antigen again in the months or years ahead. When a body responds to an invading organism, it amplifies a subpopulation of B-lymphocytes producing protein molecules that recognize and react against specific components of the foreign invaders. The molecules that are made by lymphocytes that recognize foreign molecules are called **antibodies.** The foreign molecules that stimulate antibody production are called **antigens.** We all produce antibodies on a continuous basis and they circulate throughout the

Humoral response an immune reaction resulting in the production of Ig molecules by B-lymphocytes

Antibodies a large class of proteins in the immunoglobulin family produced by B-lymphocytes

Antigens literally "antibody generators"; molecules that can be used to induce antibody production

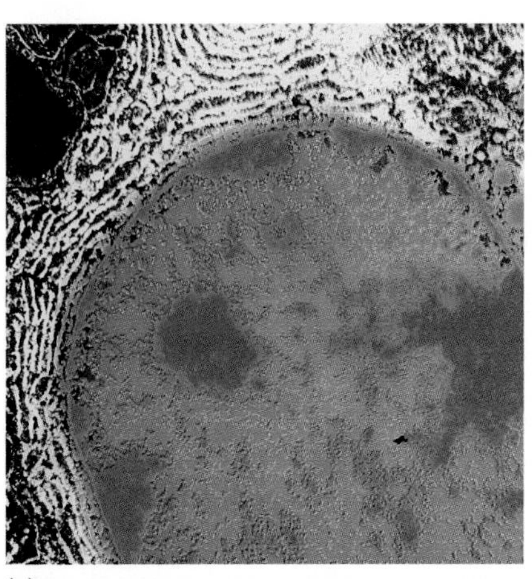
(a)

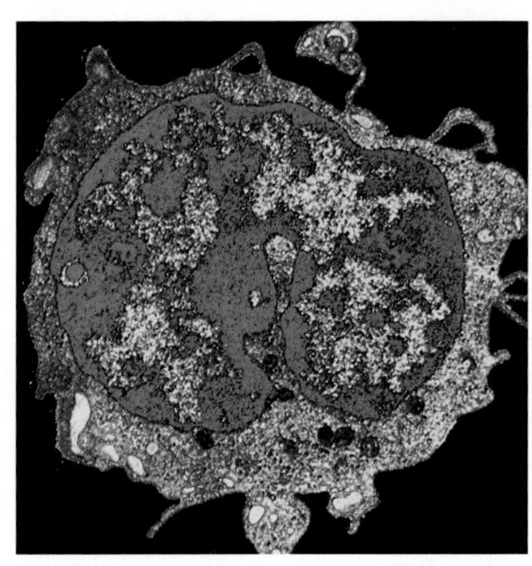
(b)

Figure 12–10 B- and T-Lymphocytes
Lymphocytes are remarkably similar when observed as resting cells (a). Upon activation, B-cells form plasma cells and secrete antibodies, and T-cells (b) become activated for cell-mediated lysis (killer cells) or helper functions.

body, along with the memory cells capable of supplying plasma B-cells to produce more. Antibodies are one of the major classes of molecules found in our blood and are classified as *immunoglobulins*.

Immunoglobulins There are five types of immunoglobulins or Ig molecules: IgA, IgD, IgE, IgG, and IgM. The IgG molecule is the best known of these and has a unique molecular structure involving the association of *heavy* and *light polypeptide chains* (Figure 12–11). The subunit chains combine to form the binding sites for antigen recognition, as well as other effector functions. One of the most important effector functions of antibodies is in the initiation of **complement-mediated cell lysis** (Figure 12–12). The set of proteins known as *complement* are present in the blood and they assemble on the target cell surface and create a hole in the cell, killing it. The complement proteins assemble in a sequence to an antibody bound to a foreign cell surface. Factors C5b, C6, C7, and C8 interact to establish a site at which multiple C9 units integrate with the plasma membrane to form an aqueous transmembrane channel (Figure 12–12).

Each type of immunoglobulin is involved in a different aspect of human immunity to agents of disease (Table 12–2). The IgA molecule is secreted by exocrine glands throughout the body and is particularly important in protecting mucous membranes. The IgE molecule is found on basophils and mast cells and mediates responses to allergens. This will be discussed in more detail in the sections on hypersensitivity and anaphylaxis. The IgM molecule is the largest of the

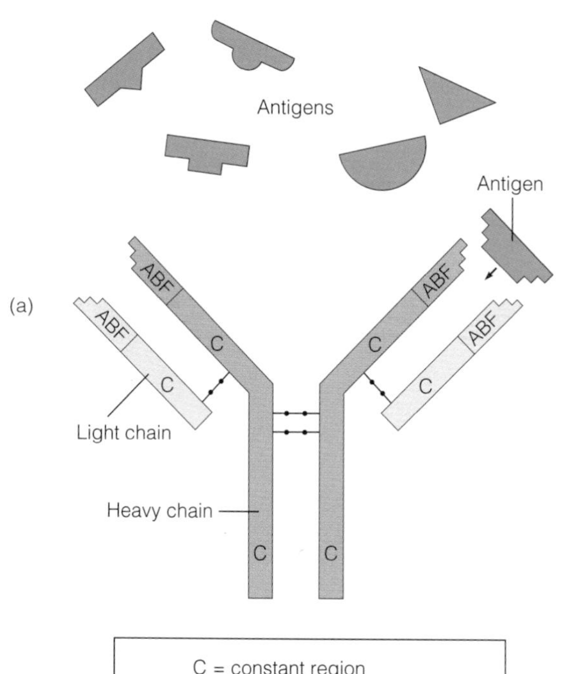

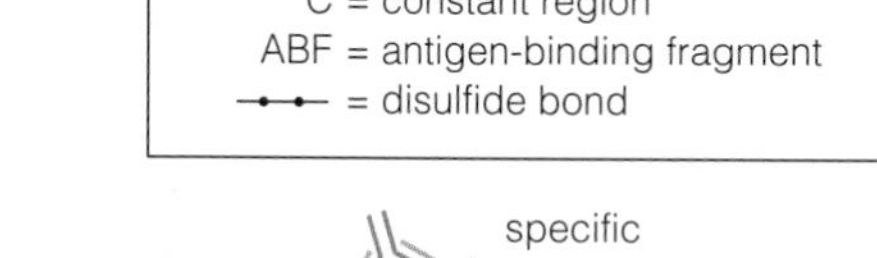

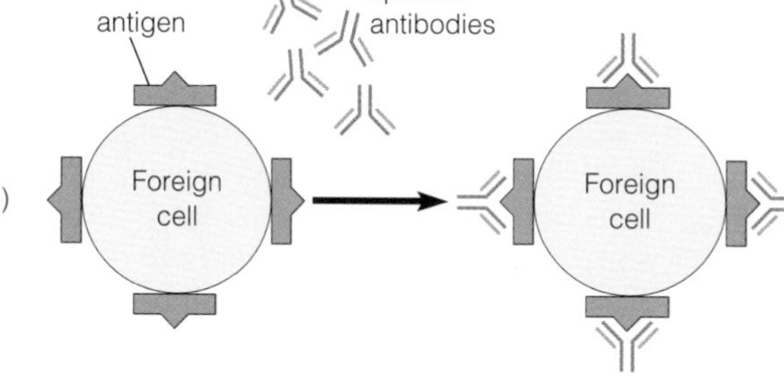

Figure 12–11 An Antibody Molecule and Its Antigen-Binding Sites

Antibodies are composed of four chains—two heavy, two light. The antigen-binding sites are composed of parts of heavy and light chains (a). Binding of antibodies to a foreign cell mark it for destruction by the immune system.

Figure 12–12 Complement and Cell Lysis

The assembly of complement proteins begins with binding of complement to antibodies (not shown) and finishes with the sequential assembly of complement factors C5b, C6, C7, C8, and C9 into a channel or hole in the plasma membrane. The foreign cell shown here has a transmembrane aqueous channel composed of complement proteins C8 and C9, which makes it "leaky."

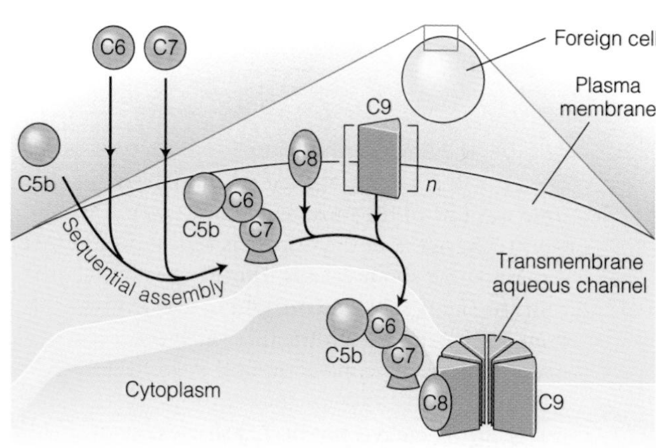

Complement-mediated cell lysis a process by which Ig molecules target foreign cells for destruction using a series of specialized complement proteins

Table 12-2 Antibodies

CLASS	LOCATION	FUNCTION
IgG	Interstitial fluid, plasma	Bind bacteria, viruses, and toxins
IgA	Exocrine gland secretion (saliva)	Bind bacteria, fungus, viruses on mucous membranes
IgM	Blood plasma	Bind and agglutinate bacteria
IgD	Surface of B-cells	Unknown
IgE	Exocrine gland secretions (on surface of basophils)	Produce allergic reactions

immunoglobulin molecules and is carried in the blood plasma along with its relative the IgG molecule. Both types of immunoglobulins react against bacteria, and IgG molecules react against viruses and toxins as well. The IgD molecule is found on the surfaces of B-lymphocytes, but no function has yet been ascribed to it.

Different populations of B-lymphocytes produce the five Igs on a continuous basis. From the point of view of human health, it is important to be exposed to a wide diversity of antigens for the specific purpose of letting the body make antibodies against them. This kind of response is called **acquired immunity,** because it is the body itself that is building up an arsenal of molecular weapons against foreign materials [Figure 12–13(a)]. As will be described later, one of the means of bringing about acquired immunity is through *vaccination,* which is a common way to develop antibodies against childhood and adult disease agents such as viruses causing measles, mumps, polio, and hepatitis, as well as infectious bacteria and their toxins. The efficacy of vaccination is perhaps most clearly observed in the case of smallpox. There has not been a reported case of smallpox anywhere in the world in decades (Biosite 12–1).There is also another kind of immunity, called **passive immunity,** which arises from the transfer of antibodies between one person and another (Figure 12–13[b]). Transfer of antibodies provides short-term protection, but under some circumstances this is of great importance. For example, much of a newborn baby's resistance to disease comes from the transfer of IgG antibodies from mother to child through the placenta *in utero* and neonatally during breast-feeding. Mother's milk is rich in nutrients and rich in antibodies. Of particular importance are IgG and IgA, which are absent from the newborn baby's repertoire of defenses because the immune sys-

Acquired immunity immunity resulting from exposure to an antigen

Passive immunity immunity resulting from acquisition of preformed Ig molecules from another individual

BIO site

12–1

THE MILKMAID'S TALE

In 1796, the English physician Edward Jenner developed the first vaccine against smallpox, which at the time was one of the most feared diseases in the world. The agent of the disease is a virus, but Jenner did not know about viruses. During the eighteenth century, smallpox killed one out of every three children in England within the first three years of their lives. Millions of people around the world have been killed by this disease over the centuries. However, not today. There has not been a single case of smallpox reported in the world for over a decade, thanks to Edward Jenner and the milkmaids.

How did all this come about? It seems that in Jenner's time there was a second viral disease related to smallpox called cowpox. If one contracted cowpox, one was protected against the far more devastating smallpox virus. The ingenious Jenner took material from a pustule of cowpox on the wrist of a milkmaid (whose name was Sarah Nelms) and rubbed it onto the skin of a young boy. A few weeks later he deliberately exposed the boy to smallpox. The boy never developed the disease. He was protected, just as the milkmaid was. This protection is a form of immunization, which was called a vaccination, from the Latin for cow, *vacca.* The reason this protection is possible is because the cowpox virus (vaccinia) is closely related antigenically to the smallpox virus (variola), such that the antigens are cross reactive with antibodies in the blood of vaccinated individuals. The milkmaid's tale proved a valuable model for scientists around the world, and in recent times people can be immunized against many previously deadly or debilitating diseases such as polio, measles, and mumps.

tem is too immature. IgG helps protect against a variety of childhood viral infections, and IgA is normally secreted by exocrine glands serving to trap antigens in the gut. If a newborn infant has the opportunity to utilize its mother's IgA passively until it can form its own, many allergies to foods and other ingested materials can be avoided.

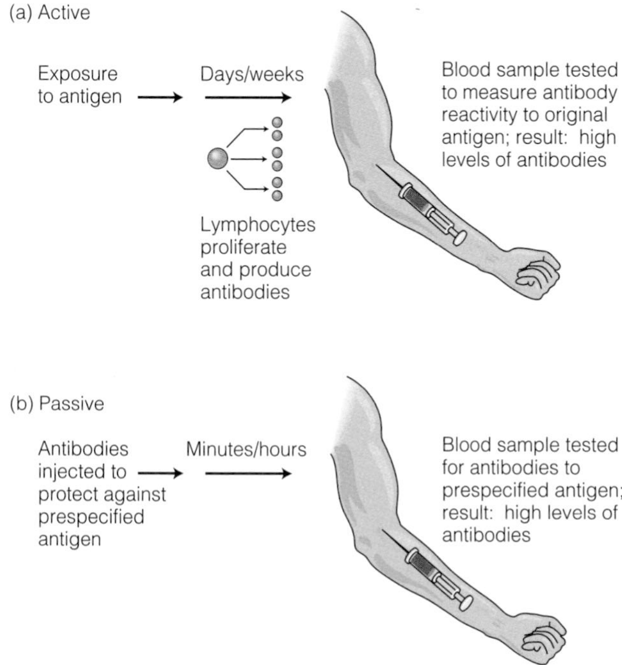

Figure 12–13 Active and Passive Immunity
Active immunity is one in which the body forms its own antibodies against an antigen (a). Passive immunity is one in which antibodies are imported from an outside source (such as a breast-feeding mother) into the body to protect it against a prespecified antigen or antigens (b).

T-Lymphocytes and Cell-Mediated Responses

As stated earlier, lymphocytes are represented by two types of cells—B-cells, which have been described, and T-cells, which will be described now (Figure 12–14). There are three main types of T-cells. The first are the *cytotoxic or killer T-cells,* which participate in the direct destruction of foreign cells or virus-infected cells. The second are the *helper T-cells,* which interact with B-cells, macrophages, and other T-cells to properly initiate and control the immune response. The third type is the *memory T-cells,* which are formed during the initial stimulation of killer T-cells and supplies a population of latent T-cells for subsequent exposures to the antigen in months or years to come. T-cells produce recognition molecules on their cell surfaces but do not secrete circulating antibodies. Instead, the molecules stick out of the surface of the cell, just waiting to encounter and bind to the correct antigen. Killer T-cells are phagocytic and very efficient in directly destroying foreign cells that enter the body. They participate in cell-mediated lysis. T-cells are particularly important in killing cancer cells and in rejection of foreign cells and tissues, as in the case of organ transplants (see Histocompatibility later in this chapter).

T-lymphocytes undergo the final stages of their differentiation in the thymus gland (Figure 12–14). It is in this environment that the proper growth factors and hormones are available for T-cells to mature. The thymus gland is large and very active in newborn humans and undergoes shrinkage and degeneration as we age and become adults. All the T-cells that we have mature early in life. This has been shown to be the case in situations in which the thymus is surgically removed. In adults, loss of the gland leaves the immune system largely unaffected. In newborns and children, absence or removal of the gland may mean that the individual lacks T-cells altogether. Without the T-cells a body cannot survive the infections that inevitably occur. It should be pointed out that in HIV infections one

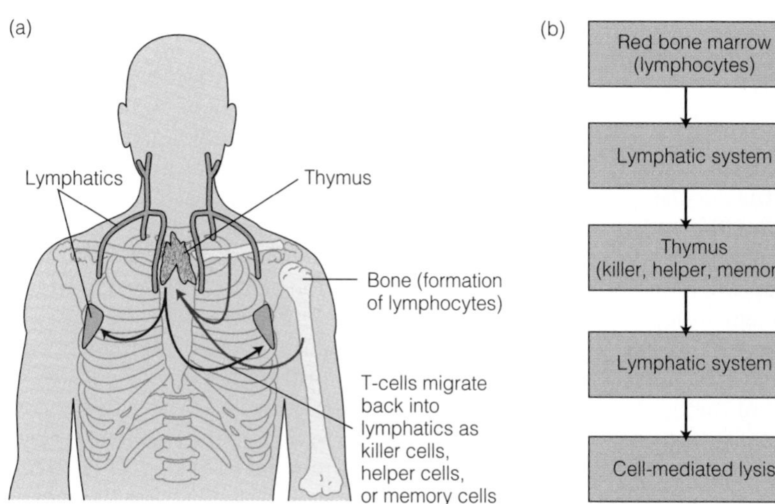

Figure 12–14 T-cells and Their Development
The course of lymphocytes destined to become T-cells is indicated in (a). A flowchart of the process is presented in (b).

of the targets of the virus in humans is T-cells (see Chapter 16 on sexually transmitted diseases). In full blown AIDS, the T-cell count of an individual is too low to sustain effective defenses against microbial and viral infections. It is to these infections that individuals with AIDS succumb.

Killer T-cells are also very important in fighting virus infections. However, they do not directly bind viruses; they bind to virally infected cells. How do they know a cell is infected? Viruses enter cells by endocytosis and then take over the cell's metabolic machinery. The virus uses its own genetic material to make mRNA to produce proteins to make new viruses. New viruses are released by a budding process. This means that viral proteins are exposed for a time on the cell surface of the host. It is the exposure of these antigens that is recognized by the killer T-cells, which then proceed to destroy the infected cell and the viruses within. A description of the immune response would be incomplete without consideration of one more group of cells, the monocyte/macrophages.

Monocyte/Macrophage Cells

Monocytes are formed in the bone marrow and are released to circulate in the blood. In a fashion similar to that described for neutrophils and eosinophils, monocytes exit the circulation through capillaries and enter the interstitial tissues. At this point they become macrophages. Macrophages are large, highly motile cells that crawl around within tissues and on the surfaces of some types of epithelia. One of their functions is to ingest microorganisms, debris, and viruses tagged with antibodies. But the monocyte/macrophage cell has another extremely important role in the immune response. It presents antigens to the helper T-cell so that the immune response can be initiated and highly specific. The three types of white blood cells described here, B-cells, T-cells, and monocyte/macrophages work together to tailor a suitable response to any foreign antigen that enters a healthy human body.

EXPOSURE TO FOREIGN ANTIGENS

When a foreign antigen enters the body, it is exposed to the numerous immune surveillance cells within the circulatory system. During this initial phase of exposure, the body begins its response. Out of the billions of lymphocytes within the body, only a few are inherently capable of recognizing and interacting with a particular foreign antigen, but this process is complex and involves several steps. In the first step, a macrophage encounters and binds an antigen, ingests it, and then displays a small part of the original molecule back onto its own surface. The macrophage in this case is called the *presenting cell,* and the cell to which it presents the modified antigen is the helper T-cell. The helper T-cell stimulates B-lymphocytes and T-lymphocytes, which divide and differentiate. The activated B-cells produce a humoral response, and the killer T-cells utilize cell-mediated responses that bring about the destruction of any cells bearing the appropriate foreign antigens on their surfaces (Figure 12–15).

The recognition of a molecule as foreign is based on chemical composition, such as amino acid sequence in a protein or the sugar groups on a glycoprotein or lipid. Activated lymphocytes involved in an immune response may be reactive to different parts of the same foreign molecule. Remember that the macrophage presents only small sections of the original antigen, so a single large antigen may provide several different small antigens for the helper T-cells to encounter. The specific small part of an antigen that is recognized by a particular T-lymphocyte or antibody is called an *epitope.* This can be related to the fable of the three blind men and the elephant (Figure 12–16). Each man is asked to describe the creature, based on what he feels with his hands. If one man is holding the tail, his description of the elephant becomes "a long, thin, flexible creature." If one man is holding a leg, his description may be "a thick, stumpy, treelike creature." If the last is feeling the abdomen of the elephant, he may decide "it is a giant, balloonlike creature floating just above the ground." Each is describing a different, but important part of the elephant. An antigen molecule is similar to an elephant. Different parts of the molecule are recognized by a select group of essentially "blind" lymphocytes, each of which is capable of recognizing only a part of the total antigen presented to it.

Self and Non-Self

What determines whether antibodies are made or not? Usually it depends on the body's capacity to distinguish between "self" and "non-self." We tolerate our own molecules as part of the self from which we are made, and by which we are maintained, and recognize almost all other molecules as non-self, if and when they are introduced into the body. When you think about it, non-self is an enormously large category. Imagine all the kinds of creatures in the world whose molecules are unique to them and, thus, foreign to us. There are millions of different species of animals and plants, each of which makes thousands of different types of molecules, every one of which is nonhuman. However, the ability to recognize non-self does not stop there. We recognize as different even molecules from other human beings. The rejection of a transplanted heart, or kidney, attests to the subtle differences that are recognizable between individual humans. Non-self is extremely individualized, referring to any antigen that is not a natural part of the body. With such tremendous diversity in the organic world around us, it is no wonder that the human species has evolved the potential to respond to billions of different foreign antigens even before any individual is actually exposed

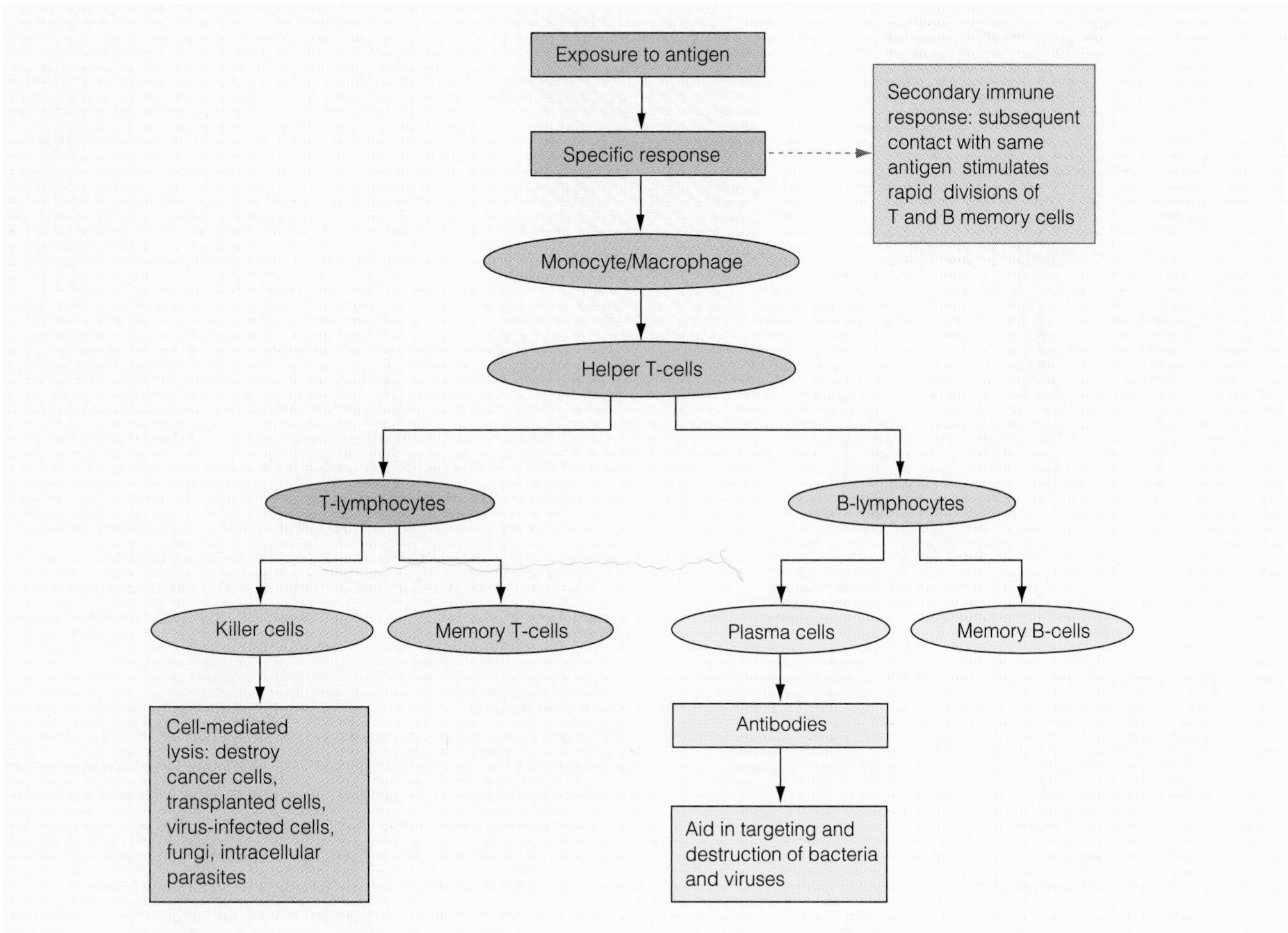

Figure 12–15 The Immune Response
Exposure of the human body to a foreign antigen results in a cascade of events that forms active T- and B-cells. Information on the antigen is transmitted by monocyte/macrophages in combination with helper T-cells. Once activated, killer T-cells directly destroy invaders, and B-cells produce antibodies that indirectly do so.

to them! It is estimated that B-lymphocytes of the human immune system are inherently capable of producing 10 billion to 20 billion antibodies with different specificities.

Selectivity and Proportional Response

One of the most important aspects of acquired immunity is its selectivity, which results in the immune system's ability to mount a proportional response. We actually make antibodies only against the antigens to which we are exposed, even if the potential exists to produce billions of different antibodies. It is not an all-or-none response in which exposure to an antigen results in antibodies against everything from A to Z. This is a waste of time, energy, and effort, and the body hoards its energy. This all-or-none response would be something akin to mobilizing an entire army every time two people get into an argument. What is needed is a precise and limited response; a response proportional to the demand. How does the immune system do this? It does it by recruiting existing lymphocytes capable of producing the precise set of antibody molecules that are specific to a particular antigen. This is called **clonal selection**

Clonal selection the matching of an antigen with a specific and limited subpopulation of lymphocytes that are stimulated to grow and differentiate

Figure 12–16 Blind Men and the Elephant
How are we to deal with descriptions of objects for which we have incomplete knowledge? The blind men all have a different idea of what an elephant is by examining only one part of the creature.

(Biosite 12–2). The *clonal selection theory* seeks to explain how lymphocytes committed to produce specific protective antibodies are selected for proliferation and differentiation.

Clonal selection depends on a process undergone by lymphocytes that is called *antigen dependent differentiation.* This is true for both B- and T-cells, although the molecules involved in maturation of each cell type are somewhat different (Figure 12–15). For example, with respect to the process described in the last section in which a helper T-cell activates other T- and B-cells, an immature B-lymphocyte has antibodies on its cell surface but is not yet capable of secreting those antibodies. When a population of lymphocytes is exposed to antigens, only the lymphocytes with the appropriate cell surface antibodies will be stimulated. They then undergo antigen dependent differentiation and become secreting cells and, in doing so, produce antibodies with the same specificity as those expressed on the cell surface. Overall, the response leads to the selective proliferation of a set of single cells, each of which produces a single type of antibody. The population of cells formed from a single cell under such circumstances is called a *clone.* Thus, the designation of this process is clonal selection. Most of these cells will become plasma cells and secrete antibodies, but a few will become memory cells. In a secondary response, instead of a few, uninitiated cells starting from scratch against a foreign antigen, the body retains a cache of memory lymphocytes. They are present in sufficient numbers to ensure a rapid and efficient response upon re-exposure to the antigen. However, it does take a significant period of time to develop an initial immune response to an antigen.

The initial response to an antigen takes days to weeks to fully develop. Eventually, the antibodies produced by the B-cells are in sufficiently high concentration in the body fluids to be able to effectively target the antigen and bind to it. This binding forms what is called an *antibody-antigen complex.* The complex can then be removed rapidly from the body by excretion through the kidney or through the phagocytic activity of macrophages, which recognize such complexes and ingest them. The second time an individual is exposed to the same antigen, the response is nearly instantaneous. Thanks to the memory T- and B-cells and any lingering circulating antibodies in the blood, the antigen and any antigen-bearing cells are immediately identified and destroyed or eliminated from the body. This is acquired immunity, and knowledge of this response has been of great benefit to humankind.

Vaccination

Vaccination is the controlled introduction of antigens into a body with the goal of establishing immunity against disease-causing organisms or toxins. Almost everyone has had a vaccination at some point. We vaccinate ourselves against the viruses that cause polio, mumps, measles, and some forms of the flu. Each of these viral antigens is readily recognized as foreign. Thus, we become immune to diseases caused by these agents. The resident army of lymphocytes and the antibodies they produce are used defensively and offensively as weapons. In the absence of a protective vaccination, a human has to contract the disease to develop an acquired immunity. This is a dangerous course to take because many of the diseases against which we are vaccinated may kill us (Biosite 12–1).

THE COURSE OF A VIRAL INFECTION

The Circumstance

To see the system in action, let us hitch a ride on a flu virus that is just about to enter the nasal cavity of a healthy human female. Someone next to her in the store has just sneezed and the spray surrounds her unexpectedly. She breaths in. Flu viruses are particularly insidious forms of quasi-life. They are not really living, but they are quite capable of interacting with living cells, taking over cellular metabolism, and affecting cells in detrimental ways. The viruses we have chosen to travel with are flu viruses that have escaped all the fixed and mobile defens-

12–2

THE THEORY OF CLONAL SELECTION

The theory of clonal selection is based on the conjecture that all the antigen-recognizing capacity of the cells of the immune system is genetically programmed in the lymphocytes. These cells do not have to be exposed to a foreign antigen to learn how to recognize the antigen; they need only be selected from the existing population of lymphocytes in the body by exposure to antigen(s) (as shown here).

Each of us has a population of lymphocytes whose individual members have the capability of responding to one of tens of billions of antigens. Exposure of the body to 10, or 100, or 1000 of these antigens selectively affects a specific subset of the lymphocytes, which then proliferate and produce the appropriate antibody. Selectivity of primed cells is the essence of the theory, and the theory has proven to be valid under a wide range of experimental testing. The human immune system is capable of generating enormous immunoglobulin diversity during the development of B- and T-lymphocytes. The total potential of the population may never be called upon, but as a hedge against the appearance of new and damaging antigens (a virus, bacterium, or toxin) the evolution of the immune system has provided an amazing mechanism to quickly and permanently acquire immunity.

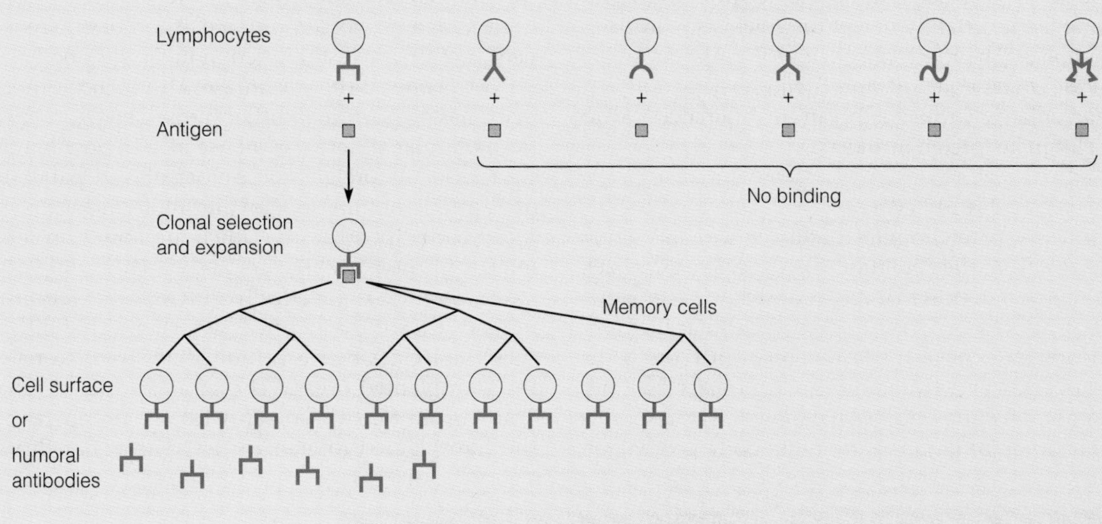

es of this woman's body and entered into the epithelial cells lining the sinuses. The last line of defense is immunity acquired through exposure to new antigens. The one thing that viruses do particularly well is multiply. If they did not multiply, our bodies would probably not mount an attack against them. This is because they probably would be present in too low a number, or for too short a time, for even our sensitive immune system to encounter and respond to them. However, they do multiply rapidly, and for a virus to produce more copies of itself, it uses resources within a cell. Unfortunately, cells are often severely damaged or destroyed by flu viruses in the process. As each cell is lysed, huge numbers of new virus particles are released into the surrounding tissues. Some of these viruses get into the infected woman's circulatory system and spread to other parts of her body. Viruses multiply so fast that her immune system cannot keep up. Within a few hours she goes from feeling just fine to the first uncomfortable symptoms of the flu.

Hot Pursuit

However, the posse is hot on the trail of the viral antigens. Macrophages and lymphocytes are responding. Macrophages take up the viruses and display their pieces on the cell surface. Helper T-cells process and pass on the molecular information. B- and T-cells are activated to divide and they begin to produce antibodies or cell surface recognition molecules. The virus infection is in full bloom over the next few days, but the woman's defenses are fighting as hard as they can. Her lymph glands are swollen with defensive activity. Lymphocytes keep dividing. They keep pumping out antibodies as well as undergoing cell surface changes. Inflammation of her tissues induces elevated body temperature. Her muscles may ache, and there is probably a general body weakness incapacitating her as well.

Catching Up

Then things begin to change. The defenses are finally catching up with the runaway viruses. The antibodies specific for the viruses are binding to a significant number of them, and macrophages are gobbling up the antibody-coated particles. The killer T-cells are recognizing the infected cells and destroying them and the viruses they contain. Antibodies are the molecular "smart weapons" of the body's defenses. Most viruses cannot long hide from them. It's similar to the paint bomb that blows up in the face of a would-be bank robber, as he or she opens the bag of loot freshly filched from the frightened bank teller. It certainly makes the thief easily identifiable to the police. The macrophages and T-lymphocytes are equivalent to cellular police. B- and T-lymphocytes, antibodies, and macrophages work together in a coordinated way to mark and eliminate foreign intruders.

Thanks for the Memory

Finally, the viruses are eliminated and the woman's body slowly recovers from the debilitating disease. She goes back to work or school and tries to forget how sick she was. However, her body does not forget the virus. Cellular memory is quite acute. The next time someone with the virus sneezes on her, whether two months or two years later, her immune system, having once been exposed to it, responds immediately and strongly. This time her body is prepared for the virus and eliminates it before it can get a foothold. Unfortunately, flu viruses mutate rapidly and change their molecular signatures (the composition of their molecules) significantly every few years. A body exposed to one form of flu may have little or no resistance to another. Around and around we go.

HISTOCOMPATIBILITY

Self versus non-self recognition has taken on a new meaning in recent years. As medical research and understanding of the immune system have become progressively more sophisticated, there has been more and more effort put into replacement of damaged or unhealthy tissues and organs with healthy ones. Healthy organs come from live donors willing to give, for instance, bone marrow-derived blood cells or one of their kidneys. Organs also are donated by the families of individuals who have recently died in accidents. The organs of the latter can be preserved for hours under the right conditions. However, the ability of a surgeon to transplant a new organ into a person does not ensure that person of its successful function in the new body.

"New body" is the key phrase. Although the organ transplant has come from another human being, there are subtle molecular differences between individuals that make the transplant foreign. These molecules are part of the elaborate family of immunoglobulin molecules, known as histocompatibility antigens, which are found on most types of cells in the body (Figure 12–17). This group of specialized molecules is genetically different in every individual human (except for identical twins), and the difference in these molecules between individuals leads to rejection of the transplanted organ. The same system of recognition that worked for the woman with the flu operates against the transplanted foreign tissues. The major difference is

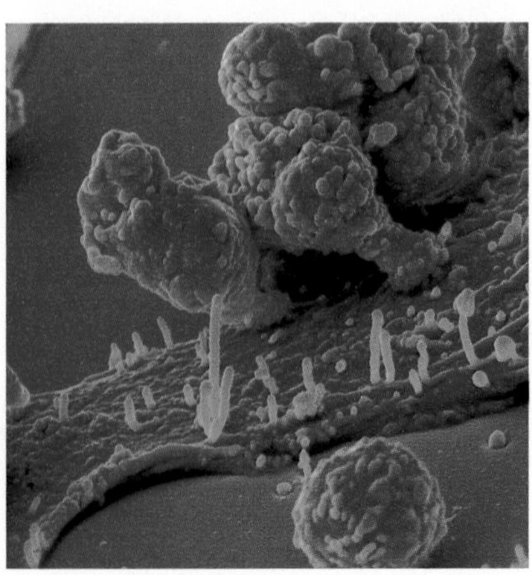

(a)

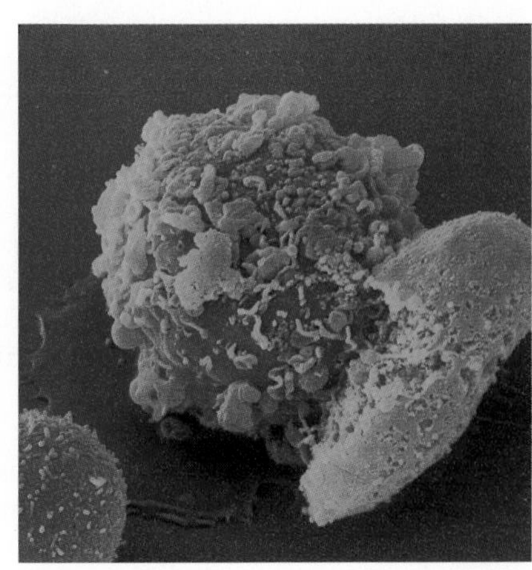

(b)

Figure 12–17 Histocompatibility and Rejection

Host T-cells recognize foreign cells and tissues and destroy them. The basis for the recognition is a group of proteins known as histocompatibility antigens, which occur in different combinations as a result of genetic differences in each individual. The T-cell in (a) binds to the foreign cell and destroys it (b).

that the lymphocytes that do the job of destruction in this case are solely the killer T-cells. Killer T-cells attack the transplanted organ and destroy it because they recognize it as different. This task may take only hours to occur or it may take days or weeks to accomplish, but it is usually inevitable. Those times that rejection does not occur appear to depend on how well the recipient and the donor are matched with respect to their histocompatibility antigens (Table 12–3). Screening individuals for histocompatibility is a life and death assessment. There are lists available to a physician and his or her patient of willing potential donors whose histocompatibility profiles are known. However, the lists for the nation are relatively small and the likelihood of a perfect match are very low.

Sometimes, to save a life, transplantation of mismatched tissues has to take place. There are some ways in which the rejection of a foreign organ can be prevented, depending on the use of drugs that suppress the immune response. As long as the immune suppressive drugs are present, the body is severely retarded in its rejection of the graft. In some cases this gives the system a chance to develop *tolerance* for the transplant. Tolerance, or the ability to tolerate self, arises during fetal development and gives the developing human body time to catalog its own natural molecular constituents and prevents what would otherwise be a sure-fire trip towards self-destruction. The problem with prolonged use of immunosuppressive drugs is that they also compromise the body's ability to fight off ordinary agents of infections or abnormal cells that may arise within the tissues, as in cancer.

AUTOIMMUNITY

Nothing is perfect. There are special conditions in which the body attacks itself using the same weapons usually reserved for foreign invasion. This is not so good for the body. Tissues and cells are selectively destroyed and functions are lost. Rheumatoid arthritis, which reflects the progressive destruction of the body's bones and joints, is at least in part an autoimmune disease. Some forms of kidney disease (nephritis) are caused by the immune system attacking the kidney's tiny filtering units, the nephrons (described in Chapter 10). The nervous system (as described in Chapter 17) is also a potential target for autoimmune destruction. Myasthenia gravis and muscular dystrophy involve a slow immunological wasting-away of the nerves and muscles, characterized by progressive physical weakness and eventual physical incapacitation. Multiple sclerosis leads to the destruction of the myelin sheath insulating neurons in the brain. The basis for such responses by the body against itself are not well understood, but their effects are disastrous.

Table 12-3 Organ or Tissue Transplants

TISSUE	HISTOCOMPATIBILITY MATCH REQUIRED?
Heart	Yes
Kidney	Yes
Skin	Yes
Bone marrow	Yes
Lens	No
Cornea	No

ALLERGY—IMMEDIATE AND DELAYED HYPERSENSITIVITY

Our cases of seasonal hayfever or sensitivity to certain foods and other substances are called **allergies** (Figure 12–18). But they also fit into the category of immune responses called **immediate hypersensitivity,** because they occur in a matter of minutes upon exposure to an allergen. The molecules involved in the process are IgE molecules produced by B-cells and bound to the surfaces of mast cells and basophils. The allergens generally stimulate local responses, such as swelling or hives. The nasal epithelium and sinuses are classic, locally responsive regions. The infamous ragweed, as well as hundreds of other pollen-based allergens, enter the nose and adhere to the moist nasal and sinus epithelia. Mast cells and basophils with bound IgE are located in the interstitial tissues of the sinuses and are the principal culprits in this response. However, other immune cells participate as well, including T-cells and the monocyte/macrophages. When allergens come in contact with a mast cell, they bind to the IgE on the cell's surface. The binding of allergens to IgE triggers the cell to release histamines, which cause an immediate and usually locally restricted swelling of tissue and dilation of blood vessels. Sufficient swelling of the connective tissue of the sinuses can cause painful pressure on bones of the face, the teeth, and jaw. Fluids are forced out of the tissues and flow from the nasal and sinus membranes. As long as an individual is exposed to the allergen, the symptoms persist. If the stimulus is removed, the response dwindles and the sinuses return to normal, eyes stop watering, and pain and swelling

Allergies a form of immune reaction involving local response (inflammation, runny nose, sneezing) to antigens such as pollen, dander, and other foreign materials

Immediate hypersensitivity a standard allergic response

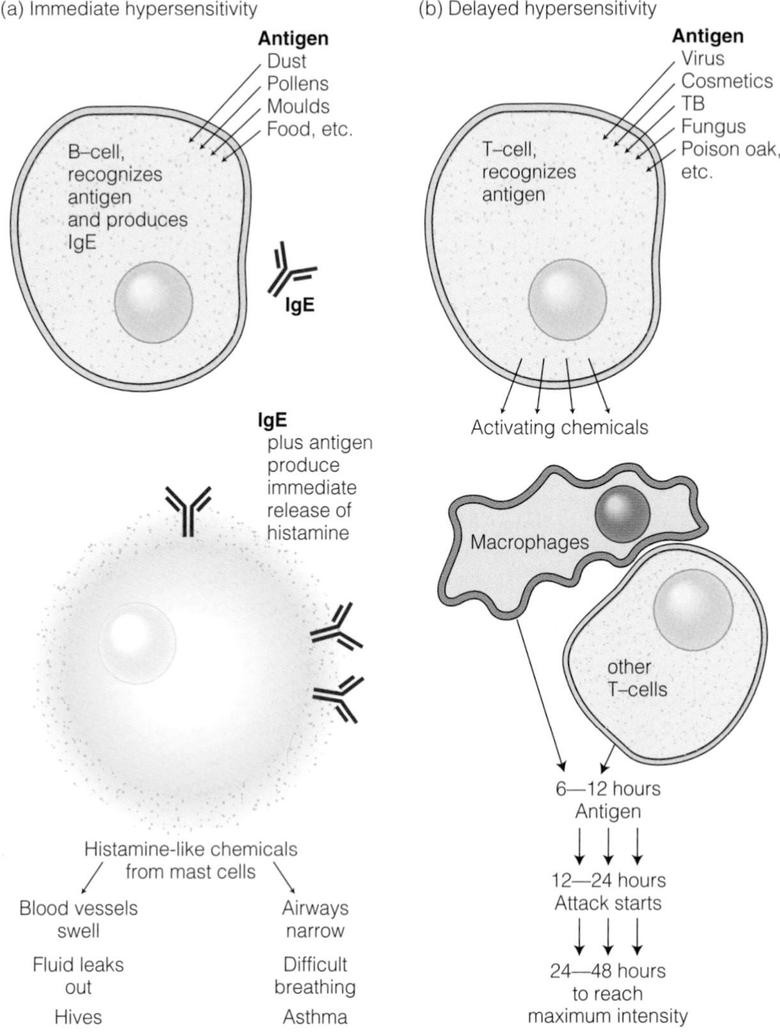

Figure 12–18 Immediate and Delayed Hypersensitivity
Immediate hypersensitivity (a) occurs in minutes and constitutes the most common form of allergic responses, as well as the more serious ones (hives, asthma). Delayed hypersensitivity (b) is mediated by T-cells and takes longer to develop.

are reduced or eliminated. It is not always possible to avoid the things to which we are allergic. Pharmaceutical science has devised a temporary way around the problem—**antihistamines.** These drugs suppress the release of histamine from mast and basophilic cells and reduce the symptoms of allergy, but they do not eliminate the underlying causes.

The immediate hypersensitivity just described is usually more of a nuisance than a danger. However, there is an extreme form of immediate hypersensitivity known as **anaphylactic shock** that can be deadly. The principle of the response is the same as in hayfever, but, instead of a local reaction, the effects of the allergen are systemic. In addition to histamine, other highly active molecules are also released. These include *leukotrienes* and *prostaglandins,* which act to constrict smooth muscles in the bronchii of the lungs. The system-wide release of histamine and other active molecules in this response may cause *hives* all over the body, or difficulty in breathing and *asthma,* as well as a severe drop in blood pressure that arrests the heart. The hives result from swelling induced by histamine's effects on capillaries and water movement. The difficulty in breathing results from the effects of histamines, prostaglandins, and leukotrienes constricting the airways. The redistribution of water from blood to tissues in the capillary bed causes blood

Antihistamines compounds whose action is to block or reduce the effects of histamines released during an allergic reaction

Anaphylactic shock drastic and immediate hypersensitive response resulting in bronchial and cardiac reactions that can be lethal, for example, allergy to penicillin or bee stings

pressure to drop and the heart to arrest or to beat arrythmically. The major causes of anaphylactic shock are bee and other insect stings and penicillin. Immediate treatment with antihistamines to block the histamine's effects, and epinephrine to open the airways and constrict blood vessels, is usually sufficient to save an individual's life.

Delayed hypersensitivity takes days or weeks to appear and is observed commonly in the human response to poison ivy and some cosmetics (Figure 12–18). It is also observed in the skin test for tuberculosis (TB). The scratching open of the skin and introduction of TB antigens produces an inflammation over several days if the body has acquired an immunity based on previous exposure to the bacterium. The degree of the reaction is a measure of the immune system's memory. This reaction is a T-cell-mediated response and is triggered when the allergen causes the T-cells to bind antigens and release highly reactive molecules called *lymphokines*. The lymphokines attract macrophages, which release their own set of active molecules and induce inflammation and itchiness in the area of exposure.

Delayed hypersensitivity a T-cell-mediated response to specific allergens such as poison oak or some cosmetics; may take several days to fully develop

Summary

There are three basic levels of defense in the human body—fixed, mobile, and acquired. The fixed defenses include the skin and mucous membranes. The acidic, oxidative, and enzymatic qualities of the secretions of mucous membranes bring about the destruction or inactivation of a variety of potentially dangerous substances and/or organisms. The other fixed defenses are part of the inherent internal protection of the body if the outer fixed defenses are breached. This includes the cells that inhabit the reticuloendothelial system, which are found localized in specialized regions of the body, such as the bone marrow, lymph nodes, spleen, and liver. The cells in these locations act as phagocytes and devour invaders. The lymphatic system is composed of a network of vessels in which many lymphatic cells mature. The thymus gland, spleen, and lymph nodes are part of this system.

The immune cavalry of mobile defenses attacks invaders where and when they enter the body. The granulocytic neutrophils and eosinophils are attracted to the site of entry by special targeting chemicals, whose gradients they follow to the source. They proceed to attack and consume the invaders and, in the process, form pus as the area is cleaned up afterwards. The monocyte/macrophage cells also form part of this phagocytic set of cell types. Basophils release chemicals, such as histamine, that cause local swelling and inflammation, in addition to causing the loosening of endothelial cells in capillaries to promote the migration of other granulocytes into areas of inflammation.

Agranulocytic cells are involved in acquired immunity. The acquired defenses are based on molecular recognition of self and non-self. The cells involved are the B- and T-lymphocytes. B-cells come in two varieties. The first are plasma cells that secrete copious amounts of antibodies. The second are the memory cells, which allow the immune system to maintain the ability to rapidly respond to the inducing antigen months or years later. T-cells come in three varieties. The first is the killer T-cell, which carries out the process of cell-mediated lysis. The second is the helper T-cell, which activates both the killer T-cell and the plasma B-cell. The third type of T-cell is the memory cell, which retains for the body the ability to respond to subsequent exposure to the same antigen. These types of lymphocytes are capable of being stimulated to divide and to differentiate to produce antibodies or other related immunoglobulin-type recognition molecules specific for use in ridding the body of foreign agents. Macrophages play an important role in presenting antigens to helper T-cells. The number of different types of molecules to which the immune system is able to respond is estimated to be in the tens of billions. Such diversity, in conjunction with selective generation of molecular and cellular "smart weapons," protects us from nearly any naturally occurring foreign substances, and some unnatural ones as well!

Questions for Critical Inquiry

1. Against what does the human body need to be defended? Why?
2. What is the outcome of vaccination? What importance does this procedure have in our society?

3. What are the physical barriers of the human body that provide the first line of defense?
4. What is histocompatibility? How does it relate to the human body's ability to recognize self and non-self? What if no perfect match can be found for an organ transplant?
5. What role do mast cells and basophils play in an allergic response? What might be the consequences of a systemic release of the contents of these cells?

Questions of Facts and Figures

5. What are the two major types of immunity? How are they related?
6. What are three basic categories of cellular and/or molecular defenses?
7. What are the classes of immunoglobulins? In what capacities do they function?
8. What are the types of granulocytes? Describe their functions.
9. B-lymphocytes produce and secrete what molecules essential to immunity?
10. What is an antigen?
11. Why is it important that development of acquired immunity be a proportional response?
12. What is autoimmunity? What diseases are associated with it?

References and Further Readings

Alberts, B., Bray, D., Lewis, J., Roff, M., Roberts, K., and Watson, J. D. (1994). *Molecular Biology of the Cell,* 3rd ed. New York: Garland Publishing.

Dwyer, J. (1988). *The Body At War.* Middlesex, England: Penguin Books.

Stalheim-Smith, A., and Fitch, G. K. (1993). *Understanding Human Anatomy and Physiology.* St. Paul, MN: West Publishing.

CHAPTER 13

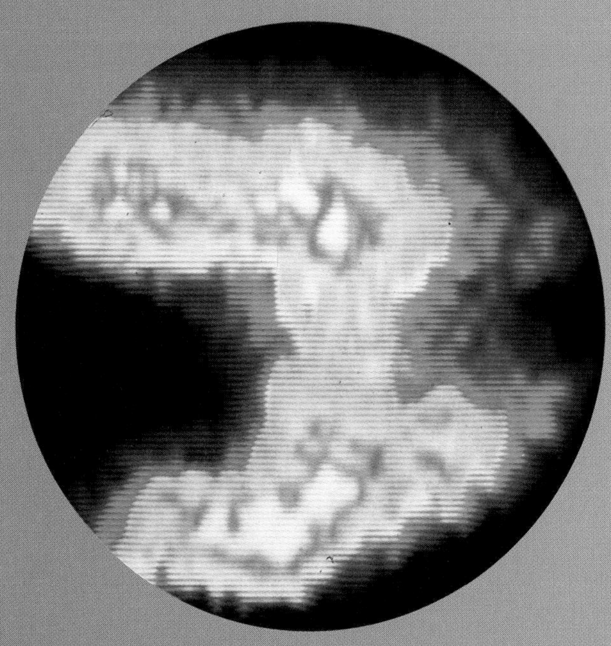

The Endocrine System

INTRODUCTION

Hormones are some of the most potent molecules found in nature. Those produced by the human body have tremendous effects on our early growth and development, as well as on adult anatomy, physiology, and behavior. To be considered a hormone, a molecule must have certain properties. It must be a diffusible substance synthesized by one cell type but produce its effect on other cells, often at great distance from the source. Hormones that control the growth of bones are formed in the brain. Hormones that control the characteristics of maleness and femaleness are formed in the gonads. Hormones that prepare us for emergency situations that demand that we either stand and fight or run away are produced in tiny adrenal glands that lie atop the kidneys. What are these hormones that affect us so expansively, and how do they work? In this chapter, we will explore the effects of the many types of hormones of the human body and the endocrine system of glands that produce them.

HORMONES AND HOMEOSTASIS

The endocrine system of glands is composed of a set of hormone-secreting organs, distributed widely throughout the human body (Figure 13–1). This system of organs carries out a major portion of the chemical coordination of metabolic activities of the human body. There are nine types of glands included in this set: the pituitary gland, hypothalamus, thyroid gland, parathyroid glands, adrenal glands, pancreas (islet of Langerhans), pineal gland, thymus gland, and gonads (testes and ovaries), (Table 13–1). Molecular communication takes place among the endocrine glands, so that their functions are often coordinated by feedback loops.

The hormones produced by the glands of the endocrine system are secreted directly into the circulatory system, where they are dispersed throughout the body. The target cells for each hormone respond through receptors either on the cell surface or within the cytoplasm of those cells, which leads to altered metabolic activity.

Through the release of hormones, the endocrine system coordinates the relatively slow-changing metabolic activity of the body. The hormones produced by the glands of the endocrine system can have positive, as well as negative, effects on their target tissues. Changes in the presence, absence, and/or relative amount of hormones and the number of hormone receptors of cells in the body may elicit a wide range of responses. As has been described in earlier chapters, the control of change, that is maintaining homeostasis, is fundamental to survival. Hormones are essential to carrying out the processes necessary to maintain homeostasis.

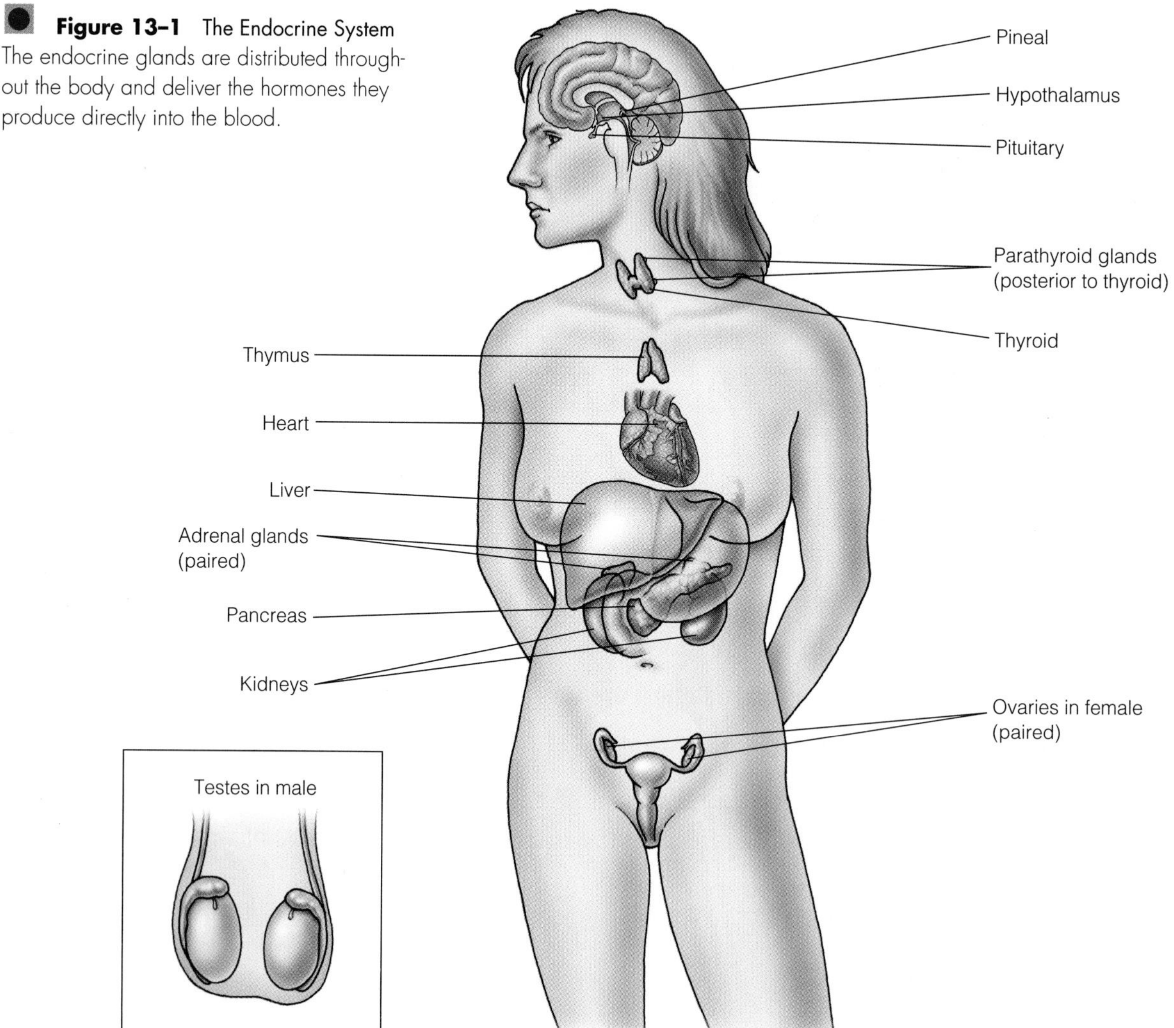

Figure 13–1 The Endocrine System
The endocrine glands are distributed throughout the body and deliver the hormones they produce directly into the blood.

Table 13-1 The Endocrine Glands

GLAND	LOCATION	HORMONE(S)	CHEMICAL CHARACTER
Pituitary			
Anterior	Brain	LH, FSH, ICSH, ACTH, TSH, prolactin	Glycoprotein
Posterior	Brain	ADH, oxytocin	Oligopeptide
Hypothalamus	Brain	Releasing factors	Glycoprotein
Thyroid	Neck	Thyroxine	Amino acid derivative
		Calcitonin	Glycoprotein
Parathyroid	Posterior thyroid	Parathyroid hormone	Glycoprotein
Adrenal			
Cortex	On top of kidney	Aldosterone, sex steroids, glucocorticoids	Steroid
Medulla	On top of kidney	Epinephrine, norepinephrine	Catecholamine
Islet of Langerhans	Pancreas	Insulin, glucagon	Polypeptide
Pineal	Brain	Melatonin	Amino acid derivative
Thymus	Sternum	Thymosin	Glycoprotein
Gonads			
Ovary	Abdominal cavity	Estrogen, progesterone	Steroid
Testis	Scrotum	Testosterone	Steroid

This chapter describes some of the basic features of the glands of the endocrine system, including the various hormones that they produce, their specific tissue targets and the general and specific effects the hormones have (Biosite 13–1). In addition, the relationships the glands have to one another and the way in which their activities are regulated will be illustrated. We will also examine what can go wrong if the glands that produce hormones do not function properly and the hormonal balances of the body are upset.

ENDOCRINE GLANDS

The nine types of endocrine glands are diverse in their sizes and structures and widespread in their locations throughout the body. From the head (hypothalamus, pituitary, and pineal glands) to the neck (thyroid, parathyroid, and thymus) and down through the torso (adrenals and pancreas) and groin (gonads), they share a connection to the circulatory system that makes them distinctly different from other glands in the body.

13–1

HORMONES

The word *hormone* comes from the Greek and means to stir up or to excite. That is clearly what these molecules do. They stir up the metabolism of the cells composing all the tissues and organs of the human body. They bring about, or modulate, the changes needed for the body to respond to changing conditions.

Hormones are the specific products of the endocrine glands and are diverse in chemical structure. For example, the anterior pituitary hormones are all glycoproteins, and those of the posterior pituitary are oligopeptides. The hormones of the adrenal cortex are steroids, as are those produced by the gonads. The hormones of the adrenal medulla are catecholamines, very like neurotransmitters of the nervous system. The thyroid hormone thyroxine is a modified amino acid and the only molecule in the body that requires iodine for function. Hormones are matched with specific receptors in target cells and together they mediate cellular activities that help tissues and organs to maintain the body's homeostasis.

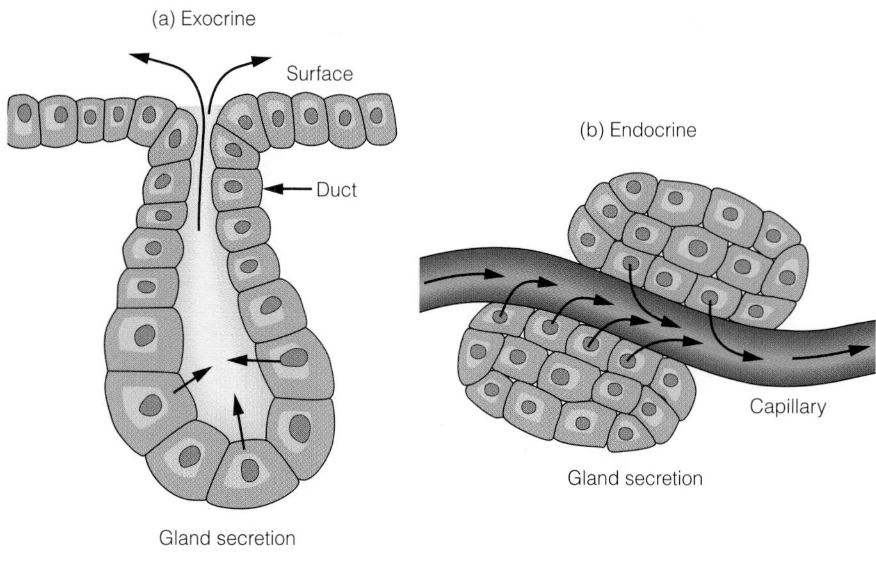

Figure 13–2 Exocrine Versus Endocrine Glands
Exocrine glands (a) secrete their products through a system of ducts onto surfaces. Endocrine glands (b) secrete their products directly into the blood for distribution throughout the body.

Endocrine Versus Exocrine

Before going into the details of the endocrine system it should be pointed out that **endocrine glands** are different from the other major type of glands found in the body, which are referred to as exocrine glands (Figure 13–2). Exocrine glands secrete the substances that they produce directly onto surfaces through a system of ducts or channels. For example, exocrine glands, such as the salivary, gastric, and intestinal glands that line the digestive tract, produce enzymes, mucus, or acid, and secrete them directly into the lumen of the gut to mix with and bring about the breakdown of food. Sweat and sebaceous glands secrete their watery or oily products, respectively, directly onto the outside surface of the skin (see Chapter 5). Exocrine secretions are local in both distribution and effect, and they are basically released onto the various surfaces of the body.

Endocrine glands are fundamentally different from exocrine glands. First, they do not have ducts but release their hormones directly into the blood. As a result, endocrine secretions are systemic in distribution and are regularly involved in controlling intermediate metabolic processes, that is, processes that affect the performance of cells and tissues. Second, hormones represent a form of molecular messages, which cells translate into information that affects gene expression, synthetic processes, and/or the release of specialized products. Cells capable of receiving and deciphering specific molecular signals are then able to carry on, alter, or stop metabolic processes in which they are involved.

Size of the Glands

The glands of the endocrine system are in general physically small, but their products are very potent. If all of the glands were collected together, they could easily be held in one hand (Figure 13–3). However, their small size belies their great importance to human health and vitality. Tiny amounts of hormones can have dramatic effects on the body. For example, the pancreatic hormone insulin stimulates many types of cells to increase the uptake of glucose. When glucose levels rise in the blood after eating, insulin is released. When glucose levels drop, insulin release is curtailed and is scarce in the blood. The most widely known insulin abnormalities are associated with the disease known as *diabetes mellitus*. In many cases, taking minute amounts of insulin by injection will temporarily correct abnormalities affecting cellular transport of sugar throughout the entire body. Without insulin, the glucose levels of the blood can become intolerably high. An individual suffering from such elevations of sugar may go into a **diabetic coma** and without proper treatment may die. Insulin is only one example of the many different types of endocrine hormones circulating in the blood. Each type has different, yet often profound, effects on the performance of the body's metabolic engines.

Types of Glands

Of the nine types of endocrine glands in the human body, some are found as diffuse groups of cells within other

Endocrine glands ductless glands secreting hormones directly into the blood

Diabetic coma a state of unconsciousness induced by abnormal carbohydrate metabolism

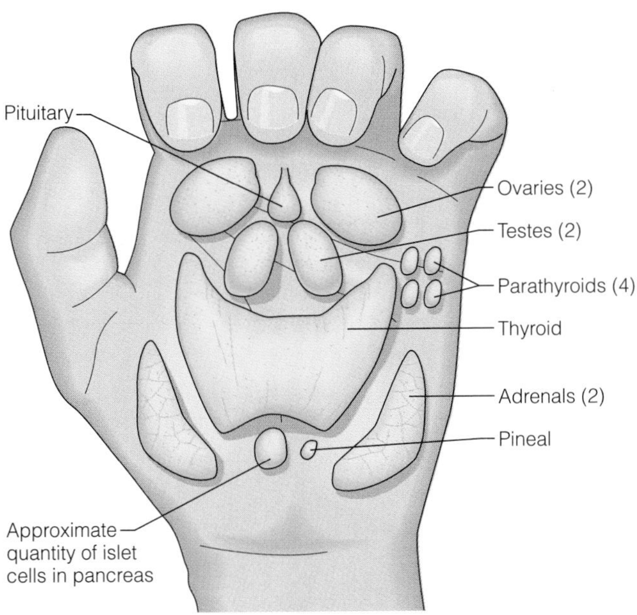

Figure 13–3 Size of Endocrine Glands
The principal endocrine glands of the body are relatively small in overall mass.

organs, such as the islets of Langerhans in the pancreas; some are distinctive, single glandular structures, such as the hypothalamus and pituitary; and some are matched pairs, such as the adrenal glands or gonads (Figure 13–1). Each type of endocrine gland produces a specific type or types of hormones, but some glands are relatively more important than others. The most important of the glands of the human body is the pituitary gland (Figure 13–4). This gland functions in conjunction with the hypothalamus and produces, or stores for release, many different and critically important hormones. The pituitary gland is central to the coordination of endocrine system activity. This does not lessen the importance of the other glands of the body but does suggest a hierarchical structure in the interrelationships among the glands.

PITUITARY GLAND

What is it that makes the pituitary gland so important? There are essentially two aspects of this gland that are vital to know: (1) the number and kinds of hormones produced by its cells, and (2) the effect(s) that these hormones have in stimulating and regulating other endocrine glands and nonendocrine target tissues. Because hormones have been equated to molecular messages, which, in turn, are translated into specific responses, it follows that the more types of hormones a gland produces, the more specific responses it can elicit and, therefore, the more important it is to the coordination of bodily functions. The control capacity of the pituitary gland is certainly a case in point. Many of the peripheral endocrine glands are affected directly by pituitary hormones, including the adrenal glands, the *gonads* (male and female), and the thyroid gland.

In humans, there are two lobes to the pituitary gland—an anterior lobe and a posterior lobe (Figure 13–4). Structurally and functionally, each lobe is quite different from the other and will be dealt with separately. For its size, which is about the size of a marble, the pituitary gland is one of the most potent tissues in the human body. Its products influence so many aspects of our physiology and behavior that we cannot survive long without it.

Origin of the Pituitary Gland

What are the hormones produced by the pituitary gland? There are six different hormones made by cells of the anterior lobe of the pituitary gland, and two released from the posterior lobe (Table 13–1). The origin of the anterior lobe is the epithelium that gives rise also to the roof of the mouth during embryonic development. This tissue grows upward on a stalk into the developing brain. The posterior lobe is derived from brain tissue itself. The outgrowing epithelial and brain tissues contact each other and form a single unit, the pituitary gland. In many types of animals there is also an intermediate lobe, but it is not evident in adult humans. The posterior lobe retains its direct connections to the brain throughout development, growth, and maturity. The connections form a bridge specifically to the hypothalamus. The anterior lobe, on the other hand, loses its epithelial stalk and becomes completely disconnected from the tissue of its origin. The pituitary gland eventually becomes located in a bony cup known as the *sella turcica,* which is part of the sphenoid bone of the skull (Figure 13–4 and see Chapter 6). The cup provides both support and protection for the gland.

The release of hormones from both lobes of the pituitary gland is regulated by the hypothalamus (Figure 13–4 and 13–5). The hypothalamus is a tiny area of the midbrain and is in anatomical proximity to the pituitary gland. In some cases, hypothalamic cell processes are directly connected with the pituitary gland. The anterior lobe of the pituitary gland contains cells within it that both produce and release hormones into the body through the circulation. The posterior lobe, on the other hand, contains only the end feet, or cellular extensions, of neural cells whose cell bodies are situated in the hypothalamus itself (Figure 13–4). The hormones that are released from the posterior pituitary are actually synthesized in the hypothalamus. They are transported along nerve cells to the posterior lobe, which is the site of their release.

In addition to the differences in origin and organization of cells that produce pituitary hormones, there are also

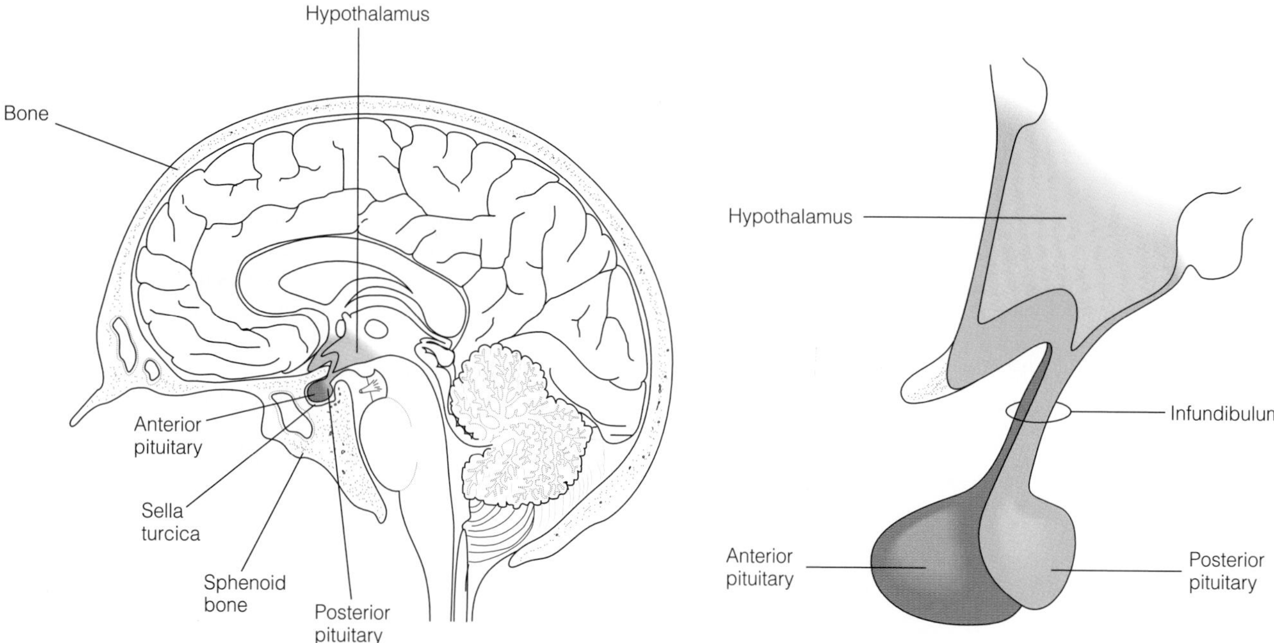

Figure 13–4 Hypothalamus and Pituitary Gland
Both the hypothalamus and pituitary glands are located in the brain (a). The posterior pituitary is an extension of the hypothalamus (b).

distinct differences in the types of hormones that are produced. All the hormones of the anterior lobe of the pituitary gland are glycoproteins or polypeptides. The two hormones of the posterior lobe are oligopeptides (Table 13–1, Figure 13–4). Each type of hormone, whether from the anterior lobe or posterior lobe of the gland, has its own unique effect on target cells and tissues.

HORMONES OF THE ANTERIOR PITUITARY

Follicle Stimulating and Luteinizing Hormones

There are two pituitary hormones that have a direct effect on the gonads. They are *follicle stimulating hormone* and *luteinizing hormone,* FSH and LH, respectively. These hormones are polypeptides and are produced and released under conditions regulated by complex negative feedback loops affecting the female reproductive cycle. They are **gonadotropic hormones.** This designation describes their target, the gonads, and their function, the stimulation of egg and sperm production. Briefly, FSH and LH are secreted in response to stimulation by hypothalamic releasing factors. In females, FSH and LH specifically target and stimulate the growth, maturation, and release of an egg(s). In the process of responding to these pituitary hormones, the ovaries themselves produce **estrogen** and **progesterone,** which feed back to the hypothalamus and inhibit the releasing factors produced there (Figure 13–5). This cycle goes around and around in mature females, as they undergo a monthly reproductive cycle.

It should be noted that FSH and LH are also made and released by the anterior lobe of the pituitary in males. FSH targets the testes in which it stimulates sperm development and testosterone production. The effect of LH in males, in whom it is called *interstitial cell stimulating hormone,* or ICSH, is on the production of testosterone by the interstitial cells of the testes. These are complex cycles, indicative of the primary role of the pituitary gland in the internal self-regulation of sexual and reproductive processes in the human body.

Gonadotropic hormones LH and FSH

Estrogen a steroid hormone produced principally by the ovaries and controlling the ovarian and menstrual cycles

Progesterone a steroid hormone produced in the ovaries and controlling the ovarian and menstrual cycles

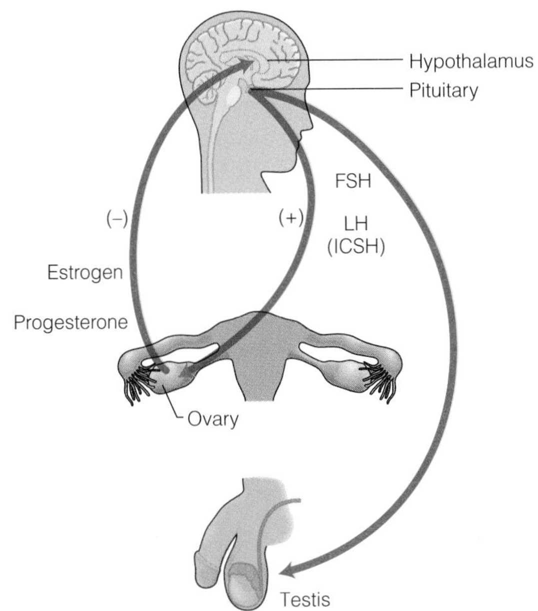

Figure 13–5 The Hypothalamus-Pituitary Axis and Gonadotropins

The hypothalamus gland controls the secretion of FSH and LH, which are hormones produced in the anterior pituitary. FSH and LH influence the activity of the ovary (or testes) and the production of estrogen and progesterone. Estrogen and progesterone inhibit the hypothalamus and prevent release of FSH and LH from the anterior pituitary.

Prolactin

There is another polypeptide hormone produced by the pituitary that is involved indirectly in processes related to reproduction and neonatal nutrition. This hormone is known as *prolactin*. Prolactin is intimately involved in stimulating the mammary glands to produce milk almost immediately after birth (see Chapter 15 for details). This hormone targets the glandular epithelium of the mammary glands. Although the level of prolactin rises during pregnancy, prolactin is strictly inhibited by a factor produced by the hypothalamus. The growth phase of development of the breasts during pregnancy is dependent on estrogen and progesterone, which are maintained at higher than normal levels throughout pregnancy, as well as corticosteroids and insulin. Together these hormones bring about glandular growth and maturation to milk-producing capacity. Prolactin is under the control of both stimulatory, and inhibitory, releasing factors produced in the hypothalamus. After the birth of the baby and the delivery of the placenta, estrogen and progesterone levels drop dramatically. Under these conditions prolactin release from the pituitary is no longer inhibited, and lactation begins. The influence of the suckling response during breast feeding is to suppress the prolactin inhibitory factor produced in the hypothalamus and, thus, to allow release of prolactin and stimulation of milk production. In addition, suckling triggers the release of hormones that stimulate milk flow.

Why aren't the effects of prolactin more obvious at times other than after birth? For instance, why not during the menstrual cycle? In this case, the high levels of estrogen and progesterone during most of the cycle play a critical role in suppressing prolactin. They help prevent prolactin release from the anterior pituitary. However, when estrogen and progesterone levels drop during the late phase of the menstrual cycle, the effects of prolactin may be felt as tenderness or sensitivity of the breasts. From the hormonal interactions described here, it is apparent that the human body requires several levels of control of hormone release to perform properly. In all likelihood, the human body has evolved multiple molecular controls and feedback systems as checks and balances to protect itself against the potential tyranny of single hormone control.

Growth Hormone

For general effects on the growth of the human body there is no pituitary hormone more influential than the growth hormone, or GH. GH is a glycoprotein composed of a chain of approximately 190 amino acids. The effects of GH on target cells is at the level of protein synthesis. The generally stimulating effect of GH suggests that cells throughout the body have receptors for the growth hormone on their surface. In some ways receptors may be thought of as the mailboxes of cells. The molecular messages arrive at the cell surface and their messages are transmitted onto the cell where they bring about specific metabolic responses.

GH and Bone Growth Normally, during human growth and development, the steady stream of GH released into the body by the pituitary gland helps maintain the continuous elongation of bones. As we mature and attain our adult size and stature, the amount of GH dwindles to maintenance levels for the body. The length of time over which relatively elevated levels of GH influence the elongation of bones determines in part how tall a person will be. There are nutritional and genetic factors involved in this as well, but, if too little GH is produced or it is produced over too short a period of time during growth and development, the human body will not attain a normal size. There is a range of values for the heights of individuals in the population of human beings that we consider normal. The variation in size reflects individual differences among human beings, differences that include genetic background, sex, nutrition, and many other intrinsic and extrinsic factors. However, if there is too little GH produced by the anterior pituitary, a condition of **pituitary dwarfism** results.

Pituitary Dwarfism There are several thousand pituitary dwarfs in the United States and, other than the fact that they are only 3-feet to 4-feet tall, they are generally well proportioned and show a normal range of intelligence. What they missed was the proper *somatotropic effects* of GH on the growth zones of the long bones of their bodies. In Chapter 6, we discussed the growth of bones and learned that growth takes place near the two ends, or epiphyses, of the long bones (Figure 13–6). GH keeps the cells in this region of the bones dividing and growing and inhibits them from undergoing differentiation, which in this case means ossification and terminal bone formation. Once ossification of the epiphyseal plates takes place, the length of a bone cannot be further extended. A pituitary dwarf lacks sufficient GH to maintain the growth of bones.

> **Pituitary dwarfism** a condition contingent upon the lack of pituitary growth hormone during childhood development

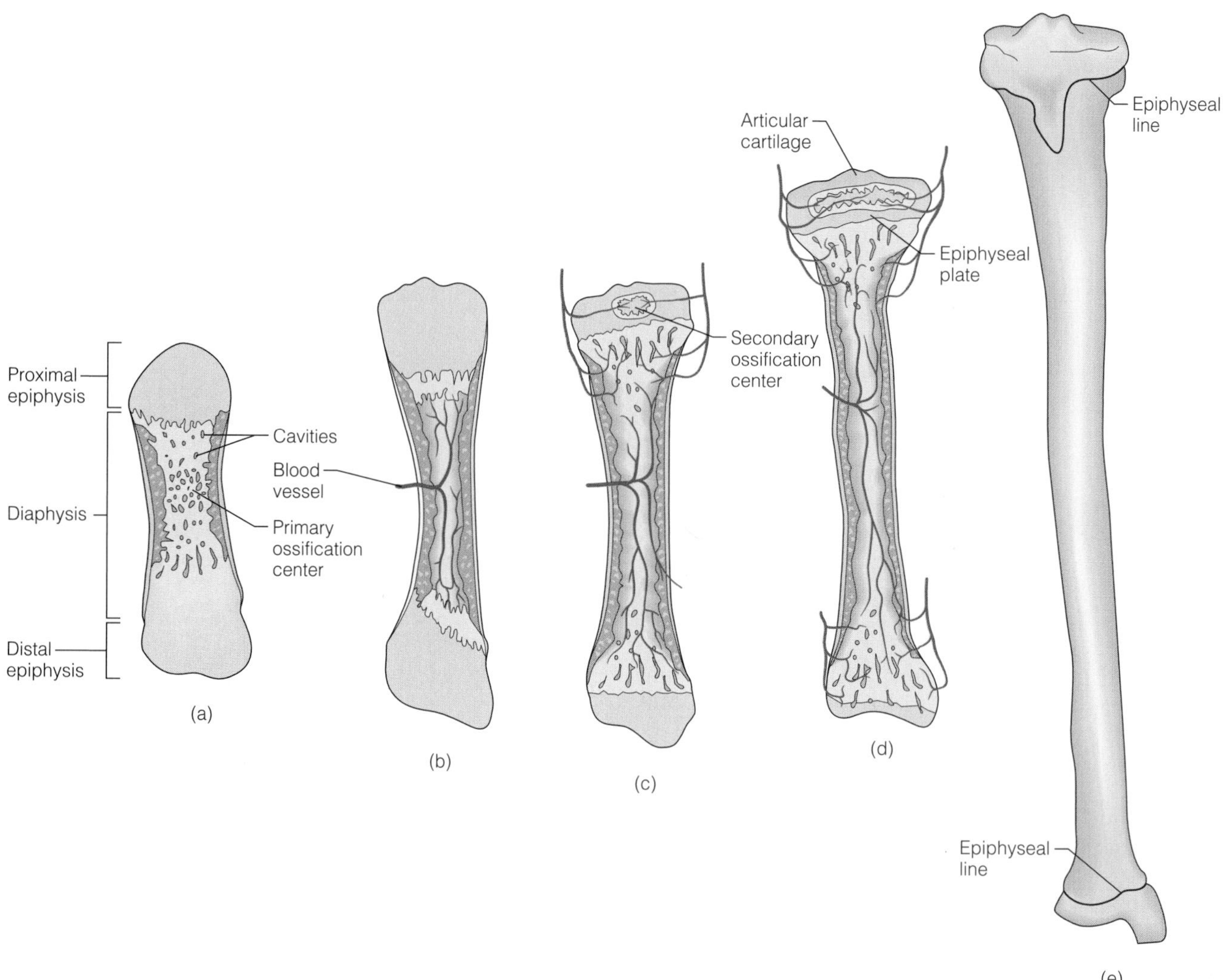

Figure 13–6 GH and Bone Growth
Pituitary growth hormone affects growth of fetal and neonatal bones along their long axis at the epiphyses. If too little GH is produced during these periods and continuing into adolescence, bones stop growing longer. If there is too much GH, the bones fail to stop growing normally.

In fact, an individual affected by this condition may attain an adult height of only one meter (Figure 13–7). The famous Tom Thumb (Charles Stratton) of circus notoriety was less than a meter tall. This condition is not necessarily an inherited one. A parent affected by pituitary dwarfism may have children who are perfectly normal in size and rate of growth. Using modern techniques of genetic engineering, much is being done to develop the technology to produce sufficient quantities of this hormone *in vitro*. For those babies that are diagnosed as having too little GH, treatment with an artificially synthesized hormone may help alleviate the condition.

Pituitary Giantism At the other end of the spectrum of size are the pituitary giants. As you might have guessed, these individuals have too much GH or GH levels are maintained too long during adolescent growth. Overproduction of GH is often caused by a tumor in the pituitary gland. Individuals subjected to this condition may grow well over 8-feet tall. For comparison, the height of an average American male is about 5-feet 10-inches tall. The tallest basketball players (who are not considered giants in the sense used here) are around 7-feet to 7.5-feet tall. Pituitary giants may measure in at nearly 9-feet tall. One of the tallest humans on record was a man named Robert Wadlow, who was 8-feet 11-inches tall. There are serious anatomical and physiological problems with relative body mass and volume when humans reach these staggering sizes. Human bones themselves are not really strong enough to support the weight of a 9-foot individual, where weights may reach 500 pounds.

Acromegaly What if the increase in GH occurs later in life, after a normal adult size has been attained and the long bones of the body can no longer grow in length? Unfortunately, elevated levels of GH in adults leads to a disfiguring condition known as **acromegaly.** Pituitary tumors are probably the culprit in this condition, as they are in giantism. Although it is true that GH cannot stimulate the elongation of the long bones of the body, it can stimulate the growth of cartilage associated with bones and joints. The increase in the size of these regions distorts the normal relationships among bones, particularly the face, hands, and feet (Figure 13–8). Other effects of the acromegalic overproduction of GH are headaches, loss of vision, and lethargy, suggesting that too much GH brings about complex physiological responses that are not nearly so obvious as its effects on expansion and growth of cartilage. Treatment for the cause of these problems, in all likelihood a tumor, is difficult because of the inaccessibility of the pituitary gland itself. It resides right in the middle of the brain. However, surgery, radiation treatment, and some drugs may be effective in alleviating the problem. Unfortunately, while it is possible to add GH into the body of a pituitary dwarf to increase the level to near normal, there is no way yet known to effectively subtract GH from the human body to reduce the excess.

Acromegaly a condition in adults resulting from overexposure to growth hormone

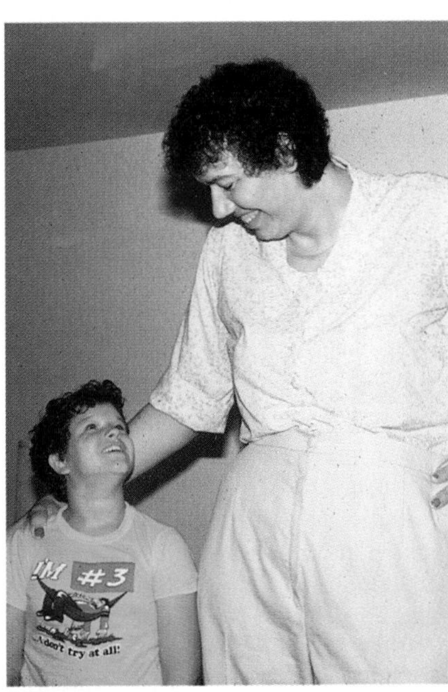

Figure 13–7 Growth Hormone (GH) Abnormalities
Too little or too much growth hormone? The woman shown standing with a young relative is nearly 8½ feet tall.

(a)

(b)

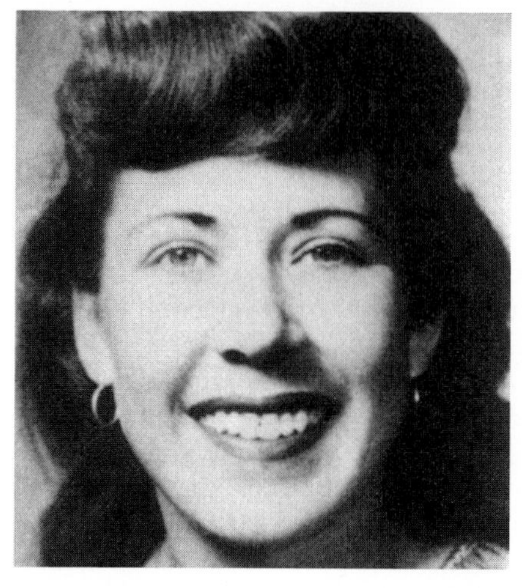

(c)

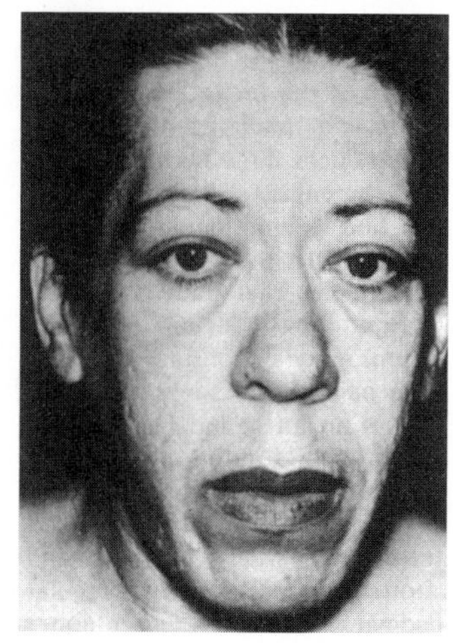

Figure 13–8 Acromegaly
This series of photographs shows the progression of acromegaly: (a) a young woman at age 20; (b) the same woman at age 24 (note changes in features); (c) the woman at age 40, showing characteristic enlargement of nose and jaw and distortion of facial features.

GH is the only anterior lobe pituitary hormone with such generalized, direct effects on the growth and development of cells of the body. As has been seen for FSH, LH (ICSH), and prolactin, these important hormones have profound local effects, but they influence the body as a whole in a more indirect way. FSH and LH specifically stimulate gonads, which in turn produce steroid hormones that have generalized body effects. The specific target for prolactin is the mammary epithelium, essentially postpartum (after birth). This local targeting is also an attribute of the remaining two hormones of the anterior lobe of the pituitary gland, thyroid stimulating hormone (TSH) and adrenocorticotropic hormone (ACTH). These two hormones and the glands they regulate will be discussed before returning to the posterior pituitary. In addition, other aspects of thyroid gland and adrenal gland structure and function will be discussed.

THYROID GLAND

Thyroid stimulating hormone (TSH) does just what its name implies, it stimulates the thyroid gland. The thyroid gland is located just below the larynx in the front of the neck and produces thyroid hormone known as thyroxine (Figure 13–9). TSH is regulated by a specific hypothalamic releasing factor, which itself is regulated by the level of thyroid hormone in the blood. In this way, a negative feedback loop is established and the pituitary-thyroid gland interaction is governed by homeostasis. The metabolic action switches to the thyroid gland after release of TSH. The thyroid gland is a fairly large gland by endocrine standards. Within its tissue are islands of cells within *follicles* that are involved in the production and storage of thyroxine (Figure 13–9). The release of thyroxine into the blood has an impact on most cell types of the body. Much as in the case for growth hormone, thyroxine influences the basic metabolism of cells. Thyroxine is even more pervasive in its action than GH and stimulates not only protein synthesis but sugar metabolism, fat metabolism, DNA and RNA synthesis, mitochondrial activity, respiratory rates, and growth.

What is the structure of thyroxine? It is very different from thyroid stimulating hormone and all of the other anterior pituitary hormones (Table 13–1). Thyroxine is a derivative of the amino acid tyrosine and contains several atoms of iodine. The incorporation of iodine is very important to the function of this hormone. Without iodine, thyroxine has none of the metabolic effects listed above and, therefore, is useless to the body. Iodine is an essential micronutrient and must come from our diet. This is why it is so important to have a stable source of dietary iodine. In the United States most of the commercially available table salt is iodized. It takes an extremely small amount of iodine to satisfy our daily requirements, less than 0.0000002 grams per day, but, if it is missing from the diet, there are serious consequences.

Goiter

One consequence of insufficient iodine is a condition known as goiter. In this condition, the thyroid gland enlarges to such an extent that the neck becomes obviously swollen (Figure 13–10). What causes this response? In the absence of iodine the thyroid gland is thrown into confusion. It senses that it must produce thyroxine to satisfy the body's needs and so goes on producing molecules similar chemically to thyroxine. But these molecules do not incorporate iodine—no iodine, no interpretable molecular message. The information flow to the cells is interrupted.

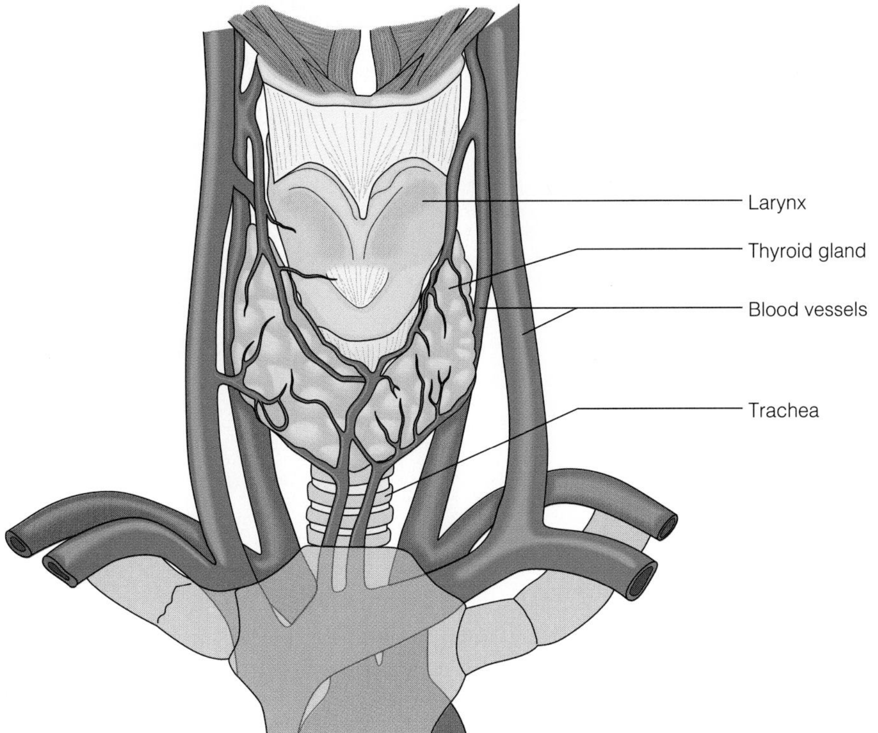

Figure 13–9 The Thyroid Gland
The thyroid gland is located overlying the larynx and trachea. Follicles within the thyroid gland produce the hormone thyroxine. The thyroid gland also produces calcitonin, a hormone involved in the regulation of calcium.

The body responds with a return message to the thyroid gland, "Where's the hormone?" Thus, the thyroid gland itself grows in mass to form more and more useless molecules. This is analogous to expanding a factory to build cars with no wheels. The demand for cars is there, but no one in the factory recognizes the wheel-less bodies. Management is confused and expands to build more of what they think are perfectly fine cars.

So it is with the thyroid gland, an essentially useless expansion of the gland to produce more hormone in an attempt to satisfy a demand that cannot be fulfilled. Simply adding iodine to the diet corrects the metabolic defect. However, this does not reduce the size of the goiter itself. The results of iodine deficiency, other than the swelling of the thyroid gland, give rise to other symptoms as well. Lethargy is one side effect. A tired, worn-out feeling relates to a general turn-down of the body's metabolic rate. This includes sluggish mental processes and a slowing of the heart beat. The pallor of the skin also changes, along with its increased dryness.

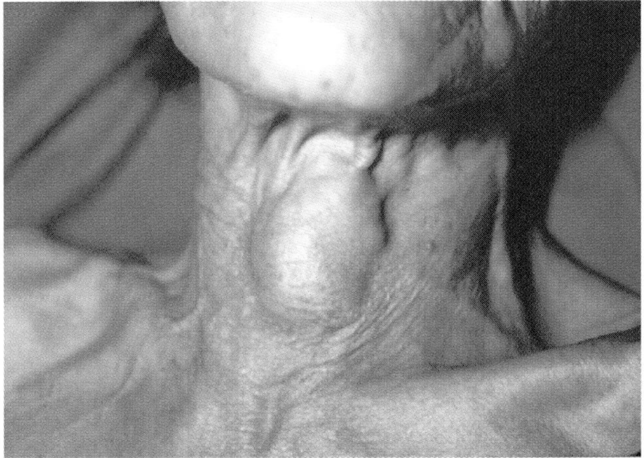

Figure 13–10 Goiter
Enlargement of the thyroid gland may result in a goiter. Goiters often arise due to lack of dietary iodine which simulates overgrowth of the gland in an effort to produce thyroxine.

Hypothyroidism and Hyperthyroidism

Myxedema There are other problems that can arise in the function of the thyroid gland that do not involve dietary iodine. *Hypothyroidism* and *hyperthyroidism* are quite common and correctable conditions associated with the under- and overproduction, respectively, of functional molecules of thyroxine. The symptoms of hypothyroidism in humans are essentially those of goiter. The difference is that a perfectly normal hormone is synthesized (so no goiterous swelling occurs) but in insufficient amounts to

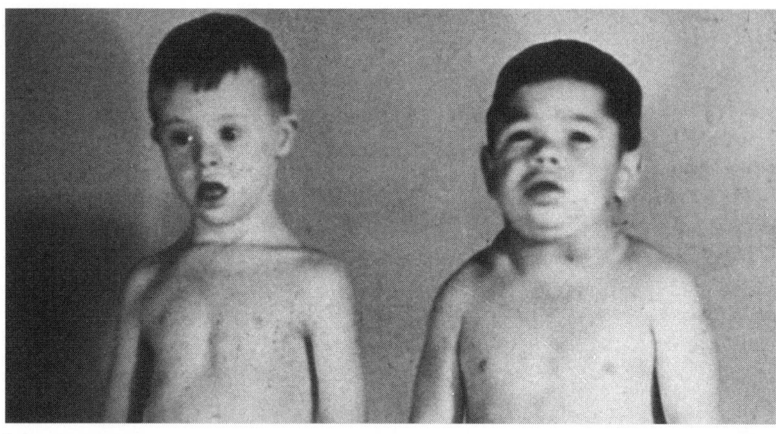
(a)

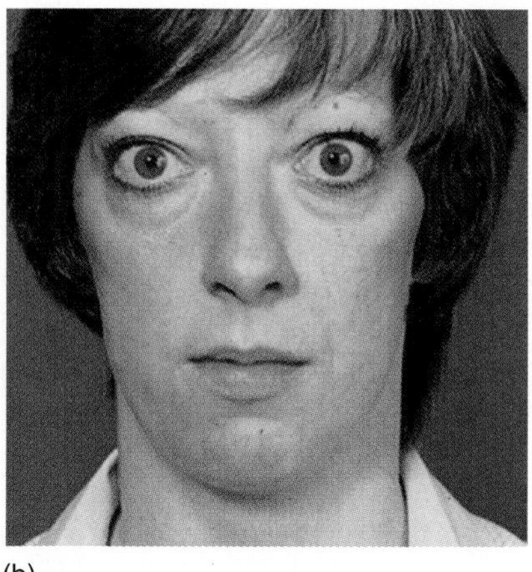
(b)

Figure 13–11 Hypothyroidism and Hyperthyroidism
Photo (a) is an example of cretinism, which is a hypothyroid condition of children. Photo (b) is an example of Graves' disease, which is a condition of adult hyperthyroidism.

maintain the metabolic engines of the body. If the deficiency is severe and persists long enough in an adult, a condition known as **myxedema** arises (Figure 13–11[a]). Myxedema is characterized by lethargy, mental dullness (that is, decreased intelligence), obesity, and circulatory and skin problems. Can this condition be corrected? Yes. Individuals who do not have enough thyroxine can be given hormones (usually tablets to be taken orally) with dramatic results. Most, if not all, symptoms can be reversed by simply controlling the amount of thyroid hormone ingested. The individual may well have to take supplements for the rest of his or her life, but what a small price to pay for relief from the dehumanizing effects of this hormone deficiency.

Cretinism Unfortunately, severe thyroxine deficiencies during fetal development, or during infancy, may have permanent debilitating effects. Such a condition is called **cretinism** and results in impairment of growth (dwarfism), intelligence, and sexual maturation. Cretins are mentally retarded and never grow much beyond the size of young children. These early abnormalities prevent subsequent sexual maturation (Figure 13–11[b]). The reason that this condition cannot be corrected by supplemental thyroxine after the symptoms have developed is because there are critical phases of growth in a fetus that are not possible to reinitiate at later stages of growth. For example, the growth of the brain depends on a nutrient-rich diet and an energy-efficient metabolism. Low thyroxine levels reduce metabolic rates in brain cells drastically and, thus, cause irreparable damage. The profound effects of this type of deficiency, as is seen to a somewhat lesser extent in the case of GH, serve to underscore the fundamental importance of the endocrine glands and their secretions.

Graves' Disease What about hyperthyroidism? What happens if we overproduce thyroid hormone? This is the flip side of the thyroid coin. Excess TH increases the body's metabolism markedly, speeds up the heart beat, enhances the appetite, and induces a state of nervousness and excitement far above normal. One form of this condition is called **Graves' disease** and results from improper regulation of thyroid hormone synthesis and release (Figure 13–11[c]). There may be an immunological component to some forms of this disease. Autoimmunity is an important factor in a number of serious disorders of the body, and antibodies against cells of the thyroid gland may trick it into continuous production of thyroxine. The hyperthyroid condition can be corrected by administration of certain kinds of drugs that interfere with synthesis of the hormone, by surgical removal of a portion of the thyroid gland, or by treatment of the thyroid with radioactive iodine. The thyroid gland concentrates radioactive iodine, and the radioactive decay of these iodine atoms destroys a portion of the cells of the thyroid tissue. Problems may occur with the surgical removal or destruction of the entire thyroid gland. This is because thyroxine is not the only important hormone produced by the thyroid gland.

Myxedema a form of hypothyroidism characterized by lethargy and mental dullness; reversible by treatment with thyroid hormones

Cretinism a form of hypothyroidism in children characterized by irreversible mental retardation

Graves' disease a form of hyperthyroidism; characterized by hyperactivity, enhanced appetite, and nervousness; also known as exophthalmic goiter

PARATHYROID GLANDS

What are these other hormones? One of the lesser known hormones, *calcitonin,* is associated with the thyroid gland itself and another hormone, *parathyroid hormone* (PTH), is associated with a set of glands known as the *parathyroid glands*. The four tiny parathyroid glands are located in pairs, embedded in the back surface of the thyroid gland (Figure 13–12). Calcitonin and PTH are responsible for regulating the levels of calcium and phosphate ions in the blood, and their synthesis is stimulated, or inhibited, by the relative level of calcium in the blood (Figure 13–13). Calcitonin and PTH are not themselves regulated by any hormone from the pituitary gland. Not all endocrine glands are tied to the apron strings of the pituitary. Calcitonin is produced by the thyroid gland proper and has as its main effect the lowering of calcium ion concentration in the blood. It does this by reducing the amount of calcium released from the bones, which are the principal storehouses of calcium in the body, and by increasing the deposition of calcium into bones (see Chapter 6). PTH, on the other hand, increases the level of calcium in the blood by recruiting calcium from the bones and aiding in calcium resorption through the kidneys and intestine into the blood.

How do these two hormones act together? They are clearly part of a hormonal on-off switch that is intimately involved in regulating calcium levels in the blood. Calcium is one of the most important ions of the body, not just because of its importance for bones but because both the contraction of muscle and the activity of nerves depend on it as well. This is a reason why maintenance of calcium balance in the body (homeostatic mechanisms) is so important. The question should perhaps be restated in terms of what hormones do as messengers. What are the targets for calcitonin and PTH?

Calcitonin inhibits cells that breakdown bone and release calcium, the osteoclasts, and stimulates bone cells that deposit calcium into new bone, the osteocytes and osteoblasts (Figure 13–13). PTH, on the other hand, stimulates osteoclast activity and targets the intestine to increase adsorption of dietary calcium. It does this in conjunction with vitamin D or calciferol. PTH also inhibits the excretion of calcium through the kidney. This balancing act between calcitonin and PTH is crucially important. Overactive parathyroid glands can cause a dramatic decrease in bone density and deposition of calcium in noncalcifying tissues.

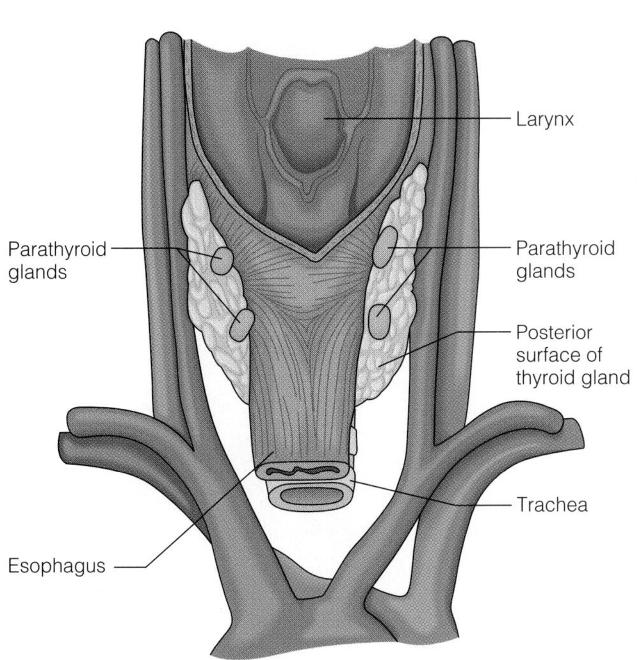

Figure 13–12 The Parathyroid Glands
The parathyroid glands are located within the posterior surface of the thyroid gland. Cells in the parathyroid glands produce parathyroid hormone, which is involved in the regulation of calcium.

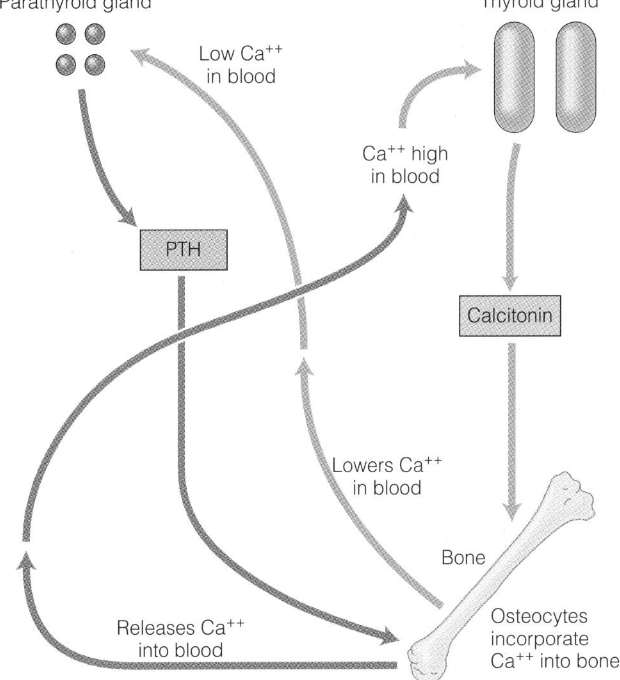

Figure 13–13 Targets of Parathyroid Hormone and Calcitonin
It is in the balance of the effects of calcitonin and parathyroid hormone that calcium is regulated in the human body. Calcitonin increases calcium incorporation into bone, and parathyroid increases calcium withdrawn from bone and taken up for excretion by the kidney.

ADRENAL GLANDS

The final hormone of the anterior lobe of the pituitary gland left to consider is **adrenocorticotropic hormone,** or ACTH. ACTH influences the production and release of hormones found in the adrenal glands, which are located atop each of the kidneys (Figure 13–14). Actually, ACTH affects only a limited part of the adrenal gland, the cortex, which is an area where several classes of steroid hormones are produced. Steroid hormones are different biochemically from any of the endocrine hormones so far discussed. Instead of being related to amino acids, either as simple derivatives (for example, thyroxine) or as chains of amino acids (for example, oligopeptides and polypeptides) steroids are related to cholesterol and are, therefore, lipid in nature (Table 13–1).

Their lipid nature determines the manner in which they affect cells. Steroids are **lipophilic,** or lipid-loving, and readily pass through the plasma membrane directly into the cytoplasm and bind to cytoplasmic receptors (Figure 13–15). With steroids all cells are targets, largely because there is no way to stop these lipophilic hormones from entering a cell. However, not all specificity and selective targeting are lost. There are several classes of cytoplasmic, steroid-binding molecules that mediate the regulatory effects of the various steroid hormones. The effects of steroids on the cells of the human body are profound and wide ranging. We will discuss the importance of the sex hormones testosterone, estrogen, and progesterone when we cover reproductive biology in Chapter 14. These hormones are made predominantly in the gonads but are also part of a much larger family of steroids that are made in the cortex of the adrenal glands as well.

Adrenocorticotropic hormone (ACTH) a hormone produced by the anterior pituitary affecting the regulation of the adrenal cortex and the synthesis of steroids

Lipophilic compounds that are soluble in fats and lipid membranes

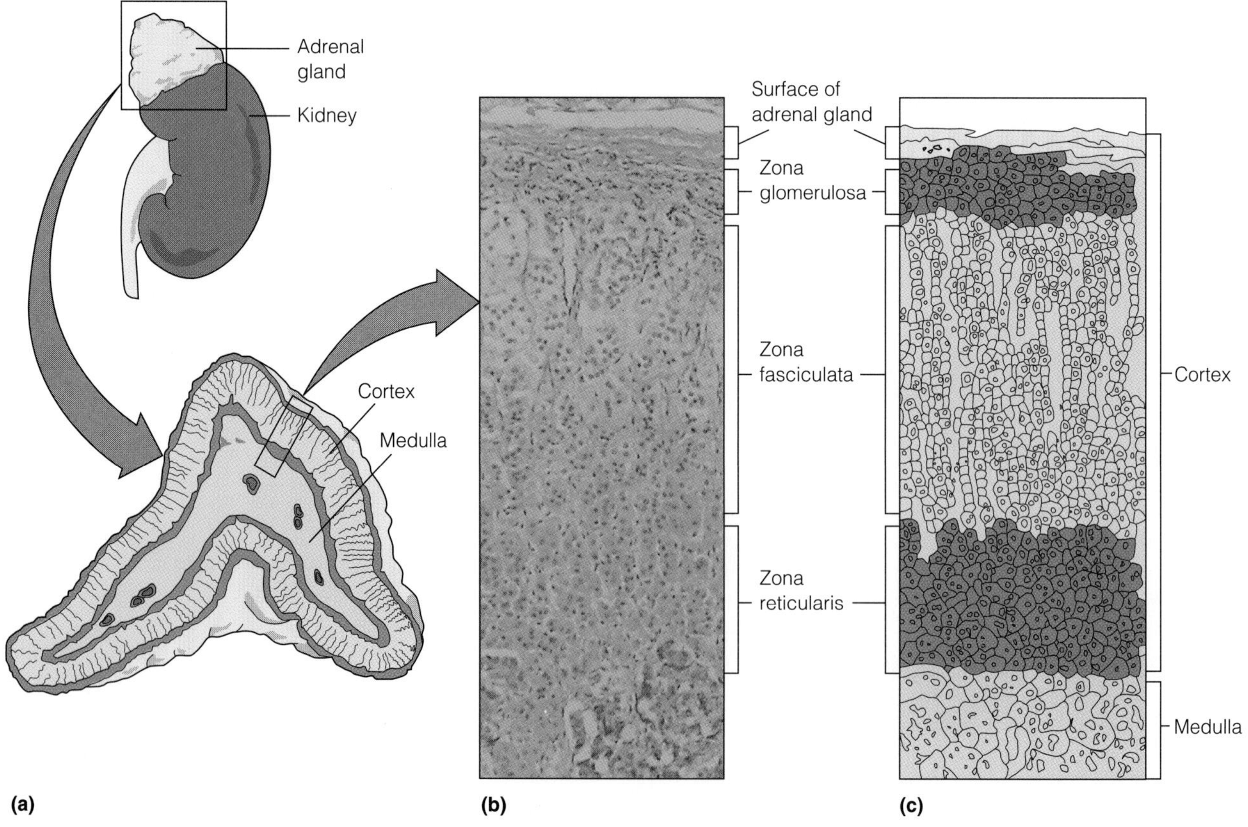

Figure 13–14 The Adrenal Glands
The adrenal gland is organized into two very different regions—the cortex and the medulla (a). The cortex is organized into layers (b) and (c), which produce steroid hormones.

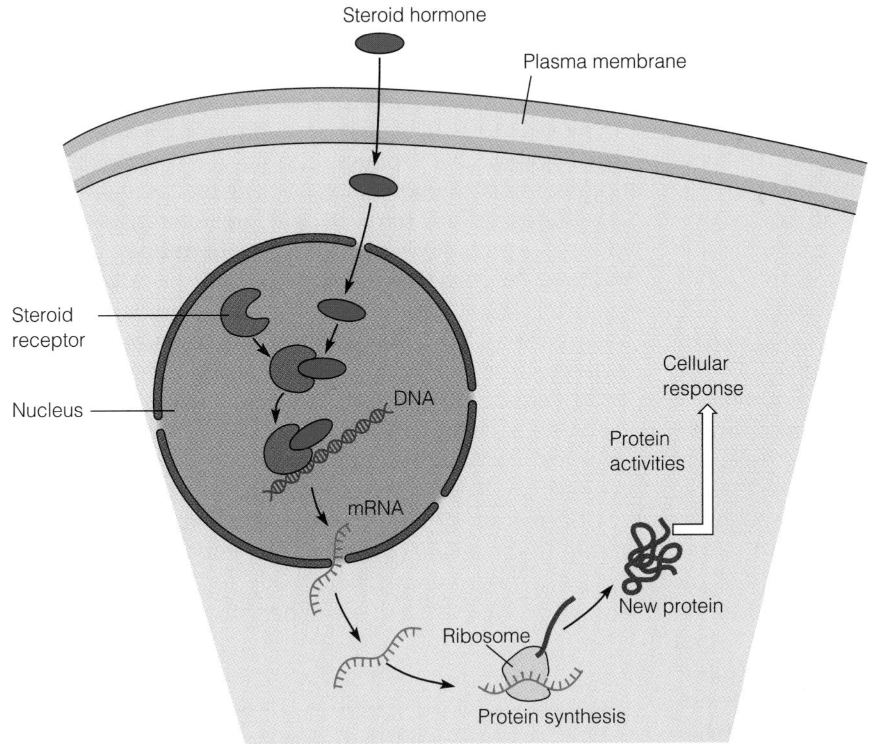

Figure 13–15 Steroid Hormones and Intracellular Receptor
Steroid hormones are lipophilic, easily passing through the plasma membrane. Steroids bind to receptors in the cytoplasm, or the nucleus (as shown), which are translocated into the nucleus and bind DNA. This binding alters transcription so new genes and proteins are expressed as part of the cellular response.

Adrenal Cortex and Steroid Hormones

The cortex is organized in layers and produces basically three different types of steroids in three different regions of the gland (Figure 13–14). The types of steroid hormones synthesized in the cells of the cortex are the *mineralocorticoids, glucocorticoids,* and *sex hormones*. They are all remarkably similar in chemical structure, but their effects on the cells of the body are distinctly different. Their names are an indication of their functions. Mineralocorticoids affect the balance of minerals (particularly sodium and potassium) in the body. Glucocorticoids affect the metabolism of sugar (glucose) by the body and help the body respond to stress, tissue damage, and inflammation. Sex hormones determine secondary sexual characteristics and libido and influence the body's general metabolism level. The synthesis of these three types of hormones from a cholesterol precursor is directly influenced by ACTH. ACTH binds to the cell surface of cortical cells and activates a special cytoplasmic messenger system that activates enzymes that are directly involved in the synthesis of steroids. Each cell that produces steroids responds differently to this signal and, so, produces specific types of hormones. The ATCH signal targets the cortical cells of the adrenal gland but does not affect all the cells in the same way.

Aldosterone It is important to consider how each of these classes of steroids functions. A major mineralocorticoid is aldosterone. Its target is the kidney, in which it affects the absorption of sodium and potassium by the kidney tubule cells (see Chapter 10). Aldosterone works in conjunction with ADH, which is an oligopeptide hormone secreted by the posterior lobe of the pituitary, in controlling the kidney's capacity to resorb water. Excessive secretion of aldosterone upsets the water balance of the body by increasing retention of sodium ions. Where sodium ions are found in higher concentration, water generally moves by osmosis. This leads to swelling and hypertension, which can be both troublesome and dangerous. Hypersecretion of aldosterone also increases the excretion of potassium. If potassium is overdepleted from the body, nerves cannot function properly. This is disastrous. In contrast, hyposecretion of aldosterone may lead to dangerously low levels of sodium ions in the body and retention of excessive amounts of potassium. It is unlikely that an individual with a severe aldosterone dysfunction could live very long. Aldosterone may be administered to individuals to correct modest insufficiencies in its production.

Glucocorticoids Glucocorticoids help regulate the breakdown of glycogen in the liver and of fat stored in adipose tissues. In addition, they also are pivotal in the conversion of cellular protein to glucose, if glycogen and fats are in low supply. The action of glucocorticoids, such as cortisol and cortisone, have a balancing effect on the action of insulin, which stimulates transport of blood sugar across plasma membranes into cells and activates its conversion to glycogen.

Cortisol and cortisone also play roles in resistance to stress. Glucose is the primary source of chemical potential energy for the body (that is, use in production of ATP), and in a stressful situation the body recruits as many energy sources as are available. Glucocorticoids are also effective as anti-inflammatory agents. They help reduce swelling and capillary dilation associated with damage to tissues. However, one can get too much of a good thing.

Excessive secretion of glucocorticoids (particularly hydrocortisone and cortisone) leads to a condition known as *Cushing's syndrome* (Figure 13–16). In this condition there is a significant and dramatic redistribution of the body's fat. A moon-shaped face, a hump on the back, and thinning of the legs are all characteristics of the disease. Although glucocorticoids are powerful anti-inflammatory drugs, they also interfere with the rate at which tissues repair themselves after damage. In Cushing's syndrome the body bruises easily and heals slowly. The reciprocal condition is even more serious. *Addison's disease* is characterized by severe reduction in secretion of glucocorticoids (Figure 13–16). This leads to low blood sugar, low blood pressure, lethargy, weight loss, muscle weakness, and mental fatigue. Untreated, Addison's disease may well be fatal.

Sex Hormones The final type of steroid molecules produced by the adrenal cortex is that of the sex hormones. Sex hormones are vitally important in establishing and maintaining the **secondary sex characteristics** of males and females. The adrenal cortex produces a minute amount of these steroids, the bulk is produced by the gonads under the influence of FSH and LH. The general class of steroids to which male sex hormones belong are the **androgens.** This includes testosterone, which is the predominant sex hormone in males. The predominant female sex hormones are estrogen and progesterone. Problems of over- or undersecretion of sex hormones can have detrimental effects on the development of secondary sex characteristics. *Hypergonadism* in males results in overdevelopment of the genitalia, pronounced muscularity, and hairiness. If hypergonadism arises prior to the onset of puberty, the individual affected may undergo not only precocious sexual maturation but be shorter in stature than expected because the growth of long bones ceases as a result of premature fusion of the epiphyseal plates.

The reverse situation is *hypogonadism.* This condition results in the failure of the male, or female, to produce sufficient sex hormones for the body to develop secondary sexual characteristics at all. When the proper level of sex

> **Secondary sex characteristics** the development and maintenance of mature features of males and females at puberty under the influence of sex steroids; for example, facial hair, breast development
>
> **Androgens** the male sex hormones, such as testosterone

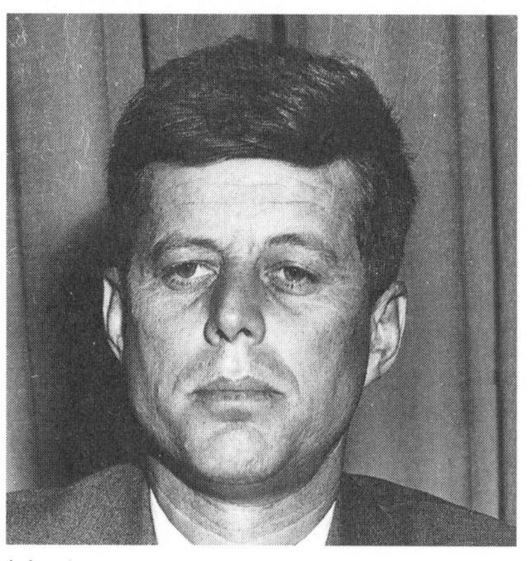
(a)

(b)

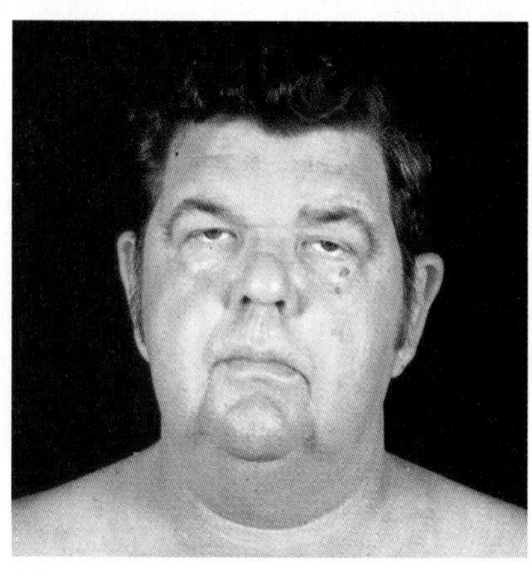
(c)

Figure 13–16 Diseases Affecting the Adrenal Cortex
Addison's disease is a serious disorder in which secretion of glucocorticoids is severely reduced (a) and (b). President Kennedy was affected by this disease and the effects of steroid treatment altered his facial appearance. Cushing's syndrome (c) is a condition in which glucocorticoids are oversecreted.

hormone production is upset in a body, or when sex hormones are introduced in excess into the body from an outside source, a wide range of body tissues are affected. For example, treatment of males with estrogen tends to feminize them. Breasts develop, the voice changes, facial hair recedes, and there is a redistribution of body fat and a reduction of muscle. Behavior may change as well. On the other hand, treatment of females with androgens, such as testosterone, increases the proportion of muscle to fat, increases strength, deepens the voice, reduces the breasts, and brings about the growth of facial hair and balding.

The Dark Side of Steroids

There is a darker side to androgens in our modern society. Because of our technical ability to synthesize steroids artificially, their normal chemical structure can be modified, and it is relatively easy to make large quantities of steroid derivatives. Some chemical modifications make the synthetic steroids hundreds of times more powerful than the natural ones. By taking these drugs an individual can significantly alter his or her metabolism. Because androgens are involved in anabolic or building-up processes, this equates to growth of bigger muscles and developing greater strength in a shorter time. In addition to the obvious physical changes in muscularity, many other tissues may be affected. The cardiovascular system may become more susceptible to atherosclerosis; the gonads may cease functioning normally, leading to sterility and/or sexual impotence; and aggressive, antisocial behavior may develop. Anabolic steroids, as with any abused substance, may severely disrupt the homeostatic mechanisms of the body and, thus, seriously undermine human physical and mental health. Fooling Mother Nature is dangerous business.

Adrenal Medulla

The innermost region of the adrenal gland is called the *adrenal medulla* (Figure 13–14). The cells in this region are of a different origin than those of the cortex and do not produce steroid hormones. Medullary cells produce a class of compounds known as *catecholamines*. These hormones are identical to neurotransmitter molecules found in the nervous system, such as *epinephrine* and *norepinephrine* (Table 13–1). The medulla is so different from the cortex that it is considered here as a separate gland. The similarity in hormones produced by medullary cells and neurons suggests that there is a close relationship between the endocrine system and the nervous system. However, there is an important distinction between the two systems. The endocrine system acts indirectly on target cells by releasing hormones into circulating blood, and the nervous system acts directly by making synaptic connections between neurons and their target cells.

The adrenal glands use the circulatory system to deliver hormones over relatively great distances. Medullary cells produce both epinephrine and norepinephrine. These hormones cause rapid and dramatic changes in the physiology and metabolism of the body in preparation for so-called fight-or-flight behaviors. The medulla secretes these hormones (principally epinephrine) into the circulatory system in times of emergency, and they nearly instantaneously cause a variety of effects that prepare the body for action. This includes constriction of arterioles and the consequent elevation of blood pressure, increased flow of blood to skeletal striated muscles, increased rate and depth of heartbeat, opening of airways in the lungs and recruitment of glucose from the breakdown of glycogen stored in the liver. The heightened physiological state prepares us for a wide range of responses to emergency situations.

The function of the hormones of the adrenal medulla would seem to be a vitally important feature of the body's defenses in times of stress, but is it? If we were to lose the function of the adrenal medulla, surprisingly, the loss would not be life threatening. Although we might be somewhat hampered in our ability to sustain the fight-or-flight response, normal activities would not be critically affected. This is most likely because of the close physiological and chemical relationship between the hormones of the adrenal medulla and those of the nervous system in general. The nervous system can "make up" to some extent for functions lost with the elimination of the function of the medullary core of the adrenal gland. By comparison, the hormones produced by the cortex of this gland are absolutely critical for survival.

The nervous system (Chapter 17) is more direct and limited to close communication between neurons and their respective targets, that is, muscles, glands, and other neurons. However, the brain may also be considered as a gland, instead of exclusively modeling it as an electrical switch board or electronic computer. A glandular nature of the nervous system is certainly reflected in the structural identity and in the similarity of the mode of action of neurotransmitters in comparison to the products of cells in the medulla of the adrenal gland.

HORMONES OF THE POSTERIOR PITUITARY

We now return to the posterior lobe of the pituitary gland, which is so different from the anterior lobe that it is here considered as a separate and distinct glandular structure. The posterior lobe of the pituitary gland originates exclusively from the nervous system. In fact, as noted earlier, the posterior lobe of the pituitary gland is composed of the cellular extensions of brain cells whose bodies are in the hypothalamus. As a result, when we consider the hormones of the posterior lobe, we are, in actuality, considering the secretion of hormones produced in the brain and shipped to a special location to be released. There are two hormones secreted from the terminal cell processes of nerves in the posterior lobe (Table 13–1). They are antidiuretic hormone (ADH) and **oxytocin.** Both are oligopeptides and both have profound effects on the target tissues influenced by them. It is fair to say that the posterior lobe of the pituitary gland is not a separate gland at all, but merely a satellite structure of the hypothalamus, which is itself an important gland. However, because of the clear and well-organized nature of the region from which the hormones are released, the posterior lobe will be considered as a separate functional region, if not a separate gland.

ADH

ADH is extremely important in controlling and regulating the balance of water in the body by the kidneys (Chapter 10). ADH is made in cells located in the hypothalamus, transported down relatively long axonic extensions and stored in the ends of these extensions in the posterior lobe of the pituitary. Changes in the osmolarity of blood fluids are detected by receptor cells in the hypothalamus, which triggers the release of ADH. A negative feedback loop is involved, as in many cases of regulation of metabolism that require on-off switches to function properly (Figure 13–17). Even as ADH elicits increased absorption of water back into the body through the kidney tubules, the quality of the body's fluids change and continued ADH release is inhibited. The balance of water and solutes are of such importance (see also aldosterone in the section on the adrenal cortex) that the continuous coordination of the expression and release of the hormone is tightly regulated. The disease **diabetes insipidus** results from the failure of the hypothalamus to produce or release ADH. If ADH is not produced and released, the body excretes water but does not resorb it. As discussed in Chapter 10, covering the urinary system, this leads to unregulated loss of water from the body. Literally gallons of dilute urine are produced each day in the face of continuous, thirst-driven consumption of water. For lack of this simple oligopeptide the entire body is thrown into homeostatic chaos. It is not surprising that this form of diabetes is lethal.

Oxytocin

The other hormone released from the posterior lobe of the pituitary gland is oxytocin. Oxytocin is also an oligopeptide hormone (Figure 13–18). It has powerful effects on the contraction of striated muscle fibers and/or muscle cells, particularly in the female body. This includes the muscular layer of the uterus known as the *myometrium* and the *myoepithelial cells* associated with the **let-down reflex** of mammary glands. Each of these target tissues has a specialized function. The regulation of oxytocin offers a rare glimpse into processes that function as positive feedback loops. For example, for the birth of a baby to occur, the myometrium must undergo persistent and sustained muscular contractions. These contractions grow continuously stronger over the period of a few hours (or many hours, if labor is prolonged) prior to birth. Oxytocin is the hormone that induces the contraction of the uterine myometrium.

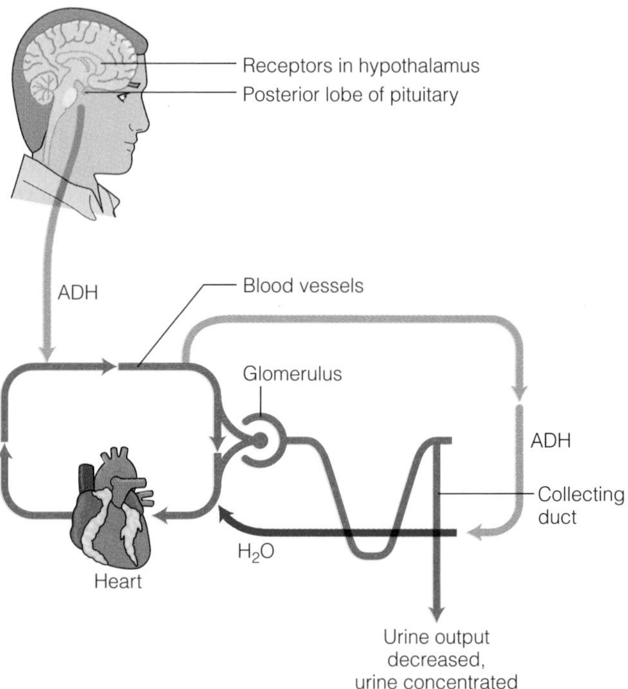

Figure 13–17 Antidiuretic Hormone (ADH) and its Action
Regulation of urine output depends on ADH. ADH controls the resorption of water across the walls of the kidney collecting ducts.

Oxytocin a hormone produced by the hypothalamus and released from the posterior pituitary

Diabetes insipidus a rare condition in which lack of ADH production results in excessive urine production

Let-down reflex a reflex in mammals in which the tactile stimulation of suckling brings about release of milk from mammary glands

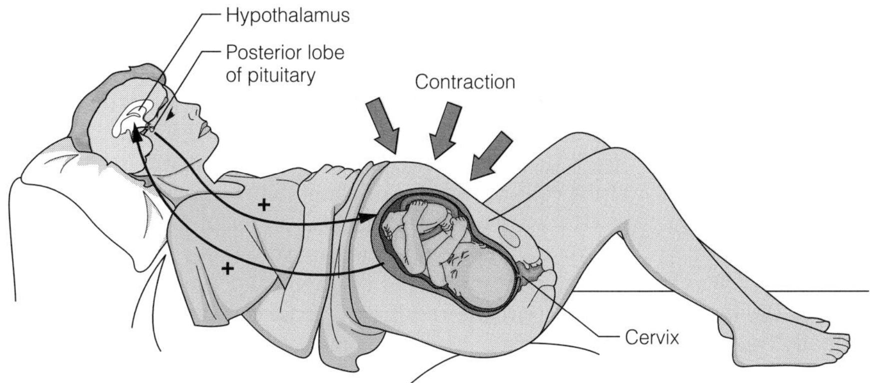

Figure 13–18 Oxytocin and Childbirth
The effect of oxytocin is to increase uterine contractions during childbirth. This positive feedback loop results in more and more oxytocin being released from the posterior pituitary as stronger and stronger contractions occur.

The first contractions serve as a signal to the soon-to-be-mother that delivery time is near.

The contractions are initially relatively weak, and the interval between them relatively long. The process depends on the uterus sending signals to the hypothalamus to release more oxytocin. The next wave of contractions is stronger. The cycle repeats itself, contractions building in strength and shortening in interval. How is this different from the effects of other hormones? Instead of the signals from the contracting uterus inhibiting the release of oxytocin, they elevate it more at each successive step. This is positive feedback, and it continues to build until the strength and duration of the contractions are great enough (take a deep breath!) to push the baby out of the uterus through the birth canal and into the waiting hands of the doctor or nurse. A negative feedback loop would never work in this situation, because it would turn itself off with every cycle of contractions. On the other hand, a positive feedback loop for hormones such as GH or ADH would be very detrimental.

This mechanism is also useful in explaining the initiation of flow of milk during breast-feeding. In this case, the effect of oxytocin on the myoepithelial cells of the mammary glands is a positive feedback. Tactile stimulation of the nipple(s) of a new mother from the sucking of the baby at her breast, sends a signal to the hypothalamus and elicits the release of oxytocin. The oxytocin causes the tiny muscle cells surrounding the mammary glands to squeeze the milk out of the gland into the ducts and out through the nipple. This let-down reflex continues to operate until the tactile stimulus is removed (Figure 13–19).

PANCREAS

The pancreas has two very different types of cells that separately constitute the functional parts of its activity. For this reason it may be thought of as a dual purpose organ. In Chapter 9, we saw that the pancreas produces a wide range of digestive proenzymes that aid in the enzymatic degradation of nutrients in the duodenum. This is an exocrine function of the pancreas. The other important function of the pancreas is endocrine in nature. In specialized regions of this organ are islands of cells that produce and release the hormones insulin and glucagon. The

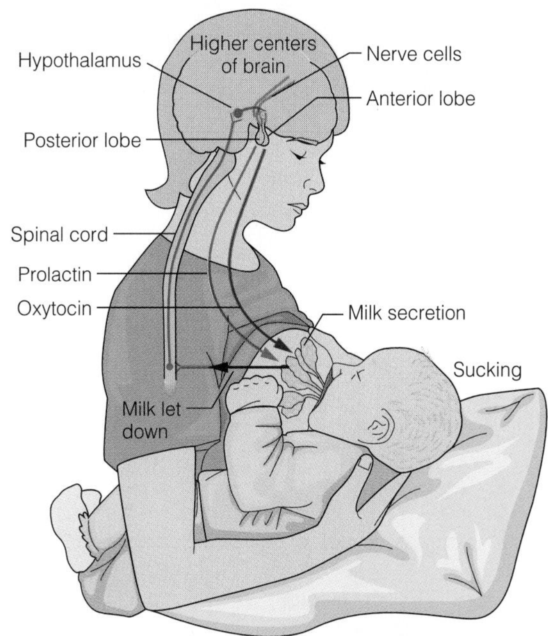

Figure 13–19 The Let-Down Reflex
The let-down reflex occurs when suckling stimulates tactile receptors that send signals to the brain to bring about the release of oxytocin. Oxytocin stimulates myoepithelial cells to squeeze the mammary gland and bring about milk secretion.

islands of cells are called the islet of Langerhans, named after the German anatomist who discovered them (Figure 13–20). They make up only about 1% of the total mass of the pancreas, but they are tremendously important in the regulation of glucose.

Alpha and Beta Cells

There are two main types of islet cells—the *alpha cells* and the *beta cells*. Insulin is produced by the beta cells and glucagon is produced by the alpha cells. Roughly 75% of the islet cells are the beta type. As discussed at the beginning of the chapter, insulin controls principally the transport of glucose into the liver and its storage in the form of the polysaccharide glycogen. Insulin also increases the transport of glucose into skeletal muscle cells, so that they have the chemical resources to make ATP and, thus, the capacity to contract. When all is functioning properly, insulin provides a convenient and efficient mediator of glucose transport and storage. However, this is only half the story of sugar metabolism. We need to consider how we normally get the glucose out of storage, when the body needs fuel. This is the job of the hormone produced by the alpha cells of the islet of Langerhans, glucagon.

Glucagon, as is insulin, is a polypeptide hormone. Unlike insulin, however, glucagon stimulates liver cells to activate enzymes involved in the breakdown of glycogen and to transport the resulting glucose molecules back into the vascular circulation. This makes glucose available for general use by cells of the body and is particularly important for the activities carried out by muscle cells and brain cells. The reciprocal relationship between alpha and beta cells, and the hormones that they produce, establishes a feed-

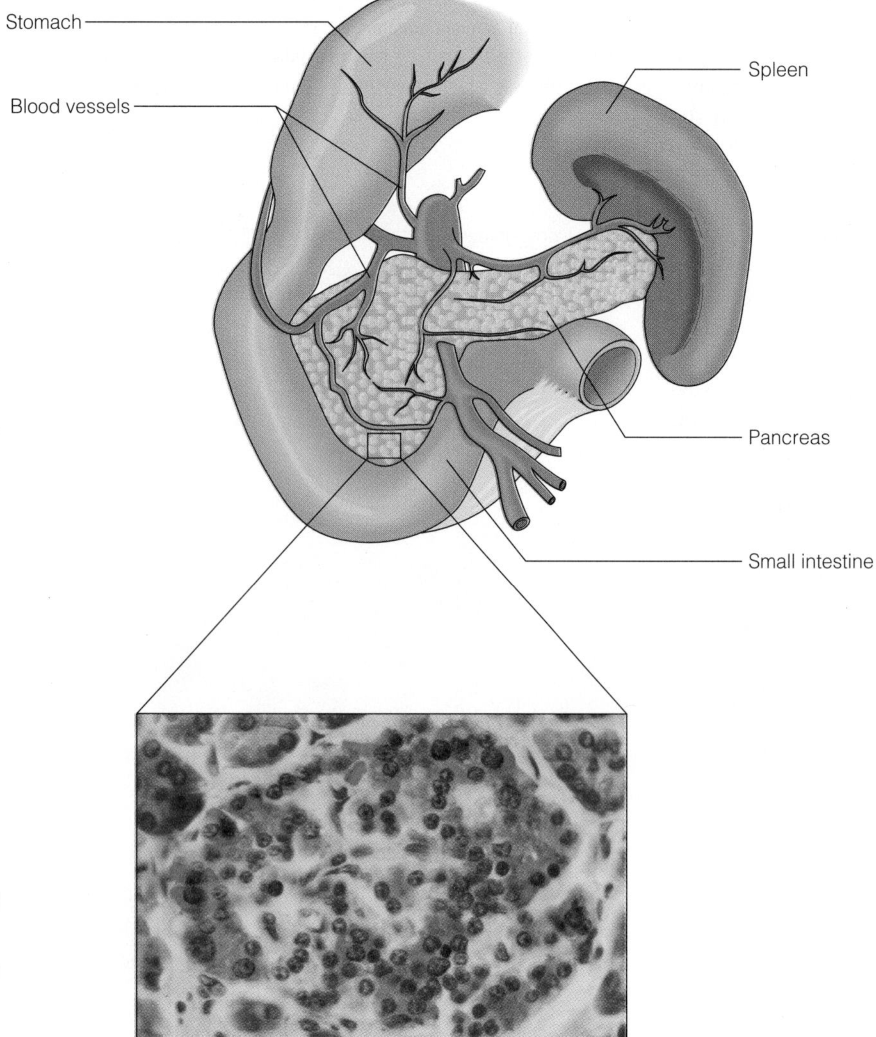

Figure 13–20 The Islet of Langerhans

The pancreas is a dual-function organ with both exocrine and endocrine activity. The endocrine function is supported by the islets of Langerhans which produce the hormones insulin and glucagon.

back loop. Insulin also has other important effects on cells, including a role in cell growth and proliferation. Insulin is one of the most studied hormones in the human body. A major reason for this is that abnormalities in the production or use of insulin are fairly common in humans (Figure 13–20). A serious human disease related to abnormalities in the production or use of these hormones is diabetes mellitus, which was described earlier in this chapter. Left untreated, this form of diabetes can lead to death.

There are two other glands or glandular tissues that are to be considered. They are important in their own ways, but less well understood in their regulation than many of the glands discussed in this chapter. They are the **pineal gland** and the **thymus gland.**

PINEAL GLAND

The pineal gland is a small island of cells on the upper surface of the third ventricle of the brain. It is sometimes referred to as the *third eye* (Figure 13–21). This latter designation arises from the fact that in many animals, such as birds and reptiles, the pineal gland is directly affected by changes in light. The thin skull bones and semitransparent skin of these animals make it possible for the pineal gland to receive and detect light in this way. In humans the pineal gland is indirectly affected by light, through the input of optic nerves from the retina. The pineal gland produces the hormone **melatonin,** which is thought to be involved in influencing *circadian rhythms.* An example of this is the jet lag we suffer from as a result of the alteration of day-night periods when we travel over several time zones. The disturbance of our activity cycles is in part a consequence of melatonin release occurring on the original schedule (home clock) and out of phase with our new environment. Another function of melatonin is regulating the secretion of steroids in the ovaries. In this way it may be involved in the timing of the onset of sexual maturation in females and in regulating the menstrual cycle. Other pineal hormones, such as **serotonin,** which subtly influence the brain and the adrenal cortex, may also be involved in these processes.

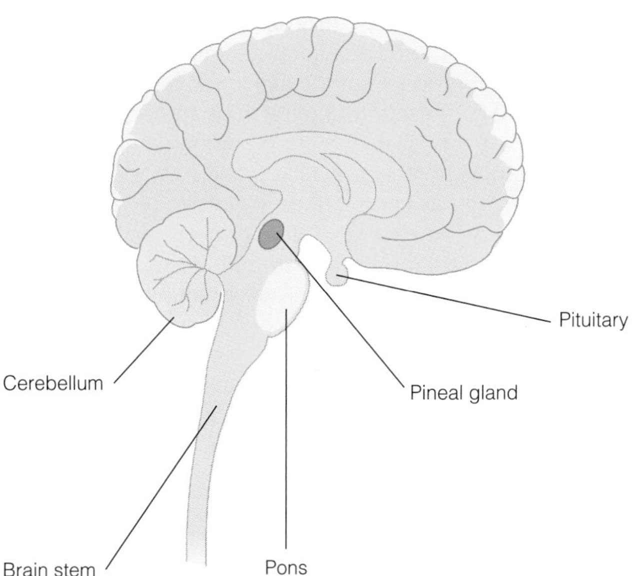

Figure 13–21 The Pineal Gland
The pineal gland is located in the center of the brain. It produces the hormone melatonin, which has influence over circadian rhythms and sexual maturation.

THYMUS GLAND

The thymus gland has an interesting developmental history. Anatomically, it is located behind the sternum just above the heart and is quite large when we are young (Figure 13–22). During infancy and early childhood, the growth and development of the thymus is active in bringing about the maturation of T-lymphocytes (see Chapter 12). The production of T-lymphocytes takes place relatively early in human growth because, as we age, the thymus undergoes degeneration or involution. During involution, the thymus shrinks and becomes inactive. Later in life the thymus generates no new T-lymphocytes. The thymus produces a hormone called **thymosin.** The effects of thymosin on T-lymphocyte differentiation seem to be dependent on both the hormone and the T-lymphocytes being present simultaneously in the thymus. This suggests that, in addition to the hormone itself, the environment of the thymus is important to the lymphocytes that colonize this "temporary" endocrine gland.

Pineal gland a gland located in the brain and secreting melatonin

Thymus gland a gland located in the region of the neck and, in childhood, the site of formation and maturation of T-lymphocytes; produces the hormone thymosin

Melatonin a hormone produced by the pineal gland; influential in establishing circadian rhythms

Serotonin an essential neurotransmitter prevalent in the brain

Thymosin a hormone essential to T-lymphocyte differentiation in the thymus gland

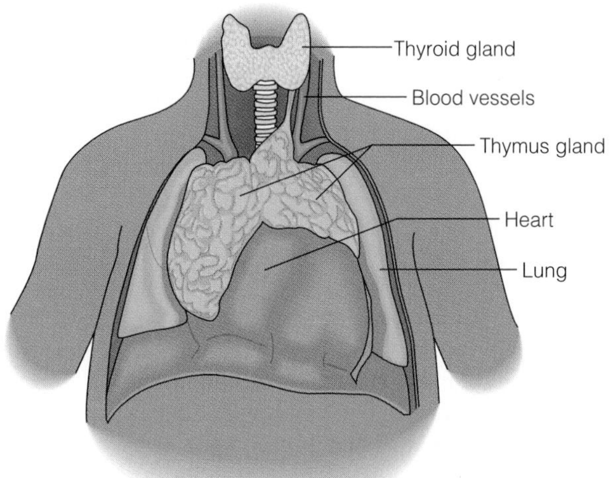

(a) Thymus in young child

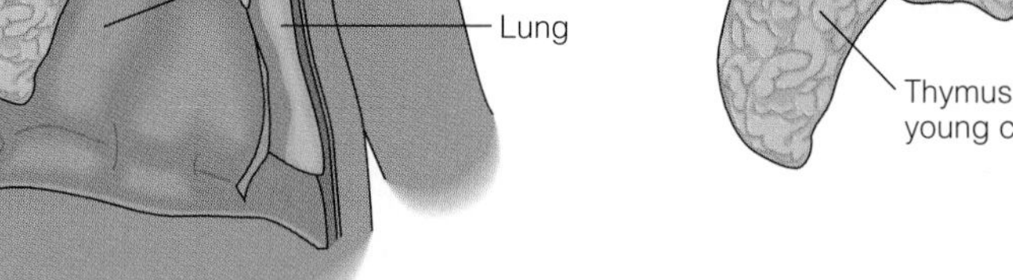

(b) Comparison of thymus in young child and adult

Figure 13–22 The Thymus Gland

The thymus gland is large and active in children and young adults (a) but undergoes atrophy and involution in adults (b). The thymus gland provides the tissue environment and the hormone (thymosin) necessary for differentiation of T-lymphocytes.

Summary

The endocrine system is a collection of many relatively small glands located in many different regions of the body. The reason for considering them together is not because they have the same effects on the tissues of the body, but because the underlying mechanism by which they operate is the same. Each of the glands releases hormone molecules directly into the vascular circulation. The consequence of precisely controlled systemic releases of hormones is to provide stimulatory signals to alter target cell functions at relatively great distances from the source. Some of the endocrine glands produce hormones whose primary effects are on the performance of other endocrine glands. The hormones of the hypothalamus, known as releasing factors, and many of the hormones released from the anterior lobe of the pituitary gland (FSH, LH, TSH, ACTH) fall into this category. However, the process by which these hormones work is self-limiting; that is, they establish metabolic homeostasis in the body through negative feedback regulation from the glands that they, themselves, initially stimulate.

The overall mass and number of cells in the human body is enormous. Most of these cells are organized into complex tissues, which in turn are built up into organs and organ systems. The endocrine system of glands is key to coordinating the metabolic activities and, thus, the survivability of all the body's cells. It does this through the subtle, and pervasive, use of hormone molecules that elicit specific metabolic and behavioral responses in their targets.

The normal processes of life depend in large part on the stability and conformity of interactions among different tissues and organs. This stability is maintained by exceedingly fine adjustments in metabolic rates and cell activities, such as muscle contractions, glandular secretion, and neurotransmission. The pendulum of hormonal effects swings slowly, and inexorably, around the set points that define the state of homeostasis. As we will see in Chapter 17, which describes the human nervous system, the faster and more direct route by which cues are delivered between nerves does not so much change the mechanisms by which signal molecules affect cells, but rather the rate and distance over which these effects occur.

Questions for Critical Inquiry

1. Which of the endocrine glands is considered the master gland? Why?
2. Too much or too little growth hormone (GH) and/or thyroid hormone (TH, thyroxine) has profound effects on human anatomy, physiology, and behavior. Describe some of the conditions associated with too much, or too little, GH and TH both during growth and development and in adulthood.

3. What types of steroids are made in the adrenal cortex? Why is this class of hormones so vitally important to homeostasis?
4. What effects do insulin and glucagon have on sugar metabolism? What can go wrong in this relationship and what are the consequences?

Questions of Facts and Figures

5. What are the differences between endocrine and exocrine glands?
6. What relationship exists between the pituitary gland and the hypothalamus?
7. What are the products of the anterior lobe of the pituitary? The posterior lobe?
8. FSH and LH are regulated by what two hormones from the ovaries?
9. What role do calcitonin and parathyroid hormone have in the homeostasis of calcium?
10. The let-down reflex is controlled by what activity? What hormone is involved?
11. What kind of problems can arise from overuse or abuse of anabolic steroids?
12. Do FSH or LH have effects on male physiology? If so, what?

References and Further Readings

Alberts, B., Bray, D., Lewis, J., Raff, M., Roberts, K., and Watson, J. O. (1994). *Molecular Biology of the Cell,* 4th ed. New York: Garland Publishing.

Sherwood, L. (1993). *Human Physiology,* 2nd ed. St. Paul, MN: West Publishing.

Stalheim-Smith, A., and Fitch, G. K. (1993). *Understanding Human Anatomy and Physiology.* St. Paul, MN: West Publishing.

CHAPTER 14

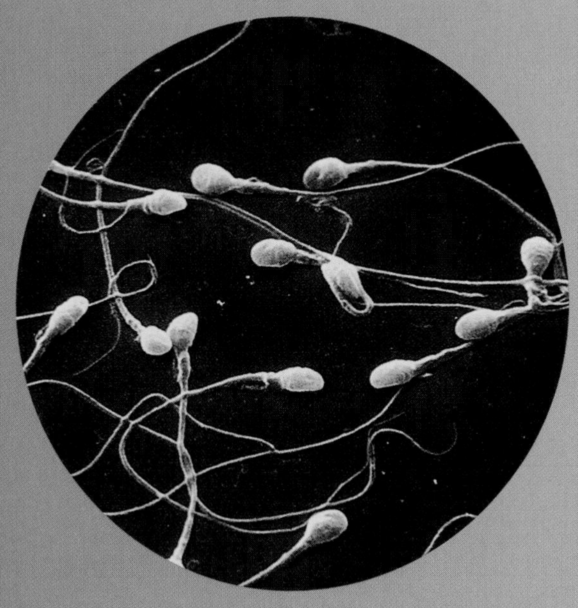

Sex and Life

INTRODUCTION

In human biology, to talk about life, we must talk about sex and reproduction. Human sexuality is a cultural phenomenon, as well as a biological one. Sexual images are all around us. Art, music, politics, and religion fill the news with controversies involving sex. Advertising firms bombard us with sexual images to lure us to buy particular kinds of clothes, cosmetics, and cars. People are alternately offended, gratified, amused, or bemused by this onslaught. Very few people in our society can resist being influenced in some fundamental way by the topic of sex. Everyone has an opinion on what is good about sex, what is bad, what is art, what is obscene or pornographic, and what is tasteful and what is not. However, everyone agrees that sex and sexual reproduction are important aspects of life and the fate of our species. This chapter deals with sexual anatomy and physiology of males and females in their reproductive capacity.

REPRODUCTIVE STRATEGIES

Different species of living organisms display reproductive strategies and behaviors very different from those of humans. Birds have unique flight patterns and aerial activities as a prelude for mating; amphibians may carry their mates along with them; and fish may swim upstream to reproduce and die. As different as these various strategies appear, they are all sexual in nature. How is sexual reproduction different from other types of reproduction? The simplest form of reproduction is asexual, as represented by reproduction of bacteria and protista, which occurs by mitosis or simple duplication (Figure 14–1). Asexual reproduction also includes the process of budding, in which a new individual is formed by directly growing out of another, as is the case for many types of yeast and for hydra. *Parthenogenesis* is also a form of asexual reproduction and occurs as a result of an egg developing without being fertilized. This happens naturally in many types of invertebrates including aphids, bees, and wasps. In each of these cases, the newly formed individual is identical genetically to its progenitor, thus, forming a clone.

Sexual reproduction also involves cell division, but a new individual arises only after the fusion of two highly specialized, genetically different cell types called gametes. A gamete is a cell specialized for sexual reproduction and is represented in males as *sperm* and in females as an *ovum* or *egg* (Figure 14–2). Thus, gametes are derived from two morphologically and functionally distinct forms of individuals composing a species such as ours. One consequence of the interaction of the two different types of human gametes is fertilization. This leads to the development and growth of an entirely new and genetically unique individual and not a clone of either one of the two parents.

External Fertilization

Many species have mechanisms of reproduction in which the gametes are released into the extracellular environment and undergo fertilization outside the organism. Sea urchins release their gametes directly into the shallow seas and tidal pools in which they live. Fertilization takes place almost immediately upon contact between sperm and egg. This is a common strategy for reproduction in animals and results in what is called **external fertilization.** For external fertilization to be effective as a mechanism for reproduction, there have to be a tremendous number of gametes (eggs and sperm) involved. In fact, hundreds of millions of gametes are released by male and female sea urchins during *spawning*.

Fish and frogs also undergo external fertilization of eggs, but usually the male and female representatives of these classes are closer to one another than sea urchins when they reproduce. Most of us are familiar with the sexual behavior of salmon. These fish are noted for traveling hundreds of miles upstream to release their gametes during spawning (Figure 14–3). They unerringly return to the loca-

External fertilization the fertilization of an egg by a sperm outside the body of the female

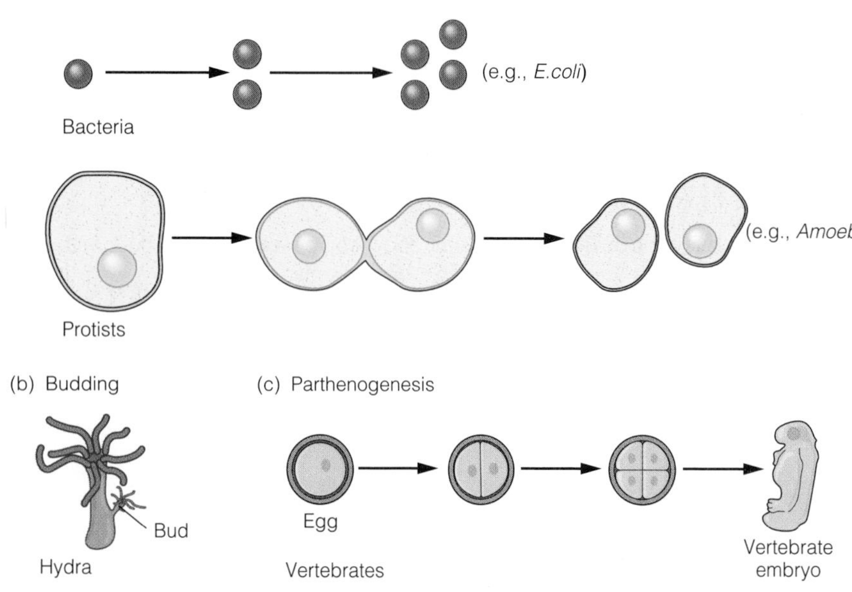

Figure 14–1 Simple Strategies for Reproduction

Simple duplication is asexual in nature (a). Bacteria and protists use this mechanism. Budding organisms (b), such as the hydra, asexually produce small complete copies (clones) of themselves that separate and become new individuals. Insect and some vertebrate eggs are capable of unfertilized development, a phenomenon known as parthenogenesis (c).

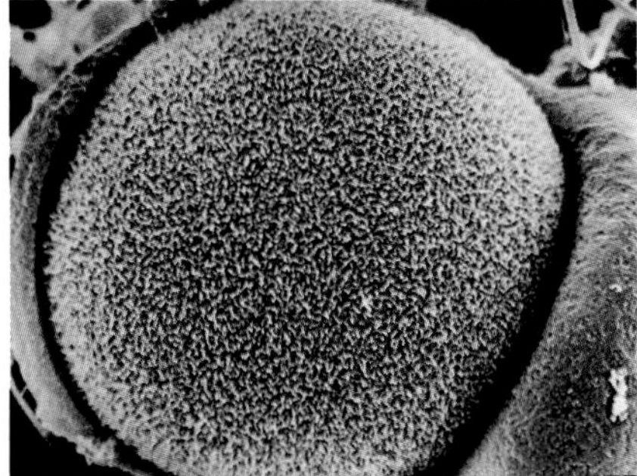

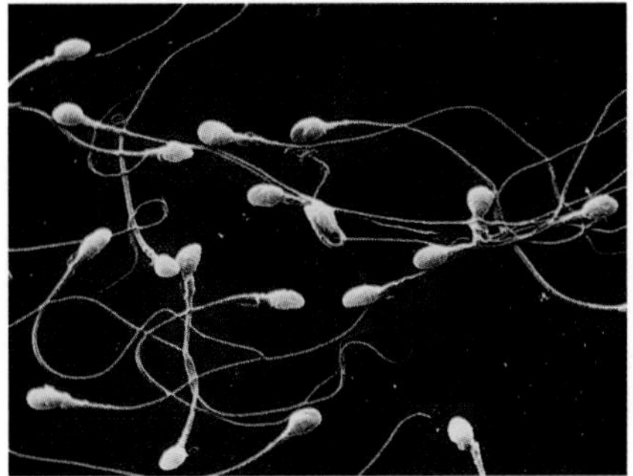

Figure 14–2 Human Gametes

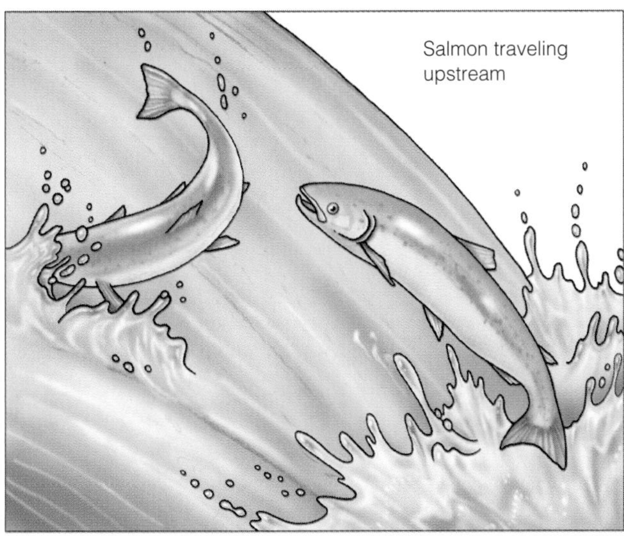

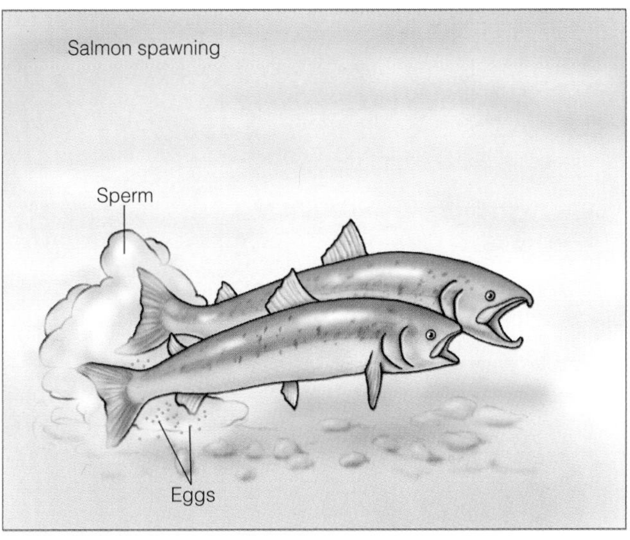

Figure 14–3 External Fertilization
Salmon hatch in freshwater streams and migrate downstream to the ocean where they mature. After several years of growth, they return to the streams of their origin to spawn. Using a sensory homing mechanism that identifies the home stream, they release gametes (egg and sperm), which are fertilized externally.

tion of their own hatching to participate in the process of reproducing the next generation. Once this arduous journey is complete, and they have discharged their gametes into the freshwater spawning sites, they die. Frogs do not display behavior quite so dramatic, but some species also have unique strategies that ensure proper and timely fertilization of eggs. A female frog may carry a smaller male partner on her back for days to continuously fertilize the eggs she produces.

Internal Fertilization

Reptiles, birds, and mammals do not release their gametes into the environment for external fertilization. These groups of animals have evolved mechanisms of **internal fertilization** (Figure 14–4). Internal fertilization occurs within the female body. Egg and sperm encounter one another in specialized organs and begin development internally. After fertilization, and depending on the species, an embryo may be enclosed within a hard-shelled, externally laid egg (birds, reptiles) or directly implanted into the highly vascular tissue of the uterus of a pregnant female (mammals). Internal fertilization is required for life on land and is a somewhat more sophisticated and complex process than that associated with external fertilization. The

Internal fertilization the fertilization of an egg by a sperm inside the female body

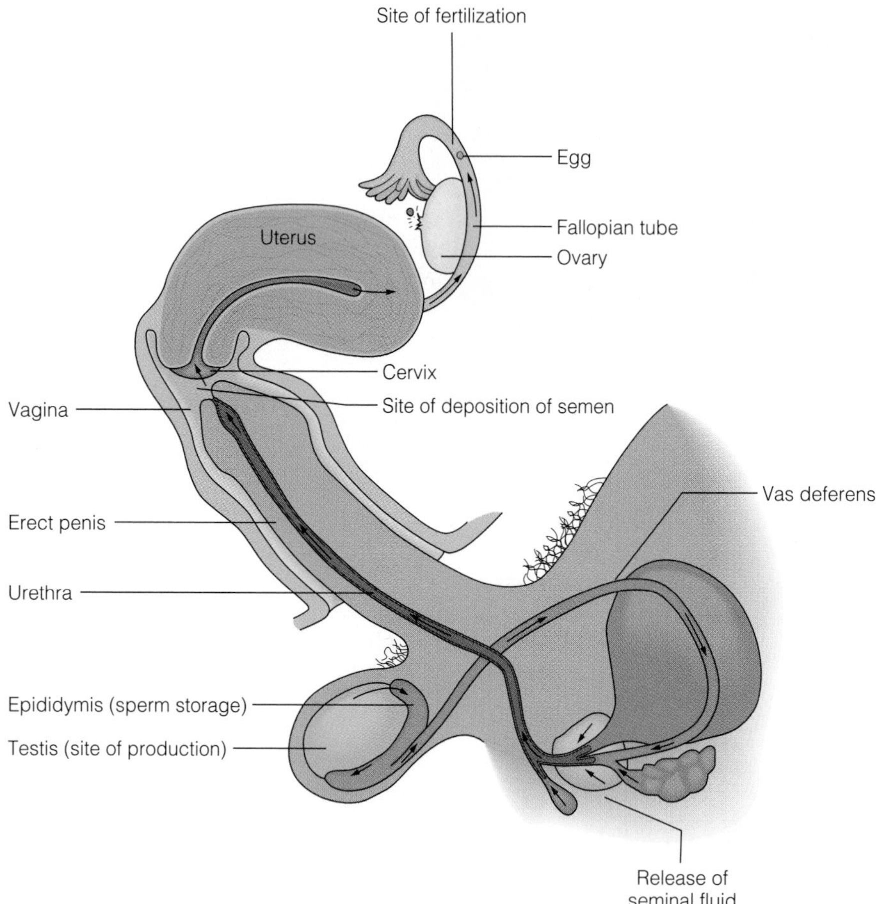

Figure 14-4 Mammalian (Human) Internal Fertilization
Internal fertilization in mammals involves penetration of the penis into the vagina and release of sperm into the female reproductive tract.

reasons for this are that internal fertilization involves more intimate interactions between individuals, utilizes specialized tissues to sustain the embryo, and relies for success on specialized behaviors that bring the parents together in the first place.

Human Encounters

A human male inseminates a female through direct and intimate sexual contact. Fertilization takes place inside the female's body. This greatly increases the chances for fertilization to occur. In fact, human females (and females of all internally fertilizing species) produce far fewer eggs than organisms that reproduce externally. However, males in all classes of vertebrates produce large numbers of sperm. When released, sperm are concentrated within the reproductive canal of the female and can survive for relatively long periods of time.

The initial development of reptile, bird, or mammal embryos is also internal. However, most reptiles and all birds package the embryos for external growth and development. Reptilian and avian embryos are surrounded by thick, wet, and viscous nutrient layers and encased in a tough, hard, and protective shell before they are pushed out from the female's body and incubated externally. In mammals, on the other hand, embryonic development is entirely internal and requires the invasion of the mother's tissues by an embryo in a process known as **implantation.** Long-term interaction between human embryonic and maternal tissues results in the formation of a *placenta,* which mediates exchange of nutrients and wastes between the two. The period of time between implantation and birth is called **gestation,** or more commonly, *pregnancy.* Gestation ranges from as little as two weeks in small mammals, such as mice, to two years in large mammals, such as

Implantation the embedding of the blastocyst into the wall of the uterus in placental animals

Gestation the average duration of time spent for a species developing *in utero* before birth (nine months in humans)

elephants. In *Homo sapiens,* gestation averages approximately nine months (an average of 265 days).

SEXUAL DIMORPHISM

Anatomically, our species has two easily differentiated sexual forms—female and male (Figure 14–5). This sexual dimorphism has been shaped by natural selection, and the sexual form and function of an individual is a permanent genetic condition. The anatomy of the male and female are complementary to allow for successful participation in the internal fertilization process needed to produce a new individual of our species. The complementarity of human genitalia, in this case the vagina in females and the penis in males, is a species-specific variation on the general structural requirements for all mammals. These organs provide a mechanical means to bring together the gametes produced by their respective reproductive organs (**ovaries** and **testes**). Biologically speaking, the ends not only justify the means, but they require them.

While the outcome of human reproduction is for the most part comparable biologically to that of other higher animals, we are unique with respect to our sexual behavior. Human sexual activity is not directed solely to reproduction. Many animals have seasonal cycles of reproductive activity, usually initiated in the spring or fall. The sexual activity of the animals during these times is driven by hormonal changes, which are responsible for ensuring the proper timing of reproductive capacity in males and

Ovaries gonads of females; producers of eggs
Testes gonads of males; producers of sperm

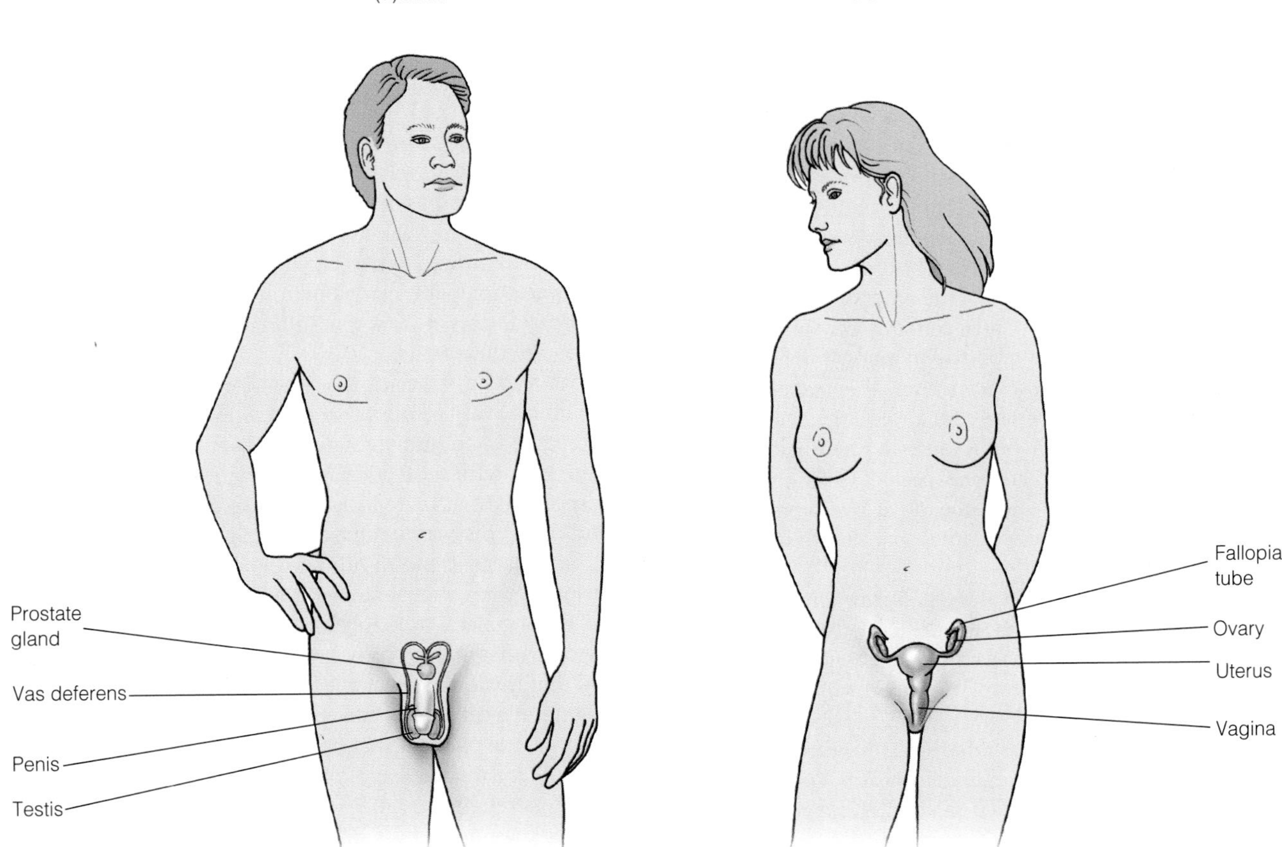

Figure 14–5 Reproductive Structures of Humans
The location of reproductive structures in human females (a) and males (b) is shown.

females and appropriate behavior in each for successful reproduction. Humans and a few other mammals, on the other hand, are fertile on a continuous basis throughout the year (a monthly cycle for humans). But the area in which we appear to be truly unique among animals is in seeking sexual activity for pleasure (Biosite 14–1).

This brief overview of sexual and asexual reproduction and behavior is aimed at demonstrating the diversity of mechanisms that animal species have evolved to produce new generations. It is also meant to show how human sexual interactions have far more than biological consequences. A living human organism is a very complicated entity with trillions of cells involved in constructing hundreds of different cell types and tissues found within an adult human being. However, no organ system is more important to the survival of the human species than the

BIO site 14–1

SEX AND PLEASURE—PROBLEMS IN PARADISE

In seeking pleasure from sexual encounters humans attempt to fulfill an ancient biological need. The evolution of sexual reproduction was probably selected through the linkage of the feelings of pleasure with coitus. However, major problems arise from conflicts between cultural and biological processes in today's world because many individuals who have sex for pleasure do so without regard for or, in many cases, ignorance of the reproductive potential of their own bodies. Clearly, females bear the brunt of the social and biological burden of sex, because they bear the children. Males are not involved in embryogenesis, except as co-initiators of the process that leads to fertilization. However, men certainly can (and should) be involved and supportive in many other ways. Pregnancy can be a dream come true and a wonderful, fulfilling experience for a woman when planned for or anticipated, but it can be a nightmare if it occurs unwanted and unexpected. Raising and nurturing a child is a full-time job. It is also probably one of the most important responsibilities we have as human beings.

Another problem associated with unrestrained sexual interactions is the spread of disease, specifically the sexually transmitted diseases. There are many different disease-causing organisms harbored in the human body that are spread by sex. Viruses and bacteria and the infections they cause can have devastating effects on human health. Diseases caused by bacteria such as syphilis and gonorrhea, while treatable with antibiotics, are nonetheless dangerous and on the rise. This topic will be discussed in more detail in Chapter 16. In the most serious situations, such as those involving human immunodeficiency virus (HIV), the consequences of ill-considered sexual activity often prove dangerous and may be fatal.

On a more positive note, it is interesting to point out that no other animals on earth face each other during sexual intercourse. Only humans do. The evolution of the human body, particularly development of an upright stance instead of remaining on all fours, required major changes in the skeleton, including changes in the angle of the pelvis and rearrangement of the internal organs of reproduction. This did not eliminate the possibility of front-to-back sexual intercourse, but uprightness probably tended to favor face-to-face encounters, which almost certainly played a role in enhancing development of more intimate, personal relationships between human males and females. Why these changes in anatomy and physiology occurred during human evolution is not known. There were, no doubt, many environmental and behavioral factors at work that influenced the natural selection of these traits.

With respect to face-to-face sex, however, it may have been advantageous in strengthening the pair bonding between human males and females to ensure long-term protection for the young, who in our species take years to grow and develop to self-sufficiency. This bonding may be thought of as part of a much larger trend in primate evolution to develop a group of behaviors that in humans we call love. It is a general principle in biology that the longer an offspring survives, the greater the likelihood that it will have a chance to reproduce and pass on traits carried by it to the next generation. Therefore, the changes that brought about the evolution of face-to-face sexual physiology and, arguably, strengthened male-female bonding may be considered as profoundly important biologically. The cultural changes stemming from these biological changes continue to evolve in terms of the structure of the human family and society.

reproductive system. It behooves us to know as much about our reproductive potential as we can. This chapter and the following one will focus on how the genetic material of males (potential fathers) and females (potential mothers) is packaged into gametes, how gametes develop and mature, and how accessory sex organs function to ensure that these specialized cells get together.

THE REPRODUCTIVE SYSTEMS OF HUMANS—GENETICS

There are 46 chromosomes found in the genome of *Homo sapiens*. The 46 are represented by 22 homologous pairs and one pair that may not be perfectly matched (Figure 14–6). This set of chromosomes and the genes of which they are composed provide all the genetic information necessary to produce a member of our species and differentiates us from other species. Each member of each identical pair of chromosomes is known as an autosomal chromosome. The remaining pair are known as sex chromosomes and are designated X and Y. In females there are two matched X chromosomes. An exception to identical pairing is observed in the sex chromosomes of males. In this case, a small Y-shaped chromosome can be identified and is mismatched with a much larger X chromosome. It is because of this profound influence on the genetic, anatomical, and physiological fate of the fertilized egg that the X and Y chromosomes are called sex chromosomes. The first part of this chapter describes the anatomy and physiology of mature human females, the second part describes that of mature human males.

THE FEMALE REPRODUCTIVE SYSTEM

Ovaries and Eggs

The female reproductive system is designed for the production of cells (eggs, oocytes, or ova) with the potential to give rise to a new, genetically distinct individual. A female's organs of reproduction are located entirely internally (Figure 14–7). Oocytes are produced by a pair of almond-shaped structures known as ovaries. The ovaries are located within the female pelvic cavity on either side of the uterus and are held in place by connective tissue and a number of special ligaments.

Oogenesis The process of gametogenesis in females is called **oogenesis,** and takes place entirely within the ovary. A human female usually produces only one mature egg each month. Oogenesis occurs for a limited number of years in the lifetime of a mature female and is part of a complex process referred to as the **ovarian cycle** (Figure 14–8). Gametogenesis in females is quite different from that found in adult males, who are capable of producing hundreds of millions of sperm a day and do so on a continuous basis throughout their lifetimes. For females, one mature egg is formed on average each month, or about twelve a year. Let us assume that a female is fertile for at least 25 years to 30 years (from her first ovulation to cessation of the ovarian cycle). During this time, she will produce approximately 300 to 360 fertilizable eggs. This number varies depending on how many pregnancies a woman has, because there is no ovulation during the nine months of human gestation, and on the potential release of multiple eggs during any one cycle.

All the primordial germ cells that an individual female has in her lifetime, initially as many as 2 million, migrate during embryogenesis from outside the embryo proper into the developing ovary. This large number of *oogonia* is reduced to a few hundred thousand after puberty. Immature ova are called *primary oocytes* and are stored in the ovaries until stimulatory signals induce one oocyte (or sometimes

Oogenesis the development of eggs within the ovary

Ovarian cycle monthly cycle of production of mature eggs from immature precursors; influenced by pituitary and ovarian hormones

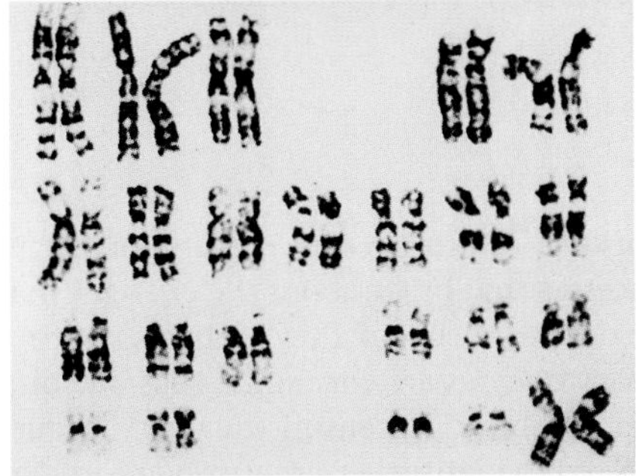

(a)

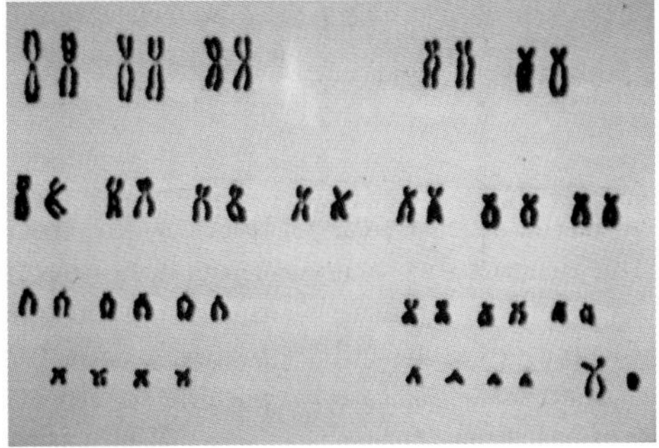

(b)

Figure 14–6 The Human Karyotype—Differences in X and Y Chromosomes

The human karyotype is a presentation of all 46 chromosomes in condensed form (a). There are 22 pairs of autosomes and 1 pair of X chromosomes in females (a). In males there are 22 pairs of autosomes and an X and a Y chromosome (b). X and Y chromosomes are called sex chromosomes.

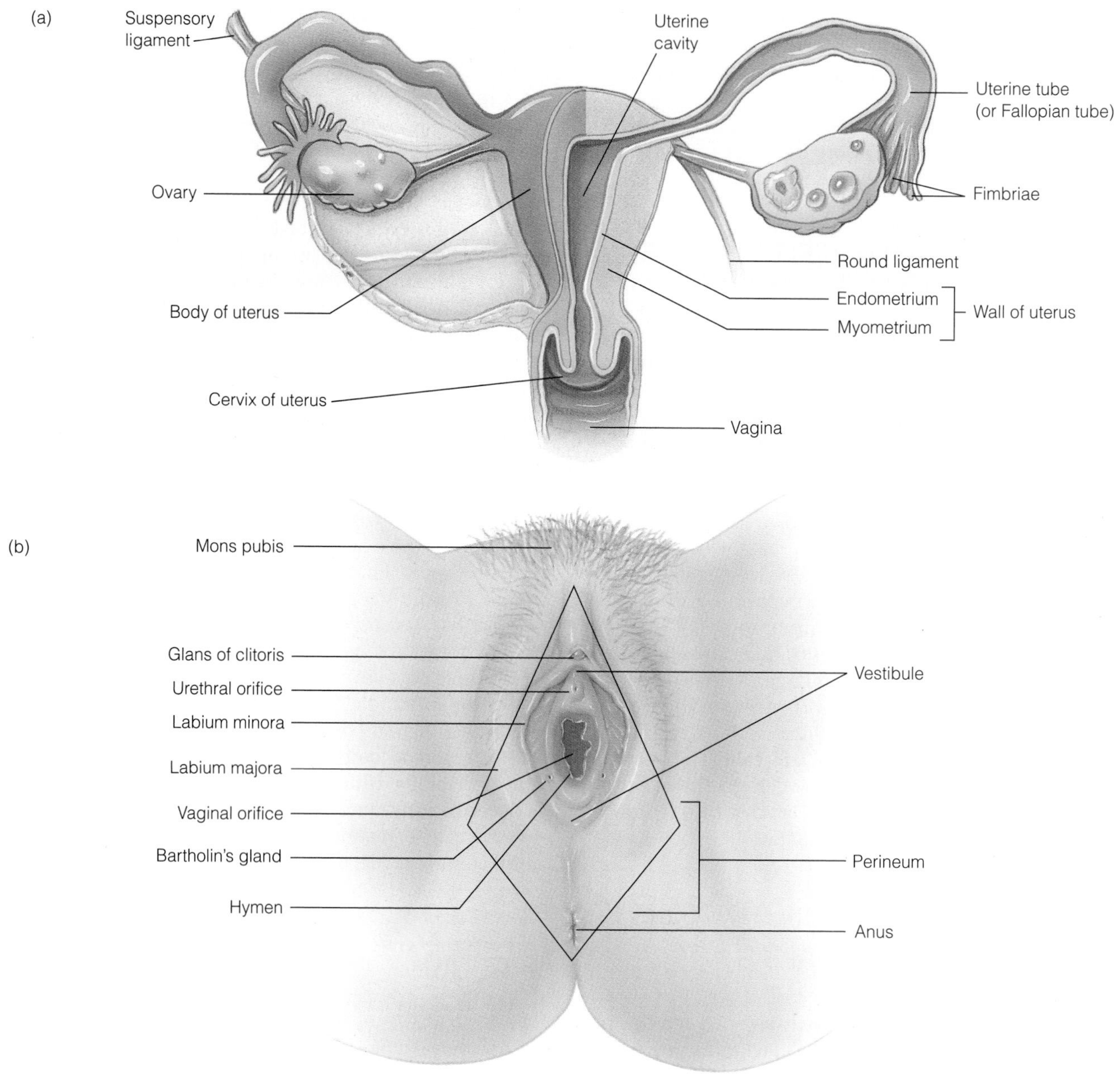

Figure 14–7 Female Reproductive System
The female reproductive system is organized internally. The ovaries, fallopian tubes, uterus, and vagina are all within the pelvis (a). The vulva is externalized and is composed of the clitoris, the labia, and the hymen.

more) to undergo maturation into a fertilizable ovum. Hormonal changes occurring during the time of onset of puberty stimulate the production of oocytes to form mature ova. This begins a continuous cycle of changes that will occur throughout the fertile years of a woman's life. Because these changes are so crucial to the success of sexual reproduction and the survival of our species, it is important to understand the processes that bring about the changes that instruct a primary oocyte to begin differentiating into a mature ovum. Primary oocytes are very small cells, of comparable size to other neighboring cells in the ovary (Figure 14–8). When stimulated to mature, a primary oocyte under-

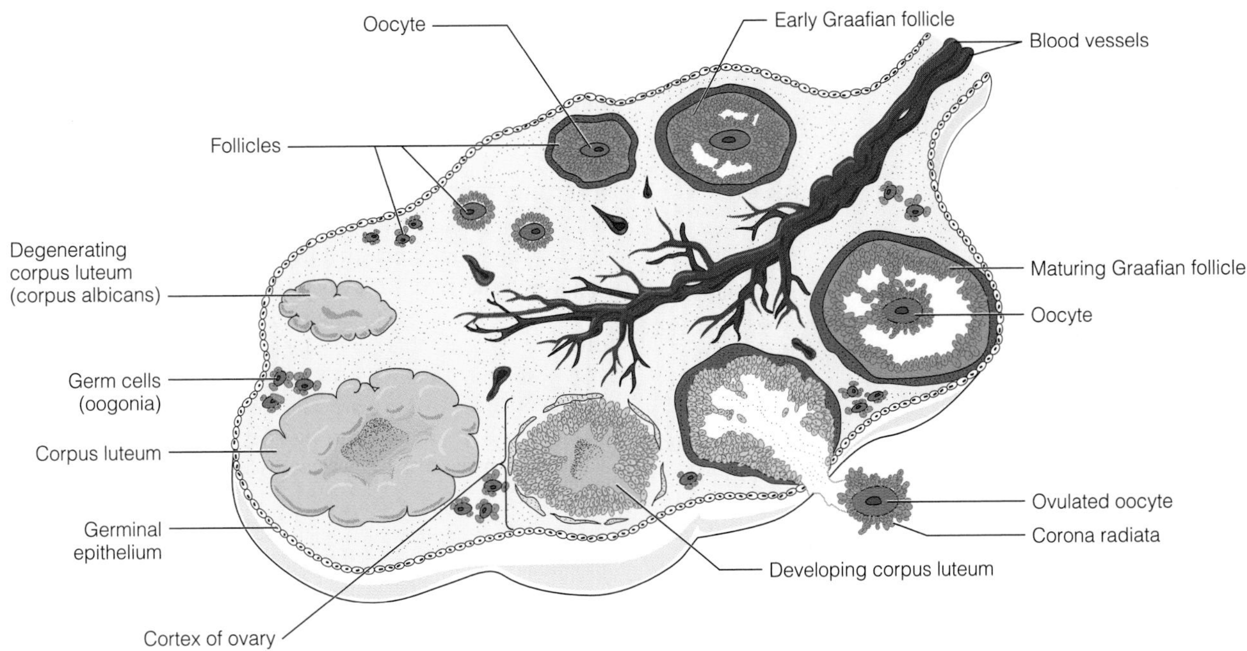

Figure 14–8 Ovary and Follicle Formation
The ovary is shown with follicles in various stages of development.

goes meiosis and develops into a *secondary oocyte.* This cell grows in size and eventually becomes a mature egg. Meiosis serves the same purpose in sperm development as it does in egg development; that is, it ensures that every sperm ends up with only one set of chromosomes, which is half the number found in somatic cells (see Chapter 3, Meiosis—Reduction Division).

Support Cells and the Follicle Oocytes are not the only cell type within the ovary. An oocyte stimulated to undergo maturation is surrounded by *support cells* in the ovary, including connective tissue, blood vessels, nerves, and epithelia. It is in conjunction with special supportive cells that the egg forms a multicellular follicle and undergoes *folliculogenesis* (Figure 14–8). The oocyte grows in size as it matures, and the follicle itself increases in size and complexity along with it. This process has been likened to the development of fruit, and a follicle may be thought of as "ripening." The ovum grows from a cell that is so small that it can be seen only with the aid of a microscope to one that can be seen with the naked eye. The human egg is still relatively small (about the size of the head of a pin), but it is enormous with respect to all other types of cells in the human body. An entire follicle, composed of an egg and its attendant supportive cells, is even larger (Biosite 14–2).

While the human oocyte cannot match the size of amphibian, reptile, and bird eggs, it does grow fairly rapidly. A human egg completes its maturation during the first two weeks of the ovarian cycle. Initially, a fluid-filled cavity develops within the follicle surrounding the oocyte. The follicle is attached to the oocyte by a simple stalk of cells. This development results in the formation of a *Graafian follicle.* Under the appropriate conditions and with the proper timing, the follicle ruptures the epithelial surface of the ovary, and the ovum is separated from the ovary. The release of a fully ripened egg occurs at the outer surface of the ovary and the egg floats freely (with its follicular cells) within the abdominal cavity of the female. The ovum and its attendant support cells move into proximity of a **Fallopian tube** and are drawn into the tube by the movement of *fimbriae.* Once the egg is safely within the Fallopian tube, it begins its migration to the **uterus** (Figure 14–9). Movement of the egg is aided by the slow beat of cilia on the surface of the epithelial cells that line the Fallopian

Fallopian tube part of the reproductive system of females providing a channel for the transport of the egg from the ovary to the uterus; also known as a uterine tube

Uterus major muscular and glandular structure of the reproductive tract of females into which the blastocyst implants and then is carried throughout pregnancy

14–2

ALL OUR EGGS IN ONE BASKET

Human ova are tiny compared to their amphibian, reptilian, and avian relatives. We need look only as far as the grocer's shelf to see the size of chicken eggs, which are sold by the dozen, or into the still waters of the pond at the park in which hundreds of easily seen frog eggs can be discovered embedded in their jelly along the bank. The biggest eggs produced by any living creature are those of the ostrich, which reach a diameter of 20 centimeters or more and may have a volume of a liter. A look inside the eggs on the grocer's shelf reveals the presence of a familiar structure, the yolk. Although the shell is commonly included in a description of an egg, the actual ovum is associated with the yolk. The rest of the egg is for external protection and nutrient storage. Therefore, the size of mammal eggs is small because they do not have to contain all the nutrients for growth and development needed by eggs that are laid, nor do they need the hard shells that protect the embryos within.

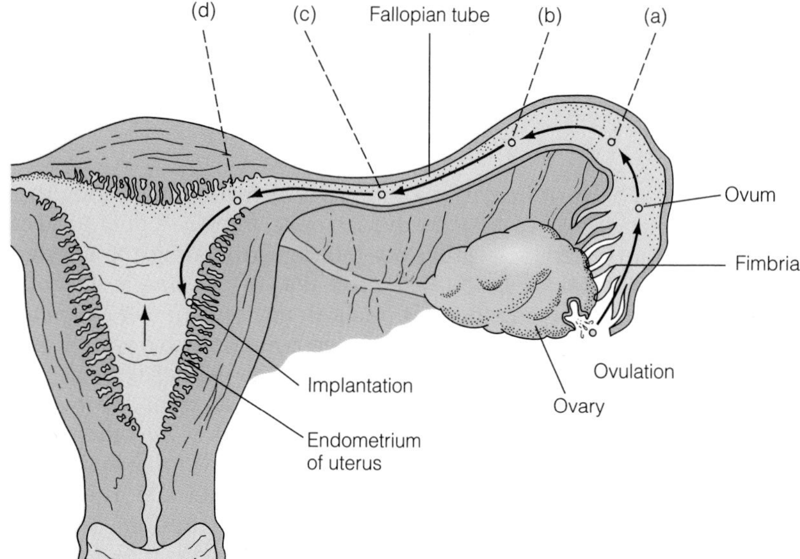

Figure 14–9 Egg Release and Migration

An ovum released following ovulation is captured by the fimbriae and moves through the Fallopian tube. If the egg is not fertilized, it will pass into the uterus and vagina and be discharged from the body. If the egg is fertilized (usually at site [a]), it will undergo cleavage (at sites [b] and [c]) and form a blastocyst (site [d]), which will implant in the uterus (for details see Figure 14–14).

tube. It is within the first part of the Fallopian tube that the egg is most likely to have a successful encounter with sperm. Success, as defined under these circumstances, is based on the probability that a fertilized egg will have time to undergo *cleavage* and be prepared to implant itself into the **endometrium** of the uterus of the mother. If the egg does not contact a sperm and remains unfertilized, it travels a long journey of elimination from the body.

Sexual Anatomy

Structurally, the female and male reproductive tracts are homologous to one another (Figures 14–5 and 14–7). Homology in this case means that the structures of both male and female reproductive systems arise from a common set of cells and tissues very early in development. A system of ducts for transport of gametes is as necessary for an egg-producing ovary as it is for a sperm-producing testis. The ducts and canals of female anatomy provide passageways for the two-way traffic of the gametes (sperm in and egg[s] out). Internally, females have two Fallopian tubes (one to serve each ovary), a single uterus, and a single vagi-

Endometrium the glandular epithelium of the uterus

na (Figure 14–7). The relatively small externalized portion of the genitalia of females is called the *vulva* and includes several fleshy, protective folds called *labia,* which enclose a region referred to as the vestibule. In the vestibule, which includes the structures and space enclosed by the labia, are situated *Bartholin's glands,* which secrete a lubricating fluid in conjunction with sexual arousal. This helps in the initial entry of the penis into the vagina during sexual intercourse (Figure 14–7). Above the labium and in the region of the bony protuberance (the pubic symphysis) along the front of the pelvic girdle is the *mons veneris.* Below the mons, but above the vagina, is a specialized, hooded organ known as the **clitoris.** The clitoris of females and the male penis are homologs and each is very sensitive to tactile stimulation during sexual arousal and intercourse. Both females and males may reach *sexual climax* or *orgasm* (Figure 14–10; see also Figure 14–19 later in this chapter). In females this involves the stimulation of muscular contractions of the vagina and uterus and results in the displacement of sperm deep within the uterus, thus, increasing the chance for an egg-sperm encounter and fertilization. *Resolution* is the return to a normal resting state following a sexual climax. The *refractory period* is the minimum time needed to recover between orgasms and is more rapid in females than in males. In some cases females may have multiple or extended orgasms with no intervening refractory period.

Hormones and the Cyclic Production of Eggs

One of the most fascinating and important research efforts made in recent times has been systematically carried out to develop an understanding of the formation of mature sex cells, in what has been referred to previously as gametogenesis. For females, this research effort has resulted in a much better understanding of the events and processes of the ovarian cycle and the periodic changes in the tissues lining the uterus, which is known as the **menstrual cycle** (Figure 14–11). These two cycles are on average about 28 days in duration for adult humans. The role of the endocrine system, particularly the hypothalamus, pituitary

Clitoris a specialized, highly receptive female genital organ homologous to the male glans penis

Menstrual cycle the periodic buildup and sloughing of the endometrium; on average, repeats every 28–29 days in conjunction with the ovarian cycle

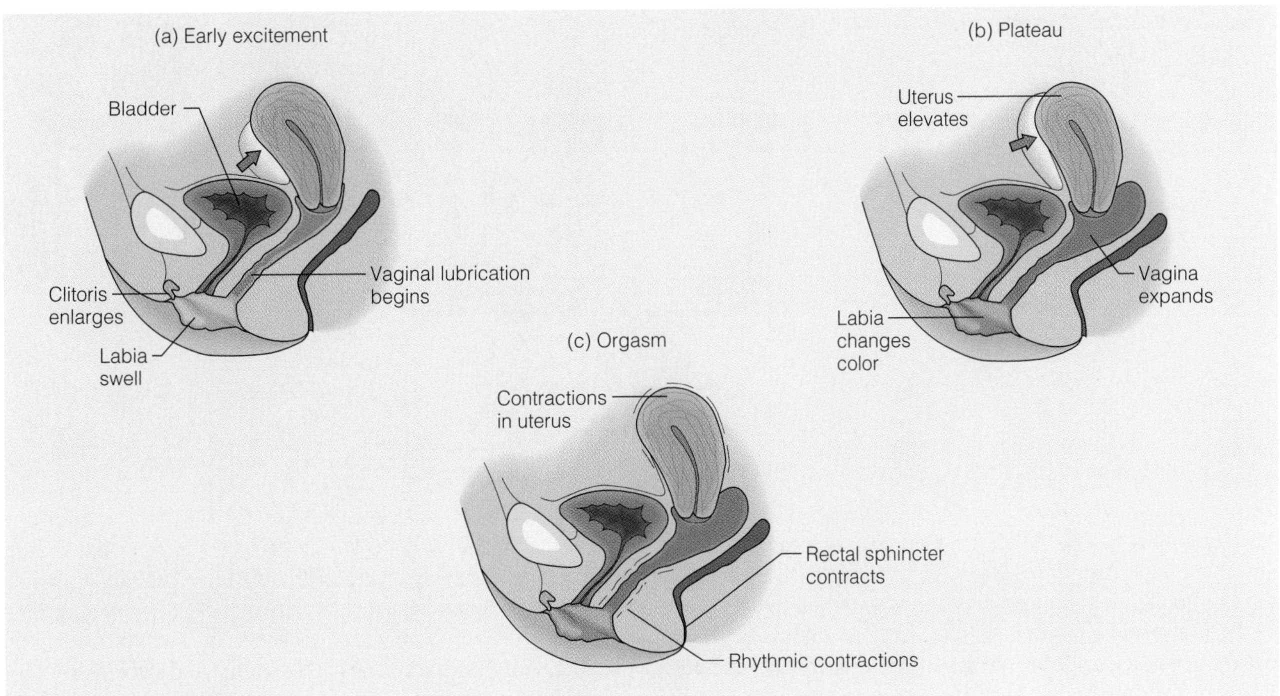

Figure 14–10 Female Orgasm
The sexual arousal and orgasm of a human female involve a complex series of muscular contractions, secretions, and changes in blood flow that lead from early excitement (a) to a plateau state (b), and to the rhythmic contractions of orgasm.

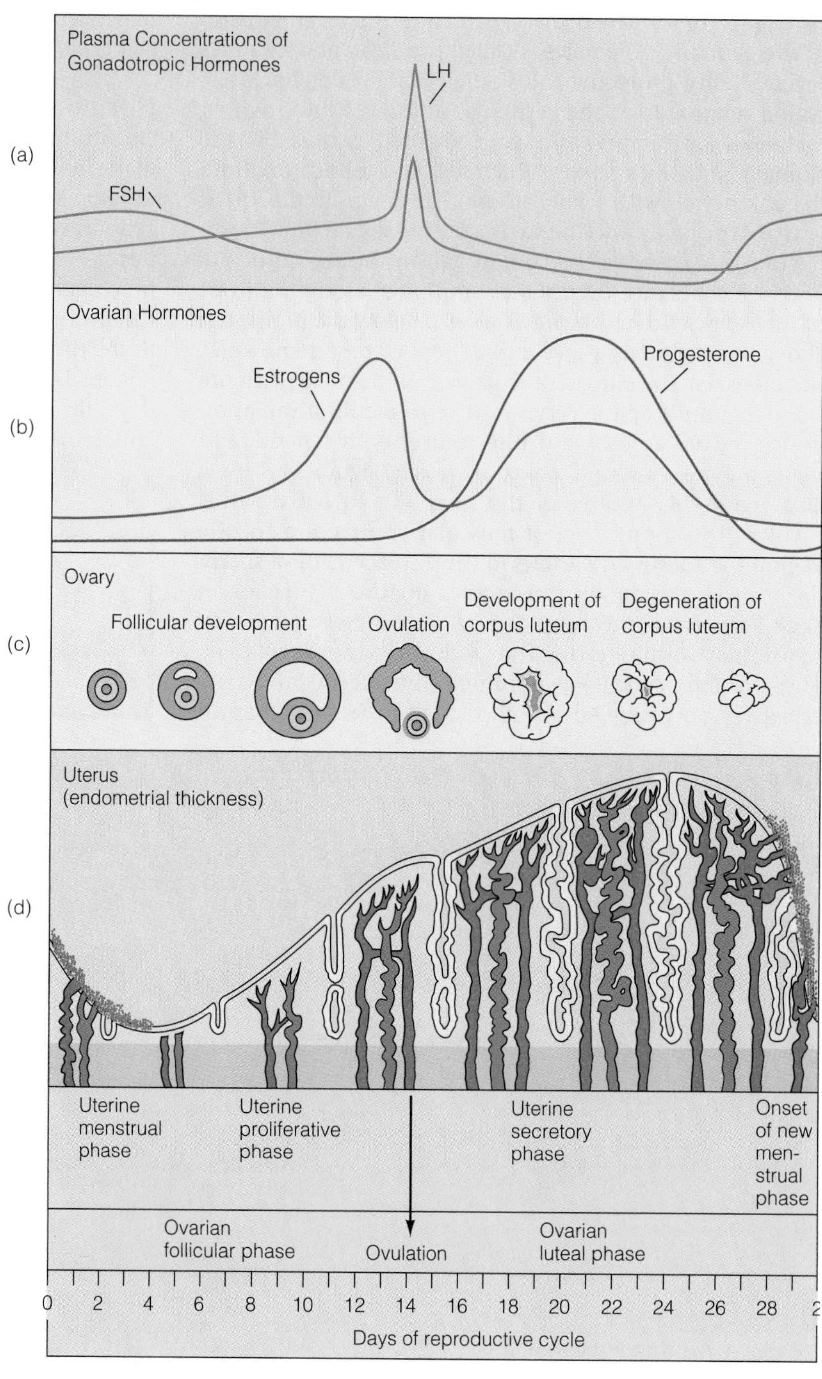

Figure 14–11 Ovarian and Menstrual Cycles
The ovarian cycle and menstrual cycle are interrelated. Gonadotropic hormones (a) and steroid hormones of the ovaries (b) together influence follicular development and luteal changes (c). Changes in the endometrium occur in response to ovarian hormones (d).

gland, and the glandular nature of the ovaries themselves, figure prominently in these processes (Figure 14–12; see also Chapter 13). In the following section, the changes in the egg, the ovary, and the uterus that occur during the ovarian and menstrual cycles are described.

Ovarian and Menstrual Cycles The ovarian and menstrual cycles are intimately interrelated and coordinated in their timing and activities. The menstrual cycle is by convention divided into several phases. The first is called the *proliferative phase,* the second is called the *ovulatory phase,* the third is the *secretory phase,* and the final stage is called *menses* (Figure 14–11). Menses is the part of the cycle during which the endometrium of the uterus is sloughed off and expelled from a woman's body. This phase of the cycle lasts, on average, three to four days depending on the individual and a variety of extenuating conditions. When this flow ceases, the cycle is ready to begin again. The changes that occur during the monthly cycle are both complicated and self-perpetuating. They provide an excellent example of homeostasis and negative feedback. In the next section, we explore the form negative feedback takes in this process.

FSH, LH, Estrogen, and Progesterone After menstruation has begun, when the bleeding associated with the loss of the endometrium is occurring, the ovary and the primary oocytes within come under the influences of two pituitary hormones known as follicle stimulating hormone, or FSH, and luteinizing hormone, or LH. The ovary itself produces two steroid hormones, estrogen and progesterone. These two sets of hormones bring about changes in their target tissues, which drive the cycle (Figure 14–13). The two pituitary hormones are considered gonadotropic hormones because they primarily influence the function of the gonads. Initially, FSH stimulates the maturation of a follicle in either one of the two ovaries. A primary oocyte undergoes meiosis and begins growing in size on its way to becoming a mature egg. The ovum is surrounded by follicle cells, which nurture and protect the egg during this period of growth. Usually only one egg is stimulated to complete its maturation during any one ovarian cycle. The growing follicle produces estrogen, which has a dramatic effect on the uterus as well as on the hypothalamus and pituitary gland.

Both the egg and the follicle grow rapidly in size and complexity. Within two weeks of the initial stimulation by FSH, the egg and its follicle are large enough to be seen with the naked eye. A fluid-filled cavity surrounds the egg, which is suspended within it by a cell stalk. This entire structure, including egg, cavity, and support cells, is a Graafian follicle. At this point in its maturation it is positioned close to the surface of the ovary (Figures 14–8 and 14–9).

The next major step in the cycle is ovulation (Figures 14–8 and 14–11). This is the time at which the egg and a portion of the follicle cells that surround it are released from the ovary. This rupturing event is triggered by an increasing concentration of estrogen, which causes an elevation in the level of the LH a day or so before ovulation actually occurs. Although the egg and part of its follicle separate from the ovary at this time, there is a sizeable follicular structure that remains behind. The luteinizing hormone is important for the development of this remnant, which becomes the *corpus luteum* (Figure 14–8). The corpus luteum in turn produces both estrogen and progesterone, which together are needed to bring about further changes in the female reproductive tract. Once the egg has been released and for several days afterwards, it is capable of being fertilized as it moves slowly down the Fallopian tube. The greatest chance for successful fertilization occurs in the upper region of the Fallopian tube. After passage through the Fallopian tube, an unfertilized egg is doomed. An egg must be fertilized and begin to divide in the Fallopian tube if it is to successfully implant in the uterus (Figure 14–14). The timing must be right, or the zygote is not capable of implantation.

At this juncture, a summary of what has happened in the cycle will be helpful (review Figures 14–11 and 14–13). We have discussed the initial two weeks of the ovarian

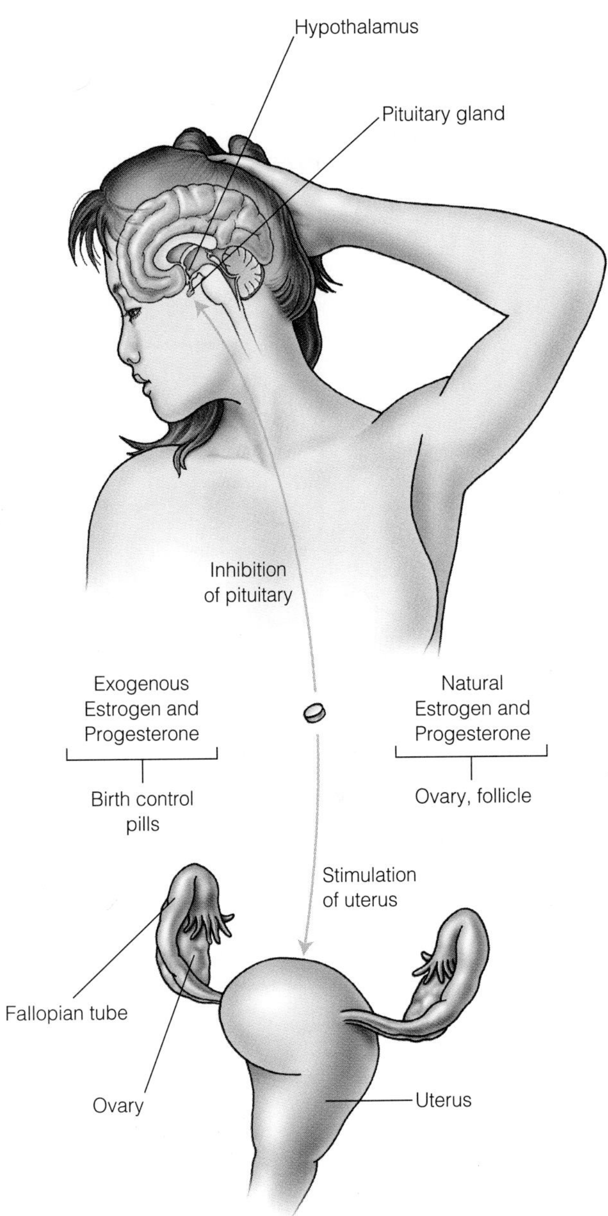

Figure 14–12 Endocrine Connections
The hypothalamus and pituitary gland influence, and are influenced by, the ovarian hormones. Feedback loops are the bases for the cyclic phenomena associated with ovarian and menstrual cycles.

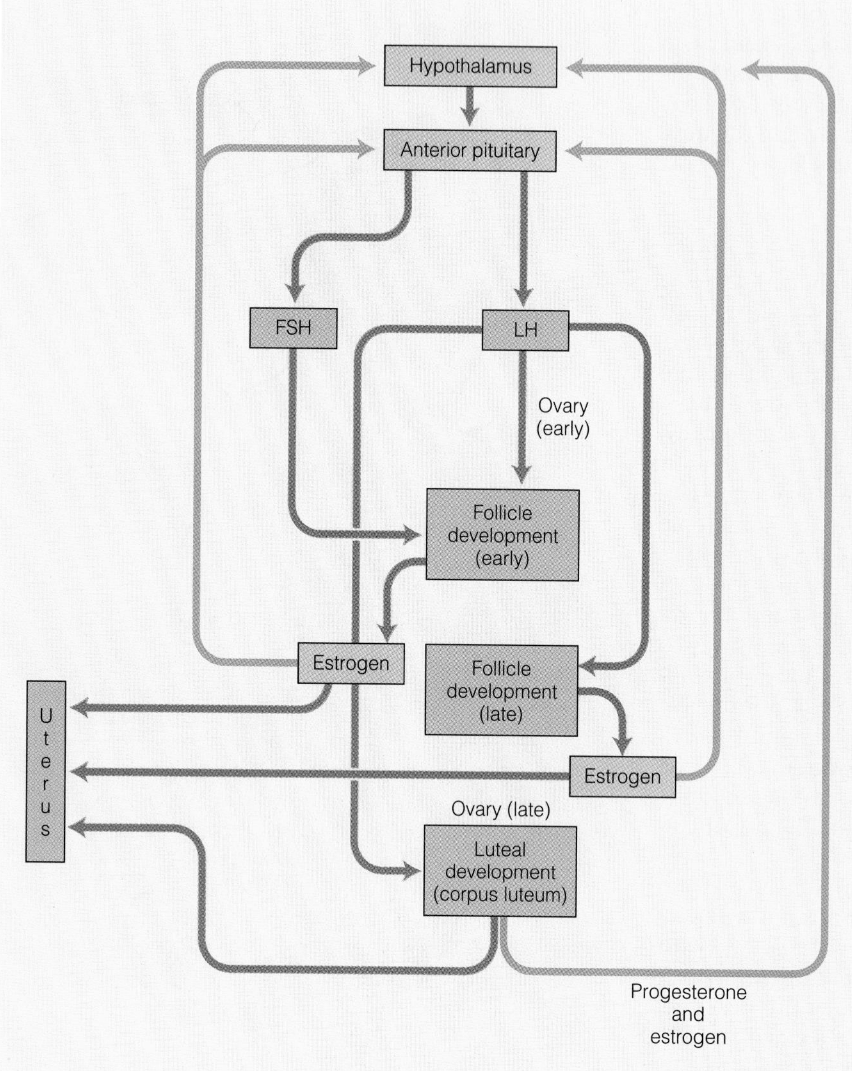

Figure 14-13 Feedback and Ovarian Cycling
The hypothalamus allows release in the anterior pituitary of FSH and LH. FSH and LH stimulate follicle formation. The ovary produces small amounts of estrogen, which feeds back to the hypothalamus and pituitary to limit FSH and LH release. Estrogen and progesterone induce changes in the uterus to prepare for implantation. The corpus luteum is the remnant of the follicle left in the ovary and produces both estrogen and progesterone.

cycle. The egg and its attendant follicle cells have been stimulated to develop by FSH. During that development they produced estrogen, which began the preparation of the uterus for potential later events. The egg and follicle have burst from the ovary under the influence of LH (and FSH) and begun their journey through the Fallopian tube. The remaining follicle within the ovary is under the influence of LH. The follicle is transformed into a corpus luteum that produces estrogen and progesterone. The rest of the process focuses on the preparation of the female body for pregnancy, and this means preparing the uterus for accepting the embryo into its nurturing folds.

Changes in the Uterus—The Endometrium

In humans, the uterus is a large muscular organ lined with a specialized epithelium known as the endometrium. The uterus is positioned anatomically just above the vagina, and overlies the urinary bladder. Access to the uterus from the vagina is through the *cervix*. While the epithelium of the reproductive tract is continuous throughout, the epithelia of the vagina, cervix, and uterus are different from one another. The uterine wall is the normal site of implantation and development of an embryo. The cervix is not well suited for implantation. The vagina has a thick, stratified epithelium that is durable and elastic enough to withstand the rigors of sexual intercourse but incapable of supporting successful implantation. However, for the uterus to be able to serve as the site for implantation, it must be prepared in advance to provide the right environment for the embryo.

The endometrium, which lines the uterus, undergoes cyclic changes during the four phases of the menstrual

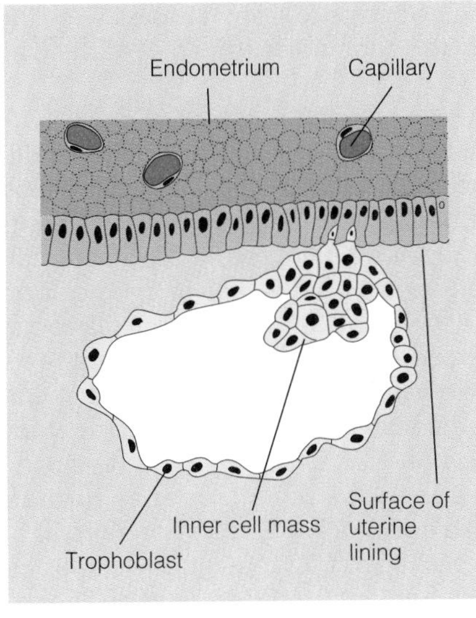

(a)

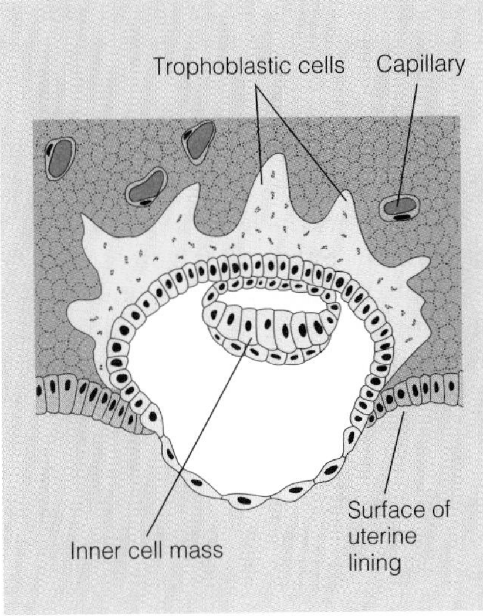

(b)

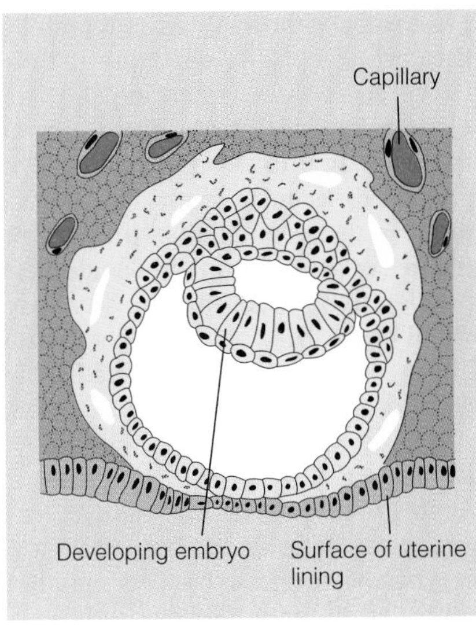

(c)

Figure 14–14 Implantation
The blastocyst in (a) contacts and attaches to the endometrium. The trophoblastic cells invade the uterus (b) and the blastocyst becomes completely embedded within a few days (c).

cycle. The phases are coordinated in time with the development of the egg and follicle, such that, if the egg is fertilized during passage through the Fallopian tube, the embryo will have a well-developed, "friendly" tissue in which to embed itself. The activities of the ovarian and menstrual cycles and the several cell and tissue types involved are functionally linked together and regulated chemically by estrogen and progesterone. The two hormones are clearly necessary for the proper function of the reproductive tract, but they are also involved in maintaining female physiological well-being in general. The endometrial tissue is a prime target for these two steroid hormones and is very sensitive to changes in their concentration in the blood. During the first two weeks of the ovarian cycle, there is a dramatic increase in the amount of estrogen produced by the ovarian follicle(s). This increase in the level of estrogen stimulates the proliferation of endometrial cells and brings about a generalized thickening of the uterine lining. This constitutes the proliferative phase.

Negative Feedback and Estrogen The elevation of estrogen levels during the first phase of the ovarian cycle is important for another reason as well. As already established, the pituitary gland initiates the development of an egg through the release of FSH. However, the pituitary gland is itself controlled by hormones from a region of the brain known as the hypothalamus. The hypothalamic hormones directly affect endocrine cell secretions of the pituitary gland. In the case of FSH, the hypothalamus stimulates the pituitary with an FSH releasing factor, which allows the pituitary to secrete FSH. It is as if the brain provides the keys to open a chemical gate so that FSH can be released. There is also an LH releasing factor. So what, you may be asking at this point, is to stop the hypothalamus-pituitary combination from stimulating egg production at all times? The answer to this question is negative feedback by way of the effects of estrogen on the hypothalamus (Figure 14–13).

The cycle may be summarized in the following way. FSH stimulates the development of an egg from the collection of oocytes stored within each ovary. The development of an egg and its follicle cells in turn brings about an increase in the production of estrogen. The estrogen then has two very different and important effects. The first is to stimulate the endometrium to thicken, and the second is to act as an inhibitory messenger to the hypothalamus to signal it to stop secreting releasing factors. Without hypothalamic releasing factors the pituitary chemical gates are closed and secretion of FSH shut off. Thus, the negative feedback loop established between brain and ovary depends on signals (follicular estrogen molecules) from the tissue originally stimulated by pituitary hormones. The product of the target (estrogen from the ovary) inhibits the original source of stimulation (FSH from the pituitary). A high level of estrogen also triggers the release of LH, setting the stage for ovulation. During the second half of the ovarian cycle, both estrogen and progesterone prevent the secretion of FSH and LH and the development of any new eggs.

LH and Ovulation As is represented in Figure 14–11, the level of LH rises sharply to bring about ovulation. However, as already mentioned, there is a second role for LH in this cycle. LH transforms the remnant of the follicle in the ovary into a corpus luteum. This structure produces progesterone and accounts for an elevated level of this hormone in the second half of the ovarian cycle. Under the influence of progesterone, the endometrium passes from the proliferative phase into the secretory phase of the menstrual cycle. This is the time of the buildup of the uterine lining into a highly vascularized, glandular epithelium capable of sustaining an implanted embryo.

Fates and Options

At this point, all the preparations for possible fertilization and implantation have been made or are in progress. The egg has been released and is moving down the Fallopian tube, the pituitary has been shut down and the uterus has been built up. During this time, perhaps a period as short as one to two days, the egg may be successfully fertilized. Thus, at this juncture there are two likely fates for an egg migrating through the Fallopian tube (see Figures 14–9 and 14–11):

1. If no fertilization occurs, the egg dies and degenerates as it passes through the reproductive tract on its way to expulsion from the body, and the final phase of the menstrual cycle, menses, occurs.
2. If fertilization of the egg occurs, the egg undergoes cleavage, passes out of the Fallopian tube into the uterus, and attaches to the endometrium. This normally leads to the implantation of the embryo into the sustaining, nurturing environment of the uterus and the beginning of a pregnancy.

Option One Starting with a description of the first option, it is clear that if the egg is not fertilized, it dies. The ramifications of this have profound effects on the reproductive system. An unfertilized egg cannot implant into the uterus. One consequence of this is the rapid breakdown of the corpus luteum. This culminates in a drastic drop in the production of estrogen and progesterone. Because the endometrium is dependent on high levels of these hormones to maintain its integrity, it begins to disintegrate. Evidence of this is apparent by day 28 or so of the cycle as the endometrium is sloughed off and begins to discharge from the body, signaling the beginning of the menstrual period.

In this scenario the cycle has come full circle. The drop in estrogen and progesterone over the final few days of the ovarian/menstrual cycle not only affects the endometrium but removes the block controlling the release of factors of the hypothalamus. This in turn unlocks the chemical gates of the pituitary gland, which once again releases FSH to the blood and ultimately the hormone reaches the ovaries. The cycle starts again.

Option Two In option two, the outcome is entirely different. The egg is fertilized in the Fallopian tube (becoming a zygote), and begins dividing (cleavage) within hours (Figure 14–9). The cells of the new zygote continue to divide and the zygote attains an early embryonic form known as the *blastocyst* (Figure 14–15). It is the blastocyst that attaches to the uterine wall. This signals the entire reproductive system to maintain itself. The implanted embryo interacts with the maternal tissue and together they begin to form a placenta. The cells involved in producing this structure synthesize a hormone known as **human chorionic gonadotropin,** or HCG. HCG is important in maintaining the corpus luteum. As a result, the corpus luteum develops further and continues to produce progesterone and estrogen.

Because the placenta develops only when implantation has occurred, HCG is a unique molecular indicator of pregnancy. In fact, the various pregnancy test kits that are available over the counter at a pharmacy or through a physician are designed to detect the presence of HCG in the urine. The placenta also produces progesterone and estrogen at high levels, thus, helping to inhibit the hypothalamus and supporting the maintenance of the uterus. As a result, no FSH is released from the pituitary gland. The absence of pituitary gonadotropins ensures that once an egg is fertilized and implanted no more eggs will be stimulated to mature and be released. This is a concise description of the beginning of a pregnancy, but much more is in store for the female body and its newly implanted embryonic partner, as will be discovered in the next chapter.

THE MALE REPRODUCTIVE SYSTEM

In studying the male reproductive system, two main features are to be considered:

1. The organs and tissues responsible for the formation of sperm (in a process called **spermatogenesis**)

Human chorionic gonadotropin (HCG) a hormone produced by the developing placenta and used as a molecular indicator of pregnancy

Spermatogenesis the formation of sperm in the testes

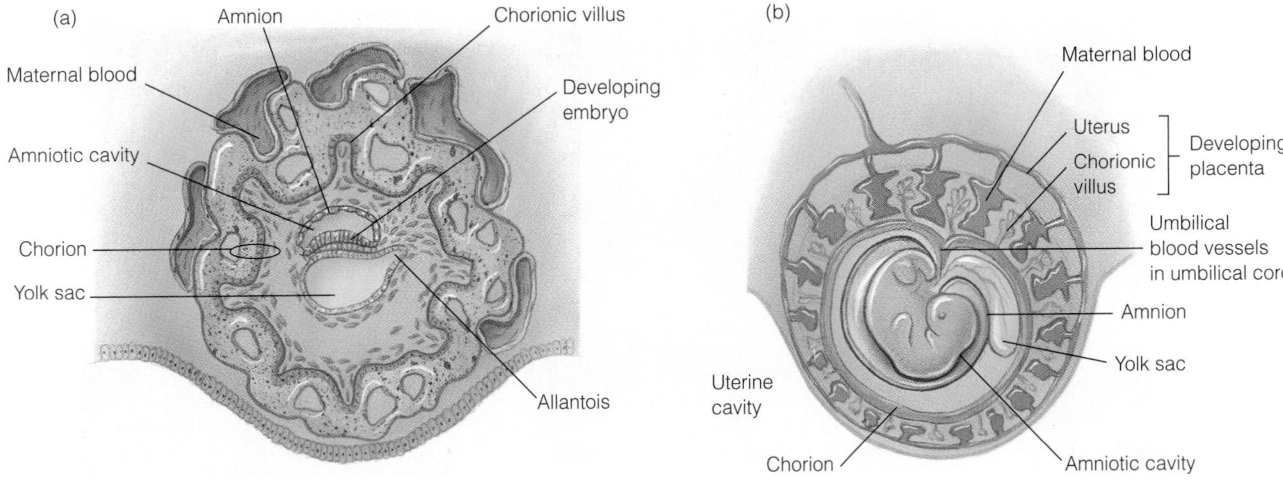

Figure 14–15 The Placenta and Umbilicus
The developing embryo is surrounded by an amnion and chorion as well as other extraembryonic structures (e.g., allantois, yolk sac) and maternal tissue (a). The embryo is connected to its mother through umbilical connections to the placenta (b). The placenta functions to exchange materials (nutrients, waste products, oxygen, carbon dioxide) between the mother and embryo.

2. The system of accessory structures, including tubes, ducts, and glands in which the sperm mature, are stored, and through which they are delivered into the female reproductive tract.

The sex-related organs of the male are for the most part externalized, but there is considerable internal organization of this system as well (Figure 14–16).

Sperm and Their Formation

The most important product of the male reproductive system is sperm. The gametes are formed in two almond-shaped structures known as the testes or testicles. The starting point for the description of the male reproductive system is spermatogenesis, the generation of sperm. The testes are contained within a protective sack called the *scrotum,* which hangs outside the body between the legs (Figure 14–16). The externalized location of the testes is important because the viability and fertility of sperm produced within them is low if the testes are at normal body temperature. In fact, the temperature of the testes is 3°–4° C below the internal temperature of the body. Sperm are among the smallest cells in the human body. Each cell has a single long flagellum or tail (Figure 14–17). They are the only flagellated cells produced in the human body (Chapter 3). Sperm grow and develop solely in the testes within a convoluted system of epithelial tubes known as the *seminiferous tubules* (Figure 14–18). There are several hundred meters of seminiferous tubules in each of the two testes. Sperm are derived from precursor cells known as *spermatogonia,* which undergo a series of developmental changes as they migrate through the epithelium of the seminiferous tubule to the inner channel or lumen of the tubule.

Development of Sperm

Seminiferous Tubules Sperm development depends on interactions with other cell types. The seminiferous tubules are a complex staging area for sperm development and maturation and there are other important nongamete cell types in the tubule (Figure 14–18[b]). One of these is the *Sertoli cell,* the other is the *Leydig cell.* Sertoli cells are very large epithelial cells that span the thickness of the wall of a seminiferous tubule. Sertoli cells wrap themselves around the developing spermatogonia in the outer edge of the tubule and help control sperm development by providing structural support, nutrients, and signals for their migration, growth, and differentiation (Figure 14–18[b]). Just outside the seminiferous tubules, in a region known as the interstitium, are found blood vessels, nerves, and the second type of specialized cell, the Leydig or interstitial cells (Figure 14–18[b]). Leydig cells produce one of the most important molecules in the male reproductive system, testosterone. The role of testosterone in sperm production is only one of many it serves in growth and development of the male human body (see Biosite 14–3).

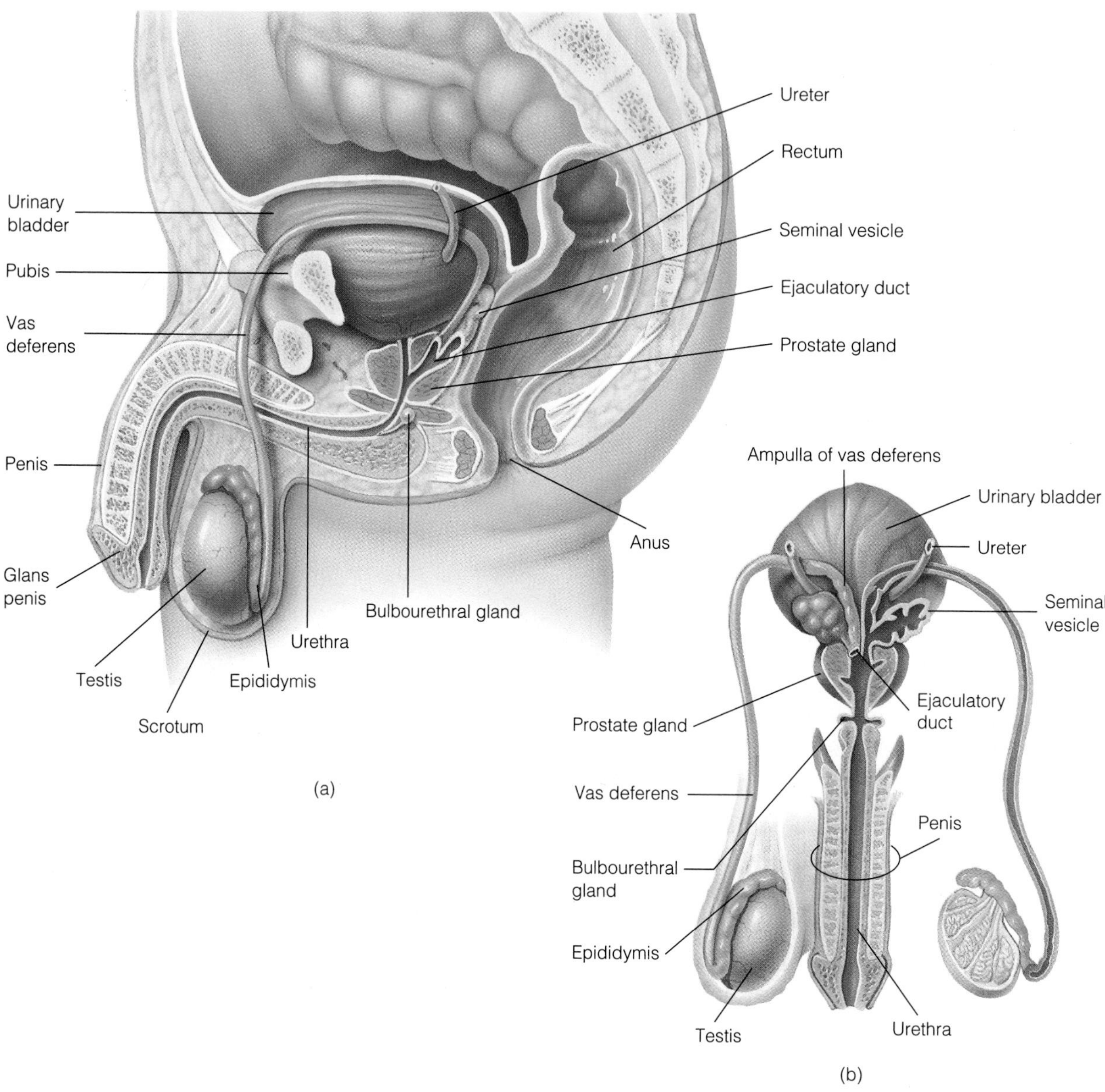

Figure 14–16 The Male Reproductive System
The male reproductive system is organized externally. The testes, epididymis, vas deferens, and penis are exposed outside the body (a). Other ducts and glands are internal and aid in the transport of semen through the internal vas and urethra (b).

Sperm development within the seminiferous tubules begins at the outside edge of the tube (Figure 14–18[c]). Sperm reach the inner surface of the tubule only after dividing and migrating through and along the network of Sertoli cells. Throughout this process developing sperm are under the influence of many factors including testosterone. However, the journey does not end when sperm are released into the lumen of the tubule. It is only the beginning of another journey. Newly released sperm have all the morphological characteristics of mature sperm, but they are not yet mature. Maturity in this case is defined as the ability to successfully fertilize an egg. Sperm removed directly from the seminiferous tubules are not capable of fertilizing an egg. The collective length of the seminiferous tubules of the testes has been estimated to be over one-half mile, so the journey of a sperm through this twisted system of tubes takes time.

The Epididymis After the journey through a seminiferous tubule, sperm enter a series of collecting tubes, which eventually combine into a single tube, the *epididymis* (Figure 14–18[a]). An epididymis can be felt as a coiled tube on the top backside of each testicle. Each epididymis is about 20-feet long when uncoiled and sperm take up to six weeks to pass through the region and mature. The great length and overall volume of the epididymis is compatible with the rate of production and release of sperm that are produced on a continuous basis in the seminiferous tubules. Hundreds of millions of sperm may be produced each day in a healthy, mature human male. In fact, a single ejaculate of semen during coitus can contain as many as 500 million sperm (compare this output with the single mature egg that is usually produced each month by an ovary in a healthy, mature human female).

For a sperm to finish its maturation and eventually have the potential to fertilize an egg, it must be exposed to the microenvironment of the epididymis. Epididymal sperm are not quite ready to fertilize an egg. They must also undergo **capacitation,** a process that prepares the sperm for what is called the *acrosome reaction* (see Figure 14–17 for the location of the acrosome). Normally, capacitation takes place in the vagina or uterus as a result of contact with substances secreted by the female reproductive tract.

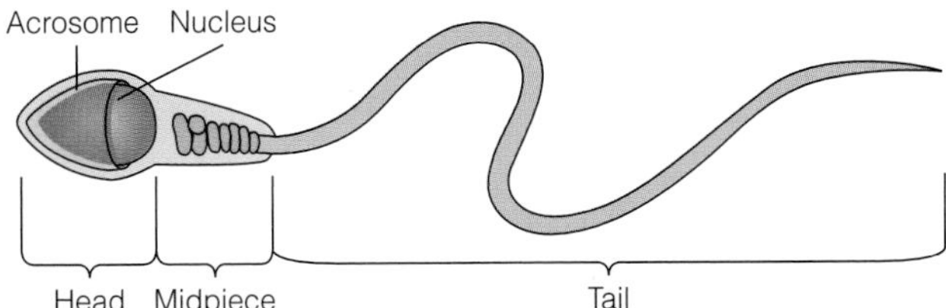

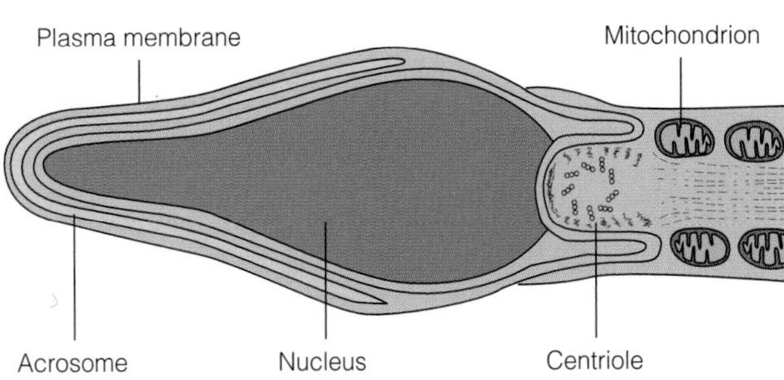

Figure 14–17 Sperm
Sperm are long, flagellated cells. They are composed of a head, midpiece, and tail. The nucleus is in the head, mitochondria is in the midpiece, and axoneme is in the tail.

The Vas Deferens From the epididymis, sperm pass into a muscular tube known as the **vas deferens** (Figure 14–16 and 14–18). The vas provides a route for sperm out of the scrotum and into the body proper. The vas follows a route up and over the bladder toward the prostate gland. The expanded distal end of the vas, a region called the *ampulla,* is a holding area for sperm prior to *ejaculation*. Ejaculation occurs during an orgasm and the sperm are forced by muscular contractions into the ejaculatory ducts, where they mix with other fluids produced by glands in other parts of the reproductive tract and form a viscous, milky substance known as **semen** (Figure 14–19). Ejaculation results in the rapid expulsion of semen from the male reproductive tract.

The Urethra From the ejaculatory ducts the sperm pass into the urethra. This transition occurs within the walnut-sized prostate gland, which is situated surrounding the junction of the ejaculatory ducts and the urethra. The urethra is also the channel through which the urinary bladder is drained. The urethra extends through the length of the penis and opens externally (Figure 14–16).

The Penis The penis is a fleshy, highly sensitive organ that has two very different roles in male physiology (Figure 14–16[b] and 14–20). The first, and most common, is its role in the elimination of urine from the body. In this case, the penis surrounds the final length of the urinary tract plumbing.

The second function of the penis is to deliver sperm into the female reproductive tract during sexual intercourse. To perform this function, the penis switches between two different anatomical states. Most of the time the penis is in a flaccid, flexible state. This is conducive to carrying out its function as part of the urinary system. However, under conditions of sexual excitement, the penis becomes rigid and erect (Figure 14–19). The purpose of such rigidity is to facilitate penetration into the female vagina. The size and structure of the erect human penis is complementary in shape and proportional in size to the human female vagina. This should come as no surprise. Natural selection for sexual dimorphism in humans has ensured that there is a

Capacitation the final step in sperm activation that allows them to fertilize an egg; occurs in the vagina

Vas deferens the tube connecting the epididymis to the urethra and carrying the sperm from the region of the scrotum into the body

Semen sperm-bearing fluid produced in testes and reproductive glands of male animals for internal fertilization

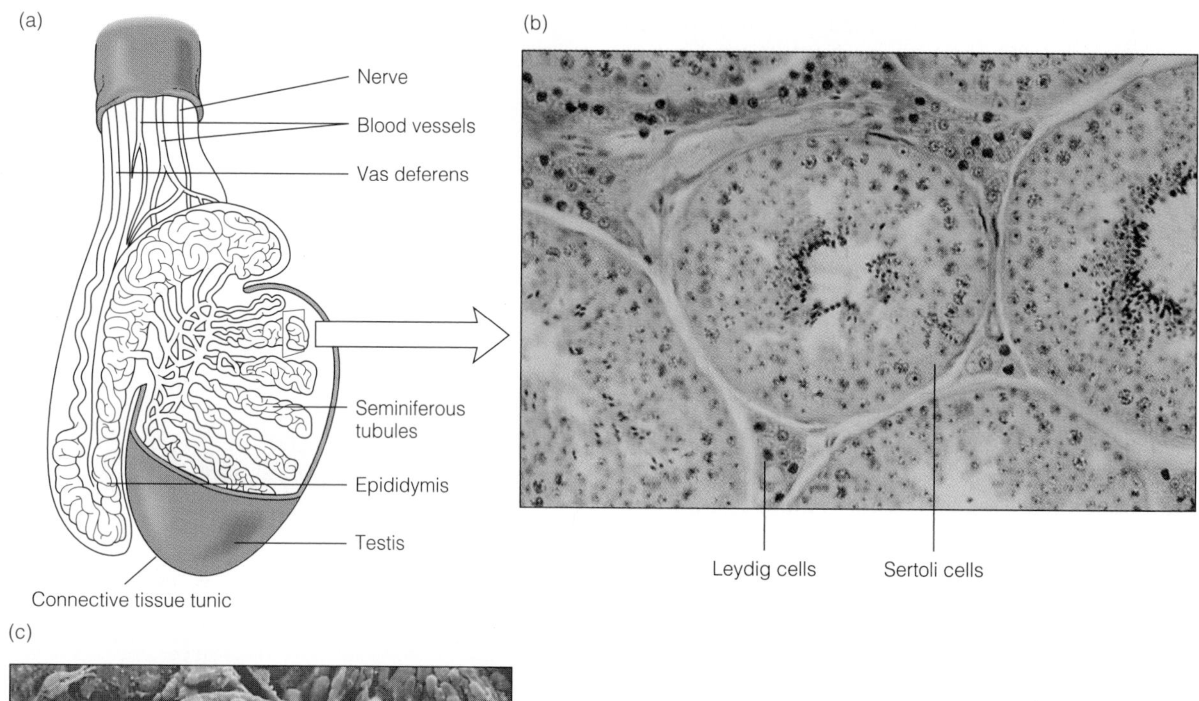

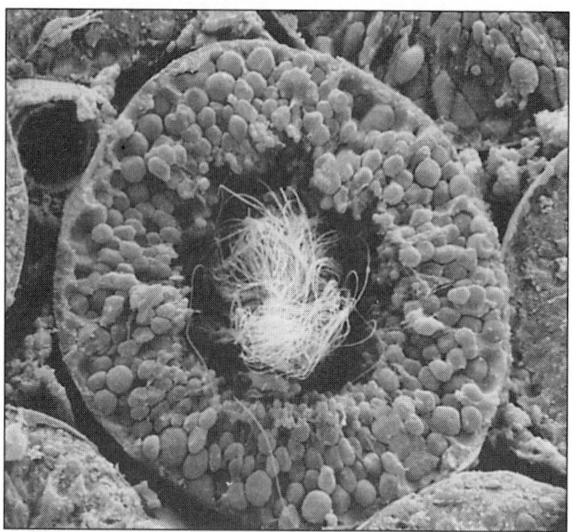

Figure 14–18 Organization of the Testes
The testes are composed of an outer layer of skin and connective tissue. Within are densely packed seminiferous tubules (a), (b), and (c). Sperm are made in the seminiferous tubules in conjunction with Sertoli cells (b). Leydig cells are found outside the seminiferous tubules. The epididymis is located along the posterior of the testis and connects to the vas deferens to allow sperm translocation.

balance in the general size and shape of these specialized tissues. This, in turn, ensures that ejaculation during coitus brings about the deposition of sperm deep within the female reproductive tract, which is needed for successful fertilization of an egg.

Erection and Ejaculation How does a penis change from a flaccid to a rigid state? To make this transition, the penis uses "captured" blood from the circulatory system. The internal structure of the penis is organized into tissue reservoirs to retain arterial blood (Figure 14–20). The two types of reservoirs are the *corpus cavernosum* and the *corpus spongiosum*. An increase in the volume of blood within the cavernous and spongy tissues of the penis allow it to become firm and erect. For this to occur, the blood vessels that normally allow the venous blood to circulate out of the spongy tissues of the penis are constricted. Arterial blood flow into the cavities, however, continues unabated. Thus, there is an unbalanced flow of blood into and out of the penis. Blood flow into the spongy and cavernous tissues of the penis in turn puts increased pressure on the veins further compressing them. This results in a net increase in blood volume within the spongy tissues of the penis, which engorge and result in an erection.

14–3

TESTOSTERONE AND SEX

Testosterone is a steroid and fundamentally important in the formation of sperm and in the development of secondary sex characteristics in males. Females also produce testosterone but at a level many times less than that produced in males. The physiological effects of testosterone on males encompass structural and functional influences on the entire reproductive system, on the proportion of muscle mass in the body (males have more muscle mass than females), on the growth of facial and body hair, and many more aspects of behavior. For better or for worse, high levels of testosterone in males also correlates with aggressiveness and dominance behavior. This is not just a characteristic of humans, but true of all mammals, birds, and reptiles.

It is likely that the disparity in steroid levels in males and females was naturally selected as a survival characteristic in our ancestors. The biological importance of aggressive behavior was perhaps more clearly defined in the role of males as principal protectors (fighting) and providers (hunting) of family and/or tribe. The strength and aggressiveness needed to carry out this role were genetically built into the growth and maturation of the male. This aggressiveness is still in evidence today. It certainly cannot be turned off just because we are now surrounded with an organized and extensive social structure. Patterns of growth and development for modern mammals, including humans, that were laid down over tens of millions of years of evolution are not easily set aside. Testosterone, as well as other related steroid hormones, are of central importance in influencing the growth, health, and behavior of vertebrate organisms of all types.

Females normally produce low levels of testosterone. Abnormalities in testosterone production in females can and do result in development of male physical characteristics, that is, thick facial hair, deepened voice, increased muscularity. Addition of testosterone and related steroids to the diet dramatically affects the human body. Many scandals in world-class athletics have been caused by the use of performance enhancing drugs, among which testosterone-based steroids are prominent.

Ejaculation is facilitated when the penis is erect. The tubes, ducts and glands within which sperm and semen are produced and stored are surrounded by muscles. The rhythmic contractions of these muscles, controlled by the nervous system during a sexual climax, result in the forced displacement of the fluids within the entire system of tubes, ducts, and glands and their expulsion through the urethra (Figure 14–19). The penis returns to its flaccid state following resolution of sexual excitement, usually after ejaculation, at which time blood flow and the physiological state of the entire reproductive tract returns to normal.

Glands and the Production of Semen

There are a number of specialized glands associated with the reproductive tract in males (Figure 14–16). Each one produces its own type of fluid secretion, which, taken together with the sperm, constitutes semen. Semen is a thick, milky nutrient-rich fluid with a pH of about 7.4, making it slightly alkaline. Semen lubricates the tissues, protects and provides sperm with sustenance, particularly after they are introduced into the female reproductive tract. The prostate gland, which surrounds the urethra just underneath the urinary bladder, is very important in the production of semen. The secretion of this gland is an alkaline fluid that affects the ability of sperm to move effectively (motility). The volume of secretion of this gland makes up about 25% of the entire volume of semen.

There are some serious medical problems that can develop in the prostate gland. As males get older, particularly beyond the age of 50, the gland may grow abnormally large and constrict the flow of urine from the bladder. This growth may result from infection, a number of inflammatory conditions known collectively as *prostatitis,* or cancer. In many cases emergency procedures (catheterization), or surgery, are needed to correct the problem of restricted flow. Imagine the feeling of your bladder filling up but not being able to release the urine. This is painful as well as dangerous. A ruptured bladder can be lethal. Periodic, routine physical examination of the prostate by a physician as males attain middle age and beyond can help detect problems early and prevent them from becoming serious. In recent years, prostate cancer has become a serious concern for human male health. This cancer may go undetected for years and, when it is diagnosed, it may be too late to do

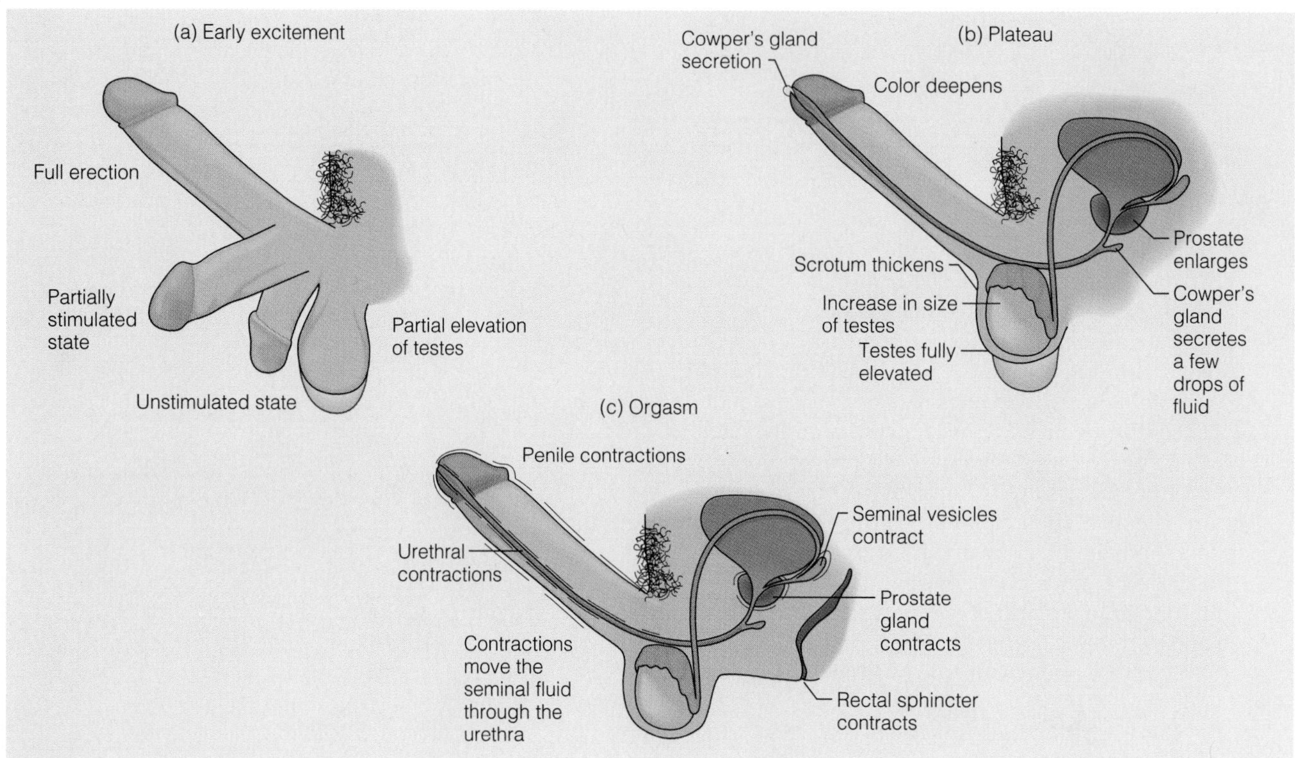

Figure 14–19 Male Orgasm
The sexual arousal and orgasm of a human male involves a complex series of muscular contractions, changes in blood flow, and secretions. In (a) excitement brings about a capture of blood in cavernous and spongy tissues, which brings about erection. Further stimulation prepares the penis for ejaculation (b) (c) in which muscle contractions move seminal fluid through the urethra.

anything about it. Most men are expected to develop this condition as they age into their 70s and 80s. This is generally not a concern because this form of cancer grows very slowly. However, there is a portion of males in our society, particularly black males, who in their 40s or 50s develop prostate cancer. This form is very aggressive and lethal. A new test called a PSA test (or prostate serum antigen test) is available through physicians to detect increases in the level of these antigens, which are observed in developing cancer. Early detection can save lives.

Seminal Vesicles There are two *seminal vesicles* in the male reproductive tract. They are irregular in shape and about 4 cm–5 cm in length (Figure 14–16). The vesicles are located at the base of the back of the bladder and produce an alkaline fluid rich in the sugar fructose. The alkaline pH of the fluids produced by all the accessory glands of the male reproductive system serves to help neutralize the normally acidic environment of the vagina. Sperm are very sensitive to acid. Between 50% and 60% of the volume of the semen is provided by secretions from the seminal vesicles. Sperm are held in the reservoir of the ampulla of the vas.

The Bulbourethral Glands The *bulbourethral glands* (also known as Cowper's glands) are pea-sized, paired glands that are found in a region just under the prostate gland (Figure 14–16). The ducts of the glands open into the urethra and secrete lubricants as well as substances that neutralize any residue of urine left in the urethra through which sperm must pass. Sperm entering the vagina and uterus are protected by the fluids in which they are suspended.

In brief summary, the male reproductive system is a complex assembly of glands and ducts that serve to produce, preserve, and deliver sperm into the female reproductive system during sexual intercourse. Sperm are produced in the seminiferous tubules of the testes and each one is potentially capable of fertilizing an egg. They are transported through the epididymis and the vas deferens to holding areas. Glandular secretions are mixed with the cells to form semen. The most important attribute of the reproductive system is the production of gametes. These cells are found nowhere else in the body and represent the genetic link to future generations. Little wonder that this system of reproductive organs, glands, and ducts is so important to our species.

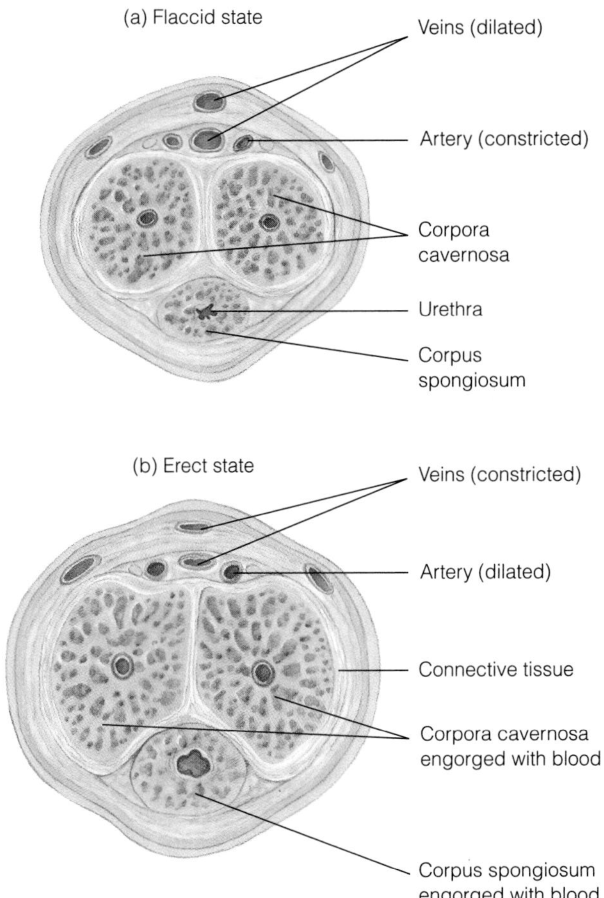

Figure 14–20 Cross Section of Human Penis
In the erect state, corpora cavernosa and corpus spongiosum fill with blood (b) constricting veins and dilating arteries. For an anatomical reference compare with Figure 14–19(a).

REGULATING FERTILITY

Hormonal Intervention and the Ovarian Cycle

Research in the physiology of human fertility over the last few decades has increased our understanding of many of the basic processes described in this chapter. Much of this work has been simply descriptive, but in the case of human fertility there have been practical applications as well. The most important of these has been the introduction of useful and safe methods of molecular contraception. The most common type of molecular contraception is **birth control pills.** Birth control pills are composed of steroid hormones related to natural forms of progesterone and estrogen (Figure 14–21). Modest to low levels of these steroids mimic pregnancy and, therefore, suppress the release of pituitary FSH and the stimulation of follicles in the ovary (as depicted in Figure 14–8). From the previous description of the ovarian cycle, it is clear that sex hormones play a key role in negative feedback, directly controlling the hypothalamus and indirectly suppressing gonadotropic hormones of the pituitary gland. If there are no primary oocytes stimulated to mature, there is no ovulation and, thus, no eggs present in the reproductive tract to be fertilized by sperm.

The timing of ingestion of the steroid hormones can be controlled, so that their withdrawal for several days at the end of a cycle dramatically reduces their levels in the body. This in turn triggers menstrual flow. The ingestion of hormones on the proper schedule following menses once again prevents the release of FSH. When properly used, this method is very effective (99%+) in preventing pregnancy. This is because there is never an egg released to be the target for sperm. In addition, the method is also very popular because it is easily reversed by simply not taking the pills.

One problem that may be encountered in this method is mistakes in taking the pills. Such errors can lead to pregnancy. Most birth control pills contain relatively low amounts of hormones, so as to not disturb other functions of the body. Missing as few as one or two tablets can allow FSH to be released and start the maturation of an egg. Abstaining from sexual intercourse or using other mechanical means of preventing the successful encounter of an egg and sperm should be employed under these circumstances (Figure 14–22).

Other Methods of Birth Control—Contraception

Abstinence Abstinence from having sex at all is a 100% effective method of preventing pregnancy. While this is an absolutely foolproof method, it is not a very realistic one. Sexual relationships are a natural part of our lives and represent our emotional needs for companionship and love, even when there is no desire to procreate. It is fine to talk about controlling oneself (or controlling others) by invoking abstinence as a method, but the evidence of the need for consenting sexual interactions in our society suggests rather that we should look toward methods that allow for safe, nonreproductive sexual interactions that protect the participants from some of the consequences of their actions.

Rhythm The rhythm method is a natural technique, which takes into consideration the timing of ovulation. If sexual intercourse is avoided during the time of egg release

Birth control pills any of a number of combinations of estrogen and progestins in pill form used to control the ovarian cycle

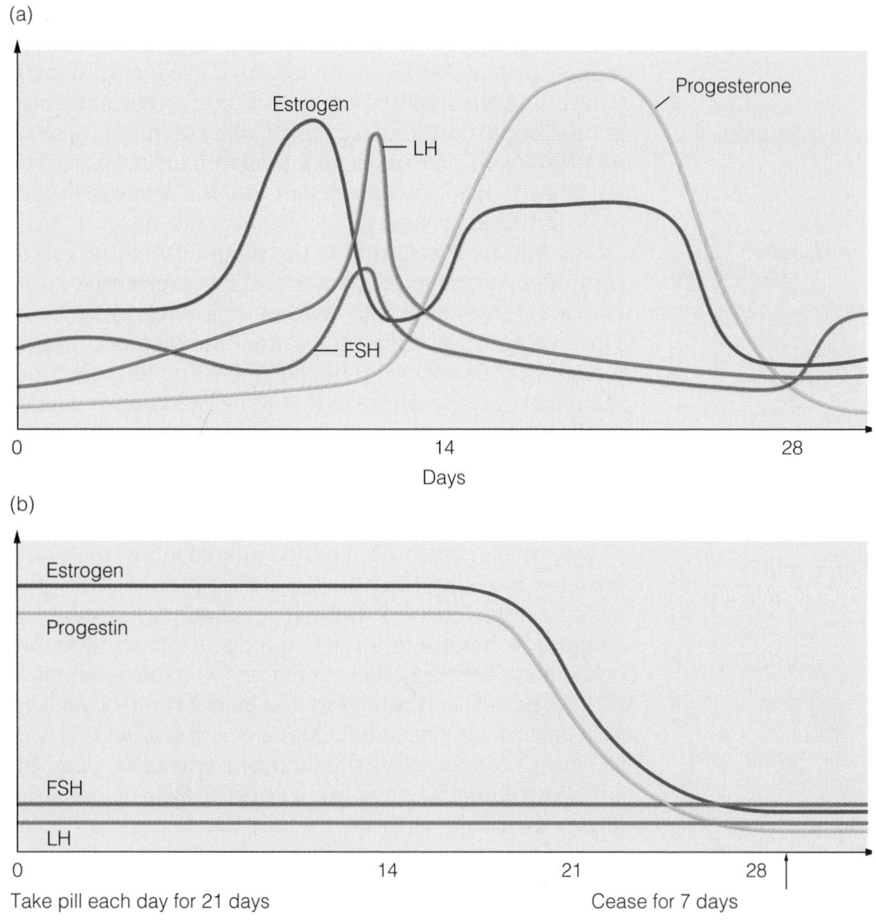

Figure 14–21 Regulating Fertility—The Birth Control Pill
The normal ovarian cycle is depicted in (a). Cyclic variation in FSH, LH, estrogen, and progesterone occur naturally. Oral contraceptives (b) contain estrogen and progestins that suppress release of FSH and LH so no follicle is formed. Cessation at day 21 induces menses.

and the passage of the egg through the Fallopian tube, the chances of fertilization and pregnancy drop dramatically. For humans this means a few days during which the egg makes this journey. A problem with the rhythm method is knowing exactly when ovulation is going to occur. There is considerable variability among females as to the progression of events during a cycle, and a day one way or the other can be extremely important. Another psychologically inherent problem with this method is controlling sexual desire, which may or may not conform to the schedule of ovulation. For these reasons this method has serious drawbacks for long-term family planning.

Condoms Methods that are considerably more reliable for protection against fertilization than rhythm involve the use of barriers to the entry of sperm into the female reproductive tract (Figure 14–22). Historically for males, and more recently for females, condoms have been used as form-fitting barriers that prevent direct sexual contact. Condoms take the form of thin sheaths made of latex or natural membranes that fit snugly over the penis, or form-fit into the vagina and over the labia and block sperm from entering the vagina (Figure 14–22 [a] and [b]). Condoms are relatively inexpensive and, in addition to preventing fertilization, they are also valuable in preventing transmission of diseases. For this reason they are also called *prophylactics,* which means guarding from or preventing disease.

The Diaphragm Another type of device that keeps sperm and eggs from encountering one another is called a *diaphragm* (Figure 14–22[c]). This device takes the form of a tightly fitting membrane that covers the female cervix and prevents sperm from moving into the uterus and upper regions of the reproductive tract. This barrier is individually fitted for a woman by a physician and has to be inserted manually prior to having sexual intercourse (though it may be worn on a continuous basis). In many cases, in addition to a diaphragm, a woman may use contraceptive foam or jelly, to ensure not only the blockage of sperm but also the killing of them.

The availability and use of male/female condoms, diaphragms, and spermicidal chemicals has proven to be

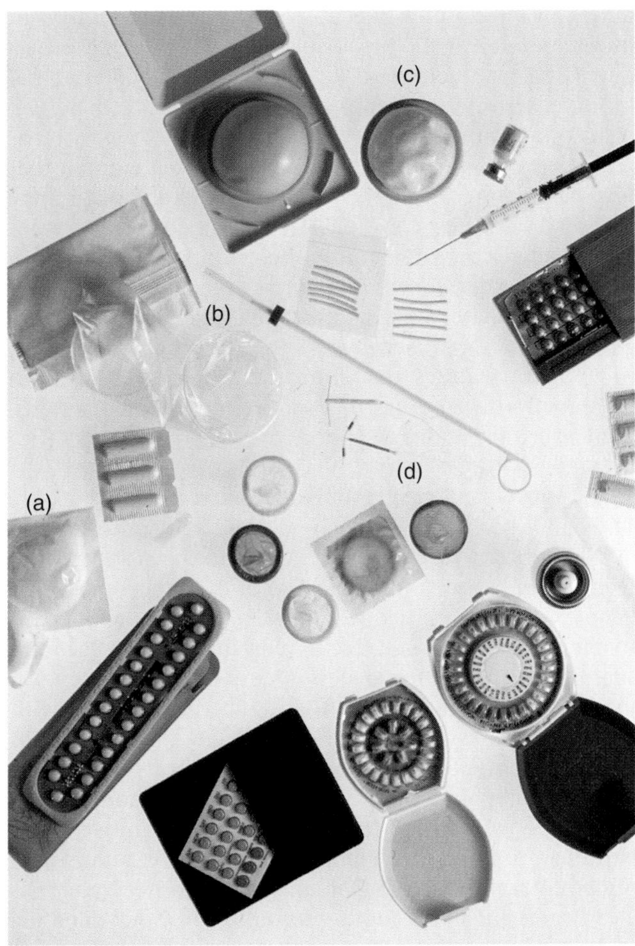

Figure 14–22 Mechanical Methods of Contraception
Mechanical methods of contraception prevent sperm and egg from encountering one another (a), (b), and (c) or interfering with sperm viability or implantation of a blastocyst (d).

of great value in preventing unwanted pregnancy. Though often uncomfortable, or unwieldy to use, these devices and compounds have proven to be very effective. When everything works well, the failure rate of these methods is less than 1%. Spermicidal compounds of a variety of types and systems of delivery are available on the market today. It is important to note that the spermicidal compound nonoxynol-9 is thought to prevent the entrance of human immunodeficiency virus into cells of the human body and so is considered prophylactic against infections that might lead to AIDS (covered in Chapter 16).

IUDs Another method to prevent pregnancy involves the use of an intrauterine device or IUD (Figure 14–22 [d]). These devices generally do not prevent sperm from contacting the egg but rather interfere with sperm function and survival and with implantation of the embryo. The shape and composition of these devices vary greatly, as does the degree of difficulty that women have had in using them. A significant occurrence of expulsion of the devices (15%–20%) adds more difficulty than ease to their use. There has been a trend away from the use of these devices in recent years, so there are only a few types of IUDs available. Part of this trend is because of fairly common and serious side effects that have shown up in a substantial number of women who have used this method of birth control. This includes bleeding, infection, and sterility in a significant number of women. When IUDs work, their protection rating is quite high, but the potential problems associated with their use leaves in question whether they are worth the risk.

Vasectomy and Tubal Ligation Along with abstinence, the only other sure way to prevent pregnancy is to disconnect the testes and/or ovaries from the rest of the reproductive tract (Figure 14–23). This does not mean removal of the gonads but rather the cutting of the tubes through which gametes migrate from the gonads. For males this generally means severing the vas deferens and tying off the ends. Known as a **vasectomy,** this is a relatively simple surgical procedure that entails tying off and/or cutting the vas deferens within the scrotum (Figure 14–23[a]). Sperm are still produced within the testes, but they simply have no route through which to enter any other part of the reproductive tract. It is of great physiological and psychological importance that the testes be retained in males. These organs produce testosterone, which maintains the secondary sexual characteristics of a male. The cutting of the vas deferens results in little or no impairment of sexual desire or performance.

In females, the surgical procedure is somewhat more complicated because a female's gonads, the ovaries, are internally located. However, the principle of separation is the same. The Fallopian tubes are severed or ligated (tied off) to prevent an egg from migrating into the uterus (Figure 14–23[b]). This also prevents sperm from passing upward towards the ovary and eliminates the possibility of gametes contacting each other. This procedure is known as a **tubal ligation** and is 100% effective. Again, as in the case of the testes, the physiological functions of the ovaries

Vasectomy a surgical procedure in which the vas deferens is cut and sealed to prevent transport of sperm

Tubal ligation a surgical procedure in which the Fallopian tubes are ligated and cut to prevent transport of eggs

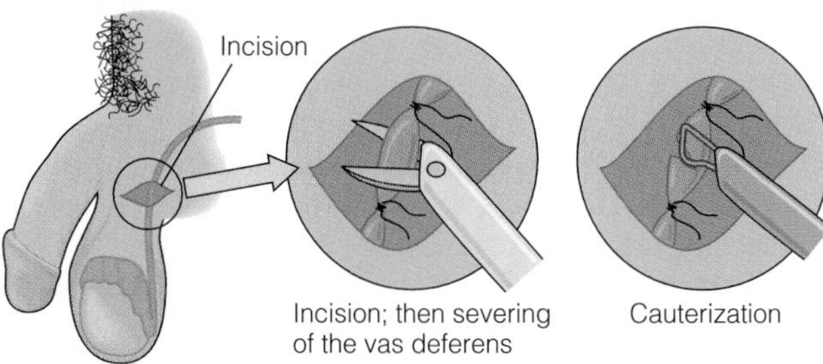

(b) Tubal ligation

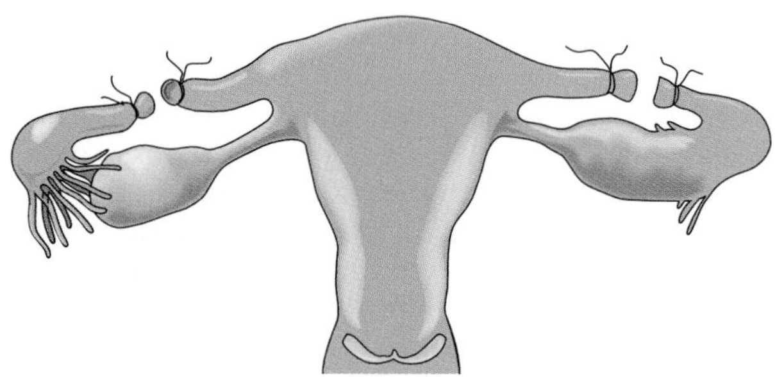

Figure 14–23 Vasectomy and Tubal Ligation
Surgical methods of contraception result in the separation of the tubes connecting the gonads to the rest of the reproductive tract in males (vasectomy) and females (tubal ligation).

are important because they produce estrogen and progesterone, which are necessary for the good health and well-being of the female body.

Chemical Intervention There are also a number of controversial techniques that involve the chemical induction of an abortion of an implanted embryo. One of them involves the drug RU 486. This "morning after" method requires an injection or injections of the drug and results in interference with the action of progesterone. One consequence of treatment with RU 486 is that the endometrial lining of the uterus cannot be maintained and, therefore, disintegrates and is expelled from the body. Along with the endometrium, of course, goes any implanted embryo that may be present. The method is generally safe, though there are some side effects possible. This method is readily available in a few countries around the world, including France where it was developed. It is not yet generally available in the United States but is now being tested for approval. Other similar approaches are being developed and all of them involve drugs that interfere with progesterone in combination with drugs that induce uterine contractions, such as prostaglandins.

Ethics and Basic Scientific Research

The major problem with many of the birth control methods described here, and particularly with the morning after techniques, is not a biological one; it is an ethical one. People who are opposed to surgical abortions are opposed to chemical ones as well. Techniques that prevent eggs and sperm from interacting are less offensive to most people on moral grounds. Techniques that putatively interfere with implantation, as represented by IUDs, are questionable in the minds of people opposed to abortions, because an unimplanted blastocyst is actually a new, and unique human genetic entity, though in a very primitive form. However, an implanted embryo may well have attained an uncomfortable level of humanness, and the embryo is clearly growing in the mother's womb. The moral dilemma is that the freedom of a woman to choose not to be pregnant, with all the emotional and psychological complications that implies, and the right of the unborn child to be given a chance to come into the world are clearly in conflict. There are over 1.5 million surgical abortions performed in the United States each year. The introduction of much simpler, safer, and more private chemical procedures complicates the issue of abortions even more, particularly in the United States.

Summary

In this chapter, we have focused on the anatomical and physiological differences and similarities in the reproductive systems of human males and females. The production of gametes, ova and sperm, and their migrations through their respective reproductive tracts reveal an intricate interrelationship between reproductive organs and glands. The principal consequence of an egg and sperm encountering one another is fertilization of the egg. This leads to the formation of a zygote, a new human being.

In the twentieth century, what has been found is that the more we have learned about the anatomy, physiology, and biochemistry of gametogenesis and fertilization, the more we have been able to control it. Basic research in human reproductive physiology has led to molecular and mechanical methods of controlling fertility and safely preventing fertilization.

However, it should not be forgotten that the issues of controlling fertility and terminating a pregnancy strike deep into the foundations of people's moral and ethical sensibilities and have grave political consequences. Although few would deny the pervasive influence that the discoveries of scientists and physicians have made in this area, the uncertainty in the larger world is in how to apply them.

Questions for Critical Inquiry

1. Why would we, as humans, want to regulate our fertility?
2. What types of reproductive strategies are used by animals? What are some advantages and disadvantages of each type?
3. What is the genetic basis for the difference between human males and females?

Questions of Facts and Figures

4. In what tissue are sperm produced? Where is this tissue located? What cells do the developing sperm interact with during their maturation?
5. What hormones are involved in the ovarian cycle? What are their sources and what cells and tissues do they affect?
6. Describe the pathway through which a sperm migrates after release from the epithelium of the seminiferous tubule.
7. What is the outcome of a vasectomy? A tubal ligation?
8. Where does capacitation of human sperm take place? What reaction is triggered when sperm get in close contact with an egg?
9. What is the composition of most birth control pills and how do they affect the female reproductive system?
10. What are the effects of estrogen and progesterone on the endometrium?

References and Further Readings

Cummings, M. (1996). *Biology: Science and Life.* St. Paul, MN: West Publishing.

King, B. M., Camp, C. J., and Downey, A. M. (1991). *Human Sexuality Today.* Englewood Cliffs, NJ: Prentice Hall.

Stalheim-Smith, A., and Fitch, G. K. (1993). *Understanding Human Anatomy and Physiology.* St. Paul, MN: West Publishing.

Wilson, E. O. (1978). *On Human Aggression.* Cambridge, MA: Harvard University Press.

CHAPTER 15

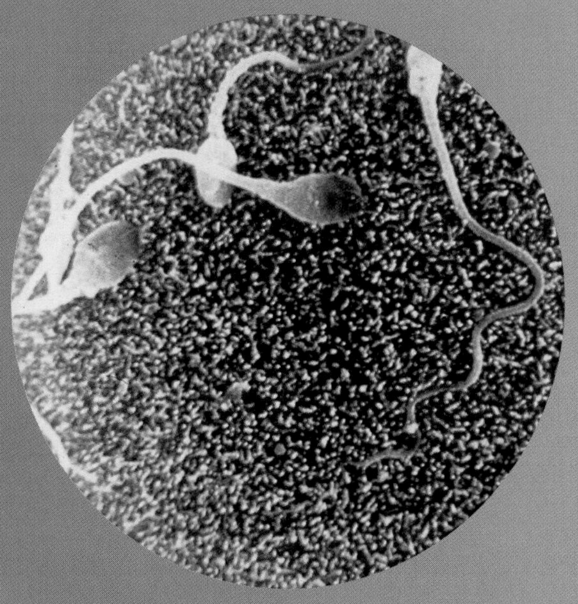

The Changing Form of New Life

INTRODUCTION

The formation of eggs and sperm through meiosis and their subsequent maturation into functional forms are a prelude to their combining to produce a new and unique individual. Fertilization of an egg begins a process of growth and development that does not cease throughout the lifetime of that individual. Certainly, the rates of growth and the appearance of new features are different at different times in the process. Anyone who has observed a pregnancy (or been pregnant) is familiar with outward changes in the mother and wonders about the hidden changes that are taking place in the fetus. Likewise, neonatal growth, childhood growth, puberty, and maturation into an adult each have their own special characteristics. In this chapter we will follow, in broad terms, the course of life from the instant of conception to the moment of our passing—a journey of being and becoming that we all take.

FERTILIZATION AND THE INSTANT OF CONCEPTION

Cell Recognition and Its Outcome

The modern view of human embryonic development begins with an equal genetic partnership between the male and female gametes. As was discussed in Chapter 14, each of the two sexes provides an egg or sperm, which contain half the genetic endowment necessary to make a new human being (Figure 15–1). Sperm and eggs are specialized for their roles in reproduction. However, it should be kept in mind that the gametes have only the potential for making a new individual; they are not in-and-of themselves the new individual.

Sperm, Eggs, and Zygotes

Human sperm are tiny, extremely long, flagellated cells that must be introduced into the female reproductive tract in great numbers in order to have a chance to fertilize an egg. Generally, a single, large egg is produced and released by a female each month as part of her ovarian cycle (Figure 15–1; review Chapter 14). From an array of millions, only one sperm will ultimately fertilize an egg. The manner in which a new individual forms from a single cell is called **morphogenesis,** literally the genesis of shape. Shape changes may occur by locally increasing cell numbers through cell proliferation, increasing individual cell size through cell growth, repositioning cells through migration, and eliminating cells through controlled or programmed cell death. Morphogenesis is a complex, multistep process that is initiated with fertilization and continues to different extents throughout embryonic, fetal, and neonatal life. Sustaining life is a precarious undertaking at any stage of development, but it is most susceptible to disruption during the early stages. If anything goes wrong at any one of the early steps of morphogenesis, the entire process of embryonic development may fail. There are basically three phases to the early stages of life, focusing around the time of fertilization and zygote formation (Figure 15–2):

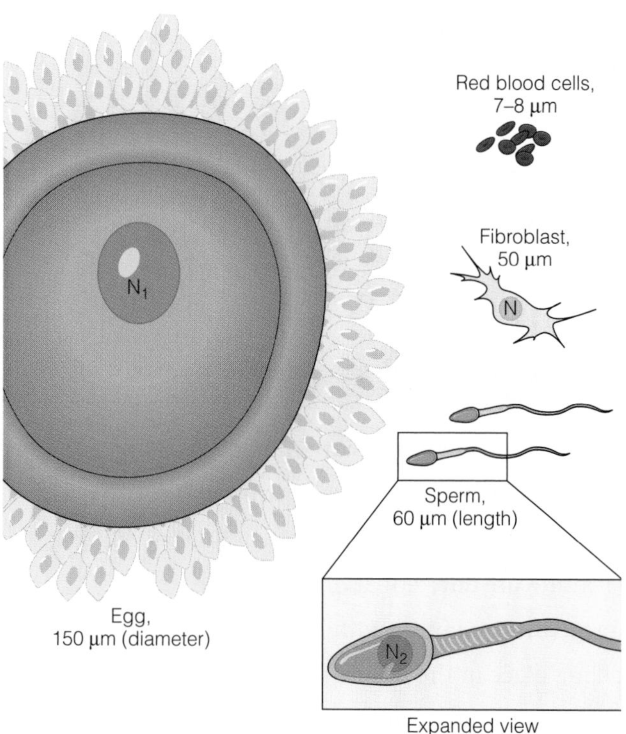

Figure 15–1 The Sizes of Human Sex Cells
Human sperm are approximately 60 μm in length but contain only a small volume of cytoplasm. A human egg is 120 μm–150 μm in diameter and contains several thousand times the volume of a sperm. Other cell types are shown for comparison. Nucleus of egg (N_1) and sperm (N_2) result from meiosis and contain a haploid set of chromosomes.

1. Cell-to-cell contact and adhesion (sperm meets egg)
2. Chromosomal association (fusion, egg and sperm chromosomes intertwine) and zygote formation
3. Cell division and cleavage (dividing up the territory).

Sperm Meets Egg The actual contact of an egg and a sperm in phase 1 is preceded by much movement and migration of cells within the female reproductive tract (see Chapter 14). First, we will consider how an egg and a sperm actually get together and then how, within the microenvironment of the reproductive tract, a sperm cell may recognize and adhere to an egg cell when they encounter one another (Biosite 15–1).

Motile human sperm are introduced into the female vagina as a result of coitus. Most, if not all, the sperm undergo capacitation in the vagina or uterus and become fully capable of fertilizing an ovum. They actively move through the cervix, into the uterus, and, eventually, into the Fallopian tubes. Some sperm do not make it out of the vagina, others travel in the wrong direction, but some do make it into deeper areas of the reproductive tract. What is clearly different about this set of circumstances, and the story of swimmers in Biosite 15–1, is that there is no realization of failure for a sperm that does not fertilize an egg and no feelings of success by the one sperm that does fer-

Morphogenesis the development of form or shape

tilize an egg. However, the idea of traveling blindly over relatively great distances and randomly encountering an ovum is similar in spirit to the swimmer encountering the prize island and gives some feeling for the size, scale, and nature of the situation.

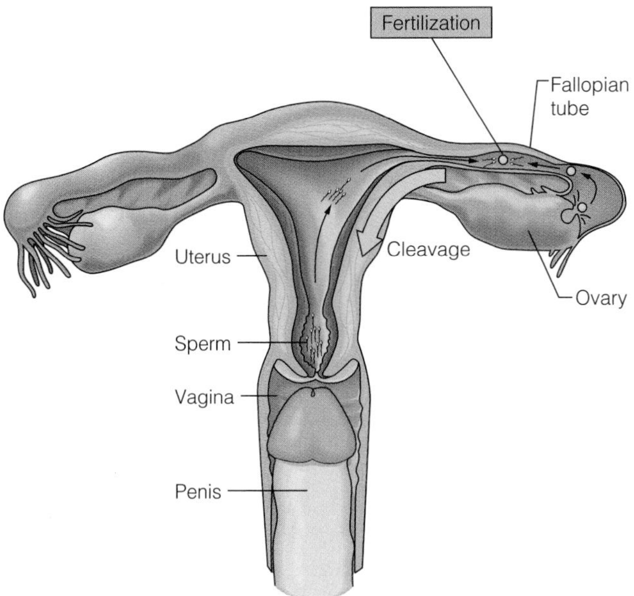

Figure 15–2 The Timing and Location of Fertilization
Following ovulation the ovum travels through the Fallopian tube. Introduction of sperm during this time has the greatest chance to successfully fertilize the egg. Fertilization usually takes place in the upper half of the Fallopian tube and subsequent cleavage results in a blastocyst that implants within five to six days afterward.

Randomness There is an inherent randomness to the process of fertilization, an uncertainty as to which (if any) of the sperm will actually make contact with the egg at the right time and in the right place. In fact, the number of sperm that eventually make it into the vicinity of the egg in the Fallopian tube is small. Although there are hundreds of millions of sperm suspended in the original semen, by the time they reach the upper region of the Fallopian tube there may be only a few hundred, or a few dozen, left to encounter an egg. The sperm must first overcome the barrier imposed by the follicle cells that surround the ovum. This barrier is made up of follicular cells (collectively known as the *corona radiata*) and a jelly layer (the *zona pellucida*) (Figure 15–3). These barriers help protect and sustain the egg as it moves through the Fallopian tube.

Capacitated sperm are stimulated to release enzymes by the acrosome reaction. Each sperm contains within it a collection of hydrolytic enzymes that are located inside a special vesicle at the tip, known as the acrosome (see Chapter 14). Acrosomal enzymes are released under the appropriate conditions, that is, during the encounter with the outer barriers of the follicle. These enzymes dissolve a path through the corona and the zona to the surface of the egg. This is when initial, direct contact may be made between egg and sperm (Figure 15–3 and 15–4). The presence of many sperm surrounding the egg and the releasing of their acrosomal enzymes as a group helps hasten the breakdown of the corona and zona so one of the sperm can contact and adhere to the egg plasma membrane. Ultimately, the first sperm to contact the egg cell surface is the one to initiate the process of fertilization.

Contact and Adhesion What are the details of the sperm-egg encounter? The first phase is simply physical

15–1

BLIND SWIMMERS

Imagine being prepared to swim blindfolded with thousands of males in a big lake with lots of small islands situated two to three miles off shore. The goal of the group of swimmers is to get to a particular, very special island in this archipelago and to claim it. The prize is the island. The swim takes a lot of energy and by the time the swimmers get out as far as the islands many of them are already out of the race or very tired. However, one is lucky and ends up targeting the correct island. Many of the swimmers dropped out early in the race or swam the wrong direction. As the lucky swimmer emerges onto the shore and removes the blindfold, he realizes that he has managed to be the first to reach the right island. The prize is won! This short story is about individuals, the nature of competition, and chances for success. It can also apply to sperm as they work their way through the female reproductive tract in quest of the egg.

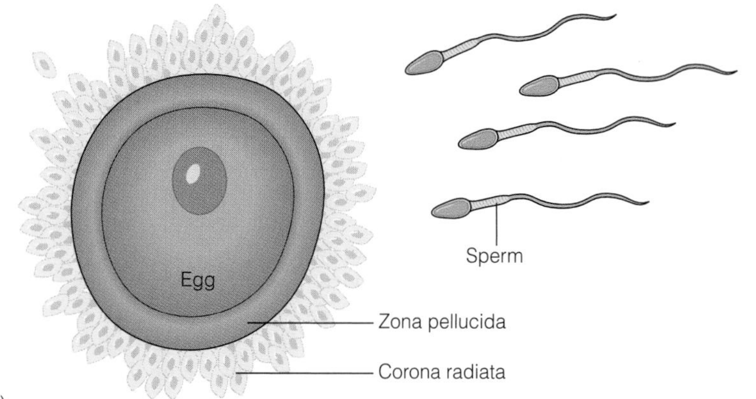

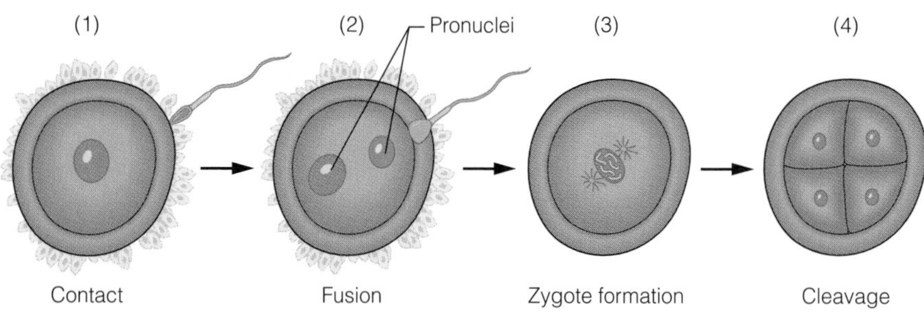

Figure 15–3 Fertilization and Cleavage
Sperm are confronted with the surrounding layers of the egg (corona radiate and zona pellucida) through which they must penetrate to contact the egg surface. In (b), contact (b1), fusion (b2), and zygote formation (b3) begin the process of cleavage (b4).

contact and then adhesion. However, even though the sperm has contacted the egg, fertilization has yet to occur. Stable attachment is extremely important. What if the sperm falls off? Another sperm may well take its place. Only one sperm can successfully fertilize an egg and, as a reflection of the randomness of the process, it really does not matter which one. As described, a major problem that sperm encounter on their way to the egg is the barrier that surrounds the egg itself.

The attachment of a sperm to the surface of an egg depends on molecular interactions between molecules at the surface of the two cells. They interact through a kind of molecular lock and key mechanism. The sperm has molecular keys that insert into special molecular lock receptors on the surface of the egg (Figure 15–5). This interaction is "sticky" and results in the adhesion of the two cells together. This cell-to-cell binding is the culmination of phase 1.

Block to Polyspermy During phase 2, there is a rapid and dramatic chemical change that prevents any other sperm from adhering to the egg cell surface and repeating the attachment process for itself. This is called the **block to polyspermy,** and it is very effective in preventing a second encounter. The block to polyspermy is a biochemical change in the surface membrane of the egg that occurs within seconds after the sperm has attached to the egg surface. The block spreads instantly, as a wave, over the entire egg surface. The door is locked on any uninvited extra guests. Actually, this is a fundamentally important response on the part of an egg. If more than one sperm were to simultaneously fertilize an egg (a phenomenon known as polyspermy), there would be far too many chromosomes in the egg for it to develop normally.

Cell Fusion The next step in the encounter of these two cells requires the formation of a channel between the egg and the sperm, so that the sperm *pronucleus* can migrate into the egg cytoplasm. This is akin to the fusion of oil droplets on the surface of water, except that this process requires the plasma membranes of the two cells to fuse and become as one (Figure 15–3). Once this has occurred, the pronucleus of the sperm and the pronucleus of the egg in-

Block to polyspermy a reaction of the egg to prevent more than one sperm from participating in fertilization

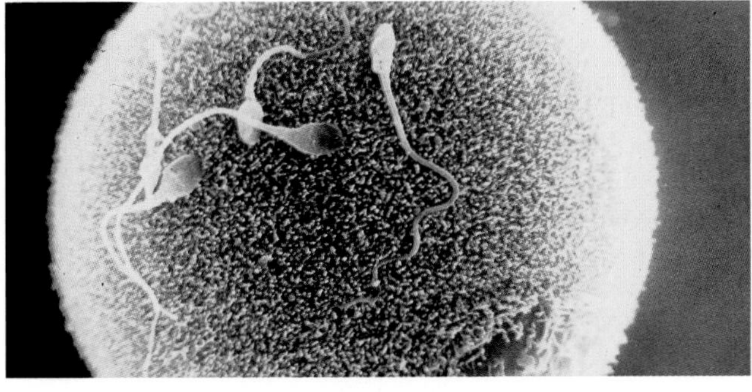

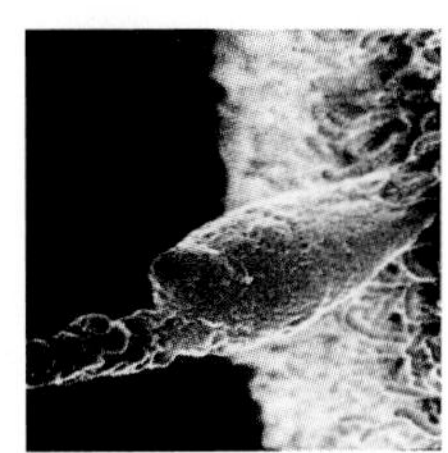

Figure 15–4 Contact Between Egg and Sperm
In (a) many sperm attempt to penetrate the membrane of an ovum. Only one will be successful. Part (b) is a close-up of sperm on the surface of an ovum.

tegrate with each other to form the nucleus of what is then considered the zygote. Cell fusion brings the chromosome count of the fertilized egg to 46 or 23 pairs, which is the appropriate number for our species.

This progression of events takes some time, particularly unraveling the chromosomes from the pronuclei and organizing them in the new nucleus so they can begin to undergo cell division or cleavage. It is perhaps worth envisioning this process (with some poetic license) as a delicate chromosomal dance in which the long, slender molecules of life, conjoined by chance are unwound in a pas de deux that creates an entirely new nucleus with a combination of genetic information that makes a unique new individual. Then, the work of the new individual starts. Once the nucleus is established, the cell is no longer an egg, nor a sperm; it is a zygote.

The facts presented in describing this process of zygote formation were not always known. In the past there were many imaginative ideas about how an embryo or zygote was formed. Biosite 15–2 discusses some of the historical and sometimes hysterical concepts.

Cleavage The third phase of the process is initiated as the single-cell zygote, and then it begins to divide (Figure 15–6). The initial cell divisions of a zygote are called cleavage and, as more and more cells arise through mitosis, the embryo begins to undergo self-organization. Keep in mind that the average size of a cell in the human body is far smaller than that of a zygote. How does the egg, which is relatively large, reduce the size and volume of its daughter cells to the size of normal human somatic cells? It is the process of cleavage that subdivides the zygote cytoplasm

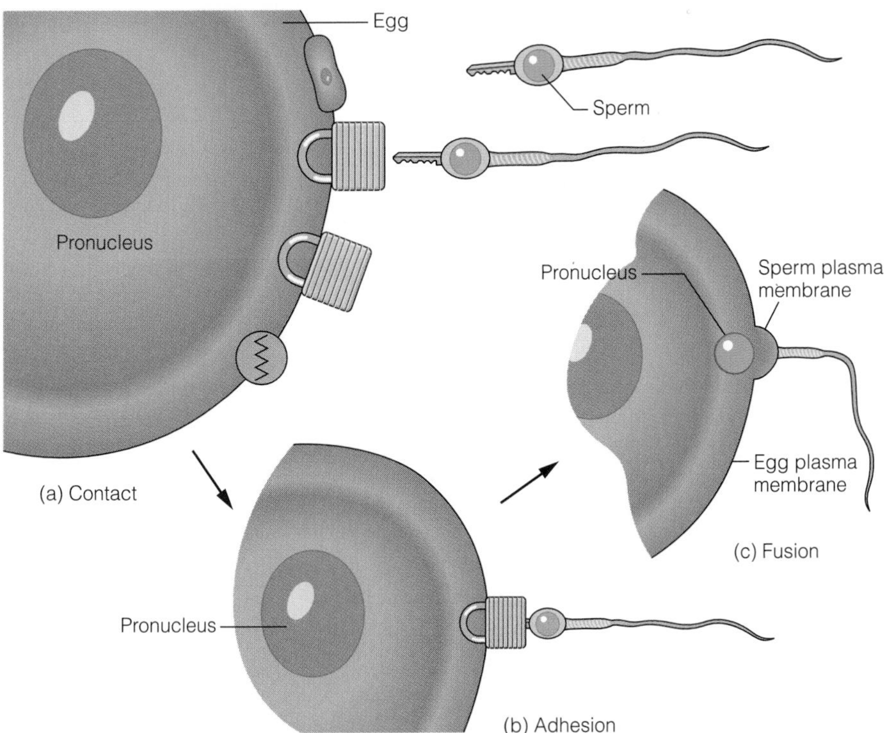

Figure 15–5 Locks and Keys
Contact between an egg and a sperm involves binding of sperm to receptors on the egg surface. The lock-and-key analogy shown in (a) represents receptors (egg) as locks and sperm molecules as keys. Once a sperm has unlocked the egg no other sperm can do so (b). Fusion (c) brings about release of the sperm pronucleus into the egg cytoplasm.

15–2

PREFORMATIONISTS AND EPIGENETICISTS

Long ago it was believed by some that the embryo was preformed and simply got bigger as development proceeded. The proponents of this notion were called *preformationists* (appropriately), and they had some very interesting ideas about how humans developed. One of the most innovative of these ideas was that of the *homunculus*. This concept had as a core assumption that a tiny, fully formed human being was packaged inside each gamete. Not only that! It was also believed that within each homunculus there existed gametes that also contained homunculi. The logic of this particular view has a biblical connotation as well. If each female in each generation contained the seeds of future generations, like a Russian doll with smaller and smaller dolls within itself, then a look back in time would lead directly to Adam and Eve! This famous couple must have had all the generations of humanity within them in the form of smaller and smaller homunculi right from the beginning. As a scientific proposal this idea is useless, but, as support for creationist views, it fit the times.

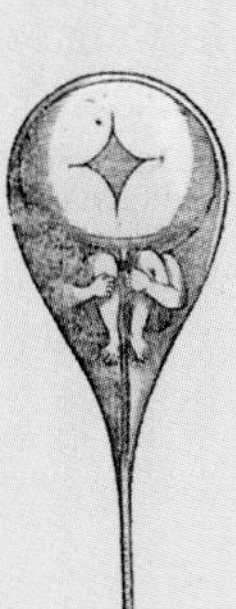

A second possibility with respect to the process of embryonic development suggested that characteristic anatomical structures arise from initially formless, apparently unorganized materials within the embryo. In this case the parts were to be assembled, as if they were the pieces of an unassembled bicycle, following a set of special instructions. This self-assembly process was called *epigenesis* and its proponents were the epigeneticists. As it turns out, this second view prevailed, with some modifications. As more powerful experimental methods became available during the nineteenth and twentieth centuries and the images of homunculi within cells were shattered, a new era arose in the science of life that compelled us to explain how the orderly assembly of indistinct, apparently amorphous, unorganized materials within cells could be brought about to form such a complicated and well-organized entity as a human being. Even today in the modern biotechnical world, while we do know quite a bit, we are still far from a complete understanding of how such a complex process might occur.

and reduces the size and volume of the newly formed embryonic cells at each division. As the number of cells in the embryo goes up, their average size goes down.

Where Does Cleavage Start? The start of cleavage for the zygote usually begins while it is still in a Fallopian tube (see Figures 15–2 and 15–6). This is where successful encounters between an egg and the sperm are most likely to take place. Prior to fertilization, but following ovulation, the egg is gently wafted down from the ovary by the action of cilia on the surface of epithelial cells in the Fallopian tube. Sperm, on the other hand, are working hard to move against the flow by the action of their flagella. The two cells usually encounter one another somewhere in the middle region of the Fallopian tube. Fertilization takes place in this warm, moist environment and the new zygote continues the journey down the Fallopian tube, dividing as it goes.

Embryogenesis Prior to Implantation

There are several stages of development that describe the state and shape of the new embryo. After the first few cleavages of the zygote, the embryo looks as if it were a solid ball

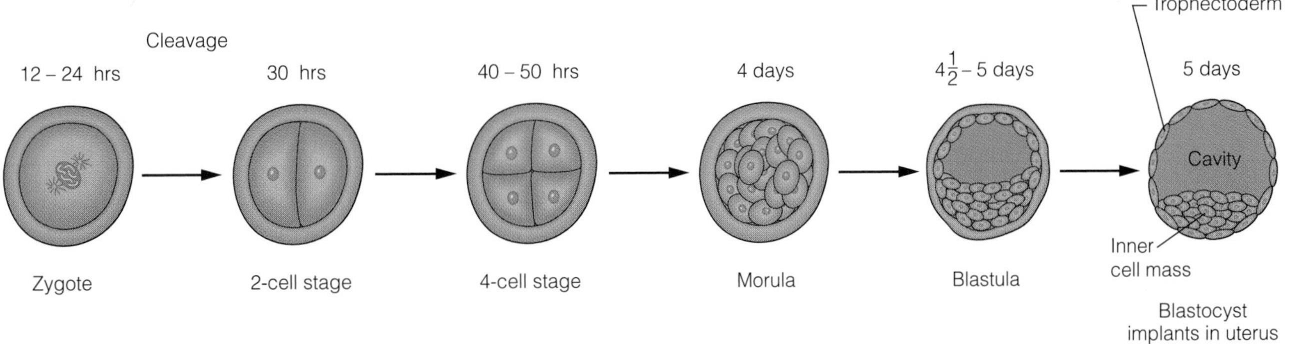

Figure 15–6 Cleavage and Early Development
Cleavage of the zygote gives rise to a multicellular structure within a few days. The morula is formed by four days, followed by the blastula and the blastocyst within five days.

of cells (Figure 15–6). This solid ball of cells is called a **morula.** The journey of the embryo through the Fallopian tube continues. The morula cells keep dividing, but cells within the morula separate from one another to form a cavity, a stage of development known as the **blastula.** The journey of the embryo continues. With the formation of an internal cavity and the separation of the **inner cell mass** and the **trophectoderm** from one another, the embryo becomes a blastocyst. At this stage in embryogenesis the embryo is capable of attaching to the endometrial lining of the uterus. Several days have passed by this time (usually three to five days) since fertilization, and the blastocyst has usually finished its journey down the Fallopian tube and entered the uterus. Arrival of the blastocyst within the uterus signifies the end of the journey. The endometrium of the uterus is the site in which the embryo implants and begins its intimate association with mother. As has already been pointed out, implantation is the real beginning of pregnancy.

er and can begin receiving nutrients. A woman in this state may not even suspect she is pregnant yet.

As mentioned earlier, cells on the outside of a blastocyst are different from the cells on the inside (Figure 15–7[a]). The cells on the outside of the blastocyst are the trophectoderm cells. They form what will become the extraembryonic layers, which are the layers surrounding and enclosing an embryo (Figure 15–7). They are not destined to form any part of the embryo proper. However, they are extremely important to the survival of the embryo. The extraembryonic cells will share with the mother's tissues the job of forming a placenta, which connects the mother and the baby through a filtered vascular circulation. The trophectoderm will also form the chorion and amnion, which are membranous, fluid-filled sacs in which the embryo resides. For the next several months the embryo will float within these membranes in a sea of warm and protective fluid. The cells inside the blastocyst, composing the inner cell mass, form the embryo itself (Figures 15–6 and 15–7).

IMPLANTATION

Implantation of the blastocyst into the wall of the uterus is a complex and multistep process. If implantation occurs in the wrong location, serious problems may arise (Biosite 15–3). Once the initial adhesive contact has been established between the trophectoderm cells on the outside of the blastocyst and the endometrial lining of the uterus the embryo begins to embed itself into the endometrial lining (Figure 15–7). This process of embedding starts immediately, so that the embryo is protected from the possibility of being physically dislodged by the activity of the new moth-

Morula a stage of the cleavage of the egg in which the embryo is a solid ball of cells

Blastula a stage of the cleavage of the egg in which the embryo is a ball of cells with a cavity within it

Inner cell mass the part of a blastocyst that will give rise to the embryo

Trophectoderm the part of the blastocyst that will give rise to the extraembryonic layers and part of the placenta

15–3

ECTOPY—BEING IN THE WRONG PLACE AT THE RIGHT TIME

Implantation can be confusing to the new embryo. The adhesive capacity of a blastocyst to attach and implant itself in the uterus is quite high. This is both good and bad. The best place to implant is, of course, the uterus, optimally in the upper half, and this is normally the case. However, because of its adhesive properties, the blastocyst can attach to just about any of the tissues in the reproductive tract and even to sites outside of it. The attachment and implantation of an embryo at sites outside the appropriate region (that is, the uterus) leads to an ectopic pregnancy. Embryos have been known to implant in the region of the cervix, in a Fallopian tube (known as a tubal pregnancy), on the surface of an ovary, and even as far away as the epithelium lining of the abdominal cavity (see figure).

The possibility of implantation in the abdominal cavity requires a bit of an explanation. The ovaries and their Fallopian tubes are not directly connected with one another. There is an open, unconnected space between them in the abdominal cavity through which the egg must initially "float" in order to be captured and begin moving down the Fallopian tube. To prevent escape from the proper path, there are a series of fingerlike projections, called fimbriae, arising from the end of the Fallopian tube nearest the ovary. The motion of the cilia on the cells of the fimbriae help pull the egg into the Fallopian tube. Sometimes, however, an egg moves away from the fimbriae and into the abdominal cavity. If, by chance, the egg was fertilized before actually entering the tube, or was somehow pushed back out of the tube following fertilization, it might end up migrating into the abdominal cavity and attaching to the peritoneum. Because the blastocyst is naturally sticky, it is fully capable of adhering to the peritoneal membrane and trying to implant.

Ectopic attachment of the blastocyst in the abdomen is the beginning of a real problem. The mother's body usually cannot support such growth because sites outside the uterus are not really prepared to receive and provide nutrients to an embryo.

Generally, an ectopic pregnancy does not last long. In such cases a natural abortion occurs. This usually means that the embryonic cells are resorbed by the mother's body and leave little sign of their passing. Furthermore, areas outside the uterus are not prepared to grow rapidly in conjunction with the embryo even if gestation should continue. Abdominal pregnancy can be dangerous for both mother and child because the fragility of the connection between them can lead to internal bleeding and hemorrhage at later stages in embryonic or fetal development.

Implantation within the reproductive tract but outside the uterus can be hazardous as well. A tubal pregnancy can cause rupture of the Fallopian tube and internal bleeding. Implantation in the region of the cervix is equally problematic, as is implantation at or around an ovary. Structural distortion of internal anatomy with the magnitude of growth of a fetus can interfere with normal physiological functions, such as restriction of vascular circulation, nerve inputs, tissue growth, and gas exchange. All in all, ectopic pregnancies are problematic, dangerous, and rarely go to term. Most ectopic implants that do occur probably self-abort. However, there are cases in which an ectopically located fetus develops to term and is delivered normally. In the case of an abdominal pregnancy, a surgical procedure akin to a cesarean section must be performed.

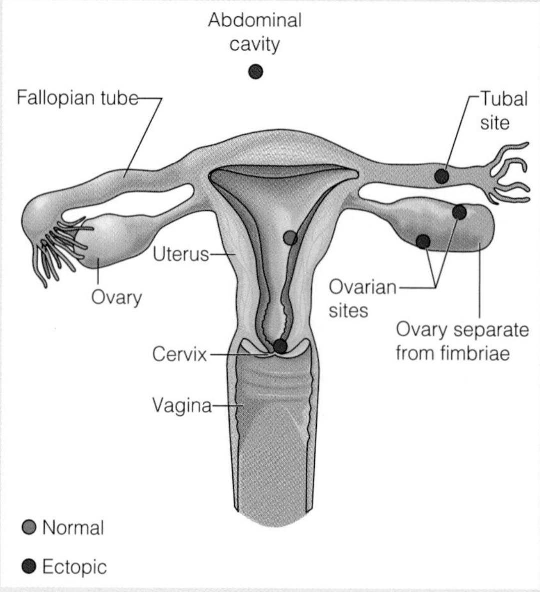

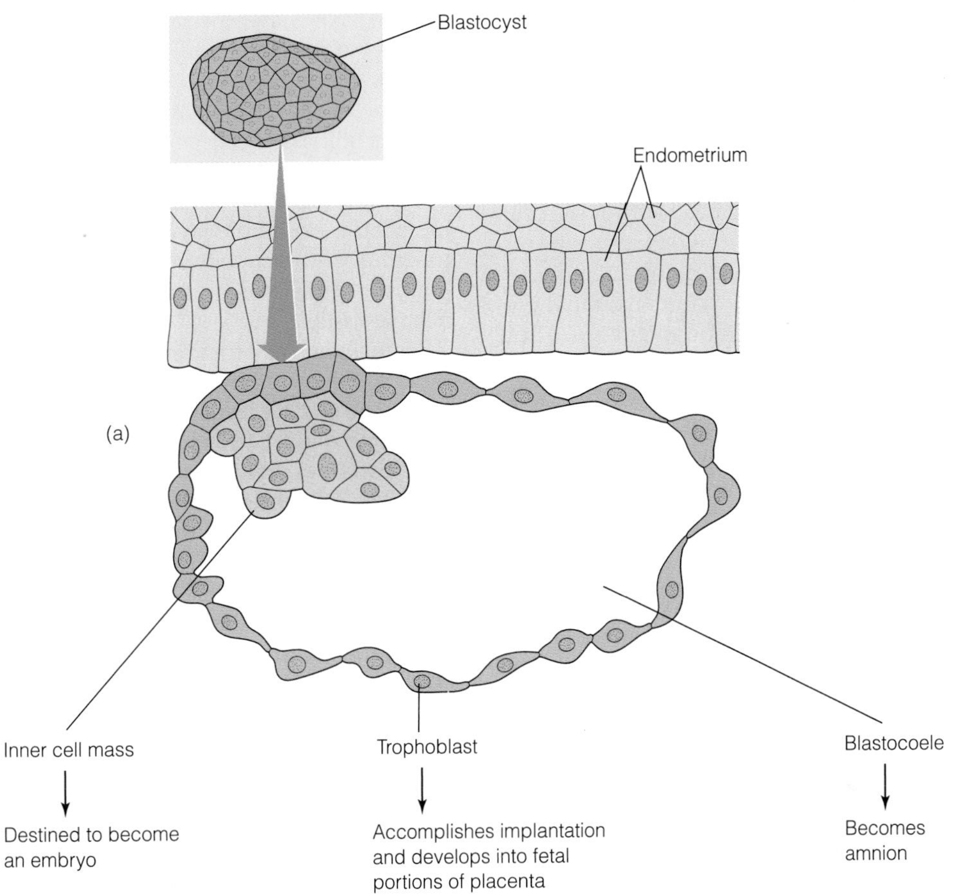

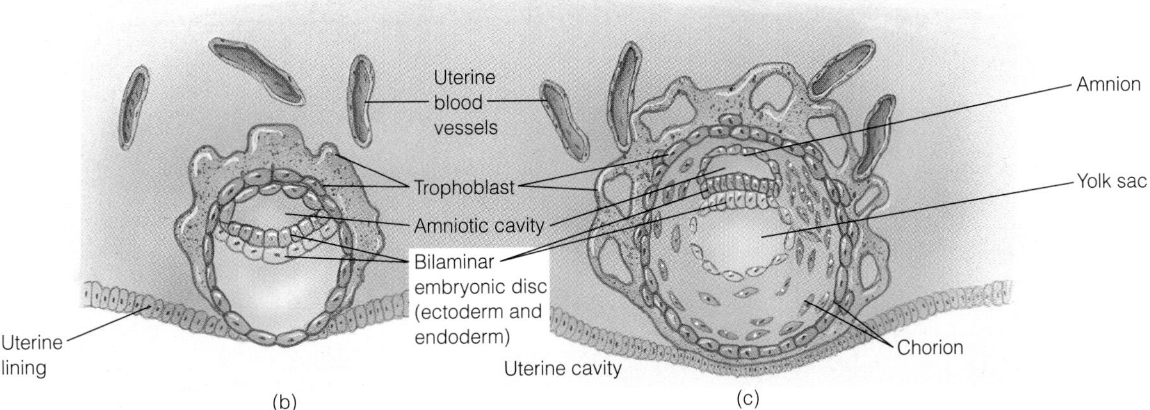

Figure 15-7 Implantation of the Blastocyst
The blastocyst attaches to the endometrium (a) and embeds itself into the uterine lining (b) and (c).

POSTIMPLANTATION DEVELOPMENT

Gastrulation

The mother-child union is an intimate one, yet each individual, adult and embryo, retains independence from the other. The embedding of the embryo deep within the uterine wall ensures its safety and a continuous source of sustenance. Protected within the extraembryonic membranes, the cells of the inner cell mass begin to undergo rapid proliferation and changes in cell shape and migration. These processes are part of the ongoing and continuous changes that shape the embryo during morphogenesis. One of the most important processes of morphogenesis occurring during the early part of embryogenesis is called **gastrulation** (Figure 15–8). In this and many of the succeeding steps of growth and development all animals from fish to humans share common characteristics (Figure 15–9). The main importance of gastrulation is to establish bilateral or two-sided symmetry and to establish distinct layers of cells within the embryo. For cell movement and migration to occur, cells lose their adhesion to one another, separate and travel along the midline of the embryo and migrate inward and laterally to establish layers within and along the length and breadth of the embryo. When this stage is completed, the embryo has a front and back, a top and bottom, sides and cavities within that are composed of three distinct cell layers.

The Primary Germ Cells—Ectoderm, Mesoderm, and Endoderm The three distinct cell layers are known as the **primary germ cell layers.** The layers are called ectoderm, endoderm, and mesoderm (Figure 15–8), and cells in each layer have special fates (Table 15–1). *Ecto* means outside, so the ectoderm is the cell layer found on the outside of the embryo. The ectodermal cells will eventually give rise to the epidermal cells of the skin, pigment cells, and the billions of cells of the nervous system. *Endo* means inside, so the endodermal layer of cells form what will be the future digestive tract, liver, pancreas, and lungs. The mesoderm, or middle layer, is positioned between the ectoderm and the endoderm. Mesodermal cells are destined to form the skeleton, muscles, and most of the body's connective tissue.

Humans and all other vertebrates attain three-dimensional form based on bilateral symmetry. As development proceeds, the embryo generates many paired structures along the length of the body axis. *Homo sapiens* has two arms, two legs, two eyes, two ears as well as many other paired anatomical features. However, some structures that may not appear to be paired in adults are in fact constructed of cells originally derived from opposite sides of the embryo. For instance, the palate forms from the fusion of two matched plates of tissue that meet at the midline of the roof of the mouth. Failure of the two plates to come together correctly and completely leads to cleft palate, which in later life may impose an impediment for normal speech (Figure 15–10).

Gastrulation the period in embryonic development in which the cells of the primary germ layers separate from one another

Primary germ cell layers established during gastrulation and composed of ectoderm, mesoderm, and endoderm

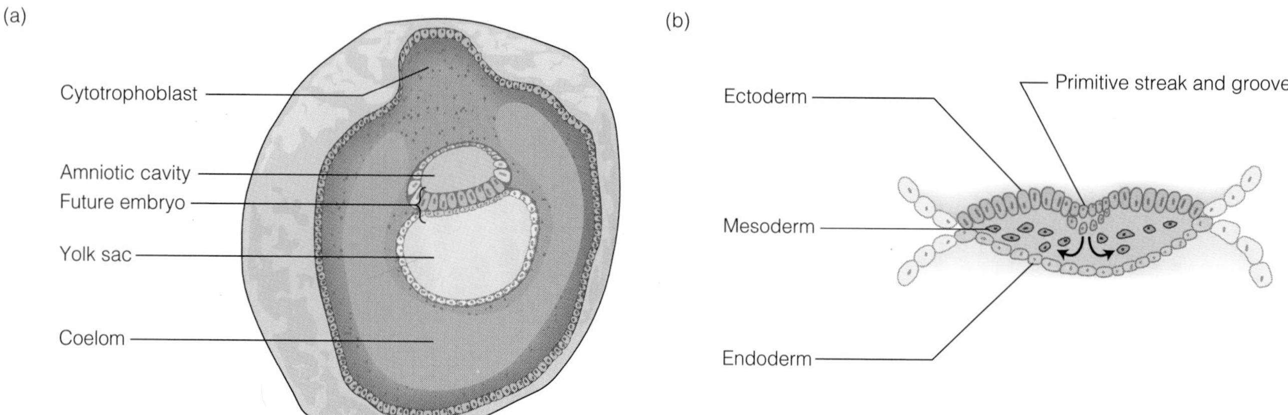

Figure 15–8 Gastrulation and Primary Germ Cell Layers
The human blastocyst in (a) is a two-layered (colored) structure. During gastrulation (b), cells move from the future ectoderm into the space between the ectoderm and endoderm. These cells are the mesoderm.

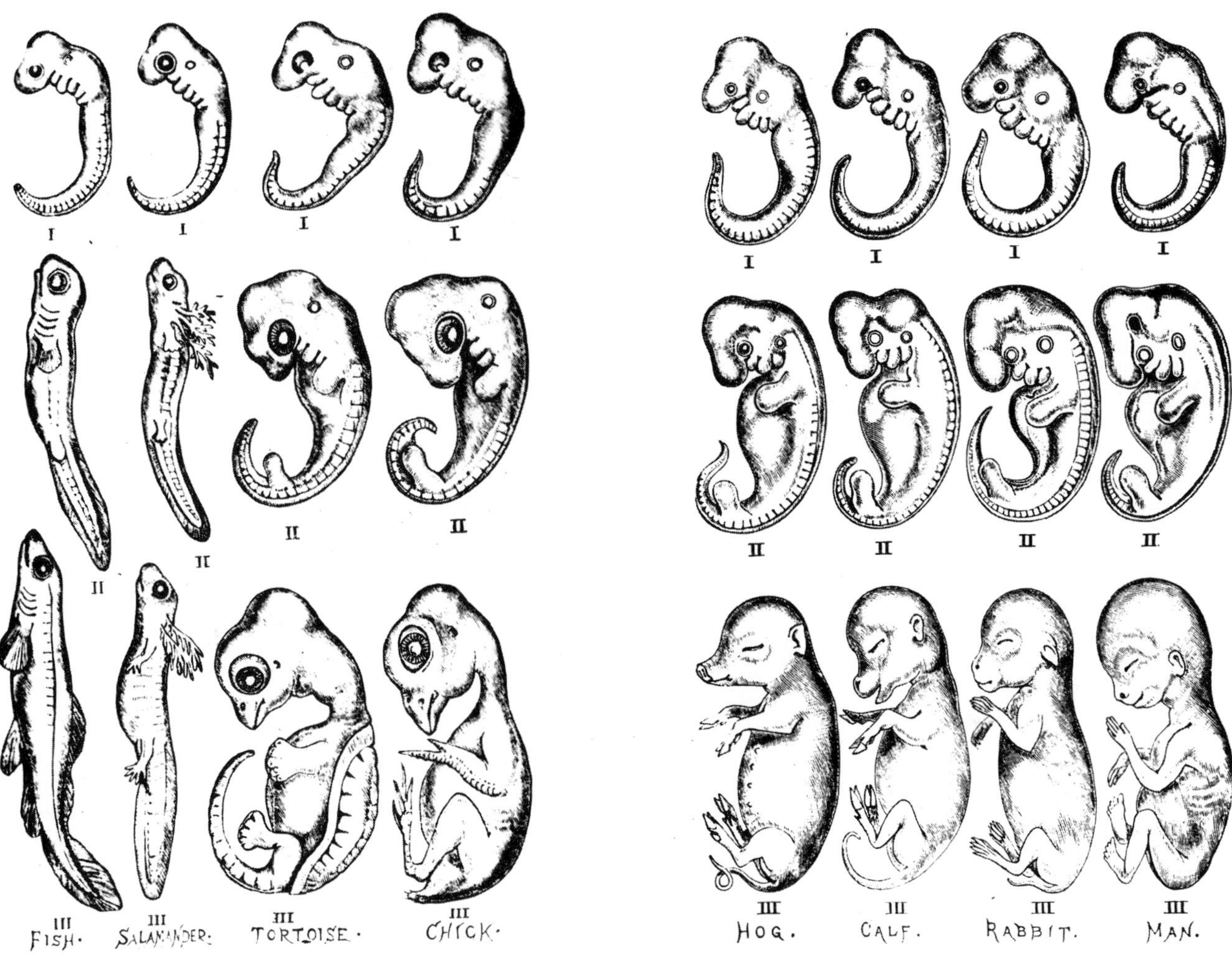

Figure 15–9 Comparison of the Embryogenesis of Selected Animals
The early stages of embryonic development in vertebrate animals are very similar (a). The later stages (b) (c) show divergence of form that characterizes each species. (From E. Haeckle, 1874. Courtesy of the Bodleian Library, Oxford)

Neurulation

Following gastrulation, the development of the primordial nervous system occurs. This process is called **neurulation** and proceeds sequentially from the head region to the tail. Neurulation is an elegant process and one that is essential (Figure 15–11). Prior to neurulation, the movement of cells in the gastrula has largely been as individuals, similar to a crowd of people walking towards the same destination. During neurulation, this changes in such a way that cells tend to retain their connections to one another while still undergoing morphogenesis. The ectoderm on the top surface of the embryo folds over as a sheet of cells and forms a tube. This is called the *neural tube* and it zips itself together right down the middle of the back of the embryo from head to tail.

Neurulation the period of embryonic development in which the neural tube forms and the basic framework of the nervous system is established

Table 15-1 Fates of Embryonic Ectoderm, Mesoderm, and Endoderm

ECTODERM	MESODERM	ENDODERM
Brain	Muscles	Stomach
Spinal cord	Bones of body	Liver
Skin—epidermis	Skin—dermis	Lungs
Bones of head	Kidney	Pancreas
Cranial nerves	Tendons	Intestines
Pigment cells	Ligaments	Esophagus

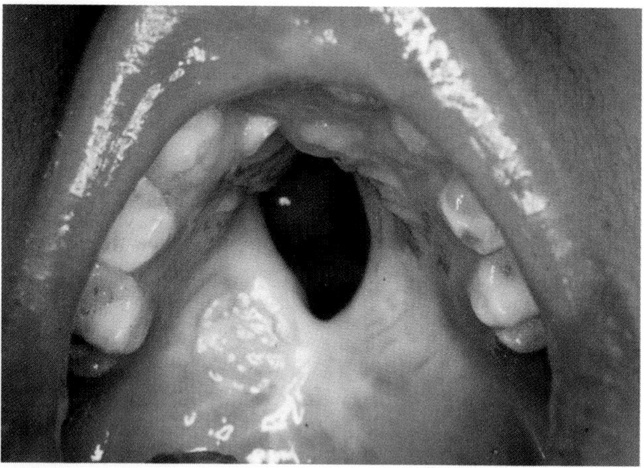

Figure 15-10 Bilateral Symmetry and the Cleft Palate
Many of the structures of the human body form along the midline. If their development is altered, as in the case of the formation of the palate, abnormalities may occur.

Observing the movement of cells during neurulation and the formation of the neural tube, you can appreciate how early variation seen among different types of animals actually begins. During the process of neural folding and zipping, the tube sinks below the surface and is covered by a layer of ectoderm that will become the skin. Once inside the embryo the neural tube is surrounded by mesoderm. The mesoderm begins to organize itself around the newly formed nervous system and eventually participates in the formation of the bones of the cranium and vertebrae that enclose and protect the brain and spinal cord. These developmental processes are completed within days to weeks depending on the species of animal. As mentioned previously the outcome of gastrulation and neurulation are observed to be very similar in each of the tiny, well-organized embryos of humans, fish, frogs, and other vertebrates (Figure 15–9).

Embryonic Period

The embryonic period of development for humans is completed within approximately two months. During this period, the appearance and position of all major tissue types and organs are established (Figure 15–12). This process is called **organogenesis.** The brain and spinal cord are formed, as are the heart and somites (which will become muscle and bones of the back). The future arms and legs

Organogenesis the period of time during embryogenesis in which the rudiments of all organs of the adult body are formed

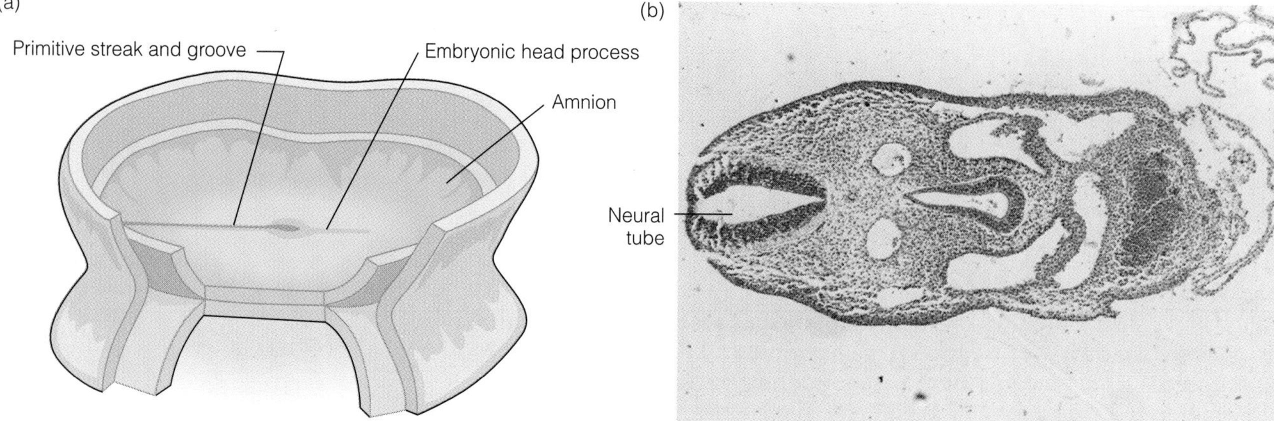

Figure 15-11 Neurulation—The Beginning of the Nervous System
After gastrulation has occurred, the nervous system forms through a process of neurulation. This involves the formation of a groove (a) and, eventually, a simple tube (b). The tube differentiates into a complex structure with the head at one end and the spinal cord extending from it.

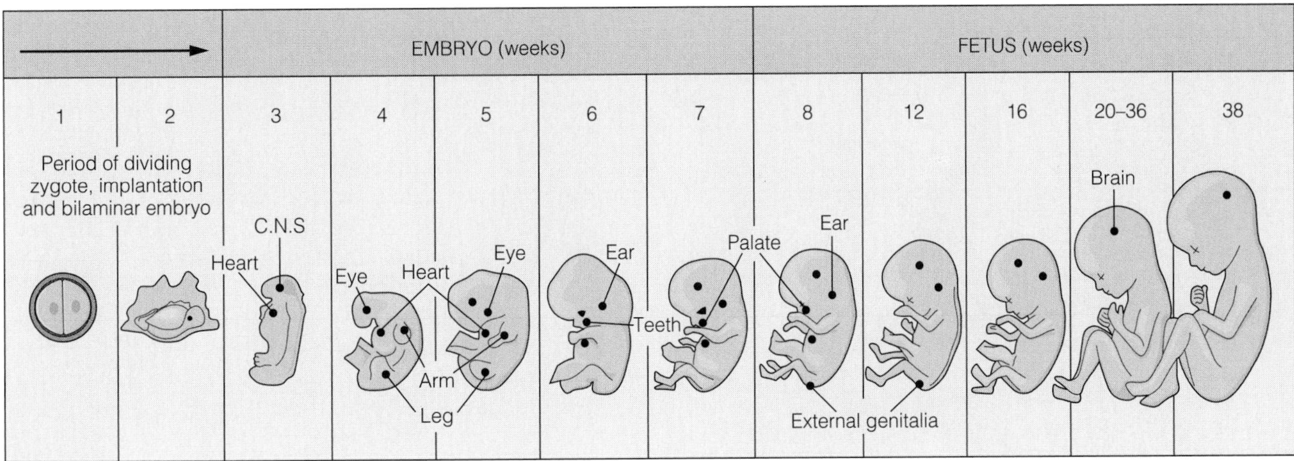

Figure 15–12 Organogenesis
Organogenesis refers to the formation of organs and organ systems in the embryo. Most organs (called organ rudiments) are formed within the first eight weeks of development. Growth shown in this figure is not to scale.

occur as limb buds during this stage and the skeleton, while present, is a cartilaginous model of the bony structure that it will later be. Eyes, ears, lungs, gonads, and liver are all present. These structures are, of course, incompletely formed at this early stage, but what is important to note is that the pattern upon which later growth and maturation will occur is established during the first few weeks of embryogenesis. We pick up the story of further changes in tissues and organs at the point of transition of a human embryo to the next phase of its development, the **fetus.**

FETAL DEVELOPMENT

The beginning of fetal development is considered to be marked by the time at which the cartilaginous skeleton of the embryo begins to ossify into bone. The conversion of cartilage to bone begins to occur at approximately two months of gestation. The fetal stage is characterized by the continued rapid growth and development of the tissues and organs initially formed during embryogenesis (Figure 15–12). The fetus at seven to eight weeks weighs approximately 4 g–5 g and is 2 cm–4 cm in length. For some idea of these dimensions, this is about the length and weight of a small paperclip. The changes that occur during early fetal development give rise to more and more clearly defined human features. We actually begin to look human. Changes in the shape of the head bring the eyes into a position such that they are directed forward, the nose is well formed, limbs lengthen and differentiate, ossification of the skeleton begins, the mouth and digestive tract are formed, and the external genitalia are visible. We get bigger and more active as time goes by and, eventually, we make our presence felt by movements within the uterus.

What Happens to Mom Along the Way?

Endometrium The fetus is not the only one growing during the process of its development. The mother's body is also reshaping itself to accommodate the fetus (Figure 15–13). As has already been discussed, the uterus was prepared for the invasion of the blastocyst during the ovarian cycle. The thick, highly vascularized endometrium initially surrounds the implanting embryo and buries it deep within the walls of the uterus. The trophectoderm of the blastocyst plays a key role in the initial attachment, but it is also important in the formation of the placenta (Figure 15–13).

Placenta The placenta is formed of cells from both the fetus and the mother. This is a very important organ, even though it is only a temporary one, because the fetus is contained within a fluid-filled envelope (the amnionic sac), does not have lungs to breath, nor eyes to see, nor ears to hear, and is certainly incapable of digesting food in any manner familiar to us as adults. Mom has to do it all. Not only does she provide oxygen and nutrients through her own lungs, digestive system, and circulation, she also gets

Fetus the stage of development following embryogenesis and characterized initially by the formation of an ossifying skeleton from a cartilaginous model

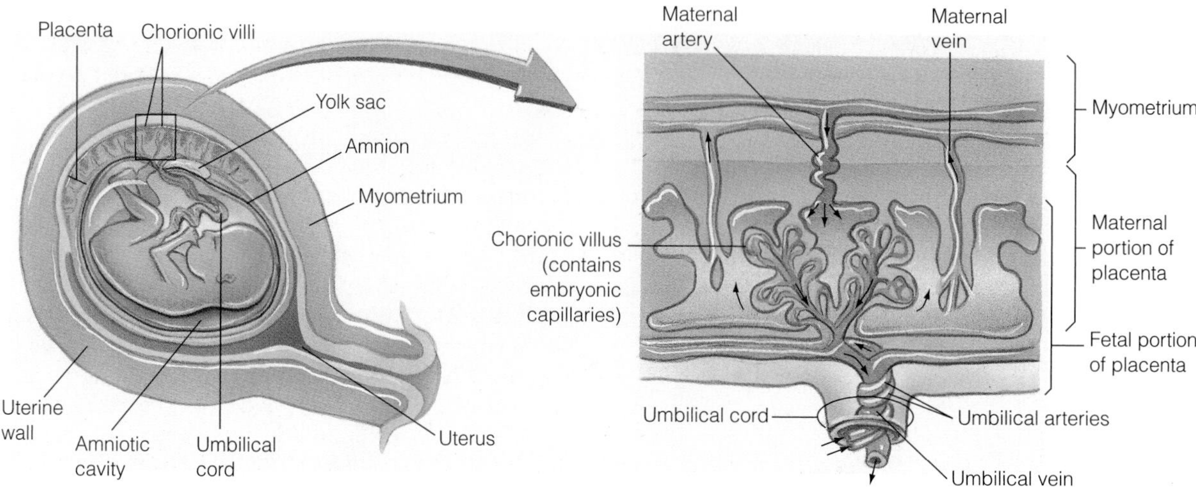

Figure 15–13 The Uterus and Placenta
The embryo (and later the fetus) is surrounded by an amnion and connected to the uterus through umbilical connections to the placenta (a). The placenta is a combination of maternal and embryonic/fetal tissues (b) and supports a great amount of blood flow to bring about exchange of materials between the mother and offspring.

rid of the metabolic wastes that the growing fetus produces and shares her antibodies to protect the fetus from disease—in with the good and out with the bad. All this exchange is mediated by the placenta (Figure 15–14).

Myometrium A mother's body undergoes obvious shape changes during pregnancy (Figure 15–15). She must make room for the baby in other ways as well. The uterus must grow to accommodate the changes in fetal size and to prepare for birth. The muscular layers of the uterus, called the myometrium, grow tremendously during pregnancy (Figure 15–13). The birth of a child requires great effort on the part of the mother and the push that eventually expels the newborn child out of the womb comes from the contraction of the muscles of the uterus. Pound for pound the uterus is probably the strongest muscular organ in the human body, male or female.

Pelvis Other changes in shape during pregnancy involve the loosening and restructuring of the connective tissue of the pelvis. The region of contact between the pubic bones of the pelvic girdle, known as the pubic symphysis, softens and becomes flexible and extendable (Figure 15–15). This provides a capacity for expansion that helps in the widening of the birth canal for passage of the baby during delivery. The largest and most difficult part of the fetus to push through the birth canal is the first to come out, the head.

The mother's body also undergoes dramatic physiological and biochemical changes. Relatively high levels of estrogen and progesterone are maintained throughout pregnancy and suppress FSH and the initiation of the ovarian cycle. The sources of estrogen and progesterone are the placenta, the ovaries, and for the first part of gestation, the corpus luteum. These hormones also aid in maintaining the essentially "double" metabolism of the mother-child union. In addition, these hormones have a stimulatory effect on the development of the mammary glands. This is in anticipation of production of milk for breast feeding. Along with estrogen and progesterone, functional development of the glandular epithelium of the breast also requires the pituitary hormone prolactin, whose concentration in the body increases at the end of pregnancy and rises dramatically right after birth in conjunction with lactation (Table 15–2). The gestation of humans takes an average of 265 days (9 months). This is by no means a fixed schedule. A baby, or babies in the case of multiple births, may come early or arrive late depending on many different factors, including the age and health of the mother, previous pregnancies, prenatal nutrition, disease, and size and developmental state of the offspring.

To summarize some of the important aspects of human embryonic and fetal development covered so far, a fertilized egg undergoes rapid cleavage to form a morula, as it is gently wafted down the Fallopian tube. The cells of the morula undergo a dramatic change in organization, forming a cavity within a blastula, thus, establishing an inside and an outside. This collection of cells continues to organize around a slightly eccentric cavity to become a blastocyst, with trophoectoderm cells on the outside, and an inner cell mass that is destined to form the embryo proper.

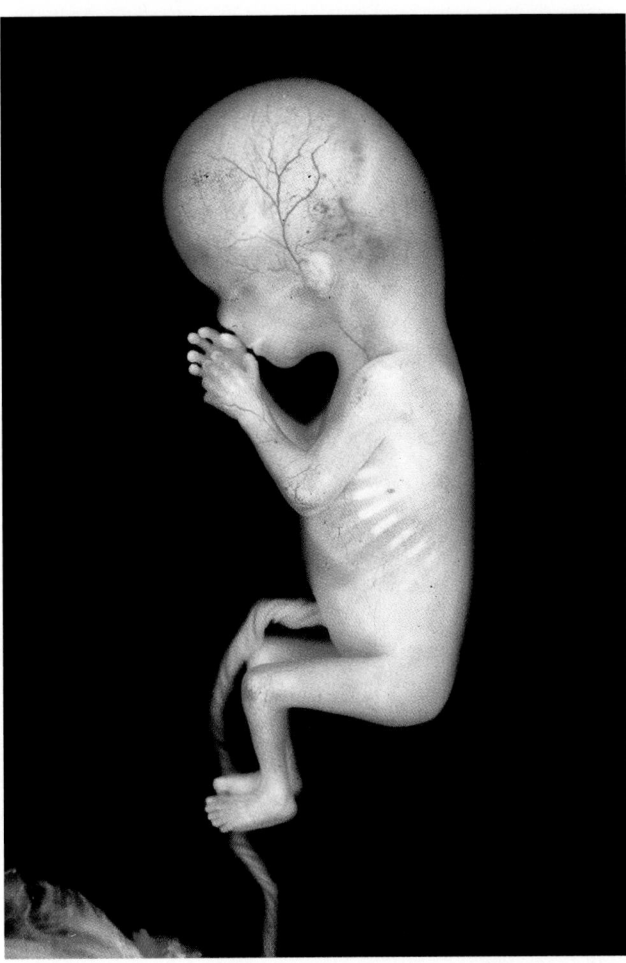

Figure 15–14 The Mother-Child Connection
The fetus is connected to the mother through the placenta.

Upon entering the uterus, the blastocyst quickly adheres to the endometrium and begins embedding itself. The blastocyst implants deeply into the folds of the uterine wall, and, in concert with the mother's tissues, forms a placenta. The human embryo is called a fetus after seven to eight weeks of development, a time at which the skeleton begins to change from cartilage to bone. Growth of the fetus continues at a rapid rate as it fills and stretches the mother's uterus (Figure 15–16).

MATURATION OF THE FETUS AND BIRTH

How much does a fetus actually grow and what problems may arise with changes in the pattern of growth and development? The average size of a child at birth ranges between 48 cm and 52 cm in length and 2.7 kg and 3.6 kg in weight. Even with all the feedback systems and regulation of growth and development, many babies are born considerably outside this range. Smaller babies, or low birth weight (*LBW*) babies, have a lower survival rate than normal-sized babies. Factors affecting LBW include the age of the mother (young teenage mothers have smaller babies), general health (proper nutrition is a necessity for proper fetal growth), and disease (bacterial and viral infections; metabolic disorders, such as diabetes). Larger-than-average sized babies do not suffer the same problems as LBW babies, but mothers who deliver such babies may have more anatomical difficulty than those who deliver average-sized babies. A child that is more physically mature and well developed has a better chance for survival and thriving. However, there are also limits on how developed a child can be at birth, particularly with regard to development of the brain and head.

Birth Canal

Simply stated, to be born in a natural fashion (not surgically removed as in a **cesarean section**), a baby's head must fit through the mother's birth canal (Figure 15–17). The head is the largest structure that has to pass through this canal. If the head is too big, the bones and tissues of the mother's pelvis may not be able to expand enough to accommodate it. As you can imagine, this would be a serious problem. Yet, we know that the more mature and well developed the baby is, the better the chance for its survival. There has to be a balance struck between the size of the mother's skeleton and the size of the baby. But what is it?

To balance out these two factors, matching parental anatomy to baby size, we as humans end up being born in a developmentally immature state and almost totally helpless. All the organ systems of the newborn's body are in this state, but the brain, in particular, is unfinished. Why might this be? If we stayed in the womb longer in order to develop more fully, it would mean that we would grow more and our heads would become larger as brain size increased. A larger-than-average human fetal skull may be too big to pass through the average-sized birth canal, however. Historically, lacking effective surgical procedures, this would have been like signing a mother's death warrant. So, a factor in the natural selection for our small size and immaturity at birth is that it ensures our mother's survival of the birthing event!

Cesarean section a surgical procedure in which the uterus is opened by an incision through the abdominal wall and the baby is removed

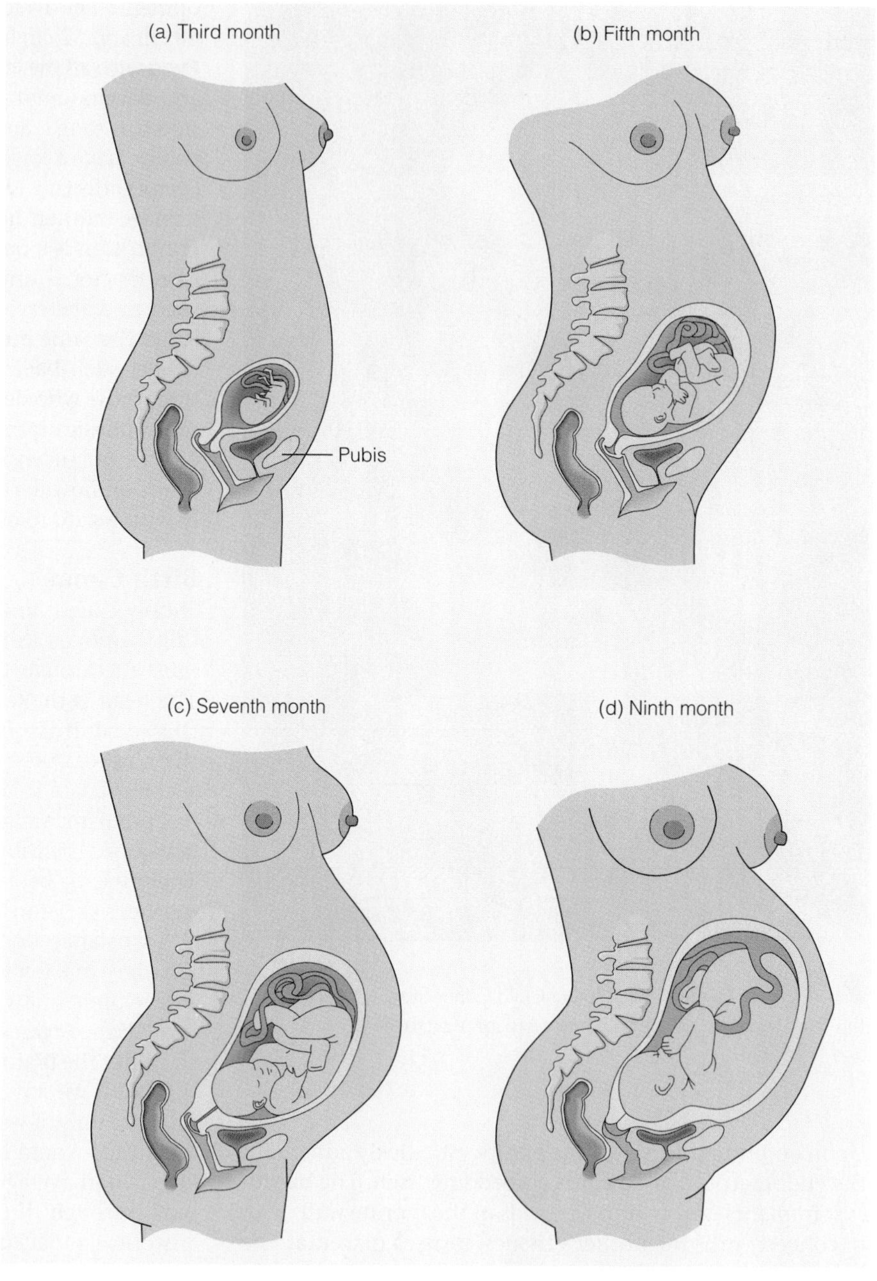

Figure 15–15 Changes in Mom's Shape
A mother's body undergoes changes to accommodate the developing fetus.

Birth

Birth is the final step in the long series of events initiated with coitus and fertilization. Just prior to birth, the program of growth and development has gone as far as it can within the uterus. An average baby is nearly 50 cm long and weighs 3.5 kg. Additional space in the uterus is taken up by the extraembryonic membranes, amniotic fluid, the placenta and the umbilical cord. All the space in the uterus has been used up. There is nowhere to go but out. The uterus is as much as 30 times its regular mass and stretched to its limits. Now, it is prepared to begin the muscular contractions that will push the enclosed baby out into the world. Let us follow the activities that an imaginary expectant American mother may go through on the day of birth.

The Clock Starts on the Day's Events

4:00 A.M. Her water breaks. This is not an unusual occurrence. It involves the breakdown of the chorion and amnion so that the amnionic fluid drains from the uterus. This is a clear indication that the time has arrived (and the baby is about to also!). Call the doctor. The doctor asks if there are any contractions yet; if no, he or she says not to worry and to head to the hospital at morning light. The term *contractions* refers to the rhythmic, periodic muscular activity of the uterus. The uterus is the driving force behind birth. Once contractions begin, they will progress at intervals that occur closer and closer together until delivery.

Table 15–2 Physiological and Anatomical Changes During Pregnancy

METABOLISM AND CELLS	HORMONES	ORGANS
General increase in metabolic rate, appetite, and body temperature Increase in the number of white blood cells Enhanced calcium, phosphate, and iron regulation	Elevated levels of estrogen, progesterone Suppression of FSH and LH Appearance of chorionic gonadotropin Increase in eyrthropoeitin	Increase in kidney function Increase in heart size Softening of pubic symphysis Increase in mass of uterus Increase in skin

6:30 A.M.–9:30 A.M. Once checked into the hospital, the mother-to-be is monitored and made comfortable. Contractions start; first, they are 5 min apart, then 4 min apart, 3 min apart, 2 min apart, 1 min apart. She's getting ever closer to that fateful moment. The doctor checks the cervix for dilation. This means that the cervical region, which normally demarcates the vagina from the uterus temporarily disappears, a process known as **effacement,** and the uterus and vagina are established as a single channel, the birth canal.

9:35 A.M. Dilation is complete and the contractions are very close to one another. The soon-to-be mother, with the support of her husband, decided much earlier in the pregnancy to have natural childbirth. She's learned to control her breathing and focus her energy. Her husband helps in this process. She's on a special gurney, draped in sterile

Effacement the period during labor in which the uterus and vagina establish a temporary birth canal

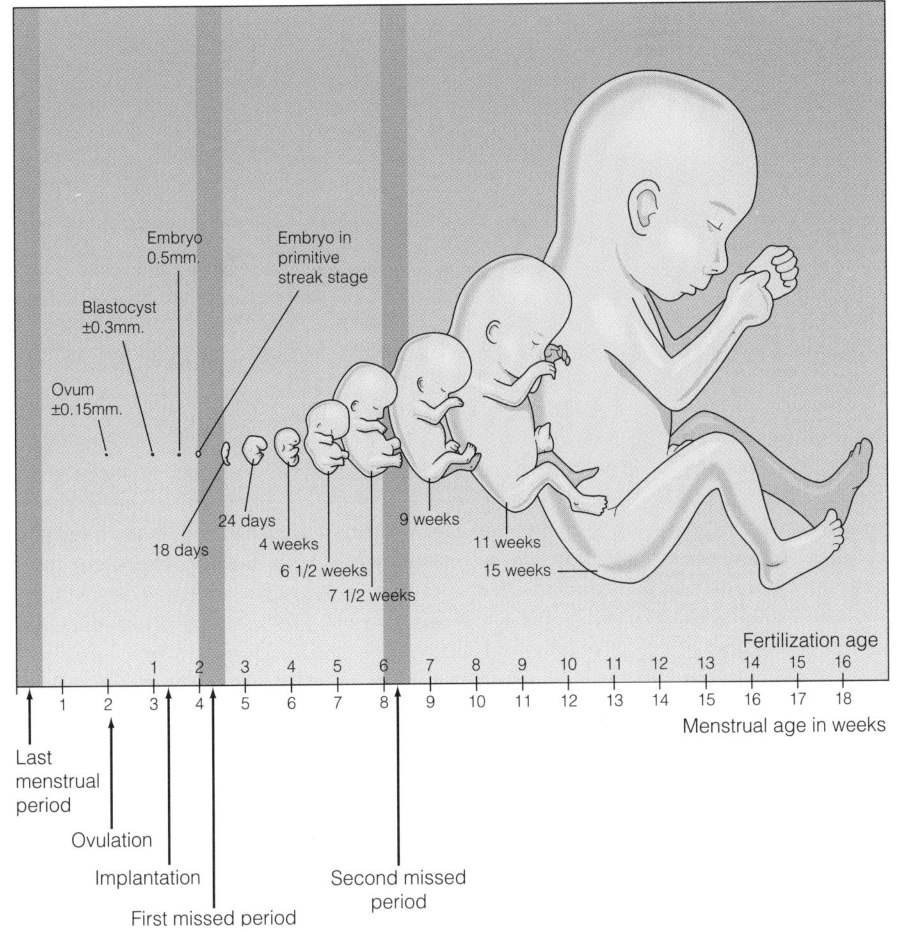

Figure 15–16 Relative Size of Embryo and Fetus During Development

The size and mass of the embryo change enormously during development—from a pinpoint at fertilization to a 3 kg newborn more than a billion times more massive.

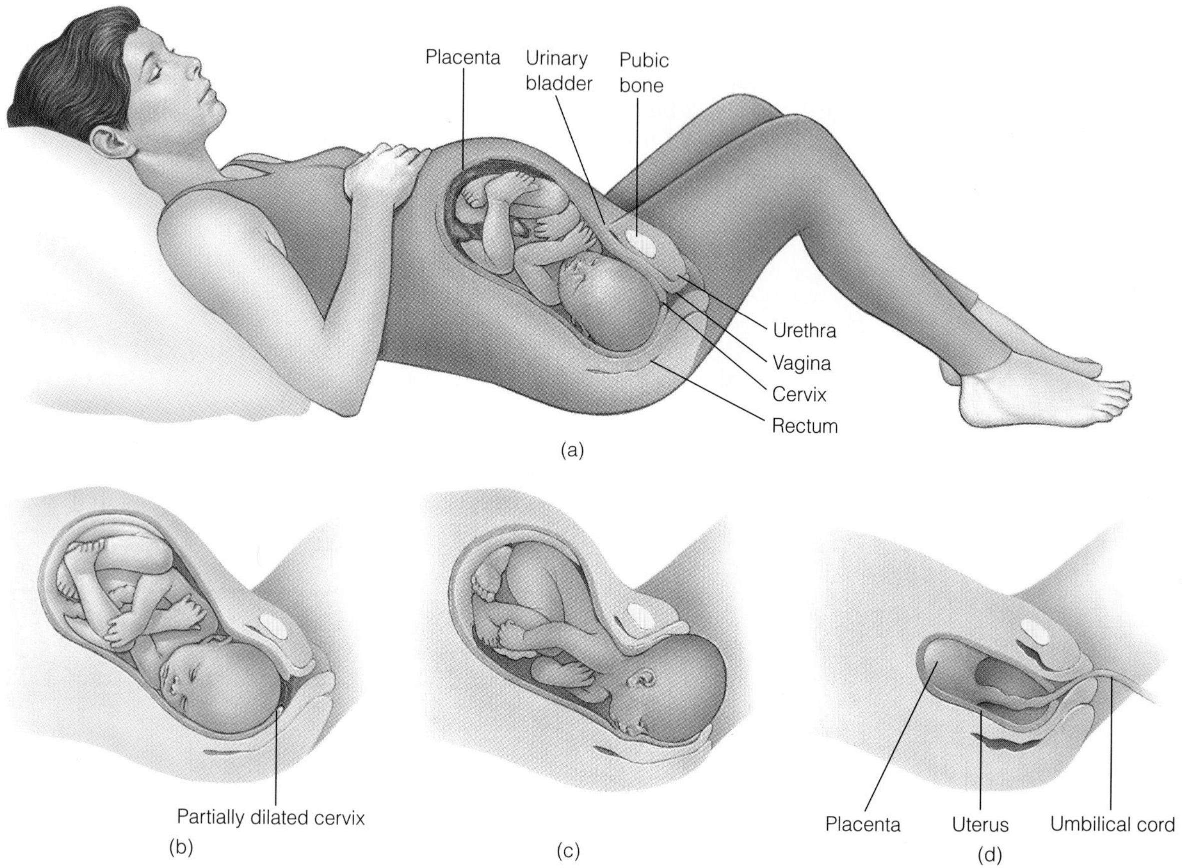

Figure 15–17 Birth and the Birth Canal
During labor, the mother's body changes under the pressure of muscle contractions (a). The cervix dilates (b) and the uterus and vagina form a single birth canal (c). The afterbirth (placenta, umbilicus) is delivered following childbirth.

cloth, has her legs elevated and begins to push the baby out of her body.

10:10 A.M. A quick and successful delivery of the neonate and, shortly thereafter, the placenta. A healthy new baby, an exhausted but exhilarated mother, a proud father, and a satisfied team of doctors, nurses and/or midwives.

Births occur every second of every day somewhere around the world. Not all mothers have the benefits of clean, well-staffed hospitals; many deliver babies perfectly well alone and in primitive settings, but the succession of events is pretty much the same and the result is inevitable. Stripped of the human story line, the steps in the process of birth are simple and straightforward:

1. Contractions of the uterus
2. Dilation and effacement of the cervix and establishment of a temporary birth canal (vagina and uterus)
3. Delivery of the baby; amnionic sac breaks at, if not before, this time
4. Delivery of the placenta (afterbirth).

The control of contractions of the muscles of the uterus is linked to a special hormone released from the posterior lobe of the pituitary gland. This hormone is an oligopeptide known as oxytocin (see Chapter 13). Oxytocin has a profound effect on the muscles of the uterus. The sustained contractions of this muscle are like no other muscular activity in the body. The contractions are involuntary in nature and normally do not stop until the baby and the afterbirth have been delivered. The regulation of this activity falls into the category of a positive feedback loop (Figure 15–18). This is reasonably rare in biological systems because, unlike the more prevalent negative feedback loop, there is no way to stop it once it gets going. It builds on itself, as in pushing someone on a swing. A little push each

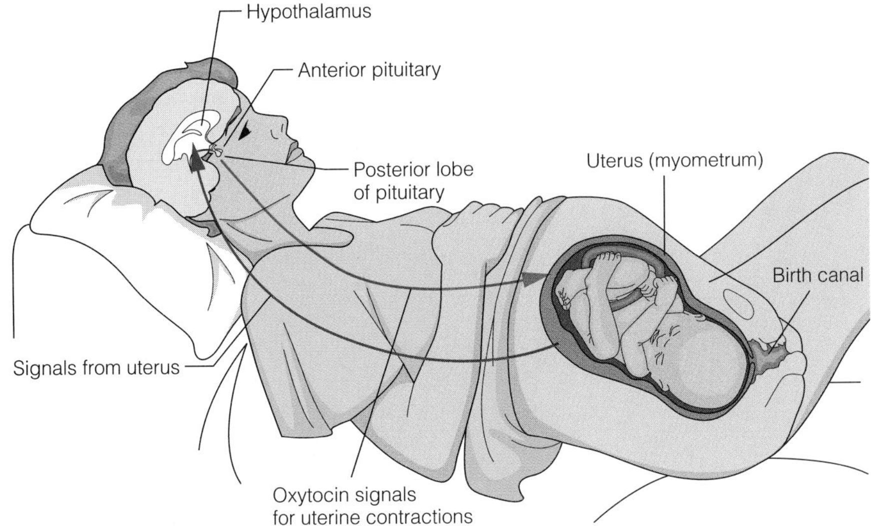

Figure 15–18 Oxytocin and Delivery
The posterior pituitary and the uterus form a positive feedback loop during labor. Oxytocin induces uterine contractions, which feed back to the pituitary to release more oxytocin.

time the swing returns causes it to travel out farther the next time. Eventually, the swing gets out of control and the rider may fall out of the seat or get whiplash; so it is with birth and the push of uterine contractions. They stop only when the job is finished.

Mammary Gland Development and the Let-Down Reflex

Prolactin and Oxytocin As discussed earlier, the mammary glands undergo significant development and maturation during pregnancy under the influence of estrogen, progesterone, and, eventually, prolactin (Figure 15–19). Prolactin brings about the production of mother's milk, which is a rich mixture of proteins, fats, and carbohydrates sufficient to satisfy the nutritional requirements of the newborn baby (Table 15–3). Each breast has eight to ten mammary glands, which drain into a duct that exits the breast through the nipple (Figure 15–19). These glands are poised to produce milk continuously and await only the proper set of signals. **Parturition,** or separation of mother and child at birth, initiates this process in the presence of prolactin. Mother's milk also provides the baby with passive immunity, because it contains antibody molecules of the mother's immune system that are shared with her baby. The stimulation of this milk production begins in earnest after birth as the level of sex steroids drops in the mother's body with the loss of the placenta. The elevated level of prolactin is the key to this synthesis.

A short time after the delivery of the baby, oxytocin is once again called upon to control the stimulation of tiny muscles in what is known as let-down reflex. This response initiates the flow of milk during breast feeding. Oxytocin has its effect on specialized cells that are the "gate-keepers" of the dam holding back the flow of milk from the glands. These cells are called myoepithelial cells, and they surround the mammary glands. The myo prefix suggests a musclelike epithelial cell. Oxytocin causes these cells to contract, just as if they were tiny smooth muscle cells, thus, forcing the stored milk of the glands into the ducts that carry it to the nipple and out of the body (Table 15–3). What causes oxytocin to be released in the first place? Oxytocin is released from the posterior lobe of the pituitary gland in response to stimulation of the area around the nipple by the suckling baby. The sensory nerves from this region influence the hypothalamus, which in turn controls oxytocin release from the pituitary. This stimulus-response cycle continues for months, or even years in some cases, as the child grows and develops. Mother's milk is rich in proteins, fats, and carbohydrates. During the first few days the mammary secretions are watery and rich in protective proteins such as antibodies (Table 15–3). This early form of milk is called colostrum.

The journey of two individuals, bound together anatomically, physiologically, and emotionally from fertilization to birth, has been a fairly long one. The months of internal growth and development of both the mother and child culminated in the emergence of a new and fragile human being into the external world. The female body responded dramatically to the changes induced by the implantation of the tiny blastocyst, which itself journeyed down the reproductive tract to find its place in the uterus. The responses to hormones prepared the female body to allow

Parturition the act of giving birth to a baby

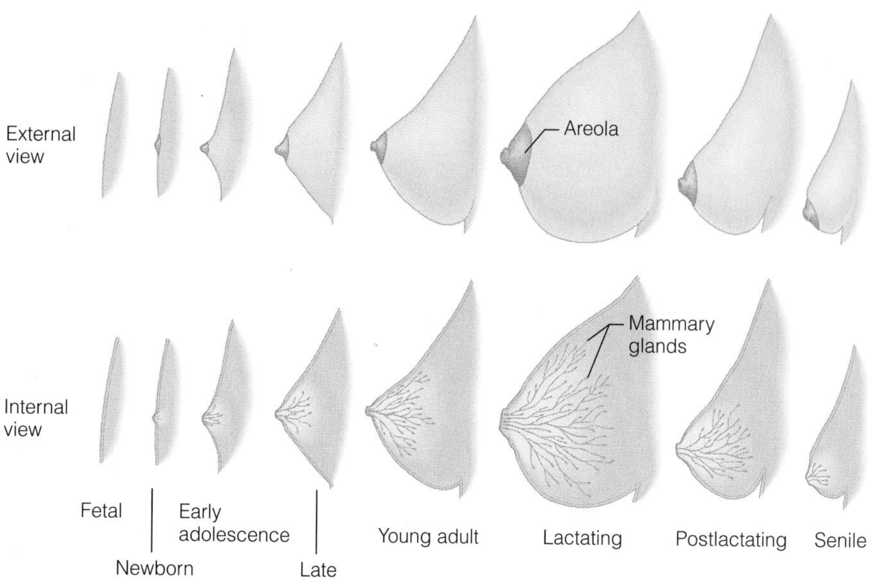

Figure 15–19 Breast Development, Prolactin, and Milk Production
Female breast development involves both increases in size and in degree of glandular differentiation. Prolactin stimulates mammary glands to produce milk shortly after delivery of a baby and maintains lactation during the period in which the newborn is breast-feeding.

for implantation, protection, nutrition, and gas exchange with the temporary passenger she carried. These physiological and anatomical changes also provided for the establishment of mechanisms for delivery of the baby at the appropriate time and for the subsequent nourishment of the newborn. However, the journey of life is just beginning for this child, as it did for each of us, and many more significant changes in growth and development are yet in store.

BEYOND BIRTH

The life *in utero,* discussed earlier in this chapter, is a hidden phase of growth and development. Growth and development outside the mother is better known and composes the rest of the story. In fact, it is clear that for humans nine months of embryonic and fetal growth is not enough to ensure independent long-term survival of an individual immediately after birth. As we emerge from the warm protective womb, where nearly all our physical and biochemical needs are met by our mother's body, we are thrust almost completely helpless into an outside world that is extremely hostile (Figure 15–17). We cannot see well, we can barely move, and we are not able to effectively regulate our body temperature. We need protection from the environment, as well as a very special diet. Our relatively tiny, fragile new bodies are not prepared to survive alone. We need our parents to protect and nurture us not for just a short, transitory period (weeks or months) but for more than a decade. The complex set of behaviors of adults involved in protecting and providing for their young (a family) allows the initial period of immaturity in offspring to be sustained much longer in humans than in any other species on Earth. Nine months of gestation can be viewed as a minimum time in which sufficient growth, development, and differentiation has taken place in the fetus to ensure survival in the outside world.

Mechanisms of Growth

Human growth depends on many factors, both genetic and environmental, and it takes a great deal of time for physical maturation to be fully and satisfactorily attained. As a generalization, growth of a living organism occurs by four basic processes (Figure 15–20):

1. An increase in the number of cells
2. An increase in the size of cells
3. Specialization of cells
4. Programmed cell death.

Table 15–3 Composition of Human Mother's Milk

ONE TO FIVE DAYS POSTPARTUM COLOSTRUM	FIVE DAYS POSTPARTUM MILK
Low in triglyceride fat and sugar (lactose)	High in triglyceride fat
Rich in protein (lactoferrin and immunoglobulins) that protect newborn against infection and stimulate natural immunity	High in lactose High in protein Rich in vitamins and the minerals calcium and phosphate

Note: Breast-feeding stimulates production of oxytocin that may help in returning the mother's uterus to normal (a process called involution).

The first process is called cell proliferation, the second process is called hypertrophy, and the third is called differentiation. The fourth mechanism, programmed cell death, requires a bit more explanation because it is important in eliminating cells that have fully differentiated and served the purposes for which they were intended. For example, the cells of the epidermis that migrate from the proliferative basal layer undergo differentiation, followed by a programmed death, and, eventually, are sloughed off (see Chapter 5). If this did not happen in epidermal cells, as well as in a host of cell types throughout the organism (liver, blood cells, intestinal cells) in a timely and controlled manner, our bodies would accumulate more and more cells, grow to a larger and larger size, and eventually succumb to the crush of the weight we would carry.

All the processes and mechanisms by which we grow at the cellular level are regulated by internally derived factors such as hormones, nutrients, and constraints on metabolism. They are also influenced by external factors such as temperature, amount and quality of food in the diet, hygiene, and diseases.

In terms of overall growth and development, there is no question that the rate of growth is greatest *in utero.* If you imagine the diameter of a fertilized human egg to be that of a ping-pong ball, then, by the time we are born our total growth would bring that diameter to roughly one mile, easily a factor of 50,000. In terms of mass, we increase from egg to fetus by more than a billion times. By comparison, during that part of growing up with which we are most familiar (that is, after birth), we may increase in mass 15–30 times and in length by a factor of only 3–4. However, as important as embryonic development is, most of our lives (99%, if one lives to be 75 years old) and certainly all that we remember is spent outside the womb. It is this growing up and maturing, this expansion from a newborn baby to adult, that we focus on in the last part of this chapter.

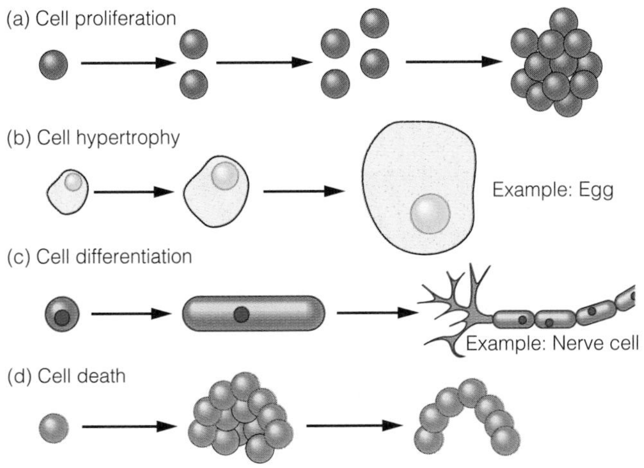

(a) Cell proliferation

(b) Cell hypertrophy

Example: Egg

(c) Cell differentiation

Example: Nerve cell

(d) Cell death

Figure 15-20 Mechanisms of Growth and Change
The principal mechanisms of growth are (a) cell proliferation, (b) hypertrophy (or increase in cell size), and (c) cell differentiation (change in cell characteristics). Examples of (b) and (c) are egg development and nerve development, respectively. In addition, programmed cell death plays a role in growth by selectively eliminating cells in specific locations. The formation of fingers occurs by the death of cells between the digits.

Genes and the Environment Human growth and development occur in stages, with rapid initial growth and then a slowing as we reach maturity. We never stop growing during our lifetimes. Males and females grow at different rates during childhood, with females generally maturing somewhat more rapidly than males, but males catch up later (Figure 15–21). Males are on average larger than females and have different proportions of muscle and fatty tissues. The interplay of genetics and a variety of factors in the environment are the major influences on the extent of this growth. If we were to plot human growth on a graph, it would take the form of an S-shaped curve, with lags in growth at some stages, rapid growth at others, and a general plateau as we reach maturity (Figure 15–21).

Within the context of the overall growth of the human body, there are periods of great importance to the growth of particular organs and organ systems. For example, the human brain finishes growing by early childhood (Figure 15–22). The heart grows rapidly in the first few years of life, slows for a time, and then grows rapidly as we attain adulthood. The growth of the body is reflected in the growth of its parts. A good diet with protein-rich nutrients is particularly important early in life, when organs such as the brain are growing rapidly. The absence of proper nutrition during

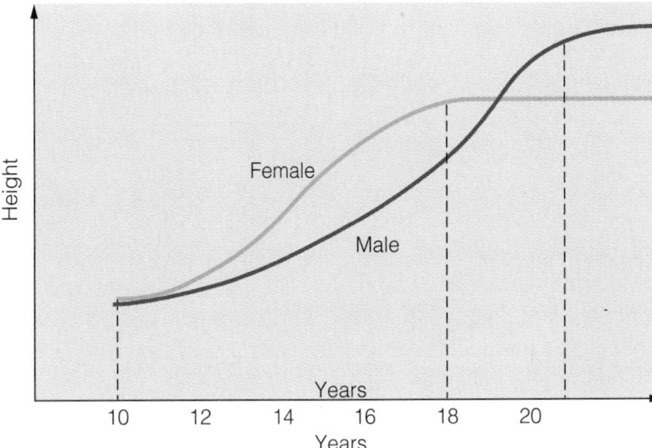

Figure 15-21 Relative Growth of Females and Males
Females and males grow at different rates at different periods of development. Females grow more rapidly following the onset of puberty, but their growth levels off earlier than males.

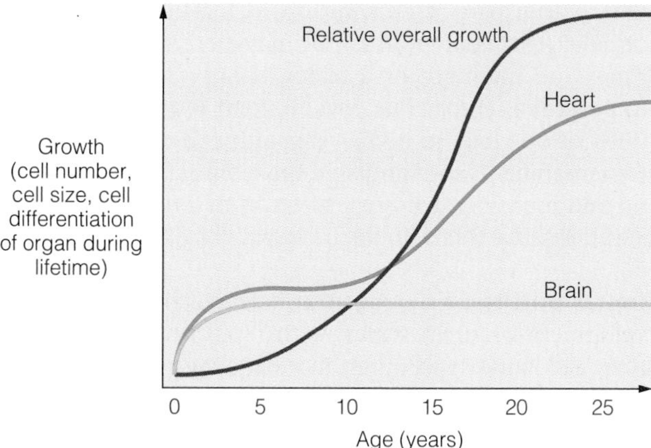

Figure 15–22 Human Growth Curve
The growth of the human body occurs at different rates at different times. The brain grows early to a maximum cell number. In contrast the heart undergoes early (5 to 6 years) and late (10 to 20 years) growth spurts. The overall rate is represented by an S-shaped curve.

this critical growth period can have severe and life-long consequences, as is observed in children growing up under conditions of famine (Figure 15–23). Sufficient intake of protein and total calories are needed to sustain the rapid growth of children. This is usually provided by mother's milk but, in famine, that milk may be inadequate or the child may be weaned too early. In the face of low protein in the diet, children suffer from a condition known as *kwashiorkor.* When a child's diet is both low in protein and low in total calories, a condition known as *marasmus* develops. Symptoms of both these conditions are bloated bodies, edema, lethargy, mental retardation, and diarrhea. If such conditions are sustained, death is common within the first year or two of life.

The initial growth of the body is predominantly under genetic control. This is particularly true during embryogenesis and fetal growth but extends into the neonatal period as well. These stages represent predominantly the unfolding of the programmed plan for growth built into the human genome. Later in life, many more factors in the environment are brought to bear on the growth of the body including availability or consumption of nutrients, exercise, effects of childhood diseases, climate, lifestyle, and timing of the onset of puberty. The overall growth of an individual (that is, rate and extent) is a balance of inherent genetic factors and the action of the environment on them.

AGING AND SENESCENCE

The negative aspect to the general pattern of positive growth and maturation is *aging* and **senescence.** As a human body ages, its cells and tissues accumulate irreparable damage and metabolic waste and become less and less capable of maintaining functional integrity. Aging and senescence are both defined as growing old and losing function. Aging is the passage of time for an individual in relation to the commencement of function, that is, birth. We all grow old. We cannot arrest time nor halt irreversible

Senescence aging but with the connotation of loss of function or deterioration of the body

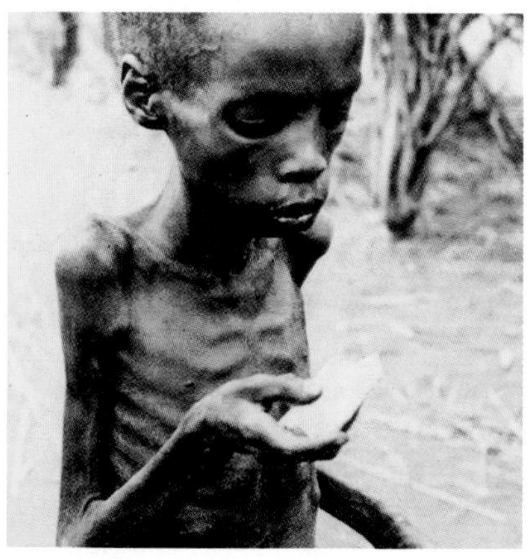

Figure 15–23 Children of Famine
(a) Marasmus and its ultimate consequence (b). Kwashiorkor in a little boy suffering from protein deficiency (c).

change. Senescence also means to grow old but has the added connotation of the breakdown and/or deterioration of the body's cells, tissues, and organs. This may come about through the accumulation of detrimental metabolic products, inefficiency of repair of damaged tissues, or genetically programmed changes in cell viability, all of which combine to negatively affect body functions. It is senescence of the human body, not aging alone, to which we eventually succumb.

The chance for things to go wrong (or at least not to be repaired quickly) in cell and tissue functions increases with age. The process(es) of growth that occurred during development and brought us to maturity after birth are modified once we have attained it. The capacity to maintain the body's tissue, part of the overall process of homeostasis, becomes less and less efficient with age. In addition, some of the tissues that grow and develop during childhood cease growing altogether as they mature. For example, nerve cells, which proliferated so dramatically during fetal and neonatal development, actually die at a greater rate than the rate at which they are replaced starting at five or six years of age (Figure 15–22). We end up with fewer neurons than when we started. Another example is heart and skeletal striated muscle. When cells composing these tissues die or are destroyed by accident or disease, they are replaced with general connective tissue, rather than new muscle cells. This is because muscle cells and fibers cannot regenerate themselves in adults. How do human cells and tissues get to the point where they cannot repair or replace themselves? How does this occur in other animals? There are a number of theories that attempt to account for the mechanisms of aging and senescence.

Life Expectancy and Life Span

During the last century, there has been a dramatic increase in human longevity, the length of time we live, particularly in technologically advanced societies. This does not mean that people in the past did not live to old age. For example consider the following two populations:

Population 1	Population 2
5 people die at birth	10 people die at age 60
5 people die at age 100	
Average life expectancy = 50	Average life expectancy = 60

It is clear from this example that population 1 has a significant number of very old people but also a significant number who died very young. Population 2 lived uniformly longer, but did not attain very old age. In dealing with life expectancy we must be careful to consider all the statistics. With this in mind let us consider the last 100 years in the United States. In the 1890s the average life expectancy was in the range of 40 years. Life expectancies for a human being in the late 1990s range up to almost 80 years. The average for males is approximately 71 years and that of females is 79 years. What is the basis for this substantial increase in life expectancy? To what factors can we attribute a doubling of our lifetimes in less than 100 years?

There are several reasons why we live longer than ever before in human history. Among them are advances in medicine and treatment of diseases, particularly those advances that have influenced child mortality. A principal role must be given to the discovery and development of antibiotics. Without them, even what are minor infections by today's standards, would be lethal, as they often were even 50 years ago. Control and eradication of disease agents and their vectors (for example, malarial parasites and their host mosquitos) have been instrumental in limiting rampant spread of vector-borne diseases. Better nutrition, both with respect to the quantity of food available for consumption and the quality of the food in terms of nutrient content, also figures into the equation of extending our life expectancy. New technologies of all kinds add to the formula as well. For example, better and cheaper construction materials for homes means better protection against the outside environment; better plumbing, sewage control, and reclamation systems eliminate, or neutralize, hazardous waste; and water purification systems ensure safe water supplies. Another important factor is the development of new attitudes and awareness about personal hygiene and care for ourselves. The twentieth century has seen so many advances in medical, technological, and health-related areas that it is difficult to ascribe to any one of them a specific percentage of the overall effect.

The two terms that we have discussed here are *life span* and *life expectancy.* These terms should be clearly differentiated from one another. Life span is the number of years that any one of us actually lives. Life expectancy is the number of years that any individual can expect to live based on the average of all the life spans of a population at a given time (historically and in the present). This does not mean that any individual is guaranteed to live up to that expectation. Because members of *Homo sapiens* in some societies have doubled their life expectancy in the last century, how much more might be expected? It is impossible to set absolute limits, of course, but, if one assumes that the oldest people on Earth today represent a target age for life expectancy in the future, say by the 2090s, an individual born at that time might well expect to live 100 years.

To return to the last century, there is an interesting phenomenon that goes hand in hand with human longevity of the times. Because men and women of the 1800s had life expectancies of around 40 years, they did not live long enough in most cases to develop many of the health problems that are prevalent in older members of our population today. Certainly, the main causes of death in the United States were not the same then as they are today. Today, we die most frequently from heart disease and cancer and much less often from infectious diseases (at least in developed countries). Viral and bacterial infections, parasites, lack of proper nutrition, serious accidents (for which there was inadequate medical technology and expertise), and child mortality were considerably more important factors in determining longevity of individuals in the population then than now. Some physiological processes that we commonly recognize today were virtually unknown a century and more ago. **Menopause** is one such process.

On average, women who grew up in the last century and the early part of this century did not live long enough to undergo all the changes we associate with menopause. Menopausal changes usually begin for a woman in her mid-forties and can go on for several years. The profound changes in the reproductive cycles of women have long-term and persistent effects on their physiological, emotional, and behavioral well-being. One of the most serious challenges in dealing with aging and senescence is the emotional component that enters the equation as lifetime and life expectancy increase. In this sense we must try to ensure that the quality of life stays high as the length of life is extended.

Theories of Aging and The End Game

As most people who play chess will tell you, chess is a very complex and demanding game. The game is described as having different stages and different strategies for beginning, middle, and end. Chess is a case of a game imitating life. Life is infinitely more complex than chess, but life has been presented in this chapter as having a beginning, a middle, and an end. It is now time to deal with some aspects of the end game of life. Eventually the processes of senescence, as an inevitable checkmate, catches up to all of us. Senescence, however is not a simple or straightforward phenomenon. It represents many different types of structural and functional changes in the cells and tissues of the body with many different causes. There are a number of hypotheses and theories of aging that attempt to account for the timing and extent of degradative changes in the human body and culminate in the cessation of life.

The Internal Clock One hypothesis suggests that an internal molecular clock is built into the cells of our body. There are two main possibilities. One is an internal timing mechanism for initiating or programming cell death. The other is a specific signal from somewhere in the body that triggers the onset of cell death. As discussed in conjunction with the description of the brain, there appears to be a time in postnatal development during which proliferation of neurons ceases and loss of nerve cells is greater than gain. This starts quite early in human development, perhaps in the fifth or sixth year of life. The neurons that are produced in prodigious numbers in the fetus and young child begin dying in relatively small numbers but are not replaced by new neurons. There is little or no reserve of undifferentiated neurons, and mature neurons do not replicate. This inability to replicate and divide reflects the fact that the differentiation and morphogenesis of nerve cells is intimately associated with their functional organization as a network. Cell division requires significant changes in cell shape, whereas the organization of the nerve network, for example, the brain, requires stable, long-range structural associations.

The number of neurons that die each day in our adult life is estimated to be relatively few, perhaps in the thousands. Because there are tens of billions of neurons in the brain, this amounts to a fraction of a percent loss each year (0.001% per year). Are there "death" genes that somehow inactivate essential metabolic pathways or cellular process such as a cell's ability to divide? There is growing evidence that such genes exist and affect processes such as programmed cell death. However, their relationship to longevity in humans is not yet known.

Mutations The cells of the body are continuously exposed to environmental agents that can cause mutations (Table 15–4). Mutations are permanent, inherited changes in the genetic information coded by the DNA of chromosomes. When mutations affect individual cells, they are called *somatic mutations*. These alterations occur randomly in the DNA molecules of a cell's chromosomes and some of these changes are expected to be more significant than oth-

Table 15–4 Mutagens and Mutation

MUTAGEN	ALTERATIONS LEADING MUTATIONS
Radiation	
Ultraviolet (UV)	Alter chemical structure of DNA
X-rays	Break the strands of DNA
Chemicals	
Heterocyclic hydrocarbons (anthracenes, benzopyrenes)	Alter DNA; convert nucleotides and DNA sequences in genes
Viruses	
RNA type (retrovirus)	Alter gene regulation; act indirectly in humans
DNA type (Epstein-Barr and herpes)	Lead to transformation of cells and development of cancer

Menopause a time in the life of aging females in which ovulation and menstruation cease (onset from 35 years old to 50 years old)

ers. Given sufficient time (years or decades), changes affecting one or more genes in a cell could alter the ability of that cell or its progeny to properly perform their functions. The cumulative effects of such changes in the vast population of cells of the body could be a factor that helps seal our doom. For example, cells have molecular repair mechanisms to correct chemical alterations in DNA (the source of mutations). Imagine a mutation that affects a cell's ability to repair DNA damage. This reduced capacity could contribute significantly to limiting cell survival. The cumulative effect of this type of damage to many cells in the body over a lifetime may eventually manifest itself in the decline and death of the organism.

The Immune System Connection The immune system is attuned to the body in a very special way. The molecules produced by cells that compose the immune system are capable of detecting differences between that which is self and that which is non-self, that is, foreign materials that enter the body. This line of defense protects us against all kinds of disease agents (viruses and bacteria) and toxins that could otherwise overwhelm us. The immunological theory of senescence suggests that at some point in an individual's life the immune system becomes less efficient and so offers less protection from invading organisms and the damage they do. Even more devastating to a body, the immune system may go awry and begin to attack the cells and tissues of the host (review Chapter 12). This autoimmune response can inflict a level of internal self-destruction that leads to serious disability and death. There are a number of well-known diseases in which the body's immune system aggressively attacks and destroys the body's own tissues. Rheumatoid arthritis, multiple sclerosis, glomerular nephritis, myasthenia gravis, and scleroderma all have autoimmune components to them. Even though such lethal abnormal functions of the immune system are relatively rare, the potential is there in all of us. Sublethal, or low-level, effects of damage brought about by the immune system may add to the burden of repair that is normally needed to maintain homeostasis.

Bridges Between Molecules

The molecules found outside of cells in the extracellular matrix are extremely important to the structure and function of the tissues and organs of the body. Bridges between such extracellular molecules provide the basis for the theory of cross-linking, which suggests that, if and when extracellular molecules are abnormally linked together, they may become inactive, or an impediment to other functions, or useless in the jobs they should routinely perform. The buildup of waste products in the body may promote the formation of these abnormal bridges or cross-linking alterations. One important class of these metabolites is known as *free radicals*. These materials are highly chemically reactive and wreak havoc with molecular structure and function. They may directly participate in cross-linking reactions. A metaphor that is commonly used to describe desperate actions is "burning our bridges behind us." The idea that building bridges may lead to our demise is, therefore, somewhat ironic.

Some ordinarily important and structurally benign molecules, when increased in abundance, can affect molecular cross-linking. Glucose can react with collagen molecules to bring about covalent cross-links. These linkages in turn bring about changes in the texture and quality of the connective tissues of the body. Collagen is one of the most prevalent and important extracellular molecules found in animals. Alterations in its structure can directly and indirectly affect circulation of blood, muscle movement and strength, and elasticity of connective tissues. Diabetics, whose glucose metabolism is altered, have a very high level of glucose in their blood and body tissues. They also have a higher proportion of altered collagen than nondiabetics in a comparable age group. The cross-linking and senescence of the body are correlated in this case. In juvenile diabetes, these damaging changes begin early in life. The life expectancy of severe diabetics is greatly reduced.

There are many other theories of aging and senescence. This is a very active and important area of research. However, for all the work being done, there is still very little known about the specific details of how aging and senescence occur.

DEATH

The end of life is death. The critically important organ systems (that is, the brain, lungs, heart) of the body shut down, and cells and tissues cease to function. The attributes that we associate with living organisms disappear. But what is death? How is it determined that a body has died? What criteria do we use to define this end state? To attempt to answer these questions, we need not only a biological approach but a legal and ethical one as well. The legal definition of death has changed in recent years. As medicine has advanced and scientists have learned more about the physiology of the human body, its repair, and maintenance, we as a society have had to confront our own success. Death is no longer simply defined as when the heart stops beating or when an individual stops breathing. The apparent moment of our passing may well be put off. Many people whose hearts have stopped and whose breathing has ceased have survived these interruptions and gone on to live decades more of rich, rewarding lives. So how do we define death? Death in the modern, medically oriented Western world is determined by three basic criteria:

1. The absence of brainwaves. This phenomenon is measured by an electroencephalogram (EEG). Failure of the brain to generate waves is called **brain death.**
2. Lack of breathing and heartbeat. The absence of both these activities represents loss of spontaneous muscular movement. Specifically, this refers to lack of contraction of the diaphragm to expand the lungs and lack of contraction of the cardiac muscle to pump the blood.
3. No reflex responses to external stimuli. Fixed and dilated pupils are a prime example. If the pupil fails to contract in response to light directed upon it, the autonomic nervous system is no longer functioning: a clear sign of death.

Fulfilling any one of these criteria of death may not be sufficient. Some conditions warrant continued medical intervention even though death may be apparent. Certainly, the cessation of breathing and heartbeat in an otherwise healthy individual have proven to be an insufficient reason to curtail emergency treatment. In some rare, but important, instances individuals who have fallen through the thin ice of winter-frozen lakes and whose bodies have not been recovered for 20 or 30 minutes have been resuscitated. How does this happen? Apparently, the intense cold of the water slows down the metabolic reactions of the body and prevents all the oxygen in the brain from being used up rapidly. This establishes a temporary state approaching suspended animation, which persists for much longer than would be expected. Ordinarily, deprivation of oxygen to the brain for even a few minutes leads to massive cell death and irreparable damage. Overdoses of some classes of drugs, such as barbiturates, can depress the EEG so that it appears flat. In many cases, individuals in this "flatliner" condition survive for hours and ultimately recover. As with most things in life, it would appear that in death also there are exceptions to the rules. There is much yet to be learned about the limits of both life and death.

Brain death one of the criteria of death in which normal electrical activity of the brain ceases

Summary

This chapter has provided a travelogue for a journey from the instant of conception to the moment of death—a lifetime. The growth and development of the cells, tissues, and organs of the human body depend on basic processes of cell proliferation, cell growth, and cell differentiation and specialization. This allows the cells of a newly fertilized zygote to initially establish themselves and their organization, then flourish, and ultimately to maintain the structure and function of tissues and organs in adults. The maintenance of the body is the province of a complex set of mechanisms known as homeostasis. Many of the mechanisms of homeostasis are known and at least partially understood. Scientists have provided the means to help the human body extend its longevity. These advances were brought about through changes in medical practices, discovery and use of antibiotics, institution of environmental controls (elimination of many disease vectors, water quality regulations, and sewage processing), promotion of awareness of the importance of personal hygiene and establishment of criteria for nutrition and diet. Today, we can expect (and usually demand!) to live longer and healthier lives than at any time in history.

Questions for Critical Inquiry

1. What have been some of the social, political, and environmental ramifications of the increased life expectancy of humans?
2. What conditions reflect the lack of sufficient protein and/or total calories in the diet of a growing child? What critical period(s) in development is affected by this lack? What is the fate of these individuals?
3. How has the definition of death changed in the last few decades? What are the personal, family, and social implications of these changes?

Questions of Facts and Figures

4. What are the primary cell layers? What tissues and organs do cells from these layers give rise to?
5. What are some of the major changes that take place in the female body during pregnancy?

6. How is fertilization of an egg by more than one sperm prevented?
7. What is the birth canal? How is it formed?
8. Where does cleavage usually begin? Why is it important that it start in this region?
9. What are the basic cellular mechanisms of growth?
10. Name and explain the theories of aging.
11. What is the function of the placenta?
12. What defines the difference between the human embryo and fetus? When does this change occur?

References and Further Readings

Alberts, B., Bray, D., Lewis, J., Roff, M., Roberts, K., and Watson, J. D. (1994). *Molecular Biology of the Cell,* 3rd ed. New York: Garland Publishing.

Balinsky, B. I. (1970). *Introduction to Embryology,* 3rd ed. Philadelphia: W. B. Saunders.

Cummings, M. R. (1996). *Biology: Science and Life.* St. Paul, MN: West Publishing.

Oppenheimer, S. B., and Lefevre, G. (1980). *Introduction to Embryonic Development,* 2nd ed. Boston: Allyn and Bacon.

Sherwood, L. (1993). *Human Physiology,* 2nd ed. St. Paul, MN: West Publishing.

Stalheim-Smith, A., and Fitch, G. K. (1993). *Understanding Human Anatomy and Physiology.* St. Paul, MN: West Publishing.

CHAPTER 16

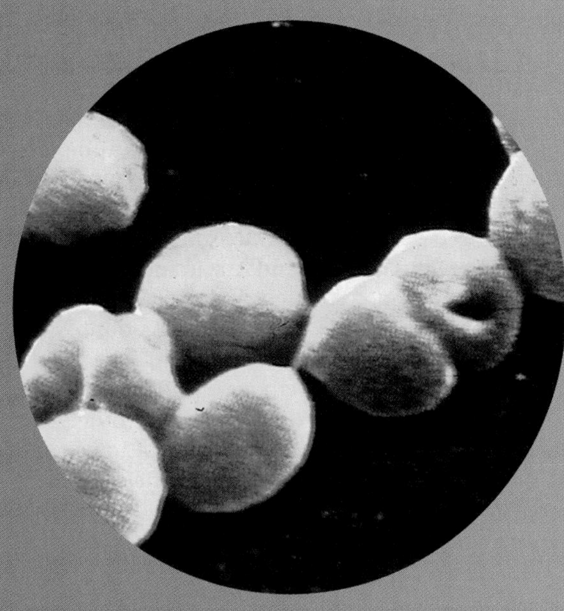

Sexually Transmitted Diseases

INTRODUCTION

It is not as if sexual transmission of disease popped up in the last few decades. Human diseases transmitted by sexual interactions have probably always existed in human (and prehuman) populations. They are certainly described in the historical records of the millennia. Individuals in many well-known civilizations of antiquity are described in the writings of those times as having symptoms of diseases that fit our modern descriptions of sexually transmitted diseases. For example, an outbreak of herpes may have been responsible for the banning of public kissing in ancient Rome. Europe was affected by an epidemic of syphilis during the early sixteenth century, possibly brought back from the new world by Columbus and his crew. Columbus himself died of syphilis while in prison. Although he may have been imprisoned because he failed to bring enough gold back from the newly rediscovered Americas to satisfy the Spanish crown, perhaps he brought back another "treasure" much more costly in terms of human health. In this chapter, some of the principal agents of sexually transmitted diseases (STDs), old and new, will be described. In addition, the effects of these diseases on individuals who are infected with them and the ramifications of these diseases in the health of our society will be explored.

GLOBAL IMPACT OF SEXUAL REPRODUCTION

We understand better today than at any time in our history the physiological implications and consequences of our sexual activities. Sexual interactions are a perfectly natural and necessary part of human biology—no sex, no babies, no next generation. Knowledge of the sexual physiology of males and females works in our favor because with understanding comes a certain degree of control of these organ systems. By being able to control our fertility, for example, we are provided a measure of choice in planning a family. Uncontrolled, however, our reproductive potential is enormous and, in some ways, disruptive of the quality of our lives. In some areas of the world, particularly in the developing nations, this negative potential has already been realized (Figure 16–1). The resulting overpopulation of some of these regions has led to serious social, economic, and health problems (Figure 16–1b). The most dire consequences of overpopulation are those coupled with agricultural disasters (for example, prolonged drought). The wide-spread famines and hunger in many regions of Africa and Asia in the last decades are testimony to the tenuous existence of millions of people (Figure 16–2). These problems in world population, widespread famine, economic hardship and their relationship to sex and reproductive potential perhaps do not strike home for most of us. They are things that happen to others; it's "their" problem. However, with the recent appearance and worldwide spread of the human immunodeficiency virus (HIV), there is a new awareness of a dark side to unrestrained

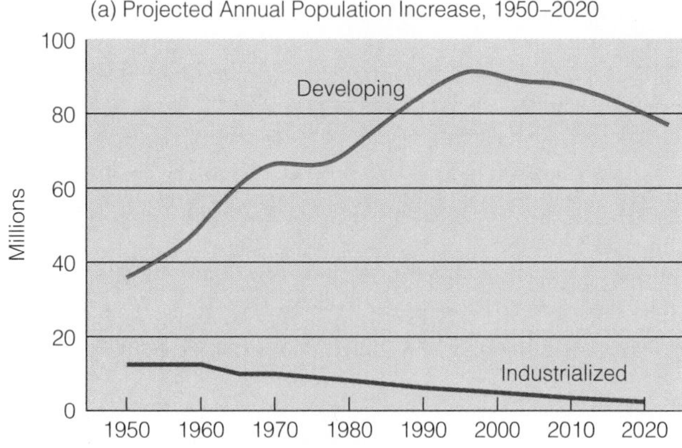

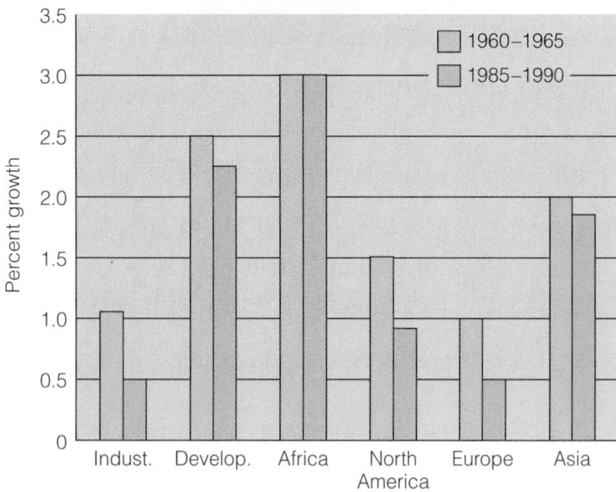

Figure 16–1 Population Growth
(a) Projected annual population increase, 1950–2020
(b) Average annual population growth rates, 1960–1965 and 1985–1990
Source: United Nations Populations Division, *World Population Prospects 1991* (United Nations, New York, 1991), pp. 228–231.

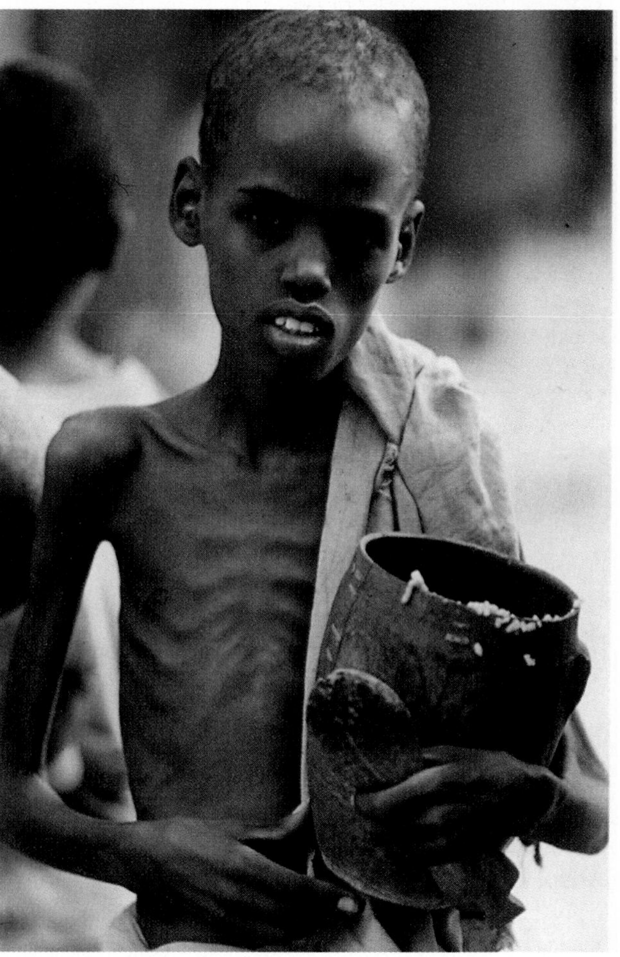

Figure 16–2 Starvation
For many, the world is not a kind place. Starvation faces over a billion people on Earth each day.

or ill-considered sexual interactions, which affects us personally with respect to our health and psychologically creates an atmosphere of fear and uncertainty in human relations. The unforgiving aspect of some sexual interactions is felt in the rapid rise and prevalence of sexually transmitted diseases. It's not "their" problem—it's "our" problem.

GENERAL KNOWLEDGE

Surveys of teenagers and young adults indicate that general knowledge of sexually transmitted diseases is low, even though the individuals surveyed thought that they were well informed on such issues. Knowledge of the diseases, their history and how they are transmitted, how they are treated, and what happens if they are ignored, is fundamental to maintaining good health. Such knowledge is also fundamental for developing ways to avoid contracting these diseases in the first place. Safe sex depends on knowing what the risks are before, not after, sexual encounters. The agents of sexually transmitted diseases fall into a number of categories, including bacteria, fungi, and protozoa as well as the quasi-living viruses.

BACTERIA AND STDS

Among the most common sexually transmitted diseases are **syphilis, gonorrhea,** and **chlamydia.** Each type of disease is caused by a different organism. A fragile, spiral-shaped bacterium known as *Treponema pallidum* is the agent of syphilis. The gonococcal bacterium *Nesseria gonorrheae* causes gonorrhea, and the bacterium *Chlamydia trachomatis* is responsible for chlamydia (Figure 16–3). Each of these types of bacteria grow well in seminal and vaginal fluids and are exchanged during sex between an infected individual and his or her partner. If you have one type of STD, it does not protect against getting another, and if you have had any one of these diseases, it does not protect against reinfection. Each type of bacteria has its own set of symptoms and each causes progressively more damage as the infection progresses.

Syphilis

Syphilis is generally transmitted during coitus, but it can also be transmitted during oral sex or anal sex. Thus, it can infect the lips, mouth, throat, and areas around the anus as well as the genital tract. There were approximately 100,000 new cases of syphilis in the United States each year at the beginning of the 1990s. This is contrasted with lower rates from previous decades (7,000 in the late 1950s), which, in turn, reflected the social conditions and the first availability and effective use of antibiotics. In general, bacteria grow wherever they are provided the right conditions for their survival, which in the case of organisms that are transmitted sexually, is generally on or in moist, protected areas of the body. The organisms

> **Syphilis** a disease caused by infection with the spirochete *Treponema pallidum*
>
> **Gonorrhea** a disease caused by infection with the organism *Nesseria gonorrheae*
>
> **Chlamydia** a disease caused by infection with the organism *Chlamydia trachomatis*

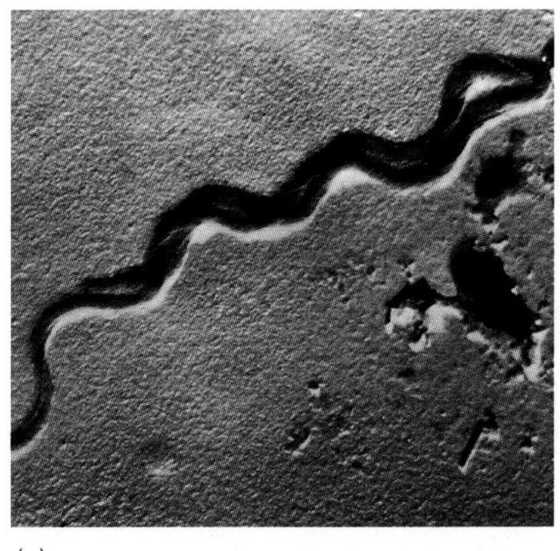
(a)

(b)

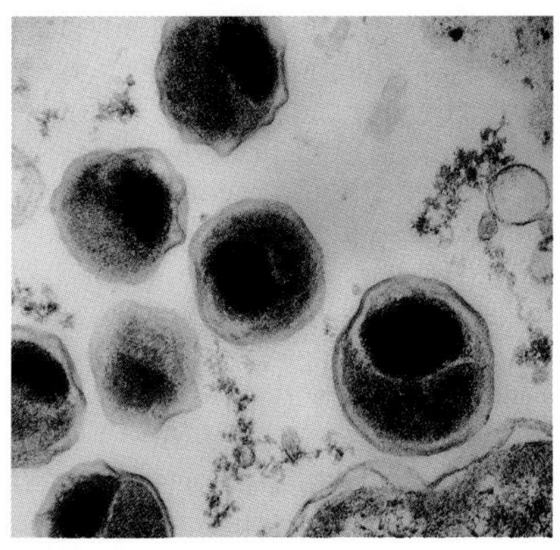
(c)

Figure 16–3 Bacteria and STD
(a) *Treponema pallidum,* agent of syphilis
(b) *Nesseria gonorrheae,* agent of gonorrhea
(c) *Chlamydia trachomatis,* agent of chlamydia
The three organisms shown are common causes of STDs affecting millions of people worldwide.

that cause syphilis, *Treponema pallidum,* do not survive exposure to oxygen-rich conditions outside the body. This makes infection by disease organisms from toilet seats or shared dishes highly unlikely. There are three stages of a syphilis infection: primary, secondary and tertiary. The stages progress from early relatively mild infection to massive organ destruction, mental instability, and death (Figure 16–4).

Primary Stage During the primary stage, symptoms appear typically within two weeks to a month. In males, the first signs of the infection are sores on the penis, scrotum, or site of initial entry of the bacterium into the body (Figure 16–4a). These sores ulcerate and form what are called *chancres,* which are generally painless. For males, it is difficult not to notice such changes on their genitalia. However, for females, who may develop these sores on the vagina or cervix, an infection may go unnoticed (see Figure 16–4b). Antibiotics, when given during the first stage of the infection have been the most effective treatment for syphilis. *Penicillin* and related antibiotics are the drugs of choice for this treatment, and the infection can be completely eliminated with little or no permanent damage. If the disease is left untreated, the sores will disappear within a few weeks and the pathogen will spread to other tissues of the body.

It should be pointed out that *natural resistance* to antibiotics by pathogenic bacteria, such as syphilis and other STD agents, has been on the rise in recent years and may in part account for the increase in reported incidence of these diseases. Resistance to antibiotics arises through mutation of individuals in the bacterial population. A mutation may alter the genetic capacity of the bacterium in such a way that it can more effectively circumvent the effects of the antibiotic. It may not initially have full resistance, but it may well have a distinct selective advantage over its counterparts. This allows it to grow and dominate the population. This process of mutation and selection is continual and may eventually give rise to fully resistant strains of bacteria.

Secondary Stage The second stage of syphilis is considerably more serious and bothersome. A new set of symptoms may show up within days of the first stage, or the disease may remain latent for several months. When the symptoms do appear, they take the form of a rash over part or all of the body (Figure 16–4c). Some forms of the rash will develop raised, fluid-filled bumps. The fluid of these bumps is filled with pathogenic bacteria and the individual is highly contagious under these circumstances. There are other symptoms as well, including sore throat, fever, sore joints, appetite loss, and headaches. However, these symptoms are common to many other kinds of diseases, from colds to flu, and will disappear within a few weeks to months even if left untreated.

Tertiary Stage At this juncture in the disease, there may be a long period of **latency.** The pathogens have embedded themselves into many tissues of the body, including the brain, spinal cord, heart, blood vessels, and bones. In most cases, infected individuals are not contagious during the latent phase of the disease, and in 50% or more of the cases, the disease will not develop further. However, the percentage of spontaneous recovery is small. For many infected people, there is a final and destructive stage of the disease—the tertiary stage. In the worst cases, there is massive destruction of tissues and organs when the disease reemerges from latency. Emergence from latency may occur perhaps as early as 3 to 5 years or as long as 30 to 40 years after the initial infection. The damage can range from ulcers on the skin and internal organs to fatal destruction of the heart and blood vessels. Damage to the brain and spinal cord can be so extensive as to result in mental illness, paralysis, and ultimately death.

This is not a pretty picture and does not need to happen if one is aware of the signs of this disease and acts upon recognition. It is difficult to be sure that an infection is syphilis without blood tests or samples of the fluid from chancres. The infection is most easily treated in the first and second stages. It should be noted that the presence of syphilis, gonorrhea, and/or chlamydial infections increases the susceptibility of an individual for other infections with pathogens, including HIV.

Treatment in the tertiary stage of an infection is more difficult. Higher doses of antibiotics are needed, often for longer periods of time. The resistance of some strains of the pathogen to antibiotics further complicates the problem. As pointed out earlier, there has been a general and progressive rise in the resistance of syphilis strains treated in the infected population in recent years. It should be noted that the damage done to internal organs and tissues prior to treatment is not necessarily undone when the infection is eliminated. Some of the damage of syphilis is even extended into the next generation. Unborn children can be infected during pregnancy. If the mother has a primary or secondary stage infection during her pregnancy, the child usually does not survive. However, even if the baby is born and survives, it is subject to *congenital syphilis* (see Figure 16–4d). Blindness, internal organ abnormalities, and mental retardation are likely outcomes. What a legacy to pass on to a child! Safe sex and STD awareness are beneficial to all.

Latency a delay in the expression of a disease until an undetermined future time

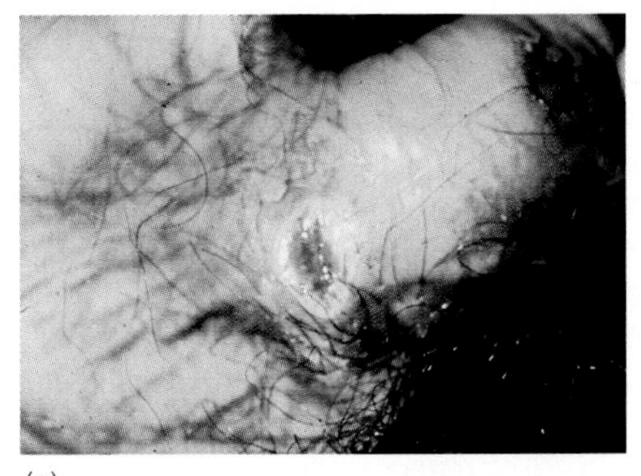

(a)

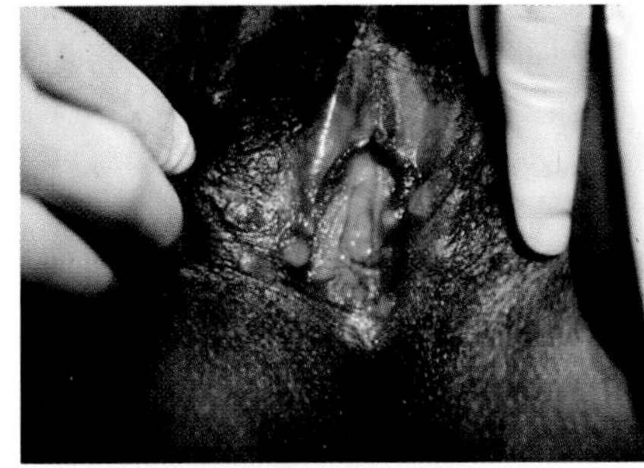

(b)

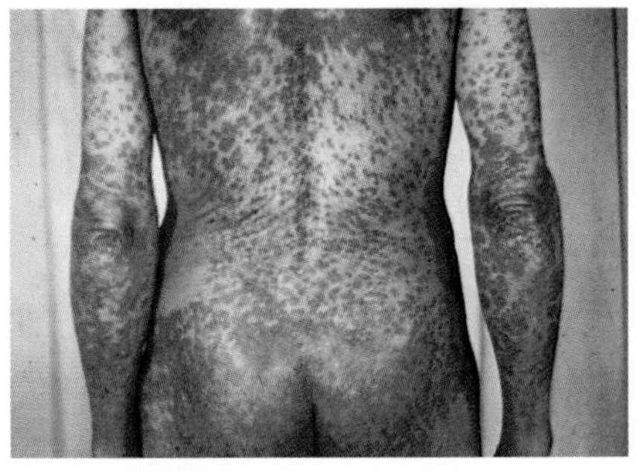

(c)

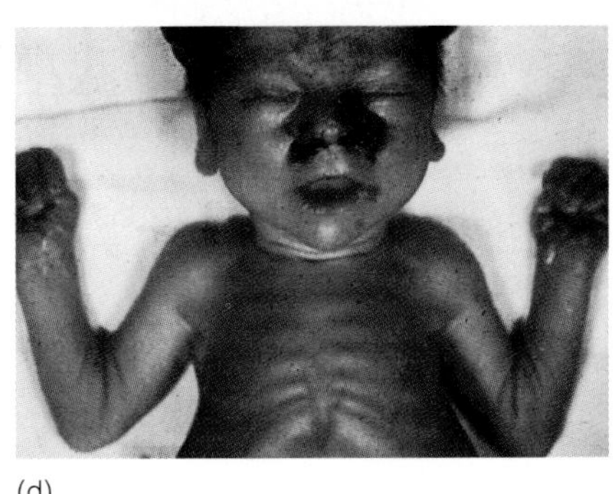

(d)

Figure 16–4 Stages of Syphilis
The first—primary stage of syphilis is shown in (a) for a male and in (b) for a female. The second stage of syphilis is shown in (c) as a rash over the entire body. In (d) this baby was born with congenital syphilis.

Gonorrhea

Gonorrhea is a more common STD than syphilis. More than a million cases are reported each year. According to epidemiologists at the Centers for Disease Control (CDC), this is probably a gross underestimate of the actual number of cases. Gonorrhea is also a generally less damaging disease to the infected individual. It affects mainly the urogenital structures of the body, although it is known to cause more serious disorders if other tissues and organs are infected. As with syphilis, the bacteria responsible for gonorrhea grow well in the moist, protective mucous membranes of the urethra, vagina, rectum, eyes, mouth, and throat. Gonorrhea pathogens are also sensitive to environmental conditions and do not survive exposure outside the body for very long. Gonorrhea may be one of the oldest known agents of STD. Symptoms of a disease described in the Old Testament of the Bible and by physicians in ancient times appear to fit those observed for gonorrhea today.

Symptoms The symptoms of gonorrhea include puslike discharges from the urethra, vagina, or other affected mucous membrane. The word "gonorrhea" means seed flow in Greek, a reference to the urogenital source of the fluids that drip from tissues associated with reproduction. The infection in males causes painful urination, which is a clear sign of the disease. Five percent to 20% of cases in males are asymptomatic (showing no symptoms). The disease can spread throughout the reproductive system, painfully infecting the prostate gland, seminal vesicles, epididymis, and bladder. Damage to reproductive tissues can lead to scarring and obstruction, which lead to sterility.

Asymptomatic Cases Females are generally **asymptomatic.** A female may not know she has contracted the disease unless her infected male partner tells her. The gonorrhea pathogen initially infects the cervix. An abnormal and/or irritating discharge may be noted in as few as 20% of those infected. As in the case of asymptomatic males, treatment may not be sought because there is no sign of disease. This leads to two problems—one social, the other personal. The social dimension of the problem is that if a female does not know she has the disease, she may infect others. The personal aspect is that unknown to her until too late, the disease can spread to her uterus and Fallopian tubes. The Fallopian tubes can become swollen and painful, leading to fever and abdominal pain. This is a condition known as *pelvic inflammatory disease (PID),* which is damaging to the reproductive tract and may lead to sterility.

Asymptomatic infected but not showing symptoms

Gonorrhea pathogens do not normally enter the blood, so the bacteria are not spread to a developing fetus except at delivery when the new baby's eyes may be exposed and infected in the birth canal. Most states in the United States require by law precautionary measures to prevent this type of infection and, so, physicians put silver nitrate drops in the eyes of all newborns. However, the absence of bacteria in the blood means that there can be no simple and effective blood test for the disease. Usually, a doctor has to visually identify these organisms under a microscope or grow cells in culture that have been isolated from the urethra of males or cervix of females to ascertain the presence of this pathogen.

Treatment Treatment of gonorrhea also requires the use of antibiotics. Penicillin has been the most effective curative in this regard. However, as in the case of syphilis, highly resistant strains of *Neisseria* have arisen that are not killed by penicillin. How did these bacteria become resistant? In the case of *Neisseria,* resistant strains of these bacteria have evolved an enzyme that inactivates or circumvents the effects of penicillin molecules. The enzyme is known as *penicillinase* and, regardless of how high the dosage of penicillin may be, the bacteria are unaffected. This causes great fear in the health care community because it suggests that the time-tested, inexpensive treatment of gonorrhea may soon be ineffective and that further increases in infection rates will result. Combinations of antibiotics with compounds that inhibit the bacterial penicillinase are proving to be effective. Eliminating or circumventing the penicillinase once again makes many of these organisms susceptible to relatively inexpensive treatment.

Chlamydia

Another disturbing fact in the complex world of sexually transmitted diseases is that some STD pathogens come in pairs. That is, the presence and identification of one type of pathogen in an infected individual is a strong indicator of the presence of another. This is the case with gonorrhea. Up to 50% of those infected with gonorrhea are also infected with chlamydia. Chlamydia is caused by a bacterium and is even more commonly an agent of infection than is *Neisseria* (Figure 16–3c). Estimates from the CDC are that there are 4 million to 5 million new cases of chlamydia each year. An additional complication is that chlamydial pathogens are not susceptible to penicillin at all and require another class of antibiotics known as tetracyclines. As a precaution, most physicians treating a known case of gonorrhea will routinely give both penicillin and tetracycline.

Symptoms Chlamydia is less well known among young people today than it should be. Aside from the fact that it is several times more prevalent than gonorrhea, it is also more dangerous. This is because it is far more likely to damage reproductive organs and lead to sterility or ectopic pregnancy (see Chapter 15). The symptoms of chlamydia are the same as gonorrhea, if not initially milder. Inflammation of the urethra, discharge, and painful urination in males is characteristic and usually prompts a doctor's visit. Females are much less likely to know they even have the disease and, so, less likely to seek treatment immediately. Estimates of the prevalence of chlamydia suggest that nearly one-fifth of sexually active teenagers are infected, one-third of college students, and a significant percentage of pregnant women. These high numbers, in turn, reflect the epidemic proportions of chlamydial infections in the U.S. population.

Other Bacterial Agents of Disease

There are a number of other types of bacteria that are agents of sexually transmitted disease. Many of these are rare in the United States but are found commonly in the tropical latitudes. *Chancroid,* which causes painful sores on the genitalia, is seen in only a few thousand new cases each year. *Granuloma inguinale* spreads from a painless sore to surrounding areas and may lead to extensive damage and even death if not treated. Fewer than 100 cases a year are treated in the United States. *Shigellosis,* caused by the bacterium in the genus *Shigella,* is not technically an STD but leads to diarrhea, pain, and fever in those infected. The source of the infecting bacterium is usually feces.

VIRUSES AND STDS

The distinct advantage of having a bacterial STD (if there is any!) is that they can be effectively treated with antibiotics. This is not the case for the other major class of STD-causing pathogens. Viruses are different in many ways from bacteria, but, for the purposes of discussion here, the main differences are that bacteria live *with* cells of the host and viruses live *in* the cells of the host and that antibiotics do not affect the viruses. There are four groups of viruses that will concern us here. They are **herpes viruses, human papilloma viruses, hepatitis viruses,** and **human immunodeficiency virus (HIV)** (Figure 16–5). The last of these is clearly the most serious and threatening to

Herpes viruses DNA viruses; two types, I and II, both of which may be transmitted sexually

Human papilloma viruses DNA viruses; many types, some associated with warts, condyloma, and cervical cancer

Hepatitis viruses DNA viruses; three main types: A (infectious hepatitis), B (serum hepatitis) and C

Human immunodeficiency virus (HIV) an RNA virus that infects human cells and destroys the function of T-cells

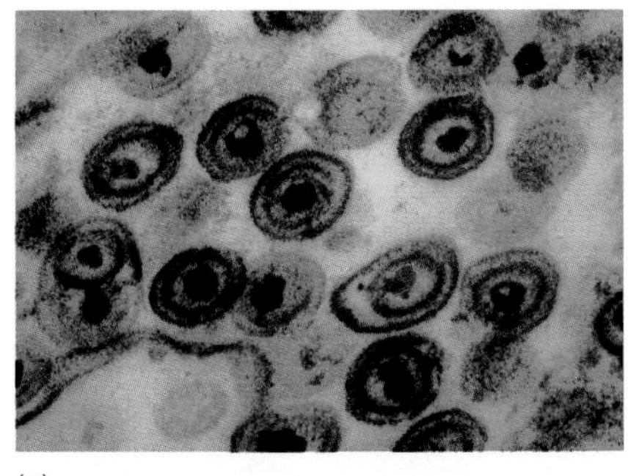
(a)

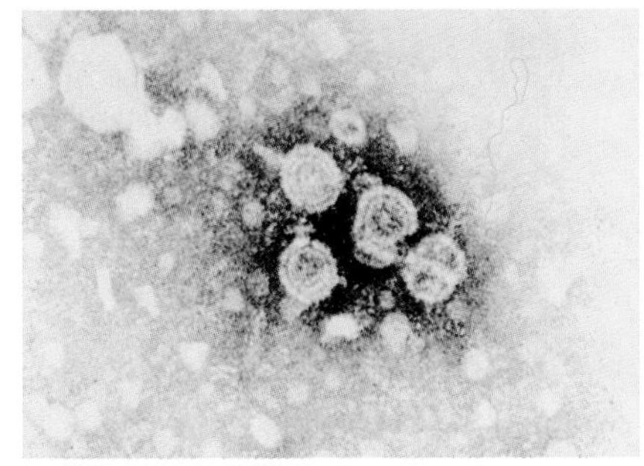
(b)

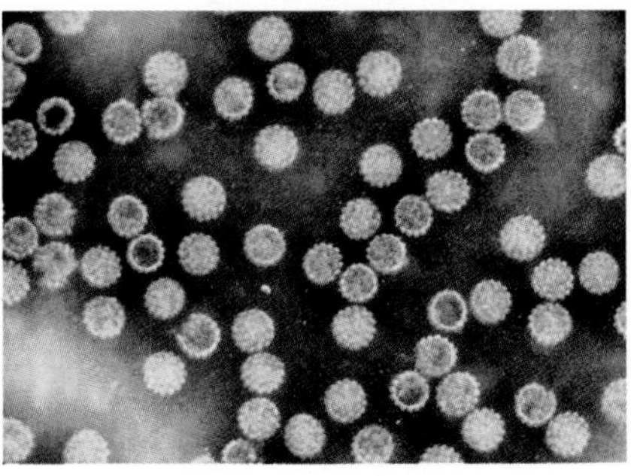
(c)

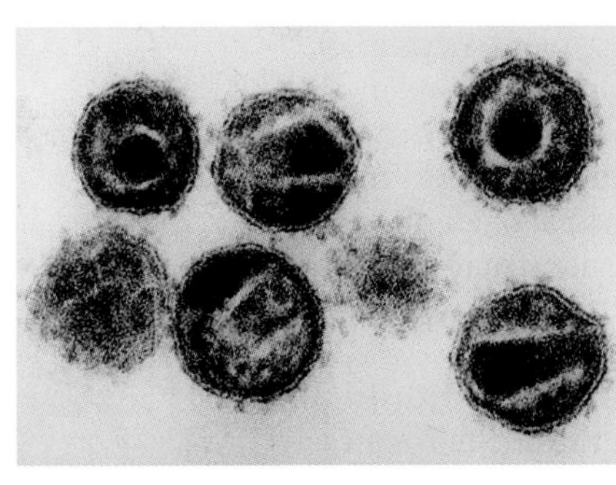
(d)

Figure 16–5 Viruses and STD
Four types of viruses are commonly associated with STD; (a) herpes viruses, (b) hepatitis viruses, (c) human papilloma viruses, and (d) human immunodeficiency viruses (HIV).

human health and survival. In its apparently inevitable and fatal form, HIV may cause a whole set of symptoms, or a syndrome. This set of symptoms is known as **AIDS (acquired immune deficiency syndrome).** Viruses associated with sexually transmitted diseases generally cannot be completely eliminated from the cells of the body once they have been infected. Humans are host to the viruses for the remainder of their lifetimes.

Herpes

We do not hear as much about herpes infections in the media as we used to. There are probably two main reasons for this. The first is that concern for HIV and AIDS has come to dominate our attention, and the second is that drugs for the symptomatic treatment of herpes infections have been developed that bring much relief to those who suffer from them. What is herpes? Actually, we should ask what are herpes, because there are two related viruses that cause similar clinical manifestations. They are called herpes simplex I and herpes simplex II. These two forms of herpes viruses are also related to the types of viruses that cause chicken pox, infectious mononucleosis, and shingles.

Symptoms Both types of herpes viruses cause painful blisters (fever blisters, cold sores) around the mouth and blisters and ulcerations on and around the genitalia. It used to be thought that the two types were site specific; that is, herpes simplex virus type I (as the virus is named) was associated with the oral lesions, and that herpes simplex virus type II was associated with the genital lesions. This led to the belief that oral herpes could only be passed by kissing and genital herpes could only be passed by sexual intercourse. This has proven not to be the case. Either type of virus can infect either region, as well as other parts of the body. The basis for this change in the distribution of the two herpes viruses is thought to be the increased popularity of oral-genital sexual activity. Although this may have played a part in the change, it is also known that individuals with oral herpes can, in fact, infect other parts of themselves including the genitals and the eyes.

The Cure? There is no known cure for herpes. Once infected, we carry the virus with us for the rest of our lives. What is important to point out in this age of HIV infections and AIDS is that herpes is not fatal, though there can be serious complications. The symptoms are irritating and painful, but even when left untreated, they disappear with time. The symptoms of a primary herpes attack occur within two days to three weeks of the initial infection and last two to three weeks. Attacks last longer in females than

AIDS (acquired immune deficiency syndrome) a multitude of symptoms including weight loss, T-cell destruction, rare infections, and cancer, associated with the advanced stages of an HIV infection

Figure 16–6 Testing Human Blood for Antibodies (Western Blot)
Blood samples are taken from individuals never infected or exposed to a virus (such as herpes or HIV) or from an individual exposed or infected by a virus (a). Following preparation of the serum, viral proteins separated from one another by size (b) are tested to determine whether antibodies have been made against the virus. As shown here, if antibodies recognize proteins at 41 and 24, the test is positive for HIV.

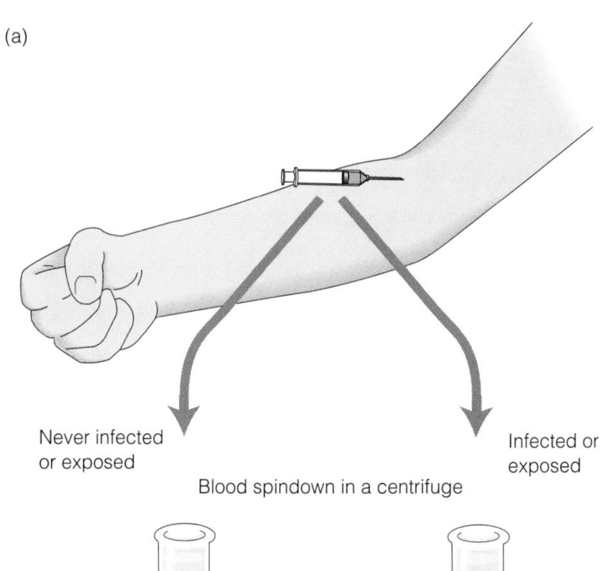

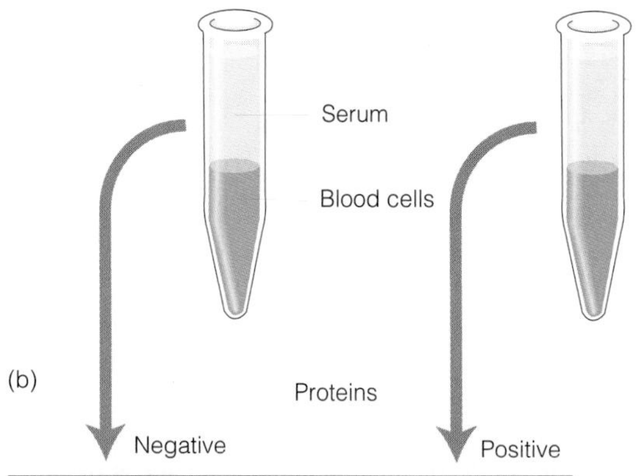

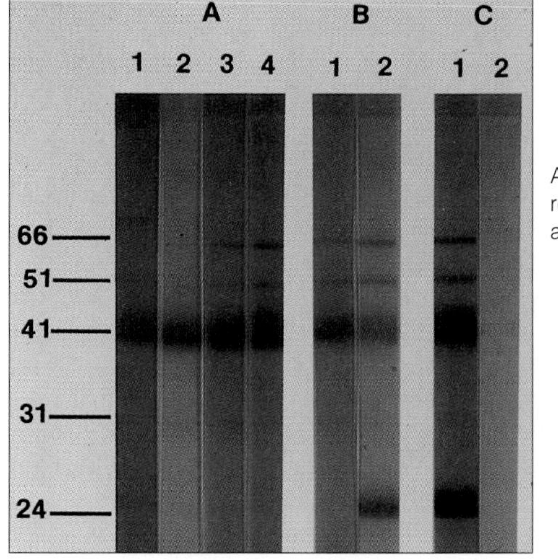

in males. Subsequent, recurrent attacks usually last about a week and are generally less aggravated than the initial attack. This difference in the strength of subsequent outbreaks may have to do with the fact that our immune system is primed for fighting the virus released during an attack only after the first exposure to the virus. Herpes outbreaks may arise recurrently as a result of fatigue, physical or emotional stress, illness, overexposure to sunlight, menstruation, and possibly some dietary factors.

Detecting the Presence of the Virus and Determining the Numbers How many people in the United States harbor the herpes virus? One way to estimate the number of people in the United States with herpes is to measure the antibodies in blood (called a Western blot test). The presence of antibodies against herpes viruses in the serum (called **seropositive**) indicates exposure to the virus at some time during the life of that individual (Figure 16–6). The number of people in the United States that are seropositive for herpes is staggering. It is estimated that more than 100,000,000 people in the United States have been infected with herpes and that 25,000,000 have shown the symptoms of infection. All infected individuals do not have attacks, suggesting that there may be a genetic component to the different levels of resistance to the viruses in different individuals.

Serious Problems with Herpes Infections A number of potentially serious problems may arise as the result of a herpes infection. The herpes virus can infect cuts and scrapes anywhere on the body. Failure to follow certain rules of hygiene, such as washing after using the bathroom and not touching blisters, can lead to the spread of the infection, particularly to the eyes. A herpes infection of the eye is called *herpes keratitis* and is the leading cause of blindness in the United States. Contact lens users need to be especially careful. Herpes encephalitis and herpes meningitis, affecting the brain and the membranes surrounding the central nervous system (the meninges), respectively, are rare but dangerous. The risk of cervical cancer is as much as eight times greater than normal if women are infected with herpes. Because cervical cancer is one of the most frequent types of cancers encountered by women (second or third in

Seropositive having antibodies in one's serum reactive against a virus; an indication of exposure to the virus

incidence), it is clearly important to have a doctor monitor the physiological status of the cervical mucosa more often than the normal yearly interval (Figure 16–7a).

Neonatal Herpes Perhaps the most devastating form of herpes is the neonatal form of the infection. Pregnant women with active herpes at the time of delivery are quite likely to pass on the infection to the child. Within three weeks of birth, the primary infection will occur and is often fatal to the baby who has an immature immune system with essentially no defense against the rampant spread of infection. Half of all babies infected in this way have serious and permanent neurological problems or succumb to death. In the face of these statistics, most doctors choose to deliver babies of mothers with genital herpes by cesarean section.

Treatment of Herpes Infections Remember that there is at present no cure for herpes. However, there is a treatment available that lessens the severity of an outbreak. The drug acyclovir, applied as an ointment to the affected area or taken orally as a tablet, reduces the symptoms and shortens the duration of infection. In some cases of recurrent attacks, the onset of an outbreak of herpes can be anticipated by an individual before the symptoms actually appear. Prompt use of acyclovir in these cases helps forestall the severity of the subsequent outbreak. In some cases of frequent recurrent attacks, the drug must be taken on a continual basis. The length of time that the drug can be taken safely is under study. In general, as more data on side effects are obtained, the length of time over which the drug can be used is being extended. For those who take acyclovir continually to suppress outbreaks, a withdrawal from the drug will result in a return to recurrent outbreaks.

Herpes carries with it a stigma. The disease often induces feelings of guilt and self-deprecation in those infected and avoidance and fear in those who find out about a partner being infected. How do we deal with personal and intimate relations with others when it is known that one of the individuals involved in the relationship has herpes? Is there a "safe" time in which sexual relations can occur? Clearly, honesty is the best policy. Telling your partner that you have herpes before you have sex is not only fair but it gives that individual a chance to decide just how committed he or she is to the relationship. Talking about herpes also gives you a chance to demystify the disease. Basically, what needs to be known is that most individuals are contagious only during an overt attack and that use of a condom protects against infection. The viral baggage carried by those who are infected need not result in psychological and emotional distress and the end of meaningful relations between people. For more detailed information on herpes (and other STDs), there are books and periodicals listed at the end of this chapter.

Human Immunodeficiency Virus (HIV) and Acquired Immune Deficiency Syndrome (AIDS)

The symptoms expressed by individuals with AIDS are in great part the consequence of the infection of cells of the immune system with a virus known as human immunodeficiency virus or HIV (Figure 16–5d). A prime target for the HIV in most cases is T-cells, a subgroup of lymphocytes that provide cellular immunological defense for the body (see Chapter 12). The virus is also known to invade other types of cells as well. The targeted destruction of immune system cells is double trouble because it makes the body more susceptible to, and eventually defenseless against, a variety of other infectious diseases and cancer. This breakdown in defense unleashes the avalanche of disorders observed in full-blown AIDS. T-cells are the key to protecting the body against foreign invaders and abnormal cells from within. It is the constellation of fatal diseases that apparently arises as a result of the viral destruction of T-cells that constitutes AIDS.

Who Discovered HIV? French scientists at the Pasteur Institute in Paris led by Luc Montaigner were the first to isolate the virus in 1984. American scientists at the National Institutes of Health also isolated and characterized the virus during this time. Today, research on AIDS dominates the health research carried out in many countries around the world. Such an effort has been mandated because the spread of the virus by sexual contact has already reached epidemic proportions in some countries and the problem continues to grow. In our cosmopolitan world, where foreign lands are only a few hours away by jet, disease agents know no national boundaries.

What Are the Symptoms of AIDS? AIDS was first diagnosed in the early 1980s in men who had rare infections or cancers related to the suppression of the immune system. Respiratory infections involving the fungus *Pneumocystis carinii* and a cancer known as Kaposi's sarcoma were found in an alarming number of cases. Other symptoms of the disease include loss of appetite, diarrhea, loss of weight, fevers and night sweats, swollen lymph nodes, and abnormal blood counts. The first cases of the disease appeared to be limited to homosexual males, establishing the belief that it was a problem for the homosexual community alone. This is not the case, however, because the virus is spread by heterosexual interactions and other means as well. As our understanding of the disease has evolved, the groups at most risk are homosexual or bisexual males, drug users of both sexes who share needles, and children who are born to HIV-infected mothers. However, infection resulting from heterosexual interactions, particularly in females, is on the rise (Figure 16–8 and Table 16–1).

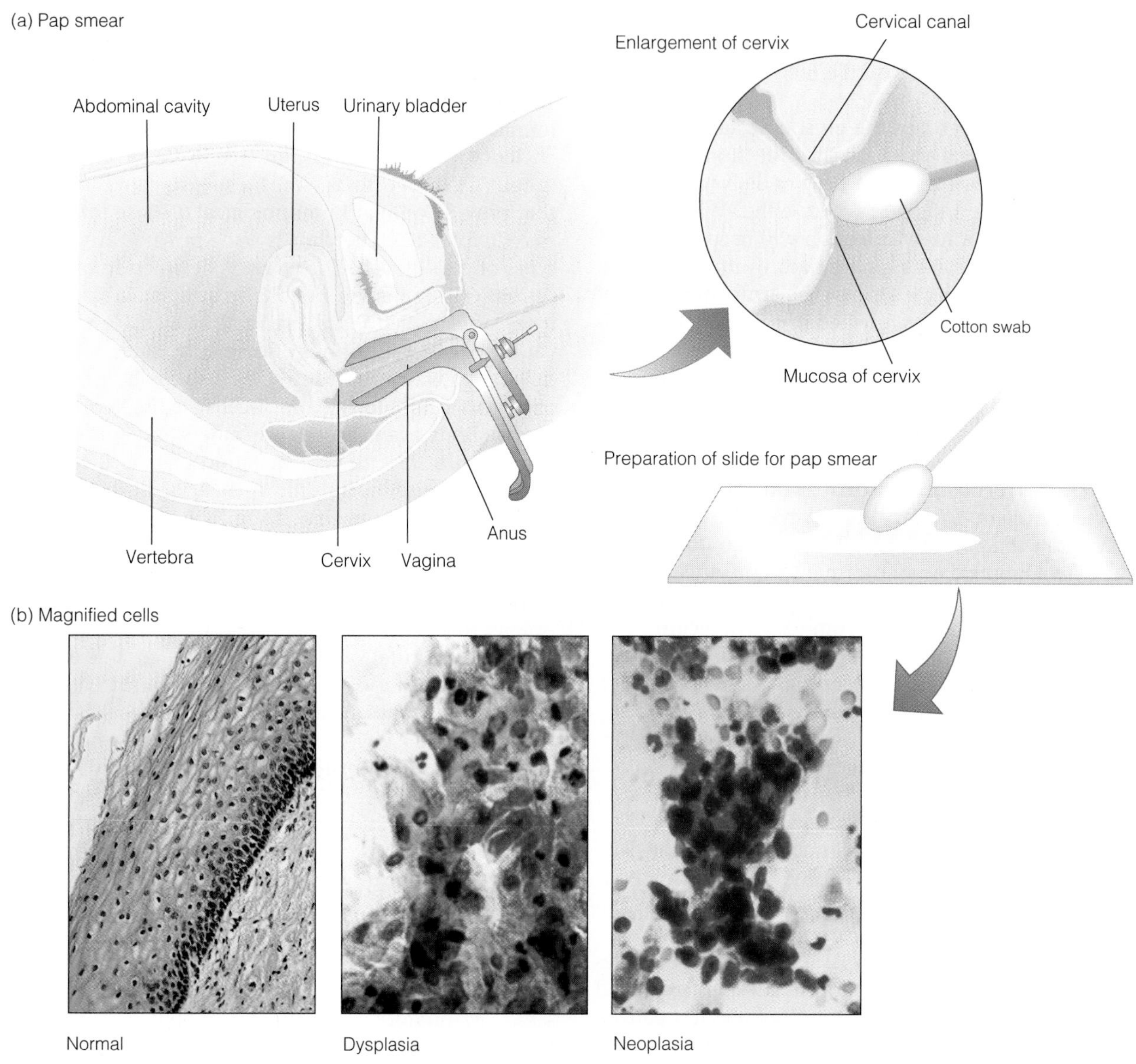

Figure 16–7 A Pap Smear
In (a) preparation of a pap smear (using the Papanicolaou technique) is shown. Cells from the cervix are applied to a slide for observation under a microscope. In (b) normal cervical tissue, dysplasia of cells (a pre-cancerous state), and neoplasia are shown.

What Are the Numbers? Opinions differ depending on how data is analyzed, but between 800,000 and 1,200,000 Americans are believed to be infected with HIV. This means that these individuals may not yet be showing the symptoms but are seropositive for the virus. There were over 227,000 deaths resulting from AIDS reported by the early 1990s. For many reasons, ranging from the right to privacy to protection against social ostracism, not all cases of AIDS are reported as such, suggesting that there may be many more with AIDS than the number listed here. The predictions for the increase in the cumulative number of cases of AIDS based on U.S. Public Health Service statistics is depicted in Figure 16–8.

The average period of time during which a person who is HIV positive may actually develop AIDS is five to six years. However, the disease has not been around long enough to make such estimates with absolute certainty. This does not mean that if an HIV positive individual does not develop the disease within five to six years that individual is safe. It may well be that certain individuals, based on a variety of as yet unknown factors, will have a decade or more in delayed onset. Reports from the international meetings on HIV and AIDS held in Japan during 1994 suggest that a key to finding a cure for the infection may be revealed in studying the resistance of individuals who have been HIV positive for 10–15 years, yet show no sign of AIDS. As we learn more about the virus and the disease and attempt to develop rational treatments to arrest or delay its full-blown effects, HIV may become manageable.

The Origin of HIV There have been many stories put forward to explain the origins of AIDS, ranging from the then Soviet Union blaming the CIA to claims that it must be God's revenge against homosexuals. These stories serve only to promote hatred, fear, and blame among and between people by those who wish to satisfy their own self-serving ends. The appearance and evolution of viruses that kill human beings is well known throughout history and includes polio, influenza, and smallpox. These viruses, which are not considered to be sexually transmitted, have led to the death and disfigurement of millions of people throughout history. Although it may never be known exactly where, and under what circumstances, HIV arose, the general belief is that it initially developed in central Africa. AIDS is certainly in epidemic proportions in central Africa today and is rampant in the heterosexual population. The disease is called Slim in that part of the world, which reflects the weight loss associated with AIDS symptoms, and it has been on record as occurring since the 1970s (Figure 16–9). Serum from blood samples taken as far back as the late 1950s and early 1960s has been shown to be HIV positive. The route out of Africa probably was established with cultural exchanges between developing central African nations of the 1970s and Haiti. The popularity of Haiti as a vacation spot for Americans made it an ideal setting for exchanging much more than culture. The spread of the disease to the United States was initiated in the male homosexual population, particularly among individuals that had frequent and promiscuous sexual encounters.

Transmission of HIV Infection Because this is a chapter on sexually transmitted diseases, it is appropriate that HIV is described principally as an infective agent passed between individuals during intimate sexual contact. In fact, the virus is spread specifically by exchange of infected bodily fluids, among which semen and vaginal secretions are implicated. The virus can also be spread by exchange of blood or blood products. This is reflected in the high incidence of AIDS among drug users who share needles. It is also observed in the rare, but significant, occurrence of infection from contaminated blood used in a

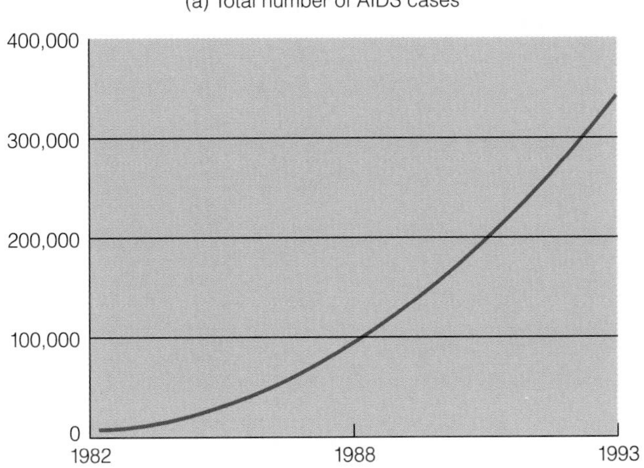

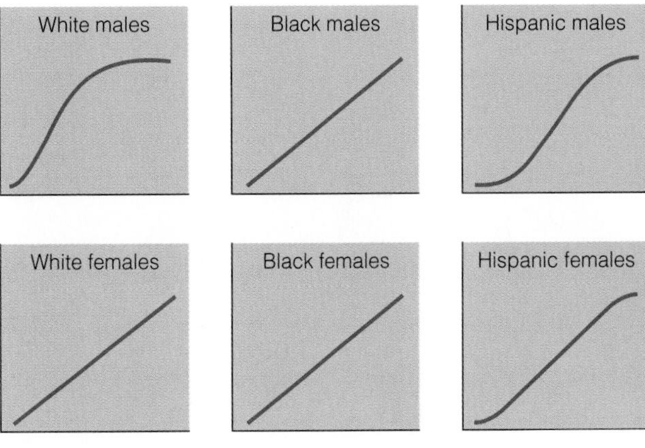

Figure 16–8 Incidence and Distribution of AIDS in the United States

The total number of AIDS cases increases dramatically each year (a). The rates between groups is different (b). For example, compare the rates of new cases reported for white males versus white females.

Source: Adapted from Rosenberg, P. (1995) *Science* 270: 1372-1375.

blood transfusion or from blood products used in other blood disorders such as hemophilia. The great tennis player Arthur Ashe suffered from the first type of occurrence and Ryan White suffered from the second (Figure 16–10). The story of Ryan White is one of great courage and perseverance in the face of social ostracism. The worry of many people with respect to Ryan White was that the infection was more generally contagious than reported and that, if he were allowed to attend public school, their children would come down with the disease. The fear of many, even in the face of overwhelming evidence against it, was that HIV could be transmitted by

Table 16–1 Percentage of AIDS by Exposure Categories, 1985–1989

EXPOSURE CATEGORY	BEFORE 1985	1985	1986	1987	1988	1989
Adult Male						
Homosexual/bisexual contact	69	72	71	70	64	63
I.V. drug user	15	15	15	14	21	20
Homosexual and I.V. drug user	10	8	8	8	7	7
Hemophiliac/Coagulation disorder	1	1	1	1	1	1
Heterosexual contact	2	2	2	2	2	3
Transfusion	1	1	2	2	2	2
Undetermined	2	2	2	3	3	5
Adult Female						
I.V. drug user	58	53	49	49	53	50
Hemophiliac/Coagulation disorder	<1	1	<1	<1	<1	<1
Heterosexual contact	24	27	34	31	30	32
Transfusion	8	11	10	13	11	8
Undetermined	10	9	6	6	6	9
Children (<13 years)						
Hemophiliac/Coagulation disorder	4	6	6	7	7	4
Transfusion	11	14	13	14	11	6
Mother with AIDS or at risk for HIV infection	78	78	74	77	79	85
Undetermined	6	0	2	3	4	4

Source: Centers for Disease Control

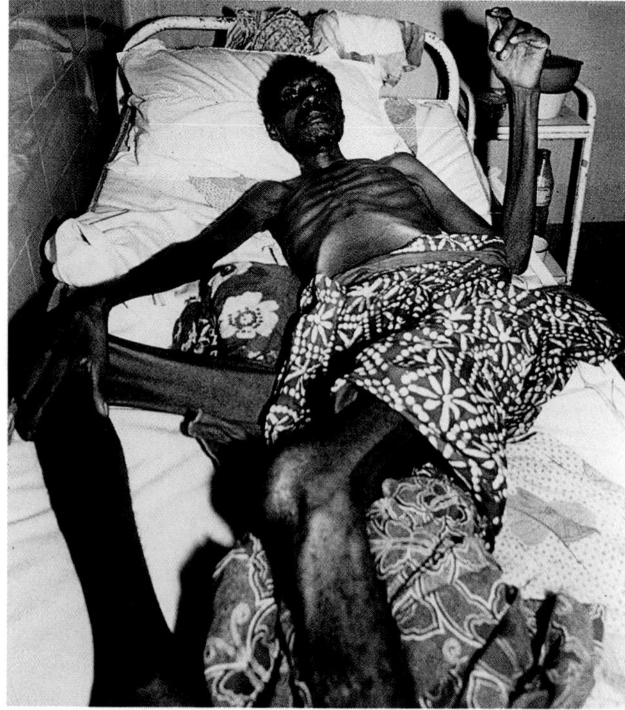

Figure 16–9 Slim in Africa
The occurrence of AIDS (called Slim in many African countries) is high. Estimates are that HIV infection among the population may be as much as 30%.

casual contacts, such as touching, kissing, or sharing a glass used by a person with the virus. However, on sober reflection, if HIV were so easily transmitted, then we would expect to see many more nonsexual and nonblood-related incidences. There is no known case of a family member of an AIDS patient living at home developing AIDS, even though they have routinely touched, hugged, kissed, and shared food and kitchen utensils with that infected person. In addition, HIV infection does not appear to occur as a result of mosquitos or other *insect vectors.* If it did, the transmission of the disease likely would be correlated with areas of greatest insect vector activity. There is as yet no such correlation. All this suggests that the HIV is a difficult virus to transmit effectively, other than by the most intimate of contacts or fluid exchanges.

The Blood Supply of a Nation In those cases where the HIV is transmitted to individuals not involved in risky behaviors, it is usually traceable to contaminated blood used for transfusion, to blood products used for other treatments, as in the case of Ryan White and others like him with blood diseases, or to infection suspected of resulting from procedures performed by an HIV-positive individual on a patient or client. The probability of HIV infection resulting from a transfusion of blood supplies of the United States is presently estimated to be less than 1 in about 30,000. The contamination that has been traced to blood supplies has had a very negative effect on donation of blood. Too many people have the idea that giving blood

(a)  

(b) 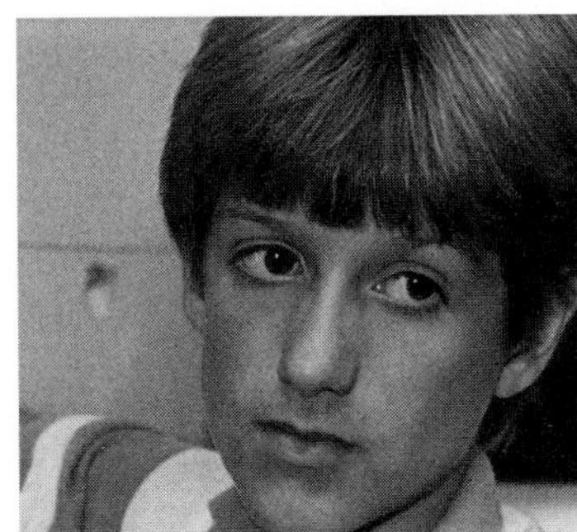

Figure 16–10 Victims of AIDS
(a) Arthur Ashe
(b) Ryan White
Some individuals infected with HIV and showing full-blown AIDS acquired the infection from contaminated blood or blood products. Tennis star Arthur Ashe and student Ryan White contracted HIV in this way. Both died from complications resulting from AIDS.

will somehow expose them to possible infection by HIV. Nothing could be further from the truth. Giving blood under the sterile conditions and well-regulated environment of a donation center does not expose a person to blood other than his or her own. The long-term effect of not donating blood is to deplete the blood supply of the country, and this is having serious consequences for people who need healthy, safe blood for medical emergencies.

AIDS Treatment The first thing that needs to be established is whether an individual has been exposed to HIV. This may seem obvious, but because of the variability in the development time of AIDS, a person may not know whether they were infected until they actually begin to show symptoms. On the other hand, knowledge that an individual is infected may prevent him or her from passing on the infection to future partners. This knowledge is also important because recent evidence suggests that early treatment may delay the onset of full-blown AIDS by years. If there are any doubts about a person's status in this regard, it is better to know than not to know. The two ways in which an individual is routinely tested for HIV are by an immunological test known as an *ELISA* (enzyme-linked immunosorbent assay) and/or a Western blot (Figure 16–6). These tests detect the presence of antibodies against HIV or viral proteins in the blood. The presence of such antibodies, as in the case of herpes, indicates that a person has been exposed to the virus. This does not mean that the person will develop AIDS. One of the problems in serum testing is that it may take weeks to months to develop antibodies against the virus. In some cases, it may take years to develop a detectable level of response. This means that it is possible for a person to test negative but actually be HIV positive. An individual is fully infective under these ambiguous conditions.

At this time, there is no cure for an HIV infection. However, progress in treatment is being made. There are a number of drugs that have proven directly effective against the virus and others that are effective against AIDS-related conditions. The drugs presently being used are azidothymidine (AZT) and dideoxyinosine (ddl), and, most recently, thiocytidine (3TC) has been added to the arsenal. These compounds are powerful inhibitors of viral replication and so prevent the spread of the disease. Unfortunately, they also have serious side effects. AZT can induce severe anemia and, in many cases, the individuals being treated are so weakened by the disease that their bodies cannot handle the treatment. However, ddl is often used for patients who cannot tolerate AZT, which has proven to be beneficial to many who otherwise could not be treated. The use of these drugs extends the life expectancy of AIDS victims and often increases their general health. An antiviral drug that has been available for some years, alpha interferon, is also used in conjunction with AZT or ddl to inhibit viral spread. There is so little known about how this virus operates once it has infected the various types of cells within the body that each patient is essentially a testing ground for the right combination and level of treatment of the antiviral drugs.

A new type of HIV inhibitor, known as a protease inhibitor, has been developed as a consequence of the detailed study of the replication cycle of the HIV virus. One of the crucial steps in viral development involves the precise cutting of large proteins into smaller, active forms of polypeptides used to assemble the virus. The enzyme that cuts the protein is called a protease and scientists have discovered a protease inhibitor that blocks this step. The newest approach to treatment for the HIV infection and AIDS includes the use of this protease inhibitor. Another advance in treatment that has recently been made is the use of a combination of antiviral drugs. The use of AZT, 3TC, and the protease inhibitor has proven remarkably effective in suppressing virus production and may be valuable in preventing or halting the spread of the virus in newly infected individuals. One of the advantages of suppressing virus replication is that it reduces the viral load on the body and gives the immune system of the HIV-infected individual a chance to recover and naturally combat more effectively the viral, bacterial, and fungal infections associated with AIDS.

Two other drugs are used to treat the symptoms of AIDS. An aerosol form of the drug pentamidine has proven effective in treating the pneumonia associated with AIDS, and *ganciclovir* and *saquinavir* have been used to prevent eye infections that can lead to blindness.

Ultimately, the most efficacious way to halt the spread of the disease will be to develop a vaccine against HIV. As in the cases of polio, measles, and influenza, vaccines that induce a natural acquired immunity to the viruses have proven the best deterrent. Unfortunately, the HIV is constantly mutating so as to avoid the immune surveillance of the body's natural defenses (see Chapter 14). Hope for a vaccine is not lost, however. Major research efforts continue in order to discover ways of preventing this disease through vaccination.

The Costs The cost of full-blown AIDS in terms of the individual are inestimable. Pain, guilt, and emotional trauma from loss of support from friends and family are devastating. The suicide rate for individuals who are diagnosed HIV positive is 30 to 40 times the norm. These people are often social outcasts, as fear and disgust from the general population are focused against them. We should all keep in mind that it is the disease that is frightening, not the people who have it.

The cost to society is also very high. It is estimated that to treat and care for an individual with AIDS, particularly in the terminal phases of the disease, the average hospital bill would be $70,000 to $80,000. This puts an enormous burden on the health care system of this, and any other, country that has to deal with thousands of terminally ill AIDS patients. The number of individuals who will contract HIV infections and eventually manifest AIDS is increasing, especially if we consider that there is reason to believe that every HIV-positive individual will develop the syndrome. A large part of the federal budget for the U.S. Public Health Service is already targeted for AIDS research.

There are a number of excellent sources of information on AIDS and its prevention and treatment. Table 16–2 provides some of them. To avoid becoming one of the statistics, use good sense in your sexual activities. If you are not willing to abstain from having sex, as most of us are not, then be concerned about who your partners are and who their partners have been. Always use safeguards such as condoms (synthetic latex type only, as natural membranes may allow virus-size particles to pass through) and spermicides containing the compound nonoxynol-9. Educate yourself and your partner. As with herpes, once a person is infected, HIV becomes part of his or her permanent genetic baggage. Unlike herpes, HIV will kill you.

Table 16–2 AIDS Hotlines and Resources

Public Health Service AIDS Hotline 1-800-342-2437 (look for your local hotline agency) AIDS information from state departments of health

Alabama—205-261-5131	Montana—406-444-4740
Alaska—907-561-4406	Nebraska—402-471-2937
Arizona—602-255-1203	Nevada—702-885-4988, 885-5948
Arkansas—501-661-2395	New Hampshire—603-271-4487
California—916-445-0553	New Jersey—609-588-3520
Colorado—303-331-8320	New Mexico—505-984-0911
Connecticut—203-549-6789	New York—518-473-0641
Delaware—302-995-8422	North Carolina—919-733-3419
District of Columbia—202-332-AIDS	North Dakota—701-224-2378
Florida—904-488-2905	Ohio—614-466-4643
Georgia—800-342-2437	Oklahoma—405-271-4061
Hawaii—808-735-5303	Oregon—503-229-5792
Idaho—208-334-5944	Pennsylvania—717-787-3350
Illinois—312-871-5696	Rhode Island—401-277-2362
Indiana—317-663-8406	South Carolina—803-734-5482
Iowa—515-281-5424	South Dakota—605-773-3364
Kansas—913-862-9360	Tennessee—615-741-7247
Kentucky—502-564-4478	Texas—512-458-7504
Louisiana—504-342-6711	Utah—801-538-6191
Maine—207-289-3747	Vermont—802-863-7240
Maryland—301-945-AIDS	Virginia—804-786-6267
Massachusetts—617-727-0368	Washington—206-361-2914
Michigan—517-335-8371	West Virginia—304-348-5358
Minnesota—612-623-5414	Wisconsin—608-267-3583
Mississippi—601-354-6660	Wyoming—307-777-7953
Missouri—816-353-9902	

Minority Task Force on AIDS 212-749-1214
National crisis line run by the National Gay Task Force 800-221-7044
National Council of Churches/AIDS Task Force 212-870-2421
National Association of People with AIDS 202-483-7979
National AIDS Testing 1-800-356-2437
National AIDS Prevention 1-800-872-8378
National Mobilization Against AIDS (a national clearinghouse for all kinds of AIDS service organizations): 1012 14th St., N.W., Suite 601, Washington, D.C. 20005. 202-347-0390
For a 20-minute educational film on the transmission of AIDS entitled "Sex, Drugs, and AIDS," write to ODN Productions, Suite 304, 74 Varick St., New York, NY 10013.

Hepatitis and Human Papilloma Virus

Hepatitis There are five forms of viral hepatitis (Figure 16–5[b]). The best known of these are Hepatitis A (known as infectious hepatitis), Hepatitis B (known as serum hepatitis), and Hepatitis C. They all cause inflammation of the liver, but Hepatitis B, in particular, can lead to serious damage to that organ and possibly death. Hepatitis A is spread by contaminated feces and, so, sexual contacts involving the anus can lead to infection. However, this form of hepatitis can also be spread by other nonsexual means including drinking or eating contaminated water and food. This has been a problem for suppliers of seafood and harvesters of shellfish when the areas in which these commercial organisms grow has been contaminated with human refuse.

Hepatitis B is transmitted by contaminated blood and body fluids such as saliva and genital fluids. This form of the disease is extremely prevalent both in the United States and worldwide. As many as 200,000,000 people may be infected. Hepatitis B can cause severe liver damage, such as cirrhosis and cancer. It is spread predominantly by sexual contacts, with the incidence in homosexual males accounting for 50% of the cases in the United States. Sharing needles for drug use is also a source of infection. Pregnant mothers can pass the infection on to their babies. The availability of effective vaccines to Hepatitis A and B is also providing excellent prophylaxis against infections and is particularly important to protect health care professionals and others who have a high probability of being exposed to the virus.

Hepatitis C is a recently discovered form of the disease. As in its alphabetical cohorts, it will probably be shown to be spread predominantly through sex and sharing needles. The symptoms can range from nonexistent to serious and fatal.

Papilloma Viruses Human papilloma viruses (HPV) are the cause of what have been described as venereal warts. The designation "venereal" is derived from Venus, the Olympian goddess of love, although once contracted, this disease would hardly be considered loving in nature. The viruses are spread by sexual intercourse and other forms of sexual contact and are highly contagious. They can cause large lumpy outgrowths in many areas of the body but are most common on the moist areas of genitalia and anus. However, the virus need not result in the formation of these obvious outgrowths. They can infect cells without accompanying growths, and they are just as infective even if their effects cannot be visibly observed. The most aggravating problem with this highly prevalent disease occurs for females. Infection with certain forms of the virus (particularly in those forms designated HPV 16 and HPV 18) increase the chance for the development of cervical cancer. Males are not safe from this transformation, however, as the incidence of cancer of the penis also goes up in association with the infection. As in the case of all the viruses discussed in this chapter, there is no cure for HPV. However, knowledge of the symptoms and care in their management with your physician can protect against the potentially dire outcome of their presence in your body.

OTHER DISEASES TRANSMITTED BY SEXUAL ACTIVITY

In addition to bacteria and viruses, there are other types of organisms that cause human health problems that fall into the category of STD. The protozoan *Trichomonas vaginalis* grows well in the vagina and urethra of women and causes a disease known as *trichomoniasis* (Figure 16–11a). Unlike the bacterial pathogens we have discussed, this protozoan can survive outside the genital tract quite well in urine or water for an extended time. This means that the disease can be "caught" from a wet toilet seat, or other moist medium, with which the body may come in contact. Infection in females results in a foul-smelling, frothy discharge from the vagina that itches constantly. However, other than smell and itchiness there is little or no damage done to the

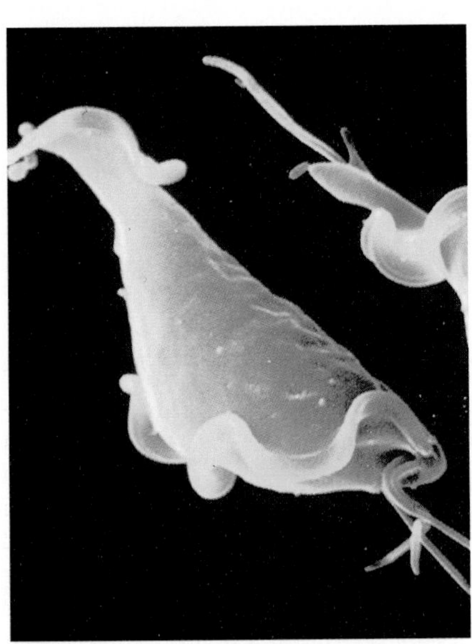

(a)

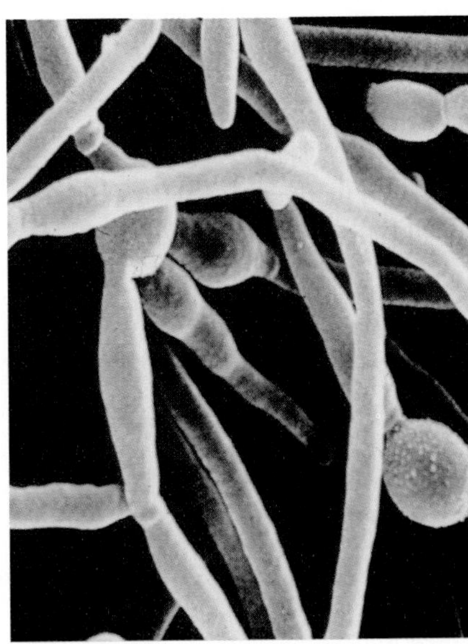

(b)

Figure 16–11 Other Sex-Related Diseases
The protozoan *Trichomonas vaginalis* (a) and the fungus *Candida albicans* (b) are also agents of disease transmitted by sexual activity.

reproductive tract. Trich, as it is called, can be passed to a male during intercourse, but males tend to be asymptomatic. Males who do not know that they have the infection may become carriers and, unknowingly, pass it on to new partners.

The yeast **Candida albicans** is another infectious agent that is considered to be sex related (Figure 16–11b). Infection from this organism is very common in women but is transmitted by sex only in a small percentage of incidences. This species of fungus is a normal inhabitant of the vagina and the real problem emerges when the conditions within the vagina change so as to induce abnormal growth of the organism. Antibiotic treatments (for other unrelated infections) may alter the microbial population of the vagina, or changes in hormones may trigger the overgrowth. The symptoms are a malodorous, intensely itching white, thick discharge from the vagina. The form of the disease in which the fungus grows on the tongue and linings of the mouth is called *thrush.* Effective treatment is usually in the form of antifungal compounds prescribed by a physician.

A DIVERSITY OF FORMS

There are many types of organisms that can be passed sexually from one person to another. Some of the most serious of those are discussed here. Many of the agents of sexually transmitted or sex-related diseases can be treated easily and effectively with various kinds of antimicrobial compounds. Even the more serious bacterial diseases, such as syphilis and gonorrhea, can be cured with little or no permanent damage if treated in a timely fashion. "A timely fashion" requires that a person have a working knowledge of symptoms and an awareness of significant changes in personal bodily functions associated with these diseases.

The viral STDs are not so easily dealt with. They persist in a human body for a lifetime as residents within cells and tissues. Their damage may be done by interfering with the normal function of many different kinds of cells, as in the eruptive sores in skin and eyes caused by herpes simplex viruses, or by targeting specific cell types (that is, T-lymphocytes for HIV), or even by increasing the chance for cancerous transformation of epithelial cells as with human papilloma viruses.

RULES TO LIVE BY

No one wants a sexually transmitted disease. The sure way to avoid the most dangerous forms of these diseases is to abstain from sex altogether. Barring conversion to celibacy and a monastic lifestyle or similar behavior, there are recommendations that have been made that may be useful in avoiding contracting these types of infections.

1. Limit the number of sex partners that you have or will have. Do not sign your own death warrant by having indiscriminate sex or by having sex with intravenous drug users or prostitutes. Remember, when you have sex with someone, you are in some ways having sex with everyone that person has had sex with. Monogamous sex with a faithful partner is a good way to ensure safety and pleasure for both of you.
2. Use condoms and spermicidal compounds. When in doubt about the health status of an individual, either you or your partner should use a condom to provide a barrier to exchange of, or contact with, bodily fluids.
3. Examine your partner for symptoms of an STD. Be aware of the symptoms of various STDs. This may be futile in the case of potential partners who are asymptomatic, but it is better to be wary than not. Wash genitals thoroughly after sex, and urinate to flush organisms from the urethra. Be aware of changes in your own body and make regular visits to your doctor. Ask specifically about tests or examinations for STDs.

Sex is a perfectly natural human biological activity. Ensuring that sex is safe in a less-than-perfect world is a personal and social necessity.

Summary

Descriptions of diseases transmitted by sexual interactions have been reported for thousands of years, though clearly no understanding of the nature of the agents was known. Today, we know that sexually transmitted diseases, or STDs, are caused by a variety of microbial agents. These include bacteria, viruses, fungi, and protozoans. The bacterial infections of syphilis, gonorrhea, and chlamydia are particularly prevalent in human societies and are increasing. Bacterial infections are treatable by antibiotics, though resistance of bacteria to such drugs is on the rise.

Viruses are not so easily treated. The herpes simplex group, papilloma viruses, human immunodeficiency virus (HIV), and hepatitis viruses are all permanent residents of the cells of the human body once they have infected it. In addition to the overt diseases caused by these infective agents, they may predispose an individual for cancer.

Among this group of infective agents, HIV is the deadliest of the viruses listed, and infections may lead to AIDS, or acquired immunodeficiency disease syndrome, and, ultimately, to the demise of the immune system. This subjects the individual to lethal infections by other microbial or viral agents.

Fungi and protozoans also cause infections that can be transmitted through sexual interactions. The yeast infections caused by *Candida albicans,* which are manifested mostly in women, are irritating but not deadly. This is also true of the protozoan infection known as trichomoniasis, whose itching and vaginal discharge do little or no damage to the reproductive system.

Knowledge of these diseases is fairly low in our society, particularly among the most sexually active groups of teenagers and young adults. Knowledge of the diseases, the agents of transmission, the means of transmission, and the treatments is of fundamental importance for personal health and for the health of all.

Questions for Critical Inquiry

1. Where is the human immunodeficiency virus thought to have originated? Why is it important to know?
2. What does it mean to be asymptomatic with respect to an STD? Why is it a problem?
3. Is the blood supply of the United States safe? How safe? Why (with respect to agents of STDs) is it safe to give blood?
4. What rules do you apply to yourself with regard to sex?

Questions of Facts and Figures

5. What species of bacterium is the agent of syphilis? What are the stages of this disease and how effective is treatment at each stage?
6. Name four viruses associated with sexually transmitted diseases. How are they treated? Can they be cured?
7. Are any of the viruses associated with STDs also associated with cancer? Which ones?
8. What types of organisms, other than bacteria and viruses, are associated with diseases transmitted by sexual contacts?
9. What is AIDS?
10. Is HIV always contracted through sex? If not, what other ways are there?

References and Further Readings

Centers for Disease Control (1994). Morbidity and Mortality Weekly Report 43:826.

Cowley, G. (1993). "Bad News on Two AIDS Fronts." *Newsweek* 122 (Aug 2):62–63.

King, B. M., Camp, C. J., and Downey, A. M. (1991). *Human Sexuality Today.* Englewood Cliffs, NJ: Prentice Hall.

Rosenberg, P. (1995). "Scope of the AIDS Epidemic in the United States." *Science* 270:1372–1375.

United Nations Population Division (1991). *World Population Prospects 1991.* New York: United Nations.

U.S. Department of Commerce, Bureau of Census, (1995). *The National Data Book; Statistical Abstracts of the United States.* Washington, DC: USGPO.

CHAPTER 17

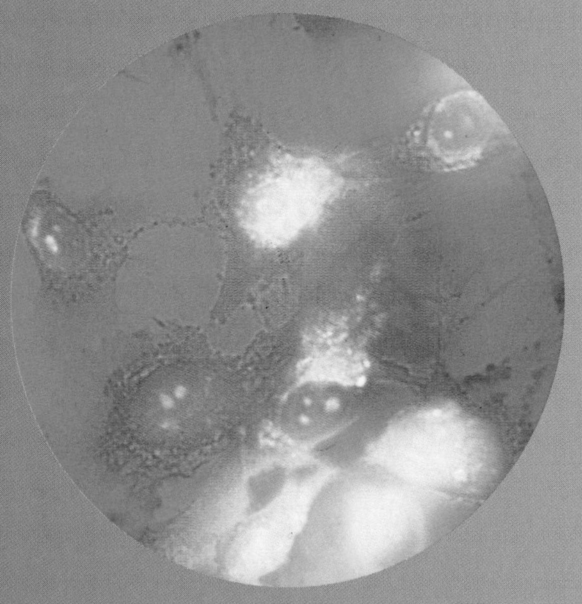

The Nervous System

Between the idea and the reality
Between the motion and the act
Falls the shadow.

T. S. Eliot, "The Hollow Men"

INTRODUCTION

The most important integration of form and function in the human body is mediated by the nervous system. How is it that we can learn to play the piano or swing a baseball bat and actually hit the ball (some of us) being hurled toward us by a pitcher? How do we learn to speak, read, and write? These are activities that many of us take for granted in the ordinary course of events in our lives. At an even deeper level, we might ask how an artist creates a visual image or a poet captures a sense of action in a metaphor. The Cubist images of Pablo Picasso ask us to suspend our disbelief of what is real and what is not. The lines of poetry written by T. S. Eliot, in "The Hollow Men", suggest great uncertainty about how we actually accomplish our intended actions. Yet, these artistic urges and accomplishments arise from the same source as simply speaking, reading, and writing. In fact, all our perceptions and actions, from the simple to the sublime, require the most intricate and integrative organ system yet produced by Nature, the human nervous system.

NERVOUS SYSTEM VERSUS ENDOCRINE SYSTEM

As briefly alluded to in Chapter 13 in describing the functions of the endocrine system, a primary difference between the endocrine system and the nervous system is the rate at which signals are received by cells and acted upon. The diffusion of endocrine hormones through the circulatory system and tissue fluids takes only a few seconds to occur, but the cellular response to such signals may take hours or days to fully develop. The nervous system is much faster (Figure 17–1). A fair temporal comparison of the two systems might be that in the endocrine system we mail letters and wait days for a response, while in the nervous system we pick up the phone and dial direct. The principal cells of the nervous system have special properties of excitability and signal transmission that make intercellular communication much more rapid and cellular response times almost instantaneous. The signals that are sent along the network of cells that make up the nervous system are electrochemical impulses propagated along the plasma membrane of neurons by the movement of ions. The rate

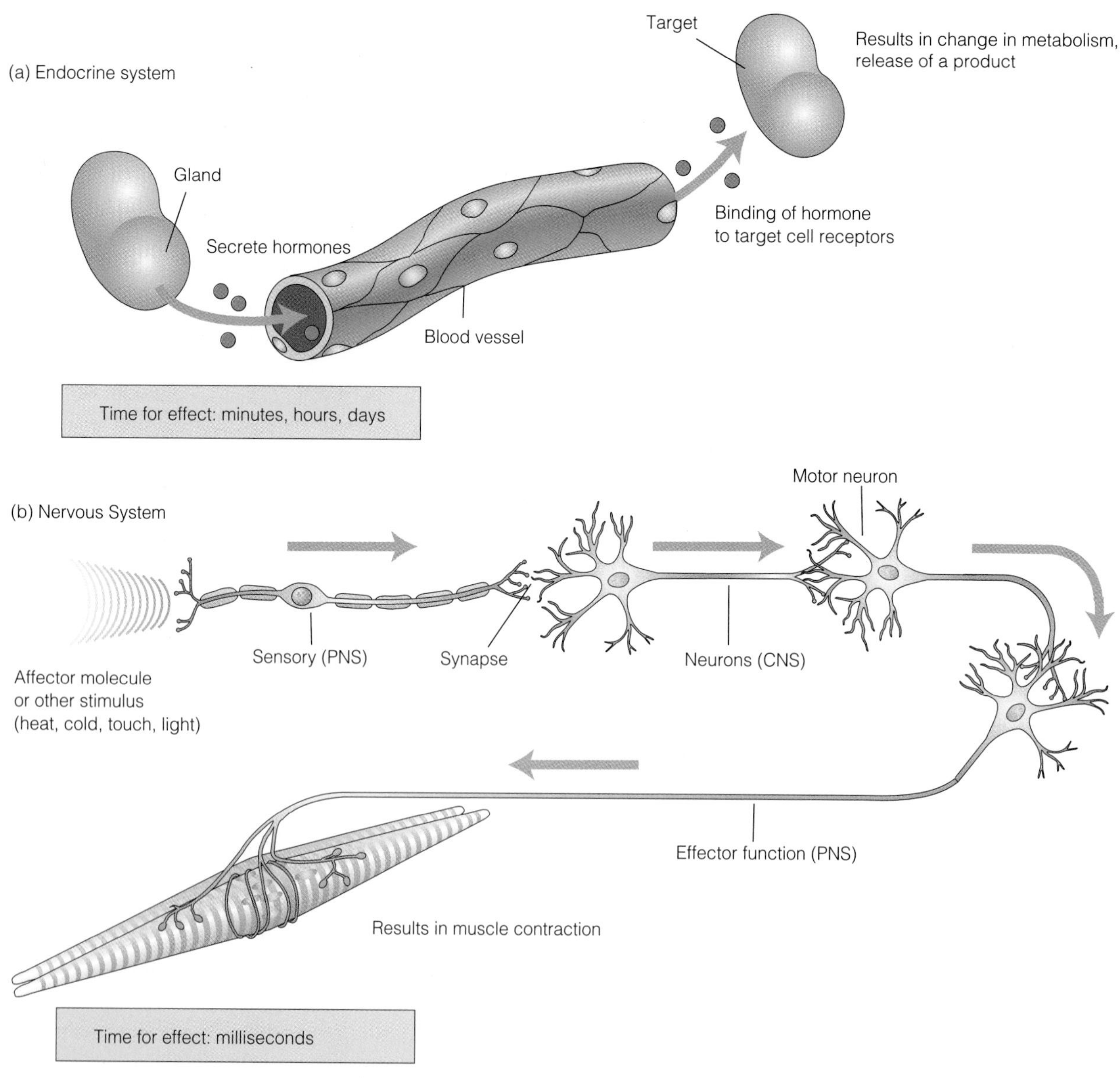

Figure 17–1 Comparison Between Endocrine System and Nervous System
The endocrine (a) and nervous (b) systems differ with respect to the mechanisms by which they affect target tissues.

at which these signals travel is on the order of 100 meters a second. This means that a signal from the brain to move a finger may take only 10 milliseconds! The international telephone system might well be thought of as a model of connectivity and communication to help envision the nature of the human nervous system, a difference being that the velocity of impulses along fiber optic transmission lines is near the speed of light.

The first part of this chapter will focus on the cells and tissues that compose the human nervous system. This will provide a foundation for understanding the function of these cells and tissues in signal transmission, which will be the focus of the second part of the chapter. The means by which cell structure and function combine in the nervous system are of fundamental importance not only for understanding the mechanisms that underlie our abilities to move, see, and speak but also for understanding complex human behaviors and the nature of our unique consciousness.

ANATOMY OF THE NERVOUS SYSTEM

Neurons

The basic structural unit of the nervous system is the neuron. Neurons have a number of morphological characteristics that make them easy to recognize (Figure 17–2). The most significant characteristics are the axon, the cell body or soma, and the dendrites. Neurons may be extremely long and thin and interconnect among themselves and with many different types of cells in the body. For example, the neurons that originate in the spinal cord (known as **motor neurons**) may be one to two meters in length in humans and connect with muscle fibers in distant parts of the body, such as those controlling the hands and feet. Motor neurons found in larger animals, such as elephants or giraffes, may be many meters in length.

Axon The longest and thinnest cellular structure of a neuron is the axon. An axon is an extension of the cell surface that arises from the cell body and connects to target sites of muscles and glands. There is only one axon for each neuron. In the major nerves of the body, thousands of axons may be bundled together and distributed through the body before branching off to reach their target sites (Figure 17–2).

Cell Body The cell body, or soma, of a neuron is its central structural feature (Figure 17–2). It contains the nucleus and most of the metabolic machinery including endoplasmic reticulum, ribosomes, mRNA, tRNA, and the Golgi apparatus. Materials synthesized in the soma are transported to the end of the axon by an elaborate system of neurofilaments and microtubules. Transport of materials also occurs from the end of an axon to the cell body. Thus, the neuron is capable of recycling useful molecules to maintain its functional capacities.

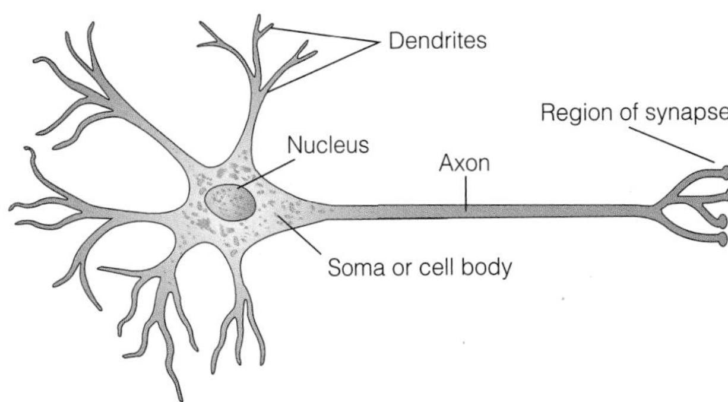

Figure 17–2 A Neuron
A neuron has three basic morphological features: the cell body (or soma), the dendrites, and an axon.

Dendrites Another type of extension associated with a neuron is known as the dendrite (Figure 17–2). The origin of the word "dendrite" is tree and dendrites often form multiple branches from the soma. In fact, neurons may have many dendrites. Dendrites are shorter than axons and remain in the vicinity of the soma. These branchlike extensions interact with axons from other neurons to establish the neural networks needed for effective function of the nervous system.

To briefly summarize, a neuron has three basic parts—an axon, a soma, and dendrites. Most organelles and metabolic functions are associated with the soma. The connections between neurons are established at the ends of axons through the formation of synapses, as will be discussed later in the chapter.

Morphology of Neurons

The number of branches that a neuron possesses is a key criterion for its categorization. Neurons come basically in three forms—*unipolar, bipolar,* and *multipolar* (Figure 17–3).

Unipolar Neurons A unipolar neuron (Figure 17–3) is quite rare in the human body but is best approximated by the shape of sensory neurons found in spinal ganglia. A spinal ganglion is a group of nerve cells whose cell bodies are clustered together and whose extensions pass out of the cluster and connect to distant target sites. The shape of unipolar neurons reflects the fact that the axon and dendrite of each neuron arise from the soma together and run in opposite directions. Because these are sensory neurons, the dendrite is located in the periphery and the axon is targeted toward the central nervous system.

Motor neurons neurons involved in stimulating muscle cell contraction

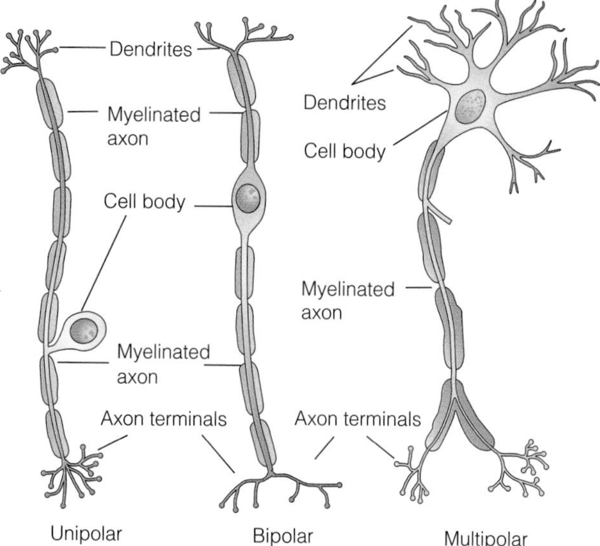

Figure 17–3 Three Types of Neurons
The three basic types of neurons, based on morphology, are shown here as unipolar, bipolar, and multipolar.

Bipolar Neurons The second type of neuron is known as a bipolar cell (Figure 17–3). In this case, the cell has a clearly defined single axon and a single dendrite. An example of this type of nerve cell is found in the **neural retina** of the eye (Figure 17–4). Retinal bipolar cells are involved in making connections between the neurons that respond to light (the **photoreceptors**) and the cells (*ganglion cells*) that carry the information about light stimulation to the brain (see Chapter 18 for more details). The individual axons and dendrites are considerably shorter than the combined extension of the unipolar cell, which serves to point out that it is not the length nor the specific shape of a neuron that is of greatest importance but how it provides connections for proper signal transmission. The most revealing definition of a neuron is, therefore, a functional one; a neuron is a cell that is capable of generating and propagating nerve impulses and delivering those signals to target cells. Bipolar and unipolar cells act as nerve cell bridges to form a signal conduit to the brain.

Multipolar Neurons The final category of neurons is the most prevalent in the human body. Multipolar cells have an axon and multiple dendrites (Figure 17–3). The longest cells of the human body are multipolar neurons, such as the motor neurons of the spinal cord. The extreme length of the axon of a motor neuron provides an uninterrupted route along which a **nerve impulse** may travel to its target. In addition to the length of an axon that is necessary to establish uninterrupted routes to a target, the diameter of an axon determines the velocity of the nerve impulse—the larger the diameter, the faster the nerve impulse can travel along it. Axons of large caliber can carry impulses at transmission rates of 100 meters per second or more. Dendrites of a multipolar cell by comparison are much shorter than the axon. The cell body of a motor

> **Neural retina** complex layers of nerve cells in the eye
>
> **Photoreceptors** light-sensitive cells in the neural retina; rods and cones
>
> **Nerve impulse** the electrical signal carried along the length of a neuron

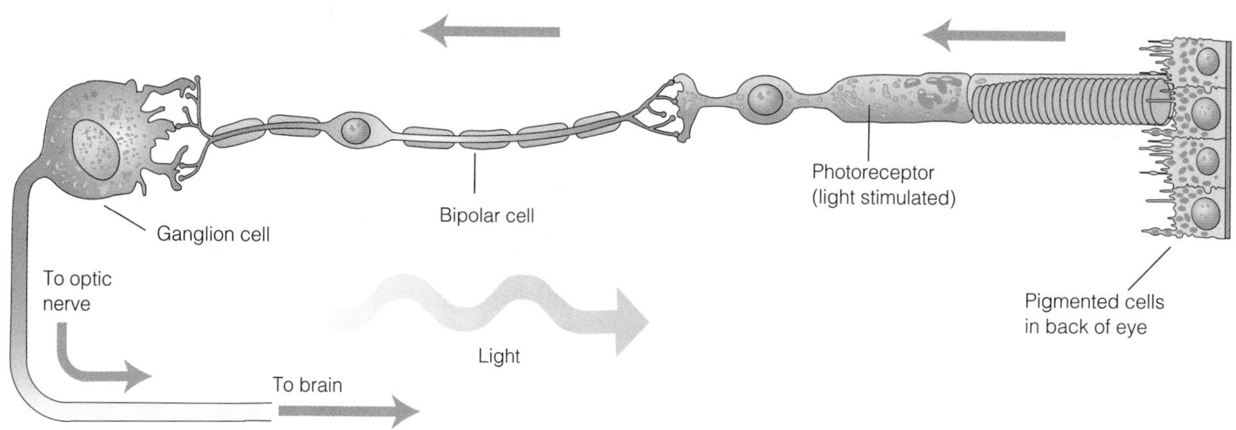

Figure 17–4 Connection of Photoreceptor Cell to Bipolar Cell to Ganglion Cell in the Retina
The connection between a photoreceptor cell, a bipolar cell, and a ganglion cell provides a pathway for light stimuli to be passed to the brain.

neuron and its dendrites are located within the protective structure of the vertebrae.

What might be the basis for having so many dendrites? The answer to this question is connections. For the nervous system to run an efficient signaling service for the body, there must be an extensive "interconnectedness" of the cells that generate and carry the signals. A single connection between neurons is not sufficient to maintain the function of this network. In fact, it is not uncommon for a single neuron to be simultaneously connected with hundreds or thousands of other neurons. The magnitude of this interconnection reaches a zenith in the *cerebellum,* where a single *Purkinje cell* may have associations with 25,000 other neurons. Purkinje cells are multipolar neurons with tremendously elaborate branches of dendrites arising from the soma.

Glia

Before going further into a description of the anatomy of the nervous system, let us consider another group of cells found in association with neurons, the glia. Glia are not neurons. They are not excitable and do not have the capacity to generate and propagate nerve impulses. However, glia (the word means glue) are important and often essential structural elements of the nervous system. They provide the adhesion that holds the neurons together and the insulation that prevents short circuits between neurons. They have the capacity to provide neurons with nutrients and neurotransmitter precursor molecules. The complementary properties of glial cells to those of neurons establishes a nervous system in which electrically active cells are stabilized in the flexible grip of a living glue and insulated from one another so short circuits are prevented. There are many times more glial cells in the nervous system than neurons, attesting to their importance as structural and functional cellular elements.

Types of Glial Cells

There are several different types of glial cells: Schwann, oligodendroglia, astrocytes, and *microglia* (Figure 17–5). Schwann cells are found exclusively in the **peripheral nervous system (PNS),** and oligodendroglia, astrocytes,

Peripheral nervous system (PNS) nerves outside the protection of the skeleton

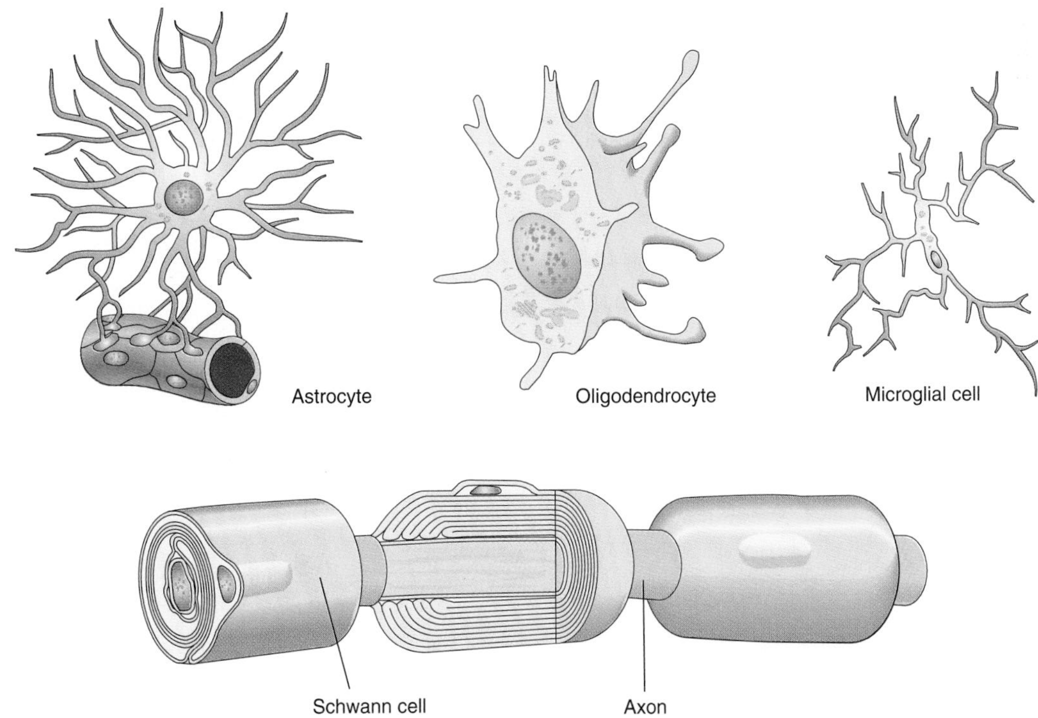

Figure 17–5 Glial Cells
There are four basic types of glial cells. Astrocytes and oligodendrocytes are found in the central nervous system (CNS). Schwann cells are found in the peripheral nervous system (PNS). Microglia are phagocytic cells found in the CNS.

and microglia cells are found only in the **central nervous system (CNS).**

Schwann Cells The Schwann cell is named after the famous nineteenth-century anatomist Theodor Schwann who discovered it. This unique cell wraps itself around the axons of motor and sensory neurons in layer after layer of its own plasma membrane (Figure 17–6). These layers are referred to as *myelin.* The process is called *myelination* and is of critical importance in insulating the neurons of the PNS. By the time a Schwann cell has finished wrapping a neuron, it looks a lot like a jelly roll. This snugly fitting, multilayered membrane seals off the neuron from potential interference from outside. This includes preventing lateral contacts with other neurons, as well as isolation from ions and molecules found in interstitial fluids of tissues through which nerves must pass on their way to distant targets. Because axons in the PNS are tremendously long, it may take hundreds of thousands of Schwann cells to effectively myelinate them. The wrapping is not completely uniform along the length of a myelinated axon. There are nodes apparent where one Schwann cell ends and the next begins. These nodes are called the *nodes of Ranvier,* and they are very important in increasing the velocity of conduction of a nerve impulse.

Oligodendroglial Cells Myelination also occurs in the CNS. However, instead of Schwann cells being involved, the job is accomplished by another type of glial cell known as an oligodendroglial cell. As imposing as this name appears, it has a simple derivation. "Oligo" means many, "dendro" means treelike, that is, with branches, and "glia" refers to glue. What we have then, is a glial cell with several branches (Figure 17–7). Oligodendroglial cells are also known as oligodendrocytes and are able to myelinate axons, but not exactly as described for the Schwann cell. A Schwann cell interacts with only one axon, and an oligodendroglial cell is capable of interacting with the axons of several different neurons simultaneously. This allows for some complicated arrangements among glia and neurons but is structurally important for holding the organization of the nervous system in place. Not all neurons are fully myelinated. However, this does not mean that the axons are not insulated by glia. Unmyelinated axons are simply those whose wrapping by oligodendrocytes is less tight fitting and extensive than the jelly roll configuration (Figure 17–7).

If something goes wrong with the process of myelination or with the stability of the myelin sheath, serious structural and functional problems may arise. For example, the disease multiple sclerosis (MS) is an autoimmune condition in which the human body's immune system attacks itself. The target of the attack is myelin. Destruction of the myelin sheath gives rise to short circuits between neurons, which in turn interferes with the transmission of nerve impulses along axons.

Central nervous system (CNS) nerves within the skull and vertebrae; protected by enclosure within the skeleton

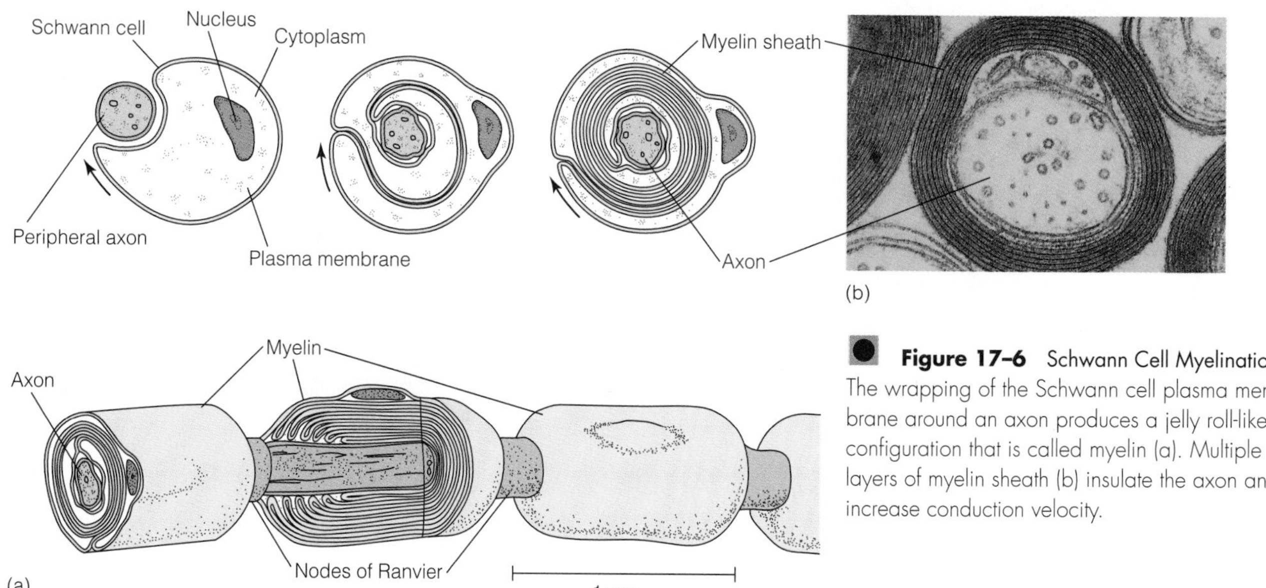

Figure 17–6 Schwann Cell Myelination The wrapping of the Schwann cell plasma membrane around an axon produces a jelly roll-like configuration that is called myelin (a). Multiple layers of myelin sheath (b) insulate the axon and increase conduction velocity.

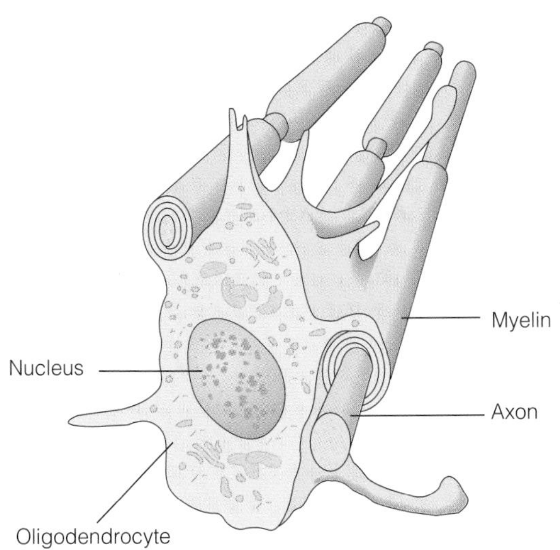

Figure 17–7 Myelination of the Central Nervous System—Oligodendrocytes

One oligodendrocyte provides myelin sheath for several axons in the CNS.

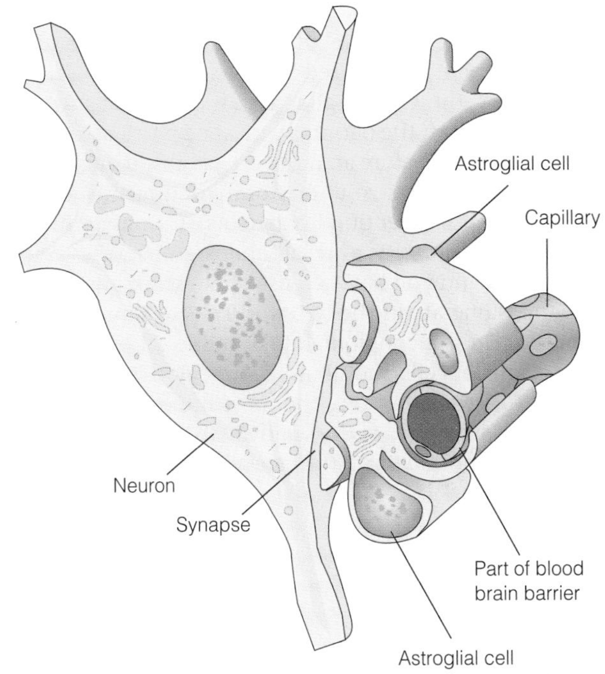

Figure 17–8 Astrocytes

Astroglial cells are found in the CNS (a) and help form the blood-brain barrier (b).

Astroglial Cells The other major type of glial cell in the CNS is the astrocyte. These star-shaped cells (Figure 17–8) have a distinctly different role in the structure of the CNS from their cousins the oligodendrocytes. Astrocytes help to establish a blood-brain vascular barrier, as well as provide nutrients and maintain connectivity of neurons. Important small molecules such as water, oxygen, carbon dioxide, and glucose easily pass through the blood-brain barrier, but other types of molecules such as antibodies, blood proteins, and many antibiotics do not (Figure 17–8). Astrocytes do not form myelin sheaths. They are structural elements and associate with the basement membranes surrounding capillary blood vessels to restrict passage of many classes of compounds into the brain proper.

A comparison of astrocyte and oligodendroglia provides the following picture of their structural roles in the CNS. Oligodendrocytes surround neurons with insulation in a process known as myelination. Myelin is a product of the plasma membranes of these cells wrapping around the axons of several different neurons. This prevents interneuronal contact. Astrocytes wrap their plasma membrane around blood vessels, particularly capillaries, and help prevent the diffusion of many types of molecules from blood to brain. Both types of glia provide a living glue that supports the structural cohesiveness and organization of the billions of neurons in the CNS. Schwann cells myelinate axons in the PNS and, unlike oligodendroglia, are associated with only one axon.

Microglial Cells The fourth type of glial cells are the microglia (Figure 17–5). Strictly speaking, this cell is probably not a glial cell in the same sense as astrocytes and oligodendrocytes. The microglia are scavengers. They migrate within the brain and collect and eliminate debris and/or foreign material that may have entered the CNS. Their activities are similar to that of macrophages in the rest of the body. In fact, the microglia are related more to the family of phagocytic cells than to the glue-insulator cells that hold the CNS and PNS together.

HIGHER ORDER STRUCTURES

In general, the cells composing the nervous system serve three basic needs: (1) collecting information, (2) integrating information, and (3) acting on information. These needs and the processes used to fulfill them are of a continuous nature; that is, they are occurring all the time. Nerve cells are necessary to maintain constant control of body functions, particularly those associated with homeostasis. Using the cells integrated into the nervous system in this way, the human body is able to act (or react) to changes in external and internal environments and to optimize the chances for survival of the individual.

The Brain

The brain acts as the center of control for the entire human nervous system. It is housed in the cranium and is a principal part of the CNS (Figure 17–9). Nearly all incoming information from the periphery must be processed in some way by the brain before an individual is able to act or react. Almost every action, voluntary or involuntary, is initiated by the brain (an exception is **reflexes**). Combine this control with the capacity to recall and use memories of past experiences and you have a sense of how humans and other animals operate in the real world (see Chapter 19). The human brain is a conglomerate of many different specialized regions of cells and fibers. Each region is dedicated to specific functions. Among the most important and easily recognized structures of the brain are the *cerebrum,* the *thalamus* and hypothalamus, the cerebellum, the *midbrain,* the *medulla oblongata,* and the *pons* (Figure 17–9). The extension of the CNS of the head into the spinal cord is reflected in changes in organization but not in continuity of connections of nerve cells, as will be discussed later in the chapter.

The Cerebrum and Its Hemispheres

Sulci and Gyri The cerebrum is the seat of memories and consciousness. It is what most of us think of when we use

> **Reflexes** innate and automatic neuromuscular response to internal or external stimulus

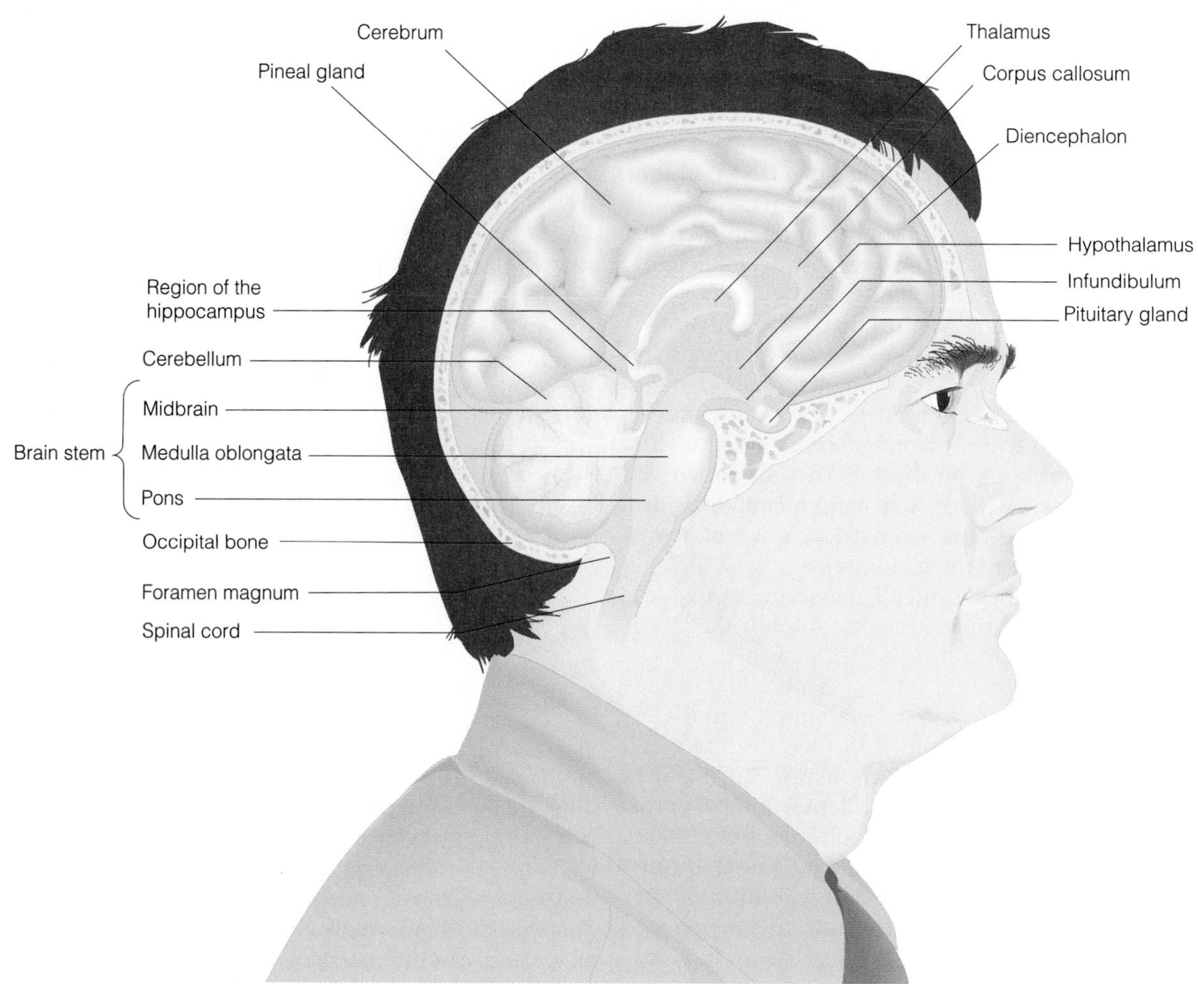

Figure 17–9 Brain and Brain Stem

the term "brain" (Figure 17–10). The cerebrum is actually two distinct interconnected structures within the cranium. These structures are shaped like hemispheres or domes. Each hemisphere is extensively folded upon itself, giving the characteristic surface features of fissures and folds (**gyri**) with grooves (**sulci**) between them, as observed in diagrams of the brain (Figure 17–10). The connection between the two cerebral hemispheres is called the *corpus callosum* and represents hundreds of millions of nerve fibers crossing between the left and right hemispheres (Figure 17–9).

Ventricles The cerebral hemispheres develop from the front part of the brain (the forebrain) during embryogenesis. The tissue substance of the cerebrum and the rest of the brain surrounds a set of interconnected cavities known as *ventricles*. Each cerebral hemisphere has its own ventricle. These two cavities communicate with the much smaller third, fourth, and fifth ventricles and, eventually, with the *central canal* of the spinal cord (Figure 17–9). The ventricles are filled with a lymphlike fluid called **cerebrospinal fluid (CSF),** which circulates throughout the ventricular system. CSF is produced by tissues known as choroid plexuses. CSF is also found within the spaces of the layered membranes that surround the brain. These membranes are called the **meninges** (Figure 17–11).

Meninges In addition to being encased within the cranium and vertebrae, the brain and the spinal cord are surrounded by a series of layers of membranes. These membranes are known as the meninges. There are three layers of meninges—the *dura mater,* the *arachnoid mater,* and the *pia mater* (Figure 17–11). The dura mater is a tough layer that protects the outermost surface of the brain. The pia mater is a more delicate layer that covers the folds of the brain. The arachnoid layer and the subarachnoid space lie between the dura and the pia. It is the subarachnoid space

Gyri regions of outfolding of the cerebrum

Sulci regions of infolding of the cerebrum

Cerebrospinal fluid (CFS) the fluid filling the ventricles and central canal and surrounding the brain and spinal cord

Meninges several layers of membranes immediately surrounding the brain and spinal cord

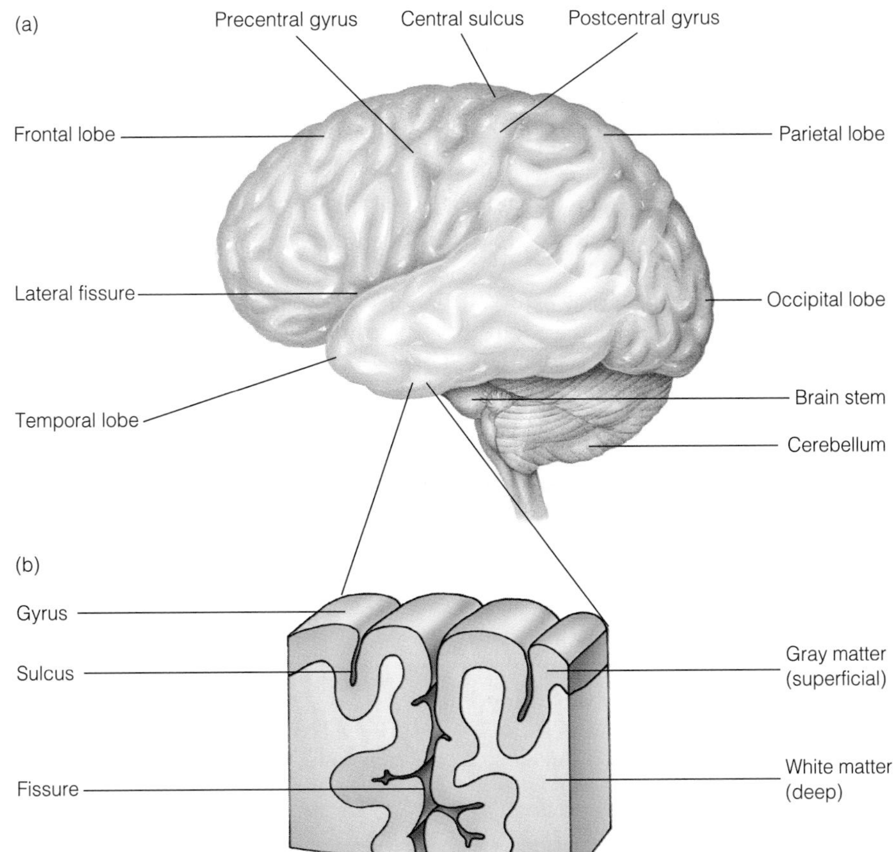

Figure 17–10 Organization of Lobes in the Cerebrum

Each cerebral hemisphere is divided into four lobes (a). The lobes are folded into a series of sulci and gyri (b). The organization of nerve cells is in gray matter (cell bodies) and white matter (fibers), which in the cerebrum are superficial and deep, respectively.

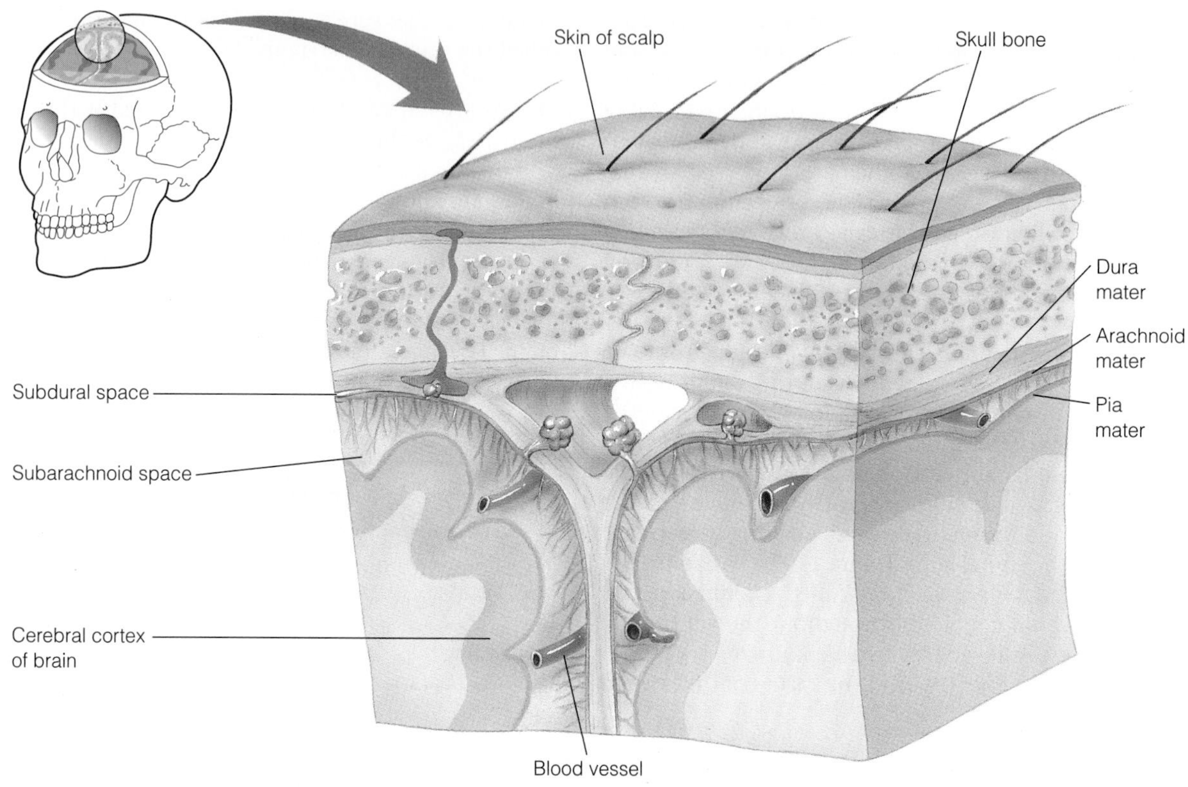

Figure 17–11 Meninges
The surface of the brain is covered with meningeal membranes of three types—dura mater, arachnoid mater, and pia mater.

that is filled with cerebrospinal fluid. The CSF in this space helps protect the brain by cushioning the effects of blows to the head or even simple movements of the head. The brain basically floats within the cavity of the cranium supported by a watery cushion.

The floating-brain plan is efficient in reducing vibrations and contact with bones, which might otherwise disrupt brain tissue. Without CSF and the system of spaces in which it flows, even nodding your head would be painful and damaging. Boxing is probably the single most dramatic example of a legitimate activity demonstrating the degree of abuse the head can take without obvious effects. Unfortunately, the cumulative effect of blows to the head, although not obvious as a result of one fight, may have later effects on a boxer's life.

Lobes There are a number of lobes of the cerebrum. They include the *frontal, parietal, occipital,* and *temporal* lobes (Figure 17–10). Each lobe has functional attributes as well. As examples, the occipital lobe is primarily involved in the processing of visual information, the temporal lobe serves in sound associations, the frontal lobe is involved with primary motor responses, and the parietal lobe is a region of sensory input and integration.

Gray and White Matter There are two distinctive layers of tissues in the cerebrum called the gray matter and the white matter. Gray matter is found in the outer or superficial layer of the brain and contains the neuronal cell bodies (Figure 17–12). Gray matter is organized in compartments or collections of cells in the CNS. For example, there are a number of islands of gray matter within the cerebral hemispheres. They are referred to as **nuclei.** Similar collections of cells in the PNS are called **ganglia.** If gray matter is more sheetlike in appearance, it is referred to as a *lamina,* as found in the cerebral cortex.

Nuclei (nucleus) regions of gray matter (cell bodies) in the central nervous system; for example, the basal nuclei

Ganglia structures containing a high density of cells, usually in the peripheral nervous system; for example, spinal ganglia

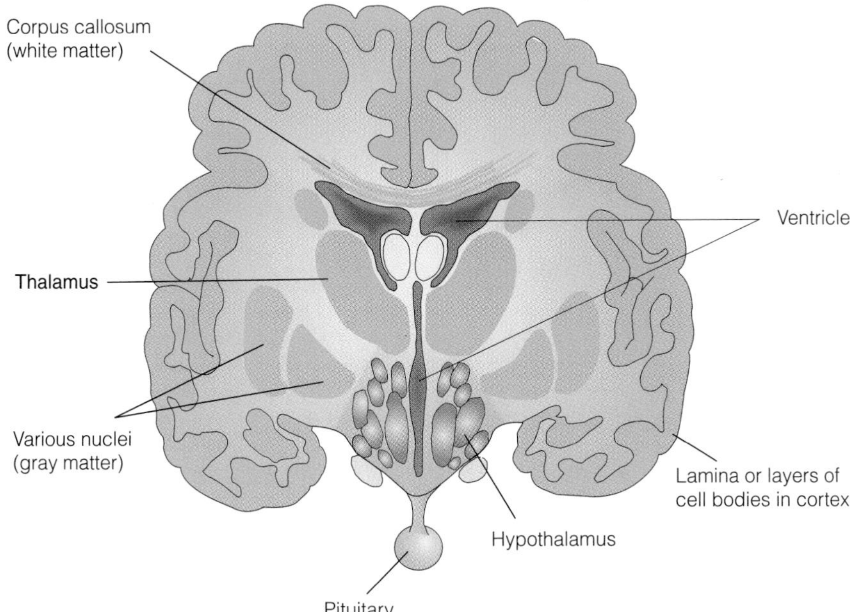

Figure 17–12 The Distribution of Gray Matter and White Matter
Gray matter is composed of layers (lamina) or islands (nuclei) of cell bodies located at the surface of the brain and within it. White matter forms the fiber interconnections made of axons that allow regions of the brain to communicate with one another.

The white matter is a deeper underlying layer, which represents the axonal fibers emanating from the soma. The white matter is made up predominantly of myelinated axons. Myelination gives a glistening, white appearance to the nerve fibers. The dense accumulation of individual axons into bundles results in tracts of nerve fibers within all regions of the central nervous system.

Thalamus and Hypothalamus The thalamus is a large oval region containing many ganglia or nuclei (Figure 17–9). These nuclei participate in most of the sensory input to the brain, including vision, taste, and hearing. The hypothalamus is a tiny region of the brain that is very important in controlling many processes involved in maintaining homeostasis. The region in which the hypothalamus is found is below and around the third ventricle. As discussed in Chapter 13, the hypothalamus is important in regulating the release of pituitary hormones from the pituitary gland. In addition, the hypothalamus is involved in heart rate regulation, smooth muscle contraction in the viscera, glandular secretions, feelings of rage and aggression, and regulation of body temperature, to name but a few.

Cerebellum The cerebellum is a large, open clam-shaped structure with deep folds similar to those observed in the cerebrum. The cerebellum is in the posterior region of the brain and separate from the overlying cerebrum (Figure 17–13). The cerebellum is in charge of coordination of bodily movements and combining the information of many sources to make this possible. Eye-hand coordination, reading and playing music, equilibrium and balance, and maintenance of posture are all examples of activities in which the cerebellum plays a role.

Midbrain The midbrain surrounds the region between the third and the fourth ventricles. Nuclei in the midbrain control eye movements, head movements, and a number of reflexes. It is the main connection area for interconnections between the upper and lower regions of the brain (Figure 17–9).

Pons The pons, as its Latin derivation implies, is a bridge connecting the spinal cord with the brain. There are a number of nuclei in the pons, including those for cranial nerves (discussed in a later section) involved directly in controlling chewing, receiving and processing facial sensations, regulating eye movement, and salivation. Signals from the *vestibule* and *semicircular canals* in the inner ears, which concern equilibrium (Figure 17–9), are initially received in subregions of the pons.

Medulla Oblongata The medulla oblongata of the brain is a direct continuation of the top of the spinal cord (Figure 17–9). It lies just above the foramen magnum and just below the pons. The white matter of the medulla oblongata contains all the nerve tracts that ascend into or descend out of the brain. An important aspect is the *reticular formation*. This region functions in arousal and consciousness. Other aspects of bodily function controlled by the medulla are breathing, dilation and constriction of blood

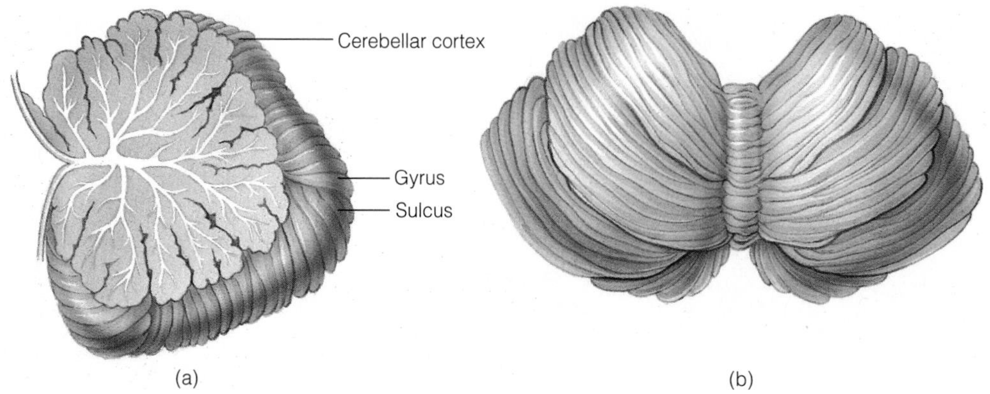

Figure 17-13 Cerebellum
The cerebellum is folded into gyri and sulci (a). The cerebellum has two symmetrical hemispheres (b) that communicate with the cerebrum to coordinate body movement.

vessels, and the coordination of swallowing, sneezing, and coughing.

Cranial Nerves There are 12 pairs of cranial nerves that arise from different regions of the brain or brain stem and control peripheral targets of muscles and glands or are involved in receiving and distributing input of information from the senses (Figure 17–14). The cranial nerves form part of the involuntary nervous system and control the parasympathetic pathways that regulate the function of many organs. In addition, some of the cranial nerves are also part of the somatic nervous system, which conveys signals from the CNS to the voluntary muscles of the skeleton.

The cranial nerves have a diverse set of targets and functions. Many of the nerves serve different parts of the same organ or structure and therefore are highly integrated in their performance. For example, the optic nerve forms synapses with neurons that send impulses to the cranial nerves III, IV, and VI that control muscles involved in eye movements. The vagus nerve (X) connects to the heart and helps regulate the rate of the heartbeat.

The Spinal Cord

The portal to the spinal cord is through the foramen magnum in the *occipital bone* of the skull (Figure 17–9 and see Chapter 6). There is complete continuity with the nerve tracts that run through the brain stem, particularly the medulla, and the nerve tracts that run through the spinal cord. The spinal cord is found within a canal formed by the articulated vertebrae sitting one atop the other to form a spinal column (Figure 17–15). Protection of the spinal cord is also afforded by the meninges, which cover both the brain and the cord, and cerebrospinal fluid that cushions the delicate nerve tissue.

A cross section of the spinal cord reveals it to contain an internal butterfly-shaped gray matter and a surrounding region of white matter (Figure 17–16). This is the reverse of the arrangement found in the cerebrum in which the gray layer is on the outside and the fiber tracts are on the inside. The spinal cord has a front and back (anterior and posterior, respectively) that serve distinct functions. The posterior region is basically sensory in function, and the anterior region contains the cell bodies of motor neurons that control muscle and glandular activity. The spinal cord has 31 pairs of nerves that extend into and out of the vertebral column (Figure 17–17). The bilaterally paired nerves are named for the region of the spine from which they arise—8 cervical, 12 thoracic, 5 lumbar, 5 sacral, and 1 coccygeal. The spinal nerves form large bundles that have both sensory and motor neurons contained in them.

Ascending and Descending Fibers In addition to the input from the peripheral nervous system to the cord and out to the periphery, there are tracts of fibers that ascend (to the brain or higher regions of the cord) or descend (from the brain or within the cord to lower levels). The neurons that serve the function of ascending or descending nerve tracts among and between levels provide a vast population of interconnecting, or *internuncial neurons.* This ensures that information from the periphery gets to all the sites in the central nervous system that are required in order to integrate function throughout the body. For example, sensory information from the bottom of the foot informs us of an irritating stone in the shoe, which in turn leads to conscious actions to remove it. In addition to consciously acting to remove an irritant, be it a stone, a thorn, or a pesky insect, the body also has a means to respond automatically and unconsciously (Figure 17–18). These reactions are known as spinal reflexes.

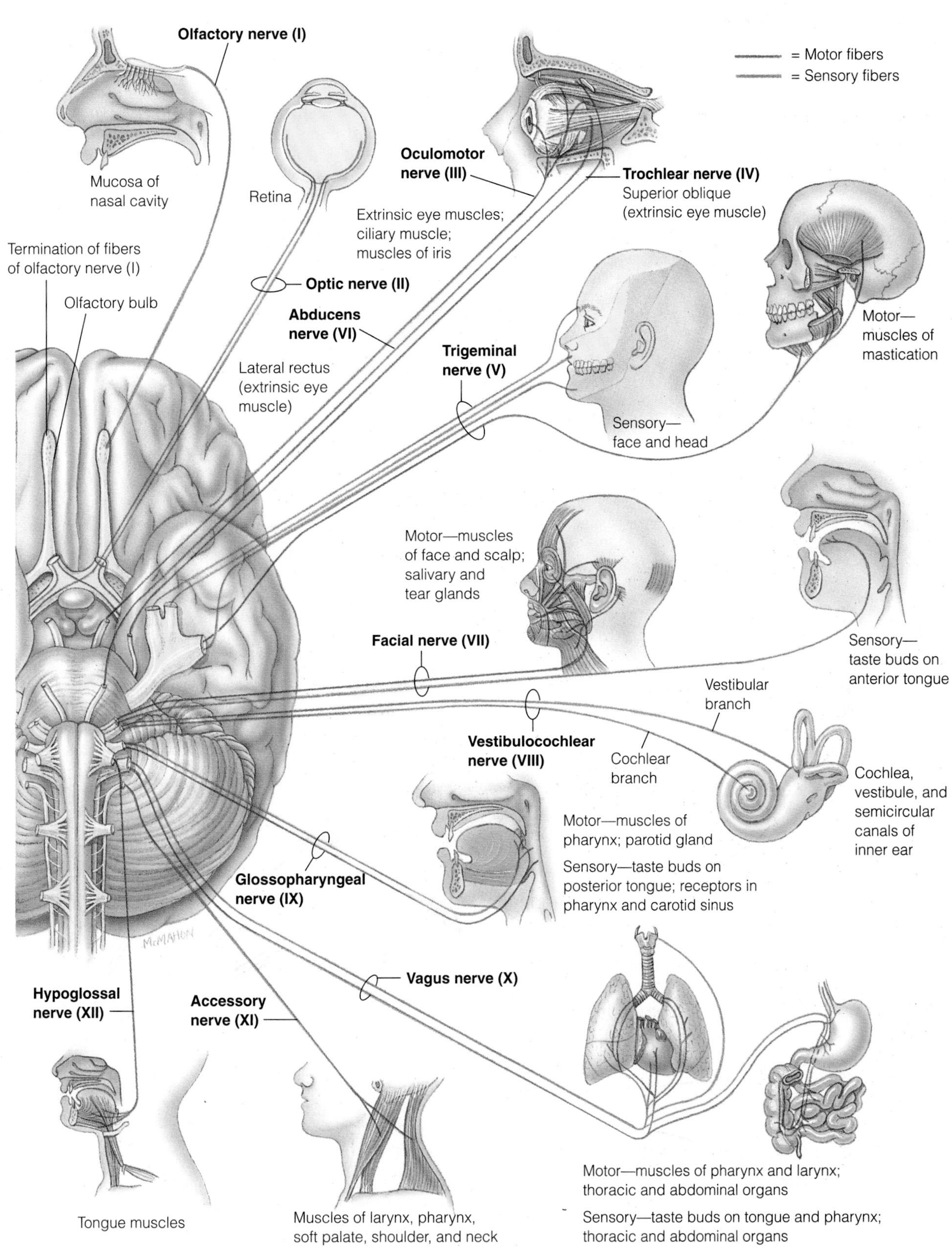

Figure 17–14 Cranial Nerves

There are 12 pairs of cranial nerves (I–XII) whose functions are diverse in controlling parasympathetic activities of the nervous system. The nerves have both motor and sensory functions.

The Reflex Arc There are situations in which the body needs to respond rapidly and without the hesitation of conscious consideration. The information that arrives in the posterior portion of the spinal cord is similar to passengers on a plane arriving at a busy airport. Not all the passengers on that plane go to the same gate for their next flight. They disperse to different gates and different destinations. The signals arriving in the spinal cord are bound for many different regions of the CNS; some regions are local, some are more distant. Connections may go up the spinal cord along ascending pathways or down the spinal cord along descending pathways. Connections may cross over to the opposite side of the cord (called *contralateral connections*) or stay on the same side (*ipsilateral connections*) or they may go directly from the posterior spinal cord to the anterior and activate motor neurons. This activation of motor neurons brings about muscle contractions that allow withdrawal of the foot shown in Figure 17–18.

This withdrawal reflex provides one of the simplest complete circuits in the human nervous system, a **reflex arc.** Most of us have vivid memories of our reflexes at work. For example, if a hand is placed on a hot burner on the stove, heat receptors in the skin begin to send frantic signals. The rate of generation of these impulses carries a message of "too hot" as it races to the posterior spinal cord. There, the impulses are passed directly to a motor neuron in the anterior region of the cord, which in turn directs a message back to various muscles in the arm to contract immediately. This contraction jerks the hand away from the hot spot before

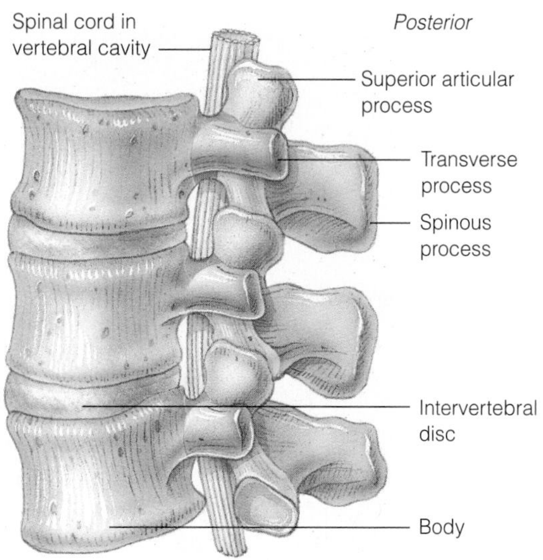

Figure 17–15 Vertebrae
The vertebrae are stacked vertically and separated by intervertebral discs. The spinal cord passes through the vertebral cavity.

> **Reflex arc** nerve connections made through the spinal cord

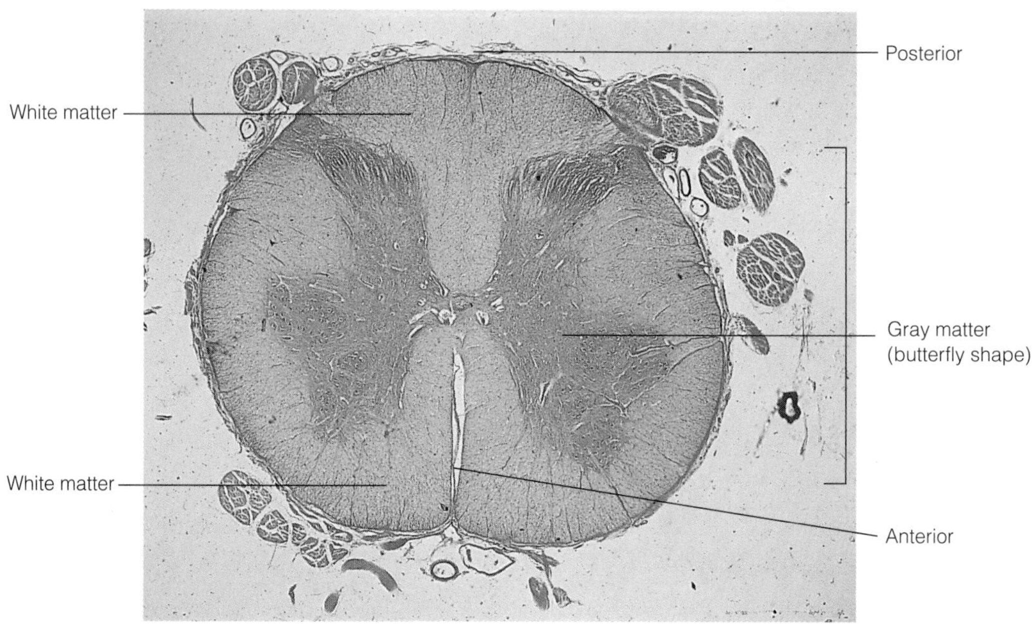

Figure 17–16 The Spinal Cord
The spinal cord is organized into white and gray matter. The gray matter is deep within the cord surrounded by fibers (see Figure 17–12 for comparison).

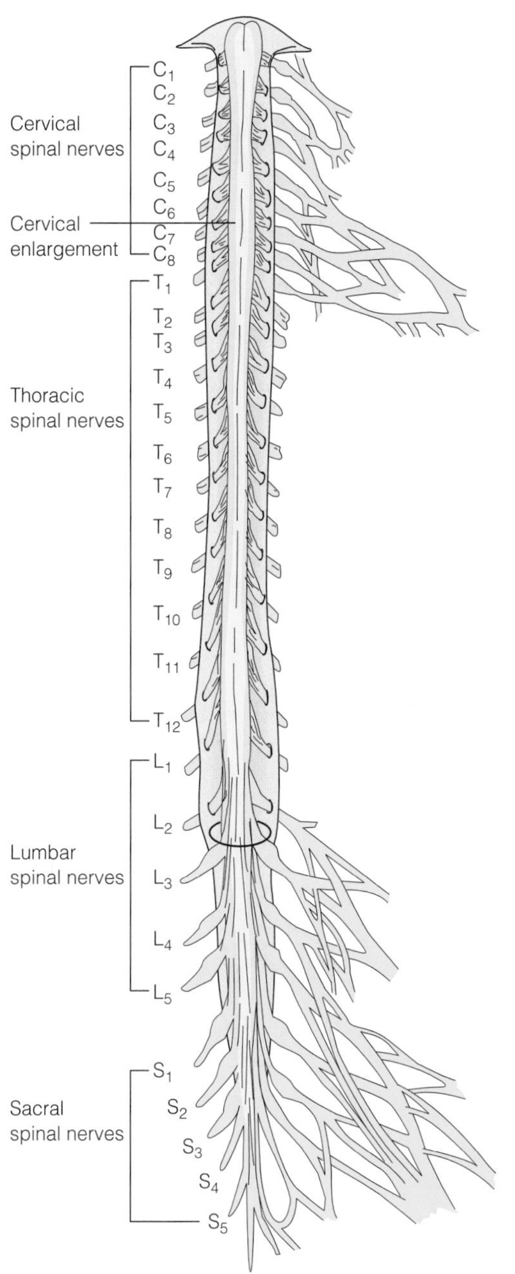

Figure 17–17 Spinal Cord and Nerves
The spinal cord exits the vertebral column to connect to peripheral targets. The cervical, thoracic, lumbar, and sacral spinal nerves are shown. (Vertebrae are designed C_1 through S_5).

the individual even has a chance to think about it. Thinking about it would require that the signal get to the brain along ascending fibers, be integrated, and return a responsive signal along descending fibers to the appropriate motor neurons in the spinal cord—the more distance the signal has to travel, the more time it takes to respond. The reflex arc is the shortest distance possible in the nervous system to respond to dangerous situations in which damage is occurring. In a more benign example, anyone who has had a physical examination knows that when the doctor or nurse taps the area just below your knee with his or her mallet, there is an automatic leg kick (called the *patellar response*) (Figure 17–19). The hammer blow stimulates receptors in the tendons of the knee, which sends an impulse through a reflex arc. If a body is functioning normally, the pattern of response is fixed. However, if something is wrong with or interrupts the circuit, as in the case of nerve damage from an accident or disease, it is immediately apparent.

Organizing a Reflex Action Reflexes are not just fast, they are accurate. To account for the accuracy of a response, the manner in which the neurons are connected must be considered. Implied in the functional responses of reflexes or sensory inputs is a map of the connections from the surface and interior of the body that precisely displays the origins of the incoming signals and the outputs from motor neurons to their targets. Signals arising in specific regions of the body do not necessarily go right back to the site of origin of the stimulus, although in the case of the hot burner they do. A more complex response would occur, for instance, if while walking along, you stepped on a nail.

First, the immediate response to such a painful signal is to jump back. In such a circumstance, for the motor neuron output to be useful, the response must occur in conjunction with muscle groups on both sides of the body and at different levels in the spinal cord. To pull your foot away from the nail, both contralateral and ipsilateral connections must be made. Balancing and pushing off with the opposite leg is coordinated with the lifting of the affected foot (Figure 17–18).

Second, the sensory neurons in the spinal cord interact with internuncial neurons ascending to the brain. Thus, information about the events in the periphery are integrated by regions of the nervous system capable of processing that information. When sensory information arrives in the brain, we are aware only then of what happened and exactly where on or in the body it occurred.

All sensory information eventually ends up being processed in the brain. A sensory neural network of regions on and in the body is connected to specific regions or areas in the brain that further control the behavioral responses to augment the reflex reaction. How the raw data from the sensory network gets to the brain and is processed and integrated is not precisely known. There are clearly well-mapped routes to and from the brain to handle specific types of signals, and within the brain itself are specific regions with special capabilities for processing selected types of sensory raw data. However, as in the lines of poetry beginning this chapter, between the input and the response "falls the shadow."

Figure 17–18 Withdrawal Reflex
The withdrawal reflex shown here results from a reflex arc and establishes the means to coordinate right leg flexion (movement away from a source of pain) and left leg extension to remove weight from the right leg. Information is sent both ipsilaterally and contralaterally.

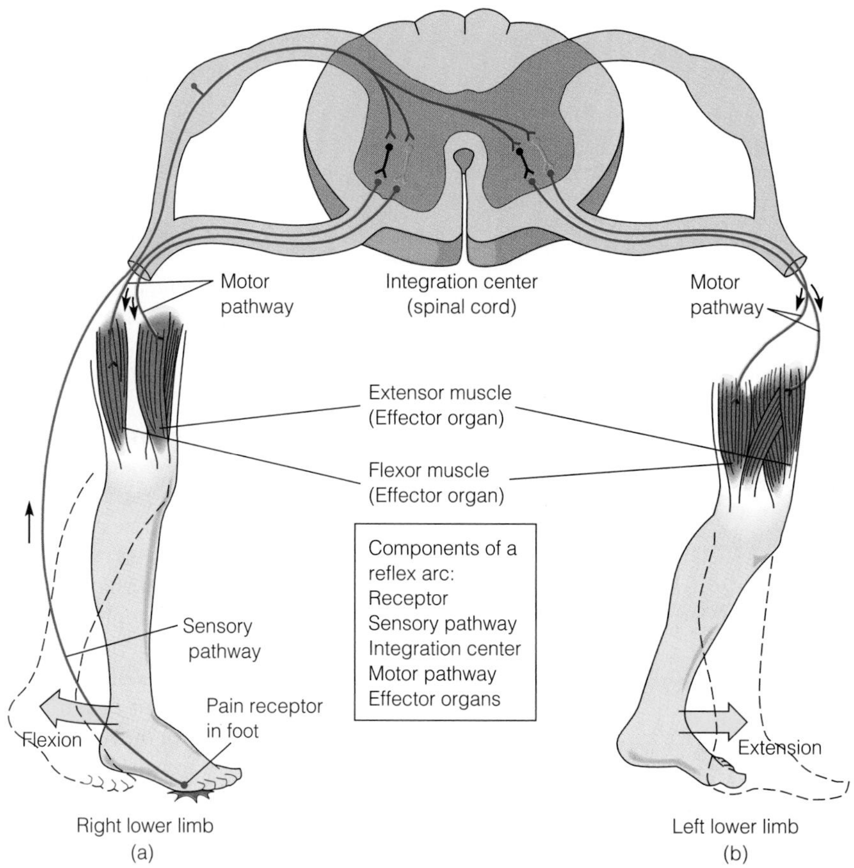

Motor pathway
Integration center (spinal cord)
Motor pathway
Extensor muscle (Effector organ)
Flexor muscle (Effector organ)
Components of a reflex arc:
Receptor
Sensory pathway
Integration center
Motor pathway
Effector organs
Sensory pathway
Pain receptor in foot
Flexion
Extension
Right lower limb (a)
Left lower limb (b)

Figure 17–19 Patellar Reflex Arc
The patellar reflex involves a simple reflex arc (arrows). It results from the activation of stretch receptors in the patellar tendon and stimulation of muscles of the upper leg.

Extensor muscle of knee (Quadriceps femoris)
Patellar tendon
Motor neuron
Muscle spindle
Reflex arc

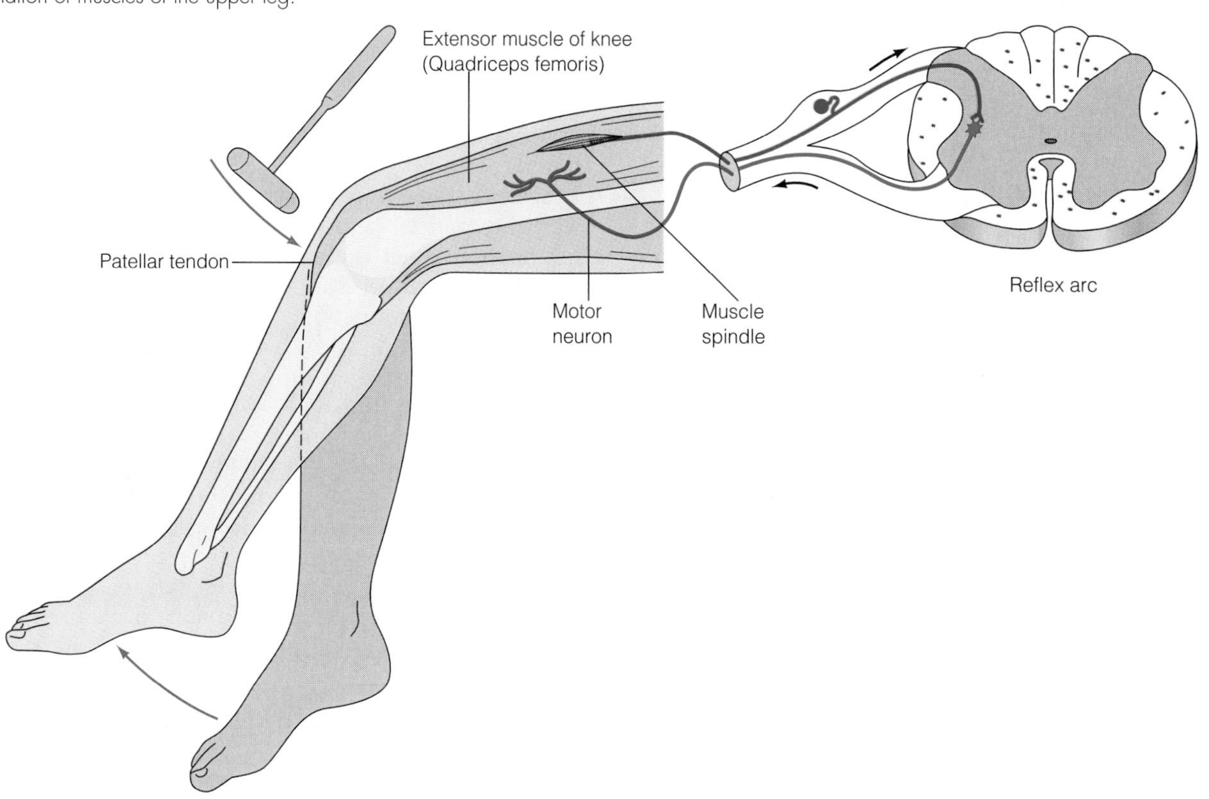

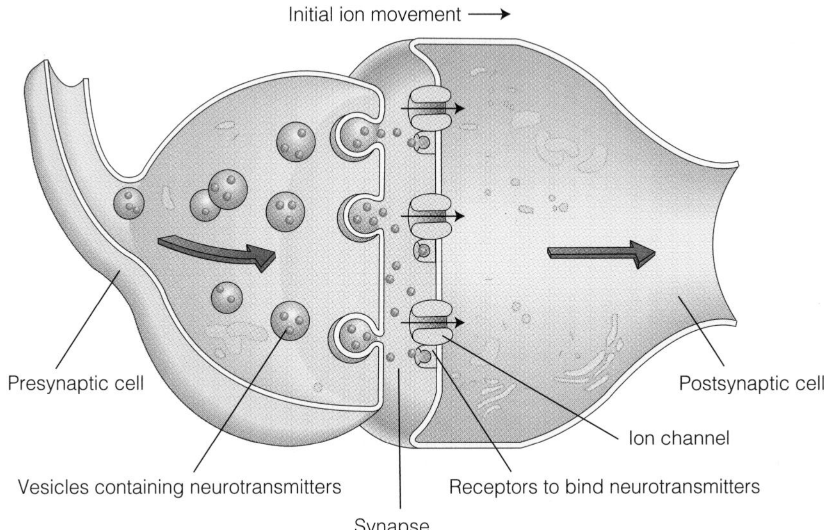

Figure 17–20 Pre- and Postsynaptic Cells

A presynaptic nerve cell releases neurotransmitters into a synapse. The neurotransmitters bind to receptors on the postsynaptic nerve cell and open channels that allow movement of ions across the membrane to generate an action potential.

PHYSIOLOGY OF THE NERVOUS SYSTEM

In its most general description, the function of the nervous system is to distribute signals that can be consciously and unconsciously interpreted as information about the environment. This capability is of paramount importance for human survival. The fact that this information is based on nerve impulses with an underlying basis in biochemical reactions is very intriguing. The impulses travel rapidly and in only one direction over an integrated network of billions of specialized cells. What does it mean to say that the nervous system functions in one direction? How is a signal generated in the network? What molecules are involved? How does a signal move through a cell? What allows the nerve impulse to get from one cell to the next in the system? These questions all relate to the performance of cells in the nervous system, the direction of signals, and to the structural features of neurons and glia described earlier.

Unidirectionality

Inherent to the organization of the nervous system, impulses and information travel only in one direction. Sensory signals arising in the periphery (for example, touching a surface) are transmitted to the central nervous system along sensory neurons to the spinal cord and on to the brain. Integration of the information takes place in the brain and subsequent responses are relayed in the form of *effector functions* in contraction of muscles and/or secretion of glands. Considering all the billions of cells involved and all the trillions of connections that they make with their neighbors, what is the basis for transfer in only one direction along nerves? The answer lies in the structural polarity of the nerve cells themselves. Nerve cells communicate with one another through synapses (Figure 17–20). Synapses are tiny gaps between neurons whose ends are asymmetrically organized. Synapses commonly occur between the axon of one cell and the dendrites or cell body of another. There is a presynaptic cell and a postsynaptic cell (Figure 17–20). The transfer of a signal (signal transmission) from one cell to another occurs as a result of stimulatory molecules known as neurotransmitters. Neurotransmitters are released from a presynaptic cell and affect a postsynaptic cell, not the other way around (Figure 17–21). This is the basis of unidirectionality. Such asymmetrical connections avoid the confusion of having to transmit nerve impulses in both directions simultaneously along the same nerve cell. One might imagine that the capacity to send nerve impulses in opposite directions might result in canceling one another out. Such a situation for the regulation of the contraction of the muscles involved in body movement, digestion, heartbeat, and breathing would be potentially a disaster.

Nerve Impulses, Neurotransmitters, and Synapses

Nerve Impulse Nerve impulses are the basic signaling units in the nervous system. Their generation depends on the structure of the nerve cell plasma membrane and its ability to transport ions from outside the cell to inside and from inside to outside (Figure 17–22). The plasma membrane of a nerve cell is normally a barrier to the passage of ions such as sodium (Na^+) and potassium (K^+). The buildup of sodium ions outside a cell by active transport (see Chapter 3) and their exclusion from passive movement back inside establishes an electric potential of approximately −70 millivolts (mV) (Figure 17–23). This is called the cell's resting potential. The basis for this voltage potential is the difference between isolated plus- and minus-charged ions. There are many more plus charges outside the cell than inside. In the case of the nerve cell membrane, there are special ion channels that can be opened or closed

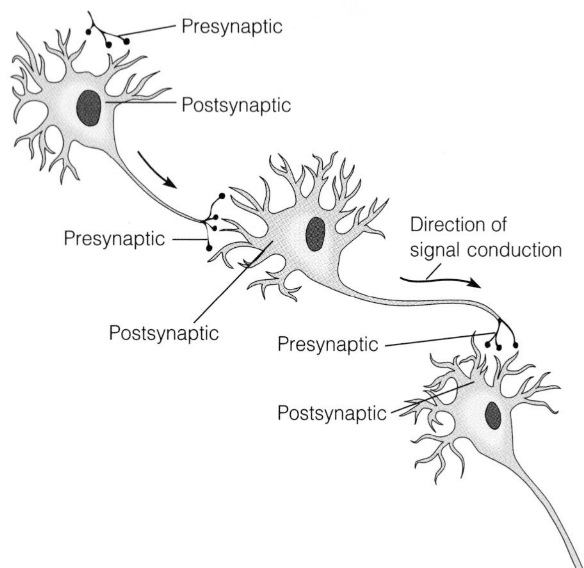

Figure 17–21 Synaptic Connections
A series of neurons is connected by synapses. A neuron may be presynaptic at one synapse and postsynaptic at another, different synapse.

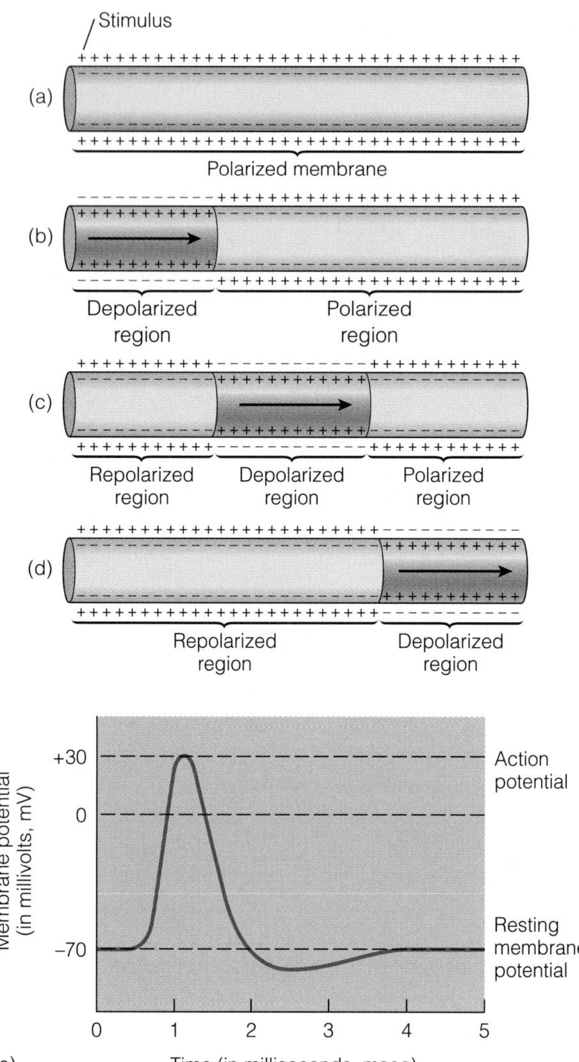

Figure 17–22 Propagation of a Nerve Impulse
The nerve impulse is propagated by changes in permeability to sodium ions (Na^+) (a). As ions move in (b), the membrane is depolarized, then repolarized (c) and (d). The voltage change across the membrane is relatively large and rapid (e).

under special circumstances, which change the voltage potential across the cell membrane (Figure 17–23). When the channels are open, sodium ions flow in and change the membrane voltage potential, which rises above zero (+30 mv). This opening creates an impulse that migrates or propagates rapidly from the point of origin to regions farther away along the length of the cell (Figure 17–22). As the channels open in new regions, they close in those that were initially opened. Sodium ions are actively pumped back out into the extracellular environment almost immediately after they have entered. In this way, the nerve impulse moves along the entire length of a neuron even as the neuron recovers in its wake (Figure 17–23). The changes in membrane channels that allow the rapid transport of ions along the length of a cell from dendrite to cell body to axon takes place in milliseconds.

Neurotransmitters and the Synapse What is it that disturbs the nerve cell membrane in the first place and opens the sodium channels? Part of the answer is neurotransmitters (Figure 17–20 and Table 17–1). Neurotransmitters are produced predominantly in neurons, with the exceptions of epinephrine and norepinephrine, which are also produced in abundance by the adrenal medulla. The effects of neurotransmitters are restricted primarily to other neurons through their release at synapses, but other cell and tissue types also are affected. Such types include muscles through neuromuscular interactions mediated by acetylcholine and neuroglandular interactions mediated by epinephrine and norepinephrine.

Neurotransmitters are small but potent molecules that are released into the synaptic cleft between two nerve cells (or between a neuron and another type of target cell) and bind to membrane receptors (Figure 17–20). The neurotransmitter molecules are stored in vesicles in the presynaptic cells, released into the synapse, and bind to

(a)

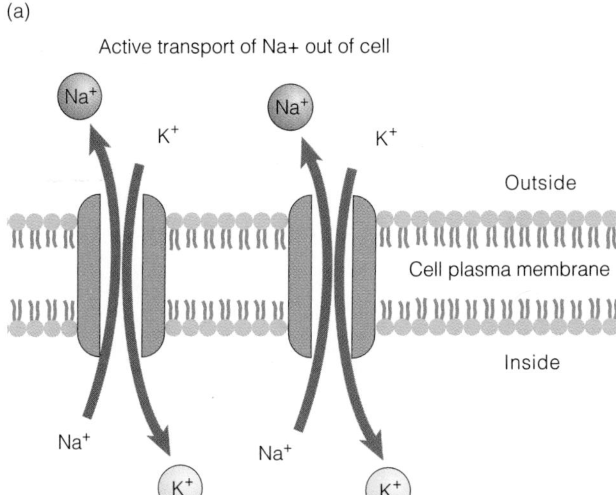

(b) Resting potential -70mV

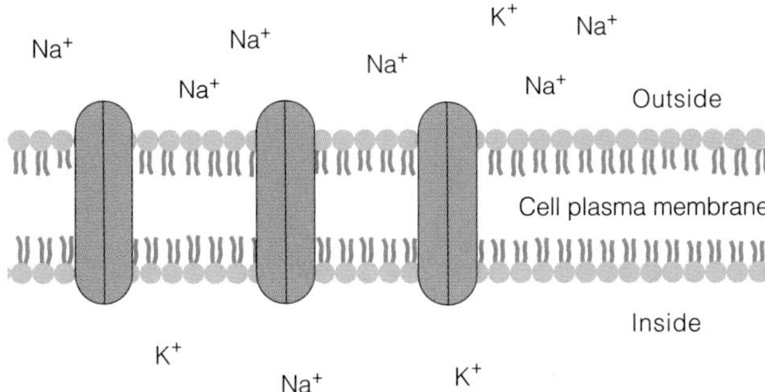

Figure 17–23 Ion Transport and Membrane Potential
Under resting conditions, a neuron pumps Na ions out of the cell and pumps K ions in (a). This establishes a resting membrane potential of approximately −70 millivolts, which is used to drive a nerve impulse (see Figure 17–22).

receptors on the postsynaptic neuron (Figure 17–20). The binding of the neurotransmitter to its receptor opens ion channels in the postsynaptic neuron and an **action potential,** or a nerve impulse, is initiated (Figure 17–22 and 17–23). As pointed out previously, the synapse is a tiny gap between cells, often no more than 20 nanometers (nm) in width. The diffusion of neurotransmitters across this gap takes only microseconds to occur.

Timing of Impulses The other part of the answer to channel opening is exemplified in the action of sensory receptors. Disturbances of the various types of sensory receptors throughout the body elicit dynamic changes in

Table 17–1 Neurotransmitters

NAME	TISSUE SOURCE	CELL OR TISSUE TYPES AFFECTED
Acetylcholine	Nerve	Neurons and muscle
Dopamine	Nerve	Neurons
Norepinephrine	Nerve, adrenal medulla	Neurons, glands, muscle
Epinephrine	Nerve, adrenal medulla	Neurons, glands, muscle
Serotonin	Nerve	Neurons
Glycine	Nerve	Neurons
Glutamate	Nerve	Neurons
Asparate	Nerve	Neurons
GABA (gamma-aminobutyric acid)	Nerve	Neurons

membrane potential that are then propagated between cells by neurotransmitters (Figure 17–24). For example, the stimulation by light of photoreceptors in the retina of the eye sends signals to other types of neurons in the retina via neurotransmitters, which in turn are then transmitted to cells in the brain. Similarly, the binding of a sugar molecule to a taste receptor on the tongue stimulates a change in membrane potential that is propagated to other neurons in the taste centers of the brain via neurotransmitter release at synapses. The underlying mechanisms eliciting nerve impulses throughout the nervous system are based on sensory cell responsiveness and neurotransmitters.

A nerve cell does not transmit impulses all the time nor should it. Because the binding of neurotransmitters initiates a nerve impulse, there must be a way to regulate them in order for nerve impulses to stop being initiated. For example, the neurotransmitter acetylcholine is released into synapses between nerves or between a nerve and a muscle fiber. Acetylcholine binds to its receptor and initiates a nerve impulse or a muscle contraction (Figure 17–25). In the space of the synapse are also enzymes that are capable of inactivating acetylcholine. The enzyme *acetylcholine esterase* serves this purpose, eliminating excess acetylcholine and stopping the generation of nerve impulses. What is turned on by neurotransmitters in the synapse is turned off very quickly by the action of the enzyme.

There are specific classes of molecules produced by other organisms that act as inhibitors of neural transmission in humans and other mammals. Their interference can be fatal. *Botulin toxin* (from the bacterium *Clostridium botulinum*) is a

> **Action potential** the form of the electrical signal carried by a neuron

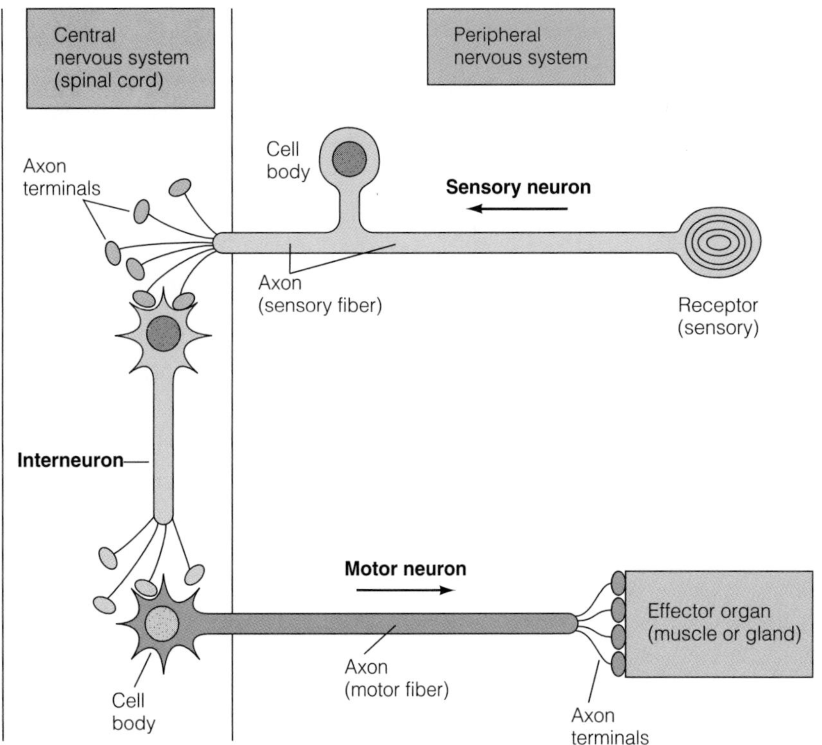

Figure 17–24 The Scheme of Nerve Connections

The sensory receptor produces a nerve impulse that passes from the periphery to the central nervous system. Interneurons relay that signal to other regions of the CNS as well as to motor neurons of the spinal cord. The motor neuron sends a signal to an appropriate target and elicits a response. The system is one way as depicted and represents the general plan by which information is distributed in the nervous system.

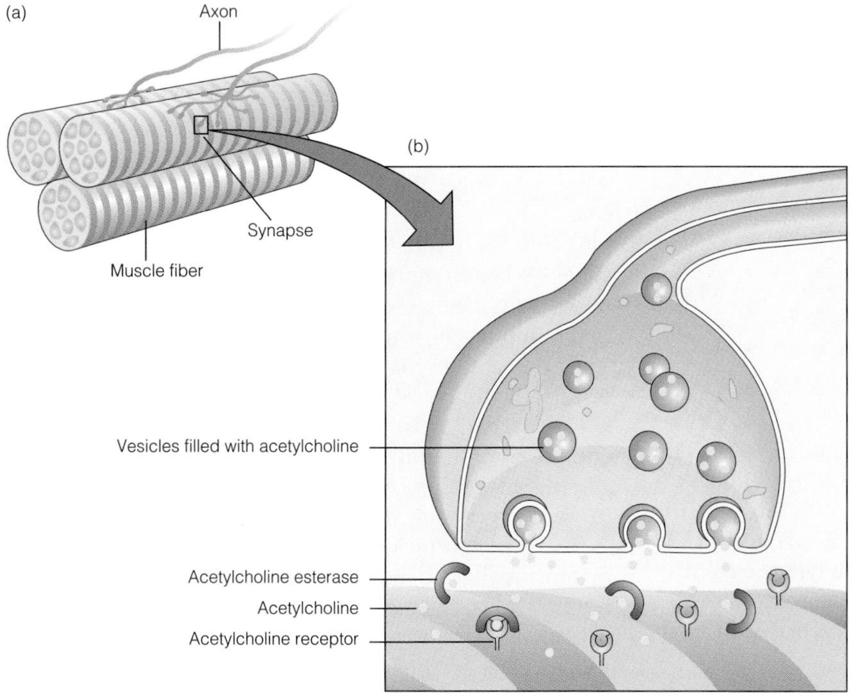

Figure 17–25 The Neuromuscular Junction

Muscle fiber in (a) is innervated by an axon that forms synapses on its surface. In (b) acetylcholine is released from the presynaptic cell and binds to acetylcholine receptors on the postsynaptic muscle, which then contracts. Acetylcholine esterase deactivates excess acetylcholine, allowing that muscle to relax.

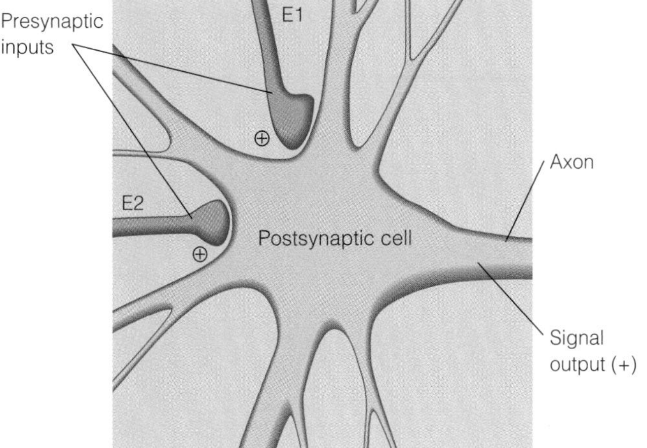

Figure 17–26 Summation
Summation is the process by which both excitatory (+) and inhibitory (−) multiple presynaptic inputs are evaluated by a postsynaptic cell. As shown, two (+) inputs result in a (+) output.

case in point. This toxin prevents release of acetylcholine from its receptor in the synapse and, therefore, interferes with the relaxation of muscles (Figure 17–25). The tetanus toxin (from *Clostridium tetani*) blocks inhibitory impulses so that muscles stay contracted. This unrelenting contraction is called *tetany* and if it occurs in conjunction with the functioning of the diaphragm, for instance, breathing is prevented.

Excitatory and Inhibitory Synapses In addition to biochemical changes in response to neurotransmitters bound to receptors as a way of turning off or interfering with nerve impulses, there is also indirect inhibition of the responsiveness of neurons to excitatory neurotransmitters. An individual neuron may have more than one synapse. In fact, some cells may have hundreds or thousands of synapses. The synapses function in basically one of two ways, excitation, as described above, or inhibition. The binding of inhibitory neurotransmitters at synapses can prevent a neuron or target cell from responding to excitatory stimulation. This is observed in the opposition of the sympathetic and parasympathetic nervous systems, in which the former is **excitatory** (using the neurotransmitter epinephrine) and the latter is **inhibitory** (using acetylcholine). A nerve cell is somehow able to sum up (called *summation*) the excitatory and inhibitory signals and determine a net result (Figure 17–26). If the number of excitatory signals is greater than the inhibitory signals, a nerve impulse is generated. If the reverse occurs, the nerve cell is prevented from responding. Thus, signals of on and off for nerve cells are determined by a complex system of checks and balances.

It should be pointed out that an excitatory response is an "all-or-none response." A nerve cell that produces an impulse does so without regard to the strength of the force that initiated it. The perception of the strength of signals from a sensory nerve cell in the hand to the brain depends on the rate at which the sensory receptor produces impulses. In the hot stove example, the heat caused the receptors to fire rapidly enough to invoke a reflexive reaction.

Threshold The *threshold* for the initiation of a nerve impulse is very important. It represents the level of stimulation needed to bring about the firing of a nerve by the minimum amount of neurotransmitters or by the lowest level of some external physical signal such as low light, or faint sound. If the nervous system were overly sensitive to stimuli capable of inducing a nerve impulse, there would be no discrimination as to the intensity of the stimulus to which a nerve would respond with an action potential. Thus, any level of stimulus would generate an action potential. The threshold of a neuron is a measure of the sensitivity it has to the amplitude of a signal eliciting a response (or to the net excitatory stimulation with regard to summation), and/or the rate at which the stimulation occurs.

DAMAGE AND DISORDERS OF THE NERVOUS SYSTEM

Damage

There are a number of kinds of damage that affect the nervous system and many types of clinical neurological disorders. Accidental destruction of nerve tissue is very serious. Trauma to the brain can be fatal, as can loss of cerebrospinal fluid due to puncture of the membranes enclosing the CNS. Severing or *transection* of the spinal cord can result in death (at the level of the upper cervical vertebrae). *Quadriplegia* results from partial transection of the cervical spinal cord, leaving both sets of limbs impaired (Figure 17–27[b]). *Paraplegia,* or paralysis of the lower extremities, results from partial severing of the spinal cord between the cervical and lumbar segments (Figure 17–27[a]). Nerve severing in the extremities (such as hands) can result in loss of function. In general, nerves do not regenerate, and damage of the types described here results in permanent disability.

Excitatory the quality of signals between nerve cells that stimulate an action potential

Inhibitory the quality of signals between nerve cells that prevent an action potential

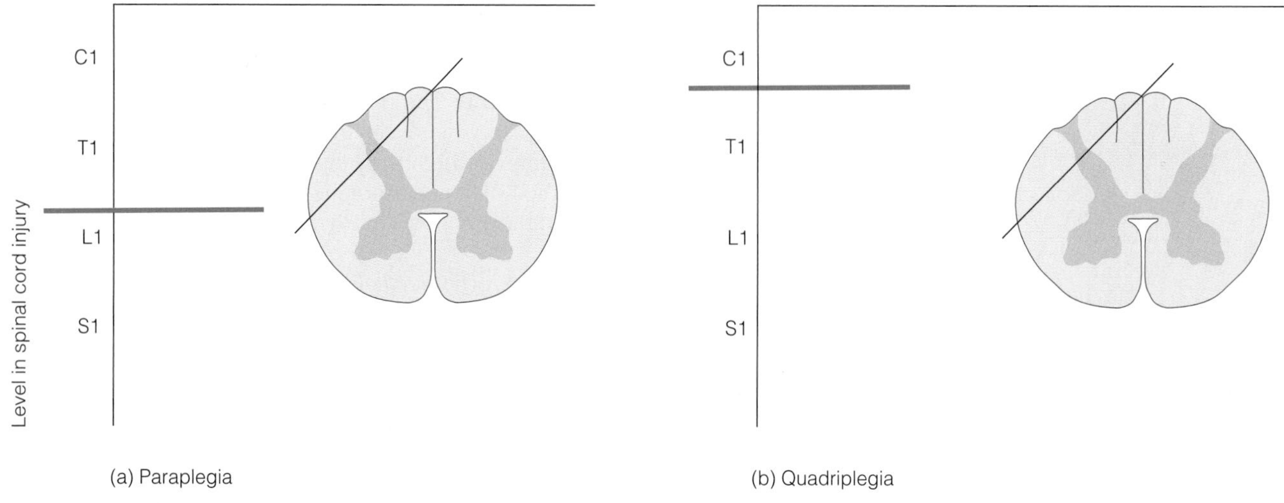

Figure 17–27 Paraplegia and Quadriplegia
The damage done in partial transection of the spinal cord results in loss of function. In (a), damage to the cord in the thoracic region results in loss of normal function of lower extremities. In (b), damage to the cord in the cervical region results in loss of function of both arms and legs.

Disorders

Neuritis and Neuralgia *Neuritis* is the inflammation of a nerve or nerves and results from nerve damage (for example, from a blow or contusion). Such inflammation also arises from dietary deficiencies, poisons, and heavy metals, such as lead. *Neuralgia* is pain along the entire length of a nerve. The causes are often unknown. Damage due to disintegration of neural tissue in the brain or from the buildup of toxic metabolic products can occur.

Parkinson's, Alzheimer's, and Huntington's Diseases *Parkinson's disease* results from the destruction of a region of the brain known as the *basal nuclei. Alzheimer's disease* occurs progressively as a result of the buildup of fibers and plaque in a region of the brain known as the hippocampus (Figure 17–9). A genetically inherited disease that develops late in life is *Huntington's disease.* This dominant genetic defect results in loss of brain mass and progressive deterioration of personality as well (see Chapter 20). Lethargy, loss of motor control, and memory loss characterize these diseases.

Other Problems Multiple sclerosis is the immune system attacking its own myelin sheaths and forming hardened scar tissue that interferes with motor control. Myasthenia gravis results also from autoimmunity, which leads to inhibition and destruction of acetylcholine receptors and the cells that display them, particularly at neuromuscular junctions.

Aneurysms and **embolisms** in the brain can cause strokes, which leave individuals with loss of motor control. The lack of oxygen from the blockage of arteries to the brain rapidly kills neurons. Infections of the meninges (**meningitis**) can be serious and debilitating. Some forms of meningitis are fatal. The sexually transmitted disease syphilis, in its tertiary stage, can cause massive destruction of the brain and spinal cord tissue as well as many other tissues and organs in the human body.

It is clear that damage or deterioration of the nervous system may be life threatening. The origin of the damage can be viral, bacterial, genetic, immunological or physical trauma, but the outcome is often severe, and the long-term prognosis is a significant lessening of the quality of life.

Embolism obstruction of blood flow by a detached blood clot

Meningitis inflammation and infection of the meninges

Summary

This chapter has been concerned with the types of cells that compose the human nervous system, the organization of those cells, and their functions. The principal cell type of the nervous system is the neuron. Neurons are categorized into three main morphological types—unipolar, bipolar, and multipolar cells. Each of these types has the unique feature of being able to generate nerve impulses, establish synapses, and stimulate impulses in other nerve cells through the release of neurotransmitters. The other major type of cells found in the nervous system are the glia. The glia are involved in the process of myelination, which insulates neurons, and the establishment of the blood-brain barrier, which prevents passage of large molecules from the vascular circulation into the brain proper. Glia have other more general structural and nutritional roles in the nervous system as well.

The coverage of the anatomy of the brain includes a description of the central nervous system and the peripheral nervous system. The cerebrum, thalamus, hypothalamus, midbrain, pons, medulla oblongata, and spinal cord are major regions of the brain that are organized to receive, integrate, and act on external and internal stimuli. The cranial and spinal nerves interact together in the functioning of the parasympathetic and sympathetic nervous systems. The sympathetic and parasympathetic nerves have positive and negative influences on the activity of their target organs. This influence is based on the excitatory neurotransmitter epinephrine and the inhibitory neurotransmitter acetylcholine. Various outcomes of damage or diseases affecting nerves clearly show how important an intact nervous system is to homeostasis of the human body.

The most rapid and all-encompassing integration of form and function in the human body is mediated by the nervous system. Although human science and technology have greatly expanded our knowledge of the nervous system, there is much more to be discovered in terms of understanding nerve cell differentiation, nerve cell function, and nerve cell interactions during embryonic growth and development. Advances in each of these areas will help to establish a better understanding of human behavior.

Questions for Critical Inquiry

1. Comparison of the nervous system with the endocrine system indicates degrees of similarities as well as differences. What are the similarities between the two? What are the differences?
2. In what sense is the nervous system unidirectional? What is the structural basis for it? Why is it important that the signals not pass along a single nerve cell in both directions?
3. The number of synapses that a single neuron may have is enormous. How might a decrease in the number of synapses affect the brain? How might an increase affect the brain?
4. What is the blood-brain barrier? What cells are involved in its formation? Why is it important that such a barrier exist? What disadvantage might there be to its presence?

Questions of Facts and Figures

5. How are the CNS and the PNS different from one another?
6. What are sulci and gyri? What do such features of the brain represent in terms of functions?
7. What are the major cellular elements of the human nervous system?
8. What are the cranial nerves? How do they function?
9. Schwann cells and oligodendroglia cells produce what insulating material? Where are these cells located? Do they have exactly the same characteristics?
10. What are the meninges? How are they organized?
11. The ventricles of the brain are filled with fluid. Name the fluid and recount how it protects the brain.
12. How does acetylcholine operate as a neurotransmitter? How is its presence in the synapse regulated?
13. What is gray matter? What is white matter? How are they different in the brain and spinal cord?

References and Further Readings

Edelman, G. M. (1988). *Topobiology.* New York: Basic Books.

Franklin, J. (1987). *Molecules of the Mind.* New York: Atheneum.

Noback, C. R., and Demarest, R. J. (1977). *The Nervous System,* 2nd ed. New York: McGraw-Hill.

Sherwood, L. (1993). *Human Physiology,* 2nd ed. St. Paul, MN: West Publishing.

Stalheim-Smith, A., and Fitch, G. K. (1993). *Understanding Human Anatomy and Physiology.* St. Paul, MN: West Publishing.

CHAPTER 18

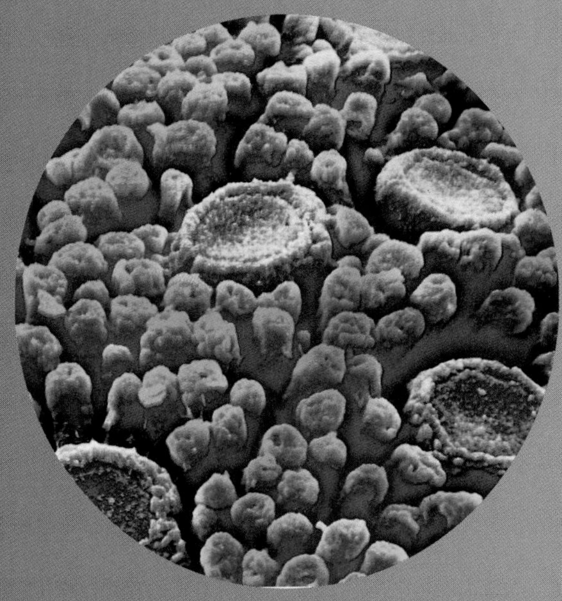

The Senses

INTRODUCTION

The structure and function of the nervous system provides the anatomical framework in which to integrate and act upon signals generated within us as well as signals arising in the external environment. But, what are the receptors and the nature of the stimuli that are to be integrated? We perceive a spectrum of electromagnetic energies, as well as sound waves, the binding of a variety of types of molecules and ions, changes in heat and cold, and contact with physical objects. Reception of these stimuli depends on the activity of several specialized cells, which may compose complex tissue types and organs. These specialized parts are connected to the nervous system and are commonly known as the senses. This chapter deals with the cells and organs of sense and the stimuli to which they respond.

INFORMATION FROM THE SENSES

Sense organs are the antennae of the body. Through their capacity to convert stimuli into nerve impulses, they help inform us of the presence and intensity of specific types of stimuli on a continual basis. Ultimately, these stimuli give rise to nerve impulses, or patterns of nerve impulses, that are sent to the brain for interpretation and integration. For example, what we perceive as seeing or hearing depends on the way the brain organizes and interprets nerve impulses from the eyes and ears. In essence, the activation of the senses by external stimuli is the first step in the information conduit to the brain. Sensory receptors also receive and send signals that inform us of internal conditions, for example, the position of the body and its appendages in time and space. A specific group of tactile sensory cells is also responsible for signaling bodily contact with the outside world. Regardless of the source or nature of the stimuli, whether light, music, perfume, or a stone in your shoe, the receptor types involved inform us of the sights, sounds, smell, and feel of the world around us.

SIX GROUPS OF SENSES

Six main groups of senses will be described in this chapter (Figure 18–1). The first four are the special senses and include the reception of light (vision) and sound (hearing), and the capacity to discriminate chemically among different types of molecules and solutes (taste) and (smell). The fifth group is known as the general senses. The general senses help determine our perception of the shape, texture, and heat content of objects or surfaces (the touch-temperature complex). The sixth group of receptors provides

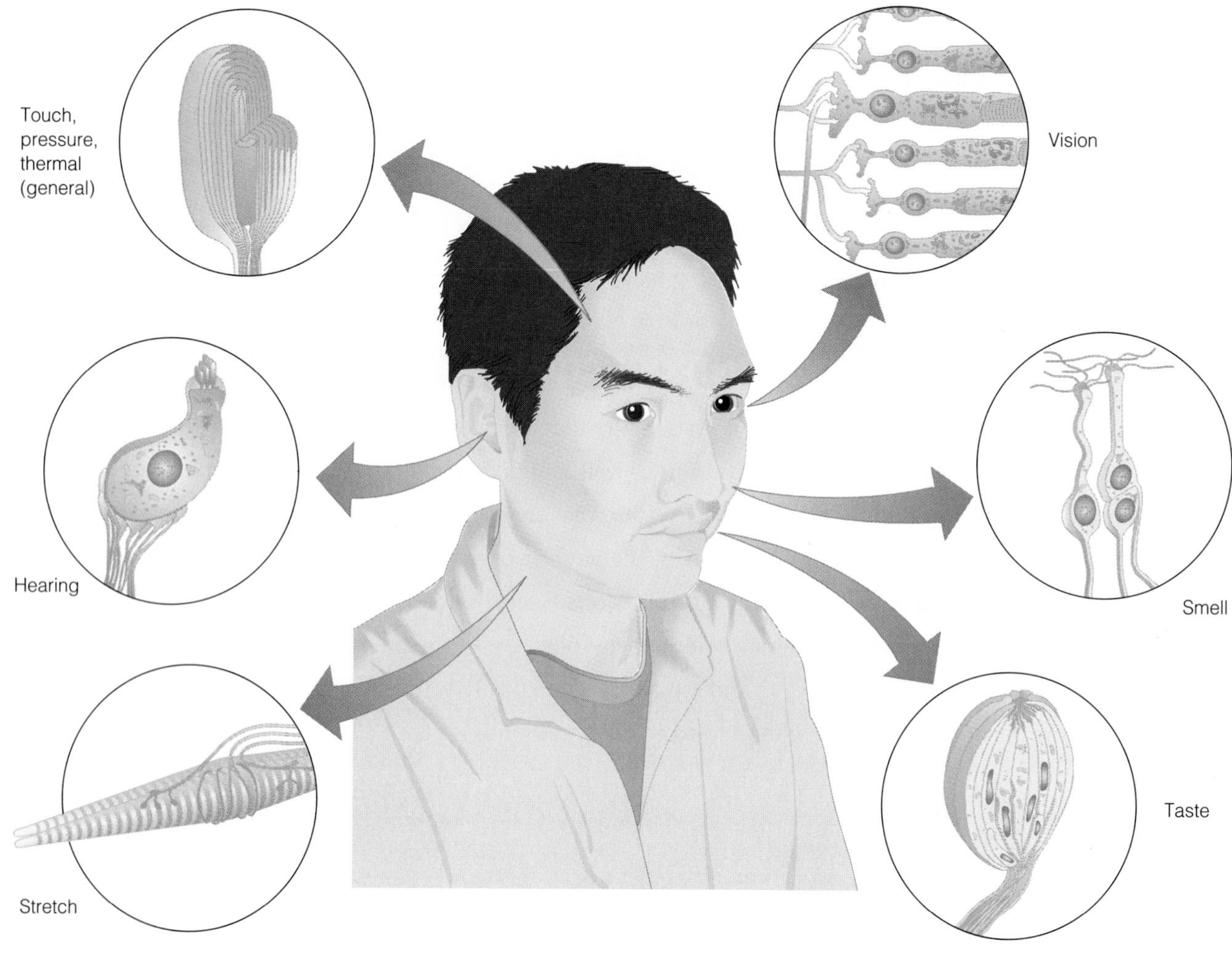

Figure 18–1 The Six Types of Senses

a means to monitor events inside the body and includes **stretch receptors** in our muscles and tendons. These receptors continually send signals from which the brain interprets the relative position of the body and the state of muscle tension. In this way, muscle tone can be automatically maintained so we will not suffer from tetany, which is a state of continuous muscle contraction, or slackness, the muscle state of complete relaxation. The contraction and balance of tone in antagonistic and synergistic muscles is continually adjusted in response to signals generated within muscles, tendons, and joints (see Chapter 7). This continual adjustment allows the body to accommodate to changes in position and levels of intensity of activity.

Eyes and Ears

The eye is able to absorb specific wavelengths of light and, in doing so, to generate nerve impulses. These nerve impulses act as signals to the brain and are interpreted in terms of patterns of dark and light as well as color and intensity. The design of the ear provides a means to capture and convert a specific range of vibrations in the air (or other conducting medium) and to translate them into nerve impulses that are interpreted by the brain as having a specific frequency (pitch), complexity, and loudness. Because of their importance and complexity, the eyes and ears will be discussed more fully in later sections of the chapter.

Tongue and Nose

Taste buds are cells specialized for discriminating among types of molecules based on their chemical composition (Figure 18–2, also see Chapter 12). Taste buds are located predominantly on the tongue but occur in other areas of the mouth, including the palate and pharynx. In the moist environment of the mouth, many of the chemicals in food are dissolved and find their way into individual taste buds, where they may generate signals interpreted as bitter, sweet, sour, and salty. From this collection of taste sensations, in combination with our sense of smell, flavors are discerned. The perception of smell, or **olfaction,** is much the same as that of taste, and, in fact, they are related in function (Figure 18–3). The receptors of smell are specialized cells of

> **Stretch receptors** receptors in muscle, epithelium, or connective tissue sensitive to tissue distortion
>
> **Olfaction** the act of smelling

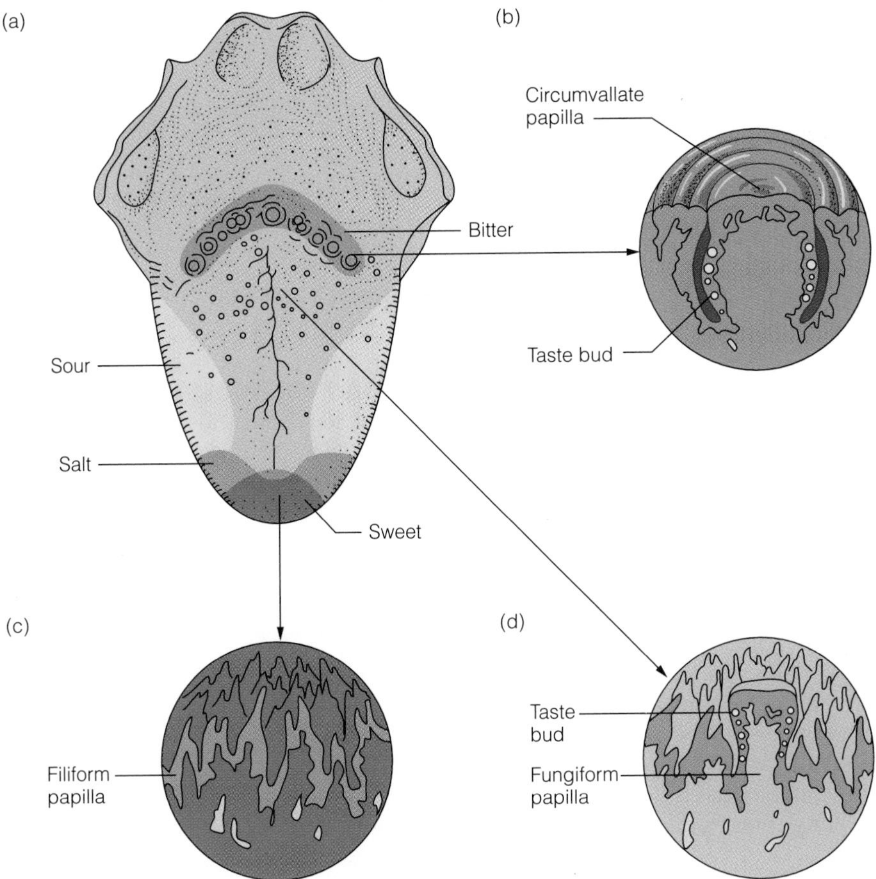

Figure 18–2 The Tongue and Taste Buds

The tongue (a) is covered with specialized structures known as papillae, (b), (c), and (d). Taste buds (b) and (d) are formed on some types of papillae.

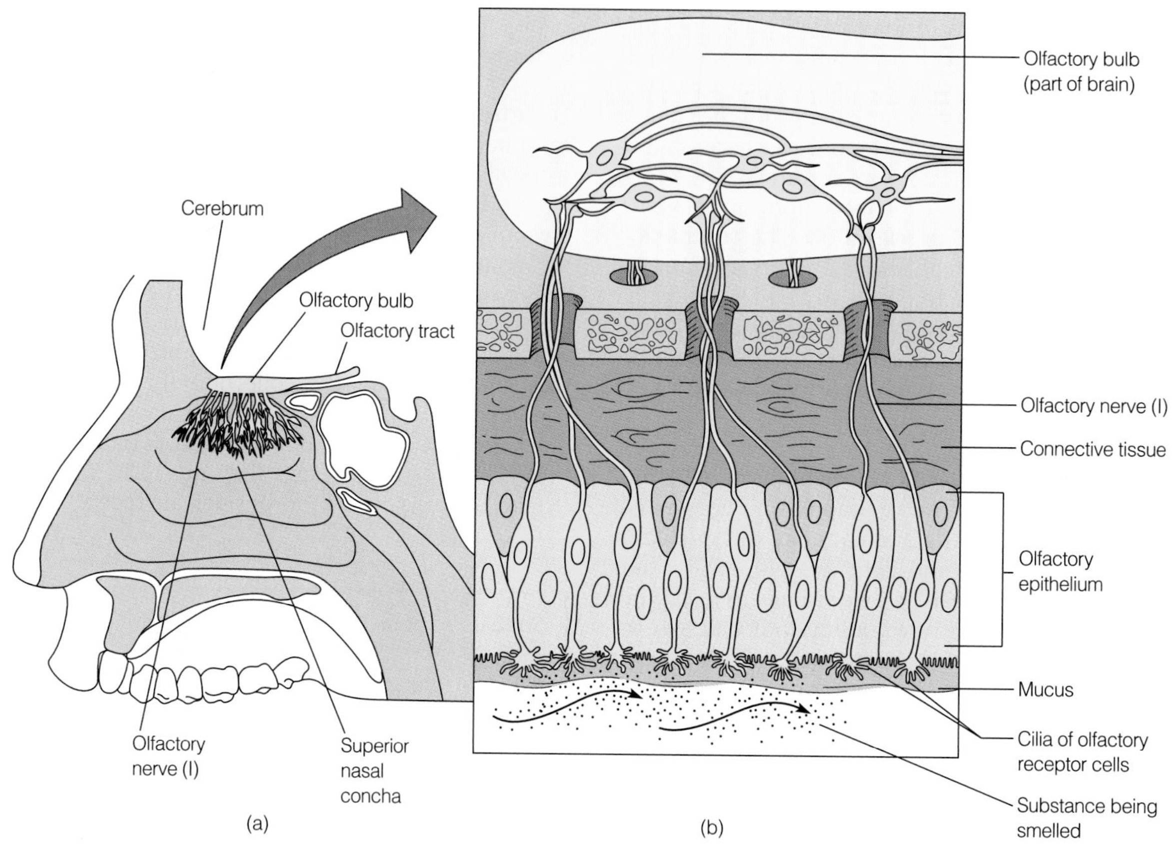

Figure 18–3 Olfaction
Chemical receptors of the olfactory nerve (a) capture odorants, are stimulated, and send nerve impulses to the brain (b). The receptors are located in the olfactory bulb of the olfactory epithelium (b) in the upper regions of the nasal cavity (a).

the olfactory epithelium located in the mucous membrane covering the roof of the nasal cavity. Airborne chemicals are dissolved in mucus clinging to the olfactory epithelium and, thus, gain access to receptors. Later in this chapter, taste and smell will be covered in more detail.

Skin and Internal Tissues

The skin contains a wide range of specialized cells and tiny structures used for sensory reception. Receptors of skin fall into the category of **general senses.** These senses allow us to perceive objects by touch and pressure all over the surface of the body and to perceive sensations of hot and cold. Signals from the specific and general senses are involved in establishing levels of pain and pleasure and in determining conditions of danger and/or potential damage to the body. The responses to sensory input can be emotionally or psychologically overpowering at times. "Stop to smell the roses" is so sensually pleasurable that it has come to be used as a metaphor for taking time from our busy lives to enjoy the world around us.

RECEPTORS OF THE GENERAL SENSES

Mechanoreceptors and Thermoreceptors

The general structures of sense are those associated with the *touch-temperature complex.* They are the **mechanoreceptors** and the **thermoreceptors.** These receptors are single cells or small collections of cells that are located primarily in the surface of the skin but may occur in other regions of the body as well (Figure 18–4). They are considered simple in type to differentiate them from the structurally more complex special types discussed later in the

General senses senses of touch, pressure, and heat
Mechanoreceptors receptors sensitive to touch and pressure
Thermoreceptors receptors sensitive to heat (hot and cold)

chapter. As was pointed out in Chapter 5, the skin is made up of two types of tissues—epithelial (the epidermis) and connective (the dermis). Sensory cells and sense organs of the skin occur in the epidermis and dermis over the entire surface of the body. However, there is considerable variation in the number of receptors in different regions. For example, although there are receptors responsive to light touch within every square centimeter of the body surface, the highest density areas are on the gripping areas of the skin of the thumb and fingers. A region with one of the lowest receptor concentrations is the back. Why might this be the case? If the greater numbers of receptors per area equates to greater sensitivity in discriminating the form or other physical qualities of objects, then there is clear functional value to having high densities of receptors on the fingers, with which we manipulate objects in the world around us.

Free Nerve Endings

There are a number of different types of sensory cells in this general sense group, and they are organized at different depths within the skin. This discussion begins with the most superficial of these, the *free nerve endings* (Figure 18–4). The free nerve endings, or naked nerve endings as they are also called, are essentially modified dendrites and are located within the epidermis proper. Free nerve endings are also

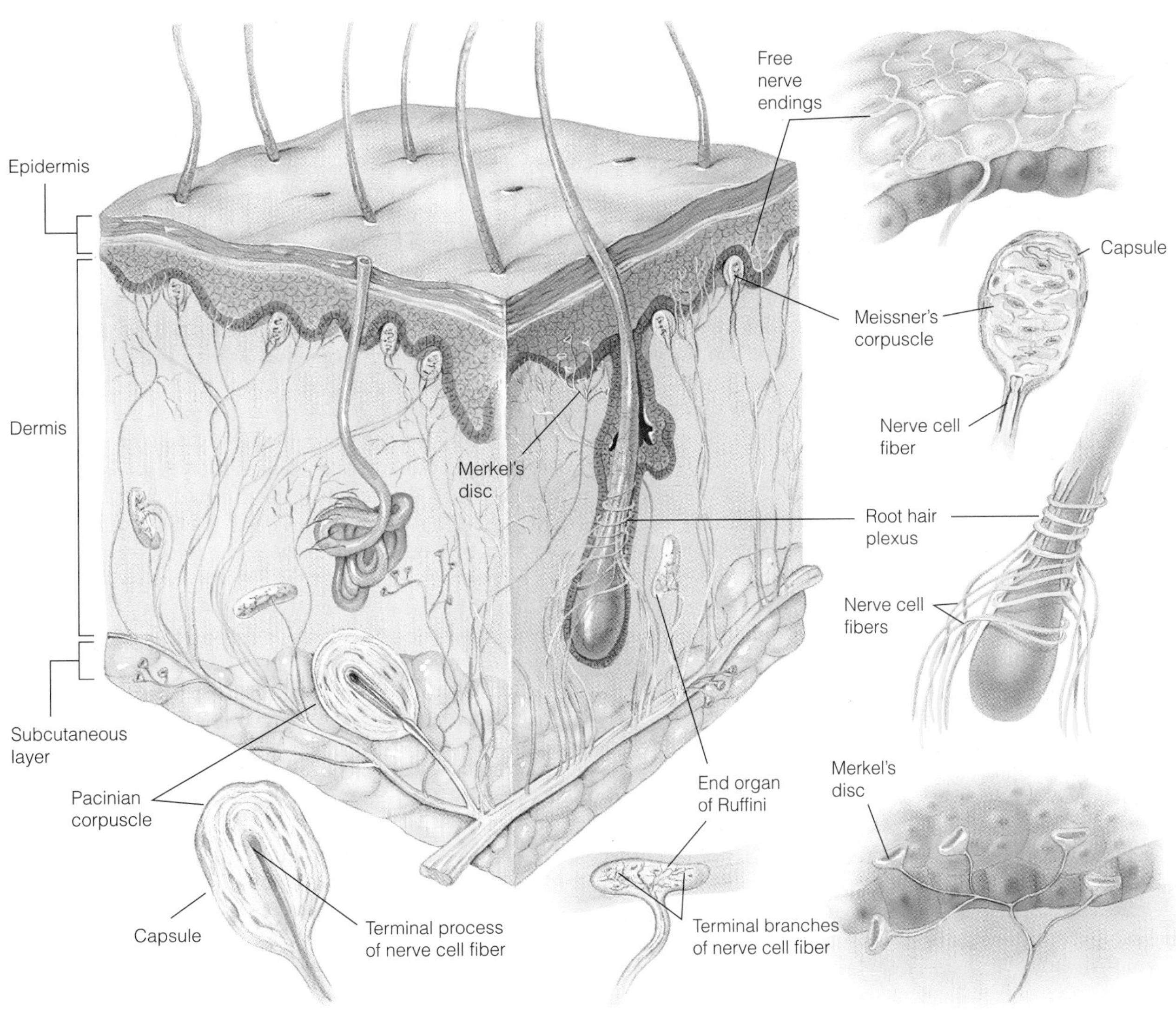

Figure 18–4 General Senses
The general senses include mechanoreceptors and thermoreceptors of the skin and other tissues. They are diverse in form and function.

wrapped around the bases of hair follicles and respond to hair movement. As you probably know from experience, these nerve endings are exquisitely sensitive to touch. For us to feel anything by touch, whether with fingers or along the back, there must be a distortion or change in the resting state of the nerve. The distortion induced by contact between the surface of the skin and an object elicits a firing of a nerve cell. So, what we monitor during touch sensations is the degree to which nerve endings are distorted by contact. Free nerve endings are among the most sensitive receptors.

Naked nerve endings also generate signals interpreted by the brain as pain. When the distortion of the skin is sufficiently large, or when it results in tissue damage, the rate at which these nerve endings fire increases dramatically. If you pick up a sharp-pointed object or scrape hard against a rough surface, this sends signals that are interpreted by the brain as pain. These sensations are uncomfortable and often elicit immediate reactions. Pain may induce a reflex response that results in rapid release or pulling away from the source of painful stimulation.

The Hair Root Plexus

The requirement for change in a resting state holds true for nerve cells involved in all the senses, regardless of type. The threshold for initiating a nerve impulse in a free nerve ending correlates to the amount of distortion it undergoes, which in turn depends on the degree to which the skin is affected by contact with an object surface. A minimum distortion of the skin may occur in picking up a small object, such as a postage stamp, somewhat more in holding a glass of water, and a lot more in lifting heavy boxes. Naked nerve endings also surround the base of a hair follicle and form part of what is called the *hair root plexus* (Figure 18–4). The importance of such structures is that they extend the range of our sense of touch farther than what we might ordinarily consider as contact with the skin. For example, the movement of a hair occasioned by our brushing against a surface, or by a breeze blowing over the skin, activates nerve cells.

Merkel's, Meissner's, and Pacinian Receptors

In addition to naked nerve endings, there are also more complex sensory organs associated with touch. Two tiny receptor structures associated with light touch are called *Merkel's discs* and *Meissner's corpuscles.* A class of receptors associated with deeper pressure is called *Pacinian corpuscles* (Figure 18–4). These several types of discs and corpuscles are named for the biologists who discovered or described them first. However, it was not clear at the time of their initial discoveries exactly what role (if any) these organs played in sensory reception. Today, we have a much more complete understanding of their nature, but there are still aspects to their function yet to be discovered.

Merkel's discs are located in the deep epidermis and consist of formations of dendrites organized among epidermal cells. In this case, distortion of the epidermis elicits nerve firings in the dendrites of these structures. Meissner's corpuscles are generally located very close to the junction between the dermis and epidermis in a region of the dermis known as a papilla. They are responsive to light touch and are abundant in areas of the body that are particularly sensitive to these sensations, such as the fingertips and lips. In general, when the position of the sensory receptors of the skin is deep, more distortion of the skin is necessary for them to be activated. This provides the potential for a graded response to varying degrees of local pressure exerted on the skin.

The deepest of the sensory receptors of the skin are the Pacinian corpuscles. These onion-shaped structures are located predominantly in the dermis or in deeper subcutaneous tissue, around joints and tendons, around muscles, and in visceral organs. This uniquely structured receptor is composed of a nerve ending that is encapsulated by several concentric layers of cells. The system of touch reception is dependent on a diversity of types and a variety of locations of mechanoreceptors. The signals they send provide not only an awareness of location of stimulation on or in the body but also the potential for a quantitative assessment of the intensity of tactile stimulation.

The Receptors of Krause and Ruffini

Reception of touch or pressure are not the only modes of reception in the skin. Two other receptors, the *end bulbs of Krause* and *Ruffini's corpuscles* are thermoreceptors (Figure 18–4). They are generally associated with the perception of cold and heat, respectively. Each of these types of receptors is located in the dermis, relatively close to the epidermal-dermal junction. How they are activated by heat or cold is not known. The two types of receptors are probably also capable of responding to *tactile stimuli.* When considered in this regard, they are probably functionally similar to Meissner's corpuscles, which are sensitive to light touch. The end bulbs of Krause and Ruffini's corpuscles are both encapsulated receptors and, so, bear structural similarity to Meissner's corpuscles. In fact, there is a controversy among neurobiologists working in the field as to how exactly these three types of receptors may be related to one another.

Interoceptors and Proprioceptors

The receptors just discussed are predominantly involved in transducing signals from outside the body to supply information about the environment around us. As such, they are considered to be in a category known as **exteroceptors.** However, there are also two classes of receptors that provide

Exteroceptors receptors for stimuli from outside the body

us with information about the internal environment. They are called **interoceptors** and **proprioceptors.**

Interoceptors Many interoceptors are located in blood vessels and organs throughout the body and, as in the case of exteroceptors, act as antennae for stimuli or cues from internal locales. Such receptors respond to changes in the internal environment of the body, such as those that are involved in monitoring carbon dioxide and oxygen levels in the blood or the level of ions and nutrients. Signals from these receptors elicit behaviors that result in changes in breathing, drinking, and eating. They also send signals of changes in blood pressure. The integration of mechanical and thermal information, including signals of pain, contact, pressure, hot and cold, with chemical information in the blood takes place in the brain. It is in the brain that integration results in a continual series of voluntary and/or involuntary responses which control body functions. This epitomizes the meaning of homeostasis. The receptor systems of the body send a continual flow of information to the brain, which, in turn, provides the adaptive responses needed to change and adjust the body accordingly.

Proprioceptors There are internal receptors in muscles, tendons, and joints that provide the brain with an awareness of the general state of muscle contraction, tension on tendons, and positional information on the orientation of joints and bones (Figure 18–5). This provides a sense of the spatial position of the body at every instant in time. These body-awareness signals are called proprioception and are mediated by proprioceptors. Collectively, proprioceptors are involved in what is a very subtle and continual process. Proprioception also involves information from receptors located in the inner ear. These receptors are associated with mechanisms necessary to maintain balance and equilibrium (see the section on sound reception and sense of balance later in this chapter). Much of the information from proprioceptors is targeted specifically to the region of the brain known as the cerebellum, which, in turn, plays a major role in coordinating the activities of skeletal muscles throughout the body in the effort to maintain balance.

The proprioceptive system allows us to sense how we are oriented in the space around us. In this way, we can perform many actions even if we are in total darkness. We can stand, move, stop, know up from down, get dressed, lie down, and still retain a "feel" for where we are. Test yourself right now. Stand up and tightly close your eyes. Lift your arms up and stretch them out to the right and left. Close each hand into a fist, but leave your index fingers extended. Keep your eyes closed. Now, bring your extended arms around in front of you and touch your fingertips together. You should be able to perform this simple routine without much difficulty. This is proprioception at work. It is, in some ways, an inner vision of the body as a whole, a sense of each part working together in an ever-changing pattern of balancing movements. Proprioception is dancing in the dark, diving for the

Interoceptors receptors for stimuli from inside the body
Proprioceptors specialized receptors providing body awareness information

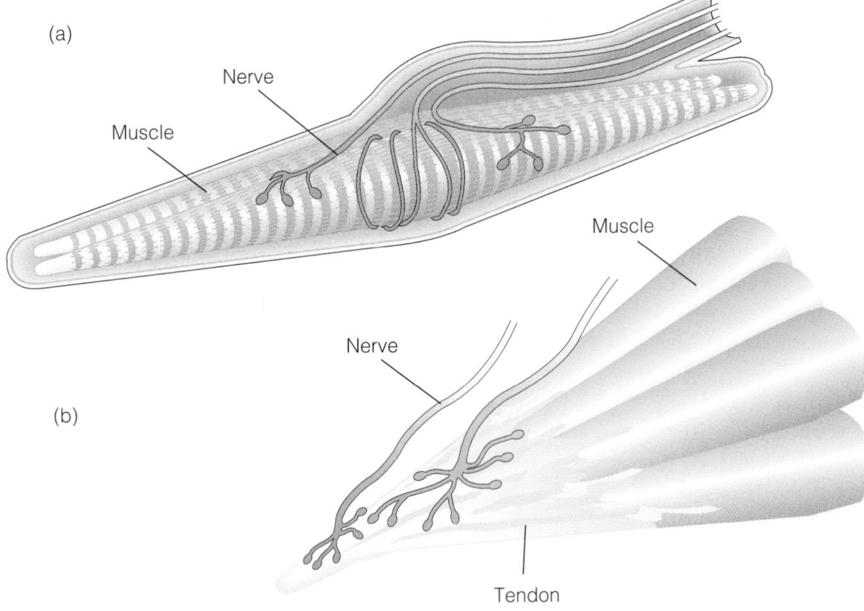

Figure 18–5 Proprioception
Internal receptors in muscles, tendons, and joints are important in maintaining awareness of the body. Neuromuscular (a) and neurotendinous (b) receptors help carry out this process.

catch, swimming in the sea, and knowing you will not fall down when you climb the stairs.

TASTE AND SMELL

The senses of taste and smell are closely related to one another. They both depend on special neuronal and non-neuronal cells distributed in particular locations in the mouth and nasal cavity. The cells involved in receiving signals translated as taste or smell have chemoreceptors on their surfaces (Figure 18–2).

Taste

The chemoreceptors of taste (also known as *gustatory receptors*) are located in **taste buds** (Figure 18–2). Most of our taste buds are located on the tongue, but some are found in other locations of the oral cavity. Taste buds are oval-shaped arrangements of cells that are integrated into the mucous epithelium of the mouth and tongue. Taste buds contain two types of cells—*gustatory cells* and *support cells.* The capsule outside of the taste bud is formed by support cells, and the cells that they support are the gustatory cells (Figure 18–2). Several gustatory cells are in each taste bud, and each of these cells has a hairlike process on the end that extends toward, or into, the *taste pore.* The hairlike processes of a gustatory cell have surface receptors for various classes of chemical compounds. The activation of these cells by the binding of chemicals to the receptors gives rise to the taste sensations of bitter, sweet, sour, and salty (Figure 18–2). The signals arising in the gustatory cells stimulate the dendritic ends of nerve cells that are wrapped around the taste bud. These nerve impulses are sent to the brain by way of the cranial nerves and target the association centers of the cerebral cortex.

Smell

The sense of smell works much the same way as the sense of taste. In this case, the chemoreceptors for smell (called *olfactory receptors*) are located in the mucous epithelium lining the roof of the nasal cavity. The signals generated by the binding of *odorant molecules* to the olfactory receptors are sent along nerves that pass through the ethmoid bone of the skull and directly to the frontal lobe of the cerebrum (Figure 18–3). The olfactory cells, as do the gustatory cells of the taste bud, have hairlike dendritic processes and special supporting cells.

It is essential that the environment in which receptors function be moist. The substances that are detected by taste and smell receptors must be dissolved. This is clearly the case for taste because we moisten our food with saliva as we chew, but is less obvious for smell. The sense of smell is very sensitive. It takes only a few molecules of an odorant to stimulate the olfactory cells to produce a nerve impulse. This sensitivity is thousands of times greater than is observed for taste reception. However, under the constant bombardment of a specific type of odor molecule, the sense of smell can become dulled rapidly. This phenomenon is called **habituation** and results in the cessation of signals generated by a receptor in the continuous presence of an otherwise appropriate stimulus. For example, the smell of perfume in the cosmetic area of a department store is initially intense but tapers off rapidly (within a minute or two) with persistent exposure.

What about situations in which we temporarily lose the ability to smell? We have all experienced instances in which this may have occurred, perhaps when we had a cold or the flu. In these cases, the sense of smell is greatly diminished or temporarily eliminated by excessive mucous secretions and/or inflammation and swelling of the olfactory epithelium. This blocks access of molecules to receptors. What happens if the odor molecules cannot reach their targets? Along with the loss of smell, one also loses the sense of taste, even if the gustatory receptors are working perfectly. This suggests that much of what is called taste may actually be made possible only in association with the sense of smell. Have you ever held your nose closed while taking a spoonful of unpleasant tasting medicine? The principle of combining taste and smell is involved as we try to keep the smell of the vile stuff from intensifying the bad taste (Figures 18–2 and 18–3).

VISION AND THE EYE

The eyes are our windows on the world. All mammals have two eyes, surrounded by bone and covered by layers of connective tissue. The eyes are directed in their orientation by muscles and covered intermittently for protection with the special skin of the eyelids. In humans, the eyes are positioned in the front of the face and their fields of view overlap extensively (Figure 18–6). Specialized cells in the *retina,* called photoreceptors, are activated by light. These cells are able to discriminate among colors and intensity and provide what is arguably one of the most important sensory systems in our species. The general senses, whose extero-, intero-, and proprioceptors maintain the body's homeostasis and sense of self-awareness, are probably more important, but eyes provide the means to see beyond ourselves and into the world around us.

Taste buds structures in which taste receptors are located

Habituation losing sensitivity to a smell due to overexposure

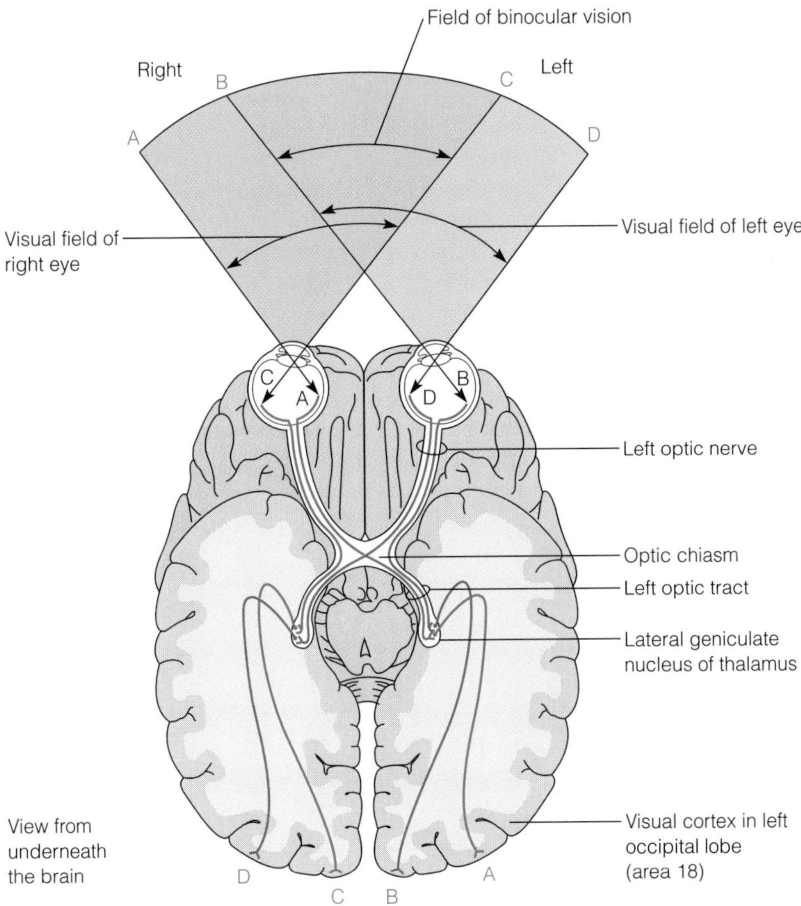

Figure 18–6 Field of View
One of the properties of binocular vision is that the eyes have overlapping fields of view. The right and left eyes capture light (photoreceptor cells) and send signals to the brain along the optic nerves.

The Structure of the Eye

The shape of the human eye is basically spherical, with connection to the brain through the optic nerves (Figure 18–6). Because the eyes are derived directly from the brain (Biosite 18–1), the retinas have an unusually complicated and intricate organization (Figure 18–7). Only the retina has cells with the capability to transduce light to nerve impulses. However, the other layers of the eye are very important. A retina is surrounded by layers of connective tissue, the *sclera* and the *choroid.* Light impinges on the retina through several transparent layers, including the *cornea, lens,* and **vitreous humor.** The cornea and lens help focus incoming light on photoreceptor cells at the back of the eye. The vitreous humor is the transparent jelly that fills the central part of the eyeball and helps maintain its spherical shape. Human eyes are fitted into bony sockets known as orbits, which are located in the front of the head (see Figure 7–1). There are several small extrinsic muscles that attach the eyes to the connective tissue of the sclera and to the bones of the face. This provides a means for controlling the direction of the gaze without moving the head. A detailed description of these features and others is presented, starting from the outside of the eyeball and moving inward.

Sclera and Cornea The outside of the eyeball is surrounded by the sclera and the cornea (Figure 18–7[a]). The sclera is a tough, fibrous connective tissue layer that protects the structural integrity of the eye. It covers most of the eye and is a white opalescent color. The white of the eye, as the sclera is commonly called, gives way in the front of the eye to a transparent layer called the cornea. It is through the transparent cornea that light first passes and is focused as it enters the eye.

Choroid The second layer of the eyeball is the choroid. The choroid layer is also composed of connective tissue, but it is not as tough and fibrous as the sclera. The choroid

Vitreous humor the clear viscous content of the center of the eyeball

18–1

THE DEVELOPMENT OF THE EYE

The eye develops during the first few weeks in human embryonic development. It arises from an outgrowth of the brain, which forms an optic vesicle that contacts the prospective epidermis. The embryo is so young at this point in its development that the surface layer is not yet called skin; it is called ectoderm. The optic vesicle undergoes an invagination that will form the optic cup, and the ectoderm develops into a lens. This process of lens formation is called lens induction, and if it fails to occur, abnormal or lensless eyes are formed. The optic cup gives rise to the neural retina and is surrounded by connective tissue that will become the sclera and the choroid. The part of the sclera overlying the lens will form the cornea (see figure below). The complexity of the neural retina results in part from the fact that it derives from an outgrowth of the brain. The multilayering of cells in the neural retina and the interactions among the cells in different layers ensure that much of the processing of light-generated nerve impulses takes place in the neural retina before being sent to the brain.

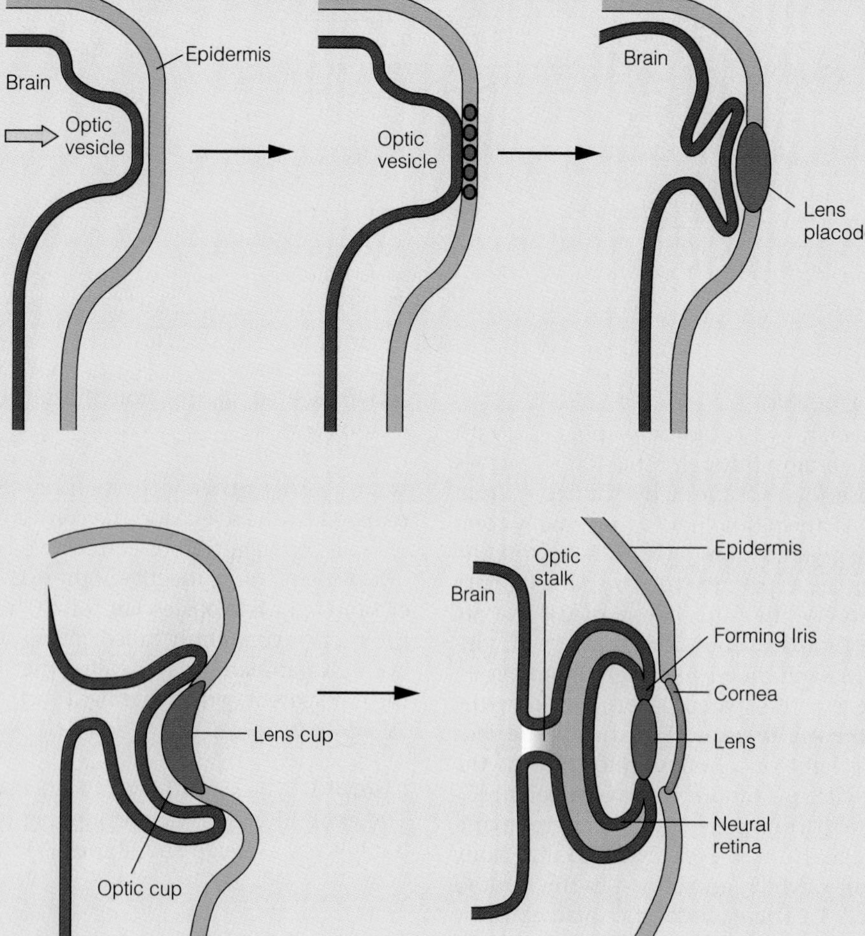

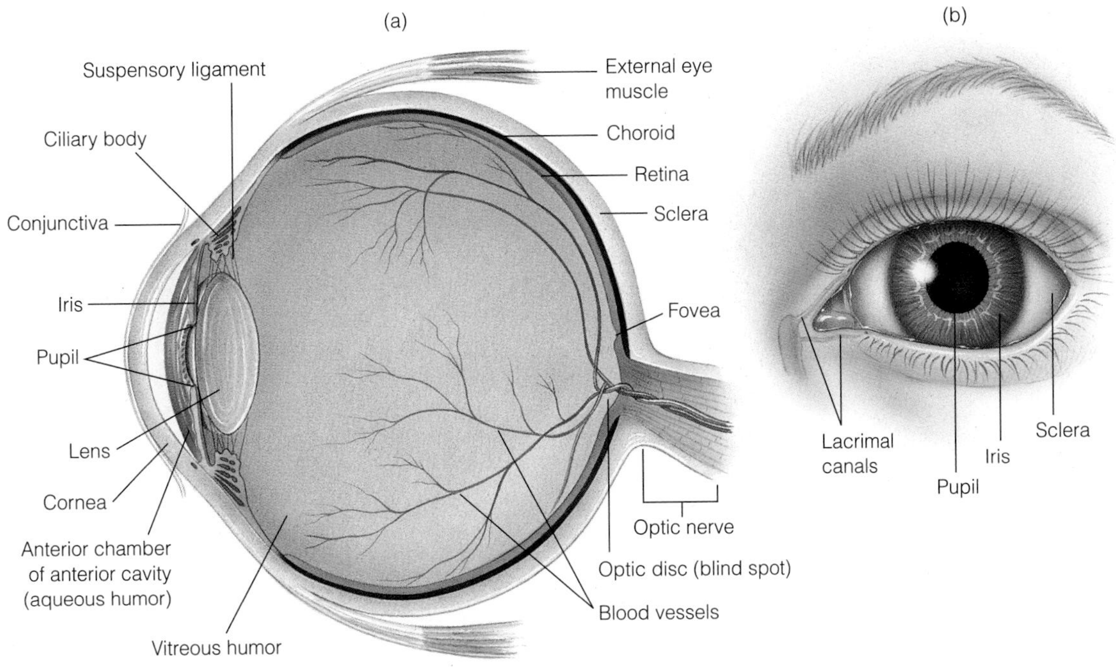

Figure 18–7 The Eye
The eye is a multilayered structure with the capacity to capture light (retina) and send signals on light and dark to the brain. The retina is supported by a nutritive choroid and a protective sclera. Transparent structures (cornea, lens, vitreous humor) allow light to enter the eye.

layer is, however, rich in blood vessels. These vessels are needed to bring oxygen and nutrients to the eye, particularly the photoreceptor cells of the retina. The choroid layer also forms the *iris,* which is generally a darkly pigmented, donut-shaped structure that restricts the amount of light entering the eye. The iris automatically dilates or constricts to regulate the amount of light that enters the eye. The *pupil* is the open space left in the donut-shaped iris (Figure 18–7[b]). For example, entering a darkened room triggers the iris to dilate, thus, letting more light in. Conversely, entering a brightly lit room, or walking outside on a bright sunny day, brings about an immediate constriction to reduce excessive light. The choroid layer also forms what is called the *ciliary body,* which changes the shape of the lens so that the eye can focus on objects near or far away. Behind the cornea and in front of the iris is a special *anterior chamber* filled with a watery fluid called the *aqueous humor.* The aqueous humor supplies nutrients to structures in the anterior part of the eye with which it is in contact and applies pressure against the outer surface to help keep the cornea in proper shape.

The Lens The lens is formed during the development of the eye from cells of the prospective embryonic epidermis. As the future eye, or *optic vesicle,* grows out from the brain during development (Biosite 18–1), it contacts the overlying future epidermis of the embryo. This region will become the lens. The lens is a thick, flexible, transparent disk-shaped structure composed of cells that produce proteins called **crystallins.** The lens is the principal structure by which an image is focused. In adults, the lens of the eye is similar in shape to the lens of a camera and performs essentially the same function. As will be seen, however, the way in which the lens of an eye focuses the light is different from that of a camera.

The Neural Retina The inner layer of the eye is the retina (Figure 18–7[a]). This is the part of the eye that is derived directly from the brain. There are several cell layers within the retina and each has its own special function in generating or relaying nerve impulses as signals to the brain (Figure 18–8). The signals from the eye to the brain are integrated there to produce vision. The outermost layer

Crystallins clear proteins of the lens

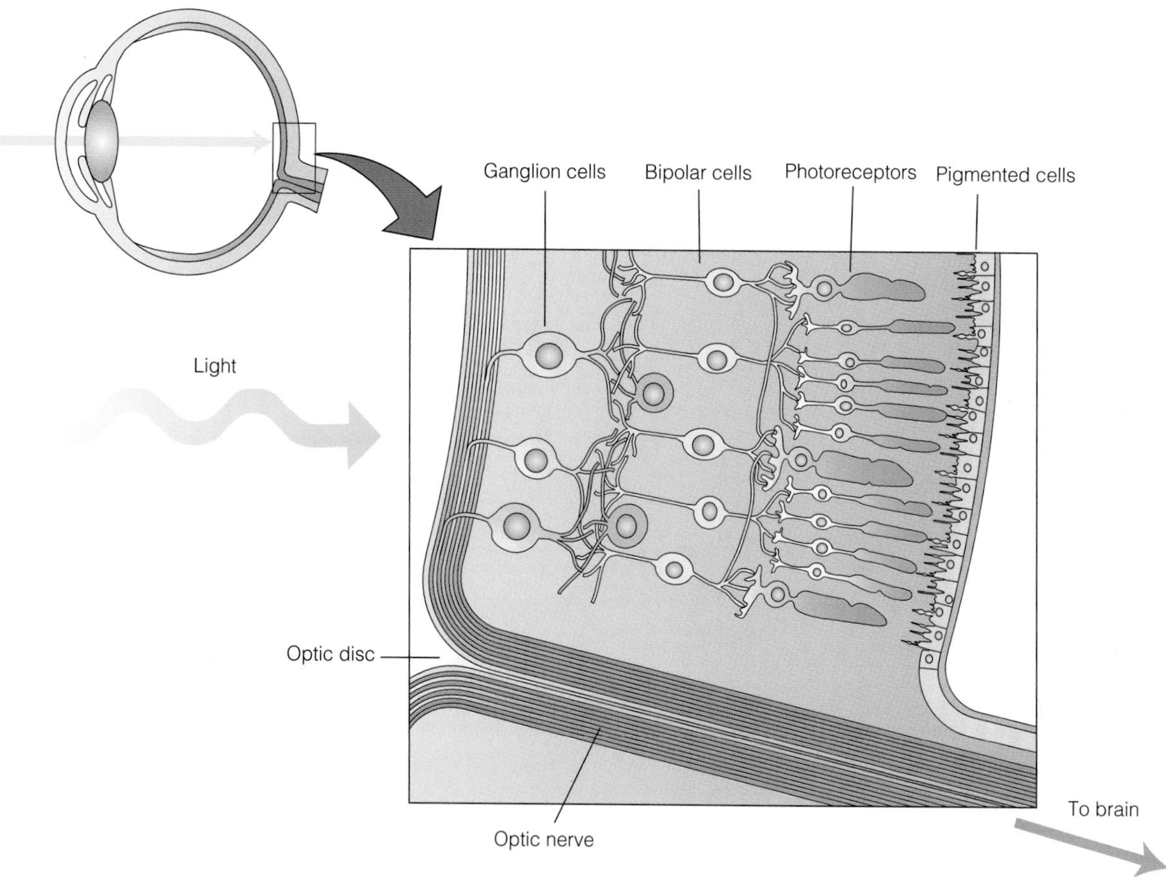

Figure 18–8 The Retina and Light
The retina is multilayered. The principal cells used in the relay of light information are the photoreceptors, bipolar cells, and ganglions. Light passes through the thickness and stimulates the photoreceptor cells. The photoreceptors send nerve impulses back to the ganglion cell layer.

of the retina is a nonneuronal layer of pigmented epithelial cells. Beneath the pigmented layer, and in a series of deeper layers, are the photoreceptor cells, bipolar cells and ganglion cells. Ganglion cells mark the interface between the retina and the centrally located vitreous humor.

Photoreceptor Cells—Rods and Cones

The keys to light reception are the photoreceptor cells. There are two types found in humans—*rods* and *cones* (Figure 18–9). Rods have evolved to respond to light intensity and, essentially, are responsible for black and white images. Cones have evolved to respond to different wavelengths of light (Biosite 18–2) and, hence, provide color vision. Rods and cones get their names from their shapes, but it is their functional differences that are most important to vision. Rods allow us to perceive the presence or absence of light, that is, black and white and shades of gray. Cones allow us to perceive color.

The structure of a photoreceptor cell is quite complex. Each cell has an inner segment, an outer segment, a cell body, and an axon (Figure 18–9). The inner segment contains mitochondria, which provide the cell with the energy needed to function. An axon arises from the photoreceptor cell body at the opposite side of the cell from the outer and inner segments and extends inward to connect to the bipolar and amacrine cells. The molecules that absorb light are found in the outer segments. The structure of the outer segment is characterized by a stack of membrane plates, and the molecules responsible for absorbing light are found within the stacks of membranes. In rods, these molecules are called **visual purple,** or rhodopsin. The outer seg-

Visual purple light-absorbing pigment of rods; also known as rhodopsin

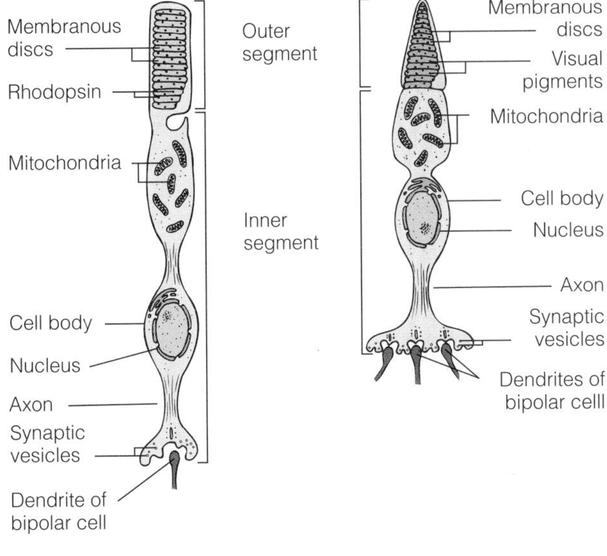

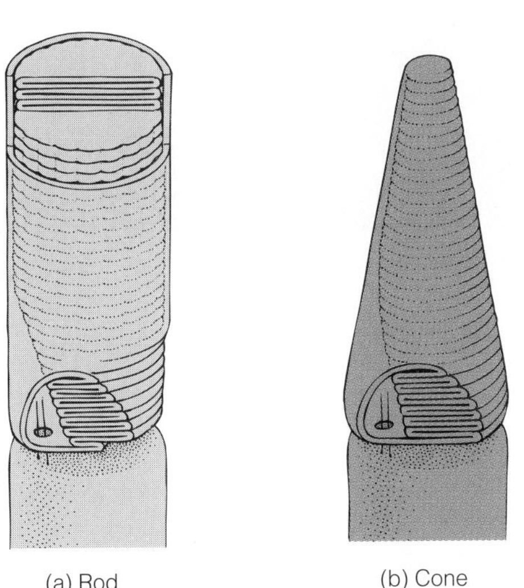

Figure 18–9 Rods and Cones
Rods (a) and cones (b) are the two types of photoreceptors found in the retina. Rods are sensitive to the intensity of light, and cones are sensitive to color (wavelength) of light.

ments are surrounded by surface extensions of the pigmented epithelial cells of the choroid that help protect and maintain them (Biosite 18–3).

There are three types of cones in the human eye, each type corresponding to the reception of a particular color. There are red cones, green cones, and blue cones. Each one has pigments that act to absorb a particular wavelength of light. The absorption of red, green, or blue light alters the pigments of a cone and helps generate a signal that is sent to the brain through the retina cell network. These pigments, as are rhodopsin, are continually regenerated. The brain sorts out the signals from the different types of cones and provides us with a full range of color perception. There is an interesting anomaly in color perception, which is relatively common in human beings. If cones of a certain color type are missing or in low relative abundance in the retina, a person may be color blind. The most common form of color blindness is in red-green wavelengths, which suggests that red cones are missing or malfunctioning. For a person with red-green color blindness, red and green appear to be the same color. Color blindness is a sex-linked hereditary disorder and is found mostly among males.

The Path of Light

How does this sequence of layers, starting with the pigmented cells and dependent on functions of the rods and cones, provide for efficient reception of light and mechanisms for passage of visual information to the brain? To understand this process, let us follow the path of light as it passes through the various cellular layers of the eye on its way to the retina and then on to the brain (Figures 18–7 and 18–8). Light arrives at the eye from a source or sources outside the body. The source can be direct, as we perceive the stars in the night sky and the reading lamp in our room, but it can also be indirect, as a reflection of light off the surface of an object. Whatever its source, the light first passes through and is focused by the curved cornea. Light then traverses the anterior chamber, which is filled with the transparent aqueous humor. The partially focused light slips past the iris-ringed pupil and through the lens. The lens completes the focusing of light, which then passes through the large central chamber of the eye, which contains the vitreous humor. The word "vitreous" means glass, and, in fact, the vitreous humor is almost perfectly clear. The focused light then falls on the retina itself, passing through its entire thickness (Figure 18–8). Having traversed the thickness of the retina, light strikes the pigmented epithelium and the photoreceptor cells that are closely associated with that layer. The function of the pigmented layer, and, to some extent, the choroid layer behind it, is to absorb stray light. This prevents the light from being reflected and interfering with light reception by the photoreceptor cells. The photoreceptor cell layer is in close contact with the pigmented epithelium.

If we assume that a photoreceptor excited by light will generate a nerve impulse, then that nerve impulse must have a pathway to leave the retina and arrive in the brain. It is through the activities of the brain that we actually see. Photoreceptor cells are connected by synapses to bipolar cells, which in turn stretch across the thickness of the retina and contact the ganglion cells on the inner surface. There are two horizontally arranged cell layers in the retina—the amacrine cell layer and the horizontal cell layer.

BIO site

18–2

THE QUALITY OF LIGHT

Light is electromagnetic radiation and travels approximately 300,000,000 meters per second. Light has a dual nature, one as waves and the other as particles. It is the interaction of photons with light-sensitive molecules found in the photoreceptor cells of the visual system that allows them to convert light energy to chemical energy. This conversion produces an action potential and stimulates a nerve impulse. The spectrum of electromagnetic radiation that is technically described as light is very wide and includes light waves such as X rays, microwaves, and radiowaves, none of which we can see with our eyes. The only part of the spectrum that we can see is called visible light. The wavelengths of visible light occur between 400 nm and 700 nm, roughly between violet-blue light and red light, which are essentially the colors of the rainbow or of white light split by refraction through a prism (see accompanying figure). It is the visible light wavelengths to which the photoreceptor cells in the human retina are able to respond. Knowledge of these facts about light, in general, and the response capabilities of the photoreceptor cells to specific wavelengths of light provides us with a sense of the limits of visual perception. We see only part of the spectrum of light and, therefore, many aspects of the world are invisible to us.

They help process nerve impulses to produce better images. Amacrine cells are found between the photoreceptors and the bipolar cells. Horizontal cells reside in the region between the bipolar cells and the ganglion cells. Ganglion cells are the innermost layer of cells in the retina. The axons from the ganglion cells are collected into a bundle of nerve fibers and exit the eye through the optic disc. The nerves pass into the brain along the *optic nerve* through a passage in the bone of the orbit. Interactions among retinal cells in all layers are very important to the quality of visual information. In fact, unlike any other sensory system, the cells of the retina actually process the signals generated by light stimuli before they are sent to the brain. In this way, the retina can be thought of as a functional extension of the central nervous system.

Visual Acuity

Rods and cones are distributed differently in the retina. Rods are concentrated in the periphery of the retina, and the highest concentration of cones is found in a central area of the retina known as the *fovea* (Figure 18–7). Of the millions of photoreceptor cells found in the human retina, only 4%–5% are cones. However, the cone-rich fovea is the area of greatest **visual acuity.** This means that the sharpest visual image we can have of an object comes as a result of focusing the image on the fovea. In fact, it is at this site that the lens naturally focuses images of objects onto the retina. In the case of the fovea, more cone cells per unit area means a higher-resolution color image.

The pattern and intensity of light that is projected onto photoreceptor cells from an outside source results in the generation of impulses in some rods and cones and suppression of impulses in others. This on-or-off process in an array of millions of photoreceptors results in a potentially continuously changing pattern of nerve impulses to the brain. These impulses are integrated in the cerebrum and establish an image depicting in great detail the original object or source of light in the field of view.

Visual acuity a quality of vision determined by the sharpness of the image perceived

18–3

PIGMENTS AND VISION

Rhodopsin is extremely sensitive to light. When a solution containing isolated, purified rhodopsin molecules is exposed to the proper wavelengths of light, it fades from purple to white. This reaction is called photobleaching, and when this bleaching reaction occurs inside the outer segment of a rod, there is a nerve impulse produced. This signal is relayed through the neural retinal cell network, out along the optic nerve and into the brain. One part of the rhodopsin molecule is similar chemically to vitamin A. In fact, vitamin A is known to be necessary for the proper function of the visual system. A deficiency of vitamin A in the diet can lead to decreased sensitivity to low light and to a condition known as night blindness. Using vitamin A supplied from the diet, molecules of rhodopsin are continually, and rapidly, regenerated in photoreceptors and are used over and over to transduce the energy of light to the neural signals needed for vision.

The Shape of the Eye

The lens is the principal, dynamic, focusing structure of the eye. Light initially passes through the cornea and is bent, but it is the lens that is responsible for focusing the image of an object onto the curved surface of the retina. The shape of the eye is crucial for good vision. If the eye deviates from an optimal spherical shape, incoming light cannot be focused properly by the lens and vision will be impaired (Figures 18–7, 18–10 and 18–11). The lens is particularly important with respect to focusing objects near to us. To focus the light from an object, the lens must change shape. How does this occur? The lens is attached to the ciliary body by tiny ligaments. The smooth muscles of the ciliary body contract or relax, and, in so doing, change the shape of the lens (Figure 18–10). This process of focusing is called **accommodation** and provides a mechanism for the eye to focus on objects very near to us, as well as those far away. To compare the eye to a camera, the equivalent focusing procedure in a camera involves changing the position of the lens with respect to the focal plane of the film, as opposed to changing the shape of the lens (Figure 18–10).

One consequence of light passing through the lens is that the image cast on the retina is inverted. You may be asking yourself if that is the case, why don't we see the world upside down? The nerve impulses generated by the light image cast on the retina are sent to the brain and are passed along several interceding neurons into a region of the occipital lobe of the cerebrum known as area 18 (Figure 18–6). The impulses are distributed to the cells of this region in such a way that they are reinverted, and we perceive the world in its proper orientation.

Myopia and Hyperopia As mentioned earlier, good vision requires proper shape of the eye and proper structure of the cornea and lens. Changes in these features of the eye lead to the most common defects in vision. Among these are **myopia, hyperopia, astigmatism,** and *cataracts* (Figure 18–11). Myopia is a result of the lengthening of the eyeball along the optical axis. Thus, when an object is relatively far away, it leads to an image that is actually focused in front of the retina and, so, appears blurry. Hyperopia results from a shortening of the eye along the optical axis such that the image of objects close at hand is ideally focused only along an imaginary curve behind the eye. The distance at which an object becomes blurry, or out of focus, depends on how badly misshapen the eyeball is. The degree to which an individual is myopic or hyperopic is measured in an eye test and the evaluation of your vision is rated on a special scale. We have all heard that a person who does not need glasses has normal, 20/20, vision. This means that the person can see and resolve a set of letters or numbers of a standard size on a chart while standing 20 feet away. A rating of 20/50 means that an individual must be 20 feet away to read letters of a size that a person with normal vision can read at 50 feet away. Myopia and hyperopia can be corrected with glass or plastic lenses (glasses or contacts) added in the path of light to the eyes (Figure 18–11[b] and [d]).

Astigmatism Astigmatism is a consequence of irregular curvature of the cornea or lens (Figure 18–11[e]). A flat spot

Accommodation (vision) the mechanism by which objects are brought into focus by the lens

Myopia nearsightedness

Hyperopia farsightedness

Astigmatism an irregular curvature of the cornea causing distortion of vision

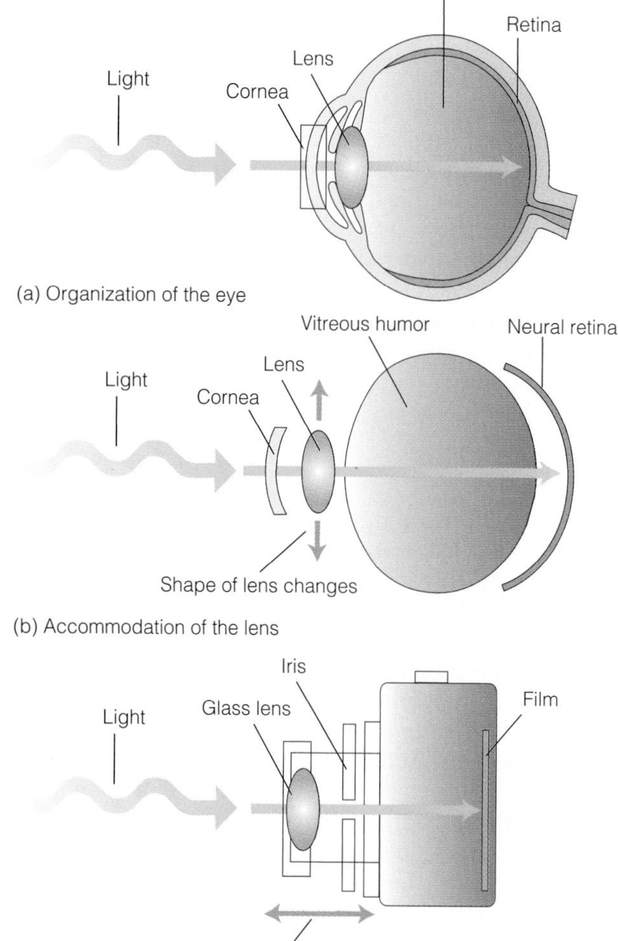

Figure 18–10 Focusing an Image—The Eye as Camera
The image of objects we see are focused by the lens using a process of accommodation (a). The shape change of the lens during accommodation (b) is similar in outcome to moving the lens of a camera back and forth (c).

on the cornea, for instance, results in a local distortion of the refraction of light, which is maintained as a distortion of the pattern of light onto the retina. This type of defect is usually present from birth and does not progress and get worse, as in the case of myopia or hyperopia. Continual changes in the shape of our eyes keep us buying new glasses every few years (or less!). Astigmatism can also be corrected with appropriately shaped glass or plastic lenses.

Cataracts Another problem that affects vision is the formation of cataracts. Cataracts are cloudy-appearing defects in the human lens. They usually develop in older individuals and represent cumulative biochemical changes that result in decreased transparency of the lens. Cloudy areas within the lens block the light on its passage to the retina. How does this damage occur? A lifetime of exposure to radiation from the sun, particularly UV radiation, is one cause. UV radiation brings about chemical changes in the proteins that make up the bulk of the lens. These proteins are generally crystal clear (hence, their name crystallins), and their denaturation seriously and progressively reduces the transparency of the lens and eventually leads to blindness.

Field of View

With all the structures and mechanisms in place in the human visual system, what do we see when we turn our gaze on the outside world? We see primarily sources of light and objects that reflect light. These primary and secondary sources of light provide a representation of the objects and light sources occurring in our field of view. Because we have two eyes, we actually have two fields of view (Figure 18–6). The extrinsic muscles of each eye control the tracking, or object-following, capacity of both eyes simultaneously and in a highly coordinated manner. This precise coordination of both eyes results in what is called *convergence* and keeps the image of objects properly focused on the fovea of each retina. The visual fields of each eye overlap to some degree, as a result of the eyes being positioned directly on the front of the face. It is the incomplete overlapping of these two fields that helps us establish a sense of depth of visual perception. Perceiving an object from two different positions, even locations as close together as our eyes, provides a basis for geometric triangulation (Figure 18–6). This triangulation, in turn, allows for an assessment of the relative distance between us and an object.

Depth Perception

Depth perception is a very important attribute for human vision because we require the coordination of eyes and hands in many of the activities in which we are involved. The surgeon and the watchmaker are renowned for the exquisite eye-hand control they develop in their professions. This coordination has to occur in three-dimensional space to be useful. Each of us can reach out and touch an object in space without hesitation, in part because of visual field overlap and the sense of depth it imparts. Many other species have eyes located on the sides of their heads and have little or no visual field overlap. For example, a horse cannot see an appropriately sized object directly in front of its head. Thus, the blinders on a racehorse are to prevent it from seeing and being distracted by the animals next to it. Because the horse cannot see directly in front of itself, the jockey steers the animal in its race to the finish. There are other cues that help humans determine depth in visual space, such as the relative size of objects (for exam-

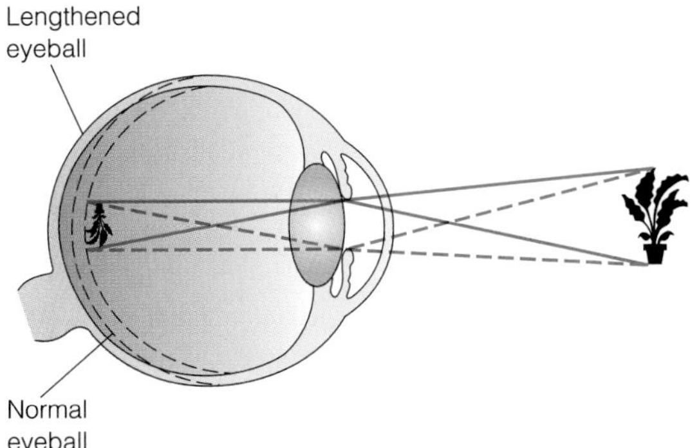

(a) Myopia (nearsightedness) uncorrected

(b) Myopia (nearsightedness) corrected

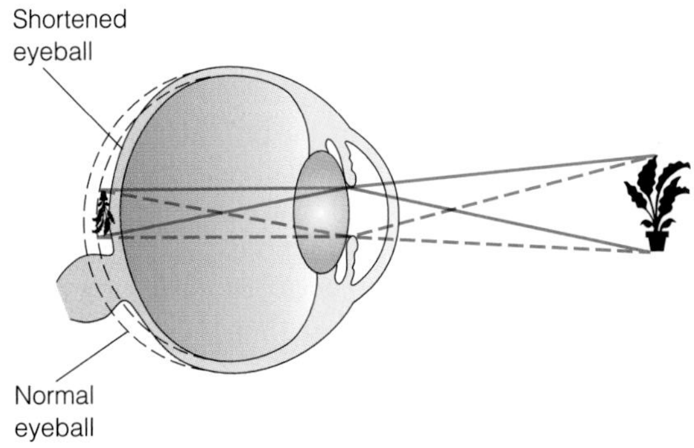

(c) Hyperopia (farsightedness) uncorrected

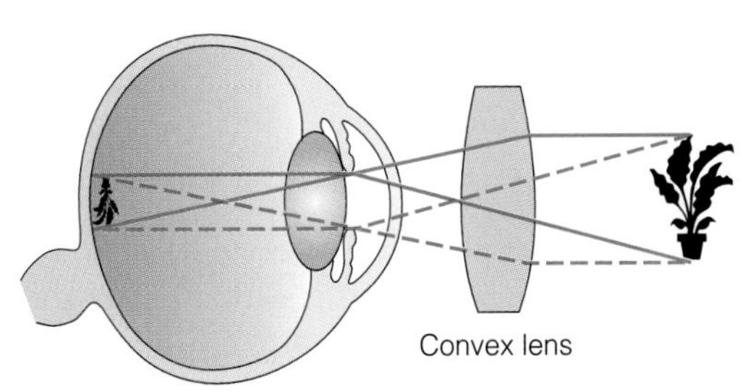

(d) Hyperopia (farsightedness) corrected

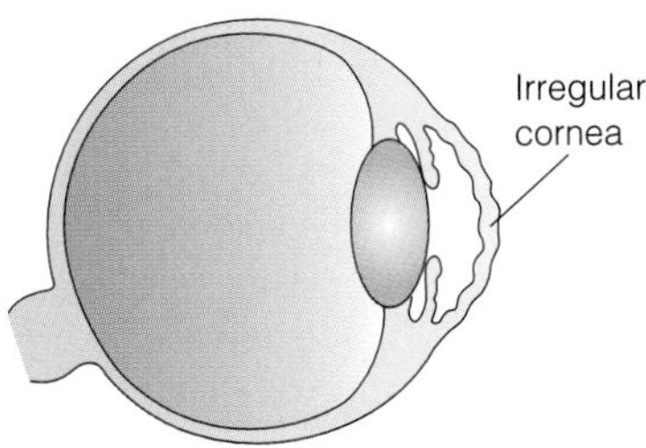

(e) Astigmatism from irregular cornea

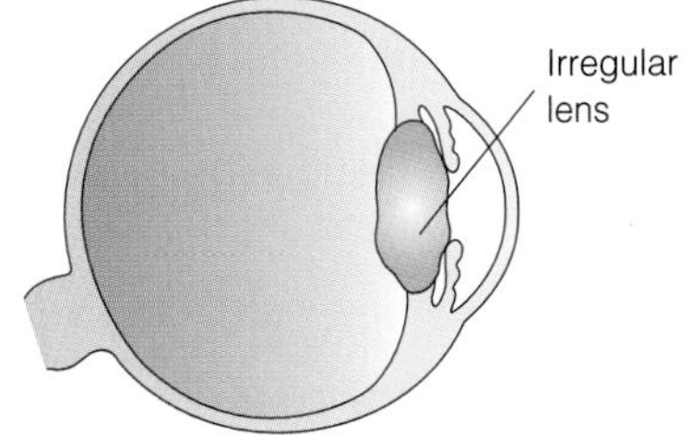

(f) Astigmatism from irregular lens

Figure 18-11 Myopia, Hyperopia, and Astigmatism
Myopia is nearsightedness and is a consequence of an elongated eyeball (a). Hyperopia is farsightedness and is a consequence of a foreshortened eye (c). Both conditions can be corrected with lenses (b) and (d). Astigmatism results from an irregular cornea or lens surface (e) and (f).

ple, houses are bigger than cars, cars are bigger than people), but in many cases, this evaluation requires experience to be of much use. We can be fooled by unexpected object position and relative size differences in the world around us (Figure 18–12).

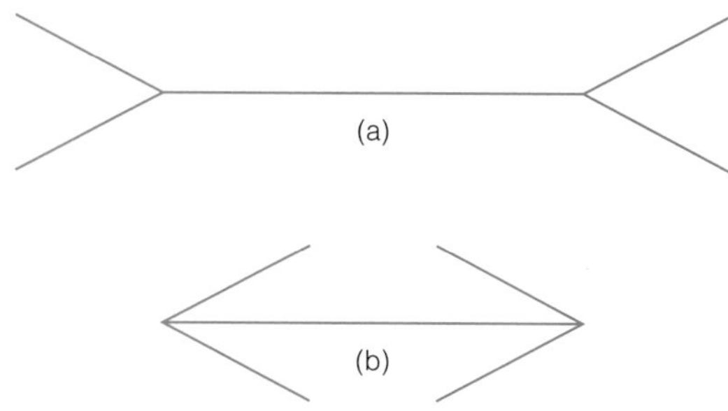

Figure 18–12 Optical Illusion
Which line is the longer straight line, (a) or (b)?

SOUND RECEPTION AND THE SENSE OF BALANCE

The ears of humans are located on exactly opposite sides of the head. As was the case for the eyes, this separation provides the means to establish the direction from which sounds originate. The ear has three regions, or compartments, that serve different functions in hearing and the sense of balance. They are known as the *external ear,* the *middle ear,* and the **inner ear** (Figure 18–13). The external and middle parts of the ear serve acoustical and mechanical functions. Structures located in the inner ear contain phonoreceptors and proprioceptors for maintaining body balance and equilibrium.

External Ear

The outer ear consists of a fleshy, somewhat rigid structure on the side of the head and a canal that enters the skull (Figure 18–13). The external structures are called *pinna* and are composed of cartilage covered with skin. The shape of the pinna helps direct sound waves down the *external auditory canal,* which passes into the temporal bone of the skull. The rest of the ear resides within bone. All regions of the ear are subject to infection. Outer-ear infections, commonly known as swimmer's ear, are readily controlled. Middle- and inner-ear infections are more complicated and more dangerous. At the end of the external auditory canal is the eardrum or **tympanic membrane** (Figure 18–13), which separates the outer ear from the middle ear. As sound reaches the tympanic membrane, it induces vibrations that are transferred to the tiny bones of the middle ear.

Middle Ear

The middle ear has within it the three smallest bones, or ossicles, found in the human body. They are known commonly as the *hammer* (or malleus), *anvil* (or incus), and *stirrup* (or stapes) (Figure 18–13). The hammer is directly attached to the tympanic membrane and moves in response to its vibrations. The ossicles serve to amplify the vibrations 10 to 20 times and make human hearing very sensitive. The movements of the hammer are passed on to the anvil and stirrup. The stirrup is directly attached to the inner ear and moves rhythmically against an opening known as the *oval window.*

The middle ear essentially is sealed at both ends between the tympanic membrane and the oval window. To avoid problems with differences in air pressure, there is a tube, known as the **eustachian tube,** that connects the middle ear to the pharynx. The opening and closing of the tube allows the pressure in the middle ear to be equalized with atmospheric pressure. You can feel the pressure changes in the ear when taking off in an airplane or scuba diving. Yawning or chewing gum usually helps open the eustachian tube and allows the pressure to balance. In the event that the tube does not open, serious and painful distress or damage to the eardrum can occur. Scuba diving with a cold, hay fever, or an ear infection that blocks the eustachian tube can be disastrous and lead to a ruptured tympanic membrane.

Infections of the middle ear (often referred to as *otitis media*) are caused by bacteria that migrate through or are forced up the eustachian tube. Such infections are common in babies and young children in whom the tube is incomplete in its development. The most serious of these infections may lead to meningitis, which is an infection of the membranes that surround the brain.

Inner Ear

The inner ear contains the sensory receptors of the ear and the nerve connection that will lead to the brain. There are two sensory organs surrounded by bone in the inner ear and they are interconnected (Figures 18–14 and 18–15). The space within the bones of the skull in which the inner ear resides is called the *bony labyrinth.* The regions within the

Inner ear innermost part of the ear embedded in the bones of the skull; includes structures involved in hearing and balance

Tympanic membrane the eardrum

Eustachian tube the canal connecting the pharynx to the middle ear

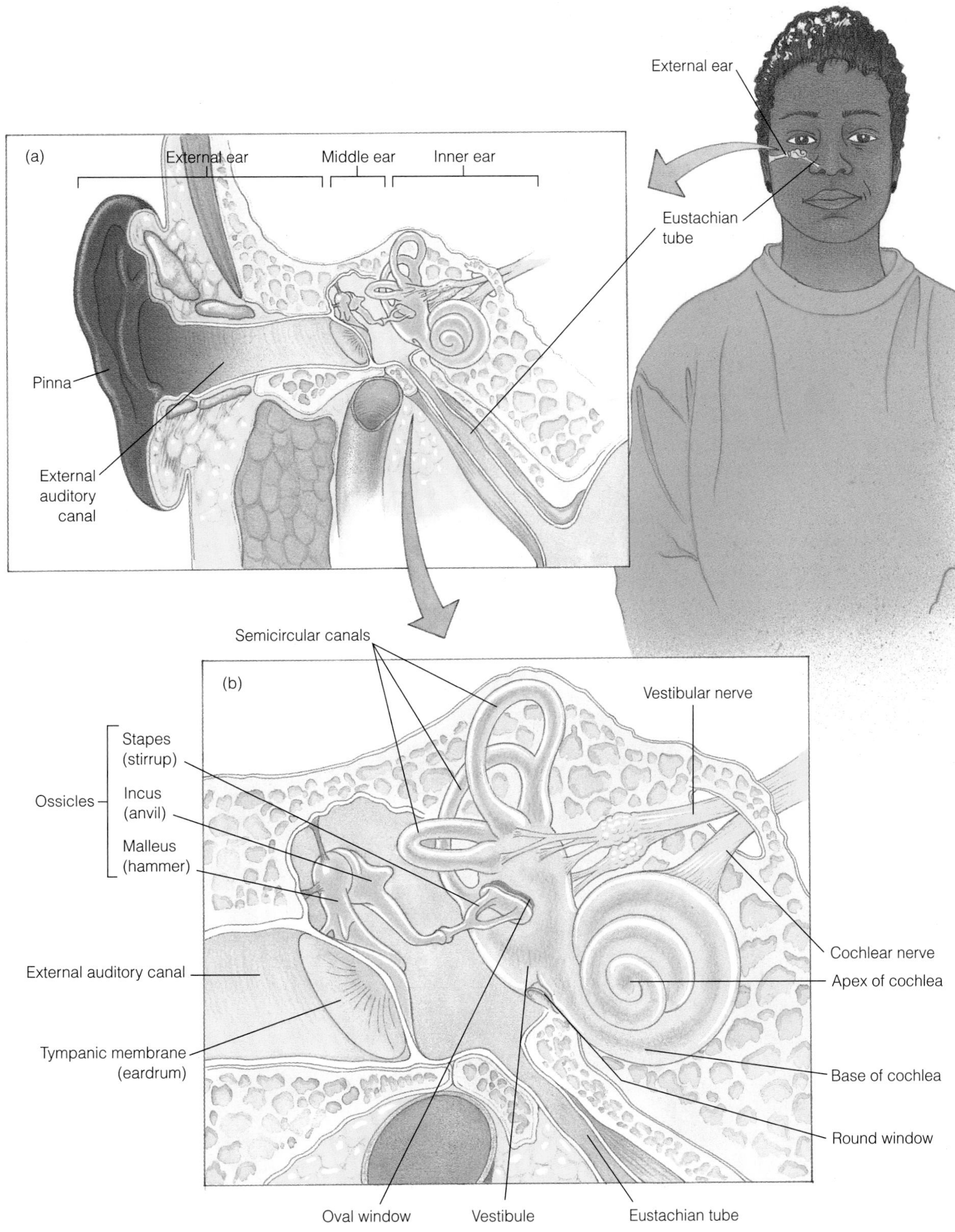

Figure 18–13 The Ear
The ear has three regions—external, middle, and inner.

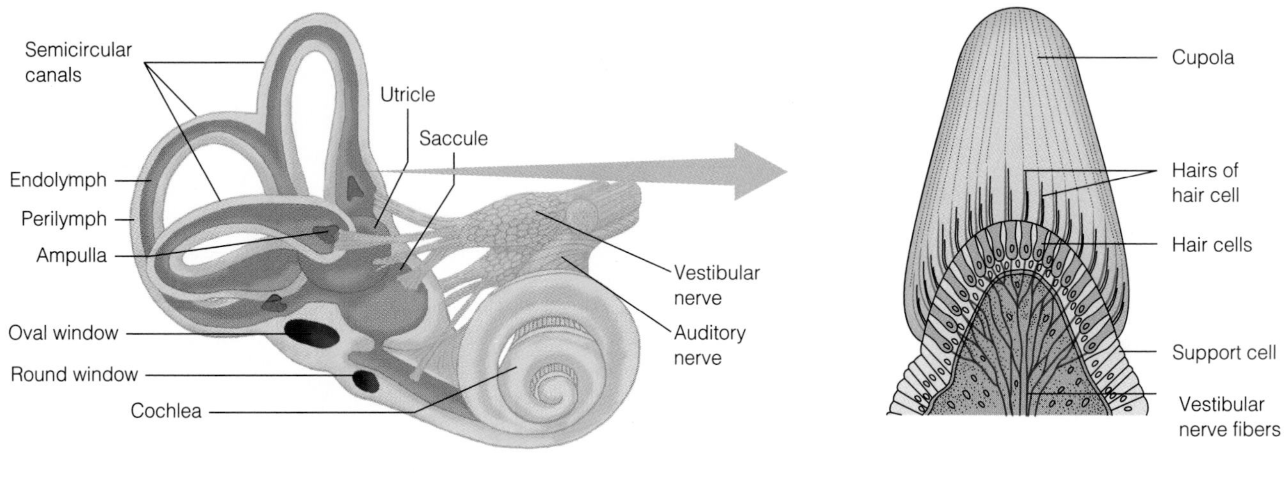

Figure 18–14 The Vestibular Apparatus—Dynamic Equilibrium
The vestibular apparatus (a) contains the structures and receptors involved in dynamic equilibrium and static balance. At the base of each semicircular canal is a receptor-rich structure known as the ampulla (b). Each ampulla contains a structure known as a cupola, which has hair cells that are sensitive to the movement of endolymph.

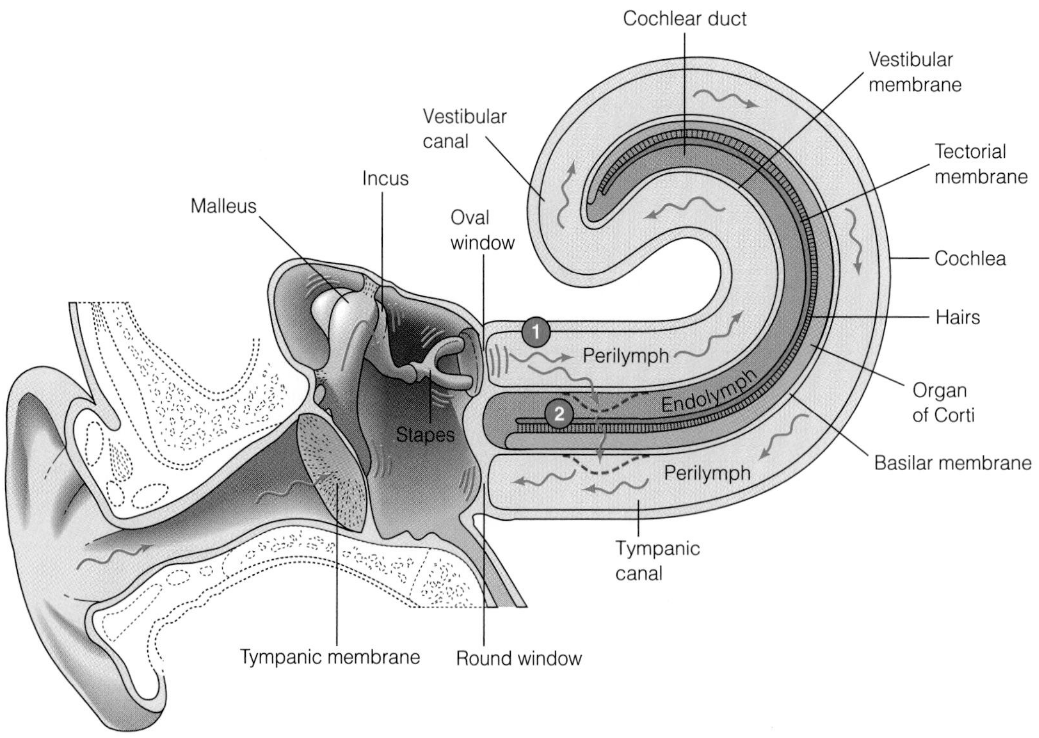

Figure 18–15 The Cochlea
The cochlea in this figure is unrolled to show that the vestibular canal and the tympanic canal are part of the same canal system. The vestibular canal starts at the oval window. The tympanic canal ends at the round window.

labyrinth are divided into the cochlea, the vestibule, and the semicircular canals. The *cochlea* is a spiral-shaped structure containing the phonoreceptors and, thus, the means to hear. The semicircular canals and vestibule contain a complex structure known as the *vestibular apparatus,* which houses the mechanoreceptors involved in **static balance** and **dynamic equilibrium.** The consequences of infections of the inner ear are dangerous because they can lead to serious problems in the control of balance and loss of hearing. The first part of the discussion of the ear will focus on the vestibular apparatus and how we maintain our balance.

Dynamic Equilibrium and the Semicircular Canals

The mechanoreceptors for dynamic equilibrium are located in three *semicircular canals,* and those for static balance are located in two cavernous structures known as the *utricle* and *saccule* (Figure 18–14[a]). Together, the semicircular canals and the combination of utricle and saccule form the vestibular apparatus. With respect to dynamic equilibrium, the receptor cells of the semicircular canals are located in the base of each of the three loops in an area known as the *ampulla* (Figure 18–14[a]). These receptors have long, hairlike extensions that are embedded in a viscous structure known as the *cupola* (Figure 18–14[b]). The movement of the fluid within the canals (called *endolymph*) deflects the hair cells, which, in turn, generate nerve impulses that are sent to the brain. Movement of the head causes movement in the endolymph, which affects the receptor hairs and is translated into signals to the brain concerning the rate and degree of movement that occurred. A fundamental feature of the semicircular canals is that they are arranged roughly at right angles to each other. This means that movement of the head in any direction (up, down, back, forth, left, or right) is detected by receptors in one or more of the three canals. This provides a 3-D representation of movement. The relaying of this 3-D information to the cerebellum allows the body to react in a pattern of muscle activities that maintain spatial and bodily balance and equilibrium. The exquisite control of the dynamic movements of the human body afforded by this system of position detectors is observed in both the graceful pirouettes of the ballerina and the driving, twisting, slam-dunk of the basketball player.

Static Balance and the Vestibule

The vestibule is connected to the semicircular canals but contains a functionally separate sensory system. The vestibule is composed of two chambers, the saccule and the utricle, which contain the receptor cells for static balance (Figure 18–14). The receptor cells are themselves housed within tiny structures called *maculae.* Static receptors are similar in type to the receptor cells of the semicircular canals in that they have hairlike processes and are embedded in a viscous cap. However, tiny mineral crystals called **otoliths,** or ear stones, are also embedded in the maculae. These stones give greater mass to the viscous cap and are responsive to accelerating movement and to gravity. The receptors of the utricle and saccule are oriented differently from one another to provide vertical and horizontal spatial information. A good example of the function of these receptors is observed in bending the head forward. In this case, the otoliths of the maculae are deflected downward and the gelatinous material in which they are embedded stimulates the mechanoreceptors to send nerve impulses to the brain. Thus, the proprioceptive system, knowing where the head is spatially and in what direction it is moving, can make the proper adjustments in the tension on muscles and tendons and through them, position bones so that we may maintain balance.

Hearing and the Cochlea

Hair Cells, Hairs, and Membranes The other neural system of the inner ear is the cochlea (Figures 18–13 and 18–14). This is the structure where receptors are found, which are capable of responding to vibrations transmitted to the inner ear from the bones of the middle ear. The third and final bone of the middle ear, the stirrup, attaches to the oval window, thus, bringing the amplified vibrations of the tympanic membrane to the doorstep of the vestibule. The oval window is a hole in the base of the vestibule and provides a pathway into the *vestibular* and *tympanic canals* that surround the *cochlear duct* (Figures 18–14 and 18–15). These two canals are actually part of a tube that spirals to an apex and then returns to the base of the cochlea at the *round window.* The entire spiraling canal is filled with a fluid known as *perilymph.* The perilymph vibrates in response to the pattern of pulsations on the oval window. The vibrations are similar to compression waves occurring in water when it is disturbed by sound. The cochlear duct has located within it two specialized membranes known as the *vestibular membrane* and *basilar membrane* (Figure 18–16). The duct is filled with endolymph. The vibrations of the perilymph are transmitted to the endolymph and distort the shape of the basilar membrane. Thus, it may be valuable to think of the phonoreceptors in the cochlear duct as detecting the shape of sound.

Static balance balance associated with the inner ear and receptors in the utricle and saccule

Dynamic equilibrium balance associated with the function of receptors located in the semicircular canals

Otoliths calcified "ear stones" associated with detection of movement and position of the head

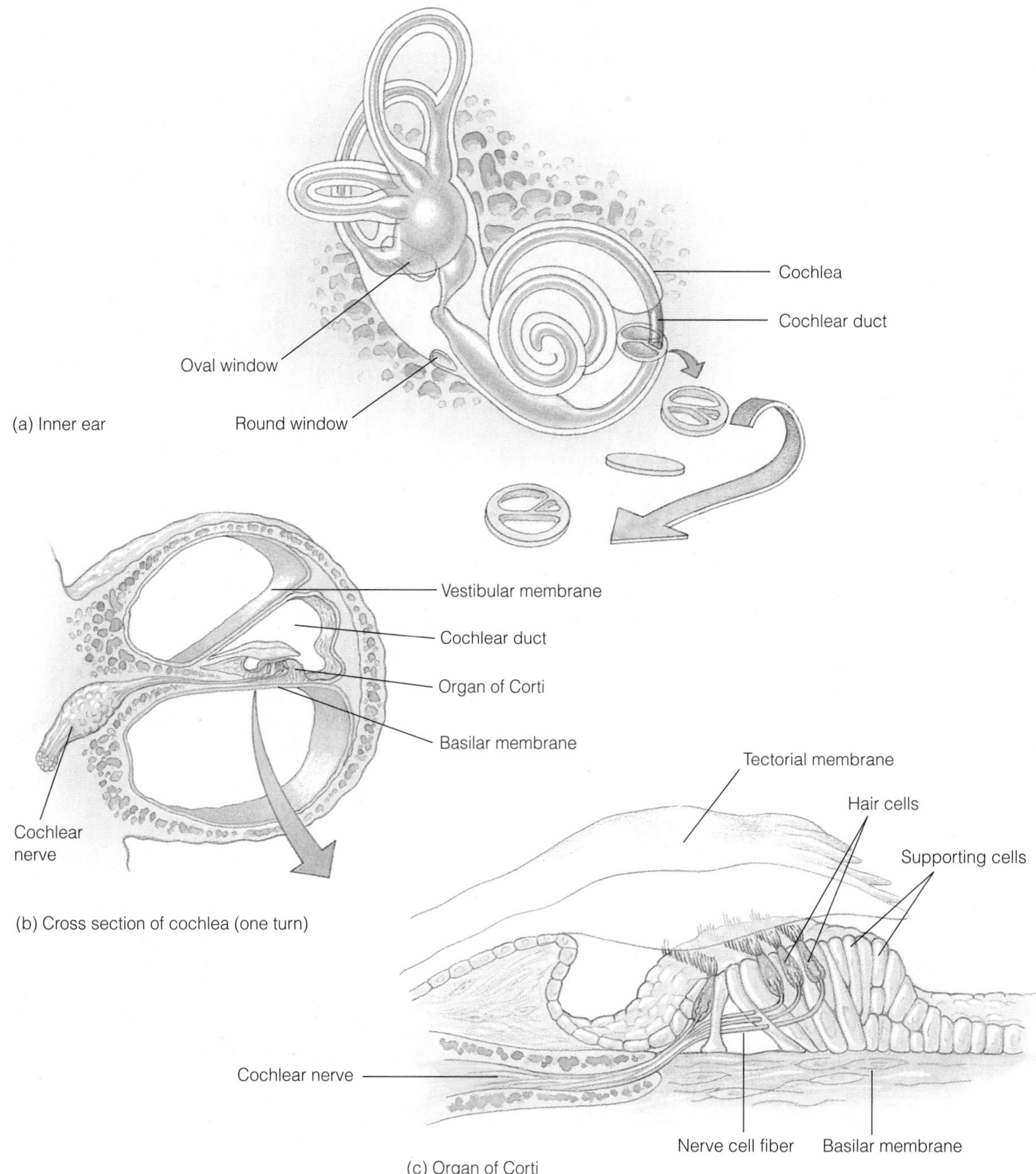

Figure 18–16 Organ of Corti
The cochlea is sectioned to observe inside it (a). The organ of Corti has hairlike phonoreceptor cells embedded in the overlying tectorial membrane (c). The organ of Corti rests on the basilar membrane (c). Nerve impulses that are perceived as sound enter the brain along the cochlear nerve.

Organ of Corti The phonoreceptor cells are located in a structure of the cochlea called the **organ of Corti.** The receptor cells have hairlike extensions that are embedded in an overlying layer known as the *tectorial membrane,* which holds the hairs in place. When the hairs of the hair cells move, nerve impulses are generated. The organ of Corti rests upon the basilar membrane (Figure 18–16[c]). Thus, the hair cells in the organ of Corti and the underlying and overlying membranes with which they associate combine to create a system of extreme sensitivity to vibrations. The deflection of the hair cells is brought about by compressing the oval window, setting up vibrations in the perilymph, and transmitting the pressure waves to the endolymph and surrounding cochlear duct membranes. These vibrations distort the basilar membrane, which evokes nerve impulses by deflecting sensory hairs. Signals are sent directly to the temporal lobe of the cerebrum where auditory information is integrated and perceived as sound.

Discrimination of Sounds

The ear is capable of detecting sound over a wide range of pitch and loudness. How do we discriminate between different sound intensities? How are high frequencies differentiated from low frequencies? The ability to discriminate between frequencies of vibrations results from the fact that phonoreceptor cells in different regions of the organ of Corti respond in a graded fashion to different frequencies of vibrations. The region nearest the oval window is receptive to high frequencies, and the region of the apex is receptive to low frequencies. The basis for this frequency response is the distribution of a series of fibers that form part of the basilar membrane. The shortest fibers are found at the base of the cochlear duct, and the longest are found at the apex. Functionally, as vibrations are transmitted through the eardrum to the cochlear duct, the short wavelengths maximally vibrate the short fibers, and the long wavelengths maximally affect the long fibers. This results in the discrimination of pitch within specific regions of the cochlea (Figure 18–16).

The full range of acoustic frequencies over which humans are able to hear covers roughly 20 cycles per second (or 20 hertz) to 20,000 cycles per second (20 kilohertz). If the sound is too loud, it can be painful and/or damaging. A measure of the loudness or *amplitude* of sound can be established using the decibel system (Table 18–1). The decibel system represents a measure of exponential increases or decreases in sound amplitude. If we consider the sound of breathing to be 10 on an arbitrary scale, then increases in the sound level by thousands or millions of times can be assessed relative to it. The decibel system allows for a simplified representation of the sound level to avoid awkwardly large numbers, such as 1,000,000,000,000 associated with jet engines (Table 18–1). This enormous number can be simply presented as 120 (1 followed by 12 zeros). Anyone who has been close to a jet liner at takeoff can appreciate the difference in loudness relative to breathing and the inherent danger of such high amplitude sound to the ear.

A pleasing outcome of the structure and responsiveness of the ear to vibrations is that we can hear many types and qualities of sounds singly or simultaneously. Human hearing is extraordinarily sensitive and discriminating. This range of sensitivity and range of responsiveness to sound, complex or simple, is easily demonstrated if we consider our ability to hear the bold sound of music performed by each type of instrument in a full orchestra and the irritating drip of water in the bathroom sink that disturbs us as we try to fall asleep at night.

Table 18-1 Range of Human Hearing—Frequency and Amplitude

(a) Frequency Range

20 hertz to 20,000 hertz[1]

(b) Amplitude

RELATIVE INTENSITY OF SOUND	DECIBELS[2]	LEVEL OF SOUND
10	10	Breathing
100,000	50	Quiet conversation
10,000,000	70	Traffic on a highway
10,000,000,000	100	Operation of a jackhammer
1,000,000,000,000	120	Close to a jet takeoff
10,000,000,000,000	130	Rupture of an eardrum

[1]hertz = 1 cycle per second

[2]decibel = common unit of sound. Note that decibels are exponential numbers increasing in powers of 10. For example, 100,000 has 1 followed by 5 zeros. Such a number may be represented as $1 + 10^5$ in scientific notation or as 50 in the decibel system.

Organ of Corti a cochlear structure in which hairlike phonoreceptors are located

Summary

The sense cells and organs of the human body are our antennae to the world. They set the range and limits of our perceptions. The general, or so-called simple senses, include exteroceptors, interoceptors, and proprioceptors. These are, for the most part, distributed in the skin, viscera, muscles, tendons, and joints of the body. There are also several special sense organs whose function requires elaborate and complex structures. These organs include the eyes and ears. A review of what kind of stimuli our senses actually receive indicates that sensory receptors fall into several distinct classes—mechanoreceptors, chemoreceptors, thermoreceptors, photoreceptors, and phonoreceptors. These correspond to touch or tactile sensations, to taste and smell, to heat and cold sensations, and to light and sound, respectively.

Sensory cells and organs are exquisitely responsive to changes in the environment. On a conscious level, this is of paramount importance in establishing what is going on around us and within us. On a subconscious level, many of the senses are intimately linked to the body's ability to continually maintain physiological equilibrium and, so, are homeostatic in nature. However, as sensitive as many of these receptors are, they are also limited in their capacities to respond to stimuli. Touch that is too light is not felt. Light waves outside the visible range of the photoreceptors in the human eye are not seen. Acoustical vibrations of too high a frequency are not heard. Ultimately, the brain integrates the signals provided by our senses, limited though they be, into information and experiences that form the basis for behavior and decision making, which are necessary for carrying out actions in the world in which we live.

Questions for Critical Inquiry

1. What is the nature of the general senses? Where are the receptors of these senses located? How would their loss affect you?
2. How is the eye similar to a camera? How is it different? What kinds of damage or disease to the eye might be expected to lead to blindness?
3. What is the relationship between the interoceptors and proprioceptors and homeostasis? What would happen if the ability to receive signals from these receptors was lost?
4. How is the organ of Corti organized within the cochlea? How does it respond to different frequencies of sound waves? How might they be damaged? What would happen to the quality of hearing?
5. Is the human eye sensitive to all wavelengths of light? If not, to which wavelengths is it sensitive? What would it be like to be sensitive to all wavelengths of light?

Questions of Facts and Figures

6. How are taste and smell related to one another? Which is more sensitive?
7. What is static balance? What structures are involved in maintaining static balance?
8. The oval and round windows are located in what part of the ear? What are their roles in hearing?
9. What are the principal layers of the eye?
10. What are the cell layers of the retina. How are they organized? What cells are involved in the reception of light?
11. What is the role of the pigmented epithelium?
12. Where is the concentration of cones the highest in the human retina?
13. What are the bones of the middle ear and what is their function?
14. What is the difference between myopia and hyperopia? How can they be corrected?

References and Further Readings

Noback, C.R., and Demarest, R.J. (1977). *The Nervous System,* 2nd ed. New York: McGraw-Hill.

Sherwood, L. (1993). *Human Physiology,* 2nd ed. St. Paul, MN: West Publishing.

Stalheim-Smith, A. and Fitch, G.K. (1993). *Understanding Human Anatomy and Physiology*. St. Paul, MN: West Publishing.

CHAPTER 19

Human Behavior and Learning

INTRODUCTION

Behavior is a reflection of how an organism works out an interactive relationship with the environment in which it lives. As with overall biodiversity, there is tremendous diversity in behavior. From an evolutionary viewpoint, Nature is the testing ground for genotypic and phenotypic variations in behavior among individuals in a population, just as it is for structure and function. A slight change in a heritable trait affecting behavior or affecting the outcome of that behavior may allow an organism to compete more effectively for resources (food, water, mates) than others of its species. Humans demonstrate some of the most complex behaviors of any living organisms. A major part of human behavior is learning. *Homo sapiens* has developed language and symbols to communicate thoughts and feelings, arts, and sciences to reveal the rules of nature, as well as laws and ethical systems to help guide civilization's advance. In this chapter, we explore the basis for many types of behavior in an effort to understand how we successfully adapt to the ever-changing environment around us.

INSTINCTUAL VERSUS LEARNED BEHAVIOR

Instinctual behaviors are genetically predetermined. They do not require any experience or learning in order to occur. Instinct is linked to programmed performance, so any enhancement of a part of the capacity of the organism to respond to a set of stimuli, such as genetic changes affecting muscle function, skeletal modification, or hormone production and reception, may play a determining role in the success of that trait. Instinctual behaviors will be discussed in this chapter and compared to behaviors that are learned. Learning builds on the genetic framework of innate behaviors, but learning is more complex than instinct. This is because the behaviors associated with learning are subject to extensive modification by experience. Both innate and learned behaviors require elaborate and interactive neuro-endocrine systems.

Instinctual Behavior

Under what conditions are instincts likely to be considered valuable to an organism? It is easy to imagine that if an environment provides a stable or recurrent source of a particular stimulus, such as light, gravity, or shapes, then instinctual, genetically predetermined behaviors associated with these stimuli may be favored. Instinctual reactions are predetermined genotypically to reduce variability in the response of individuals within populations to set signals. This means that they can respond or communicate effectively without having to learn. For example, in baby birds, the response to the color, shape, and movement of their parents' beaks is inborn (Figure 19–1). A change in response to such an important, and specific, stimulus might be detrimental if performance (feeding) needs to follow rigid and precise steps for which there is no time to learn. Suckling behavior in human babies is also innate (Figure 19–1, a). Presented with a mother's breast, the newborn child begins the sucking action necessary to stimulate release of milk from the mammary glands.

Learning

On the other hand, learning from experience may be advantageous when stimuli are variable, unstable, or unpredictable. The ability to respond to changes in environmental stimuli requires continual input of new and modifying experiences in order to be appropriate and adaptive. A rat in an artificially contrived maze learns by *trial and error* which route to follow to succeed in obtaining food or water (Figure 19–2). The rat learns the path through the maze and remembers it. There is clearly no way to build such specific knowledge into the rat's genes. It is through trial and error and the experiences that are built up from them that the rat learns. The responses to the environment are interpreted as beneficial or detrimental based on success or failure in the test maze. How different is this from the environment in which a rodent lives naturally? Learning where food and water may be, where a safe haven is to be found, and where to go when danger arises are clearly important to survival.

Memory

Learning develops from experience, and underlying the ability to learn is *memory*. Memory is the brain's record of experiences and includes the mechanisms by which memories are organized and made accessible within the brain. Memories of experiences may, by these mechanisms, be recalled or recollected. If an organism does not remember previous experiences, then decisions based on its present circumstances may be fatally flawed. For example, the outcome of a small animal not remembering a predator's shape or smell may be lethal. Learning requires memory, and survival in complex and ever-changing environments requires learning.

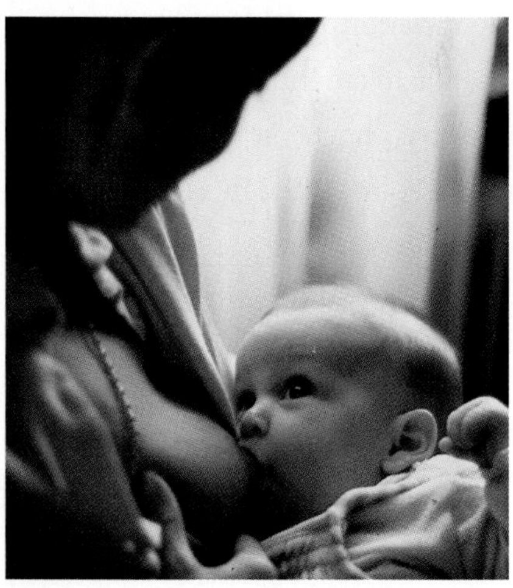
(a)

(b)

Figure 19–1 Instinctual Behavior Instinctive behaviors, including (a) sucking in human infants during breast feeding and (b) gaping mouths of baby birds, occur in anticipation of feeding and in response to sensory cues or releasers in the environment.

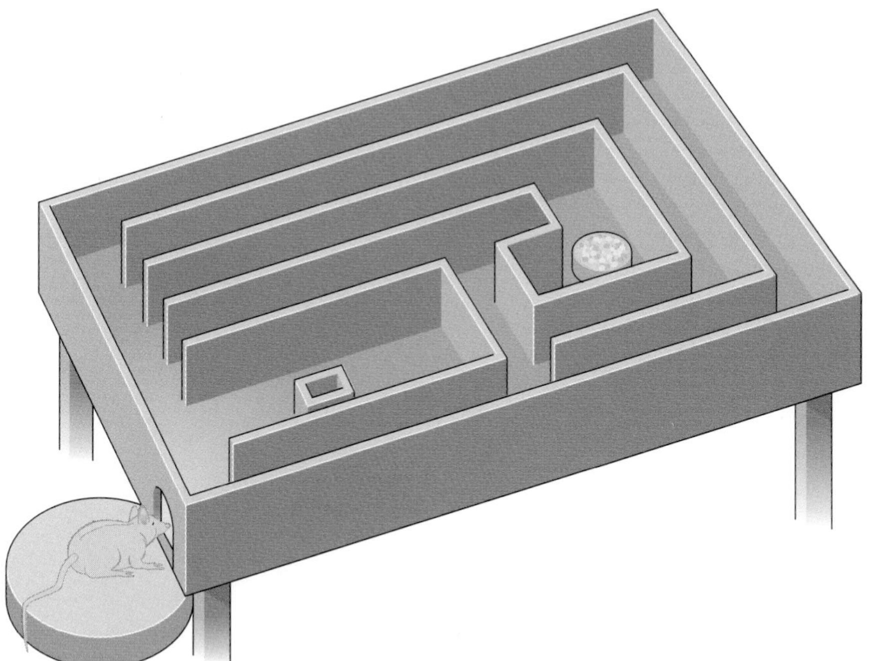

Figure 19–2 Trial and Error Confronted with a maze, a rat will explore the environment and through trial and error discover the food. Subsequent trips through the maze meet with more rapid success until the path is learned and the rat goes directly to the food.

VIEWS OF BEHAVIOR

A Mechanistic View

Studies of behavior fall into two broad categories, mechanistic and operational. The first category is mechanistic because it involves behavior that is describable in terms of biochemical and/or cellular mechanisms. The mechanistic view of behavior, thus, involves a basic understanding of how the endocrine and/or nervous systems function in conjunction with processing and responding to sensory information. The mechanistic view includes establishing how hormones and neurotransmitters work and the function of receptors in synapses that regulate fluxes of ions into and out of cells during a nerve impulse. Sensory nerves convert physical and/or chemical stimuli into nerve impulses and interconnect to the massive integrating framework of nerve cells within the brain (see Chapters 17 and 18 for review). It is in the brain that sensory information elicits a preprogrammed, instinctual response or activates a learned behavior that, in turn, depends on specific recollection and interpretation of memories of previous experiences. Both patterns of response (instinctual and learned) affect the activities of the body in specific ways.

An Operational View

The second category of studies of behavior depends on observations of the end result of all the underlying cellular and biochemical activity, that is, the behavior itself. The overall behavior is correlated to the initiating stimulus in the environment, without regard to precisely which biochemical or cellular pathways are involved. This second approach is taken by psychologists, who study human behavior, and ethologists, who study the behavior of animals in the wild.

The Combination

The combination of biochemical/cellular approaches, involving studies of the structure and function of the nervous system, and the whole-organism psychological approach has proven to be most useful in the quest to understand animal behavior, particularly human behavior. In fact, the use of animals in behavioral research has resulted in many discoveries applicable to humans that otherwise would not have been possible to make. It is important to realize, however, that even detailed knowledge of how the nervous system is organized and functions has not yet proven to be sufficient to explain how behaviors arise and persist in the human repertoire. The simplest way to define animal behavior is to refer to it as a response or set of responses to a specific set of environmental stimuli. However, behavior does not exist as an isolated event. Many different behaviors are integrated simultaneously in response to a spectrum of many different stimuli. This feature of behavior, the blending together of many actions and reactions in a response to multiple stimuli or cues in the world around us, makes the study of human behavior complicated and fascinating.

SIMPLE AND COMPLEX—THE NATURE OF BEHAVIOR

The discussion of behavior begins by describing and defining it in two different categories—simple and complex. For example, growth or movement responses of organisms toward light, or against the pull of gravity, or to the presence or absence of nutrients in the environment fall into the simple category. Instinctual behaviors, such as feeding activities in birds and learning in humans, are considered to be complex. The most complex behaviors are those observed in higher vertebrates, whose large brains and elaborate sensory systems provide a tremendous capacity to receive and process information about the internal and external environment.

Simple Behavior

Tropism No behavior is really simple, because nothing living is simple. However, some behaviors are less complex than others. The simplest types of behaviors are found in organisms that are not able to move around freely throughout most of their life cycles. Such organisms are called *sedentary* and include classes of organisms within all the Kingdoms. For example, a behavior that involves a change in the direction of growth toward or away from a continuously present (or recurring) stimulus is called a **tropism.** The growth of a plant toward sunlight is a form of *phototropism.* Light-responsive growth of the stem on the side opposite the light results in reorienting of the leaves (Figure 19–3). If you have a plant on your window sill at home, try turning it away from the sun and observe how the stem and leaves will redirect themselves to the source of light. Tropisms reflect relatively slow metabolic changes that eventually result in morphological or structural changes in the organism.

The root of that same potted plant grows downward into the earth in which it is embedded. This growth is in response to gravity and is called *geotropism,* or gravitropism. Interestingly, at the same time a root grows down in positive response to gravity, the stem of the plant is negatively gravitropic, and grows up and away from the attraction of Earth's gravity. To put this into a behavioral context, such responses are extremely important in the germination of seeds, which must send a shoot up through the ground to attain the sunlight needed for further development, while sending a root downward for stability and acquisition of nutrients.

As immobile as most plants are, there are a number of species that have specialized structures that react quite rapidly to stimulation. For example, the Venus flytrap lays open its leaves as an invitation to insects (Figure 19–4). Under these conditions, when a fly alights on the moist, sticky surface of the spread leaves, the leaves are stimulated to close rapidly and form a cage from which the insect prisoner cannot escape. A slow process of digestion begins as the plant converts the prey to organic soup. Even with this unique behavior the Venus flytrap is earthbound and not equipped to actually move itself away from its fixed location.

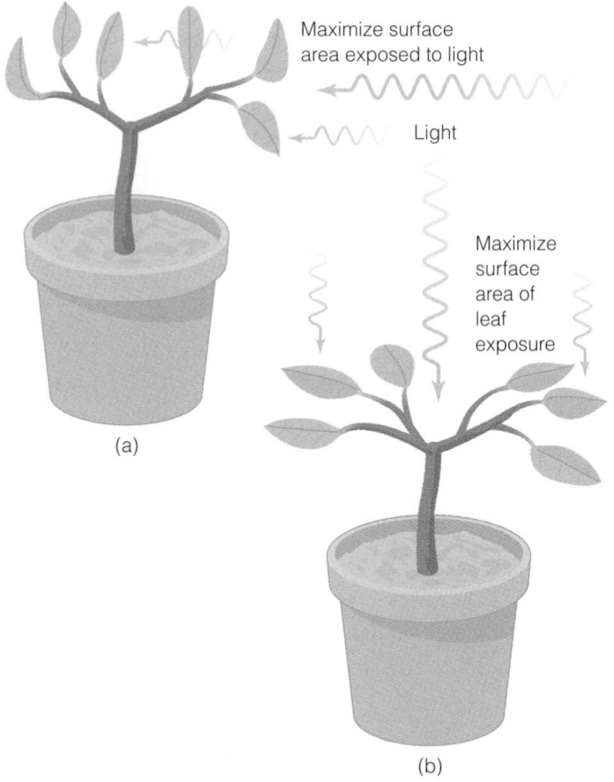

Figure 19–3 Tropism
The reorientation of leaves (a) to maximize direct exposure to incoming light (b) results from growth of the plant in response to light stimulus, a response known as phototropism.

Taxis A **taxis,** (pl. taxes) on the other hand, is a relatively rapid movement of a whole, mobile organism in response to a stimulus or stimuli in the environment in which it lives. For example, the movement of a marine unicellular organism toward the surface of the sea may be stimulated by sunlight from above (Figure 19–5). These organisms have flagella, which provides them a means to move up and down the water column in response to light. Active movement of an organism in the direction of the source of the

Tropism growth toward a stimulus
Taxis (pl. taxes) movement toward a stimulus

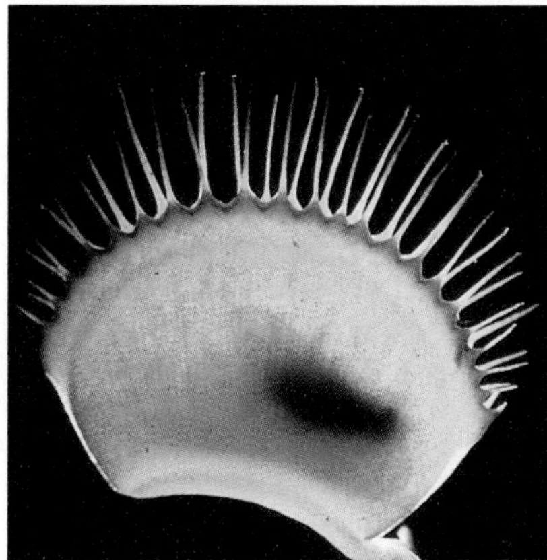

Figure 19–4 Venus Flytrap
A Venus-flytrap plant (a) has the capacity to capture prey. The leaf is sensitive to touch and responds rapidly to the presence of a fly (b) by closing and (c) by trapping its prey.

stimulus is called a positive taxis. Active movement away from the source is referred to as a negative taxis. Response to light is a *phototaxis*. Migration of an organism, such as a bacterium or protozoan, along a concentration gradient of nutrients in the environment is a *chemotaxis*. The movement of an organism along a gradient of heat is a *thermotaxis*. These types of responses involve receptors and chemical processes that are sensitive to the intensity or concentration of the stimulus and elicit behaviors resulting in orientation of the organism to the gradient. This may translate into a movement toward or away from the source. Light, temperature, and dissolved chemical substances in air or water are, thus, cues for positive as well as negative taxes. These responses may seem to be simple, perhaps because we observe them in single-celled organisms, such as a bacterium or a protozoan, but the details of the behavior of these organisms are very complex and have yet to be completely understood (Biosite 19–1).

These examples of relatively simple forms of behavior have been used to point out how diverse what we call behavior really is. Every action that any living organism takes in responding to stimuli or cues in the environment in which it lives can be considered behavior. This is true regardless of whether it involves change in the pattern of growth (tropism) or change in the direction of movement to follow a physical or chemical trail (taxis) to a source of heat, light, or food.

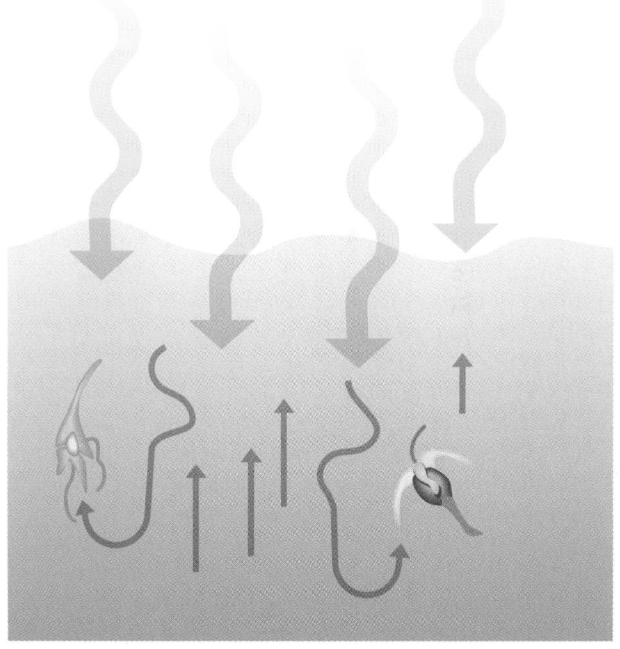

Figure 19–5 Taxis
The movement of flagellated phytoplankton in surface waters of the ocean is in response to light. This is called phototaxis and involves rapid movement of the entire organism.

Complex Behavior

Multicellular organisms retain many of these simple behaviors in the activities of the individual cells composing them. This may be because of the inherent qualities of cellular life; however, for the multicellular organism, it is also necessary to add mechanisms to coordinate the activities of cells and the tissues they compose into coherent patterns of action. Natural selection clearly places a premium on behavioral adaptations that endow an organism with the widest range of chances for success within the changing environments to which they are exposed.

19–1

PREDATORY BEHAVIOR—FOLLOWING CHEMICAL TRAILS

Many types of protozoa are predatory. Predatory organisms that track their prey often depend on a chemical trail to do so. The voracious protozoan *Didinium* targets its prey in this way and in an unparalleled act of consumption, engulfs and devours an organism that may be nearly its own size (see accompanying figure). Likewise, an amoeba uses a similar tracking technique, but instead of being free swimming as the *Didinium* is, it crawls along a surface in search of the source of a chemical attractant secreted by its prey. The protean amoeba wraps itself around its food and ingests it by phagocytosis. The behavior of these and many other types of single-celled organisms is highly adaptive, and the mechanisms they use have been very successful throughout the history of life on Earth. In many cases, a single drop of pond water will contain a microcosm of organisms so diverse and varied in their sizes, shapes, and behaviors that one wonders whether "simple" is really the right word to use in describing them.

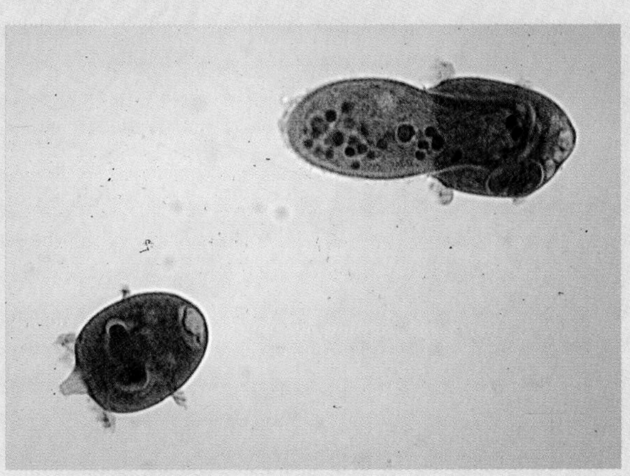

Complexity of form requires complexity in function, and complex behavior requires both.

Reflexes First on the list of complex behaviors for higher organisms are reflexes. Reflexes are the simplest of behaviors in higher organisms. They do not require consciously thinking about the act in order for it to occur and are fixed in the way in which they are carried out. A reflex involves several types of cells working together. Reflex pathways are anatomically built into the cellular circuitry of the human nervous system. They occur in conjunction with muscle contraction and control actions in a single part of the body or in a few parts simultaneously. The hand on a hot stove is removed by retraction of the arm. Striking the tendon below the knee with a mallet results in the kicking of only the leg that was struck. Stepping on a nail results not only in the movement of the foot but also in a reorientation of the body to balance on the opposite leg to gain leverage and pull away from the source of painful stimulation (Figure 19–6 and see Chapter 17 for more details). Sequences of reflexive behavior in lower organisms may make up a large part of their behavioral repertoire. Reflexes are a form of **stereotyped behavior.** This means that they occur in a predictable and repeatable fashion in response to the same or similar stimuli. Reflexes are highly adaptive and may themselves be integrated into complicated sequences referred to as **fixed-action patterns.**

Fixed-Action Patterns Some reflexive responses to stimuli are so complex that they may be thought of as sequences of reflexes or as fixed-action patterns. Interestingly, once such a sequence of reflexive behaviors is initiated, it goes through all the steps to completion. In birds, displays of feathers, nest-building movements, and some attack-and-escape movements are fixed-action patterns. An amusing example of the fixity of such behavior is observed in the egg retrieval movement of the greylag goose (Figure 19–7). If by chance (or human intervention) an egg rolls away from a goose's nest, the mother goose extends her neck and head toward the moving egg and places the surface of her lower beak against it. She then carefully pulls her head back and rolls the egg into the nest. However, if the egg is removed as she is pulling it back toward the nest, she will not stop the retrieval movement to wonder where the egg went, but rather she finishes the action as if the egg were there. The pattern of response is fixed and is programmed to go to

Stereotyped behavior predictable patterns of reflexive behavior

Fixed-action patterns fixed sequences of reflexes composing a complicated behavior

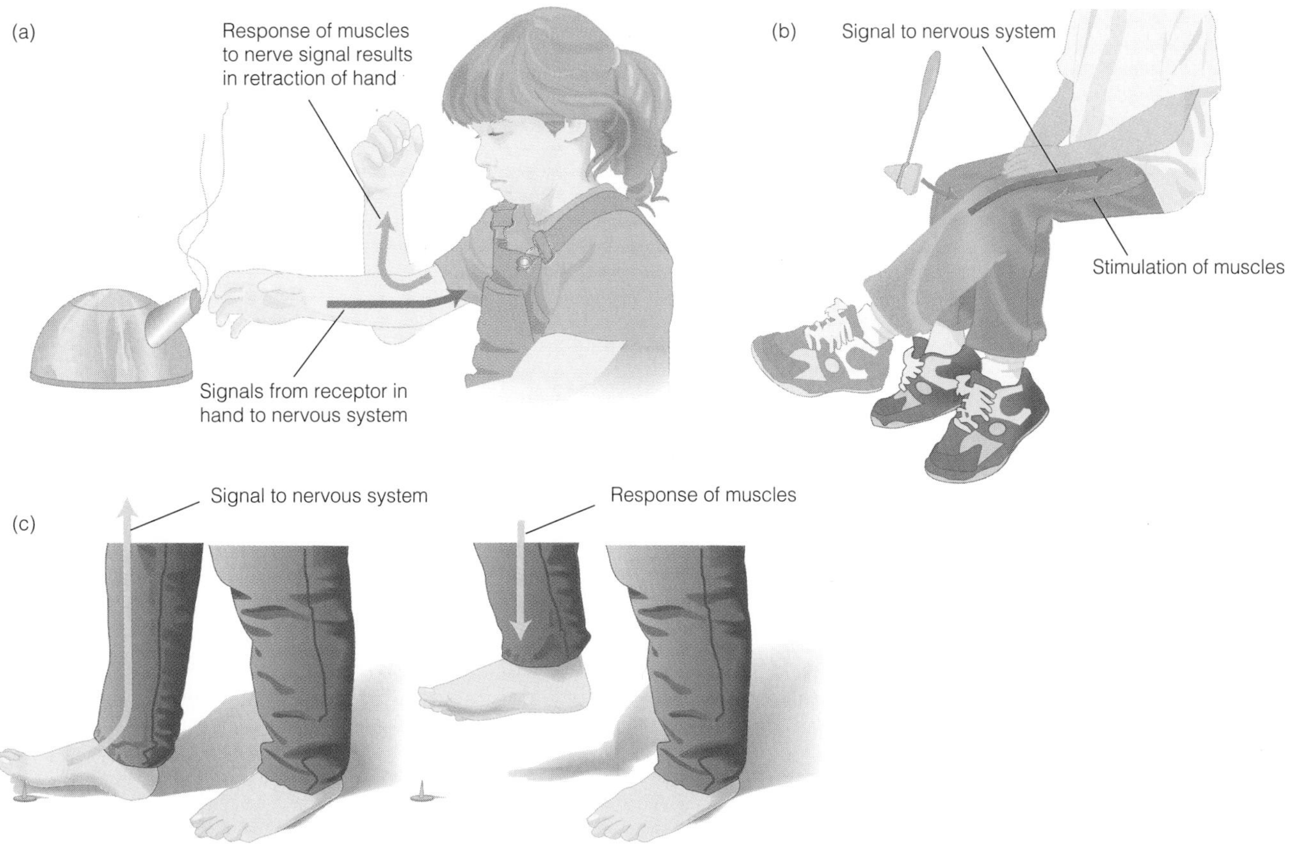

Figure 19-6 Reflexes
Responses of higher organisims to external stimuli that occur without conscious thought are reflexes. The stimuli may be from sources of pain (a), (b), and/or simple activation of stretch receptors in the knee tendon (c).

completion. This whole series of actions has been described as being akin to flushing a toilet. Once the handle is depressed and the valve is opened, the toilet goes through the entire process of emptying and filling up again to complete the cycle.

THE RELATIONSHIP BETWEEN INSTINCTUAL BEHAVIOR AND LEARNING

To what extent is behavior instinctual and to what extent not? This is a fascinating question because it focuses on the contrast between two kinds of behavior that are characteristic of all higher vertebrates. Instinct and learning are related to one another. In fact, learning may have evolved in animals as the capacity to modify innate behavior. This would certainly have been advantageous, because learning provides a way to modify responses to stimuli based on recalling previous experience(s). As evidence for this relationship, innate behaviors themselves are known to change during an organism's lifetime.

Instinctive behaviors by their nature are very stereotyped. Such behaviors are genetically programmed. They result in patterns of responses to stimuli arising in the environment with which organisms have not had previous experience. However, instinctual responses can get better and/or more accurate with maturation and experience. For example, a newborn human child is instinctively able to grip a finger that is offered by a parent or other individual. The grip is really quite strong—strong enough that the baby can be lifted up in midair without falling. This is just the kind of behavior needed to hold on to Mom if she needed to take flight in the face of an enemy or predator. However, the baby outgrows this instinctual response as it learns to interpret danger on its own. Suckling behavior in

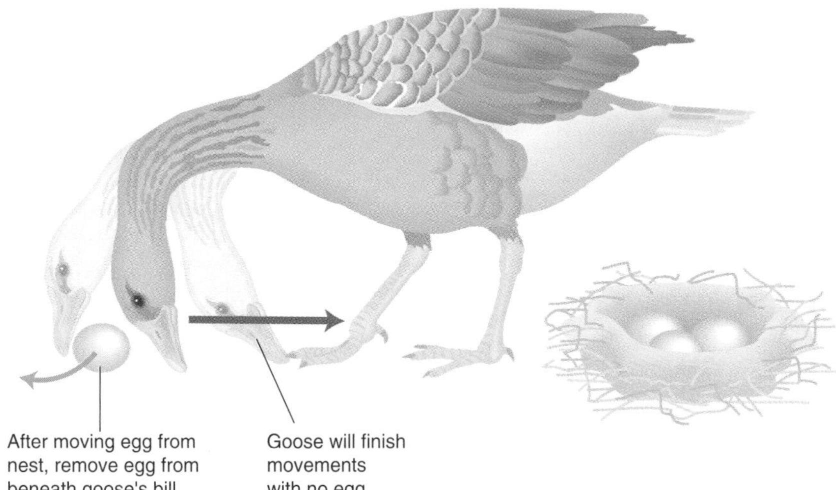

Figure 19–7 Fixed-Action Pattern
The retrieval of an egg by the greylag goose involves (a) a series of coordinated reflexive actions. If the retrieved egg is removed from under the mother goose's bill, the goose will finish the action in the egg's absence. The set of actions is fixed in its occurrence.

humans and other mammals is also instinctual. As the baby's mouth contacts its mother's breast, it begins to suck. This stimulates the let-down reflex in the mammary glands and milk begins to flow (see Chapter 15 for review). The baby soon learns that when hungry, it can initiate contact with the intention of suckling and feeding. It is fair to assume that these types of behavior are anatomically built into the cellular connections of the nervous system prior to birth and are based on the genetic blueprint for embryonic and fetal development. This suggests that the fetal nervous system establishes a cellular framework for learning. The neural framework ensures that the behavior of the newborn infant requires only the correct stimulation or cue in order to be released. A **releaser** or key stimulus is that stimulus or aspect of the environment that causes an individual to perform a preset behavior (Biosite 19–2).

Instinctual behaviors are quite common in most classes of vertebrates but are not all that common in humans. Clearly, gripping and suckling are instinctual, but human behaviors are for the most part learned (Table 19–1). In different branches of the animal kingdom, including other kinds of mammals, birds, reptiles, fishes, and all classes of invertebrates, a greater portion of their behaviors is based on instinct. Learning can and does occur in many of these animals but does not predominate. The middle ground on which instinct and learning overlap is an interesting terrain to explore.

The Song of Birds

Bird songs are probably one of the most intriguing examples of instinctive behavior and learning. In many songbirds, the particular song that an individual sings is species specific. This specificity increases the probability of proper mate selection and virtually assures a successful courtship will take place. In general, only organisms of the same species can produce viable offspring. How do birds acquire the ability to sing their song? Experiments involving some species of songbirds have shown that when birds are hatched in isolation and grown to maturity in the absence of others of their kind, they have an instinctive ability to perform the song of their species correctly. The genetic program of a bird's song somehow is preset in the neuroanatomy of the brain. This is not to say, however, that all bird song is instinctual and that birds cannot learn and embellish their songs through experience. In fact, individuals in some species do require hearing the song from their mature relatives. The juvenile birds may not sing that song until the next spring, but, once having been exposed to it, they remember it.

What happens if a bird is misled? Young birds exposed initially to the songs of different species may not learn their own. Birds are capable of learning new songs and/or variations on previously learned songs and adding them to their repertoires. Anyone who has ever listened to a mockingbird knows how versatile they are in their mimicry. A parrot can even form sounds that mimic the human voice, "Polly wanna cracker?"

Assuming that instinctive behaviors are built into a particular species by the expression of genes during the formation of the nervous system, what then is learning? A simple way of defining learning is to call it the modification of a behavior based on a response to experience. A bird's song may be initially instinctive, but it can be, and

Releaser an aspect of the environment that acts as a cue to stimulate a particular behavior; also known as a key stimulus

BIO site

19–2

RELEASERS

Releasers come in many different forms, most of which are not obvious. In the case of birds that have just hatched, it is important to the survival of the young hatchlings that they specifically respond to the parent birds returning to the nest. When the parents come laden with food to the edge of the nest, the hatchlings must be able to recognize and respond appropriately to the parents in order to be fed. The classic studies of Nikko Tinbergen provide a vivid example of releasers and what is actually recognized by the young herring gulls when their parents lean over the nest to feed them. Is it the parents' eyes? The shape of their heads? The color of feathers? No. What these young birds recognize in the parents is the beak (see accompanying figure). This seems reasonable because the food is carried in the beak and the parent's beak extends first and farthest over the edge of the nest.

Tinbergen went a step further. He mimicked the parental gulls' beaks with models to see what aspects of the beak could actually trigger the feeding response; in a word, what was the releaser?. He made and presented different shapes of fake beaks with different patterns of colors on the artificial beaks. It was in the pattern of color and shape that he found the answer to his question. The young birds would respond well to a model that matched the parents' beaks, which normally have a single red spot near the tip, but they would respond even better to a pattern of three red lines drawn around the circumference of a pencil-shaped shaft (see [b] in accompanying figure). Why? The answer is not yet known, but striped sticks have yet to replace Mom and Dad.

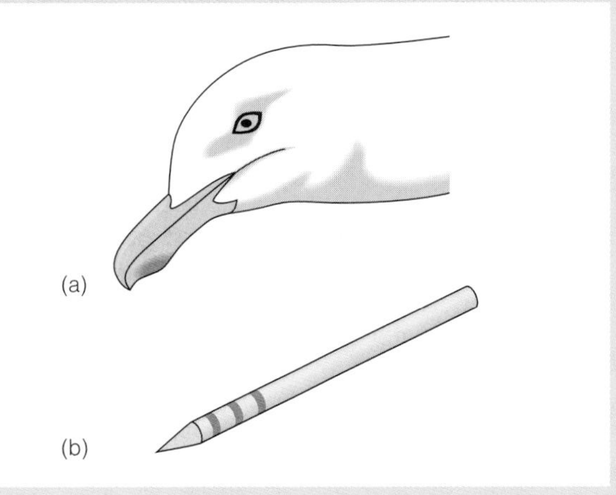

is, modified through experience. Each year as springtime blooms, so do the voices of birds. An individual bird does not lose its genetically endowed pattern, but its song may change subtly with each new season. A bird does learn. In contrast to the predominance of instinct in birds, humans learn almost exclusively by experience. This is one of the major attributes of being human. Members of our species have a tremendous capacity to learn and, in so doing, are able to modify the way they behave, as does a baby with respect to suckling. The question that now arises is how do we learn?

Table 19-1 Human Behaviors

INSTINCTIVE	LEARNED
Suckling	Social interactions
Head movements to find breast	Language
Grasping with hands and feet	Insight
Smiling	Abstraction
Response to mother's voice (a possible form of imprinting)	

WAYS OF LEARNING

One way to learn is to know what to ignore. This we do quite effectively by habituation. Habituation, as a way of learning, depends on the loss of responsiveness to the continuous presence of a stimulus that has little or no value to an organism. A common example of the use of the term sensual habituation occurs with regard to the sense of smell. Within a matter of minutes, our sense of smell habituates or becomes null with respect to odor molecules that may have been initially repugnant. We simply cease to perceive them. The behavioral form of habituation is observed in animals that are confronted with a stimulus that has no meaning to them. For example, in experiments designed to test this phenomenon, snails will stop moving and retract into their shells in response to a physical disturbance around them. Tapping on a glass plate over which a snail is crawling will initially elicit the retraction response. If the stimulus is repeated at intervals after the snail has reemerged, the snail will be less and less affected by it and, eventually, will ignore it. As a result of learning that no consequences are associated with a stimulus, the snail becomes habituated to that specific aspect of its environment. Many types of organisms use this simplest form of learning to exclude processing and responding to sensory information from stimuli that have little or no importance to survival. It saves energy and effort to just ignore them.

Imprinting

A more complicated and enigmatic form of learning is **imprinting.** Imprinting is perhaps one of the most rapid, yet transient, capacities for learning observed in higher animals. It occurs only during a brief period of time, immediately after birth or hatching, but may determine behavior that can last a lifetime. Sight is the primary sensory mode through which imprinting takes place. Basically, the newborn animal attaches itself to the first object it perceives in its field of view, usually one of its parents. This takes place quite quickly and, once established, is set for life. There is a crucial period of imprinting, which, once passed, does not occur again. In those animals that imprint on their parent(s), depriving them of association with the parents during the crucial period may be detrimental biologically and emotionally.

Figure 19–8 Precocity
Many ground-dwelling birds, such as quail, are hatched in a highly developed state. Within a short time, they are capable of walking and migrating efficiently. This behavior is considered precocious, as opposed to the situation observed in Figure 19–1(b) in which young birds are less well developed and nearly helpless (altricial) at hatching.

Precocial and Altricial Under what circumstances might imprinting be valuable? Many types of animals are able to travel almost immediately after they are born. These animals are called **precocial** (as opposed to animals that are delayed in this capability, which are called **altricial**). Anyone who has watched a newly hatched chick has seen it struggle out of its shell, dry off, stand up, and walk around within a matter of minutes (Figure 19–8). The initial few minutes are safe in the confines of an incubator in a school classroom or at home, but in the wild, this is one of the most dangerous periods in any animal's life. Newly hatched wild birds must be able to recognize one or both of their parents and to follow them unhesitatingly to safety. The ability to do this is imprinting.

Our present understanding of imprinting resulted from an ingeniously contrived encounter between ethologist Konrad Lorenz and a number of his young feathered friends. Lorenz had many animal houseguests throughout his long and illustrious career in the behavioral sciences. He believed that the only way to gain an understanding of the behavior of animals was to study them unfettered by the bars and cages of zoos. He had a house full of birds, reptiles, amphibians, and mammals, all with freedom to move in, on, over, under, and through it as they wished. This made for some very interesting encounters.

One of the most important of these encounters occurred when Lorenz took a group of hatchling geese and presented himself as the first moving object in their field of view. What resulted was the imprinting of a gaggle of geese on Lorenz himself. They followed him wherever he went, as if he were their parent (Figure 19–9). In many subsequent trials of this phenomenon, it has been shown that an animal prefers its imprinted parent, who or whatever it may be, more than members of its own species. Imprinting does

Imprinting rapid learning occurring in the young of some species, which determines associations that may last a lifetime

Precocial the young of animals that are ready to act independently immediately after birth or hatching

Altricial the young of animals that are born or hatched in a dependent state

not even require that the moving object be living, just moving. This was shown when a newly hatched chick imprinted on a moving model electric train. The discovery of imprinting in animals prompted a search for such behavior in humans. It does appear that human infants imprint on their mothers. This has been shown in both a positive and a negative way. When an imprinted baby is denied access to its mother, even for a short time during the first year of life, he or she may develop serious behavioral retardation. If separation continues for an extended period (months), behavioral damage may be irreparable.

The mechanism of imprinting is not known. It is clearly a function of the genetically developed structure of the neural network of the brain and mediated through the sense of sight. A permanent record of the initial visual encounter of offspring and parent is somehow stored in memory, never to be forgotten. As in the cases of imprinting on false parents (animate or inanimate) by newborn animals or the response of hatchlings to the fake herring gull beak with red-colored rings as recounted in Biosite 19–2, imprinting provides us with a picture of how complicated the organization of the brain is with respect to instinct, learning, and memory.

Figure 19–9 Imprinting
Imprinting is the rapid learning that takes place in newly born or hatched animals. Organisms display this behavior in recognizing and following their parents, who are usually the first objects to come into their view. Konrad Lorenz substituted himself for a mother goose and, thus, became a mother himself.

Classical Conditioning—Pavlov's Dogs

Among the most famous experiments on learning were those carried out by the great Russian behaviorist and physiologist, Ivan Pavlov. He developed the concept of what is now called *classical conditioning*. Conditioning is the key word in this behavior. Pavlov was interested in the physiology of the salivation response that occurred in a dog given food (Figure 19–10). This behavior is instinctual in dogs and involves a reflexive nerve pathway—sensing food starts the secretion of saliva. What Pavlov was interested in was whether there was a way in which that response could be transferred to another unrelated or *neutral stimulus*. How tightly linked to a particular stimulus was an innate behavior such as salivation?

Experimentally, Pavlov found that every time he exposed a hungry dog to food it salivated. What Pavlov added to this environment was the ringing of a bell when the food was

Figure 19–10 Classical Conditioning
Classical conditioning involves the association of a physiological response (salivation) with an anticipated occurrence (feeding). This response can be "conditioned" and transferred to a neutral stimulus (bell) under specific experimental and natural circumstances. Dr. Pavlov is pictured (center, white beard) with his staff and his dog.

offered. Over and over, he repeated this simple procedure. The final test was to ring the bell alone with no food present and see what happened. Any guesses? The dog salivated. The canine was now conditioned to respond to the new stimulus by associating it with the arrival or consumption of food. Without the costimulation, the dog would never have responded to the bell by salivating. Such a response was simply not in the dog's natural repertoire. Pavlov and many other scientists over the years tried many different types of neutral stimuli, and each was shown to elicit salivation, if they too were conditioned initially with the natural releaser, food.

Extinction What happens if the food is not given to the conditioned animal after the bell is rung? In this case, the conditioned response fades quickly. Without food, the dog's appetite is not satisfied and the unnatural occurrence of salivation elicited by the sound of the bell is lost. When there is no consistent reward associated with a conditioned response, it undergoes *extinction*. This reaction in animals under experimental conditions relates to human behaviors in several ways.

First, humans can also be conditioned in this fashion. This may be evidenced by the effectiveness of advertisements that attempt to link desirable traits (youth and beauty) or lifestyles (rich and famous) with the use of particular products. Ask yourself the next time you go shopping why you choose to buy a specific brand of clothing, food, or drink. Are you conditioned? As in the case of that first Russian dog, classical conditioning is an effective way to transfer a response from a natural to an unnatural stimulus.

Second, forgetting or ignoring memories or experiences that are no longer of functional use is a valuable trait. It allows us to eliminate responses to useless or inappropriate stimuli or cues. More important, this trait allows new adaptive learning and memories to take their place.

Operant Conditioning—The Choices of Rats

A more complicated type of learning results from what is called *operant conditioning* or *instrumental conditioning*. In this type of learning, an animal (human or otherwise) acts in response to stimuli under a controlled set of conditions. Experiments verifying that this type of learning takes place require the direct participation of the subject, who must operate within the test system. Operant conditioning was discovered and developed conceptually by a number of scientists, the most notable being B. F. Skinner, after whom the famous testing cage for rats was named (Figure 19–11). Operant conditioning is different from classical conditioning, in that the animal response is independent of outside influence. There are no bells and whistles to act as neutral stimuli. For example, if a rat is placed in a cage equipped with various kinds of mechanical or electrical devices that can be activated by the test animal, then his or her behavior in that environment can be observed and recorded. If one of the devices is a bar that, when pushed, delivers food pellets into the cage, it is only a matter of time until the rat chances to push on the bar and is rewarded with a pellet of food. It takes only a few trials for the animal to learn this important lesson and to repeat the performance as often as it wants food. The same is true for avoidance behavior. For example, if pressing a bar elicits a mild shock, the rat will cease to press it. It learns to avoid painful stimuli even more quickly than stimuli for which it is rewarded. If food and shock are administered simultaneously, however, the rat has a dilemma. It has to make a choice. The animal is eventually driven by hunger to press the bar. Survival is often an overriding compulsion. The word "driven" is particularly important in this context. What does it mean to be driven to do something? **Drives,** by definition, are specific patterns of behavior that are controlled by needs that are ultimately associated with survival of an organism.

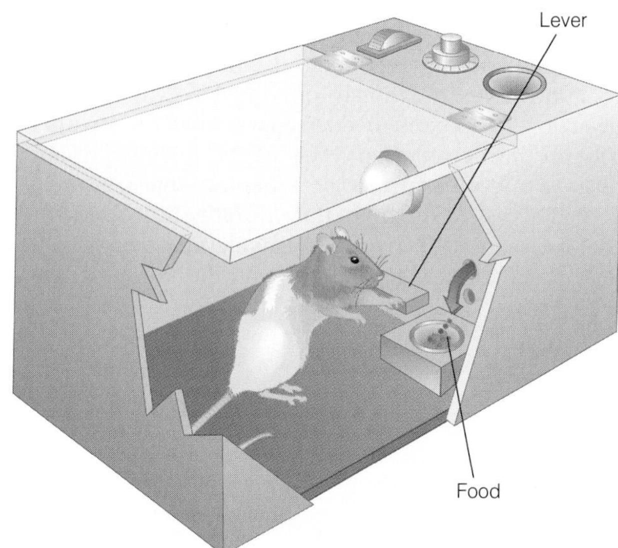

Figure 19–11 Operant Conditioning
In a Skinner box, a rat presses a lever that delivers food into a cup. The rat learns this behavior by operating in the environment.

Drives behaviors controlled by basic needs, for example, thirst and drinking

Drives Drives come from within, built into the nervous system and linked to the far-reaching systems of homeostasis that sustain and balance bodily functions. Terms that are often applied to this type of activity are called appetites or **appetitive behaviors,** and consumption, or *consummatory acts*. Drives result from internal chemical changes in the state of the body, which trigger the nervous system to initiate complex behavioral responses. Feeding is a consequence of the drive to satisfy hunger. Drinking is the consequence of a drive to satisfy thirst. Appetitive behaviors and consummatory acts are linked together. Appetitive drives are generally transient behaviors occurring in response to stimuli that arise periodically and are quenched by an act of consumption. These types of behaviors are goal-oriented and are satisfied only when the goal is achieved. When you get hungry or thirsty, it's hard to resist the desire to eat or drink. The body (through changes in blood chemistry) is communicating its nutritional needs to the brain. The nervous system, in turn, directs the body to the fulfillment of those needs through goal-oriented behavior.

MEMORY AND LEARNING

All types of learning depend on stored experiences in the form of memories and on the ability to recall those memories and use them to make informed decisions. When confronted with a new and unusual environment, an organism may test that environment to ascertain what it is like. Such testing involves exploration of the scene, manipulation of objects in that setting, and cautious consideration of unknown aspects of the environment. Some types of responses to the unknown may be correct and successful; others may be inappropriate and failures. This process is called trial and error and is the basis for both classical and operant conditioning. Trial and error is probably the most common way in which we all learn.

Accumulating and retaining experiences as memories is no simple matter. It is not as if memories are automatically and permanently recorded in the brain and open to easy recall. There are *short-term* and *long-term memory* processes. Each has its own special characteristics and very little is known about how such processes work. In some neurological disorders, the brain appears unable to convert an experience to a permanent memory trace (also called an **engram**). Another possibility is the memory itself is stored in the brain, but we have lost the ability to recall or recollect it. In this second case, a person is commonly thought to have **amnesia,** which is the inability to remember specific events of the past. Amnesia can be induced by damage to the brain, perhaps a blow to the head, or by neurological disorders that interfere with *associative processes*.

Amnesia

In *retrograde amnesia,* the most recent memories of an individual are more likely to be lost. Depending on the severity of the brain damage, years of memories may be suddenly unretrievable. This is usually a temporary state, and as memories return, they come back in the order in which they were originally experienced from oldest to newest. This suggests that humans organize memories in the brain in a special way, somehow associated with the order of their original occurrence. It is probably useful to think of memory as a process, with a beginning and an end, and that the process can be interfered with. Thus, the process of memory falls into the two categories mentioned previously. The first is called long-term memory, such as we associate with remembering family and old friends. The second is called short-term memory, such as we associate with trying to remember a phone number or the answer to a question on a test for which we studied late last night for the first time.

Short-term and Long-term Memories

There are disorders affecting stabilization of short-term memories, but the neurological basis for them is not well understood. A person with a serious short-term memory problem may have to relearn things every time they are to be done. These individuals often get confused when asked to recall a sequence of words or numbers they were exposed to just seconds or minutes before. Simply turning away from this person for a moment, after an initial introduction, may require you to reintroduce yourself because the person does not remember meeting you. To explain this odd behavior and to come to grips with the mechanisms by which experiences are transformed into memories, neuroscientists can only speculate on the nature of short-term and long-term memory processes.

For humans, memories are our link to the past and our passage to the future. Memories provide a framework that makes sense out of experiences that we can call up and assess to make decisions. The philosopher George Santayana remarked once that "those who cannot remember the past are condemned to repeat it." This seems true for individuals as well, which is attested to in the memory-deficient person in the preceding example. We cannot learn without memory. We may respond to stimuli and be conscious of our actions, but for those responses to be of any lasting value, they must be recorded somewhere in the brain and be available for recall.

Appetitive behaviors drives to fulfill a basic need; also known as appetites

Engram a permanent memory trace

Amnesia the loss of memory

Short-term memory is the ability to recall information that you just learned, a name, a phone number hurriedly acquired from a new acquaintance, or some of the facts that you just learned in the last paragraph. As everyone has experienced, with the passage of time, the ability to recall this new information in detail diminishes. It is this transient holding of information that we call short-term memory. If you don't use it, you lose it.

The more permanent storage of retrievable information is called long-term memory. The mechanisms involved in the process of long-term storage and recollection are a mystery, but they undoubtedly involve systematic and permanent changes in the organization of the nervous system. These changes in the nervous system may reflect to some extent changing patterns of electrical and biochemical activity or the number of synapses in already existing nerve pathways in the brain.

The organization of the brain and central nervous system arises during early embryonic development, setting up major anatomical structures such as the cerebrum, cerebellum, spinal cord, and all the functional nerve pathways. Instinctive behavioral potential is built into such an elaborate structural network. The modification of those pathways, including those that affect instincts, may occur by learning if the mechanisms involved in memory storage provide for their modification. How the structural and/or biochemical features of the brain relate to processes of memory and recall are certainly problems whose resolutions lie in the future.

Storage of Memories

Where are memories stored? Is there a special region of the brain where we file our experiences? Are they placed in little cellular compartments with proper names and dates and places? It does not appear that this is the case. Memories are stored throughout the grey matter of the cerebral cortex. Much of the evidence for this is the fact that damage to specific regions of the brain does not necessarily lead to loss of specific types of memories. The cerebral hemispheres in humans are committed predominantly to association and processing of sensory and stored information. The associative areas of the brain integrate sensory input so as to give meaning to the signals from photoreceptors, phonoreceptors, chemoreceptors, mechanoreceptors, and thermoreceptors. The ability to consciously compare new sensory input with existing information held in memory or to compare two or more sets of memories with no new sensory input is what gives humans the exceptional quality of **insight.**

Insight Learning

Insight learning is used to solve many types of problems, real and hypothetical, that require logic and intuition. It is not so much the mechanisms by which a memory is recorded (it is assumed that all mammals have similar facilities) but the way in which humans are able to associate and link together those experiences that underlie insight. Humans have developed systems of languages to communicate ideas, we have developed mathematics and science to explain the nature of reality, and we have systems of laws to govern our actions, not only as individuals but as societies.

BIOLOGY AND SOCIETY

Social organization is an important element in the development of individual behavior. Individuals learn from other individuals. Children learn from parents, siblings, teachers, friends, and enemies. The ability to communicate between and among individuals and to cooperate collectively has determined in large part the evolution of society. Human beings are exceptionally good at learning by trial and error, and this has been crucial in establishing successful societies.

However, *Homo sapiens* is not the only group of social animals on the planet. In fact, many higher animals form groups of characteristic size and composition, some small, some large. They all have unique aspects yet share certain basic features necessary to maintain society. The troupe of baboons is a functional social unit, as is the pack of wolves and the herd of bison. Honeybees have one of the most highly structured societies of any invertebrates known. Their key to social success is organization and communication. Bees have a divided society with jobs and duties clearly defined. There is one queen per hive, a few fertile males called drones, and a preponderance of workers. Ninety-nine plus percent of the population of a hive are workers. They are all females and may live only a few weeks. The worker bees pass through a characteristic series of genetically determined behavioral changes in which they go from hive-tending nurses that feed the queen and larvae to wax-producing builders who maintain the comb and, eventually, to foraging dancers who not only search for and find food but communicate that information to others in the hive through an elegant waggle dance (Figure 19–12). Bee society is group oriented. A fundamental difference between insect societies and vertebrate societies is in the recognition of other members of the group either as individuals or simply as part of the whole. We recognize each other as separate entities; bees may not recognize themselves as such. This presents us with the interesting and provocative possibility that an insect society, such as a hive of bees, should be viewed as a single organism.

Insight ability to connect experiences to generate ideas

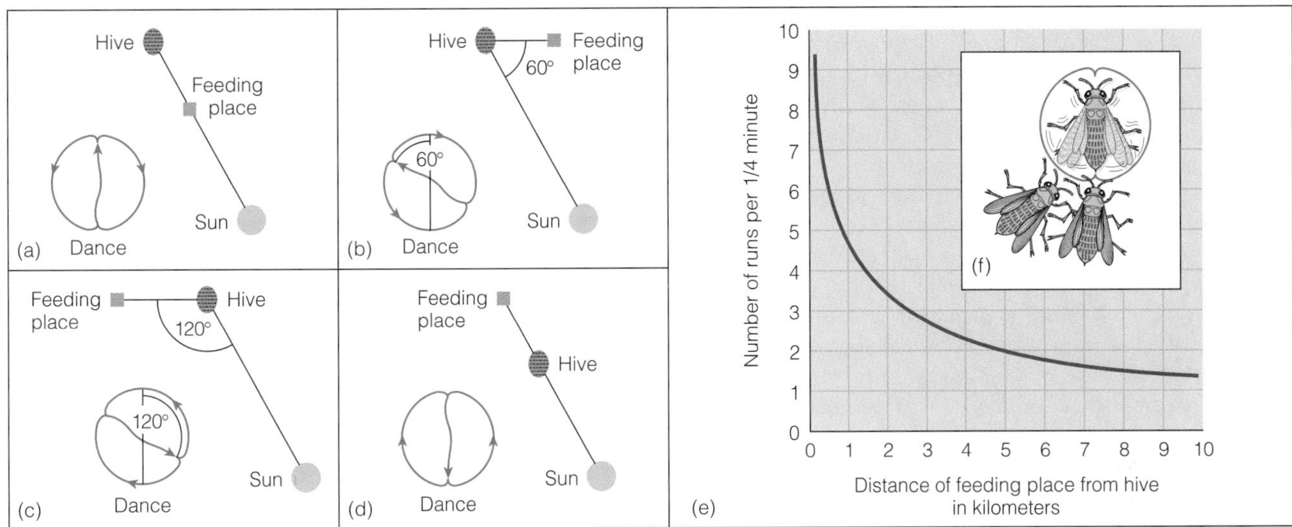

Figure 19–12 The Dance of Bees
The dance of bees informs of the direction and distance of the feeding place from the hive (a)–(d) relative to the sun. The distance from the hive is indicated by the number of turns (e) and (f) per minute in the dance. The direction is determined by the angle of the dance.

Vertebrate societies are much less strictly inclusive. They are organized to allow for diversity in behavior, learning, and recognition of individuals. That is not to say that vertebrate societies are not highly structured, but rather that the dynamics of such societies are based on individual interactions rather than programmed, instinctive "social organism" performances. Vertebrates set up dominance structures or *hierarchies*. This means that one or a few individuals are able to control the actions of the group to their own specifications. The dominant member of such a group is usually a male, and he establishes his dominance through conflict with other males vying for the same position. This conflict is often ritualized, so that a weaker male does not die from fighting but rather becomes subservient. The dominant male gets his choice of mates, food, and just about anything else he wants. He also plays a major role in protecting and defending the group from outside aggression. This often involves establishing and/or preserving a territory. *Territoriality* occurs in the smallest of units or families as well as in much larger groups. We humans have even established extremely complex laws of ownership and property, but they come down to the same territorial imperatives that apply to wild animal populations.

Territoriality

Birds and other animals aggressively defend their mating grounds, their nests, and the foraging areas around them to various extents, depending on the species (Figure 19–13). The area in which an animal forages for food is called the *home range*. In many cases, the defended territory will be

(a)

(b)

Figure 19–13 Territory and Defense
Territory is established by animals for mating, nesting, and food supply. Threat displays in males are a prelude to combat, as observed with herring gulls (a) squabbling over territory. A female snowy owl (b) protects her nest territory against encroachment. Most vertebrates aggressively establish and maintain their territories.

smaller than the home range. In some cases, they are the same. In others, such as troops of gorillas, there is little territoriality at all and these huge primates travel unhurriedly through their home range at a peaceful measured pace. Many animals mark their territories, and, familiar to all of us, dogs and cats are among them. You may notice when you walk your dog or watch someone else walk a dog that the animal seems to stop constantly to urinate or sniff around the various objects that it inspects. Why is it that a dog relieves itself so often in so short a time? For the most part, they are not simply urinating but are chemically marking their own territories and checking out the chemical marks of other dogs. What humans do with fence posts and survey lines other animals do with territorial behaviors that range from physical displays of aggression to the distribution of molecular signposts in the environment through chemicals produced in their own bodies.

There is much to be gained by increasing the number of individuals in a group and maintaining strict social order. Protection from predators, proximity to other individuals for breeding, safety in rearing and training young, and various divisions of labor such as hunting and foraging are all part of the advantage of numbers and an unambiguous social structure. The actions and interactions of individuals in social groups are often beneficial to the group as a whole. To keep the group intact, it is necessary for new members or young animals to learn how to behave properly. Individuals learn, directly or indirectly, from other individuals in the group.

Learning—Vertical and Horizontal

Human societies are the most complicated forms of organization of all animal groups. Humans have an enormously complex social and historical structure. We span continents and millennia with our civilizations. For each individual, understanding how society is organized and works requires much learning. In general, we as a group learn and pass on what we learn to others by direct communication among individuals. This is a form of horizontal transmission of information and depends on parents, peers, and teachers to facilitate learning in the young. Parents also pass on the potential to learn in a vertical manner, that is, through the genome transmitted from one generation to the next. We need both genes (nature) and the environment in which to exercise the potentialities they provide (nurture) to survive both as individuals and as individuals in society.

LEARNING TO LEARN

An important part of the underlying framework of human behavior, which depends to so great an extent on experience, is in learning to learn. It is potentially dangerous to genetically "fix" the behavioral capacity of an organism prior to an actual experience. In the ever-changing world in which we live, the ability of a human to alter his or her behavior with changing conditions and to recall previous experiences in making such decisions, has translated into a successful method of adapting.

Our lives are filled with events and processes of maturation that are common to all of us. The events of birth, puberty, marriage, and death are common to all peoples and all cultures. We also have experiences that are limited to smaller groups to which we belong, either by birth or through other affiliations. Cultural traditions, religion, and associations in social or political contexts provide a more narrow range of experiences. Many experiences we have are unique to us alone. Affection for a particular pet, influence of a mentor, a serious illness, or involvement in a catastrophic event mark us in very specific ways. We learn from all these experiences, and we, in turn, participate in events that affect the experiences and learning of others.

Growth and Change

As we grow and develop, we change. This is fairly obvious with respect to anatomy and physiology during the first dozen years of our lives. During this early period of development, we grow rapidly and approach puberty. The relative rates of growth of infants and children at certain stages of life are critical to the development of specific organs (see Chapter 15 for review). This is true for behavior as well. We do not suddenly emerge in the full flower of human reasoning ability and intelligence. We go through stages of brain and intellectual development, which we hope bring us to a state of mature intelligence.

The development of the brain controls many of the changing behavioral patterns observed during the period of growing up. The total number of neurons present in the human brain is maximal within a few years after birth. Estimates for this number are placed at 100 billion to 200 billion, or so, neurons. The general organization of the central nervous system is in place at birth, though not all connections in the neural network have been made. This is important, because making and maintaining intercellular connections among nerve cells requires that these cells lose the capacity to divide. Relatively few or no new neurons are produced in the human brain after the first four to five years of growth.

Stages of Brain Development and Behavior

Although most of the cells that will ever be in the brain are there at a young age, human behavior appears to follow a pattern of change that suggests that different parts of the brain mature at different times during growth (Table 19–2). Early behavior is controlled by the evolutionarily older parts of the brain as they mature during development. During the

Table 19–2 Developmental Stages

STAGE	AGE
Brain development (neurogenesis)	Embryonic to 1 to 2 months
Organization of brain (synaptogenesis)	2 months to birth
Brain stem (arousal/awareness)	Birth to 6 months
Higher brain levels	6 months through childhood and puberty
Associative centers (attention span increase, memory expansion, enhanced motor control)	
Cerebral hemispheres (ideation, abstraction)	Adult

first few weeks and months after birth, we have a general sense of awareness and arousal. These states are controlled by the reticular system and the thalamus, particularly the hypothalamus. Evolutionarily speaking, these are considered ancient regions because they are found in all vertebrates from fish to humans and govern similar kinds of regulatory and instinctual behavioral responses (see Chapter 13 and 17).

As development and growth of the individual human being proceeds, control of behavior passes upward to the associative centers of the cerebral hemispheres. As we grow during subsequent months and years, we develop an attention span and increasingly use memory to connect events. Short-term memory skills increase rapidly in the first year and long-term memory continues to develop and improve throughout childhood and on. Control of muscular activity and the integration of the senses also change rapidly during this period. We are able to accomplish sophisticated eye-hand coordination, balancing and locomotion, control of urinary and anal sphincters, control of the voice box, and associations of sounds (words) with objects and people in the world around us, all within the first year or two of life. "Mama" and "Dada" are relatively simple sounds to form with the lips, tongue, and palate, and their persistent occurrence in the child's environment is virtually ensured by their continual use by proud parents!

Last but not least, humans develop the capacity to generate ideas and produce abstract thoughts. Although the timing of behavioral changes in individuals is by no means rigidly determined, we all tend to follow roughly the same stages. Significant variation from them is often reason for concern. For example, childhood **dyslexia,** which is a sensory perceptual disorder, may affect the perception of the orientation and order of symbols such that they are incorrectly perceived. This leads to serious reading problems, because a letter *b* may be perceived as *d* or the word *was* may be perceived as *saw.* These problems become apparent only when the child has to learn to read, usually at the beginning of grade school, and are often interpreted by teachers as simply reflecting a slow or retarded individual. Even when there are no functional abnormalities involved, it is important to consider the capabilities of children as they develop in terms of how and what to teach at different stages of their physical and mental development, or what to expect in their responses to exposure to concepts that may be "over their heads." Why waste time and energy on demanding an understanding of abstract concepts if the child is not physiologically and behaviorally prepared to benefit the most from being presented with them? These are controversial issues because all of us, scientists and nonscientists alike, are still trying to understand not only how but when we learn.

Play

Recent research on brain activity in children has shown that one of the most important learning behaviors for children is **play.** Play involves curiosity and exploration, as well as interactions between individuals. Curiosity is the interest in things around us. Curiosity is seeking out new situations and exploring new environments. The exploratory behavior of primates and carnivores is very important to them in learning about the features of a new or unfamiliar territory. It is during play that children learn how to socially interact with one another without suffering from the potentially serious consequences often encountered in the real world. Play behavior may extend for years and even into adulthood. We continue to play throughout our lifetimes and, as a result, are able to resolve conflicts within the rules of a game and, thereby, limit fatal aggression. It is clear that games imitate life and are a reasonably safe way of learning and socializing, though all too often people seem to get the relationship reversed (Figure 19–14).

RECOGNITION AND RECALL

New experiences and memories of them are stored during activities such as play and trial-and-error investigations. Access to memories of such experiences is valuable in subsequent decision making. How do we retrieve information from storage in the brain? Although certainly related to one another, recognition and recall are actually two different ways in which the memory of an experience can be drawn from the memory bank. The difference between them is in the stimulus used to evoke a memory. Recognition occurs when the memory of an experience is stimulated by repeating

Dyslexia a disturbance of the ability to read

Play behavior involving exploration and interactions that function in learning and social development

(a)

(b)

Figure 19–14 Play
Play is one of the most important ways in which animals learn the rules of their species or societies without suffering the lethal consequences that might arise if they did not develop them in a protected environment. In (a), young girls compete in a pick-up game of basketball. In (b), young baboons play on the rocks in the protection of their troupe.

the experience, as in remembering the style of a shirt or dress when you see it again. We all have had occasion to say, "I'll know it when I see it." Recall is the retrieval of a memory without repeating or simulating the experience. The first is a sensory stimulation of a memory. The second is mental manipulation of information, a "flash of insight" that comes from thinking about an event or problem without having to experience it again. Recall is the key to the processes of abstraction and reasoning, which, in turn, form a basis for what we would in general call **intelligence.**

INTELLIGENCE

What is intelligence? Although there is no simple definition for intelligence, it is related to the way we use information. Intelligence is a measure of what we learn and what we know, as well as how we apply what we know to our activities. To make sense of what we learn, we need a process of organization. New learning has to be associated in some systematic way with what we already know. Intelligence tests often confront us with questions and problems we may have never encountered before. Those who do well on such tests have usually been exposed to similar types of problems and have previously learned how to approach their resolution. Learning to learn is important. This is why the measure of intelligence registered on some forms of standardized tests must be viewed cautiously, as it may reveal only a part of a person's intelligence if that individual's experiences are different from the norm. How many people have you talked with who have had mediocre results in standardized testing, yet who were fully capable of carrying on an intelligent and insightful conversation? Ethnic background, educational background, experiences, and behavioral development all play a part in whether one does well on standardized tests. However, success in life may have little to do with how well a person scores on a test and much to do with how one learns and adapts to the environments around them.

ASSIMILATION AND ACCOMMODATION

Learning is a continual process, mediated by neurological mechanisms that allow for reception and integration of stimuli from the environment. Whether at play, at work, or interacting in other ways in the world around us, we are constantly responding and adapting to changes in that environment through *assimilation* and *accommodation.* Assimilation is the process by which new data are organized in the brain, and accommodation is the process by which old experiences are replaced or supplemented by new. Complex human behaviors, such as learning, arise from the sum total of experiences in life, past and present, internal and external. This is true to some extent for all animals, whether their responses to the stimuli are rigidly prescribed by instincts or open-ended and continually modifiable as with human learning. What differentiates behavior and learning among animals quantitatively and qualitatively is the degree to which it is modifiable by experience. A combination of an enormously complex brain, a massive network of neuronal connections, and genotypically prescribed mechanisms to receive, store, and alter information underlies human intelligence. Humans are unique in this capacity because we are the only ones with an acute and often abstract awareness of ourselves and our relationship to the universe around us.

Intelligence what an organism knows and how it applies that knowledge to its activities

Summary

Why do we act the way we do? In this chapter, the discussion focused on descriptions of the basis for simple and complex behaviors. To the extent that any behavior can be called simple, behaviors that involve changes in growth in response to forces or stimuli, such as gravity and light, are called tropisms. Those involving the movement of an entire organism toward or away from a stimulus are called taxes. Reflexes are responses of a part or parts of an organism to stimuli. For example, the knee jerk response in humans is a reflex, which requires no conscious thought in order to occur, only the proper connections of sensory and motor nerves and muscles. In more complex animals, and certainly in all vertebrates, there is a well-developed network of nerves that mediates both the reception of stimuli and the response to them. Genes control the development and maintenance of the nervous system and, therefore, provide a genotypic framework in which behavioral responses may occur.

Complex behaviors involve instinctual and learned responses to stimuli. Behaviors that do not require previous experience to be performed are called instinctive behaviors. Bird songs, animal feeding behaviors, and the egg-rolling behavior of the greylag goose provide instances of instinctive behaviors, the stimuli that release them, and the patterns of their performance (fixed-action patterns). In contrast to instinct, learning occurs through the capacity to modify behavior. Learning requires memory. Memory is both the process by which an experience is stored in the brain (in a manner that is not yet understood) and the manner in which it is retrieved. There are two types or stages of memory, short term and long term, and the ability to recall events depends on where the specific memory of it is stored. Habituation (learning to ignore selected stimuli), classical conditioning (learning to associate two different stimuli), and operant conditioning (learning through trial and error) are all means by which we interact and develop patterns of behavior in conjunction with and suitable for survival in the world around us.

Intelligence is a measure of reasoning ability and is adaptive in that it allows for assimilation of data (systematic organization of information into existing patterns) and accommodation (changing the patterns of that organization). Recognition is the ability to retrieve the memory of an experience by external repetition of that experience or a similar one. Recall is the ability to internally retrieve a memory of an experience without the reoccurrence of an event and is essential to abstract reasoning.

A society is an organization of individuals into a large collection or group. Some species recognize themselves not so much as individuals but as part of a whole (for example, a hive of honeybees). Others are clearly composed of independent individuals whose interactions depend on communication and rules of engagement in order to function together (humans). Play and curiosity/exploration represent activities used by individuals to learn the rules of their societies and of the world around them.

Questions for Critical Inquiry

1. What is instinctual behavior? How is it related to the structure of the nervous system? How might it be related to learning?
2. How are memory and learning related to one another?
3. What is imprinting? Under what circumstances would this type of learning be of paramount importance? When would it be a hindrance?
4. How are brain development and learning related to one another? How is this important in teaching or training children?

Questions of Facts and Figures

5. What is a fixed-action pattern? Provide an example.
6. What is a conditioned response? What brings about extinction of a conditioned response?
7. What is the difference between short-term memory and long-term memory?
8. Differentiate between recognition and recall.
9. What are two forms of simple behavior?
10. What is the main difference between classical conditioning and operant conditioning? Use examples of each to explain.

References and Further Readings

Cummings, M. R. (1996). *Biology: Science and Life*. St. Paul, MN: West Publishing.

Lewontin, R. C. (1991). *Biology as Ideology*. New York: Harper Perennial.

Lorenz, K. (1966). *On Aggression*. New York: Harcourt Brace & World.

Wilson, E. O. (1978). *On Human Nature*. Cambridge, MA: Harvard University Press.

CHAPTER 20

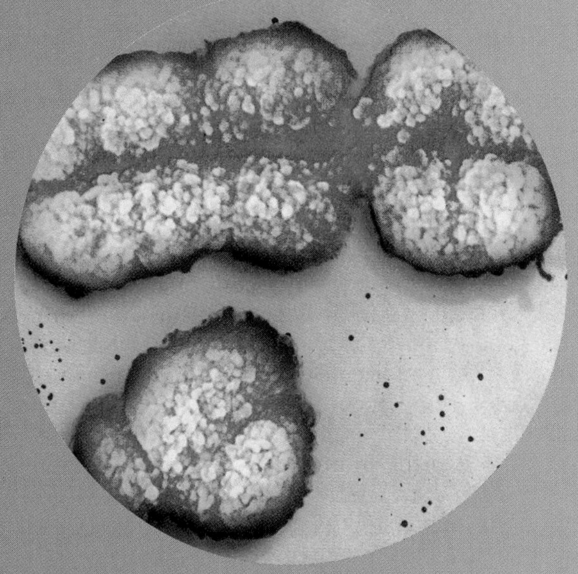

Genetics and Human Heredity

INTRODUCTION

The deciphering of codes found in nature has been an important endeavor in the development of science. Physicists have uncovered the structure of matter in the combination of quarks and have provided us with a coherent view of the cosmos from inside atoms to inside stars. Chemists have deciphered the periodic relationships among elements, the key to understanding chemical bonds and molecular structure. Biologists have cracked the genetic code and propelled us into the age of genetic engineering and molecular technologies capable of altering the code of life. This chapter presents the subject of genetics against the backdrop of human heredity and the nature of the structure and function of DNA.

THE CODE OF LIFE

The genetic code is encrypted in the sequence of nucleotide bases in the nucleic acids of DNA (see Chapter 2 for details). The discovery of the structure of DNA in 1953 as a double helix was the biological equivalent of uncovering the Rosetta stone and the decipherment of the hieroglyphics of ancient Egypt (Biosite 20–1). Within a few years of discovering the structure of DNA, the code of life was deciphered, and with it scientists have been able to make sense of the inheritance of traits and the appearance of combinations of those traits in individuals. This knowledge has proven to be applicable not only in *Homo sapiens* but in all living organisms. This chapter will focus on genes, the units of inheritance, and the means by which they are encoded and transmitted from one generation to the next. The description of the kind of traits encoded by genes will include how mutations in them occur and how they may alter cell and organismal growth and development.

GENES—UNITS OF INHERITANCE

Genes are the fundamental units of inheritance in all living organisms. Genes serve the same basic function in single-cell organisms, such as a bacteria, as they do in complex multicellular organisms, such as *Homo sapiens*. The information in a gene is encoded in the sequence of nucleotides found in DNA (Figure 20–1). A gene represents a complete molecular code specifying the synthesis of an RNA molecule. RNA is related to DNA in containing nucleotides that retain the genetic information of the original DNA molecule. However, RNA molecules are much shorter than the DNA molecules from which they are derived and contain only a small subset of the original information. The process by which RNA is made from DNA is called transcription and takes place in the nucleus. Transcription represents the first stage of the expression of a gene into a useful cellular product. RNA is used as a messenger during the process of

20–1

THE ROSETTA STONE—DECIPHERING CODES

In 1799, a soldier in Napoleon Bonaparte's army in Egypt uncovered a stone that would prove to be an important link in deciphering the written language of ancient Egypt. The Rosetta stone contained a message inscribed in different languages, one being Greek, which was understood, and another, Egyptian hieroglyphics, which was not. Breaking the code of the glyphs opened a window to the ancient world and has provided us with a rich and rewarding view of human history. Breaking the genetic code has opened genes to manipulation and alteration using newly developed techniques in molecular biology and genetic engineering. The ability to identify, isolate, sequence, and alter a gene underlies the burgeoning of biotechnology industries during the 1980s and 1990s. Correcting a disease, or condition, by changing the gene responsible for it may be accomplished at two levels—the somatic cell level and the germ cell level.

Gene therapy of somatic cells targets the source of a problem in the body of an individual. Gene therapy is being carried out by the National Institutes of Health in their studies of an immune deficiency known to be caused by the lack of a functional enzyme called adenosine deaminase. Researchers have isolated human bone marrow cells from affected children, introduced normal genes, and then reintroduced the altered cells back into the marrow to proliferate. The results are still coming in, but it appears that the technique works.

Research targeting the alteration of genes in the germ cells is not being carried out on humans at this time. Unlike the situation with individuals, in which the somatic cells are affected, manipulation of genes in gametes means that the changes are heritable, thus, changing all future generations derived from that individual. At present, too little is known about the long-term effects of genes introduced into the gene pool. For example, we need to know whether introduced genes will disrupt the action of other genes and whether they will alter phenotypes in unsuspected ways that would be deleterious to future generations. Even the best intentions for correcting a lethal genetic disorder are not enough, as yet, to override the need for caution in tampering with the code of life.

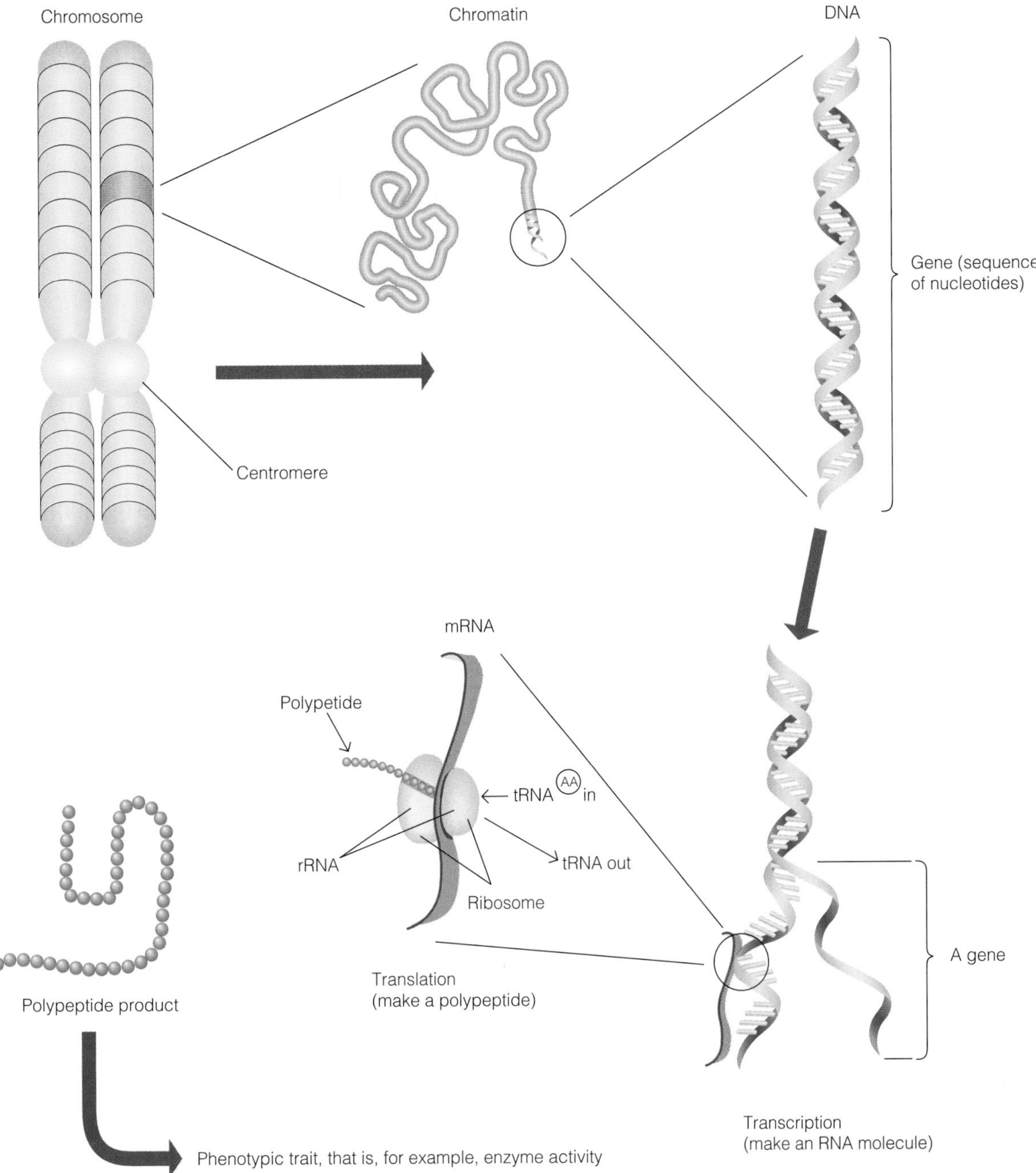

Figure 20–1 Chromosomes and Genes
There is a hierarchical structure to the organization of genes. Genes are the units of heredity and, generally, are involved in translational processes (synthesis of proteins). Linear sequences of genes are organized into chromatin, which makes up the structure of the chromosomes.

translation in the cytoplasm of a cell, which results in the synthesis of a protein. Proteins, in turn, provide structural materials for the growth and assembly of cells and the means to carry out and regulate all the cellular biochemical reactions needed for life (see Chapter 2).

From a structural point of view, human traits such as the color of the eyes, the shape of the nose, and the length of the fingers result from the expression of genes into proteins. For example, cells synthesize and secrete protein hormones that regulate how and when genes are expressed. This control helps organize cells and leads to the differentiation of cells whose own products produce bone, cartilage, and skin. The processes of gene expression are quite complicated, because a cell requires the products of thousands of genes simultaneously, at different levels of concentration, and in specific sequences to support growth and development of an organism throughout its life.

Chromosomes

Genes in human DNA are organized sequentially in long linear arrays. Many thousands of genes are linked together in a single structural unit known as a **chromosome** (Figure 20–2). Because genes may contain thousands of nucleotides, and chromosomes may contain thousands of genes, chromosomes are extremely large, intricate, and elegant structures. What does it take to be a chromosome? A chromosome must have three basic characteristics to serve effectively as a carrier of genetic information (Figure 20–3). First, it must have the capacity to be duplicated or undergo **replication.** For this to occur, specific sequences of DNA known as *origins of replication* must be present in the DNA. These sequences interact with special proteins that initiate DNA synthesis. Second, the ends of the chromosomes must be protected. This is accomplished by specific sequences of DNA at each end of every chromosome, known as *telomeres*. Third, each chromosome must have sequences of DNA that can be used to connect the chromosome to the microtubules of the spindle apparatus (see Figure 20–3 and Chapter 2 for details). The region of the chromosome called the centromere serves this function and allows for the controlled, evenly balanced separation of chromosomes into the two daughter cells during mitosis.

Thus, it is apparent that not all of the DNA in chromosomes is transcribed and translated into RNAs and proteins, respectively. Noncoding sequences of DNA are also thought to support the structure of a chromosome and to act as spacers between regions of DNA that contain meaningful coding sequences. There are 46 chromosomes contained in the nucleus of human cells, and each has two telomeres, one centromere, and many origins of replication (Figure 20–2). These chromosomes are organized within a nucleus, which is itself too small to be seen without the aid of a microscope. However, the length of all the chromosomes, stretched out end to end, is estimated at almost two meters!

The 46 chromosomes in humans are represented by 23 identical pairs, called *homologous pairs*. Formation of gametes during meiosis results in the parental homologous pairs separating from one another and recombining during fertilization to form a new and unique individual (see Chapter 3 for details of meiosis and Chapter 14 for fertilization). The size and morphology of pairs of human chromosomes have been established by displaying them in their condensed form as isolated from cells undergoing mitosis (Figure 20–3). The visualization of pairs of condensed chromosomes is called a **karyotype.** The chromosomes in humans form 23 matched sets, except for one pair in males, the X and Y chromosomes. As we will discuss later in this chapter, the X and Y chromosomes determine the sex of an individual.

Chromosome a single DNA molecule associated with proteins which is found in the nucleus of a cell and contains hereditary information represented by genes

Replication the duplication of DNA during the S phase of the cell cycle

Karyotype the representation of condensed chromosomes with pairs aligned

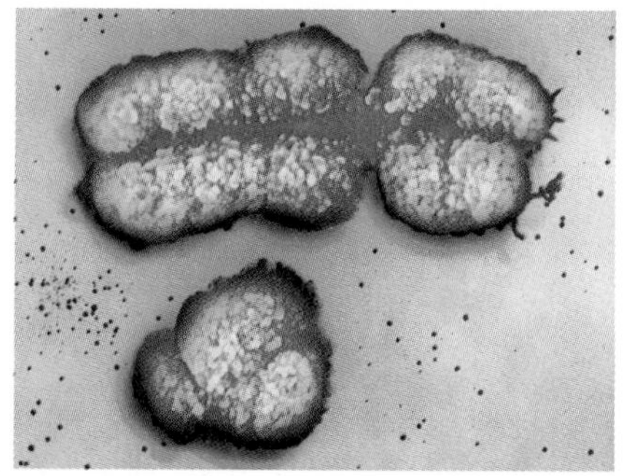
(a)

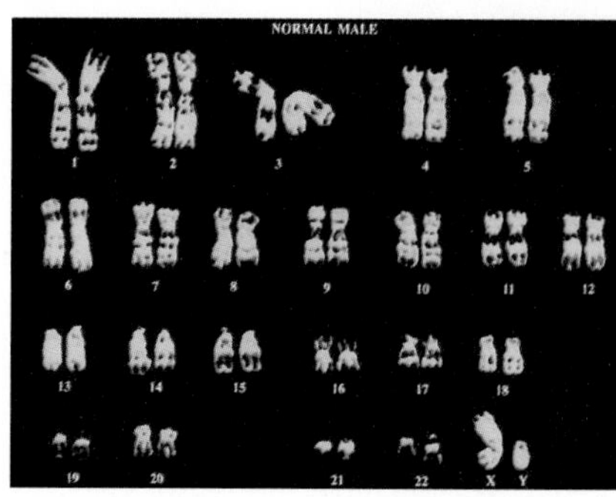

(b)

Figure 20–2 Human Chromosomes

The X and Y chromosomes (a) are part of a set of chromosomes (b) that are representative of the human genome. Matched pairs of chromosomes displayed in this way are called a karyotype.

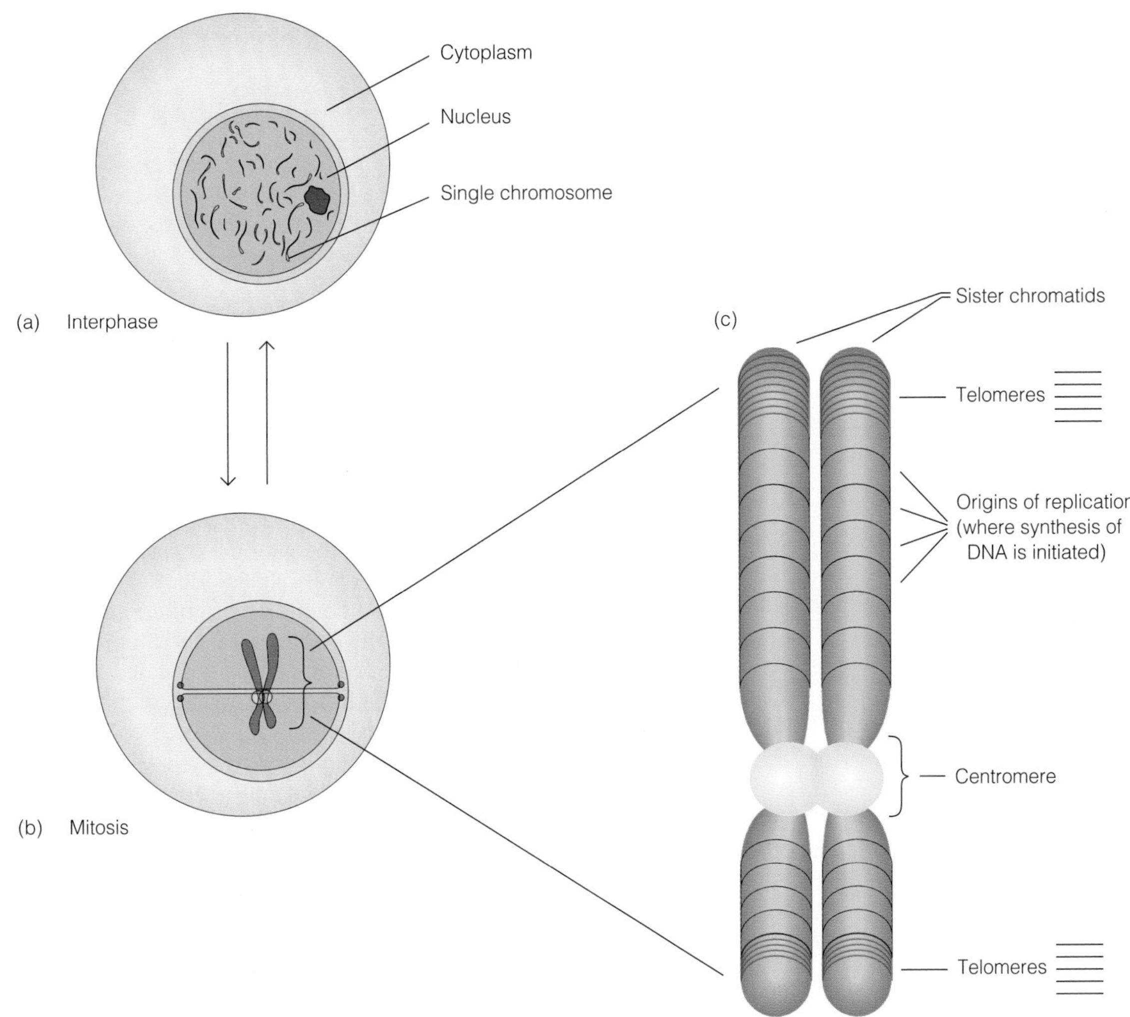

Figure 20–3 Essentials of a Chromosome
The appearance of chromosomes changes during the cell cycle (a) and (b). Interphase chromosomes are spread out in the nucleus. Chromosomes are condensed in mitosis. The essential features of a chromosome do not change with changing appearances (c). Telomeres protect the ends of a chromosome, centromeres provide attachment points between sister chromatids and for the attachment of microtubules, and the origins of replication provide initiation sites for DNA synthesis.

Chromatin

As described in the previous section, each human chromosome is composed of thousands of interconnected genes arranged in specific linear sequences (Figure 20–4). There is, however, more to a chromosome than simply DNA. The DNA of a chromosome is found in association with a variety of special proteins known as histones, which surround, protect, and organize DNA in the nucleus. There are also a large number of nonhistone proteins that interact with DNA and are involved in regulating and controlling gene expression, replication, and transcription. This complex combination of DNA and protein is called chromatin.

The Genome

Each of the chromosomes of humans contains a tremendous number of genes. This brings up the question of how many genes an organism needs to reproduce itself and to grow and develop (Table 20–1). Estimates suggest that a bacterium, which is the simplest type of organism, may

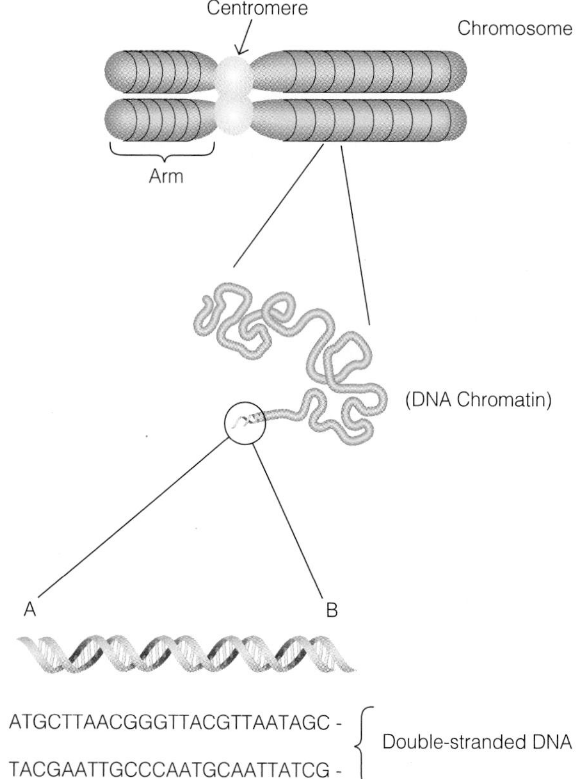

Figure 20–4 Nucleotide Sequence in a Gene
The sequence of DNA representing part of a gene is shown to be a sequence of nucleotides (ATG . . . TAC . . .) with a start point (A) and an end point (B). The gene is integrated in a linear array with other genes that compose one arm of a chromosome.

need as few as a thousand genes, but a human being requires upwards of 50,000 to 60,000. The full set of genes contained in the full collection of chromosomes of an individual is called the genome. The genome contains all information needed as a blueprint for life, both at the level of the individual and for a species. The genome is subtly different and unique in every individual of the same species. For example,members of *Homo sapiens* are clearly similar anatomically and physiologically, but detailed analysis of genes shows that they are not genetically identical. Genomes are vastly different among distant species, for instance, between a human and a bird or an amphibian (Table 20–1). However, the basis for storing and using the coded information in all Kingdoms of life is the same—DNA in genes. Fifty thousand to 60,000 genes is a staggering number to be contained and used from within the nucleus of every human cell. The linkage of genes into linear chromosomes was an essential step in the evolution of the genome in humans and all other eucaryotic organisms.

Table 20–1 Size of Genomes

ORGANISM	NUMBER OF NUCLEOTIDE PAIRS	NUMBER OF GENES
Bacteria (*E. coli*)	4.7 million	1,000
Yeast	10 million	2,000–3,000
Plants	100 million to 100 billion	10,000
Amphibians	600 million to 90 billion	10,000–50,000
Birds	1 billion	10,000–50,000
Reptiles	2 billion	50,000
Humans	3 billion	50,000–60,000

Where are genes located along a chromosome? Are they organized in a specific pattern or are they all mixed up? The location of a gene along the chromosome is known as its **locus** (Figure 20–5). The locus (pl. loci) of genes is always the same in the genome of a species, and the order of arrangement of genes is identical on homologous chromosomes. Much effort has gone into establishing the loci of genes onto a map of the human chromosome (Biosite 20–2). Specific genes are located along the length of a chromosome in specific arrangements, such that gene A always precedes gene B, B precedes C, and so forth. This has been firmly established by molecular genetic analyses. In terms of genetic traits and inheritance, genes that are located on the same chromosome are considered to be *linked genes* and are inherited together (Figure 20–5). However, as the focus of the next section will show, there may be some differences in the exact form of genes found at any particular locus. The genes on nonhomologous chromosomes are not physically linked and are inherited independently from one another.

Alleles

Up to this point, we have been concerned with what genes are and how they are organized along the set of chromosomes that constitute our genome. However, genes at the same locus on homologous chromosomes are not necessarily identical. There are differences in the nucleotide sequences of genes at a particular locus between the pairs. These differences suggest that a gene, say gene A, may come in more than one form, yet still be considered gene A. Altered forms of a single gene are called **alleles** (Figure 20–6). Allelic differences arise by mutation, which bring about changes in nucleotides

Locus the position along a chromosome at which a gene (or allele) is located

Alleles altered forms of the same gene, only one of which occurs at a locus

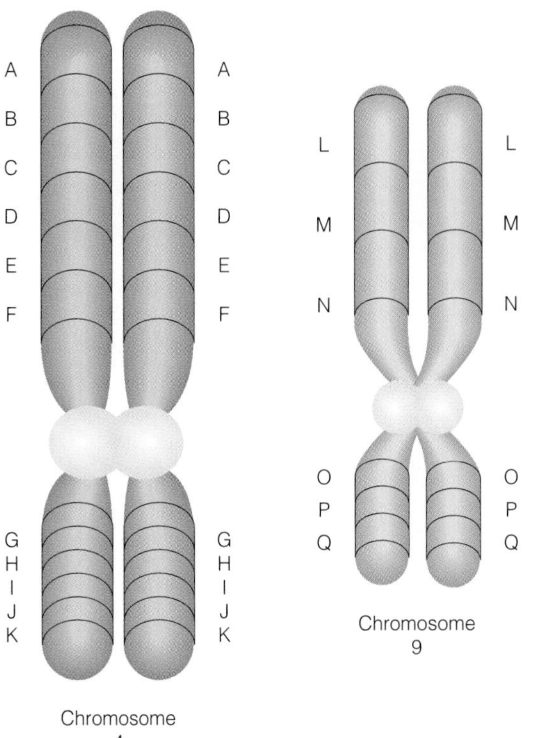

Figure 20–5 Order and Linkage of Genes
Different genes, A–Q, are located on different chromosomes. Chromosome 4 in this example has genes A–K, which are considered linked. Chromosome 9 has linked genes L–Q. Genes on different chromosomes are nonlinked. The order of genes along normal chromosomes is always the same. Thus, each gene has a specific locus.

within the gene sequence. Considering that a single gene may contain thousands of nucleotides, a single nucleotide altered by mutation is only slightly different structurally from the original sequence. However, from a functional point of view, allelic forms of a gene may result in extreme changes in the expression of a trait. This includes the failure of a trait to be expressed at all or the product to be faulty. In many cases, mutations result in little or no observable change in the structure and function of a gene product. Mutation of a gene may change a trait, for example, by affecting the ability of a protein product of that gene to function as an enzyme. If the nonfunctional enzyme in question were involved in the production of melanin pigments during embryonic development, the individual would end up with little or no skin and eye color. Such alleles are well known and have been genetically analyzed in individuals who express a trait known as *albinism* (Figure 20–7).

DOMINANCE AND RECESSIVENESS

What are other possibilities for differences in alleles between homologous chromosomes? An allelic difference between parents at a specific locus may provide a zygote with two different forms of the gene. If the products of such alleles are different, then one possibility is that the traits resulting from expression of one allele may dominate the other. A *dominant gene* is one that leads to the expression of a specific trait regardless of what other allele may be matched with it. A *recessive gene* is one that is not expressed as a trait when present with a dominant allele. Under these

BIO site

20–2

THE GENE POOL

The gene pool of a species is the complete collection of the genes represented in the individuals of an interbreeding population. Within the population, the number of alleles for genes exceeds the capacity of any one individual to have them all represented. The genome of a species represents the generalized genetic composition of a species, that is, the specific types and arrangements of genes organized within the chromosomes that are necessary for growth, development, and reproduction. At any one gene locus, however, there may be one of many different possible alleles. This increases the genetic diversity of a species and is important in its long-range evolutionary survival. Some alleles are more prevalent than others and, thus, their appearance in the population is greater. For example, the gene pool of *Homo sapiens* contains a number of alleles for the ABO blood group, but these alleles are not equally represented in the population. In the United States, there are more O-type individuals than AB types. However, it should be kept in mind that within subgroups of humans, the prevalence of a specific allele may change dramatically, based on the population's geographic distribution or national origin.

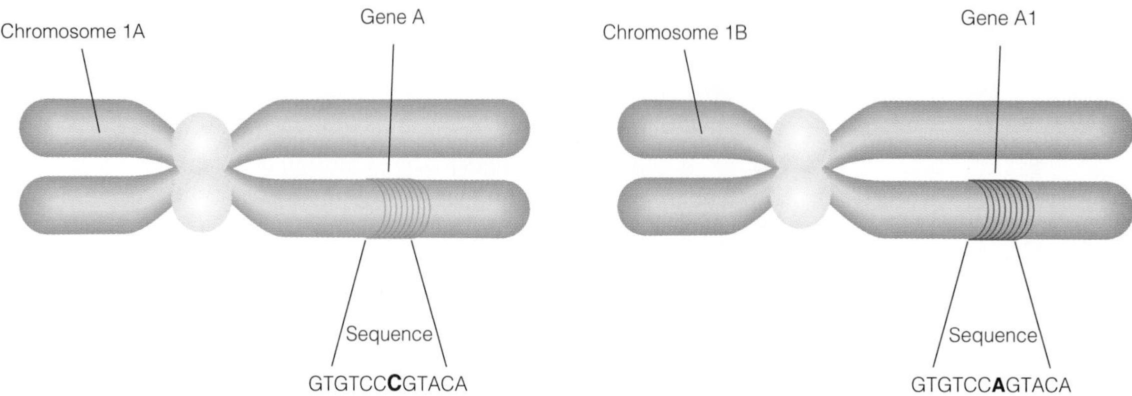

Figure 20–6 Alleles
Chromosomes 1_A and 1_B form a homologous pair. At the locus for gene A, different forms of that gene (alleles) may occur. The difference observed is in a substitution of "T" for "C" in gene A_1. The product of gene A is normal and functions as it should, but the product of gene A_1 may work poorly or not at all.

Figure 20–7 Albinism
Lack of an enzyme needed to make melanin results in albinism. The allelic form of the normal gene for the enzyme is recessive and only occurs when the trait is homozygous. Among the children of a village in Cameroon is one albino child.

conditions, recessive genes are essentially silent, but, as we will see, they are not forgotten (Figure 20–8).

If the exact same allele is inherited from each parent, the individual is considered to be **homozygous** for that gene. Both genes produce the exact same product and are expressed as precisely the same trait. If the alleles are different from one another, the individual is **heterozygous.** If one allele is dominant over the other, its product will be expressed in the heterozygous individual and may completely mask the presence or activity of a recessive trait. Some genes are incompletely dominant, showing signs of their presence in combination with other alleles. Other genes are neither dominant nor recessive, and so traits of both are observed simultaneously in the cells and tissues in which they are expressed. This is called *codominance.* Some traits depend on the relative contributions of many genes, a commonly encountered situation referred to as *polygeny* (Figure 20–9). Different combinations of alleles of many different genes affecting polygenic inheritance may influence traits that show continual gradations in human growth and development, such as height or skin color.

GENOTYPE AND PHENOTYPE

The genome of a species is a complete set of information needed for the growth and development of individuals in that *population.* The specific information that is encoded in the chromosomes of any one individual of that species is called the **genotype.** This is a unique set of genes and does not change during the lifetime of the organism. The physical traits associated with genes of a specific genotype, such

Homozygous a condition in which the alleles on homologous chromosomes are identical

Heterozygous a condition in which the alleles on homologous chromosomes are not identical

Genotype the set of genes carried by a particular individual

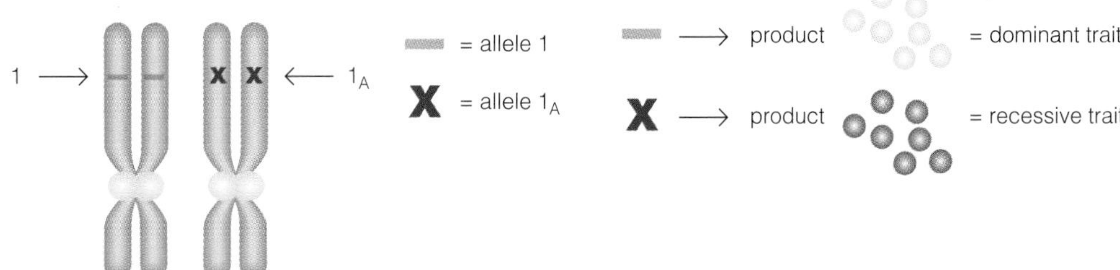

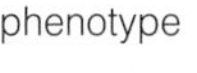

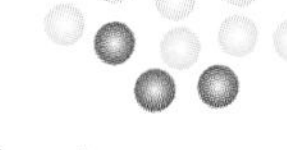

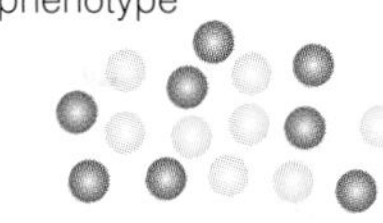

Figure 20-8 Dominant Versus Recessive
Allele 1 and allele 1_A have different quality products, one of which dominates the other when both are present in the cell.

as tallness or shortness, blue eyes or brown eyes, and other physical characteristics, are called the **phenotype.** The phenotypic characteristics of an individual result from the action or expression of genes associated with a specific genotype. In molecular terms, the phenotype results from the transcription and translation of genetic information into proteins, or sets of proteins, involved in a physical and/or behavioral trait.

REALIZATION

What we inherit from our parents is analogous in many ways to a blueprint for a new house. However, the plans are not the house itself. We are not simply a collection of all the genes in the genome, but rather the outcome of their expression. We inherit genetic potential, to be realized over time during the formation of a new individual. A fertilized egg is not a human being, but it does have the potential to become one. Right from the start, a new zygote is unique. It has all the information encoded in the DNA within its nucleus to become a functional human male or female. As previously pointed out, each and every species has a different and unique set of genetic information coded into its genome, and each individual has a specific combination of alleles that compose its genotype.

Barring any obstacles to growth and development, the genomic DNA is decoded gradually and in an organized manner to form an individual of a particular species and that species alone. The mouse genome is different from the chicken

Phenotype the traits of genes as expressed in a particular individual

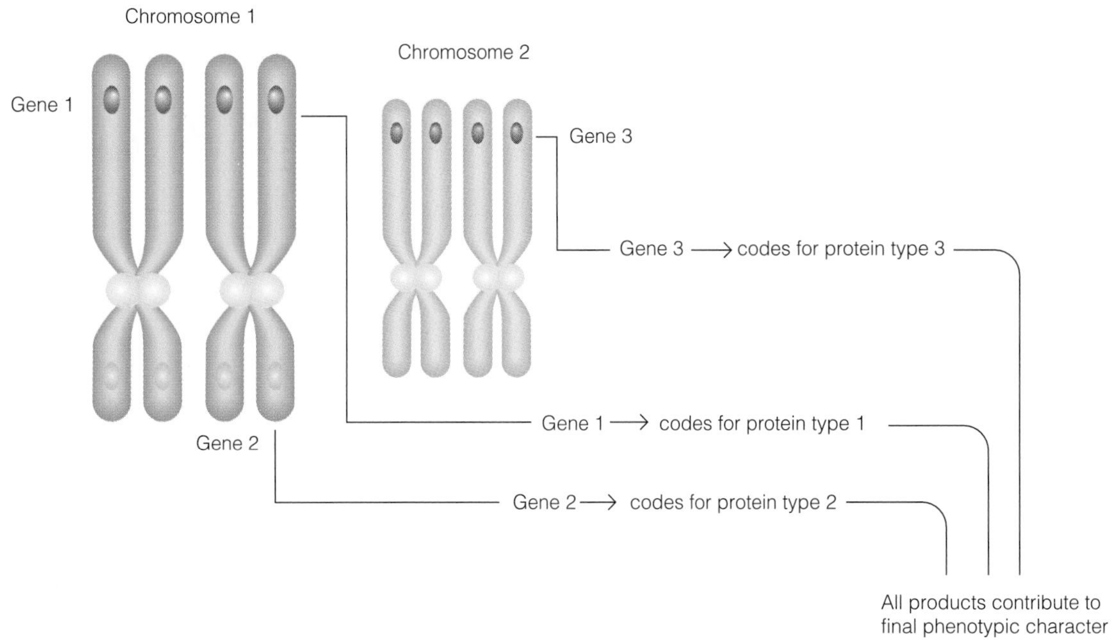

Figure 20–9 Polygeny
Polygenic traits such as height, intelligence, and skin color involve the products of many genes. Variation of alleles adds further to the diverse character of polygenic traits.

genome, and both are very different from the genome of a pine tree. You can not get a mouse from the genetic information for a pine tree. However, as different as the genomes for these organisms are from one another, the structure and the composition of their genetic information are based on the same coding mechanisms and use the same molecular principles for their expression. DNA is the Rosetta stone of life.

CHANCE AND NECESSITY

Genetic variability arises in two different ways. The first and most common is through a process of *recombination* that involves exchange of parts of chromosomes during crossing over. Crossing over occurs during meiosis, when homologous chromosomes exchange equivalent parts of themselves. When this occurs reciprocally between the chromosomes, the alleles of one chromosome are replaced with equivalent but nonidentical alleles of the other. This process provides the opportunity to mix the alleles into unique combinations in gametes and enhance the chance for success of the new individual.

Mutations, on the other hand, induce inheritable changes in the genetic information of somatic and/or germ cells. Changes in DNA can be passed on through mitosis to daughter cells within the body, and changes in germ cell DNA can be passed on by meiosis to gametes and then, following fertilization, to a new generation of individuals. Changes in the sequence of nucleotide molecules in DNA are discussed here, and exchanges of parts of chromosomes involving more problematic translocations are discussed in a later section of the chapter.

Mutation

Mutations are represented by changes in the nucleotide sequence of a gene, such as in the allelic differences discussed earlier. These types of mutations are called *point mutations* because they affect a single nucleotide in the sequence of DNA. A *substitution* of one nucleotide for another alters only a single codon (in this case, affecting a single amino acid) and may or may not significantly alter the functional polypeptide product. Functional alleles of a gene usually arise by substitutions. *Deletion* of a single nucleotide in a gene alters the sequence that follows it and changes the way that the transcription machinery of the cell reads the new sets of triplet sequence. This is called a *frameshift mutation* (see Chapter 2 for review of DNA). This alters RNA and the polypeptide products to a greater or lesser extent depending on whether the deletion occurs at the beginning of a gene or near the end. The same is true of adding or inserting a nucleotide. *Insertion* also alters the reading frame of the code and produces a new sequence in transcription and translation. Nonfunctional alleles are more likely to arise with deletions or insertions.

What causes mutations? Any phenomenon that alters the chemistry of nucleotides may lead to a mutation. Exposure to X rays and ultraviolet (UV) light can alter the structure of nucleotides in DNA (Table 20–2). The human body is continually bombarded with high-energy radiation, and uncorrected changes in the DNA can occur. Many chemicals that we may ingest, inhale, or otherwise are exposed to in the environments in which we live and work are also capable of altering DNA. Many different types of organic and inorganic materials comprise classes of compounds that induce mutations. They are known collectively as **mutagens.** The list of these mutagenic chemicals is long indeed (Table 20–2). However, although it is well known that mutagens can alter nucleotides and change the genetic code, we can never know precisely which cell will be affected or where in the DNA of that cell a change may occur. For this reason, mutations are considered to be random in occurrence.

Table 20–2 Effects of Agents That Alter DNA—Mutagens

AGENT	EFFECT(S)
X rays	Breaks in DNA strands
UV radiation	Chemical alterations of bases in DNA
Chemicals*	Chemical alterations in DNA

*Some common chemical mutagens: Ethylmethane sulfonate, anthracenes, epoxides, benzene, nitrogen mustard, safrole, aflatoxin, methylnitrosourea, dimethylsulfate

Natural Selection

Change in the genetic code by mutation is a perfectly natural phenomenon. Such changes in and of themselves are neither good nor bad, though the outcome of these changes can have serious effects on an individual. Without a continual occurrence of mutations in the genetic material, there would be little chance for the changes that we associate with evolution. If there is no mechanism for altering genes, there can be no change in the phenotype. It is at the level of the performance and survival of the individual that natural selection works. One outcome of genetic change is that the attributes and capacities of individuals for survival or reproductive success are altered. Once changes in the genetic information have taken place, they are subjected to "testing" by natural selection in the environment in which that organism lives. Successful competition may depend on changes that give even the slightest advantage. However, not all genetic change enhances individual survivability and success. Some genetic changes are neutral with respect to selection. Furthermore, as will be presented later in the chapter, certain types of mutations in DNA are disastrous.

Once changes have occurred in the DNA by mutation, the uncertainty of the change is gone. The new DNA sequence either encodes a product that works (regardless of how well) or it does not. Chance occurrence is replaced by necessity for function. The change in a gene is now a stable, heritable part of the genome, and the product coded for by that gene must serve its bearer by performing successfully in the environment in which that organism lives. The rate of mutational change at a specific locus is on the order of 10^{-6} mutations per gene per cell division. This means that one mutation of a gene is likely to take place on average every 1,000 cell divisions. Therefore, because we have 50,000–60,000 genes in the genome, 50–60 mutations may occur in a cell's DNA during any cell division cycle.

Domestication and Artificial Selection

Homo sapiens realized fairly late in its history how to breed animals and plants for domestic use. Domestication of species by humans brought about rapid changes in genotypes and phenotypes of individuals in the populations of those organisms. Domestication represents the outcome of artificial selection. Both in the past and today, animals and plants are bred to express certain qualities. There is great practical value in such breeding. It has worked quite well in terms of selecting animals and plants that grow faster, are hardier and tastier, or are more compatible with a human sense of beauty. Until the last century, nothing was known about how such changes actually came about. The idea of "blending" traits has been used throughout history, but as intuitively pleasing as this concept was, it was not useful in explaining the mechanisms of inheritance. In the middle of the nineteenth century, and then again in the beginning of the twentieth century, all this changed. The beginning of modern genetics is traced to the 1850s and 1860s in a small monastic garden in an abbey in what was then part of the Austro-Hungarian Empire. The scientist involved, Gregor Mendel, was by profession a monk, and his work was, for the most part, little noted nor long remembered by his contemporaries.

MENDELIAN GENETICS—GREGOR MENDEL AND HIS PEAS

Gregor Mendel grew thousands of pea plants in his garden, and each season he would collect pollen and cross-fertilize

Mutagens chemical, biological, radiative agents responsible for causing mutations

various plants of both pure and mixed varieties and analyze their offspring for the inheritance of specific traits. He did this for many generations of plants and was able to establish a set of rules of inheritance. The underlying significance of Mendel's work in modern times is that he approached the analysis of plant inheritance using mathematical principles. As will be presented shortly, Mendel grew and counted thousands of plants over many generations and used statistics to support his conclusions. The rules he formulated have stood the test of time, even though they had to be rediscovered by scientists a generation after Mendel's death. A review of Mendel's experiments will establish a framework in which to understand human inheritance.

LAWS OF INHERITANCE

Segregation

Mendel was not just insightful; he was lucky. He happened to choose an organism to study with genes that were inherited independently of one another and could, therefore, be observed in all possible combinations. The most important outcome of Mendel's work was the establishment of two sets of rules that later became known as Mendel's laws of inheritance. The first was his *law of segregation* and the second was the *law of independent assortment*. In the first law, he determined that the units of inheritance came in pairs (Mendel did not call the units of inheritance genes, he called them "factors") and that the pairs separate during the formation of gametes and are brought back together at fertilization. This is what we now know to be the result of meiosis and the separation of genes (alleles) carried on homologous chromosomes (see discussion of meiosis in Chapter 3). When Mendel's factors separated from one another during meiosis, each remained intact and capable of reappearing, in the appropriate allelic combinations, in subsequent generations. Certain traits in these plants did not appear at all from one generation to the next, but they were not lost; they were simply carried along silently in a heterozygous state. These findings provided the underpinning for the concept of recessiveness and dominance.

Independent Assortment

Mendel's second law states that different genes segregate independently of one another. For instance, a trait that affects the shape of seeds is inherited independently from the trait that determines the color of flowers. In modern terms, this suggests the location of the traits that Mendel examined were genes on different chromosomes. This is where Mendel was lucky. He did not know about genes or chromosomes. All the traits he chose to examine were in fact on separate chromosomes. Genes that are on the same chromosomes do not behave in this manner, because they are physically linked. Such linked genes are inherited as a unit because they are on the same chromosome. Let's review how Mendel worked all this out in the isolation of his monastic garden.

TRAITS AND THEIR PASSAGE

The color of flowers, length of stems, and shapes of seeds are all characteristics that are easy to see and evaluate in garden peas. Mendel chose these and several other traits and followed the patterns of their inheritance over many plant generations (Table 20–3). Pea plants have both male and female sexual organs and can self-fertilize and, so, perpetuate themselves. Mendel bred and selected plants for tallness and shortness, as well as for phenotypic characteristics of seed surfaces known as round (smooth) or wrinkled (Figure 20–10), which we will use throughout this section

Table 20–3 Traits in Peas Examined by Gregor Mendel

DOMINANT	RECESSIVE
Tall	Short
Round seed	Wrinkled seed
Yellow seed	Green seed
Green pod	Yellow pod
Gray seed coat	White seed coat
Inflated pod	Pinched pod
Flowers axial on stem	Flowers terminal on stem

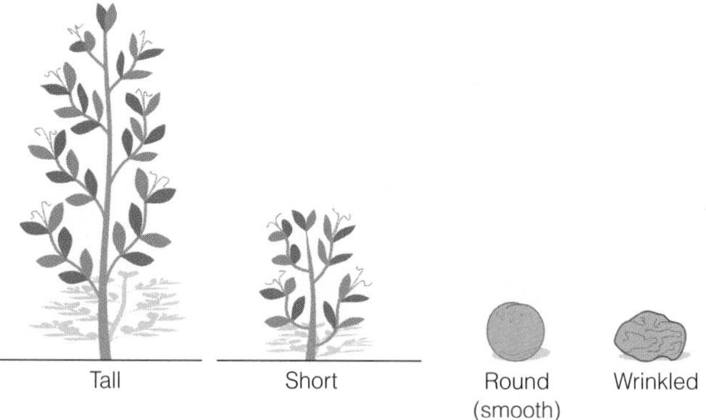

Figure 20–10 Mendelian Characteristics
The Mendelian characteristics described for pea plants throughout this chapter are allelic differences resulting in tall or short plants, or in round (smooth) or wrinkled seeds.

as examples. In terms of the dominance or recessiveness of this set of traits, what is known is that tall dominates short, and round dominates wrinkled. With these character types (and those listed in Table 20–3), Mendel was able to establish the general rules of inheritance and in one of the greatest achievements of science, to discover a valuable key to understanding the genetics of all organisms.

The F_1 Generation

When plants that produce only round seeds are crossed with (that is pollenated) others of like type, the outcome is always a new generation of plants with round seeds. This is also true for tall and short plants (Figure 20–11). The plants that are bred together are called the **parental generation (P),** and their offspring are called the first filial generation, or the **F_1 generation** (Figure 20–11). Continual inbreeding of these plants always results in tall plants. The same is true for short plants. They breed true. However, when pure-breeding tall plants are crossed with pure-breeding short plants, the F_1 progeny produced are all tall. Why is this? Segregation of genes is the key. Visualization of the results makes the explanation easier. When the breeding involves only a single factor difference between two plants, it is called a *monohybrid cross*. A standard way to represent such a crossbreeding is to use a device known as a Punnett square (Figure 20–12).

The Punnett Square

The Punnett square allows us to organize the relationship between the factors and to get a sense of the probabilities associated with the generation of specific genotypes and phenotypes. The tall plant crossed with the short plant (represented as tall × short) can be interpreted as the union of gametes of the male and the gametes of the female, resulting in a generation of new individuals (Figure 20–12). In this case, the alleles for tallness are different in each of the parents, and traits are distributed uniformly to each of the F_1 offspring. Notice, however, that unlike the parents,

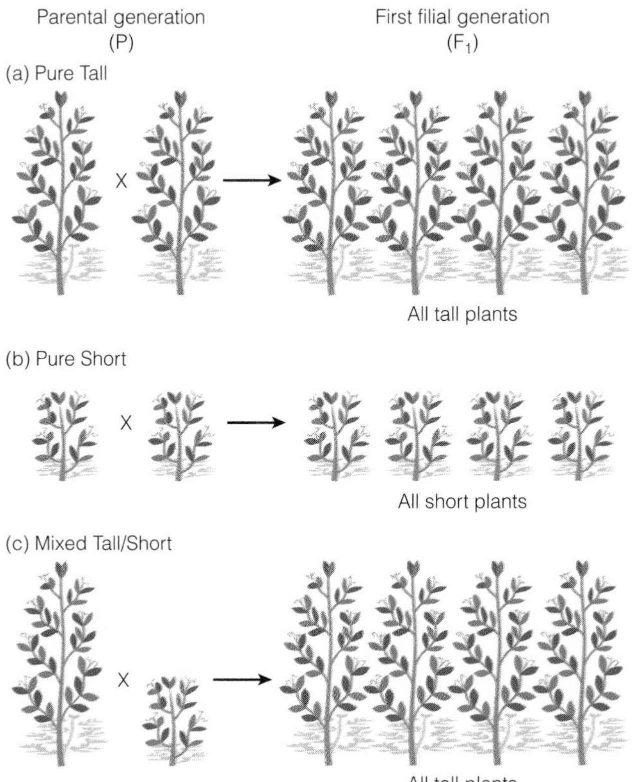

Figure 20–11 P, F_1, and F_2 Generations
In (a), pure tall parental plants (P) are crossed (mated) to give a pure tall F_1 generation. In (b), pure short parental plants produce a pure short F_1 generation. In (c), tall cross short plants result in all tall plants in F_1. $F_1 \times F_1$ results in F_2 generation.

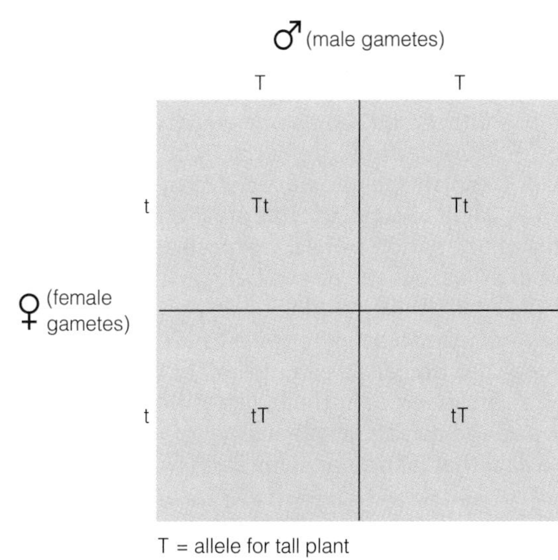

Figure 20–12 Punnett Square and a Monohybrid Cross
This cross involves that presented in Figure 20–11(c) and here takes the form of TT (pure tall) × tt (pure short) → tT (mixed tall). Segregation of single chromosomes during meiosis results in only *one* allele being present in any one gamete.

Parental generation (P) called P, the progenitors of subsequent filial (F) generations

F_1 generation the first filial (F) generation of siblings

the offspring are heterozygous for these traits. Because the tall trait dominates short, the individuals of all the F_1 offspring are tall. One advantage in using plants in these studies is that a large number of seeds are produced at the time of each fertilization and they grow fairly rapidly. This allows the population to be dealt with on a statistical basis. Mendel was the first to quantitatively assess genetic data to establish probabilities of outcomes.

The next step is to cross individuals from the F_1 generation to produce an **F_2 generation**. Because all of the F_1 generation are heterozygous, each of the F_1 individuals produces gametes that have either a chromosome with a dominant tall allele (capital letter T), or a recessive short allele (lower case t). The Punnett square is now filled with three different letter combinations representing F_2 genotypes: TT, Tt, and tt (Figure 20–13). The ratio of genotypes is expected to be 1:2:1 (Tt being twice as likely as tt or TT). Something very interesting occurred in this F_2 generation. The previously absent short phenotype (tt) has re-emerged from its silence in the F_1 generation, where it was obviously not lost but only dominated by the presence of T. tt appears in only one square of the Punnett diagram.

Large numbers of offspring are useful in establishing the probabilities. One square of tt represents the probability of obtaining a short plant in a cross between parents heterozygous for this trait. If only one plant were formed, the chance for it to be short would be 1 in 4. If this cross-produced 100 plants, approximately 25 of them would be short and 75 tall. In actuality, this process is not perfect, so there may be variation from the expected number in natural populations, but the more plants involved, the closer the results come to expectations. Mendel grew thousands of plants and used statistics to come to his conclusions. He concluded that the idealized ratios of 1:2:1 for genotypes and 3:1 for phenotypes in the F_1 generation represented a natural and predictable phenomenon. It was on the basis of these data that he proposed his first law.

A Test Cross

Let's do a bit of practical detective work. How can we differentiate between a TT individual and a Tt individual? Certainly not by the phenotype, because T is dominant and plants of these genotypes are all tall. In fact, the ideal phenotypic ratio for a cross between heterozygous parents is predicted to be 3 to 1, tall to short. One way to determine whether a particular plant is homozygous or heterozygous for a dominant trait is to carry out what is called a *test cross* (Figure 20–14). A test cross involves a plant of unknown genotype (is it *TT or Tt?*) and a plant that is known to be homozygous for the recessive trait in question. How can so simple a test be so valuable? Because the outcome of such a cross results in a 1:1 ratio of tall plants and short plants in the next generation if the unknown plant is a heterozygote and all tall plants if the unknown genotype is homozygous. With one test cross, the original unknown genotype can be established.

A Double Cross

For Mendel to establish his second law, that of independent assortment, he needed to show that different traits could appear together, yet be inherited independently of one another. The Punnett square can easily accommodate

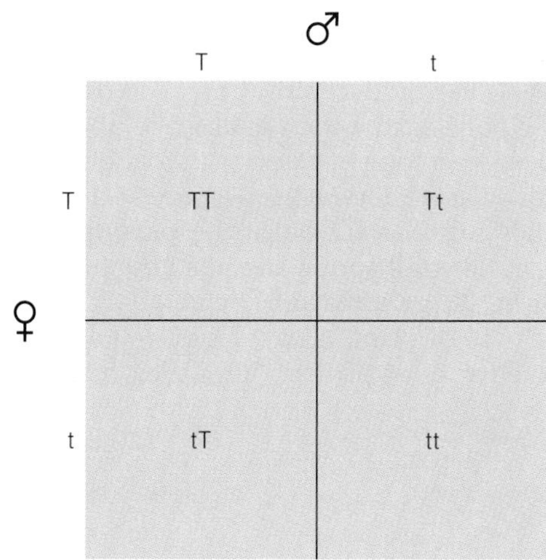

Figure 20–13 Crossing Heterozygotes (F_1 + F_1)
The cross between siblings of the F_1 generation results in three different genotypes and two different phenotypes. They are

Genotypes: TT, pure tall; Tt, mixed tall; tt, pure short. The ratio among types is 1:2:1 (1TT: 2Tt:1tt)
Phenotypes: Tall and Short. The ratio among types is 3:1 (1TT + 2Tt:1tt).

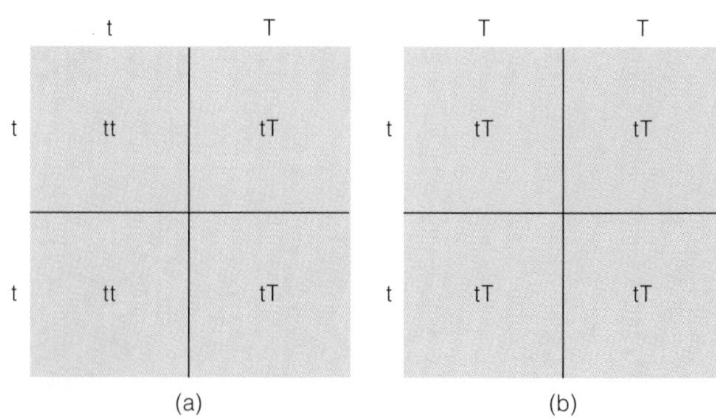

Figure 20–14 Test Cross
A test cross is used to ascertain the genotype of an unknown plant type. If the unknown is mixed (tT), then the ratio of tall to short is 1:1 as in (a). If the unknown is pure (TT), then all offspring are tall, as in (b).

F_2 generation the second filial generation (offspring of siblings in F_1)

this type of test, which is called a double or *dihybrid cross*. Suppose that there is a pure-breeding tall plant that produces round seeds, and it is crossed with a short plant with wrinkled seeds (Figure 20–15). If each plant is homozygous for the traits, the starting genotypes are *RR TT* for tall/round (where *R* is the gene for round seeds), and *rr tt* for short/wrinkled. The F_1 generation will be solely *Rr Tt* heterozygotes that are all tall and possess round seeds (Figure 20–16). If members of the F_1 generation are crossed (*Rr Tt* × *Rr Tt*), an assortment of phenotypes and genotypes reflecting all possible combinations among the traits in question is obtained. The predicted ratio of the four possible phenotypes in a dihybrid cross is 9:3:3:1, with nine tall plants with round seeds, three tall plants with wrinkled seeds, three short plants with round seeds, and one short plant with wrinkled seeds. This ratio reflects the probability for obtaining an individual with one of the four possible phenotypes. The genotypes are more diverse, appearing in a ratio of 1:1:2:2:4:2:2:1:1. As Mendel's second law predicts, the ratio of the two individual traits affecting height and seed shape are independent of one another and occur with exactly the same probability as when they are examined in monohybrid crosses.

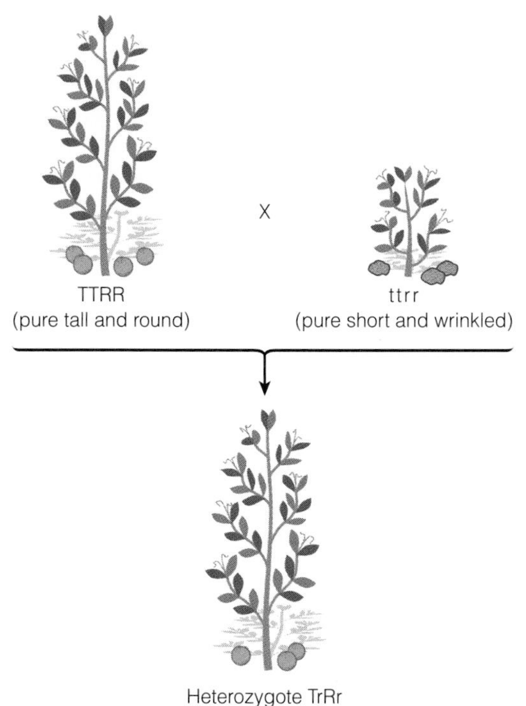

Figure 20–15 Double Cross (Dihybrid Cross)
The cross of plants with respect to two nonlinked traits is shown when pure tall and round-seeded plants are crossed with pure short and wrinkled-seeded plants. All F_1 are tall with round seeds.

Pedigrees and Family Trees

A pedigree is a history of interrelationships among individuals in families over many generations. It is also an accounting of different traits and characteristics that appear in individuals of those families. Just about everyone has wondered how they are related to other members of their family and how each individual is similar to, and yet different from, all the others. Most of us have some knowledge of our ancestors, how they looked, how they behaved, and where they came from. Such relationships can be plotted out, and different branches of the family can be followed to see who was married to whom, how many children they had, and where we personally fit in the scheme of these relationships. In other words, we can construct a family tree. A cousin may be an albino, an uncle may have curly hair, and a niece may have blue eyes instead of brown eyes as the rest of her family (Figure 20–17).

Another kind of diagram useful in relating patterns of inheritance of traits to individuals in families, or species, is one that represents the presence of a selected gene on a model chromosome. This allows us to follow a specific gene during meiosis and combination with its homolog in a zygote. Because a specific gene is found only on a specific chromosome, we can simplify matters by representing only one chromosome pair at a time (Figure 20–18). If we re-examine a Mendelian cross between a round-seeded plant and a wrinkle-seeded plant with this simplification in mind, all alleles are shown to occur at the same chromosomal

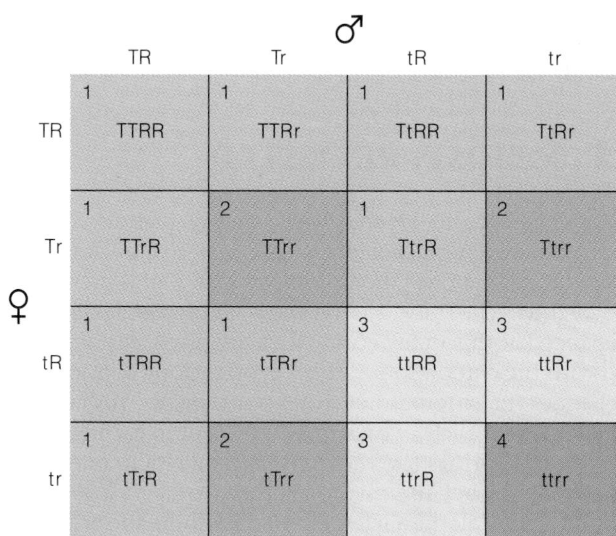

Figure 20–16 Dihybrid Cross ($F_1 \times F_1$)
This Punnett square is less complicated than it appears. Each square represents the combination of traits (alleles) arising from segregation of the alleles during gamete formation. T dominates t and R dominates r. The phenotypic ratio is 9^1 tall/round: 3^2 tall/wrinkled: 3^3 short/round: 1^4 short/wrinkled. Colors of squares indicate the phenotype.

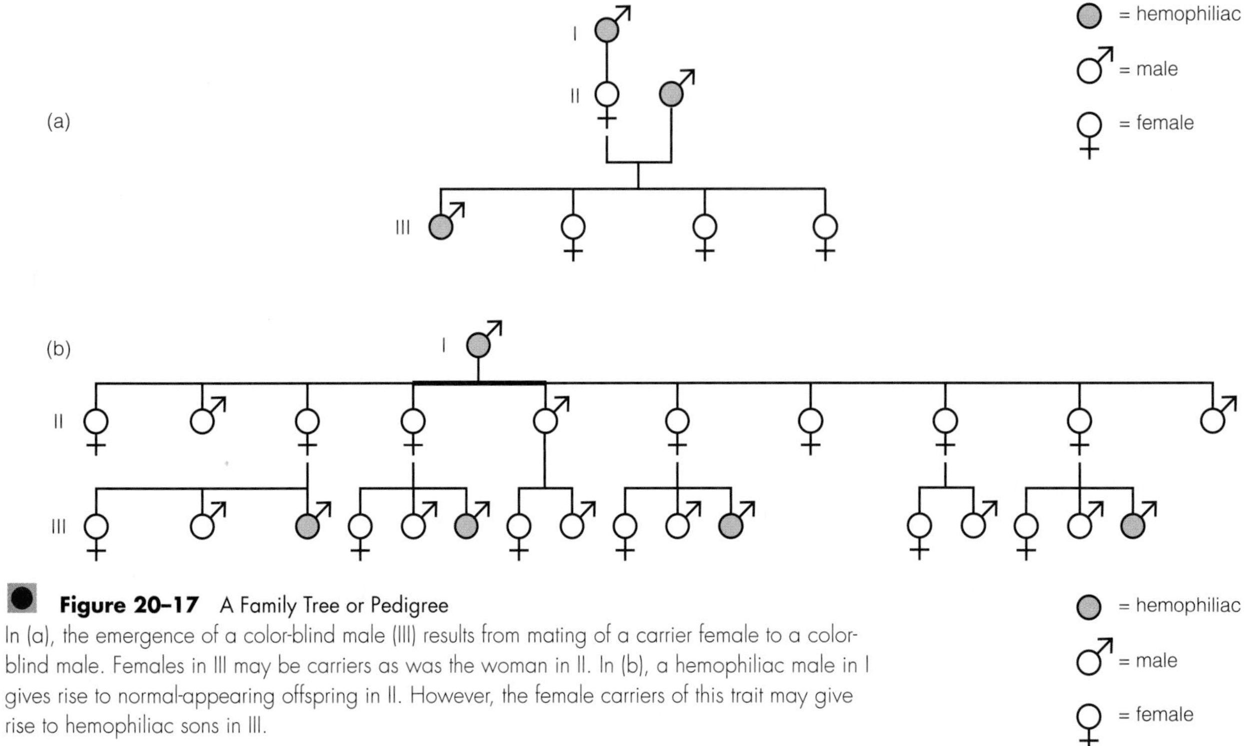

Figure 20–17 A Family Tree or Pedigree
In (a), the emergence of a color-blind male (III) results from mating of a carrier female to a color-blind male. Females in III may be carriers as was the woman in II. In (b), a hemophiliac male in I gives rise to normal-appearing offspring in II. However, the female carriers of this trait may give rise to hemophiliac sons in III.

location on homologous chromosomes, and a recessive trait on one or both chromosomes may be indicated by specific marks (× and ·). In the F_1 generation, all individuals are tall, even though they carry a recessive allele. In the F_2 generation, this genetic trait can re-emerge as a pure trait (homozygous recessive and dominant) and as heterozygous (Figure 20–18).

MEIOSIS REVISITED

With reference to Figure 20–18, during meiosis (I), the homologous chromosomes bearing recessive alleles separate (Ia), so that one chromosome of each pair is present in a daughter cell. The alleles on those chromosomes are also separated at this time because they are physically associated with the chromosome. The second meiotic division (Ib) separates the chromosome at its centromere. This ensures that only one allele of each gene is present in each gamete. Fertilization brings together individual alleles in one combination (II), but the many gametes represent a random assortment of all the alleles, thus, establishing the probabilities manifested in the crosses Mendel carried out. Formation of gametes from the F_1 heterozygotes results in the Punnett distribution presented in Figure 20–12. Of the four possible outcomes of this cross, one is the homozygous dominant, two are heterozygotes, and one is the homozygous recessive. If you think about genes being distributed in this way, that is, as homologous chromosomes separating and then dividing at the centromere during meiosis, then it does not matter what the trait actually is but rather how the process of segregation occurs. What is of significance for the genotype is that a specific allele is responsible for all, or part, of a trait. Once the combination of alleles in a zygote is set, the phenotype emerges as the expression of the alleles, depending on whether they are dominant/recessive, incompletely dominant, or codominant in quality.

In humans, the outcome of the expression of homozygous recessive traits may be quite striking. Nearly everyone has seen an individual who appears to have no coloration (Figure 20–7). Albino individuals are completely devoid of pigmentation in hair, skin, and eyes. The albino trait results from the occurrence of a pair of recessive alleles for a gene that normally codes for an enzyme involved in the synthesis of melanin from the amino acid tyrosine. As in the case of recessive traits in Mendel's peas, each of the human parents of an albino child must contribute a recessive allele to their albino offspring. In this case, if each parent is heterozygous for the trait, they may be unaware that they carry recessive genes. Based on probability alone, the chance for a heterozygotic pair of people having an albino child is 1 in 4. The actual incidence of albinism in human populations is probably around 1 in 10,000 or so, and the number of heterozygotic carriers in the population is

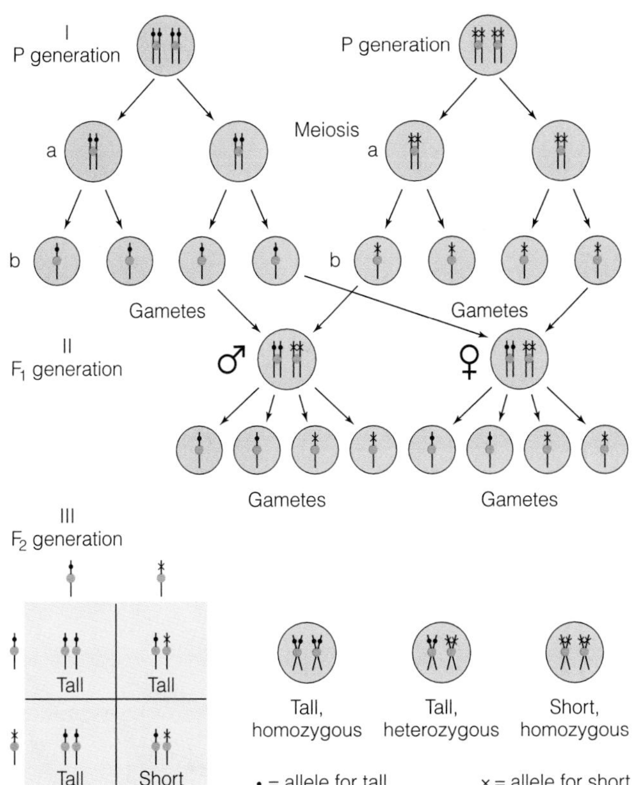

Figure 20–18 Locus on a Chromosome

The locus of the gene can have a dominant or recessive allele. Pure parents give rise to mixed F_1. What is crucial in this diagram is that the particular allele at a chromosomal location and the linkage or lack of linkage determine which traits will occur in the F_1 and F_2 generations.

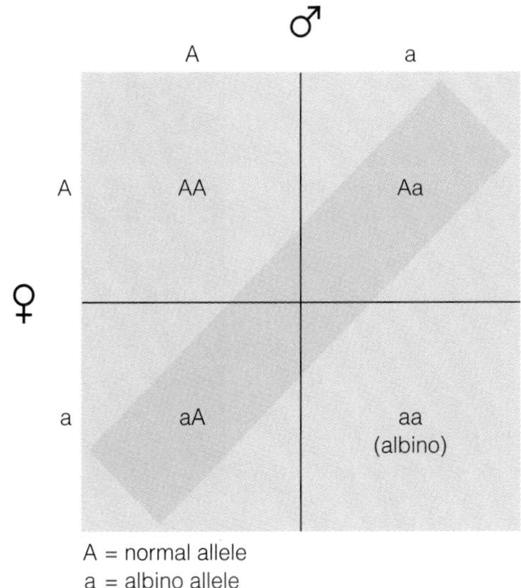

Figure 20–19 Carriers of Hidden Traits

Approximately 1 in 200 in the human population are carriers of the albino gene. Fifty percent of the offspring of an affected individual and a normal individual will be carriers of the recessive trait but will not express it. In a cross between carriers, there is a 25% chance of producing an albino child.

approximately 1 in 200. Although an albino individual may have some health problems in dealing with sunlight or exposure to UV radiation, he or she is normal in other ways, including fertility. When, or if, they marry, albino individuals will, in all likelihood, marry an individual that does not have the defective gene. However, all the offspring of that marriage will be heterozygotes. These individuals will be silent carriers of the recessive gene until a marriage to another heterozygote reveals its presence in their offspring (Figure 20–19).

The Dark Side of Recessive Genes

The fact that recessive genes are carried by individuals without their knowing it has a dark side. The effects of a recessive trait may be lethal. For example, diseases such as *cystic fibrosis* and *phenylketonuria* can result in death of offspring during infancy or early childhood. Both are recessive traits. These inborn errors of metabolism, as many such defects are described, continue to be carried as recessive alleles in the human population simply because the heterozygous state is phenotypically normal and has little effect on health. Defective genes are often carried in families and, so, have a higher probability of being carried in the siblings over many generations. For a couple who have had an albino child, there is a 50% chance that any child they have will be a carrier (Figure 20–19). This means that three-fourths of the siblings either manifest the abnormality or carry it hidden in their genes.

Consanguinity

If we extend this argument to include the inheritance of lethal recessive genes, we can see, based on genetic principles, why marriage of close relatives is dangerous. Close relatives are more likely to share the same genes (good genes and bad genes) than unrelated individuals. Intermarriage of family members, known as *consanguineous marriages,* closer than third cousins is illegal in the United States, but in other places and during other times, such intrafamily marriages were frequent. In the nineteenth century, the royal houses of Great Britain and other European monarchies often intermarried for political as well as aristocratic reasons. One historically important biological consequence of marriage and sexual unions within the royal families was the spread of hemophilia (Figure 20–20).

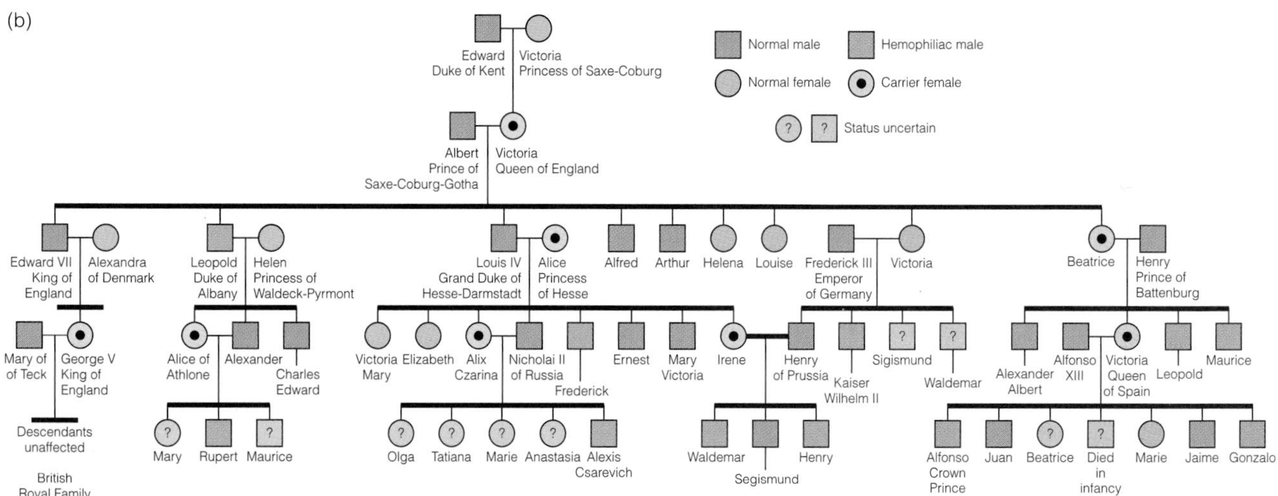

Figure 20–20 The Royal Family and Hemophilia
The incidence of hemophiliac males in the royal families of Europe is thought to originate in genes carried by Queen Victoria, seated front and center in (a). The pedigree (b) shows carriers and those expressing the disease.

In species other than *Homo sapiens,* this selective breeding is practiced with regularity. It is called **inbreeding,** a term not politely applied to humans. Inbreeding occurs for many kinds of domestic animals, from dogs and cats to sheep and cattle, as well as in agriculturally important plants. Domestication involves the artificial selection of desirable traits in animals (or plants) and breeding males and females that best reflect them in the population. The traits selected may be hair color, hair length, milk production, hunting ability, or innumerable other qualities. As different as they may appear today, all breeds of dogs were originally derived from wolf ancestors. Unfortunately, as farmers, ranchers, and pet breeders selected for desirable traits, the inbreeding schemes that they used also increased the probability for undesirable traits to occur, much to the detriment of the health and viability of inbred stocks. Breeding programs now include the intermingling of genes from individuals outside the inbred population, a process

Inbreeding matings between closely related individuals, used in domestication of animals and plants

referred to as **outbreeding.** Modern genetics has provided insights into the pluses and minuses of inbreeding and the benefits of outbreeding.

Dominant Inheritance

Another way in which traits may be passed on is through dominant genes. Obviously, the probability of phenotypic expression of a trait, whether normal or defective, changes dramatically if it is represented by a dominant gene. An example of this type of inheritance is observed in individuals who have *syndactyly,* which is characterized by differing degrees of fusion of fingers and/or toes (Figure 20–21). The cellular basis for this abnormality is the failure of digits to separate properly during fetal development. In an individual who is heterozygous for the defect, one expects that marriage and mating with a normal individual would result in a 50% probability of an offspring having the defect. Mating of two affected heterozygous individuals would result in a 75% chance of passing on the defect.

A different twist to the inheritance of dominant defective genes is observed in a disease known as Huntington's disease, which is a serious and terminal neurological disorder. The twist is that the defect does not show up in the individuals who carry the trait until they are adults. The symptoms of the disease usually do not appear until a person is in their late thirties. Plenty of time to have a family. The legacy that a heterozygote for the Huntington gene passes on to his or her children is a 50% chance of a late-arising, ceaselessly progressive lethal disease. In the last few years, geneticists and molecular biologists have developed a way of determining whether an individual has this defect long before the symptoms are expressed. The technique involves isolating DNA from the cells of a potentially affected individual and using special molecular techniques to identify the Huntington gene.

Incomplete Dominance

Not all genes can be categorized as dominant or recessive. As a result, heterozygotes having genes that are incompletely dominant show traits that are intermediate between the homozygous recessive state and the homozygous dominant state (Figure 20–22). For an abnormal trait or condition, the degree of abnormality is determined by the dose of a recessive gene. Incomplete dominance results in a heterozygote individual (with one abnormal allele and one normal allele) who may be mildly affected, and a homozygote (with the double recessive) who is severely affected. Abnormalities of this type are exemplified by two serious blood diseases, one known as **sickle-cell disease** and the other as *thalassemia.*

Sickle-Cell Disease The problem associated with sickle-cell disease is severe hemolytic anemia, that is, massive destruction of red blood cells. This trait is common among black populations in Africa and probably arose only within the last few thousand years. Today, 4%–5% of all children in many areas of Africa are afflicted with sickle-cell anemia. The

	S	s
S	SS syndactyl	Ss syndactyl
s	sS syndactyl	ss Normal

(a)

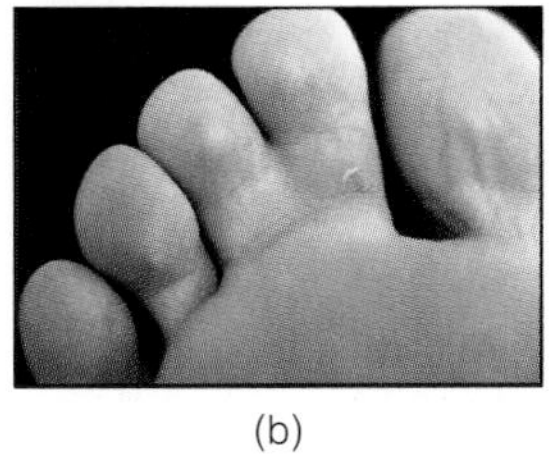

(b)

Figure 20–21 Dominant Inheritance
Unlike recessive traits, dominant traits appear whenever the dominant allele is present (a). For example, the Punnett square for the dominant abnormal trait for fused digits, or syndactyly (represented by S), is shown in (a). The recessive normal trait is represented by s. The effect of this dominant trait is shown in (b).

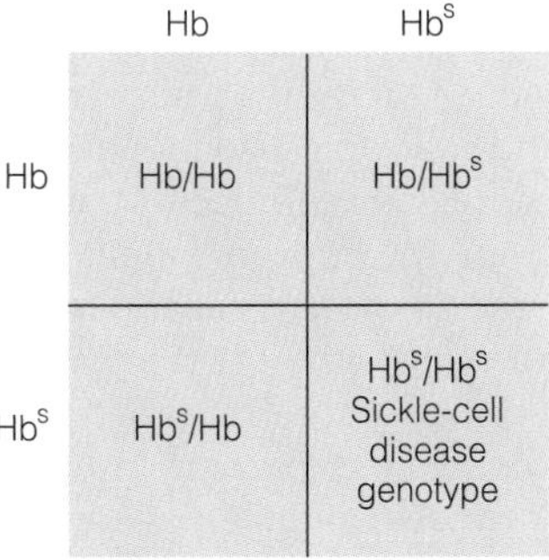

(a)

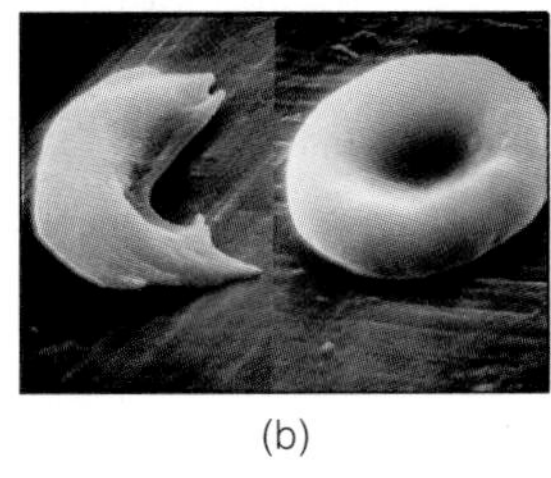

(b)

Figure 20–22 Incomplete Dominance
Sickle-cell disease is an incompletely dominant trait. Hb is the normal allele for hemoglobin and Hb^s is the sickle allele. Hb/Hb is normal and Hb^s/Hb^s is severely abnormal, which results in changes in cell shape, see (b). Hb^s/Hb is intermediate in abnormality and represents carriers of the sickle-cell trait.

Outbreeding purposeful matings between unrelated individuals, used in domestication of animals and plants

Sickle-cell disease a recessive genetic disorder in which hemoglobin is abnormal and cell shape is altered

disease got its name from the characteristic change of red blood cells from a normal biconcave shape to half-moon shape that occurs when cells are deprived of oxygen. The drastic change in cell shape results in blocked blood vessels, cell lysis, joint pain, ulcerations, fever, and other symptoms that are serious enough to eventually cause death, usually in early childhood. The sickle-cell gene encodes a defective protein molecule called hemoglobin-S. The molecule differs from normal hemoglobin-A by a single amino acid change in the protein. Even though the mutational change in the globin gene occurred fairly recently, it has persisted in the population. As will be seen, the phenotype of incomplete dominance has consequences far beyond lethality.

Sickle-Cell Trait The symptoms described above and, therefore, the designation sickle-cell disease apply to an individual who is homozygous for the defective genes. On the other hand, what about the heterozygote? How are individuals with a single dose of the gene affected? Heterozygotes express what is called the *sickle-cell trait.* Thirty percent to 40% of their hemoglobin is S-type, but the rest is normal. The defect is not lethal in heterozygotes, though extreme conditions of oxygen deprivation (high altitude, physical overexertion) will cause some sickling of red blood cells (Figure 20–22). If this gene is so lethal, how is it that it has been maintained at such a high frequency in the populations of people of African ancestry? The conjecture is that in the heterozygous state, this gene must be advantageous to the individuals bearing it. Into the picture comes the scourge of the tropics, *malaria.* Malaria is a parasitic disease endemic to the tropics and subtropical zones and is a major cause of death in those regions. Apparently, what fixed the hemoglobin-S gene in the African population (and in populations of other regions where malaria is endemic) was that when it was carried as a trait (a single dose of the gene), it protected individuals from malaria infections, somehow preventing the malarial parasite from effectively invading red blood cells. The double dose of the gene was deadly (and continues to be), but the natural selective advantage gained by individuals heterozygous for the trait, through malaria resistance, more than balances the negative potential of homozygous lethality.

Thalassemia Another intermediate or incompletely dominant trait is thalassemia, a disorder of blood that is common in Italy and the areas surrounding the Mediterranean Sea. "Thalassa" means sea, and so this disease was named for the region in which it was first identified and studied. In homozygotes who carry the genes for this defect, the outcome is predictably bleak. They usually die within the first few years of their lives from progressive hemolytic anemia. This form of the condition is referred to as *thalassemia major.* Changes in the bones are also observed, as are enlargement of the spleen and liver and a yellowing of the skin. In the heterozygous condition, the gene is not lethal, but its effects can be observed in altered blood cells, skeletal changes, and yellowing of skin. This milder condition is called *thalassemia minor* and as in the case of the sickle-cell trait, was probably initially associated with resistance to blood parasites, perhaps malaria. Having expanded in incidence within the population for reasons that no longer exist in Italy (malaria has been eradicated), the trait is perpetuated in the population because having a single dose of the gene is not detrimental. On the contrary, from a cultural point of view, the trait may have actually made the women who carried it more physically attractive to men. This may have resulted from the delicate bone structure of these women and the appreciation for the rareness of such perceived beauty. This, in turn, may have led to competition for this limited resource that was both beautiful and potentially lethal.

As described, there are genes that are completely dominant, such that a single dose of the gene results in a phenotype identical to that of a double dose, and there are genes that are incompletely dominant, in which case, a single dose of the gene produces a milder form of the defect or disease. There is also another condition in which allelic genes do not dominate one another but are coexpressed. These genes show the characteristics of codominance.

Codominance

So far in this discussion, we have limited consideration of alleles to only two alleles at a particular locus on a chromosome, one being completely or incompletely dominant over the other. However, many genes have more than two forms and, thus, many alleles. These are called "multiple alleles." Because an individual can have no more than two alleles for a particular gene, a larger pool of alleles must be present in the population (Biosite 20–2). The genome of a species includes all the allelic variation at all the loci on all the chromosomes and, thus, constitutes a large pool of genes. The chance of getting a particular set of alleles in a randomly mating population depends on their frequency in the population. Combining this idea of multiple alleles in the gene pool with codominance, let us consider human blood group antigens.

ABO Blood Groups The blood group antigens of the ABO group are glycolipid molecules occurring on the surface of human red blood cells. The most commonly recognized blood types in humans are called A, B, AB, and O. Each of us expresses only one of these blood types, and it is important to know what type you are. For example, transfusion of blood from one individual to another can save a life, but transfusion of the wrong blood type can be life threatening. We will see why this is true after discussing the nature of the blood groups.

ABO Antigens What are the genetics of this blood group? First of all, there are three main alleles I^O, I^A, and I^B (actually, there are many more than this because each of the main alleles has variants as well). Any individual can express only two of the alleles, one from each parent (Figure 20–23). If an individual has type A blood, that person is either AO or AA. In this case, A dominates O, and the phenotype is A. If a person has type B blood, that individual is either BO or BB in genotype. In this instance, B dominates O. If an individual has type AB blood, then both alleles are equally expressed or codominant. The genotype of that individual is AB, the same as is the phenotype. O is recessive to A and B, but A and B are codominant to each other. The type O phenotype is expressed only when the OO genotype occurs, as in a homozygous recessive trait. The existence of multiple alleles points out to us that the genes carried in an individual represent only one of many possible combinations of a large and ever-changing pool of human genes.

ABO Antibodies Why is knowledge of human blood types so important? Certainly, from a genetic point of view, it offers insight into the structure and expression of the human genome. However, it is the clinical application of this knowledge that saves lives. There is a real danger in blood transfusion if blood types are not matched. Why? In addition to having antigens on the surface of red blood cells, the serum of that individual contains antibodies against the antigen or antigens that are not represented on their cell surfaces (Table 20–4). For example, the serum of a blood type A woman contains antibodies that will react against type B red blood cells. This means that if she needed a transfusion of blood for an emergency operation, she could not accept blood from a type B individual. The reverse of this is true as well. The blood type B person could not tolerate type A blood because that person's serum contains antibodies against the A antigen. The result of mismatched transfusions is clumping together of donor cells, blockage of blood vessels, and cell lysis.

(a)

♀ \ ♂	I^O	I^O
I^A	I^AI^O Type A	I^AI^O Type A
I^O	I^OI^O Type O	I^OI^O Type O

(b)

♀ \ ♂	I^O	I^B
I^O	I^OI^O O-Type	I^OI^B B-Type
I^A	I^AI^O A-Type	I^AI^B AB-Type

Figure 20–23 Codominance
In (a), a type O male mates with a type A female who is heterozygous for I^A and I^O. In (b), a type B male heterozygote mates with a type A female heterozygote. I^AI^B are codominant alleles and both appear as antigens on red blood cells.

Table 20–4 Blood Group Antigens and Antiserum

ANTIGEN	ANTISERUM
A	Anti-B
B	Anti-A
AB	None
O	Anti-A and anti-B

Genotypes for the ABO group: A = I^A/I^A or I^A/I^O, B = I^B/I^B or I^B/I^O, AB = I^A/I^B, and O = I^O/I^O

What about the serum of AB and OO individuals? In the case of an AB individual, which is a rather rare blood type regardless of a person's geographic or ethnic origin, there are no antibodies to either A or B in the blood (Table 20–5). As you might have guessed, this individual can receive blood from all blood groups and is considered a **universal acceptor.** An O blood type person, a very common type worldwide, has serum that contains both anti-A and anti-B antibodies and, so, can receive blood only from other O type individuals. However, the blood cells themselves have no A or B antigens and, therefore, are not recognized by anti-A or anti-B antibodies. Thus, the type O blood cells are tolerated by individuals with any combination of the ABO gene products. O type individuals are known as **universal donors.**

Rh Factors The genetics of ABO blood groups provide insight into the complexity of inheritance and the importance of knowing how such traits manifest themselves. There are many more blood types known. Each is similar in that they are cell surface antigens, and each has associated antibodies in the serum. A particularly well-known type is referred to as the Rh type. This group of antigens was discovered in rhesus monkeys (hence the Rh designation) and has proven to be important during human embryonic and fetal development. A mother who is Rh negative (that is, does not have the antigen on her red blood cell surfaces) who mates with a man who is Rh positive may have a child who is Rh positive. Should the mother be exposed to that child's blood cells during pregnancy, she may develop an

Universal acceptor an individual with AB type blood
Universal donor an individual with O type blood

Table 20–5 Blood Groups and Populations (%)

GEOGRAPHIC ORIGIN	A	O	B	AB	Rh (NEG)	Rh (POS)
European	30–40	40–50	8–12	1–6	15	85
Native American	2	98	0	0	0	100
Asian	30	30	30	10	0	100
African	20–30	40–50	30–40	1–10	0	100
Pacific Islands	—	—	—	—	0	100

immune response against the Rh factor (that is, antibodies to Rh in her blood). In subsequent pregnancies, those antibodies may affect the fetus she carries if that baby is Rh positive. The situation is dangerous for the fetus but can be treated in consultation with a physician.

Anthropology Blood groups have been valuable in determining not only the basis for clinical safety of blood transfusion but also the genetic relationships among populations of peoples around the world. The frequency of occurrence of specific blood group alleles, or the occurrence of rare alleles in different human populations, has been used to show that such groups of peoples may be closely, or distantly, related and for how long they may have been isolated from one another.

Paternity Blood groups have also been used as evidence in paternity/identity lawsuits. The claim of a woman that a certain man is the father of her child can be disproved if the blood type of the baby is not possible from any combination of the mother's and putative father's blood types. For instance, an AB child cannot arise from the mating of an A type mother (either AA or AO) and an O type father (OO) (Figure 20–23). However, because so many individuals share a specific blood type, even if a potential parental blood group combination makes it possible for the child to have arisen from the sexual union of a particular man and woman, it does not prove that he is the father.

Polygenic Inheritance

To simplify the principles of genetic inheritance, we have focused on traits that are represented by genes that differ from one another in terms of alleles. In many cases, this one-to-one correspondence between an allele and the resulting phenotype does not hold. As mentioned earlier (see Figure 20–9), traits such as tallness, skin color, and intelligence appear to be determined by many different genes acting together. This type of inheritance is very difficult to analyze. Not only do we have to consider the many different genes involved in an adult but all the genes that participated in a process that was carried on throughout growth and development. You do not get tall suddenly at 21 years old. These kinds of traits are not all or none. A population of individuals is neither tall nor short, black nor white but rather shows gradations with respect to those traits and many other traits. This is called polygenic inheritance. Based on random mating, the gene pool of humanity provides the genetic diversity needed to affect the distribution of traits in this way. In addition, there is a component of environment involved in the degree to which polygenetic potential is actually achieved. For example, a nutritional deficiency in protein throughout early growth of an individual may prevent attainment of a tall stature, even if genes for tallness are present in the right combination.

SEX CHROMOSOMES

Sex-Linked Traits

The X and Y chromosomes are considered the human sex chromosomes (Figure 20–2). Their presence in the zygote, and in all cells that arise from it, determines (with rare exceptions) what sex we will be. Two X chromosomes in a fertilized egg lead to the development of a female. The combination of an X and a Y chromosome leads to the development of a male. The X chromosome is quite a bit larger than the Y chromosome and carries a number of well-characterized genes expressed as distinctive phenotypic traits (Figure 20–24). These are called **sex-linked traits.** Genes affecting color blindness and hemophilia are both found on the X chromosome and, so, are sex linked. Although most of the genes of the human genome are carried on the 44 autosomes, many essential genes are carried on the X chromosome. The absence of X chromosomes is lethal. This is not true of the Y chromosome.

Very few traits are carried on the Y chromosome (Figure 20–24). Because of this, we know very little about how the Y chromosome influences sex determination, although recent research has led to the discovery of a gene

Sex-linked traits genes occurring on the X chromosome

Figure 20–24 Traits Carried on X and Y Chromosomes

X CHROMOSOME	Y CHROMOSOME
Color blindness	Maleness (SRY)
Hemophilia	Ear hairs
Femaleness (alone or multiple copies)	
Albinism	
Duchene muscular dystrophy	
Cleft palate	
Lesch-Nyhan syndrome	
Diabetes insipidus	
Ichthyosis	
Fragile X, mental retardation	

The X chromosome carries thousands of genes, a few of which are listed. The Y chromosome carries very few genes, two of which are listed. One is known as SRY and determines maleness. The other carries the trait for long ear hairs.

whose product is necessary for normal male development. However, what is generally meant when we talk about sex-linked genes is X-linked genes, and there are some very interesting phenomena associated with genes carried on the X chromosome.

Color Blindness Color blindness is a recessive, X-linked trait. There are far more color-blind males than there are color-blind females. Why should this be the case? Because the trait is carried on the X chromosome and males have only one X chromosome. If we examine crosses between color-blind males and normal females, we observe that all of their children are normal in phenotype, but that each of the female offspring carries the trait (Figure 20–25). Because color blindness is a recessive trait, its phenotypic expression is masked in females by the normal dominant gene for eye color on the homologous X chromosome. However, if a female from the F_1 generation described above marries a normal male, the probability is that half of their male children will be color blind (Figure 20–25[b]). This is because to be a male there must be an X and a Y chromosome present. The Y chromosome carries no compensating gene for eye color, so the only genetic information that any male will receive is the information on the X chromosome received from his mother. This condition in males, in which there is no offsetting normal gene present on the diminutive Y chromosome, results in what is called a hemizygous state. Furthermore, because the mother can provide only X chromosomes to the possible X-Y combination, there is a 50% chance that an X chromosome with a recessive gene will be inherited. It is also true that there is a 50% chance for female offspring to inherit the gene, but this is masked by the normal X chromosome derived from the father. The only time that a color-blind female can arise is if a female carrier and a color-blind male have children together (Figure 20–25[c]). Under those conditions, there is a 50% chance for both males and females to be color blind. Color blindness is relatively common and not fatal, unless a color-blind person forgets the position of the red and green lights in traffic signals! Other X-linked traits are far less benign.

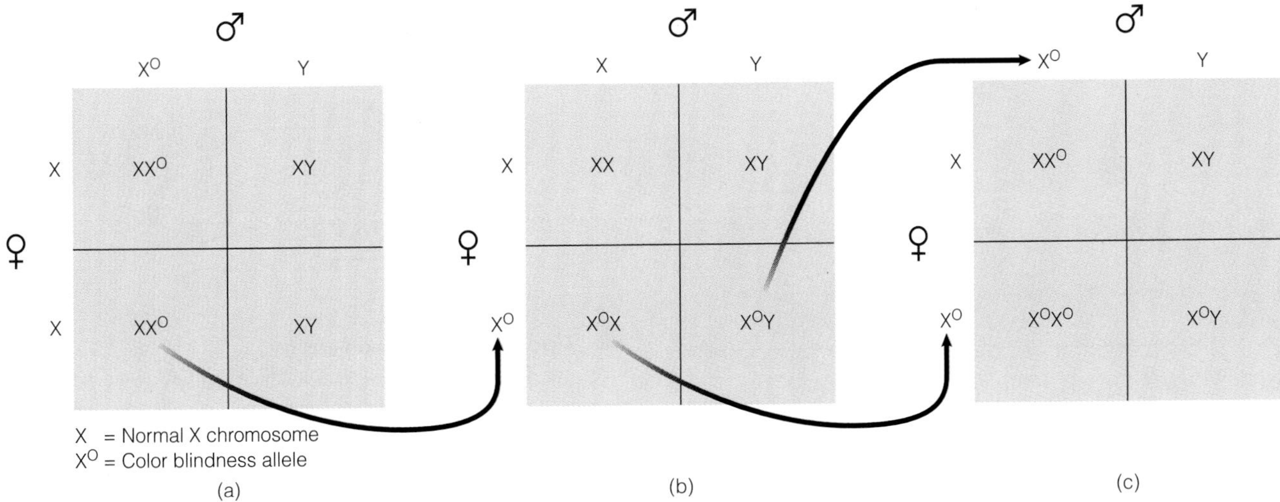

Figure 20–25 Sex-Linked Traits
In (a), a color-blind male passes his X chromosome on to his daughters, who are carriers. In (b), a carrier female passes on the X to her son, who is color blind. In (c), a color-blind female is rare and requires that a color-blind male mate with a female carrier.

Hemophilia If we consider a recessive gene for hemophilia, as we have the gene for color blindness, the distribution of the two traits is identical. The difference is lethality. Hemophiliacs cannot stop bleeding, because their blood will not clot. A male who cannot stop bleeding due to a missing clotting factor(s) in the blood is far worse off than a male who cannot differentiate green from red. Females are carriers of the trait but do not themselves express it. The genes for hemophilia and color blindness (that is, sex-linked genes) are generally passed from mother to son and from father to daughter. Cases of female hemophilia are very rare because they require a hemophiliac male (assuming he survives to sexual maturity) to mate with a mature carrier female, a rare event. Assuming that this combination of individuals actually did occur, there would be only a 25% chance of the trait showing up in females produced from such matings. The occurrence of a hemophiliac female would most likely arise from a consanguineous marriage.

Sex-Limited and Sex-Influenced Traits

In addition to sex-linked traits, there are also **sex-influenced** and **sex-limited traits.** In these cases, the genes involved are on autosomal chromosomes, but their effects are different when expressed in males and females. For example, a well-known sex-influenced trait is *pattern baldness,* or alopecia, which is much more common in males than in females. The gene affecting alopecia acts as a dominant gene in males and a recessive gene in females. As a result, the transmission and expression of sex-influenced genes can occur from mother to son and father to son (which does not happen in sex-linked traits) and appears in females only as a double recessive (Figure 20–26). Sex-limited traits appear to be exclusively associated with either male or female secondary sexual characteristics and probably reflect differences in the way that hormones are produced and utilized in males and females.

♀ \ ♂	P	p
P	PP Normal	Pp Alopecia
p	pP Alopecia	pp Alopecia

P = Normal
p = Alopecia (pattern baldness)
(a) Male

♀ \ ♂	P	p
P	PP Normal	Pp Normal
p	pP Normal	pp Alopecia

(b) Female

Figure 20–26 A Sex-Influenced Trait—Pattern Baldness
In males, p acts as a dominant trait (colored squares in [a]) and in females (b) as a recessive allele.

HUMAN CHROMOSOMAL DISORDERS

Nondisjunction

Although the process by which chromosomes are distributed during meiosis to forming gametes is remarkably accurate and reproducible, it is not perfect. Occasionally, the separation of homologous chromosomes does not happen as planned (Figure 20–27). Instead of the chromosome pair separating and segregating during meiosis, they stay together, and both chromosomes migrate into one of the gametes generated by cell division and, thus, are missing from the

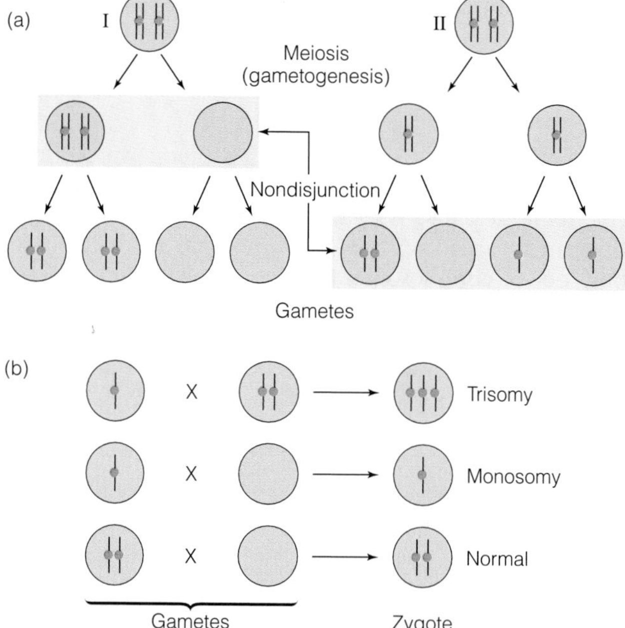

Figure 20–27 Nondisjunction
In (a), meiosis leads to a reduction in chromosomes during gametogenesis. If a chromosome fails to segregate (I and II) either at the first (I) or second (II) meiotic division, some gametes will have a missing chromosome. The possibilities in the F_1 generation are shown in (b).

Sex-influenced traits traits whose expression is influenced by the sex of the individual; affects genes on autosomes

Sex-limited traits traits arising from autosomal genes occurring only in individuals of one sex or the other

other. This failure of a chromosome to separate is called **nondisjunction,** and it can occur at either the first or the second meiotic division (Figure 20–27[a]). Nondisjunction may result in the zygote having three, instead of two, chromosomes of a particular type. Having three of one type of chromosome is called **trisomy,** and it can lead to serious problems of growth and development.

Down's Syndrome Nondisjunction of most autosomal chromosomes is, in fact, lethal during early development. Almost half the chromosomal abnormalities associated with spontaneous abortions are autosomal trisomies. The few types that are not lethal early in development have profound phenotypic effects on the individuals who carry and express them. If an individual has three copies of chromosome 21 (called *trisomy-21*), there may be alterations in general body morphology, changes in behavior, and reduced intellectual capacity. This affliction is called Down's syndrome (Figure 20–28). The incidence of Down's syndrome is approximately 1 in 600 live births, a fairly high rate for a congenital anomaly involving trisomic distribution of chromosomes. Statistically, the incidence goes up as the age of the potential mother goes up. Younger women (under 30) have a low probability of giving birth to a child with Down's syndrome (1 in 2000 or so), but women over 40 have a ten times greater chance for having a Down's baby.

Other Trisomies Trisomy can occur with respect to any chromosome. Very few are actually observed, suggesting that the great majority of them are lethal early in development and are spontaneously aborted. Other autosomal trisomies that are observed at birth include *trisomy-13* (Patau's syndrome) and *trisomy-18* (Edward's syndrome). Survival of individuals with these combinations of chromosomes is limited to a few weeks or months. However, individuals with Down's syndrome may live for many years and mature. They are usually gentle and caring individuals, who as a group show widely different physical and mental capabilities, suggesting that the effects of trisomy-21 are

> **Nondisjunction** the failure of chromosomes to segregate into separate daughter cells during meiosis
>
> **Trisomy** the presence of three chromosomes of one type in cells

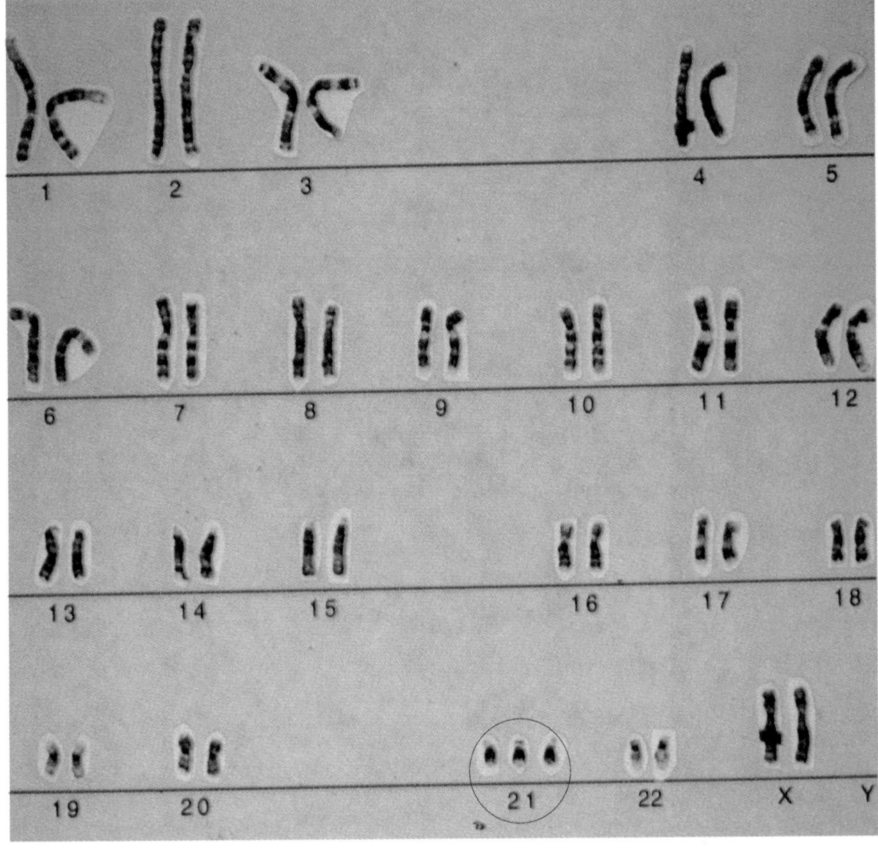

Figure 20–28 Trisomy 21—Down's Syndrome
In (a), the karyotype of trisomy 21 (circled) is shown. The human outcomes of this chromosome imbalance due to nondisjunction are mental retardation, shortened life span, and morphological changes in body shape (b).

influenced or altered by other genes. We tend to be harsh judges of the inadequacies of others, but what we need is not pity nor disgust but understanding and empathy. The random occurrence of nondisjunction and the addition of just one extra dose of genes from one extra chromosome of the 46 chromosomes that we normally bear in each of the cells of our body are enough to upset the genetic balance required for normal development. There, but for good fortune, might be any of us.

Nondisjunction and Sex Chromosomes

Klinefelter's Syndrome The nondisjunction of sex chromosomes provides a somewhat less lethal picture of the effects of genes and chromosome dosage. Extra sex chromosomes have variable effects, depending on the ratios in which they occur. In an XXY male individual, one of the parent's gametes underwent nondisjunction of the X chromosome. The XXY genotype has 47 chromosomes and leads to what is called **Klinefelter's syndrome** (Figure 20–29). These individuals are outwardly fairly normal males, but their testes fail to mature and often lack sperm-producing capacity. Mild mental deficiency is observed in some, but most have normal capacity.

Turner's Syndrome **Turner's syndrome** is the result of an XO female genotype. The absence of a Y chromosome makes this individual a phenotypic female. In this case, an egg received no X chromosome during meiosis, so an X-chromosome-bearing sperm from the father provided it. Instead of having an extra chromosome, these individuals are missing one and have only 45 (Figure 20–30). External genitalia of individuals with Turner's syndrome do not develop to maturity, gonads are abnormal (though eggs are present), and mental retardation is common. The most obvious physical features of these females is their short stature, bordering on dwarfism, and a webbed neck.

Neither Klinefelter's nor Turner's syndrome is lethal, unlike the situation for most autosomal abnormalities.

Klinefelter's syndrome a condition arising in phenotypic males from the presence of two X chromosomes and one Y chromosome

Turner's syndrome a condition arising in phenotypic females from the presence of only one X chromosome

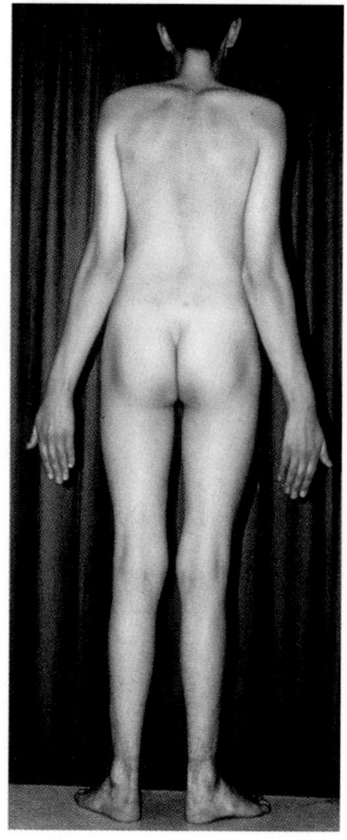

(a)

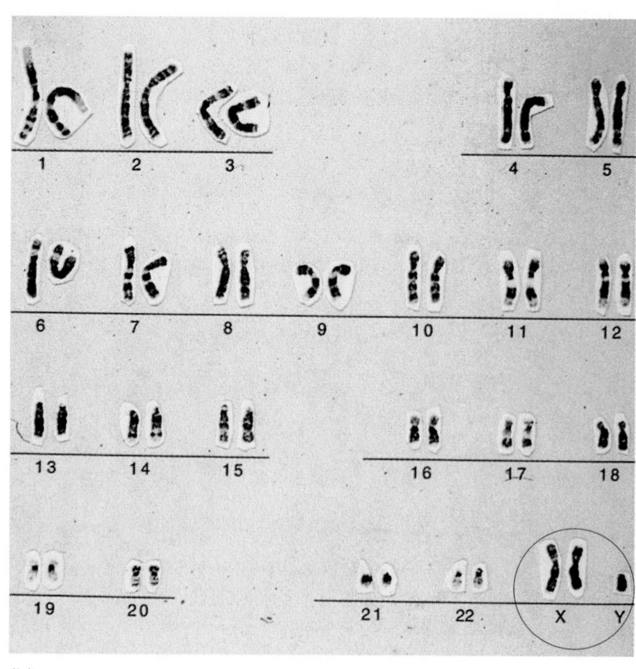

(b)

Figure 20–29 Klinefelter's Syndrome
In (a), an individual with Klinefelter's syndrome is shown. In the karyotype shown in (b), the double X and Y chromosomes are circled.

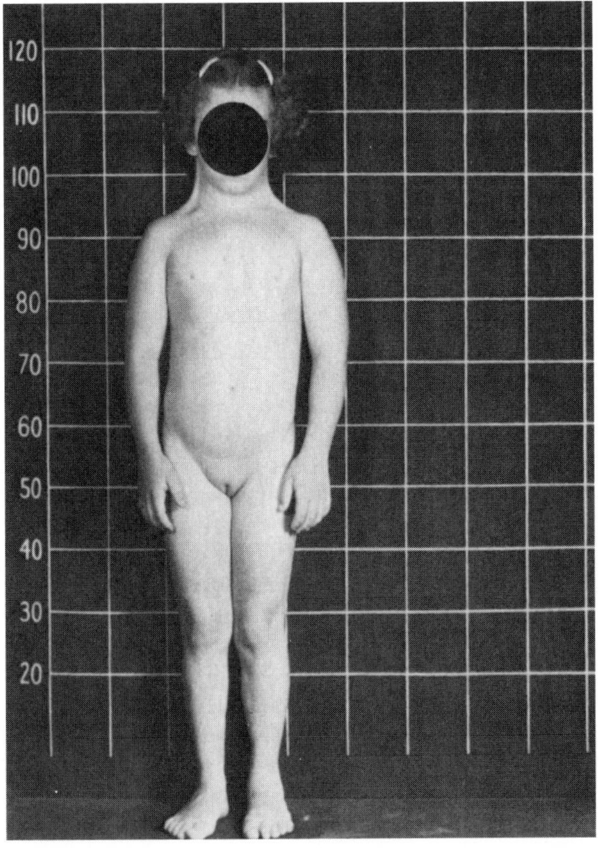

(a)

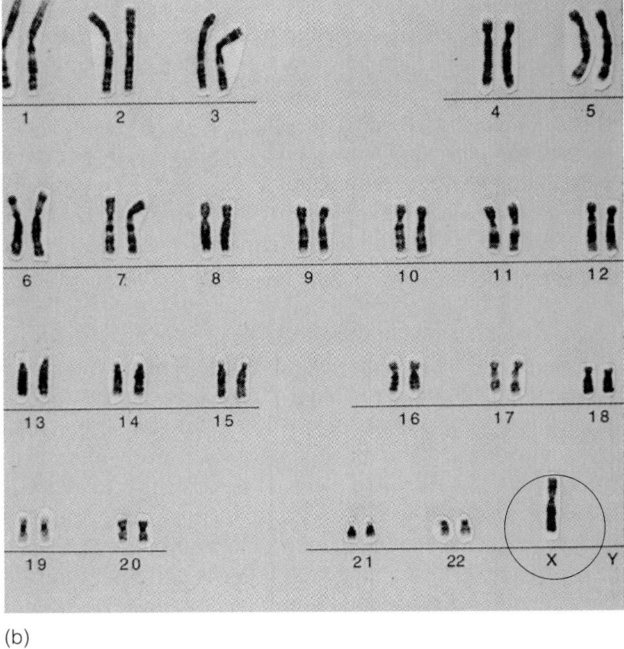

(b)

Figure 20–30 Turner's Syndrome
In (a), an individual with Turner's syndrome is shown (scale is in feet). In the karyotype shown in (b), there is only one X chromosome (circled) and no Y chromosome.

However, the YO genotype is lethal. No individual has ever been described in which there was no X chromosome. This strongly suggests that the genes carried on the X chromosome must be present in at least one dose for survival of an individual. There are many cases in which the dosage of X and Y increases without any obvious effect. XXX females display no outward signs of abnormalities, though breasts may not develop normally, external genitalia may be immature, and subnormal mentality is more likely than in the general population. XYY males, the so-called supermale, have been suspected of being highly aggressive and even to be more likely to participate in criminal behavior. The criminal potential of XYY males has not proven to be the case. Individuals with the XYY genotype are generally taller than average and may be relatively infertile.

More skewed ratios of sex chromosomes have been observed for both males and females. XXXY, XXXXY, and XXYY males tend to have Klinefelter-like symptoms, with more profoundly abnormal physical and mental defects. XXXX females show more exaggerated traits of the same type observed in XXX females. It is not easy to determine the exact number of chromosomes in polysomic individuals. The way in which the number of X chromosomes is determined in many cases of the presence of multiple X chromosomes is through the identification of the Barr body (Figure 20–31). This dark-staining region of the nucleus of normal 46-chromosome females is actually a single, highly condensed X chromosome. X chromosomes tend to look much like some of the smaller autosomes, particularly chromosomes numbered 6–8, but because an X chromosome is potentially identifiable as a Barr body that shows up on microscopic examination, they can often be unambiguously identified. The general rule is that the number of Barr bodies in a cell is equal to one less than the total number of X chromosomes. This is because only one of the X chromosomes found in normal females is thought to be active, and the others, however many that may be , are inactive. An active chromosome is one whose genes are being expressed and whose products are used in the structure and function of cells.

CHROMOSOME BREAKS, REARRANGEMENTS, AND TRANSLOCATIONS

There are other ways in which the distribution of genetic material can be altered. Genetic alteration occurs through the breaking off and loss of a piece of one chromosome, or in the rearrangement of the orientation of a piece of a chromosome (through breakage and reattachment) to the same chromosome, or by attaching the broken-off piece to a nonhomologous chromosome (Figure 20–32). The first condition is called a deletion. The second is called an *inversion* because the piece of chromosome involved is flipped over, thus, changing the arrangement and order of genes on the chromosome map. The third type of change, involving attachment of a broken piece of chromosome to a different chromosome, is called a **translocation** (Figure 20–32).

Deletion and Inversion

The deletion of genes can be, and usually is, lethal. As in nondisjunction of autosomal chromosomes, the loss of large pieces of genetic information is incompatible with growth and development. Inversions are more complicated, however. In this case, there is no loss in total genetic information, but there is a change in the arrangement of that information in the chromosome. Sometimes the position of genes relative to one another is extremely important. Imagine changing the order of towns along a highway, so that they are no longer in the order ABCD, but rather DCBA. If it is important to stop at A before B, or C before D, then there may be serious trouble in delivering goods and services. This is often true for the order of genes as well, because products and activities of genes may be

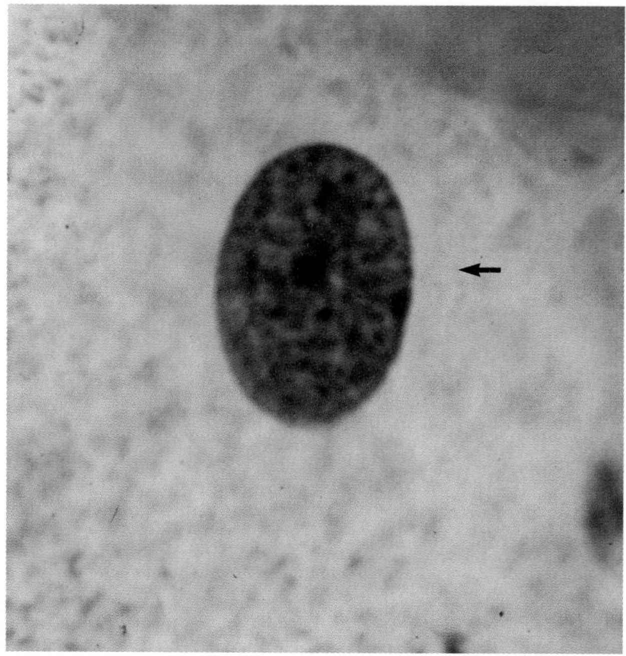

Figure 20–31 The Barr Body
The nucleus from a female cell shows one Barr body (arrow), which is a condensed and inactive X chromosome.

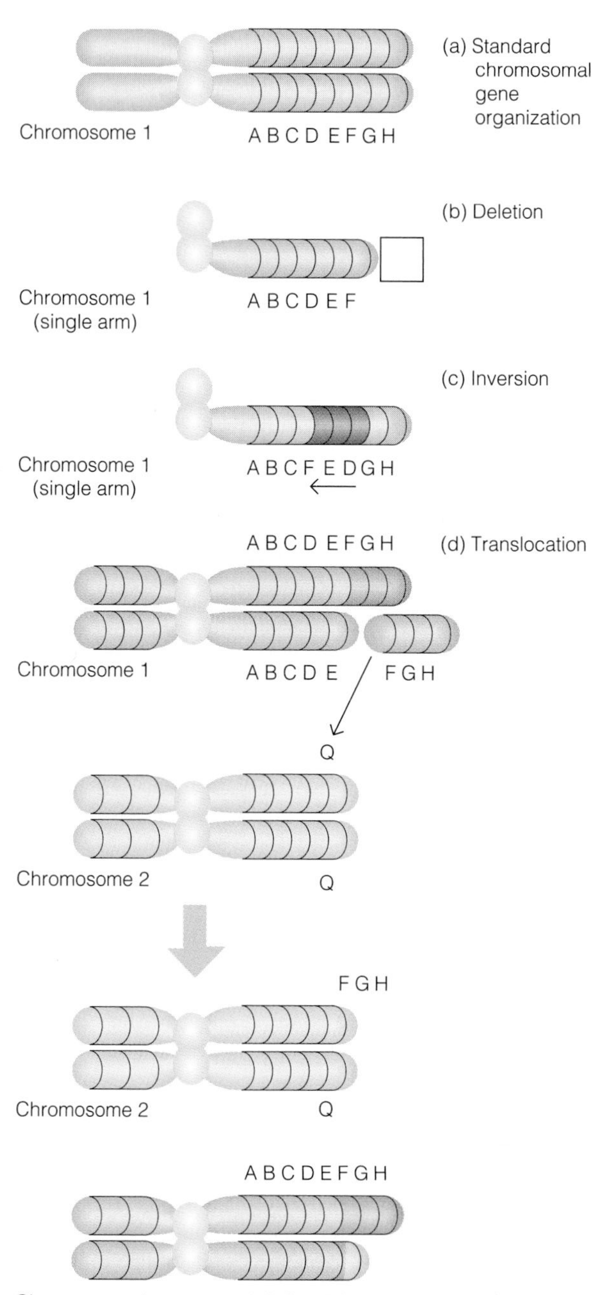

Figure 20–32 Large-Scale Changes in Chromosomes
In (a), the normal or original organization of genes is shown for chromosome 1 (gene region A through H). In (b), the gene region G and H are deleted or lost from the chromosome. Genetic information is lost. In (c), the gene region DEF is inverted so that the order of genes changes to FED. Genetic information is not lost but may be altered by position effects. In (d), gene regions from chromosome 1 (FGH) and chromosome 2 (Q) are reciprocally or evenly exchanged. No information is lost, but position effects may alter gene expression. Translocation can also occur without reciprocity.

ordered in such a way that they affect the phenotype of an individual. The sequence of gene expression may be related to the sequence of genes and is of fundamental importance in organismal growth and development.

Translocation

Likewise, translocations do not change the total number of chromosomes or the total number of genes, but they do alter the amount of DNA in a particular chromosome and have what are called **position effects.** In some cases, if the translocation of a piece of one chromosome is balanced by the reciprocal translocation of another, everything works out normally. However, in many cases, this is not what happens. Another form of Down's syndrome involves the reciprocal translocation of a large piece of chromosome 21 to chromosome 14 during meiosis. This translocation results in the possibility of passing on extra genes from chromosome 21, the very chromosome that in trisomic individuals leads to Down's syndrome. The problem in this case is that carriers of this translocation appear normal, but have a much higher chance of passing the trait on to their offspring than can be accounted for in the expected incidence of nondisjunction. Translocation of a piece of a chromosome is a permanent change in the parental genotype, not the result of an altered fate in the separation of chromosomes into a single gamete of an otherwise genetically normal parent. This means that the probability of more than one Down's child produced by the same parents may occur irrespective of the age of the mother. Establishing the karyotype of the mother and father should be carried out in conjunction with genetic counseling, which may be vital to parents considering whether to have children or not.

Translocation the movement of a piece of one chromosome onto another chromosome

Position effects changes in the expression of genes based on their position relative to other genes in the chromosome

Summary

The sequence of nucleotides in DNA encodes genetic information. The basic unit of genetic information is the gene. Genes are located in sequences along chromosomes, and all the genes of all the chromosomes found in the nucleus of cells constitute a genome and an individual's genotype. Chromosomes are macromolecular complexes of DNA, histones, and nonhistone proteins. In humans, there are two sets of chromosomes (22 homologous pairs and the sex chromosomes) in each cell, one member of which is derived from the mother, the other from the father. Genes code for products (RNA and proteins) that have effects on the expression of traits, which are manifested as the phenotype of the individual. Different forms of the same gene are called alleles. Alleles of a given gene may be completely dominant or recessive, others incompletely dominant, and still others may be expressed in a state of codominance.

Sex in humans is determined predominantly by the presence of an XX or XY pair of chromosomes. Sex-influenced and sex-limited traits are carried on autosomes, but differentially expressed depending on the sex of the individual. Many of the characteristics that are phenotypically expressed in humans depend on the action of many different genes. These polygenic traits include growth (height), pigment production and distribution (the color of the eyes, skin, and hair), and intelligence. Not to be forgotten, however, is the important influence of the environment on the expression of genetic potential.

Mutations in human DNA occur at a low but fairly constant rate. Mutations are random with respect to which genes may be affected. The agents of mutations include X rays, UV light, and a long list of chemical substances known as mutagens. Although mutations may occur in a gene, that gene can still be expressed and translated into a protein. However, the protein may not function as well as the normal nonmutant product. The products of genes in an organism determine the overall performance (and potential survival) of that individual in its environment. Mutations generally are detrimental because their outcome is proteins whose function is compromised.

The Punnett square, pedigrees, and chromosomal diagrams provide ways to visualize the organization and distribution of genes and the outcome of genetic crosses. With these means, we can assess the probability of the occurrence of normal and abnormal traits. Meiosis separates and distributes homologous chromosomes into gametes. Fertilization of an egg by a sperm brings about reestablishment of a full set of genes in a new individual and, thus, the inheritance of traits into the next generation.

Questions for Critical Inquiry

1. What are the units of inheritance? Of what class of molecules are they composed? Do all other organisms use the same encoding structures as humans?
2. What types of characteristics result from polygenic inheritance? How might such traits be affected by environmental factors?
3. Describe the phenomenon of nondisjunction and provide examples of how it influences growth and development. Use examples involving the nondisjunction of autosomes and of sex chromosomes. What are the individual and social consequences of the failure of chromosomes to properly separate?
4. Does substitution of a single nucleotide in a gene sequence necessarily alter the function of that gene? How might these kinds of changes relate to allelic differences in a gene?

Questions of Facts and Figures

5. Into what structures are genes organized within the nucleus of a cell? What are the molecular components of this structure and what is the combination called?
6. What are alleles? Are there limitations on their number?
7. What are sex-linked genes? What traits are associated with them and under what circumstances do they appear?
8. Differentiate between a dominant and recessive trait. Using Mendel's peas as a model, how would you test for whether an individual was homozygous or heterozygous for a dominant trait?
9. What is codominance? How is this shown in the expression of the antigens of the ABO blood group?
10. What is a gene pool? Why is it an important aspect of the genetic composition of a species at the population level?
11. With respect to chromosomal structure, what is (a) A translocation? (b) An inversion? (c) A deletion?
12. What is a point mutation? How do deletions and insertions affect the reading frame of a gene?

References and Further Readings

Berg, P., and Singer, M. (1992). *Dealing with Genes.* Mill Valley, CA: University Science Books.

Cummings, M. R. (1996). *Biology: Science and Life.* St. Paul, MN: West Publishing.

Jones, S., (1993). *The Language of Genes.* New York: Anchor Books.

Lewontin, R. C. (1993). *Biology as Ideology.* New York: Harper Perennial.

Roberts, J. A. F. (1973). *An Introduction to Medical Genetics.* Oxford: Oxford Press.

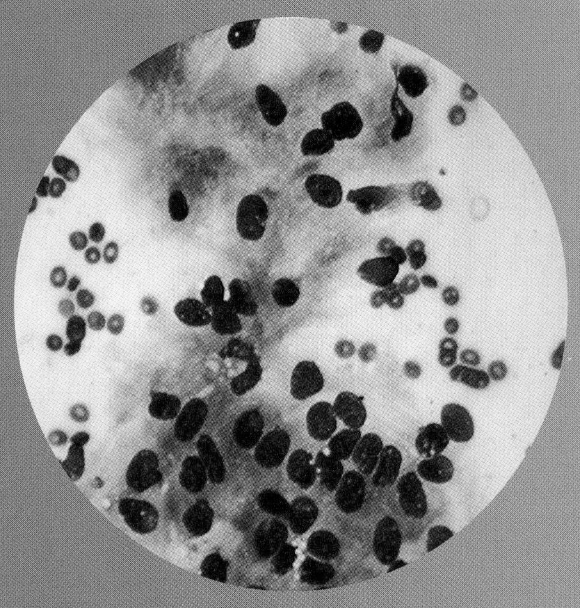

Cancer

INTRODUCTION

Cancer is one of the most frightening words to be heard when a physician presents it as part of a clinical diagnosis. The word conjures up images of pain and wasting away of the human body. Although most forms of cancer are successfully treated if they are discovered early enough, those types that are not successfully treated have devastating effects not only on the human body but on the human spirit. What is this disease we call cancer? What are its causes? How are cancers treated? With these questions in mind, this chapter explores the nature of abnormal cell growth that is the underlying basis for the formation of tumors and the development of cancer.

TUMORS AND CANCER

Tumors occur in all tissue types in the human body. Tumor cells arise initially from normal cells that are transformed by exposure to a variety of external agents, known as **carcinogens.** The induction by these agents is the first step of many toward a state of uncontrolled cellular growth. It is the abnormal proliferation of cells that gives rise to tumors. Tumors grow within their tissues of origin or may spread to other regions of the body. Tumors that remain at the site of origin and do not invade surrounding tissues are called **benign tumors.** Benign tumors arise with a fairly high frequency in humans and other animals, but because they are localized, usually they present little risk. Growths such as warts on the skin fall into this category.

Tumors that spread to other regions of the body and grow are called **malignant tumors** and are the cause of cancer. The spread of cancerous tumor cells occurs by two means—*invasion* and metastasis (Figure 21–1). Initial growth of transformed cells forms a localized mass that is encapsulated or confined by host tissue (Figure 21–1[a] and [b]). At this point, the tumor is benign. Invasiveness is the property of a tumor cell that allows it to escape the confines of the capsule and penetrate into surrounding tissues (Figure 21–1[c]). Tumors of this type are classified as malignant. Malignant cells within tumors are also referred to as *neoplastic,* and the tumors they form are referred to as **neoplasias.** Continued invasion of tissues allows tumor cells to penetrate through the blood vessels and to be distrib-

Carcinogens chemicals or other agents that cause cancer

Benign tumors tumors whose growth is restricted to one area and which remain encapsulated

Malignant tumors tumors whose growth is unrestricted and which eventually invade surrounding tissues or spread by metastasis

Neoplasias tumor growths resulting from neoplastic cells

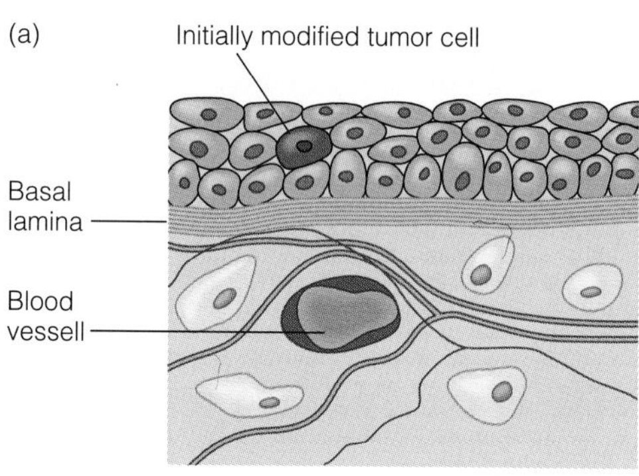

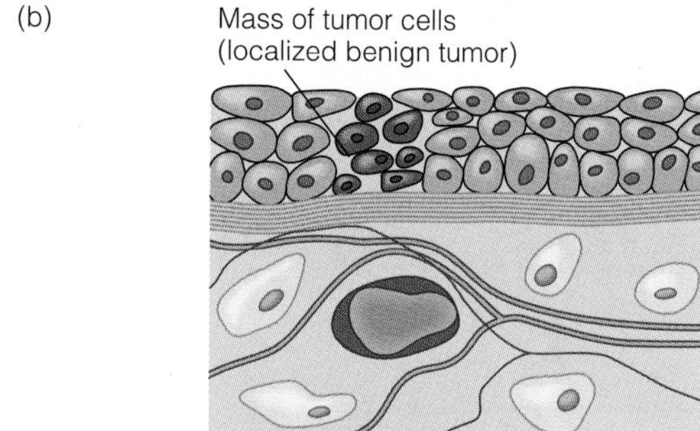

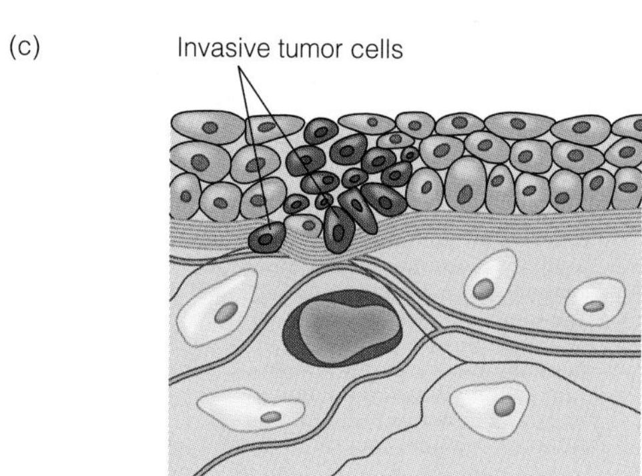

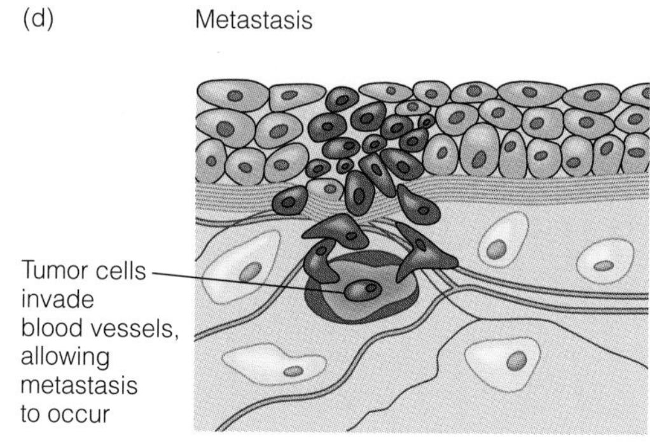

Figure 21–1 Invasiveness and Metastasis
Tumors arise from a single cell (a). The cells grow to form a mass that is encapsulated (b) and are considered benign. Cells that penetrate into other tissue compartments (c) are invasive, and those that penetrate blood vessels and disperse are metastatic (d).

uted to the rest of the body by the circulatory system. This process is called metastasis (Figure 21–1[d]).

TYPES OF CANCER

There are a variety of types of cancer, more than a hundred have been named and characterized (Figure 21–2). The major types of tumors of epithelial origin are called **epitheliomas** or carcinomas. The suffix "oma" added to the end of a named tissue or cell type refers to the fact that it is transformed and cancerous. Many of the most malignant cancers have their origins in epithelial tissues. For example, Figure 21–2 shows cancers of the human lung and colon and other epithelial types (for examples of skin cancers see Chapter 5). Most cancer of the breast, cervix, and prostate gland are carcinomas also. Sarcomas are cancers of connective tissue, that is, bone, cartilage, and may involve muscle as well. Subtypes of tumors arising in bone are called *osteomas;* those arising in cartilage are *chondromas.* Transformation of the blood-forming tissues are called hematopoietic cancers and include lymphomas and myelomas and many forms of *leukemia.* The cancers of the nervous system are *neuromas, neuroblastomas,* and *gliomas.* One of the most deadly cancers for children is called *retinoblastoma,* and, as will be discussed later in this chapter, its expression is linked to the abnormal function of specific genes that suppress cancer development.

CELLS AND TRANSFORMATION

The structural and functional roles of tissues depend in part on whether the cells in those tissues proliferate and are renewable, as in epithelia, or do not proliferate and are nonrenewable, such as neurons. The cells of different tissue types of the human body undergo cell division at different rates, fast or slow, or are arrested in their ability to divide. The epithelial cells of the small intestine divide every day or two, epidermal cells divide every few days, and liver cells divide only rarely. Mature neurons in the brain do not divide at all. The regulation of the proliferation of all types of cells depends on factors that control the cell cycle. As presented in more detail in Chapter 3, the cell cycle is divided into four parts—G_1, S, G_2, and M. The cell cycle occurs in conjunction with a series of biochemical events that drive a cell to initiate the synthesis of chromosomal DNA and to prepare itself to divide. Each step in the cell cycle is required for progression to the next step.

There are a number of **checkpoints** in the cycle at which internal biochemical mechanisms assess the state of the cell before allowing it to go on in the cycle (Figure 21–3). The checkpoints occur in late G_1, in late G_2, and in M. The G_1 checkpoint ensures that the cell is prepared to synthesize DNA. The G_2 checkpoint ensures that the DNA is intact and undamaged and that the cell is an appropriate size. It would not do to have the cell divide too soon and produce cells too small to survive. The M-phase checkpoint ensures that all the chromosomes in metaphase of mitosis are lined up on the metaphase plate and ready to separate. Problems in regulating passage through checkpoints are associated with cancer.

> **Epitheliomas** malignant tumors of epithelial origin; also known as carcinomas
>
> **Checkpoints** times during the cell cycle that are important in determining normal progression in the cycle and initiating mechanisms to repair damage to DNA

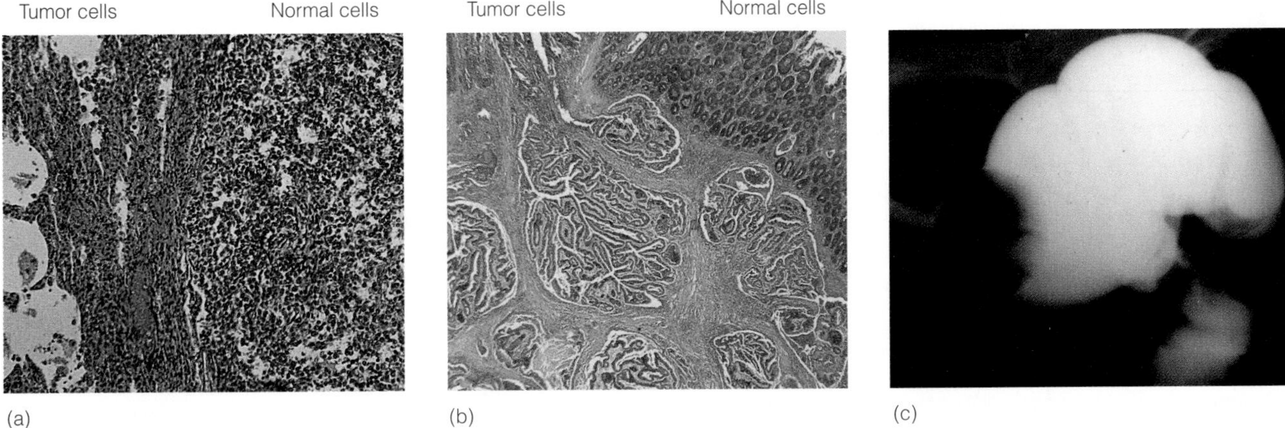

Figure 21–2 Carcinomas—Cancers of Epithelia
In (a), a lung cancer is shown with tumor and normal cells. In (b), a cancer of the colon is shown along with an X ray of a colon tumor (c).

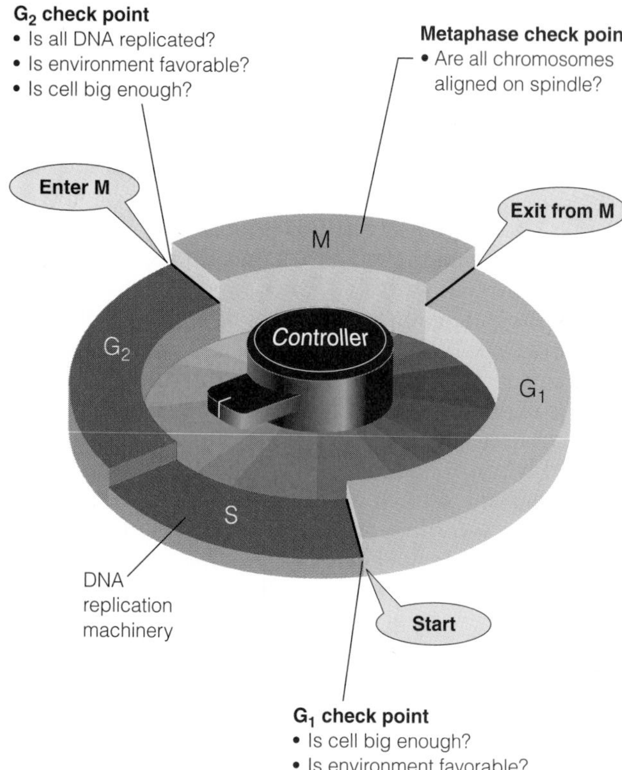

Figure 21–3 Checkpoints
Checkpoints in the cell cycle are important in regulating passage of cells from one cell division (mitosis, M) to another. The cell asks questions of itself in several phases before progressing to the G_1, G_2, and M checkpoints.

During the embryonic stage of human development, the rate of cell division is very high for all cells because they increase the mass of the new individual. The general rate of cell division drops during fetal growth and eventually slows further, or stops, for many cell types during childhood. In adulthood, the patterns of proliferation, or lack of proliferation, depend on a complex set of factors tied to the way cells function in the body, the number of cells needed to maintain tissue and organ structure, and the conditions under which the cell cycle is regulated to control cell proliferation or cell differentiation. Sometimes the control of a cell's normal pattern of cell division and differentiation is disrupted and it begins to divide and grow inappropriately with respect to its role in the structure and function of a tissue. These are transformed cells and they pass the characteristics of rapid proliferation and growth down to their descendants. The rapid growth of these cells results in a cellular mass that is often great enough to displace normal tissue and interfere with normal function. The invasiveness and metastasis of cells from such tumors leads to cancer. What does it mean for cells to undergo transformation? What kind of properties do transformed cells have? These are very important questions to answer for long-term human health, and cancer researchers have been on the trail toward their discovery for decades.

PROPERTIES OF TUMOR CELLS ASSOCIATED WITH MALIGNANCY

Many possible factors are involved in changes in cells that bring about transformation. Among the possibilities are changes in growth factors and hormones, both of which may be altered in their functions to influence cell growth and division (Figure 21–4). This pathway of influence may also be affected by cell surface receptors for hormones and growth factors. The cell may get false or misleading signals from receptors. There are also intracellular proteins that affect gene expression by binding to DNA in the nucleus. Such proteins alter the expression of otherwise normal genes resulting in uncontrolled cell growth. Many of the products of abnormal gene expression affect the cell cycle. It is the alteration of what are normally well-regulated genes controlling cell growth that result in normal cells going awry. Abnormal genes or genes expressed abnormally are called oncogenes or cancer-causing genes (Figure 21–4).

Cancer is generally a disease of older individuals because cancer develops in a *multistep process* which will be discussed in detail later. The first step is exposure to an agent capable of inducing transformation (a carcinogen or **initiator**). The second step is penetration of a carcinogen into the cell and alteration of the sequence of nucleotides in one of the two DNA strands, uncorrected damage to DNA, or breakage of the DNA. The third step, and possibly many subsequent steps, often involves a second mutational event and/or the enhancement of the changes in the presence of other factors called **promoters.** Cancer is not what has been termed a single-hit phenomenon, which means that more than one altering event may have to take place for a cell to be stably transformed. The likelihood of two hits affecting the same or a related region of DNA in a single cell is very low, which is why it may take a lifetime to develop some cancers. For example, skin cancer arises from long-term exposure to sunlight over many years. The changes that take place in epidermal cells and/or melanocytes during the tanning of youth may come to deadly fruition decades later. Lung cancers may arise 20 to 30 years after an

Initiator a compound or agent that alters or damages DNA; a carcinogen

Promoters compounds or agents that allow the changes induced by events of initiation to proceed

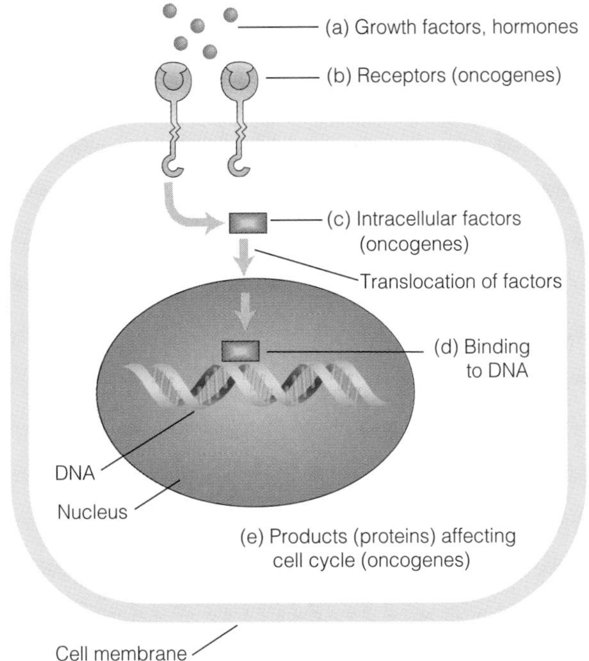

Figure 21–4 Factors in Cancer Cell Changes
Many factors are involved in cell changes to a cancerous cell (a process of transformation). Changes in growth factors (a) or their receptors (b) are important. Intracellular factors (c) and DNA binding molecules (d) affect the way the cell behaves. The products (e) of genes influenced by (d) may also affect oncogenesis. Oncogenes are normal cellular genes or gene products that have gone awry.

individual takes up smoking. Even fairly rapidly arising leukemias take four to five years to develop (Figure 21–5). The histological changes in a tissue in which cancer cells are growing are distinctive (Figure 21–6). Tumors take time to show up; the number of cells needed to detect a tumor by X ray is 10 million to 100 million (Figure 21–6[e]).

Rate of Cell Division

Once the cells have been stably transformed, they acquire a number of special characteristics. One of the key characteristics of malignant cancer cells is the rapid and uncontrolled rate at which they divide. When normal cells are cultured *in vitro,* the cells interact with one another and cease dividing as they reach confluence (Figure 21–7[b]). This is called *contact inhibition* of cell growth. The state of arrested division is called *quiescence* and reflects what happens normally to cells in the body—they are in contact with their neighbors and either do not divide or slow the rate of their proliferation to match the tissue's needs. Malignant cancer cells grown under the identical culture conditions do not stop dividing as they contact their neighbors and continue to pile up on top of one another, as if there was no contact between the cells at all (Figure 21–7[a]). This behavior is observed in the body as tumor cells grow, invade surrounding tissues, and undergo metastasis.

Cell Adhesion and Stickiness

The property of adhesion of cells is important in establishing relationships with neighboring cells and molecules of the extracellular matrix. In culture, normal cells attach to the glass, or plastic, substratum of a culture dish and stick strongly to it. They will not divide if they do not attach and flatten out on the substratum. Malignant cells under the same conditions attach poorly and do not stick well to the substratum, yet proliferate rapidly. This is referred to as the loss of dependence on anchorage for cell growth (Figure 21–7[a]). This alteration in adhesion correlates very well with the capacity of malignant cells to form tumors.

Cell Shape

Cell shape is an important characteristic in assessing transformation of malignant cells. When comparing cells of the same origin, one normal and one malignantly transformed, the shape of the cells is observed to be very different (Figure 21–7). Normal cells tend to be organized side to side, as well as stretched out over the surface of the substratum (Figure 21–7[b]). Malignant cells are often rounded-up, an indication that they are not attaching well and are disorganized with respect to their neighbors (Figure 21–7[a]). It is clear that cell shape is related to cell stickiness and the ability to adhere to other cells. The abnormal shape of cells is also observed in cancerous tumors in the body, where neoplastic cells are often large, distended, and poorly organized with respect to the normal surrounding tissues.

Cell Metabolism

The metabolic rate of malignant cells is higher than their normal counterparts. One reason for this metabolic increase is that transformed cells are much more efficient in the transport of nutrients than normal cells. This is particularly evident with respect to the transport of glucose. Through the glycolytic pathway and the Krebs cycle (see Chapter 2), glucose supplies tumor cells with sufficient energy (ATP) needed to maintain rapid growth. Because tumor cells are more efficient in transport of nutrients, they require less available nutrients than normal cells of similar type. In addition, tumor cells may grow independently from growth factors essential to normal cell growth. These attributes, enhanced transport efficiency and independence from growth factors, allow them to grow under conditions that inhibit or restrict the growth of normal cells.

Cell DNA and Chromosomes

The DNA of malignant tumor cells is altered by transformation. This represents mutations in the DNA, which are inherited by all subsequent generations of transformed cells. A cancer arises from a single transformed cell. Thus,

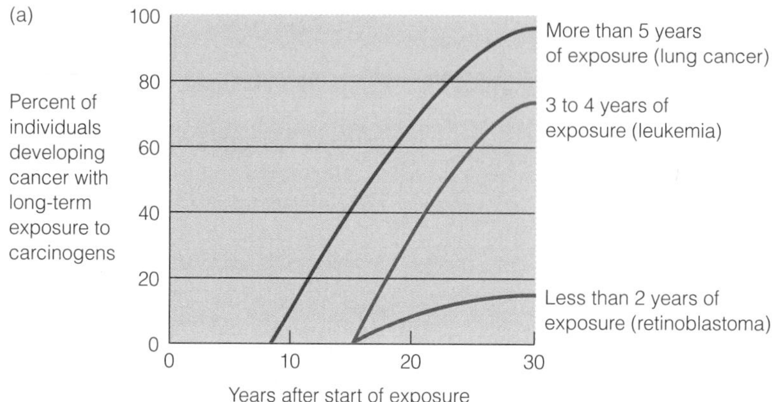

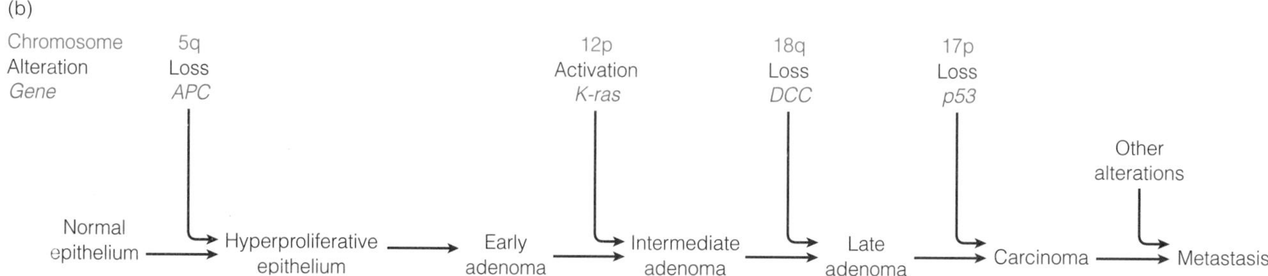

Figure 21–5 Cancer Takes Time to Develop
In general, exposure to carcinogens (initiators) does not immediately lead to cancer. It may take years for cancer to develop even with continual exposure (a). Cancer develops as a multistep process, as indicated in (b), where changes in chromosome 5 lead to changes in genes on chromosome 12, 18, and 17, that in turn lead to development of a metastatic carcinoma.

the tumor is a clone of cancer cells. These changes are progressive, with early generations of tumor cells less abnormal than later generations. As stated previously, these changes may take years to develop. In many cases, the number of chromosomes in human tumor cells is usually slightly greater than the 46 normally found in the human genome. This altered number of chromosomes is called **aneuploidy** and occurs in conjunction with alterations in the morphology of chromosomes. Translocations and fragmentation of chromosomes are hallmarks of malignantly transformed cells.

Altered Cell Cycles

As discussed earlier, the phase of the cycle in which the cell physically separates chromosomes and divides is mitosis. Mitosis leads into a period of interphase in which a cell prepares to replicate its DNA and, then, to divide again, or withdraw from the cell cycle and stop dividing (Figure 21–3). During the G_1 stage of the cell cycle, a cell makes a major decision to stop dividing or to go through another cycle. Figure 21–4 shows there are a number of types of molecules that regulate this process, including protein growth factors (a), cell surface receptors (b), and intracellular proteins that interact with and regulate the expression of DNA (c). Alteration in the normal action of any of these factors may lead to uncontrolled cell growth. Some factors in the checkpoints of the cell cycle may be involved in the loss of a cell's ability to stop cell division. The genes for these proteins, when altered by mutation through the action of carcinogens, are likely to lead to the development of cancer. This process is known as **oncogenesis,** and genes for these types of growth-controlling proteins, as noted earlier, are called cellular **oncogenes.**

Tumor Suppressor Genes

In addition to oncogenes, which promote the transformation process, there are also a number of genes whose prod-

Aneuploidy abnormalities in the number of chromosomes found in cells

Oncogenesis the development and progression of cancer

Oncogenes the cellular genes whose abnormal expression is linked to the development of cancer

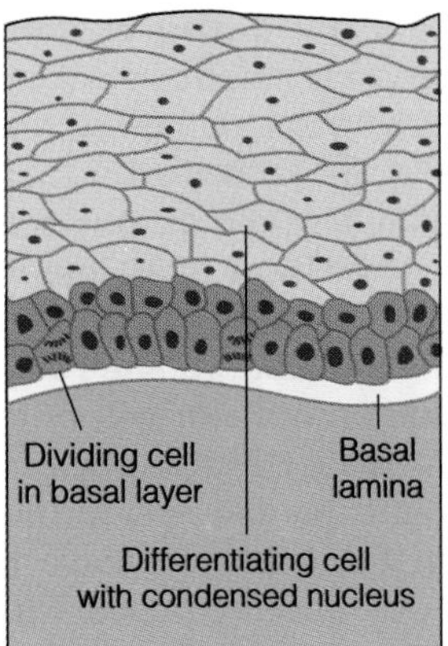

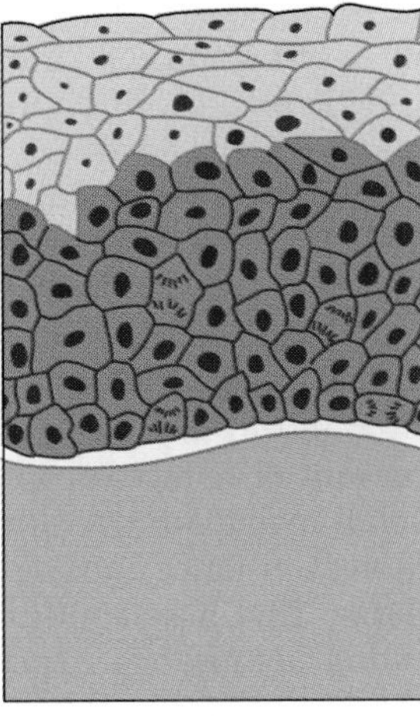

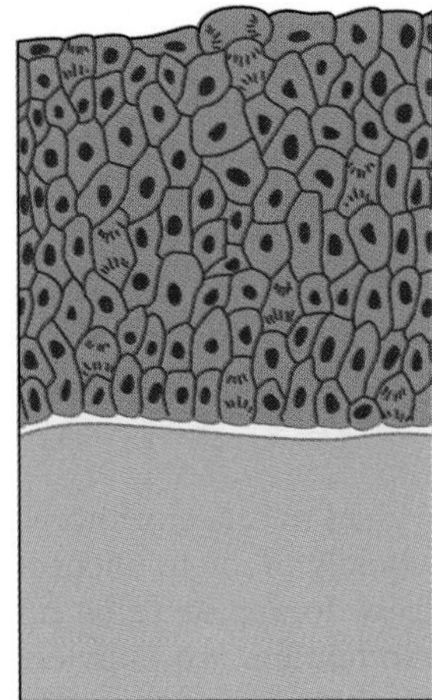

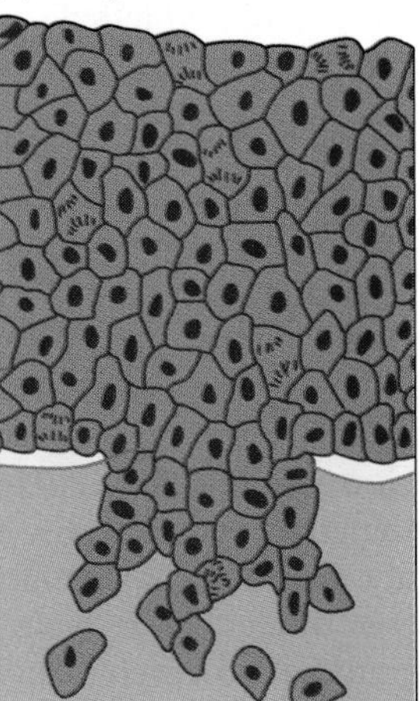

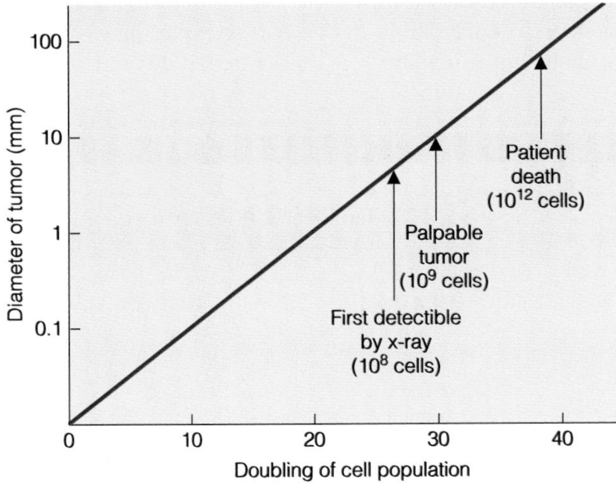

Figure 21–6 Histological Changes in a Cancer over Time The changes in an epithelium to a malignant carcinoma are shown in (a) through (d). In (e), the first appearance of a tumor on X ray involves 10 million to 100 million cells.

ucts interfere with, or suppress, transformation. These proteins are known as **anti-oncogenes** and have only recently been discovered. The role of anti-oncogenes is in correcting the changes or damage in DNA that are introduced by mutagenesis and carcinogenesis. If the damage cannot be corrected, anti-oncogene products bring about a programmed cell death, or **apoptosis.** It was mentioned previously that the development of retinoblastoma in young children is a genetic problem. In fact, the problem for these children is that they have only a single functional copy of a tumor-suppressing gene, which is altered after a single-hit mutational event. This mutation leads to the inability of the target cells to suppress cell transformation. The affected individuals do not start out with cancer, but because they cannot effectively suppress it after losing the anti-oncogene, they have a very high likelihood of

Anti-oncogenes the cellular genes whose expression suppresses the initiation of cancer

Apoptosis programmed cell death

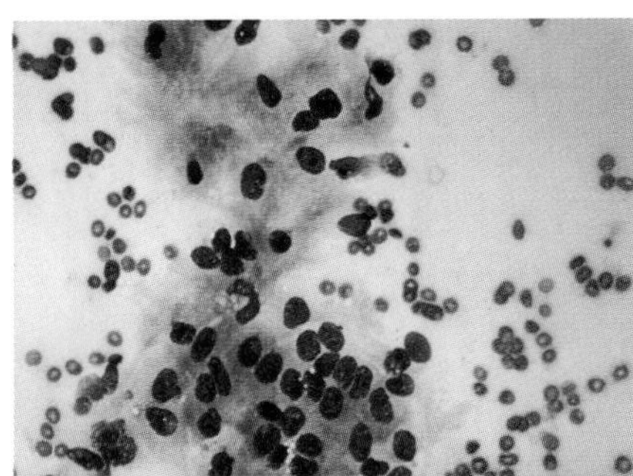

(a)

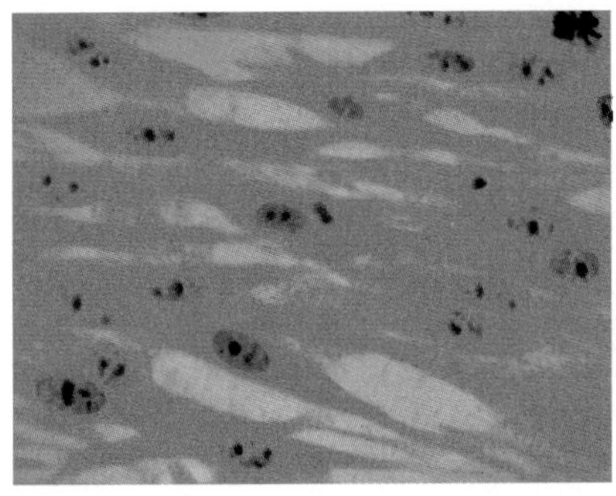

(b)

Figure 21–7 Cancer Cell Shape and Growth
In general, transformed or cancerous cells are more round (a) than their normal counterparts (b). The growth of cancer cells is not inhibited by contact with their neighbors as is the case with normal cells. The adhesion of cells to each other and to the substratum helps determine their shape.

oncogenesis. This particular cancer is very aggressive, metastatic, and lethal.

ASSESSING RISKS

There are a number of risk factors associated with cancer. Among them are genetic predisposition, as described above, and a variety of factors in the environment that induce mutations in DNA.

Genetic Predisposition

Genetics can predispose an individual to cancer. However, not all genetic backgrounds are equally prone to oncogenesis, though some types of cancers do appear to be programmed to develop and to run in families. An individual from a family with a background of cancer, such as colon cancer or breast cancer, may have a significantly higher risk of developing that cancer than the population at large. It should be kept in mind that this does not mean that a person with such a genetic predisposition will develop that cancer, it means only that the chances are greater.

Environmental Factors

There are agents in the environment that have the capacity to transform cells and cause cancer. With respect to the genetic predisposition for cancer, many of these environmental agents determine whether such cancers will occur. Sunlight, cigarette smoke, air and water pollution, and common constituents of the diet are known to initiate and/or promote cancer. It should be pointed out that agents that cause or initiate cancer and those that promote cancer may be different. There is a synergism between the two, and in many cases, the absence of a promoter may prevent or reduce the chance that an initiator will bring about malignant transformation. To be effective, the promoter substance must be consistently present after the initiator. This has ramifications in the habits we have in eating, because a large percentage of cancers correlate with the diet (see the following section of this chapter on Nutrition and Cancer).

The relationship of cancer to environmental factors is also observed in the worldwide distribution of cancers and their prevalence (Table 21–1). For example, many cancers occur in high incidence in the United States but occur in

Table 21–1 Worldwide Incidence of Common Cancers

	HIGH-INCIDENCE POPULATION		LOW-INCIDENCE POPULATION	
SITE OF ORIGIN OF CANCER	LOCATION	INCIDENCE*	LOCATION	INCIDENCE*
Lung	USA (New Orleans, blacks)	110	India (Madras)	5.8
Breast	Hawaii (Hawaiians)	94	Israel (non-Jews)	14.0
Prostate	USA (Atlanta, blacks)	91	China (Tianjin)	1.3
Uterine cervix	Brazil (Recife)	83	Israel (non-Jews)	3.0
Stomach	Japan (Nagasaki)	82	Kuwait (Kuwaitis)	3.7
Liver	China (Shanghai)	34	Canada (Nova Scotia)	0.7
Colon	USA (Connecticut, whites)	34	India (Madras)	1.8
Melanoma	Australia (Queensland)	31	Japan (Osaka)	0.2
Nasopharynx	Hong Kong	30	UK (southwestern)	0.3
Esophagus	France (Calvados)	30	Romania (urban Cluj)	1.1
Bladder	Switzerland (Basel)	28	India (Nagpur)	1.7
Uterus	USA (San Francisco Bay Area, whites)	26	India (Nagpur)	1.2
Ovary	New Zealand (Polynesian Islanders)	26	Kuwait (Kuwaitis)	3.3
Rectum	Israel (European and USA born)	23	Kuwait (Kuwaitis)	3.0
Larynx	Brazil (São Paulo)	18	Japan (rural Miyagi)	2.1
Pancreas	USA (Los Angeles, Koreans)	16	India (Poona)	1.5
Lip	Canada (Newfoundland)	15	Japan (Osaka)	0.1
Kidney	Canada (NWT and Yukon)	15	India (Poona)	0.7
Oral cavity	France (Bas-Rhin)	14	India (Poona)	0.4
Leukemia	Canada (Ontario)	12	India (Nagpur)	2.2
Testis	Switzerland (urban Vaud)	10	China (Tianjin)	0.6

*Incidence = number of new cases per year per 100,000 population, adjusted for a standardized population age distribution (to eliminate effects due merely to differences of population age distribution). Figures for cancers of breast, uterine cervix, uterus, and ovary are for women; other figures are for men.

Adapted from V. T. DeVita, S. Hellman, and S. A. Rosenberg (eds.), *Cancer: Principles and Practice of Oncology*, 4th ed. Philadelphia: Lippincott, 1993; based on data from C. Muir et al. *Cancer Incidence in Five Continents*, Vol. 5. Lyon: International Agency for Research on Cancer, 1987.

low incidence in other countries, and vice versa (compare Table 21–2 with Table 21–1). What environmental factors may be involved in cancer is more complex than simply pointing to known carcinogens, such as the components of cigarette smoke in lung cancer. Not all smokers develop cancer. Not all people with lung cancer necessarily have been smokers. The time over which an individual is exposed to carcinogens is directly related to his or her chance for developing cancer. The general health of individuals, the state of their immune systems, and the types of diseases or infections they have had may all correlate to cancer incidence under circumstances in which they occur together with smoking. Migration of people from one country or geographical location to another often leads within a generation or two to the acquisition of cancer rates characteristic of the host country or region. The roles of genetics, environment, culture, and lifestyle in the propensity to develop cancer are not known, but it is probably reasonable to assume that 70%–80% of cancers are avoidable.

CAUSES OF CANCER

Any factor that brings about the transformation of a normal cell into a cancer cell is considered a carcinogen. There are three main categories of carcinogens—physical, chemical, and biological.

Physical Agents of Carcinogenesis

Classic forms of carcinogens are X rays and ultraviolet light. The effects of exposure to these wavelengths of light is damage to molecules of all types. However, the most important of these damaging effects is on DNA. DNA may be chemically changed, altered in structure, or even physically destroyed. Alterations in DNA by UV radiation result in chemical changes in nucleotides in one of the nucleic acid strands (Figure 21–8[a]). This leads to functional mutations, which, in turn, may affect the growth of cells by direct or indirect effects on oncogenes. X rays and other

Table 21–2 Types of Cancer and Incidence in the United States, 1993

TYPE OF CANCER	NEW CASES PER YEAR		DEATHS PER YEAR	
Total cancers	1,170,000		528,300	
Cancers of epithelia: carcinomas	992,700	(85%)	417,175	(79%)
Oral cavity and pharynx	29,800	(3%)	7,700	(1%)
Digestive organs (total)	236,900	(20%)	120,325	(23%)
Colon and rectum	152,000	(13%)	57,000	(11%)
Pancreas	27,700	(2%)	25,000	(5%)
Stomach	24,000	(2%)	13,600	(3%)
Liver and biliary system	15,800	(1%)	12,600	(2%)
Respiratory system (total)	187,100	(16%)	154,200	(29%)
Lung	170,000	(15%)	149,000	(28%)
Breast	183,000	(16%)	46,300	(9%)
Skin (total)	(>700,000)*		9,100	(2%)
Malignant melanoma	32,000	(3%)	6,800	(1%)
Reproductive tract (total)	244,400	(21%)	59,950	(11%)
Prostate gland	165,000	(14%)	35,000	(7%)
Ovary	22,000	(2%)	13,300	(3%)
Uterine cervix	13,500	(1%)	4,400	(1%)
Uterus (endometrium)	31,000	(3%)	5,700	(1%)
Urinary organs (total)	79,500	(7%)	20,800	(4%)
Bladder	52,300	(4%)	9,900	(2%)
Cancers of the hemopoietic and immune system: leukemias and lymphomas	93,000	(8%)	50,000	(9%)
Cancers of central nervous system and eye: gliomas, retinoblastoma, etc.	18,250	(2%)	12,350	(2%)
Cancers of connective tissues, muscles, and vasculature: sarcomas	8,000	(1%)	4,150	(1%)
All other cancers + unspecified sites	57,050	(5%)	43,425	(8%)

*Nonmelanoma skin cancers are not included in total of all cancers, since almost all are cured easily and many go unrecorded.

In the world as a whole, the five most common cancers are those of the lung, stomach, breast, colon/rectum, and uterine cervix, and the total number of new cancer cases per year is just over 6 million. Note that only about half the number of people who develop cancer die of it.

(Data for USA from American Cancer Society, *Cancer Facts and Figures, 1993*.)

(a)

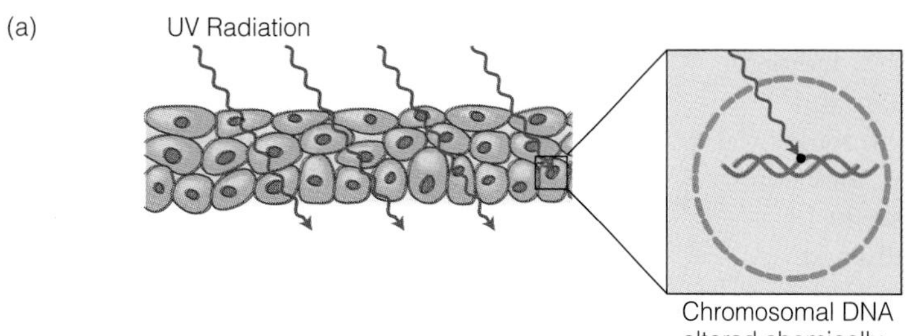

(b)

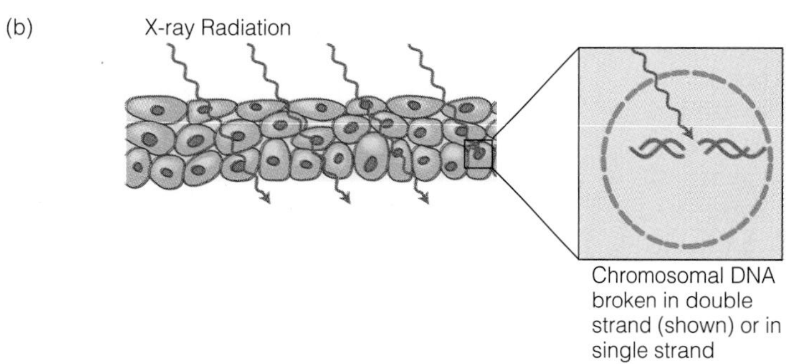

Figure 21–8 Effects of Radiation UV radiation (a) can bring about point mutations, chemical alterations of purines or pyrimidines (thymidine dimers), and failure to repair damaged DNA. X rays (b) cause breaks in DNA that are difficult or impossible to repair. Both types of damage may lead to cancer.

forms of ionizing radiation cause multiple breaks in the DNA that are difficult or impossible for the cell to repair (Figure 21–8[b]). This type of radiation is known to cause human leukemias. These cancers of hematopoietic cells were prevalent among the survivors of the atomic bomb blasts that destroyed Hiroshima and Nagasaki during World War II.

Chemical Carcinogens

Chemical carcinogenesis was first discovered in studies of animals whose skins were painted with substances being tested for the ability to induce cancer. The compounds in this class have a wide range of chemical structures (Figure 21–9). A key discovery in chemical carcinogenesis was that not all the compounds are directly involved in carcinogenesis but rather must undergo chemical changes in the cell to have effects. There are two groups of chemical carcinogens—those that act directly and those that act indirectly (Figure 21–9). *Direct-acting compounds* react directly with DNA. *Indirect-acting compounds* must be metabolized before they are capable of reacting with DNA. It is interesting to note that bacteria of the human gut often produce carcinogens as a by-product of their metabolisms. Many of these compounds are glycosides, or sugar-containing molecules, and result from the enzymes that react with perfectly harmless carbohydrates that we ingest (see also the section Nutrition and Cancer).

There is an ingenious and simple method known as the **Ames test** (Biosite 21–1) used to determine whether a compound is potentially a carcinogen or not. It involves the use of bacteria and their susceptibility to undergo carcinogen-induced changes in growth. All new compounds formulated for human consumption are assessed using the Ames test and must pass it before they can be sold.

Biological Agents of Carcinogenesis

The principal biological agents of carcinogenesis are viruses (Table 21–3). There are two types of viruses that are known to cause cancer. The RNA viruses, called *retroviruses,* and the *DNA tumor viruses.* Retroviruses make a special copy of DNA (cDNA) from their RNA and integrate it into the host cell genome. Integration of viral cDNA is a random process and involves cutting the host DNA, inserting the viral sequence(s), and then closing up, or ligating, the cut. Once integrated, these genes take over a portion of the cell's metabolic machinery to make new viruses. If the integration of viral genes should occur in, or near, cellular genes associated with regulating growth and proliferation (known as proto-oncogenes), the cellular genes may be captured and incorporated into new viruses. The altered viruses, when released, may go on to infect other cells. In this way, a retrovirus may carry an altered proto-oncogene

Ames test a test using bacteria to assess mutagenicity and carcinogenicity of new compounds

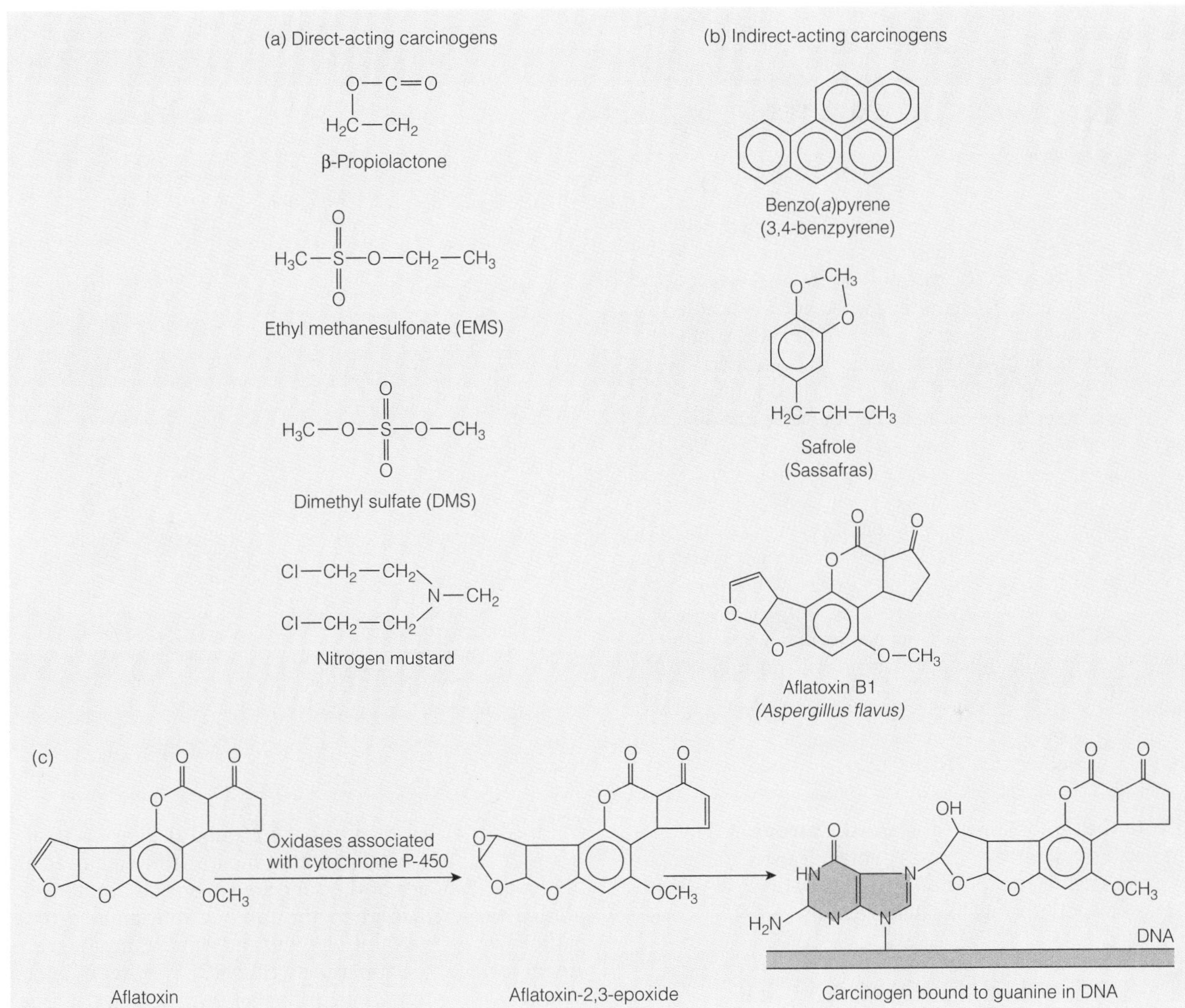

Figure 21–9 Some Chemical Carcinogens
Chemical carcinogens are either direct acting (a) or indirect acting (b). Indirect-acting compounds, such as aflatoxin (c), are altered chemically by the cells to an active form (aflatoxin 2, 3 epoxide). It is the activated compound that binds to DNA and initiates formation of cancer cells.

into a new host cell, where it is changed sufficiently to function as an oncogene and to transform the infected cell.

DNA tumor viruses have a different mechanism for transforming animal cells; they contain tumor genes that when integrated into the host genome, directly alter the host cell. The copies of DNA and cDNA of viruses are a permanent part of the host genome and are inherited in every subsequent generation of cells. Tumors in humans caused by RNA and DNA viruses are rare. However, the herpeslike DNA virus called the *Epstein-Barr virus* may play a direct role in *Burkitt's lymphoma* and in some *nasopharyngeal cancers*. The human immunodeficiency virus (HIV) may be associated indirectly with the development of cancer. An HIV infection reduces the immune functions of the human body so that the Epstein-Barr virus, or a related DNA virus, may be expressed in cells and give rise to cancer. This is the

BIO site

21–1

THE AMES TEST

The Ames test is a test to assess the mutagenicity of chemical compounds. Most compounds that are mutagens are also carcinogens, so the test is very important in determining whether a molecule is cancer causing. The test itself uses a bacterial strain of *Salmonella* that has been genetically engineered to be highly susceptible to mutation. The organisms require the presence of an amino acid, histidine, in order to grow (see diagram). Because of the susceptibility to mutation, exposure of the strain to potential mutagens and carcinogens may cause a genetic reversion of the bacteria to a condition in which they do not require histidine. If the bacteria are exposed to a carcinogen and then grown in a growth medium that does not contain histidine, only the revertants will grow. This test is applied to all new chemical compounds, food additives, drugs, and antibiotics that are offered for public consumption.

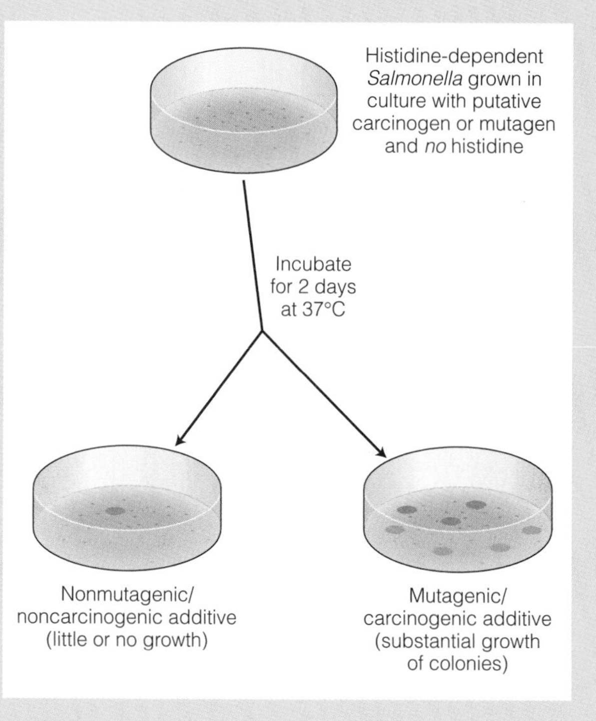

case for the tumor known as **Kaposi's sarcoma.** The present limited evidence of viruses in major cancers of humans does not exclude the possibility that they may yet prove to play an important role in carcinogenesis.

TREATMENT OF CANCER

There are basically three ways to treat cancerous tumors—remove them surgically, destroy them with ionizing radiation, or destroy the cells involved in the cancer with chemical compounds that kill rapidly dividing cells. The success of these approaches depends on a number of factors, including the type of cancer, the stage of development of the cancer, and its location in the body. These factors constitute a major part of the pathology of cancer and provide a physician with the information needed to make sound judgments on treatment. This information is considered clinical and should be provided to a patient.

Surgery

Success at surgically removing tumors depends on the accessibility of the cancer and the degree to which it is encapsulated, or confined, by host tissue. If the cancer is well confined, the chance of success is good (Figure 21–10[a]). How does a physician know that he or she has removed all the tissue? Samples of the cancerous tissue, and putatively normal tissue adjacent to the tumor, are examined by a pathologist to determine whether and where cancerous cells are present. Tumors are usually classified as either *in situ* or invasive. *In situ* tumors are those confined within a normal tissue capsule. Invasive tumors have penetrated the tissues around the tumor. Therefore, we can say, the less invasiveness, the better. The size of a tumor is also important. Small tumors (less than 2 cm or so) are an indication of early discovery. The **prognosis,** or prospect of recovery, is better when the tumor is smaller.

Tumors are graded with respect to the degree of abnormality of the cells. A commonly used grading system contains three grades—I, II, and III. A grade I tumor is well differentiated and the prognosis is generally good.

Kaposi's sarcoma a rare cancer of connective tissue associated with HIV suppression of the human immune system
Prognosis with respect to disease, the prospect of recovery

Table 21-3 Viruses and Cancer

VIRUS	ASSOCIATED TUMORS	AREAS OF HIGH INCIDENCE
DNA viruses		
Papovavirus family		
Papillomavirus (many distinct strains)	Warts (benign)	World wide
	Carcinoma of uterine cervix	World wide
Hepadnavirus family		
Hepatitis-B virus	Liver cancer (hepatocellular carcinoma)	Southeast Asia, tropical Africa
Herpesvirus family		
Epstein-Barr virus	Burkitt's lymphoma (cancer of B lymphocytes)	West Africa, Papua New Guinea
	Nasopharyngeal carcinoma	Southern China, Greenland (Inuit)
RNA viruses		
Retrovirus family		
Human T-cell leukemia virus type I (HTLV-I)	Adult T-cell leukemia/lymphoma	Japan (Kyushu), West Indies
Human immuno-deficiency virus (HIV-1, the AIDS virus)	Kaposi's sarcoma (cancer of endothelial cells of blood vessels or lymphatics)	Central Africa

Differentiated, in this case, means that the cells are fairly normal in morphology and histology. A grade II tumor is moderately differentiated, cells are less normal in appearance, but prospects are still good. The majority of cancers are grade II at the time of surgery. Grade III tumors are poorly differentiated, with cells grossly abnormal in appearance and in histological organization. Grade III tumors are aggressive, and the prognosis is less favorable.

One of the characteristics that a pathologist is asked to examine is the nature of the margin or edge of the tumor. This is an important structural feature of the tumor and helps the surgeon decide whether he or she has removed all the cancerous growth. This test is done while the surgeon is in the operating room. A biopsy of the tissue in question is frozen and cut into thin sections. The borders of the section are painted with India ink. If the border of the tumor is right at the edge of the ink line, the margin is said to be involved, and more tissue needs to be removed. If the ink is well away from the tumor, the margin is considered clear and the surgeon closes up.

In most cases, the lymph node status of the specimen is also ascertained. This means examining the lymph nodes for the presence of cancer cells. In cases such as breast cancer or colon cancer, there are many lymph nodes present in the fatty tissues that may surround a tumor. Evidence of tumor cells in the lymph nodes means that the cancer has spread. The spread of tumor cells occurs by metastasis, and microscopic tumor foci may be beyond the capacity of surgical procedures to find and remove them.

Radiation Therapy

Another means of eliminating cancer is to destroy it with radiation. This technique takes advantage of the powerful ionizing potential of radioactivity targeted at the site of the tumor. In many cases, radioactive materials can be placed in proximity to tumors that are inoperable (Figure 21–10[b]). The radiation destroys the tumor as well as the normal tissue surrounding it—the larger the area irradiated under these circumstances, the greater the damage to normal tissue. Radiation treatments usually occur in a sequence of exposures

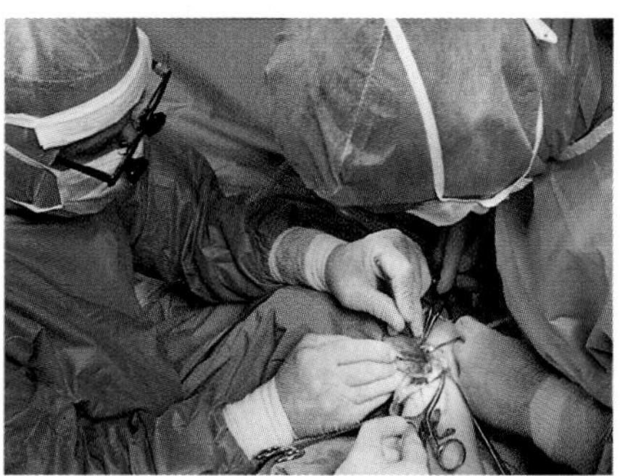

(a)

(b)

Figure 21-10 Treatment of Cancer

In these examples of cancer treatment, surgery seeks to remove obstructive cancerous tissue from bone to relieve pressure on nerves (a). Radiation treatment seeks to eliminate cancer cells that are not obvious and may be missed by surgery (b). The illuminated discs on the patient's chest indicate areas of treatment.

21–2

NATURAL COMPOUNDS AND CANCER

It is well known that naturally formed compounds from plants and animals are often of great medicinal value. This is the case for chemotherapeutic agents as well. The vinca alkaloids, including vincristine and vinblastine, are derived from the periwinkle plant (*Vinca*) and are known to interfere with microtubules. Colchicine and colcemid are derived from the crocus plant (*Colchicum autumnale*) and are also very powerful inhibitors of the assembly of microtubules. The recently discovered natural compound taxol is derived from the bark of the Pacific yew tree (*Taxus*) and is very effective in stabilizing microtubules. The stabilization of microtubules by taxol prevents them from undergoing the changes necessary to separate the chromosomes during anaphase of mitosis. This interference, and that of vinca alkaloids and colchicine, eventually alters the cell sufficiently to kill it. These compounds are preferentially effective against cancer cells because cancer cells divide so rapidly. However, they also interfere with normal cells that are dividing as well.

to limit the dose at any one time. However, the systemic effects of radiation treatment may be very debilitating. The individual undergoing treatment may lose hair, be nauseated, suffer from diarrhea, and have digestive tract problems.

As an example of a specific treatment, there are two methods of radiation used to treat prostate cancer (see Chapter 14)—external beam therapy and radioactive isotope implantation. The former involves an X-ray beam delivered to the site of the cancer by a large X-ray machine. The treatment schedule may be daily for seven or eight weeks. In the implant method, radioactive particles are seeded into the prostate gland permanently, thus, killing all cells in close proximity. The implantation technique has the advantage of being site specific. Implanting of the seeds may involve the use of ultrasound techniques to target the specific sites involved.

Chemotherapy

Chemotherapy uses drugs to kill cancer cells. The drugs are taken orally or by injection, but in either case, they are systemic in their distribution and effects within the body. Chemotherapy is used in conjunction with local treatment (surgery and/or radiation) to make sure that cells that may have spread outside the area of direct treatment do not escape destruction. This approach takes advantage of the fact that malignant tumor cells divide rapidly and, so, are more susceptible to the effects of inhibitors of cell division. Chemotherapeutic drugs specifically affect cell division and, thus, selectively kill tumor cells.

What molecules or structures are involved in cell division? One particularly important set of cytoskeletal structures involved are the microtubules. As covered in the section on cell division in Chapter 3, the spindle apparatus of a dividing cell is composed principally of microtubules. Chromosomes are attached to the tubules and migrate to the poles of a dividing cell during anaphase. If chromosomes do not separate, the cell dies. Thus, chemical compounds that interfere with the function of microtubules are particularly useful as therapeutic agents. The classes of compounds known as vinca alkaloids and colchicine derivatives prevent the assembly of microtubules and interfere with the separation of chromosomes into the daughter cells. Another natural product, a drug known as taxol, interferes with microtubules by preventing their disassembly (Biosite 21–2). In both cases, cells exposed to these compounds are unable to divide and, thus, die. These compounds and others have been used with success to treat cancer and help kill tumor cells that may have escaped elimination through surgery or radiation.

Unfortunately, these drugs do not differentiate between tumor cells that are dividing and normal cells that are dividing. This leads to the death of cells that are actively dividing as a normal part of the body's processes, including the epithelial cells lining the digestive tract and mouth and hair follicle cells in the skin. Nausea, bleeding of the gums, and loss of hair are all consequences of the drug therapy. In determining the level of treatment in chemotherapy, a physician seeks to balance the short-term bad effects on normal tissues with the long-term benefits of eliminating cancer cells. It is important to note that other approaches to selectively eliminating cancer cells are being explored. One

Chemotherapy the use of chemical compounds to kill cancer cells

of the properties of malignant cancer cells is that they are highly undifferentiated. This means that they do not look or act the same as the cell types from which they originally arose. If drugs could be found that stimulated cancer cells to differentiate or undergo a course of programmed cell death, they then would be less prolific and could potentially be subjected to normal tissue regulation. Such approaches to cancer therapy would in principle be far less damaging to the body of the person being treated because uninvolved cells would not have to be killed to conquer the cancer.

NUTRITION AND CANCER

Fats, Alcohol, and Cured Meats

Estimates from the National Cancer Institute suggest that at least 35% of all cancers have a nutritional cause. For women, it may be as high as 50%. The basis for these high percentages is that poor nutrition affects the function of the immune system, and it is immunity that protects us against cancer. In addition, some people believe that the foods we eat may contain carcinogens in the form of food additives and pesticides. There is little evidence for these factors in the foods grown and prepared for consumption in the United States. However, there are correlations with the consumption of fats, alcohol, and pickled or salt-cured foods and cancer. Both initiators and promoters of cancer may be in, or derived from, the foods we eat.

Fat consumption and storage have been implicated in the promotion of cancer, and a number of types of cancer are specifically associated with obesity. These include colorectal cancer, breast cancer, stomach and kidney cancer, and uterine cancer. In rat animal model systems used to study obesity, linoleic acid and *omega-6 fatty acid* were shown to enhance cancer development. How does this research relate to the human diet? The average American derives 40% of his or her calories from fat. The American Cancer Society recommends 30% for adults, so reduction in consumption is warranted. However, caution should be taken with children under two years of age, who should not be on fat-free diets because they need cholesterol for development of the brain and central nervous system.

Cancers of the mouth and throat are associated with the consumption of alcohol. Red wine and beer are implicated in this in part because they may contain carcinogens. Urethane in wine and nitrosamines in beer are known carcinogens. As described previously, the action of bacteria of the gut may also produce carcinogenic glycosides. Red wine is more carcinogenic than white wine in this regard.

Stomach cancer in many parts of the world is correlated with the consumption of large amounts of salt-cured, smoked, or pickled meats. The basis for this level of effects is the presence of nitrosamines in these preparations, or the conversion of nitrates and nitrites used for preservation and curing into nitrosamines by the action of our metabolism.

A balancing factor in the promotion of cancer is found in the consumption of fresh fruits and vegetables. The fiber of fruits and vegetables helps to move food through the digestive tract. This prevents the buildup of materials in the colon that may promote cancer and, generally, increases the efficiency with which absorption of nutrients takes place. Antioxidant molecules, such as vitamins C and E, which are present in fruits and vegetables, may aid in preventing cancer by inactivating the highly reactive oxygen and hydroxyl free radicals that can damage cells and initiate cancer.

Nutrition is an important part of health, both in supplying energy and molecular building blocks for growth and in maintaining the body in its fight to prevent the initiation and promotion of cancer. There are a number of general recommendations from the American Cancer Society that will help protect us against cancer.

Recommendations

1. Avoid obesity.
2. Eat a varied diet. This will help make eating more enjoyable, as well as avoid constant exposure to foods that may have specific carcinogens.
3. Eat plenty of fruits and vegetables.
4. Eat high-fiber foods. This includes whole-grain cereals, vegetables, and fruits.
5. Reduce the intake of fats.
6. Limit consumption of alcohol. Cancer is correlated with heavy drinking and exacerbated by smoking.
7. Avoid excessive intake of salted, cured, or pickled foods. The nitrites used to preserve and prepare such foods is a precursor for conversion to nitrosamine.

These recommendations establish the basis for good health, long life, and protection against one of the most dreaded classes of disease known to humanity, cancer.

Summary

This chapter has dealt with the nature, causes, and treatment of cancer. Cancers are basically of two types—benign and malignant. Benign cancers are encapsulated and localized, and malignant cancers are invasive and metastatic. There is a variety of types of cancers, including carcinomas, sarcomas, lymphomas, neuromas, and gliomas, to name but a few. The cells of malignant cancers are altered in many ways, the principal among them being that they grow rapidly and out of control. However, there are a number of other characteristics that are associated with cancer cells, including changes in adhesiveness, shape, metabolism, abnormalities in DNA and chromosomes, and specific alterations in the cell cycle. The genes involved in cancer transformation are known as oncogenes. In addition to oncogenes, there are tumor-suppressor genes, known as anti-oncogenes, that help prevent the development of cancer. Mutations in these genes may result in aggressive, highly metastatic cancers, such as retinoblastoma.

There are a number of factors that influence the potential for initiation and promotion of cancer—genetics, environment, and nutrition. The environmental factors include a number of agents that cause or initiate cancer. These agents are called carcinogens. The three types of carcinogens are physical, chemical, and biological or genetic. Physical agents are X rays and UV light; among the chemical agents are heterocyclic hydrocarbons and a variety of other chemical compounds. Potential chemical carcinogens can be assessed using the Ames test. Viruses are biological agents of carcinogenesis. Both RNA and DNA viruses are implicated directly or indirectly in some forms of cancer.

The treatment of cancer involves surgery, radiation therapy, and chemotherapy. The selection of a specific form of therapy depends on the type of cancer, the stage of development, and the location in the body. A description of pathological changes sheds light on how a physician might establish the nature and extent of a cancerous growth and what to expect should you have to consider clinical information provided by a physician about you or a member of your family.

Nutrition and cancer are closely related. The things we eat may both initiate and promote cancer. Fats, alcohol, and cured foods have been implicated in the promotion of cancer, and fruits and vegetables have been found beneficial in preventing cancer. As a result, the American Cancer Society publishes dietary recommendations for the prevention of cancer.

Questions for Critical Inquiry

1. What is cancer? How does it affect the lives of the people who develop it? How has it affected your life?
2. What is a carcinogen? What forms do the agents of carcinogenesis take? What effects do they have? Where do they come from?
3. Does the predisposition for cancer run in families? Does it mean that those individuals will get cancer?
4. Why does excessive tanning when we are young generally result in skin cancer only when we are decades older? Why is cancer in general considered a disease of the old?

Questions of Facts and Figures

5. Differentiate between a benign tumor and a malignant one.
6. How is nutrition related to cancer? What factors in a diet are promoters of cancer?
7. What are some of the main properties of tumor cells?
8. What are three ways of treating cancer? Under what conditions might they be used separately? Together?
9. What does a tumor-suppressor protein do? What happens if there is a mutation in the gene coding for such a protein?
10. What is an oncogene? What is a proto-oncogene? How are they related?

References and Further Readings

Alberts, B., Bray, D., Lewis, J., Roff, M., Roberts, K., and Watson, J. D. (1994). *Molecular Biology of the Cell,* 3rd ed. New York: Garland Publishing.

Berg, P., and Singer, M. (1992). *Dealing with Genes.* Mill Valley, CA: University Science Books.

Sherwood, L. (1993). *Human Physiology,* 2nd ed. St. Paul, MN: West Publishing.

Stalheim-Smith, A., and Fitch, G. K. (1993). *Understanding Human Anatomy and Physiology.* St. Paul, MN: West Publishing.

CHAPTER 22

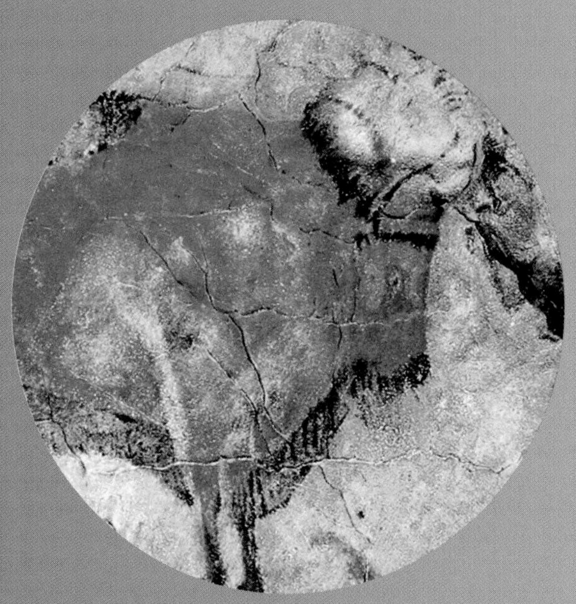

Evolution and Human History

Descent with Modification
Mechanisms of Evolution
Mechanisms for Genetic Variation
Speciation
Geological and Biological Clocks—A Changing Earth
Primates and Human Evolution
Migration and Isolation
Evolution and Macroevolution
Summary
Questions for Critical Inquiry
Questions of Facts and Figures
References and Further Readings

INTRODUCTION

Most people have an interest in the genealogy of their family. However, keeping track of one's relatives, particularly over many generations, is difficult and complicated. People move to new jobs and lose touch, divorce breaks up family units and replaces them with new ones, children are born in distant places, college and schooling open doors to new opportunities that carry family members far from their roots, and there are a hundred other ways by which individuals disperse throughout the world. In this chapter, we will consider a much larger and more complex framework of genealogy—the history of *Homo sapiens* and its ancestors going back to the beginnings of life on Earth. How did our species come to exist in its present form in this time and place? What were the changes on Earth that allowed life to start and evolve? What were the mechanisms at work to bring about these changes? The road to these discoveries leads us back hundreds of millions of years and involves some of the most profound scientific insights ever proposed.

22–1

LAMARCK AND DARWIN

Charles Darwin was not the first individual to be concerned with evolution, nor was he the last, but modern evolutionary biologists have stood on Darwin's shoulders to extend the view of biology, while the ideas of others, such as Lamarck, have been left by the wayside. There were certainly many thinkers from past centuries whose ideas held sway for different periods of time, but none of them was completely divorced from the idea of creation of life by an outside force and the unchangeability, or fixity, of the species created. According to some biblical scholars, the entire world was created approximately 6000 years ago, and life forms have not changed fundamentally since. In this view, Man was considered very special, made by a Creator in His own image, and for some higher purpose. Other religions, past and present, have equally complex creation mythologies. All these stories portray the world as being created and life arising by design and intervention from an outside force, a Creator.

These views have proven to be inadequate to explain the diversity and continual change observed in the world around us. Darwin discovered and expounded on the roles of variation and natural selection in the process of evolution, as did his contemporary Alfred Wallace. Using the work of Charles Lyell on **uniformitarianism,** which suggests that geological processes occurring today are the same as they have been throughout time, and Thomas Malthus, who provided the framework concepts of supply and demand, Darwin presented us all with a new view of life.

DESCENT WITH MODIFICATION

The scientific view that developed in the eighteenth and nineteenth centuries held that Earth was much older than Bible scholars supposed and that all life, including human life, descended by modification from simpler forms that were no longer in existence. There are many who added to the knowledge of life and its origin and evolution but none more than the British naturalist Charles Darwin. Darwin's ideas owe much to scientists of previous generations, as well as to those of his own (Biosite 22–1). It was Darwin who developed the **theory of evolution** in its modern form and articulated the general mechanisms by which it must occur. With this in mind, the focus of this section is on the mechanisms of evolution—variation and natural selection—which will provide a framework in which to view the evolution of humankind (Table 22–1).

Table 22–1 Recognizable Forms of Life from the Paleozoic

TAXON	COMMON NAME
Porifera	Sponges
Cnidaria	Corals
Arthropoda	Trilobites, insects
Mollusca	Clams
Annelida	Worms
Chordata	Fish
Plantae	Ferns, cycads

MECHANISMS OF EVOLUTION

Lamarck and Inheritance of Acquired Characteristics

Jean Baptiste Lamarck lived during the latter half of the eighteenth and early part of the nineteenth centuries, a generation before Darwin. He was an eminent scientist of his times and an influential intellectual. Lamarck's ideas in biology were revolutionary and were to have a great influence on Darwin, a young British naturalist. Why was Lamarck's thinking on evolution so important? It was important because he was the first to clearly set forth a self-consistent and biologically feasible mechanism to account for inheritable changes in organisms and to consider how change over time (evolution) might give rise to new species. Lamarck published his theory the year that Darwin was born, in 1809. Darwin would later become well aware of Lamarck's work.

Theory of evolution the theory proposed by Charles Darwin to explain change and the origin of species through variation and natural selection

Uniformitarianism a hypothesis that states that geological processes occur at similar rates throughout Earth's history

The Neck of the Giraffe Lamarck's theory proposed that the physical characteristics acquired during the lifetime of an individual were passed on to its offspring (Figure 22–1). A classic example of the operation of this kind of mechanism focuses on the giraffe, but the sequence of changes observed in the fossil record for any organism could be substituted just as easily. This includes how the zebra got its stripes or the elephant got its trunk. How did a giraffe's neck come to be so long? According to Lamarck's hypothesis, the ancestors of the modern giraffe worked to reach higher and higher into the branches of trees to gather their diet of leaves and their necks, through constant use, became longer. An individual who had acquired a slightly longer neck passed this feature directly on to its offspring in the next generation. Why would the ancestors of the giraffe need to stretch in the first place? The basis for stretching was linked to the environment in which the giraffe lived. Drought, destruction, or overbrowsing of leaves on the lower branches of trees would have forced at least some giraffes to stretch their necks to survive. It was a matter of competition for resources. Succeeding generations would have to feed higher and higher in the branches also, resulting in individuals with incrementally longer necks. Conditions in the environment, then, induced changes in behavior of giraffes, which altered their anatomy and influenced their heredity. This was a simple, straightforward process linking heritable genetic change with traits acquired within one's lifetime. Thus, over many generations, giraffes evolved long necks—simple, elegant, but wrong.

Darwin and Descent

The biological basis for inheritance of traits had to have another explanation; otherwise, all you would have to do to get tailless mice or featherless chickens would be to cut the tails or pluck the feathers off the parents prior to mating. Are children of weight lifters and gymnasts genetically endowed with their parents' strength and agility? Of course not, at least not without considerable work during their own lifetimes. Experiments and observations on the potential for inheritance of an acquired characteristic have been carried out by many investigators, and the conclusion is always the same—no passage of acquired traits from parents to offspring (Figure 22–2).

Diversity and Variation If evolution is not determined by inheritance of acquired traits, then what? Darwin took a different path toward the principles involved. His detailed studies on a variety of organisms throughout his lifetime, and during his youthful travels around the world in the 1830s on the British survey ship HMS *Beagle,* gave him a unique perspective on the diversity of life on a global scale. The term **biodiversity** is often applied to this all-inclusive grouping of organisms. The survey ship traveled extensively along the east and west coasts of South America,

> **Biodiversity** the wide range of structure and function of life forms adapted to specific environments

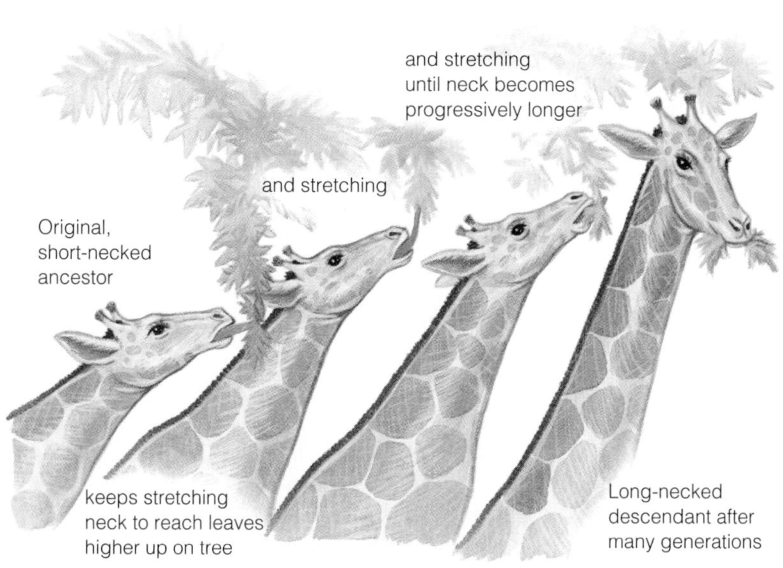

Figure 22–1 Inheritance of Acquired Characteristics
Lamarck postulated that stretching of the neck during a giraffe's lifetime brought about an elongation. The trait once acquired could be passed on to descendants who would have longer necks.

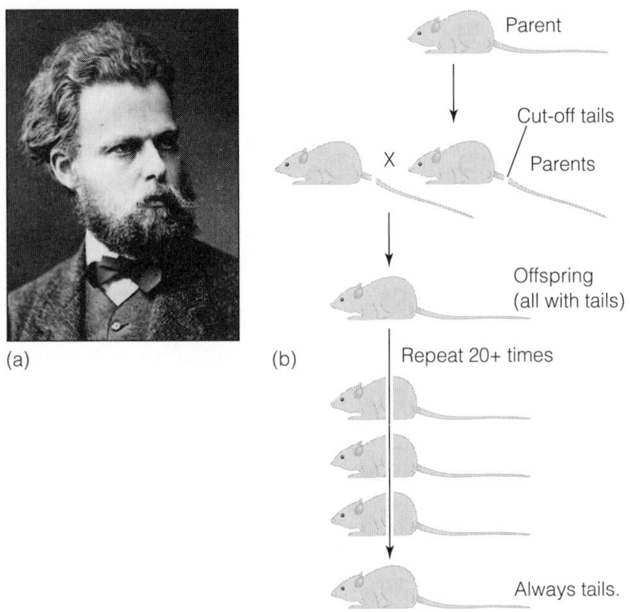

Figure 22–2 The Mouse's Tale
August Weismann (a) tested Lamarck's hypothesis of inheritance of acquired traits by cutting off the tails of more than 20 generations of mice (b). Not once did "tailless" parents produce tailless offspring.

the Galapagos Islands, New Zealand, Tasmania, Australia, and other lands during the five-year world tour. Darwin recorded in his journal all he saw and thought about on the trip. He noted the variations, as well as the similarities, among species and observed how animals and plants in different regions, on different islands, and on different continents adapted to the environments in which they lived.

Diversity Darwin wondered at the immense diversity of life past and present, because he had also discovered fossils of species of animals no longer living on Earth. If all types of life were created within the last few thousand years, where were living examples of these fossil organisms whose remains were buried deep within ancient sediments? Because Earth was no longer a mysterious and undiscovered place, there were no hidden lands in which such organisms might still exist. These fossil discoveries whispered of an Earth of great antiquity, with all forms of life subject to great change over long periods of time. Darwin believed that time—vast expanses of time—was needed for speciation. His idea was that new species arise by descent, with modification from old ones.

Variation and Natural Selection The mechanisms of Darwinian evolution can be distilled down to two concepts—**variation** and **natural selection.** Variation refers to the differences among individuals in a population, and natural selection refers to the forces that act upon those differences. Darwin noticed, as did many other naturalists, that species produced many more offspring in a given generation than survived. This suggested that some of the individuals must be better suited to survive in the environments in which they lived than others of that species. This was a "population" approach to evolution because it focused on variation among individuals in a population as the key to species success over time. A population is a collection of individuals of the same species that interact together as a group and interbreed to produce viable and fertile offspring. As pointed out in Chapter 20, recombination between homologous chromosomes and mutation in DNA sequence give rise to allelic forms of genes in different combinations in a population and, thus, provide a means for introducing variation into a species.

Throughout his studies, Darwin consistently observed that individuals within a population showed variation in their morphological characteristics, that is, no two organisms were exactly alike. Excluding sexual differences for the moment, some individuals were bigger, some faster, some had different patterns of color on their surfaces, and some were more cunning than others. As discussed in Chapter 20, the numbers and kinds of these characteristics reflect the complex phenotypic expression of thousands of genes present in unique genotypes of individual organisms. A population shares a common pool of genes with many alleles, not all of which are expressed in any one individual. Although unknown to Darwin in his lifetime, variation in genotype ultimately reflects the differences observed in phenotype. Phenotypic differences in individuals are acted upon by the environment—"tested", as it were—and the most successful not only survive but are more likely to pass on their trait(s). **Fitness** of an organism is measured in the number of offspring it produces and, ultimately, the success of those progeny based on their genetic endowment, that is, the longer an organism survives, the more chance it has to reproduce (Biosite 22–2).

MECHANISMS FOR GENETIC VARIATION

How are variations introduced into the gene pool of a population? They are generated by mutations, chromosomal

Variation the genotypic and phenotypic differences observed in individuals of a single species

Natural selection the environmental forces acting on genetic variation in individuals to determine fitness

Fitness a measure of reproductive success

BIO site

22–2

BIOTIC POTENTIAL

Part of the means by which selection operates is through competition for resources. Because the resources of Earth are limited, competition can be intense. If resources were unlimited, there would be no constraints on the growth of any (and all) populations of organisms, which would grow exponentially, and this would be a reflection of biotic potential. However, the normal growth of a population follows a different course (an S-shaped curve, see accompanying figure), which reflects the limited resources available to a growing population. This simple curve does not include the effects of the growth of other species, which also compete for the same, or similar, resources.

Thomas Malthus wrote a book dealing with the problem of growth at the turn of the nineteenth century and suggested that the limit on growth, including human growth, was tied to the amount, availability, and increase in production of resources—supply and demand. The specific problem of growth at the time Malthus wrote was that agricultural production was expanding at a slower rate than the demands placed on that production by growing human populations. Darwin was familiar with the thesis of the book and was struck by its applicability to natural selection and the evolution of species.

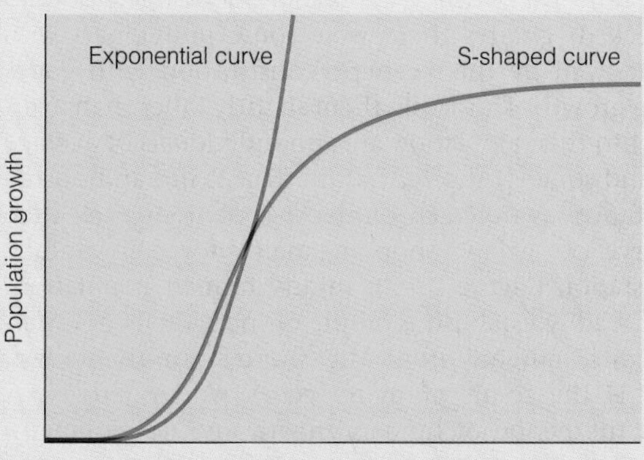

rearrangements, and recombination of genes during meiosis (see Chapter 20). The altered genes of an affected individual would have a chance to intermix with the genes of other members of the population through sexual reproduction. This assumes that an individual with the new form of a gene is able to compete effectively with his or her counterparts. Ultimately, the particular set of genes (new and/or old) that an individual has, and expresses, determines the capacity of that individual to compete in and adapt to the environment in which it lives.

A Hypothetical Situation—Selection at Work

If the environment changes over time, as it does continuously, then the genetic capacities of some members of the population may either provide a selective advantage or prove to be a disadvantage. For example, imagine a trait that increases the number of hair follicles and, thus, the thickness of fur on an organism's body. If the climate of a region changes and average temperature declines, it might be an advantage to have thicker fur. However, if the climate changes and the average temperature increases, the thick fur might be a distinct disadvantage. Survival is linked to external conditions, and competitiveness is determined by how those conditions affect the anatomy, physiology, and behavior of an individual. An advantage in thermal regulation in the cold could well increase the likelihood of survival. Longer survival results in more chances to mate and reproduce, which leads to more individuals with the hair coat trait and, thus, an increase in the number of individuals carrying and expressing the trait in the population. This would also lead to an increase in the prevalence of that gene in the gene pool of the population.

In summary, natural selection acts on the capacities of individuals to perform successfully in the environments in which they live. These capacities are called **adaptations.** Competition among individuals is often keen for the limited resources that limit population growth; therefore, any advantage gained by an adaptation of one individual over another may be invaluable. Mutation and recombination introduce changes in the genotypes of individuals, which, if functionally useful, may become an important or fixed part of the gene pool of a species. It is the variations expressed in phenotypes of individuals on which natural selection acts.

Giraffes Revisited

Returning to the problem of the evolution of the giraffe's long neck, how do Darwinian concepts of variation and

Adaptations genetic characteristics related to fitness

natural selection solve the problem that was unable to be answered using Lamarck's hypothesis? In terms of variation, assume that within a population of ancient (and short-necked) giraffes there were some individuals who were endowed by their genetic constitution with genes affecting growth. This made them slightly taller than average. Within this population are also individuals of average height and some of shorter stature. This is not an unusual circumstance, as noted in human populations each time we observe a crowd in a shopping mall, a football game, or a train station (Figure 22–3). In any human population, you can readily establish a profile of individuals of different heights, ranging from the shortest to the tallest. Tallness is the result of many genes working together through the period of human growth and development. This is called polygenic inheritance. With respect to a population of giraffes, who also have genetic differences affecting tallness, imagine that a climatic change results in a sudden and persistent reduction in the availability of food.

Normally, there would have been plenty of leaves on low and intermediate branches to feed the entire population of ancestral giraffes. However, because of environmental change, the trees bear fewer leaves, and those that are present are higher up on the branches. Perhaps the rainy season came late, or there was a series of long dry summers. Because taller individuals in the population were able to reach higher into the trees for leaves than shorter animals, the chance for success in obtaining food clearly was greater for taller individuals. Note that the taller animals did not acquire a longer neck through stretching it, but rather acquired it by chance, as a result of recombinations of genes already present in the gene pool or introduction of new beneficial alleles by mutation. Add to this the fact that starvation selectively eliminated giraffes that were too short to obtain a sufficient amount of food. What was left? What was left was a population of generally taller giraffes. Survival translated to increased longevity, and longevity resulted in increased opportunities to breed. More individuals with genes for tallness were subsequently included in the pool of genes found in the ancient population of giraffes.

Although the focus of our attention has been on the long neck of a giraffe, these animals have traits that affect the length of their legs as well. The length of the legs, like the length of a neck, is also under selective pressure by the environment and reflects part of the complex, polygenic basis for tallness. It is genotypic variability expressed as differences in phenotype, not the physical characteristics acquired through experience or work, that determines the trait that is passed on to an offspring. The process of change is gradual in a species and occurs over many generations. The sequence of changes as described here may well have been what resulted in the evolution of modern giraffes with long necks and long legs.

In review, individuals with genes that made them taller than other members of the population were selected by the environmental conditions that imposed restrictions on individuals who did not express the tallness trait(s). Tall survivors had more opportunity to breed and left more offspring who carried those traits. This is natural selection in a nutshell. Through variation in the natural genetic endowment of individuals in a population, environmental selective pressures bring about the elimination of some individuals and an increase in the representation of others.

Extend the idea of evolutionary change over the enormous span of the ages of life on Earth. Earth and all the life on it, now and in the past, is subject to ceaseless change. However, if our view encompasses only a very short period of time, many aspects of the world do not appear to change at all, giving us a false sense of their permanence. For example, mountains do not crumble overnight, and seas do not dry up over a weekend. However, when viewed over long spans of time, natural change is inevitable. Mountains do indeed erode away to hills and plains, and ancient seas become land. Even within a human lifetime, a pond may evolve into a meadow, or an abandoned field will reestablish forest growth.

(a)

(b)
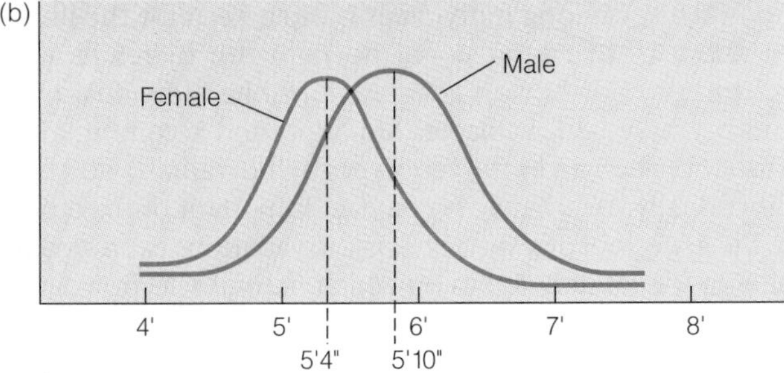

Figure 22–3 Variation—Stature
Human stature is a polygenic trait. There is a continual gradation of heights among men and women.

SPECIATION

Evolution occurs in the context of random changes arising in individuals within a population. As more and more of these changes accumulate, a subpopulation of a species may be so altered that it behaves differently and can no longer interact with its former relatives. One criterion for defining a species is the ability of a group of similar individuals to interbreed and produce viable offspring. Changes that lead to one set of related organisms losing those capacities usually take a long time to occur. The appearance of a new species is measured over thousands or millions of years and often results in the extinction of older forms. The outcome of this process of speciation was suggested by Darwin in his observations of finches on the Galapagos Islands (Figure 22–4).

Darwin favored the idea of **gradualism** in evolution, which suggests that changes that give rise to new species are slow and accumulate such that the final form of a species goes through a series of changes represented by intermediate stages to get there. Small changes may offer small advantages. The accumulation of these changes over time establishes intermediate types. However, natural selection need not operate in a gradual manner. Natural selection may give rise to rapid evolution of organisms and to new species in discrete steps resulting from what is called

> **Gradualism** a hypothesis that evolutionary change and the appearance of new species occur slowly and incrementally

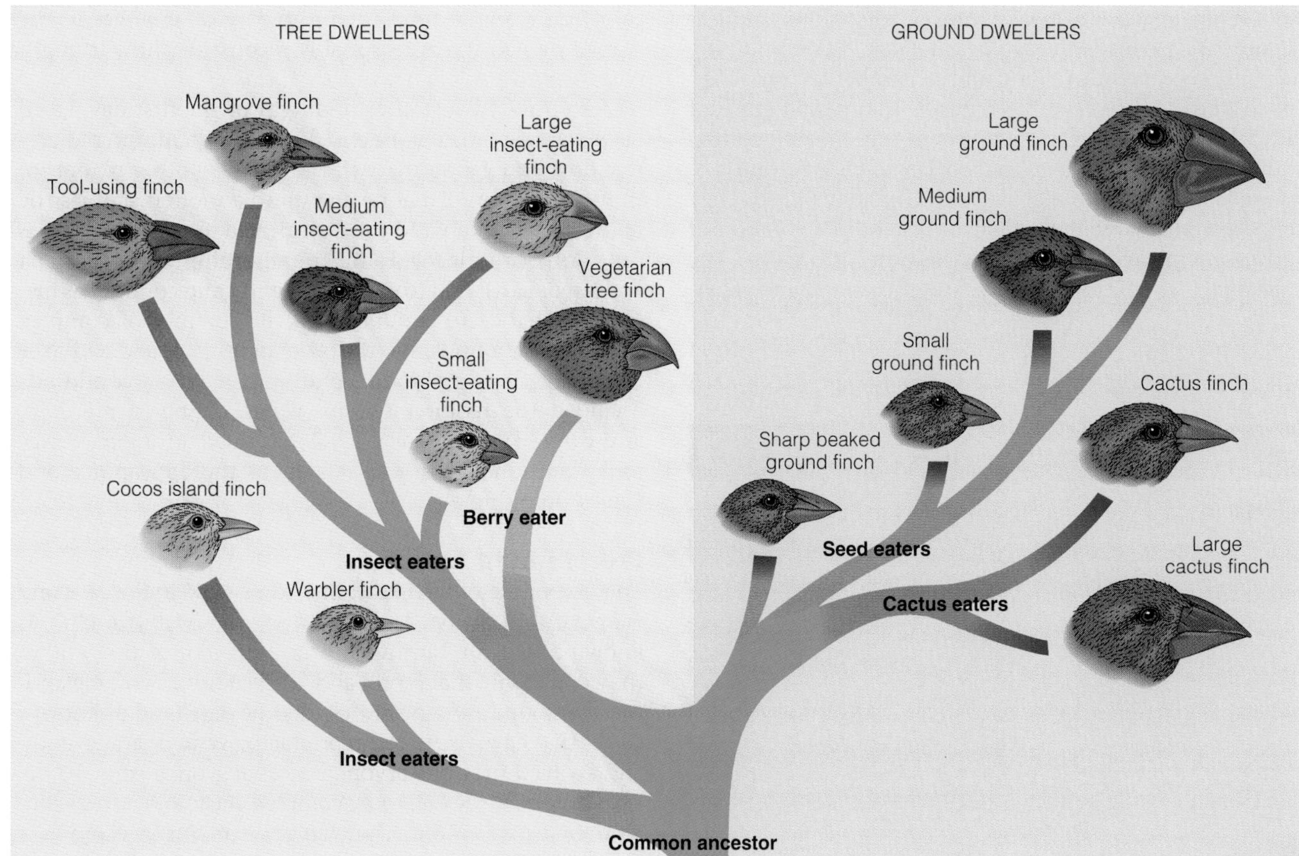

Figure 22–4 Speciation
When Charles Darwin visited the Galapagos Islands, he found related types of birds well adapted to their environments. The six species on the right branches are ground dwellers, and those in the left branches are tree dwellers. The shapes of bills vary depending on diet. Darwin concluded that all the types shown evolved from a common ancestor.

punctuated equilibrium. Whatever mechanisms bring about change, it is clear that of the millions of species presently living, there have been tens of millions that came before and are now extinct. As will be presented in the next section, *Homo sapiens* arose in a continual line from ancestors that no longer walk the earth.

GEOLOGICAL AND BIOLOGICAL CLOCKS—A CHANGING EARTH

Continents and oceans, glaciers and deserts have come and gone as part of the pattern of ceaseless alteration of the evolving surface of Earth. Earth has been in rotation around the Sun for 4.5 billion years. The span over which changes in the planet have occurred is called geological time and is inconceivably long. The geological clock started ticking when Earth was formed, but a second clock started about 3.5 billion years ago and it moves to a different beat (Figure 22–5). This is the biological clock. Three and one-half billion years is also a long time, and much that has happened to the structure of Earth during that period makes it difficult to know exactly when, where, and how life started and began to evolve. The two processes of change, one geological and the other biological, are intimately linked. They are parts of the overall physical and chemical evolution of Earth; one process is inorganic, the other is organic. The appearance of new forms of life, and the underlying genetic changes observed in them over time, is the basis for the study of biological evolution. The effects of the activity of living things on Earth is also an integral part of the process of geological change.

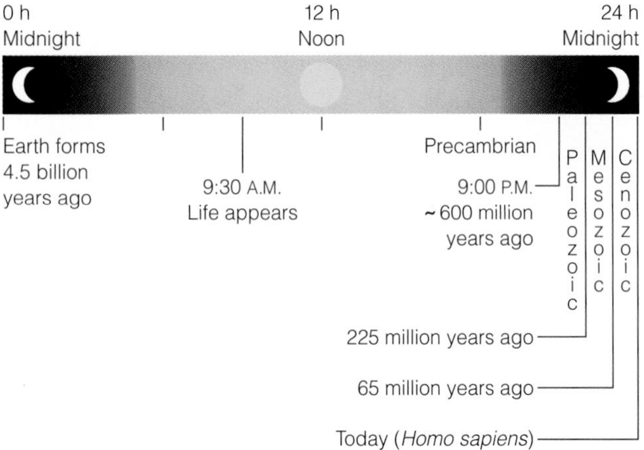

Figure 22–5 Geological and Biological Time
The time line shows that at zero (0) hours Earth formed as a planet approximately 4.5 billion years ago. Life began around 9:30 A.M. in the morning and present forms arose at 9:00 P.M. that night. Each second in this 24 h clock is 50,000 years. Using this scale, *Homo sapiens* as a species evolved at approximately 11:59:50 P.M., nearly midnight of the same day.

One goal of this chapter is to present a view of the evolution of life in general and to see how human beings have evolved within that framework. There were not always humans on Earth. In fact, we show up on the scene only very recently on the geological and biological clocks. As a way of giving shape to the concept of geological time and the biological clock, consider all of Earth's history compressed into a single day, a mere 24 hours. Under such a time crunch, each second on the clock is equal to roughly 50,000 years (Figure 22–5). The entire written history of humankind encompasses perhaps the last 5000 to 6000 years or so, and the presence of our species, *Homo sapiens,* has occurred within the last 500,000 years or less. Ten seconds, starting just before midnight, that's all we've had! With this perspective on the scale of time and the enormity of the scope of physical and biological changes in the world, we will try to reasonably reconstruct how it might have been when life began.

Zero time. At just after midnight of the first day, Earth began to take form, condensed by gravity out of cold cosmic dust and gas along with the Sun and other planets of the solar system. Earth heated up to a molten state as the matter from which it formed continued to condense. It was not until around 9:30 in the morning that life first appeared. Earth had to cool sufficiently and conditions had to arise that were conducive to the stability of complex organic compounds, such as proteins and nucleic acids (see Chapter 1 on prebiotic conditions). These conditions included the requirement for water, dissolved minerals, and gases. The biological clock has overlaid upon its face four traditional divisions related to the appearance and evolution of life.

Precambrian Era

The first phase of life began in ancient seas and oceans and evolved during what is called the *Precambrian era.* Little is known about life during this long interval of time, other than conjecture that it probably derived from a complex primordial soup through various intermediate (and unknown) stages that gave rise to self-reproducing systems and eventually evolved into procaryotic unicellular organisms. In the later phases of the era, more complex eucaryotic multicellular organisms evolved. They left few remains, because, it is

Punctuated equilibrium a hypothesis that evolutionary change and the appearance of new species take place rapidly

suspected, they were soft bodied and not easily preserved as fossils. Those remains that have been found are often bizarre jellyfish-like organisms and worms that have no known counterparts among modern groups of animals. Other remains show characteristics clearly similar to modern forms of life. This period of early evolving life started 3 billion to 3.5 billion years ago, midmorning on our geological clock, and ended approximately 570 million years ago at about 9:00 P.M., using compressed clock time. The Precambrian era ended rather abruptly, in what is called the **Cambrian explosion** (see Chapter 1, Biosite 1–1). Remnants from the explosion of those diverse life forms are with us today, preserved in the rich fossil beds of the **Paleozoic era,** known as the era of old life.

Paleozoic Era

It was not until organisms developed "hard parts," such as mineralized shells or exoskeletons capable of being preserved indefinitely in sedimentary rocks, or occurred in sufficient numbers in the shallow seas that individuals consistently left evidence as a fossil record. Fossils are the modified remains or impressions of the remains of previously existing life (Figure 22–6). Why it is that many different kinds of organisms suddenly evolved in this way is a hotly contested issue among paleontologists, who study the fossil record for clues of what may have happened to trigger such an explosion of new forms of life. Whatever the cause may have been, however, it is clear that the basic design of animals changed dramatically around 570 million to 600 million years ago. The Cambrian explosion of animal life marked the end of the Precambrian era and the beginning of the Paleozoic era. Many different life forms observed today had recognizable ancestors in those ancient seas (Table 22–1).

The Paleozoic era started in the mid-evening of evolving Earth and was a time of tremendous expansion of the variety and diversity of forms of life. All the major phyla had their beginnings during this time (Table 22–1). In terms of human ancestors, our predecessors clearly began with the evolution of chordates (phylum Chordata), which attained bilateral symmetry along an axis specified by a primitive structure known as the notochord (see Chapter 15). An example of relationships among chordates is represented in Figure 22–7, which shows more and more specific characteristics or traits are selected from related groups. In Figure 22–7, across the top row of vertebrate organisms are representatives of reptiles (turtle), fishes, amphibians (salamander), birds, and mammals. At each successive row, further distinctions are made to class (Mammal), order (Carnivore), family (Canidae), and, finally, genus and species. *Canis latrans* is the coyote but it is related to all other vertebrates, which, in turn, are all included in the phylum Chordata.

A notochord is formed during the early part of human embryonic development, a reminder that we share a long history of anatomical organization and evolution with other forms of life. From an evolutionary point of view, the early chordates gave rise to the vertebrates (subphylum Vertebrata), which have an internal skeleton. The first vertebrates were fishes, and they filled the seas during the early and middle Paleozoic era. They became so abundant that the Devonian period of the Paleozoic era has come to be known as the *Age of Fishes.* From fish evolved amphibians, who were adapted to a semiaquatic lifestyle that served them well, in what was to be the organismal transition to land. The evolution of land plants was also occurring during this era. The first reptiles arose late in the Paleozoic era and continued to evolve into the next era, the **Mesozoic,** as did birds and mammals. The time line clock is still ticking and the crunch of world time puts us at just about 11 P.M. Only slightly more than an hour left in the day of the evolution of life on Earth.

Figure 22–6 Fossil Remains
These days, stories about the discovery of new dinosaur remains are common. The buried remains of an allosaurus (a) and a stegosaurus (b) had their bones converted to silicates and were preserved under successive layers of sediments. Uplifting and erosion of the land exposed the bones as shown here and revealed the remains of long-dead and extinct species of the past.

Cambrian explosion the sudden and dramatic appearance of most modern phyla; approximately 650 million years ago during the Cambrian period of the Paleozoic era

Paleozoic era the most ancient era of life; from 650 million years ago until 225 million years ago

Mesozoic era the middle era following the Paleozoic between 225 million and 65 million years ago

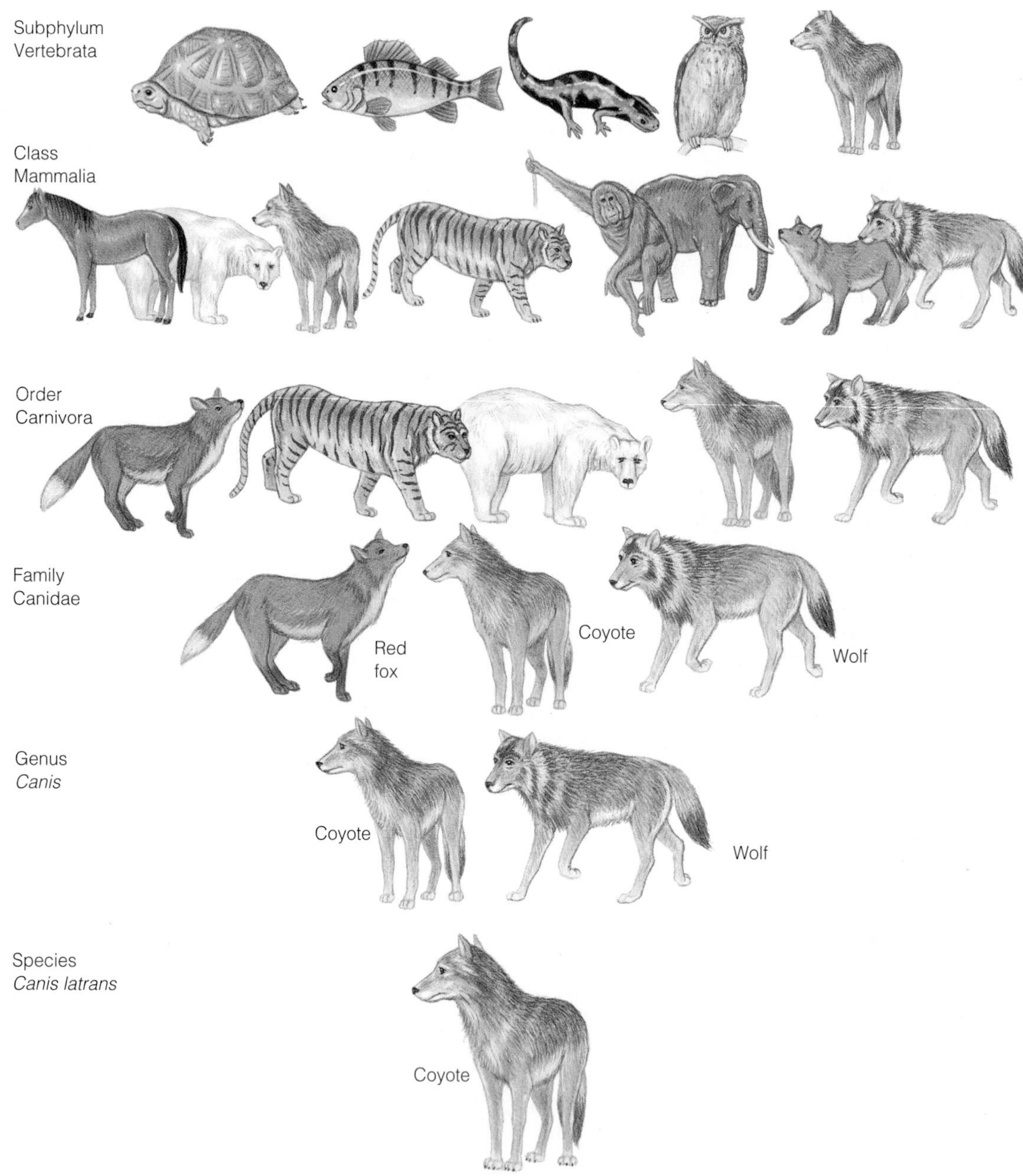

Figure 22–7 Phylum Chordata

The phylum Chordata has within it the subphylum Vertebrata, which contains a number of classes including mammals. Follow the line of the coyote to observe that similarities increase among species of the order Carnivora, family Canidae, genus *Canis*, and species *Canis latrans*.

Mesozoic Era

The end of the Paleozoic era, as with the end of each era, was marked by the extinction of many types of organisms and great changes in the lands and waters of Earth. The **Permian extinction,** as this particular crisis is called, dealt a death blow to many of the species characteristic of the Paleozoic era, including trilobites, some fishes (placoderms), most corals and brachiopods, some reptiles (polycosaurs), and all but one group of ammonites. In short, between 70% and 90% of all marine species became extinct.

Many of these organisms may seem strange and unfamiliar by today's standards, but they existed for hundreds of millions of years (much longer than *Homo sapiens* has been on Earth) before losing out in the struggle to survive. The Mesozoic era started approximately 225 million years ago and lasted 160 million years. This is about three-quarters of an hour on the clock. During this time, the single, gigantic continent of **Pangea,** which had coalesced during the previous era, began to break up into many smaller continents (Figure 22–8). This was brought about by geological forces operating deep within the planet. The separation of the enormous contiguous surface plates of Earth occurred as a result of the plates essentially "floating" on the molten magma beneath them in Earth's core. The movement of these plates is called continental drift. The outcome of this slow geological process, which amounts to the movement of the United States and Canada (the North American plate) approximately 1 cm per year, brought about the separation of the great land masses of Earth. Today, 225 million years later, many land masses once connected to one another have attained positions on opposite sides of Earth. The forces that drive the movement of the plates (called tectonic plates) continue to operate today. In another 100 million years, the continents may once again jam themselves together into a new supercontinental land mass.

Permian extinction a time of dying out of numerous species at the end of the Paleozoic era

Pangea a supercontinent including all land masses presently called continents that existed approximately 200 million years ago

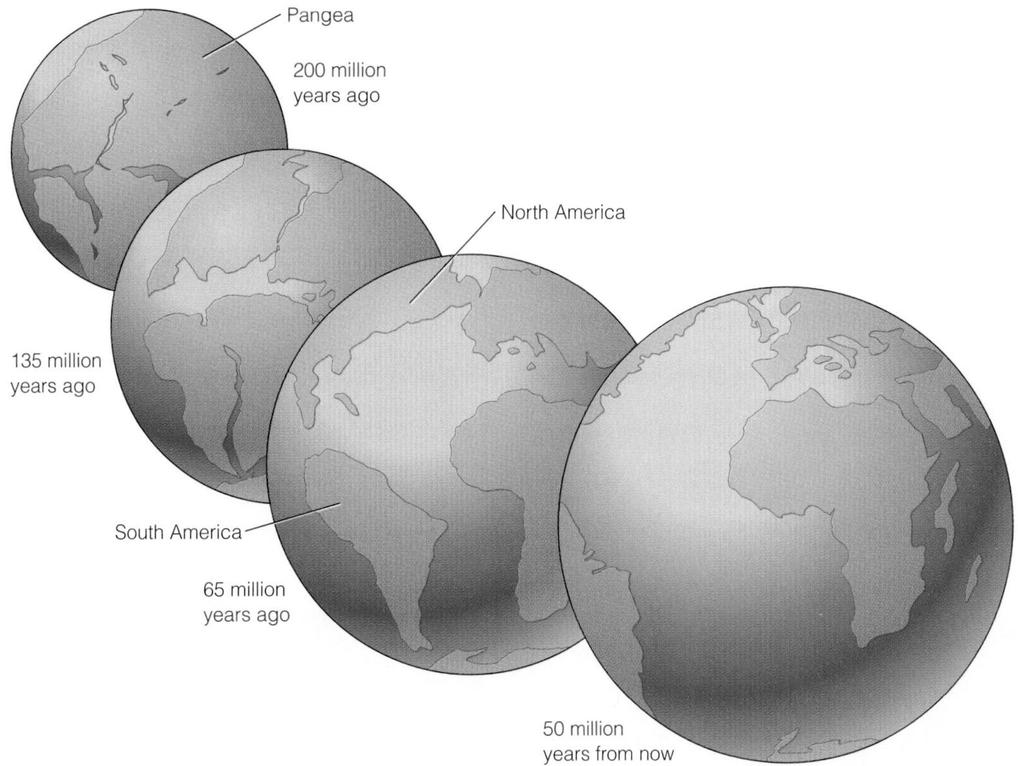

Figure 22–8 The Breakup of Continents
The continents began to form from the breakup of the supercontinent of Pangea 200 million years ago. The separation of land masses occurs through the process of continental drift.

Primitive Mammals With the beginning of the Mesozoic, the first dinosaurs appeared and along with them our relatives, the primitive mammals. The Mesozoic era was the zenith of reptilian life. Reptiles were enormously successful. In world clock time, this was quite literally their finest hour. They occupied habitats of all descriptions on land, in the water, and in the air. The best known of the reptiles were the dinosaurs, who arose to dominate the large-animal niches of the *Jurassic* and *Cretaceous periods* (Figure 22–9). However, many other types of reptiles evolved as well, including turtles, lizards, and snakes. Birds first appeared during the Mesozoic and may have arisen from or along with the dinosaurs. At the end of the Mesozoic era, there was another major extinction. This one is called the *Cretaceous extinction* and includes the demise of the dinosaurs, others of the giant reptiles, many kinds of clams, the ammonites (those that survived the Permian extinction!), and a large group of invertebrates. What organisms persisted through this time of extinction? Many different types survived, but of particular importance to this discussion were mammals.

Cenozoic Era

It is now 20 minutes to midnight on the clock of rapid evolution. After the fall of the reigning dinosaurs 65 million years ago, the organisms that survived the extinction had to attain a new balance with the environments in which they lived. Whereas dinosaurs and reptiles had once been the large-animal masters of Mesozoic Earth, now smaller animals were provided an opportunity to expand and diversify into **habitats** and **niches** that were open to them. The new era is called the **Cenozoic,** and it is the one in which we presently live. The animals that would come to replace the reptiles and dominate the large-organism niches of this era were the **mammals** (Figure 22–10). It is fair to say, however, that the real masters of the Cenozoic era, based on number and diversity, are the *arthropods* (that is, insects, spiders, crabs) who compose over 80% of all species in existence today. One of the greatest advantages that mammals and birds have over reptiles, amphibians, and fish is that they are endothermic; that is, mammals and birds are able to maintain a constant body temperature over a wide range of environmental heat conditions. This was probably a factor in their surviving the Cretaceous extinction, which many think may have been caused by a sudden (by geological time standards), and as yet unexplained, lowering of the average temperature of the surface of Earth. During the early Cenozoic era, 60 million years ago, the first primitive **primates** (order Primates) appeared. With primates, the story of the descent of humans and their monkey and ape relatives begins.

PRIMATES AND HUMAN EVOLUTION

The primitive primates of 60 million years ago gave rise to the prosimians, monkeys, apes, and humans of today. People have argued bitterly over this lineage, particularly since the days of Darwin. Did humans descend from apes? The simple answer is no. Humans and apes descended separately from a common ancestor, who is no longer in existence. The family tree of primates has many branches, and, as with any family tree, the branches can be traced back to a main trunk (Figure 22–11). The primitive primate from which we were derived no longer exists. We did not descend *from* apes, we descended *with* apes. We are related to other primates, but we have followed a very different path in our descent from a distant, primordial common relative.

The First Primates

The first primates were probably similar to the insect-eating prosimians of today. Prosimians are our most distant pri-

Habitats the environments in which organisms live

Niches the role that organisms play in their habitats

Cenozoic era the most recent era of life, from 65 million years ago to the present, during the early part of which mammals flourished and forest primates appeared

Mammals warm-blooded organisms having hair and mammary glands

Primates order of mammals to which *Homo sapiens* belongs

(a)

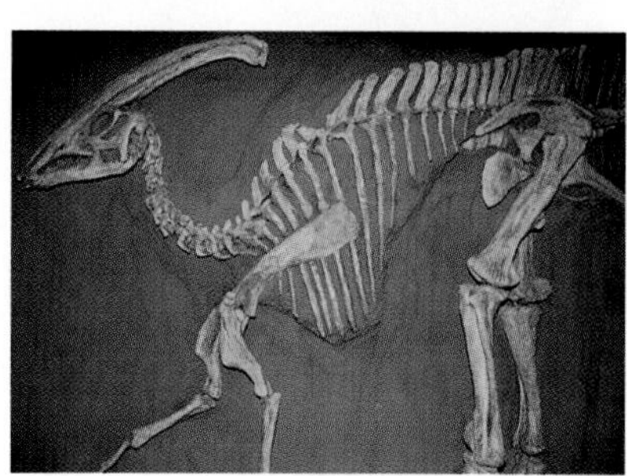
(b)

Figure 22–9 Mesozoic Era
The Mesozoic era produced some of the most gigantic and terrifying of animals. The allosaurus shown here (a) with a stegosaurus was probably a predator/scavenger of tremendous speed and ferocity. The duck-billed dinosaur (b) was probably a mild-mannered plant-eater.

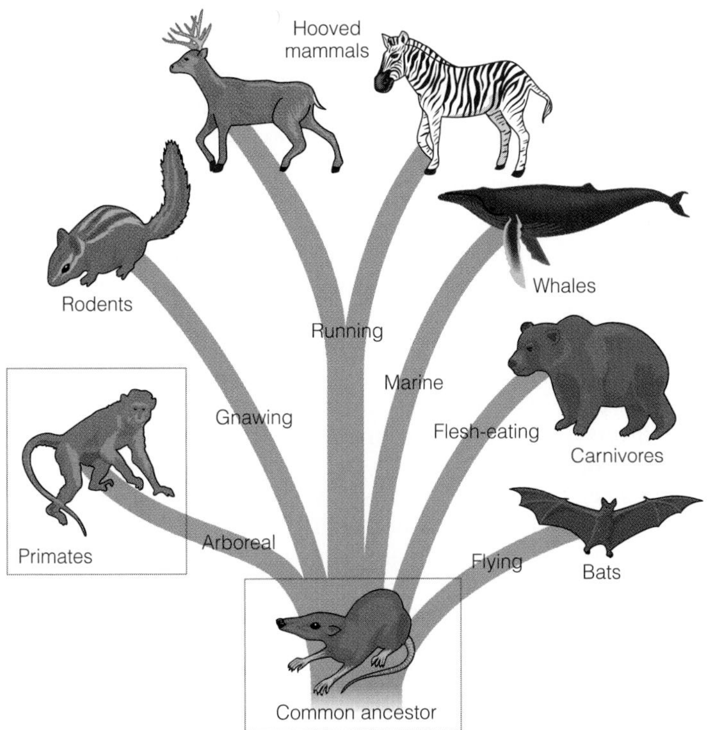

Figure 22–10 Cenozoic Era The Cenozoic era saw the rise of mammals from small ancestral mammals that survived the Mesozoic extinctions. Primates (box) were among the groups of mammals to evolve and will be considered in detail in conjunction with human evolution.

Figure 22–11 Ascent of Humans *Dryopithecus* appeared approximately 25 million years ago and was the first hominoid. About 10 million years ago, *Ramapithecus* gave rise to the great apes and to human ancestors, *Australopithecus afarensis*. A number of side groups evolved, including the extinct australopithecines (*A. robustus*, *A. africanus*, *A. boisei*). *Homo erectus* preceded *Homo sapiens*.

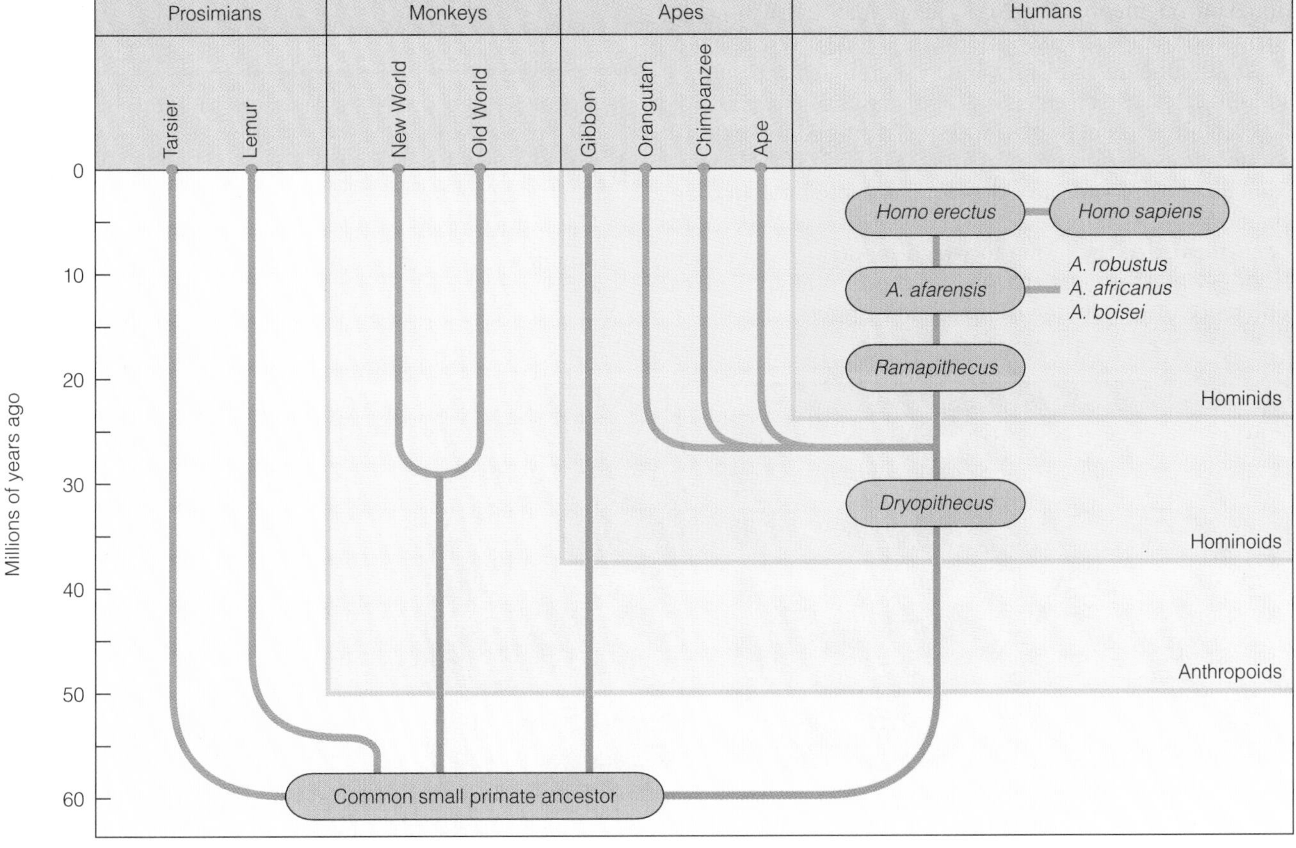

mate relatives. These organisms, including *lemurs* and *tarsiers,* have attributes that reflect generalized primate characteristics and were the first branch of the primate part of the tree of evolutionary divergence (Figure 22–12). Their eyes face forward, their hands can grasp with opposable thumbs and forefingers, and some species have flattened fingernails instead of claws. The present distribution of prosimian species is limited to tropical and subtropical Africa, Madagascar, and Asia, but 50 million years ago, they were found all over the world.

Anthropoidea

Monkeys, apes, and humans fall into the category of primates known as **anthropoids** (suborder Anthropoidea). All members of this group have a full set of flexible fingers, relatively large brains (reaching a maximum in humans), forward-facing eyes, and depth perception based on **stereoscopy.** The specific ancestors of this group are obscure, but probably arose about 20 million years ago during the Miocene period, in the form of a group of organisms called *dryopithecines* (Figure 22–11). Dryopithecines were hominoids, which means, among other attributes, that they were tailless after birth and had spinal curvature that suggested adaptation to partial or complete **bipedalism.** They were apelike creatures, who ranged in size from a small modern monkey to a gorilla. Much of their relationship to modern humans and apes is based on partial skulls and jaw fragments. The Miocene primates had a small brain and their face shape and teeth patterns were apelike.

At this juncture in history, a mere six minutes ago on evolution's clock, the divergence of apes began. *Ramapithecus* is an even more recent human ancestor, arising 8 million to 10 million years ago in India and Africa (Figure 22–11). The few skull and jaw pieces that exist are even more modern than their dryopithecine ancestors. *Ramapithecus* is considered to be a **hominid,** the group to which modern humans also belong, while the great apes are considered hominoids. The difference in hominid and hominoid distinctions is in the degree to which the traits of both relate to present-day humans. Hominids share more characteristics with modern humans (skull size, jaw, erect stature, bipedalism) than hominoids do. The split to the great apes may have occurred slightly before *Ramapithecus* appeared. However, it may be that the hominoids split off later than suggested in Figure 22–11.

It is interesting to consider detectives of human history at work. Based on dental structure, the eating habits of *Ramapithecus* are thought of as very humanlike. However, based on biochemical differences in blood proteins in present-day apes and humans, *Ramapithecus* is considered a side road from the highway to humanity. There is still much to solve in human ancestry. Fossil evidence has led to the conjecture that our early relatives, whatever they may have been, had attributes that included living on the ground, having an erect posture, and walking on two legs. It is much later, perhaps only 5 million years ago, that our closest primate relatives, the chimpanzees and gorillas, diverged from the human line (Figure 22–13). There is considerable controversy over this late date for separation of our human ancestors, but all the evidence taken together supports it. The steps that we are to take next in human evolution are the ones that make us most different from all other primates. These include evolution of the increase in size and complexity of the brain, formation of the hand, and development of a persistent erect stature.

Anthropoids primates of the suborder Anthropoidea to which humans and the great apes belong

Stereoscopy binocular vision that establishes a sense of depth perception

Bipedalism walking on two legs

Hominid a designation describing humankind and its ancestors starting with *Ramapithecus*

(a)

(b)

Figure 22–12 Prosimians
Shown in (a) is a Philippine tarsier and in (b) a ringtailed lemur. These species are limited largely to equatorial regions and are abundant on the island of Madagascar.

(a) (b) (c) (d)

Figure 22–13 Our Hominoid Relatives
The closest of our living hominoid relatives are the (a) chimpanzee, (b) gorilla, (c) orangutan, and (d) gibbon.

Australopithecines

This phase of human evolution took place predominantly in the savannas of eastern and southern Africa, beginning about 4 million years ago. The clock shows just less than two minutes before midnight, but better late than never for brain development! *Australopithecus afarensis (A. afarensis)* was a relatively new inhabitant of the fertile eastern plains of Africa at that time (Figure 22–14). They were ground dwellers, had a large brain as indicated by the structure of the head, stood erect, and left footprints in the 3.5-million-year-old layer of volcanic ash in present-day Ethiopia. Australopithecines were approximately 4 feet tall and may have weighed 40 to 50 pounds. The most famous of this group was **Lucy,** a diminutive female, whose bones compose one of the most complete skeletons of its type ever discovered (Figure 22–15). She did not stand alone. Entire families, or tribes, of individuals have also been discovered, and this strengthens the view that these individuals were smart and social. Many other features of the skeletal remains of australopithecines suggest a clear line to modern *Homo sapiens,* but our australopithecine ancestors were not there quite yet (Figure 22–16).

Why would the kind of characteristics observed in *Dryopithecus, Ramapithecus,* and *Australopithecus* have been selected in the first place? The evolution of a large brain would certainly have been important in reasoning and in developing tools. This evolutionary tendency to increase brain size would have benefited from the evolution of hands that were not only highly manipulative but continually, or

Lucy a member of the species *Australopithecus afarensis (A. afarensis),* a famous early hominid ancestor to humankind

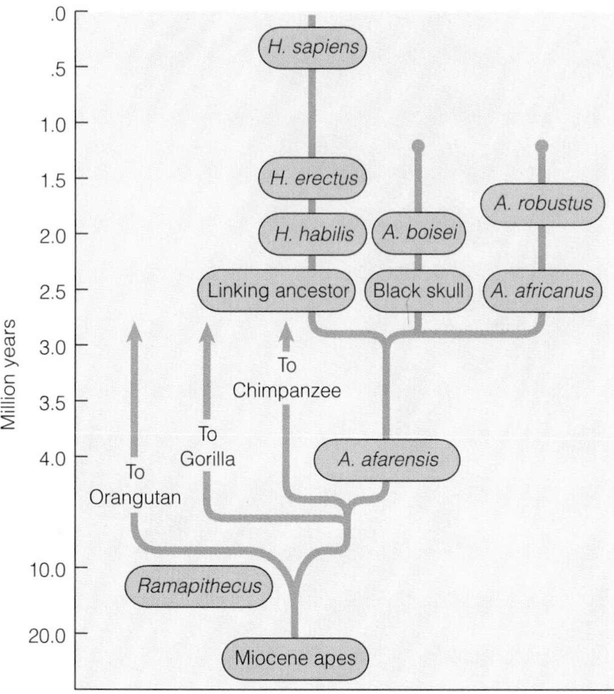

Figure 22–14 Evolution of the genus *Homo*

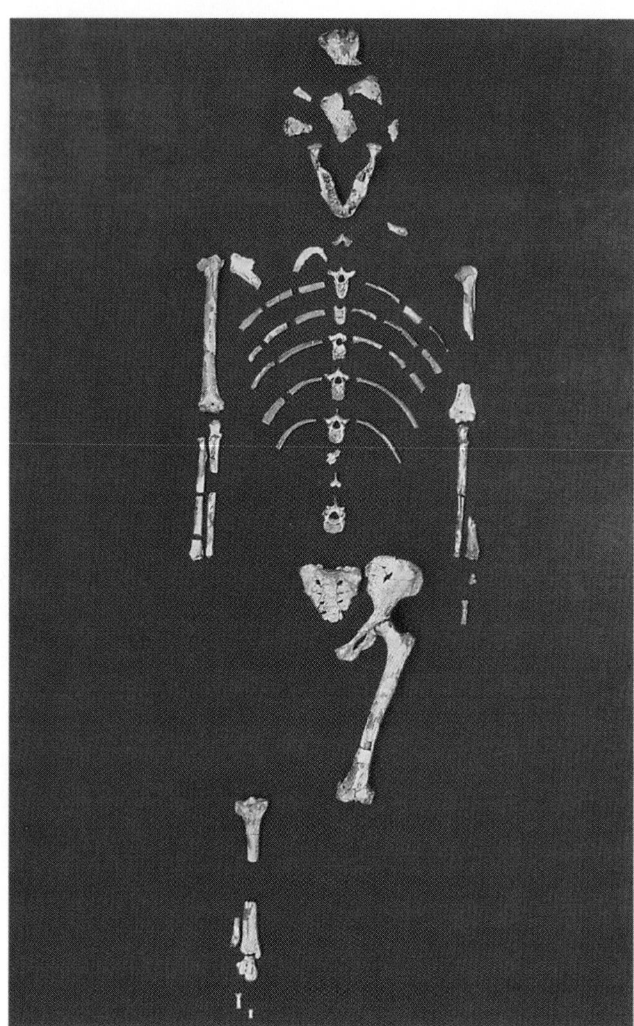

Figure 22–15 Lucy
The australopithecines were on a direct route to evolution of the genus *Homo*. Lucy (a) is *Australopithecus afarensis;* her remains are among the best of any remains from 3.5 million years ago in present-day Ethiopia.

at least predominantly, free from the need to balance the body as a quadruped. Hands that can utilize and adapt materials that the mind envisions for weapons or tools would clearly enhance success in many activities. An advantage gained by erect posture was the ability to see greater distances over the tall savanna grasses. Also, the behaviors associated with the excellent vision of primates would have been further enhanced by fully erect posture. The abilities to see well with less obstruction and to estimate distance would certainly have been important for success in both hunting for food and identifying enemies. Depth perception based on the spacing of front-facing eyes also would have been essential and had already evolved in primates.

An additional possibility, with respect to erect posture, is that uprightness evolved because it allowed for more efficient thermal regulation. Standing up exposes less of the surface area of the body to the intense radiation of the tropical and subtropical sun. This exposure of body surface from standing erect may have occurred in conjunction with the evolution of sweat glands, which could have been instrumental in allowing our ancestors to thermoregulate and, thus, to be active during the day. Most carnivores, who lack sweat glands, tuck themselves into the shade and await the cool of evening. Human evolution was not as simple as this makes it sound. Hindsight seems to give an inevitability to the importance of the changes described above, which in reality does not or did not exist. Clearly, a multitude of selective forces were involved in the changes that led to the continual success of hominids. The forces involved are complex and interactive with respect to natural variation in the adaptations of organisms to the environment. It is the change in the phenotype of individuals, gradual or rapid, subtle or obvious, that is tested by nature.

The Genus *Homo*

Time is getting short. The genus *Homo* appeared with less than one-half minute left on the clock. The first of the species of the genus *Homo* was *Homo habilis (H. habilis),* a descendant of *A. afarensis,* about 2 million years ago. There are several other species of australopithecines that also arose from *A. afarensis,* including *A. africanus, A . boisei,* and

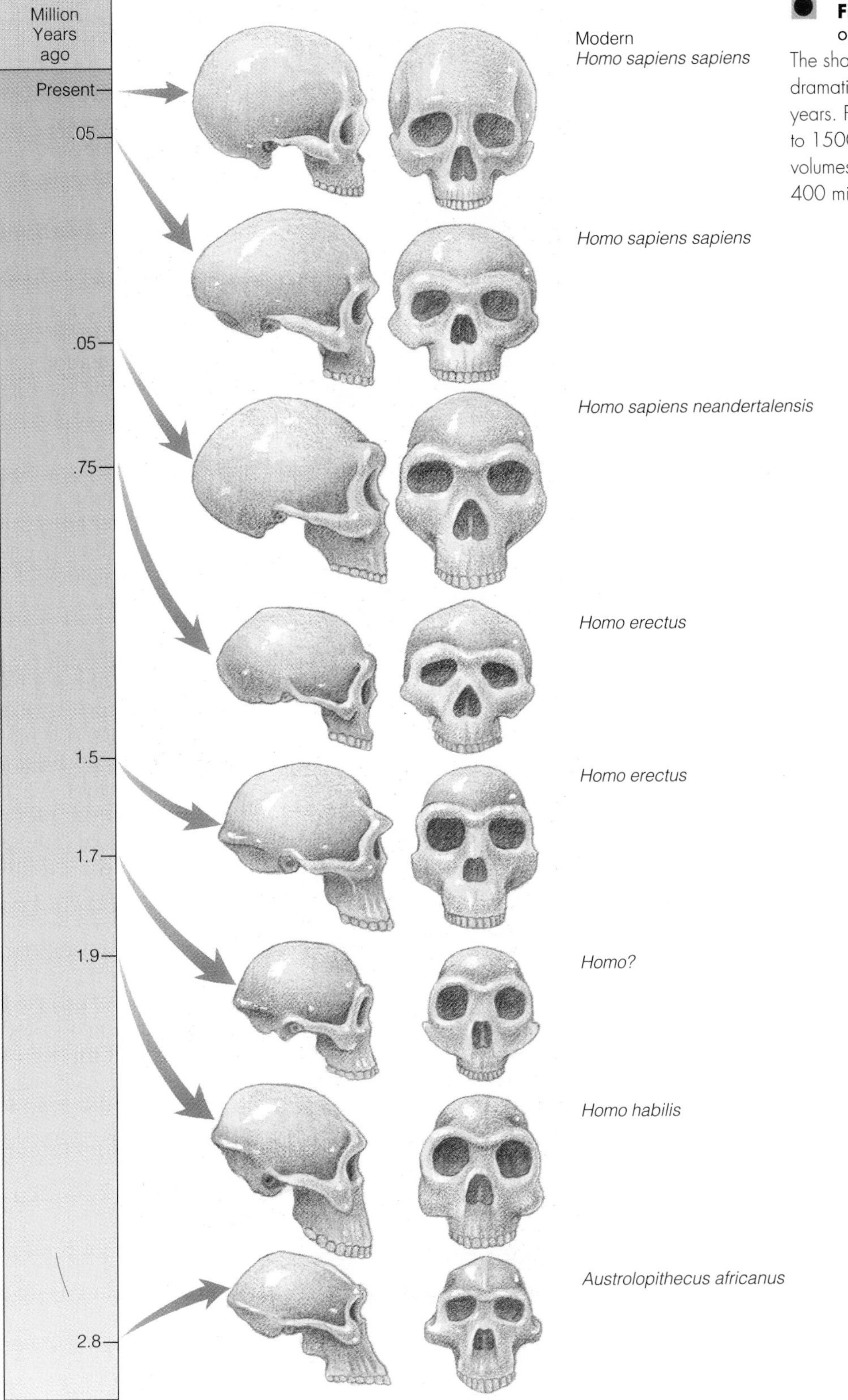

Figure 22-16 Recent Ancestors of Modern Humans
The shape of hominid skulls has changed dramatically over the last 2.8 million years. Present cranial volumes are 1400 to 1500 milliliters. Australopithecines had volumes estimated to have been 350 to 400 milliliters.

A. robustus, but they became extinct by approximately 1 million years ago (Figure 22–16). *Homo habilis,* as derived from Latin, means "able man" because this species was named for its ability to make and use tools. The discovery of tools with fossil remains of *H. habilis* indicates sophisticated use of materials from the environment in which this species lived. Stone tools, which were the first types to be produced, provided a way to kill, cut, crush, and manipulate prey and other materials. It takes a large, complex brain to conceive of tools and their use.

The next steps in our evolution, definable in the emergence of the species *Homo erectus,* saw an increase in brain size to near modern human capacity and a doubling of that observed in *H. habilis* and *A. afarensis*. *H. erectus* arose about 1.5 million years ago and persisted until 500,000 years ago. During that time, the species migrated to Europe and Asia, hunted, used fire, ate meat, and presumably had a complex social structure as reflected in the building of camps and occupation of caves (Figure 22–17). *H. erectus* did well in its environment but was eventually displaced by *Homo sapiens*.

Homo Sapiens *Homo sapiens* arose within the last 500,000 years and has proven to be one of the best-adapted and most widely distributed species on the face of Earth. Our early ancestors succeeded very well during and between periods of glaciation, which occurred periodically. Remnants of glaciers exist today at very high latitudes (polar) and altitudes (mountains) (Figure 22–18). These ancestors inhabited sites over an enormous area of Europe and Asia. One of the groups was the **Neanderthals,** named after a valley in Germany where the first fossil remains were discovered. Neanderthal humans were a relatively short-lived subgroup of *H. sapiens,* occurring within the last 75,000 years, and were the epitome of the stereotypical "caveman." They were good hunters, stocky and strongly built, and had large jaws and strong teeth (undoubtedly needed for their diet). They also produced some art, made tools, and buried their dead. They became extinct about 30,000 years ago.

The other group was the **Cro-Magnon** humans. These humans were more lightly built than their Neanderthal cousins but were apparently smarter. They made more sophisticated tools of bone and ivory and were organized into tribes with complex social structure and rituals associated with nature, such as formal burials. These two groups probably overlapped to some extent, though the fossil record is not clear on this issue. Missing pieces of the fossil record at many steps in the evolution of humans have made for controversy as to the exact chain of events. However, Cro-Magnon humans left an indelible legacy of art for us to ponder and appreciate. Deep within caves in Europe are the symbolic works of human artists no less inspired or talented than their modern counterparts (Figure 22–19). With such demonstrations of art and ritual as have been discovered, we can see the clear path for the emergence of modern humans.

MIGRATION AND ISOLATION

Migration

However, the story of modern humans involves one more factor—*migration* and **geographical isolation.** During the last ice age, about 20,000 years ago, groups of *H. sapiens* migrated over land on ice bridges from Asia to the Americas. The original pioneers of America came through Alaska to a new continent and not directly from Europe. With the recession of the glaciers 10,000 to 15,000 years ago, we come to the time of recorded history, or at least some physical traces of events and activities of humans. The stories of great floods early in human history may in some cases reflect the mythological transformation of oral histories of peoples who were affected by the melting of mile-thick glaciers and mountain ice pack. As the huge masses of ice melted and receded, the waters filled up the oceans and seas. Water once again separated the continents and isolated the migrant peoples from their origins in Asia, Africa, and Europe for the next 15,000 years. The coming of modern Europeans to North and South America within the last 1,000 years ultimately led to the demise of the descendants of many of the original pioneers.

Isolation

The separation of populations of people by physical barriers, such as mountains, rivers, and seas, represents what is called geographical isolation. All species are affected by physical separation in direct and indirect ways. First and foremost, individuals in the separated populations are prevented from interbreeding. This means that the available gene pool of the species is potentially reduced. The gene pool is the total complement of genes carried collectively by individuals in a population. By interbreeding among many individuals, new combinations of genes can be passed on to their offspring. Individuals expressing unique sets of genes are acted upon by the forces of natural selection. Recombination of genes associated with gamete formation (meiosis) is an important way in which variation may be introduced into a population (see Chapter 20). Unlike mutation, variation introduced by recombination does not increase the number of alleles of a gene. However, it greatly expands the possible mix of different genes individuals may receive from their parents.

Neanderthals a subgroup of *Homo sapiens*

Cro-Magnon a subgroup of *Homo sapiens* from which modern humans evolved

Geographical isolation a separation of members of a species by physical barriers, for example, a mountain range or body of water

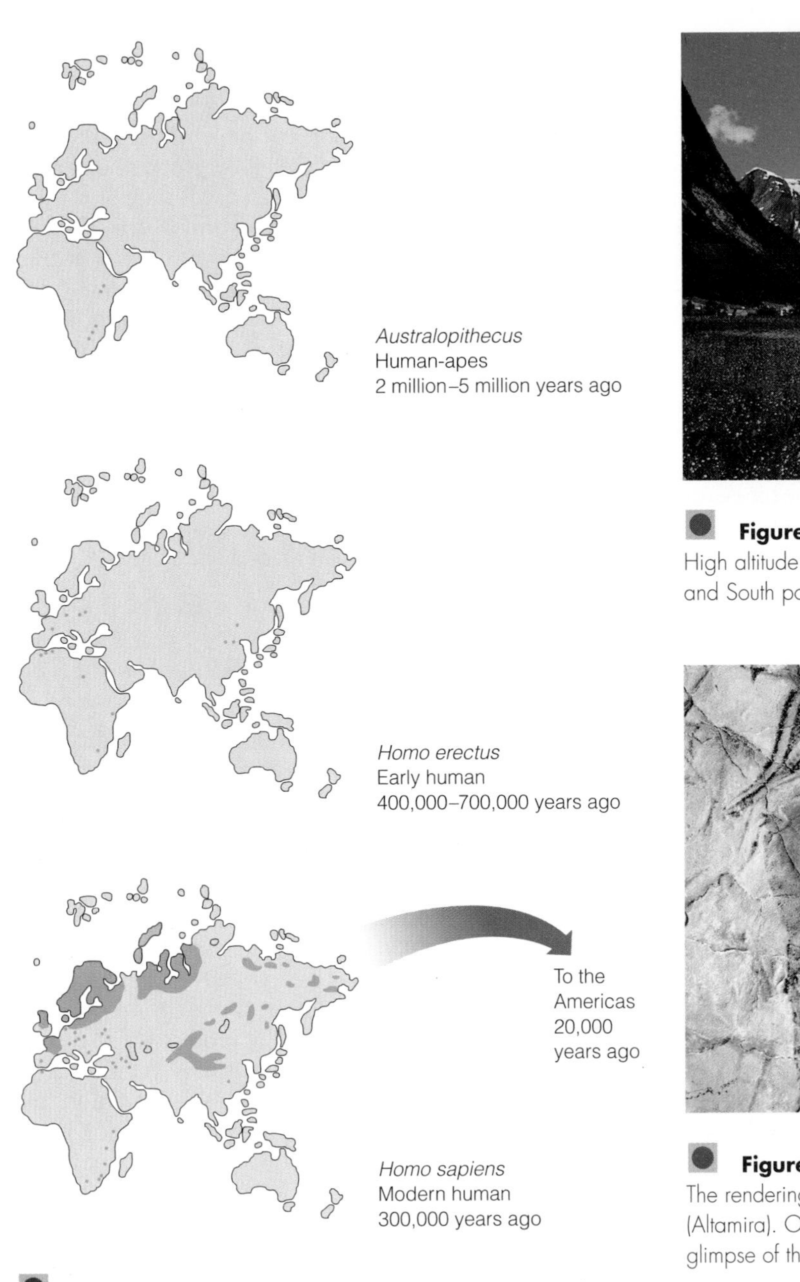

Figure 22–17 Geographical distribution of *Australopithecus, Homo erectus,* and *Homo sapiens*

Figure 22–18 Glacial Remnants
High altitude glaciers remain today in mountain valleys. The North and South polar regions also have huge, permanent glaciers.

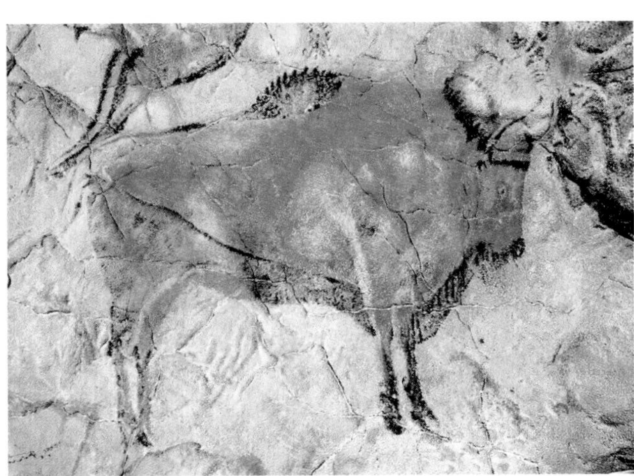

Figure 22–19 Cave Art
The rendering of a horse was discovered in a cave in Spain (Altamira). Other caves in Spain and France have provided a glimpse of the skills and technical expertise of early humans.

Randomness in the choice of a mate is one of the advantages of sexual reproduction. For humans, it results in the chance for continual mixing and flow of genes in the population. The diversity of variants depends on the size of the population and the complexity of the gene pool. By isolating one group from another in different environments, each separate group will evolve somewhat differently over time under the influence of natural selection. Individuals of such groups are still members of the same species, but continued separation and introduction of variation through mutation may well lead to what are called *subspecies* or *races*. This idea underlies Darwin's concept of gradualism, as the basis for forming new species. Combine the physical separation of subpopulations through continental drift, recession of glaciers, the disappearance of land bridges, and other geophysical phenomena on Earth's surface with mutations, recombinational variation, and natural selection and you can begin to see how important Darwin's ideas about the origins of the diversification and evolution of life really are.

Genetic Drift

What kind of circumstances may be involved in altering the phenotypes of populations? If a separated group of individuals of a species is quite small, the chances are that many alleles of the original gene pool will be absent. The result is quite different from the outcome of natural selection acting on individuals of the entire population. This process of separation and isolation is called **genetic drift** and may account for a fairly rapid evolution of new species. In a normal, large population of a species, the frequency of genes is quite stable. With genetic drift, chance governs which individuals may be isolated and, thus, what genes they carry. How might this process occur? Imagine a flock of birds flying along the coast during a bad storm. A small group of birds within the flock is blown off course and swept out over the sea. When the storm ends, the birds find themselves on an island far off the mainland. They have no sense of how to get back, assuming they would desire to do so. This is an accidental isolation of a subgroup of a species, which as a separate group has a relatively limited gene pool. The island turns out to be a fairly nice place, but it is not exactly the same as their previous habitat. The new inhabitants of the island are essentially starting a subpopulation with a unique set of genes, the products of which can be acted upon by natural selection. Because of this, the subpopulation will have a chance to change in ways that were not possible for the original mainland population.

Humans and Geographical Isolation

Geographical isolation has influenced human evolution as well. The migration and isolation of humans from Africa to Europe and to Asia, as well as to the North and South American continents, has led to distinctive changes in human phenotypes. There are three main groups or races of humans commonly recognized, though many more are possible to construct, depending on the criteria used. The three major races are the negroids, caucasoids, and mongoloids. The first is largely African in origin, the second European, and the third Asian. The distinguishing difference among the races is skin color (black, white, and yellow, respectively). However, there are many other differences as well, particularly in facial features, such as eye shape, and in biochemical differences, such as frequency of blood types. The concept of human races is difficult to sort out. A great many factors are involved. Biological, racial differences reflect the natural and expected buildup of variations in genes or in gene frequencies due to thousands of years of separation of potentially interbreeding populations. Different environments provide different selective pressures. Phenotypic and genotypic differences among individuals of isolated populations arise from selective pressures unique to the environments in which they live.

Race and Status

The problem with respect to humans in particular is that the concept of race has negative social connotations. Variation in the color of skin or shape of eyes, resulting from evolutionary changes affected by geographical isolation, has all too often been equated with superiority or inferiority and lesser or greater status and worth. Those in positions of political or economic power often feel their exalted positions are a reflection of inherited characteristics that make them inherently better or smarter. In the world of "survival of the fittest," such thinking is called **social Darwinism,** since it equates social status and economic success with genetic superiority. Throughout history, differences in the way people look or behave have been used against them. You need consider only the institution of slavery prevalent in the United States during the eighteenth and nineteenth centuries to see the far-reaching effects of perceptions of social inequality.

EVOLUTION AND MACROEVOLUTION

There is no doubt that evolution occurs. The evidence for evolutionary change is all around us, from erosion and weathering of Earth's surface to the presence of the fossilization of organisms trapped in sediments laid down millions of years ago. These changes require long periods of time to occur, a fact which fits well within the time frame needed for the evolution of life. What is less certain are the mechanisms involved in bringing about specific organismal changes and long-range changes in biodiversity. The general mechanisms by which evolution is thought to occur are theoretically sound, but because the processes are so complex and interactive, the precise details have yet to be completely worked out. As we have discussed, mutation and genetic recombination are known to lead to variation in the genotypes and phenotypes of individuals in a population. The fitness associated with those variations among individuals establishes a basis for competition for resources and mates. Fitness means a longer life and more opportunities to mate and produce offspring. Add to this competition equation the ever-changing influence of patterns of climate and environment, and the basis for the effects of geograph-

Genetic drift a significant change in gene frequencies in a population, resulting from random causes, usually associated with geographical or reproductive isolation

Social Darwinism the application of Darwinian principles to the inherent status of human groups

ical isolation and genetic drift come into sharp focus. We are all part of a complex set of molecular, behavioral, organismal, and environmental interactions that drive the evolutionary process. The argument that the theory of evolution has been disproven or is in serious doubt is similar to saying that the world is flat or that the Sun rotates around Earth. Evolution is a fact. New ideas about the mechanisms of evolution continue to be put forth, both because of how important it is to know how it works and because of the interest we have in knowing how we got here.

Macroevolution

Evolution of a species, or **microevolution,** may occur by a slow accumulation of small changes in the genome of a population, which in turn changes the genotypes and phenotypes of individuals. Alternatively, evolution may occur relatively rapidly, or both. The life history of a species is a family tree. Little is known about how a new species arises. However, once a species is established, it may maintain itself for long periods of time.

Evolution above the species level is called **macroevolution.** Even less is known about how macroevolution works, but it is related to events such as mass extinctions (for example, the Permian extinction), or rapid expansion of diversity of entire new groups of organisms, as observed in the Cambrian explosion and the ensuing Paleozoic era, when nearly all the major groups of modern organisms appeared. Macroevolution theory attempts to explain how this diversity arose and how natural selection works on the entire, ever-changing ensemble of organisms inhabiting this planet.

Gradualism and Preadaptation

One of the problems confronting evolutionary biologists is how complex structures, whose all-or-none function depends on the interaction of many parts, could evolve gradually. For instance, how did the vertebrate eye evolve? An eye is one of the most complex structures produced in the history of life and in vertebrates is derived from complex interactions involving the skin and the brain during embryonic development. It could not have arisen as a complete structure and cannot function without all the parts. To function in vision, an eye must have a precise shape and completeness to it (see Chapter 18). It must have a cornea, a lens, a vitreous humor, and a retina that connects to the brain. Ten percent of an eye is not enough. If eyes evolved gradually from a simple pre-eye form to the complex functional units we know today, the predecessors to the modern vertebrate eye must have had a somewhat different and more simple function than vision. The principle of *preadaptation* provides a clue for the evolution of this complex structure.

What if the predecessor to the eye did not originally evolve to be used for vision? At many points in the evolution of multicellular organisms, variation among individuals in a population lead to more complex capacities as a way to increase fitness. Perhaps the ability of some cells to respond to light provided an advantage because an organism could orient positively with respect to light. This phototrophic ability might have been important in allowing an organism to stay near the surface of water for feeding or acquiring oxygen. A subsequent step may have been the ability to track movement of objects in the presence of light. This would require a considerably more complex structure but could not have arisen had there not first been the ability to detect light. These changes eventually gave rise to species with full vision and complex eyes. The point to be made here is that preadaptation, in this case light detection, was a relatively simple function that later was useful to build upon. Preadaptation is potentially very important in the evolution of many kinds of traits for which today there is an all-or-none quality.

It should be noted that the evolution of such adaptations as vision is not limited to vertebrates. Invertebrates have undergone analogous but separate functional evolution. The eyes of invertebrates are dramatically different in structure from those of vertebrates, yet they serve the same purpose. The eyes of insects and mammals are considered analogous to one another. This means that although the eyes of this group are functionally similar (allowing for vision), they are structurally dissimilar (in cell types and embryonic origins). This is an example of *convergent evolution,* in which the final functional form of a system arising from natural selection is similar, but the structure and means by which it evolved are genetically and developmentally entirely different (Figure 22–20). Convergent evolution is observed in many features of organisms evolving in geographic isolation in similar environments and is reflected in the properties of **analogy** and **homology** (Figure 22–21). The eyes of mammals, on the other hand, are considered homologous to each other because they are similar genetically, developmentally, and structurally. Eyes

Microevolution the changes in appearance of a species over generations

Macroevolution the large-scale changes (origins and extinctions) of species and higher taxonomic levels over geological time

Analogy functionally related structures of dissimilar genetic or developmental origin; for example, wings in flies and bats

Homology structures arising from genetically and developmentally related processes; for example, forelimbs in vertebrates

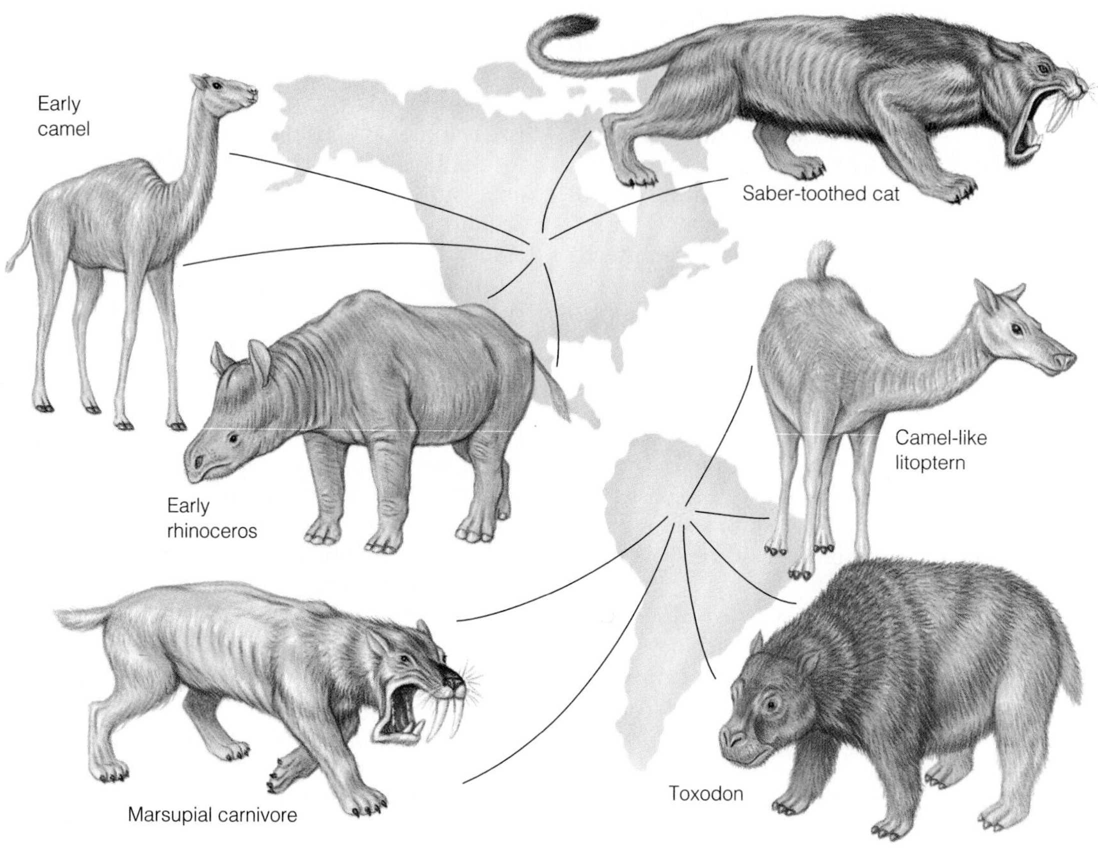

Figure 22–20 Convergent Evolution
Convergent evolution is the development of similar phenotypic features in animals separated on different isolated continents.

in all vertebrates are considered to have arisen by *parallel evolution* (Figure 22–22).

Consider also the function and structure of limbs, wings, and lungs in vertebrates and invertebrates as further evidence of convergence of functional adaptations and examples of homology and analogy. Limbs in terrestrial vertebrates arise from groups of cells that organize along the long axis of the body during embryogenesis. Limbs in invertebrates, particularly insects, arise from clusters of cells within larvae that have no clear relation to the adult insect body axis. The relationship between limbs in these two groups is one of analogy (Figure 22–21). Wings in terrestrial vertebrates are derived from forelimbs, and so are homologous structures in birds and mammals (bats). Wings in insects, as in the case of legs, are derived from clusters of larval cells showing little or no relationships to vertebrates. Thus, wings in flies and birds are considered analogous to one another. Lungs in mammals are specialized structures deriving from endodermal cells in the embryo. The equivalent, analogous air-delivery system in insects is represented by a series of *tracheae* that form passageways from the exoskeleton to the internal tissues of the body.

Human history is but a small episode in the general and gigantic process of evolutionary change over time and space. It is only one of many patterns of diversification of life on Earth. It is assumed that 3 billion to 3.5 billion years ago life began as simple organisms in ancient seas. What such organisms were like and how they arose is a mystery. During the last several hundred million years, life forms have evolved into tremendously complex and diverse types. The forces involved in bringing about change are continual, and the evolution of organisms from simple to complex seems a natural progression. However, evolution should not necessarily be thought of as progressive. Progress implies purpose, and the basis for evolution is ultimately chance and necessity. Variation and selection allow one phenotype in a population to succeed and survive longer than another and to produce more genotypically similar offspring. There is no clear-cut underlying purpose to the evolution of life. What purpose we have as human beings, we create for ourselves.

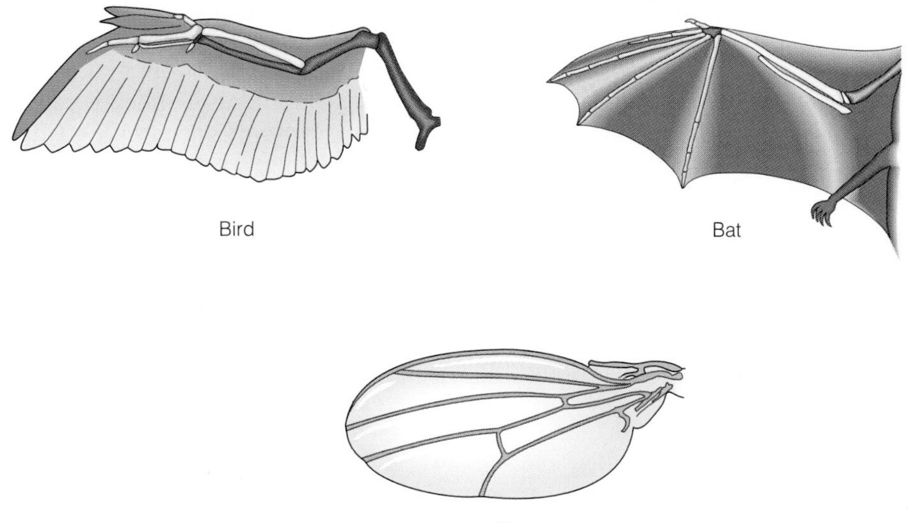

Figure 22–21 Homology and Analogy
The wings of a fly, bird, and bat are functionally similar—all are used in flight. However, the embryological origin of a fly's wing is from nonlimb larval tissues, which is entirely different from that of a bird or bat. Organs that are functionally similar but arise from dissimilar structures during development are considered analogous. The embryonic origin of wings in birds and bats is similar, both arising from the forelimb during development. As such, they are considered homologous.

Figure 22–22 Parallel Evolution
The kangaroo rat and jerboa are closely related species that have evolved similar features independently. This is a form of parallel evolution.

Summary

Charles Darwin's ideas have profoundly influenced the way we think about the diversity of life on Earth. He was the first to propose a workable theory of evolution of species. Evolution is the process of change in species over time. The time required in the case of evolution on Earth is measured in billions of years. Evolutionary change is brought about by altering the genetic blueprint for making an individual and then competitively testing individuals expressing those variations for success in the environments in which they live. Alterations in the genetic blueprint arise from mutation and, particularly in higher organisms which reproduce sexually, from the recombination of genes. In any population, there is a gene pool that represents an assemblage of all the genes (the genome) in a species, but not all genes (alleles) of the pool are found in any one individual. Unique combinations of alleles result from the mating of individuals, for whom mate selection is largely a random process. The constant mixing of alleles in the interbreeding population prevents genetic drift. Genetic drift occurs in populations in which individuals with a limited set of genes of the gene pool are accidentally separated from the population to which they initially belonged. A few birds being isolated on an island by storm is an example of such an accident. It is highly unlikely that natural selection could have acted in such a manner as to isolate or select for that particular set of genes. Geographical isolation leads to reproductive isolation and potentially to the evolution of new species.

The evolution of life on Earth has been compressed in this chapter into a single day for purposes of comparison and has involved both geological and biological clocks. Human evolution has been presented as a continual process, whose origins could be traced to chordates and vertebrates from the Paleozoic era. Particular emphasis has been placed on the descent of humans from their primate ancestors beginning 60 million years ago. A plausible sequence of events suggests the relationships we have to monkeys, chimpanzees, and apes. It is clear that humans *did not* descend from apes, but rather underwent evolution along with apes. *Homo sapiens* arose within the last 500,000 years, a mere 10 seconds before our chapter's biological clock struck midnight at the end of evolution's day.

Earth has been in existence for approximately 4.5 billion years, and life has been evolving for 3 billion to 3.5 billion years. Earth is unique in this solar system and perhaps in this region of the galaxy. It is the only planet with resident life. The conditions for the initiation of life have been obscured by time and remain controversial, but there is little doubt that the conditions needed to maintain life and to allow its evolution are inherent to the land, waters, and air encompassing Earth. The conditions on Earth have established a sphere of life, the biosphere, which is discussed in the next chapter.

Questions for Critical Inquiry

1. Why is it incorrect to say that humans are descended from apes?
2. What are some mechanisms for generating genetic variation in a species?
3. Compare and contrast Lamarckian and Darwinian views on the evolution of the neck of giraffes. What is the major problem with Lamarck's hypothesis explaining how the neck evolved?

Questions of Facts and Figures

4. What are some of the major factors affecting evolution?
5. What are the major eras of life? When did life arise on Earth? What form(s) did early life likely take?
6. When did the first mammals appear? The first primates? The first humans?
7. Why was it important for Darwin to have gone on a five-year, round-the-world survey on the HMS *Beagle?*
8. In the evolution of complex characteristics, such as the evolution of the eye, what role does preadaptation play?
9. What effect does genetic drift have on the gene pool?
10. What is the basis for the evolution of differences associated with human races?

References and Further Readings

Bowler, P. J. (1983). *Evolution: The History of an Idea.* Berkeley, CA: University of California Press.

Cummings, M. R. (1996). *Biology: Science and Life.* St. Paul, MN: West Publishing.

Monroe, J. S., and Wicander, R. (1994). *The Changing Earth.* St. Paul, MN: West Publishing.

Patterson, C., ed. (1987). *Molecules and Morphology in Evolution: Conflict or Compromise.* Cambridge: Cambridge University Press.

Raup, D. M., and Jablonski, D., Eds. (1986). *Patterns and Processes in the History of Life.* Berlin: Springer-Verlag.

CHAPTER 23

The Biosphere

INTRODUCTION

A view of life at the largest scale must include a consideration of its diversity and distribution over the entire Earth. The surface of Earth is composed of realms of solid (the crust), liquid (the oceans), and gas (the air), each of which plays a special role in supporting life. In fact, these layers have such special attributes that Earth is unique among all known planets in having provided both the means for initiation of life and conditions to sustain its evolution. The biodiversity of the planet is so vast that even though millions of species have been categorized to date, they may represent only a small percentage of the estimated total. Major inquiries for us in this chapter focus on what types of environments support life, what types of life evolve within them, and how the balance of life is endangered by pollution. The study of the interactions of organisms in their environments is called ecology, and the series of planet-spanning regions in which life abounds are called ecosystems. The ecosystems taken together form an elaborate, interactive, and delicately balanced sphere of life known as the biosphere. The characterization of ecosystems and the balance of life and energy within them concern us in this chapter.

THE SURFACE OF EARTH

The surface of Earth is composed of three fairly easily defined layers known as the *lithosphere,* the *hydrosphere,* and the atmosphere (Figure 23–1). Each of these layers is crucial to the composition of the different ecosystems and to the life within them. The lithosphere is the solid crust of Earth, the hydrosphere is the water that fills the vast basins of oceans and seas, and the atmosphere is the envelope of air that blankets Earth. In addition, each of the layers is affected by radiation from the Sun, which, as will be discussed later, provides the primary source of energy for life. The interfaces of lithosphere, hydrosphere, and atmosphere form the biosphere, which is a relatively thin but essential zone of life at Earth's surface. Almost all living organisms are found within a few meters down into the earth or water and a few dozen meters above the ground in the air.

The Lithosphere

The lithosphere is the solid crust of Earth, the ground upon which we stand, the mountains we climb, and the basins over which the waters of the world flow (Figure 23–1). The composition of the continental crust is mineral in nature, with an abundance of the elements silicon, oxygen, aluminum, calcium, potassium, sodium, and iron making up more than 90% of the mass (Table 23–1). There are three major types of rocks found in the crust—*igneous, sedimentary,* and *metamorphic.*

Igneous rocks have as their source the molten lava pushed up through the crust during volcanic activity or cooled slowly below the surface. As the molten materials cool on or near the surface of Earth, they form a variety of rock types including pumice, obsidian, granite, and basalt. Mountain building and volcanic activity are often closely related. For example, it is granite and basalt that compose

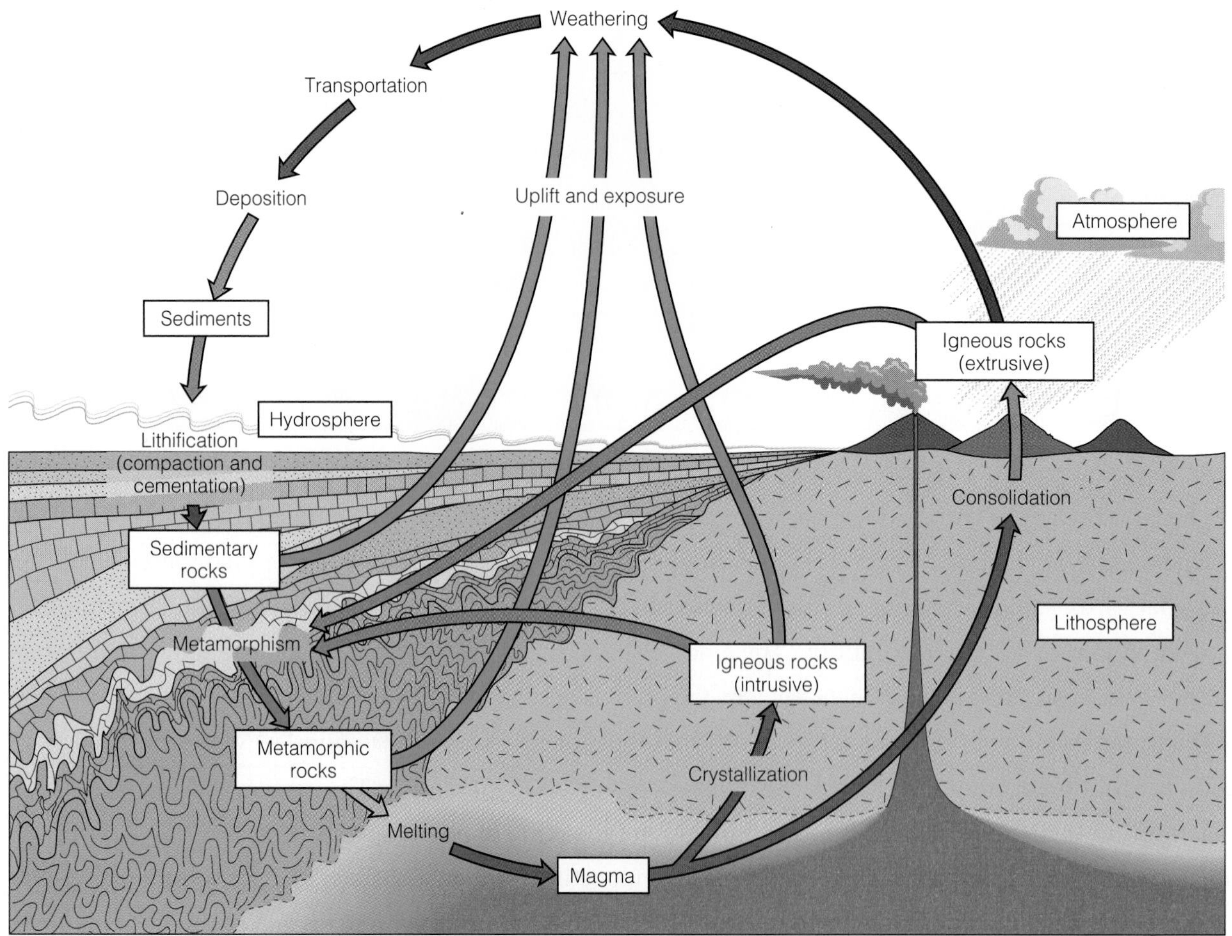

Figure 23–1 Lithosphere, Hydrosphere, and Atmosphere
The lithosphere, hydrosphere, and atmosphere are interrelated. The rocks of Earth undergo cycles of formation and transformation.

Table 23-1 Abundance of Some Chemical Elements in Earth's Crust

ELEMENT (CHEMICAL SYMBOL)	RELATIVE ABUNDANCE (%)
Oxygen (O)	62.5
Silicon (Si)	21.2
Aluminum (Al)	6.47
Sodium (Na)	2.64
Calcium (Ca)	1.94
Iron (Fe)	1.92
Magnesium (Mg)	1.84
Phosphorus (P)	1.42
Carbon (C)	0.08
Nitrogen (N)	0.0001

the Rocky mountains of the central United States and the Sierra Nevada of the West. The lithosphere overlies a hot, molten core, which in some regions may be as close as 5 to 6 miles beneath the crust (Figure 23–1).

Sedimentary rocks form from the erosion and breakdown of igneous and metamorphic rocks. This erosion gives rise to progressively smaller and smaller pieces of rock through the action of wind, rain, friction, and cycles of freezing and thawing. For example, *sand* and *silt* are carried by rivers to lakes and seas, where they are deposited as sediments, compacted, and eventually form hard thick layers. Sedimentary rock also forms from the mineralized remains of dead aquatic organisms that sink down and collect into thick layers on the bottoms of lakes and seas. The sedimentary rocks of Earth are built up layer upon layer over millions of years and contain the remains of once living organisms as part of the fossil record.

Metamorphic rocks arise when igneous or sedimentary materials are subjected to great heat and pressure deep within the crust. As a result, they do not melt completely but do undergo chemical changes. Because such rocks are severely altered by heat and pressure, fossil remains are destroyed. A good example of a metamorphic rock type is marble, in which the flow of melted rock can be seen. The interactions between the lithosphere, the hydrosphere, and the atmosphere have occurred over vast stretches of geological time and are extremely complex.

How do these three layers of Earth interact? Much of the lithosphere is porous, as is a sponge, and a large percentage of Earth's fresh waters are contained within it. It is interesting and important to note that only 0.01% of Earth's water is available as fresh water. The lithosphere acts as a gigantic filter system, which helps clean and purify the waters. These are the waters that we and the majority of living terrestrial organisms consume and count on for life. Many people rightfully worry that ground pollution from human activities may become so extensive that it will alter the capacity of Earth to carry out this vital purifying role. There have been many instances of ground water contamination that have led to poisoning of public and private water supplies. If the quality of water is compromised, the quality of all life also is compromised.

The Hydrosphere

The hydrosphere is the water that fills the oceans, seas, rivers, and lakes. More than 99.9% of this water is the saltwater of oceans and seas (Figure 23–2). Earth's huge masses of water wash away most of the debris, waste, poison, and toxins that are put into them by human activity. Water is the great diluter of contaminants. In addition, organisms of many types living in oceans, seas, and lakes detoxify much of the waste through their metabolic activities. However, humans today in many environments are adding more contaminants than can be effectively diluted or efficiently detoxified in many regions of the world. Contaminants are produced during industrial manufacturing and processing, as well as from activities such as agriculture and mining. The lithosphere and hydrosphere combine to protect the water users of Earth from hazardous materials, but their ability to do so is not limitless.

The Atmosphere

The third partner of the biosphere is the atmosphere. The atmosphere is the envelope of gases that surrounds Earth. It

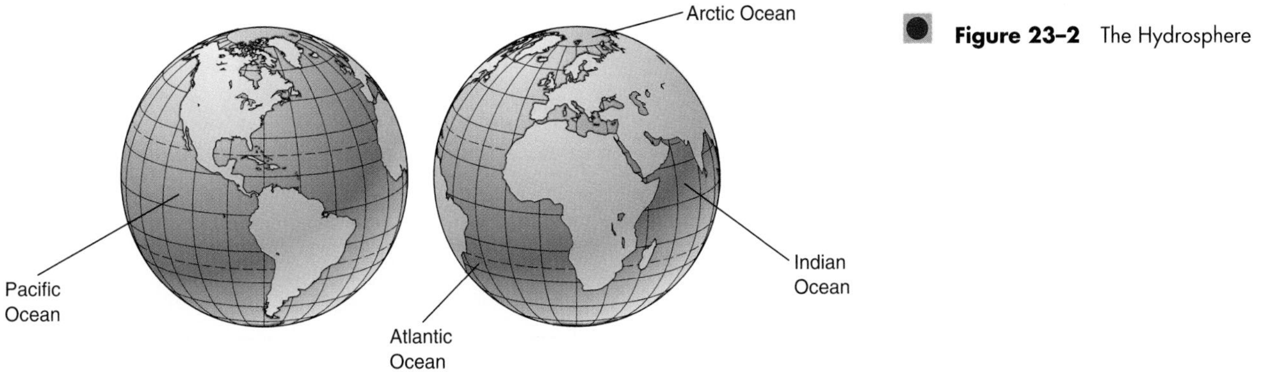

Figure 23–2 The Hydrosphere

is several miles thick and is rich in the gases that are of preeminent importance to life—oxygen, carbon dioxide, and nitrogen (Figure 23–3). Oxygen makes up 20% of the atmosphere and carbon dioxide makes up about 0.03%–0.04%. The difference is made up of nitrogen. As previously discussed in Chapter 8, oxygen is consumed by animals in order to efficiently run their metabolic processes, and carbon dioxide is produced as a waste product. Carbon dioxide is converted by plants into complex carbohydrates (that is, sugars and cellulose) utilizing the energy of the Sun, which is captured by photosynthesis. Atmospheric carbon dioxide is also important in regulating the temperature of Earth's surface, as we will see later in this chapter in a discussion of the **greenhouse effect.**

Contamination of the atmosphere with excessive carbon dioxide and other airborne compounds is a major concern for all of us. The cars and industries of modern human civilizations fill the air with pollutants, which diffuse readily within the atmosphere. The volume of Earth's atmosphere is huge and can withstand and dilute many dangerous gaseous and particulate wastes released into it. However, as is the case for the lithosphere and hydrosphere, the atmosphere does not have unlimited capacity and is not immune to overloading. Air, as water, knows no international boundaries. The winds of the world blow where they will and carry with them the wastes of human industry. We all breathe from the same reservoir of air and, if it is despoiled, we all share a common fate.

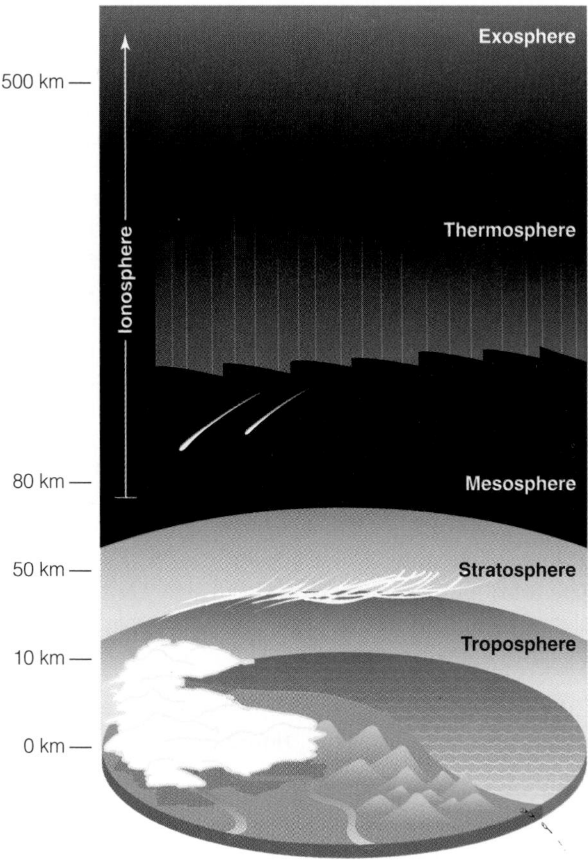

Figure 23–3
The atmosphere consists of approximately 79% nitrogen, 20% oxygen, small amounts of carbon dioxide, ozone, and water. The average surface temperature of the planet is 15°C. The region of the atmosphere nearest Earth's surface is the troposphere.

THE DISTANT SUN

The Sun is a distant and essential benefactor of Earth. It is the single most important source of energy for life. However, it is critical that the distance between Earth and Sun be what it is, an average of 93 million miles. If it were much closer, Earth would probably have been too hot for life to evolve and persist. Our planetary neighbor Venus is nearer to the Sun and has a most inhospitable atmosphere. Venus has an atmosphere composed in part of sulfuric acid and high levels of carbon dioxide, with an average surface temperature range of 460°C–475°C. Not much life as we know it can exist under such dire conditions. Thus, a location closer to the Sun than Earth is clearly a disadvantage.

Being farther away is no more advantageous. Mars is the next planet out from the Sun. At 135 million miles away, it is a cold and austere red planet with average temperatures well below 0°C and very little atmosphere at all. The unmanned American Viking probe to the planet was able to collect and analyze samples of soil for the presence or remnants of life. All the tests were negative. Recent discoveries of putative organic materials from a meteorite whose origin was the surface of Mars have led some scientists to suggest that there was once life on Mars. This is still a controversial issue among planetary scientists, but it is of tremendous biological importance in establishing the origins of life, which may well have occurred in other places in our solar system and beyond. There are clearly no complex life forms on Mars and therefore not much chance for a war of the worlds!

What makes Earth uniquely suited to support life is that the Sun provides enough energy to heat our world to an average temperature of approximately 15°C–20°C, ranging from record lows well below 0°C (at the poles) to a high of around 55°C (in the deserts of North Africa). Within the aver-

Greenhouse effect natural effect of Earth's atmosphere on the heat retained from solar radiation

age range of temperatures on Earth, the molecule H_2O is in its liquid form, water. Water is the molecular key to life. If water becomes too hot, it boils away as vapor, as on Venus. If it is too cold, it solidifies permanently to ice, as on Mars and more distant planets in our solar system. So far in the exploration of space, only Earth has this remarkable fluid, water, and only on Earth has life occurred and evolved. Recent discoveries on Jupiter's moon Europa suggest water may exist under a thick layer of ice. It would be reasonable to assume that if water were present on Mars or Europa at some point in their histories and on other planets circling stars such as our Sun, life would have a chance to start and evolve.

The biosphere blankets the surface of Earth and is roughly 12 to 13 miles thick. This estimate includes a measure from the deepest trenches of the ocean (the Mariannas Trench in the South Pacific, nearly 7 miles deep) to the highest peaks of the Himalayas (Mt. Everest, roughly 6 miles up) (Figure 23–4). Relative to Earth's size, the biosphere appears about as thick as the peel on an apple. It provides a thin but effective buffer against the endless night of the universe, with just the right combination of solids, liquids, and gases for life to have initiated, to have evolved, and to flourish.

SPECIES, POPULATIONS, COMMUNITIES, AND ECOSYSTEMS

There is great diversity of form and behavior of life on Earth. There is also great similarity among the diverse types. Functionally, all organisms need to acquire food to eat and water to drink. Animals in particular must also provide for adequate means of defense of self and young and establish proper circumstances for reproduction. The difference between any two species, and by extension all forms of life, is that chance genetic and phenotypic variations in ancestral forms occur regularly and have been selected for by natural selection (see Chapter 22 for details). The fittest of these organisms go on to reproduce and establish relatively stable groups that are well adapted for their environments. The study of the relationships between extant organisms and their environments is known as **ecology.** The individual organism within a species is the basic biological unit of both evolution and ecology, but an individual cannot survive alone. It must be part of a group called a population. A population is a collective of socially interactive individuals of a species that shares a common gene pool (Figure 23–5, and see Chapter 20). Another way of stating this is that a population is a group of organisms of the same species that interact and breed together to produce viable and fertile offspring. As discussed in Chapter 22, different populations of the same species may exist in isolation from one another as in geographical isolation, but these populations do not normally have a chance to interbreed, although they can if brought together again.

The next level of interaction involves populations of different species living together in the same environment. This level of complexity is called a **community** (Figures 23–6 and 23–7). The concept of community encompasses both the species and the physical environment in which they live. The interaction of communities of organisms in different types of environments within a region is the basis for larger units of ecological interactions known as **ecosystems.** Ecosystems are the large-scale divisions of the biosphere (Figure 23–8), and the major types of terrestrial ecosystems are called **biomes.**

Different types of ecosystems exist within the biosphere (Figure 23–8). Many may be familiar to you; some you may have grown up in. Within those ecosystems are myriad communities whose compositions are determined by climatic, geographical, and biological factors. The balance in numbers and types of organisms in an ecosystem is dynamic. Different types of plants compete for space, light, water, and nutrients. Local changes in the availability of these resources exert profound effects on interacting populations. Different types of animals also compete for space and resources. How successfully an individual competes with

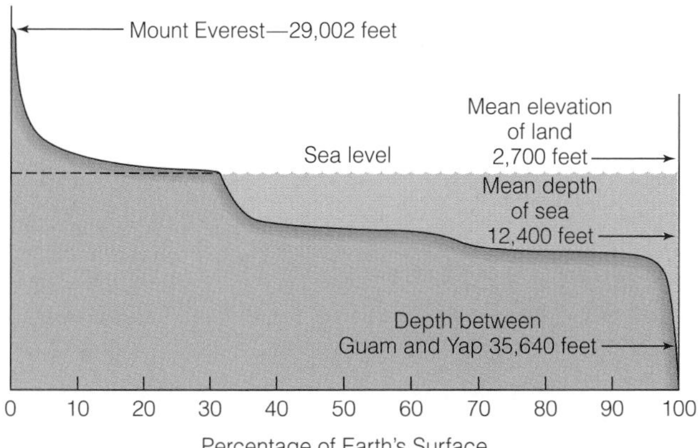

Figure 23–4 Earth's Highs and Lows
The highest point on the surface of Earth is Mt. Everest and the lowest is the depths of the South Pacific Ocean. The total thickness in which life may exist is, thus, approximately 65,000 feet or 12–13 miles.

Ecology the study of the environment and the living organisms within it

Community assemblage of populations living in a particular habitat or environment

Ecosystems the large-scale divisions of the biosphere

Biomes terrestrial or land-based ecosystems

(a)

(b)

(c)

(d)

Figure 23–5 Populations
Populations are groups of the same species living and breeding together in a habitat. This applies to all plants (a) and animals (b), (c), and (d), as well as all other sexually reproducing groups of organisms (a) redwoods; (b) elephants; (c) king penguins; (d) humans.

(a)

(b)

Figure 23–6 Terrestrial Community
A community includes an environment and all the populations that live within it. A prairie community is shown with a prairie dog (a) and various types of grasses and trees (b) filling niches within it.

Figure 23–7 Marine Community
Marine communities are as complex as their terrestrial counterparts. The coral reef has a variety of species of coral, as well as fish, sponges, and other marine organisms.

other individuals of its own and other species determines not only who will survive but in what numbers and in what distribution. **Competition** is not just about absolute survival of the fittest but also about the relative success of the moderately well adapted. There are no perfect adaptations, only ones that work more or less successfully than others.

Habitat and Niche

The area or region in which an organism lives is its habitat. What an organism does in that environment is called its niche. A niche, in human terms, is similar to a job. It is the ecological role that an organism plays in the community, particularly with regard to the acquisition of resources and energy. Many different species may live in the same habitat, but each has a different role to play. Fish, frogs, snakes, insects, and plants are all physically present at the same address in and around, say, a pond, but what each species does in that habitat is different. The snake hunts frogs (and many other prey), the fish eats insect larvae and frog eggs, the frog eats insects, and on and on. The essence of the community concept is that it represents the dynamic interactions of species within the environments in which they live.

RELATIONSHIPS WITHIN ECOSYSTEMS

Within all ecosystems, specific types of relationships are always found. Most relationships have to do with energy use or energy transfer, which is necessary to have life at all. The energy of life is the energy that is built into molecules, or chemical energy. However, the second law of thermodynamics suggests that only certain forms of energy are usable for supporting living things. One of the problems for any living system is energy acquisition. Another is that usable energy is depleted as the system is operated, so energy runs downhill. Once energy reaches the bottom of the hill, it is no longer available to do work. Batteries must be recharged, gasoline must be refilled in the tank, and the metal springs of toys must be wound up. It takes energy to do work. This one-way street for energy is the general rule for the universe. It can be summed up by saying that although the total energy in the universe does not change, the level of useful energy does. Useful energy in the case of ecology means energy that can be utilized by individuals to support their activities. For Earth and all the life on it, the Sun provides an essentially endless source of energy.

With the Sun as the source, the major service that must be performed in any ecosystem is to trap the light energy of the Sun and transform it to forms of chemical energy useful to living organisms. The underlying mechanism responsible for converting light energy to chemical energy is photosynthesis and involves the molecule **chlorophyll** (Figure 23–9). As a result of this capacity, chlorophyll is probably the single most important molecule for life on Earth. It is at the beginning of the chain of food sources. Plants produce chemical energy using solely inorganic ingredients (water and carbon dioxide) in the process of photosynthesis. As a result of this autonomy from organic molecules, plants are considered to be *autotrophic* (see also Chapter 1).

Competition the activities of organisms within a community for shared resources

Chlorophyll the molecule of plants that captures radiant energy

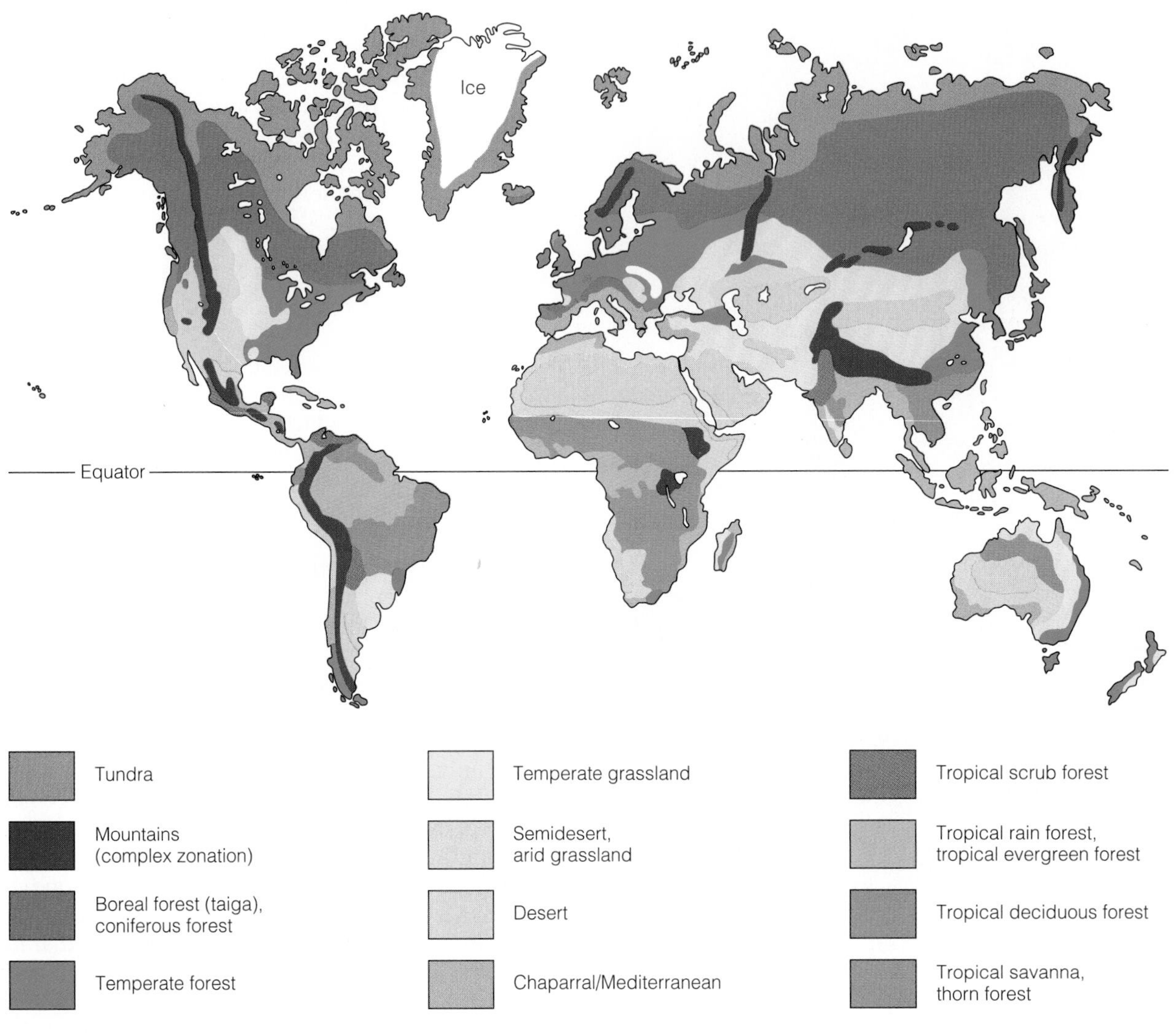

Figure 23–8 Terrestrial Ecosystems—Biomes

The distribution of ecosystems is in large part determined by their global position. The warmth and wetness of the equatorial region (around 0° on the map) is contrasted in the extreme with the coldness and dryness of the tundra and polar (arctic and antarctic) ice at latitudes above 60°. (As will be shown in Figure 23–31, there are also effects of altitude on the distribution of ecosystems.)

Based on a wide diversity of plants to perform this primary energy-capturing and energy-converting service, an ecosystem can be established. All the rest of the major forms of life ultimately obtain their energy directly or indirectly from plants. Animals consume plants to obtain energy. Carbohydrates, proteins, and fats are all organic materials that are produced by plants and obtained by animals by consuming them. Organisms that depend on an external organic source of nutrients for life are called *heterotrophic*. Animals that consume plants are often themselves consumed by other animals. Thus, energy stored in molecules is transferred among several different types of consumers. Animals that eat only plants are called **primary consumers,** and we give them the name herbivores.

Primary consumers organisms that eat plants; also known as herbivores

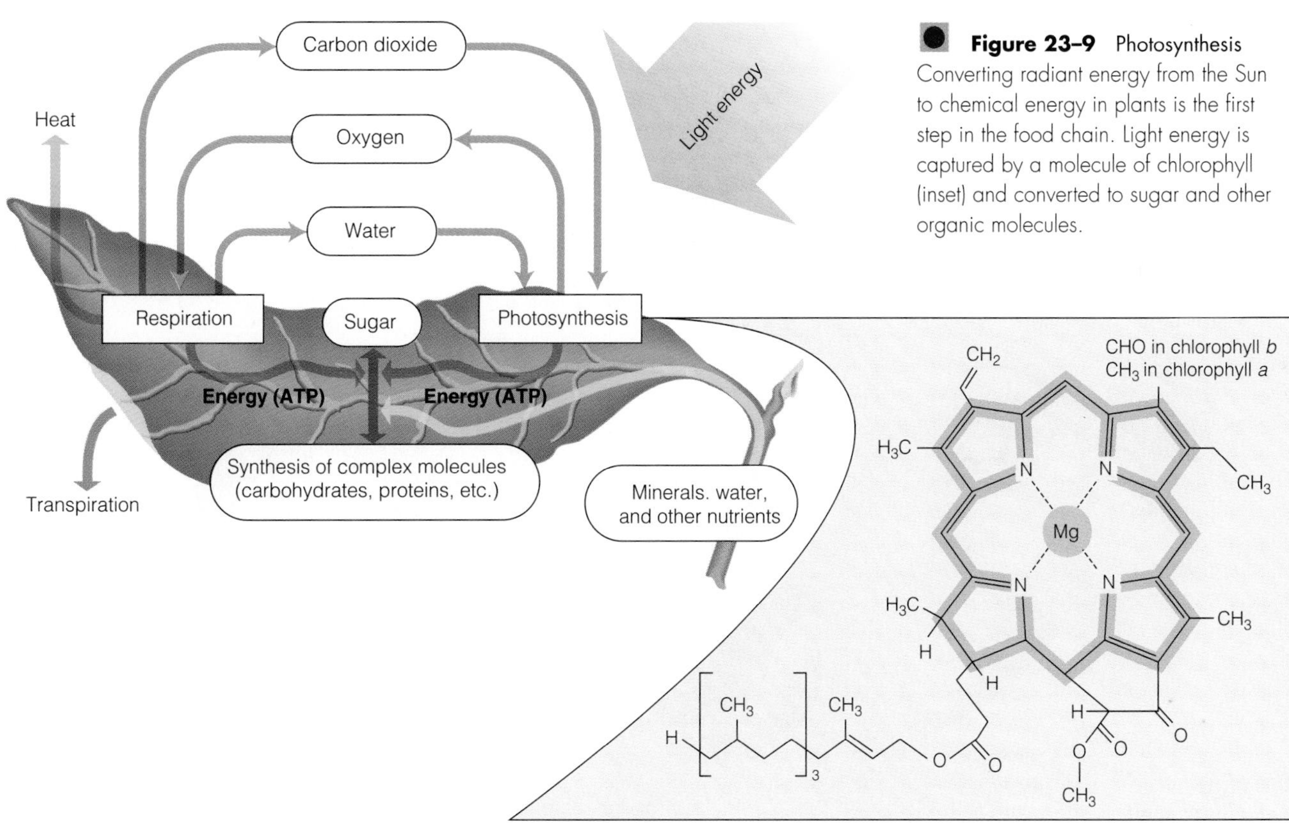

Figure 23–9 Photosynthesis Converting radiant energy from the Sun to chemical energy in plants is the first step in the food chain. Light energy is captured by a molecule of chlorophyll (inset) and converted to sugar and other organic molecules.

Animals that eat both plants and other animals are **secondary consumers,** and we name them omnivores. Animals that eat only other animals are **tertiary consumers,** and we name them carnivores (Figure 23–10).

Producers and Consumers

The simplest view of these categories of energy production and consumption is that organisms can be divided into two groups—producers and consumers. Producers are plants. They capture the Sun's radiant energy and convert it to chemical energy. All other forms of life, except for a few special types of photosynthetic or chemosynthetic bacteria, are consumers. One aspect of complexity in ecological systems arises as we consider networks of interactions among populations and communities of producers and consumers in the ecosystems of Earth. There are chains and webs of interactions between producers and consumers that diversify the energy economy of the biosphere. Plants are the first link in the chain.

Food Chains and Food Webs

In addition to the interactions between producers and consumers, one other group is necessary for the success of life—the decomposers and saprophytes. The major types of organisms that decompose organic materials are bacteria and fungi. Thousands of species of these two Kingdoms of organisms produce an array of enzymes capable of catalyzing the breakdown of organic molecules into less complex chemical forms. These less complex forms of organic materials are then available for other organisms (producers and consumers) to use. Without decomposition of complex molecules into simple compounds, there is no rapid way to recycle the components of dead consumers and dead producers (Figure 23–11).

Consider what might happen if no decomposition of an organism occurred after its death. For example, what would happen to all the leaves after they fell each autumn? Assuming that the trees and bushes from which leaves fell kept growing and producing new leaves, each season would add a layer of undecomposed leaves to the ground. The layers would get deeper and deeper until they were so thick that they no longer had anywhere to fall

Secondary consumers organisms that eat plants and other animals; also known as omnivores

Tertiary consumers organisms that eat primarily other animals (meat eaters); also known as carnivores

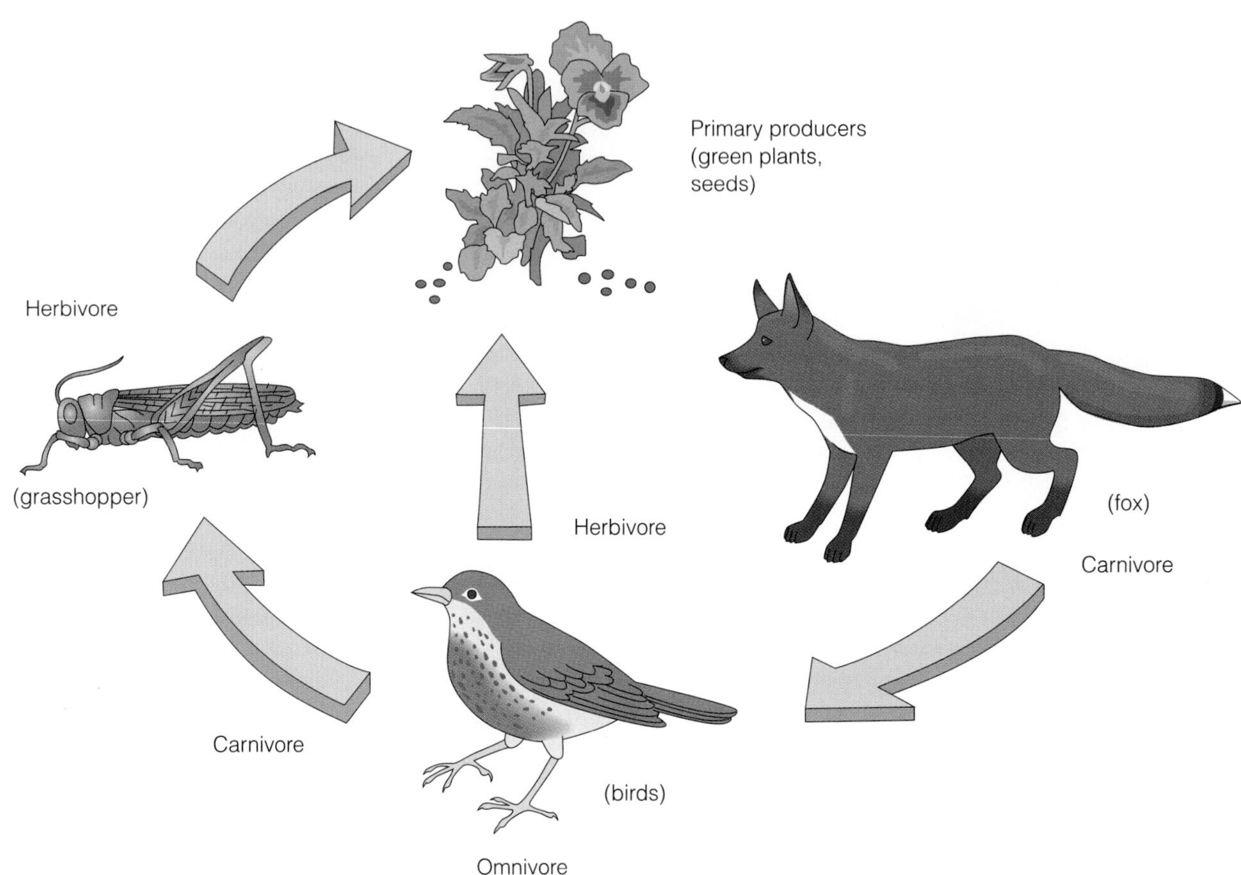

Figure 23–10 Herbivore, Omnivore, and Carnivore
Herbivores use plants as their source of food energy. As shown here, the grasshopper eats the plant. Omnivores, such as the bird depicted, use both plants and plant products, such as seeds, as well as animals as sources of food. Carnivores are predominantly meat eaters. The fox shown here consumes smaller animals including birds, reptiles, and other mammals.

(Figure 23–11[d]). In reality, if there was no decomposition of the leaves on the ground, the trees and bushes would deplete the soil of needed nutrients and become ever more deeply buried. Whole forests would disappear under the piles of leaves! Eventually, burial would bring about the cessation of photosynthesis for lack of light, growth would stop, and the plants would eventually die. Obviously this does not happen because decomposers turn the leaves to mulch and return nutrients to the soil. They decompose the fallen tree, the dead squirrel, the expired fish, and all the other creatures of the world that die. The rapid turnover and recycling of organic and inorganic compounds is healthy and essential for sustaining life in the biosphere.

A Food Chain What is a food chain? A **food chain** is a series of links between producers and consumers, plants and animals, represented in a specific, linear relationship (Figure 23–12). For instance, the grass along the edge of a pond may be eaten by an insect. The insect, in turn, may be consumed by a frog, which is eaten by a snake, who is devoured by a hawk. When the hawk dies, its body is consumed by scavengers and the remains are decomposed by microorganisms, thus, returning nutrients to the soil from which the grass grew. The links of the chain connect a producer to a series of consumers whose fates are all intertwined. Each level of feeding in this chain is called a **trophic level,** and, as will be seen, the energy transferred

Food chain simple linear relationship between primary, secondary, and tertiary consumers

Trophic level the feeding level of an organism in a food chain or food web

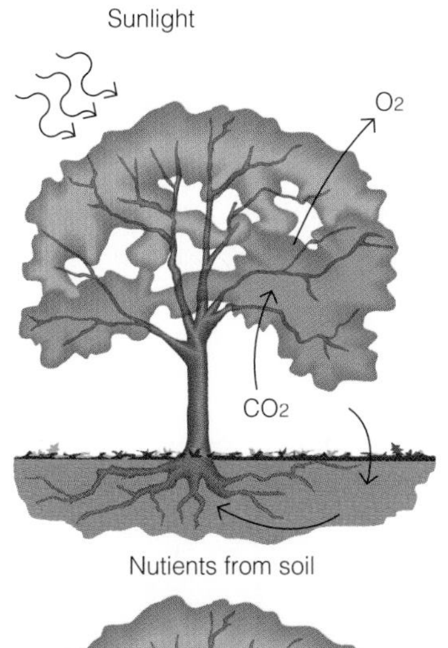

Figure 23-11 Decomposition
What if no decomposition occurred? Over time, leaf fall (a) would build up (b)–(d) until it was thick and choking.

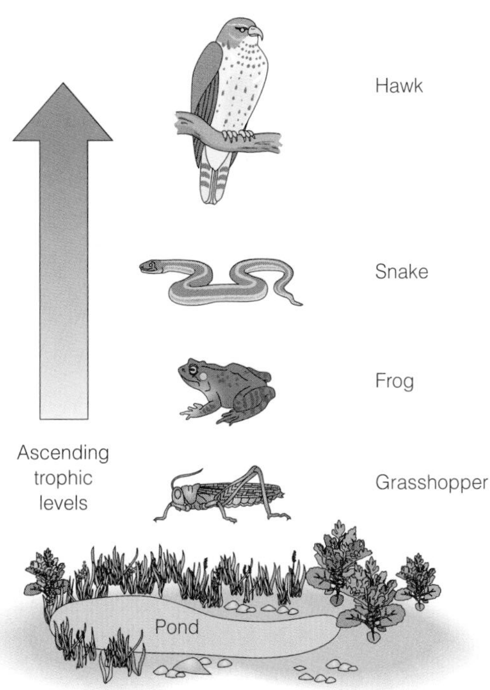

Figure 23-12 Food Chain
The food chain represents relationships between consumers and producers. The food chain is named as such because it establishes a chain of trophic level activities. The hawk consumes the snake, which eats the frog which, in turn, devours the insect that ingests plants. Eventually, the hawk dies and is consumed by scavengers and degraded by bacteria and other microorganisms.

from one trophic level to the next is very important to the energy economy of the entire ecosystem.

A Food Web One of the most important attributes of a community to understand is that the organisms inhabiting it are multifaceted in their activities and behaviors. This means that they may carry out a variety of activities that sustain them in the environments in which they live. Acquisition of food is a constant demand for organisms of all types, and there are many different sources of food available to each organism. Unlike a food chain, which seeks to represent a specific linear relationship among a specific set of organisms at different trophic levels, a food web takes into consideration the more generalized sources of food for each organism in a community. These sources of food may be used by each of many different types of organisms and, thus, a competition for resources is established (Figure 23–13). In addition, roles carried out by organisms may change when considering two trophic levels in the web. For example, the mouse in Figure 23–13 is a predator in one level of the food web and prey in another.

Thus, a food web is much more complex than a food chain and, in fact, might be thought of as interconnecting a number of existing or possible food chains. The complexity of a food web can be nearly limitless at a broad scale of analysis. A food web interlinks producers and consumers together in a vast interactive array.

More accurately than a chain, a web portrays the relationships among populations of organisms found in a community (see the arrows in Figure 23–13). As you might

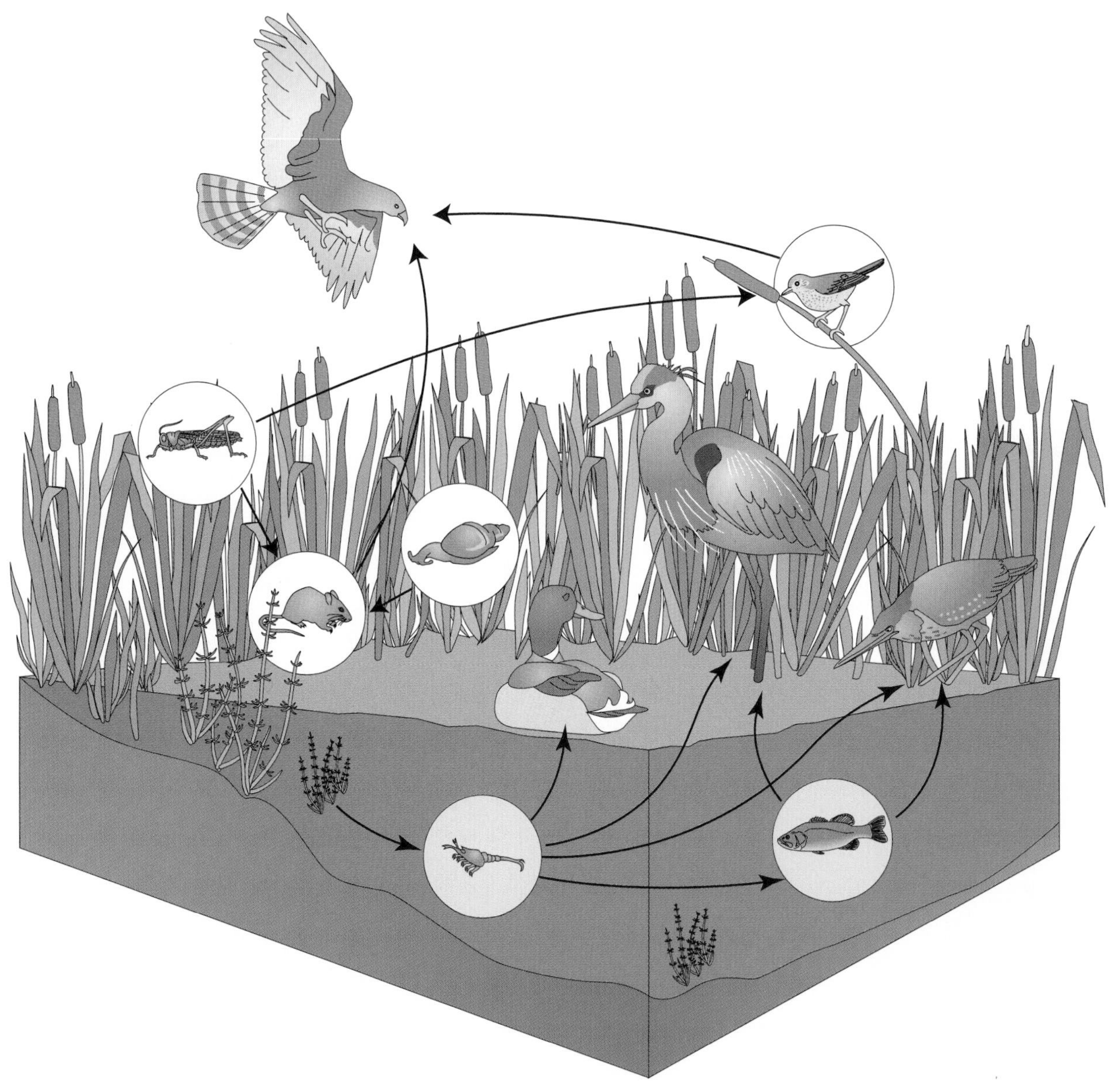

Figure 23–13 Food Web
The food web represents more complex relationships among consumers and producers in a community. As indicated by the network of arrows, consumers at all trophic levels have more than one food source. Designation of an organism as a predator or prey is determined by the trophic level examined. For example, the mouse (as predator) that eats a grasshopper or snail is a secondary consumer but may fall prey to a hawk, which is a tertiary consumer.

imagine, there are no bold lines of demarcation among communities; they often overlap. Some species are parts of many different communities. For example, the hawk that feeds on organisms at the edge of the pond may also hunt in the adjoining forest or along the distant shoreline of a large lake. Food webs reach across community boundaries and often beyond specific types of ecosystems, eventually including a vast array of environments and organisms, including and entangling even humankind. Depending on the trophic level of the food web under consideration and the species observed, some organisms, such as the mouse, may be alternatively considered as **predator** or **prey.**

CYCLES OF RESOURCE USE

Earth is essentially a closed system. The supply of resources on and within the planet may have to be used over and over again—recycling is necessary at the global level. The fundamental resources are the elements themselves, including carbon, oxygen, nitrogen, hydrogen, sulfur, and phosphorus, as well as many more complicated compounds composed of these elements. These materials pass from being incorporated into living things to being part of nonlife, organic to inorganic and back again in an unending and relentless cycle.

There are some exceptions to this closed-system rule. Some light gases, such as hydrogen and helium, can escape the gravitational attraction of Earth and diffuse off into outer space. Some matter, in the form of meteorites and ionized particles, can make its way from outer space and enter the biosphere. However, these outgoing and incoming materials are a very small portion of the balance of matter on Earth. Some of the main cycles of elements and resources are briefly described in the following sections.

The Carbon Cycle

Carbon is the single most important element for life and is found in all types of organic molecules. Most of these types of compounds are produced only by synthetic processes associated with living things. The carbon cycle (Figure 23–14) is intimately related to the activities of plants and animals. For example, carbon dioxide in the air is captured by plants during photosynthesis, and the carbon is incorporated into

> **Predator** an organism that captures and consumes other organisms (prey)
>
> **Prey** the organism captured and consumed by another organism (predator)

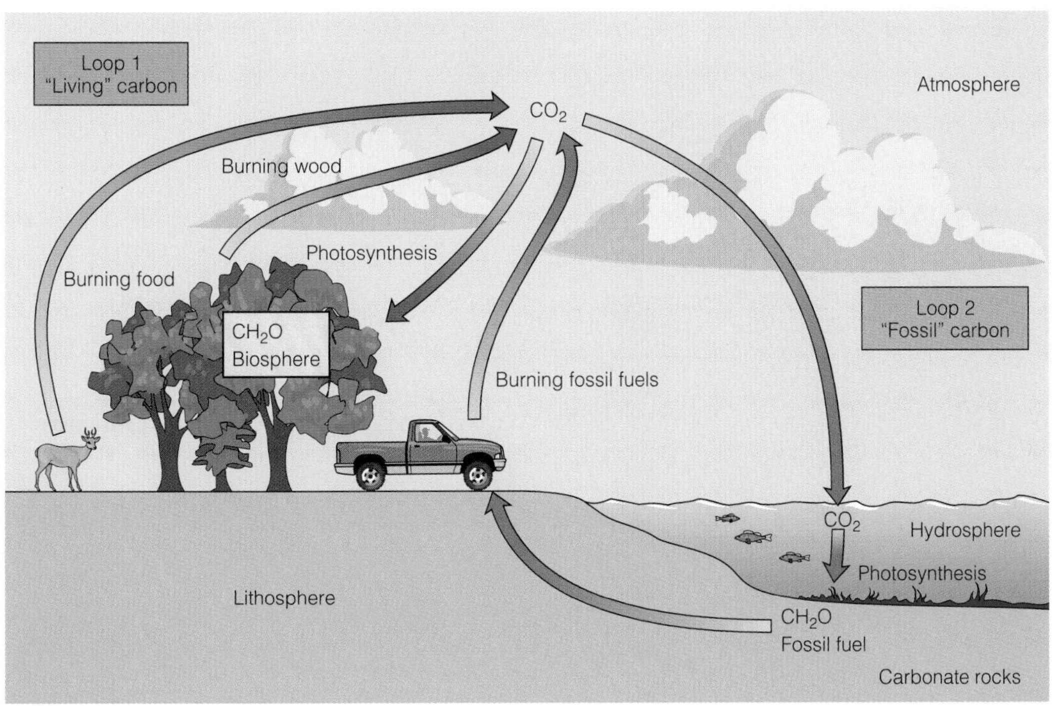

Figure 23–14 Carbon Cycle
Carbon is cycled in many forms from organic materials in animals and plants to the carbon dioxide released in the burning of fossil fuels.

carbohydrates and other compounds during this process. Some of the compounds are used for the energy metabolism of plants and some are stored or used structurally in the form of complex carbohydrates, such as starch or cellulose. It is the stored organic compounds that become available for consumption by herbivores and omnivores. Carbohydrates are energy-rich molecules and their oxidation by the biochemical reactions involved in the complex metabolic pathways of consumers results in the extraction of the chemical energy stored in the bonds between atoms in the molecules. The end products of carbohydrate oxidation are carbon dioxide and water, which are released into the atmosphere.

Several other avenues in the carbon cycle are also possible. Decomposition of dead animals and plants provides organic materials for the growth of new plants. In addition, organic materials can be converted to fossil fuels over millions of years of geological time. Natural gas, oil, and coal are all derived from once living organisms. Organic molecules are converted to highly combustible hydrocarbons and are burned to provide sources of energy for homes, cars, and industry. Combustion returns carbon dioxide to the atmosphere where plants once again can capture and convert the carbon to organic compounds. Because the driving force for this cycle is the nearly limitless energy from the sun, individual carbon atoms themselves may be reused continuously in an essentially endless process. Some of the very carbon atoms from which the dinosaurs were made are incorporated in your body at this very moment.

The Oxygen Cycle

Oxygen is a key ingredient for animal and plant metabolism. Oxygen in the atmosphere is there entirely as a result of the action of living organisms. A billion years of photosynthesis by microorganisms in the seas was needed to establish an oxygen-rich atmosphere for the planet, and this process continues to be utilized to maintain it (Figure 23–15). Although trees, bushes, grasses, and their terrestrial relatives produce a significant portion of the world's oxygen, most is still produced and released into the atmosphere by plant life, known as phytoplankton, in

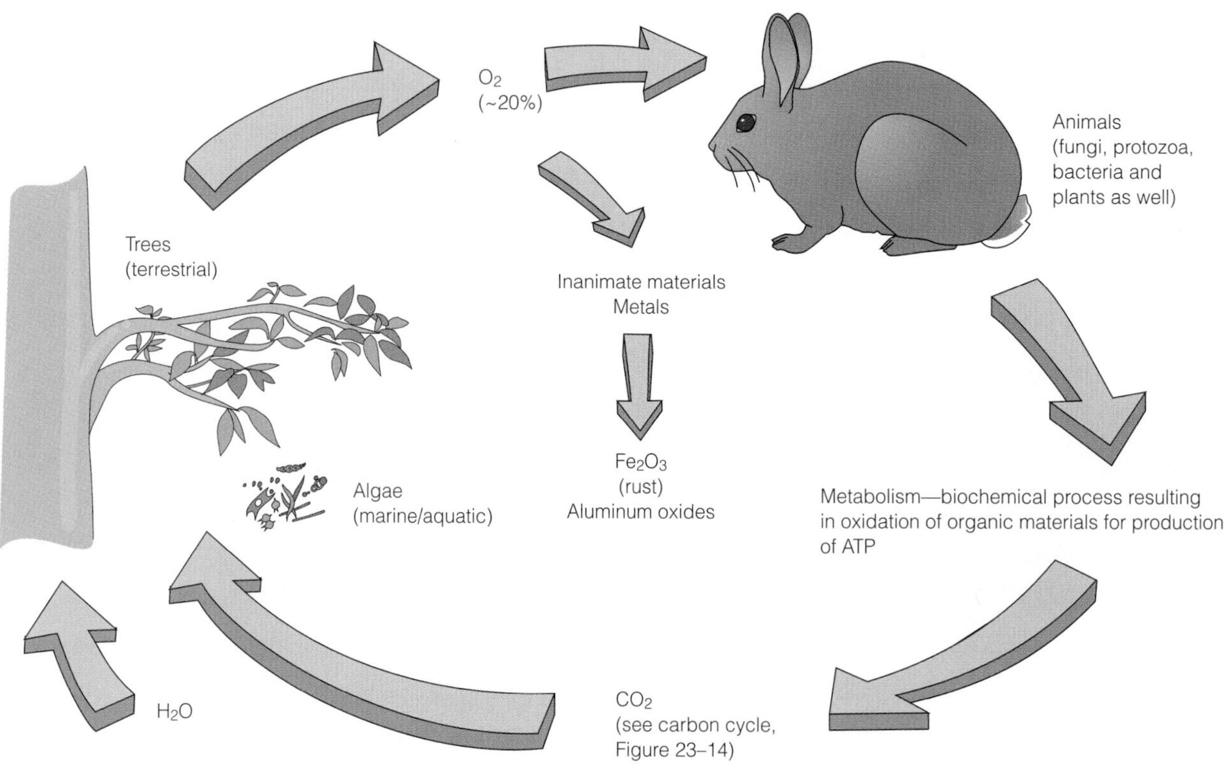

Figure 23–15 Oxygen Cycle
Oxygen is produced by plants as a by-product of photosynthesis. Oxygen is used by animals to oxidize organic materials and, thus, produce carbon dioxide. Carbon dioxide is used during photosynthesis to produce organic materials.

the waters of the world. These organisms live in the seas and oceans within a few meters of the surface where light can reach them. Estimates are that 70% of the oxygen we breathe is derived from these microscopic organisms.

The oxygen that plants produce is split from water during reactions associated with photosynthesis. The oxygen is then available for use by animals, where it is combined with carbon and hydrogen to produce carbon dioxide, water, and energy. Oxygen also reacts chemically with many other elements. Iron rusts and silver tarnishes as a result of reactions with oxygen, a process known as oxidation. In fact, much of the planetary oxygen is bound up in inorganic materials. Clays, mica, and the innumerable grains of sand on a beach are composed of silicon dioxides. These compounds do not readily give up oxygen and are not part of the oxygen cycle described here. However, they do provide us with an excellent solid surface upon which to carry out most of the activities of our lives.

The Water Cycle

The combination of hydrogen and oxygen in water is the single most important compound found in the biosphere (Figure 23–16). Hydrogen is a diatomic gas and is not prevalent in the atmosphere, but its relationship to oxygen in water makes it one of the most common elements in the biosphere. Water is chemically very stable and readily cycles through plants and animals, as well as through soil and air. As a measure of its importance to us, reflect on the fact that the human body is composed of 65%–70% water.

Consider a molecule of water as it travels through a simple cycle through air, soil, and plants and back again. This particular molecule evaporates off the surface of a lake in Minnesota on a warm day in late summer (Figure 23–16[a]). It rises in the atmosphere as a vapor and is blown west by the prevailing winds. The warm rising vapor cools as it gets higher and higher in the atmosphere. The confrontation of the warm moist rising air and the cold air above causes the water vapor to begin to condense and form clouds. The clouds get thick in the late afternoon. A storm is brewing (Figure 23–16[d]).

Tiny bits of dust in the upper air provide a nucleating center to form drops of water when it rains. Molecules of water come plummeting down as water drops from the sky and splash onto the ground. Water is absorbed into the earth and works its way down through the porous dirt until it comes in contact with a plant root. The root absorbs the molecule and transfers it to the complex vascular system of an enormous elm tree (Figure 23–16[e]). The storm ceases its incessant precipitation, the clouds break up, and the Sun comes back out to warm the atmosphere.

Trees use their leaves as a surface from which to evaporate water and in the process (called *transpiration*) draw more water molecules through the plant's vascular system (Figure 23–17). The original water molecule works its way up inside the elm and is released into the air. Back it goes into the sky, and when it falls again as rain, it may plunge back into the lake, or onto another plot of land. Variations on this theme of the water cycle have occurred in endless succession since life on Earth began.

The Nitrogen Cycle

Nitrogen is the other major gaseous component of air. It makes up approximately 79%-80% of the gas we breathe, and although not directly involved in metabolic pathways, as are oxygen and carbon dioxide, it helps make breathing possible. Nitrogen is abundant in the atmosphere, but it is far less prevalent in the many other forms useful for life such as nitrates and nitrites. Nitrogen plays a role in a number of complicated biospheric cycles associated with volcanic eruptions and the decomposition of organic materials by bacteria and other decomposers. Nitrogen is also present in coal and oil. When burned, the nitrogen forms oxides that react with water in the atmosphere to produce acids.

In some types of plants, special nitrogen-fixing bacteria are found in the roots. These bacteria help convert nitrogen gas to compounds usable by living organisms, notably in the form of nitrates and nitrites. Plants convert nitrates and nitrites to nitrogen-containing organic compounds that are essential to life. Such important compounds as proteins and DNA contain nitrogen (amino acids and purines and pyrimidines), as do the metabolic wastes of animals, such as ammonia, uric acid, and urea. These waste products recycle through the environment to supply essential nutrients for new growth. As discussed in Chapter 10, urea is the principal nitrogenous waste product of human metabolism and is eliminated from the body as urine through the action of the kidneys and urinary system. The nitrogenous waste of animals becomes the substrate for the many microorganisms that use and recycle it.

The Sulfur and Phosphorus Cycles

Sulfur and phosphorus are also essential elements for life. *Sulfur* reacts with oxygen to form sulfates and sulfites that are, in turn, utilized by plants and animals to produce organic materials, particularly sulfated carbohydrates. Sulfur is also found in proteins, specifically in two amino acids, methionine and cysteine. In the absence of oxygen, sulfur may form hydrogen sulfide. This gas, which smells of rotten eggs and decay, accounts in part for the strong fetid smell of marshes and swamps where oxygen is in short supply. Sulfur oxides arise during the combustion of coal and oil in which sulfur is present to varying degrees. A problem caused by sulfur oxides (and nitrogen oxides from the same fossil fuel sources) is that they combine with water in the atmosphere to form acids that are harmful to living organisms. Sulfuric acid and nitric acid are powerful acids and have been implicated in the destruction of lakes and forests by acid rain.

Phosphorus is an essential component of DNA and RNA. As in the cases of sulfur and nitrogen, it must be converted to an

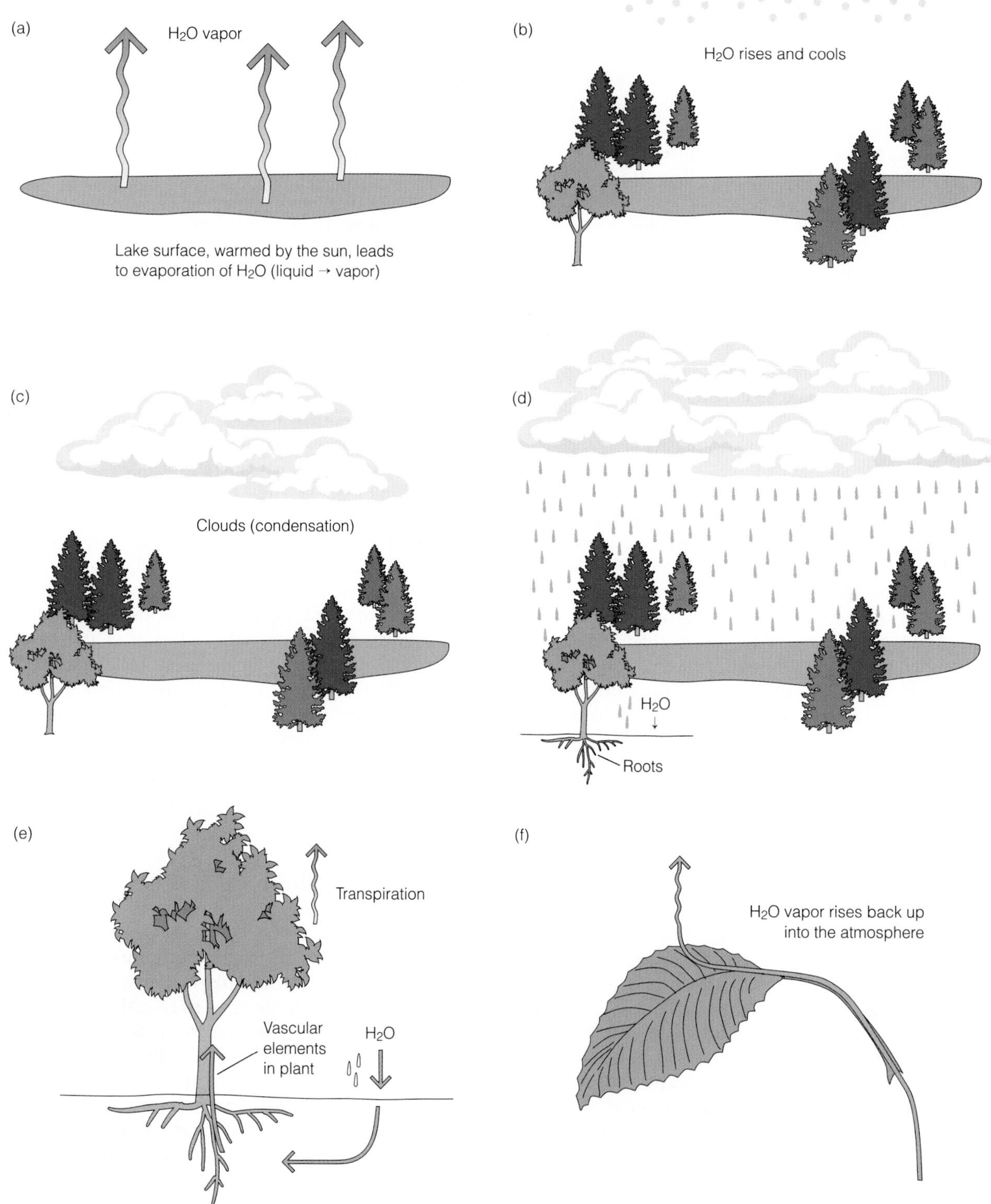

Figure 23–16 Water Cycle

The water cycle involves changes in state from liquid to vapor and back again. Occasionally, water is frozen to ice and is then essentially unavailable to cycle.

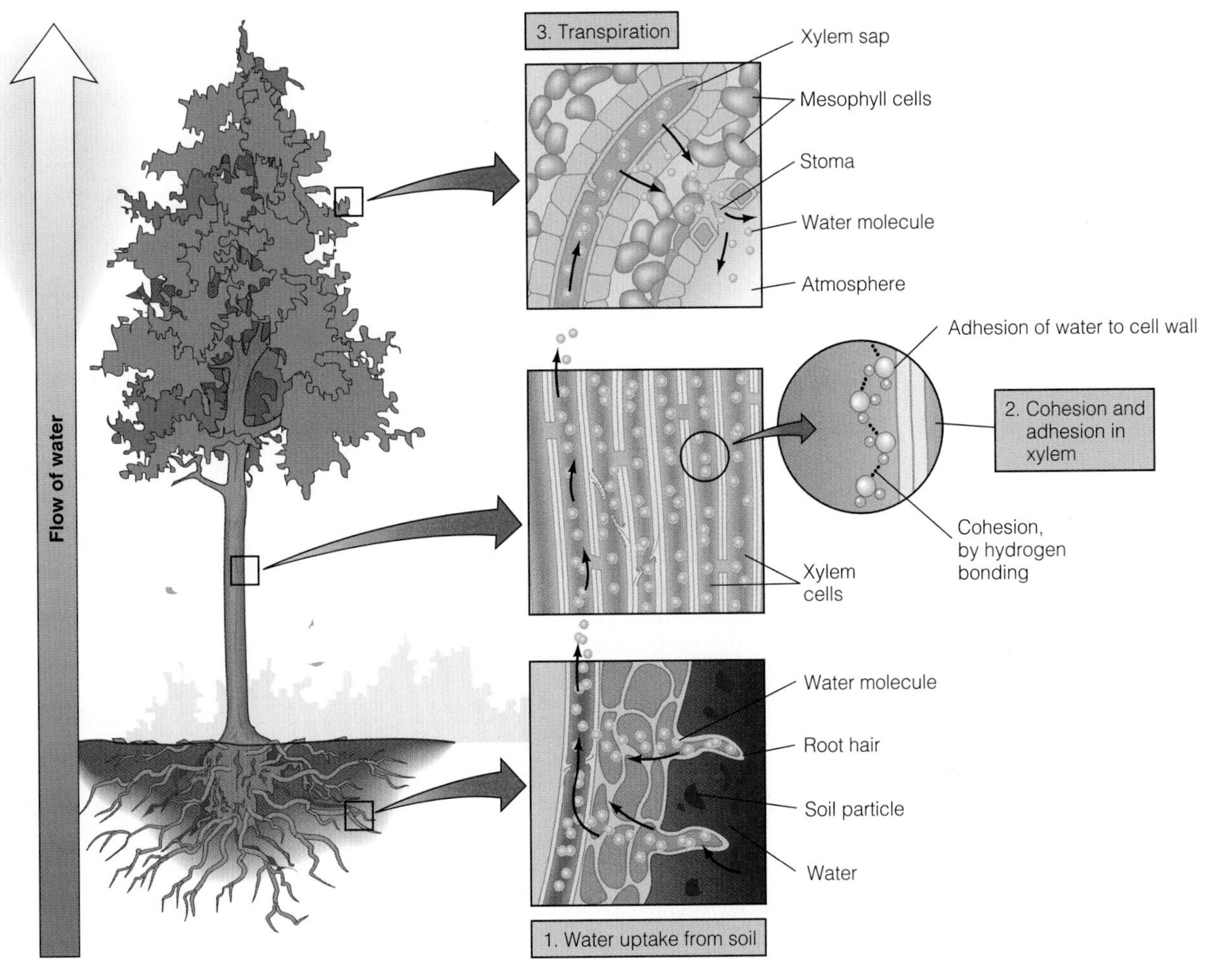

Figure 23–17 Transpiration
Transpiration is the process by which water is moved from roots to leaves in a plant. Xylem is the special tissue in plants through which water moves.

organic form usable by cells. The form that phosphorus takes, called *phosphate,* is necessary for animal and plant growth. Phosphates are absorbed readily by plants and incorporated into plant organic materials and eventually through the food web into animals. The cycle of phosphate from decaying organic material in the soil and back into plants and animals is an important one. There are substantial mineral phosphate deposits found on Earth that are often utilized in the manufacture of fertilizer for agriculture. The use of phosphate-rich fertilizers often disrupts the natural phosphorus cycle and introduces excessive amounts of phosphates into the ecosystem through runoff from the land. This may, in turn, lead to adverse ecological effects and destruction of life in ponds, lakes, and streams through a process known as **eutrophication.** Eutrophication results from an increase in nutrients, which, in turn, leads to massive overgrowth of a dominant species. Excessive nutrients increase respiration of organisms which use up available dissolved oxygen. This leads to anaerobic conditions and the death of many lake species. The phosphate runoff from the washing machines of homes and laundromats can be devastating to life in both fields and streams, so much so that in many areas phosphate-free detergents are used in washing clothes and dishes.

Eutrophication a rapid increase in nutrient status of a body of water; may lead to excessive respiration and oxygen depletion

It is clear that many of the problems we have in the use of the land, air, and water of Earth are related to our ignorance of the way in which these cycles work. We often blunder into the use of certain methods without thinking about their long-term consequences. Phosphate fertilizers and detergents are effective and convenient. Sulfur and nitrogen oxides and other atmospheric pollutants are simply by-products of our voracious consumption of fossil fuels to power our industries, homes, and cars. A discussion of these two aspects of the problem of pollution and its potential long-term effects is presented later in this chapter (The Greenhouse Effect, The Hole in the Ozone Layer).

Cycles within Cycles

There are many other cycles of elements and compounds in nature, including those involving micronutrients, such as iodine and iron, that play important roles in cellular activities. Iodine is essential for production of thyroid hormones, and iron is required for forming functional hemoglobin to carry oxygen in our blood. These many and diverse cycles are not described in detail here but suffice it to say that nearly all aspects of recycling in the biosphere are linked at some level (Figure 23–18). In fact, the cycles of individual elements and compounds may be viewed as part of a collective of cycles of exchange between living organisms and among the diversity of organisms and their physical environments. Each separate part of these processes is a cycle within a cycle that principally includes carbon, hydrogen, oxygen, nitrogen, sulfur, and phosphorus transforming from organic forms to inorganic forms and back again. Oxygen, carbon dioxide, water, ammonia, sulfates, and phosphates are all inorganic molecules. Yet, carbon from carbon dioxide is used directly by plants to produce carbohydrates, and oxygen in animals is used to form covalent bonds to oxidize organic compounds during the production of cellular energy.

The biological and nonbiological activities carried out by elements and compounds in the lithosphere, hydrosphere, and atmosphere are each integral parts of the sphere of life. These activities help bring about the chemical and physical transformation of matter from inorganic to organic and back again in an endless cycle.

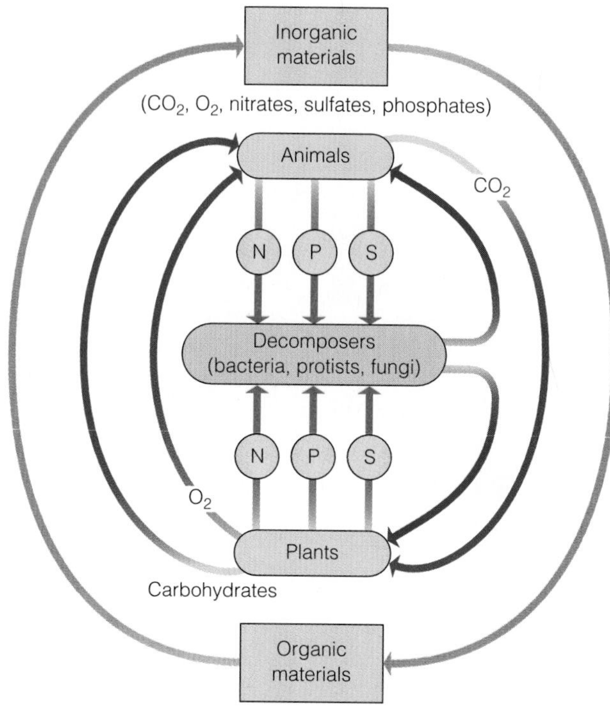

Figure 23–18 Cycles within Cycles
Cycles of elements, ions, and compounds are ultimately interrelated. Plants produce carbohydrates and oxygen (O_2), as well as many other organic materials containing nitrogen (N), phosphorus (P), and sulfur (S). These are principally found in proteins, nucleic acids, lipids, and carbohydrates. Animals utilize the oxygen and carbohydrates produced by plants to produce cellular energy and carbon dioxide (CO_2). Carbon dioxide is used by plants to synthesize carbohydrates. The production of waste products and the decomposition of both plants and animals by bacteria, protists, and fungi provide the opportunity to break down organic compounds into inorganic forms of phosphates (rich in P), nitrates and nitrites (rich in N), and sulfates (rich in S), which can be used by animals and plants in their growth and development. Thus, the overall process is one in which transformation of materials occurs from inorganic to organic forms and back again in an endless, interconnected set of cycles.

THE ECONOMICS OF ECOSYSTEMS

In general, when we think of economics, we think of the study of the production, distribution, and consumption of goods and services by and for humans. However, it is easy to see that this kind of study also applies to the flow of energy in an ecosystem. Primary production of energy is the province of plants, through the capture and conversion of radiant energy to chemical energy. Distribution of chemical energy occurs through the network of consumers and decomposers whose activities are interconnected in the food chains and food webs of an ecosystem. Modest gains and major losses occur at each step in the process of distribution and consumption. Energy does not cycle; it travels a one-way, downhill street. To depict energy transfer among living organisms on an economic balance sheet eases the understanding of trophic levels.

Figure 23–19 represents the chemical energy produced and stored in plants and starts with the assumption that this equals 100% of the initial energy of an ecosystem. What happens to the energy as it works its way through the various trophic levels? When an insect eats a plant, only part of the energy stored in the plant is available for the animal's metabolic demands. Estimates suggest that an average of 10% of the original energy of the plant is usable by the primary consumer. When a bird eats an insect, again only 10% of the chemical energy of the prey is transferred to the predator. The balance sheet reflects a 10 to 1 ratio of total energy to useful energy at each level of consumption, or approximately a 10% efficiency rate. This 10% rule applies generally to every trophic level encountered. A hawk eating the bird that fed on the insect who ate the plant receives only 0.1% of the energy originally available in the plant. Clearly, energy runs downhill at a rapid rate.

Because of the 10% rule, in each succeeding link of a food chain a consumer must eat 10 times more prey to survive. As a result, a community always has many fewer consumers than producers. The quantity (mass) of organisms and their products at each trophic level in a community is called the **biomass.** The shaded areas of Figure 23–19 labeled as producers and primary consumers (herbivores) indicate that the bulk of the biomass in any particular environment is in these components. Less and less biomass is encountered in successive trophic levels (secondary and tertiary consumers). Biomass ultimately represents organisms and their products. At the highest levels, a small percentage of biomass is represented by very few carnivores, represented as the thin line at the top of the diagram (Figure 23-19). Eagles, hawks, bears, wolves, and humans fit into this category. None of these organisms has natural predators (although humans are known to kill the carnivores named, including themselves).

A serious problem is associated with the passage of materials and energy between trophic levels. In the contemporary world, many toxic pollutants accompany energy molecules and nutrients in their flow through animals and plants in food chains and food webs. A pollutant may be taken up directly by any animal or plant at any level in the ecosystem, but in some cases, the pollutants follow an even more destructive course—they accumulate to higher concentrations in animals at each succeeding trophic level. This is called **biological magnification.**

Biological Magnification

Starting at the lowest level of a food chain, let us follow a toxic substance such as a pesticide through the trophic infrastructure and consider the consequences. Let us suppose that the pesticide is sprayed each week on the grass and bushes of all the homes in a residential area in order to keep the plants from being eaten by insects. In low doses, this compound has little or no effect on humans or pets.

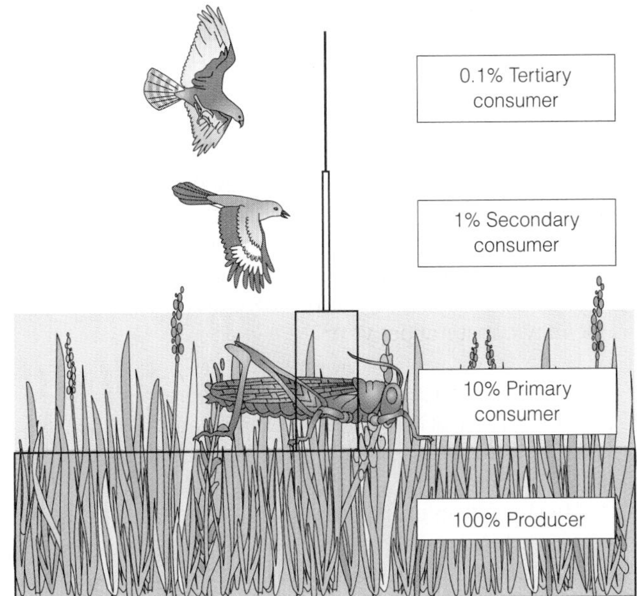

Figure 23–19 Energy Transfer
In general, energy transfer between trophic levels is about 10% efficient. If the producer contains 100% of the energy initially, only 10% of that energy is useful to the primary consumer. This 10% rule holds for energy transfer at all trophic levels.

However, there are warnings on the label concerning overexposure. The homes are built around a beautiful lake, which, in turn, is part of a natural ecosystem for the as-yet-undeveloped areas nearby. Some small percentage of the pesticide runs off into the lake and is taken up and concentrated by microorganisms and other inhabitants of the aquatic environment. For purposes of simplifying the arithmetic, assume that each microorganism takes up 100 units of pesticide in its lifetime. A hundred microorganisms are eaten by insect larvae and the toxin is concentrated within the larval tissues. The larvae, in turn, are eaten by small fish of various species. The small fish eat, on average, 100 larvae before they are eaten by medium-sized fish. In this series of events, the pesticide is never excreted nor detoxified by the host organism; it simply accumulates to higher and higher concentrations at each trophic level.

Biomass the amount of organisms and their products in the environment

Biological magnification the accumulation of a contaminant through trophic levels

Each week, a medium-sized fish, which ate 100 small fish, is eaten by the largest species of fish in the lake. The largest fish have no predators in the lake and each eats 100 or more of the medium-sized fish during its years in residence in the lake.

In this scenario, the ecosystem of the lake has a forest around it occupied by a family of eagles. Once each week for a year an eagle swoops down out of the sky and captures a large fish from the lake. What great luck for the eagle! But is it? The arithmetic of biological magnification reveals a rough estimate of how many units of the pesticide the eagle may consume per year:

$$100 \text{ units per microorganism} \times 100 \text{ microorganisms} \times 100 \text{ larvae} \times 100 \text{ small fish} \times 100 \text{ medium-sized fish} \times 50 \text{ large fish} \times 1 \text{ eagle} \times 52 = 26{,}000{,}000{,}000{,}000{,}000$$

Twenty-six trillion units of the pesticide accumulate in the body of the bird over a year of living, hunting, and feeding on fish from the pesticide-contaminated lake. Quite possibly, the poison concentrates within one organ, such as the liver, or one tissue type, such as fatty tissue. Perhaps, it is not so surprising that the eagle gets ill and dies or that embryos in the eggs that the female eagle lays do not develop or that the shells of the eggs are too thin and break before hatching (Figure 23–20). Unfortunately, pesticides that would be safe for humans and other vertebrates, if accumulated only to low levels by direct consumption, may concentrate through biological magnification to levels that have devastating effects on organisms at the highest trophic levels. Keep in mind that eagles are not the only predators of the fish in the lake. What might happen to people in the neighborhood who fish regularly in the lake and take home their catch for dinner?

Figure 23–20 Biological Magnification
Through biological magnification, a pesticide such as DDT interferes with eggshell formation. Note the crushed egg of the brown pelican made fragile by exposure of the mother to DDT.

Eating Plants versus Eating Animals

According to the 10% rule, one consequence of energy transfer efficiency has serious ramifications with respect to what we eat. The human diet generally consists of a mixture of fat, protein, and carbohydrate. In some parts of the world, most of the dietary intake is carbohydrates from plants, notably through one of the types of grasses such as wheat, corn, or rice. In others, particularly in the United States and Western industrial nations, there is much more animal protein in the diet. Cattle, pigs, sheep, chickens, and many other animals are grown for their meat to satisfy our dietary desires.

When humans are considered as primary consumers, eating grains and plant products exclusively, then they obtain 10% of the energy of the plants. Plants represent 100% of the available energy because they directly transform the energy of light to organic molecules. If, on the other hand, humans eat beef, pork, or chicken, which feed on plants, then they are considered as secondary consumers. In this case, humans get only 10% of the energy stored in the meat and, thus, only 1% of the original plant-stored energy. Application of the 10% rule suggests that it requires 10 times more biomass of plants fed to animals to provide humans with the energy they could have gotten had they eaten plants or plant products in the first place.

With the human population of the world increasing at a geometric rate, it is becoming inefficient and expensive to eat animal flesh as a primary source of nutrients and energy. The agricultural resources available per person are declining. At some point in the future, the 10% rule will apply directly to us and our consumption. We will not be able to afford the cost of eating animal meat, in terms of both price (dollars per pound) and energy efficiency (biomass consumed between trophic levels). As a result, the average American diet will change. Conjure up an image of a dinner plate filled with food a hundred years from now. What portion of the meal do you think might be the meat of secondary consumers?

TYPES OF ECOSYSTEMS

The biosphere is a self-enclosed system and all life ultimately interacts within it. However, the biosphere is very complex. We are therefore forced to divide it up locally into smaller units to study the principles by which it is organized and operates. Ecosystems are the largest subunits of the biosphere that provide relatively clearcut divisions suitable for study.

There are many types of ecosystems (Figure 23–8), and each has its own particular physical, climatic, and biological characteristics. Some ecosystems are more hospitable and habitable than others for humans and, so, are more directly known to us. Others are very inhospitable to human life, and we live there only in small numbers or with great technological effort. The hydrosphere has aquatic ecosystems in both fresh and marine waters, including lakes, rivers, and oceans. The lithosphere has a great variety of terrestrial ecosystems called biomes, including deserts, grasslands, tundra, rain forests, and chaparral. There is overlap between terrestrial and aquatic ecosystems in regions where they meet, as found along the intertidal zones of oceans and seas and on the shores of lakes and streams.

Aquatic Ecosystems

Lakes and Ponds *Lakes* and *ponds* are usually freshwater systems that contain standing water and complex animal and plant life (Figure 23–21). A pond is a much smaller entity than a lake. A pond is shallow enough so that plants can be rooted in the bottom. It is different from wetlands, as might occur in an **estuary** or in a *swamp,* in that it is limited in size as well as depth (usually less than 6 feet). Lakes range in size and depth from small and relatively shallow to thousands of square miles in area and great depth. The Great Lakes of the northern United States and southern Canada are among the largest lakes in the world. The deepest lake in the world is Lake Baikal in Russia. It is over a mile deep and contains 20% of Earth's fresh lake water. Relatively speaking, fresh water is a fairly rare commodity on the surface of Earth, though there is a huge reserve stored in the ground. The greatest reserve on the planetary surface is held inaccessible and frozen in polar ice.

Lakes are formed in several ways. During periods of glaciation in Earth's geological history, glaciers plowed deep furrows in the lithosphere as they advanced, and as they receded, the melting waters filled the massive furrows. That is how the Great Lakes formed. Geological activity such as mountain building also forms lakes. Further, erosion of land by moving water (streams and rivers) can create lakes and ponds. Water seeks the lowest level as it

Estuary a type of shallow wetlands, where the tide meets a river current

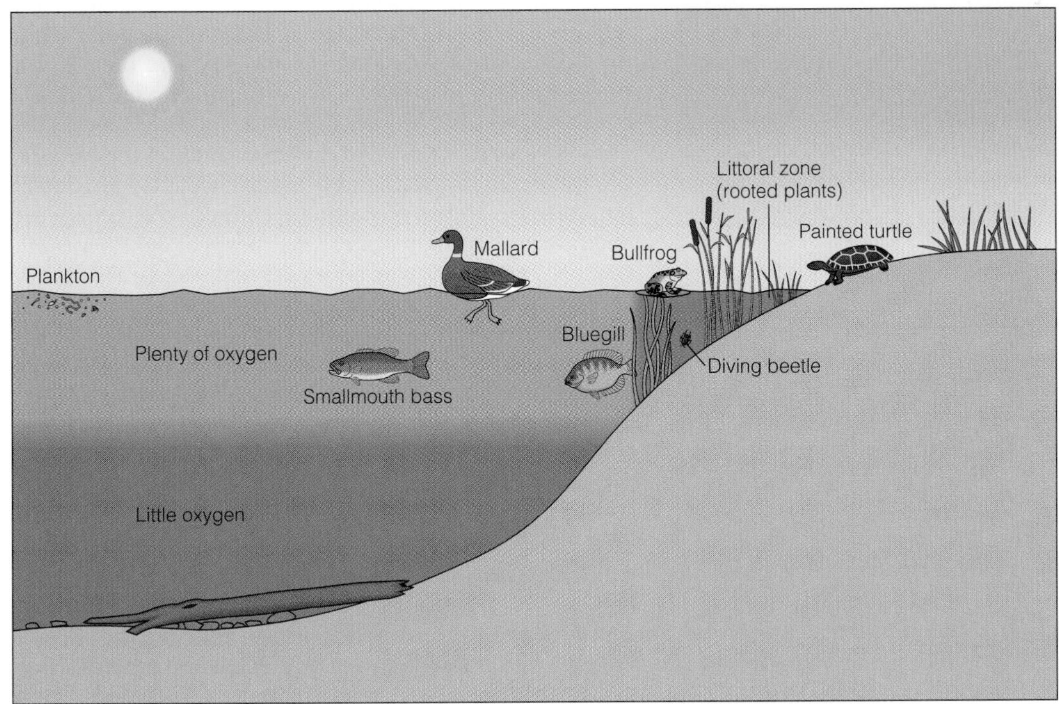

Figure 23–21 Lake
Lakes are bodies of fresh water, usually too deep for plants to be rooted on their bottoms. Lakes are teeming with life from bacteria and protista to fish and birds.

follows the course that gravity and landforms set for it. A tremendous variety of life is associated with lakes and ponds, including microorganisms, plants, and animals. Fluctuations in light, oxygen, and temperature during different times of the year bring about periodic changes in the environmental conditions of lakes and ponds.

Rivers and Moving Water Rivers and streams are moving water. They cut and erode Earth as the waters within them ceaselessly flow over dirt and stone (Figure 23–22). Many streams form as water moves downhill under the influence of gravity from snowmelt. Eventually (and often transiently), these waters collect in low places in the terrain over which they move and form lakes and ponds. Overspill of these waters continues to find a path that eventually leads downward to the sea.

Figure 23–22 Rivers
The Colorado River in the Grand Canyon of Arizona is fast moving. Erosion from the movement of water has formed the canyon over millions of years.

Rapidly moving or **lotic** water tends to be more highly oxygenated than slow-moving or **lentic** water. Ordinarily, oxygen dissolves in water rather poorly. Lakes and ponds generally have much less oxygen per unit volume than rivers and streams, and fast-moving waters tumbling over an irregular streambed have the highest level of oxygen. This is because the surface area for absorption of oxygen is greatly increased by the breakup of the smooth surface of water into atomized droplets. Another aspect of rapidly moving water in streams is that most of the animal life in streams and rivers lives out of the direct current under rocks and pebbles or in slow-moving areas where the inhabitants are not easily displaced. There are exceptions, of course, because many vertebrates and invertebrates thrive in the direct flow of water in rivers and streams. Many have developed specific adaptations that enable them to strongly adhere to surfaces, alter their resistance to flow, or swim powerfully upstream.

Oceans The greatest mass of water on Earth is in the oceans (Figures 23–2 and 23–4). Seventy percent of the surface of Earth is covered by oceans and because of their depth and volume, they offer 300 times more space for living things than does the atmosphere. Whereas only three major phyla dominate the land (vascular plants, arthropods, and vertebrates), all major phyla are represented in the ocean. Life on Earth originated 3.5 billion years ago in the oceans (review Chapters 1 and 22), and only within the last 500 million years have organisms moved onto land.

There are several named oceans, including the Atlantic, the Pacific, the Indian, and the Arctic (recall Figure 23–2), but all these bodies of water are connected and they ebb and flow unceasingly against the shores of continents and islands. There are great ocean currents, such as the *Gulf Stream,* that have profound effects on the climate and weather. Did you ever wonder how London, which is relatively high in latitude, stays pleasant or at least livable throughout the year (Figure 23–23)? The warm waters of the Gulf Stream flow to the British shores and moderate what would be at that latitude a far colder and more inclement climate.

The oceans abound with life, and most of this life is located in the first hundred meters of its depth, where light can penetrate. However, even on the bottom of the deepest valleys of the sea, near thermal vents, life clings, swims, or crawls around in complex associations (Figure 23–24). Phytoplankton abounds in surface waters and produces most of Earth's oxygen. The food chains and food webs of the

Lotic water the fast-moving water of a river or stream
Lentic water the slow-moving water in a river or stream

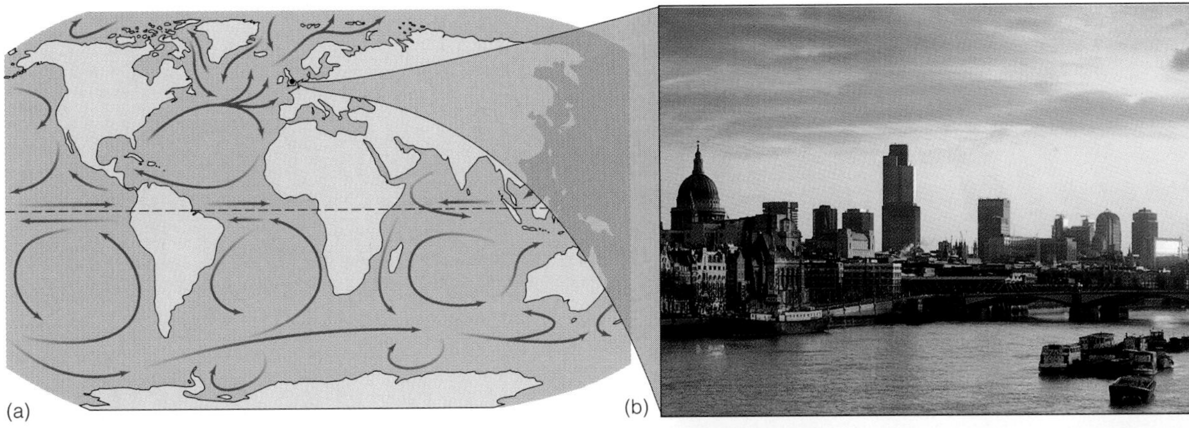

Figure 23–23 The Gulf Stream and London
The ocean currents (a) circulate water around the world. The warm waters of the Gulf Stream make the climate in London (b) much different from other regions at the same latitude.

Figure 23–24 Thermal Vents
Life at 2500 meters below the surface, near ocean thermal vents, is fairly abundant and diverse. Crabs, clams, and tube worms can be identified here.

oceans and seas are as complex as those on land and overlap the land-based chains and webs along the millions of miles of shoreline where oceans, rivers, land, and tide converge, such as along the ocean beaches of the coast (Figure 23–25). These areas of convergence are complex and often form intertidal zones of great diversity. The animals and plants within these zones must be able to tolerate changes as dramatic as wet to dry and back again with the tides.

Terrestrial Ecosystems—Biomes

The terrestrial ecosystems, as those of the ocean, blend together continually at their edges. A quick trip across the United States from east to west offers a sketch of the diversity of the biomes and their blending together. Each can be seen to contain certain distinctive types of plants and can be described as forested or grassy, but do not let these simplifications fool you. Every biome contains a vast interacting web of hundreds of thousands of species of animals, plants, protists, fungi, and bacteria. The largest percentage of the life forms on Earth are too small to see.

Rain Forest and Jungle For the purposes of this trip across the United States, let us begin at the tip of the Florida peninsula, so we can include a tropical forest in our travels. *Tropical rain forests* are found extensively in other parts of the world, as a belt around Earth at the equator (Figure 23–26). The diversity of life in tropical rain forests is tremendous. There are more species of plants and animals and organisms of other Kingdoms in tropical rain forests than in any other terrestrial ecosystem. This biome is wet (heavy rainfall throughout the year), humid, and hot (an average of nearly 26°C) and delicately balanced ecologically.

Rain forests are constantly in the news in recent years because they are being destroyed by the activities of an ever-increasing human population. For example, the Amazon rain forest is being destroyed at the rate of thousands of acres a day and converted to agricultural use. The delicacy by which this type of ecosystem maintains itself precludes recovery to its natural state after being razed. When a rain forest is destroyed, it is replaced by *jungle*. Jungle is a tangle of low-growing plants that invade the denuded areas of a forest. A jungle is not nearly as complex an ecosystem as a rain forest and cannot sustain the diversity of animal and plant life. Rain forests are only sparsely inhabited by humans, and those at the tip of Florida are no exception.

Temperate Deciduous Forests After a short drive up the Florida peninsula, we find ourselves in a vast region of the eastern United States called *temperate deciduous forest*

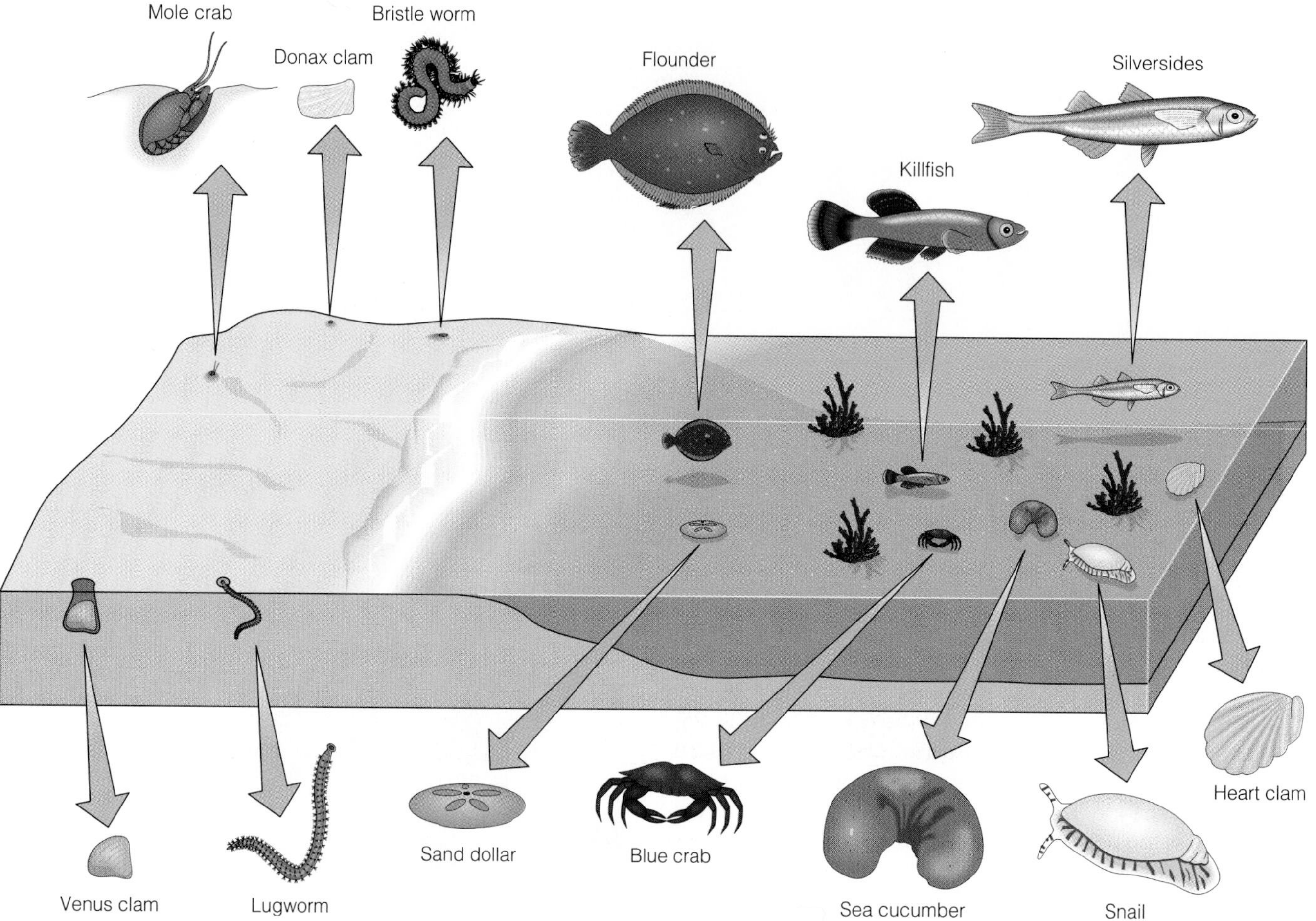

Figure 23–25 Beaches
Beaches are a commonly enjoyed environment. A closer examination of the beach and near-shore waters reveals a great diversity of species. This environment bridges two different types of ecosystems, terrestrial and aquatic, and constitutes a zone that changes with the tides (intertidal zone).

(Figure 23–27). Most biomes are determined by the climate and the types of plants that grow in them. There are north-south variations in the composition of the plant life present in temperate deciduous forests in the United States. *Northern deciduous forests* range from southern Canada to Appalachian North Carolina and Tennessee and are filled with hemlock, white pine, and mixed hardwood trees. Further south, beech, sugar maple, basswood, birch, red oak, and white pine are the predominant species. In Florida, such forests are dominated by magnolia and live oak.

Savanna and Grasslands A left turn in Georgia and we are headed west. The temperate deciduous forest continues to be the dominant ecosystem all the way to the Mississippi river. There the environment begins to change again. We transit into what was at one time the *savanna* (Figure 23–28), and then into the great *grasslands* of the Midwest (the natural transition between forest and grassland is called savanna). Not much savanna is left in the United States because of land development and the destruction of that biome. However, had we come with the first immigrants to North America several hundred years ago, we would have found it. In savannas, trees thin out and grasses dominate. Earth's preeminent savannas are found throughout central and southern Africa. Vast expanses of short trees and grasslands in this region were probably home to the first humans over 3 million years ago (see Chapter 22).

Plains and Prairies Grasslands in North America come in two main types—*plains* and *prairies* (Figure 23–29). Because of changes introduced by farming and agricultural development over the last 150 years in the United States,

Figure 23–26 Tropical Rain Forest Biome
The center photo is a view from the floor of a tropical rain forest. Tropical rain forests are the most biologically diverse environments on Earth. Organisms of many kinds abound: (a) toucan, (b) orchid, (c) butterfly, (d) howler monkey, and many other larger and smaller organisms.

Figure 23–27 Temperate Deciduous Forest
Found over a vast region of the eastern United States, deciduous forests are composed of hemlock, white pine, beech, birch, and red oak, plus many other types of hardwood trees.

Figure 23–28 Savannas
Not much of the original savanna remains in the United States, but in Africa the vast savannas support a great diversity of species, many familiar to us all, including elephants, lions, zebras, gazelles, birds, and acacia trees.

there is little left of the original prairies and plains. Prairies are characterized by greater average rainfall and taller grasses than plains. Plains not only have less rainfall but they have shorter, more sparsely distributed species of grasses. Traveling east to west across the middle of the United States, we pass through the remnants of those regions where prairie and plains grasses once flourished.

The prairies are now the breadbasket of America. Wheat and corn are the agricultural crops of human choice, but natural grasses that grew there were represented in many varieties long before the farmer appeared. Those original grasses fed huge herds of animals, notably bison, which were nearly brought to extinction in the nineteenth century. Prairies, as did bison, once flourished on the North American continent before the present stock of humans arrived.

In general, plains have shorter grasses and more sparse vegetation, than found in the prairies. But the plains too have been altered by agriculture. Overfarming and drought have had catastrophic effects on the western plains. The infamous Dust Bowl of the 1930s, which provided the background for John Steinbeck's book *The Grapes of Wrath*, resulted from poor farming techniques and severe climatic conditions. The dry, wrathful winds simply blew away the topsoil and left the region desolate for decades.

Coniferous Forests and Taiga We now approach the major mountain range of the central United States known as the Rockies. As we go up in altitude in an effort to pass over the gigantic peaks, the ecosystem begins to transform once again. This time the conversion is into a biome of the coniferous forest or, as it is also called, *Taiga* (Figure 23–30).

Coniferous forests are dominated by spruce, fir, and pine trees. Mountains provide an interesting gradation of ecosystems as we go up in altitude. The lower elevations have Douglas fir and ponderosa pine, with an undergrowth of grasses and shrubs. Higher up on the slopes of mountains grows what is called a *subalpine forest,* with fir trees and bristlecone pines. This phenomenon is called **altitudinal zonation** and occurs as we move to higher and higher altitudes on the sides of mountains until eventually there is little or no life existing in those environments at all (Figure 23–31). The changes observed in ecosystems with respect to altitude reflect similar changes seen in ecosystems at different latitudes. This similarity between the ecological outcomes of the type of changes that occur, altitudinal and latitudinal, reflect the similarity in conditions present in those environments even though they are not equivalent geographically.

Figure 23–29 Grasslands—Prairie
Little of the original grassland is left in the United States. Prairies are characterized by tall grasses and more average rainfall than the other grassland biome, the plains.

Alpine Tundra Higher yet in altitude on the sides of mountains, plant life begins to thin out and then cease to grow. Above the so-called **tree line,** the conditions are too harsh for the growth of trees in forests. The land is too cold and too windy and there is too little soil. The top of the Rockies is solid igneous rock, from which the mountains

Altitudinal zonation the changes in biomes that occur with changing altitude, as up the side of mountains

Tree line the zone in altitudinal zonation after which no trees grow, alpine tundra

Figure 23–30 Taiga
Taiga, or coniferous forests, are found in northern latitudes or at higher altitudes, as in the Rocky Mountains. Spruce, fir, and pine trees provide habitats for owls, bobcats, and hares.

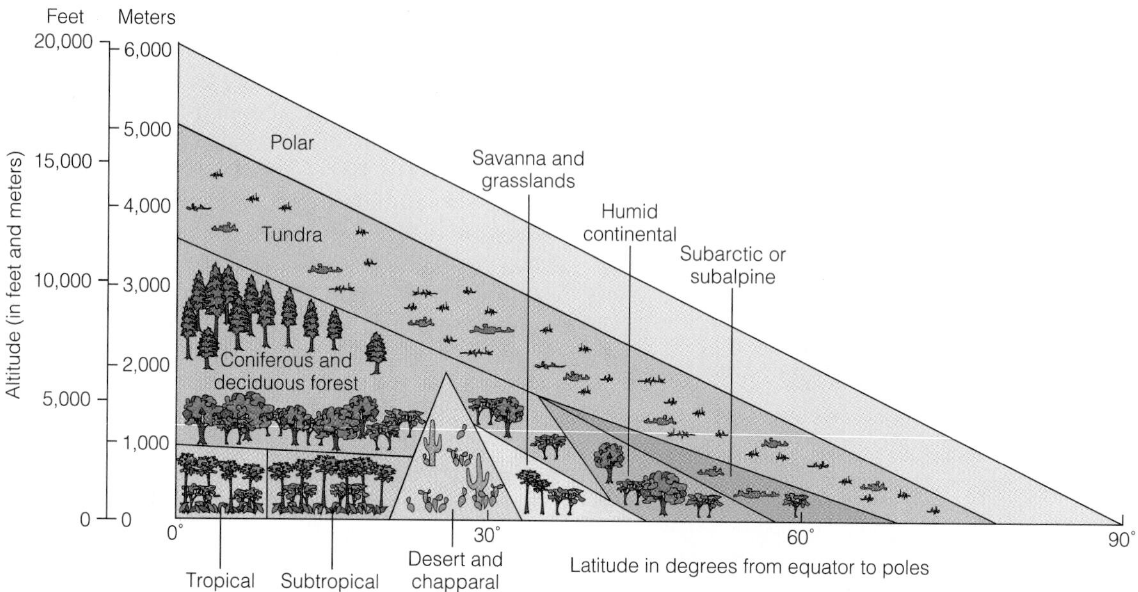

Figure 23–31 The Effects of Altitude and Latitude on Ecosystem Type

As you move from the equator (0° on the horizontal scale) to the poles (90°), the ecosystems change from tropical, desert, and temperate to tundra and polar types. A similar set of changes occurs when moving up in altitude, along the sides of mountains. This similarity between latitudinal and altitudinal changes in ecosystems reflects responses of living organisms to prevailing temperature and climatic conditions regardless of how they are established.

got their name, and is permanently under a cap of ice and snow. This is the transition zone to the *alpine tundra. Tundra* is usually associated with the arctic region (located at higher latitudes), where a short relatively warm season provides ponds and bogs for the few types of migrating birds that nest there in spring and for the hearty mammals who reside there year round (Figure 23–32). Arctic winters are harsh and unforgiving. An alpine tundra occurs at high altitudes in mountain ranges around the world. The Sierra Nevada, the Rockies, and the Appalachian mountains all display altitudinal variation in ecosystems.

Desert Coming down out of the mountains and the coniferous forest biome, we feel the warm dry air of the desert. Deserts are *arid* and, in the hottest types, appear relatively sparse with respect to the diversity of life (Figure 23–33). Annual precipitation is often less than 10 inches to 12 inches a year, which may come in a seasonal burst that leaves the rest of the year parched. Animals and plants have to complete their life cycles rapidly in order to survive. Special adaptations to preserve water are also necessary. Animals seek the refuge of shade during the day and are often **nocturnal,** that is, active at night. Human development in the great Southwest of the United States has depended on the importation of water, which establishes a strange oasislike quality to the cities of this region, such as Phoenix and Las Vegas. Much of Southern California is desert, spotted with towns and cities where water has made them green. The desert of the southwestern United States extends down through much of northwestern Mexico as well.

Chaparral As we approach the west coast of the United States, the desert biome gives way to the final terrestrial ecosystem we will consider in our travels, the *chaparral* (Figure 23–34). Chaparral is a semiarid ecosystem also known as *mediterranean* because similar conditions are found in regions bordering the Mediterranean Sea. The plants of the chaparral are composed predominantly of broad leaf and evergreen shrubs and dwarf trees (those under 7 feet to 8 feet tall). California laurel, sage, and manzanita abound, and interspersed among this dense foliage are live oak trees. Winters are generally mild and wet with rainfall of 10 inches to 20 inches, and summers are relatively long, dry, and hot, often reaching over 100°F.

Nocturnal at night, particularly in reference to the timing of animal activities and behavior

(a)

(b)

Figure 23–32 Tundra
Life adapts to the cold. Throughout most of the year, tundra is frozen. As observed in (a), the arctic fox shown is adapted to the winter-white and cold by its thick white fur. The lengthy cold season occurs because of the extreme latitude of the ecosystem. However, during the brief spring and summer, the tundra is home to the caribou (b) and blooms with flowers and grasses.

Figure 23–33 Deserts
Deserts come in many types and are different in various parts of the world. They all share the hot, dry environment that receives very little rain (often less than 10–12 inches per year). The Sonoran desert of Arizona is shown. Animals and plants have adaptations that allow them to survive, including behavioral adaptations that make them active nocturnally.

Figure 23–34 Chaparral
Chaparral is a pyroclimax community. Low, thick shrubby plants and sparse large trees are characteristic of this ecosystem. Chaparral is subject to mild wet winters and periodic fires that germinate the seeds dropped by the plants in this community and keep the species presence relatively stable.

Succession and Climax Communities Chaparral has a fascinating ecology. The chaparral community forms one link in a chain of changing ecosystems that can occur in succession, one following the other when there is destruction of the environment from fire or human land use. This sequence of changes is, in fact, called **succession** and occurs in all types of communities when they have been significantly disturbed. For example, when a fire destroys a forest, the recovery of the area affected occurs in stages

Succession the progressive changes in plant species in disturbed or denuded habitats

(Figure 23–35). The forest does not simply grow back directly where it once stood. Initial plant growth is in the form of grasses and shrubs. These fast-growing pioneering species help to hold the forest soils together against erosion and offer low cover and food for small animals. This is followed by more complex, slower growing plants (including young trees) and the migration of larger animals into the region. The final stage of succession in the case described here is called the **climax community,** which represents the original forest. It may take decades or centuries for this succession of ecological changes to transpire, or it may not happen at all as in the case of rain forests destroyed by human activities.

A Pyroclimax Community What is interesting about a chaparral community is that it is not a climax community. The final stage of succession for this ecosystem, of which chaparral is the penultimate form,is called *coastal sage*. The coastal sage community is composed of widely dispersed oak trees, with grasses and sagebrush. What is it that maintains the chaparral in this arrested stage of succession? The answer is fire. Without fire, chaparral gives way to the climax community, coastal sage. However, fire is quite common in the regions in which chaparral establishes itself. In times before the immigration of Europeans and the build up of cities and their suburbs, fire was frequently ignited by lightning and spread freely over the land. The chaparral is extremely dense in foliage and highly volatile, so any fire within it spreads rapidly with quick, high heat. Thus, chaparral is called a *pyroclimax community*. This means that for a chaparral community to be maintained as the predominant ecological unit, the plants must be burned down every few years. This works well for the plants of the chaparral because their seeds, which they drop in huge numbers

Climax community the final type of community in a succession

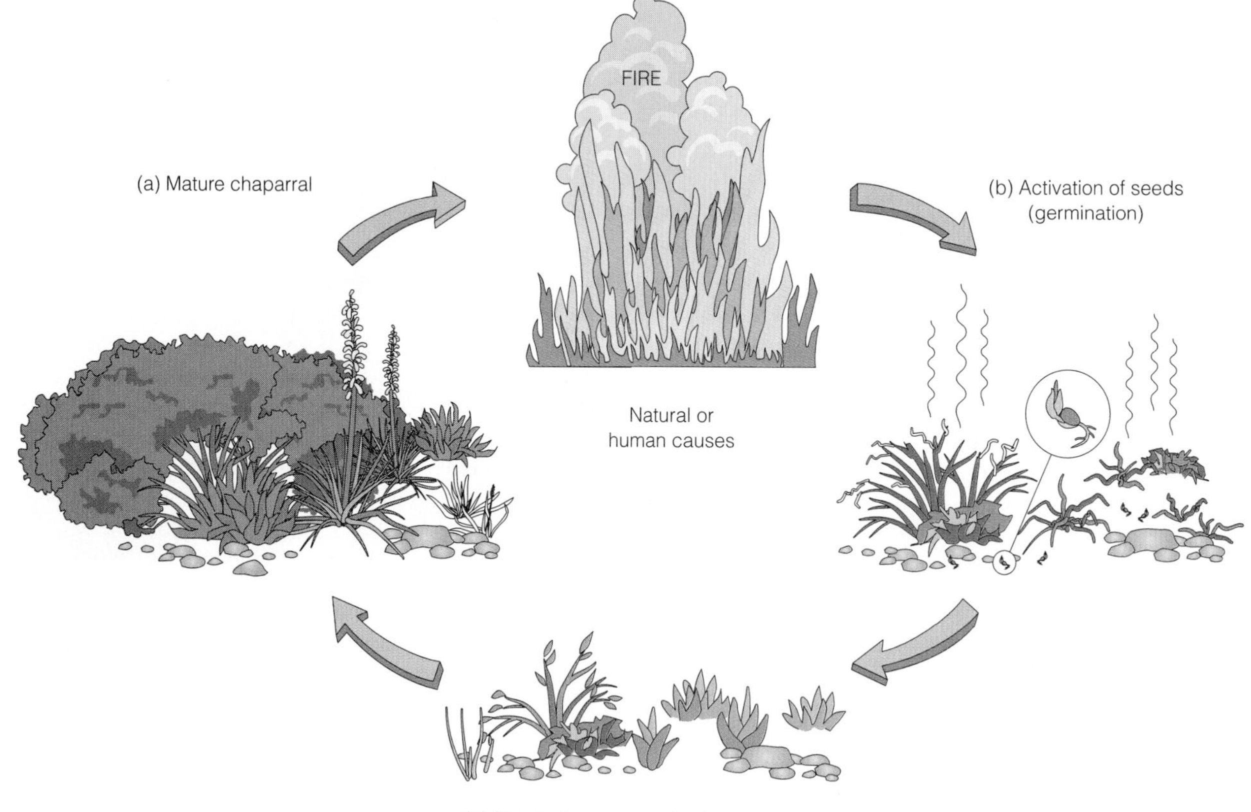

Figure 23–35 Succession in a Pyroclimax Community
A pyroclimax community such as chaparral (a) requires fire to activate seeds of the species to grow (b). The burn is usually intense and rapid, not enough to kill everything but enough to clear the ground for new growth of grasses, shrubs, and evergreens.

each year, cannot germinate unless they are exposed briefly to high temperature. Thus, the dense scrub of the chaparral, which burns hot and fast, is perfectly poised to regenerate itself following a fire. A chaparral community is like the mythological phoenix, which is famous for arising anew out of its own ashes.

This ecological cycle offers an interesting lesson for the ever-expanding growth of civilization. The southwest coast of California is densely populated. A large portion of the population lives in areas that are of the chaparral type. In many cases, families have built their homes right in the middle of a pyroclimax community. If this ecosystem is not subjected to fire every few years, it will be replaced by a coastal sage ecosystem, which is the climax community. It is ironic that the only way to retain the chaparral, and the natural beauty of the area that was the reason for building there in the first place, is to burn it down. Unfortunately, along with the plants of the chaparral, a good many homes that are built there to enjoy its rugged natural beauty are burned down also.

The trip through the terrestrial biomes of the United States has been a simple and descriptive one. There is great complexity in each and every ecosystem and much yet to be learned about them. An understanding of how ecological communities develop, evolve, and sustain themselves may ultimately help us live better and more wisely within nature instead of outside it.

POLLUTION AND POLLUTANTS

A **pollutant** is any agent that when released or accumulated in an environment is detrimental to the quality of life or property (Table 23–2). This is an open-ended definition because some natural materials that are not normally considered pollutants can become pollutants if they occur at high concentrations. Carbon dioxide is not considered a pollutant until it occurs in excess, a situation in which too much of a good thing is bad. Carbon dioxide is essential for life on Earth, as the oxygen and carbon cycles described previously have shown, but a high concentration of carbon dioxide is lethal. Even modest increases in the amount of carbon dioxide in the atmosphere may have long-range effects on the world's climate. The same is true of heat, which is necessary for life but in excess becomes *thermal pollution.* Similarly, light (for example, streetlights or stadium lights) can disrupt ecosystems and is, thus, *light pollution. Noise pollution* is another example, as anyone who lives near an airport or industrial area can attest. Still another example, of a conditional pollutant is ozone, which is a relative of oxygen. Ozone in the upper layers of the atmosphere protects life from the dangerous radiations from the sun. Yet, ozone in the air over cities such as Mexico City or Los Angeles is a dangerous pollutant.

The word "pollution" is used to describe processes as well as particular types of substances. Pollution is the production and release of detrimental agents into the environment. Pollution of the biosphere is predominantly a consequence of human actions, although certain types of natural disasters can cause extensive pollution, such as forest fires caused by lightning and volcanic activity that release dust and gas into the atmosphere. However, these are usually minor compared with the types and amounts of pollution that human activities produce. We alter the earth, the water, and the air with a wondrous and worrisome variety of gases, acids, particles, metals, chemicals, radioactivity, and refuse (Table 23–2). Many of these materials are of our own creation and there is often no way to naturally detoxify them. Others are taken up into the food web and ascend through trophic levels to do their damage in unsuspected ways (recall Figure 23–20 and the fate of the eagles around the lake). Only in the past few decades have the problems of pollution effectively reached the general public, and legislation and regulatory action have been taken.

Table 23–2 Pollutants

TYPE	EFFECT
Carbon dioxide	Metabolic poison, greenhouse gas
Carbon monoxide	Metabolic poison
Ozone	Irritant, oxidizer, greenhouse gas
Hydrocarbons	Carcinogens
Sulfur oxides	Acid rain
Nitrogen oxides	Acid rain, photochemical smog
Particulate	Irritants in lungs, eyes
Radioactive materials	Carcinogens

The Shock of the Future

From an economic point of view, a scenario of the future may proceed as follows. The increase in human population worldwide will open new markets for the sale of products. Greater productivity and sales will yield greater profits. Greater productivity will mean more jobs, and more jobs mean more money for more people. More money means more spending, which increases demand for new products and increases productivity. This sounds good to most people, and it is. The problem with this vision of growth is that it does not take into consideration the limits on natural resources and the inevitable polluting, by-products of industry and manufacturing.

The alternative, pessimistic view is that populations will increase, productivity will increase, and waste will increase.

Pollutant an agent whose presence in an environment is detrimental to life and the quality of life

The waste will be released into the environment, potentially lowering the quality of life, in fact, endangering life. There will be a rapid depleting of nonrenewable natural resources, such as fossil fuels and minerals, and a severe strain on renewable sources, such as trees and wild game. Products will become more expensive and it will cost more to live. Workers will demand higher wages, and on and on. Not a pretty picture. Where does this type of cycle of rampant growth and consumption stop? How many people can Earth provide for? How should the resources of Earth be used? Consider the population of the world in the year 2000 to be 6+ billion. By the year 2030, at the present rate of population growth, that number will be 10 to 12 billion. Because much of the world population is at present starving or sustaining itself below levels of nutrition set forth by the World Health Organization (WHO, an agency of the United Nations), what is the world going to be like in the middle of the next century?

These questions are unresolved at present, but their resolution will require people in all facets of our society to work together. Most people, whether scientists, politicians, or people in business, want to see the world stay the beautiful life-sustaining place that it is. We all want to pass on a healthy world to our children and to their children. However, there are clear signs that the edges of nature are frayed by human activity. Lakes and streams are polluted and undrinkable, air in big cities is often unbreathable, and the arable land for farming and agriculture is receding. Biological reality offers a harsh vision of the future for humans.

The problems we face in degraded ecological systems constitute a crisis of the environment. Most problems of contaminated air and water are local problems, or at least these are perceived to be local. However, there are problems of national and international importance that have consistently broken through our awareness threshold and are reported often in the national and local news media. This is because of the potential widespread disaster associated with their unchallenged progression. Two environmental problems will be discussed here to describe the potential global magnitude of their effects should they continue to worsen. They are the increase in Earth's atmospheric greenhouse effect and the reduction or thinning of the stratospheric *ozone layer.*

The Greenhouse Effect

If you have ever parked a car in the open for an hour with the windows rolled up on a bright, hot summer day, you know what the greenhouse effect is. When you return to the car and open the door, you can feel the heat that has accumulated inside. The heat flows out to meet you as you move to get in. How does this overheating occur? The primary agents for this effect on your car are sunlight and glass. Sunlight is radiation and carries energy. The glass of the windows acts as a filter. It allows some of the high-energy light to pass through, but prevents lower-energy heat from escaping back out. The sunlight passes through the windows and heats up everything inside the car. Once the heat is generated, most of it has nowhere to go; it cannot get back out through the glass. The only thing that it can do is get hotter and hotter in the car. You open the door and it feels as if it is a furnace inside.

Earth's atmosphere acts in some ways as a thin pane of glass curving over the entire globe (Figure 23–36). The energy of the sunlight travels through the atmosphere to the surface, heating up everything it impinges on and, thus, reducing its energy. The atmosphere acts as a filter. Some of the light is reflected off the surface of Earth and bounces back out into space. This type of reflection of sunlight from the surfaces of other planets in the solar system allows us to see them. Venus is the brightest object in the night sky, other than the moon, for this reason. However, much of the heat generated by sunlight on Earth does not escape; it is trapped near the surface by the atmosphere and as a result, warms the planet enough to make it habitable. It is this heat that keeps Earth at the appropriate temperature to sustain the fluidity of water. The atmosphere is important also because without it there would be a serious radiation problem for Earth. Too much dangerous high-energy radiation (and fast-moving particles) would impinge on the surface and too much heat would escape back out into space.

What is it in the atmosphere that causes this greenhouse effect? The principal ingredient is carbon dioxide,

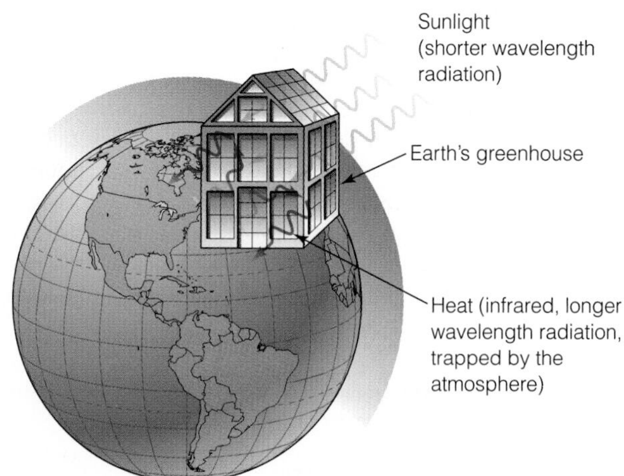

Figure 23–36 Earth as Greenhouse
The greenhouse effect is a natural phenomenon needed to maintain thermal balance of Earth's biosphere. Sunlight (short wavelength radiation) enters the atmosphere, passes to the surface, and is re-emitted (longer wavelength radiation). Some of the longer wavelength light is absorbed and is not reflected, thus, warming the surface.

although other greenhouse gases are important as well (Table 23–2). Carbon dioxide helps retain the heat generated by sunlight on the surface of Earth. Carbon dioxide makes up a very small percentage of the atmosphere (approximately 0.03%–0.04%). The percentage may be small, but carbon dioxide is very important. If Earth had much less than this concentration of carbon dioxide, more heat energy would escape back into space, leaving the planet colder. If Earth had a higher concentration of carbon dioxide, more heat would be retained and the surface regions would heat up. The problem we face today is an increase in carbon dioxide in the atmosphere.

How does this occur? It happens principally through the burning of fossil fuels. When oil, coal, and natural gas are burned, the major end product of their combustion is carbon dioxide. The industries of the world operate on fossil fuels, and fossil fuels will continue to be our major source of energy for the next century or so. This being the case, carbon dioxide levels should rise steadily from the combustion of the hydrocarbons that constitute the chemical components of these fuels. The question is how much this will increase the surface temperature of the planet in the long run. Some scientists say there will be very little effect; some say the effects will be dramatic and devastating. Others take the middle ground and predict substantial and progressive changes, but with less of a sense of doom and gloom.

What are the long-term consequences of a few degrees increase in the average temperature of the atmosphere? One consequence may be that a significant portion of polar ice will melt, raising the level of the oceans. This would immerse low areas along the coastlines of all continents and islands and change the geography of Earth's terrestrial land masses.

Another possibility is that significant climatic changes may occur. **Climate** is a regional phenomenon establishing the long-range patterns of precipitation, temperature range, humidity, and other weather-related processes. Thus, global warming would potentially alter rain patterns and lead to desertification in some regions and new agricultural capacity in others. Areas that are marginal today may become arid and useless tomorrow or vice versa. This kind of global change has underlying political ramifications. The loss of arable lands for farming could disturb agricultural economics among nations, such that exporters of today become importers of tomorrow. The next century may see global food wars based on competitive needs for arable land and usable water.

A third possibility is that plants will grow faster and be healthier. In this case, more carbon dioxide and more efficient photosynthesis could lead to the production of a greater mass of plant materials. This would mean an increase in plant foodstuffs, including complex carbohydrates, fats, and proteins. One drawback to this growth change is that the number and types of organisms associated with plants probably would be altered also. This includes plant pathogens (viruses, bacteria, fungi) as well as insect populations feeding on plants. The ratios of these organisms could change significantly. Such changes would modify all the terrestrial ecosystems. No significant change in one part of the biosphere can occur in isolation from the others. An environmental domino effect, one change leading to others, could ensue. This, in turn, would alter in unpredicted and unsuspected ways many aspects of the biosphere.

The drastic planetary changes alluded to here and suggested to be the result of increases in greenhouse gases such as carbon dioxide are not foregone conclusions. However, they are serious food for thought. How humans use the resources of the world today will determine what kind of world future generations will inherit.

The Hole in the Ozone Layer

Ozone is a Dr. Jekyll and Mr. Hyde. Under one set of conditions, it is of vital importance as a radiation-filtering ozone layer in the stratosphere (15 km–50 km above Earth's surface) and perfectly safe (Figure 23–37). Under another set of conditions closer to the surface (0 m–1000 m), it is a pollutant and is very dangerous to human health. What is this molecule with two entirely different environmental personalities? Ozone is a molecule composed of three oxygen atoms and under normal conditions, is a gas. It is produced by high-energy reactions involving the combination of regular oxygen, which has two oxygen atoms and a split form of the oxygen molecule that has only one oxygen atom and an unpaired electron called a **free radical** (Figure 23–37). Ozone has a sharp, distinctive odor that can be sensed by smell during lightning storms and around electrical equipment. It is formed during combustion of hydrocarbons, as is the case in automobile emissions, and as a result of the effects of sunlight on nitrogen oxides and subsequent interactions with oxygen (Figure 23–37). This is called a **photochemical reaction.** Ozone is chemically highly reactive and can damage cells and tissues of the human body and other animate and inanimate materials.

The presence of ozone in the air at the surface of Earth is predominantly a consequence of industrial activity and

Climate the long-range patterns of weather and precipitation

Free radical highly reactive form of molecule usually associated with light-induced breaking of covalent bonds.

Photochemical reaction the effects of sunlight on molecules in the atmosphere, for example, formation of ozone at surface levels of the operation of motor vehicles. In sunbelt cities such as Los Angeles, with millions of cars operating all the time and atmospheric conditions capable of trapping pollutants within *inversion layers,* ozone alerts are a fairly common occurrence. The advice of health officials to people during these alerts is to stay inside and avoid heavy exertion. Little children (who tend to be very active) and older adults with respiratory problems are the most seriously compromised groups. This is the Mr. Hyde form of ozone. The Dr. Jekyll form is part of a more complicated, globally important process protecting life on Earth.

Ozone is found in a layer that envelops Earth several kilometers above Earth's surface. It is formed in a region where incoming high-energy radiation encounters oxygen molecules at the fringe of the atmosphere. Ozone formation is a cyclic process in which O_3 is converted to O_2 and $O^\cdot$, and O_2 and $O^\cdot$ are then converted back to O_3 (Figure 23–37). The $O^\cdot$ symbol indicates the appearance of an oxygen free radical. Free radicals have an unpaired electron and are highly reactive chemically. They exist only a short time before they react with other molecules. This is the case for the combination of $O^\cdot$ and O_2 into O_3. Ozone molecules specifically absorb high-energy ultraviolet radiation and convert that energy to the formation of $O^\cdot$ and O_2. This prevents UV radiation from passing unhindered to the surface of the planet and interacting with molecules of living organisms. UV radiation is dangerous and damaging to all forms of life. At high intensities, it destroys molecules and even at low intensity, it can chemically alter DNA and proteins. It causes mutations, cancerous transformation of cells, and cataracts in the eyes. The incidence of skin cancers in humans is correlated with exposure of the surface of the body to the damaging radiation of sunlight accumulated over a lifetime. The deep, dark sensuous tan of summer that sunbathers go to such trouble to develop leaves a sinister legacy later in life.

The ozone layer does a great job protecting life at the surface from the dangerous wavelengths of radiation emitted from the Sun. What would happen if the ozone layer began to disappear? A few years ago, this idea would have been used as the basis for a science fiction novel. How could anything we did on Earth affect a layer 15 km up in the atmosphere? Truth is often stranger than fiction and in this case, considerably more dangerous. In the 1980s, atmospheric scientists began to notice changes in the ozone layer (Figure 23–38). There was a drop in the concentration of ozone in a region over Antarctica. A hole was developing in the ozone layer and each year the hole got bigger. What was causing this to occur? What were the ramifications of this trend? Several teams of scientists started working on the problem.

The origin of the problem presently confronting the ozone layer lies in the past. One of the most important contributions to organic chemistry during the 1940s and 1950s was the discovery of the means to synthesize **chlorofluorocarbons.** The most prominent among the fluorocarbons synthesized was a compound called Freon-12.

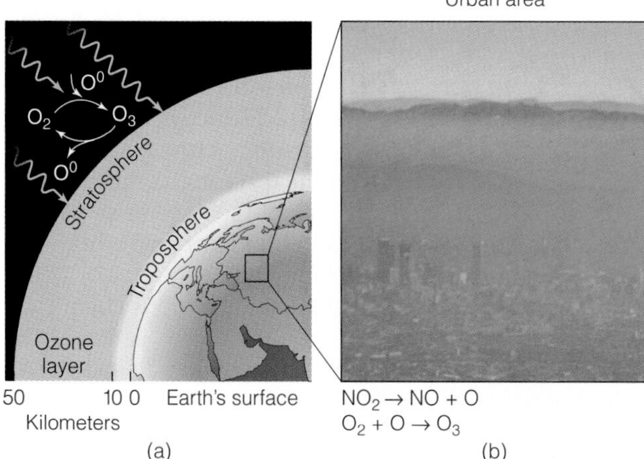

Figure 23–37 Ozone—High and Low
The stratosphere contains the ozone layer, which protects Earth's surface from high-energy radiation. Light energy is absorbed by O_3, which splits to O_2 and the oxygen frees radical O, then recombines to O_3 again. In (b), O_3 at ground level is dangerous and combines with other gases to form smog. In addition to the O_3 cycle occurring in the stratosphere, O_3 is also produced at Earth's surface through the light-induced breakdown of nitrogen dioxide (NO_2 to nitric oxide [NO] and free radical oxygen [O]). The formation of ozone then occurs when molecular oxygen (O_2) reacts with $O^\cdot$ to form O_3.

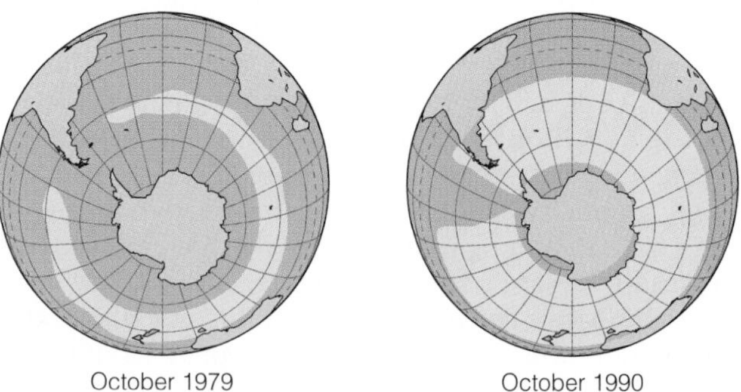

Figure 23–38 The Ozone Hole
The ozone layer reportedly has become thin, particularly over the South Pole (shown). The suspected cause of this phenomenon is chlorofluorocarbons released into the atmosphere. Dramatic changes took place in the ozone layer throughout the 1980s.

Chlorofluorocarbons compounds rich in chlorine and fluorine such that they interfere with the production of ozone in the stratosphere

Freon has a very low boiling point (−28°C), is chemically inert, long-lived, and harmless to humans and has proved to have many useful applications. It is the major coolant compound for the compressors of our home and car air conditioners, as well as refrigerators and freezers, though steps are being taken to eliminate its use. Freon gas is (or was) used as a propellant for canned paints, deodorants, and hairsprays that overflowed the shelves of hardware stores, markets, and pharmacies around the country and around the world. It seemed the ideal compound for our emerging high-technology civilization. In addition, Freon is a very stable molecule. It takes decades, perhaps centuries, to break down chemically. Slowly but surely, decades of human use released thousands of tons of Freon into the atmosphere and it worked its way up to the most rarefied heights, including the stratospheric ozone layer.

No one knew at the time it was produced that Freon interferes with the synthesis of ozone. It is now known that Freon prevents the cycle of chemical reactions that regenerates ozone from oxygen. The initial effects of this began to show up in the atmosphere over the South Pole and, more recently, a similar depletion has been detected in the atmosphere over the North Pole. Because Freon is so incredibly stable chemically, it is thought that the molecules presently in the atmosphere will persist for decades and continue to interfere with ozone production. There is not a thing that we can do about it even if we stop manufacturing and using Freon from this day on.

What happens if the ozone layer is depleted over inhabited areas of the world? The main consequence is that more UV radiation will penetrate to Earth's surface. All the destructive effects of this radiation observed under present atmospheric conditions will be multiplied. The incidence of skin cancers will increase because mutation rates will increase. Cataracts and eye damage will be more prevalent. Plants and animals of all kinds will be affected as well as humans. How would agricultural plants survive? Should we start selecting for hybrid plants that can survive more intense solar radiation? The amount of radiation will also potentially increase the heat introduced into the biosphere. This, in conjunction with increases in atmospheric carbon dioxide, could accelerate the rate of increase in the average temperature of the planet over future decades and centuries and make Earth a very different place than it is today.

Much of what is presented in this section of the chapter is speculative. It is not known what the long-term effects of increases in carbon dioxide and/or the depletion of ozone will be. However, we need to act now to prevent the possible disasters spelled out here. Pollution of the outer atmosphere has been emphasized here, but many other pollution problems, some in your own neighborhood, complicate the picture of the future of Earth. The following pollution problems demand our attention: the occurrence of acid rain, the buildup of pesticides in soil and sediments, the occurrence of heavy metals (such as mercury and cadmium) in sediments, and the storage and containment of radioactive waste. In addition, thermal pollution of water at nuclear reactor sites (used in cooling and then returned to the environment) and noise pollution in urban and industrial areas constitute problems that need reasonable and long-term solutions.

An awareness of the major environmental problems, just enumerated, that face us today can be our first step toward bringing into balance the consequences of human activities with the natural forces of nature that sustain the planet itself and the delicate biosphere that surrounds it. First, you have to know you have a problem. Then, you face the problem and work on solving it.

Summary

The lithosphere, hydrosphere, and atmosphere are parts of an overarching sphere of life known as the biosphere. Earth is unique among planets of the solar system in supporting life. One of the main reasons it can do this is that the energy of the Sun is sufficient to adequately warm (but not scorch) Earth, and another reason is that water is present in great abundance. The Sun is the ultimate source of energy for life on Earth because energy runs downhill on a one-way street and has to be supplied on a continuous basis.

The capture of radiant energy from the Sun through the process of photosynthesis is the province of plants. In this regard, plants are autotrophs, making all their own organic molecules using simple ingredients including carbon dioxide, water, and sunlight. All plants produce oxygen. Most of the atmospheric oxygen comes from the phytoplankton of the sea, which share with terrestrial plants the green pigment chlorophyll necessary to absorb light. Plants convert light energy to chemical energy in the form of complex carbohydrates. Animals consume plants and may themselves be consumed by other animals. Decomposers break down complex organic materials built up in plants and animals into simple compounds that are recycled within the biosphere. There are complex cycles controlling the availability and reuse of all biologically important elements and chemical compounds, because Earth is a closed system with limitations on its natural resources.

Populations are groups of individuals of the same species that interact and interbreed. Communities are groups of populations interacting in an environment. The interactions among individuals in these communities establish food chains and food webs. Plants are the fundamental basis for establishing a biological framework for community structure. Plants are primary producers and the first link in the food chain. The products of plants are passed directly or indirectly to consumers (herbivores, omnivores, and carnivores), but the overall efficiency of energy between links in the food chain is relatively low. The 10% rule expresses the general efficiency with which energy is transferred between trophic levels in a food chain or food web. As omnivores, we should be aware that this translates into a tenfold decrease in efficiency when comparing the overall energetics of eating plants versus eating animals that feed on plants.

Biological amplification is a process of accumulation in which there is successive increase in concentration of chemicals in animals at different trophic levels. Concentration of toxic chemicals in animals at the highest trophic levels can be lethal and can cause disturbances throughout the ecosystem in which they occur.

Habitats represent the addresses of organisms. The roles that species play in the habitats in which they live are called niches. The largest units of organization in the biosphere are called ecosystems or biomes. The study of the interactions of living things with the environments inhabited by them is called ecology. The boundaries of all types of aquatic and terrestrial ecosystems are not clear-cut but appear, for example, as intertidal zones and gradations such as the changes in ecosystem types that occur as altitudinal variations up the side of a mountain. There are myriad interactions of organisms across the arbitrary borders of ecosystems, borders that grade into one another over the face of the planet. All of life on Earth, including *Homo sapiens,* is interconnected and interactive.

Questions for Critical Inquiry

1. What is ecology? Why should its study be important to us?
2. What is the basis for the greenhouse effect? How does Earth's greenhouse effect operate? Is it necessary for life to exist? What will happen if it is enhanced?
3. What is the function of the ozone layer? How is it being altered? What are the long-term consequences of these changes?
4. What is biological magnification? How might it affect organisms at the highest trophic levels? How might it affect you?
5. If you were isolated on a desert island with six chickens (and the means to cook them) and 50 pounds of corn, which would you eat first and why?

Questions of Facts and Figures

6. What are the three main layers of the biosphere? How do they interact?
7. What is the difference between a population and a community?
8. What is pollution?
9. What is the 10% rule and why is it important in the economics of ecosystems?
10. What are some of the major ecosystems found in the United States? How can you tell when you are leaving one and entering another?
11. Describe a food chain. How is a food chain different from a food web?
12. What is the ultimate source of energy for life on Earth? How is it trapped for use by living organisms?
13. What is a producer? A consumer?
14. What is the difference between an autotroph and a heterotroph?
15. Why are decomposers important to a community?
16. What is a niche? A habitat?

References and Further Readings

Cummings, M. R. (1996). *Biology: Science and Life.* St. Paul, MN: West Publishing.

Monroe, J. S., and Wicander, R. (1994). *The Changing Earth.* St. Paul, MN: West Publishing.

APPENDIX

Kingdoms of Life

KINGDOM MONERA

The monerans are single-celled or unicellular organisms that lack definitive internal structures. The lack of specialized internal structures, such as a nucleus, provides another way to classify them, in this case, as procaryotes. The monerans are microscopic in size, have specialized cell walls, and include what are commonly known as the bacteria and blue-green algae (also called the cyanobacteria) (Figure 1–8). Some species of cyanobacteria have the capacity to photosynthesize. There are thousands of species in this Kingdom, including many commonly known to cause diseases in humans, but it is expected that this is only the tip of the biological iceberg in terms of their actual numbers. There may be thousands more species of Monera yet to be discovered.

KINGDOM PROTISTA

Protista organisms, like the monerans, are unicellular. The distinctive internal cellular organization, however, that includes a nucleus and specialized organelles differs from that of the monerans. For this reason, they are classified as eucaryotes. In addition, the protists are larger than the monerans. Some species of protists are capable of photosynthesis, such as unicellular algae, and are, therefore, **autotrophs;** others, lacking the ability to make their own food, must obtain sustenance and energy from materials in the environments in which they live. These are called **heterotrophs.**

Protistan taxonomy is somewhat controversial. This is because some organisms are both unicellular and photosynthetic and, therefore, could fit into Kingdom Protista or Kingdom Plantae. Some protists are funguslike (slime molds) supporting the contention that each of the major multicellular Kingdoms had its origin in a particular group of unicellular predecessors (see Figure A–1). The key criteria for organisms in this Kingdom are the requirements that they be unicellular and live in an aqueous environment. In addition many of these species are often highly motile (Figure 1–8).

KINGDOM FUNGI

Fungi are eucaryotes and for the most part are multicellular in structure. Common unicellular forms are represented by a large class of fungi known as yeast (Biosite A–1). All fungi are heterotrophs and feed by absorption of nutrients from their environments. In most cases fungi secrete

Autotroph an organism that produces all its cellular constituents from inorganic materials and, thus, grows independently of exogenous organic carbon sources; for example, plants

Heterotroph an organism whose growth is dependent on an exogenous source of organic compounds; for example, animals

Figure A–1 Broad, General Interrelationships of Kingdoms

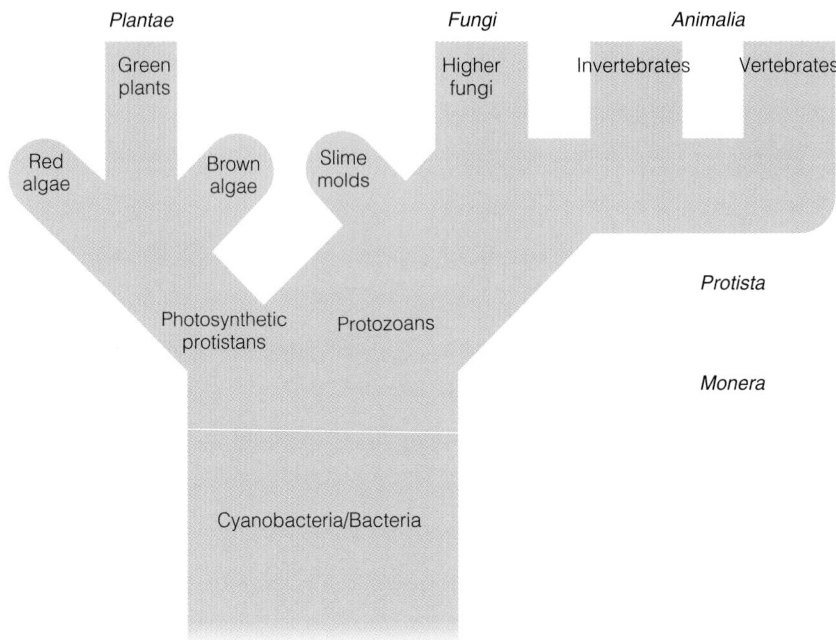

enzymes that digest and break down dead materials, so they are also classified as **saprophytes.** Fungi have specialized cell walls and unique reproductive structures that give them their often familiar forms (Figure 1–8).

KINGDOM PLANTAE

Organisms in the Kingdom Plantae are all multicellular, photosynthetic, and, therefore, autotrophs. They contain specialized organelles, called **chloroplasts,** in which the reactions of photosynthesis are carried out. Plants are eucaryotes and multicellular in structure and their protistan relatives are unicellular. Higher plants have a multitude of complicated cell and tissue types, and some forms are the most massive and long-lived organisms on Earth (for example, the mammoth redwood trees of California). There are many subdivisions of this Kingdom ranging from red and brown algae and mosses to various classes of vascular plants familiar to us such as ferns, flowering plants, and coniferous trees (Figure 1–8).

KINGDOM ANIMALIA

All animals are multicellular eucaryotes and heterotrophic. Even the simplest of animals has specialized cell types and some form of motility. The organization of the Kingdom Animalia will be used to show how organisms within a kingdom are further subdivided, categorized, and named.

Subdivisions of Animals

While a Kingdom is the largest most inclusive category of similarity, it is rather useless unless there is further division into smaller and more exclusive categories. Below the level of Kingdom are many subcategories that order the arrangement and relationships among organisms into groups that share more and more traits. This hierarchy is arranged in descending order as follows:

Kingdom
Phylum
Class
Order
Family
Genus
Species
(as well as subcategories, such as subphylum, subclass, etc.).

Saprophyte an organism that obtains organic compounds needed for its growth from absorption of materials from dead and decaying organisms; for example, many types of bacteria and fungi

Chloroplast the chlorphyll-containing organelle found in green plants that is responsible for photosynthesis

YEAST AND THE PROCESS OF FERMENTATION

The basis for industrial fermentation is the enzymatic breakdown or degradation of organic molecules by microorganisms. It is a process in which complex compounds are broken down into simpler organic products. Some of these products are important commercially, such as alcohol and acetic acid. Louis Pasteur proved that microorganisms were involved in this type of fermentation in the 1860s, but it was not until 1878 that the term enzyme was first used.

Fermentation and anaerobic respiration are synonymous. The classic example of commercially valuable fermentation is that brought about by yeast, particularly *Saccharomyces cerevisiae*. The fermentation carried out by this species results in the formation of carbon dioxide and ethanol from glucose and provides the basis for production of alcoholic beverages. However, many microorganisms and other types of cells are capable of fermenting glucose to a variety of different compounds. Bacteria convert glucose to acetic acid in the formation of vinegar, and the cells of our own body are capable of producing lactic acid by anaerobic respiration.

Dividing the Kingdom Animalia into smaller categories, the major phyla of animals are the Porifera (sponges), Cnidaria (coral and jellyfish), Ctenophora (sea combs), Annelida (segmented worms), Mollusca (clams and snails), Arthropoda (insects and spiders), Echinodermata (sea urchins and starfish), and Chordata (mammals, birds, reptiles, amphibians, and fishes). With respect to phylum Chordata, one might ask, how do humans fit into the hierarchy and naming categories established in the Linnean system? Humans are in the

Kingdom:	Animalia,
Phylum:	Chordata,
Subphylum:	Vertebrata,
Class:	Mammalia,
Order:	Primates,
Family:	Hominidae
Genus and Species:	*Homo sapiens*

For the specific naming of all individual species, both the genus and species designation must be used. *Homo sapiens* is the scientific name for the present population of humans existing on Earth. There have been others before us, including *Homo erectus* and *Homo habilis,* but they as species have long ceased to exist. There are many other familiar examples of Latin binomial combinations of common animals, including *Felis leo,* for the lion; *Canis lupus,* for the wolf; *Pan troglodytes,* for the chimpanzee; and *Gallus gallus,* for the chicken. Using the Linnean classification system as a key to establishing relations among organisms, each of those listed above are related to humans because each is in the Kingdom Animalia and the phylum Chordata. However, the chimpanzee is our closest relative, because we share similarities at the level of order. We are both primates. Our next closest kin are the felines and canines, because we share with them similarities in structure and function at the level of class. All are mammals. Birds are in the class Aves and quite distant taxonomically from us, but we do share similarities at the level of subphylum. We are all vertebrates. This organization of life connects all species living (and extinct).

Metric and English Equivalencies

LENGTH

1 km = 1 kilometer = 1000 meters = 0.621 miles
1 mile = 5280 feet = 1.609 kilometers
1 m = 1 meter = 100 centimeters = 39.37 inches
1 cm = 1 centimeter = 0.01 meter = 0.394 inches
1 inch = 2.54 cm
1 mm = 1 millimeter = 0.001 meter = 0.039 inches

MASS

1 kg = 1 kilogram = 1000 grams = 2.205 pounds
1 lb = 1 pound = 453.6 grams
1 oz = 0.0625 lb = 28.35 grams

VELOCITY

1 m/s = 100 cm^5 = 3.28 ft/s = 2.24 mph
1 km/s = 2240 mph
1 km/h = 0.621 mph

TIME

1 d = 1 day = 24 h = 1440 minutes = 86,400 seconds
1 yr = 1 year = 3.16×10^7 seconds

On Numbers and Scientific Notation

DECIMAL	EXPONENTIAL SCIENTIFIC NOTATION	METRIC UNIT PREFIX (GRAM, SECOND, LITER)
0.000000000001	10^{-12}	pico
0.000000001	10^{-9}	nano
0.000001	10^{-6}	micro
0.001	10^{-3}	milli
0.01	10^{-2}	centi
0.1	10^{-1}	deci
1	10^{0}	no prefix
10	10^{1}	deka
100	10^{2}	hecta
1000	10^{3}	kilo
1,000,000	10^{6}	mega
1,000,000,000	10^{9}	giga
1,000,000,000,000	10^{12}	tera

GLOSSARY

A-bands anisotropic zone of a sarcomere; region of overlap of actin and myosin on either side of the M-line

Acclimation adapting to a new environment or set of conditions

Accommodation (learning) the process by which memories of old experiences are replaced or supplemented by new ones

Accommodation (vision) the mechanism by which objects are brought into focus by the lens

Acetic acid mild organic acid found commonly in vinegar

Acetylcholine esterase an enzyme that degrades and inactivates acetylcholine

Acetylcholine neurotransmitter used between nerve and muscle

Achilles tendon attaches gastrocnemius muscle to the heel

Acidosis a condition of blood or body fluids in which the pH becomes abnormally acidic

Acne condition of skin in which sebaceous glands are infected and inflamed

Acquired immunity immunity resulting from exposure to an antigen

Acromegaly a condition in adults resulting from overexposure to growth hormone

Acrosome reaction the release of the contents of the acrosome, an enzyme-filled organelle, when a sperm contacts the outer barriers of material surrounding an egg

Actin actin protein composing microfilaments

Action potential the form of the electrical signal carried by a neuron

Activation energy the energy needed to initiate a chemical reaction

Active transport cellular transport processes requiring ATP

Adam's apple the cartilage of the larynx, which protrudes in human males

Adaptations genetic characteristics related to fitness

Addison's disease a condition characterized by lethargy, weight loss, and mental fatigue resulting from reduced production of corticosteroids

Adductor muscles muscles of the pelvis and inner side of the upper leg

Adenoids lymphatic tissue located in the back of the region of the pharynx

Adenosine diphosphate (ADP) a signal molecule used to activate platelets during thrombosis

Adhesive junctions specialized regions of the cell surface that provide attachment between cells

Adipose tissue fatty storage tissues of the body

Adrenal glands endocrine glands located atop the kidneys; produce steroids and catecholamines such as adrenaline

Adrenal medulla innermost region of the adrenal gland; produces the catecholamines epinephrine and norepinephrine

Adrenocorticotropic hormone (ACTH) a hormone produced by the anterior pituitary affecting the regulation of the adrenal cortex and the synthesis of steroids

Aerobic cellular respiration energy-producing reactions requiring oxygen

Age of Fishes the Devonian period of the Paleozoic era when fish were a dominant life form

Aging the passage of time; simply, getting older

Agranulocytes a class of white blood cells containing no obvious granules

AIDS (acquired immune deficiency syndrome) a multitude of symptoms including weight loss, T-cell destruction, rare infections, and cancer, associated with the advanced stages of an HIV infection

Albinism an inherited genetic condition in which melanin is not produced

Albumin a serum protein produced by the liver

Aldosterone steroid hormone produced by the adrenal glands; regulates sodium and potassium in the kidney

Alkalosis a condition of blood or body fluids in which the pH becomes abnormally basic

Alleles altered forms of the same gene, only one of which occurs at a locus

Allergens materials or agents that elicit an allergic immune response

Allergies a form of immune reaction involving local response (inflammation, runny nose, sneezing) to antigens such as pollen, dander, and other foreign materials

Alpha cells glucagon-producing cells found in islets of Langerhans

Alpine tundra an altitudinal zone in which snow is constant and life is limited

Altitudinal zonation the changes in biomes that occur with changing altitude, as up the side of mountains

Altricial the young of animals that are born or hatched in a dependent state

Alveoli grapelike clusters of cavities in which gas exchange takes place in the lungs

Ames test a test using bacteria to assess mutagenicity and carcinogenicity of new compounds

Amnesia the loss of memory

Amnion or amnionic sac *see* extraembryonic tissues

Amphipathic chemical property of a molecule in which both positive and negative charges exist simultaneously under certain conditions

Amplitude loudness

Ampulla (1) the region of sperm storage; (2) the structure at the base of the semicircular canals housing receptors

Anaerobic cellular respiration energy-producing reaction occurring in the absence of oxygen

Analogy functionally related structures of dissimilar genetic or developmental origin; for example, wings in flies and bats

Anaphase third stage of mitosis; chromosomes separate at centromere

Anaphylactic shock drastic and immediate hypersensitive response resulting in bronchial and cardiac reactions that can be lethal; for example, allergy to penicillin or bee stings

Anatomical of or related to bodily structure

Androgens the male sex hormones, such as testosterone

Aneuploidy abnormalities in the number of chromosomes found in cells

Aneurysm a bulge or outpocketing of a blood vessel resulting from weakness or damage

Angina pectoris the reference pain of a heart attack that is felt in the left arm

Angular movements movements that change the angle between bones; for example, the elbow

Anions negatively charged ions

Anisotropic able to polarize light

Antagonism working against one another

Anterior chamber the fluid-filled space between the cornea and the iris

Anthropoids primates of the suborder Anthropoidea to which humans and the great apes belong

Antibodies a large class of proteins in the immunoglobulin family produced by B-lymphocytes

Antibody-antigen complex the results of an antibody binding to its antigen, which is recognized by the immune system and destroyed or degraded

Anticoagulant a compound that interferes with blood clotting

Antidiuretic hormone (ADH) a peptide hormone of the posterior pituitary; regulates water movement through the nephron

Antigen dependent differentiation the growth and differentiation of lymphocytes in the absence of any antigens

Antigens literally "antibody generators"; molecules that can be used to induce antibody production

Antihistamines compounds whose action is to block or reduce the effects of histamines released during an allergic reaction

Anti-oncogenes the cellular genes whose expression suppresses the initiation of cancer

Antioxidant a molecule that retards or prevents oxidation

Antiport mechanism of cellular transport involving exchange in both directions across the cell membrane

Anvil a bone of the middle ear; also called incus

Aorta the largest arterial structure of the body; exit from the heart to systemic circulation

Aortic arch the arterial structure from which arise a number of large arteries carrying blood from the heart to the body

Aortic semilunar valve the valve controlling exit of blood from the heart to the aorta and the rest of the body

Apocrine sweat glands exocrine gland producing sweat and a thick secretion

Apoptosis programmed cell death

Appendicitis infection and inflammation of the appendix

Appendix the blind end of the cecum, an outpocketing of the intestine

Appetitive behaviors drives to fulfill a basic need; also known as appetites

Aqueous humor the fluid in the anterior chamber

Arachnoid mater the middle layer of the meninges

Areolar tissues loose connective tissue

Arid dry, as applied to a desert

Arteries vascular elements of the circulatory system carrying blood away from the heart; usually oxygenated

Arterioles small arteries

Arthritis an inflammation of a joint or joints

Arthropods segmented organisms with chitinous exoskeletons; includes insects, spiders, crabs, shrimp, and even trilobites (extinct)

Ascending aorta a region of the main artery of the body (aorta) as it leaves the heart

Ascending colon the first segment of the large intestine that rises from the lower to upper abdomen

Asexual reproduction a form of reproduction not requiring specialized germ cells

Assimilation the manner in which new information is organized in the brain

Associative processes brain functions in which memories can be recalled or recollected

Asthma an allergy-related condition in which bronchoconstriction makes breathing difficult and results in wheezing and coughing

Astigmatism an irregular curvature of the cornea causing distortion of vision

Astrocytes class of glial cells; part of the blood-brain barrier

Asymmetrical an unbalanced organization

Asymptomatic infected but not showing symptoms

Atlas the uppermost cervical vertebra, supports the skull

Atmosphere the specific group of gases surrounding Earth that compose the air we breathe

ATP adenosine triphosphate; a nucleoside triphosphate utilized as a common "energy currency" of all cells

ATP'ase enzyme capable of hydrolysing phosphate group off ATP. Used to produce the energy needed for muscle contraction

Atrioventricular node also known as AV node, propagates signal from SA node in pace of heart beat

Atrium the upper chamber of vertebrate heart that passes blood to ventricle

Atrophy a wasting away

Autoimmunity abnormal condition in which the body's immune system attacks and destroys the body or its parts

Autotroph an organism that produces all its cellular constituents from inorganic materials and, thus, grows independently of oxygenous organic carbon sources; for example, plants

Autotrophic state of self-sufficiency in organisms that require only simple compounds, such as CO_2 and H_2O, to live

Axis the second vertebra upon which the atlas rotates

Axon extension from the surface of a neuron

Axoneme microtubular structure of cilia and flagella

Bacteriophages a class of viruses that infect bacterial cells

Ball-and-socket joints joints involving a ball-ended bone that fits into a deep cup-shaped receiving bone

Barometric pressure the force of the weight of air on a particular region or location; measured by the height of a column of mercury

Bartholin's glands located in the vestibule, these glands secrete lubricating fluids associated with sexual arousal

Basal cell carcinoma malignant cancer cell arising from basal cells of the epidermis

Basal lamina an assemblage of extracellular matrix materials found in conjunction with epithelia (*see* basement membrane)

Basal nuclei islands of gray matter in the cerebrum controlling motor activity and interconnections between brain regions

Base opposite of an acid; hydrogen ion acceptor

Basement membrane extracellular matrix organized to support an epithelium

Basilar membrane one of two specialized membranes surrounding the cochlear duct

Basophils a rare type of granulocyte

Benign tumors tumors whose growth is restricted to one area and which remain encapsulated

Beta cells insulin-producing cells found in islets of Langerhans

Biceps brachii muscle of the front of the upper arm

Bicuspid valve a valve with two flaps occurring between the left atrium and ventricle; also called the mitral valve

Bicuspids relatively flat, bilobed teeth for chewing

Bidirectional in two directions

Bile a compound produced by the liver from degraded red blood cells and used to emulsify fats

Bile duct a common channel for bile formed from the hepatic ducts

Bile salts emulsifying molecules formed by the liver by degradation of hemoglobin

Bilirubin a pigment in bile

Binary fission reproduction of a cell by division into two equal parts

Biochemistry the chemistry of the products of living organisms

Biodiversity the wide range of structure and function of life forms adapted to specific environments

Biofeedback control of unconscious or involuntary bodily processes through conscious thought

Biological magnification the accumulation of a contaminant through trophic levels

Biomass the amount of organisms and their products in the environment

Biomes terrestrial or land-based ecosystems

Biosphere the zone formed by a combination of atmosphere, lithosphere, and hydrosphere in which life exists on Earth

Bipedalism walking on two-legs

Bipolar neuron a neuron with separate axon and dendrite

Birth control pills any of a number of combinations of estrogen and progestins in pill form used to control the ovarian cycle

Bite the alignment of the upper and lower jaws, particularly with respect to teeth

Blastocyst a multicellular embryonic structure that implants into the uterus

Blastula a stage of the cleavage of the egg in which the embryo is a ball of cells with a cavity within it

Block to polyspermy a reaction of the egg to prevent more than one sperm from participating in fertilization

Bolus softened, chewed-up food ready to swallow

Botulin toxin neurotoxin produced by the bacterium *Clostridium botulinum*

Bowman's capsule a double-walled structure surrounding the glomerulus and into which the filtrate of blood flows

Brachial artery major artery of the arm from which blood pressure is commonly measured

Brachialis muscle of the upper arm

Brachioradialis muscle that flexes the forearm

Brain death one of the criteria of death in which normal electrical activity of the brain ceases

Brain stem the region of the brain connecting the cerebrum with the spinal cord

Bronchioles tiny branches of the bronchus that lead to the alveoli

Bronchus a branch-off of the trachea that enters the lungs

Brownian motion random movement of microscopic particles influenced by kinetic energy of the molecules composing the liquid or gas in which they are suspended or dissolved

Budding reproduction of a new individual by outgrowth from the parent organism

Buffers compounds capable of neutralizing solutions of both acids and bases

Bulbourethral glands small glands of the male reproductive system that secrete lubricating fluids

Burkitt's lymphoma a cancer of lymphoid cells associated with Epstein-Barr virus

Bursitis a condition of inflammation and pain of a bursa, a fluid-filled membrane associated with synovial joints

Calcification the process by which calcium is utilized to form mineralized materials, such as bone

Calcitonin a hormone produced by the thyroid gland affecting calcium regulation

Calcium phosphate the major mineral component of bone

Callous connective tissue mass associated with initial bone repair

Calorie a unit of heat energy

Calyx a narrow cavity formed at the upper ends of the renal pelvis; associated with collection and movement of urine from the papillae

Cambrian explosion the sudden and dramatic appearance of most modern phyla; approximately 650 million years ago during the Cambrian period of the Paleozoic era

Canaliculi meaning little canals; open connections between lacunae within an osteon

Canaliculi small canals connecting cellular domains within bone

Canines relatively pointed tearing teeth

Capacitation the final step in sperm activation that allows them to fertilize an egg; occurs in the vagina

Capillaries the smallest vascular elements of the circulatory system

Capsule the connective tissue sheath covering the outside of each kidney

Carbon dioxide (CO_2) a gas; waste product of human cell metabolism; 0.03%–0.04% of Earth's atmosphere

Carcinogens chemicals or other agents that cause cancer

Carcinomas malignant cancers of epithelial origin

Cardiac myocytes muscle cells in the heart

Carotid artery the artery delivering blood to the head and brain

Carpals the bones of the wrist

Cartilage model embryonic skeleton prior to ossification

Cartilagenous joint tight fitting, slightly moving joint as in a symphysis

Catabolism the breakdown of complex molecules in living organisms

Catalyst an agent that increases the rate of a chemical reaction without itself undergoing permanent change

Cataracts cloudy areas within the human lens that alter transparency

Catecholamines a class of molecules often utilized as neurotransmitters; for example, epinephrine

Catheterization the insertion of a tube into the urethra with the intent of draining the urinary bladder

Cations positively charged ions

Cecum an outpocketing of intestine at junction of ileum and colon

Cell body especially related to the central region of a neuron from which axons and dendrites arise and in which the nucleus is located

Cell cycle the repetitious process by which a cell prepares for and completes division

Cell fusion combining of two or more cells together within a shared plasma membrane

Cell respiration energy-forming activities in a cell that require oxygen

Cellular respiration energy-producing reactions in a cell

Cellulose fibrous polysaccharide produced by plants and indigestible in vertebrates

Cenozoic era the most recent era of life, from 65 million years ago to the present, during the early part of which mammals flourished and forest primates appeared

Central canal fluid-filled tube within the spinal cord

Central nervous system (CNS) nerves within the skull and vertebrae; protected by enclosure within the skeleton

Centromere region of a chromosome at which it separates during anaphase; attachment site for spindle microtubules

Cerebellum the region of the brain controlling body movement

Cerebrospinal fluid (CFS) the fluid filling the ventricles and central canal and surrounding the brain and spinal cord

Cerebrum the largest part of the human forebrain; formed of two hemispheres; seat of consciousness

Cervical of, or referring to, the region of the neck; the first 7 vertebrae

Cervix the region of transition between the uterus and the vagina

Cesarean section a surgical procedure in which the uterus is opened by an incision through the abdominal wall and the baby is removed

Chancres ulcers of the skin formed in the primary stage of syphilis

Chancroid a tropical disease that may be passed through sexual interactions and leads to painful sores on the genitalia

Chaparral a biome known as a pyroclimax community, located in mediterranean environments such as those found in the Pacific southwest

Checkpoints times during the cell cycle that are important in determining normal progression in the cycle and initiating mechanisms to repair damage to DNA

Chemotaxis movement toward a chemical attractant, usually along a concentration gradient of a diffusible substance

Chemotherapy the use of chemical compounds to kill cancer cells

Chlamydia a disease caused by infection with the organism *Chlamydia trachomatis*

Chlorofluorcarbons compounds rich in chlorine and fluorine such that they interfere with the production of ozone in the stratosphere

Chlorophyll the molecule of plants that captures radiant energy

Chloroplast the chlorphyll-containing organelle found in green plants that is responsible for photosynthesis

Cholera toxin a deadly toxin produced by the bacteria *Vibrio cholera,* which causes massive dehydration through water loss from the intestines

Cholesterol common lipid substance in mammals; part of the sterol family

Chondrocytes cartilage forming cells

Chondromas tumors of cartilage

Chorion *see* extraembryonic tissues

Choroid inner connective tissue layer of the eye, rich in blood vessels

Chromatin DNA and associated proteins in the nucleus

Chromosome a single DNA molecule associated with proteins which is found in the nucleus of a cell and contains hereditary information represented by genes

Chyle milky emulsified fats absorbed from small intestine into lacteals

Chyme the semisolid material resulting from partial digestion of food in the stomach

Cilia cell surface extensions surrounding an axoneme; involved in cell movement

Ciliary body a part of the choroid layer; helps change shape of the lens during accommodation

Circadian rhythms an aspect of the biological clock that controls a cycle of physiological activities on a 24-hour basis; for example, sleep patterns and some hormone secretions

Circumductive movements rotational movements; for example, arm movement around shoulder

Circumvallate papillae mound-shaped papillae with deep surrounding creases found on the back third of the tongue in a V-shaped pattern

Cis-region region of the Golgi apparatus closest in proximity to the endoplasmic reticulum

Cisternal space volume within the membranes of an organelle such as the endoplasmic reticulum or Golgi apparatus

Classical conditioning behavioral modification using the methods of Ivan Pavlov

Cleavage the first divisions of the fertilized egg or zygote

Climate the long-range patterns of weather and precipitation

Climax community the final type of community in a succession

Clitoris a specialized, highly receptive female genital organ homologous to the male glans penis

Clonal selection the matching of an antigen with a specific and limited subpopulation of lymphocytes that are stimulated to grow and differentiate

Clonal selection theory the theory that explains how clonal selection works at the molecular and cellular level

Clone a population of cells (or organisms) derived from a single parental cell

Coastal sage the climax community associated with chaparral

Coccyx the tail bone in humans; fused vertebrae at the base of the spine

Cochlea the spiral-shaped structure of the inner ear having phonoreceptors

Cochlear duct the central duct of the cochlea

Codominance a condition in which the expression of each of two different alleles is observed simultaneously

Collagen fibrous extracellular matrix protein found as major component of connective tissues

Collecting tubules tubes into which distal convoluted tubules of many nephrons terminate; collect urine for transport through the pyramids and papillae

Colon the large intestine

Coma a persistent abnormal state of unconsciousness

Community assemblage of populations living in a particular habitat or environment

Compact bone dense arrays of osteons

Competition the activities of organisms within a community for shared resources

Complement a family of proteins, some of which recognize Ig molecules and others of which bring about cell lysis of foreign cells

Complement-mediated cell lysis a process by which Ig molecules target foreign cells for destruction using a series of specialized complement proteins

Complex carbohydrates polysaccharides including those that can be digested (starch) as well as those that cannot (cellulose)

Concentration gradient difference in the concentration of a substance from one region or compartment to another

Condensation reactions chemical reactions in which water molecules are incorporated into the final product(s)

Conduction relating to exchange of heat between materials in direct contact

Cone color–sensitive photoreceptor cell

Congenital syphilis a lethal syphilis infection of newborns, arising from an infected mother

Connective tissue proper fibrous type connective tissues

Consanguineous marriages marriages between close relatives

Constitutive condition a condition in which the expression of a trait is continuous and not controlled by external stimulation, e.g., pigmentation of skin

Consummatory acts fulfillment of an appetite

Contact inhibition the effect of cell-to-cell contact on restricting cell growth

Continental drift slow movement of crustal land masses over the surface of Earth

Contralateral connections nerve connections that cross to the opposite side of the body

Convection transfer of heat by circulation of gas or fluid

Convergence the proper coordination of orientation of both eyes on an object

Convergent evolution the evolution of phenotypically similar organisms or parts of organisms that are not genetically or developmentally related

Core body heat the heat of the central portion of the human body; in humans maintained at approximately 38° C.

Cornea transparent outer layer of the eye overlying the region of the pupil

Corneocytes dead cells of the outer layers of the epidermis

Corona radiata the layer of support cells around the egg released from the ovary (part of the original follicle)

Coronary arteries major arteries that supply oxygen to the heart

Corpus callosum the region of dense nerve fiber connections between cerebral hemispheres

Corpus cavernosum a specialized tissue of the penis capable of being engorged with blood

Corpus luteum a specialized hormone-secreting structure resulting from the influence of LH

Corpus spongiosum the spongy tissue of the penis capable of being filled with and retaining blood

Cortex a layer of kidney underlying the capsule containing nephrons

Cotransport cellular transport involving simultaneous translocation of two substances across a membrane

Covalent bonds chemical bonds between atoms in which electrons are shared

Cranium subset of bones of the skull that surround the brain

Cretaceous extinction a time of dying and extinction of great numbers of species at the end of the Mesozoic era

Cretaceous period the final period of the Mesozoic era

Cretinism a form of hypothyroidism in children characterized by irreversible mental retardation

Cro-Magnon a subgroup of *Homo sapiens* from which modern humans evolved

Crossing-over exchange of DNA between homologous chromosomes usually during meiosis

Crystallins clear proteins of the lens

Cupola the viscous structure into which receptors in the ampulla are embedded

Cushing's syndrome a condition characterized by obesity and muscle loss due to overproduction of corticosteroids

Cuspids relatively flat single-lobed teeth for chewing

Cyclic AMP an intracellular messenger molecule

Cystic fibrosis an inherited condition affecting mucus secretion

Cytokinesis physical cleavage of one cell into two

Cytoplasm the viscous fluid found within a cell, excluding the nucleus

Cytoskeleton protein polymers produced in the cytoplasm of cells; involved in cell shape, motility, and contraction

Defecation the act of releasing or expelling feces from the body

Dehydration loss of water; drying out

Dehydrogenation removal of hydrogen atoms, specifically removal of hydrogen in fatty acids to form carbon–carbon double bonds and less saturated fats

Delayed hypersensitivity a T-cell-mediated response to specific allergens such as poison oak or some cosmetics; may take several days to fully develop

Deletion the loss of a nucleotide in a gene, altering the sequence, which leads to a frameshift mutation

Deltoid muscle of the shoulder

Dendrites extensions arising from a nerve cell body

Dengue fever a viral infection carried by mosquitos to humans and characterized by headache, severe joint pain, and rash

Dental caries bacteria-induced deterioration of teeth; cavities

Dermis the component of skin that underlies the epidermis and is connective in character

Descending aorta the branch of the aorta delivering blood to the lower parts of the body

Descending colon the third segment of the large intestine descending from the upper to lower abdomen

Desmosome symmetrical adhesive junction formed between cells

Detoxification processes used by cells to make toxic compounds less toxic

Diabetes insipidus a rare condition in which lack of ADH production results in excessive urine production

Diabetes mellitus a disease in which the metabolism of glucose is disrupted by failure to produce or respond to the hormone insulin

Diabetic coma a state of unconsciousness induced by abnormal carbohydrate metabolism

Dialysis procedure in which blood is cleansed in an artificial kidney machine

Dialysis membrane specialized membrane through which blood is filtered during dialysis

Diaphragm muscle at the interface of the thoracic and abdominal cavities; involved in breathing

Diaphragm with respect to birth control, a device that fits over the cervix and prevents egg-sperm encounters

Diarrhea a condition in which the feces are watery and without form

Diastole the period in a heartbeat during which both atria and ventricles are relaxed

Diffusion movement of a substance from an area of higher concentration to one of lower concentration

Dihybrid cross a mating involving analysis of two independent traits

Diploid the full complement of 46 chromosomes (23 pairs in humans) Also known as the 2N number of chromosomes

Direct-acting compounds carcinogenic chemicals that themselves alter DNA during initiation

Disaccharides covalent combinations of two monosaccharides

Distal convoluted tubule follows the loop of Henle and continues the job of resorption of nutrients, especially water

Distal end the end farthest away from the long axis of the body

Distributing arteries arteries of midsize that distribute blood to all regions of the body; the commonly named arteries, such as the carotid or brachial arteries

DNA deoxyribonucleic acid; the genetic material

DNA tumor viruses viruses whose genomes are based on DNA and that may be the cause of or involved in some types of cancer

Dominant gene a gene or allele whose expression is observed and masks the expression of other alleles

Drives behaviors controlled by basic needs, for example, thirst and drinking

Dryopithecines Miocene ancestors of hominoids

Duodenum the first part of the small intestine; receives chyme from the stomach and enzymes and bile from the pancreas and liver, respectively

Dura mater the outermost and toughest layer of the meninges

Dynamic equilibrium balance associated with the function of receptors located in the semicircular canals

Dyslexia a disturbance of the ability to read

Ebola virus biological agent of a lethal disease in which hemorrhaging occurs in all organs of the body; vector unknown

Eccrine sweat glands exocrine glands producing watery secretion of sweat commonly distributed over the surface of the entire body

Ecology the study of the environment and the living organisms within it

Ecosystems the large-scale divisions of the biosphere

Effacement the period during labor in which the uterus and vagina establish a temporary birth canal

Effector functions the responses or behaviors that result from a reflex; for example, the contraction of a muscle or secretion from a gland

Ejaculation the forceful expulsion of semen from the male reproductive tract during sexual arousal

Elastic arteries arteries invested with elastic connective tissue to allow expansion and contraction under the pressure of blood flow

Elastin elastic extracellular matrix protein found in many types of connective tissue

Electrolytes the water soluble charge-bearing molecules and ions found in bodily fluids

ELISA immunological technique used to detect antigens or antibodies in blood and other tissues

Ellipsoidal joint elliptical bone surfaces that mediate back and forth movements

Embolism obstruction of blood flow by a detached blood clot

Embolus a bulging out of a blood vessel from weakened wall structure

Emphysema irreversible and progressive deterioration of the lungs

Emulsifiers molecules that disperse large fat globules into smaller lipid vesicles

End bulbs of Krause a class of thermal receptors associated with sensitivity to cold

Endocardium the epithelial lining of the chambers of the heart

Endocrine glands ductless glands secreting hormones directly into the blood

Endocytosis active, energy-requiring transport process involving invagination and vesiculation of the plasma membrane

Endolymph fluid within the semicircular canals and cochlear duct

Endometrium the glandular epithelium of the uterus

Endomysium delicate connective tissue surrounding muscle fibers

Endosome vesicle arising from endocytosis

Endosteum the layer of cells lining the inner cavities of bone

Endothelium the epithelial lining of blood vessels

Endothermic warm-blooded; maintenance of constant body temperature

Engram a permanent memory trace

Entropy in thermodynamics, the tendency towards disorder

Eosinophils a common type of granulocyte

Epicardial membranes membrane layers surrounding the heart

Epicardium the epithelial layer surrounding the heart

Epidermis specialized epithelium of skin; overlies dermis

Epididymis the tube connecting the seminiferous tubules to the vas deferens

Epigenesis originally, a counterargument to the preformationists that stated that the embryo developed from unformed parts, more like the modern concept

Epiglottis the flap that covers the glottis during swallowing

Epimysium connective tissue surrounding an entire muscle or muscle group

Epinephrine and norepinephrine *see* adrenal medulla

Epiphyseal growth plate growth zone in lengthening bones

Epithelia a group of tissues composed of tightly adherent cells that form barriers to the movement of materials into and out of the body

Epitheliomas malignant tumors of epithelial origin; also known as carcinomas

Epitope a part of a molecule against which antibodies are made; one protein may have several epitopes

Epstein-Barr virus DNA virus that may play a role in Burkitt's lymphoma

Equilibrium state of balance between opposing forces or actions

Erythrocytes the most prevalent formed elements of blood; also called red blood cells

Erythropoietin a hormone stimulating the production of erythrocytes

Esophagus tube connecting the region of the pharynx to the stomach

Essential amino acids amino acids that are required for protein synthesis but not synthesized by the human body

Essential hypertension high blood pressure for which there is no specific cause

Essential nutrients those nutrients we cannot synthesize but without which we cannot live

Estrogen a steroid hormone produced principally by the ovaries and controlling the ovarian and menstrual cycles

Estuary a type of shallow wetlands, where the tide meets a river current

Eucaryotic organisms a single or multicellular organism whose cells have a nucleus

Eustachian tube the canal connecting the pharynx to the middle ear

Eutrophication a rapid increase in nutrient status of a body of water; may lead to excessive respiration and oxygen depletion

Evaporation change of state of a substance from liquid to vapor

Excitatory the quality of signals between nerve cells that stimulate an action potential

Excrete to eliminate waste from a cell, tissue, or organ

Exobiology the study of life outside or beyond Earth

Exocrine glands glands with ducts that deliver secretions to the surface of an epithelium

Exocytosis active, energy-requiring transport process involving evagination or budding of plasma membrane

Expiration breathing out, exhaling

Expirational muscles muscles used to exhale

Expiratory reserve the volume of air that can be exhaled beyond the level of an ordinary breath

Extension to move or separate parts of the body from one another; for example, arm extension

External auditory canal the part of the external ear leading to the tympanic membrane

External ear the outer part of the ear, including pinna and auditory canal

External fertilization the fertilization of an egg by a sperm outside the body of the female

External oblique overlie the side of the body in the abdominal region

External urethral sphincter one of two sphincters that control flow of urine from the urinary bladder; under voluntary control

Exteroceptors receptors for stimuli from outside the body

Extinction the loss of effectiveness of neutral stimulus in the absence of reinforcement

Extracellular matrix (ECM) a complex, multicomponent material produced by surrounding cells; composed predominantly of proteins and carbohydrates

Extraembryonic tissues the membranous tissues that surround the embryo but are not part of it, e.g., amnion, chorion, and yolk sac (from which early red blood cells migrate)

F_1 generation the first filial (F) generation of siblings

F_2 generation the second filial generation (offspring of siblings in F_1)

Facilitated diffusion use of specialized membrane channels to make the transport of materials into and out of cells easier; no cellular energy is required

Fallopian tube part of the reproductive system of females providing a channel for the transport of the egg from the ovary to the uterus; also known as a uterine tube

False ribs the ribs indirectly connected to the sternum via cartilagenous connections to the true ribs

Feces solid waste products of digestion

Feedback loop the effect of an end-product of a process to alter the process

Femur the bone of the upper leg

Fertilization the interaction of germ cells to form a new individual by sexual reproduction

Fetus the stage of development following embryogenesis and characterized initially by the formation of an ossifying skeleton from a cartilaginous model

Fiber indigestible materials in a diet that help in movement and absorption of nutrients in the intestinal tract, for example, cellulose

Fibrin a protein of blood involved in the clotting of blood

Fibrinogen inactive precursor protein to fibrin

Fibroblast common connective tissue cell type, secretes ECM molecules

Fibrocyte a mature, less active form of fibroblast

Fibrosis formation of fibrous scar tissue based on growth of connective tissue cells

Fibrous joint tight fitting, nonmoving joints, as in a suture

Fibrous pericardium the tough outer layer of the pericardium

Fibula one of the two bones of the lower leg

Filiform papillae slender, filament-like elevations on the tongue

Filtration the action of separating components of the blood on the basis of size

Fimbriae fingerlike projections of the Fallopian tube (nearest the ovary) that help capture the egg upon release from the surface of the ovary

First degree burns damage to skin from heat directly affecting only the epidermis

Fitness a measure of reproductive success

Fixed-action patterns fixed sequences of reflexes composing a complicated behavior

Flagella extensions of specialized cells (such as sperm) containing axonemes

Flat bones thin or platelike bones

Flexion to move or bend parts of the body by muscle contraction; for example, arm flexion

Floating ribs the ribs with no connection to the sternum or to other ribs

Follicle stimulating hormone (FSH) a hormone produced by the anterior pituitary, affecting the ovarian cycle

Follicles with respect to the thyroid gland, the cellular structures in which thyroxine is produced

Folliculogenesis the formation of a follicle during the ovarian cycle

Fontanels meaning tiny fountains; regions between bones of the cranium that have yet to form sutures

Food chain simple linear relationship between primary, secondary, and tertiary consumers

Foramen magnum meaning large hole; the hole in the cranium through which the spinal cord enters the brain

Formed elements the cellular components of blood, including red and white blood cells and platelets

Fossil record the imprint or mineralized remains of previously living things preserved in geological sediments

Fovea the central region of the retina with the highest concentration of cones

Frameshift mutation altered reading of triplet genetic code, for example, due to deletion

Free nerve endings a class of receptors for touch and pain

Free radicals highly reactive forms of molecules that damage cellular components

Frontal lobe the most anterior and upper part of each cerebral hemisphere

Fulcrum fixed position at which a lever operates

Fungiform papillae mushroom-shaped elevations of the front part of the tongue

Gametes cells specifically associated with sexual reproduction

Ganciclovir and saquinavir antivirals used to prevent eye infections characteristic of AIDS

Ganglia structures containing a high density of cells, usually in the peripheral nervous system; for example, spinal ganglia

Ganglion cells neurons forming part of the optic nerve

Gap junction an intercellular junction that establishes a direct connection or channel between two adjoining cells

Gastrocnemius muscle of the backside of the lower leg (or calf); flexes the leg

Gastroesophageal sphincter a valvelike flap that closes off the opening between the stomach and the esophagus to prevent reflux; also known as the cardiac sphincter

Gastrulation the period in embryonic development in which the cells of the primary germ layers separate from one another

Gene genetic unit of inheritance; composed of DNA

Gene pool full complement of genes (alleles) found in a population of interbreeding organisms (a species)

General senses senses of touch, pressure, and heat

Genetic drift a significant change in gene frequencies in a population resulting from random causes, usually associated with geographical or reproductive isolation

Genome the full set of genes found in the chromosomes of an individual, which specify the development, growth, and maintenance of that individual

Genotype the set of genes carried by a particular individual

Geographical isolation a separation of members of a species by physical barriers, for example, a mountain range or body of water

Geometric progression a doubling of values at each succeeding step in a sequence

Geotropism growth under the influence of gravity; also gravitropism

Germinative cell layer in skin, the deepest living cell layer from which cells of other layers are derived

Gestation the average duration of time spent for a species developing *in utero* before birth (nine months in humans)

Gingiva the gums of the mouth

Gingivitis disease-resulting destruction or deterioration of the gums

Glia a class of cells found in nerve tissue; plays a supportive, structural role in nervous system

Gliding joints joints that allow sliding movements in two dimensions between bones

Gliding movements movements associated with gliding joints; for example, the wrist and ankle

Gliomas tumors arising from glial cells

Glomerulus an array of arterioles through which blood is filtered into Bowman's capsule

Glottis the entrance to the larynx

Glucagon a hormone formed in the pancreas affecting glycogen breakdown in cells

Glucocorticoids hormones of the adrenal cortex affecting carbohydrate metabolism; for example, cortisone

Gluteus maximus principal muscles of the buttocks

Glycerol a three-carbon compound used as a backbone for the formation of triglycerides and phospholipids

Glycolipid a combination of lipid and carbohydrates

Glycolysis metabolic pathways within a cell converting glucose to pyruvate

Glycoprotein a combination of lipid and carbohydrates

Glycosylated with sugars covalently linked to a molecule, as in the case of glycoproteins or glycolipids

Goblet cells mucous gland cells of the stomach

Goiters growths of the thyroid gland resulting from insufficient levels of dietary iodine

Gonadotropic hormones LH and FSH

Gonads the testes and ovaries

Gonorrhea a disease caused by infection with the organism *Nesseria gonorrheae*

Graafian follicle a mature follicle with egg and full complement of support cells

Gradualism a hypothesis that evolutionary change and the appearance of new species occur slowly and incrementally

Granulocytes a class of white blood cells containing obvious cytoplasmic granules

Granuloma inguinale an infection that spreads from relatively painless sores, but if untreated, may be lethal

Grasslands biomes in which tall or short grasses predominate

Graves' disease a form of hyperthyroidism; characterized by hyperactivity, enhanced appetite, and nervousness; also known as exophthalamic goiter

Greenhouse effect natural effect of Earth's atmosphere on the heat retained from solar radiation

Growth hormone the pituitary endocrine hormone affecting bone growth

Gulf Stream the ocean current running along the eastern coast of the United States and toward northwestern Europe

Gustatory cells cells found in taste buds; receptors of taste

Gustatory receptors taste receptors

Gyri regions of outfolding of the cerebrum

Habitats the environments in which organisms live

Habituation losing sensitivity to a smell due to overexposure

Hair root plexus nerve endings wrapped around the hair base and sensitive to movement

Hammer a bone of the middle ear; also called malleus

Haploid one complete set of chromosomes represented by one member of each pair of a diploid set, 23 in humans. This set is also called the IN number of chromosomes

Haversian system/osteon in compact bone, a canal surrounded by concentric rings of ossified material in which osteocytes are found

Heavy and light polypeptide chains the subunits of Ig molecules

Helper T-cells lymphocytes involved in signaling other immune cells during the initiation of an immune response

Hemoglobin a protein complex involved in binding oxygen in red blood cells

Hemophilia a condition in which blood clotting does not occur due to a factor(s) missing in the cascade of clotting reactions

Hemopoietic cells progenitor cells of blood cells

Heparin a molecule of the circulatory system that prevents the spread of a clot during thrombosis

Hepatic ducts channels from the liver to the duodenum delivering bile

Hepatitis viruses DNA viruses; three main types: A (infectious hepatitis), B (serum hepatitis) and C

Herpes keratitis a herpes virus infection of the eye; a leading cause of blindness in the United States

Herpes viruses DNA viruses; two types, I and II, both of which may be transmitted sexually

Hertz units of frequency, in cycles per second

Heterotroph an organism whose growth is dependent on an exogenous source of organic compounds; for example, animals

Heterotrophic a state in which organisms require organic materials produced by other organisms to live

Heterozygous a condition in which the alleles on homologous chromosomes are not identical

Hexoses six-carbon sugars, i.e., glucose

Hierarchies systems of dominance in vertebrate groups

Hilum the region of exit of the renal pelvis from the kidney

Hinge joints joints that allow movement between bones in a single plane

Histamine a compound produced and released by basophils during an allergic response

Histocompatibility the mutual tolerance among tissues of two individuals that allows transplanting; tissue differences establish the molecular basis for tissue rejection

Histology the study of tissue structure

Histone proteins special proteins associated with DNA in chromosomes

Hives an elevated, inflamed, and bumpy, itchy skin that results from many types of allergies; also known as urticaria

Home range the area in which an animal forages for food

Homeostasis the tendency for a system or organism to reach a state of physiological equilibrium

Hominid a designation describing humankind and its ancestors starting with *Ramapithecus*

Homologous equivalent anatomical or genetic elements arising from the same or shared origins

Homologous pairs the two chromosomes in the nucleus with identical or nearly identical order of gene loci that pair up during meiosis

Homology structures arising from genetically and developmentally related processes; for example, forelimbs in vertebrates

Homo sapiens the name of a species within the genus Hominidiae; usually characterized as a grouping of bipedal primates; employed as a term for the human species

Homozygous a condition in which the alleles on homologous chromosomes are identical

Homunculus one of the preformationists' ideas, a fully formed embryo in a sperm

Hormone a substance produced by endocrine glands and secreted directly into bloodstream, which elicits specific responses from target cells

Human chorionic gonadotropin (HCG) a hormone produced by the developing placenta and used as a molecular indicator of pregnancy

Human immunodeficiency virus (HIV) an RNA virus that infects human cells and destroys the function of T-cells

Human papilloma viruses DNA viruses; many types, some associated with warts, condyloma, and cervical cancer

Humoral antibodies Ig molecules produced and secreted by B-lymphocytes

Humoral response an immune reaction resulting in the production of Ig molecules by B-lymphocytes

Huntington's disease a genetic disorder appearing in adults leading to brain damage and mental aberrations

Hydrogen bond attractive bond between molecules by positive and negative charged regions, as in the case of interacting hydrogen and oxygen in water molecules

Hydrolysis reactions chemical reactions in which water molecules are formed as one of the final products

Hydrophilic "water loving"

Hydrophobic interactions "water hating"; interactions between molecules to the exclusion of water

Hydrophobic property of molecules that are not soluble in water

Hydrosphere the fresh and marine waters of Earth

Hyperbaric conditions of high pressure, usually associated with depth in water

Hypergonadism an excessive production of male or female sex steroids

Hyperopia farsightedness

Hyperproliferation excessive cell division; usually associated with increase in tissue or organ size

Hypertension high blood pressure

Hyperthermia overheating of a body above normal limits

Hyperthyroidism a condition arising from the overproduction of thyroxine

Hypertonic solution a solution containing a higher concentration of solutes relative to physiological norm

Hypertrophy increase of cell size to accommodate growth in tissue or organ size

Hypogonadism an underproduction of male or female sex steroids

Hypothalamus region of the brain involved in controlling involuntary body functions, e.g., thermal regulation

Hypothermia loss of heat from a body beyond normal limits

Hypothyroidism a condition arising from the underproduction of thyroxine

Hypotonic solution a solution containing a lower concentration of solutes relative to physiological norm

I-bands isotropic zone of a sarcomere; region of actin filaments between Z lines and A-bands; disappear during contraction

Identical twins two individuals arising from a single fertilized egg; genetically identical

Igneous rock rock formed from molten materials below Earth's crust

Ileocecal valve valve involved in regulating movement of material from the ileum to the colon

Ileum the third and final part of the small intestine, transitional to large intestine

Immediate hypersensitivity a standard allergic response

Immiscible phase that which cannot be mixed; often refers to lack of solubility of a compound in water

Immunoglobulins antibodies and related proteins found in vascular and lymphatic circulation and on cell surfaces of immune cells; five types are found in humans: IgA, IgD, IgE, IgG, and IgM

Implantation the embedding of the blastocyst into the wall of the uterus in placental animals

Imprinting rapid learning occurring in the young of some species, which determines associations that may last a lifetime

Inbreeding matings between closely related individuals, used in domestication of animals and plants

Incisors relatively sharp-edged cutting teeth

Incontinence inability to control the urethral sphincters and, thus, to control urine flow from the urinary bladder

Indirect-acting compounds carcinogenic compounds that are modified by cell metabolism before acting as initiators

Ingestion intake of food, eating

Inhibitory the quality of signals between nerve cells that prevent an action potential

Initiator a compound or agent that alters or damages DNA; a carcinogen

Inner cell mass the part of a blastocyst that will give rise to the embryo

Inner ear innermost part of the ear embedded in the bones of the skull; includes structures involved in hearing and balance

Insect vector an insect (such as a mosquito) that transmits a virus to another organism by bite or sting

Insertion addition of a nucleotide in a gene, altering the sequence, which leads to a frameshift mutation

Insertions sites on bone at which tendons attach to allow muscle contraction to move bones around a joint

Insight ability to connect experiences to generate ideas

Inspirational muscles muscles used to inhale

Inspirational reserve the volume of air that can be inhaled above the level of an ordinary breath

Insulin a hormone formed in the pancreas affecting glucose uptake by cells

Integral proteins proteins that are embedded in a cell membrane and exposed on both sides of the phospholipid bilayer

Intelligence what an organism knows and how it applies that knowledge to its activities

Intercalated discs junctional complexes containing desmosomes and gap junctions; prevalent in cardiac muscle

Intercostal muscles muscles between the ribs

Intermediate-sized filaments family of filaments composing the cytoskeleton; 8 nm–11 nm in diameter

Internal fertilization the fertilization of an egg by a sperm inside the female body

Internal naris opening in the back of the nasal cavity to allow air flow to pharynx

Internal oblique underlie the external oblique muscles

Internal urethral sphincter one of two sphincters, or constrictors, that control flow of urine from the urinary bladder; under involuntary control

Internuncial neurons neurons that connect a series of neurons into a network; also known as interneurons

Interoceptors receptors for stimuli from inside the body

Interphase portion of the cell cycle in which the cell is not dividing

Interstitial cell stimulating hormone (ICSH) a hormone produced in the anterior pituitary affecting testosterone production in Leydig cells of the testes

Intervertebral discs cushioning pads of cartilaginous material occurring between vertebrae

Intracellular parasites viruses or some classes of monerans that live inside cells and generally destroy cells

Invasion growth of tumors into surrounding tissues

Inversion a change in the order or arrangement of a piece of a chromosome

Inversion layer a layer in which pollutants are trapped close to the surface by overlying atmospheric layers

Ipsilateral connections nerve connections that stay on one side of the body

Iris pigmented structure surrounding the pupil of the eye

Irregular bones bones of undefined shape

Ischemia a condition in which blood flow is restricted and low or no oxygen is delivered to an organ

Isometric nonuniform contraction of increasing tension against a resistance

Isotonic solution a solution containing solutes matching physiological concentrations; of or relating to conditions of equal tension or tonicity

Isotonic uniform contraction of a muscle during movement of an object

Isotropic nonpolarizing

Jejunum the second part of the small intestine

Joints regions of attachment between bones

Jungle a biome composed of a tangle of low-growing plants that invade denuded forest areas

Jurassic period the time of the giant dinosaurs during the Mesozoic era

Kaposi's sarcoma a rare cancer of connective tissue associated with HIV suppression of the human immune system

Karyokinesis separation of chromosomes during mitosis; specifically during anaphase

Karyotype the representation of condensed chromosomes with pairs aligned

Keratinized stratified epithelium the epidermis of skin

Keratinocytes cells producing keratin in the epidermis

Keratins a class of proteins forming intermediate-sized filaments and bundles; the structural proteins of keratinocytes and many other types of epithelial cells

Kidney stones mineralized structures formed in the kidney that may block urine flow, damage delicate epithelia, and cause pain; also known as renal calculi

Killer T-cells thymus-derived lymphocytes (T) capable of directly destroying foreign or virus infected cells

Kinetic energy the energy of motion

Kingdoms the most inclusive category of taxonomy; five Kingdoms generally recognized

Klinefelter's syndrome a condition arising in phenotypic males from the presence of two X chromosomes and one Y chromosome

Krebs cycle pathway of sequential enzymatic reactions occurring in mitochondria utilizing acetyl-CoA to generate carbon dioxide and ATP

Kupffer cells phagocytic cells of the liver sinusoids

Kwashiorkor a condition resulting from a diet in which insufficient protein is consumed; in children, leads to small size, abnormal bone growth, and mental retardation

Labia fleshy, protective folds of the vulva

Lacrimal glands tear-producing glands

Lacteal lymphatic vessels involved in fat uptake from the small intestine

Lacunae small spaces within bone and cartilage in which cells (osteocytes or chondrocytes) reside

Lakes large bodies of fresh water

Lamella concentric ring structures of osteons

Lamina layers of cells and fibers in the cerebrum

Langerhans cells immune reactive cells of the skin; derived from bone marrow cells

Language a form of communication

Laryngitis inflammation or infection of the vocal cords or larynx

Larynx the region above the trachea in which sounds are made; also known as the voice box

Latency a delay in the expression of a disease until an undetermined future time

Latissimus dorsi large muscle that covers the back and side

Law of independent assortment genes on nonhomologous chromosomes segregate independently from one another

Law of segregation alleles on homologous chromosomes separate from one another during meiosis

LBW (low birth weight) factors physiological and anatomical factors that influence the birth weight of children, including the mother's age, general health, and history of disease and drug use

Lemurs primitive placental primates known as prosimians

Lens a specialized clear structure for focusing light

Lentic water the slow-moving water in a river or stream

Let-down reflex a reflex in mammals in which the tactile stimulation of suckling brings about release of milk from mammary glands

Leukemia a cancer of blood-forming tissue

Leukotrienes powerful and often damaging chemicals produced by cells during an immune reaction

Leverage the action of a lever to facilitate movement

Leydig cells the cells located outside the seminiferous tubules (the interstitium) that produce testosterone under the influence of ICSH

Ligaments tough, resilient connective tissue connecting bones

Light pollution excessive and disturbing light in an environment; for example, city lights at night

Linea alba vertical connective tissue separating halves of the rectus abdominous

Lingual tonsils tonsils associated with the base of the tongue

Linked genes the genes located on the same chromosome

Linoleic acid an unsaturated fatty acid found in triglycerides and essential in human diet

Linolenic acid an unsaturated fatty acid essential in the human diet; related to linoleic acid

Lipases lipid-specific enzymes

Lipids class of hydrophobic molecules composing fats, waxes, and sterols

Lipophilic compounds that are soluble in fats and lipid membranes

Lithosphere the solid materials of Earth's surface

Lithotripter literally stone smasher; a machine that uses vibrations to destroy certain types of kidney stones

Lockjaw the results of exposure to the toxin of the bacterium *Clostridium tetani;* tetanus of voluntary muscles, especially the jaw

Locus the position along a chromosome at which a gene (or allele) is located

Long bones bones with one significantly longer dimension

Long-term memory permanent storage of experiences

Loop of Henle an extended region following the proximal convoluted tubule in which urine is concentrated

Lotic water the fast-moving water of a river or stream

Lucy a member of the species ***Australopithecus afarensis*** (*A. afarensis*), a famous early hominid ancestor to humankind

Lumbar of or referring to the 5 vertebrae in the region of the lower back, immediately below the thoracic vertebrae

Lumen a hollow space within blood vessels (or other tubular structures) through which blood flows

Lumen of endoplasmic reticulum (ER) space or volume contained within the membranes of the ER

Lungs organs used in the exchange of gases between the body and the atmosphere

Luteinizing hormone (LH) a hormone produced by the anterior pituitary, affecting ovulation and corpus luteum formation

Lymph the fluid of the lymphatic system; similar to serum

Lymphatics the system of lymph vessels, nodes, and organs

Lymphokines molecules produced by some immune cells that are involved in either stimulating other immune cell types or in directly destroying target foreign cells

Lymphoma cancerous lymphocytes forming solid tumors, usually in lymph nodes

Lysis breakdown or disintegration of substances or cells

Lysozyme an enzyme of tears with antibacterial action

Macroevolution the large-scale changes (origins and extinctions) of species and higher taxonomic levels over geological time

Macrophages phagocytic cells of monocyte origin; part of the agranulocyte class

Maculae specialized structures containing receptors for motion in the utricle and saccule

Malaria a parasitic disease of the blood

Malignant tumors tumors whose growth is unrestricted and which eventually invade surrounding tissues or spread by metastasis

Mammals warm-blooded organisms having hair and mammary glands

Mandible bone of the lower jaw

Marasmus a condition arising when a child's diet is low both in protein and in total calories; results in bloating, edema, lethargy, diarrhea, and mental retardation

Mastication the action of chewing

Matrix of or relating to the innermost compartment of a mitochondrion

Maxilla bone of the upper jaw

Mechanoreceptors receptors sensitive to touch and pressure

Medial region of or relating to the middle plates of a Golgi apparatus

Mediastinum space in which the heart resides surrounded by the left lung

Mediterranean describes a semiarid ecosystem composed predominantly of chaparral

Medulla a layer of kidney underlying the cortex containing nephrons and collecting ducts

Medulla oblongata the most posterior part of the human brain; origin of several cranial nerves

Medullary cavities the inner cavities of bones

Megakaryocyte platelet-producing giant cell found in red marrow

Meiosis reduction division of cells (particularly in formation of gametes) in which chromosome number is reduced by half (*see* haploid)

Meissner's corpuscles a class of receptors of light touch

Melanin brown/black pigment of skin produced by melanocytes

Melanocytes pigment-producing cells of the body that produce melanin

Melanomas malignant cancer cells derived from the transformation of a melanocyte

Melanosome organelle found in melanocytes in which melanin is stored; passed from a melanocyte to epidermal cells

Melatonin a hormone produced by the pineal gland; influential in establishing circadian rhythms

Memory cell a type of immature lymphocyte that has the capacity to differentiate into a mature lymphocyte

Memory T-cells *see* memory cells

Memory the brain's record of experiences

Meninges several layers of membranes immediately surrounding the brain and spinal cord

Meningitis inflammation and infection of the meninges

Menopause a time in the life of aging females in which ovulation and menstruation cease (onset from 35 years old to 50 years old)

Menses the sloughing of the endometrium in the absence of implantation; usually lasts 3–4 days

Menstrual cycle the periodic buildup and sloughing of the endometrium; on average, repeats every 28–29 days in conjunction with the ovarian cycle

Merkel's discs a class of receptors of light touch

Mesentery membranous structures of the peritoneal cavity attaching intestines to the abdominal wall

Mesoderm middle layer; embryonic layer of cells from which muscle arises

Mesozoic era the middle era following the Paleozoic between 225 million and 65 million years ago

Messenger RNA one of three major types of RNA; contains genetic information used in translation

Metabolism the totality of chemical reactions taking place within an organism

Metacarpals the bones connecting the carpals to the fingers (phalanges)

Metamorphic rock rock formed under high heat and pressure but not completely melted

Metaphase plate alignment via spindle fibers of condensed chromosomes across an equatorial plane of a cell during metaphase

Metaphase second stage of mitosis in which condensed chromosomes are fully attached to spindle fibers

Metastasis break up of a tumor and spread of cancer cells to other parts of the body via the circulatory system

Metatarsals the bones connecting the ankle to the toes (phalanges)

Meteorology study of weather and atmospheric conditions that give rise to it

Microevolution the changes in appearance of a species over generations

Microfilaments ubiquitous components of the cytoskeleton of eucaryotic cells; composed of actin; 5 nm–6 nm in diameter

Microglia a type of phagocytic glial cell of the CNS

Microtubules ubiquitous components of the cytoskeleton of eucaryotic cells; composed of tubulin; 22 nm–25 nm in diameter

Microvilli fine, filamentous extensions of the cells lining the intestines and other bodily sites

Midbrain the region of the brain connecting cerebrum to brainstem

Middle ear the area between the external and inner ear, includes tympanic membrane and three small bones

Migration the translocation from one geographical location to another

Mineralocorticoids hormones of the adrenal cortex affecting mineral balance of the body; for example, aldosterone

Mitosis phase of cell cycle in which chromosomes are divided equally into two daughter cells; consists of four stages: prophase, metaphase, anaphase, and telophase

M line central region of a sarcomere; rich in myosin

Monohybrid cross a mating involving analysis of a single trait

Monosaccharides single simple sugar

Mons veneris literally, the mound of Venus; a fatty protuberance over the pubic symphysis in females

Morphogen a molecule that influences growth and morphogenesis of an organism

Morphogenesis the development of form or shape

Morula a stage of the cleavage of the egg in which the embryo is a solid ball of cells

Motor neurons neurons involved in stimulating muscle cell contraction

Multiple alleles many forms of a single gene in a population

Multiple sclerosis degenerative disease of the nervous system; involves destruction of myelin surrounding neurons

Multipolar neuron a neuron with a single axon and multiple dendrites

Multistep process with reference to cancer, the stepped progression of normal cells to cancerous cells by means of initiation and promotion, which may take years to occur

Muscular dystrophy a hereditary, degenerative muscle disease

Mutagens chemical, biological, radiative agents responsible for causing mutations

Myasthenia gravis progressive disease manifested as weakness of voluntary muscles

Myelin the insulating material around neurons composed of modified glial cell membranes

Myelination the process of forming myelin

Myeloma cancerous tumor of the bone marrow

Myoblasts immature muscle cells, particularly those fated to fuse and form muscle fibers

Myocardial infarction a blockage of blood flow to the myocardium; commonly associated with heart attacks

Myocardium the muscular layer of the heart

Myocyte a mature muscle cell

Myoepithelial cells contractile epithelial cells that are stimulated by oxytocin to squeeze mammary glands to bring about the secretion of milk

Myofibrils assemblages of myofilaments within a myofiber

Myofilament–based cytoskeleton cytoskeleton of muscle cells; involved in contraction

Myogenesis the development or genesis of muscles

Myometrium the muscle layer of the uterus

Myopathy the destruction or degeneration of muscle

Myopia nearsightedness

Myosin filament–forming protein of muscle and nonmuscle cells; a principal structural element of a sarcomere

Myotonia spasm (and change of tone) of one or more muscles

Myotubes form from the cell fusion of myoblasts during embryonic development

Myxedema a form of hypothyroidism characterized by lethargy and mental dullness; reversible by treatment with thyroid hormones

N-acetylsalicylic acid aspirin, a weak acid

Nasal septum connective tissue separating the two sides of the nasal cavity

Nasopharyngeal cancer cancer of the nose and mouth associated with Epstein-Barr virus infection

Natural resistance resistance in bacteria to antibiotics resulting from mutation and selection

Natural selection the environmental forces acting on genetic variation in individuals to determine fitness

Neanderthals a subgroup of *Homo sapiens*

Negative feedback condition in which variance from a set point induces changes in a system that returns that system to the set point

Neoplasias tumor growths resulting from neoplastic cells

Neoplastic the most malignant stage of cancer-cell development

Nephritis a disease associated with the kidney function, particularly with respect to the nephron

Nephron the functional unit of the kidney wherein blood is filtered and nutrients are resorbed

Nerve impulse the electrical signal carried along the length of a neuron

Neural retina complex layers of nerve cells in the eye

Neural tube the embryonic structure arising during neurulation from which the nervous system will develop

Neuralgia pain in the nerves

Neuritis an inflammation affecting nerves

Neuroblastomas tumors of immature, fast-growing neuronal cell types, highly malignant

Neuromas tumors of the nerve

Neuromuscular junction synapse between nerve cell and muscle cell

Neurons electrically excitable cell type of nerve tissue

Neurotransmitters chemical signal molecules involved in regulating the electrical activity of nerve and muscle

Neurulation the period of embryonic development in which the neural tube forms and the basic framework of the nervous system is established

Neutral stimulus in classical conditioning, the transfer from a proper stimulus to an unrelated or neutral one

Neutrophils the most prevalent type of granulocyte

Niches the role that organisms play in their habitats

Nitrogen (N) natural element number 7; diatomic molecule (N_2); nearly 80% of Earth's atmosphere

Nitroglycerin a drug used to dilate arteries in the heart to allow blood flow to the myocardium

Nocturnal at night, particularly in reference to the timing of animal activities and behavior

Nodes of Ranvier regions in which the myelin sheath of a Schwann cell is absent

Noise pollution excessive amount and type of sound in an environment; for example, jet takeoff over a neighborhood

Nondisjunction failure of a chromosome to separate during mitosis or meiosis and enter one of the two daughter cells

Northern deciduous forest a biome characterized by forests of hemlock, pine, and mixed hardwoods

Nostril opening in the nose to allow entrance of air; also known as the external naris

Nucleases DNA- and RNA-degrading enzymes

Nuclei (nucleus) regions of gray matter (cell bodies) in the central nervous system; for example, the basal nuclei

Nucleic acid DNA and RNA; composed of covalently linked nucleotides

Nucleoside a purine or pyrimidine base with covalently linked ribose or deoxyribose sugar

Nucleotide a nucleoside with one or more phosphate groups covalently linked to the ribose or deoxyribose sugar

Nutrients substances from which we derive sustenance; consist of six classes: water, fat, protein, carbohydrates, minerals, and vitamins

Occipital bone the bone at the base of the skull through which the foramen magnum penetrates to allow the spinal cord to enter the cranium

Occipital lobe the most posterior and lower part of each cerebral hemisphere

Occlusion a blockage of flow in the circulatory system

Occlusive junctions form barriers to movement of substances between cells; tight junctions

Odorants molecules or substances that interact with nerve receptors and that we perceive as smells

Olfaction the act of smelling

Olfactory epithelium specialized epithelium located along the upper surface of the nasal cavity in which nerve receptors for odorants are found

Olfactory receptor cells nerve cells capable of binding odorants; located in the olfactory epithelium

Oligodendrocytes glial cells responsible for myelination in the central nervous system

Omega-6 fatty acid a form of lipid consumed in the diet that has been shown under experimental conditions to enhance development of cancer

Oncogenes the cellular genes whose abnormal expression is linked to the development of cancer

Oncogenesis the development and progression of cancer

Oogenesis the development of eggs within the ovary

Oogonia immature precursor cells of ova

Operant conditioning learning by direct participation of the subject; also known as instrumental conditioning

Opposition working against one another

Optic nerve the bundle of axons exiting the eye to connect with the brain

Optic vesicle embryonic structure leading to formation of the eye

Organ of Corti a cochlear structure in which hairlike phonoreceptors are located

Organogenesis the period of time during embryogenesis in which the rudiments of all organs of the adult body are formed

Origins of replication locations in DNA at which replication is initiated

Origins sites on bone at which tendons attach and stabilize muscles

Osmolarity the property of a solution determined by the concentration of ions dissolved in it

Osmosis the movement or diffusion of water from an area of higher concentration or activity to a lower one across a semipermeable membrane barrier

Osmotic balance the equilibrium in the kidney that regulates the movement of water

Osmotic pressure the force or pressure produced by osmosis

Ossification the process of bone formation

Osteoblasts immature bone-forming cells

Osteoclasts cells involved in the breakdown of bone

Osteocytes mature bone-forming cells found within cavities of bone

Osteoma cancer of bone

Osteomalacia a softening of bones in adults from lack of sufficient vitamin D

Osteomas tumors of bone

Osteons small oval structural units repeated within compact bone, composed of a central canal surrounded by concentric rings of bone

Osteoporosis a disease state in which bones become brittle from loss of organic materials; such bones are particularly susceptible to breakage

Otitis media a common infection of the middle ear

Otoliths calcified "ear stones" associated with detection of movement and position of the head

Outbreeding purposeful matings between unrelated individuals, used in domestication of animals and plants

Ova (egg) female gamete

Oval window the opening into the inner ear against which the stirrup moves

Ovarian cycle monthly cycle of production of mature eggs from immature precursors; influenced by pituitary and ovarian hormones

Ovaries gonads of females; producers of eggs

Ovulatory phase the thickening of the endometrium around the time of ovulation

Oxygen (O) natural element number 8; diatomic molecule (O_2) required for human cellular metabolism; 20% of Earth's atmosphere

Oxytocin a hormone produced by the hypothalamus and released from the posterior pituitary

Ozone 0_3, a highly reactive form of oxygen

Ozone layer a stratospheric layer (15 km above Earth's surface) in which there is an abundance of ozone; responsible for absorbing high energy radiation

Pacemaker (1) the sinoatrial node of the heart; (2) an implanted electrical device to control the beat of the heart

Pacinian corpuscles a class of receptors of deep pressure

Palate the upper surface of the mouth

Palatine tonsils tonsils associated with the palate

Paleozoic era the most ancient era of life; from 650 million years ago until 225 million years ago

Pancreatic ducts channels from the pancreas to the duodenum delivering digestive enzymes

Pangea a supercontinent including all land masses presently called continents that existed approximately 200 million years ago

Papillae (1) found at the end of each renal pyramid; involved in emptying urine into the cavity of a calyx; (2) specialized raised areas of the tongue and mouth

Parallel evolution the evolution of similar structures in organisms genetically and developmentally related

Paranasal sinuses mucus-lined spaces within several bones of the face

Paraplegia paralysis of lower body including both legs

Parasites organisms living within other organisms obtaining nutrients at the host's expense; eosinophils function to destroy many types of parasites

Parasympathetic nerves nerves arising from the brain and brain stem (cranial nerves) and lower spine that control organ function; generally inhibitory

Parathyroid glands four small glands embedded in the back of the thyroid gland and producing parathyroid hormone

Parathyroid hormone a hormone produced by parathyroid glands affecting calcium regulation

Parental generation called P, the progenitors of subsequent filial (F) generations

Parietal lobe the posterior and upper part of each cerebral hemisphere

Parietal pleura the layer of pleural membrane nearest the lungs

Parkinson's disease a chronic, progressive nervous disorder resulting in muscle weakness and tremor

Parotid glands glands of the mouth that produce saliva (salivary glands)

Parthenogenesis the development of an egg into an individual in the absence of fertilization

Parturition the act of giving birth to a baby

Passive immunity immunity resulting from acquisition of preformed Ig molecules from another individual

Patella the kneecap

Patellar response kicking reflex of the leg

Pathology the study of damage, disease, or abnormal growth of tissues

Pattern baldness a sex-influenced trait much more common in males than females; also called alopecia

Pectoralis major a principal muscle of the chest

Pedicels footlike parts of specialized cells surrounding the glomerulus that allow passage of blood into Bowman's capsule

Pelvic inflammatory disease (PID) an infection of the urogenital tract often resulting in irreparable damage to the reproductive system

Penicillin an antibiotic derived from fungus

Penicillinase the enzyme that inactivates penicillin.

Penis the male organ of copulation; fleshy structure through which the urethra passes

Pepsin a stomach enzyme

Peptidases peptide-degrading enzymes

Peptones water soluble products resulting from the partial hydrolysis of protein

Pericardial cavity the space enclosing the heart

Pericardial sac a double-walled membrane bounding the pericardial cavity; also known as the pericardium

Pericarditis inflammation of the pericardium

Perilymph the fluid found within the vestibular and tympanic canal

Perimysium connective tissue surrounding small muscle groups

Periosteum the layer of cells surrounding bone from which new bone cells are derived

Peripheral nervous system (PNS) nerves outside the protection of the skeleton

Peripheral proteins proteins associated with one or the other surface of a cell membrane

Peristalsis waves of muscular contractions that move materials within a hollow tubular structure or organ, e.g., material movement within the digestive tract

Peritonitis an infection or inflammation of the peritoneal membrane lining the abdominal cavity

Peritubular capillaries capillaries that surround the nephrons (particularly the proximal and distal convoluted tubules) and exchange oxygen and carbon dioxide

Permeases proteins involved in cell transport (*see* facilitated diffusion)

Permian extinction a time of dying out of numerous species at the end of the Paleozoic era

Peroneus muscles muscles of the lower leg; flex the foot

Phagocytosis active, energy-requiring cellular transport process (cell eating); used to bring large molecules or particles into a cell

Phagosome an intracellular vesicle composed of endosome(s) and lysosome(s)

Phalanges the bones of fingers and toes

Pharynx region of the mouth and throat that extends from the nasal cavity to the larynx and esophagus

Phenotype the traits of genes as expressed in a particular individual

Phenylketonuria an inherited condition in which the amino acid phenylalanine cannot be metabolized properly; an inborn error of metabolism

Phosphate a form of oxidized phosphorus; used in detergents

Phosphorus an element that cycles in the environment and is biologically important in nucleic acids

Photochemical reaction the effects of sunlight on molecules in the atmosphere, for example, formation of ozone at surface levels

Photoreceptors light-sensitive cells in the neural retina; rods and cones

Photosynthesis capture and conversion of light energy to chemical energy

Phototaxis movement toward light

Phototropism growth toward a source of light

Physiological of or related to bodily functions

Pia mater the innermost and most delicate of the meninges

Pineal gland a gland located in the brain and secreting melatonin

Pinna the outer, fleshy structure of the ear that directs sound to the auditory canal

Pinocytosis active, energy-requiring cellular transport process (cell drinking); used to bring fluids and dissolved substances into a cell

Pitch the frequency of vibrations of a sound

Pituitary dwarfism a condition contingent upon the lack of pituitary growth hormone during childhood development

Pituitary gland the primary endocrine gland; consists of two lobes, anterior and posterior, that secrete a variety of protein or peptide hormones

Pivot joints joints between bones involving a rounded end on one bone and a depression on the other

Placenta temporary structure derived from the uterine wall and the implanted blastocyst and serving to support and nurture the developing embryo and fetus

Plains a subtype of grasslands with relatively short, sparse grasses and limited rainfall

Plaque fatty materials deposited along the surface of blood vessels, often partially or completely occluding the flow of blood

Plasma membrane complex phospholipid bilayer membrane composing the barrier surrounding and defining a cell

Plasma the nonformed or fluid material of blood

Play behavior involving exploration and interactions that function in learning and social development

Pleura the membranes surrounding the lungs

Pleurisy an infection and/or inflammation of the pleura

Pneumonia infection of the lungs

Point mutations alterations of a single nucleotide in a gene by mutation

Pollutant an agent whose presence in an environment is detrimental to life and the quality of life

Polygenic inheritance traits involving the products of many genes

Polygeny the influence of many genes on a trait, with variations in combinations of genes leading to a range of variance in a population, for example, height in humans

Polyps saclike projecting outgrowths of the intestinal tract; may be tumorous

Polysaccharides many monosaccharides covalently linked together to form long chains, i.e., glycogen or cellulose

Ponds small bodies of water shallow enough for plants to be rooted

Pons the region of the brain that connects the cerebrum to the cerebellum

Population a group of organisms of the same species sharing a particular habitat that interbreed to produce viable offspring

Position effects changes in the expression of genes based on their position relative to other genes in the chromosome

Positive feedback a condition in which variance from a set point induces further variance of that system from a set point

Potential energy the energy of state or position, as a rock sitting at the top of a hill or an unconnected charged battery

Prairie a subtype of grasslands with relatively long, dense grasses and more rainfall than plains

Preadaptation the evolution of a physical feature for one function initially and for another, different, or more complicated function in later generations

Precambrian era the time before the appearance of most modern phyla, approximately 650 million years ago

Precapillary sphincters regions of arterioles in which smooth muscle cells may contract to control blood flow

Precocial the young of animals that are ready to act independently immediately after birth or hatching

Predator an organism that captures and consumes other organisms (prey)

Preformationists those who hold a historical belief that the embryo was fully formed in either the egg or the sperm and only needed to grow larger

Pregnancy the condition of a female following implantation of the blastocyst and lasting until birth

Presenting cell a macrophage that has processed a foreign antigen and interacts with helper T-cells and other T- and B-lymphocytes

Prey the organism captured and consumed by another organism (predator)

Primary consumers organisms that eat plants; also known as herbivores

Primary germ cell layers established during gastrulation and composed of ectoderm, mesoderm, and endoderm

Primary oocytes egg precursor cells having undergone the first meiotic division

Primates order of mammals to which *Homo sapiens* belongs

Probability functions mathematical terms used to predict statistical likelihood of events or spatial distributions

Procaryotic organisms unicellular organisms with no definable nucleus

Proenzymes an inactive form of enzymes

Progesterone a steroid hormone produced in the ovaries and controlling the ovarian and menstrual cycles

Prognosis with respect to disease, the prospect of recovery

Prolactin a hormone of the anterior pituitary affecting lactation

Proliferative phase refers to the replacement of the endometrium following the end of the last menstrual period or menses

Promoters compounds or agents that allow the changes induced by events of initiation to proceed

Pronator teres muscle of the forearm to turn arm inward and palm back

Pronucleus a term used to describe the nuclei of the egg and sperm before they integrate to form the complete nucleus of the zygote

Prophase the first stage of mitosis

Prophylactics protective sheaths of natural or synthetic materials that fit tightly over the penis or into the vagina; used to prevent fertilization and disease

Proprioceptors specialized receptors providing body awareness information

Prostaglandins fatty-acid-derived hormones released by cells during an allergic response and affecting blood pressure, bronchodilation, and smooth muscle contraction

Prostate gland a male reproductive system accessory gland; produces seminal fluid; surrounds the urethra at its exit from the urinary bladder

Prostatitis an inflammation or infection of the prostate gland

Proteases protein-degrading enzymes

Proteoses small water-soluble peptides resulting from the breakdown of proteins

Prothrombin an inactive form of thrombin

Protoplasm viscous fluid and all components of a cell contained within the plasma membrane, including the nucleus (*see* cytoplasm)

Proximal convoluted tubule tubule immediately following Bowman's capsule and responsible for resorption of nutrients

Proximal end the end nearest the long axis of the body

Pseudostratified epithelium specialized epithelium that appears to be stratified but in which all cells are directly attached to the basement membrane

Psoas major muscle of the pelvic girdle and upper leg

Ptosis a sagging or falling of an organ from its origin site

Pubic symphysis joint between the pubic bones in the pelvic girdle known to be altered during late pregnancy to accommodate birth

Pulmonary artery the artery directing deoxygenated blood from heart to lungs

Pulmonary circulation the flow of blood from the right ventricle through lung circulation and returning to the left atrium

Pulmonary circulatory system blood vessels leading to and away from the lungs; associated with the right side of the heart

Pulmonary edema buildup of watery fluids in the lungs; associated with poor heart function

Pulmonary semilunar valve the valve-controlling exit of blood from the heart to pulmonary circulation

Pulmonary vein the vein directing oxygenated blood from the lungs to the heart

Punctuated equilibrium a hypothesis that evolutionary change and the appearance of new species take place rapidly

Pupil the variably sized opening into the eye, surrounded by the iris

Purkinje cell a multipolar neuron in the cerebellum

Purkinje fibers fibers associated with the distribution of electrical signals in the heart to control the heartbeat

P-wave the initial electrical activity at the beginning of a heartbeat

Pyramids structures of the kidney that project into the renal pelvis and are associated with a papilla and a calyx; involved in urine collection

Pyroclimax community a community of organisms, principally plants, whose regeneration depends on activation of their seeds by the heat of fires

Pyrophosphate groups the terminal two phosphate groups linked to ATP (and other nucleotides)

Pyruvic acid (pyruvate) three-carbon intermediate molecule in energy metabolism of a cell

QRS complex a series of waveforms representing the contraction and relaxation of the heart during the middle part of a heartbeat

Quadriplegia paralysis of both legs and both arms

Quadrupeds four-legged animals

Qualitative relating to quality or kind

Quantitative relating to quantity or amount

Quasi-life forms viruses and intracellular parasites that cannot reproduce or function outside a host

Quiescence a state of inactivity of cells in which there is an arrest of cell division

Radiation energy emitted in the form of waves and particles (i.e., lightwaves and photons)

Ramapithecus an ancient ancestor of hominids

Recessive gene a gene or allele whose expression is masked

Recombination refers to equivalent exchange of parts of homologous chromosomes, particularly during meiosis

Rectum the final region of the digestive tract in which solid wastes are retained for later expulsion

Rectus abdominus stomach muscle

Rectus femoris muscle of the front upper leg, the kicking muscle

Red marrow blood forming tissue within the medullary cavity

Reducing atmosphere conditions in which oxygen is absent from an environment

Reflex arc nerve connections made through the spinal cord

Reflexes innate and automatic neuromuscular response to internal or external stimulus

Reflux movement of material from the stomach into the esophagus

Refractory period the time after orgasm an individual needs to recover before being able to orgasm again; longer in males than in females

Releaser an aspect of the environment that acts as a cue to stimulate a particular behavior; also known as a key stimulus

Remodeling with respect to bone, the constant turnover of old bone and its replacement with new bone

Renal arteries the large arteries that split off the aorta to supply the left and right kidneys with blood under high pressure

Renal pelvis the final cavity of a kidney to which urine flows from the pyramids and papillae and from which the ureters emerge

Renal vein the large vein that carries deoxygenated blood away from the kidneys

Replication the duplication of DNA during the S phase of the cell cycle

Residual volume the volume of air that cannot be exhaled; maintains the expansion of the lungs

Resolution the time after an orgasm in which the body returns to a normal resting state

Respiratory bronchioles bronchioles immediately adjacent to alveoli

Reticular formation a region of the brain associated with arousal

Reticuloendothelial system macrophages and phagocytic cells circulating in blood, connective tissue, and capillaries capable of removing and destroying foreign cells

Retina light receptive portion of the eye; also called the neural retina

Retinoblastomas tumors of retinal neural cells, highly malignant

Retrograde amnesia condition of amnesia in which most recent memories are lost

Retroviruses RNA viruses implicated in some types of cancer

Rheumatism a disease characterized by inflammation and pain of the joints

Rheumatoid arthritis a form of arthritis caused by an autoimmune reaction

Ribosomes intracellular structures involved in the synthesis of protein during the process of translation

Ribs long, flat bones enclosing the thorax in vertebrates

Rickets a condition in which lack of vitamin D in children results in soft bones

RNA ribonucleic acid; formed during transcription; a carrier of specific genetic information

Rod light–sensitive photoreceptor cell

Rotational movements movement around an axis; shaking your head "no" is an example

Round window the hole in the base of the cochlea at the opposite end of the canal system from the oval window

Ruffini's corpuscles a class of thermal receptors associated with sensitivity to heat

Rugae the deep folds in the stomach lining that allow for volume expansion

Saccule one of the chambers housing receptors of static balance

Sacrum fused vertebrae forming the back of the pelvis

Saddle joints joints in which both interacting bones are shaped in the manner of a saddle

Saliva watery fluid produced by glands and secreted into the mouth

Salivary amylase an enzyme produced by salivary glands that digests starches

Sand and silt forms of sediments of different sizes

Saprophyte an organism that obtains organic compounds needed for its growth from absorption of materials from dead and decaying organisms; for example, many types of bacteria and fungi

Sarcolemma name given to the plasma membrane of a muscle fiber

Sarcomas tumors of muscle and connective tissues

Sarcomeres units of contraction in some types of muscle cells; composed principally of actin and myosin

Sarcoplasm name given the cytoplasm of a muscle fiber

Sarcoplasmic reticulum name given to the endoplasmic reticulum of a muscle fiber

Savanna a tall grassy ecosystem transitioning between forest and plains

Scab the hardened material resulting from the formation of a blood clot

Schwann cells glial cells responsible for myelination in the peripheral nervous system

Sclera the tough outer layer of connective tissue of the eyeball

Scrotum the saclike structure in which the testes are located

Scuba diving swimming under water with scuba gear; scuba is short for *s*elf-*c*ontained *u*nderwater *b*reathing *a*pparatus

Sebaceous glands exocrine glands of skin that produce oily secretion released into hair follicle

Sebum oily secretion of sebaceous gland

Second degree burns damage to skin from heat affecting both the epidermis and underlying dermis

Secondary consumers organisms that eat plants and other animals; also known as omnivores

Secondary oocytes egg precursor having undergone the second meiotic division

Secondary sex characteristics the development and maintenance of mature features of males and females at puberty under the influence of sex steroids; for example, facial hair, breast development

Secretory phase the buildup of glands within the endometrium and secretion of fluids in preparation for implantation

Sedentary permanently attached, nonmigrating

Sedimentary rock rock formed from the accumulation and compaction of material eroded from igneous and metamorphic rock

Segmentation a breaking-up of material into small masses in the digestive tract to facilitate movement

Sella turcica the part of the sphenoid bone in which the pituitary gland is located

Semen sperm-bearing fluid produced in testes and reproductive glands of male animals for internal fertilization

Semicircular canals a set of three canals in the inner ear associated with dynamic balance

Semiconservative replication synthesis of DNA in which one strand of a double-stranded molecule is used to copy and make a new strand

Seminal fluids a viscous mixture of sperm and secretions produced by the male reproductive accessory glands

Seminal vesicles the glands of the male reproductive system producing seminal fluids

Seminiferous tubules the specialized epithelium of the long thin tubules in which sperm undergo development and differentiation

Senescence aging but with the connotation of loss of function or deterioration of the body

Septum the internal wall of the heart, separating chambers

Seropositive having antibodies in one's serum reactive against a virus; an indication of exposure to the virus

Serotonin an essential neurotransmitter prevalent in the brain

Serous pericardium the delicate, lubricating, inner layer of the pericardium

Serratus anterior muscles of the back and side of the thorax

Sertoli cells the epithelial cells of the the seminiferous tubules that support and nurture the developing sperm

Serum the fluid material of the blood minus the factors involved in blood clotting

Sex hormones hormones produced in the gonads and the adrenal cortex; for example, testosterone

Sex-influenced traits traits whose expression is influenced by the sex of the individual; affects genes on autosomes

Sex-limited traits traits arising from autosomal genes occurring only in individuals of one sex or the other

Sex-linked traits genes occurring on the X chromosome

Sexual climax (orgasm) in females, a highly aroused sexual state in which uterine muscle contractions pull semen into the uterus; in males, sexual arousal that results in muscle contractions associated with ejaculation of semen

Sexual reproduction a form of reproduction involving interaction of germ cells

Shigellosis a disease caused by infection with the organism *Shigella dysenteriae;* bacterial dysentery

Short bones bones with roughly similar lengths in all dimensions

Short-term memory temporary storage in recallable memory of information from an experience

Shoulder blades large flat bones of the upper back; also known as the scapulae

Sickle-cell disease a recessive genetic disorder in which hemoglobin is abnormal and cell shape is altered

Sickle-cell trait the recessive gene carried by individuals heterozygous for sickle hemoglobin trait

Sieving the process by which components of the blood pass through the pedicels or are retained in the glomerular arterioles

Sigmoid colon the S-shaped region of the colon just above the rectum, also known as the sigmoid flexure

Sinoatrial node also known as SA node; controls pace of the heart beat

Skull bones of the head

Sliding-filament model model used to explain the interaction of myosin and actin during the shortening of sarco-meres in muscle contraction

Social Darwinism the application of Darwinian principles to the inherent status of human groups

Soleus muscle of lower leg; flexes the foot

Somatic cells from the Greek soma or body; cells of the body excluding the gametes

Somatic mutations mutations in the DNA of general body cells (not including germ cells)

Somatotropic effects literally, body growth effects; applied to hormones involved in growth and development

Spawning a form of external fertilization; for example, as observed in fish

Special connective tissue designation for connective tissues composing bone, cartilage, and blood

Species a population of genetically similar organisms that is reproductively isolated, produces viable and fertile offspring, and shares an evolutionary history

Sperm male gamete

Spermatogenesis the formation of sperm in the testes

Spermatogonia the precursor cells of sperm

Sphygmomanometer instrument used to measure blood pressure

Spinalis muscles muscles attached to vertebrae

Spindle apparatus assembly of microtubules used to translocate chromosomes to the poles during anaphase

Spindle fibers microtubules found in the spindle apparatus

Spleen the lymphatic organ in which red blood cells are held in reserve; location for development of early forms of red blood cells

Spongy bone bone riddled with open spaces lacking osteons

Squamous cell carcinomas malignant cancer cells arising from the transformation of squamous or elongated cells of the epidermis

Staghorn calculus massive mineralized layer precipitated along the inner surface of the renal pelvis and calyces

Static balance balance associated with the inner ear and receptors in the utricle and saccule

Stereoscopy binocular vision that establishes a sense of depth perception

Stereotyped behavior predictable patterns of reflexive behavior

Sternum bone plate in the center of the thorax to which the ribs are attached

Sterols a class of lipids with multi-ring structure to which steroids and cholesterol belong

Stirrup a bone of the middle ear; also called stapes

Stomach a principal organ of digestion

Stress physiological state resulting from factors that tend to alter an existing equilibrium

Stretch receptors receptors in muscle, epithelium, or connective tissue sensitive to tissue distortion

Stroke the loss of brain function (particularly motor control) due to oxygen deprivation caused by blockage of blood flow

Subalpine forest a high altitude forest of spruce and bristlecone pines, where life is sparse

Subclavian vein major vein of the thorax delivering materials from the lymphatic system to the blood

Sublingual glands glands located under the tongue that produce saliva (salivary glands)

Submandibular glands glands of the lower jaw that produce saliva (salivary glands)

Subspecies a category of types of a species, usually geographically separated from one another; also called a race

Substitution the replacement of one nucleotide for another in a gene sequence

Succession the progressive changes in plant species in disturbed or denuded habitats

Sulci regions of infolding of the cerebrum

Sulfur an element that cycles in the environment and is biologically prominent in proteins and proteoglycans; also found in coal

Summation the balancing of inhibitory and excitatory signals to affect activity of a nerve cell

Superior vena cava the large vein returning blood to the right side of the heart

Suppinator muscle of forearm to turn arm outward and palm forward

Support cells (1) cells found in a taste bud that provide structural support for gustatory cells; (2) with respect to the ovarian follicle, cells that associate and nurture the developing oocyte

Surface tension the property of liquids that makes them resistant to penetration at their surface due to the cohesive interactions of the molecules that compose them

Surfactant a wetting agent or detergent that reduces surface tension

Suture fibrous joint particularly evident between the bones of the cranium

Swamp a type of boggy wetlands in which land is saturated with or covered by water

Sweat glands exocrine glands secreting sweat or other substances (*see* apocrine and eccrine sweat gland)

Symbiosis living together of two dissimilar organisms in a mutually beneficial relationship

Symbiotic relationship a relationship in which two independent entities coexist together to their mutual benefit

Symmetry division of structure into exactly similar parts with respect to shape, size, and position

Sympathetic nerves nerves arising from spinal ganglia that control organ function; generally excitatory

Symphysis slightly moveable joint type exemplified by the connection of bones of the pubis

Symport transport channel in plasma membrane

Synapse junction between excitable cells; a gap across which neurotransmitters diffuse between nerve cells or between nerve cells and muscle cells or fibers

Synapsis exchange of parts of chromosomes during recombination events of meiosis

Syndactyly fused digits; failure of fingers and/or toes to separate during early development

Synergism working together

Synovial cavity space between bones having synovial joints

Synovial fluid slippery, lubricating fluid occurring in synovial cavities of the moveable joints

Synovial joint a fully moveable joint

Syphilis a disease caused by infection with the spirochete *Treponema pallidum*

Systemic circulatory system blood vessels leading to the entire body other than the lungs and back to the heart; associated with the left side of the heart

Systole the period in a heartbeat during which both atria and ventricles are contracting

Tactile stimuli stimulation of nerves arising from touch

Taiga high latitude or altitude forests of pines and firs; also known as coniferous forest

Tarsals the bones of the ankle

Tarsiers prosimians, related to lemurs

Taste buds specialized chemoreceptors in supporting structures located on fungiform and circumvallate papillae

Taste pore channel through which gustatory cells send hairlike processes containing taste receptors

Taxis (pl. taxes) movement toward a stimulus

Taxonomy the study of the classification of living and extinct organisms

Tectonic plates large areas of Earth's crust undergoing constant deformation

Tectorial membrane the layer overlying the organ of Corti in which hairlike phonoreceptors are embedded

Telomeres the ends of a chromosome

Telophase the final stage of mitosis

Temperate deciduous forest a biome characterized by mixed hardwood trees and distributed widely in higher latitudes

Temporal lobe the lateral part of each cerebral hemisphere

Tendinitis a condition of inflammation and pain in tendons

Tendinous intersections or cross-tendinous insertions of the rectus abdominus

Tendons tough, resilient connective tissues connecting muscle to bone

Teratogen a molecule or agent that causes deformities in development and growth of organisms

Teres major muscle of the back and upper arm

Territoriality the behavior used to establish and defend an area or volume of a habitat for breeding or feeding

Tertiary consumers organisms that eat primarily other animals (meat eaters); also known as carnivores

Test cross a mating used to determine genotype of individual

Testes gonads of males; producers of sperm

Testosterone steroid hormone, associated with development of male secondary sexual characteristics

Tetanus continued state of muscle contraction resulting from continuous stimulation of a muscle

Tetany a state of continuous contraction of muscle

Thalamus a part of the forebrain; relays sensory information from spinal cord

Thalassemia a genetic disorder affecting hemoglobin and blood

Thalassemia major a severe condition arising in individuals homozygous for the thalassemia gene

Thalassemia minor a mild condition arising in individuals heterozygous for the thalassemia trait

The bends a condition associated with formation of gas bubbles within blood vessels as a result of decreased pressure from rising too rapidly from great depths in water

Theory of evolution the theory proposed by Charles Darwin to explain change and the origin of species through variation and natural selection

Thermal pollution excessive heat added to an environment, for example, water heated during nuclear reactor cooling

Thermoreceptors receptors sensitive to heat (hot and cold)

Thermotaxis movement toward a source of heat

Third degree burns severe damage to skin and underlying tissues (connective tissue, muscle) from heat

Third eye an ancient name for the pineal gland

Thoracic of, or referring to, the cavity enclosed by the ribs in the upper torso; the set of 12 vertebrae immediately below the cervical vertebrae

Threshold the minimum level of stimulation needed to stimulate an action potential

Thrombin the enzyme used to activate fibrinogen to fibrin during thrombosis

Thrombocyte another name for platelets

Thrombosis the process of blood clotting; involves a cascade of complex interactions

Thrombus a blood clot

Thrush a form of yeast infection (*Candida albicans*) occurring in the mouth

Thymosin a hormone essential to T-lymphocyte differentiation in the thymus gland

Thymus gland a gland located in the region of the neck and, in childhood, the site of formation and maturation of T-lymphocytes; produces the hormone thymosin

Thyroid gland an endocrine gland located anterior to the trachea and involved in the production of growth and other hormones

Thyroxine the growth hormone produced by the thyroid gland

Tibia one of the two bones of the lower leg; shins

Tibialis muscles muscles of the lower leg; flex the foot

Tidal volume the volume of air exchanged in an ordinary breath of air

Tissue rejection the failure of a host to accept a donated organ or tissue

Tissues specialized assemblages of cell types—epithelia, connective tissue, muscle, nerve

Tolerance the capacity of the immune system of an individual to recognize "self" antigens and not destroy or react against them

Tongue a flexible, flattened muscular structure of the mouth used in eating, swallowing, and speech

Tonicity refers to the concentration of solutes in a solution

Tonsillectomy surgical removal of the tonsils

Tonsillitis infection of the tonsils

Tonsils lymphatic tissue located in the throat

Topologically distinct the configuration of compartments within cells, especially the distinction between cytoplasmic and noncytoplasmic space

Total capacity the total volume of air that the lungs can hold

Touch-temperature complex the set of sensory receptors primarily associated with skin, for example, mechanoreceptors

Trabeculae plate- or barlike structures in spongy bone

Trachea the wind-pipe, a cartilagenous tube connecting throat to lungs

Tracheae passageways in arthropods that lead from the exoskeleton to internal tissues of the body

Traits genetically inherited characteristics

Trans region of or relating to the distal or most distant plates of a Golgi apparatus

Transcription the synthesis of RNA from a DNA template

Transection a cut through the spinal cord

Transitional epithelium specialized epithelium whose cells undergo shape change (flattening) during stretching, as in the urinary bladder

Translation the synthesis of protein from an RNA template held in a ribosome

Translocation the movement of a piece of one chromosome onto another chromosome

Transpiration evaporative process used by plants to move water through vascular elements (xylem)

Transverse colon the second segment of large intestine crossing from left to right in the upper abdomen

Trapezius a muscle of the upper back

Tree line the zone in altitudinal zonation after which no trees grow, alpine tundra

Trial and error a means of learning by experience

Triceps brachii muscle of the back of the upper arm

Trichomoniasis the condition resulting from infection with the protozoan *Trichomonas vaginalis*

Tricuspid valve a valve with three membranous sections to prevent backflow of blood between chambers of the heart

Triglycerides a class of lipids characterized by linkage of three fatty acids to a glycerol molecule

Trisomy the presence of three chromosomes of one type in cells

Trisomy-13 three copies of chromosome 13; known as Patau's syndrome

Trisomy-18 three copies of chromosome 18; known as Edward's syndrome

Trisomy-21 a congenital anomaly in which three copies of chromosome 21 are present in all cells of the body; known as Down's syndrome

Trophectoderm the part of the blastocyst that will give rise to the extraembryonic layers and part of the placenta

Trophic level the feeding level of an organism in a food chain or food web

Tropical rain forests biomes of dense growth and biodiversity, generally equatorial

Tropism orientation by an organism or part of an organism to external stimulation

Tropomyosin protein associated with myosin; involved in controlling muscle contraction

Troponin calcium-binding protein involved in controlling muscle contraction

True ribs ribs connected directly to the sternum

T-tubules tiny, deeply penetrating invaginations of the sarcolemma

Tubal ligation a surgical procedure in which the Fallopian tubes are ligated and cut to prevent transport of eggs

Tubulin protein composing microtubules

Tundra a latitudinal zone associated with the arctic; cold and unforgiving

Tunics connective tissue coverings of blood vessels

Turner's syndrome a condition arising in phenotypic females from the presence of only one X chromosome

T-wave the final electrical activity at the end of a heartbeat

Tympanic canal part of the fluid-filled channel system of the cochlea

Tympanic membrane the eardrum

Ulcer an open wound, particularly in the digestive tract, in which the epithelial lining is eroded

Unidirectional in one direction

Uniformatarianism a hypothesis that states that geological processes occur at similar rates throughout Earth's history

Unilateral effects affecting only one side of the body, as in a stroke; for example, damage to the left side of the brain affects the right side of the body

Unipolar neuron a neuron with a single extension combining axon and dendrite

Universal acceptor an individual with AB type blood

Universal donor an individual with O type blood

Uremia a toxic buildup of urea in blood

Ureter tubes that exit the kidneys and deliver urine to the urinary bladder

Urethra the tube through which urine flows to exit the body; shorter in females than males

Urinary bladder the organ in which urine is stored prior to elimination from the body

Urine excreted waste product in mammals containing nitrogen and produced in kidneys and eliminated through the urinary system

Urogenital source arising from the urinary and reproductive systems

Uterus major muscular and glandular structure of the reproductive tract of females into which the blastocyst implants and then is carried throughout pregnancy

Utricle one of the chambers housing receptors of static balance

Uvula a flaplike extension of the palate at the entrance to the pharynx

Vaccination purposeful exposure to an antigen (particularly a noninfectious or killed form of virus) to stimulate acquired immunity

Vagina a sheathlike internal structure of the female reproductive system leading to uterus

Vagus nerve the tenth cranial nerve; involved in control of the heart beat

Variation the genotypic and phenotypic differences observed in individuals of a single species

Varicose veins abnormally swollen veins that have broken away from the connective tissues that surround them and protrude outward on a surface

Varicosities any abnormally swollen or dilated structures

Vas deferens the tube connecting the epididymis to the urethra and carrying the sperm from the region of the scrotum into the body

Vasectomy a surgical procedure in which the vas deferens is cut and sealed to prevent transport of sperm

Vastus muscles muscles of the upper leg that insert on the patella and help extend the leg

Vectors carriers of infectious disease agents between natural host and organisms pathologically infected

Ventricle (1) the lower chamber of vertebrate heart that pumps blood out of the heart; (2) one of five interconnected, fluid-filled cavities within the brain

Venules tiny veins arising from capillaries

Vertebrae 26 bones of the spinal column

Vestibular apparatus the triple-loop structure and cavernous regions of the inner ear used in maintaining balance

Vestibular canal part of the fluid-filled channel system of the cochlea

Vestibular membrane one of two specialized membranes surrounding the cochlear duct

Vestibule (1) region of the external genitalia above the vagina; (2) the region of the nasal cavity immediately behind the nostrils; (3) with respect to the inner ear, the region associated with balance

Villi folds in the epithelium lining the intestines

Visceral pleura the layer of pleural membranes furthest from the lungs

Viscosity the thickness or thinness reference of a fluid and its ability to flow

Visual acuity a quality of vision determined by the sharpness of the image perceived

Visual purple light-absorbing pigment of rods; also known as rhodopsin

Vitamin D hormone needed for bone formation; also known as calciferol

Vitamin K a vitamin produced by bacteria in the intestines and important in blood clotting

Vitreous humor the clear viscous content of the center of the eyeball

Vocal cords structures located within the larynx that vibrate to form sounds

Voltage potential specifically the potential difference in charge across the plasma membrane of a cell

Vomiting the forceful ejection of materials in the stomach back up the esophagus and out of the body

Vulva the externalized genitalia of females

Walking pneumonia a viral form of pneumonia

Weather local conditions of the atmosphere; such as, rain, wind, hot, and cold

Weight the attraction of mass to the Earth resulting from gravity

Working volume the volume of exchangeable air in the lungs

Yellow fever lethal viral disease carried by mosquitoes to humans

Yellow marrow fatty tissue within the medullary cavity with no blood cell-forming capacity

Yolk sac *see* extraembryonic tissues

Z lines also known as Z discs; regions at the ends of a sarcomere that separate sarcomeres from one another

Zona pellucida the layer of jelly surrounding the egg released from the ovary (part of the original follicle)

Zygote product of fusion of two gametes; a new genetically distinct individual of a sexually reproducing species

INDEX

A

B

C

D

E

F

G

N

O

P

T

Illustration Credits
Jon Clark, Deborah Cowder, Hespenheide Design, Carlyn Iverson, Stan Maddock, Sandra McMahon, Randy Miyake, Elizabeth Morales-Denney, Laurie O'Keefe, Precision Graphics, Rolin Graphics, Tech Graphics, Cyndie C. H. Wooley

Photo Credits
Chapter 1 opener, page 1 ©Earth Imaging/Tony Stone Images
Fig. 1.2b, page 4 ©Earth Imaging/Tony Stone Images
Fig. 1.8a left, page 12 ©S. Lowry/Univ. Ulster/Tony Stone Images
Fig. 1.8a right, page 12 ©Paul Johnson/BPS/Tony Stone Images
Fig. 1.8b left, page 12 ©Richard L. Carlton/Visuals Unlimited
Fig. 1.8b right, page 12 ©Phil Degginger/Tony Stone Images
Fig. 1.8c left, page 12 ©Robert Brons/BPS/Tony Stone Images
Fig. 1.8c right, page 12 ©Davis M. Philips/Visuals Unlimited
Fig. 1.8d left, page 12 ©Larry Ulrich/Tony Stone Images
Fig. 1.8d right, page 12 ©Carole Elies/Tony Stone Images
Fig. 1.8e left, page 12 ©Jeff Lepore/Photo Researchers, Inc.
Fig. 1.8e right, page 12 ©David Grossman/Photo Researchers, Inc.

Chapter 2 opener, page 17 ©Andrew Syred/Tony Stone Images
Fig. 2.5b, page 23 ©Andrew Syred/Tony Stone Images

Chapter 3 opener, page 51 ©Andrew Syred/Tony Stone Images
Fig. 3.7b, page 59 ©Don W. Fawcett/Visuals Unlimited
Fig. 3.11a, page 62 ©M. Schliwa/Visuals Unlimited
Fig. 3.11b, page 62 ©Loren Knapp
Fig. 3.11c, page 62 ©M. Schliwa/Visuals Unlimited
Fig. 3.17b, page 69 ©Andrew Syred/Tony Stone Images
Fig. 3.17c, page 69 ©David M. Philips/Visuals Unlimited
Fig. 3.17d, page 69 ©David M. Phillips/Visuals Unlimited

Chapter 4 opener, page 81 ©Robert Brons/BPS/Tony Stone Images
Fig. 4.1a, page 82 ©Fred E. Hossler/Visuals Unlimited
Fig. 4.1b, page 82 ©David M. Philips/Visuals Unlimited
Fig. 4.1c, page 82 ©Charles W. Stratton/Visuals Unlimited
Fig. 4.1d, page 82 ©Karl Aufderheide/Visuals Unlimited
Fig. 4.1e, page 82 ©T. E. Adams/Visuals Unlimited
Fig. 4.3b, page 84 ©G. W. Willis/BPS/Tony Stone Images
Fig. 4.8a, page 90 ©Cabisco/Visuals Unlimited
Fig. 4.8b, page 90 ©Cabisco/Visuals Unlimited
Fig. 4.8c, page 90 ©Robert Brons/BPS/Tony Stone Images
Fig. 4.8d, page 90 ©Fred E. Hossler/Visuals Unlimited
Fig. 4.8e, page 90 ©Biophoto Associates/Photo Researchers
Fig. 4.8f, page 90 ©Makio Murayama/BPS/Tony Stone Images
Fig. 4.8g, page 90 ©Cabisco/Visuals Unlimited
Fig. 4.9a, page 91 ©David M. Phillips/Visuals Unlimited
Fig. 4.10, page 92 ©Bill Beatty/Visuals Unlimited
Fig. 4.12a, page 93 ©Bruce Iverson/Visuals Unlimited
Fig. 4.12b, page 93 ©Fred Hossler/Visuals Unlimited
Fig. 4.15a, page 95 ©John D. Cunningham/Visuals Unlimited
Fig. 4.19, page 99 ©Fred Hossler/Visuals Unlimited

Chpater 5 opener, page 101 ©James Darell/Tony Stone Images
Fig. 5.2a, page 103 ©James Balog/Tony Stone Images
Fig. 5.2b, page 103 ©Tim Davis/Tony Stone Images
Fig. 5.2c, page 103 ©Stephen Frink/Tony Stone Images
Fig. 5.2d, page 103 ©Christoph Burki/Tony Stone Images
Fig. 5.2e, page 103 ©Tim Davis/Tony Stone Images
Fig. 5.13a, page 115 ©Ken Greer/Visuals Unlimited
Fig. 5.13b, page 115 ©Ken Greer/Visuals Unlimited
Fig. 5.13c, page 115 ©Ken Greer/Visuals Unlimited
Fig. 5.15a, page 117 ©Art Wolfe/Tony Stone Images
Fig. 5.15b, page 117 ©Tony Stone Worldwide
Fig. 5.15c, page 117 ©Bruce Berg/Visuals Unlimited
Fig. 5.15d, page 117 ©James Darell/Tony Stone Images
Fig. 5.15e, page 117 ©Bruce Berg/Visuals Unlimited

Chapter 6 opener, page 119 ©SIV/Visuals Unlimited
Fig. 6.1, page 120 ©John M. G. Ross/Photo Researchers, Inc.
Fig. 6.3a, page 122 ©Michael Newman/PhotoEdit
Fig. 6.3b, page 122 ©Biophoto Associates/Science Source/Photo Researchers, Inc.
Fig. 6.3c, page 122 ©Jim Stevenson/Science Photo Library/Photo Researchers, Inc.
Fig. 6.10, page 128 ©SIU/Visuals Unlimited

Chapter 7 opener, page 145 ©David M. Phillips/Visuals Unlimited
Table 7.1a, page 173 ©Philippe Plailly/Science Photo Library/Photo Researchers, Inc.
Table 7.1b, page 173 ©Biophoto Associates/Science Source/Photo Researchers, Inc.
Table 7.1c, page 173 ©Will and Deni McIntyre/Tony Stone Images, Inc.
Table 7.1d, page 173 ©Sue Ford/Science Photo Library/Photo Researchers, Inc.
Table 7.1e, page 173 ©CNRI/Science Photo Library/Photo Researchers, Inc.

Chapter 8 opener, page 177 ©David Phillips/Visuals Unlimited

Chapter 9 opener, page 201 ©S. Ito–D. Fawcett/Visuals Unlimited
Fig. 9.12e, page 213 ©G. Shih-R. Kessel/Visuals Unlimited
Fig. 9.20a, page 222 ©Max Listgarten/Visuals Unlimited
Fig. 9.20b, page 222 ©M. I. Walker/Photo Researchers, Inc.
Fig. 9.20c, page 222 ©SIV/Visuals Unlimited

Chapter 10 opener, page 233 ©Fred E. Hossler/Visuals Unlimited
Fig. 10.6a, page 238 ©Fred E. Hossler/Visuals Unlimited
Fig. 10.6b, page 238 ©Michael Webb/Visuals Unlimited
Fig. 10.6c, page 238 ©Fred E. Hossler/Visuals Unlimited
Fig. 10.11b, page 247 ©David Joel/Tony Stone Images
Fig. 10.15a, page 250 ©SIU/Visuals Unlimited
Fig. 10.15b, page 250 ©Dr. E. Walker/Science Photo Library/Photo Researchers
Fig. 10.16, page 251 ©SIV/Visuals Unlimited
Fig. 10.17a, page 251 ©Hank Morgan/Photo Researchers
Fig. 10.17b, page 251 ©Will and Demi McIntyre/Science Source Photo Researchers

Chapter 11 opener, page 255 ©Science VU/Visuals Unlimited
Fig. 11.7b, page 263 ©SIV/Visuals Unlimited
Fig. 11.7c, page 263 ©David Leah/Photo Researchers
Fig. 11.10, page 265 ©SIV/Visuals Unlimited

Chapter 12 opener, page 281 ©Science Photo Library/Photo Researchers, Inc.
Fig. 12.3a left, page 283 ©Fred Hossler/Visuals Unlimited
Fig. 12.3a middle, page 283 ©Custom Medical Stock Photo
Fig. 12.3a right, page 283 ©Custom Medical Stock Photo
Fig. 12.3b top, page 283 ©David M. Phillips/Visuals Unlimited
Fig. 12.3b bottom left, page 283 ©George B. Chapman/Visuals Unlimited
Fig. 12.3b bottom right, page 283 ©David M. Phillips/Visuals Unlimited
Fig. 12.10a, page 288 ©Science Photo Library/Photo Researchers, Inc.
Fig. 12.10b, page 288 ©David M. Phillips/Photo Researchers, Inc.
Fig. 12.17a, page 296 ©Boehringer Ingelheim Science VU/Visuals Unlimited
Fig. 12.17b, page 296 ©Boehringer Ingelheim Science VU/Visuals Unlimited

Chapter 13 opener, page 301 ©SIU/Visuals Unlimited
Fig. 13.7a, page 309 ©Richard Hutchings/Photo Researchers, Inc.
Fig. 13.7b, page 309 ©Bettina Cirone/Photo Researchers, Inc.
Fig. 13.8a,b,c, page 310 ©Courtesy of Milton Crane, Loma Linda University Medical School
Fig. 13.10, page 311 ©Ken Greer/Visuals Unlimited
Fig. 13.11a, page 312 ©Lester V. Bergman/Corbis
Fig. 13.11b, page 312 ©Biophoto Associates/Science Source/Photo Researchers, Inc.
Fig. 13.14b, page 314 ©John D. Cunningham/Visuals Unlimited
Fig. 13.16c, page 316 ©Science VU/Visuals Unlimited
Fig. 13.16a, page 316 ©UPI/Corbis-Bettmann
Fig. 13.16b, page 316 ©UPI/Corbis-Bettmann
Fig. 13.20b, page 320 ©Cabisco/Visuals Unlimited

Chapter 14 opener, page 325 ©David M. Phillips/Visuals Unlimited
Fig. 14.2a, page 327 ©David M. Phillips/Visuals Unlimited
Fig. 14.2b, page 327 ©David M. Phillips/Visuals Unlimited
Fig. 14.6a, page 331 ©Science VU/Visuals Unlimited

Fig. 14.6b, page 331 ©L. Lisco, D. Fawcett/Visuals Unlimited
Fig. 14.18b, page 344 ©Elizabeth Walker, Associate Professor, Department of Anatomy, School of Medicine, West Virginia University
Fig. 14.18c, page 344 ©From Tissues and Organs: A Text-Atlas of Scanning Electron Microscopy by Richard G. Kessel and Randy Kardon. Copyright ©1979 by W.H. Freeman and Company
Fig. 14.22, page 349 ©Charles Thatcher/Tony Stone Images

Chapter 15 opener, page 353 ©David M. Phillips/Visuals Unlimited
Fig. 15.4a, page 357 ©David M. Phillips/Visuals Unlimited
Fig. 15.4b, page 357 ©David M. Phillips/Visuals Unlimited
Fig. 15.10, page 364 ©Biophoto Associates/Science Source/Photo Researchers, Inc.
Fig. 15.11b, page 364 ©John D. Cunningham/Visuals Unlimited
Fig. 15.14, page 367 ©Cabisco/Visuals Unlimited
Biosite 15.2 left, page 358 ©National Library of Medicine/ Science Photo Library/Photo Researchers, Inc.
Biosite 15.2 right, page 358 ©Charles and Josette Lenars/Corbis
Fig. 15.23a, page 374 ©Science VU-WHO, A. Vollan/Visuals Unlimited
Fig. 15.23b, page 374 ©Forest W. Buchanan/Visuals Unlimited
Fig. 15.23c , page 374 ©Nancy D. McKenna

Chapter 16 opener, page 381 ©CNRI/Science Photo Library/Photo Researchers
Fig. 16.2, page 382 ©Reyters/Corbis-Bettman
Fig. 16.3a, page 383 ©Science VU/Visuals Unlimited
Fig. 16.3b, page 383 ©CNRI/Science Photo Library/Photo Researchers
Fig. 16.3c, page 383 ©David M. Phillips/Visuals Unlimited
Fig. 16.4a, page 385 ©Ken Greer/Visuals Unlimited
Fig. 16.4b, page 385 ©Ken Greer/Visuals Unlimited
Fig. 16.4c, page 385 ©Science VU/Visuals Unlimited
Fig. 16.4d, page 385 ©Science VU/Visuals Unlimited
Fig. 16.5a, page 387 ©Veronika Burmeister/ Visuals Unlimited
Fig. 16.5b, page 387 ©Science VU/Visuals Unlimited
Fig. 16.5c, page 387 ©Science VU/Visuals Unlimited
Fig. 16.5d, page 387 ©Hans Gelderblom/Visuals Unlimited
Fig. 16.6b, page 388 ©M. G. Sarngadharan and Philip D. Markham/Science VU/Visuals Unlimited
Fig. 16.7b, page 390 ©Science Photo Library/Photo Researchers
Fig. 16.7c, page 390 ©Cabisco/Visuals Unlimited
Fig. 16.7d, page 390 ©Cabisco/Visuals Unlimited
Fig. 16.9, page 392 ©Reuters/Corbis-Bettmann
Fig. 16.10a, page 393 ©UPI/Corbis-Bettman
Fig. 16.10b, page 393 ©UPI/Corbis-Bettman
Fig. 16.11a, page 395 ©David M. Phillips/Visuals Unlimited
Fig. 16.11b, page 395 ©David M. Phillips/Visuals Unlimited

Chapter 17 opener, page 399 ©SIU/Visuals Unlimited

Chapter 18 opener, page 423 ©Omikron/Photo Researchers
Biosite 18.2, page 436 ©James L. Amos/Corbis

Chapter 19 opener, page 447 ©David Young Wolff/Tony Stone Images
Fig. 19.1a, page 448 ©Ursula Markus/Photo Researchers, Inc.
Fig. 19.1b, page 448 ©Gregory Scott/Photo Researchers, Inc.
Fig. 19.4a, page 451 ©David Sieren/Visuals Unlimited
Fig. 19.4b, page 451 ©N. ET Perennou/Photo Researchers, Inc.
Fig. 19.4c, page 451 ©Dr. Jeremy Burgess/Science Photo Library Photo Researchers, Inc.
Fig. 19.8, page 456 ©Barry Lewis/Tony Stone Worldwide
Fig. 19.9, page 457 ©Thomas McAvoy, Life Magazine ©Time Warner, Inc.
Fig. 19.10, page 457 ©Corbis-Bettman
Biosite 19.1, page 452 ©Stan Elems/Visuals Unlimited
Fig. 19.13a, page 461 ©A. Morris/Visuals Unlimited
Fig. 19.13b, page 461 ©Art Wolfe/Tony Stone Images
Fig. 19.14a, page 464 ©David Young Wolff/Tony Stone Images
Fig. 19.14b, page 464 ©Jeanne Drake/Tony Stone Worldwide

Chapter 20 opener, page 467 ©Biophoto Associates/Science Source/Photo Researchers
2a, page 470 ©Biophoto Associates/Science Source/Photo Researchers
Fig. 20.2b, page 470 ©Biophoto Associates/Science Source/Photo Researchers
Fig. 20.7, page 474 ©Jane Thomas/Visuals Unlimited
Fig. 20.20a, page 484 ©Mary Evans' Picture Library/Photo Researchers, Inc.
Fig. 20.21b, page 485 ©Keith/Custom Medical Stock Photo
Fig. 20.22b, page 485 ©Stanley Flegler/Visuals Unlimited
Fig. 20.28a, page 491 ©Valerie Lindgren/Visuals Unlimited
Fig. 20.28b, page 491 ©Gary Parker/Science Photo Library
Fig. 20.29a, page 492 ©NMSB/Custom Medical Stock Photo
Fig. 20.29b, page 492 ©Valerie Lindgren/Visuals Unlimited
Fig. 20.30a, page 493 ©Reproduced with permission from John Money, *Sex Errors of the Body* ©1968, The Johns Hopkins University Press.
Fig. 20.30b, page 493 ©Valerie Lindgren/Visuals Unlimited
Fig. 20.31, page 494 ©George Wilder/Visuals Unlimited

Chapter 21 opener, page 497 ©SIU/Visuals Unlimited
Fig. 21.2a, page 499 ©John D. Cunningham/Visuals Unlimited
Fig. 21.2b, page 499 ©Cabisco/Visuals Unlimited
Fig. 21.2c, page 499 ©SIU/Visuals Unlimited
Fig. 21.7, page 503 ©Don W. Fawcett/Visuals Unlimited
Fig. 21.10a, page 509 ©Hank Morgan/Science Source/Photo Researchers
Fig. 21.10b, page 509 ©Martin Dohrn/Science Photo Library/Photo Researchers

Chapter 22 opener, page 513 ©Robert Frerck/Tony Stone Images
Fig. 22.2b, page 516 ©Corbis-Bettmann
Fig. 22.3a, page 518 ©Don Spiro/Tony Stone Images
Fig. 22.9a, page 524 ©Jim Steinberg/Photo Researchers
Fig. 22.9b, page 524 ©Kevin Schafer/Tony Stone Images
Fig. 22.12a, page 526 ©Tom McHugh/Photo Researchers
Fig. 22.12b, page 526 ©M. Loup/Jacana/Photo Researchers
Fig. 22.13a, page 527 ©Ken Lucas/Visuals Unlimited
Fig. 22.13b, page 527 ©James Balog/Tony Stone Worldwide
Fig. 22.13c, page 527 ©Tim Davis/Tony Stone Images
Fig. 22.13d, page 527 ©Joe McDonald/Visuals Unlimited
Fig. 22.15a, page 528 ©Institute of Human Origins
Fig. 22.18, page 531 ©Charles Preitner/Visuals Unlimited
Fig. 22.19, page 531 ©Robert Frerck/Tony Stone Images

Chapter 23 opener, page 537 ©Earth Imagery/Tony Stone Images
Fig. 23.5a, page 542 ©Brooking Tatum/Visuals Unlimited
Fig. 23.5b, page 542 ©Art Wolfe/Tony Stone Images
Fig. 23.5c, page 542 ©Art Wolfe/Tony Stone Images
Fig. 23.5d, page 542 ©Jim Pickerell/Tony Stone Images
Fig. 23.6a, page 542 ©François Gohier/Photo Researchers
Fig. 23.6b, page 542 ©Rod Planck/Photo Researchers
Fig. 23.7, page 543 ©Stuart Westmorland/Tony Stone Images
Fig. 23.20, page 556 ©L. Kiff/Science VU/Visuals Unlimited
Fig. 23.22, page 558 ©Charles Preitner/Visuals Unlimited
Fig. 23.23b, page 559 ©Ian Shaw/Tony Stone Images
Fig. 23.24, page 559 ©WHOI, D. Foster/Visuals Unlimited
Fig. 23.26a, page 561 ©Tim Davis/Photo Researchers
Fig. 23.26b, page 561 ©Kjell B. Sandved/Visuals Unlimited
Fig. 23.26c, page 561 ©Leonide Principe/Photo Researchers
Fig. 23.26d, page 561 ©William Grenfell/Visuals Unlimited
Fig. 23.26, page 561 ©Gregory Dimijian/Photo Researchers
Fig. 23.27, page 561 ©Bill Beatty/Visuals Unlimited
Fig. 23.28a, page 562 ©David L. Pearson/Visuals Unlimited
Fig. 23.28b, page 562 ©Daniel J. Cox/Tony Stone Images
Fig. 23.28c, page 562 ©Barbara Gerlach/Visuals Unlimited
Fig. 23.29, page 563 ©Ron Spomer/Visuals Unlimited
Fig. 23.30, page 563 ©Charlie Ott/Photo Researchers
Fig. 23.30a, page 563 ©David E. Myers/Tony Stone Images
Fig. 23.30b, page 563 ©Joe McDonald/Visuals Unlimited
Fig. 23.30c, page 563 ©Tom J. Ulrich/Visuals Unlimited
Fig. 23.32a, page 565 ©S. McCutcheon/Visuals Unlimited
Fig. 23.32b, page 565 ©Jim Zipp/Photo Researchers
Fig. 23.33, page 565 ©Arthur Morris/Visuals Unlimited
Fig. 23.34, page 565 ©Verna R. Johnston/Photo Researchers
Fig. 23.37b, page 570 ©Frank Hanna/Visuals Unlimited